Harmonic Analysis

A Comprehensive Course in Analysis, Part 3

Harmonic Analysis

A Comprehensive Course in Analysis, Part 3

Barry Simon

AMERICAN MATHEMATICAL SOCIETY

Providence, Rhode Island

2010 *Mathematics Subject Classification.* Primary 26-01, 31-01, 46-01;
Secondary 30H10, 30H35, 42C40, 42B20, 46E35.

For additional information and updates on this book, visit
www.ams.org/bookpages/simon

Library of Congress Cataloging-in-Publication Data

Simon, Barry, 1946–
    Harmonic analysis / Barry Simon.
        pages cm. — (A comprehensive course in analysis ; part 3)
    Includes bibliographical references and indexes.
    ISBN 978-1-4704-1102-2 (alk. paper)
    1. Mathematical analysis—Textbooks. 2. Harmonic analysis—Textbooks. I. Title.

QA300.S5275   2015
515′.2433—dc23

                                                                                2015024457

To the memory of Cherie Galvez

extraordinary secretary, talented helper, caring person

and to the memory of my mentors,
Ed Nelson (1932-2014) and Arthur Wightman (1922-2013)

who not only taught me Mathematics
but taught me how to be a mathematician

Contents

Preface to the Series

Young men should prove theorems, old men should write books.

—*Freeman Dyson*, quoting G. H. Hardy[1]

Reed–Simon[2] starts with "Mathematics has its roots in numerology, geometry, and physics." This puts into context the division of mathematics into algebra, geometry/topology, and analysis. There are, of course, other areas of mathematics, and a division between parts of mathematics can be artificial. But almost universally, we require our graduate students to take courses in these three areas.

This five-volume series began and, to some extent, remains a set of texts for a basic graduate analysis course. In part it reflects Caltech's three-terms-per-year schedule and the actual courses I've taught in the past. Much of the contents of Parts 1 and 2 (Part 2 is in two volumes, Part 2A and Part 2B) are common to virtually all such courses: point set topology, measure spaces, Hilbert and Banach spaces, distribution theory, and the Fourier transform, complex analysis including the Riemann mapping and Hadamard product theorems. Parts 3 and 4 are made up of material that you'll find in some, but not all, courses—on the one hand, Part 3 on maximal functions and H^p-spaces; on the other hand, Part 4 on the spectral theorem for bounded self-adjoint operators on a Hilbert space and det and trace, again for Hilbert space operators. Parts 3 and 4 reflect the two halves of the third term of Caltech's course.

---

[1]Interview with D. J. Albers, The College Mathematics Journal, **25**, no. 1, January 1994.

[2]M. Reed and B. Simon, *Methods of Modern Mathematical Physics, I: Functional Analysis*, Academic Press, New York, 1972.

While there is, of course, overlap between these books and other texts, there are some places where we differ, at least from many:

(a) By having a unified approach to both real and complex analysis, we are able to use notions like contour integrals as Stietljes integrals that cross the barrier.

(b) We include some topics that are not standard, although I am surprised they are not. For example, while discussing maximal functions, I present Garcia's proof of the maximal (and so, Birkhoff) ergodic theorem.

(c) These books are written to be keepers—the idea is that, for many students, this may be the last analysis course they take, so I've tried to write in a way that these books will be useful as a reference. For this reason, I've included "bonus" chapters and sections—material that I do not expect to be included in the course. This has several advantages. First, in a slightly longer course, the instructor has an option of extra topics to include. Second, there is some flexibility—for an instructor who can't imagine a complex analysis course without a proof of the prime number theorem, it is possible to replace all or part of the (non-bonus) chapter on elliptic functions with the last four sections of the bonus chapter on analytic number theory. Third, it is certainly possible to take all the material in, say, Part 2, to turn it into a two-term course. Most importantly, the bonus material is there for the reader to peruse long after the formal course is over.

(d) I have long collected "best" proofs and over the years learned a number of ones that are not the standard textbook proofs. In this regard, modern technology has been a boon. Thanks to Google books and the Caltech library, I've been able to discover some proofs that I hadn't learned before. Examples of things that I'm especially fond of are Bernstein polynomials to get the classical Weierstrass approximation theorem, von Neumann's proof of the Lebesgue decomposition and Radon–Nikodym theorems, the Hermite expansion treatment of Fourier transform, Landau's proof of the Hadamard factorization theorem, Wielandt's theorem on the functional equation for $\Gamma(z)$, and Newman's proof of the prime number theorem. Each of these appears in at least some monographs, but they are not nearly as widespread as they deserve to be.

(e) I've tried to distinguish between central results and interesting asides and to indicate when an interesting aside is going to come up again later. In particular, all chapters, except those on preliminaries, have a listing of "Big Notions and Theorems" at their start. I wish that this attempt to differentiate between the essential and the less essential didn't make this book different, but alas, too many texts are monotone listings of theorems and proofs.

(f) I've included copious "Notes and Historical Remarks" at the end of each section. These notes illuminate and extend, and they (and the Problems) allow us to cover more material than would otherwise be possible. The history is there to enliven the discussion and to emphasize to students that mathematicians are real people and that "may you live in interesting times" is truly a curse. Any discussion of the history of real analysis is depressing because of the number of lives ended by the Nazis. Any discussion of nineteenth-century mathematics makes one appreciate medical progress, contemplating Abel, Riemann, and Stieltjes. I feel knowing that Picard was Hermite's son-in-law spices up the study of his theorem.

On the subject of history, there are three cautions. First, I am not a professional historian and almost none of the history discussed here is based on original sources. I have relied at times—horrors!—on information on the Internet. I have tried for accuracy but I'm sure there are errors, some that would make a real historian wince.

A second caution concerns looking at the history assuming the mathematics we now know. Especially when concepts are new, they may be poorly understood or viewed from a perspective quite different from the one here. Looking at the wonderful history of nineteenth-century complex analysis by Bottazzini–Grey[3] will illustrate this more clearly than these brief notes can.

The third caution concerns naming theorems. Here, the reader needs to bear in mind Arnol'd's principle:[4] *If a notion bears a personal name, then that name is not the name of the discoverer* (and the related Berry principle: *The Arnol'd principle is applicable to itself*). To see the applicability of Berry's principle, I note that in the wider world, Arnol'd's principle is called "Stigler's law of eponymy." Stigler[5] named this in 1980, pointing out it was really discovered by Merton. In 1972, Kennedy[6] named Boyer's law *Mathematical formulas and theorems are usually not named after their original discoverers* after Boyer's book.[7] Already in 1956, Newman[8] quoted the early twentieth-century philosopher and logician A. N. Whitehead as saying: "Everything of importance has been said before by somebody who

---

[3]U. Bottazzini and J. Gray, *Hidden Harmony—Geometric Fantasies. The Rise of Complex Function Theory*, Springer, New York, 2013.

[4]V. I. Arnol'd, *On teaching mathematics*, available online at `http://pauli.uni-muenster.de/~munsteg/arnold.html`.

[5]S. M. Stigler, *Stigler's law of eponymy*, Trans. New York Acad. Sci. **39** (1980), 147–158.

[6]H. C. Kennedy, *Classroom notes: Who discovered Boyer's law?*, Amer. Math. Monthly **79** (1972), 66–67.

[7]C. B. Boyer, *A History of Mathematics*, Wiley, New York, 1968.

[8]J. R. Newman, *The World of Mathematics*, Simon & Schuster, New York, 1956.

did not discover it." The main reason to give a name to a theorem is to have a convenient way to refer to that theorem. I usually try to follow common usage (even when I know Arnol'd's principle applies).

I have resisted the temptation of some text writers to rename things to set the record straight. For example, there is a small group who have attempted to replace "WKB approximation" by "Liouville–Green approximation", with valid historical justification (see the Notes to Section 15.5 of Part 2B). But if I gave a talk and said I was about to use the Liouville–Green approximation, I'd get blank stares from many who would instantly know what I meant by the WKB approximation. And, of course, those who try to change the name also know what WKB is! Names are mainly for shorthand, not history.

These books have a wide variety of problems, in line with a multiplicity of uses. The serious reader should at least skim them since there is often interesting supplementary material covered there.

Similarly, these books have a much larger bibliography than is standard, partly because of the historical references (many of which are available online and a pleasure to read) and partly because the Notes introduce lots of peripheral topics and places for further reading. But the reader shouldn't consider for a moment that these are intended to be comprehensive—that would be impossible in a subject as broad as that considered in these volumes.

These books differ from many modern texts by focusing a little more on special functions than is standard. In much of the nineteenth century, the theory of special functions was considered a central pillar of analysis. They are now out of favor—too much so—although one can see some signs of the pendulum swinging back. They are still mainly peripheral but appear often in Part 2 and a few times in Parts 1, 3, and 4.

These books are intended for a second course in analysis, but in most places, it is really previous exposure being helpful rather than required. Beyond the basic calculus, the one topic that the reader is expected to have seen is metric space theory and the construction of the reals as completion of the rationals (or by some other means, such as Dedekind cuts).

Initially, I picked "A Course in Analysis" as the title for this series as an homage to Goursat's *Cours d'Analyse*,[9] a classic text (also translated into English) of the early twentieth century (a literal translation would be

---

[9]E. Goursat, *A Course in Mathematical Analysis: Vol. 1: Derivatives and Differentials, Definite Integrals, Expansion in Series, Applications to Geometry. Vol. 2, Part 1: Functions of a Complex Variable. Vol. 2, Part 2: Differential Equations. Vol. 3, Part 1: Variation of Solutions. Partial Differential Equations of the Second Order. Vol. 3, Part 2: Integral Equations. Calculus of Variations*, Dover Publications, New York, 1959 and 1964; French original, 1905.

"of Analysis" but "in" sounds better). As I studied the history, I learned that this was a standard French title, especially associated with École Polytechnique. There are nineteenth-century versions by Cauchy and Jordan and twentieth-century versions by de la Vallée Poussin and Choquet. So this is a well-used title. The publisher suggested adding "Comprehensive", which seems appropriate.

It is a pleasure to thank many people who helped improve these texts. About 80% was TEXed by my superb secretary of almost 25 years, Cherie Galvez. Cherie was an extraordinary person—the secret weapon to my productivity. Not only was she technically strong and able to keep my tasks organized but also her people skills made coping with bureaucracy of all kinds easier. She managed to wind up a confidant and counselor for many of Caltech's mathematics students. Unfortunately, in May 2012, she was diagnosed with lung cancer, which she and chemotherapy valiantly fought. In July 2013, she passed away. I am dedicating these books to her memory.

During the second half of the preparation of this series of books, we also lost Arthur Wightman and Ed Nelson. Arthur was my advisor and was responsible for the topic of my first major paper—perturbation theory for the anharmonic oscillator. Ed had an enormous influence on me, both via the techniques I use and in how I approach being a mathematician. In particular, he taught me all about closed quadratic forms, motivating the methodology of my thesis. I am also dedicating these works to their memory.

After Cherie entered hospice, Sergei Gel'fand, the AMS publisher, helped me find Alice Peters to complete the TEXing of the manuscript. Her experience in mathematical publishing (she is the "A" of A K Peters Publishing) meant she did much more, for which I am grateful.

This set of books has about 150 figures which I think considerably add to their usefulness. About half were produced by Mamikon Mnatsakanian, a talented astrophysicist and wizard with Adobe Illustrator. The other half, mainly function plots, were produced by my former Ph.D. student and teacher extraordinaire Mihai Stoiciu (used with permission) using Mathematica. There are a few additional figures from Wikipedia (mainly under WikiCommons license) and a hyperbolic tiling of Douglas Dunham, used with permission. I appreciate the help I got with these figures.

Over the five-year period that I wrote this book and, in particular, during its beta-testing as a text in over a half-dozen institutions, I received feedback and corrections from many people. In particular, I should like to thank (with apologies to those who were inadvertently left off): Tom Alberts, Michael Barany, Jacob Christiansen, Percy Deift, Tal Einav, German Enciso, Alexander Eremenko, Rupert Frank, Fritz Gesztesy, Jeremy Gray,

Leonard Gross, Chris Heil, Mourad Ismail, Svetlana Jitomirskaya, Bill Johnson, Rowan Killip, John Klauder, Seung Yeop Lee, Milivoje Lukic, Andre Martinez-Finkelshtein, Chris Marx, Alex Poltoratski, Eric Rains, Lorenzo Sadun, Ed Saff, Misha Sodin, Dan Stroock, Benji Weiss, Valentin Zagrebnov, and Maxim Zinchenko.

Much of these books was written at the tables of the Hebrew University Mathematics Library. I'd like to thank Yoram Last for his invitation and Naavah Levin for the hospitality of the library and for her invaluable help.

This series has a Facebook page. I welcome feedback, questions, and comments. The page is at `www.facebook.com/simon.analysis`.

Even if these books have later editions, I will try to keep theorem and equation numbers constant in case readers use them in their papers.

Finally, analysis is a wonderful and beautiful subject. I hope the reader has as much fun using these books as I had writing them.

Preface to Part 3

I don't have a succinct definition of harmonic analysis or perhaps I have too many. One possibility is that harmonic analysis is what harmonic analysts do. There is an active group of mathematicians, many of them students of or grandstudents of Calderón or Zygmund, who have come to be called harmonic analysts and much of this volume concerns their work or the precursors to that work. One problem with this definition is that, in recent years, this group has branched out to cover certain parts of nonlinear PDE's and combinatorial number theory.

Another approach to a definition is to associate harmonic analysis with "hard analysis," a term introduced by Hardy, who also used "soft analysis" as a pejorative for analysis as the study of abstract infinite-dimensional spaces. There is a dividing line between the use of abstraction, which dominated the analysis of the first half of the twentieth century, and analysis which relies more on inequalities, which regained control in the second half. And there is some truth to the idea that Part 1 in this series of books is more on soft analysis and Part 3 on hard, but, in the end, both parts have many elements of both abstraction and estimates.

Perhaps the best description of this part is that it should really be called "More Real Analysis." With the exception of Chapter 5 on H^p-spaces, any chapter would fit with Part 1—indeed, Chapter 4, which could be called "More Fourier Analysis," started out in Part 1 until I decided to move it here.

The topics that should be in any graduate analysis course and often are, are the results on Hardy–Littlewood maximal functions and the Lebesgue differentiation theorem in Chapter 2, the very basics of harmonic and sub-harmonic functions, something about H^p-spaces and about Sobolev inequalities.

The other topics are exceedingly useful but are less often in courses, including those at Caltech. Especially in light of Calderón's discovery of its essential equivalence to the Hardy–Littlewood theorem, the maximal ergodic theorem should be taught. And wavelets have earned a place, as well. In any event, there are lots of useful devices to add to our students' toolkits.

Chapter 1

Preliminaries

Begin, be bold, and venture to be wise.

—*Horace (65 BC–8 BC)*[1]

This volume could be named "more real analysis" and many of the chapters complete and extend ones in Part 1. Accordingly, we assume the reader is familiar with the basic topics in that part. We emphasize some of the more significant themes in Section 1.2 below. In many places, complex analysis plays a role—in particular, Chapter 5 is a kind of synthesis of ideas from Parts 1 and 2. We suppose that the reader knows the basics of analytic function theory as found, for example, in Chapters 2 and 3 of Part 2A. Section 1.3 below spells out some of the more advanced topics in complex analysis, all found in more detail in Part 2A. We will also occasionally need results from operator theory, including the spectral theorem. This is the topic of Chapter 5 of Part 4.

1.1. Notation and Terminology

A foolish consistency is the hobgoblin of little minds ... Is it so bad, then, to be misunderstood? Pythagoras was misunderstood, and Socrates, and Jesus, and Luther, and Copernicus, and Galileo, and Newton, and every pure and wise spirit that ever took flesh. To be great is to be misunderstood.

—*Ralph Waldo Emerson* [**243**]

For a real number a, we will use the terms positive and strictly positive for $a \geq 0$ and $a > 0$, respectively. It is not so much that we find nonnegative bad, but the phrase "monotone nondecreasing" for $x > y \Rightarrow f(x) \geq f(y)$ is

---

[1]This translation is from "The Danger of Procrastination" in A. Cowley, *Cowley Essays*, Scribner, New York, 1869; available online at Google Books as a free ebook.

downright confusing so we use "monotone increasing" and "strictly monotone increasing" and then, for consistency, "positive" and "strictly positive."
Similarly for matrices, we use "positive definite" and "strictly positive definite" where others might use "positive semi-definite" and "positive definite."

Basic rings and fields.

$$\mathbb{R} = real\ numbers \qquad \mathbb{Q} = rationals \qquad \mathbb{Z} = integers$$

$$\mathbb{C} = complex\ numbers = \{x + iy \mid x, y \in \mathbb{R}\}$$

with their sums and products. For $z = x + iy \in \mathbb{C}$, we use $\operatorname{Re} z = x$,
$\operatorname{Im} z = y$. $|z| = (x^2 + y^2)^{1/2}$.

Products. $X^n = n$-tuples of points in X with induced vector space and/or
additive structure; in particular, $\mathbb{R}^n$, $\mathbb{Q}^n$, $\mathbb{Z}^n$, $\mathbb{C}^n$.

Subsets of $\mathbb{C}$.

- $\mathbb{C}_+ =$ upper half-plane $= \{z \mid \operatorname{Im} z > 0\}$;
  $\mathbb{H}_+ =$ right half-plane $= \{z \mid \operatorname{Re} z > 0\}$
- $\mathbb{Z}_+ = \{n \in \mathbb{Z} \mid n > 0\} = \{1, 2, 3, \dots\}$;
  $\mathbb{N} = \{0\} \cup \mathbb{Z}_+ = \{0, 1, 2, \dots\}$
- $\mathbb{D} = \{z \in \mathbb{C} \mid |z| < 1\}$; $\partial\mathbb{D} = \{z \in \mathbb{C} \mid |z| = 1\}$
- $\mathbb{D}_\delta(z_0) = \{z \in \mathbb{C} \mid |z - z_0| < \delta\}$ for $z_0 \in \mathbb{C}$, $\delta > 0$

Miscellaneous Terms.

- For $x \in \mathbb{R}$, $[x] =$ greatest integer less than x, that is, $[x] \in \mathbb{Z}$, $[x] \le x < [x] + 1$
- $\{x\} = x - [x] =$ fractional part of x
- $\sharp(A) =$ number of elements in a set A
- $\operatorname{Ran} f =$ range of a function f
- $\log(z) =$ natural logarithm, that is, logarithm to base e; if z is complex,
  put the cut along $(-\infty, 0]$, i.e., $\log(z) = \log(|z|) + i \arg(z)$ with $-\pi < \arg(z) \le \pi$
- For sets A, B subsets of X, $A \cup B =$ union, $A \cap B =$ intersection, $A^c =$ complement of A in X, $A \setminus B = A \cap B^c$, $A \triangle B = (A \setminus B) \cup (B \setminus A)$
- For matrices, M_{12} means row one, column two
- $f \upharpoonright K = restriction$ of a function to K, a subset of the domain of f

One-Point Compactification Notation.

- $X_\infty =$ one-point compactification of a locally compact space
- $C_\infty(X) =$ continuous functions vanishing at infinity

- $C(X_\infty)$ = continuous functions with a limit at infinity (= $C_\infty(X)$ + $\{\alpha \mathbb{1}\}$)

Given a set, S, a *relation*, R, is a subset of $S \times S$. For $x, y \in S$, we write xRy if $(x, y) \in R$ and $\sim xRy$ if $(x, y) \notin R$. A relation is called

$$reflexive \Leftrightarrow \forall \, x \in S, \text{ we have } xRx$$

$$symmetric \Leftrightarrow \forall \, x, y \in S, \; xRy \Rightarrow yRx$$

$$transitive \Leftrightarrow \forall \, x, y, z \in S, \; xRy \,\&\, yRz \Rightarrow xRz$$

An *equivalence relation* is a relation that is reflexive, symmetric, and transitive.

If R is an equivalence relation on S, the *equivalence class*, $[x]$, of $x \in S$ is

$$[x] = \{y \in S \mid xRy\} \tag{1.1.1}$$

By the properties of equivalence relations,

$$\forall \, x, y \in S, \quad \text{either } [x] = [y] \text{ or } [x] \cap [y] = \emptyset \tag{1.1.2}$$

The family of equivalence classes in S is denoted S/R.

We use "big oh, litttle oh" notation, e.g., $f(x) = O(|x|)$ or $f(x) = o(1)$. If the reader is not familiar with it, see Section 1.4 of Part 2A.

Banach Space Notation.

- X = Banach space, $\quad X^*$ = dual space
- $\sigma(X, Y)$ = Y-weak topology (Y are linear functionals acting on X, $\sigma(X, Y)$ is the weakest topology on X in which $x \mapsto \langle y, x \rangle$ is continuous for each y).
- $\mathcal{H}$ = Hilbert space (always complex; $\langle \cdot, \cdot \rangle$ is antilinear in the *first* factor.
- $\mathcal{L}(X)$ = bounded linear transformations from X to X
- $A \in \mathcal{L}(X), \quad A^t \in \mathcal{L}(X^*)$ by $(A^t \ell)(x) = \ell(Ax)$
- $A \in \mathcal{L}(\mathcal{H}), \quad A^* \in \mathcal{L}(H)$ by $\langle A^* \varphi, \psi \rangle = \langle \varphi, A\psi \rangle$.

1.2. Some Results for Real Analysis

We will suppose the reader is familiar with the language of measure theory (Chapters 4 and 8 of Part 1) and topological vector spaces, especially Banach and Hilbert spaces (Chapters 5 and 3 of Part 1) and of countably normed and Fréchet spaces (Section 6.1 of Part 1). Occasionally, we'll need some of the basic tools of Banach space theory, specifically the Hahn–Banach (Section 5.5 of Part 1) and Banach–Alaoglu theorem (Section 5.8 of Part 1).

1.2.1. L^p and $\mathcal{S}(\mathbb{R}^\nu)$ Space. Let (M, Σ, μ) be a σ-finite measure space. Measure theory is developed in Chapter 4 of Part 1, first for Baire measures on compact sets, and then for σ-finite measures on σ-compact, locally compact spaces although there is discussion of measure theory on Polish spaces and even on general σ-algebras. One forms for $1 \le p < \infty$

$$L^p(M, d\mu) = \{\text{Baire functions}, f \mid \int |f(x)|^p \, d\mu < \infty\} \tag{1.2.1}$$

if $p < \infty$ and bounded Baire functions if $p = \infty$. More precisely, elements of L^p are equivalence classes of Baire functions, where $f \approx g \Leftrightarrow \mu(\{x \mid f(x) \ne g(x)\}) = 0$. This equivalence implies that

$$\|f\|_p = \left(\int |f(x)|^p \, d\mu(x) \right)^{\frac{1}{p}} \tag{1.2.2}$$

is strictly positive on nonzero functions. That $\|\cdot\|_p$ is a norm is an important result known as *Minkowski's inequality*,

$$\|f + g\|_p \le \|f\|_p + \|g\|_p \tag{1.2.3}$$

(1.2.2) is the definition if $p < \infty$. $\|f\|_\infty$ is the essential sup norm, i.e., the smallest α, so $\mu(\{x \mid |f(x)| > \alpha\}) = 0$. If $\mu(M) < \infty$, $\|f\|_\infty = \lim_{p \uparrow \infty} \|f\|_p$.

It is a fundamental result that L^p is a Banach space and that L^p is separable for one $p \in [1, \infty)$, if and only if it is for all $p \in [1, \infty)$. Another fundamental result is *Hölder's inequality*, if

$$1 \le p, q \le \infty \text{ and } \frac{1}{p} + \frac{1}{q} = 1 \tag{1.2.4}$$

then, $f \in L^p$, $g \in L^q \Rightarrow fg \in L^1$ and

$$\|fg\|_1 \le \|f\|_p \|g\|_q \tag{1.2.5}$$

More generally,

$$1 \le p, q, r \le \infty, \quad p^{-1} + q^{-1} = r^{-1},$$
$$f \in L^p, g \in L^q \Rightarrow fg \in L^r \text{ and } \|fg\|_r \le \|f\|_p \|g\|_q \tag{1.2.6}$$

In particular, writing $|f| = |f|^{1-\theta} |f|^\theta$, this implies for $p_0 < p < p_1$, we have that $L^p \subset L^{p_0} \cap L^p$ and if

$$p^{-1} = (1 - \theta)p_0^{-1} + \theta p_1^{-1} \tag{1.2.7}$$

then

$$\|f\|_p \le \|f\|_{p_0}^{1-\theta} \|f\|_{p_1}^\theta \tag{1.2.8}$$

For $1 \le p < \infty$, we have a theorem of Riesz that

$$\left(L^p(M, d\mu) \right)^* = L^q(M, d\mu) \tag{1.2.9}$$

where p and q are related by (1.2.4)—called dual indices. This duality is realized via associating $g \in L^q$ with the linear functional on L^p,

$$L_g(f) = \int f(x)\, g(x) d\mu(x) \tag{1.2.10}$$

L_g is well defined and gives a bounded linear functional with $\|L_g\|_{(L^p)^*} = \|g\|_q$ by Hölder's inequality (and the fact that $f = |g|^q g^{-1}$ gives equality in Hölder's inequality). The key part of Riesz' theorem is that every $L \in (L^p)^*$ is an L_g. This implies that for $1 < p < \infty$, L^p is reflexive. Except for very special (M, μ), L^1 is not reflexive.

Another useful general inequality is *Markov's inequality* for $p < \infty$ and $f \in L^p(M, d\mu)$ and $\alpha \in (0, \infty)$

$$\mu(\{x \mid |f(x)| > \alpha\}) \leq \alpha^{-p} \|f\|^p \tag{1.2.11}$$

For $(M, \mu) = (\mathbb{R}^\nu, d^\nu x)$, we have *Young's inequality*

$$\|f * g\|_r \leq \|f\|_p \|g\|_q \tag{1.2.12}$$

where

$$(f * g)(x) = \int f(x - y)\, g(y)\, d^\nu y \tag{1.2.13}$$

is the convolution and $1 \leq p, q, r \leq \infty$

$$p^{-1} + q^{-1} = 1 + r^{-1} \tag{1.2.14}$$

Included in the inequality is that for a.e. x, the integral is absolutely convergent.

Three important convergence theorems for L^p, $1 \leq p < \infty$ are: if $f_n \geq 0$ and measurable, then (Fatou's lemma)

$$\int \liminf_n f_n(x)\, d\mu(x) \leq \liminf \int f_n(x)\, d\mu(x) \tag{1.2.15}$$

If $f_n(x) \to f(x)$ for a.e. x and there is $g \in L^p(M, d\mu)$ ($p < \infty$) so that $|f_n(x)| \leq g(x)$ for all n and a.e. x, then $|f_n - f|_p \to 0$ (dominated convergence theorem). If $0 \leq f_n \leq f_{n+1}$ pointwise a.e., then (monotone convergence theorem) $f_\infty(x) = \sup_n f_n(x)$ has $\|f_\infty\|_p = \lim_{n \to \infty} \|f_n\|_p$ (where if $\int |f_\infty(x)|^p\, d\mu(x) = \infty$, we write $\|f_\infty\|_p = \infty$) and if $\sup_n \|f_n\|_p < \infty$, then $\|f_\infty - f_n\|_p \to 0$. These theorems hold for sequences and may fail if $\{f_n\}_{n=1}^\infty$ is replaced by a net $\{f_\alpha\}_{\alpha \in \mathcal{I}}$, of functions.

Two other spaces that enter heavily in this volume are $\mathcal{S}(\mathbb{R}^\nu)$, the *Schwartz space*, and $C_0^\infty(\mathbb{R}^\nu)$, the C^∞-functions of compact support. A multi-index is a set of ν numbers $\alpha_1, \ldots, \alpha_\nu$ in $\{0, 1, 2, \ldots\}$ with $|\alpha| = \sum_{j=1}^\nu |\alpha_j|$. For two ν-component multi-indices α, β, we set

$$\|f\|_{\alpha,\beta} = \|x^\alpha D^\beta f\|_\infty \tag{1.2.16}$$

with

$$D^\beta f = \partial^{|\beta|} f / \partial^{\beta_1} x_1 \ldots \partial^{\beta_\nu} x_\nu \qquad (1.2.17)$$

$\mathcal{S}(\mathbb{R}^\nu)$ is the set of functions on $\mathbb{R}^\nu$ which are C^∞ with $\|f\|_{\alpha,\beta} < \infty$ for all α, β.

$$\rho(f, g) = \sum_{\alpha, \beta} \min(\|f - g\|_{\alpha,\beta}, 2^{-|\alpha| - |\beta|}) \qquad (1.2.18)$$

defines a metric on $\mathcal{S}$ in which it is complete—so $\mathcal{S}$ is a Fréchet space, i.e., a complete linear metric space.

The dual of $\mathcal{S}$, denoted $\mathcal{S}'(\mathbb{R}^\nu)$ is the set of *tempered distributions*. Chapter 6 of Part 1 has extensive discussion of the structure of distributions. In particular, if $\tau \in \mathcal{S}'(\mathbb{R}^\nu)$, one defines its derivatives by

$$(D^\alpha \tau)(f) = (-1)^{|\alpha|} \tau(D^\alpha f) \qquad (1.2.19)$$

so integration by parts holds by definition. It is a basic fact that for every distribution τ, there are finitely many polynomially bounded continuous functions $g_1, \ldots g_\ell$ and multi-indices, $\beta_1, \ldots, \beta_\ell$ so that $\tau = \sum_{j=1}^\ell D^{\beta_j} g_j$.

$C_0^\infty(\mathbb{R}^\nu)$ is the set of C^∞-functions of compact support. As discussed in Chapter 9 of Part 1, it has a natural nonmetrizable topology. Its dual, $\mathcal{D}(\mathbb{R}^\nu)$, the *ordinary distributions* can be described as follows: Given $\tau \in \mathcal{D}$ and $R < \infty$, there exists C_R and N_R so that

$$|\tau(f)| \le C_R \sum_{|\alpha| \le N_R} \|D^\alpha f\|_\infty \qquad (1.2.20)$$

for all $f \in C_0^\infty(\mathbb{R}^\nu)$ with $\mathrm{supp}(f) \subset \{x \mid |x| \le R\}$. As $R \to \infty$ both N_R and C_R can grow at arbitrary rates.

1.2.2. Fourier Series and Transforms. Fourier analysis is a critical tool because it turns differentiation into multiplication. Indeed, joint analysis in phase space, the combination of the original and the Fourier variables will be the subject of a whole chapter, Chapter 4, of this book. In Part 1, we saw two guises for Fourier analysis: series (in Chapter 3, mainly Section 3.5) and transform (in Chapter 6, especially Sections 6.2, 6.4, 6.5, and 6.6)—put differently for functions on $[0, 2\pi]$ or on $\partial\mathbb{D} = \{z \in \mathbb{C} \mid |z| = 1\}$ and functions on $\mathbb{R}$ and $\mathbb{R}^\nu$.

$\{\varphi_n(e^{i\theta}) \equiv e^{in\theta}\}_{n=-\infty}^\infty$ is an orthonormal basis for $L^2(\partial\mathbb{D}, \frac{d\theta}{2\pi})$ so any $f \in L^2(\partial\mathbb{D}, \frac{d\theta}{2\pi})$ has a formal expansion

$$f(e^{i\theta}) = \sum_{n=-\infty}^\infty \langle \varphi_n, f \rangle \, \varphi_n(e^{i\theta}) \qquad (1.2.21)$$

We sometimes denote $\langle \varphi_n, f \rangle$ as $f_n^\sharp$. If

$$f_N(e^{i\theta}) = \sum_{n=-N}^{N} \langle \varphi_n, f \rangle \, \varphi_n(e^{i\theta}) \tag{1.2.22}$$

the abstract theory of expansions in orthonormal basis in a Hilbert space implies that

$$\|f - f_N\|_{L^2(\partial \mathbb{D}, \frac{d\theta}{2\pi})} \to 0 \tag{1.2.23}$$

Theorem 3.5.6 (Fejér's Theorem) of Part 1 says that for any continuous $f \in C(\partial \mathbb{D})$,

$$C_N(f) \equiv \frac{1}{N} \sum_{j=0}^{N-1} f_j \tag{1.2.24}$$

converges uniformly to f. This actually implies that $\{\varphi_n\}_{n=-\infty}^{\infty}$ is a complete set (that it is orthonormal is an elementary calculation). Fejér's theorem depends on a direct formula for C_N, namely

$$(C_N f)(e^{i\theta}) = \int_0^{2\pi} F_N(\theta - \psi) \, f(e^{i\psi}) \, \frac{d\psi}{2\pi} \tag{1.2.25}$$

$$F_N(\eta) = \frac{1}{N} \left[\frac{\sin(\frac{N\eta}{2})}{\sin(\frac{\eta}{2})} \right]^2 \tag{1.2.26}$$

and considerations we turn to in Section 1.2.4 below. F_N is called the *Fejér kernel*.

A major theme in this book will be pointwise convergence a.e. but, while it is true that for $f \in L^p(\partial \mathbb{D}, \frac{d\theta}{2\pi})$, $1 < p < \infty$, $f_N(e^{i\theta}) \to f(e^{i\theta})$ for a.e. θ, we will not develop the tools that prove it. We will, however, prove in Section 2.11 that if $\sum_{n=-\infty}^{\infty} |\langle \varphi_n, f \rangle|^p < \infty$ for some $p \in [1, 2)$, then $f_N \to f$ for a.e. θ, and we'll prove in Section 5.8 that if $1 < p < \infty$, and $f \in L^p(\partial \mathbb{D}, \frac{d\theta}{2\pi})$, then $\|f_N - f\|_p \to 0$. We note that these things are subtle; there are f in $L^p(\partial \mathbb{D}, \frac{d\theta}{2\pi})$ for $p = 1$ or $p = \infty$ (distinct for these two values) where $\|f_N\|_p \to \infty$ and there are $f \in L^1(\partial \mathbb{D}, \frac{d\theta}{2\pi})$ where f_N diverges everywhere (see Problem 10 of Section 5.4 of Part 1 for the former and Notes to Section 3.5 of Part 1 for references to the latter).

One special use for $\mathcal{S}(\mathbb{R}^\nu)$ is in the study of the Fourier transform. For $f \in \mathcal{S}(\mathbb{R}^\nu)$, we define the Fourier transform, $\widehat{f}$, and inverse Fourier transform, $\widecheck{f}$, by

$$\widehat{f} = (2\pi)^{-\nu/2} \int e^{-ik\cdot x} \, f(x) \, d^\nu x \tag{1.2.27}$$

$$\widecheck{f}(x) = (2\pi)^{-\nu/2} \int e^{ik\cdot x} \, f(k) \, d^\nu k \tag{1.2.28}$$

Then the *Fourier inversion formula* says that for $f \in \mathcal{S}(\mathbb{R}^\nu)$

$$\check{\hat{f}} = \hat{\check{f}} = f \tag{1.2.29}$$

and the *Plancharel formula* that

$$\|\hat{f}\|_{L^2} = \|f\|_{L^2} \tag{1.2.30}$$

Chapter 6 of Part 1 has several proofs of these basic facts. We note also that

$$\widehat{f * g} = (2\pi)^{\nu/2} \hat{f} \hat{g}; \quad \widehat{fg} = (2\pi)^{-\nu/2} \hat{f} * \hat{g} \tag{1.2.31}$$

Fourier analysis will be a recurring theme of this volume and in Part 4 (Section 6.9), we'll extend the Fourier transform from $\partial \mathbb{D}$, $\mathbb{Z}$, and $\mathbb{R}^\nu$ to arbitrary locally compact abelian groups.

1.2.3. Segal–Bargmann and Zak Transforms. In Sections 4.4 and 4.5, we'll need two transforms on L^2-functions closely related to Fourier transforms and presented in Chapter 6 of Part 1. Because they are not as widely known, we discuss them briefly here for the case $\mathbb{R}^\nu$, $\nu = 1$. The first, the Segal–Bargmann transform is closely related to the very useful Hermite basis for $L^2(\mathbb{R}^\nu, d^\nu x)$. We begin with the case $\nu = 1$. φ_0 is the function

$$\varphi_0(x) = \pi^{-\frac{1}{4}} \exp\left(-\frac{1}{2} x^2\right) \tag{1.2.32}$$

which has $\|\varphi_0\|_{L^2(\mathbb{R}, dx)} = 1$.

We define differential operators

$$A = \frac{1}{\sqrt{2}}\left(x + \frac{d}{dx}\right), \quad A^\dagger = \frac{1}{\sqrt{2}}\left(x - \frac{d}{dx}\right) \tag{1.2.33}$$

so $A\varphi_0 = 0$. One defines

$$\varphi_n = (n!)^{-1/2} \left(A^\dagger\right)^n \varphi_0 \tag{1.2.34}$$

Theorem 6.4.3 of Part 1 proves that $\{\varphi_n\}_{n=0}^\infty$ is an orthonormal basis of $L^2(\mathbb{R}, dx)$. Each φ_n is a polynomial of degree n times φ_0. Moreover, by Theorem 6.4.4 of Part 1, $f \in L^2(\mathbb{R}, dx)$ lies in $\mathcal{S}(\mathbb{R})$ if and only if for all ℓ, there is C_ℓ so

$$(1+n)^\ell |\langle \varphi_n, f \rangle| \leq C_\ell \tag{1.2.35}$$

for all n and $\tau \in \mathcal{S}'(\mathbb{R})$ if and only if

$$|\tau(\varphi_n)| \leq C_{\ell_0} (1+n)^{\ell_0}$$

for some fixed ℓ_0.

The *Segal–Bargmann transform* of $f \in L^2(\mathbb{R}, dx)$ is defined by

$$(\mathbb{B}f)(z) = \pi^{-\frac{1}{4}} \int A_z(x) f(x) dx \tag{1.2.36}$$

$$A_z(x) = \exp(-\tfrac{1}{2}(z^2 + x^2) + \sqrt{2}\, xz) \tag{1.2.37}$$

$\mathbb{B}$ is viewed as a map from $L^2(\mathbb{R}, dx)$ to $L^2(\mathbb{C}, \pi^{-1} \exp(-|z|^2) \, d^2z)$. Then, by direct calculation for $f \in \mathcal{S}(\mathbb{R})$ (see Theorem 6.4.11 and 6.4.12 of Part 1)

$$\mathbb{B}\varphi_0 = \mathbb{1}, \quad \mathbb{B}A^\dagger f = z\mathbb{B}f, \quad \mathbb{B}Af = \frac{d}{dz}(\mathbb{B}f) \qquad (1.2.38)$$

In particular, this implies that

$$\mathbb{B}\varphi_n = (n!)^{-1/2} z^n \qquad (1.2.39)$$

which is an orthonormal family in $L^2(\mathbb{C}, \pi^{-1} \exp(-|z|^2) d^2z)$. The span of $\{\mathbb{B}\varphi_n\}_{n=0}^\infty$ is called *Fock space* and consists precisely of the elements of L^2 which are entire analytic functions. For $\tau \in \mathcal{S}'(\mathbb{R})$, $\mathbb{B}$ can be defined by (1.2.36) since $A_z(\,\cdot\,) \in \mathcal{S}(\mathbb{R})$.

The *Zak transform* is defined by

$$(Zf)(x, k) = \sum_{j \in \mathbb{Z}} f(x + j) \, e^{-2\pi i jk} \qquad (1.2.40)$$

defined for functions f on $\mathbb{R}$ to functions Zf on $\mathbb{R}^2$ (or on $\Delta = [0,1] \times [0,1]$). For $f \in \mathcal{S}(\mathbb{R})$, this sum is convergent and defines a function on $\mathbb{R}^2$ which is C^∞ and obeys

$$\varphi(x + 1, k) = e^{2\pi i k} \varphi(x, k), \quad \varphi(x, k + 1) = \varphi(x, k) \qquad (1.2.41)$$

Alternatively, if $f \in L^2(\mathbb{R}, dx)$, $\{f(x + j)\}_{j \in \mathbb{Z}}$ is in ℓ^2 for a.e. x and (1.2.40) can be interpreted as a Fourier series in k so $(Zf)(x, k)$ is defined for a.e. x and k and is L^2 in k on $[0, 1]$. It is a simple fact (Problem 13 of Section 6.2 of Part 1) that Z is an isometry of $L^2(\mathbb{R}, dx)$ and $L^2(\Delta, d^2x)$.

1.2.4. Approximate Identities. A major theme is this volume, especially in Chapter 2, will be when a sequence, f_n, of functions converges pointwise a.e. One situation involves a family of linear operators, T_n, so $f_n = T_n f$, and we'll want to know a priori convergence for nice f's. In this regard, the notion of approximate identity, described in Section 3.5 of Part 1, is useful. We describe it here for $\partial \mathbb{D}$ but similar ideas work on $\mathbb{R}^\nu$. A sequence $\{g_N\}_{N=1}^\infty$ on $\partial \mathbb{D}$ is called an *approximate identity* if it obeys

(i) $g_N(e^{i\theta}) \geq 0$, all N, all θ

(ii) $\displaystyle \int_0^{2\pi} g_N(e^{i\theta}) \frac{d\theta}{2\pi} = 1$ \qquad (1.2.42)

(iii) For all $\varepsilon > 0$,

$$\lim_{N \to \infty} \int_{\varepsilon < \theta < 2\pi - \varepsilon} g_N(e^{i\theta}) \frac{d\theta}{2\pi} = 0 \qquad (1.2.43)$$

The basic fact (Theorem 3.5.11 of Part 1) is that if g_N is an approximate identity and $f \in C(\partial \mathbb{D})$, then $g_N * f \to f$ uniformly.

The Fejér kernel C_N of (1.2.26) is an approximate identity so for $f \in C(\partial\mathbb{D})$, $C_N * f \to f$ uniformly.

1.2.5. Probability Basics. Probability language and intuition is useful for many subjects in this volume—in particular, ergodic theory and martingales can be viewed as part of probability theory—they are the subject of Section 2.6–2.10. The basics can be found in Chapter 7 of Part 1; here we want to summarize a few ideas.

Pointwise convergence a.e. doesn't involve points although it seems to! For if f_n is a sequence of functions

$$\{x \mid \limsup f_n(x) > f(x)\} = \bigcup_{k=1}^{\infty} \bigcap_{m=1}^{\infty} \bigcup_{n=m}^{\infty} \{x \mid f_n(x) > f(x) + \frac{1}{k}\} \quad (1.2.44)$$

and $f_n(x) \to f(x)$ if and only if $\limsup f_n(x) \le f(x)$ and $\limsup(-f_n(x)) \le -f(x)$, so pointwise convergence a.e. is only a function of the measures of some measurable sets.

If $f_1, \ldots f_\ell$ is a family of measurable functions on a probability measure space, (M, Σ, μ), the joint probability distribution $d\nu_{f_1, \ldots f_\ell}$ is the unique measure on $\mathbb{R}^\ell$ with

$$\int F(y_1, \ldots, y_\ell) d\nu(y) = \int F(f_1(x), \ldots, f_\ell(x)) \, d\mu(x) \quad (1.2.45)$$

for all bounded continuous functions, F, of compact support.

In particular,

$$d\nu_f([a, b]) = \mu(\{x \mid a \le f(x) < b\}) \quad (1.2.46)$$

is the *probability distribution* of f.

A family $\{f_j\}_{j=1}^{\infty}$ of measurable functions are called *independent random variables* if for any $j_1, \ldots j_\ell$, we have

$$d\nu_{f_{j_1}, \ldots, f_{j_\ell}} = d\nu_{f_{j_1}} \otimes d\nu_{f_{j_2}} \otimes \cdots \otimes d\nu_{f_{j_\ell}} \quad (1.2.47)$$

is a product measure. If $d\nu_{f_j}$ is j independent, we say they are identically distributed.

The *strong law of large numbers* says under suitable conditions, if $\{f_j\}_{j=1}^{\infty}$ are independent, identically distributed random variables, then for a.e. x

$$\frac{1}{n} \sum_{j=1}^{n} f_j(x) \to \mathbb{E}(f) \equiv \int f(x) \, d\mu(x) \quad (1.2.48)$$

In Theorem 7.2.5 of Part 1, we proved this if $\mathbb{E}(f_1^2) < \infty$. The optimal result only requires $\mathbb{E}(|f_1|) < \infty$ and is proven in Problem 16 of Section 2.10 of this volume. It also follows from the Birkhoff ergodic theorem (see the remark after Theorem 2.7.2 below).

1.2.6. Equilibrium Measures. In Sections 5.9 and 6.8 of Part 1, we proved existence and uniqueness of potential theory equilibrium measures. We'll need these objects in Chapter 3 so we sketch the basic facts here.

The free (Coulomb) Green's function on $\mathbb{R}^\nu$, $\nu \geq 2$, is defined by:

$$G_0(x) = \left[(\nu - 2)\sigma_\nu\right]^{-1} |x|^{-(\nu-2)} \quad (\nu \geq 3)$$
$$= -\frac{1}{2\pi} \log(|x - y|) \quad (\nu = 2) \tag{1.2.49}$$

where $\sigma_\nu = \frac{2\pi^{\nu/2}}{\Gamma(\nu/2)}$ is the area of the unit sphere $\mathbb{S}^{\nu-1}$ in $\mathbb{R}^\nu$. Given a compact set, $\mathfrak{e} \subset \mathbb{R}^\nu$, and $\mu \in \mathcal{M}_{+,1}(\mathfrak{e})$, a probability measure on $\mathfrak{e}$, we define

$$\mathcal{E}(\mu) = \frac{1}{2} \int G_0(x - y) \, d\mu(x) \, d\mu(y) \tag{1.2.50}$$

Notice that $\mathcal{E}(\mu)$ takes values in $\mathbb{R} \cup \{\infty\}$. In $\nu \geq 3$, the integral defining $\mathcal{E}(\mu)$ is positive but might be $+\infty$. In $\nu = 2$, $G_0(x-y) \geq \frac{1}{2\pi} \log(\max\{|x-y| \mid x, y \in \mathfrak{e}\}) > -\infty$ so $\mathcal{E}(\mu)$ can diverge to $+\infty$ but otherwise is finite. If $\mathcal{E}(\mu) = \infty$ for all $\mu \in \mathcal{M}_{+,1}(\mathfrak{e})$, then we say $\mathfrak{e}$ has *zero capacity*. This happens if $\mathfrak{e}$ is very small, e.g., a finite set of points but also, for example, on $\mathbb{R}^2$, $C(\mathfrak{e}) = 0 \Rightarrow \mathfrak{e}$ has zero Hausdorff dimension. Otherwise,

Theorem 1.2.1. *If $\mathfrak{e}$ does not have zero capacity, then there exists a unique probability measure, ρ_e, called the* equilibrium measure *of $\mathfrak{e}$ with*

$$\mathcal{E}(\rho_e) = \inf_{\mu \in \mathcal{M}_{+,1}(\mathfrak{e})} \mathcal{E}(\mu) \tag{1.2.51}$$

Existence is Theorem 5.9.3 of Part 1. It follows from the fact that $\mu \mapsto \mathcal{E}(\mu)$ is weakly lower semi-continuous and the compactness of $\mathcal{M}_{+,1}(\mathfrak{e})$. Uniqueness if $\nu \geq 3$ is Corollary 6.8.5 of Part 1 and if $\nu = 2$, Problem 6 of Section 6.9 of Part 1. Uniqueness follows from

$$\mathcal{E}\left(\frac{1}{2}(\mu_1 + \mu_2)\right) < \frac{1}{2}\mathcal{E}(\mu_1) + \frac{1}{2}\mathcal{E}(\mu_2) \tag{1.2.52}$$

if $\mu_1 \neq \mu_2$.

The *capacity* of $\mathfrak{e}$, $C(\mathfrak{e})$, is defined by

$$G_0(C(\mathfrak{e})) = \mathcal{E}(\rho_e) \tag{1.2.53}$$

When discussing $\mathbb{C}$ (thought of as $\mathbb{R}^2$), it is a common convention to define $\mathcal{E}$ by by (1.2.50) but without $(4\pi)^{-1}$ in front of $\log(|x - y|)$. We'll shift convention in the middle of Chapter 3.

1.3. Some Results from Complex Analysis

We suppose that the reader is familiar with the basics of complex analysis as described in Chapter 2 and Section 3.1 of Part 2A. In this section, we discuss some of the more specialized aspects of complex analysis that play a prominent role in some parts of this volume. We will not limit our use of results from Part 2A to just the material in this section. In particular, we use the Phragmém–Lindelöf method of Section 5.1 of Part 2A in several places and the Weierstrass elliptic functions of Section 10.4 of Part 2A once.

1.3.1. Jensen's Formula and the Poisson–Jensen Formula. In Section 9.8 of Part 2A, we proved (Theorem 9.8.1 of Part 2A) that if f is analytic in a neighborhood of $\overline{\mathbb{D}_R(0)} = \{z \mid |z| \leq R\}$, then f has finitely many zeros, counting multiplicity, $\{z_j\}_{j=1}^N$ in $\mathbb{D}_R(0)$ and if $f(0) \neq 0$, one has *Jensen's formula,*

$$\int_0^{2\pi} \log|f(Re^{i\theta})| \, \frac{d\theta}{2\pi} = \log|f(0)| + \sum_{j=1}^N \log\left(\frac{R}{|z_j|}\right) \qquad (1.3.1)$$

The proof is easy. One introduces the *Blaschke factor,* $b(z, w)$, for $z, w \in \mathbb{D}$ by

$$b(z, w) = \begin{cases} z & \text{if } w = 0 \\ \frac{|w|}{w} \frac{w-z}{1-\bar{w}z}, & \text{otherwise} \end{cases} \qquad (1.3.2)$$

This is analytic in a neighborhood of $\overline{\mathbb{D}}$, vanishes to order 1 precisely at $z = w$ and obeys $|b(e^{i\theta}, w)| = 1$. (While not needed here, we will need in Section 1.3.2, the additional fact that $b(0, w) \geq 0$ and > 0 unless $w = 0$ and in that case $b'(0, w) > 0$. These properties determine b.)

To prove (1.3.1), one considers only $R = 1$ which suffices by scaling. In that case, one defines

$$g(z) = f(z)/(\prod_{j=1}^N b(z, z_j)) \qquad (1.3.3)$$

If f has no zeros on $\partial\mathbb{D}$, g is analytic and nonvanishing in a neighborhood of $\overline{\mathbb{D}}$, so $\log(g(z))$ is analytic there. By the real part of a Cauchy integral formula:

$$\log(|g(0)|) = \operatorname{Re} \log(g(0))$$

$$= \int_0^{2\pi} \operatorname{Re}[\log(g(e^{i\theta}))] \, \frac{d\theta}{2\pi} \qquad (1.3.4)$$

Taking into account that $|b(0, z_j)| = |z_j|$ and $|b(e^{i\theta}, z_j)| = 1$, this is exactly (1.2.3). If f has a zero on $\partial\mathbb{D}$, one proves the result for $R = 1 - \varepsilon$ and takes $\varepsilon \downarrow 0$.

The complex Poisson formula (see Section 1.3.3) relates $\log(g(z))$ to $\log(|g(e^{i\theta})|)$ in the above analysis and yields the *Poisson–Jensen formula* (Theorem 9.8.2 of Part 2A): under the above hypotheses on f, for $z \in \mathbb{D}_R(0)$:

$$f(z) = \frac{f(0)}{|f(0)|} \prod_{j=1}^{N} b\left(\frac{z}{R}, \frac{z_j}{R}\right) \exp\left(\int \frac{Re^{i\theta} + z}{Re^{i\theta} - z} \log|f(Re^{i\theta})| \frac{d\theta}{2\pi}\right) \qquad (1.3.5)$$

In Section 9.10 of Part 2A and even more Chapter 17 of Part 2B, these results are used to analyze the relation of the growth of entire analytic functions to the set of zeros of f.

1.3.2. Blaschke Products. Section 9.9 of Part 2A discusses infinite products $\prod_{j=1}^{\infty} b(z, z_j)$ given an infinite set, $\{z_j\}_{j=1}^{\infty}$ of zeros in $\mathbb{D}$. A key consideration is the *Blaschke condition*:

$$\sum_{j=1}^{\infty} (1 - |z_j|) < \infty \qquad (1.3.6)$$

One half of the theory of such products is: if f is in the *Nevanlinna space*, N, i.e., f is analytic on $\mathbb{D}$ and

$$\sup_{0 \le r < 1} \int \log_+ |f(re^{i\theta})| \frac{d\theta}{2\pi} < \infty \qquad (1.3.7)$$

then the zeros of f obey (1.3.6).

To see this, we can replace $f(z)$ by $f(z)/z^\ell$ and so suppose $f(0) \neq 0$. Thus, by Jensen's formula for $f(rz)$, one has

$$\sum_{j \,|\, |z_j| < r} \log\left(\frac{r}{|z_j|}\right) + \log|f(0)| \le \int_0^{2\pi} \log_+ |f(re^{i\theta})| \frac{d\theta}{2\pi} \qquad (1.3.8)$$

Since $\log(\frac{r}{|z_j|}) \ge 0$, we can restrict to $j \in [1, \ldots, k]$ and take $r \uparrow 1$ to see that for each k

$$\prod_{j=1}^{k} |z_j| \ge |f(0)| \exp\left(-\sup_{0 \le r < 1} \int \log_+ |f(re^{i\theta})| \frac{d\theta}{2\pi}\right) \qquad (1.3.9)$$

Now take $k \to \infty$ and see that (1.3.7) implies $\prod_{j=1}^{\infty} |z_j| > 0$ which is equivalent to (1.3.6) ! We note that if f is a bounded analytic function in $\mathbb{D}$ (called $f \in H^\infty(\mathbb{D})$), then $f \in N$, so (1.3.6) holds for the zeros of $f \in H^\infty(\mathbb{D})$.

For the other direction, one has

Theorem 1.3.1 ($\equiv$ Theorem 9.9.4 of Part 2A). *Let $\{z_j\}_{j=1}^{\infty}$ be a sequence in $\mathbb{D}$. Then $\lim_{N \to \infty} \prod_{j=1}^{N} b(z, z_j)$ converges uniformly on compacts. If*

(1.3.5) *holds, the product is an absolutely convergent product and the limit*

$$B_\infty(z, \{z_j\}_{j=1}^\infty) = \prod_{j=1}^\infty b(z, z_j) \qquad (1.3.10)$$

called a Blaschke product, *vanishes exactly at* $\{z_j\}_{j=1}^\infty$ *(counting multiplicity). If (1.3.5) fails, the limit is identically zero.*

To prove this, one first proves the estimate

$$|1 - b(z, w)| \leq \frac{(1 - |w|)(1 + |z|)}{1 - |z|} \qquad (1.3.11)$$

for all $z, w \in \mathbb{D}$.

Blaschke products will play a major role in the theory of H^p spaces—which is the topic of Chapter 5 of this volume.

1.3.3. Real and Complex Poisson Kernel. Let f be a function continuous on $\overline{\mathbb{D}}$ and analytic on $\mathbb{D}$. Then, one has the *complex Poisson representation*

$$f(z) = i \operatorname{Im} f(0) + \int_0^{2\pi} \frac{e^{i\theta} + z}{e^{i\theta} - z} \operatorname{Re} f(e^{i\theta}) \frac{d\theta}{2\pi} \qquad (1.3.12)$$

for all $z \in \mathbb{D}$. Section 5.3 of Part 2A has four proofs, one in the text and three in Problems 1–3. One depends on the fact that

$$\frac{e^{i\theta} + z}{e^{i\theta} - z} = 1 + 2 \sum_{n=1}^\infty e^{-in\theta} z^n \qquad (1.3.13)$$

and the formulae

$$\operatorname{Re} f(0) = \int_0^{2\pi} \operatorname{Re} f(e^{i\theta}) \frac{d\theta}{2\pi} \qquad (1.3.14)$$

(by the Cauchy integral formula) plus $f(z) = \sum_{n=0}^\infty a_n z^n$ implies

$$a_n = \int_0^{2\pi} f(e^{i\theta}) e^{-in\theta} \frac{d\theta}{2\pi}, \quad 0 = \int \overline{f(e^{i\theta})} e^{-in\theta} \frac{d\theta}{2\pi} \qquad (1.3.15)$$

Once one has (1.3.12) taking real parts, for $0 \leq r < 1$, one gets the (real) Poisson formula

$$\operatorname{Re} f(re^{i\psi}) = \int_0^{2\pi} P_r(\psi, \theta) \operatorname{Re} f(e^{i\theta}) \frac{d\theta}{2\pi} \qquad (1.3.16)$$

where

$$P_r(\psi, \theta) = \frac{1 - r^2}{1 + r^2 - 2r \cos(\theta - \psi)} = \operatorname{Re}\left[\frac{e^{i\theta} + re^{i\psi}}{e^{i\theta} - re^{i\psi}}\right] \qquad (1.3.17)$$

Using $f = \operatorname{Re} f - i \operatorname{Re}(if)$, one gets from (1.3.17) that

$$f(re^{i\psi}) = \int_0^{2\pi} P_r(\psi, \theta) f(e^{i\theta}) \frac{d\theta}{2\pi} \qquad (1.3.18)$$

Generalizations of (1.3.16) to higher dimensions are an important part of our analysis of harmonic functions in Chapter 3. It is natural to ask when functions analytic on $\mathbb{D}$ have some kind of boundary values on $\partial\mathbb{D}$ for which these representations hold. This is a theme of Chapter 5.

1.3.4. Montel and Vitali Theorems. In Section 3.1, we'll prove analogs for harmonic functions on $\mathbb{R}^\nu$ of these results from analytic function theory.

Theorem 1.3.2 (Montel's Theorem). *Let $\Omega \subset \mathbb{C}$ be a domain (open connected set). Let $\{f_n\}_{n=1}^\infty$ be a sequence of analytic functions so that for each compact $K \subset \Omega$, we have $\sup_{z\in K;n}|f_n(z)| < \infty$. Then there is a subsequence converging on compact subsets of Ω.*

Remarks. 1. This result and variants are found in Section 6.2 of Part 2A.

2. The proof uses a Cauchy estimate to show the f_n are locally uniformly equicontinuous and then appeals to the Ascoli–Arzelà theorem.

3. This can be rephrased as a compactness result.

Theorem 1.3.3 (Vitali Convergence Theorem). *Let Ω, $\{f_n\}_{n=1}^\infty$ obey the hypotheses of Theorem 1.3.2. Let B be an open set inside Ω for which f_n converges pointwise. Then, there is an analytic function f on Ω so $f_n \to f$ uniformly on compacts.*

Remarks. 1. This result is also found in Section 6.2 of Part 2A.

2. This is a corollary of Montel's theorem. If g is the pointwise limit on B, any two limits of subsequence limits guaranteed by Montel's theorem must equal g in B so must be equal on Ω. Thus, every subsequence has a convergent subsequence converging to the analytic continuation of g to Ω.

3. In $\mathbb{C}$, B need not be open but any set with a limit point in Ω, but in higher dimensions, the requirements on B are complicated, so we state it for B open to extend to Theorem 3.1.24 below.

1.3.5. Riesz–Thorin Interpolation. Using complex analytic techniques, Section 5.2 of Part 2A proves (Theorem 5.2.2 of Part 2A):

Theorem 1.3.4 (The Riesz–Thorin Theorem). *Let (Ω, μ) be a measure space. Let $p_0, p_1, q_0, q_1 \in [1, \infty]$ and define, for $t \in (0,1)$,*

$$p_t^{-1} = (1-t)p_0^{-1} + tp_1^{-1}, \qquad q_t^{-1} = (1-t)q_0^{-1} + tq_1^{-1} \qquad (1.3.19)$$

Let T be a linear transformation from $L^{p_0} \cap L^{p_1}$ to $L^{q_0} \cap L^{q_1}$ with

$$\|Tf\|_{q_j} \leq C_j\|f\|_{p_j} \qquad (1.3.20)$$

for $j = 0, 1$ and all $f \in L^{p_0} \cap L^{p_1}$. Then for all $t \in (0,1)$ and $f \in L^{p_0} \cap L^{p_1}$,

$$\|Tf\|_{q_t} \leq C_t\|f\|_{p_t} \qquad (1.3.21)$$

where

$$C_t = C_0^{1-t} C_1^t \qquad (1.3.22)$$

In Section 6.1 below (and in special cases, Theorem 2.2.10), a different interpolation theorem is proven. The method is different from the complex variable proof used for Theorem 1.3.4. Thus, in distinction to the "complex method" used for the Riesz–Thorin theorem, the technique of Section 6.1 is sometimes called the "real method."

1.4. Green's Theorem

In our study of harmonic functions in Chapter 3, an important role will be played by Green's theorem, a result from elementary vector calculus that we want to state.

Definition. Let $k \geq 2$. A C^k-*hypersurface* in $\mathbb{R}^\nu$ is the set $S = \{x \mid F(x) = 0\}$ where $F : \mathbb{R}^\nu \to \mathbb{R}$ is a C^k function with $\vec{\nabla} F(x) \neq 0$ for all $x \in S$. We will call $\Omega = \{x \mid F(x) > 0\}$ the *region bounded by* S if Ω has compact closure. We then call S a *compact hypersurface* bounding a compact region.

By the implicit function theorem (see Theorem 5.12.10 of Part 1), S is a $\nu - 1$ dimensional C^{k-1} manifold. For, if $\frac{\partial F}{dx_\nu} \neq 0$ at $x^{(0)}$, we think of $F : \mathbb{R}^{\nu-1} \times \mathbb{R} \to \mathbb{R}$ and find $h : \{y \in \mathbb{R}^{\nu-1} \mid |y - (x_1^{(0)}, \ldots, x_{\nu-1}^{(0)})| < \delta\} \to \mathbb{R}$ so that $F(y, h(y)) = 0$ and use $(x_1, \ldots, x_{\nu-1})$ as local coordinate near $x^{(0)}$. At any point $x \in S$,

$$n_{x_0} \equiv -\vec{\nabla} F(x_0) / |\vec{\nabla} F(x_0)| \qquad (1.4.1)$$

is called the outward pointing normal. It is orthogonal to the tangent plane to S at x_0.

Green's theorem says that:

Theorem 1.4.1 (Green's Theorem). *Let S be a compact C^2-hypersurface with Ω the region bounded by S. Let f, g be C^2-functions on Ω so that f, g and $D^\alpha f$, $D^\alpha g$ have continuous extensions to $\overline{\Omega} = \Omega \cup S$ for all multi-indices, α, with $|\alpha| \leq 2$. Then*

$$\int_\Omega \left[f(x)(\Delta g)(x) - g(x)(\Delta f)(x) \right] d^\nu x = \int_S \left(f(y) \frac{\partial g(y)}{\partial n} - g(y) \frac{\partial f(y)}{\partial n} \right) d\sigma(y) \qquad (1.4.2)$$

Remarks. 1. $\frac{\partial h}{\partial n} = \vec{n} \cdot \vec{\nabla} h$ is the outward pointing derivative.

2. $d\sigma$ is a Euclidian surface area. Locally if Euclidian distance in the normal direction is x_n, then $d\sigma dx_n$ is $d^\nu x$ the usual volume. As discussed after Example 15.3.12 of Part 2B, if $F(x_1, \ldots, x_{\nu-1}, h(x_1, \ldots, x_{\nu-1})) = 0$ locally, then $d\sigma = (1 + |\nabla h|^2)^{1/2} dx_1 \ldots dx_{\nu-1}$.

One deduces Green's theorem from Gauss' theorem:

Theorem 1.4.2 (Gauss' Theorem). *Let S be a compact C^2-hypersurface with Ω the region bounded by S. Let $\vec{E}(x)$ be an $\mathbb{R}^\nu$-valued C^1-function on Ω so that for $i = 1, \ldots, \nu$, ∇E_i has a continuous extension to $\overline{\Omega} = \Omega \cup S$. Then,*

$$\int_\Omega (\vec{\nabla} \cdot \vec{E})(x) d^\nu x = \int_S (\vec{n} \cdot \vec{E})(y) \, d\sigma(y) \tag{1.4.3}$$

Here, $\vec{\nabla} \cdot \vec{E} = \sum_{i=1}^\nu \frac{\partial E_i}{\partial x_i}$. Gauss' theorem implies Green's theorem. We take

$$\vec{E}(x) = f(x)(\vec{\nabla} g)(x) - g(x)\vec{\nabla} f(x) \tag{1.4.4}$$

noting that $\vec{\nabla} \cdot \vec{E} = f\Delta g - g\Delta f$ and $n \cdot \vec{E} = f\frac{\partial g}{\partial n} - g\frac{\partial f}{\partial n}$.

We will not provide a formal proof of Gauss' theorem (see the Notes) but remark that it is a special case of the abstract Stokes' theorem (Theorem 1.5.2 of Part 2A) if we take ω to be

$$\omega = \sum_{i=1}^\nu (-1)^{i-1} E_i(x) \, dx_1 \wedge \ldots \wedge dx_{i-1} \wedge dx_{i+1} \wedge \ldots \wedge dx_\nu \tag{1.4.5}$$

so $d\omega = \vec{\nabla} \cdot \vec{E} d^\nu x$. We also note that it is a simple consequence of the fundamental theorem of calculus. If $E_i(x) = 0$ for $i = 2, \ldots, \nu$, we note that for fixed $x_2, \ldots, x_\nu$, if $\Omega \cap \{(x_1, x_2, \ldots, x_\nu) \mid x_1 \in \mathbb{R}\} = \{(x_1, \ldots, x_\nu) \mid x_1 \in (x_1^-, x_1^+)\}$, then

$$\int_{x_1^-}^{x_1^+} \frac{\partial E_1}{\partial x_1} dx_1 = E_1(x_1^+, x_2, \ldots, x_\nu) - E_1(x_1^-, x_2, \ldots, x_\nu) \tag{1.4.6}$$

Now integrate $dx_2 \ldots dx_\nu$. The left side is $\int (\partial_1 E_1 / \partial x_1) d^\nu x$. On the right, we use the fact that $d\sigma(y) = n_1^{-1} dx_2 \ldots dx_\nu$ so $E_1 dx_2 \ldots dx_\nu = E_1 n_1 \, d\sigma$. One can repeat this for the other components and handle the case where the intersection is multiple intervals.

Notes and Historical Remarks. It is appropriate to name the theorems after Gauss and Green but not Stokes (see the Notes to Section 1.5 of Part 2A).

For a proof of Gauss' theorem (aka the divergence theorem), see Callahan [**136**], Courant–John [**174**], or Montiel–Ros [**583**].

Chapter 2

Pointwise Convergence Almost Everywhere

Nowadays, there are only three really great English mathematicians: Hardy, Littlewood and Hardy–Littlewood.

—*Harald Bohr*, in a lecture on his 60th birthday [90][1]

Big Notions and Theorems: Maximal Functions, Maximal Inequalities ⇒ a.e. Convergence, Distribution Function, Equimeasurable, Decreasing Rearrangement, Weak-L^p, Calderón Norm, Marcinkiewicz Interpolation Theorem, Accumulation Function, Hardy–Littlewood Maximal Theorem, Vitali Covering Lemma, Croft–Garsia Covering Lemma, Besicovitch Covering Lemma, Lebesgue Differentiation Theorem, Pointwise Approximate Identities, Nontangential Limits, Comparison of Measures, Stieltjes and Hilbert Transforms of Measures, Theorem of de Vallée Poussin, Ergodic Averages, von Neumann Ergodic Theorem, Maximal Ergodic Theorem, Maximal Ergodic Inequality, Birkhoff Ergodic Theorem, Calderón's Hardy–Littlewood Maximal ⇒ Maximal Ergodic, Unique Ergodicity, Strong Law of Large Numbers, Normal Numbers, Weyl's Equidistribution Theorem, Irrational Flow on the Circle, Benford's Law, Equidistribution of $P(n)$ mod 1, Skew Shift, Gauss Map, Gauss Measure, Gauss–Kuzmin Distribution, Gauss–Kuzmin–Wirsing Operator, Ergodicity of Gauss Map, Khinchin's Theorem, Lévy's Theorem, Hyperbolic Geodesic Flow, Hopf's Geodesic Theorem, Kingman Subadditive Ergodic Theorem, Lyapunov Behavior, Furstenberg–Kesten Theorem, Ruelle–Oseledec Theorem, Martingales, Sub/Supermartingales, Random Series, Kolmogorov Theorem, Doob's Inequality, Lévy 0-1 Law, Kolmogorov 0-1 Law, Tail Algebra, Doob Decomposition, Martingale Convergence Theorem, Christ–Kiselev Maximal Theorem, Pointwise Convergence of Fourier Transforms

---

[1]quoting a colleague from earlier in the century. Harald was Niels' brother.

We have often seen situations where $\{f_n\}_{n=1}^\infty$ and f lie in some $L^p(X, d\mu)$ space and $f_n \to f$ in L^p-norm. Here we will focus on when one can say $f_n(x) \to f(x)$ for μ-a.e. x. Since elements of L^p are only defined a.e., it makes no sense to say "at all x." But even when it might (e.g., all functions are continuous, so they have natural representatives), a.e. is often the correct thing to aim for (for example, if $f_n(x) = \max(1 - nx, 0)$ for $x \in [0, 1]$, $f_n \to 0$ in all L^p, $1 \le p < \infty$, $f_n(x) \to 0$ for $x \ne 0$, but $f_n(0) \equiv 1$).

From one point of view, this may seem like much ado about nothing. It follows from the argument in the proof of the Riesz–Fischer theorem that if, say, $p = 1$ and

$$\sum_{n=1}^\infty \|g_{n+1} - g_n\|_1 < \infty \tag{2.0.1}$$

then $g_n(x)$ converges to its L^1 limit at a.e. x. And any L^1-norm convergent sequence has subsequences obeying (2.0.1); indeed, every subsequence has such a subsubsequence. But mere subsequence convergence is not so satisfactory. In each of the signature theorems mentioned below, imagine settling for a subsequence and you'll see that the result is much less cogent.

If a numerical sequence has the property that for some x_∞, any subsequence has a subsubsequence converging to x_∞, then the original sequence converges to x_∞—so one might think a.e. pointwise convergence is always true. The issue is that each subsubsequence converges except on a set of measure zero, and there are uncountably many subsequences. The reader should analyze this issue in the following:

Example 2.0.1. For $n = 0, 1, 2, \ldots$ and $j = 0, 1, \ldots, 2^n - 1$, define on $[0, 1]$,

$$f_{2^n+j}(x) = \begin{cases} 0, & x \in [\frac{j}{2^n}, \frac{j+1}{2^n}] \\ 1, & x < \frac{j}{2^n} \text{ or } x > \frac{j+1}{2^n} \end{cases} \tag{2.0.2}$$

Then $\{f_m\}_{m=1}^\infty$ converges to $f \equiv 1$ in any L^p-norm, $p < \infty$, but one has pointwise convergence at no x in $[0, 1]$. □

This example proves that pointwise convergence is not automatic and is quite subtle. To get a flavor of the results we are talking about, consider the following results proven in this chapter:

Theorem 2.0.2 ($\equiv$ Theorem 2.4.1). *Let* $f \in L^1(0, 1; dx)$. *Then for a.e.* x,

$$\lim_{\varepsilon \downarrow 0} \frac{1}{\pm\varepsilon} \int_0^{x\pm\varepsilon} f(y) \, dy = f(x) \tag{2.0.3}$$

This is called Lebesgue's theorem on differentiation of integrals since it says that $x \mapsto \int_0^x f(y) \, dy$ is differentiable at a.e. point. It provides an L^1-version of the fundamental theorem of calculus. It is Theorem 4.15.4 of Part 1.

Theorem 2.0.3 ($\equiv$ Theorem 2.4.6). *Let f be analytic on $\mathbb{D}$ and obey*

$$\sup_{0 \leq r < 1} \int |f(re^{i\theta})|^2 \frac{d\theta}{2\pi} < \infty \tag{2.0.4}$$

Then for Lebesgue a.e. θ, we have $\lim_{r \uparrow 1} f(re^{i\theta})$ exists and defines an L^2-function.

Functions obeying (2.0.4) are said to lie in $H^2(\mathbb{D})$; we'll discuss them in Chapter 5.

Theorem 2.0.4 ($\equiv$ Theorem 2.7.4). *For $j = 0, 1, \ldots, 9$ and $x \in [0, 1]$, let $n_j(x, n)$ be the function "of the first n digits in the decimal expansion of x, how many equal j." Then for Lebesgue a.e. x and all j, we have*

$$\lim_{n \to \infty} \left(\frac{n_j(x, n)}{n} \right) = \frac{1}{10} \tag{2.0.5}$$

If you think about the probabilistic model of iidrv's we discussed in Chapter 7 of Part 1, you'll realize that this theorem is related to a sharp form of the law of large numbers; see Section 7.2 of Part 1.

Theorem 2.0.5 ($\equiv$ Theorem 2.11.5). *Let $f \in L^2([0, 1], dx)$ have Fourier coefficients obeying*

$$\sum_{n=-\infty}^{\infty} |f_n^\sharp|^p < \infty \tag{2.0.6}$$

for some $p \in [1, 2)$. Then its classical Fourier series converges pointwise a.e. to f.

Classical Fourier series are discussed in Section 3.5 of Part 1. Theorem 2.0.5 is true for $p = 2$ (i.e., any $f \in L^2$) but is a much subtler result and requires tools that go beyond this book.

Most of the pointwise convergence theorems we use will involve a sequence, S_n, of operators for a Banach space, X, into an L^p space. The key will be an estimate that implies $\mu(\{x \mid \sup_n |S_n f(x)| > \lambda\})$ goes to zero as $\lambda \to \infty$ uniformly on the unit ball of X. This together with pointwise convergence on a dense set will imply a.e. pointwise convergence.

Section 2.1 will study how estimates on

$$\sup_n |S_n f(x)| \equiv S^* f \tag{2.0.7}$$

the *maximal function* of $\{S_n\}$ and dense pointwise convergence implies a.e. pointwise convergence in general. Having discussed

$$m_g(\lambda) = \{x \mid |g(x)| > \lambda\} \tag{2.0.8}$$

for the maximal function, Section 2.2 has an aside on this distribution function. Section 2.3 discusses the maximal function associated to derivatives of integrals (the Hardy–Littlewood maximal function) and applies it to several analytically important situations in Section 2.4 and 2.5. Sections 2.6–2.9 discuss ergodic averages, their maximal functions, a resulting pointwise convergence theorem (Birkhoff ergodic theorem), and applications. Sections 2.10 and 2.11 discuss two further pointwise convergence theorems.

2.1. The Magic of Maximal Functions

In this section, we'll provide the engine that will drive the rest of the chapter. We'll have a family of linear maps (we'll also have a result for positive subadditive functions), T_α, from a Banach space, X, to measurable functions on a set, Ω, with a distinguished measure, μ.

The functions will be real- or complex-valued in most applications (although there is a result for Banach space-valued functions; see Problem 2). While we've only fully discussed integration theory in the locally compact or σ-compact case, integrals play no role, so we'll only need Ω to possess a sigma algebra, Σ, with a countably additive set function, μ. We'll refer to (Ω, Σ, μ) as a *measure space*.

The indices α in T_α will lie in a directed set, I, so T_α is a net. We are going to be interested in something like $\sup_\alpha \|T_\alpha x\|$. For this to be measurable, we'll consider only three paradigmal situations:

(H1) $I = \{1, 2, \dots\}$ with $\alpha \succ \beta$ in the usual sense.

(H2) I is the rationals in $(0, 1]$ with $\alpha \succ \beta$ if $\alpha < \beta$.

(H3) I is $(0, 1]$ with $\alpha \succ \beta$ if $\alpha < \beta$. In this case, we'll need $\alpha \mapsto T_\alpha x(\omega)$ to be continuous in α for each fixed x and a.e. ω. If that holds, we have that $\sup_{\alpha \in Q, \alpha < \beta} |T_\alpha x(\omega) - y_0| = \sup_{\alpha < \beta} |T_\alpha x(\omega) - y|$ (for a.e. ω) and so the lim sup's are the same, and one only needs that the limits exist for $\alpha \in \mathbb{Q}$ and we are reduced to case (H2).

Typically, (H3) happens when Ω is a topological space, each $T_\alpha x$ is continuous for a dense set of x, and $\alpha \mapsto T_\alpha x$ uniformly continuous in $\alpha \in (0, 1]$. We use $\mathbb{K}$ for $\mathbb{R}$ or $\mathbb{C}$ for the space of potential values of $T_\alpha x(\omega)$.

For each $x \in X$ and $\alpha \in I$, we'll suppose $T_\alpha x$ is a measurable $\mathbb{K}$-valued function on Ω, so that for any $C \in (0, \infty)$,

$$\mu\{\omega \in \Omega \mid |(T_\alpha x)(\omega)| > C\} < \infty \qquad (2.1.1)$$

Since we suppose that the maps T_α are linear for all $x, y \in X$, pointwise for a.e. $\omega \in \Omega$,

$$|T_\alpha(x + y)(\omega)| \le |T_\alpha x(\omega)| + |T_\alpha y(\omega)| \qquad (2.1.2)$$

We define the *maximal function* $T^* x$ on Ω by

$$(T^* x)(\omega) = \sup_\alpha |(T_\alpha x)(\omega)| \qquad (2.1.3)$$

Theorem 2.1.1 (Magic of Maximal Functions). *In the above setup, define for each $C, \delta > 0$,*

$$M(\delta, C) = \sup_{\substack{x \in X \\ \|x\| \le \delta}} \mu\{\omega \mid T^*x(\omega) > C\} \qquad (2.1.4)$$

Suppose that

(i) *X has a dense set, X_0, so that for any $x \in X_0$, $(T_\alpha x)(\omega)$ has a limit for a.e. $\omega \in \Omega$.*

(ii) *For all $C > 0$,*

$$\lim_{\delta \downarrow 0} M(\delta, C) = 0 \qquad (2.1.5)$$

Then for every $x \in X$, $(T_\alpha x)(\omega)$ has a measurable limit, $(T_\infty x)(\omega)$, for a.e. $\omega \in \Omega$. T_∞ is a linear map of X to $\mathbb{K}$-valued measurable functions. Moreover, if $\|x_n - x\| \to 0$, then $(T_\infty x_n)(\omega) \to (T_\infty x)(\omega)$ in measure, that is, for any $\varepsilon > 0$,

$$\lim_{n \to \infty} \mu(\{\omega \mid |(T_\infty x_n)(\omega) - (T_\infty x)(\omega)| > \varepsilon\}) = 0 \qquad (2.1.6)$$

Remark. The same argument shows that (if (i) is dropped) the set of x for which $(T_\alpha x)(\omega)$ has a limit for a.e. ω is closed in X.

Proof. Since limits can be obtained through sequences, measurability is immediate, and if the limit exists, T_∞ is obviously linear. We'll also suppose $\mathbb{K} = \mathbb{R}$; one can then handle $\mathbb{K} = \mathbb{C}$ by looking at $\operatorname{Re}(T_\alpha x)$ and $\operatorname{Im}(T_\alpha x)$.

Define

$$(T_+ x)(\omega) = \limsup_\alpha (T_\alpha x)(\omega)$$
$$(T_- x)(\omega) = \liminf_\alpha (T_\alpha x)(\omega) \qquad (2.1.7)$$
$$(\delta T) = T_+ - T_-$$

If we prove for all x that $T_+ x(\omega) = T_- x(\omega)$ for a.e. ω, then the limit exists.

By linearity,

$$T_+(x) = T_+((x - y) + y) \le T_+(x - y) + T_+(y) \qquad (2.1.8)$$

and similarly,

$$T_-(x) \ge T_-(x - y) + T_-(y) \qquad (2.1.9)$$

Thus, if $T_+(y) = T_-(y)$ (i.e., if $\lim T_\alpha y$ exists),

$$\delta T(x) \le \delta T(x - y) \qquad (2.1.10)$$

(in fact, by interchanging x and $x - y$ and y and $-y$, we have equality).

Pick $x_n \to x$ in X with $x_n \in X_0$. Note that $|T_\pm| \leq T^*$, so $\delta T \leq 2T^*$. By the above, for any $\varepsilon > 0$,

$$\mu(\{\omega \mid (\delta T)(x)(\omega) > \varepsilon\}) \leq \mu(\{\omega \mid \delta T(x - x_n)(\omega) > \varepsilon\})$$
$$\leq \mu(\{\omega \mid T^*(x - x_n)(\omega) > \varepsilon/2\})$$
$$\leq M(\|x - x_n\|, \varepsilon/2) \tag{2.1.11}$$

By taking $n \to \infty$ and using (2.1.5), we conclude that, for every ε,

$$\mu(\{\mid (\delta T)(x)(\omega) > \varepsilon\}) = 0 \tag{2.1.12}$$

Taking $\varepsilon = 1, \frac{1}{2}, \frac{1}{3}, \ldots$, we see for a.e. ω that $(\delta T)(x)(\omega) = 0$, that is, the limit exists. $\qquad \square$

The above argument can be modified to allow linearity to be replaced by subadditivity, that is, $T_\alpha x$ real-valued and

$$T_\alpha(x + y) \leq T_\alpha x + T_\alpha y \tag{2.1.13}$$

Theorem 2.1.2. *The conclusion of Theorem 2.1.1 holds if T_α linear is replaced by (2.1.14).*

Proof. We have that

$$T_+(x) \leq T_+(y) + T_+(x - y), \qquad T_-(x) \geq T_-(y) - T_+(y - x) \tag{2.1.14}$$

so

$$\delta T(x) \leq T^*(x - y) + T^*(y - x) \tag{2.1.15}$$

Thus, $\delta T(x)(\omega) > \varepsilon$ implies either $T^*(x - y)(\omega) > \varepsilon/2$ or $T^*(y - x) > \varepsilon/2$, so (2.1.11) becomes

$$\mu(\{\omega \mid (\delta T)(x)(\omega) > \varepsilon\}) \leq 2M(\|x - x_n\|, \varepsilon/2) \tag{2.1.16}$$

$\qquad \square$

The key therefore is (2.1.5). In many cases, T_α is linear, so $T^*(\lambda x) = |\lambda|T^*(x)$ for any $\lambda \in \mathbb{K}$ ($= \mathbb{R}$ or $\mathbb{C}$). This means $M(\delta, C) = M(1, \delta C)$ and (2.1.5) is essentially equivalent to a bound of form

$$\mu(\{\omega \mid T^*x(\omega) > C\}) \leq Q\left(\frac{\|x\|}{C}\right) \tag{2.1.17}$$

where Q is a function obeying $\lim_{\delta \downarrow 0} Q(\delta) = 0$. A bound like (2.1.17) is called a *maximal inequality*.

Notes and Historical Remarks. Theorem 2.1.1 (in a special case) is due to Banach in 1926 [**34**]. He stated the result as: When (i) holds, the set of x for which $(T_\alpha x)(\omega)$ has a limit for a.e. ω is a closed subspace of X. Our proof implies this more general result.

There is even more magic in maximal functions than we've seen. Banach, using ideas related to his work on the uniform boundedness principle, proved that in most cases when $\mu(X) < \infty$, (2.1.5) is implied by the condition that for all $x \in X$, $(T^*x)(\omega) < \infty$ for a.e. ω (see Problem 1 for a Baire category proof of this).

This says that a.e. pointwise limits of $T_n x$ imply a maximal inequality on T^*x. So, in some sense, the approach in this chapter is the only one. While this result of Banach is illuminating, we won't be able to use it to get maximal inequalities indirectly since the only way we have to prove that $(T^*x)(\omega) < \infty$ for a.e. ω is to first prove a maximal inequality!

Stein [**775**] (see also Burkholder [**122**] and de Guzmán [**205**]) has even shown that if $X = L^p(\Omega, d\mu)$, T_n bounded from L^p to L^p, $\mu(X) < \infty$, and $1 \leq p \leq 2$, a pointwise convergence result implies a maximal function inequality of the form

$$\mu(\{\alpha \mid (T^*f)(\omega) > \lambda\}) \leq C\lambda^{-p}\|f\|_p^p$$

For these results, $\mu(X) < \infty$ is critical. If $T_n \colon L^1(\mathbb{R}, dx) \to L^1(\mathbb{R}, dx)$ by

$$(T_n f)(x) = \chi_{[n,n+1]}(x) \int f(y)\, dy$$

then $(T^*f)(x) = \chi_{[1,\infty)}(x) \int f(y)\, dy$ and the measures of sets where $(T^*f)(\omega) > \lambda$ can be infinite.

Section 7.1 of Part 1 discusses pointwise a.e. convergence.

Problems

1. This will prove a result of Banach [**34**] that a.e. finiteness of the maximal function for all $x \in X$ implies a result like (2.1.5). You'll need the Baire category theorem (see Section 5.4 of Part 1). You should suppose that you are in the sequence case and for some p in $[1, \infty)$, each T_n is a bounded linear transformation of X, a Banach space, to $L^p(\Omega, d\mu)$, where $\mu(\Omega) < \infty$. Suppose also for all $x \in X$, T^*x is a.e. finite.

   (a) Prove that $x \mapsto \sup_{j=1,\ldots,n} |(T_j x)(\omega)| \equiv (T^{*,n}x)(\omega)$ is a continuous map of X to L^p.

   (b) Prove that for each $\varepsilon > 0$,

   $$C_{n,m} = \{x \mid \mu(\{\omega \mid (T^{*,n}x)(\omega) > m\}) \leq \varepsilon\}$$

   is closed in X.

   (c) Prove that for each $\varepsilon > 0$,

   $$C_m = \{x \mid \mu(\{\omega \mid (T^*x)(\omega) > m\}) \leq \varepsilon\}$$

   is closed in X.

(d) Prove that for some $m, x_0 \in X$, $r \in (0, \infty)$,

$$\|x - x_0\| < r \Rightarrow x \in C_m$$

(*Hint*: Baire category theorem.)

(e) Prove that

$$\|x\| < 1 \Rightarrow \mu\left(\left\{\omega \,\Big|\, (T^*x)(\omega) > \frac{2m}{r}\right\}\right) \leq 2\varepsilon$$

(f) Conclude that for M given by (2.1.4),

$$\lim_{C \to \infty} M(1, C) = 0$$

and then that (2.1.5) holds.

Note. As the example in the Notes shows, this result can be false if $\mu(X) = \infty$.

2. This problem will extend Theorem 2.1.1 (the linear case) to the situation where $T_\alpha x$ is a Bochner measurable Banach space-valued function (see Section 4.18 of Part 1) with values in the same Banach space, Y. For a net $\{y_\alpha\}_{\alpha \in I}$ in Y, define (AV for asymptotic variation)

$$\mathrm{AV}(\{y_\alpha\}) = \inf_\beta \left[\sup_{\alpha > \beta} \|y_\alpha - y_\beta\|\right] \qquad (2.1.18)$$

(a) Prove a net y_α is convergent if and only if $\mathrm{AV}(\{y_\alpha\}) = 0$.

(b) Prove that for two nets $\{y_\alpha\}_{\alpha \in I}$, $\{z_\alpha\}_{\alpha \in I}$ with the same index set that

$$|\mathrm{AV}(\{y_\alpha\}_{\alpha \in I}) - \mathrm{AV}(\{z_\alpha\}_{\alpha \in I})| \leq \mathrm{AV}(\{y_\alpha - z_\alpha\}_{\alpha \in I}) \qquad (2.1.19)$$

(c) Prove that if $\{T_\alpha\}_{\alpha \in I}$ are linear maps from a Banach space, X, to Y-valued functions, and if $T_\alpha y(\omega)$ has a limit, then

$$|\mathrm{AV}(\{(T_\alpha x)(\omega)\})| \leq T^*(x - y)(\omega) \qquad (2.1.20)$$

(d) Prove a Banach space-valued analog of Theorem 2.1.1.

2.2. Distribution Functions, Weak-L^1, and Interpolation

The hero of this section is the *distribution function*, $m_g(t)$, of a measurable function, g, on a measure space, (X, Σ, μ):

$$m_g(t) = \mu(\{x \mid |g(x)| > t\}) \qquad (2.2.1)$$

Many of the maximal function bounds we'll prove later take the form (where Mg is some kind of maximal function)

$$m_{Mg}(t) \leq \frac{C\|g\|_1}{t} \qquad (2.2.2)$$

and a main goal here will be to show that if one also has a bound like

$$\|Mg\|_\infty \le \|g\|_\infty \tag{2.2.3}$$

then for all $p \in (1, \infty)$,

$$\|Mg\|_p \le C_p \|g\|_p \tag{2.2.4}$$

a special case of the Marcinkiewicz interpolation theorem, a second theme of this section.

The connection to L^p-norms is that, as we'll see,

$$\|g\|_p^p = p \int t^{p-1} m_g(t)\, dt \tag{2.2.5}$$

so a bound like (2.2.2) implies that g at worst fails to lie in L^1 by logarithmic terms (i.e., $\int_\varepsilon^{\varepsilon^{-1}} t^{-1}\, dt$ only goes to ∞ as $2\log\varepsilon$). We will introduce the term weak-L^1 for such functions.

Proposition 2.2.1. *If m_g given by (2.2.1) is finite for all $t > 0$, then m_g, as a function from $(0, \infty)$ to $[0, \infty)$, obeys*

(a) *m_g is monotone decreasing, that is, $t > s \Rightarrow m_g(t) \le m_g(s)$.*
(b) *m_g is continuous from the right, that is,*

$$\lim_{t \downarrow t_0} m_g(t) = m_g(t_0) \tag{2.2.6}$$

If $\mu(\{x \mid |g(x)| = \infty\}) = 0$ (e.g., if g is everywhere finite), then

(c) $\lim_{t \to \infty} m_g(t) = 0$ \hfill (2.2.7)

Remarks. 1. A function obeying (a)–(c) will be called rcm for *right continuous monotone*.

2. We see later (see Lemma 2.2.5) that every rcm function, f, is the distribution function of a function g on $(0, \infty)$ with dx as measure. Indeed (see Problem 1), if X is nonatomic and $\mu(X) \ge \lim_{t \downarrow 0} f(t)$, f is the distribution function of a g on X.

Proof. (a) is obvious. By the monotonicity, to conclude (b), it suffices to prove that as $n \to \infty$,

$$m_g\left(t_0 + \frac{1}{n}\right) \to m_g(t_0) \tag{2.2.8}$$

If $A_s = \{x \mid |g(x)| > s\}$, then $\bigcup_{n=1}^\infty A_{t_0 + \frac{1}{n}} = A_{t_0}$ and $A_{t_0 + \frac{1}{n+1}} \supset A_{t_0 + \frac{1}{n}}$, so (2.2.8) follows from the monotone convergence theorem. The proof of (c) is similar. □

Note that, as above, for any t_0, by the monotone convergence theorem,

$$\lim_{t \uparrow t_0} m_f(t) = \mu(\{x \mid f(x) \ge t_0\})$$
$$= m_f(t_0) + \mu(\{x \mid f(x) = t_0\}) \tag{2.2.9}$$

so points of discontinuity are precisely t's for which $\mu(\{x \mid f(x) = t_0\}) \neq 0$. This is, of course, a countable set. In particular, in (2.2.6), one can take the limit through points of continuity.

Suppose that F is a monotone function on $[0, \infty)$ with $\lim_{t \downarrow 0} F(t) = F(0) = 0$. Then there is a Stieltjes measure, $d\nu$, on $[0, \infty)$ with $\nu(\{0\}) = 0$ and

$$F(t) = \nu([0, t)) \tag{2.2.10}$$

Of course, if F is C^1, then

$$d\nu = F'(t)\, dt \tag{2.2.11}$$

The connection of m_g to $\|g\|_p$ is seen in:

Theorem 2.2.2. *Let F be a monotone function on $[0, \infty)$, with $d\nu$ given by (2.2.10). Then*

$$\int F(|f(x)|)\, d\mu(x) = \int m_f(t)\, d\nu(x) \tag{2.2.12}$$

In particular, if F is monotone and C^1, we have

$$\int F(|f(x)|)\, d\mu(x) = \int m_f(t) F'(t)\, dt \tag{2.2.13}$$

and, as a special case for $0 < p < \infty$,

$$\int |f(x)|^p\, d\mu(x) = p \int t^{p-1} m_f(t)\, dt \tag{2.2.14}$$

Remark. These equalities are intended in the sense that if one side is finite, both are, and one has equality.

Proof. The special cases are immediate from (2.2.12), so we need only prove (2.2.12). In $X \times [0, \infty)$, consider the product measure, $d\mu \otimes d\nu$. Let $G \colon X \times [0, \infty) \to [0, 1]$ be the characteristic function of $\{(x, s) \mid s < |f(x)|\} = A$. Then

$$\int \left(\int G(x, s)\, d\nu(s) \right) d\mu(x) = \int \left(\int_0^{|f(x)|} d\nu(s) \right) d\mu(x)$$

$$= \int F(|f(x)|)\, d\mu(x) \tag{2.2.15}$$

while

$$\int \left(\int G(x, s)\, d\mu(x) \right) d\nu(s) = \int \mu(\{x \mid |f(x)| > s\})\, d\nu(s)$$

$$= \int m_f(s)\, d\nu(s) \tag{2.2.16}$$

Fubini's theorem (see Theorem 4.11.6 of Part 1) implies (2.2.12). $\qquad\square$

By taking μ to be a point mass at $x \in X$ (i.e., measures with $\mu(A) = 1$ (respectively, 0) if $x \in A$ (respectively, $x \notin A$)) and $p = 1$ in (2.2.14), we obtain

Corollary 2.2.3 (Wedding-Cake Representation). *For f any bounded measurable function on (X, Σ, μ), a measure space, one has*

$$|f(x)| = \int_0^\infty \chi_{\{y \mid |f(y)| > \alpha\}}(x) \, d\alpha \qquad (2.2.17)$$

The name comes from the way f is built by piling layers on one another just like a tiered cake. Problem 16 shows how, despite its simplicity, (2.2.17) provides a useful rearrangement inequality quickly.

By decomposing signed measures as a difference, we get results for F which are C^1 but not monotone (or are locally of bounded variation) such as

Corollary 2.2.4. *If F is C^1 on $[0, \infty)$ with $F(0) = 0$, if $G(s) = \int_0^s |F'(t)| \, dt$, and if*

$$\int G(|f(x)|) \, d\mu(x) < \infty \qquad (2.2.18)$$

then both sides of (2.2.12) are finite and (2.2.12) holds.

Problem 2 and the Notes discuss a function, $Q_f(t)$, determined by m_f and called the accumulation function, so that (2.2.12) has an analog for convex functions in terms of F''.

Definition. Two functions f on (X, Σ, μ) and g on $(Y, \widetilde{\Sigma}, \nu)$ are called *equimeasurable* if and only if for all $t \in (0, \infty)$, $m_f(t) = m_g(t)$.

Clearly, by (2.2.14), if f and g are equimeasurable, then $\|f\|_{L^p(X, d\mu)} = \|g\|_{L^p(Y, d\nu)}$ for any $p \in [1, \infty)$. Recall that a function on $(0, \infty)$ is called rcm if it is monotone decreasing and (2.2.6), (2.2.7) hold.

Definition. Let f be a function on (X, μ), a measure space so that $m_f(t) < \infty$ for all $t \in (0, \infty)$. $f^* \colon (0, \infty) \to [0, \infty)$ is the unique rcm function which is equimeasurable with f. It is called the *decreasing rearrangement* of f.

This definition has an implicit existence and uniqueness result which we'll see shortly. For the next few results, it will be convenient to use $m(f)$ rather than m_f.

Lemma 2.2.5. *If g is a rcm on $(0, \infty)$, then $m(m(g)) = g$.*

Proof. We want to describe the operation of going from the graph of a rcm, g, to the graph of $m(g)$ (see Figure 2.2.1). We claim (Problem 3) that you get the graph of $m(g)$ by reflecting about the line $y = x$, with rules to handle two special cases: vertical blank parts of the graph of g (where g

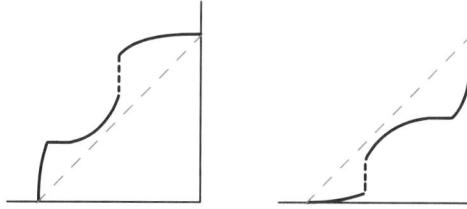

Figure 2.2.1. Construction of the graph of $m(g)$.

is discontinuous) are filled in as horizontal straight lines after the flip and horizontal pieces of the graph of g become points of discontinuity, where the value is determined by continuity from the right. This operation, when repeated, returns to the original which says $m(m(g)) = g$.   □

It is useful for understanding this lemma to consider two special cases. If g is strictly monotone, continuous, and $\lim_{t\downarrow 0} g(t) = \infty$, then so is m_g, and they are compositional inverses of each other. If $g_{\lambda,\alpha}(x) = \lambda\chi_{[0,\alpha)}(x)$, with $\chi_{[0,\alpha)}$ the characteristic function of $[0,\alpha)$, then $m(g_{\lambda,\alpha}) = g_{\alpha,\lambda}$.

Theorem 2.2.6. *If $f\colon X \to \mathbb{R}$ (or $\mathbb{C}$) is a measurable function on a space, X, with measure, μ, then*

$$f^* = m_{m_f} \tag{2.2.19}$$

is the decreasing rearrangement, that is, the unique rcm equimeasurable with f.

Remark. An equivalent formula (Problem 4), given the proof of the lemma, is

$$f^*(t) = \inf\{\alpha \mid m_f(\alpha) \le t\} \tag{2.2.20}$$

Proof. Define f^* by (2.2.19). Since m_f is an rcm, the lemma implies $m_{f^*} = m_f$, that is, f and f^* are equimeasurable and, as a distribution function, f^* is an rcm. This proves existence.

If g is an rcm and g is equimeasurable with f, then $m_g = m_f$, so $f^* = m(m_g) = g$ by the lemma. This proves uniqueness.   □

Next we turn to weak-L^p spaces. As we noted, we'll often be interested in bounds of the form (2.2.2). Thus, it is natural to consider $\sup_{t>0}(t\, m_g(t))$, or more generally:

Definition. If g is a measurable function on (X, μ) with values in $\mathbb{R}$ (or $\mathbb{C}$), we define, for $p \in [1, \infty)$,

$$\|g\|^*_{p,w} = \left[\sup_{t>0} t^p m_g(t)\right]^{1/p} \tag{2.2.21}$$

The set of g with $\|g\|_{p,w}^* < \infty$ is called *weak-L^p*, written L_w^p. Thus,

$$m_f(t) \le \frac{(\|f\|_{p,w}^*)^p}{t^p} \tag{2.2.22}$$

The name partly comes from:

Proposition 2.2.7. (a) $L^p \subset L_w^p$, *and if* $f \in L^p$,

$$\|f\|_{p,w}^* \le \|f\|_p \tag{2.2.23}$$

(b) *If* $1 \le p_1 < p < p_2 < \infty$ *and* $f \in L_w^{p_1} \cap L_w^{p_2}$, *then* $f \in L^p$ *and*

$$\|f\|_p^p \le \left(\frac{p}{p_2 - p}\right)\|f\|_{p_2,w}^{p_2} + \left(\frac{p}{p - p_1}\right)\|f\|_{p_1,w}^{p_1} \tag{2.2.24}$$

Proof. (a) $\mu\{x \mid |f(x)| \ge t\}t^p \le \|f\|_p^p$ for all t. Taking a sup over t proves (2.2.23).

(b) By (2.2.22) and (2.2.14),

$$\|f\|_p^p \le p \int_0^1 t^{p-1-p_1}\|f\|_{p_1,w}^{p_1}\, dt + p \int_1^\infty t^{p-1-p_2}\|f\|_{p_2,w}^{p_2}\, dt \tag{2.2.25}$$

which leads to (2.2.24). $\qquad\qquad\qquad\qquad\qquad\qquad\qquad\qquad\qquad\quad\square$

Remark. (2.2.24) is not optimal. One should replace 1 in the integrals in (2.2.25) by β and optimize; see Problem 5.

Example 2.2.8. Let $1 \le p < \infty$. Suppose X has disjoint sets A, B with $\mu(A) = \mu(B) = 1$. Let $f = \chi_A + 2^{-1/p}\chi_B$ and $g = \chi_B + 2^{-1/p}\chi_A$. Then

$$m_f(t) = m_g(t) = \begin{cases} 0, & t \ge 1 \\ 1, & 1 > t \ge 2^{-1/p} \\ 2, & 0 \le t < 2^{-1/p} \end{cases} \tag{2.2.26}$$

so $\|f\|_{p,w}^* = \|g\|_{p,w}^* = 1$. And we have that

$$m_{f+g}(t) = \begin{cases} 0, & t \ge 1 + 2^{-1/p} \\ 2, & 0 \le t < 1 + 2^{-1/p} \end{cases} \tag{2.2.27}$$

so $\|f + g\|_{p,w}^* = 1 + 2^{1/p} > 2 = \|f\|_{p,w}^* + \|g\|_{p,w}^*$, so $\|\cdot\|_{p,w}^*$ is not a norm. $\quad\square$

However, L_w^p is a vector space:

Proposition 2.2.9. *We have that*

$$\|\lambda f\|_{p,w} = |\lambda|\,\|f\|_{p,w} \tag{2.2.28}$$

and that

$$\|f + g\|_{p,w}^* \le 2(\|f\|_{p,w}^* + \|g\|_{p,w}^*) \tag{2.2.29}$$

In particular, L_w^p is a vector space.

Proof. (2.2.28) is immediate from $m_{\lambda f}(t) = m_f(\lambda^{-1}t)$. As for (2.2.29), we note if $|f(x) + g(x)| > t$, then either $|f(x)| > t/2$ or $|g(x)| > t/2$, so

$$m_{f+g}(t) \leq m_f(t/2) + m_g(t/2) \tag{2.2.30}$$

which implies

$$(\|f + g\|_{p,w}^*)^p \leq 2^p \left[(\|f\|_{p,w}^*)^p + (\|g\|_{p,w}^*)^p\right] \tag{2.2.31}$$

(2.2.29) then follows from the fact (Problem 6), $a, b \geq 0$, $p \geq 1$ implies that

$$(a + b)^{1/p} \leq a^{1/p} + b^{1/p} \tag{2.2.32}$$

$\square$

Remark. More generally than (2.2.30), it is useful to have

$$m_{f_1 + \cdots + f_n}(t) \leq \sum_{j=1}^{n} m_{f_j}(t/n) \tag{2.2.33}$$

While $\|\cdot\|_{p,w}^*$ is not a norm, for $p > 1$, one can define the *Calderón norm* by

$$\|f\|_{p,w} = \sup_{\{E \subset X \mid 0 < \mu(E) < \infty\}} \left[\mu(E)^{-1 + \frac{1}{p}} \int_E |f(x)| \, dx\right] \tag{2.2.34}$$

show that $\|\cdot\|_{p,w}$ is a norm, that (Problem 7)

$$\|f\|_{p,w}^* \leq \|f\|_{p,w} \leq \tfrac{p}{p-1} \|f\|_{p,w}^* \tag{2.2.35}$$

and that L_w^p is complete in this norm.

The main theorem in this section is:

Theorem 2.2.10 (Marcinkiewicz Interpolation Theorem). *Let (X, μ), (Y, ν) be σ-finite measure spaces. Let $1 \leq p_0 < p_1 \leq \infty$. Let T be a subadditive map (i.e., $|T(f + g)| \leq |Tf| + |Tg|$ pointwise) from $B_0(X)$, the bounded measurable functions on X supported on a set $S \subset X$ with $\mu(S) < \infty$, to the measurable functions on Y, so that for all $f \in B_0(X)$,*

$$\|Tf\|_{p_j,w} \leq C_j \|f\|_{p_j}, \quad j = 0, 1 \tag{2.2.36}$$

Then for $f \in B_0(X)$ and $p_0 < p < p_1$, we have that

$$\|Tf\|_p \leq C_p \|f\|_p \tag{2.2.37}$$

$$C_p = 2C_1^\theta C_0^{1-\theta} \left[\frac{p(p_1 - p_0)}{(p - p_0)(p_1 - p)}\right]^{1/p} \tag{2.2.38}$$

if $p_1 < \infty$ and

$$\theta = \left(\frac{1}{p_0} - \frac{1}{p}\right) \Big/ \left(\frac{1}{p_0} - \frac{1}{p_1}\right) \tag{2.2.39}$$

$$C_p = 2C_0^{p_0/p} C_1^{1 - p_0/p} \left[\frac{p}{p - p_0}\right]^{1/p} \tag{2.2.40}$$

if $p_1 = \infty$.

Remarks. 1. If $p_1 = \infty$, we interpret $\|\cdot\|_{p_1,w}$ as just $\|\cdot\|_\infty$.

2. This is a very special case of Marcinkiewicz's theorem; see the discussion in Section 6.1.

3. Thus, T extends to a map of $L^p(X, d\mu)$ to $L^p(Y, d\nu)$ for $p_0 < p < p_1$.

4. Since $\|g\|_{p,w}^* \leq \|g\|_p$, (2.2.37) implies a distribution function inequality

$$\mu(\{x \mid |(Tf)(x)| > \alpha\}) \leq \frac{(C_p)^p \|f\|_p^p}{\alpha^p} \tag{2.2.41}$$

Proof. We'll discuss the case $p_1 < \infty$, leaving $p_1 = \infty$ to Problem 8. For each $\alpha \in (0, \infty)$, define

$$f_{>\alpha}(x) = \begin{cases} f(x) & \text{if } |f(x)| > \alpha \\ 0 & \text{if } |f(x)| \leq \alpha \end{cases} \tag{2.2.42}$$

and

$$f_{\leq \alpha} = f - f_{>\alpha} \tag{2.2.43}$$

Since T is subadditive,

$$|Tf(x)| \leq |(Tf_{>\alpha})(x)| + |(Tf_{\leq \alpha})(x)| \tag{2.2.44}$$

Fix $\delta \in (0, \infty)$ to be chosen later. Since $w, y \geq 0$ and $w + y \geq \alpha\lambda$ implies that either $w > \frac{\alpha\lambda}{2}$ or $y \geq \frac{\alpha\lambda}{2}$, we have that

$$m_{Tf}(\lambda\alpha) \leq m_{Tf_{>\alpha}}(\lambda\alpha/2) + m_{Tf_{\leq\alpha}}(\lambda\alpha/2) \tag{2.2.45}$$

so

$$\|Tf\|_p^p = p \int_0^\infty (\lambda\alpha)^{p-1} m_{Tf}(\lambda\alpha) d(\lambda\alpha) \tag{2.2.46}$$

$$\leq I_0 + I_1 \tag{2.2.47}$$

where

$$I_0 = p\lambda^p \int_0^\infty \alpha^{p-1} m_{Tf_{>\alpha}}(\lambda\alpha/2) d\alpha,$$

$$I_1 = p\lambda^p \int_0^\infty \alpha^{p-1} m_{Tf_{\leq\alpha}}(\lambda\alpha/2) d\alpha \tag{2.2.48}$$

Since $m_{Tf_{>\alpha}}(s) \leq C_0^{p_0} \|f_{>\alpha}\|_{p_0}^{p_0} s^{-p_0}$

$$I_0 \leq C_0^{p_0} p 2^{p_0} \lambda^{p-p_0} \int_0^\infty \alpha^{p-p_0-1} \left[\int_{\{x \mid |f(x)| > \alpha\}} |f(x)|^{p_0} d\mu(x) \right] d\alpha \tag{2.2.49}$$

$$= C_0^{p_0} p 2^{p_0} \lambda^{p-p_0} \int |f(x)|^{p_0} \left(\int_0^{|f(x)|} \alpha^{p-p_0-1} d\alpha \right) d\mu(x)$$

$$= \frac{C_0^{p_0} p}{p - p_0} 2^{p_0} \lambda^{p-p_0} \|f\|_p^p \tag{2.2.50}$$

since $p > p_0$ implies the lower limit of the integral is 0.

In the same way (using $p_1 > p$, so that $\int_{|f(x)|}^{\infty} \alpha^{p-p_1-1} d\alpha = \frac{1}{p_1-p}|f(x)|^{p-p_1}$), we find that

$$I_1 \leq \frac{C_1^{p_1} p}{p_1 - p} 2^{p_1} \lambda^{p-p_1} \|f\|_p^p \tag{2.2.51}$$

If $\psi = (p - p_0)/(p_1 - p_0)$, we find the optimum choice of λ to minimize the sum of (2.2.50) and (2.2.51) is

$$\lambda = 2\left(\frac{C_1^{p_1}}{C_0^{p_0}}\right)^{1/(p_1-p_0)} \tag{2.2.52}$$

Then the constant in (2.2.50) is $2^p\, p\, C_1^{p_1 \psi}\, C_0^{p_0(1-\psi)}\, (p - p_0)^{-1}$ and in (2.2.53) $2^p\, p\, C_1^{p_1 \psi}\, C_0^{p_0(1-\psi)}\, (p_1 - p)^{-1}$. Adding these and taking a pth root, we get that C_p is $2\left[p(p_1-p_0)/(p-p_0)(p_1-p)\right]^{1/p} C_1^{p_1\psi/p} C_0^{p_0(1-\psi)/p}$. Finally, we compute

$$p_1\psi/p = \frac{p_1}{p}\frac{p - p_0}{p_1 - p_0} = \frac{p_0^{-1} - p^{-1}}{p_1^{-1} - p_0^{-1}} = \theta$$

and similarly

$$p_0(1 - \psi)/p = 1 - \theta$$

so that $C_p = 2\left[p(p_1 - p_0)/(p - p_0)(p_1 - p)\right]^{1/p} C_1^{\theta} C_0^{1-\theta}$ $\square$

This same idea can be used to prove (Problem 9) that if the hypotheses of Theorem 2.2.10 hold with $p_0 = 1$, $\mu(X) < \infty$, and

$$\int |f(x)|(1 + \log_+|f(x)|)\, d\mu(x) < \infty \tag{2.2.53}$$

(where $\log_+(y) = \max(\log(y), 0)$), then $Tf \in L^1$.

As a final result, we want to prove continuity results for m_f in f. We begin with a formal definition of convergence in measure, a notion we've used (as "convergence in probability" for probability measures) in Part 1; see Section 7.1 of that Part.

Definition. If f_n, f are measurable functions on (X, Σ, μ), we say f_n converges *in measure* to f (written $f_n \xrightarrow{\mu} f$) if and only if for every $t > 0$,

$$m_{f-f_n}(t) \to 0$$

as $n \to \infty$.

Theorem 2.2.11. *If either*

(i) $f_n \to f$ in $\|\cdot\|_{L^p(X,d\mu)}$ for some $p \in [1, \infty]$ or
(ii) $\mu(X) < \infty$ and $f_n(x) \to f(x)$ for a.e. x,

then $f_n \xrightarrow{\mu} f$.

Proof. (i) If $p = \infty$ and $\|f - f_n\|_\infty < t_0$, then $m_{f-f_n}(t_0) = 0$. If $p < \infty$,

$$\mu(\{x \mid |f_n(x) - f(x)| > t\}) \le t^{-p}\|f - f_n\|_p^p \qquad (2.2.54)$$

so $\|f - f_n\|_p \to 0 \Rightarrow m_{f-f_n}(t) \to 0$.

(ii) Fix $t > 0$. Let χ_n be the characteristic function of $\{x \mid |f(x) - f_n(x)| > t\}$. By hypothesis, $\chi_n(x) \to 0$ for a.e. x. Since $|\chi_n| \le 1$ and $1 \in L^1$ (on account of $\mu(X) < \infty$), $\|\chi_n\|_1 \to 0$ by the dominated convergence theorem. $\qquad\square$

Example 2.2.12. This example shows that (ii) can fail if $\mu(X) = \infty$. Let $X = \mathbb{R}$, and f_n the characteristic function of $[n, n+1]$. Then $f_n(x) \to 0$ for all x, but f_n does not converge to 0 in measure for $d\mu = dx$. $\qquad\square$

Theorem 2.2.13. *Let* $f_n \xrightarrow{\mu} f$. *Then*

(a) *For all* t_0,

$$m_f(t_0) \le \liminf_{n \to \infty} m_{f_n}(t_0) \qquad (2.2.55)$$

(b) *If* m_f *is continuous at* t_0, *then*

$$m_f(t_0) = \lim_{n \to \infty} m_{f_n}(t_0) \qquad (2.2.56)$$

Proof. Let $t < s$. Then for any g and h,

$$\mu(\{x \mid |h(x)| > s\}) \le \mu(\{x \mid |g(x)| > t\}) + \mu(\{x \mid |h(x) - g(x)| > s - t\}) \qquad (2.2.57)$$

since if $|g(x)| \le t$ and $|h(x) - g(x)| \le s - t$, then $|h(x)| \le s$. It follows that for any $\delta > 0$,

$$m_f(t_0 + \delta) \le m_{f_n}(t_0) + m_{f-f_n}(\delta) \qquad (2.2.58)$$
$$m_{f_n}(t_0) \le m_f(t_0 - \delta) + m_{f-f_n}(\delta) \qquad (2.2.59)$$

Since $m_{f-f_n}(\delta) \to 0$, we conclude for all $\delta > 0$,

$$m_f(t_0 + \delta) \le \liminf_{n \to \infty} m_{f_n}(t_0) \qquad (2.2.60)$$
$$m_f(t_0 - \delta) \ge \limsup_{n \to \infty} m_{f_n}(t_0) \qquad (2.2.61)$$

By (2.2.6), (2.2.60) implies (2.2.55). If m_f is continuous at t_0, $\lim_{\delta \downarrow 0} m_f(t_0 - \delta) = m_f(t_0)$, so (2.2.61) implies $m_f(t_0) \ge \limsup m_{f_n}(t_0)$. This plus (2.2.55) implies (2.2.56). $\qquad\square$

Notes and Historical Remarks. The earliest appearance of estimates of the form (2.2.2) (where Mg is not a maximal function but a conjugate harmonic function; see the Notes to Section 5.7) is in Kolmogorov [**480**] in 1925.

In Lieb–Loss [533], (2.2.17) is called the layer-cake representation. Because the layers in the cake in (2.2.17) can be unequal in size, I prefer wedding cake.

The formal definition of weak-L^p goes back to Lorentz in 1950 [546, 547]. He defined spaces $M(\alpha)$, $0 < \alpha < 1$. $L_w^p((0,1), dx)$ is his $M(\alpha)$ if $\alpha = 1 - p^{-1}$. His work [546] has the norm (2.2.34), named after later work of Calderón [128, 130].

The Calderón norm is expressed in terms of

$$f^{**}(t) = \sup\left[\frac{1}{t}\int_E |f(x)|d\mu(x) \mid \mu(E) \le t\right] \tag{2.2.62}$$

Indeed

$$\|f\|_{p.w} = \sup_t \left[t^{1/p} f^{**}(t)\right] \tag{2.2.63}$$

A little thought, using the fact that f and f^* are equimeasurable, shows that

$$f^{**}(t) = \frac{1}{t}\int_0^t f^*(s)ds \tag{2.2.64}$$

This implies (because f^* is monotone decreasing) that

$$f^*(t) \le f^{**}(t) \tag{2.2.65}$$

For more on rearrangements (spherical rearrangements as well as f^*), see the books of Kawohl [455] and Kesavan [463].

The usefulness of running averages of the symmetric rearrangements goes back to Muirhead [591] in 1902 so f^{**} is sometimes called the *Muirhead maximal function*. The key reason f^{**} is useful is that unlike f^*, it is subadditive, i.e.,

$$(f + g)^{**}(t) \le f^{**}(t) + g^{**}(t) \tag{2.2.66}$$

as is immediate from (2.2.62) (Problem 19). This is why $\|f\|_{p,w}$ is a norm.

Section 6.1 has a discussion of the history of the full-blown Marcinkiewicz theorem, which says if T is subadditive and bounded from L^{p_j} to $L_w^{q_j}$, $j = 0, 1$, and $p_0 < p_1$, $p_j \le q_j$, then T is bounded from L^{p_θ} to L^{q_θ}, where $\theta \in (0,1)$, $p_\theta^{-1} = (1-\theta)p_0^{-1} + \theta p_1^{-1}$, $q_\theta^{-1} = (1-\theta)q_0^{-1} + \theta q_1^{-1}$ (see Theorems 6.0.2 and 6.1.6).

The use of $L^1 \log L$ to get boundedness to L^1 as in Problem 9 goes back to Zygmund [875] for conjugate harmonic functions, although he did not get his results by estimates of distribution functions. $L^1 \log L$ is an example of Orlicz space, a generalization of L^p, where $F(y) = |y|^p$ is replaced by certain more general convex functions. The theory goes back to Orlicz [620]. For monographs on the subject, see [756, Ch. 2] or [489, 867].

The function f^* for maximal functions goes back to Hardy–Littlewood [360] (see also the book of Hardy–Littlewood–Pólya [363]). They noted the functional inverse of m_f is equimeasurable with f. Its use to define L^p_w is due to Lorentz [**546, 547**].

The *accumulation function* of f is defined

$$Q_f(t) \equiv \int_t^\infty m_f(s)\, ds = \int (|f(x)| - t)_+\, d\mu(x) \qquad (2.2.67)$$

where $y_+ = \max(y, 0)$. The second equality comes from (2.2.12). It is a convex function of t. If F is a C^2 convex function with $F(0) = 0$ and $F'(0) \geq 0$, then (Problem 2)

$$\int F(|f(x)|)\, d\mu(x) = F'(0)Q_f(0) + \int_0^\infty Q_f(t)F''(t)\, dt \qquad (2.2.68)$$

and, in particular, if $p > 1$,

$$\|f\|_p^p = \int_0^\infty p(p-1)t^{p-2}Q_f(t)\, dt$$

If $\mu(X) < \infty$, convergence in measure is given by a metric (see Problem 12).

Problems

1. A σ-finite measure, μ, on (X, Σ) is called nonatomic if and only if for any $A \subset X$, with $A \in \Sigma$, and $\alpha \in (0, \mu(A))$, there is $B \subset A$, with $B \in \Sigma$, and $\mu(B) = \alpha$.

   (a) Let X be a locally compact, σ-compact metric space. Prove a Baire measure is nonatomic if and only if $\mu(\{x\}) = 0$ for all x.

   (b) Let g be a rcm and (X, Σ, μ) a nonatomic measure space with $\mu(X) \geq \lim_{t\downarrow 0} g(t)$. Prove there exist sets $\{A_t\}_{t\in[0,\infty)}$ in Σ so that (i) $t > s \Rightarrow A_t \subset A_s$; (ii) $A_t = \bigcup_{n=1}^\infty A_{t+\frac{1}{n}}$ for any t; (iii) $\mu(A_t) = g(t)$. (*Hint*: Do it first inductively for $t = k/2^n$.)

   (c) Prove there is a measurable function, f, from X to $[0, \infty)$ with $\{x \mid f(x) > t\} = A_t$ so $m_f = g$. (*Hint*: Look at the proof of Urysohn's lemma in Theorem 2.2.4 of Part 1.)

   Remark. The definition of nonatomic is *not* standard. The usual one is weaker and says only that if $\mu(A) > 0$, there is $B \in \Sigma$ with $0 < \mu(B) < \mu(A)$. It is an interesting exercise to show this implies our stronger-looking definition.

2. Prove (2.2.68).

3. Prove the graph construction relating the graphs of g and m_g described in the proof of Lemma 2.2.5.

4. Prove (2.2.20).

5. Improve (2.2.24) to

$$\|f\|_p \leq A(\|f\|^*_{p_1,w})^\alpha (\|f\|^*_{p_2,w})^{1-\alpha} \qquad (2.2.69)$$

where

$$A = \left(\frac{p}{p - p_1} + \frac{p}{p_2 - p}\right)^{1/p}, \qquad \alpha = \frac{p^{-1} - p_2^{-1}}{p_1^{-1} - p_2^{-1}} \qquad (2.2.70)$$

(*Hint*: Replace 1 in (2.2.25) by β and optimize in β.)

6. Prove (2.2.32) for $a, b \geq 0$, $p \geq 1$. (*Hint*: In ℓ^p, consider $x = (a^{1/p}, 0, 0, \dots)$, $y = (0, b^{1/p}, 0, \dots)$.)

7. This problem will prove (2.2.35).

   (a) If $E_t = \{x \mid |f(x)| > t\}$, prove $\int_{E_t} |f(x)| \, d\mu(x) \geq t m_f(t)$ and then that $\mu(E_t)^{-1+\frac{1}{p}} \int_{E_t} |f(x)| \, d\mu(x) \geq t m_f(t)^{1/p}$. Conclude that $\|f\|^*_{p,w} \leq \|f\|_{p,w}$.

   (b) Given any E and f, let $g = f\chi_E$. Prove that

   $$m_g(t) \leq \min(m_f(t), \mu(E)) \leq \min((\|f\|^*_{p,w} t^{-1})^p, \mu(E)) \qquad (2.2.71)$$

   (c) Using (2.2.71), prove that

   $$\int_0^\infty m_g(t) \, dt \leq \frac{p}{p-1} \mu(E)^{1-\frac{1}{p}} \|f\|^*_{p,w} \qquad (2.2.72)$$

   and conclude that $\|f\|_{p,w} \leq p(p-1)^{-1} \|f\|^*_{p,w}$.

8. This problem will lead the reader through the proof of Theorem 2.2.10 when $p_1 = \infty$. Suppose first that $C_1 = 1$.

   (a) Prove that

   $$|(Tf)| \leq |Tf_{>\alpha\{2\}}| + \alpha/2 \qquad (2.2.73)$$

   (b) Prove that

   $$m_{Tf}(\alpha) \leq m_{Tf_{>\alpha/2}}(\alpha/2) \qquad (2.2.74)$$

   (c) Prove that

   $$m_{Tf}(\alpha) \leq C_0^{p_0} 2^{p_0} \alpha^{-p_0} \|f_{>\alpha/2}\|_{p_0}^{p_0} \qquad (2.2.75)$$

   (d) Prove that

   $$\|Tf\|_p^p \leq C_0^{p_0} 2^{p_0} \int_0^\infty p \alpha^{p-p_0-1} \left[\int_{\{x||f(x)|>\alpha/2\}} |f(x)|^{p_0} d\mu x\right] d\alpha \qquad (2.2.76)$$

   (e) Prove that

   $$\|Tf\|_p^p \leq C_0^{p_0} 2^p p(p - p_0)^{-1} \|f\|_p^p \qquad (2.2.77)$$

   and conclude (2.2.40) when $C_1 = 1$.

   (f) Prove the result for $C_1 \neq 0$. (Hint: Look at T/C_1.)

9. This problem will prove that if the hypotheses of Theorem 2.2.10 hold with $p_0 = 1, p_1 = \infty$ and (2.2.53) holds, then

$$\int_1^\infty m_{Tf}(t)\, dt < \infty \qquad (2.2.78)$$

so if $\mu(X) < \infty$, $Tf \in L^1$.

(a) Prove $\int_0^\infty m_{Tf_{\leq t}}(t)\, dt \leq C\|f\|_1$.

(b) Prove that

$$\int_{1/2}^\infty m_{Tf_{>t}}(t)\, dt \leq C\|f\|_1 + C \int_{1/2}^\infty t^{-1}\left(\int_t^\infty m_f(s)\, ds\right) dt \qquad (2.2.79)$$

and conclude the integral on the left of (2.2.79) is finite.

(c) Prove (2.2.78).

10. A sequence f_n is called *Cauchy in measure* if $\forall \delta > 0, \forall \varepsilon > 0, \exists N$ so that $n, m \geq N \Rightarrow$

$$\mu(\{x \mid |f_n(x) - f_m(x)| > \delta\}) < \varepsilon \qquad (2.2.80)$$

(a) Given such a sequence, prove that there is a subsequence g_j so that

$$\mu(\{x \mid |g_{j+1}(x) - g_j(x)| \geq 2^{-j}\}) \leq 2^{-j} \qquad (2.2.81)$$

(b) If the set in (2.2.81) is called A_j, show if $x \in \bigcap_{j=J}^\infty (X \setminus A_j)$, then

$$\lim_{j \to \infty} g_j(x) \equiv f(x) \qquad (2.2.82)$$

exists.

(c) Conclude that (2.2.82) holds for μ a.e. x.

(d) Show that g_j converges in measure to f and then that f_n converges to f in measure.

11. Let $f_n \xrightarrow{\mu} f$ and suppose for some $g \in L^p(X, d\mu)$, $|f_n(x)| \leq g(x)$ for all x. Prove that $\|f_n - f\|_p \to 0$. (*Hint*: Use Problem 10 and the fact that if M is a metric space, m_n a sequence in m, $m_\infty \in M$ fixed, and every subsequence has a subsubsequence converging to m_∞, then $m_n \to m_\infty$ in M.)

12. Suppose $\mu(X) < \infty$. For any measurable functions f, g on X, define

$$\rho(f, g) = \int \min(1, |f(x) - g(x)|)\, d\mu(x) \qquad (2.2.83)$$

(a) For $t < 1$, prove that $m_{f-g}(t) \leq t^{-1}\rho(f, g)$.

(b) For $t < 1$, prove that $\rho(f, g) \leq t\mu(X) + m_{f-g}(t)$.

(c) Prove $f_n \xrightarrow{\mu} f$ if and only if $\rho(f_n, f) \to 0$.

(d) Prove that the metric space is complete (see Problem 10).

Remark. That convergence in measure is given by a metric goes back at least to Fréchet [**287**].

13. Let $f_n \xrightarrow{\mu} f$, $g_n \xrightarrow{\mu} g$.

    (a) Prove that $f_n + g_n \xrightarrow{\mu} f + g$.

    (b) If $\mu(X)$ is finite, prove that $f_n g_n \xrightarrow{\mu} fg$.

    (c) Find f_n, f, g_n, g on $\mathbb{R}$ so $f_n g_n$ does not converge to fg in measure.

14. (Hölder's inequality for weak-L^p) If $1 \leq p, q, r < \infty$ and $p^{-1} + q^{-1} = r^{-1}$ and if $f \in L_w^p$, $g \in L_w^q$, prove that $fg \in L_w^*$ and

$$\|fg\|_{r,w}^* \leq r^{-1/r} p^{1/p} q^{1/q} \|f\|_{p,w}^* \|g\|_{q,w}^*$$

    (*Hint*: Use $m_{fg}(t) \leq m_f(t/s) + m_g(s)$ and solve the minimization problem for $x^p + y^q$ subject to $xy = 1/t$.)

15. Prove that $\|\sum_{j=1}^n f_j\|_{p,w}^* \leq n(\sum_{j=1}^n \|f_j\|_{p,w}^*)$.

16. (a) Suppose (X, Σ, μ), (Y, Σ', ν) are two measure spaces, $A, B \in \Sigma$, $A', B' \in \Sigma'$, and $\mu(A) = \nu(A')$, $\mu(B) = \nu(B')$. Suppose also either $B' \subset A'$ or $A' \subset B'$. Prove that $\mu(A \cap B) \leq \nu(A' \cap B')$.

    (b) If f, g are measurable functions on X and f^*, g^* their decreasing rearrangements, prove that

$$\left| \int f(x) g(x) \, d\mu \right| \leq \int_0^\infty f^*(t) g^*(t) \, dt \tag{2.2.84}$$

    (*Hint*: Use the wedding-cake representation and part (a) of this problem.)

17. (a) If A, B are balls in $\mathbb{R}^\nu$ centered at 0, prove that $\mathbf{x} \mapsto |(A + \mathbf{x}) \cap B|$ is spherically symmetric in x and decreasing as $|x|$ increases.

    (b) Let f and g be two functions on $\mathbb{R}^\nu$ with

$$|x| \geq |y| \Rightarrow 0 \leq f(x) \leq f(y) \tag{2.2.85}$$

    and similarly for g (these are called spherically symmetric decreasing functions); notice that (2.2.85) implies $|x| = |y| \Rightarrow f(x) = f(y)$. Prove that $f * g$ is spherically symmetric and decreasing. (*Hint*: Use part (a) and the wedding-cake representation.)

18. Prove that
    (a) $|f| \geq |g| \Rightarrow m_f \geq m_g$
    (b) for λ positive, $m_{\lambda f}(t) = m_f(t/\lambda)$
    (c) $m_{f+g}(t_1 + t_2) \leq m_f(t_1) + m_g(t_2)$ $\hspace{2em}$ (2.2.86)
    (d) $m_{\max(f,g)}(t) \leq m_f(t) + m_g(t)$ $\hspace{2em}$ (2.2.87)

19. Prove that

    (a) $|f| \geq |g| \Rightarrow f^* \geq g^*$ (2.2.88)

    (b) For λ positive, $(\lambda f)^* = \lambda f^*$ (2.2.89)

    (c) $(f + g)^*(t_1 + t_2) \leq f^*(t_1) + g^*(t_2)$ (2.2.90)

    (d) $m_f(f^*(t)) \leq t$;   when does equality fail?

    (e) $f^*(m_f(t)) \leq t$;   when does equality fail?

    Hint: You'll need to use mainly (2.2.20) but occasionally (2.2.19). You'll need (d) to prove (e). In that case, define

    $$t_0 = m_{f \cdot g}(f^*(t_1) + g^*(t_2)) \qquad (2.2.91)$$

    and prove that

    $$t_0 \leq m_f(f^*(t_1)) + m_g(g^*(t_2)) \leq t_1 + t_2 \qquad (2.2.92)$$

    Then use monotonicity of $(f + g)^*$.

20. Prove that (f^{**} given by (2.2.62)/(2.2.64))

    (a) $|f| \geq |g| \Rightarrow f^{**} \geq g^{**}$ (2.2.93)

    (b) For λ positive, $(\lambda f)^{**} = \lambda f^{**}$ (2.2.94)

    (c) $(f + g)^{**}(t) \leq f^{**}(t) + g^{**}(t)$ (2.2.95)

Remark. You'll need to use both (2.2.62) and (2.2.64). f^{**} is useful precisely because (2.2.95) holds.

2.3. The Hardy–Littlewood Maximal Inequality

The main object of study in this section is the *Hardy–Littlewood maximal function* defined by

$$(M_{\mathrm{HL}}f)(x) = \sup_r \frac{1}{|B_r(x)|} \int_{B_r(x)} |f(y)| \, d^\nu y \qquad (2.3.1)$$

for functions f, locally L^1 on $\mathbb{R}^\nu$ (i.e., $\int_{B_r(0)} |f(x)| \, d^\nu x < \infty$ for all r). Here $B_r(x)$ is the ball of radius r centered at x. We have in mind the ball in the Euclidean norm but, as we'll note, the inequality below with the same constant and proof holds for balls in any norm, so, for example, taking $\|x\| = \max_{j=1,\dots,\nu}|x_j|$, we get a maximal function based on "hypercubes." Our main result in the section is:

Theorem 2.3.1 (Hardy–Littlewood Maximal Inequality). *For any $f \in L^1(\mathbb{R}^\nu)$, we have ($|\cdot| = $ Lebesgue measure)*

$$|\{x \mid M_{\mathrm{HL}}f(x) > \alpha\}| \leq \frac{3^\nu}{\alpha} \|f\|_{L^1(\mathbb{R}^\nu, d^\nu x)} \qquad (2.3.2)$$

Remarks. 1. The 3^ν constant is not optimal (indeed, in $\nu = 1$, we'll get a bound with 2 rather than 3 below) and the bound with any constant C is given this name.

2. If f is a nonzero function, one has (Problem 1) that

$$(M_{\mathrm{HL}}f)(x) \geq C|x|^{-\nu} \tag{2.3.3}$$

so $M_{\mathrm{HL}}f$ is never in $L^1(\mathbb{R}^\nu, d^\nu x)$.

3. It is obvious that $\|M_{\mathrm{HL}}f\|_\infty \leq \|f\|_\infty$ and the M_{HL} is subadditive. Thus, by Theorem 2.2.10 and the fact that M_{HL} is subadditive,

$$\|M_{\mathrm{HL}}f\|_p \leq C_{p,\nu}\|f\|_p \tag{2.3.4}$$

for $1 < p < \infty$.

When $\nu = 1$, there are three other maximal functions often considered:

$$(M_{\mathrm{L}}f)(x) = \sup_{a>0} \frac{1}{a} \int_{x-a}^{x} |f(y)|\, dy \tag{2.3.5}$$

$$(M_{\mathrm{R}}f)(x) = \sup_{a>0} \frac{1}{a} \int_{x}^{x+a} |f(y)|\, dy \tag{2.3.6}$$

$$(M_U f)(x) = \sup_{a<x<b} \frac{1}{b-a} \int_{a}^{b} |f(y)|\, dy \tag{2.3.7}$$

(L, R, and U are for left, right, and uncentered, respectively.)

One has a number of inequalities among them (Problem 2):

$$M_{\mathrm{HL}} \leq M_U, \quad M_U = \max(M_{\mathrm{L}}, M_{\mathrm{R}}), \quad M_{\mathrm{HL}} \leq \tfrac{1}{2}(M_{\mathrm{L}} + M_{\mathrm{R}}), \\ M_{\mathrm{L}} \leq 2M_{\mathrm{HL}}, \quad M_{\mathrm{R}} \leq 2M_{\mathrm{HL}}, \quad M_U \leq 2M_{\mathrm{HL}} \tag{2.3.8}$$

that show, except for constants, Hardy–Littlewood inequalities for any of these imply them for all (see Problem 18 of Section 2.2). Different proofs of Theorem 2.3.1 prove maximal inequalities directly on different of these. Indeed, the two proofs in the text directly control M_{HL} and M_U, respectively, and the one in the problems for M_{L} and M_{R}.

To understand why we'll need a "covering lemma," let's start trying to prove (2.3.2). If $M_{\mathrm{HL}}f(x) > \alpha$, there is an r_x so that

$$|B_{r_x}(x)| \leq \alpha^{-1} \int_{B_r(x)} |f(y)|\, d^\nu y \tag{2.3.9}$$

Pretend the $B_{r_x}(x)$ are all disjoint. Since $E_\alpha = \{x \mid M_{\mathrm{HL}}f(x) > \alpha\}$ is covered by the $B_{r_x}(x)$, we would have

$$|E_\alpha| \leq \sum_x |B_{r_x}(x)| \leq \alpha^{-1} \int_{\cup B_{r_x}(x)} |f(y)|\, d^\nu y \leq \alpha^{-1} \int |f(y)|\, d^\nu y$$

Of course, the $B_{r_x}(x)$ are uncountable and certainly not disjoint. This would work, for example, if we could find a disjoint subfamily which still covered E_α. Even that is too much to ask, but if we can find a disjoint family that cover a substantial fraction of E_α, we'd get a result—and that is what the following covering lemma asserts.

Theorem 2.3.2 (Vitali Covering Lemma). *Let $\{B_{r_j}(x_j)\}_{j=1}^k$ be a finite set of open balls in a metric space, X. Then there exists a subset $S \subset \{1,\ldots,k\}$ so that*

(i) $j, \ell \in S, \; j \neq \ell \Rightarrow B_{r_j}(x_j) \cap B_{r_\ell}(x_\ell) = \emptyset$

(ii) $\displaystyle \bigcup_{j=1}^k B_{r_j}(x_j) \subset \bigcup_{j \in S} B_{3r_j}(x_j)$ $\hfill (2.3.10)$

In particular, if $X = \mathbb{R}^\nu$ and the metric comes from a norm on X,

$$\left| \bigcup_{j=1}^k B_{r_j}(x_j) \right| \leq 3^\nu \sum_{j \in S} |B_{r_j}(x_j)| \qquad (2.3.11)$$

Proof. By reordering the balls if need be, we can suppose $r_1 \geq r_2 \geq \cdots \geq r_k$. Construct S inductively as follows: $1 \in S$. If $B_{r_{j+1}}(x_{j+1}) \cap [\bigcup_{\ell \leq j, \ell \in S} B_{r_\ell}(x_\ell)] = \emptyset$, put $j+1 \in S$. Otherwise, $j+1 \notin S$. By construction, the $\{B_j(x_j)\}_{j \in S}$ are disjoint.

Suppose $j+1 \notin S$. Then for some $\ell \leq j$, $B_{r_{j+1}}(x_{j+1}) \cap B_{r_\ell}(x_\ell) \neq \emptyset$. This implies $|x_{j+1} - x_\ell| \leq r_{j+1} + r_\ell$ and thus, since $r_\ell \geq r_{j+1}$ (see Figure 2.3.1), $B_{r_{j+1}}(x_{j+1}) \subset B_{2r_{j+1}+r_\ell}(x_\ell) \subset B_{3r_\ell}(x_\ell)$. Thus, inductively $B_{r_j}(x_j) \subset$ RHS of (2.3.10), so (2.3.10) holds.

The reader will prove in Problem 3 that in $\mathbb{R}^\nu$, in the metric defined by any norm

$$|B_{\lambda r}(x)| = \lambda^\nu |B_r(x)| \qquad (2.3.12)$$

for all $\lambda, r \in (0, \infty)$ and $x \in \mathbb{R}^\nu$. Thus, (2.3.10) implies (2.3.11). $\hfill \square$

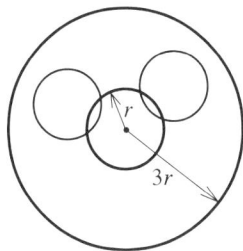

Figure 2.3.1. Graphical proof of (2.3.10).

Proof of Theorem 2.3.1. Let

$$E_\alpha = \{x \mid (M_{\mathrm{HL}}f)(x) > \alpha\} \tag{2.3.13}$$

For each $x \in E_\alpha$, there is r_x so that (2.3.9) holds. Clearly, $E_\alpha \subset \bigcup_x B_{r_x}(x)$. Let K be compact with $K \subset E_\alpha$. Since $\{B_{r_x}(x)\}_{x \in E_\alpha}$ is an open cover of K, we can find $x_1, \ldots, x_k$ so

$$K \subset \bigcup_{j=1}^k B_{r_{x_j}}(x_j) \subset \bigcup_{j \in S} B_{3r_{x_j}}(x_j) \tag{2.3.14}$$

where, if $j, \ell \in S$, $j \neq \ell$, we have $B_{r_{x_j}}(x_j) \cap B_{r_{x_\ell}}(x_\ell) = \emptyset$. The second inclusion is by the Vitali covering lemma.

Thus,

$$|K| \leq 3^\nu \sum_{j \in S} |B_{r_{x_j}}(x_j)| \tag{2.3.15}$$

$$\leq 3^\nu \alpha^{-1} \sum_{j \in S} \int_{B_{r_{x_j}}(x_j)} |f(y)| \, d^\nu y \tag{2.3.16}$$

$$= 3^\nu \alpha^{-1} \int_{\cup_{j \in S} B_{r_{x_j}}(x_j)} |f(y)| \, d^\nu y \tag{2.3.17}$$

$$\leq 3^\nu \alpha^{-1} \|f\|_1 \tag{2.3.18}$$

Here (2.3.15) holds by (2.3.14) and $|B_{3r}(x)| = 3^\nu |B_r(x)|$, (2.3.16) comes from (2.3.9), and (2.3.17) from the disjointness of the $\{B_{r_j}(x_j)\}_{j \in J}$. Since $|\cdot|$ is a regular measure, $|E_\alpha| = \sup_{K \subset E_\alpha, K \text{ compact}} |K|$ and we have (2.3.2). $\qquad \square$

Remarks. 1. In $\nu = 1$, if one uses (2.3.8) and Theorem 2.3.1, one gets weak inequalities for $M_{\mathrm{R}}, M_{\mathrm{L}}, M_{\mathrm{U}}$ with a 6 as the constant. However, by following the proof of the theorem, one can get 3. As we'll discuss in the Notes, the optimal constant for M_{R} and M_{L} is 1 and for M_{U} is 2.

2. With no change in proof, we can replace f by a finite measure, μ, defining

$$(M_{\mathrm{HL}}\mu)(x) = \sup_r \frac{1}{|B_r(x)|} |\mu|(B_r(x)) \tag{2.3.19}$$

We will also want an analog of Theorem 2.3.1 for $d^\nu x$ replaced by a general finite measure on $\mathbb{R}$:

Theorem 2.3.3. *Let μ be a finite Baire measure on $\mathbb{R}$. Let $f \in L^1(\mathbb{R}, d\mu)$. Define*

$$M_{\mathrm{U}}^{(\mu)}(f)(x) = \sup_{a < x < b} \frac{1}{\mu((a,b))} \int_{(a,b)} |f(y)| \, d\mu(y) \tag{2.3.20}$$

Then

$$\mu(\{x \mid (M_{\mathrm{U}}^{(\mu)}f)(x) > \alpha\}) \leq \frac{2}{\alpha} \|f\|_{L^1(\mathbb{R}, d\mu)} \tag{2.3.21}$$

Remark. This result implies the same result for any Baire measure on $\mathbb{R}$; for if $\mu_R(A) = \mu(A \cap [-R, R])$, μ_R is finite. If $f \in L^1(\mathbb{R}, d\mu)$ has compact support, then $(2.3.20)/(2.3.21)$ for μ_R implies the same for μ, and then a density argument yields $(2.3.21)$ for all $f \in L^1(\mathbb{R}, d\mu)$.

Notice that the constant 2 is better than 3 when $d\mu = dx$! There is an analog of this for $\mathbb{R}^\nu$, but the required Besicovitch covering lemma is somewhat more involved and will only use the $\nu = 1$ result, so we settle for describing the Besicovitch result in the Notes.

The Vitali lemma uses the translation invariance and scaling covariance of dx. We'll instead use the following:

Theorem 2.3.4 (Croft–Garsia Covering Lemma). *Let $\{I_j\}_{j=1}^k$ be a finite set of intervals in $\mathbb{R}$. Then there exist $S_1, S_2 \subset \{1, \dots, k\}$ so that for $m = 1, 2$,*

(i) $j, \ell \in S_m,\ j \neq \ell \Rightarrow I_j \cap I_\ell = \emptyset$ \hfill (2.3.22)

(ii) $\displaystyle\bigcup_{j=1}^k I_j = \bigcup_{j \in S_1 \cup S_2} I_j$ \hfill (2.3.23)

Proof. We first claim there exists $J \subset \{1, \dots, k\}$ so that

$$\bigcup_{j=1}^k I_j = \bigcup_{j \in J} I_j \tag{2.3.24}$$

and so that J has property P: for all $\ell \in J$, I_ℓ is not a subset of $\bigcup_{j \in P, j \neq \ell} I_j$. For if $\{1, \dots, k\}$ doesn't have property P, pick ℓ so I_ℓ is a subset of $\bigcup_{j \neq \ell} I_j$ and drop it. If $\{1, \dots, k\} \setminus \{\ell\}$ doesn't have property P, drop an $I_{\ell'}$ from the collection. Keep doing it until the remaining collection has property P or only one interval is left.

Order the indices $j_1, \dots, j_n$ in J so $\inf I_{j_\ell} \leq \inf I_{j_{\ell+1}}$. Let

$$I_{j_\ell} = (\alpha_\ell, \beta_\ell) \tag{2.3.25}$$

By the choice of ordering,

$$\alpha_1 \leq \alpha_2 \leq \cdots \leq \alpha_n \tag{2.3.26}$$

If $\beta_\ell \geq \beta_{\ell+1}$, $I_{j_{\ell+1}} \subset I_{j_\ell}$, violating property P, so

$$\beta_1 \leq \beta_2 \leq \cdots \leq \beta_n \tag{2.3.27}$$

If $I_{j_\ell} \cap I_{j_{\ell+2}} \neq \emptyset$, then $I_{j_\ell} \cup I_{j_{\ell+2}} = (\alpha_\ell, \beta_{\ell+2}) \supset (\alpha_{\ell+1}, \beta_{\ell+1}) = I_{j_{\ell+1}}$, violating property P. We conclude, for $\ell = 1, 2, \dots, n-2$, $I_{j_\ell} \cap I_{j_{\ell+2}} = \emptyset$. Thus, if $J_1 = \{j_1, j_3, j_5, \dots\}$, $J_2 = \{j_2, j_4, j_6, \dots\}$, we have $(2.3.22)$. $\qquad\square$

Proof of Theorem 2.3.3. Given the Croft–Garsia lemma, the proof is essentially identical to Theorem 2.3.1. Let $E_\alpha = \{x \mid M_U^{(\mu)} f(x) > \alpha\}$. Let $K \subset E_\alpha$ be compact. By definition of E_α and compactness of K, we can find intervals $\{I_j\}_{j=1}^k$ so that

$$K \subset \bigcup_{j=1}^k I_j \tag{2.3.28}$$

$$\mu(I_j) \leq \alpha^{-1} \int_{I_j} |f(y)| \, d\mu(y) \tag{2.3.29}$$

Apply the Croft–Garsia lemma. Then for $m = 1, 2$,

$$\mu\left(\bigcup_{j \in J_m} I_j \right) \leq \alpha^{-1} \int_{\cup_{j \in J_m} I_j} |f(y)| \leq \alpha^{-1} \|f\|_1 \tag{2.3.30}$$

where we used (2.2.40) and disjointness of the intervals in a single J_m.

By (2.3.28),

$$\mu(K) \leq 2\alpha^{-1} \|f\|_1 \tag{2.3.31}$$

By inner regularity, (2.3.21) holds. $\square$

Notes and Historical Remarks.   The one-dimensional case of Theorem 2.3.1 appeared in 1930 in Hardy–Littlewood [360]. They were not interested in pointwise convergence but rather asked (and answered) the question: If u is harmonic in $\mathbb{D}$, continuous on $\overline{\mathbb{D}}$, can $\int_0^{2\pi} \sup_{0 < r < 1} |u(re^{i\theta})|^p \frac{d\theta}{2\pi}$ be bounded by a multiple of $\int_0^{2\pi} |u(e^{i\theta})|^p \frac{d\theta}{2\pi}$? Their method of proof relied on first proving a discrete version using combinatorical arguments which they translated into cricket scores (Hardy was a big fan of cricket), remarking that "The arguments used are indeed mostly of the type which are intuitive to a student of cricket averages."

Their work prompted two 1932 papers by F. Riesz [688, 689] which pushed their idea: one [688] found the alternate sunrise lemma proof that we discuss below, and the other [689] used the maximal inequality to recover Lebesgue's theorem on differentiation of Lebesgue integrals, as we will see in the next section.

It should be emphasized that while maximal functions appeared earlier, for example, in work by Banach [34] and Hardy himself [354], it was this work of Hardy–Littlewood and Riesz that captured the imagination of analysts and placed the subject in the center of analytic research. It should also be noted that when this work was done, Littlewood was forty-five and Hardy and Riesz over fifty.

Godfrey Harold Hardy (1877–1947) was a British mathematician credited for moving English analysis from a focus only on applied subjects (think

Stokes, Kelvin, Barnes, Whittaker, Watson) into the modern era. He was educated at Cambridge and was there from 1896–1919 and 1931–47, spending the period between as Savilian Professor at Oxford. One reason he left Cambridge was unhappiness over Russell's dismissal for pacifism during the First World War. He was happy at Oxford but returned to Cambridge in part because, a lifelong bachelor, he would be able to keep his rooms at Trinity College, Cambridge after he reached retirement age, while Oxford did not have such a policy. Hardy was known for his love of cricket and, indeed, their proof of the maximal inequality makes reference to methods of scoring cricket.

Hardy is known for a wide variety of works in analysis and number theory, including his famous books on number theory and on divergent series. His students include Mary Cartwright, Rado, Titchmarsh, and, of course, Ramanujan, which Hardy described as "the one romantic incident in my life" and indeed, one of the great dramas in mathematics.

John Edensor Littlewood (1885–1977) was born in England but spent part of his youth in South Africa. (His father, educated at Cambridge, had moved there to be headmaster of a school.) In 1900, he returned to England, first to Saint Paul's and then Trinity College, Cambridge where, except for three post-degree years in Manchester, he spent his entire career. He is known for work in analysis, number theory, and nonlinear differential equations. His students include Chowla, Ingham, Spencer, Swinnerton-Dyer, and Verblunsky.

Both Hardy and Littlewood were vivid characters and anecdotes about each abound. The Hardy–Littlewood collaboration, spanning thirty-five years, was legendary. Because he suffered from depression, Littlewood did very little traveling outside Cambridge (until his later years when he was helped by medication), while Hardy was well-traveled. There is a persistent story that one mathematician (in some versions Wiener, in others Landau) upon meeting Littlewood exclaimed: "Oh, so you really exist. I thought that 'Littlewood' was just a pseudonym that Hardy put on his weaker papers." [**488**].

Riesz approached the Hardy–Littlewood inequalities via a particular lemma that leads to an *equality* for M_L and M_R. The lemma says:

Theorem 2.3.5 (F. Riesz's Sunrise Lemma). *Let* $g \colon \mathbb{R} \to \mathbb{R}$ *be continuous and obey* $\lim_{x \to \pm\infty} g(x) = \pm\infty$. *Let*

$$E = \{x \in \mathbb{R} \mid \exists\, y < x \text{ with } g(y) > g(x)\} \qquad (2.3.32)$$

Then E *is open and every connected component* (a, b) *of* E *is bounded and has* $g(a) = g(b)$.

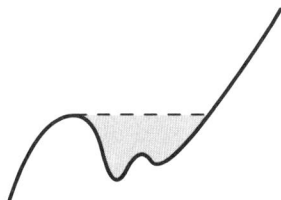

Figure 2.3.2. The sunrise lemma.

The name comes from Figure 2.3.2 and the notion that if the sun shines from the left, E is the set of points in the shade (at sunrise). The reader will prove this in Problem 4 and use it to prove a remarkable equality (Problem 5):

Theorem 2.3.6 (Riesz Maximal Equality). *Let $f \in L^1(\mathbb{R})$. Then (M_{L} given by (2.3.6)) for all $\alpha > 0$,*

$$\alpha|\{x \mid (M_{\mathrm{L}}f)(x) > \alpha\}| = \int_{\{x\mid(M_{\mathrm{L}}f)(x)>\alpha\}} |f(y)| \, dy \qquad (2.3.33)$$

This (and similar results hold for M_{R}), of course, implies that

$$m_{M_{\mathrm{L}}f}(\alpha) \leq \alpha^{-1}\|f\|_1 \qquad (2.3.34)$$

and proves (2.3.21) because of (2.3.8) and Problem 18 of Section 2.2.

Besides the covering lemma proofs and the Riesz sunshine lemma proof, we should mention that we'll later see such close connections to some of the other maximal inequalities of this chapter that they provide alternate proofs. In Section 2.6, we'll show the one-dimensional Hardy–Littlewood maximal inequality is essentially equivalent to the maximal ergodic inequality; in particular, Problem 20 of Section 2.6 shows how to go from the maximal ergodic inequality to the Hardy–Littlewood inequality and Theorem 2.6.15 goes in the other direction. In Section 2.10, we'll see that the Doob maximal inequality for martingales implies the one-dimensional Hardy–Littlewood inequality; see Theorem 2.10.19.

By using a suitable norm, one sees that (2.3.2) holds for maximal functions defined by rectangles of a fixed shape. However, if one wants to take the sup over all rectangles containing x, (2.3.2) is false (for any constant replacing 3^ν). For Busemann–Feller [**125**] noted that Lebesgue's theorem may fail for arbitrary rectangles. Instead, there is a bound if $\frac{3^\nu}{\alpha} \int |f(x)| \, d^\nu x$ is replaced by $\frac{C_\nu}{\alpha} \int |f(x)|(1 + \log_+(\frac{|f(x)|}{\alpha}))^{\nu-1} \, d^\nu x$. This is sometimes called the *strong maximal theorem*. It was proven by Jessen et al. [**429**], who iterated the Hardy–Littlewood theorem for $\nu = 1$ (a geometric proof is in Cordoba–Fefferman [**171**]). An important extension of maximal inequalities for sequences of functions is due to Fefferman–Stein [**270**].

The optimal constant in $\nu = 1$ for M_{U} is 2. We've proven that it is a bound and it is easy (Problem 6) to see it is optimal. For M_{HL}, the optimal constant is more subtle, only determined in 2003 by Melas [563]. It is, quite remarkably, $(11 + \sqrt{61})/12$. The optimal constant, C_ν, for the weak-L^1 inequality in dimension ν is not known, but it is much better than the 3^ν one gets from the covering argument. Stein–Strömberg [782] have proven $C_\nu \leq S\nu$ for a constant S. They also have a ν-independent bound, $\|M_{\mathrm{HL}}f\|_p \leq S_p\|f\|_p$, for any $p \in (1, \infty)$.

By using compactness and inner regularity, we only needed to deal with finite covers. One can instead deal with infinite covers and avoid the need for regularity (which, of course, holds for all measures!). For example, the associated Vitali theorem says that if $\bigcup_{\alpha \in I} B_{r_\alpha}(x_\alpha)$ has finite volume, there is, for each $\varepsilon > 0$, a countable subset of disjoint balls among the $B_{r_\alpha}(x_\alpha)$ so that (2.3.10) holds with 3 replaced by $3 + \varepsilon$. The reader will prove this in Problem 8.

The *spherical maximal function* (called *circular maximal function* if $\nu = 2$) is defined for f on $\mathbb{R}^\nu$ by

$$(M_S f)(x) = \sup_{R > 0} \int_{S^{\nu-1}} f(x + R\omega)\,d^{\nu-1}\omega \qquad (2.3.35)$$

where $d^{\nu-1}\omega$ is the normalized rotation invariant measure on the sphere. It is a theorem of Stein [778] that for $\nu \geq 3$

$$\|M_S f\|_p \leq C_{p,\nu}\|f\|_p, \quad p > \nu/(\nu - 1) \qquad (2.3.36)$$

It is easy to see (Problem 7) that this fails if $p \leq \nu/(\nu - 1)$. $\nu = 2$ is harder because $p = 2$ is not included in that case and it was only ten years later that Bourgain [102] proved that case. See Mockenhaupt et al. [579] for a simpler proof of Bourgain's result and applications of these inequalities and Tao [801] for an exposition of the proof of Stein's Theorem.

The use of covering lemmas to get maximal inequalities goes back to Wiener [856] in work on maximal ergodic theorems (see Theorem 2.6.11). Wiener was also the first one to state Hardy–Littlewood inequalities on $\mathbb{R}^\nu$, $\nu \geq 2$. The Vitali covering theorem (as the assertion of a subset of disjoint balls obeying (2.3.11) with a ν-dependent constant replacing 3^ν) is due to Vitali [835] in work intended to prove differentiability in dimension 2 (see Theorem 2.4.1). The elegant and simple argument of proving there is a subset of disjoint balls which, when blown up, cover $\bigcup_j B_j(x_j)$ is due to Banach [32, 33].

The simple argument in Theorem 2.3.4 is independently from Croft [178] and Garsia [308]. Garsia says it goes back to W. H. Young but gives no reference (other than to say he learned of the Young attribution from Bishop).

It is a result of "Besicovitch type," as we'll see, and it should be noted that it was known for a long time before [**178, 308**] that the Besicovitch constant in dimension 1 is 2 (although the Croft–Garsia result is more general than the Besicovitch lemma).

The $\mathbb{R}^\nu$ analogs of Theorem 2.3.3 rely on a covering result called the *Besicovitch covering lemma*. A *Besicovitch cover* of $E \subset \mathbb{R}^\nu$ is a covering by balls $\{B_{r_\lambda}(x)\}_{x \in E}$, one for every $x \in E$. The Besicovitch covering lemma asserts there is a constant b_ν for each ν, so that given any Besicovitch cover, there are countable subsets, $S_1, \dots, S_{b_\nu}$ of E so that

(i) $x, y \in S_j$, $x \neq y \Rightarrow B_{r_{x_i}}(x_i) \cap B_{r_{x_j}} = \emptyset$.

(ii) $E \subset \bigcup_{j=1}^{b_\nu} \bigcup_{x \in S_j} B_{r_x}(x)$.

This was proven for $\nu = 2$ by Besicovitch [**64**] and for general ν by Morse [**588**]. For a textbook proof, see DiBenedetto [**221**] or Kuttler [**498**].

The Croft–Garsia covering lemma implies that the optimal b_1 is 2. In general, the optimal b_ν is not known, but there is an upper bound by 5^ν. See [**294, 793**] for a discussion of these issues and for other references.

Problems

1. (a) In ν dimensions, if $f \in L^1(\mathbb{R}^\nu)$ is supported in $\{x \mid |x| \leq R\}$, prove that (where $\tau_\nu = $ volume of $B_1(0)$)

$$(M_{\mathrm{HL}} f)(x) \geq \tau_\nu(|x| + R)^{-\nu} \int |f(x)| \, d^\nu x \qquad (2.3.37)$$

(b) For any $f \in L^1(\mathbb{R}^\nu)$, prove that (2.3.3) holds.

(c) In one dimension, if f has compact support, prove that

$$\lim_{x \to \pm\infty} |x|(M_{\mathrm{HL}} f)(x) = \tfrac{1}{2} \|f\|_1, \quad \lim_{x \to \infty} |x|(M_{\mathrm{L}} f)(x) = \|f\|_1 \qquad (2.3.38)$$

$$\lim_{x \to \infty} |x|(M_{\mathrm{R}} f)(x) = \|f\|_1, \quad \lim_{x \to \infty} |x|(M_{\mathrm{U}} f)(x) = \|f\|_1 \qquad (2.3.39)$$

2. Prove (2.3.8). (*Hint*: See Problem 18 of Section 2.2.)

3. Let $B_r(x)$ be the ball of radius r about x in some norm, $\|\cdot\|$, on $\mathbb{R}^\nu$.

   (a) Let $I_\delta(\Omega)$ be the volume of the union of all hypercubes of side δ with centers in $\delta\mathbb{Z}^\nu$, whose closures are inside Ω, an open set.

   (b) Prove that $\lim_{\delta \downarrow 0} I_\delta(\Omega) = |\Omega|$.

   (c) Prove $I_{\lambda\delta}(B_{\lambda r}(0)) = \lambda^\nu I_\delta(B_r(0))$ for any $\lambda, r > 0$.

   (d) Prove that (2.3.12) holds.

4. Let $g: \mathbb{R} \to \mathbb{R}$ be continuous, with $g(x) \to \pm\infty$ as $x \to \pm\infty$. Let E be given by (2.3.32).

(a) Prove $\mathbb{R} \setminus E = \{x \mid g(y) \leq g(x) \text{ for all } y \leq x\}$. Then show that $\mathbb{R} \setminus E$ is closed and that $\mathbb{R} \setminus E$ is bounded neither from below nor from above. (*Hint*: If E is bounded below and y_0 is picked with $y_0 < x_- = \inf_{x \in E} x$ and $g(y_0) < g(x_-)$, then what can you say about a point y_1 with $g(y_1) = \sup_{x \in (-\infty, y_0]} g(x)$?)

(b) For $x_0 \in E$, let $m(x_0) = \sup_{x \leq x_0} g(x)$ and $\ell_{x_0} = \sup\{x \mid x \leq x_0 \text{ and } g(x) = m(x_0)\}$. Prove that $\ell_{x_0} \notin E$, $\ell_{x_0} < x_0$, $(\ell_{x_0}, x_0] \subset E$, and $g(\ell_{x_0}) = m(x_0)$.

(c) Let $r_{x_0} = \inf\{x \mid x \geq x_0 \text{ and } g(x) = m(x_0)\}$. Prove that $r_{x_0} \notin E$, $r_{x_0} > x_0$, $[x_0, r_{x_0}) \subset E$, and $g(r_{x_0}) > m(x_0)$.

(d) Conclude (the Riesz sunrise lemma) that E is open and any connected component of E is (a, b) with $|a|, |b| < \infty$ and $g(a) = g(b)$.

Remarks. 1. The lemma is in the paper of F. Riesz [**688**], who pointed out in a note added in proof that this simple proof was provided him by his brother, Marcel.

2. If one thinks of g as the ground level of a landscape, the sun is rising on the left with horizontal parallel rays, then E is the set of points in the shade. The lemma says that components of E are valleys below a glancing ray (see Figure 2.3.2).

5. This problem will prove the Riesz maximal equality, (2.3.33). Let $f \in L^1(\mathbb{R}, dx)$ throughout and fix $\alpha > 0$.

   (a) Let $g_\alpha(x) = \alpha x - \int_0^x |f(y)| \, dy$. Prove that g obeys the hypotheses of the sunrise lemma (see Problem 4).

   (b) Prove that $x \in E$ if and only if $(M_L f)(x) > \alpha$.

   (c) Let $I = (a, b) \subset E$ be a component of E. Prove that $\alpha |I| = \int_I |f(y)| \, dy$.

   (d) Conclude (2.3.33).

6. Prove that $\mu\{x \mid (M_X f)(x) > \alpha\} \leq C_X \, \alpha^{-1} \|f\|_1$, the optimal constant for $X = \mathrm{U}$ is $C_\mathrm{U} = 2$, and for $X = \mathrm{R}$ or L is $C_{\mathrm{R \text{ or } L}} = 1$. (*Hint*: See Problem 1(c).)

7. Let $M_S f$ be the spherical maximal function given by (2.3.35). If f is the characteristic function of the unit ball about 0, prove that
   $$|x|^{(\nu-1)} (M_S f)(x) \to C \neq 0$$
   as $|x| \to \infty$ and conclude that $M_S f \notin L^p$ if $p \leq \nu/(\nu - 1)$. Conclude that (2.3.36) fails for such values of p.

8. (a) Let $E \subset \mathbb{R}^\nu$ be covered by $\{B_{r_\alpha}(x_\alpha)\}_{\alpha \in I}$. Prove there is a countable subcover. (*Hint*: Lindelöf spaces.)

(b) Let $\{B_{r_j}(x_j)\}_{j\in\mathbb{N}}$ be a cover of a bounded set $E \subset \mathbb{R}^\nu$ with all $x_j \in E$ and $\sup_{j\in N} r_j < \infty$. (Show that there is no loss in assuming this.) Fix $\varepsilon > 0$. Define $j_1, j_2, \dots$ and $S_1, S_2, \dots \subset \mathbb{N}$ inductively by $S_1 = \mathbb{N}$, $r_{j_k} \geq (1-\varepsilon)\sup_{\ell\in S_k} r_\ell$, $S_{k+1} = \{\ell \in S_k \mid B_{r_\ell}(x_\ell) \cap B_{r_{j_k}}(x_{j_k}) = \emptyset\}$. If, after finitely many steps, $S_n = \emptyset$, stop and set $S_n = \emptyset$ for $k \geq n+1$ and $r = 0$ for j_k, $k \geq n$. Prove that $r_{j_k} \to 0$ as $k \to \infty$.

(c) Prove that $\cap S_k = \emptyset$.

(d) Prove that $E \subset \bigcup_{k=1}^\infty B_{[1+2(1-\varepsilon)^{-1}]r_{j_k}}(x_{j_k})$.

(e) Find a version of the Vitali theorem for arbitrary—rather than finite—covers by balls.

9. (a) Prove that the Vitali covering lemma is true for closed balls.

(b) Let $\varphi \in \ell^1(\mathbb{Z})$, that is, $\{\varphi_n\}_{n=-\infty}^\infty$ with $\sum_{n=-\infty}^\infty |\varphi_n| = \|\varphi\|_1 < \infty$. Define

$$(M_{\mathrm{HL}}\varphi)_n = \sup_{m=0,1,2,\dots} \frac{1}{2m+1} \sum_{j=n-m}^{n+m} |\varphi_j| \qquad (2.3.40)$$

$$(M_{\mathrm{R}}\varphi)_n = \sup_{m=0,1,\dots} \frac{1}{m+1} \sum_{j=n}^{n+m} |\varphi_j| \qquad (2.3.41)$$

Prove that

$$\#\{n \mid M_{\mathrm{HL}}\varphi > \alpha\} \leq 3\alpha^{-1}\|\varphi\|_1 \qquad (2.3.42)$$

$$\#\{n \mid M_{\mathrm{R}}\varphi > \alpha\} \leq 3\alpha^{-1}\|\varphi\|_1 \qquad (2.3.43)$$

Remarks. 1. In the same way, by using the Croft–Garsia covering lemma and Riesz sunrise lemma, one can get constants 2 and 1 rather than 3 in (2.3.42) and (2.3.43).

2. We need open sets for $\mathbb{R}$ to use compactness, but for finite sets, one can use closed covers and needs to in this case.

3. Discrete Hardy–Littlewood maximal inequalities first appeared in their original paper [**360**]. Indeed, as mentioned in the Notes, they derived the continuum result by first proving a discrete one.

2.4. Differentiation and Convolution

We have already seen that for many families of functions, $\{g_n\}_{n=1}^\infty$ on $\mathbb{R}^\nu$, we have $g_n * f \to f$ in $L^p(\mathbb{R}^\nu, d^\nu x)$ for all $f \in L^p$. Typical are the Poisson kernel and Fejér kernel (the former has a continuous parameter rather than discrete). In this section, we'll prove pointwise a.e. results. In particular, we'll prove Theorems 2.0.2 and 2.0.3 on existence of derivatives of integrals of L^1-functions and of boundary values of H^2-functions.

The key, of course, is a suitable maximal inequality and the critical realization here is that for a large class of approximate identities (see Theorem 3.5.11 of Part 1), the appropriate maximal function is dominated by a multiple of the Hardy–Littlewood maximal function. Along the way, we'll introduce a nontangential maximal function. As a warmup, we discuss derivatives of integrals, including the fundamental theorem of calculus for L^1-functions.

Theorem 2.4.1 (Lebesgue's Differentiation Theorem). *Let f be a Borel function on $\mathbb{R}^\nu$, which is in $L^1_{\text{loc}}(\mathbb{R}^\nu, d^\nu x)$, that is, for any compact $K \subset \mathbb{R}^\nu$,*

$$\int_K |f(y)| \, d^\nu y < \infty \tag{2.4.1}$$

Then for Lebesgue a.e. $x \in \mathbb{R}^\nu$,

$$\lim_{r \downarrow 0} \frac{1}{|B_r(0)|} \int_{B_r(x)} f(y) \, d^\nu y \to f(x) \tag{2.4.2}$$

Indeed, for a.e. x,

$$\lim_{r \downarrow 0} \frac{1}{|B_r(0)|} \int_{B_r(x)} |f(y) - f(x)| \, d^\nu y \to 0 \tag{2.4.3}$$

In case $\nu = 1$,

$$\lim_{\varepsilon \downarrow 0} \frac{1}{\pm \varepsilon} \int_x^{x \pm \varepsilon} f(y) \, dy \to f(x) \tag{2.4.4}$$

Remark. A point at which (2.4.3) holds is called a *Lebesgue point* of f.

Proof. We first claim it suffices to prove the result for $f \in L^1$. For then we get the result for a.e. $x \in B_R(0)$ by applying the L^1-result to $f\chi_{B_{R+1}(0)}$. Secondly, it suffices to prove (2.4.3) for it implies both

$$\lim_{r \downarrow 0} \frac{1}{|B_r(0)|} \left| \int_{B_r(x)} (f(y) - f(x)) \, d^\nu x \right| \to 0 \tag{2.4.5}$$

(which is (2.4.2)) and, if $\nu = 1$,

$$\frac{1}{2\varepsilon} \left| \int_x^{x \pm \varepsilon} (f(y) - f(x)) \, dx \right| \le \frac{1}{2\varepsilon} \int_{x-\varepsilon}^{x+\varepsilon} |f(y) - f(x)| \, dx \to 0 \tag{2.4.6}$$

which is (2.4.4).

For $f \in L^1(\mathbb{R}^\nu, d^\nu x)$, define for $r > 0$

$$(T_r f)(x) = \frac{1}{|B_r(0)|} \int_{B_r(x)} |f(y) - f(x)| \, d^\nu y \tag{2.4.7}$$

(note T_r is subadditive) and

$$(T^* f)(x) = \sup_r (T_r f)(x) \tag{2.4.8}$$

If $f \in C_0^\infty(\mathbb{R}^\nu)$, $(T_r f)(x) \to 0$ uniformly by simple uniform continuity estimates. So, by Theorem 2.1.2, we only need a maximal inequality.

Clearly,

$$(T^* f)(x) \leq (M_{\mathrm{HL}} f)(x) + |f(x)| \tag{2.4.9}$$

so

$$|\{x \mid (T^* f)(x) > \alpha\}| \leq |\{x \mid (M_{\mathrm{HL}} f)(x) > \alpha/2\}| + |\{x \mid |f(x)| > \alpha/2\}| \tag{2.4.10}$$

$$\leq \alpha^{-1}[2\,3^\nu + 2]\|f\|_1 \tag{2.4.11}$$

by Theorem 2.3.1. □

To handle convolution operators, $g \mapsto f * g$, we'll want f to be a de-
creasing function of $|x|$ or at least dominated by one.

Lemma 2.4.2. *Let f be a nonnegative function on $\mathbb{R}^\nu$ which is symmetric
decreasing, that is,*

$$|x| \geq |y| \Rightarrow 0 \leq f(x) \leq f(y) \tag{2.4.12}$$

Then for any measurable function, g, on $\mathbb{R}^\nu$,

$$|(f * g)(x)| \equiv \left| \int f(y) g(x - y)\, d^\nu y \right| \leq \|f\|_1 (M_{\mathrm{HL}} g)(x) \tag{2.4.13}$$

*More generally, for any measurable f, if its symmetric envelope, $\widetilde{f}$, is
defined by*

$$\widetilde{f}(x) = \sup_{|y| \geq |x|} |f(y)| \tag{2.4.14}$$

then

$$|(f * g)(x)| \leq \|\widetilde{f}\|_1 (M_{\mathrm{HL}} g)(x) \tag{2.4.15}$$

Proof. Since $|f| \leq \widetilde{f}$, $|f * g| \leq \widetilde{f} * |g|$, which shows that (2.4.13) implies
(2.4.15).

If f is symmetric decreasing, then $\{x \mid f(x) > \alpha\}$ is a $B_{r(\alpha)}(0)$ with
$|B_{r(\alpha)}(0)| = m_f(\alpha)$, so by the wedding-cake representation (2.2.17),

$$|f * g(x)| = \left| \int_0^\infty m_f(\alpha) \left[\frac{1}{|B_{r(\alpha)}(x)|} \int_{B_{r(\alpha)}(x)} g(y)\, d^\nu y \right] d\alpha \right| \tag{2.4.16}$$

$$\leq \int_0^\infty m_f(\alpha) (M_{\mathrm{HL}} g)(x)\, d\alpha = \|f\|_1 M_{\mathrm{HL}} g(x) \tag{2.4.17}$$

□

Theorem 2.4.3. *Let g_n be a sequence of measurable functions on $\mathbb{R}^\nu$ obeying*

(i) $\displaystyle\sup_n \|\widetilde{g}_n\|_1 < \infty$ $\tag{2.4.18}$

(ii) $\displaystyle\int g_n(x)\, d^\nu x = 1, \qquad$ *for all* n (2.4.19)

For every $r > 0$,

(iii) $\displaystyle\lim_{n\to\infty}\int_{|y|>r}|g_n(y)|\, d^\nu y = 0$ (2.4.20)

Then for any $f \in L^p(\mathbb{R}^\nu)$, $1 \le p < \infty$, $(g_n * f)(x) \to f(x)$ *for a.e.* x.

Remark. g_n need not be positive (but, of course, $\widetilde{g}_n$ is).

Proof. Let $T_n f = g_n * f$. By (ii), (iii), and the argument in Theorem 3.5.11 of Part 1, $(T_n f)(x) \to f(x)$ uniformly if $f \in C_0^\infty(\mathbb{R}^\nu)$ which is dense in L^p. Thus, by Theorem 2.1.1, we only need a maximal inequality.

By (2.4.15),
$$(T^* f)(x) = \big(\sup_n \|\widetilde{g}_n\|_1\big)(M_{\mathrm{HL}} f)(x)$$

so the Hardy–Littlewood maximal inequality (see Remark 3 after Theorem 2.2.10 and Remark 3 after Theorem 2.3.1) proves the result. $\qquad\square$

We'll often want to apply this to convolutions on the circle, $\partial\mathbb{D}$.

Theorem 2.4.4. *Let* g_n *be a sequence of functions on* $\partial\mathbb{D}$. *For* $|\theta| < \pi$, *define*
$$\widetilde{g}_n(\theta) = \sup_{\psi\in(\pi-|\theta|,\pi+|\theta|)}|g_n(e^{i\psi})| \tag{2.4.21}$$

Suppose

(i) $\displaystyle\sup_n \int_0^\pi \widetilde{g}_n(\theta)\, d\theta < \infty$ (2.4.22)

(ii) $\displaystyle\int_0^{2\pi} g_n(e^{i\theta})\,\frac{d\theta}{2\pi} = 1$ (2.4.23)

For all $\varepsilon \in (0,\pi)$,

(iii) $\displaystyle\lim_{n\to\infty}\int_\varepsilon^{2\pi-\varepsilon}|g_n(e^{i\theta})|\,\frac{d\theta}{2\pi} = 0$ (2.4.24)

Then for all $f \in L^1(\partial\mathbb{D}, \frac{d\theta}{2\pi})$ *with* $*$ *the* $\partial\mathbb{D}$ *convolution, we have*
$$(g_n * f)(e^{i\theta}) \to f(e^{i\theta}) \quad \text{for a.e. } \theta$$

Remark. In this (and also the last two theorems), the proof shows that at any Lebesgue point, $e^{i\theta}$ (and, in particular, a point where $\frac{1}{2}[f(e^{i(\theta_0+\varepsilon)}) + f(e^{i(\theta_0-\varepsilon)})] \to f(e^{i\theta})$), we have $g_n * F(e^{i\theta_0}) \to F(e^{i\theta_0})$. In particular (what is known as the Lebesgue–Fejér theorem), given the analysis in Example 2.4.8 below, Fourier series are Cesàro summable at any Lebesgue point.

Proof. Define, as functions on $\mathbb{R}$,

$$G_n(\theta) = \begin{cases} \frac{1}{2\pi} \, g_n(e^{i\theta}), & -\pi < \theta < \pi \\ 0, & |\theta| \geq \pi \end{cases} \tag{2.4.25}$$

$$F(\theta) = \begin{cases} f(e^{i\theta}), & |\theta| \leq \frac{3\pi}{2} \\ 0, & |\theta| > \frac{3\pi}{2} \end{cases} \tag{2.4.26}$$

Then Theorem 2.4.3 applies and $G_n * F \to F$ for a.e. $\theta \in \mathbb{R}$. But for $|\theta| < \pi$, $(G_n * f)(\theta) = (g_n * f)(e^{i\theta})$. $\qquad\square$

Example 2.4.5. The Poisson kernel (see Section 5.3 of Part 2A) is of the form

$$P_r(e^{i\theta}) = \frac{1 - r^2}{1 + r^2 - 2\cos\theta} \tag{2.4.27}$$

It is positive and monotone in θ for $\theta \in [0, \pi]$ and even, so $\widetilde{P}_r = P_r$ and $\int \widetilde{P}_r(e^{i\theta}) \frac{d\theta}{2\pi} = \int P_r(e^{i\theta}) \frac{d\theta}{2\pi} = 1$. Thus, except for the change from discrete n to continuous r (which is easy once the maximal function is still dominated by M_{HL}), Theorem 2.4.4 is applicable and we get Theorem 2.4.6. $\qquad\square$

Theorem 2.4.6. *Let* $f \in L^1(\partial\mathbb{D}, \frac{d\theta}{2\pi})$. *Then for a.e.* $e^{i\theta} \in \partial\mathbb{D}$,

$$\lim_{r\uparrow 1}(P_r * f)(e^{i\theta}) = f(e^{i\theta}) \tag{2.4.28}$$

Remark. This differs from Theorem 2.0.3, but one can show (see Theorem 5.2.1) that when (2.0.4) holds, there is $f \in L^2(\partial\mathbb{D}, \frac{d\theta}{2\pi}) \subset L^1(\partial\mathbb{D}, \frac{d\theta}{2\pi})$, so that $f(re^{i\theta}) = (P_r * f)(e^{i\theta})$.

There is an analog for harmonic functions in the upper half-plane, $\mathbb{C}_+$. The Poisson kernel for that case, as we'll discuss in Theorem 5.9.5, is

$$P_y(x) = \frac{1}{\pi} \frac{y}{x^2 + y^2} \tag{2.4.29}$$

Since $P_y(x) = y^{-1} P_1(x/y)$, this is an approximate identity. So if $f \in C(\mathbb{R})$, $P_y * f \to f$ uniformly on compact subsets of $\mathbb{R}$ (and if f is uniformly continuous, uniformly on $\mathbb{R}$). Moreover, since $P_y(x) = \frac{1}{\pi} \mathrm{Im}(\frac{1}{x+iy})$, P_y is harmonic on $\mathbb{C}_+$, so $g(x + iy) = (P_y * f)(x)$ is harmonic on $\mathbb{C}_+$.

Notice that $P_y(x)$ is positive, symmetric, and monotone decreasing in x, so Theorem 2.4.3 applies and we have

Theorem 2.4.7. *If* $f \in L^p(\mathbb{R})$, $1 \leq p < \infty$, *as* $y \downarrow 0$,

$$P_y * f(x) \to f(x) \tag{2.4.30}$$

for a.e. x.

Example 2.4.8. Recall that the Fejér kernel is given by

$$F_n(e^{i\theta}) = \frac{1}{n}\left[\frac{\sin(n\theta/2)}{\sin(\theta/2)}\right]^2 \tag{2.4.31}$$

This is symmetric but certainly not monotone in θ. Thus, we need to bound $\widetilde{F}_n$. We start by noting two "obvious" bounds on F_n. F_n arises as an average of Dirichlet kernels, D_j. Since $\|D_j\|_\infty = D_j(0)$, we see $\|F_n\|_\infty = F_n(0)$, that is,

$$|F_n(e^{i\theta})| \leq n, \qquad \text{all } \theta \tag{2.4.32}$$

Moreover, for $\psi \in [-\frac{\pi}{2}, \frac{\pi}{2}]$, $|\sin\psi| \geq \frac{2|\psi|}{\pi}$ (Problem 1), so

$$|F_n(e^{i\theta})| \leq \frac{\pi^2}{n\theta^2}, \qquad \text{all } \theta \tag{2.4.33}$$

$\square$

Noting that $\min(n, \frac{\pi^2}{n\theta^2})$ is monotone in θ, we conclude

$$\widetilde{F}_n(e^{i\theta}) \leq \begin{cases} n, & 0 \leq |\theta| \leq \frac{\pi}{n} \\ \frac{\pi^2}{n\theta^2}, & |\theta| \geq \frac{\pi}{n} \end{cases} \tag{2.4.34}$$

which leads to (Problem 2)

$$\int_0^\pi \widetilde{F}_n(\theta)\,d\theta \leq 1 + \pi \tag{2.4.35}$$

Therefore, Theorem 2.4.4 is applicable and we get

Theorem 2.4.9. *Let $f \in L^1(\partial\mathbb{D}, \frac{d\theta}{2\pi})$. Let $C_N(f)$ be the Cesàro average of the truncated Fourier series for f (see (3.5.7) of Part 1). Then for a.e. θ, $C_N(f)(e^{i\theta}) \to f(e^{i\theta})$.*

Finally, we turn to "nontangential" limits for boundary values of harmonic functions. We'll discuss $\mathbb{C}_+$ here and leave $\mathbb{D}$ to the Problems. For $x, x_0 \in \mathbb{R}$, $y_0 > 0$, define

$$P_{y_0,x_0}(x) = \frac{1}{\pi}\frac{y_0}{(x-x_0)^2 + y_0^2} \tag{2.4.36}$$

Then

$$\pi\widetilde{P}_{y_0,x_0}(x) = \begin{cases} \frac{1}{y_0}, & |x| \leq |x_0| \\ \frac{y_0}{(|x|-|x_0|)^2+y_0^2}, & |x| \geq |x_0| \end{cases} \tag{2.4.37}$$

so

$$\|\widetilde{P}_{y_0,x_0}\|_{L^1} = \frac{2|x_0|}{\pi y_0} + 1 \tag{2.4.38}$$

which implies, by (2.4.15),

$$g(x+iy) = (P_y g)(x) \Rightarrow |g(x+x_0+iy_0)| \leq \left[\frac{2|x_0|+\pi y_0}{\pi y_0}\right](M_{\text{HL}}g)(x) \tag{2.4.39}$$

It follows that we get a maximal inequality on each cone $|x_0| \leq \tan(\theta_0)|y_0|$ for any $\theta_0 \in \frac{\pi}{2}$. By extending Theorem 2.1.1 to allow limits for such a cone, we get:

Theorem 2.4.10 (Nontangential Limits). *Let $\theta_0 \in [0, \frac{\pi}{2})$ and $g \in L^p(\mathbb{R}, dx)$ for some $p \in [1, \infty)$. Then for a.e. $x \in \mathbb{R}$, with*

$$C_{\theta_0}(x) = \{z \in \mathbb{C}_+ \mid z \in \mathbb{C}_+, |\arg(-i(z-x))| < \theta_0\} \qquad (2.4.40)$$

and $g(u + iv) = (P_v * g)(u)$, we have

$$\lim_{\substack{|z-x| \to 0 \\ z \in C_{\theta_0}(x)}} g(z) = g(x) \qquad (2.4.41)$$

Remarks. 1. For the known dense set with pointwise convergence, we use the fact that if $g \in C_0^\infty$, the function g on $\overline{\mathbb{C}}_+$ is continuous.

2. This is called nontangential convergence since it says we have the limit so long as z does not approach x asymptotically tangentially.

A similar result is useful for $\mathbb{D}$. Rather than consider arbitrary opening angle in $(0, \frac{\pi}{2})$, we'll take angle $\frac{\pi}{4}$. Given $e^{i\theta_0} \in \partial\mathbb{D}$, let (see Figure 2.4.1)
$$N(\theta_0) = \overline{\mathbb{D}_{\frac{1}{2}}(0)} \cup \text{square of sides } \frac{\sqrt{2}}{2} \text{ with } 0 \text{ and } e^{i\theta_0} \text{ as corners.}$$
Then,

Theorem 2.4.11 (Nontangential maximal function for $\mathbb{D}$). *Let $f \in L^1(\partial\mathbb{D}, \frac{d\theta}{2\pi})$ and*

$$(M_{\text{NT}}f)(e^{i\theta}) = \sup_{\substack{z \in N(\theta) \\ z \neq e^{i\theta}}} |g(z)| \qquad (2.4.42)$$

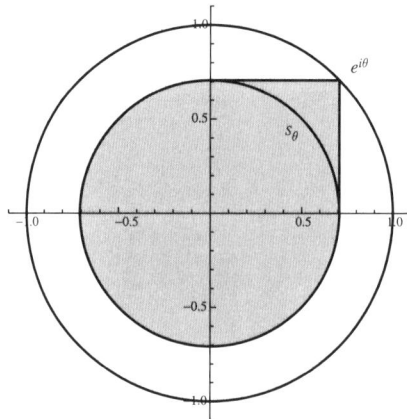

Figure 2.4.1. $N(\theta_0)$ for $\theta_0 = e^{i\pi/4}$.

where

$$g(re^{i\theta}) = \begin{cases} f(e^{i\theta}) & \text{if } r = 1 \\ \int \left[\frac{1-|r|^2}{1+r^2-2r\cos(\theta-\phi)}\right]f(e^{i\theta})\frac{d\theta}{2\pi}, & \text{otherwise} \end{cases} \qquad (2.4.43)$$

Then, there exists an f-independent constant, C, so that for all θ,

$$(M_{\text{NT}}f)(e^{i\theta}) \leq C(M_{\text{HL}}f)(e^{i\theta}) \qquad (2.4.44)$$

Since $\|f\|_1 \leq 2\pi(M_{\text{HL}}f)(e^{i\theta})$ for all θ, we get a bound on the circle. Arguments like (2.4.38) control the square.

Notes and Historical Remarks. For the case $\nu = 1$, Theorem 2.4.1 was in Lebesgue's original book on integration theory [**513**] written in 1904. He proved the general case in 1910 [**514**]. In between, Vitali [**835**] proved a result for $\nu = 2$. Lebesgue used Vitali's covering lemma in his 1910 work.

Results on the existence of a.e. nontangential limits for bounded analytic functions on $\mathbb{D}$ go back to Fatou in 1906 [**261**]. The approach via maximal functions is later—that the Hardy–Littlewood maximal function is a natural tool to get the Lebesgue differentiation theorem is a result of Riesz [**689**].

Problems

1. (a) Prove that $\tan x \geq x$ for $x \in [0, \frac{\pi}{2})$. (*Hint*: Take derivatives.)

   (b) Prove that $\frac{\sin x}{x}$ is decreasing on $[0, \frac{\pi}{2})$.

   (c) For $x \in [-\frac{\pi}{2}, \frac{\pi}{2}]$, prove that $|\sin x| \geq \frac{2|x|}{\pi}$.

2. Prove (2.4.35) from (2.4.34).

3. Let $\varphi_0(t) = 1$ on $[0, 1)$, $\varphi_0(t) = -1$ on $[1, 2)$, extend φ_0 to $\mathbb{R}^1$ so as to have period 2, and define $\varphi_n(t) = \varphi_0(2^n t)$, $n = 1, 2, 3, \ldots$. Assume that $\sum|c_n|^2 < \infty$ and prove that the series

$$\sum_{n=1}^{\infty} c_n\varphi_n(t) \qquad (2.4.45)$$

then converges for almost every t in $[0, 1]$.

Probabilistic interpretation: The series $\sum(\pm c_n)$ with random signs converges with probability 1.

Suggestion: $\{\varphi_n\}$ is orthonormal on $[0, 1]$, hence (2.4.45) is the Fourier series for the φ_n-basis of some $f \in L^2$. If $a = j \cdot 2^{-N}$, $b = (j + 1) \cdot 2^{-N}$, $a < t < b$, and $s_N = c_1\varphi_1 + \cdots + c_N\varphi_N$, then for $n > N$,

$$s_N(t) = \frac{1}{b-a}\int_a^b s_N \, dm = \frac{1}{b-a}\int_a^b s_n \, dm$$

and the last integral converges to $\int_a^b f \, dm$ as $n \to \infty$. Show that (2.4.45) converges to $f(t)$ at almost every Lebesgue point of f.

2.5. Comparison of Measures

One of our goals in this section is to prove

Theorem 2.5.1. *Let α be a monotone function on $[0,1]$. Then*
(a) *For Lebesgue a.e. $x \in [0,1]$,*

$$\lim_{\varepsilon \downarrow 0} \frac{1}{2\varepsilon} [\alpha(x+\varepsilon) - \alpha(x-\varepsilon)] \qquad (2.5.1)$$

exists and is finite. Moreover, if f is the limit in (2.5.1), $d\alpha - f(x)\,dx \equiv d\alpha_{\mathrm{sing}}$ is a measure purely singular with respect to dx.
(b) *If*

$$E = \left\{ x \,\bigg|\, \lim_{\varepsilon \downarrow 0} \frac{1}{2\varepsilon} [\alpha(x+\varepsilon) - \alpha(x-\varepsilon)] = \infty \right\} \qquad (2.5.2)$$

then $|E| = 0$ and $\alpha_{\mathrm{sing}}([0,1] \setminus E) = 0$.

Remarks. 1. Problem 1 extends this to α's of bounded variation.

2. This result (or rather its extension in Problem 1) is Theorems 4.15.4–4.15.6 of Part 1.

3. (b) is sometimes called the Theorem of de la Vallée Poussin.

In fact, we'll prove a more general result (Theorem 2.5.1 is this result for the case $d\nu = d\alpha$, $d\mu = dx$); we use $\mu(a,b)$ for $\mu([a,b])$.

Theorem 2.5.2. *Let μ, ν be two (positive) Baire measures on $\mathbb{R}$ and let*

$$d\nu = g\,d\mu + d\nu_{\mathrm{sing}} \qquad (2.5.3)$$

be the Lebesgue–Radon–Nikodym decomposition guaranteed by Theorem 4.7.6 and Problem 9 of Section 4.10 of Part 1 (so $d\nu_{\mathrm{sing}}$ is $d\mu$-singular). Then
(a) *For μ a.e. $x \in \mathbb{R}$,*

$$\lim_{\varepsilon \downarrow 0} \frac{\nu(x-\varepsilon, x+\varepsilon)}{\mu(x-\varepsilon, x+\varepsilon)} = g(x) \qquad (2.5.4)$$

(b) *For*

$$E = \left\{ x \,\bigg|\, \lim_{\varepsilon \downarrow 0} \frac{\nu(x-\varepsilon, x+\varepsilon)}{\mu(x-\varepsilon, x+\varepsilon)} = \infty \right\} \qquad (2.5.5)$$

we have that

$$\mu(E) = 0, \qquad \nu_{\mathrm{s}}(\mathbb{R} \setminus E) = 0 \qquad (2.5.6)$$

We'll give the proof under the additional hypothesis that $\mu(\mathbb{R}) < \infty$, $\nu(\mathbb{R}) < \infty$. Once one has that special case, the general Baire measure case follows by looking at μ and ν restricted to each interval $[-n, n]$ with $n \in \mathbb{Z}_+$.

We'll first prove (2.5.4) when $\nu = g\,d\mu$, then use the von Neumann trick (see the proof of Theorem 4.7.6 of Part 1) to get (2.5.4) and (2.5.6) in general, and finally translate these results to the language of Stieltjes transforms.

Lemma 2.5.3. *Let μ be a (positive) Baire measure on $\mathbb{R}$ and $g \in L^1(\mathbb{R}, d\mu)$. Then for μ a.e. x,*

$$\lim_{\varepsilon \downarrow 0} \frac{1}{\mu(x - \varepsilon, x + \varepsilon)} \int_{x-\varepsilon}^{x+\varepsilon} g(y) \, d\mu(y) = g(x) \qquad (2.5.7)$$

Proof. Let $(T_\varepsilon g)(x)$ be the function inside the limit of (2.5.7). Let g be uniformly continuous and in L^1. Then for every $x \in \mathbb{R}$,

$$|T_\varepsilon g(x) - g(x)| \leq \sup_{|y-x| \leq \varepsilon} |g(y) - g(x)| \qquad (2.5.8)$$

goes to zero, so (2.5.7) holds for a dense set of g in L^1.

Thus, we only need a maximal inequality and this is provided by Theorem 2.3.3. $\qquad \square$

Proof of Theorem 2.5.2. Let $\eta = \mu + \nu$. Then $\mu \leq \eta$, so μ is η-a.e. Thus, $d\mu = F \, d\eta$ with $0 \leq F \leq 1$. Let

$$E_0 = \{x \mid F(x) = 0\} \qquad (2.5.9)$$

By the proof of Theorem 4.7.6 of Part 1 (g given by (2.5.3)),

$$\mu(E_0) = 0, \qquad d\nu_{\mathrm{s}} = \chi_{E_0} \, d\nu, \qquad g = F^{-1} - 1 \qquad (2.5.10)$$

By Lemma 2.5.3, for η-a.e. x,

$$\lim_{\varepsilon \downarrow 0} \frac{\mu(x - \varepsilon, x + \varepsilon)}{\eta(x - \varepsilon, x + \varepsilon)} = F(x) \qquad (2.5.11)$$

Since

$$\left(\frac{\mu}{\eta} \right)^{-1} = 1 + \left(\frac{\nu}{\mu} \right) \qquad (2.5.12)$$

(2.5.11) implies for η-a.e. x,

$$\lim_{\varepsilon \downarrow 0} \frac{\nu(x - \varepsilon, x + \varepsilon)}{\mu(x - \varepsilon, x + \varepsilon)} = F(x)^{-1} - 1 \qquad (2.5.13)$$

On E_0, up to a set of η (and so μ and ν) measure zero, the limit is ∞. Thus, $E = E_0$, up to sets of μ and ν measure zero, so $\mu(E) = 0$, $\nu_{\mathrm{s}}(\mathbb{R} \setminus E) = 0$. For μ-a.e. x, $F(x) > 0$ and (2.5.13) plus (2.5.10) implies (2.5.4). $\qquad \square$

We end this section by translating these results to certain harmonic and analytic functions associated to measures on $\mathbb{R}$. We'll suppose $\mu(\mathbb{R}) < \infty$ although parts of the theory go through if $\int (1 + x^2)^{-1} \, d\mu(x) < \infty$ (see the Notes).

Definition. If μ is a finite measure on $\mathbb{R}$, its *Stieltjes transform* (also called the *Borel transform* or *Cauchy transform*), $F_\mu(z)$, is the function on $\mathbb{C}_+$ given by

$$F_\mu(z) = \int \frac{d\mu(x)}{x - z} \qquad (2.5.14)$$

Notice that

$$\operatorname{Im} F_\mu(x + iy) = y \int \frac{d\mu(s)}{(s - x)^2 + y^2} = \pi P_y * \mu \qquad (2.5.15)$$

and, in particular,

$$\operatorname{Im} F_\mu(z) > 0 \text{ on } \mathbb{C}_+ \qquad (2.5.16)$$

Thus, $|e^{i\alpha F_\mu(z)}| \leq 1$ for all $\alpha > 0$. We'll see in Theorem 5.2.1 that bounded analytic functions on $\mathbb{C}_+$ have nontangential boundary values. Picking α_1, α_2 with $\alpha_1/\alpha_2 \notin \mathbb{Q}$ shows F_μ has boundary values. In particular, one defines the *Hilbert transform* of a measure on $\mathbb{R}$ by

$$H_\mu(x) = \frac{1}{\pi} \operatorname{Re} F_\mu(x + i0) \qquad (2.5.17)$$

We'll have more to say about it in Section 5.9. Here is the translation of Theorem 2.5.1 to Stieltjes transforms:

Theorem 2.5.4. *Let μ be a finite measure on $\mathbb{R}$ and $F_\mu(z)$ its Stieltjes transform. Suppose*

$$d\mu(x) = f(x)\, dx + d\mu_s(x) \qquad (2.5.18)$$

Then

(a) *$\frac{1}{\pi} \operatorname{Im} F_\mu(x + i\varepsilon)\, dx \to d\mu$ in the sense that for any $g \in C_0(\mathbb{R})$, the continuous functions of compact support*

$$\lim_{\varepsilon \downarrow 0} \frac{1}{\pi} \int \operatorname{Im} F_\mu(x + i\varepsilon) g(x)\, dx = \int g(x)\, d\mu(x) \qquad (2.5.19)$$

(b) *For Lebesgue a.e. x,*

$$\lim_{\varepsilon \downarrow 0} \frac{1}{\pi} \operatorname{Im} F_\mu(x + i\varepsilon) = f(x) \qquad (2.5.20)$$

(c) *For a set $E \subset \mathbb{R}$ with $|E| = 0$ and $\mu_s(\mathbb{R} \setminus E) = 0$, we have for all $x \in E$ that*

$$\operatorname{Im} F_\mu(x + i\varepsilon) \to \infty \qquad (2.5.21)$$

(d) *For all $x_0 \in \mathbb{R}$,*

$$\lim_{\varepsilon \downarrow 0} \varepsilon \operatorname{Im} F_\mu(x_0 + i\varepsilon) = \mu(\{x_0\}) \qquad (2.5.22)$$

Proof. (a) Let $d\mu_\varepsilon$ be the measure on the left in (2.5.19) so $\mu_\varepsilon = P_\varepsilon * \mu$. Thus, since P_ε is even,

$$\int g(x)\,d\mu_\varepsilon(x) = \iint g(x)P_\varepsilon(x-y)\,d\mu(y)\,dx$$

$$= \int (P_\varepsilon * g)(x)\,d\mu(x) \tag{2.5.23}$$

Since $P_\varepsilon * g \to g$ uniformly, we get (2.5.19).

(b), (c) If $f(x)$ is even, C^1, and monotone decreasing in $|x|$ to 0 as $|x| \to \infty$, then (this is related to the wedding-cake representation (2.2.17))

$$f(x) = \int_{|x|}^\infty [-f'(s)]\,ds = \int_0^\infty \frac{1}{2s}\chi_{[-s,s]}(x)\,Q(s)\,ds \tag{2.5.24}$$

where $Q(s) = 2s(-f'(s))$ picked so that

$$\int_{-\infty}^\infty f(x)\,dx = \int_0^\infty Q(s)\,ds \tag{2.5.25}$$

Choosing $f(x) = P_1(x)\,(= \pi^{-1}\frac{1}{x^2+1})$ and using the fact that $\int_{-\infty}^\infty P_1(x)\,dx = 1$ and $P_\delta(x) = \varepsilon^{-1}P_\varepsilon(\frac{x}{\varepsilon})$, we see that

$$\frac{1}{\pi}\operatorname{Im} F_\mu(x+i\varepsilon) = \int_0^\infty Q(s)\frac{1}{2\varepsilon s}\mu(x-\varepsilon s, x+\varepsilon s)\,ds \tag{2.5.26}$$

Thus, since $\int_0^\infty Q(s)\,ds = 1$, Theorem 2.5.1 implies (2.5.20) and (2.5.21).

(d) As $y \downarrow 0$, we have

$$\frac{y^2}{(x-x_0)^2 + y^2} \to \begin{cases} 0 & \text{if } x \neq x_0 \\ 1 & \text{if } x = x_0 \end{cases} \tag{2.5.27}$$

and $|y^2/[(x-x_0)^2 + y^2]| \le 1$, so (2.5.22) follows from the dominated convergence theorem. $\qquad\square$

Theorem 2.5.4 relied on the Cauchy and Poisson kernels $(x-z)^{-1}$ and $y[(x-s)^2 + y^2]^{-1}$ for the upper half-plane. By instead using the Cauchy and Poisson kernels $(\frac{w+z}{w-z})$ and $P_r(\theta, \varphi) = (1-r^2)/(1+r^2 - 2r\cos(\theta-\varphi))$ for the disk, one gets the following (see Problem 2):

Theorem 2.5.5. *Let μ be a finite measure on $\partial\mathbb{D}$ and $F_\mu(z)$ its Carathéodory function*

$$F_\mu(z) = \int \frac{e^{i\theta} + z}{e^{i\theta} - z}\,d\mu(\theta) \tag{2.5.28}$$

Suppose that

$$d\mu(\theta) = w(\theta)\frac{d\theta}{2\pi} + d\mu_s(\theta) \tag{2.5.29}$$

Then

(a) $\operatorname{Re} F_\mu(re^{i\theta})\frac{d\theta}{2\pi} \to d\mu(\theta)$ *weakly as* $r \uparrow 1$.

(b) *For Lebesgue almost every* θ,

$$\lim_{r\uparrow 1} \operatorname{Re} F(re^{i\theta}) = w(\theta) \tag{2.5.30}$$

(c) *For a set* $E \subset \partial\mathbb{D}$ *with* $|E| = 0$ *and* $\mu_s(\partial\mathbb{D} \setminus E) = 0$, *we have for all* $e^{i\theta} \in E$ *that as* $r \uparrow 1$,

$$\operatorname{Re} F_\mu(re^{i\theta}) \to \infty \tag{2.5.31}$$

(d) *For all* $e^{i\theta_0} \in \partial\mathbb{D}$,

$$\lim_{r\uparrow 1} \tfrac{1}{2}(1 - r) \operatorname{Re} F_\mu(re^{i\theta_0}) = \mu(\{e^{i\theta_0}\}) \tag{2.5.32}$$

Notes and Historical Remarks. The parts of this section that involve the finite limit being Radon–Nikodym derivatives go back to Lebesgue's work on differentiation. The fact that limits are infinite on the support of the singular part is a close relative of a classical theorem of de la Vallée Poussin [**206**] proven in 1915. What he showed is that for a monotone function, α, for a.e. x with respect to $dx + d\alpha$, the limit exists (although it may be infinite). This is already implied by Theorem 2.5.1, and once you know the limit exists, it is not hard to see it is infinite for α_{sing}-a.e. x. So the theorem of de la Vallée Poussin is essentially equivalent to part (b) of Theorem 2.5.1.

In the 1990s, this result (via Theorem 2.5.4(c)) became important in the analysis of singular continuous spectrum of some differential and difference operators. Amongst the spectral theory community, the standard reference for the de la Vallée Poussin theorem was Saks [**716**], whose proof is quite involved. That one could get it from a maximal inequality for $M^{(\mu)}$ and the use of the von Neumann trick is something I realized around 1995 and Poltoratski [**661**] also realized in that time period.

For F_μ to exist, one only needs $\int (1+|x|)^{-1}\, d\mu(x) < \infty$, and the harmonic function $\operatorname{Im} F_\mu$ exists if only

$$\int (1 + |x|)^{-2}\, d\mu(x) < \infty \tag{2.5.33}$$

The results in Theorem 2.5.4 are local in x, so they extend to the case where (2.5.33) holds.

What we call the Stieltjes transform is also called the *Borel transform* or the *Cauchy transform*.

Problems

1. Let α be a function of bounded variation on $[0,1]$. Write

$$d\alpha = f\,dx + d\alpha_{\text{sing},+} - d\alpha_{\text{sing},-}$$

where $d\alpha_{\text{sing},\pm}$ come from the Hahn decomposition and are mutually singular to each other and to dx. Prove an extension of Theorem 2.5.1 where (b) is replaced by

$$E_\pm = \left\{ x \,\middle|\, \lim_{\varepsilon\downarrow 0} \frac{1}{2\varepsilon}[\alpha(x+\varepsilon) - \alpha(x-\varepsilon)] = \pm\infty \right\} \qquad (2.5.34)$$

and $|E_+| = |E_-| = 0$, $\alpha_{\text{sing},\pm}(E_\mp) = 0$, $\alpha_{\pm,\text{sing}}(\mathbb{R}\setminus E_\pm) = 0$.

2. For a measure μ on $\partial\mathbb{D}$, define its *Carathéodory function*, $F_\mu(z)$, for $z \in \mathbb{D}$ by

$$F_\mu(z) = \int \frac{e^{i\theta} + z}{e^{i\theta} - z}\,d\mu(e^{i\theta})$$

Extend Theorem 2.5.4 for this situation, replacing $\operatorname{Im} F_\mu$ there by $\operatorname{Re} F_\mu$ here.

2.6. The Maximal and Birkhoff Ergodic Theorems

> The brief history of ergodic theory has had three high points, each of which evoked a flurry of activity in the field. The first was the proof by G. D. Birkhoff in 1931 of the individual ergodic theorem, which could be said to be the beginning of the mathematical theory. The second was the introduction in 1957 by A. N. Kolmogorov of entropy into ergodic theory ... The third was the proof by D. S. Ornstein in 1970 that Bernoulli systems with the same entropy are isomorphic.
>
> —J. Brown [121][2]

As we'll explain in the Notes, ergodic theory, the subject of this section, has its roots in attempts to understand the zeroth law of thermodynamics: that a macroscopic physical system quickly approaches an equilibrium state described by a few parameters, such as temperature and pressure. This is surprising because at a microscopic level, say in a gas with individual gas molecules moving rapidly, it is in any thing but a steady state.

The idea arose that the explanation is that time averages equal space averages. To be explicit, there is a constant energy phase space (the microscopic state of the system at constant energy), Ω, and a dynamics, $T_t\colon \Omega \to \Omega$, $t \in \mathbb{R}$. By Liouville's theorem, Ω has a natural probability measure, $d\mu$, that is left *invariant*, that is,

$$\mu(T_t^{-1}[A]) = \mu(A) \qquad (2.6.1)$$

---

[2]We will not discuss entropy in ergodic theory in these books and so only the first of Brown's high points.

for every measurable set, A, in Ω. Then the idea was that for reasonable functions, f, and almost every $\omega_0 \in \Omega$, we should have

$$\lim_{t \to \infty} \frac{1}{t} \int_0^t f(T_s\omega_0)\,ds = \int f(\omega)\,d\mu(\omega) \qquad (2.6.2)$$

This is normally broken into two parts:

(a) Prove that the limit on the left side of (2.6.2) exists, at least for a.e. ω_0. This will be the issue (or rather, it and its discrete analog) that will concern us in this section.

(b) Call the limit, if it exists, $E(f)(\omega_0)$. It is easy to see (Problem 4) that $E(f)$ is invariant, that is, for all $s \in \mathbb{R}$ and a.e. ω,

$$E(f)(T_s\omega) = E(f)(\omega) \qquad (2.6.3)$$

and that

$$\int E(f)(\omega)\,d\mu(\omega) = \int f(\omega)\,d\mu(\omega) \qquad (2.6.4)$$

The measure μ is called *ergodic* for the flow, T_t, if and only if

$$\mu(T_t^{-1}[A] \triangle A) = 0 \Rightarrow \mu(A) = 0 \quad \text{or} \quad \mu(A) = 1 \qquad (2.6.5)$$

In that case (Problem 5), every function, $E(f)$, obeying (2.6.3) is a.e. constant. (2.6.4) then implies (2.6.2). The issue of which measures are ergodic for which flows is a complicated one that we'll discuss a little in the Notes and in the next two sections.

As we mentioned, we have the discrete analog, $\{T^n\}_{n=0}^\infty$, and the flows, $\{T_t\}_{t \geq 0}$. Problem 7 shows how to go from discrete results to continuous and Problem 8 shows how to go in the opposite direction. We'll discuss both cases, providing two proofs of the Birkhoff theorem, one for each case. One proof (see Theorem 2.6.15) illustrates the close connection to the Hardy–Littlewood maximal inequality and, in essence, deduces the Birkhoff theorem from little more than the Hardy–Littlewood result.

We begin our discussion with some formal definitions:

Definition. A *measurable dynamical system* is a quadruple (Ω, Σ, μ, T) where (Ω, Σ, μ) is a probability measure space and $T \colon \Omega \to \Omega$ a measurable transformation with

$$\mu(T^{-1}[A]) = \mu(A) \qquad (2.6.6)$$

We'll then say that T is a *measure-preserving map*.

Remarks. 1. Invariant measures for continuous maps on compact Hausdorff spaces were discussed in Sections 5.11 and 5.12 of Part 1.

2. As usual, we'll normally only consider probability measure spaces for which $L^2(\Omega, d\mu)$ is separable.

We claim that (2.6.6) is equivalent to

$$\int f(T\omega)\,d\mu(\omega) = \int f(\omega)\,d\mu(\omega) \qquad (2.6.7)$$

for all bounded measurable functions (and from this, one easily sees that $f \in L^1(\Omega, d\mu)$ implies $f \circ T \in L^1(\Omega, d\mu)$ and (2.6.7) holds for such f). For if (2.6.7) holds and we apply it to $f(\omega) = \chi_A(\omega)$, the characteristic function of A, then $f \circ T = \chi_{T^{-1}[A]}$ and (2.6.7) for this f is (2.6.6). Conversely, if (2.6.6) holds, then (2.6.7) holds first for $\{\chi_A\}$, then for simple functions, and finally, using the dominated convergence theorem, for any bounded measurable f.

Similar to (2.6.7), we have if $f \in L^p(\Omega, d\mu)$ for any $1 \le p \le \infty$, then $f \circ T \in L^p$ and

$$\int |f(T\omega)|^p\,d\mu(\omega) = \int |f(\omega)|^p\,d\mu(\omega) \qquad (2.6.8)$$

In particular, if T is invertible and $p = 2$,

$$(Uf)(\omega) = f(T\omega) \qquad (2.6.9)$$

is a unitary operator called the *Koopman unitary*.

Here is one way to prove invariance:

Proposition 2.6.1. *Let μ be a Baire probability measure on a compact metric space, X, and $T\colon X \to X$ a continuous transformation. Let $\mathcal{U}$ be a family of open sets in X, which is a base and which is closed under finite intersection. Suppose (2.6.6) holds for all $A \in \mathcal{U}$. Then (X, Σ, μ, T) (with Σ the Baire sets) is a measurable dynamical system.*

Proof. Since

$$\mu(A_1 \cup A_2) = \mu(A_1) + \mu(A_2) - \mu(A_1 \cap A_2) \qquad (2.6.10)$$

we see inductively that (2.6.6) holds for finite unions of sets in $\mathcal{U}$ and so, since $\mu(\bigcup_{n=1}^{\infty} A_n) = \lim_{N\to\infty} \mu(\bigcup_{n=1}^{N} A_n)$ for all countable unions. Thus, (2.6.6) holds for all open sets, A.

By regularity, for any Baire set, A,

$$\mu(A) = \inf_{U \subset A} \mu(U) = \inf_{U \subset A} \mu(T^{-1}[U]) \ge \mu(T^{-1}[A])$$

Since this holds also for $X \setminus A$, we get equality, that is, (2.6.6) holds for all Baire sets, A. $\qquad\square$

Remark. In particular, if $X = \mathbb{R}$, an interval in $\mathbb{R}$, or $\partial\mathbb{D}$, it is enough to prove (2.6.6) for all intervals.

Definition. A *topological dynamical system* is a pair (X, T), where X is a metric space and $T\colon X \to X$ a continuous map.

Given such a system, define $T^*\colon \mathcal{M}_{+,1}(X) \to \mathcal{M}_{+,1}(X)$, a map on the Baire probability measures on X, by

$$\int f(x)\, d(T^*\mu)(x) = \int f(Tx)\, d\mu(x) \qquad (2.6.11)$$

for all $f \in C(X)$. This is equivalent to

$$(T^*\mu)(A) = \mu(T^{-1}[A]) \qquad (2.6.12)$$

for all Baire sets, A. μ is called an *invariant measure* for (X, T) if and only if $T^*\mu = \mu$. If Σ is the σ-field of all Baire sets, (X, Σ, μ, T) is then a measurable dynamical system. As we already noted in Section 5.12 of Part 1, $\mathcal{I}(X, T)$, the set of such invariant measures, is nonempty since any limit point of $\{\frac{1}{n}\sum_{j=0}^{n-1}(T^*)^j(\mu)\}$ is invariant and, by the weak-* compactness of $\mathcal{M}_{+,1}(X)$, there are such limit points. $\mathcal{I}(X, T)$ is a compact convex set in the weak-* topology.

Example 2.6.2 (Bernoulli Shift). Fix $p \in (0, 1)$. Let $\Omega = \{0, 1\}^{\mathbb{Z}_+}$ and $d\mu = \otimes_{n=1}^{\infty} d\lambda(\omega_j)$, where $d\lambda$ is a measure with $\lambda(\{0\}) = p$, $\lambda(\{1\}) = 1 - p$ (infinite product measures are discussed in Section 4.12 of Part 1). Let $T\colon \Omega \to \Omega$ by $T(\omega)_j = \omega_{j+1}$, that is, drop ω_1 and shift left.

We claim T is invariant for μ. For (2.6.7) holds trivially for f of the form $\prod_{j=1}^{n} f_j(\omega_j)$ with f_j a map of $\{0, 1\}$ to $\mathbb{R}$. Finite sums of such products are dense in $L^2(\Omega, d\mu)$, so (2.6.7) holds for all $f \in L^2$ and so for all bounded f. This example is called the *Bernoulli shift* and will be discussed further in Section 2.7.1. In particular, we'll prove it is ergodic. $\qquad \square$

Example 2.6.3 (Rotations). Fix $\alpha \in [0, 2\pi)$. On $\partial\mathbb{D}$, take $d\mu = \frac{d\theta}{2\pi}$ and let

$$T(e^{i\theta}) = e^{i(\theta+\alpha)} \qquad (2.6.13)$$

It is obvious for any interval $\{e^{i\theta} \mid \theta_0 < \theta < \theta_1\}$ that (2.6.6) holds, so T is invariant. This example when $\alpha/2\pi$ is irrational will be discussed further in Section 2.7.2. In particular, we'll prove it is ergodic. $\qquad \square$

Definition. A *measurable semiflow* is a quadruple, $(\Omega, \Sigma, \mu, \{T_t\}_{t \geq 0})$, where (Ω, Σ, μ) is a probability measure space, $T_t\colon \Omega \to \Omega$ a family of maps so that $(\omega, t) \to T_t(\omega)$ is a measurable function of $\Omega \times [0, \infty) \to \Omega$ and so that

$$T_0 = \mathbb{1}, \qquad T_t T_s = T_{t+s} \qquad (2.6.14)$$

If each T_t is measure-preserving, we have a *measure-preserving semiflow*. A *flow* has $\{T_t\}_{t \in \mathbb{R}}$ so that (2.6.14) still holds and we have measurability of the map from $\Omega \times \mathbb{R}$ to Ω.

For invertible maps and flows on smooth manifolds, with smooth measures, one can check measure preserving by computing Jacobians and using change of variable formulae in (2.6.7).

In an obvious way, if $\{(\Omega_j, \Sigma_j, \mu_j, T_j)\}_{j=1}^n$ are measurable dynamical systems, $T_1 \otimes \cdots \otimes T_n$ by $(T_1 \otimes \cdots \otimes T_n)(\omega_1, \ldots, \omega_n) = (T_1(\omega_1), \ldots, T_n(\omega_n))$ is a measurable dynamical system on the product space.

If $(\Omega_1, \Sigma_1, \mu_1, T_1)$ and $(\Omega_2, \Sigma_2, \mu_2, T_2)$ are measurable dynamical systems, we'll be interested in measurable maps $\phi = \Omega_2 \to \Omega_1$ obeying:

(i) $\mu_2(\phi^{-1}[A]) = \mu_1(A)$ for all $A \in \Sigma_1$

(ii) $T_1(\phi(\omega_2)) = \phi(T_2(\omega_2))$ for a.e. $\omega_2 \in \Omega_2$, that is,

$$
\begin{array}{ccc}
\Omega_2 & \xrightarrow{T_2} & \Omega_2 \\
\phi \downarrow & & \downarrow \phi \\
\Omega_1 & \xrightarrow{T_1} & \Omega_1
\end{array}
$$

commutes. In this case, we say T_2 is an *extension* of T_1 and that T_1 is a *factor* of T_2 (more properly, we should talk about quadruples rather than T_1 and T_2).

If there are sets of measure zero $S_j \subset \Omega_j$ so that $\phi \upharpoonright \Omega_2 \setminus S_2$ is a bijection to $\Omega_1 \setminus S_1$, we say the measurable dynamical systems are *isomorphic*.

Example 2.6.2 shows measure-preserving T's may not be invertible. If $\mathrm{Ran}(T)$ is measurable, they are "almost" onto, since $\mu(\Omega \setminus \mathrm{Ran}(T)) = 0$, but they may fail to be one-one; Example 2.6.2 is two-one.

If you consider the two-sided shift: $\widetilde{\Omega} = \{0,1\}^{\mathbb{Z}}$ with $\widetilde{T}(\omega)_j = \omega_{j+1}$, then $\widetilde{T}$ is a bijection. Moreover, letting ϕ be the obvious map $\phi(\omega)_j = \omega_j$ $(j = 1, 2, \ldots)$, we see that $\widetilde{T}$ is an extension of T. Problem 19 shows this is a general phenomenon—every measurable map has an invertible extension, indeed, a minimal one (see the remark after Problem 19).

Example 2.6.4 (Doubling Map). Let $\Omega = [0,1]$, $d\mu = dx$, and

$$
T(x) = \begin{cases} 2x & \text{if } x \in [0, \frac{1}{2}] \\ 2x - 1 & \text{if } x \in (\frac{1}{2}, 1] \end{cases} \tag{2.6.15}
$$

It is easy to see that $\mu(T^{-1}(I)) = \mu(I)$ for any interval, and while T is not continuous, it is except for one point and the argument of Proposition 2.6.1 extends, so T is measure-preserving. If $\phi \colon \{0,1\}^{\mathbb{Z}_+} \to [0,1]$ by $\phi(\{\omega_j\}_{j=1}^{\infty}) = \sum_{j=1}^{\infty} \omega_j / 2^j$, then ϕ is a bijection up to sets of measure zero and shows that this map is isomorphic to the Bernoulli shift for $p = \frac{1}{2}$. This example will be discussed further in Problem 2 and Section 2.8.2. $\square$

To state the ergodic theorems, we'll need the notion of conditional expectation. As a preliminary, we note the following theorem, of general interest.

Theorem 2.6.5. *Let (Ω, Σ, μ) be a probability measure space and S a linear map of the bounded measurable functions to the measurable functions. Suppose that*

(i) $f \geq 0 \Rightarrow Sf \geq 0$

(ii) $S\mathbb{1} = \mathbb{1}$ *where* $\mathbb{1}(\omega) = 1$ *for all* ω.

(iii) $\displaystyle\int (Sf)(\omega)\, d\mu(\omega) = \int f(\omega)\, d\mu(\omega)$ \hfill (2.6.16)

for all $f \in L^\infty(\Omega, d\mu)$. Then

$$\|Sf\|_p \leq \|f\|_p \tag{2.6.17}$$

for any $p \in [1, \infty]$.

Remarks. 1. All we use below is condition (i), $S\mathbb{1} \leq \mathbb{1}$, and for $f \geq 0$, $\int Sf(\omega)\, d\mu(\omega) \leq \int f(\omega)\, d\mu(\omega)$.

2. A map S obeying (i)–(iii) is called *doubly stochastic*.

3. A condition like (i) really means that $f(x) \geq 0$ for a.e. x implies $(Sf)(x) \geq 0$ for a.e. x. A careful proof going from (2.6.18) to (2.6.19) would consider only rational θ to use only countably many sets of measure zero.

Proof. By (i), S takes real-valued functions into real-valued functions. If $g \in L^\infty$, then

$$\mathrm{Re}[S(e^{i\theta} g)] = S(\mathrm{Re}(e^{i\theta} g)) \leq S|g| \tag{2.6.18}$$

so that

$$|Sg| \leq S|g| \tag{2.6.19}$$

Thus, since $|g| \leq \|g\|_\infty \mathbb{1}$,

$$|Sg| \leq \|g\|_\infty S\mathbb{1} = \|g\|_\infty \mathbb{1} \tag{2.6.20}$$

so (2.6.17) holds for $p = \infty$.

By (2.6.16) and (2.6.19), (2.6.17) holds for $p = 1$. By the Riesz–Thorin interpolation theorem (see Theorem 6.6.1 of Part 1 and/or Theorem 5.2.2 of Part 2A), we get (2.6.17) for all p. $\qquad\square$

Theorem 2.6.6. *Let (Ω, Σ, μ) be a probability measure space and $\Sigma' \subset \Sigma$ a subalgebra. Let $\mu' = \mu \restriction \Sigma'$. For any $f \in L^\infty(\Omega, d\mu)$, there is a unique function $g \in L^1(\Omega, d\mu')$ so that for all $h \in L^\infty(\Omega, d\mu')$, we have*

$$\int h(\omega) f(\omega)\, d\mu(\omega) = \int h(\omega) g(\omega)\, d\mu'(\omega) \tag{2.6.21}$$

If we write

$$g = \mathbb{E}(f \mid \Sigma') \tag{2.6.22}$$

then for $f \in L^\infty(\Omega, d\mu)$,

$$\|\mathbb{E}(f \mid \Sigma')\|_{L^p(\Omega, d\mu')} \leq \|f\|_{L^p(\Omega, d\mu)} \tag{2.6.23}$$

(so that $\mathbb{E}(\,\cdot\, \mid \Sigma')$ extend to a map of L^p to L^p).

Remarks. 1. $\mathbb{E}(\,\cdot\,\mid\Sigma')$ is called *conditional expectation*.

2. If $A\in\Sigma$ and $\Sigma'=\{\emptyset,\Omega,A,A^c\equiv\Omega\setminus A\}$, and if $f=\chi_B$ for $B\in\Sigma$, then

$$\mathbb{E}(f\mid\Sigma')(\omega)=\begin{cases}\frac{\mu(A\cap B)}{\mu(A)}, & \omega\in A\\[4pt]\frac{\mu(A^c\cap B)}{\mu(A^c)}, & \omega\in A^c\end{cases}$$

Thus, $\mathbb{E}(\,\cdot\,\mid\Sigma')$ generalizes the naive conditional probabilities.

Proof. We can view $L^p(\Omega,d\mu')$ as a subspace of $L^p(\Omega,d\mu)$ which is closed since $L^p(\Omega,d\mu')$ is complete. In particular, $\mathcal{H}'\equiv L^2(\Omega,d\mu')$ is a closed space of $\mathcal{H}\equiv L^2(\Omega,d\mu)$, so there is an orthogonal projection $P\colon\mathcal{H}\to\mathcal{H}'$ which obeys

$$\langle h,Pf\rangle=\langle h,f\rangle \tag{2.6.24}$$

for all $h\in\mathcal{H}'$ and $f\in\mathcal{H}$. Thus, $g=Pf$ satisfies (2.6.21), proving existence.

Since $\mathbb{1}\in\mathcal{H}'$, $P\mathbb{1}=\mathbb{1}$. Since P is self-adjoint, this also implies that

$$\langle\mathbb{1},Pf\rangle=\langle P\mathbb{1},f\rangle=\langle\mathbb{1},f\rangle \tag{2.6.25}$$

which is (2.6.16) for P. If $f\geq 0$ and h is the characteristic function of $\{\omega\mid -a<(Pf)(\omega)<0\}$, then the left side of (2.6.21) is nonnegative and the right nonpositive, so both sides are 0. It follows that $Pf\geq 0$. Thus, by Theorem 2.6.5, we have (2.6.23).

We leave uniqueness to Problem 9. $\qquad\square$

In the context of our measure-preserving map, T, we define $\mathcal{I}$ to be the family of subsets, A, in Σ so that

$$\mu(A\triangle T^{-1}[A])=0 \tag{2.6.26}$$

usually phrased as "$A=T[A]$ up to sets of measure zero." $\mathcal{I}$ is a Σ-algebra and the $\mathcal{I}$ measurable functions are precisely those with

$$f(T\omega)=f(\omega) \tag{2.6.27}$$

for a.e. ω. When T is invertible, it is easy to see (Problem 1) that for every $A\in\mathcal{I}$, there is $B\subset A$ so that $T^{-1}[B]=B$ and $\mu(A\setminus B)=0$.

We call μ *ergodic* for T if and only if

$$A\in\mathcal{I}\Rightarrow\mu(A)=0\quad\text{or}\quad\mu(\Omega\setminus A)=0 \tag{2.6.28}$$

That is true (Problem 5) if and only if the only functions obeying (2.6.27) are multiples of $\mathbb{1}$ (up to sets of measure zero), and then,

$$\mathbb{E}(f\mid\mathcal{I})=\left(\int f(\omega)\,d\mu(\omega)\right)\mathbb{1} \tag{2.6.29}$$

What we (and everyone else) now call ergodic was initially called *metrical transitivity*, a name one still finds in some of the older literature.

We also define, given f and T,

$$Av_n(\omega; f) = \frac{1}{n} \sum_{j=0}^{n-1} f(T^j \omega) \tag{2.6.30}$$

With this background, we can state the first of two main theorems of this section whose proof appears later:

Theorem 2.6.7 (von Neumann Ergodic Theorem). *Let (Ω, Σ, μ, T) be a measurable dynamical system. Then for every $f \in L^2(\Omega, d\mu)$, we have that, as $n \to \infty$, in $\|\cdot\|_2$,*

$$Av_n(\,\cdot\,; f) \to \mathbb{E}(f \mid \mathcal{I}) \tag{2.6.31}$$

In particular, if μ is ergodic for T and $\mu(f) = \int f(\omega) \, d\mu(\omega)$, then as $n \to \infty$,

$$\int |Av_n(\omega; f) - \mu(f)|^2 \, d\mu(\omega) \to 0 \tag{2.6.32}$$

Remarks. 1. This is also true for L^p, for any $p < \infty$ (see Problem 11).

2. This is sometimes called the *mean ergodic theorem*.

3. You'll prove a generalization of this in Problem 12.

An immediate consequence of this is that μ is ergodic if and only if for every $f, g \in L^2(X, d\mu)$,

$$\lim_{n \to \infty} \frac{1}{n} \sum_{j=0}^{n-1} [\langle f, g \circ T^j \rangle - \langle f, \mathbb{1} \rangle \langle \mathbb{1}, g \rangle] = 0 \tag{2.6.33}$$

This in turn implies that

Theorem 2.6.8. *Let (X, T) be a topological dynamical system. Then $\mu \in \mathcal{I}(X, T)$ is ergodic if and only if μ is an extreme point of $\mathcal{I}(X, T)$. In particular (by the Krein–Millman theorem), there are always ergodic measures and all invariant measures are ergodic if and only if $\mathcal{I}(X, T)$ has a unique element.*

Remarks. 1. If $\mathcal{I}(X, T)$ has a unique element, (X, T) is called *uniquely ergodic*.

2. See Theorems 2.7.7 and 2.8.7 for more on unique ergodicity.

Proof. If $\mu(B) \neq 0$, define μ_B, a probability measure, by

$$\int f(x) \, d\mu_B(x) = \mu(B)^{-1} \int f(x) \chi_B(x) \, d\mu(x) \tag{2.6.34}$$

Since B invariant implies $\chi_B(Tx) = \chi_B(x)$ for a.e. x, we see μ_B is invariant if μ is.

If $\mu \in \mathcal{I}(X,T)$ is not ergodic, then with B an invariant set with $0 < \mu(B) < 1$, we can write

$$\mu = \mu(B)\mu_B + (1 - \mu(B))\mu_{X \setminus B} \qquad (2.6.35)$$

so, since μ_B and $\mu_{X \setminus B}$ are mutually singular, they are distinct measures, and we see that μ is not an extreme point of $\mathcal{I}(X,T)$. We have thus proven $\mu \in \mathcal{E}(\mathcal{I}(X,T)) \Rightarrow \mu$ is ergodic.

For the converse, suppose μ is ergodic and for $0 < \theta < 1$ and $\mu_1, \mu_2 \in \mathcal{I}(X,T)$, we have

$$\mu = \theta\mu_1 + (1 - \theta)\mu_2 \qquad (2.6.36)$$

We want to first show that μ_1 and μ_2 are ergodic. As a preliminary, we claim (Problem 2), if (Ω, Σ, ν, T) is a measurable dynamical system, it is ergodic if and only if for all bounded measurable f, we have

$$\lim_{n \to \infty} \int |Av_n(\omega_1; f) - Av_n(\omega_2; f)|^2 \, d\nu(\omega_1) d\nu(\omega_2) = 0 \qquad (2.6.37)$$

Since $d\mu_1 \otimes d\mu_1 \leq \theta^{-2}(d\mu \otimes d\mu)$, we see (2.6.37) for μ implies it for μ_1, and thus, μ ergodic $\Rightarrow \mu_1$ ergodic.

Applying (2.6.33) to $f = g$, a real-valued, continuous function on X, for the ergodic measures μ, μ_1, μ_2 implies, with $\gamma_j = \int f(x) \, d\mu_j(x)$, that

$$(\theta\gamma_1 + (1 - \theta)\gamma_2)^2 = \theta\gamma_1^2 + (1 - \theta)\gamma_2^2 \qquad (2.6.38)$$

Since $x \to x^2$ is strictly convex, $\gamma_1 = \gamma_2$, that is, for $f \in C(X)$,

$$\int f(x) \, d\mu_1(x) = \int f(x) \, d\mu_2(x) \qquad (2.6.39)$$

Thus, $\mu_1 = \mu_2$ and so, μ is extreme, that is,

$$\mu \text{ is ergodic} \Rightarrow \mu \in \mathcal{E}(\mathcal{I}(X,T)) \qquad \square$$

We see in the Problems (Problem 22) that this result implies that any two ergodic measures are mutually singular. The second main result (also proven later) is:

Theorem 2.6.9 (Birkhoff Ergodic Theorem). *Let (Ω, Σ, μ, T) be a measurable dynamical system. Then for every $f \in L^1(\Omega, d\mu)$, we have that for a.e. ω as $n \to \infty$,*

$$Av_n(\omega; f) \to \mathbb{E}(f \mid \mathcal{I})(\omega) \qquad (2.6.40)$$

In particular, if μ is ergodic for T, then

$$Av_n(\omega; f) \to \int f(\omega) \, d\mu(\omega) \qquad (2.6.41)$$

for a.e. ω.

Remark. This is sometimes called the *individual ergodic theorem* or *pointwise ergodic theorem*.

From this, it is easy to see what happens if $f \geq 0$ but $\int f(\omega) \, d\mu(\omega) = \infty$:

Corollary 2.6.10. *Let* (Ω, Σ, μ, T) *be a measurable dynamical system and suppose* μ *is ergodic. Let* $f \geq 0$ *be measurable with*

$$\int f(\omega) \, d\mu(\omega) = \infty \tag{2.6.42}$$

Then for a.e. ω, *we have*

$$\lim_{n \to \infty} Av_n(\omega; f) = \infty \tag{2.6.43}$$

Proof. For each $m = 1, 2, \ldots$, $f_m(\omega) = \min(f(\omega), m)$ is in L^1, so by (2.6.41) and $f \geq f_m$, we have for a.e. ω that

$$\liminf_{n \to \infty} Av_n(\omega; f) \geq \int f_m(\omega) \, d\mu(\omega) \tag{2.6.44}$$

By the monotone convergence theorem, $\lim_{m \to \infty} \int f_m(\omega) \, d\mu(\omega) = \infty$, so (2.6.43) holds for a.e. ω. □

As you might expect, given what's earlier in the chapter, we'll prove the Birkhoff ergodic theorem from a suitable maximal inequality. Let

$$f^*(\omega) = \sup_n |Av_n(\omega; f)| \tag{2.6.45}$$

Theorem 2.6.11 (Maximal Ergodic Inequality). *For any* $\alpha > 0$ *and* $f \in L^1(\Omega, d\mu)$, *we have that*

$$\mu\{\omega \mid f^*(\omega) > \alpha\} \leq \alpha^{-1} \|f\|_1 \tag{2.6.46}$$

Remark. We'll prove this below using a distinct result with a similar name: Hopf–Kakutani–Yoshida maximal ergodic theorem.

The part of the machinery that requires finding a dense set with pointwise convergence a.e. is easy, and we'll need something similar for L^2-convergence. We begin by noting two obvious cases where the limit exists:

Proposition 2.6.12. (a) *Let* f *be an invariant function, that is, for a.e.* ω, *(2.6.27) holds. Then* $Av_n(\omega; f) = f(\omega)$, *so for a.e.* ω, $Av_n(\omega; f) \to f(\omega)$ *and* $Av_n(\omega; f) \to f(\omega)$ *in* L^p *if* $f \in L^p$.

(b) *Let* $g \in L^\infty(\Omega, d\mu)$ *and let* $f = g - g \circ T$. *Then for a.e.* ω, $Av_n(\omega; f) \to 0$ *and for all* p, $1 \leq p \leq \infty$, $\|Av_n(\,\cdot\,; f)\|_p \to \infty$.

Proof. (a) is trivial. As for (b), due to cancellations,

$$Av_n(\omega; f) = \frac{1}{n}[g(\omega) - g(T^n \omega)] \tag{2.6.47}$$

so $\|Av_n\|_\infty \leq 2n^{-1}\|g\|_\infty \to 0$, proving the result. □

Proposition 2.6.13. *Let Y be the set of functions $f = f_1 + f_2$, where $f_1 \in L^\infty$, f_1 is invariant, and $f_2 = g - g \circ T$ with $g \in L^\infty$. Then Y is dense in every L^p, $1 \le p \le 2$.*

Remarks. 1. This is also true for $2 < p < \infty$; see Problem 10.

2. See Theorem 2.7.7 for a discussion of when the analog is true for $C(X)$.

Proof. Let q be the dual index of p. Since $(L^p)^* = L^q$ (see Theorem 4.9.1 of Part 1), the Hahn–Banach theorem (in the form Theorem 5.5.10 of Part 1) implies that if $\overline{Y} \neq L^p$, there is $h \in L^q$, $h \neq 0$, so that

$$\int h(\omega) f(\omega) \, d\mu(\omega) = 0 \qquad (2.6.48)$$

for all invariant L^∞ f's and all f's of the form $g - g \circ T$, $g \in L^\infty$.

Thus, for $g \in L^\infty$, and then, by taking limits, for $g \in L^p$, we have

$$\int h(\omega) g(\omega) \, d\mu(\omega) = \int h(\omega) g(T\omega) \, d\mu(\omega) \qquad (2.6.49)$$

Since $p \le 2$ and μ is a probability measure, $L^q \subset L^p$ so we can take $g = \bar{h}$. By the invariance of μ, we have

$$\int |h(\omega)|^2 \, d\mu(\omega) = \int |h(T\omega)|^2 \, d\mu = \int h(\omega) \, \overline{h(T\omega)} \, d\mu(\omega) \qquad (2.6.50)$$

so

$$\int |h(\omega) - h(T\omega)|^2 \, d\mu(\omega) = 0 \qquad (2.6.51)$$

Therefore, h is invariant.

Taking limits, (2.6.48) holds for all invariant f in L^p and so, since $L^q \subset L^p$, for $f = \bar{h}$, that is,

$$\int |h(\omega)|^2 \, d\mu(\omega) = 0 \qquad (2.6.52)$$

which contradicts the fact that $h \neq 0$. $\qquad \square$

Proof of Theorem 2.6.7. $f \mapsto Av_n(\,\cdot\,; f)$ is a bounded linear operator on L^2 of norm at most 1 since $\|f \circ T^n\|_2 = \|f\|_2$. (In fact, since $Av_n(\,\cdot\,; \mathbb{1}) = \mathbb{1}$, the norm is 1.) Thus, Av_n is a family of uniformly equicontinuous functions on L^2; explicitly,

$$\|Av_n(\,\cdot\,; f) - Av_n(\,\cdot\,; g)\|_2 \le \|f - g\|_2 \qquad (2.6.53)$$

By the same argument that proves a uniform limit of continuous functions is continuous, knowing (2.6.31) for a dense set proves it for all $f \in L^2$. $\quad \square$

For $f \in L^1$, real-valued, define (no $|\cdot|$)

$$f^\sharp(\omega) = \sup_{1 \le j \le \infty} Av_n(\omega; f) \qquad (2.6.54)$$

Our first proof of Theorem 2.6.11 will depend on:

Theorem 2.6.14 (Hopf–Kakutani–Yoshida Maximal Ergodic Theorem).
Under the hypotheses of Theorem 2.6.11, we have that

$$\int_{\{\omega \mid f^\sharp(\omega) > 0\}} f(\omega)\, d\mu(\omega) \geq 0 \tag{2.6.55}$$

Remark. Bonus Section 2.9 has an alternate proof of this result.

Proof. Let $g_0 = 0$ and for $n \geq 1$, define

$$g_n = nAv_n(\,\cdot\,; f) = \sum_{j=0}^{n-1} f \circ T^j \tag{2.6.56}$$

$$h_n = \max_{0 \leq j \leq n} g_n \tag{2.6.57}$$

$$E_n = \{\omega \mid h_n(\omega) > 0\} \tag{2.6.58}$$

Then $f + g_n \circ T = g_{n+1}$, so

$$f + h_n \circ T = \max_{0 \leq j \leq n} (f + g_j \circ T) = \max_{1 \leq j \leq n+1} g_j \geq \max_{1 \leq j \leq n} g_j \tag{2.6.59}$$

If $\omega \in E_n$, $h_n(\omega) > 0$, so by (2.6.59) and $g_0 \equiv 0$, on E_n,

$$f + h_n \circ T \geq h_n \quad \text{on } E_n \tag{2.6.60}$$

Therefore,

$$\int_{E_n} f(\omega)\, d\mu(\omega) \geq \int_{E_n} h_n(\omega)\, d\mu(\omega) - \int_{E_n} (h_n \circ T)(\omega)\, d\mu(\omega) \tag{2.6.61}$$

Since $g_0 = 0$, $h_n \geq 0$, so

$$\int_{E_n} h_n(\omega)\, d\mu(\omega) = \int_\Omega h_n(\omega)\, d\mu(\omega) = \int_\Omega (h_n \circ T)(\omega)\, d\mu(\omega)$$

which implies

$$\text{RHS of (2.6.61)} = \int_{\Omega \setminus E_n} h_n(T\omega)\, d\omega \geq 0 \tag{2.6.62}$$

since $h_n \geq 0$. Thus,

$$\int \chi_{E_n}(\omega) f(\omega)\, d\mu(\omega) \geq 0 \tag{2.6.63}$$

Notice that $h_n > 0 \Leftrightarrow$ some $g_j > 0$, $j = 1, 2, \ldots, n \Leftrightarrow$ some $Av_n(\,\cdot\,; f) > 0$, so $E_n = \{\omega \mid \max_{1 \leq j \leq n} Av_j(\,\cdot\,; f) > 0\}$. Thus, if $E = \{\omega \mid f^\sharp(\omega) > 0\}$, χ_{E_n} converges pointwise to χ_E, so (2.6.63) implies (2.6.55) by the dominated convergence theorem. $\qquad\square$

Proof of Theorem 2.6.11. Since $|Av_n(\omega; f)| \leq Av_n(\omega; |f|)$, without loss we can take $f \geq 0$. For $\alpha > 0$, let $g = f - \alpha \mathbb{1}$ and apply Theorem 2.6.14 to g. If we note

$$Av_n(\omega; g) = Av_n(\omega; f) - \alpha \mathbb{1} \tag{2.6.64}$$

we see that

$$g^\sharp(\omega) > 0 \Leftrightarrow f^*(\omega) > \alpha \tag{2.6.65}$$

and (2.6.55) for g becomes

$$\int_{\{f^*(\omega) > \alpha\}} (f(\omega) - \alpha) \, d\mu(\omega) \geq 0 \tag{2.6.66}$$

so, since $f \geq 0$,

$$\alpha\mu(\{\omega \mid f^*(\omega) > \alpha\}) \leq \int_{\{f^*(\omega) > \alpha\}} f(\omega) \, d\mu(\omega) \leq \|f_1\| \qquad \square$$

Proof of Theorem 2.6.9. By Propositions 2.6.12 and 2.6.13, we have convergence a.e. for a dense set in L^1. By Theorem 2.6.11, we have a maximal inequality, so Theorem 2.6.5 implies limits for all $f \in L^1$. $\qquad \square$

Remark. Bonus Section 2.9 has a proof of Theorem 2.6.9 that goes directly from Theorem 2.6.14 without recourse to maximal functions and Theorem 2.1.1.

The rest of this section provides an alternate proof of the maximal ergodic inequality. As mentioned, there is a close relation between the Hardy–Littlewood maximal inequality and the maximal ergodic inequality. We'll suppose one knows

$$|\{x \mid (M_{\mathrm{R}}f)(x) > \alpha\}| \leq C\alpha^{-1}\|f\|_1 \tag{2.6.67}$$

where $M_{\mathrm{R}}f$ is given by (2.3.6). The Vitali covering proof gets this with $C = 6$ (or perhaps, $C = 3$; see the remark after the proof of Theorem 2.3.1), the Garsia covering proof with $C = 2$, and the Riesz sunrise lemma proof with $C = 1$.

Theorem 2.6.15. *Let $(\Omega, \Sigma, \mu, \{T_t\}_{t \geq 0})$ be a measure-preserving semiflow. For $t > 0$ and $f \in L^1(\Omega, d\mu)$, let $Av_t(\omega; f)$ be given by*

$$Av_t(\omega; f) = \frac{1}{t} \int_0^t f(T_s\omega) \, ds \tag{2.6.68}$$

and

$$f^*(\omega) = \sup_{0 < t \leq \infty} |Av_t(\omega; f)| \tag{2.6.69}$$

Then

$$\mu(\{\omega \mid f^*(\omega) > \alpha\}) \leq C\alpha^{-1}\|f\|_1 \tag{2.6.70}$$

where C is the constant in (2.6.67).

Remarks. 1. As we'll see, the proof relies quite directly on the Hardy–Littlewood maximal inequality and little more. Remarkably, we'll also see (Problem 20) that this theorem for the special case $\Omega = \partial \mathbb{D}$, $d\mu = \frac{d\theta}{2\pi}$, $T_t(e^{i\theta}) = e^{i(\theta+t)}$ implies the Hardy–Littlewood inequality!

2. There is an immediate variant of the proof which gives the discrete case, that is, Theorem 2.6.11, from the discrete Hardy–Littlewood inequality (see Problem 9 of Section 2.3).

3. As usual, a theorem like this and Theorem 2.1.1 imply a Birkhoff theorem.

Proof. Since $f^* \leq |f|^*$, we can and will suppose that $f \geq 0$. Since invariance of μ implies

$$\int_0^n \left[\int f(T_s\omega)\, d\mu(\omega) \right] ds = n\|f\|_1 \tag{2.6.71}$$

Fubini's theorem then implies that for a.e. ω and all n, $\int_0^n f(T_s\omega)\, ds < \infty$, that is, if we let

$$\widetilde{\Omega} = \{\omega \mid \forall t, \int_0^t f(T_s\omega)\, ds < \infty\} \tag{2.6.72}$$

then $\mu(\widetilde{\Omega}) = 1$. Fix $0 < S < T$ (we'll eventually take $T \to \infty$ and then $S \to \infty$) and $\omega_0 \in \widetilde{\Omega}$. Define

$$\varphi_{T,\omega_0}(s) = \begin{cases} f(T_s\omega_0), & 0 < s < T \\ 0, & s \leq 0 \text{ or } s \geq T \end{cases} \tag{2.6.73}$$

and

$$f_S^*(\omega_0) = \sup_{0 \leq t \leq S} Av_t(\omega_0; f) \tag{2.6.74}$$

Thus, if $0 \leq s \leq T - S$,

$$f_S^*(T_s\omega_0) = \sup_{0 \leq t \leq S} \frac{1}{t} \int_s^{s+t} f(T_u(\omega_0))\, du$$

$$\leq \sup_{0 \leq t < \infty} \frac{1}{t} \int_0^t \varphi_{T,\omega_0}(s+u)\, du$$

$$= (M_{\mathrm{R}}(\varphi_{T,\omega_0}))(s) \tag{2.6.75}$$

so, by (2.6.67),

$$|\{s \in [0, T-S] \mid f_S^*(T_s\omega_0) > \alpha\}| \leq C\alpha^{-1}\|\varphi_{T,\omega_0}\|_1 \tag{2.6.76}$$

Now integrate over $d\mu(\omega_0)$ using $\mu(\Omega \setminus \widetilde{\Omega}) = 0$ to see that

$$\mu \otimes ds(\{(\omega, s) \mid s \in [0, T-S], f_S^*(T_s\omega) > \alpha\}) \leq C\alpha^{-1} \int d\mu(\omega_0)\|\varphi_{T,\omega_0}\|_1 \tag{2.6.77}$$

But, by Fubini's theorem and the invariance of T_s,

$$\int d\mu(\omega_0)\|\varphi_{T,\omega_0}\|_1 = \int_0^T \int (d\mu(\omega)f(T_s\omega))\,ds = T\|f\|_1$$

By the invariance of μ, $\mu \otimes ds(\{(\omega, s) \mid s \in [0, T - S], f_S^*(T_s\omega) > \alpha\})$ is $(T - S)\mu(\{\omega \mid f_S^*(\omega) > \alpha\})$. Thus, (2.6.77) says that

$$\mu(\{\omega \mid f_S^*(\omega) > \alpha\}) \le \frac{CT\alpha^{-1}}{T - S}\|f\|_1 \qquad (2.6.78)$$

Taking $T \to \infty$, we get

$$\mu(\{\omega \mid f_S^*(\omega) > \alpha\}) \le C\alpha^{-1}\|f\|_1 \qquad (2.6.79)$$

Since $f^*(\omega) = \sup f_S^*(s)$ with f_S^* monotone increasing,

$$\{\omega \mid f^*(\omega) > \alpha\} = \lim_{S \to \infty} \{\omega \mid f_S^*(\omega) > \alpha\}$$

and (2.6.78) implies (2.6.70). $\qquad\qquad\qquad\qquad\qquad\qquad\qquad\qquad \square$

Notes and Historical Remarks.

> The important recent work of von Neumann (not yet published) shows only that there is convergence in the mean, so that (1) is not proved by him to hold for any point P, and the time-probability is not established in the usual sense for any trajectory. A direct proof of von Neumann's results (not yet published) has been obtained by E. Hopf.
> —G. D. Birkhoff, Dec. 1, 1931 [75]

> The first one actually to establish a general theorem bearing fundamentally on the quasi-ergodic hypothesis was J. v. Neumann, who ... proved what we will call the mean ergodic theorem.
> —G. D. Birkhoff and B. O. Koopman, Feb. 13, 1932 [76]

Conditional expectation was introduced by Kolmogorov [483] in his fundamental book on foundations of probability.

As you might guess, an active and broad field like ergodic theory has produced a large number of monographs, of which we mention [15, 72, 242, 348, 442, 491, 594, 630, 640, 660, 739, 759, 842].

Ergodic theory has its roots in late nineteenth-century physics. Three of the great developments in that century's physics were thermodynamics, the kinetic theory of gases, and statistical mechanics. Thermodynamics depended on the notion of macroscopic systems and described irreversible phenomena, while the kinetic theory of gases depended on microscopic dynamics which was reversible and seemed to be anything but equilibrium. James Clerk Maxwell (1831–79) and Ludwig Boltzmann (1844-1906) were the pioneers in trying to resolve these contradictions. Boltzmann, in several papers starting with [91] and including [92], introduced the idea that a dynamical system could be described by averages over constant energy surfaces (microcanonical ensembles, in the language of Gibbs) because an orbit covered the entire energy surface. In [**92**], he introduced the name "ergodic hypothesis" for this notion, from the Greek words for energy (ergon) and odos (path).

Of course, we now know, by the topological invariance of dimension, that a one-one continuous curve cannot be space filling, and it is easy to see that for smooth curves. Boltzmann's defenders will claim that he didn't mean this literally and even he mentioned exceptional periodic orbits and that the strict notion that all points were visited was a caricature of the Ehrenfests (see below). One has to understand the milieu during the time Boltzmann was working. Many scientists doubted that atoms were real (the view was shifting, but it wasn't until early in the twentieth century that the reality of atoms was the majority view) and there was neither measure theory nor topology in that era.

Among the high points between Boltzmann and the revolution of 1931–32 we discuss below are the following: Poincaré [**653**] alluded to these problems and, in particular, emphasized that the occurrence of exceptional periodic orbits prevented all initial conditions from filling the surface. Poincaré also presented his recurrence theorem [**651, 655**], which we'll discuss shortly (and in the Problems). In 1911, Paul and Tanya Ehrenfest [**241**] published a very influential encyclopedia article on the foundations of statistical mechanics, where they expressed skepticism that orbits could cover the whole energy surface and introduced what they called the quasi-ergodic hypothesis that all orbits are dense. In 1920, George David Birkhoff (1884–1944) [**74**] conjectured that if there was a single dense orbit, then except for a few exceptional ones, they are all dense. In 1928, he and Smith [**77**] introduced the notion they called metrical transitivity, what we now call ergodicity.

In early 1931, Bernard Osgood Koopman (1900–81) published a short note [**485**] that explained that measure-preserving dynamics induced unitary operators on $L^2(\Omega, d\mu)$ and suggested that the newly discovered (by Stone and von Neumann) spectral resolution and eigenvectors/eigenvalues might be significant, but he didn't do anything further with this. That said, by realizing that operator theory could be used to study the issue of the ergodic hypothesis, Koopman was the godfather of the ergodic theorems.

The year before, when Koopman realized this, he did tell von Neumann about it (see the Notes to Section 7.1 in Part 4 for a capsule biography of von Neumann) who, within a few months, found Theorem 2.6.7 [**837**]. His proof relied on the spectral theorem: as we'll see in Theorem 5.5.4 in Part 4, Theorem 2.6.7 is immediate from that theorem and the observation that if

$\omega \in \partial \mathbb{D}$, then

$$\frac{1}{n} \left(1 + \omega + \cdots + \omega^{n-1} \right) \to \begin{cases} 1 & \text{if } \omega = 1 \\ 0 & \text{if } \omega \neq 0 \end{cases} \qquad (2.6.80)$$

with the left side bounded by 1 uniformly in ω and n. von Neumann mentioned his result to Eberhard Hopf (1902–83) who found the proof we use in this section. If one knows the spectral theorem, von Neumann's proof is simple and elegant. But since that result was not yet that widely known, Hopf's proof was regarded as more accessible and simpler—in any event, it has an immediate extension to L^p (see Problem 11).

At the beginning of October 1931, von Neumann, then in Princeton, went to New York where Koopman was on the Columbia faculty and told Koopman of his result to confirm that Koopman had not found it independently. Koopman was enthusiastic and suggested that von Neumann publish his result in the *Proceedings of the National Academy of Sciences* (PNAS), where Koopman's note had appeared. At a meeting of the American Mathematical Society in New York, von Neumann told Stone and Hopf of his result (and Hopf found his proof shortly thereafter; at von Neumann's request, he submitted his proof to PNAS, requesting that his paper only appear after von Neumann's).

Still later in October, Koopman and Birkhoff came to Princeton for the opening of Fine Hall (Fine Hall does not refer to the current home of the Princeton Mathematics Department with its Tower—that was opened in 1970—but it refers to the department home for forty years which became Jones Hall when the department took the name to its new building.). There, Koopman and von Neumann told Birkhoff of von Neumann's result, knowing of Birkhoff's long interest in the quasi-ergodic hypothesis. Within six weeks, Birkhoff had the special case of Theorem 2.6.9 where the flow came from analytic differential equations on a compact analytic manifold with invariant measure. This, too, he published in PNAS [**75**].

This brings us to the priority fight, or perhaps, since it is mild as these things go, I should say priority spat. Still, the feelings engendered at the time were intense enough that thirty-five years later, when I was a graduate student at Princeton, I heard the story. It is unusual in that we have documents written contemporaneously—not only the paper of Birkhoff–Koopman [**76**] but a letter von Neumann wrote at the time to H. P. Robertson and an article by Zund [**874**] on that letter and the background.

The issue is that while Birkhoff was clearly motivated by von Neumann, who was first, Birkhoff was more senior, a member of the National Academy, and a good friend of the managing editor of the PNAS (who held the post for almost fifty years!), Harvard chemist, E. B. Wilson. And Wilson arranged for Birkhoff's paper to jump the queue and appear in the 1931 volume rather than the 1932 volume and with an earlier communication date! While Birkhoff mentioned von Neumann, the implication is that von Neumann's work was at best independent and possibly later.

For background, you need to realize that Birkhoff was a senior professor, then forty-seven years old. Koopman and Stone had been his students (both in 1926) and Hopf was his postdoc at the time (Hopf moved to MIT to be near Wiener—he's the Hopf of the Wiener–Hopf method—between submission and publication of his paper [**387**].). Von Neumann was younger (he was twenty-six when he found the theorem, and like Hopf, Koopman, and Stone, a 1926 Ph.D.), foreign and Jewish. But he was hardly a powerless postdoc—he was recognized as a wunderkind and was a protégé of Oswald Veblen (1880–1960) who had attracted von Neumann to Princeton as the Jones Professor of Mathematics (Veblen was shortly afterwards the founding head of the math department at the Institute for Advanced Study, and von Neumann one of his first appointments there). If anyone was more established in American mathematics than Birkhoff, it was Veblen—he and Birkhoff were both students of E. H. Moore at Chicago, and Veblen, who was slightly older, a postdoc when Birkhoff was a graduate student. Zund conjectures that Robertson asked von Neumann to write the letter so that Robertson in turn could show it to Veblen. In any event, it appears that Veblen, perhaps with the help of Koopman and Stone, got Birkhoff to agree to write a paper with Koopman in PNAS with nothing but a description of the history and which clearly stated von Neumann's priority. So the story had a happy ending.

As a postscript, we might note that many modern ergodic theorists may be puzzled by this disagreement. Von Neumann's result is regarded as almost trivial, while Birkhoff's is mathematically much deeper. But this ignores the importance of demonstrating what directions to pursue. Without von Neumann's setting the scene, Birkhoff is not likely to have found his theorem. Moreover, von Neumann long argued that for physics, his result suffices (see, e.g., [**838**]). There is not only truth to that but also to the fact that his result suffices for some of the mathematical applications. Moreover, as von Neumann emphasized [**838**], there is one aspect of his result that is stronger than Birkhoff's. If one defines

$$Av_n^\ell(\omega; f) = \frac{1}{n} \sum_{j=\ell}^{n+\ell-1} f(T^j\omega) \tag{2.6.81}$$

then as $n \to \infty$, in L^2, $Av_n^\ell(\cdot\,; f) \to \mathbb{E}(f \mid \mathcal{I})$ uniformly in ℓ (as can be seen by looking at either the von Neumann or Hopf proofs), but the pointwise convergence need not be uniform in ℓ.

George David Birkhoff (1884–1944) was an undergraduate at Harvard, where he was influenced by Bôcher and Osgood (a cousin of Birkhoff's later student Koopman), and a graduate student at Chicago, where he was a student of E. H. Moore but influenced by the writings of Poincaré, whose successor in mechanics he considered himself to be in later life. He received his Ph.D. in 1907 and returned to Harvard in 1912 where he spent the rest of his career, for many years as chair and for several years as dean of the faculty.

Birkhoff was the leading figure in mechanics during his career. His accomplishments included not only his ergodic theorem but the proof of a famous conjecture of Poincaré on the three-body problem as well as work on billiards, on general relativity, and on the general theory of ODEs. His students include Koopman, Langer, Morse, Stone, Walsh, Whitney, and Widder. He was notorious for his anti-semitism [**643, 675**] although as Phillips' reminiscences [**643**] make clear, he was not alone in this in the 1930s.

The full version of Theorem 2.6.9 for a measurable rather than analytic setup is due to Khinchin [**464, 466**] in 1933/34, so much so that some literature refers to the Birkhoff–Khinchin theorem. We'll follow the convention more common in the West of using just Birkhoff's name.

Most, but not all, modern approaches to the Birkhoff theorem use the maximal inequality, Theorem 2.6.11. This was first found in 1939 by Wiener [**856**], who used a variant of Birkhoff's method, a kind of covering argument. The approach via Theorem 2.6.14 is by Kakutani–Yoshida [**865**] the same year. We have added Hopf's name to this theorem because of a 1954 paper [**389**] where he found an extension to contractions on L^1 (see Problem 21). The simple proof we give of Theorem 2.6.14 is due to Garsia [**307**]. While we, following some authors, call Theorem 2.6.14 the maximal ergodic theorem, some other authors call what we call the maximal ergodic inequality by that name.

We won't attempt to describe all the many alternate proofs of the Birkhoff theorem, but we mention several. Hartmann [**366**] has a proof based on Riesz's sunrise lemma and Section 2.9 will have a proof of Keane–Peterson [**457**] whose history will be discussed in the Notes to that section. Given the close connection between the Birkhoff and Hardy–Littlewood theorems, it is not surprising that techniques for proving the Hardy–Littlewood theorem, like the sunrise lemma, can also be used for the Birkhoff theorem. Our way of understanding the relations among the theorems is Doob martingale inequality $\Rightarrow$ HL maximal inequality $\Rightarrow$ maximal ergodic inequality.

There are also covering lemma approaches; see, for example, Wiener [**856**] and Ornstein–Weiss [**622**]. The idea of using the Hardy–Littlewood theorem directly, as we do at the end of the section (Theorem 2.6.15), is due to Calderón [**131**] (motivated, in part, by his earlier work [**126**] and by Cotlar [**173**]). It is remarkable that while this idea of the essential equivalence of the Hardy–Littlewood and ergodic maximal inequalities is well known in the harmonic analysis community, it seems less well known in the ergodic theory community.

Some of the alternate proofs of the Birkhoff ergodic theorem provide additional information. For example, there is a proof of Bishop [**79**] (see also [**435, 443**]) that proves the following. Given a sequence $y_1, y_2, \ldots$ of reals and $\alpha < \beta$, we say the sequence crosses (α, β) at least N times if there exist $k_1 < j_1 < \cdots < k_N < j_N$ so that $y_{k_\ell} \leq \alpha$, $y_{j_\ell} \geq \beta$ for $\ell = 1, 2, \ldots, N$. Then Bishop proved that if (Ω, Σ, μ, T) is a measurable dynamical system, $\alpha < \beta$ are fixed, and E_N is the set of ω so that $\{\frac{1}{n} \sum_{j=0}^{n-1} f(T^j \omega)\}_{n=1}^{\infty}$ crosses (α, β) at least N times, then

$$\mu(E_N) \leq \frac{\|f\|_1}{(\beta - \alpha)N} \tag{2.6.82}$$

Since, if $\liminf y_n < \alpha < \beta < \limsup y_n$, we have that y_n crosses (α, β) infinitely often, this estimate implies the Birkhoff theorem.

In his proof of martingale convergence theorems in 1940 [**224**], Doob used upcrossing methods and proved the precursor of (2.6.82); see Problem 10 of Section 2.10.

In the next two sections and their Notes, we'll discuss examples of maps that are ergodic. Notice that because of the work of von Neumann and Birkhoff, there is a shift of the meaning of ergodic from the dense paths of the Ehrenfests to the absence of nontrivial invariant sets from the point of view of measure. That said, it follows from the Birkhoff theorem that if T is a measure-preserving map on a compact metric space with strictly positive ergodic probability measure, then almost every path is dense (see Problem 14), but the converse isn't true; see below. We emphasize that despite the name, the ergodic theorems do not require ergodicity!

The simplest context in which to discuss dense orbits is in the context of topological dynamical systems. As we saw, the ergodic measures are exactly extreme points in the compact convex set of invariant measures, a result of Wiener [**856**]. It follows that all measures are ergodic if and only if there is only a single invariant measure, in which case, T is called *uniquely ergodic*. If all orbits $\{T^n \omega\}_{n=0}^{\infty}$ are dense in Ω, we say T is *minimal*. If T is minimal, it is easy to see that any invariant measure has $\mu(A) > 0$ for all open A. Furstenberg [**295**] (see also [**449**]) has constructed a map T on the 2-torus which is minimal but not uniquely ergodic, so there are invariant measures, μ, with $\mu(A) > 0$ for all open A, all orbits dense but μ not ergodic—the promised illustration of the lack of a converse.

There is an extensive literature on Birkhoff-type theorems along subsequences of $\mathbb{Z}_+$. For example, Bourgain [**103**] has proven that if $\mathcal{P}_n$ is the first

n primes and m_n the number of such primes, then for $f \in L^2$, one has for a.e. ω that $\frac{1}{m_n} \sum_{p \in \mathcal{P}_n} f(T^p \omega)$ converges. See [103] for additional references.

In connection with ergodic theorems, we should mention:

Theorem 2.6.16 (Poincaré Recurrence Theorem [651, 655]). *Let (Ω, Σ, μ, T) be a measurable dynamical system. Let $E \in \Sigma$. Then for almost every $\omega_0 \in E$, $\{n \geq 1 \mid T^n \omega_0 \in E\}$ is infinite.*

This says that the orbit returns arbitrarily close to ω_0. It does not require ergodicity and its proof does not need the ergodic theorem (see Problem 15), although using that theorem, one can prove that the number of returns has positive density (see Problem 16). This result depends critically on finiteness of $\mu(\Omega)$. For example, on $(\mathbb{R}, dx)$, $x \to x + 1$ is nonrecurrent!

As for how long it takes to return to E, in Problem 17, the reader will prove:

Theorem 2.6.17 (Kac Return Time Theorem [436]). *Let (Ω, Σ, μ, T) be a measurable dynamical system where T is invertible and ergodic. Let $E \in \Sigma$ be such that $\mu(E) > 0$. Let*

$$E_n = \{\omega \in E \mid T^j \omega \notin E, \ j = 1, \ldots, n-1; \ T^n \omega \in E\} \qquad (2.6.83)$$

(i.e., n is the first return time). Then

$$\sum_{n=1}^{\infty} n \frac{\mu(E_n)}{\mu(E)} = \frac{1}{\mu(E)} \qquad (2.6.84)$$

Thus, the average first return time is $1/\mu(E)$, a pleasing result. Indeed, consider the standard Maxwell thought model: a box partitioned in half, filled with gas on one side. The partition is removed and the gas spreads out. Poincaré's recurrence theorem says if you look every second, eventually all the gas will return to the originally filled half, a highly surprising result! But Kac's theorem says if there are 10^{23} molecules (and there is ergodicity!), the time it takes to return will be roughly $2^{10^{23}}$ seconds—a time so large that the term astronomical doesn't do it justice.

There are two notions stronger than ergodicity that often arise. If (Ω, Σ, μ, T) is a measurable dynamical system, we say T is *mixing* (sometimes called *strongly mixing*) if and only if for all $A, B \in \Sigma$,

$$\mu(A \cap T^{-n}[B]) \to \mu(A)\mu(B) \qquad (2.6.85)$$

and *weakly mixing* if

$$\frac{1}{n} \sum_{j=0}^{n-1} |\mu(A \cap T^{-n}[B]) - \mu(A)\mu(B)| \to 0 \qquad (2.6.86)$$

Notice that if $T^{-1}[A] = A$, then, taking $A = B$, weak mixing implies $\mu(A) = \mu(A)^2$, that is, $\mu(A) = 0$ or 1, so

$$\text{mixing} \Rightarrow \text{weak mixing} \Rightarrow \text{ergodicity} \qquad (2.6.87)$$

For $f, g \in L^2(\Omega, d\mu)$, define

$$q_n(f, g) = \langle f, g \circ T^n \rangle - \langle f, 1 \rangle \langle 1, g \rangle \qquad (2.6.88)$$

Then, by approximating f, g by simple functions, or conversely, taking $f = \chi_A$, $g = \chi_B$, we see

$$\text{mixing} \Leftrightarrow \forall f, g \in L^2, \quad q_n(f, g) \to 0$$

$$\text{weak mixing} \Leftrightarrow \forall f, g \in L^2, \quad \frac{1}{n} \sum_{j=0}^{n-1} |q_j(f, g)| \to 0$$

By (2.6.33),

$$\text{ergodicity} \Leftrightarrow \frac{1}{n} \sum_{j=0}^{n-1} q_j(f, g) \to 0$$

so we see (2.6.87) once more.

These things all have Koopmanism translations, that is, expressions in terms of the Koopman unitary, U, on $L^2(\Omega, d\mu)$ given by (2.6.9). If P_1 is the projection onto multiples of 1, that is, $P_1 f = \langle 1, f \rangle 1$, then

$$\text{mixing} \Leftrightarrow \underset{n \to \infty}{\text{w-lim}}\, U^n = P_1 \qquad (2.6.89)$$

$$\text{weak mixing} \Leftrightarrow U \restriction (\text{Ran}(P_1))^\perp \text{ has no eigenvalues} \qquad (2.6.90)$$

$$\text{ergodicity} \Leftrightarrow \text{Ran}(P_1) = \ker(U - 1) \qquad (2.6.91)$$

The first and third of these are immediate. $\Rightarrow$ in (2.6.90) is easy (Problem 23). $\Leftarrow$ in (2.6.90) requires the spectral theorem and is discussed in Theorem 5.5.5 of Part 4. Problems 24, 25, and 26 explore weak mixing further.

The Birkhoff ergodic theorem has been generalized in various directions. As you might guess, given the fact that Hopf has proven Theorem 2.6.14 for contractions on L^1 and we have Theorem 2.1.1, there is a theorem of Birkhoff-type for suitable maps on L^p. Dunford–Schwartz [**233**] have proven such a theorem for a linear map R on $L^1(\Omega, d\mu)$ so long as $\|Rf\|_1 \leq \|f\|_1$, $\|Rf\|_\infty \leq \|f\|_\infty$ and $f \geq 0 \Rightarrow Rf \geq 0$.

Chacon–Ornstein [**148**] found a result where $\|Rf\|_\infty \leq \|f\|_\infty$ is dropped, but instead of looking at the limit of $\frac{1}{n} \sum_{j=0}^{n-1} (R^j f)(\omega)$, they looked at the limit of $\sum_{j=0}^{n-1} (R^j f)(\omega) / \sum_{j=0}^{n-1} (R^j g)(\omega)$. See Garsia's book [**308**] for a discussion and additional references for this subject.

Problems

1. Suppose that (Ω, Σ, μ, T) is a measurable dynamical system with T invertible and $\mu(T^{-1}[A] \triangle A) = 0$. Prove there is a $B \subset A$ with $\mu(A \setminus B) = 0$ so that $T[B] = B$. (*Hint*: Look at $\{\omega \mid T^n \omega \in A$ for all $n \in \mathbb{Z}\}$.)

2. This problem will prove that (2.6.37) is equivalent to ergodicity of ν.

   (a) If ν is ergodic, prove that (2.6.37) holds.

   (b) If $f = \chi_A$ and $\nu(A \triangle T^{-1}[A]) = 0$, prove that for each n, $Av_n(\omega; f) = \chi_A(\omega)$ and that the integral in (2.6.37) is $2[\mu(A) - \mu(A)^2]$ so that (2.6.37) implies that $\mu(A) = 0$ or $\mu(A) = 1$.

3. This will provide an alternate proof that μ ergodic implies μ is an extreme point of $\mathcal{I}(X, T)$. Suppose $\mu = \theta \mu_1 + (1 - \theta)\mu_2$ with $0 < \theta < 1$.

   (a) Prove that μ_1 is a.c. wrt μ.

   (b) If $d\mu_1 = g \, d\mu$, prove that $g \circ T = g$ for a.e. ω.

   (c) If μ is ergodic, prove that $\mu = \mu_1$.

4. Let $(\Omega, \Sigma, \mu, \{T_t\}_{t \geq 0})$ be a measure-preserving semiflow. Define

$$(\overline{A}f)(\omega_0) = \limsup_{t \to \infty} \frac{1}{t} \int_0^t f(T_s \omega_0) \, ds \qquad (2.6.92)$$

   (a) Suppose $f \in L^1(\Omega, d\mu)$. For a.e. ω_0 and all $t < \infty$, prove that $f(T_s \omega_0) \in L^1([0, t], dt)$ so that the integral in (2.6.92) exists.

   (b) For a.e. ω_0 and all s, prove that $(\overline{A}f)(T_s \omega_0) = (\overline{A}f)(\omega_0)$.

   Remark. Do not use an ergodic theorem. This result is an element of some proofs of such theorems.

5. Let $(\Omega, \Sigma, \mu, \{T_s\}_{s \geq 0})$ be a measure-preserving semiflow.

   (a) If $A \in \Sigma$ obeys $\mu(A \triangle T_s^{-1}[A]) = 0$ for all $s \geq 0$, prove that if $f = \chi_A$, then for all s, $f(T_s \omega) = f(\omega)$ for a.e. ω. In particular, if μ is not ergodic, there are functions f not a.e. constant that are invariant.

   (b) Suppose f is a real-valued function so that for all rational α, $\mu(\{\omega \mid f(\omega) > \alpha\})$ is 0 or 1 (which value can depend on α). For some α_0, prove that $f(\omega) = \alpha_0$ for almost all ω. (*Hint*: Look at $\inf\{\alpha \mid \mu(\{\omega \mid f(\omega) > \alpha\}) = 0\}$.)

   (c) If μ is ergodic and f is invariant, prove that f is a.e. constant.

6. Prove the von Neumann ergodic theorem for measure-preserving semiflows $\{T_s\}_{s \geq 0}$.

7. Let $\{T_s\}_{s\geq 0}$ be a measure-preserving semiflow.

    (a) For $n \geq 1$, prove that for $n \leq t \leq n+1$,

    $$\frac{1}{t}\int_0^t |f(T_s\omega)|\, dt \leq \frac{1}{n}\int_0^{n+1}|f(T_s\omega)|\, ds$$

    (b) If $g(\omega) = \int_0^1 |f(T_s\omega)|\, d\omega$, prove that if $T \equiv T_{s=1}$,

    $$\sup_{t\geq 1}\frac{1}{t}\int_0^s |f(T_s\omega)|\, d\omega \leq 2\sup_n \frac{1}{n}\sum_{j=0}^{n-1} g(T^j\omega) \qquad (2.6.93)$$

    (c) Prove a maximal ergodic inequality for semiflows.

    (d) Prove a Birkhoff ergodic theorem for semiflows.

8. Let (Ω, Σ, μ) be a probability measure space and T a measure-preserving map on Ω. Let $\widetilde{\Omega} = \Omega \times [0,1]$ with product σ-algebra, Σ, and define for $t \geq 0$,

    $$S_t(\omega, s) = (T^{[t+s]}\omega, \{t+s\})$$

    where $[t+s]$ is the integral part of $t+s$ and $\{t+s\}$ its fractional part. Show that $\{S_t\}$ is a measure-preserving flow for $\nu = \mu \otimes dx$ and that for $f \in L^1(\Omega, d\mu)$ and $n \in \mathbb{N}$, we have

    $$\frac{1}{n}\int_0^n g(S_t(\omega, s))\, dt = \frac{1}{n}\sum_{j=0}^{n-1} f(T^j\omega)$$

    if $g(\omega, s) = f(\omega)$. Conclude that ergodic theorems for flows imply them for maps.

    Remark. The flow $(\widetilde{\Omega}, \widetilde{\Sigma}, \nu, S_t)$ is called the *suspension* of (Ω, Σ, μ, T).

9. If (2.6.21) holds for two functions g_1 and g_2 in $L^1(\Omega, d\mu')$ and all $h \in L^\infty(\Omega, d\mu')$, prove that

    (a) $\int h(\omega)[g_1(\omega) - g_2(\omega)]\, d\mu'(\omega) = 0$

    (b) $g_1(\omega) = g_2(\omega)$ for μ' a.e. ω

10. (a) Let $1 \leq p, q < \infty$ with $p^{-1} + q^{-1} = 1$. Let $x, y \geq 0$ with $xy = \frac{x^p}{p} + \frac{y^q}{q}$. Prove that $y = x^{p-1}$.

    (b) Suppose $f \in L^p$, $g \in L^q$ for some measure space and $1 = \|f\|_p = \|g\|_q = \int f(x)g(x)\, d\mu(x)$. Prove that for a.e. x, $g(x) = |f(x)|^{p-2}\,\overline{f(x)}$.

    (c) Prove Proposition 2.6.13 for $2 < p < \infty$.

11. Prove Theorem 2.6.7 with L^2 replaced by L^p, $1 \leq p < \infty$. (*Note:* You'll need Problem 10.)

12. Let $U : \mathcal{H} \to \mathcal{H}$ be a contraction on a Hilbert space, that is, $\|U\varphi\| \leq \|\varphi\|$ for all φ. Let P be the projection onto those $\varphi \in \mathcal{H}$ with $U\varphi = \varphi$. Let $V_n = \frac{1}{n} \sum_{j=0}^{n-1} U^j$.

    (a) Prove $V_n\varphi$ has a limit in $\mathcal{H}$ as $n \to \infty$ if $P\varphi = \varphi$ and also if $\varphi = \psi - P\psi$ with $\psi \in \mathcal{H}$.

    (b) If $U^*\eta = \eta$, prove that $U\eta = \eta$. (*Hint*: Compute $\|\eta - U\eta\|^2$.)

    (c) Prove that $\{\varphi_1 + \varphi_2 \mid P\varphi_1 = \varphi_1, \varphi_2 = \psi - U\psi\}$ is dense in $\mathcal{H}$.

    (d) Prove that $V_n\varphi$ has a limit for all $\varphi \in \mathcal{H}$.

13. Let $U : L^1(X, d\mu) \to L^1(X, d\mu)$ for μ a probability measure space be a linear map that obeys (i) $f \geq 0 \Rightarrow Uf \geq 0$; (ii) $\|Uf\|_1 \leq \|f\|_1$. Prove that if $f^\sharp(x) = \sup_{n \geq 1} \frac{1}{n} \sum_{j=0}^{n-1} (T^j f)(x)$, then (2.6.55) holds.

    Remark. This is Hopf's theorem in [**389**].

14. Let (X, Σ, μ, T) be an ergodic measurable dynamical system.

    (a) If $\mu(A) > 0$, prove that for a.e. ω_0, $T^n\omega_0 \in A$ for some n.

    (b) If X is a compact metric space (and so, second countable) and μ a Baire measure with $\mu(A) > 0$ for all open A, prove that almost every $\{T^n\omega_0\}_{n=0}^\infty$ is dense in X.

15. Let (Ω, Σ, μ, T) be a measurable dynamical system. Let $A \in \Sigma$ have $\mu(A) > 0$. You may not use either ergodic theorem in this problem.

    (a) Let $B = \{\omega \in A \mid T^j(\omega) \notin A, \ j = 1, 2, \dots\}$. Prove that $B \cap T^{-j}[B] = \emptyset$ for all $j = 1, 2, \dots$.

    (b) Prove that $T^{-j}[B] \cap T^{-k}[B] = \emptyset$ for all $j \neq k$ in $\{0, 1, 2, \dots\}$.

    (c) Prove that $\mu(B) = 0$ so that for a.e. $\omega \in A$, $T^j(\omega) \in A$ for some $j = 1, 2, \dots$ (weak form of Poincaré recurrence).

    (d) For a.e. $\omega \in A$, prove that $T^j(\omega) \in A$ for infinitely many j. (*Hint*: Apply (c) to $T, T^2, T^3, \dots$.)

16. (a) Let V be a closed subspace of $L^2(\Omega, d\mu)$ with (Ω, Σ, μ) a probability measure space. Suppose $\mathbb{1} \in V$ and that if P is the orthogonal projection onto V, then $f \geq 0 \Rightarrow Pf \geq 0$. Let $A \in \Sigma$ have $\mu(A) > 0$. Prove that

$$\|P\chi_A\| \geq \mu(A) \tag{2.6.94}$$

    (*Hint*: Schwarz inequality.)

    (b) Prove $\langle \chi_A, P\chi_A \rangle \geq \mu(A)^2$.

    (c) If $B = \{\omega \in A \mid P\chi_A(\omega) \leq \alpha\}$, prove that $\mu(B) \leq \alpha$. Conclude that for a.e. $\omega \in A$, $(P\chi_A)(\omega) \neq 0$.

    (d) Under the hypotheses of Problem 15, prove that for a.e. $\omega \in A$, $\frac{1}{n}\sharp\{j \leq n - 1 \mid T^j(\omega) \in A\}$ has a limit which is nonzero.

(e) Using the von Neumann theorem, prove that, for any ε, there is an $N(\varepsilon)$ with

$$\left\| \frac{1}{N(\varepsilon)} \sum_{j=0}^{N(\varepsilon)-1} \chi_A \circ T^j - \mathbb{E}(\chi_A \mid \mathcal{I}) \right\|_2 < \varepsilon$$

(f) Prove that

$$\inf_m \frac{1}{N(\varepsilon)} \sum_{j=0}^{N(\varepsilon)-1} \mu(T^{m+j}[A] \cap A) \geq \mu(A)^2 - \varepsilon$$

(g) Prove that every set of $N(\varepsilon)$ successive ℓ's has at least one with $\mu(T^\ell[A] \cap A) \geq \mu(A)^2 - \varepsilon$.

Remark. (f) and (g) are sometimes called the *Khinchin recurrence theorem* after Khinchin [**465**].

17. This will prove the Kac return time theorem (Theorem 2.6.17). Let (Ω, Σ, μ, T) be an ergodic invertible measurable dynamical system with $E \in \Sigma$ obeying $\mu(E) > 0$. Let E_n be given by (2.6.83). Define

$$\Omega_n^{(j)} = T^j[E_n], \qquad j = 0, 1, 2, \ldots, n-1$$

(a) Prove that up to sets of measure 0, $\bigcup_{j,n} \Omega_n^{(j)} = \Omega$.

(b) Prove that up to sets of measure 0, $\bigcup_n E_n = E$.

(c) Prove that $\Omega_n^{(j)} \cap \Omega_m^{(\ell)} = \emptyset$ if $n \neq m$ or $n = m$ and $j \neq \ell$.

(d) Prove that $\mu(\bigcup_{j=0}^{n-1} \Omega_n^{(j)}) = n\mu(E_n)$.

(e) Show that $\sum_{n=1}^{\infty} n\mu(E_n) = 1$, which is (2.6.84).

18. (a) Prove a Hardy–Littlewood maximal inequality for $\ell^1(\mathbb{Z}^\nu)$.

(b) Let $T_1, \ldots, T_\nu$ be ν commuting maps of $\Omega \to \Omega$ so that for $j = 1, \ldots, \nu$, $(\Omega, \Sigma, \mu, T_j)$ is a measurable dynamical system of Ω. For all $f \in L^1(\Omega, d\mu)$ and a.e. ω_0, prove that

$$\lim_{n \to \infty} \frac{1}{n^\nu} \sum_{0 \leq k_j \leq n-1} f(T_1^{k_1} \ldots T_\nu^{k_\nu} \omega_0)$$

exists.

19. Let (Ω, Σ, μ, T) be a measurable dynamical system. Let $\widetilde{\Omega} = \{\omega \in \Omega^{\mathbb{Z}} \mid \omega_{j+1} = T\omega_j\}$. Define $\widetilde{T} \colon \widetilde{\Omega} \to \widetilde{\Omega}$ by $(\widetilde{T}\omega)_j = \omega_{j+1}$. If $A \in \Sigma$, define $\widetilde{A} = \{\omega \in \widetilde{\Omega} \mid \omega_0 \in A\}$ and set $\tilde{\mu}(\widetilde{A}) = \mu(A)$.

(a) Prove that $\{\widetilde{A}\}$ is a σ-algebra and $\tilde{\mu}$ is a measure on it.

(b) Let $\widetilde{\Sigma}$ be the smallest σ-algebra containing $\{\widetilde{A} \mid A \in \Sigma\}$ and so that $B \in \widetilde{\Sigma} \Rightarrow T^{-1}[B] \in \widetilde{\Sigma}$. Prove there is a unique extension, $\tilde{\mu}$, of $\tilde{\mu}$ on $\{\widetilde{A} \mid A \in \Sigma\}$ to $\widetilde{\Sigma}$ so that $(\widetilde{\Omega}, \widetilde{\Sigma}, \tilde{\mu}, \widetilde{T})$ is a measurable dynamical system.

(c) If $\varphi \colon \widetilde{\Omega} \to \Omega$ by $\varphi(\omega) = \omega_0$, prove that $(\widetilde{\Omega}, \widetilde{\Sigma}, \tilde{\mu}, \widetilde{T})$ is an extension of (Ω, Σ, μ, T).

(d) For the one-sided Bernoulli shift, prove that $\widetilde{T}$ is the two-sided Bernoulli shift.

Remarks. 1. Thus, every measurable dynamical system has an invertible extension.

2. $\widetilde{T}$ is a minimal invertible extension in that any other invertible extension of T is also an extension of $\widetilde{T}$.

20. Assume you have the maximal ergodic theorem in the form (2.6.70) for the space of Example 2.6.3 with $T_t(e^{i\theta}) = e^{i(\theta+t)}$.

(a) Prove that for any $f \in L^1(\partial\mathbb{D}, \frac{d\theta}{2\pi})$, if

$$(M_{\mathrm{HL}}f)(e^{i\theta}) = \sup_{0 \leq \psi \leq \pi} \frac{1}{2\psi} \int_{-\psi}^{\psi} |f(e^{i(\theta+\eta)})| \, d\eta$$

then

$$|\{\alpha \mid (M_{\mathrm{HL}}f) \geq \alpha\}| \leq 2C\alpha^{-1}\|f\|_1$$

where C is the constant in (2.6.92).

(b) From (a) and scaling, prove the Hardy–Littlewood maximal inequality for $\mathbb{R}$.

21. Let S be a linear map of $L^1(\Omega, d\mu)$ (real-valued functions) to itself with $\|Sf\|_1 \leq \|f\|_1$. Let $f^\sharp(x) = \sup_{n\geq 1} \sum_{j=0}^{n-1}(S^k f)(x)$. Prove that (2.6.55) holds. (*Hint*: Mimic the proof of Theorem 2.6.14.)

Remark. This is the theorem of Hopf [**389**] and the theorem Garsia [**307**] found the simple proof of.

22. (a) Let μ, ν be invariant measures for a topological dynamical system (X, T). Let

$$d\mu = f \, d\nu + d\gamma \qquad (2.6.95)$$

be the Lebesgue–von Neumann decomposition, with $d\gamma$ mutually singular to $d\nu$. Prove that $T^*(\gamma)$ is mutually singular to ν and then that γ is invariant.

(b) If μ and ν are ergodic, prove that either $\mu = \nu$ or else μ and ν are mutually singular. (*Hint*: Use Theorem 2.6.8.)

23. Let (Ω, Σ, μ, T) be a measurable dynamical system. Suppose there is a $g \in L^2(\Omega, d\mu)$, $\langle \mathbb{1}, g \rangle = 0$, and $g \circ T = e^{i\alpha}g$ for some $e^{i\alpha} \in \partial\mathbb{D}$. Prove that T is not weak mixing.

    Remark. The converse is true; see Theorem 5.5.5 of Part 4.

24. Prove that T is weakly mixing if and only if for all f, g orthogonal to $\mathbb{1}$, we have

$$\frac{1}{n} \sum_{j=0}^{n-1} |\langle f, g \circ T^n \rangle|^2 \to 0$$

25. Let (Ω, Σ, μ, T) be a measurable dynamical system and $T^{(2)}$ the map on $(\Omega \times \Omega, \Sigma^{(2)}, \mu \otimes \mu)$, with $T^{(2)}(\omega_1, \omega_2) = (T\omega_1, T\omega_2)$ and $\Sigma^{(2)}$ the product σ-algebra. This problem will prove that the following are equivalent: (1) $T^{(2)}$ is ergodic; (2) $T^{(2)}$ is weakly mixing; and (3) T is weakly mixing. You may assume that you know (2.6.90).

    (a) Prove that (1) $\Rightarrow$ (3). (*Hint*: Prove the contrapositive.)

    (b) Prove that (3) $\Rightarrow$ (2). ((2) $\Rightarrow$ (1) is, of course, easy, as mentioned in the Notes.)

26. Show that T is weakly mixing if and only if for any $f, g \in L^2$, there is a subset $J \subset \mathbb{Z}^+$ with $\lim_{n\to\infty} \#\{j \in J \mid j \leq n\}/n \to 1$ so that

$$\lim_{\substack{j\to\infty \\ j\in J}} \langle f, g \circ T^j \rangle = \langle f, 1 \rangle \langle 1, g \rangle$$

2.7. Applications of the Ergodic Theorems

In this section and the next, we'll present applications of the ergodic theorems. Most are to number theory because they are technically simple and require less background (other than background on topics like Fuchsian groups and continued fractions that we discussed in Part 2A). There are many other applications to geometry, differential equations, dynamical systems, and stochastic analysis. An important part of many of these applications is the proof of ergodicity of the underlying measure which allows us to identify the limit. However, there are some applications where merely the existence of the limit is useful.

2.7.1. Bernoulli Shifts: The strong law of large numbers and normal numbers.
We first want to show that the strong law of large numbers (which we proved by another means in Theorem 7.2.5 of Part 1) follows in a few lines from the Birkhoff ergodic theorem. We'll discuss the case of bounded random variables—it is easy to handle the case of finite expectation (Problem 1).

We begin with an ergodicity result. A *generalized Bernoulli shift* is the measurable dynamical system (Ω, Σ, μ, T), where $\Omega = \times_{n=1}^{\infty} X$ with the product σ-algebra, $d\mu(\omega) = \otimes_{n=1}^{\infty} d\nu(\omega_n)$, and $T(\omega)_n = \omega_{n+1}$. Here (X, Σ', ν) is a compact probability measure space with $L^2(X, d\nu)$ separable.

Theorem 2.7.1. *Any generalized Bernoulli shift is invariant and mixing, hence ergodic.*

Proof. As noted in Example 2.6.2, if $f(\omega) = F(\omega_1, \ldots, \omega_j)$ with F continuous, then $\int (f \circ T)(\omega) \, d\mu(\omega) = \int f(\omega) \, d\mu(\omega)$. Since these F's are dense in L^2, μ is invariant.

If $f(\omega) = F(\omega_1, \ldots, \omega_p)$ and $g(\omega) = G(\omega_1, \ldots, \omega_p)$, then for $k \geq p$, f and $g \circ T^k$ depend on distinct sets of variables, so

$$\int f(\omega) g(T^k \omega) \, d\mu(\omega) = \left(\int f(\omega) \, d\mu \right) \left(\int g(\omega) \, d\mu(\omega) \right) \tag{2.7.1}$$

Since these functions are dense in L^2, a simple limiting argument proves that for all $f, g \in L^2(\Omega, d\mu)$, we have

$$\lim_{k \to \infty} \int f(\omega) g(T^k \omega) \, d\mu(\omega) = \left(\int f(\omega) \, d\mu(\omega) \right) \left(\int g(\omega) \, d\mu(\omega) \right) \tag{2.7.2}$$

so T is mixing. $\square$

We can now apply the Birkhoff theorem to get the following, which implies the strong law of large numbers for any bounded iid random variables (see Section 7.2 of Part 1).

Theorem 2.7.2 (Strong Law of Large Numbers). *Let $d\lambda$ be a probability measure on $[a, b] \subset \mathbb{R}$. Let $x_1, \ldots, x_n, \ldots$ be the nth coordinate in $(\Omega \equiv \times_{n=1}^{\infty} [a, b], \Sigma \otimes_{n=1}^{\infty} d\lambda(x_j))$. Then for a.e. $x \in \Omega$,*

$$\frac{1}{n} \sum_{j=1}^{n} x_j \to \int_a^b x \, d\lambda(x) \tag{2.7.3}$$

Proof. Let (Ω, Σ, μ, T) be the Bernoulli shift. Let $f(x) = x_1$. Then $\frac{1}{n} \sum_{j=0}^{n-1} (f \circ T^j)(x)$ is the left side of (2.7.3) and $\int f(x) \, d\mu = \int_a^b x \, d\lambda(x)$, so (2.7.3) is just the Birkhoff theorem and the ergodicity of the Bernoulli shift. $\square$

Remark. Since the Birkhoff theorem applies to any $f \in L^1$, it is easy to use this proof for L^1 iidrv.

More generally, taking $A \subset [a,b]^\ell$ and $f(\omega) = \chi_A(\omega_1, \ldots, \omega_\ell)$, we get

Theorem 2.7.3. *Let* (Ω, Σ, μ) *be as in Theorem 2.7.2 and* $A \subset [a,b]^\ell$. *Then for a.e.* $\omega \in \Omega$,

$$\lim_{n\to\infty} \frac{1}{n} \#\{1 \le j \le n \mid (\omega_j, \omega_{j+1}, \ldots, \omega_{j+\ell-1}) \in A\} = \left(\bigotimes_{j=1}^\ell \lambda_j\right)(A) \quad (2.7.4)$$

Fix $b \in \{2, 3, \ldots\}$. A real number $x \in [0,1]$ is called *normal to base b* if and only if, when x is written $x = .x_1 x_2 \ldots$ as the base b decimal expansion, we have for every $\ell = 1, 2, \ldots$ and every $y_1, \ldots, y_\ell \in \{0, \ldots, b-1\}^\ell$,

$$\lim_{n\to\infty} \frac{1}{n} \#\{j \le n \mid x_j = y_1, \ldots, x_{j+\ell} = y_\ell\} = b^{-\ell} \quad (2.7.5)$$

A number is *normal* if it is normal to every base b. Since those x's with ambiguous base representations (i.e., $x = p/b^\ell$, $p \in \mathbb{Z}_+$) are not normal no matter which representation we use, the ambiguity is irrelevant for this purpose.

Thus, a number is normal to base 10 if exactly $1/10$ of its digits are 0, $1/10$ are $1, \ldots, 1/100$ of its pairs are 00, etc. Theorem 2.7.3 and the fact that there are only countably many b's and countably many finite sequences imply

Theorem 2.7.4 (Borel's Normal Number Theorem). *Put Lebesgue measure,* λ, *on* $[0,1]$. *Then a.e.* $x \in [0,1]$ *is normal.*

Proof. For each b, map $\Omega \equiv \{0, \ldots, b-1\}^{\mathbb{Z}_+} \xrightarrow{\varphi} [0,1]$ by $\varphi(\{\alpha_1, \alpha_2, \ldots\}) = \sum_{n=1}^\infty \alpha_j/b^j$. φ is a bijection after removing countable subsets from each set (where φ is $2 - 1$ due to the familiar $\ldots x999\cdots = \ldots(x+1)000$ for base 10 exception). The shift on Ω is just moving the decimal expansion, so Theorem 2.7.3 applies. $\qquad\square$

Despite the fact that almost every number is normal, it is not known if any really interesting irrationals, like π or e are normal, and it is not easy to write down any explicit normal numbers. The base 10 number $.123456789101112\ldots$ is normal to base 10. It is not known if it is normal to base b for $b \ne 10^k$. Indeed, no explicit number is known to be normal to distinct relatively prime bases p and q; see the discussion in the Notes.

2.7.2. Irrational Rotations and Weyl Equidistribution. We begin with a question of Gel'fand, which we'll answer in this subsection. Consider the powers of 2: $2, 4, 8, 16 \ldots$ and consider their leading digits. The first forty are: 2481361251248136125124813612512481361251. You might guess that this is the start of a period 10 sequence, but it is not. The leading digit of 2^{46} is 7, not 6. The apparently remarkable period 10 phenomenon is a result of the fact that $2^{10} = 1024$ is almost exactly 1000. Gel'fand's question is what is the frequency of 7 in this sequence of leading digits. If you guessed $1/9$, you'd be wrong!

Let $\alpha \in \mathbb{R} \setminus \mathbb{Q}$ be irrational. Let $\Omega = \partial\mathbb{D} = \{e^{i\theta} \mid \theta \in [0, 2\pi)\}$, $d\mu = \frac{d\theta}{2\pi}$, and $T_\alpha \omega = e^{2\pi i \alpha}\omega$, that is, $\theta \to \theta + 2\pi\alpha \mod 2\pi$. We proved $(\Omega, \Sigma, \mu, T_\alpha)$ was a measurable dynamical system in Example 2.6.3. We'll begin here by proving it is ergodic.

Theorem 2.7.5. *If α is irrational, then T_α is ergodic.*

Proof. Ergodicity is equivalent to $U_\alpha f = f$, implying $f = c\mathbb{1}$, where U_α is the Koopman unitary. Let $\varphi_n(e^{i\theta}) = e^{in\theta}$ for $n \in \mathbb{Z}$. Then $U_\alpha^* \varphi_n = e^{-2\pi i \alpha n}\varphi_n$, so

$$\langle \varphi_n, f \rangle = \langle \varphi_n, U_\alpha f \rangle = \langle U_\alpha^* \varphi_n, f \rangle = e^{2\pi i \alpha n}\langle \varphi_n, f \rangle \qquad (2.7.6)$$

Since α is irrational, $2\pi i \alpha n \notin 2\pi\mathbb{Z}$ if $n \neq 0$, so $\langle \varphi_n, f \rangle = 0$ for $n \neq 0$, that is, $f = c\mathbb{1}$. $\square$

Remarks. 1. If $\alpha \in \mathbb{Q}$, then $U_\alpha f = f$ has additional eigenfunctions, so T_α is not ergodic.

2. Since U_α has additional eigenvectors—indeed, a basis of eigenvectors—it is not weak mixing, and so, not mixing either.

Thus, if $f \in L^1(\partial\mathbb{D}, \frac{d\theta}{2\pi})$, for a.e. θ_0, we have that $Av_n(e^{i\theta_0}; f) \to \int f(e^{i\theta})\frac{d\theta}{2\pi}$ for a.e. θ_0. But more is true.

Theorem 2.7.6 (Weyl Equidistribution Theorem). *Let $f \in C(\partial\mathbb{D})$. Then $Av_n(\,\cdot\,; f)$ converges uniformly to $\int f(e^{i\theta})\frac{d\theta}{2\pi}$. In addition, if U is an open set with $|\partial U| = 0$, then $Av_n(\,\cdot\,; \chi_U) \to \frac{|U|}{2\pi}$ uniformly.*

Proof. By a simple approximation argument, it suffices to prove the first statement in the theorem for a dense set in $C(\partial\mathbb{D})$, so by linearity and the Weierstrass density theorem (see Theorem 2.4.2 of Part 1), for $\varphi_n(e^{i\theta}) = e^{in\theta}$. For $n = 0$, $Av_m(e^{i\theta}, \varphi_0) = 1$, so convergence is trivial, and for $n \neq 0$,

$$|Av_m(e^{i\theta}, \varphi_n)| = \frac{1}{m}\left|\frac{e^{in\theta + 2\pi i m n \alpha} - e^{in\theta}}{e^{2\pi i n \alpha} - 1}\right| \leq \frac{2}{m}|e^{2\pi i n \alpha} - 1|^{-1} \to 0 \qquad (2.7.7)$$

By regularity of $|\cdot|$ and Urysohn's lemma, for any U and n, there exists $f_n \in C(\partial\mathbb{D})$, with $0 \leq f_n \leq \chi_U$ and $\int f_n\frac{d\theta}{2\pi} \geq \frac{|U|}{2\pi} - \frac{1}{n}$, and similarly, there exists $g_n \in C(\partial\mathbb{D})$ with $\chi_U \leq g_n \leq 1$ and $\int g_n\frac{d\theta}{2\pi} \leq \frac{|U|}{2\pi} + \frac{1}{n}$. Thus, by the first part, there is M_n for any n, so that for all θ and all $m \geq M_n$,

$$\frac{|U|}{2\pi} - \frac{2}{n} \leq Av_m(e^{i\theta}, f_n) \leq Av_m(e^{i\theta}, \chi_U)$$

$$\leq Av_m(e^{i\theta}, g_n) \leq \frac{|U|}{2\pi} + \frac{2}{n} \tag{2.7.8}$$

proving that $Av_m(\,\cdot\,, \chi_U) - \frac{|U|}{2\pi} \to 0$ uniformly. $\square$

Remark. The argument concerning $|\partial U| = 0$ is essentially that of Theorem 4.5.7 of Part 1.

This is actually equivalent to unique ergodicity of T_α, a concept defined after Theorem 2.6.8:

Theorem 2.7.7. *Let (Ω, T) be a topological dynamical system. Then the following are equivalent:*

(1) *For all $f \in C(\Omega)$ and $\omega \in \Omega$, $Av_m(\omega; f)$ has a limit, and this limit is ω-independent.*

(2) *For all $f \in C(\Omega)$, $Av_m(\,\cdot\,; f)$ converges uniformly to a constant function, $c(f)$.*

(3) *(Ω, T) is uniquely ergodic, that is, there is a unique invariant Baire probability measure for T.*

(4) *$\{c\mathbb{1} + g - g \circ T \mid c \in \mathbb{R}, g \in C(\Omega)\}$ is dense in $C(X)$.*

Proof. We'll show $(2) \Rightarrow (1) \Rightarrow (3) \Rightarrow (4) \Rightarrow (2)$.

$(2) \Rightarrow (1)$. is trivial.

$(1) \Rightarrow (3)$.   Let $c(f)$ be the limit. By hypothesis (1), if $f \in C(\Omega)$, $Av_m(\omega; f) \to c(f)$ pointwise, and clearly, $|Av_m(\omega; f)| \leq \|f\|_\infty$. So, by the dominated convergence theorem, for any Baire probability measure, μ, $\int Av_m(\omega; f)\, d\mu(\omega) \to c(f)$. If μ is invariant, the integral of the average is $\mu(f)$, so $\int f(\omega)\, d\mu(\omega) = c(f)$ for any invariant measure. Thus, all invariant measures are equal.

$\sim(4) \Rightarrow \sim(3)$.   If $Y = \overline{\{c\mathbb{1} + g - g \circ T \mid c \in \mathbb{R}, g \in C(\Omega)\}}$ is not all of $C(\Omega)$, the Hahn–Banach theorem (see Theorem 5.5.10 of Part 1) implies that there is $\ell \neq 0$ in $C(\Omega)^*$, so $\ell([Y]) = 0$. By Theorem 4.8.11 of Part 1, there are mutually singular pointwise measures, μ_+ and μ_-, so $\ell(f) = \int f\, d\mu_+ - \int f\, d\mu_-$. Since $\ell(\mathbb{1}) = 0$, $\mu_+(\Omega) = \mu_-(\Omega)$, so since $\ell \neq 0$, we can suppose μ_+ and μ_- are both probability measures. Since $\ell(g) = \ell(g \circ T)$ for all $g \in C(\Omega)$, $\mu_+ - \mu_-$ is invariant, which (Problem 3) implies that μ_+ and μ_- are invariant. Since $\mu_+ - \mu_- \neq 0$, $\mu_+ \neq \mu_-$, that is, there are multiple invariant measures.

$(4) \Rightarrow (2)$. $Av_n(c\mathbb{1} + g - g \circ T) = c\mathbb{1} + n^{-1}(g - g \circ T^n) \to c\mathbb{1}$ as $n \to \infty$. By a density argument, for any f in the closure, $Av_n(f)$ has a constant limit. $\square$

Example 2.7.8 (Gel'fand's Question and Benford's Law). We turn now to Gel'fand's question. 2^n has leading digit d if and only if

$$d \times 10^k \leq 2^n < (d+1) \times 10^k$$

for some $k \in \mathbb{N}$. Taking base 10 logs, this holds if and only if $\{n \log_{10} 2\} \in [\log_{10} d, \log_{10}(d+1))$, where $\{\cdot\}$ is fractional part. Thus, we are rotating the circle viewed as $\{e^{2\pi i x} \mid x \in [0,1)\}$ with $\alpha = \log_{10} 2$.

We claim α is irrational, for if $\alpha = p/q$, then $10^p = 2^q$. Since $5 \nmid 2$, $p = 0$, and then, $q = 0$. Thus, Theorem 2.7.7 applies, so

$$\lim_{N \to \infty} \frac{1}{N} \#\{n \leq N \mid 2^n \text{ has leading digit } d\} = \log_{10}\left(\frac{d+1}{d}\right)$$

7 occurs with frequency about .058 and 1 with frequency about .301. This is sometimes called Benford's law, after an observation of Benford in 1938, although he was not the first (see the Notes). □

Notes and Historical Remarks. Borel's law of normal numbers goes back to a fundamental 1909 paper [**98**] where he first proved the strong law of large numbers, using the Borel–Cantelli lemma (see Section 7.2 of Part 1), not ergodic theory.

Despite the fact that almost all numbers are normal, it is hard to write down explicit ones. This is notwithstanding the fact that π and e are believed to be normal, although it hasn't even been proven that their base 10 expansions contain every digit infinitely often. There is a conjecture going back at least to Borel in 1950 [**99**] that every irrational algebraic number is normal.

Sierpinski [**738**] found an alternate proof of Borel's result on normal numbers that, in principle, should allow an algorithm for the construction of normal numbers, but it appears no explicit one is known. There are results though on explicit numbers normal to a single base b (and so to base b^ℓ but not known to be normal to any relatively prime base): Champernowne [**149**] proved .123456789101112... is normal to base 10, and then Copeland–Erdős [**167**] proved his conjecture that .2357111317..., where the primes are strung out as a decimal, is normal to base 10. Quéfflec [**666**] has a summary of the literature on this subject.

There is a lot more known about Bernoulli shifts. They are more than mixing—they are what are known as K-systems (see, e.g., [**15**]). If the space X has k-points, $x_1, \ldots, x_k$, with $\nu(x_j) = p_j$, the *entropy* of the shift is $-\sum_{j=1}^{k} p_j \log p_j$. A famous theorem of Ornstein [**621**] (see also [**442**]) says that two finite Bernoulli shifts are isomorphic if and only if they have the same entropy.

Theorem 2.7.5 has an easy extension (Problem 4): On the ℓ-torus, given $\alpha_1, \ldots, \alpha_\ell \in \mathbb{R}$, define

$$T_{\alpha_1, \ldots, \alpha_\ell}(e^{i\theta_1}, \ldots, e^{i\theta_\ell}) = (e^{i(2\pi\alpha_1 + \theta_1)}, \ldots, e^{i(2\pi\alpha_\ell + \theta_\ell)}) \qquad (2.7.9)$$

Then $T_{\alpha_1, \ldots, \alpha_\ell}$ is ergodic if and only if $(1, \alpha_1, \ldots, \alpha_\ell)$ are linearly independent over rationals, that is, with $\alpha_0 = 1$, $\sum_{j=0}^\ell q_j \alpha_j = 0$, $q_j \in \mathbb{Q} \Rightarrow q_0 = q_1 = \cdots = q_\ell = 0$. Indeed, if this condition holds, Theorem 2.7.5 extends, showing $T_{\alpha_1, \ldots, \alpha_\ell}$ is uniquely ergodic. That orbits of $T_{\vec{\alpha}}$ are dense if and only if the rational independence conditions hold is a theorem of Kronecker [492], who stated things in terms of rational approximation, not tori. For this reason, the ℓ-dimensional analog of Theorem 2.7.6 is sometimes called the Kronecker–Weyl theorem.

Unique ergodicity of irrational rotation is connected to the fact that $\frac{d\theta}{2\pi}$ is Haar measure on $\partial\mathbb{D}$ as a group under multiplication and Haar measure is unique on compact groups (Problem 8). Indeed, if G is any compact group and $\tau(h) = gh$ for all $h \in G$, then τ is uniquely ergodic if and only if $\{g^n\}_{n=-\infty}^\infty$ is dense in G (Problem 9).

Theorem 2.7.6 was discovered independently in 1909–10 by Bohl [89], Sierpinski [737], and Weyl [849]. It is especially associated with Weyl's name because of a celebrated 1916 followup [850] which proved the theorem below (see Problem 5).

Definition. A sequence $\{x_j\}_{j=1}^\infty$ in $[0, 1]$ is called *equidistributed* if and only if for all $0 \leq a < b \leq 1$,

$$\frac{1}{n} \#(j \leq N, \, x_j \in (a, b)) \to b - a \qquad (2.7.10)$$

Theorem 2.7.9 (Weyl's Equidistribution Theorem). *A sequence in $[0, 1]$ is equidistributed if and only if for all $\ell \in \mathbb{Z}$, $\ell \neq 0$, we have*

$$\lim_{N \to \infty} \frac{1}{N} \sum_{j=0}^{N-1} e^{2\pi i \ell x_j} = 0 \qquad (2.7.11)$$

This is sometimes called *Weyl's criterion* and, of course, it generalizes Theorem 2.7.6. (The Notes of Section 7.1 of Part 2A have a capsule biography.) Hardy–Littlewood [357] had related results at about the same time. Since $e^{2\pi i \ell x} = e^{2\pi i \ell \{x\}}$ if $\{x\}$ is the fractional part of x, we immediately have

Corollary 2.7.10. *A sequence $\{x_j\}_{j=1}^\infty$ in $\mathbb{R}$ has $\{\{x_j\}\}_{j=1}^\infty$ equidistributed if and only if for all $\ell \in \mathbb{Z}$, $\ell \neq 0$, we have* (2.7.11).

We saw $\{x_j\}_{j=1}^\infty$, a sequence in $[0, 1]$, is equidistributed if and only if

$$\frac{1}{N} \sum_{j=1}^N f(x_j) \to \int_0^1 f(x) \, dx \qquad (2.7.12)$$

for any continuous function, f. de Bruijn and Post [**204**] have proven a kind of converse: A fixed function, f, obeys (2.7.12) for all equidistributed sequences if and only if f is Riemann integrable.

The next section and its Notes and Problems will say a lot more about equidistribution.

A little thought about what it means to be a normal number shows the following, which links the two parts of this section:

Corollary 2.7.11. *A number, x, is normal to base b if and and only if for all $\ell \in \mathbb{Z}$, $\ell \neq 0$,*

$$\lim_{N \to \infty} \frac{1}{N} \sum_{j=0}^{N-1} e^{2\pi i \ell x b^j} = 0 \qquad (2.7.13)$$

There is a kind of dual action to the Weyl equidistribution theorem. A *Poincaré sequence* is a subset of $\mathbb{Z}^+$, $\{s_j\}_{j=1}^\infty$ with $s_1 < s_2 < \ldots$, so that for any measurable dynamical system, (Ω, Σ, μ, T), and any $B \in \Sigma$ with $\mu(B) \neq 0$, there is j so $\mu(B \cap T^{s_j}[B]) > 0$. The name comes from the fact that Poincaré's recurrence implies $\mathbb{Z}^+$ is a Poincaré sequence. In Problem 12, the reader will prove

Theorem 2.7.12 (Glasner–Weiss [**315**]). *If for every $x \in [0,1]$, one has*

$$\lim_{N \to \infty} \frac{1}{N} \sum_{j=1}^{N} e^{2\pi i s_j x} = 0 \qquad (2.7.14)$$

then $\{s_j\}_{j=1}^\infty$ is a Poincaré sequence; indeed, $\mu(B \cap T^{s_j}[B]) \neq 0$ for infinitely many j.

Because there isn't much room in one dimension, the phenomenon of unique ergodicity applies to more than rigid rotations. There is a theorem of Denjoy [**213**] (see [**14, 344**] for textbook presentations) that if T is a C^2-diffeomorphism of $\partial\mathbb{D}$ with no periodic orbits, there is a homomorphism $\varphi \colon \partial\mathbb{D} \to \partial\mathbb{D}$ so that $\varphi \circ T = T_\alpha \circ \varphi$ for some rotations of irrational angle α. In particular, T is uniquely ergodic (the theorem is usually stated in terms of irrational rotation number, but that is equivalent to no periodic orbits). Denjoy also constructed C^1-diffeomorphisms for which this conjugacy is false.

The terminology "Gel'fand's question" in Example 2.7.8 comes from the fact that I. M. Gel'fand (see the Notes to Section 6.2 of Part 4 for a capsule biography) asked and answered the question, at least according to Avez [**21**]. For a delightful, elementary, and insightful look at Gel'fand's question, see [**469**].

The distribution on $\{1, \ldots, 9\}$, $d_j = \log_{10}(\frac{j+1}{j})$ is called Benford's law after a 1938 paper [**54**] of the physicist, F. Benford, who observed that many sets of numbers—for example, the first digits of the street address of the first 342 men listed in the then current *American Men of Science*—seemed to have this distribution. In fact, illustrating Arnol'd's principle, it was found almost sixty years earlier by Newcomb [**610**].

If $k = m^a$, $n = m^b$, then $\log_k n$ is rational and it is easy to see that is the only case (Problem 10). Thus, for example, if $m \neq 10^\ell$ for any $\ell \in \mathbb{Z}$, then the leading digits of m^n, $n = 1, 2, \ldots$ are Benford-distributed.

The fact that Weyl's law holds for all initial points means $2^n \ell$ has Benford-distributed leading integers for any ℓ. It is not hard to see the leading powers of 2^n and 3^n are independent, but not of 2^n and 5^n (Problem 11).

Benford's law is what you would expect for a set of numbers equally likely to lie in $[1, 10]$ as in $[10, 100]$ and for scale-invariant distributions. It has evoked enormous interest in a wide variety of disciplines so that the online bibliography[3] is especially useful.

Benford's law is used in tests for fraud [**612**] in a discipline known as "forensic accounting." Made-up numbers have too few leading 1's and too many mid digits $(5, 6, 7)$ compared to their Benford predictions.

Problems

1. Let $\{f_j\}_{j=1}^\infty$ be iid random variables, with $\mathbb{E}(|f_1|) < \infty$. For a.e. ω, prove that $\frac{1}{n} \sum_{j=1}^n f_j(\omega) \to \mathbb{E}(f_1)$. (*Hint*: Show first there is $d\nu$ on $[0, 1]$ and $g \colon [0, 1] \to \mathbb{R}$, so $\mathbb{E}(f_1 \in (a, b)) = \nu(\{x \mid g(x) \in (a, b)\})$.)

2. Let $f \in L^2([0, 1], dx)$ and $f_n^\sharp(x) = \int e^{-2\pi i n x} f(x)\, dx$. Let T be the doubling map (2.6.15).

   (a) Prove that $(f \circ T)_k^\sharp = f_{2k}^\sharp$.

   (b) If $f \circ T = f$, prove that $f_k^\sharp = 0$ if $k \neq 0$.

   (c) Conclude that T is ergodic.

3. Let (Ω, T) be a topological dynamical system. Let $\mu \in C(\Omega)^*$. If $T^*(\mu) = \mu$, prove that $T^*(\mu_\pm) = \mu_\pm$. (*Hint*: If $\mu = \nu_+ - \nu_-$ with $\nu_\pm \geq 0$ and $\nu_\pm(\mathbb{1}) = \mu_\pm(\mathbb{1})$, then $\nu_\pm = \mu_\pm$.)

4. (a) Let $T_{\alpha_1, \ldots, \alpha_\ell}$ on $\mathbb{T}^\ell$, the ℓ torus, be given by (2.7.9). Prove that T is ergodic if and only if for no $(j_1, \ldots, j_\ell) \in \mathbb{Z}^\ell \setminus \{0\}$, $\sum_{k=1}^\ell j_k \alpha_k \in \mathbb{Z}$.

   (b) When ergodicity holds, prove an analog of Theorem 2.7.6 and so, by Theorem 2.7.7, unique ergodicity.

---

[3]www.benfordonline.net

5. This problem will prove Weyl's equidistribution theorem (Theorem 2.7.9).

   (a) If (2.7.10) holds, prove that for any function, f, on $[0, 1]$, which is a finite linear combination of $\chi_{(a,b)}$'s, one has

   $$\frac{1}{N} \sum_{j=0}^{N-1} f(x_j) \to \int f(y)\, dy \qquad (2.7.15)$$

   (b) If (2.7.15) holds for each $\chi_{(a,b)}$, prove that it holds for any $f \in C(X)$ and, in particular, for $f(x) = e^{2\pi i \ell x}$, $\ell \in \mathbb{Z}$.

   (c) By mimicking the proof of Theorem 2.7.6, show that (2.7.11) implies (2.7.10).

6. Let X be a compact metric space and (X, T) a topological dynamical system. Let μ be an invariant probability measure. Prove that μ is ergodic if and only if there is $A \subset X$ with $\mu(A) = 1$ and so that for all $f \in C(X)$ and $x \in A$, $\frac{1}{n} \sum_{j=0}^{n-1} f(T^j x) \to \int f(x)\, d\mu(x)$.

7. Let X be a compact metric space and (X, T) a topological dynamical system. Prove that T is not uniquely ergodic if and only if there is $f \in C(X)$ and some $c \neq d$ so that $\{x \mid \frac{1}{n} \sum_{j=0}^{n-1} f(T^j x) = c\}$ and $\{x \mid \frac{1}{n} \sum_{j=0}^{n-1} f(T^j x) = d\}$ are both nonempty.

8. A *topological group* is a group which is a topological space so that $(x, y) \mapsto xy^{-1}$ is a continuous function of $G \times G \to G$. Suppose you know that when G is locally compact, it has at least one measure (*Haar measure*), $d\nu_{\mathrm{L}}$, invariant under left multiplication, that is,

   $$\int f(xy)\, d\nu_{\mathrm{L}}(y) = \int f(y)\, d\nu_{\mathrm{L}}(y) \qquad (2.7.16)$$

   for all bounded Baire functions, f, and at least one right invariant measure, $d\nu_{\mathrm{R}}$, and that if $X \subset G$ is compact, that $d\nu_{\mathrm{L}}(X) < \infty$, $d\nu_{\mathrm{R}}(X) < \infty$.

   (a) If G is compact and $d\nu_{\mathrm{L}}(G) = d\nu_{\mathrm{R}}(G) = 1$, prove that $\nu_{\mathrm{L}} = \nu_{\mathrm{R}}$. (*Hint*: Consider $\int f(xy)\, d\nu_{\mathrm{L}}(y)\, d\nu_{\mathrm{R}}(x)$.)

   (b) Prove ν_{L} is the only left-invariant probability measure.

   Remark. Haar measure is constructed on any σ-compact group and its uniqueness is proven in Section 4.19 of Part 1.

9. (This problem uses Problem 8.) Prove that if G is a compact topological group and $T(x) = gx$ for a fixed $g \in G$, then
   (a) T is ergodic for Haar measure if and only if $\{g^n\}_{n \in \mathbb{Z}}$ is dense in G.
   (b) If $\{g^n\}_{n \in \mathbb{Z}}$ is dense, then T is uniquely ergodic.

   Remark. This provides another proof of Theorem 2.7.6.

10. Let $k, n \in \mathbb{Z}^+$. Prove that $\log_k n$ is rational if and only if there are $\ell, a, b \in \mathbb{Z}^+$ so $k = \ell^a$, $n = \ell^b$. (*Hint*: Look at the prime factorization of k and n.) Conclude that if $k \neq 10^\ell$, then $\{k^n\}_{n=0}^\infty$ has leading digits that are Benford-distributed.

11. (a) Prove that the leading digits of 2^n and 3^n are asymptotically independent, that is, if $\langle\!\langle x \rangle\!\rangle =$ leading digit of x, then

$$\lim_{N \to \infty} \left[\frac{1}{N} \#\{n \leq N \mid \langle\!\langle 2^n \rangle\!\rangle = d_1, \ \langle\!\langle 3^n \rangle\!\rangle = d_2\} \right.$$
$$\left. - \frac{1}{N^2} \#\{n \leq N \mid \langle\!\langle 2^n \rangle\!\rangle = d_1\} \#\{n \leq N \mid \langle\!\langle 3^n \rangle\!\rangle = d_2\} \right] = 0$$

(*Hint*: See Problem 4.)

(b) Is the same true of 2^n and 5^n?

12. This problem will prove Theorem 2.7.12. It uses the spectral theorem; see Theorem 5.5.1 of Part 4.

(a) If (2.7.14) holds, use the spectral theorem to prove for any $f \in L^2(\Omega, d\mu)$ that, with U a Koopman unitary, one has that

$$\lim_{N \to \infty} \frac{1}{N} \sum_{j=1}^{N} \langle f, U^{s_j} f \rangle = |\mathbb{E}(f \mid \mathcal{I})|$$

(b) For any f, prove that $|\mathbb{E}(f \mid \mathcal{I})|^2 \geq (\int f \, d\mu)^2$.

(c) Prove Theorem 2.7.12.

2.8. Bonus Section: More Applications of the Ergodic Theorems

Here we want to illustrate the applicability of ergodic theory ideas with three examples that span a wide gamut. The first topic requires no background beyond the last section, but the second requires results on continued fractions as discussed in Section 7.5 of Part 2A, and the third, results on hyperbolic geometry in $\mathbb{D}$ and $\mathbb{C}_+$ as discussed in Sections 7.4 and 8.3 of Part 2A and Section 12.2 of Part 2B. The first two topics are number theoretic. The third is primarily geometric although, as we'll discuss in the Notes, there are connections to number theory also.

Typical of the theorems we'll prove are the following three:

Theorem 2.8.1. *For any irrational,* β, $\{\beta n^2\}_{n=0}^\infty$ *is equidistributed* mod 1 *in* $[0, 1]$.

Remark. In fact (see Problems 3 and 4), for any $k = 1, 2, 3, \ldots$ and β irrational, $\{\beta n^k\}_{n=0}^{\infty}$ is equidistributed, and (see Problem 5) for any $\rho \notin \mathbb{Z}$, $\rho > 0$, and any β, $\{\beta n^\rho\}_{n=0}^{\infty}$ is equidistributed.

For the second, we recall that any irrational $x \in [0, 1]$ has a continued fraction expansion

$$x = \cfrac{1}{a_1(x) + \cfrac{1}{a_2(x) + \cfrac{1}{a_3(x) + \cdots}}} \tag{2.8.1}$$

where $a_n \in \mathbb{Z}^+$

$$a_1(x) = \left[\frac{1}{x}\right], \qquad Tx = \left\{\frac{1}{x}\right\}, \qquad a_n(x) = a_1(T^{n-1}(x)) \tag{2.8.2}$$

These continued fractions were discussed in Section 7.5 of Part 2A. Among other facts, we'll prove

Theorem 2.8.2 (Gauss–Kuzmin Theorem). *For Lebesgue a.e. $x \in [0, 1]$, we have that for $k = 1, 2, \ldots,$*

$$\lim_{N \to \infty} \frac{\#\{n \leq N \mid a_n(x) = k\}}{N} = d_k \equiv \log_2\left(\frac{(k+1)^2}{k(k+2)}\right) \tag{2.8.3}$$

Remarks. 1. d_k is often written

$$d_k = -\log_2\left(1 - \frac{1}{(k+1)^2}\right) \tag{2.8.4}$$

It is called the *Gauss–Kuzmin distribution.*

2. As we must, we have, by telescoping,

$$\sum_{k=1}^{\infty} d_k = \log_2\left(\prod_{k=1}^{\infty} \frac{(k+1)^2}{k(k+2)}\right) = \lim_{N \to \infty} \log_2\left(\frac{2(N+1)}{N+2}\right) = 1 \tag{2.8.5}$$

Actually, that $\sum_{k=1}^{\infty} d_k = 1$, which one might think of as required by probability and the existence of all limits, isn't automatic because there are infinitely many bins. For example, $e - 2$ has $a_n(e - 2)$ given by $[1, 2, 1, 1, 4, 1, 1, 6, \ldots, 1, 1, 2n, \ldots]$, so 1 has frequency $\frac{2}{3}$ and all other k's have zero frequency.

3. Since

$$d_k = \frac{1}{k^2} + O\left(\frac{1}{k^3}\right) \tag{2.8.6}$$

at infinity, $\sum_{k=1}^{\infty} k \, d_k = \infty$, so $\frac{1}{N} \sum_{n=1}^{N} a_n(x) \to \infty$ for a.e. x.

4. The top row of Table 2.8.1 shows the first ten values of d_k. Figure 2.8.1 shows a plot of d_k and the last four rows of Table 2.8.1 the distributions of

	1	2	3	4	5	6	7	8	9	10
GK dist	.415	.170	.093	.059	.041	.030	.023	.018	.015	.012
$\pi - 3$	.423	.162	.089	.062	.045	.027	.025	.018	.014	.012
$e - 2$	.666	.0002	0	.0002	0	.0002	0	.0002	0	.0002
γ	.426	.166	.088	.054	.043	.031	.024	.020	.012	.013
$\sin 1$	.423	.171	.091	.056	.043	.029	.022	.020	.015	.011

Table 2.8.1. The Gauss–Kuzmin distribution.

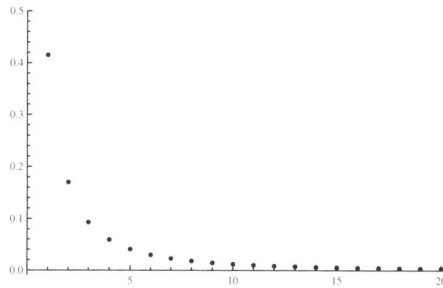

Figure 2.8.1. Plot of the *GK* distribution.

the first 5000 continued fraction a_n's for $\pi - 3$, $e - 2$, γ, $\sin 1$, computed with *Mathematica*. Notice that e is not typical. The a_n's for $e - 2$ are $1, 2, 1, 1, 4, 1, 1, 6, 1, 1, 8, \ldots$.

5. As we'll explain in the Notes and Problem 9, it isn't just the frequency of single digits that one can compute, but also of any string $k_1, k_2, k_3, \ldots, k_\ell$. For example, the frequency of $1, 2, 5$ is $\log_2\left(\frac{513}{512}\right) \cong .002815$.

The final theorem concerns geodesic flow on finite volume hyperbolic Riemann surfaces. We'll focus here on describing a special case of what we'll prove below. We start with the unit disk, $\mathbb{D}$, with the Poincaré metric of Section 12.2 of Part 2B. This Riemann metric on $\mathbb{D}$ induces a geodesic flow. The natural space, $\widetilde{\Omega}$, for this flow is described by a point $z \in \mathbb{D}$ and a unit length (in the Poincaré metric) vector in the tangent space. For any such point, z_0, and direction, there is a unique geodesic (orthocircle) through z_0 with tangent at time 0 the given direction. We can define $\widetilde{T}_t \colon \widetilde{\Omega} \to \widetilde{\Omega}$ by going along the geodesic a hyperbolic distance t and letting $\widetilde{T}_t(\omega)$ be the resulting point on the geodesic and unit tangent direction.

There is a natural measure on $\widetilde{\Omega}$ induced by the hyperbolic metric and it is invariant, but the corresponding volume is infinity. Not only is this problematic for the ideas of ergodic theory, but on the $\widetilde{\Omega}$ discussed so far, the flow is far from ergodic in an intuitive sense. The set of geodesics which start and end in $\{e^{i\theta} \mid |\theta| < \frac{\pi}{2}\}$ is large—both it and its complement have

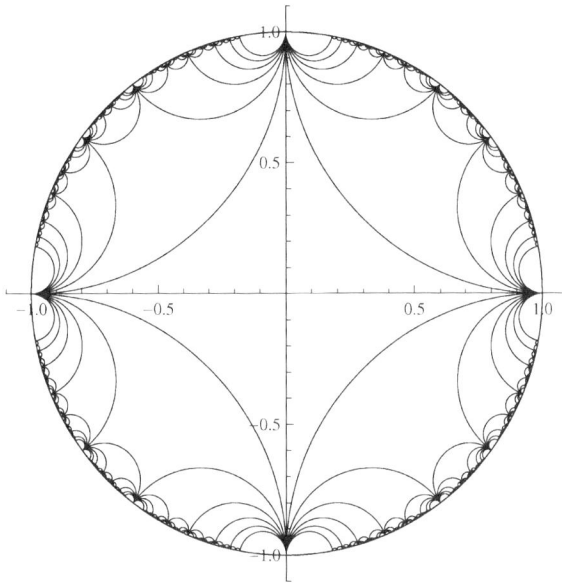

Figure 2.8.2. A fundamental domain in $\mathbb{D}$.

infinite measure. Moreover, by looking at geodesics, we easily see that none are dense.

We'll get around these issues by quotienting by a Fuchsian group. To take an explicit example, consider Figure 2.8.2: the image under the map of $\mathbb{C}_+$ to $\mathbb{D}$ of the region mapped to $\mathbb{C} \setminus \{0, 1\}$ in Section 8.3 of Part 2A. There are to basic maps, g_1 and g_2, of $\mathbb{D}$ to $\mathbb{D}$ obtained, respectively, by reflecting in the two big orthocircles in the upper half-disk and then reflecting in the real axis. g_1 and g_2 generate a discrete group, Γ, of automorphisms of $\mathbb{D}$. Figure 2.8.2 shows the images of the central region under Γ. Each orbit $\{\gamma(z) \mid \gamma \in \Gamma\}$ of z's in the central open region has one point in each open region in the picture. The set of equivalence classes under $w \sim z \Leftrightarrow \exists \gamma \in \Gamma$ so that $\gamma(w) = z$ is a Riemann surface which inherits a Riemann metric, associated volume, and geodesic flow. The volume of this Riemann surface is finite.

Ω is the unit tangent bundle of the set of equivalence classes; it is a three-(real) dimensional manifold. $\pi \colon \widetilde{\Omega} \to \Omega$ takes $z_0 \in \mathbb{D}$ into its equivalence class and maps the tangent vectors under the induced map on the tangent space. There is an induced volume on Ω which is finite. We normalized it to $\mu(\Omega) = 1$.

It is easy to see there is a unique $T_t \colon \Omega \to \Omega$ so that $\pi \widetilde{T}_t = T_t \pi$. μ is invariant for T_t.

Basically, the Riemann surface is the central open region and the top two boundaries with the top and conjugate circles associated under complex conjugation. T_t is described by taking a geodesic inside the central region. When it hits an edge, it comes back in the conjugate side. For example, the geodesic at angle $45°$ from 0, when it hits the upper-right circle, comes back in at the lower circle, remains a straight line at $-45°$, exits on the upper left, and comes in at $45°$ in the lower left. It is periodic. However, this periodic geodesic is exceptional, and one might believe most geodesics are dense (remember, only the geodesics through zero are straight lines). In fact, we'll prove:

Theorem 2.8.3. *Let Ω be the unit tangent bundle on the Riemann surface described above and μ the probability measure induced by the Riemann metric. For any $t_0 \neq 0$, $(\Omega, \Sigma, \mu, T_{t_0})$ is an ergodic measurable dynamical system.*

2.8.1. Skew Shifts and Equidistribution of $\{\beta n^2\}$. At first sight, it appears hopeless to obtain n^2 by iterating a simple map. However, if (with $\mathbb{T} = \mathbb{R}/\mathbb{Z}$, the circle as a group),

$$S_\alpha \colon \mathbb{T}^2 \to \mathbb{T}^2, \qquad S_\alpha(\theta_1, \theta_2) = (\theta_1 + \alpha, \theta_1 + \theta_2) \tag{2.8.7}$$

then, by summing an arithmetic progression,

$$S_\alpha^n\left(\frac{\alpha}{2}, 0\right) = \left(\frac{2n+1}{2}\,\alpha, \frac{\alpha}{2}\,n^2\right) \tag{2.8.8}$$

S_α is called the *skew shift*. Since we are concerned with a single orbit, ergodicity isn't enough, but unique ergodicity is. We'll prove

Theorem 2.8.4. *If α is irrational, then S_α is uniquely ergodic with invariant measure $d\mu = d\theta_1\, d\theta_2$.*

As a corollary, we have

Proof of Theorem 2.8.1. Let $\alpha = 2\beta$. If $f_\ell(\theta_1, \theta_2) = e^{i\ell\pi\theta_2}$, then by the unique ergodicity,

$$\lim_{n\to\infty} \frac{1}{N}\sum_{n=0}^{N-1} f_\ell\left(S_\alpha^n\left(\frac{\alpha}{2}, 0\right)\right) = \lim_{N\to\infty} \frac{1}{N}\sum_{n=0}^{N-1} e^{i\ell\pi\beta n^2} = \delta_{\ell 0} \tag{2.8.9}$$

since $\int f_\ell(\theta_1, \theta_2)\, d\mu = \delta_{\ell 0}$.

The Weyl equidistribution theorem in the form Corollary 2.7.10 completes the proof. $\qquad\square$

Problem 1 has an alternate elementary proof of Theorem 2.8.1; see also the Notes (in particular, Theorem 2.8.19).

We'll prove Theorem 2.8.4 in a general context that allows generalization of Theorem 2.8.1 to more general polynomials than βn^2 (see Problems 3 and 4) and exploiting a widely useful framework.

Definition. Let $(\Omega,,T)$ be a topological dynamical system. Let G be a locally compact group (i.e., locally compact space and group so that $(g,h) \mapsto gh^{-1}$ of $G \times G \to G$ is continuous). Let $\varphi_1 \colon \Omega \to G$ be a continuous function. The *group extension* defined by φ_1 is the pair $(\widetilde{\Omega}, \widetilde{T})$, where $\widetilde{\Omega} = \Omega \times G$ and

$$\widetilde{T}(\omega, g) = (T(\omega), \varphi_1(\omega)g) \tag{2.8.10}$$

If G is compact, $(\widetilde{\Omega}, \widetilde{T})$ is a toplogical dynamical system. We'll suppose Ω and G are metrizable and G is σ-compact, In particular, if G is compact, $C(\widetilde{\Omega})$ is separable.

Example 2.8.5 (Random Matrix Products). Let $G = \mathbb{GL}(n,\mathbb{C})$. If $\varphi_1(\omega) = A(\omega)$, then

$$\widetilde{T}^n(\omega, \mathbb{1}) = (T^n(\omega), A(T^{n-1}\omega)\ldots A(\omega)) \tag{2.8.11}$$

describes a random matrix product. We'll discuss this in the next section. □

Example 2.8.6 (Skew Shift and Generalized Skew Shift). If $\Omega = \mathbb{T}$, $T(\theta) = \theta + \alpha \pmod 1$, $G = \mathbb{T}$, and $\varphi_1(\theta) = \theta$, then using additive notation in $G\colon \widetilde{T}(\theta_1, \theta_2) = (\theta_1 + \alpha, \theta_1 + \theta_2)$, that is, skew shift is a group extension.

More generally, we define T_k on $\mathbb{T}^{k+1}$ inductively by $T_{k+1} = \widetilde{T}_k$ if we think of $\mathbb{T}^{k+1}$ as $\Omega \times G$ with $\Omega = \mathbb{T}^k$, $G = \mathbb{T}$, and $\varphi(\theta_1, \ldots, \theta_k) = \theta_1 + \cdots + \theta_k$. Thus,

$$T_k(\theta_1, \ldots, \theta_{k+1}) = (\theta_1 + \alpha, \theta_1 + \theta_2, \theta_1 + \theta_2 + \theta_3, \ldots, \theta_1 + \cdots + \theta_{k+1}) \tag{2.8.12}$$

This is called *generalized skew shift*. □

In a general group extension, we define

$$\varphi_n(\omega) = \varphi_1(T^{n-1}\omega)\ldots\varphi_1(\omega) \tag{2.8.13}$$

so that

$$\varphi_{n+m}(\omega) = \varphi_n(T^m\omega)\varphi_m(\omega) \tag{2.8.14}$$

A family of maps obeying (2.8.14) is called a *cocycle*. Every cocycle obeys (2.8.13). Sometimes φ_1 alone is called the cocyle.

The key to proving Theorem 2.8.4 will be

Theorem 2.8.7. *Let (Ω, T) be a uniquely ergodic topological dynamical system with η the unique invariant measure on Ω. Let G be a compact abelian group with Haar measure ν. Let $\varphi_1 \colon \Omega \to G$ be continuous. Then $\mu = \eta \otimes \nu$ is an invariant measure for $(\widetilde{\Omega}, \widetilde{T})$. Moreover, $(\widetilde{\Omega}, \widetilde{T})$ is uniquely ergodic if and only if μ is ergodic for $(\widetilde{\Omega}, \widetilde{T})$.*

Remark. If $\varphi_1(\omega) \equiv 0$, the identity in G, then any measure $\eta \otimes \tilde{\nu}$ is invariant, and if $\#(G) > 1$, μ is not ergodic. Thus, it is not automatic that μ is ergodic (although, as we'll see, it is for the skew shift).

Proof. To see that μ is invariant, note that, by the invariance of Haar measure, $\int f(T\omega, \varphi_1(\omega) + \theta)\, d\nu(\theta) = \int f(T\omega, \theta)\, d\nu(\theta)$, so

$$\int f(\widetilde{T}(\omega, \theta))\, d\eta(\omega)\, d\nu(\theta) = \int f(T\omega, \theta)\, d\eta(\omega)\, d\nu(\theta)$$

$$= \int f(\omega, \theta)\, d\eta(\omega)\, d\nu(\theta) \qquad (2.8.15)$$

Since μ is invariant, if $\widetilde{T}$ is uniquely ergodic, μ must be ergodic.

For the converse, suppose μ is ergodic and $\tilde{\mu}$ is possibly a different ergodic probability measure. Define $\pi \colon \widetilde{\Omega} \to \Omega$ by $\pi(\omega, \theta) = \omega$ and $\pi^* \colon \mathcal{M}_{+,1}(\widetilde{\Omega}) \to \mathcal{M}_+(\Omega)$ by

$$\int f(\omega)\, d(\pi^* \kappa) = \int (f \circ \pi)(\omega, \theta)\, d\kappa(\omega, \theta) \qquad (2.8.16)$$

Since $\pi(\widetilde{T}(\omega, \theta)) = T(\pi(\omega, \theta))$, we see π^* takes invariant measures to invariant measures. Since η is the unique invariant measure on Ω, we see $\pi^*(\mu) = \eta = \pi^*(\tilde{\mu})$. Applying (2.8.16) to $f = \chi_{A \times G}$, we see

$$\forall A \subset \Omega \text{ Baire}, \ \mu(A \times G) = \tilde{\mu}(A \times G) \qquad (2.8.17)$$

For $\theta_0 \in G$, let $R_{\theta_0}(\omega, \theta) = (\omega, \theta + \theta_0)$. By the form of $\widetilde{T}$, we have (since G is abelian) that

$$\widetilde{T} R_{\theta_0} = R_{\theta_0} \widetilde{T} \qquad (2.8.18)$$

Let

$$\widetilde{\Sigma} = \left\{ (\omega, \theta) \in \widetilde{\Omega} \mid \forall f \in C(\widetilde{\Omega}), \ \frac{1}{N} \sum_{j=0}^{N-1} f(\widetilde{T}^j(\omega, \theta)) \to \int f(\omega, \theta)\, d\mu(\omega, \theta) \right\}$$
$$(2.8.19)$$

Since $C(\widetilde{\Omega})$ is separable, this is a measurable set. By (2.8.19) and the invariance of μ under R_{θ_0}, we see $(\omega, \theta) \in \Sigma \Rightarrow R_{\theta_0}(\omega, \theta) \in \Sigma$, which means

$$\widetilde{\Sigma} = \pi(\widetilde{\Sigma}) \times G \qquad (2.8.20)$$

Since μ is assumed ergodic, $\mu(\widetilde{\Sigma}) = 1 \Rightarrow \eta(A) = 1$, where $A = \pi(\widetilde{\Sigma})$. Thus, by (2.8.17),

$$\tilde{\mu}(\widetilde{\Sigma}) = \tilde{\mu}(A \times G) = \mu(\Sigma) = 1 \qquad (2.8.21)$$

Since $\tilde{\mu}$ is ergodic, for every $f \in C(\widetilde{\Omega})$ and $\tilde{\mu}$-a.e. (ω, θ), $\frac{1}{N} \sum_{j=0}^{N-1} f(\widetilde{T}(\omega, \theta)) \to \int f(\omega, \theta)\, d\tilde{\mu}(\omega, \theta)$. By (2.8.21), we conclude for all $f \in C(\widetilde{\Omega})$, $\int f\, d\tilde{\mu} = \int f\, d\mu$, that is, $\mu = \tilde{\mu}$ and thus, μ is uniquely ergodic. $\qquad \square$

The following thus proves Theorem 2.8.4 and so Theorem 2.8.1:

Theorem 2.8.8. *If α is irrational and μ is the product measure $d\theta_1\, d\theta_2$ on $\mathbb{T}^2$, then μ is ergodic for the skew shift, S_α.*

Proof. We follow the proof of Theorem 2.7.5. It suffices to prove any $f \in L^2$ with $f \circ \tilde{T} = f$ is constant. For $(n_1, n_2) \in \mathbb{Z}$, let

$$\varphi_{n_1,n_2}(\theta_1, \theta_2) = \exp(2\pi i(n_1\theta_1 + n_2\theta_2)) \tag{2.8.22}$$

Let $Uf = f \circ T$. Then

$$U\varphi_{n_1,n_2}(\theta_1, \theta_2) = \exp(2\pi i(n_1(\theta_1 + \alpha) + n_2(\theta_1 + \theta_2)))$$
$$= e^{2\pi i n_1 \alpha}\varphi_{n_1+n_2,n_2}(\theta_1, \theta_2)$$

If $Uf = f$, then since U is an isometry, $U^*f = f$ and thus,

$$f^\sharp_{n_1,n_2} \equiv \langle\varphi_{n_1,n_2}, f\rangle = \langle\varphi_{n_1,n_2}, U^*f\rangle = \langle U\varphi_{n_1,n_2}, f\rangle$$
$$= e^{-2\pi i n_1 \alpha} f^\sharp_{n_1+n_2,n_2} \tag{2.8.23}$$

If $n_2 \neq 0$, for any n_1 and k, $|f^\sharp_{n_1,n_2}| = |f_{n_1+kn_2,n_2}|$, so the fact that $f^\sharp \in \ell^2$ implies $f^\sharp_{n_1,n_2} = 0$.

On the other hand, if $n_2 = 0$, (2.8.23) says $(1 - e^{-2\pi i n_1 \alpha})f^\sharp_{n_1,0} = 0$. Since α is irrational, $f^\sharp_{n_1,0} = 0$ if $n_1 \neq 0$. Thus, $f^\sharp_{n_1,n_2}$ can only be nonzero for $n_1 = n_2 = 0$, that is, f is a multiple of 1. Thus, μ is ergodic. $\square$

2.8.2. Continued Fractions: Gauss measure and the theorems of Gauss–Kuzmin, Khinchin, and Lévy. It will be convenient to define the *Gauss map* on all of $[0, 1]$ by

$$T(x) = \begin{cases} 0, & x = 0 \\ \{\frac{1}{x}\}, & x \neq 0 \end{cases} \tag{2.8.24}$$

and $a_1(x)$ on all of $[0, 1]$ by

$$a_1(x) = \begin{cases} 0, & x = 0 \\ [\frac{1}{x}], & x \neq 0 \end{cases} \tag{2.8.25}$$

and a_n by

$$a_n(x) = a_1(T^{n-1}x) \tag{2.8.26}$$

If x is irrational, x is equal to the infinite continued fraction, (2.8.1), as discussed in Section 7.5 of Part 2A. If x is rational, there is a unique N_0 so $a_n(x) = 0$ for $n \geq N_0$ and x is the finite continued fraction given by truncating (2.8.1) at $N_0 - 1$. (Moreover, if $x \neq 0, 1$, we have $a_{N_0-1}(x) \geq 2$ in this case since $Tx = 0$, $x \neq 1$ implies $a_1(x) \geq 2$, that is, our algorithm here makes a choice between the two possible choices of finite continued fraction, for example, picking $\frac{2}{3} = [1, 2]$ rather than $[1, 1, 1]$.)

We recall that if (2.8.1) is truncated by replacing $a_{n+1}(x)$ by ∞, we get a rational approximation $p_n(x)/q_n(x)$. It can be defined by the Euler–Wallis equations

$$q_n = a_n q_{n-1} + q_{n-2}, \quad p_n = a_n p_{n-1} + p_{n-2}, \quad q_0 = p_{-1} = 1, \quad q_{-1} = p_0 = 0 \tag{2.8.27}$$

(see (7.5.11) of Section 7.5 of Part 2A). It follows from $\det\left(\begin{smallmatrix} a_n & 1 \\ 1 & 0 \end{smallmatrix}\right) = -1$, induction, and $q_0 p_{-1} - q_{-1} p_0 = 1$ that

$$q_n p_{n-1} - q_{n-1} p_n = (-1)^n \tag{2.8.28}$$

By an inductive argument (see Theorem 7.5.2(a) of Section 7.5 of Part 2A),

$$q_n \geq 2^{n/2}, \qquad p_n \geq 2^{(n-1)/2} \tag{2.8.29}$$

Moreover, the expansion for $x \in (0,1)$ in terms of $a_1(x), \ldots, a_n(x)$ and $T^n(x)$ is (see (7.5.14) of Section 7.5 of Part 2A)

$$x = \frac{p_n(x) + T^n(x) p_{n-1}(x)}{q_n(x) + T^n(x) q_{n-1}(x)} \tag{2.8.30}$$

In particular, the set of x whose first n a's are $a_1, \ldots, a_n$ is given in terms of p_n, q_n solving (2.8.27) as the full interval between $(p_n + p_{n-1})/(q_n + q_{n-1})$ and p_n/q_n.

We will also need to know how good an approximation $p_n(x)/q_n(x)$ is to x, namely (7.5.24) of Section 7.5 of Part 2A,

$$\left| x - \frac{p_n(x)}{q_n(x)} \right| \leq \frac{1}{q_n(x)^2} \tag{2.8.31}$$

In the other direction, by (7.5.23) of Section 7.5 of Part 2A, if x is irrational, $(p_{n+2}(x))/(q_{n+2}(x))$ lies strictly between $p_n(x)/q_n(x)$ and x, so

$$\left| x - \frac{p_n(x)}{q_n(x)} \right| > \left| \frac{p_{n+2}(x)}{q_{n+2}(x)} - \frac{p_n(x)}{q_n(x)} \right| > 0 \tag{2.8.32}$$

Since the last difference is not zero, $|q_n p_{n+2} - p_n q_{n+2}| \geq 1$, so

$$\left| x - \frac{p_n(x)}{q_n(x)} \right| \geq \frac{1}{q_n(x) q_{n+2}(x)} \tag{2.8.33}$$

Here is our main theorem on this subject:

Theorem 2.8.9. *For Lebesgue a.e. $x \in [0,1]$, we have*

(a) (Gauss–Kuzmin Theorem) (2.8.3) *holds.*

(b) (Khinchin's Theorem)

$$\lim_{n \to \infty} (a_1(x) \ldots a_n(x))^{1/n} = \prod_{k=1}^{\infty} \left(\frac{(k+1)^2}{k(k+2)} \right)^{\log_2 k} \tag{2.8.34}$$

(c) $\displaystyle \lim_{n \to \infty} \frac{1}{n}(a_1(x) + \cdots + a_n(x)) = \infty$ $\tag{2.8.35}$

(d) (Lévy's Theorem)

$$\lim_{n\to\infty} \frac{1}{n} \log(q_n(x)) = \frac{\pi^2}{12\log 2} \tag{2.8.36}$$

(e) $\lim_{n\to\infty} \frac{1}{n} \log\left|x - \frac{p_n(x)}{q_n(x)}\right| = -\frac{\pi^2}{6\log 2}$ (2.8.37)

Remark. The right side of (2.8.34) is sometimes called *Khinchin's constant* $K \approx 2.6855$ and the exponential of the right side of (2.8.36), $\lim_{n\to\infty} q_n(x)^{1/n}$, is *Lévy's constant* $L \approx 3.2758$ (also called the *Lévy–Khinchin constant*).

We'll prove this by first finding an invariant measure of the form $f(x)\,dx$, then proving it is ergodic for T, and finally, doing the calculations to go from ergodicity to the results in the theorem. Consider T which is C^∞ on $[0,1]$ except for finitely many points $(\frac{1}{2}, \frac{1}{3}, \dots)$ of discontinuity. While $|T'| \to \infty$ as $x \to 0$, on each interval of continuity, $|T'|$ varies from n^2 to $(n+1)^2$. So on the intervals, $\sup/\inf = (1 + \frac{1}{n})^2$ is at most 4. The key to the proof of ergodicity will be that on each interval of continuity, $(T^{(\ell)})'$ has $\sup/\inf$ bounded by 4 independently of ℓ.

Notice that since T is many to one, T^{-1} has branches. $y = T^{-1}(x) \Leftrightarrow \exists n$ so $y = 1/(n+x) \equiv \varphi_n(x)$. Thus, if μ is the measure, $\mu(A) = \int_A f(x)\,dx$, then

$$\mu(T^{-1}[A]) = \sum_{n=1}^{\infty} \int_{\varphi_n(A)} f(x)\,dx \tag{2.8.38}$$

φ_n is a C^∞-bijection of $(0,1)$ to its range, so by changing variables from x to $\varphi_n(y)$, using $dx = \varphi_n'(y)\,dy$, we see

$$\mu(T^{-1}[A]) = \int_A \sum_{n=1}^{\infty} \frac{1}{(n+y)^2} f\left(\frac{1}{n+y}\right) dy \tag{2.8.39}$$

which suggests we define the *Gauss–Kuzmin–Wirsing* operator by

$$(\mathcal{G}f)(x) = \sum_{n=1}^{\infty} \frac{1}{(n+x)^2} f\left(\frac{1}{n+x}\right) \tag{2.8.40}$$

Finding invariant measures a.c. wrt dx is equivalent to finding positive solutions of $\mathcal{G}f = f$.

We'll pull the solution of this eigenvalue equation out of the air (the Notes will explain at least one plausible explanation of where Gauss found it) by noting that

$$\frac{1}{(n+x)^2} \frac{1}{(1 + \frac{1}{n+x})} = \frac{1}{(n+x)} \frac{1}{(n+x+1)} = \frac{1}{n+x} - \frac{1}{(n+x+1)} \tag{2.8.41}$$

Thus, if $f(x) = 1/(1 + x)$, the sum telescopes to

$$(\mathcal{G}f)(x) = \sum_{n=1}^{\infty} \left(\frac{1}{n+x} - \frac{1}{n+1+x} \right) = \frac{1}{1+x} \qquad (2.8.42)$$

that is, $f(x) = 1/(1 + x)$ is the required f. Since we want a probability measure, we pick c so $c \int_0^1 \frac{dx}{1+x} = c \log 2 = 1$. We have thus proven:

Proposition 2.8.10. *The measure on* $[0,1]$,

$$d\mu(x) = \frac{1}{\log 2} \frac{1}{1+x} \, dx \qquad (2.8.43)$$

is an invariant probability measure for $Tx = \{\frac{1}{x}\}$.

It is called *Gauss measure*. As a warmup for the proof of its ergodicity, we reexamine the doubling map, which is somewhat like the Gauss map in that it is many to one and C^∞ away from a countable (in this case, one!) singular point.

Example 2.8.11 (Example 2.6.4 revisited). The doubling map is given by (2.6.15). We'll provide yet another proof of its ergodicity. As with the Gauss map, there are multiple inverses. If $\varphi_0^{(1)}, \varphi_1^{(1)} \colon [0,1] \to [0,1]$ by

$$\varphi_0^{(1)}(x) = \tfrac{1}{2}\,x, \qquad \varphi_1^{(1)}(x) = \tfrac{1}{2} + \tfrac{1}{2}\,x \qquad (2.8.44)$$

then $T^{-1}[A] = \varphi_0^{(1)}[A] \cup \varphi_1^{(1)}[A]$, and for any B, $|\varphi_j^{(1)}[B]| = \frac{1}{2}|B|$, where $|\cdot|$ is Lebesgue measure.

More generally, for $n = 1, 2, \ldots$ and $j = 0, 1, \ldots, 2^n - 1$, define

$$\varphi_j^{(n)}(x) = \frac{j}{2^n} + \frac{x}{2^n} \qquad (2.8.45)$$

and see that

$$T^{-n}[A] = \bigcup_{j=1}^{2^n-1} \varphi_j^{(n)}[A], \qquad |\varphi_j^{(n)}[B]| = 2^{-n}|B|$$

In particular, if $I_j^{(n)} = \varphi_j^{(n)}[0,1]$, then

$$|T^{-n}[A] \cap \varphi_j^{(n)}[0,1]| = 2^{-n}|A| \qquad (2.8.46)$$

Now suppose A is invariant, that is, $|T^{-1}[A] \triangle A| = 0$, so $|T^{-n}[A] \triangle A| = 0$ and (2.8.46) implies

$$|A \cap \varphi_j^{(n)}[0,1]| = 2^{-n}|A| \qquad (2.8.47)$$

By Lebesgue's theorem on differentiability of integrals, we have for a.e. x,

$$\lim_{n \to \infty} 2^n \left| A \cap \left[\frac{j_n(x)}{2^n}, \frac{j_n(x)+1}{n} \right] \right| = \begin{cases} 0, & x \notin A \\ 1, & x \in A \end{cases} \qquad (2.8.48)$$

where $j_n(x)$ is defined by requiring $x \in [\frac{j_n(x)}{2^n}, \frac{j_n(x)+1}{2^n})$.

But, by (2.8.47), the limit is $|A|$. It follows that $|A| = 0$ or $|A| = 1$, proving ergodicity of dx for the doubling map. $\qquad\qquad\qquad\square$

Remark. In some ways, using the Lebesgue differentiability theorem is overkill. There is a direct way to go from (2.8.46) to $|A| = 0$ or $|A| = 1$ (see Problem 7), but conceptually, this makes the argument cleaner.

Theorem 2.8.12. *The Gauss map is ergodic for Gauss measure.*

Proof. Let $|A \triangle T^{-1}[A]| = 0$ with $|\cdot|$ Lebesgue measure. We'll prove that $|A| = 0$ or $|[0,1] \backslash A| = 0$. Since μ and $|\cdot|$ are mutually absolutely continuous, this proves ergodicity of μ.

Fix n. For $a_1, \dots, a_n \in \mathbb{Z}^+$ and $x \in [0,1]$. Let

$$\varphi_{a_1,\dots,a_n}(x) = \cfrac{1}{a_1 + \cfrac{1}{a_2 + \cdots \cfrac{1}{a_n + x}}} \tag{2.8.49}$$

This maps $[0,1]$ onto those x with $a_j(x) = a_j$ for $j = 1, 2, \dots, n$. By (2.8.30), if $\{q_j, p_j\}_{j=1}^n$ are defined by (2.8.27), then

$$\varphi_{a_1,\dots,a_n}(x) = \frac{p_n + p_{n-1}x}{q_n + q_{n-1}x} \tag{2.8.50}$$

Thus, by (2.8.28),

$$\varphi'_{a_1,\dots,a_n}(x) = \frac{(-1)^n}{(q_n + q_{n-1}x)^2} \tag{2.8.51}$$

so $(-1)^n \varphi_{a_1,\dots,a_n}$ is monotone in x. In particular,

$$I_{a_1,\dots,a_n} \equiv \varphi_{a_1,\dots,a_n}[0,1] = \begin{cases} \left[\frac{p_n+p_{n-1}}{q_n+q_{n-1}}, \frac{p_n}{q_n}\right], & n \text{ even} \\ \left[\frac{p_n}{q_n}, \frac{p_n+p_{n-1}}{q_n+q_{n-1}}\right], & n \text{ odd} \end{cases} \tag{2.8.52}$$

so that, using (2.8.28) again,

$$|I_{a_1,\dots,a_n}| = \frac{1}{q_n(q_n + q_{n-1})} \tag{2.8.53}$$

From (2.8.51), we see that

$$\frac{1}{(q_n + q_{n-1})^2} \le |\varphi'_{a_1,\dots,a_n}(x)| \le \frac{1}{q_n^2} \tag{2.8.54}$$

which implies

$$\frac{q_n}{q_n + q_{n-1}} \le |I_{a_1,\dots,a_n}|^{-1} |\varphi'_{a_1,\dots,a_n}(x)| \le \frac{q_n + q_{n-1}}{q_n} \tag{2.8.55}$$

Since $0 < q_{n-1} < q_n$, we have $(q_n + q_{n-1})/q_n < 2$ (if $a_n \gg 1$, q_{n-1}/q_n is very close to 0, and if $a_n = 1$, $a_{n-1} \gg 1$, q_{n-1}/q_n is very close to 1) and since $|\varphi[A]| = \int_A |\varphi'(x)| \, dx$, we see that

$$\tfrac{1}{2}|A| \leq |I_{a_1,\dots,a_n}|^{-1} |\varphi_{a_1,\dots,a_n}[A]| \leq 2|A| \tag{2.8.56}$$

Since $T^{-n}[A] = \cup_{a_1,\dots,a_n=1}^{\infty} \varphi_{a_1,\dots,a_n}[A]$, we see that

$$|I_{a_1,\dots,a_n}|^{-1} |T^{-n}[A] \cap I_{a_1,\dots,a_n}| \geq \tfrac{1}{2}|A| \tag{2.8.57}$$

If $|T^{-1}[A] \triangle A| = 0$, since T^{-1} preserves sets of measure zero, we have $|T^{-n}[A] \triangle A| = 0$ and thus,

$$|I_{a_1,\dots,a_n}|^{-1} |A \cap I_{a_1,\dots,a_n}| \geq \tfrac{1}{2}|A| \tag{2.8.58}$$

If $|A| \neq 0$, we see for all $x \in [0,1]$,

$$\limsup_{\substack{a,b \downarrow 0 \\ a \leq x \leq b \\ a \neq b}} |b-a|^{-1}|A \cap (a,b)| \geq \tfrac{1}{2}|A| \tag{2.8.59}$$

so, by the Lebesgue theorem, which says the limit is a.e. 0 or 1, if $|A| \neq 0$, we have that the limit is strictly positive for a.e. x, so a.e. $x \in A$, that is, $|A| = 1$. $\square$

Proof of Theorem 2.8.9. (a) Let $\chi_k(x) = \{x \mid a_1(x) = k\}$. By the Birkhoff theorem and the proven ergodicity, for a.e. x,

$$\text{LHS of } (2.8.3) = \int \chi_k \, d\mu(x) = \int_{(k+1)^{-1}}^{k^{-1}} (\log 2)^{-1} (1+x)^{-1} \, dx$$

$$= (\log 2)^{-1} \log\left(\frac{1 + k^{-1}}{1 + (k+1)^{-1}} \right)$$

$$= (\log 2)^{-1} \log\left(\frac{(k+1)^2}{k(k+2)} \right) = \log_2\left(\frac{(k+1)^2}{k(k+2)} \right) \tag{2.8.60}$$

(b) By the Birkhoff theorem and $a_n(x) = a_1(T^{n-1}x)$, for a.e. x,

$$\lim_{n \to \infty} \frac{1}{n} \sum_{j=1}^{n} \log(a_j(x)) = \int_0^1 (\log 2)^{-1} \log(a_1(x)) \frac{dx}{1+x} \tag{2.8.61}$$

$$= \sum_{n=1}^{\infty} (\log 2)^{-1} \log n \int_{(n+1)^{-1}}^{n^{-1}} \frac{dx}{1+x}$$

$$= \sum_{n=1}^{\infty} \log_2 n \log\left(\frac{(n+1)^2}{n(n+2)} \right) \tag{2.8.62}$$

which is (2.8.34).

(c) Since $\int_0^\infty a_1(x) \frac{dx}{1+x} = \sum_{k=1}^{\infty} k \, d_k = \infty$, this follows from Corollary 2.6.10.

(d) We start by noting some facts. First, an explicit integral calculated from Euler's formula for $\sum_{n=1}^{\infty} n^{-2}$ (see Problem 8),

$$\int_0^1 \log x \, \frac{dx}{1+x^2} = -\frac{\pi^2}{12} \tag{2.8.63}$$

Second, one has

$$\frac{p_n(x)}{q_n(x)} = \frac{1}{a_1 + \frac{p_{n-1}(Tx)}{q_{n-1}(Tx)}} \tag{2.8.64}$$

so that

$$p_n(x) = q_{n-1}(Tx) \tag{2.8.65}$$

as can also be seen from (2.8.27) and $p_1 = q_0$, $p_0 = q_{-1}$. Thus, since $p_1(x) = 1$,

$$\frac{1}{q_n(x)} = \frac{p_n(x)}{q_n(x)} \frac{1}{q_{n-1}(Tx)} = \prod_{j=0}^{n-1} \frac{p_{n-j}(T^j x)}{q_{n-j}(T^j x)}$$

$$= \prod_{j=0}^{n-1} \frac{1}{T^j x} \frac{p_{n-j}(T^j x)}{q_{n-j}(T^j x)} \prod_{j=0}^{n-1} (T^j x) \tag{2.8.66}$$

For any y, using (2.8.31) and (2.8.29),

$$\left| 1 - \frac{y q_m(y)}{p_m(y)} \right| = \left| \frac{q_m(y)}{p_m(y)} \right| \left| \frac{p_m(y)}{q_m(y)} - y \right| \leq \frac{1}{p_m(y) q_m(y)} \leq 2^{-(m-1)} \tag{2.8.67}$$

so uniformly in $y \in (0,1) \setminus \mathbb{Q}$ as $m \to \infty$,

$$\log\left(\frac{p_m(y)}{y q_m(y)} \right) \to 0 \tag{2.8.68}$$

From (2.8.66),

$$-\frac{1}{n} \log(q_n(x)) = \frac{1}{n} \sum_{j=1}^{n} \log\left(\frac{p_j(T^{n-j} x)}{(T^{n-j} x) q_j(T^{n-j} x)} \right) + \frac{1}{n} \sum_{j=0}^{n-1} \log(T^j x) \tag{2.8.69}$$

By the uniformity in (2.8.68), the first average goes to 0, and by the Birkhoff theorem, the second to

$$(\log 2)^{-1} \int_0^1 \log y \, \frac{dy}{1+y} = -\frac{\pi^2}{12 \log 2} \tag{2.8.70}$$

proving (2.8.36).

(e) By (2.8.31) and (2.8.33),

$$-\log(q_n(x)) - \log(q_{n+2}(x)) \leq \log\left| x - \frac{p_n(x)}{q_n(x)} \right| \leq -2 \log(q_n(x)) \tag{2.8.71}$$

so (2.8.36) implies (2.8.37).   □

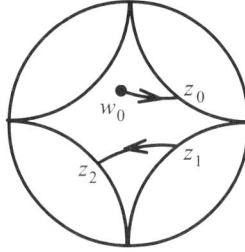

Figure 2.8.3. A geodesic in the fundamental domain.

2.8.3. Geodesic Flow on Finite Volume Hyperbolic Riemann Surfaces. The analysis of the class of examples we consider next is at the intersection of geometry, analysis, and algebra. Once we have the proper setup, the proof of ergodicity, relying on algebraic properties of the group $\mathbb{SL}(2, \mathbb{R})$, will be so slick that the underlying geometric reasons for the result might be lost. So we begin by explaining at a heuristic geometric level why ergodicity holds in the context of the example we discussed before Theorem 2.8.3.

Suppose f on $\widetilde{\Omega}$ is invariant under $\widetilde{T}_t$ for all t. Suppose also that f is uniformly continuous. For each point, z_0, on the circle, $\partial \mathbb{D}$, there is a family of geodesics "ending" at z_0. The projections of these trajectories onto $\mathbb{D}$ of these fill all of $\mathbb{D}$. We claim f is constant on these trajectories in $\widetilde{\Omega}$ for any two trajectories ending at z_0 to get close to each other since the orthocircles are tangent, so by the uniform continuity, the constant values along these trajectories must be equal.

Start at $w_0 \in \mathbb{D}$ in some direction to some z_0 on $\partial \mathbb{D}$; follow a different geodesic backwards to z_1 on $\partial \mathbb{D}$, and then return to w_0 on the geodesic from z_1. For $w_0 = 0$, see Figure 2.8.3. The return direction will be different. It is clear by considering all directions back from z_1, we'll get f is constant in all directions at w_0. This argument is consistent with our noting the flow on all of $\mathbb{D}$ has invariant sets, A, with $\mu(A) = \infty = \mu(\widetilde{\Omega} \setminus A)$ because, in this infinite measure case, χ_A is not close to a uniformly continuous function in any reasonable sense. But for the finite measure quotients we look at, we'll go beyond uniformly continuous functions.

It will be convenient to use $\mathbb{C}_+$ rather than $\mathbb{D}$, so we will—by the obvious isometries of $\mathbb{D}$ to $\mathbb{C}_+$; these results for $\mathbb{C}_+$ imply those for $\mathbb{D}$. We'll use $\mathbb{C}_+$ because the key will be to exploit the action of $\mathbb{SL}(2, \mathbb{R})$ on $\mathbb{C}_+$ to associate $\widetilde{\Omega}$ with $\mathbb{SL}(2, \mathbb{R})$. $\mathbb{SL}(2, \mathbb{R})$ is a little more intuitive than $\mathbb{SU}(1, 1)$, the automorphism group of $\mathbb{D}$. That said, some calculations are easier on $\mathbb{D}$ and we'll continue to go back and forth.

As usual, we're really interested in $\mathbb{PSL}(2, \mathbb{R}) = \mathbb{SL}(2, \mathbb{R})/\{\pm 1\}$ since if $g = \left(\begin{smallmatrix} a & b \\ c & d \end{smallmatrix} \right)$ acts on $\mathbb{C}_+$ by

$$f_g(z) = \frac{az + b}{cz + d} \tag{2.8.72}$$

then $f_g = f_h \Leftrightarrow g = \pm h$. f_g acts on $T^*(\mathbb{C}_+)$, the tangent bundle, by

$$f_g^*(z, w) = (f_g(z), (Df_g)(w)) \tag{2.8.73}$$

where

$$(Df_g)(w) = \frac{1}{(cz + d)^2} \, w \tag{2.8.74}$$

By a simple calculation (Problem 11),

$$\operatorname{Im} f_g(z) = \frac{\operatorname{Im} z}{|cz + d|^2} \tag{2.8.75}$$

so that

$$|Df_g(w)| = \frac{\operatorname{Im} f_g(z)}{\operatorname{Im} z} \, |w| \tag{2.8.76}$$

showing that Df_g is an isometry of T_z^* to $T_{f_g(z)}^*$ since the Poincaré metric on $\mathbb{C}_+$ is given by

$$\langle w, w' \rangle_z = \frac{\operatorname{Re}(\overline{w} w')}{(\operatorname{Im} y)^2} \tag{2.8.77}$$

(see (12.2.30) of Part 2B). Since f_g is an isometry, Df_g must be isometric without calculation. In particular, if $\|w\|_z = 1$, $\|Df_g(w)\|_{f_g(z)} = 1$ so f_g^* induces an isometry $h_g \colon \widetilde{\Omega} \to \widetilde{\Omega}$. Of course,

$$h_g h_{g'} = h_{gg'} \tag{2.8.78}$$

We claim that for any $\omega_0 \in \widetilde{\Omega}$, $g \to h_g(\omega_0)$ is a bijection of $\mathbb{PSL}(2, \mathbb{R})$ to $\widetilde{\Omega}$. For we need only prove the analog for $\mathbb{D}$, taking $\omega_0 \in \widetilde{\Omega}(\mathbb{D})$ to be $\omega_0 = (0, 1)$, that is, the point $0 \in \mathbb{D}$ and tangent direction along $\mathbb{R}$. If $h_g(\omega_0) = \omega_0$, then $f_g(0) = 0$, so f_g is a rotation, which must be $\mathbb{1}$ if it leaves $1 \in \partial \mathbb{D}$ invariant. Thus, $h_g(\omega_0) = h_k(\omega_0) \Rightarrow h_{k^{-1}g}(\omega_0) = \omega_0 \Rightarrow k = g$, so h is one–one. Given $\omega_1 = (z, \eta) \in \widetilde{\Omega}(\mathbb{D})$, find g_0 so $f_{g_0}(0) = z$. If $g = g_0 R_\theta$ with R_θ a rotation, $h_g(\omega_0) = (z, D_{g_0}(e^{i\theta}))$, and θ can be picked so $D_{g_0}(e^{i\theta}) = \eta$. Thus, $g \mapsto h_g(\omega_0)$ is onto.

From now on, we'll use the $\mathbb{C}_+$ model. Pick $\omega_0 = (i, i)$, that is, $z_0 = i \in \mathbb{C}_+$, and the tangent direction points up. Thus, $H \colon \mathbb{PSL}(2, \mathbb{R}) \to \widetilde{\Omega}$ by

$$H(g) = h_g(\omega_0) \tag{2.8.79}$$

is a bijection and we can, and will, henceforth associate $\widetilde{\Omega}$ with $\mathbb{PSL}(2, \mathbb{R})$.

Let $L_g \colon \mathbb{PSL}(2, \mathbb{R}) \to \mathbb{PSL}(2, \mathbb{R})$ by $L_g(k) = gk$. Then, by (2.8.78),

$$H(L_g(k)) = h_g(H(k)) \tag{2.8.80}$$

Since h_g is an isometry on $\mathbb{C}_+$, it preserves the measure, $d\tilde{\mu}$, induced by the Riemann metric on $\mathbb{C}_+$, and so, on $\tilde{\Omega}$. Thus, if this measure is pushed back to $\mathbb{PSL}(2,\mathbb{R})$ by H^{-1}, (2.8.80) says the measure is invariant under L_g, so it is Haar measure, $d\mu_H$. It is an important fact (see the Notes) that left invariant Haar measure on $\mathbb{SL}(2,\mathbb{R})$ is also right invariant.

Finally, for our preliminary study of $\tilde{\Omega}$, we want to note the action of geodesic flow on $\tilde{\Omega}$ in these group-theoretic coordinates. The geodesic through $\omega_0 = (i, i)$ is clearly a vertical straight line $\gamma(t) = iy(t)$. Since lengths of the tangent vectors are preserved and we have (2.8.76),

$$\frac{|\dot{y}(t)|^2}{y(t)^2} = 1 \tag{2.8.81}$$

We see that $\dot{y}(t) = y(t)$ and $y(t) = e^t$, $\dot{y}(t) = e^t$, that is, $T_t(\omega_0) = (e^t i, e^t i)$.

The map of $\mathbb{C}_+$ that takes i to $e^t i$ with derivative e^t is exactly $f_g(z) = e^t z$. Since g has to have determinant 1,

$$g_t = \begin{pmatrix} e^{t/2} & 0 \\ 0 & e^{-t/2} \end{pmatrix} \tag{2.8.82}$$

that is,

$$T_t(\omega_0) = H(g_t) \tag{2.8.83}$$

Since each h_k is an isometry, it maps geodesics to geodesics, that is, $h_k T_t = T_t h_k$. Thus,

$$T_t(H(k)) = T_t(h_k(\omega_0)) = h_k(T_t \omega_0) = h_k(H(g_t)) = H(kg_t)) \tag{2.8.84}$$

We summarize with:

Theorem 2.8.13. *The map, H, of (2.8.79) is a bijection from $\mathbb{PSL}(2,\mathbb{R})$ to $\tilde{\Omega}$, the unit tangent bundle over $\mathbb{C}_+$ with the Poincaré metric. H maps Haar measure on $\mathbb{PSL}(2,\mathbb{R})$ to the natural measure on $\tilde{\Omega}$ induced by the Poincaré metric. For $k, p \in \mathbb{PSL}(2,\mathbb{R})$, we have*

$$h_k(H(p)) = H(kp) \tag{2.8.85}$$

If T_t is geodesic flow on $\tilde{\Omega}$ and g_t is given by (2.8.82), then

$$T_t(H(k)) = H(kg_t) \tag{2.8.86}$$

Now let Γ be a Fuchsian group, that is, a discrete subgroup of $\mathbb{SL}(2,\mathbb{R})$ as discussed in Section 8.3 of Part 2A. For example, the picture in Figure 2.8.2 is the image in $\mathbb{D}$ associated to the action of $\Gamma(2) = \{\left(\begin{smallmatrix} a & b \\ c & d \end{smallmatrix}\right) \mid a, b, c, d \in \mathbb{Z}; a, d$ odd; b, c even$\}$. We say, for $w, z \in \mathbb{C}_+$, that $w \sim z$ if and only if $w = \gamma(z)$. If no $\gamma \neq \mathbb{1}$, $\gamma \in \Gamma$, has any fixed points, the set of equivalence classes is a Riemann surface, and the map of $\mathbb{C}_+$ to the equivalence classes is locally an analytic bijection. In that case, $\Gamma \backslash \mathbb{C}_+$, the set of equivalence classes, inherits the metric and we can continue to talk of geodesics, unit tangent bundle, etc.

In $\widetilde{\Omega}$, we say $\omega \sim \omega'$ if and only if there is $\gamma \in \Gamma$ so $h_\gamma(\omega) = \omega'$. Even if some $\gamma \in \Gamma$, $\gamma \neq \mathbb{1}$, has a fixed point, z_0, $\frac{df_\gamma}{dz}$ will be a nontrivial rotation. So even in this case, if we define $\omega \sim \omega'$ when $h_\gamma(\omega) = \omega'$, for some $\gamma \sim \Gamma$, the equivalence classes are a three-dimensional manifold.

Notice that $H(g) \sim H(k) \Leftrightarrow \exists \gamma \in \Gamma$ so $\gamma(g) = k \Leftrightarrow k = \Gamma g$. That is, the equivalence classes are just right cosets of Γ. The set of such cosets is denoted $\Gamma \setminus \mathbb{PSL}(2,\mathbb{R}) = \Omega$. It is not hard to show (Problem 12) that one can find a fundamental domain, $\Xi \subset \mathbb{PSL}(2,\mathbb{R})$, that is a Borel set, and each coset in Ω has exactly one point in Ξ. If $A \subset \Omega$ is measurable, $\mu_H(\{g \in \Xi \mid \Gamma g \in A\})$ defines a measure on Ω we will call μ.

If $\mu(\Omega) < \infty$, we call Γ a *lattice* and then normalize by $\mu(\Omega) = 1$. If Ω is compact, Γ is a lattice called a *uniform lattice*. $\Gamma(2)$ is a lattice (i.e., the central region in Figure 2.8.2), has finite volume (Problem 13), is not compact, and $\Omega = \Gamma(2) \setminus \mathbb{PSL}(2,\mathbb{R})$ is the Ω of Theorem 2.8.3. We can now state the key definition and theorem of this subsection.

Definition. Let Γ be a lattice in $\mathbb{PSL}(2,\mathbb{R})$. Let g_t be given by (2.8.82) and $\Omega = \Gamma \setminus \mathbb{PSL}(2,\mathbb{R})$. *Geodesic flow* on Ω is defined by

$$T_t(\Gamma k) = \Gamma k g_t \tag{2.8.87}$$

If no $\gamma \neq \mathbb{1}$ in Γ has a fixed point, $\Gamma \setminus \mathbb{C}_+$ is a manifold, and this is geodesic flow on that manifold. If some γ's have fixed points, $\Gamma \setminus \mathbb{C}_+$ has some singularities, but removing them leaves a manifold (which, unlike the nonsingular case, is not geodesically complete), and this is still geodesic flow. The following thus includes Theorem 2.8.3 as a special case:

Theorem 2.8.14 (Hopf's Geodesic Theorem). *If Γ is a lattice in $\mathbb{PSL}(2,\mathbb{R})$, $\Omega = \Gamma \setminus \mathbb{PSL}(2,\mathbb{R})$ with the usual probability measure, and $T = T_{t_0}$ for some t_0, then T is measure-preserving and ergodic.*

Henceforth, we'll use G to denote $\mathbb{PSL}(2,\mathbb{R})$. To begin the proof, we note unitary actions on $L^2(G, d\mu_H)$ and $L^2(\Omega, d\mu)$. For any $f \in L^2(G, d\mu_H)$ and $g \in G$, define

$$(\widetilde{V}_g f)(k) = f(kg^{-1}) \tag{2.8.88}$$

For $f \in L^2(\Omega, d\mu)$ and $g \in G$, define

$$(V_g f)(\Gamma k) = f(\Gamma k g^{-1}) \tag{2.8.89}$$

Proposition 2.8.15. *Each $\widetilde{V}_g$ and each V_g is unitary and $g \mapsto \widetilde{V}_g$ (respectively, V_g) are strongly continuous, that is, $g \mapsto \widetilde{V}_g f$ (respectively, $g \mapsto V_g f$)*

*is norm continuous for any $f \in L^2(G, d\mu_H)$ (respectively, $L^2(\Omega, d\mu)$).
Moreover,*

$$\widetilde{V}_{gh} = \widetilde{V}_g \widetilde{V}_h \qquad and \qquad V_{gh} = V_g V_h \qquad (2.8.90)$$

Remark. $\widetilde{V}_g$ is called the *right regular representation*. The inverse is put in g^{-1} for (2.8.88)/(2.8.89) to make (2.8.90) hold.

Proof. Right multiplication by g is measure-preserving on G and on Ω, so $\widetilde{V}_g$ and V_g are unitary. (2.8.90) is an elementary calculation. The inverse is needed since $V_g V_h$ first applies h^{-1} and then g^{-1} and $kh^{-1}g^{-1} = k(gh)^{-1}$.

If f is a continuous function of compact support, it is easy to see $g \mapsto \widetilde{V}_g f$ is continuous by appealing to the dominated convergence theorem. Since these f's are dense and $\|\widetilde{V}_g\| = 1$, one gets continuity for any f. The argument for V_g is the same. $\qquad\square$

In particular, $T_t = V_{g_t}$ is unitary, so measure-preserving. The proof of ergodicity in Theorem 2.8.14 will involve three steps:

(1) Consider the elements $u_s = \left(\begin{smallmatrix} 1 & s \\ 0 & 1 \end{smallmatrix}\right)$ and $\ell_s = \left(\begin{smallmatrix} 1 & 0 \\ s & 1 \end{smallmatrix}\right)$, the upper-triangular and lower-triangular subgroups in G, and the induced unitaries, $U_s = V_{u_s}$, $L_s = V_{\ell_s}$. Prove that $\{u_s, \ell_s\}_{s \in \mathbb{R}}$ generate G in that finite products are dense in G.

(2) Prove that if $T_t f = f$, then for all s, $U_s f = f = L_s f$. This will follow from the simple calculation that

$$g_t u_s g_t^{-1} = u_{se^t}, \qquad g_t \ell_s g_t^{-1} = \ell_{se^{-t}} \qquad (2.8.91)$$

This step implements the notion of distinct geodesics approaching to some point in $\partial\mathbb{D}$ exponentially fast.

(3) Prove that if $V_g f = f$ for all $g \in G$, then f is constant.

The third step seems trivial at first sight since the action of G on Ω is transitive, that is, for any k_0, $\{\Gamma k_0 g\}_{g \in G} = \Omega$, so

$$f(\Gamma k_0 g) = f(\Gamma k_0) \qquad (2.8.92)$$

for all g implies f is constant. The subtlety is that (2.8.92) only holds for a fixed g for a.e. k_0. The set of g is uncountable, so it is not evident that we can find a representative of f so (2.8.92) holds for some k_0 and a.e. g. The argument to implement step (3), with which we begin, is thus a little involved.

Proposition 2.8.16. *Let $B_1, B_2 \subset G$ be Baire sets with $\mu_H(B_1) > 0$, $\mu_H(B_2) > 0$. Then*

$$I_{B_1, B_2} \equiv \{g \mid \mu_H(B_1 \cap B_2 g^{-1}) > 0\} \qquad (2.8.93)$$

is a nonempty open set. The same is true for $B_1, B_2 \subset \Gamma$ if μ_H is replaced by μ.

Proof. We first suppose B_1, B_2 have finite μ_H measures. Then

$$\mu_H(B_1 \cap B_2 g^{-1}) = \int \chi_{B_1}(k)\chi_{B_2 g^{-1}}(k)\,d\mu(k) \qquad (2.8.94)$$

$$= \langle \chi_{B_1}, V_{g^{-1}}\chi_{B_2}\rangle \qquad (2.8.95)$$

is continuous by the continuity of V_g, so I_{B_1,B_2} is open.

Moreover, by (2.8.94),

$$\int \mu_H(B_1 \cap B_2 g)\,d\mu_H(g) = \int \left(\int \chi_{B_1}(k)\chi_{B_2 g^{-1}}(k)\,d\mu_H(k)\right) d\mu_H(g)$$

$$= \int \left(\chi_{B_1}(k)\int \chi_{B_2}(kg)\,d\mu_H(g)\right) d\mu_H(k) \quad (2.8.96)$$

$$= \mu_H(B_2)\mu_H(B_1) > 0 \qquad (2.8.97)$$

We used Fubini's theorem for positive functions in (2.8.96). Thus, I_{B_1,B_2} is nonempty.

For any B_1, B_2, since μ_H is σ-finite, we can find $\{A_1^{(j)}\}_{j=1}^\infty$ with $A_k^{(j)} \subset A_j^{(j+1)}$ and $B_k = \bigcup_{j=1}^\infty A_k^{(j)}$ so that $\mu(A_k^{(j)}) < \infty$. Clearly, by (2.8.94) and the monotone convergence theorem, $I_{B_1,B_2} = \bigcup_n I_{A_1^k,A_2^k}$, so, in general, the set is open and nonempty.

If $B_1, B_2 \subset \Omega$ and $\widetilde{B}_j = \{k \mid \Gamma k \in B_j\}$ (i.e., $\widetilde{B}_j$ is the union of the cosets in B_j), then

$$\mu(B_1 \cap B_2 g^{-1}) > 0 \Leftrightarrow \mu_H(\widetilde{B}_1 \cap \widetilde{B}_2 g^{-1}) = \infty$$

so $I_{B_1,B_2} = I_{\widetilde{B}_1,\widetilde{B}_2}$ is open and nonempty. $\qquad \square$

Theorem 2.8.17. *If $f \in L^2(\Omega, d\mu)$ and $V_g f = f$ for all g, then f is constant.*

Proof. If, for every rational α, β, $\mu(\{\omega \mid \alpha < f(\omega) < \beta\})$ is 0 or 1, it is easy to prove that f is a.e. constant. It follows that if f is not constant, we can find α so that if

$$B_1 = \{\omega \mid f(\omega) > \alpha\} \quad \text{and} \quad B_2 = \{\omega \mid f(\omega) < \alpha\} \qquad (2.8.98)$$

then $\mu(B_1) > 0$, $\mu(B_2) > 0$.

For any $g \in G$, $f(\omega g^{-1}) = f(\omega)$ for all $\omega \in \Omega_g$, a Baire set with measure 1. Let $K \subset G$ be a countable dense set and $\Omega^* = \bigcap_{g \in K} \Omega_{g^{-1}}$, which also has measure 1. Then

$$\omega \in \Omega^*, g \in K \Rightarrow f(\omega g) = f(\omega) \qquad (2.8.99)$$

and if $B_j^* = B_j \cap \Omega^*$, then

$$\mu(B_1^*) > 0, \qquad \mu(B_2^*) > 0$$

Thus, by the last proposition, $\{g \mid \mu(B_1^* \cap B_2^* g^{-1}) > 0\}$ is open and nonempty. Since K is dense, we can find $g \in K$ so $B_1^* \cap B_2^* g^{-1} \neq \emptyset$, that is, there exists $\omega \in \Omega^*$ so $\omega \in B_1$ and $\omega \in B_2 g^{-1}$, so $\omega g \in B_2$. It follows that $f(\omega) > \alpha$ and $f(\omega g) < \alpha$. This is inconsistent with $\omega \in \Omega^*$, $g \in K$, and (2.8.99). Thus, f must be constant. $\qquad \square$

Proposition 2.8.18. *Any $g \in G$ is a limit of a finite product of $u_s = \left(\begin{smallmatrix} 1 & s \\ 0 & 1 \end{smallmatrix}\right)$ and $\ell_s = \left(\begin{smallmatrix} 1 & 0 \\ s & 1 \end{smallmatrix}\right)$, $s \in \mathbb{R}$.*

Proof. We'll give a proof using the relation of Lie groups to Lie algebras (see the Notes). Problem 14 has a country bumpkin's version of the proof of this proposition.

The Lie algebra of G, call it $g = \{A \mid e^{tA} \in G\}$ for all $t \in \mathbb{R}$, is the set of real matrices with $\text{Tr}(A) = 0$. This has a basis $X_1 = \left(\begin{smallmatrix} 0 & 1 \\ 0 & 0 \end{smallmatrix}\right)$, $X_2 = \left(\begin{smallmatrix} 0 & 0 \\ 1 & 0 \end{smallmatrix}\right)$, $X_3 = \left(\begin{smallmatrix} 1 & 0 \\ 0 & -1 \end{smallmatrix}\right)$. Since $[X_1, X_2] = X_3$, the Lie algebra generated by X_1 and X_2 is all of g. This plus the theory of Lie algebras implies the result. $\qquad \square$

Next, we note, by multiplying the matrices, if

$$g_t = \begin{pmatrix} e^{t/2} & 0 \\ 0 & e^{-t/2} \end{pmatrix}, \quad u_s = \begin{pmatrix} 1 & s \\ 0 & 1 \end{pmatrix}, \quad \ell_s = \begin{pmatrix} 1 & 0 \\ s & 1 \end{pmatrix} \tag{2.8.100}$$

then

$$g_t u_s g_{-t} = u_{se^t}, \qquad g_t \ell_s g_{-t} = \ell_{se^{-t}} \tag{2.8.101}$$

In terms of FLTs, the first says that $z \to e^{-t} z \to e^{-t} z + s \to e^t(e^{-t} z + s)$ is $z \to z + se^t$.

Proof of Theorem 2.8.14. We have already noted that μ is invariant for $T = T_{t_0}$. We suppose $t_0 > 0$. We need only prove that if $V = V_{g_{t_0}}$, then

$$V f = f \Rightarrow f \text{ is constant} \tag{2.8.102}$$

If $V f = f$, then with $U_s = V_{u_s}$, by the strong continuity of $g \mapsto V_g$,

$$\|U_s f - f\| = \|V^{-n}(U_s f - f)\| = \|V^{-n} U_s f - f\|$$
$$= \|U_{se^{-nt_0}} V^{-n} f - f\| = \|U_{se^{-nt_0}} f - f\|$$
$$\to 0$$

as $n \to \infty$. Here we use $V^{-n} f = f$ and (2.8.101). Thus, $U_s f = f$ for all s. Similarly, using V^n with $n \to \infty$, $L_s f = f$. By Proposition 2.8.18 and strong continuity of $g \mapsto V_g$, $V_g f = f$ for all g. By Theorem 2.8.17, f is constant. $\qquad \square$

Notes and Historical Remarks. Theorem 2.8.1 is from Weyl [850] whose approach to the result was not via ergodic theory (which did not then exist), but via differences (see below). The approach to Weyl's theorem via skew shifts and the proof we give of unique ergodicity of group extensions is due to Furstenberg [**297**]. Earlier, Parry [**629**] had proven unique ergodicity of skew shifts. For books on equidistribution, see Kuipers–Niederreiter [**496**], Montgomery [**582**], and Rauzy [**672**].

Weyl's approach depended on

$$\left| \frac{1}{N} \sum_{j=0}^{N-1} e^{2\pi i \ell x_j} \right|^2 = \frac{1}{N^2} \sum_{j,k=0}^{N-1} e^{2\pi i \ell (x_j - x_k)} \tag{2.8.103}$$

and then changing the variables so $j = k + h$ and summing over h and k. If, for each $h \neq 0$, $x_{k+h} - x_k$ is uniformly distributed mod 1, the sums over k (for $\ell \neq 0$) with a single N^{-1} will go to zero. With some control on h-dependence, one can show for $\ell \neq 0$ that the double sum goes to zero. The idea is illustrated in Problem 1, which follows Weyl's original proof of Theorem 2.8.1.

van der Corput [**828**] found a simple trick to avoid considering infinitely many h at once and proved (see Problems 2 and 3)

Theorem 2.8.19 (van der Corput's Difference Theorem). *Let $\{x_n\}_{n=1}^{\infty}$ be a sequence in $\mathbb{R}$ so that for each $h = 1, 2, \ldots$, $\{x_{n+h} - x_n\}_{n=1}^{\infty}$ is uniformly distributed mod 1. Then $\{x_n\}_{n=1}^{\infty}$ is uniformly distributed mod 1.*

Remark. If $x_n = \alpha n$ with α irrational, $x_{n+h} - x_n = \alpha h$ is constant, and so not uniformly distributed. Thus, the converse of this theorem is not true.

With this (or Weyl's original argument), one can show (see Problem 6) if $P(x)$ is any nonconstant polynomial with at least one irrational coefficient, then $\{P(n)\}_{n=1}^{\infty}$ is uniformly distributed mod 1. One can also show directly (see Problem 5) that for $0 < \rho < 1$, $\{\beta n^\rho\}_{n=1}^{\infty}$ is uniformly distributed mod 1 for any real $\beta \neq 0$, and then Theorem 2.8.19 implies the same for all $\rho > 0$, $\rho \notin \mathbb{Z}$.

$$* \qquad * \qquad * \qquad * \qquad *$$

There is a huge literature on the dynamics of the Gauss map, including several books [**183, 377, 419, 468, 696**] (see also the presentation in [**242**]). It is known (perhaps first noted explicitly by Philipp [**641**]; see [**419**] for several proofs, Keane [**456**], and the discussion later for proofs) that the Gauss map is mixing for Gauss measure.

Let (where $|\cdot| =$ Lebesgue measure)

$$m_n(y) = |\{x \mid T^n x < y\}| \tag{2.8.104}$$

If μ is Gauss measure, $\langle \cdot , \cdot \rangle$ is the $L^2([0,1], d\mu)$ inner product and $f(x) = (\log 2)(1 + x)$, then

$$m_n(y) = \langle \chi_{[0,y]}(\cdot), f \rangle_{L^2} \to \mu([0,y]) = \log_2(1 + y) \qquad (2.8.105)$$

on account of mixing, and $\langle \chi_B, 1 \rangle = \mu(B)$, $\langle 1, f \rangle = 1$.

Remarkably, in a letter Gauss wrote to Laplace on January 20, 1812, he stated that he had been studying asymptotics of the statistics of the remainder in the continued fraction—essentially, the asymptotics of the function $m_n(y)$. He observed the limit was $\log_2(1 + y)$ (with no proof given!) and wondered the rate of convergence. On the basis of this letter (found in his complete works), one uses the names Gauss map and Gauss measure.

It is clearly a bit of magic that Gauss found $\log_2(1 + y)$. Khinchin [468] conjectures that Gauss must have derived the recursion relation

$$m_{n+1}(x) = \sum_{k=1}^{\infty} m_n\left(\frac{1}{k}\right) - m_n\left(\frac{1}{k+x}\right) \qquad (2.8.106)$$

((2.8.40) is essentially the equation for $m'(y)$) and observed that $\log(1 + y)$ solves

$$m(y) = \sum_{k=1}^{\infty} m\left(\frac{1}{k}\right) - m\left(\frac{1}{k+y}\right)$$

Keane [456] has a different conjecture for how Gauss found Gauss measure.

Results on Gauss' problem on rate of convergence as well as formal proofs of the convergence were first obtained independently by Kuzmin [499] in 1928 and Lévy [524] in 1929, who obtained $O(e^{-c\sqrt{n}})$ and $O(e^{-cn})$ errors, respectively (see below for more on this). Gauss measure also appeared in work of Khinchin [467, 468], who proved (2.8.34), and Lévy [525], who first proved (2.8.36).

As we noted, the distributional consequences of ergodicity of Gauss measure affect more than single digits. For example, if $a_1, \ldots, a_\ell$ are given and p_ℓ/q_ℓ (resp. $p_{\ell-1}/q_{\ell-1}$) are the continued fractions $[a_1, \ldots, a_\ell]$ (respectively, $[a_1, \ldots, a_{\ell-1}]$), then (Problem 9) for a.e. $x \in [0,1]$,

$$N^{-1}\{\#k \le N \mid a_{k+1}(x) = a_1, \ldots, a_{k+\ell}(x) = a_\ell\}$$
$$= \left| \log \left[\frac{(p_\ell + q_\ell)(q_\ell + q_{\ell-1})}{(q_\ell)(p_\ell + q_\ell + p_{\ell-1} + q_{\ell-1})} \right] \right| \qquad (2.8.107)$$

The results encapsulated in Theorem 2.8.9 are most easily seen using ergodicity and the Birkhoff ergodic theorem. It is striking that despite Khinchin's work on ergodic theorems shortly before his work on continued fractions quoted above, he never phrased the continued fraction results in ergodic terms. This link is often attributed to Ryll-Nardzewksi [711], although it appeared earlier in Doeblin [222]. Even earlier, in 1926, Knopp

[**476**] noted the estimates with the flavor of (2.8.65), which are the essence of ergodicity. The proof we give of ergodicity in this vein go back at least to Billingsley [**72**]. Below, in our discussion of geodesic flow, we'll discuss a different proof of ergodicity of the Gauss map.

The ultimate resolution of the Gauss problem on the rate of convergence of $m_n(y)$ to $\log_2(1+y)$ was in two papers by Wirsing [**860**] and Babenko [**26**] who analyzed the Gauss–Kuzmin–Wirsing operator, $\mathcal{G}$, of (2.8.40) (Kuzmin already analyzed this operator). They determined that in a suitable space (Babenko analyzed $\mathcal{G}$ as an operator on an H^2 space of $\{z \mid \operatorname{Re} z > -\frac{1}{2}\}$ and proved $\mathcal{G}$ compact; these notions will be discussed in Chapter 3 of Part 4), $\mathcal{G}$ had no spectrum outside a disk $\mathbb{D}_\rho(0)$ with $\rho < \lambda_2$, except for the known eigenvalue at 1 (with eigenvector $1/(1+x)$) and eigenvalue at $-\lambda_2$, where $\lambda_2 = 0.30366\ldots$. It follows that $\lim_{n\to\infty} \frac{1}{n} \log|m_n(y) - \log_2(1+y)| = \log(\lambda_2)$, that is, the exact rate of exponential convergence is known. These ideas also lead easily to a proof of mixing.

Ergodicity, or even mixing, alone is not enough to answer all statistical questions about continued fraction approximations. For example, Bosma et al. [**100**] (see also [**377**]) have found the a.e. x asymptotic fraction of n for which $q_n(x)^2|x - \frac{p_n(x)}{q_n(x)}| \leq y$ by consideration of an extension of the usual Gauss map.

$$* \qquad * \qquad * \qquad * \qquad *$$

The history of the Hopf geodesic theorem begins with the connection to continued fractions (rather than this connection being an afterthought). In 1924, Artin [**17**] wrote a paper with the title (translated to English) "A mechanical system with quasi-ergodic trajectories." The mechanical system of his paper was geodesic flow on what we would now call $\Omega = \mathbb{SL}(2,\mathbb{Z}) \setminus \mathbb{SL}(2,\mathbb{R})$, the upper half-plane with equivalence classes defined by the modular group $\mathbb{SL}(2,\mathbb{Z})$. In $\mathbb{C}_+$, he parametrized trajectories by the "endpoints at infinity," that is, real intecepts, x^+ and x^-, at $+\infty$ and $-\infty$. He showed two trajectories were the same in Ω if and only if the continued fractions for x^+ and x^-, written as a two-sided system $\ldots n_k^- \ldots n_2^- n_1^- n_1^+ n_2^+ \ldots n_k^+ \ldots$, were related by a shift. He then used that to prove most trajectories were dense in Ω. Thus, the use of the term "quasi-ergodic" which, in that era, meant dense trajectories.

After the ergodic revolution of 1931–32 spearheaded by Koopman, von Neumann, Birkhoff, and E. Hopf, the meaning of ergodic changed, and Hedlund [**371**] re-examined Artin's framework, proved ergodicity of his two-sided shift on continued fractions, and so deduced ergodicity of the geodesic flow in this case. In 1939, Hopf [**388**] (see also his Gibbs lecture [**390**]) proved Theorem 2.8.14, at least for finitely generated Γ. He didn't parameterize $\widetilde{\Omega}$ as

$\mathbb{SL}(2, \mathbb{R})$ and use group theory. Rather, in the $\mathbb{D}$ model, given a point $z_0 \in \mathbb{D}$ and a tangent direction, he looked at the geodesic from z_0 in that direction and he parametrized in terms of the two limit points, $e^{i\theta_1}, e^{i\theta_2} \in \partial\mathbb{D}$, and the geodesic distances, s, of z_0 from the midpoint of the geodesic (in the hyperbolic metric, the geodesic length is infinite and there is no midpoint—he picked the Euclidean midpoint, i.e., the point w_0 with $\arg(w_0) = \frac{1}{2}(\theta_1 + \theta_2)$ (mod π)). It turns out the measure on $\widetilde{\Omega}$ is just $\rho(\theta_1, \theta_2)\, d\theta_1\, d\theta_2\, ds$, where ρ is a function of $\theta_2 - \theta_1$. By shifting to $\mathbb{C}_+$, he found for any two geodesics, $(\theta_1, \theta_2, u), (\theta_1, \theta_2', t)$, asymptotic to the same point at ∞, there is a unique $t(u)$ so the hyperbolic distance from $(\theta_1, \theta_2, u + w), (\theta_1, \theta_2', t(u) + w)$ goes to zero as $w \to \infty$. This map is essentially the action of ℓ_s.

There is an interesting way of understanding invariance of Gauss measure and, given the connection to geodesic flow, its ergodicity. We begin by pulling a region and map out of the hat. Let

$$\Omega^\sharp = \left\{(x, \rho) \in \mathbb{R}^2 \mid x \in (0, 1),\, \rho \leq \frac{1}{1 + x}\right\} \qquad (2.8.108)$$

and $T^\sharp \colon \Omega^\sharp \to \Omega^\sharp$ by

$$T^\sharp(x, \rho) = \left(\left\{\frac{1}{x}\right\},\, x(1 - x\rho)\right) \qquad (2.8.109)$$

It can be shown (see below) that $T^\sharp$ maps $\Omega^\sharp \to \Omega^\sharp$ and is a bijection up to sets of measure 0. Since the Jacobian of the coordinate change is easily seen to be 1, $T^\sharp$ preserves two-dimensional Lebesgue measure. Projecting on x, this provides another proof the Gauss measure is preserved by $x \mapsto \{\frac{1}{x}\}$.

The map and its relevance for ergodicity comes from Einsiedler–Ward [**242**, Ch. 9]. Let C_+ be the subset of $\widetilde{\Omega}(z, w)$, where $z = ib(b \in (0, 1))$, and so that ω is pointing to the right and down (i.e., $\operatorname{Re}\omega > 0$, $\operatorname{Im}\omega < 0$). The corresponding geodesics have two intercepts on $\mathbb{R}$: one $x \in (0, 1)$ and the other $-\tilde{x}$ with $\tilde{x} \in (1, \infty)$ (see the pictures in [**242**]). The inverse diameter, $\rho = 1/(x + \tilde{x}) \in (0, (1 + x)^{-1})$. Thus, C_+ is parametrized by $\Omega^\sharp$. If $C_- = \{(z, \omega) \mid (z, -\omega) \in C_+\}$, we get the same geodesics, just pointed in the opposite direction. The point is that in $\mathbb{SL}(2, \mathbb{Z}) \setminus \mathbb{SL}(2, \mathbb{R})$, $T^\sharp$ is just the map from a point in C_+ to the next time it "hits" C_- (return map), and one can use the ergodicity of geodesic flows to get ergodicity of $T^\sharp$, and so of T.

Series [**733**] has found a way to avoid the singular points of $\mathbb{SL}(2, \mathbb{Z}) \setminus \mathbb{SL}(2, \mathbb{R})$ by discussing continued fractions in the context of $\Gamma(2) \setminus \mathbb{SL}(2, \mathbb{R})$. She also has a sequence of papers [**109, 730, 731, 732**] discussing analogs of continued fractions for other Fuchsian groups.

Any Fuchsian group, Γ, has a *limit set*, $\Lambda(\Gamma)$, the set of limit points of $\{\gamma(z_0)\}_{\gamma \in \Gamma}$. It turns out this set is independent of which z_0 in $\mathbb{D}$ or $\mathbb{C}_+$ is

Figure 2.8.4. A hyperbolic tiling with a compact fundamental domain: a computer rendition inspired by Escher's *Circle Limit III* pattern, created by Douglas Dunham, used with permission.

picked. $\Lambda(\Gamma)$ is either all of $\partial\mathbb{D}$ or $\mathbb{R}\cup\{\infty\}$ (in which case, we say Γ is *first type*) or a nowhere dense closed subset (in which case, we say Γ is *second type*). If Γ is second type, $\Gamma\setminus G$ has infinite volume. If Γ is first type and finitely generated, $\Gamma\setminus G$ has finite volume. If Γ is finitely generated in $\mathbb{D}$, its Dirichlet fundamental domain ($=\{z\in\mathbb{D}\mid |\gamma(z)|\geq |z|$ for all $\gamma\in\Gamma\}$) lies in some D_r, $r<1$, if and only if $\Gamma\setminus\mathbb{SL}(2,\mathbb{R})$ is compact and has a closure in $\mathbb{D}$ with finitely many points, all cusps if and only if $\Gamma\setminus\mathbb{SU}(1,1)$ is noncompact but finite volume. If infinitely generated, it can happen that $\Gamma\setminus G$ has finite or infinite volume. Figure 2.8.4 shows a pattern inspired by an Escher lithograph illustrating a Γ which is a uniform lattice (i.e., $\Gamma\setminus G$ is compact). For more on Fuchsian groups, see [**49, 278, 450**] or [**757**, Ch.$\tilde{9}$].

On any Lie group, G (i.e., C^∞-manifold with a group structure in which all group operations are C^∞-maps), one can construct Haar measure as follows. For any g, $L_g\colon G\to G$ by $L_g(h)=gh$ defines a C^∞-map of G to G, taking e, the identity, to g. Thus, $L_g^*\colon T_g^*(G)\to T_g^*(G)$. For any $\omega\in T_e^*(G)$, $\omega(g)\equiv L_g^*(\omega)$ defines a left invariant one-form on G, so if $\omega_1,\ldots,\omega_d$ is a basis for $T_e^*(G)$, $\omega_1(g)\wedge\cdots\wedge\omega_d(g)$ defines a left invariant measure.

For $\mathbb{GL}(\nu,\mathbb{R})$, the dimension is ν^2, and if $\{a_{ij}\}$ are the matrix elements,

$$d\mu_H(a)=\frac{\prod_{i,j=1}^{\nu}da_{ij}}{\det(A)^\nu} \qquad (2.8.110)$$

is Haar measure, since under left multiplication by B, $(a_{11}, a_{21}, \ldots, a_{\nu 1})^t$ is acted on by B as a matrix. This is true for each column, so the Jacobian is $\det(B)^\nu$ cancelled by $\det(BA)^{-\nu} = \det(B)^{-\nu} \det(A)^{-\nu}$. By noting that right multiplication just acts on the rows by B, we see that $d\mu_H$ in this case is also right invariant.

Formally, Haar measure on $\mathbb{SL}(\nu, \mathbb{R})$ is just the one for $\mathbb{GL}(\nu, \mathbb{R})$ multiplied by $\delta(\det(A) - 1)$. This means if $a_{11} \neq 0$, we can change variables from a_{11} to $\det(A)$ in (2.8.110) and get

$$d(\det(A)) \prod_{\substack{i,j \neq 1 \\ (i,j) \neq (1,1)}}^{\nu} da_{ij} \Big/ \det(A)^\nu M(A)_{11} \qquad (2.8.111)$$

where $M(A)_{11}$ is the minor, that is, $\partial \det(A)/\partial a_{11}$. Thus, for example, near points where $d \neq 0$, Haar measure for $\left(\begin{smallmatrix} a & b \\ c & d \end{smallmatrix}\right)$ in $\mathbb{SL}(2, \mathbb{R})$ is $|d|^{-1} \, db \, dc \, dd$, and where $c \neq 0$, $|c|^{-1} \, da \, db \, dd$. As with $\mathbb{GL}(\nu, \mathbb{R})$, this measure is right invariant also.

Weyl had a colorful term for the need for g^{-1} in (2.8.88), that is, in looking at $V_g V_h$, you had to first undo the g and then the h—he called it "dressing and undressing," making the analogy to putting on shoes and socks and noting that the order in which you put on/take off socks and shoes is opposite for dressing and undressing.

For a closed matrix group, G, the Lie algebra, g, is the set of matrices A so that $e^{tA} \in G$ for all t. Since

$$e^{t(A+B)} = \lim_{n \to \infty} (e^{tA/n} e^{tB/n})^n \qquad \text{(Lie product formula)} \qquad (2.8.112)$$

g is a vector space. Moreover, if $[A, B] = AB - BA$, one can also show that

$$\exp(t[A, B]) = \lim_{n \to \infty} \left(e^{A\sqrt{t/n}} e^{B\sqrt{t/n}} e^{-A\sqrt{t/n}} e^{-B\sqrt{t/n}}\right)^n \qquad (2.8.113)$$

Finally, it is a fact that any connected Lie group is a finite product of e^{tA}'s, $A \in g$ (and, if G is compact, or in many other cases including $\mathbb{SL}(\nu, \mathbb{R})$, a single e^{tA} will do). This plus (2.8.112)/(2.8.113) shows that if $X_1, \ldots, X_\ell \in g$ and $\{X_j\}_{j=1}^\ell \cup \{[X_j, X_{j_0}]\}_{j,k=1}^\ell$ span g, then any $h \in G$ is a limit of finite products of $\{e^{tX_j}\}_{t \in \mathbb{R}, \Gamma}$.

Problems

1. This will lead you through Weyl's proof that for β irrational, $\{\beta n^2\}_{n=1}^\infty$ is equidistributed mod 1.

   (a) Prove that

   $$\left| \sum_{k=1}^N e^{2\pi i \beta \ell k^2} \right|^2 = \sum_{h=-N+1}^{N-1} e^{2\pi i \beta \ell h^2} \sum_{k=\max(1,1-h)}^{\min(N,N-h)} e^{4\pi i \beta \ell h k} \qquad (2.8.114)$$

   (*Hint*: $(m+h)^2 - m^2 = 2mh + h^2$.)

(b) Prove that

$$\left| \sum_{k=1}^{N} e^{2\pi i \beta \ell k^2} \right|^2 \leq \sum_{h=-N+1}^{N-1} \min\left(N, \frac{2}{|e^{4\pi i \beta \ell h} - 1|} \right) \qquad (2.8.115)$$

(c) For any ε, prove that

$$\left| \sum_{k=1}^{N} e^{2\pi i \beta \ell k^2} \right|^2 \leq 2(2N - 2)\varepsilon^{-1}$$

$$+ N^2[N^{-1}(1 + 2\#\{1 \leq h \leq N - 1 \mid |e^{4\pi i \beta \ell h} - 1| \leq \varepsilon\})] \qquad (2.8.116)$$

(d) For any ε, prove that

$$\limsup N^{-2} \left| \sum_{k=1}^{N} e^{2\pi i \beta \ell k^2} \right|^2 \leq 2\left[\frac{\varepsilon}{\pi} + O(\varepsilon^2) \right] \qquad (2.8.117)$$

so that $\{\beta k^2\}_{k=1}^{\infty}$ is equidistributed mod 1.

Remark. With more effort and some Diophantine approximation, one can show $|\sum_{k=1}^{N} e^{2\pi i \beta \ell k^2}|$ is $O(N^{1+\varepsilon})$ for all $\varepsilon > 0$. See Montgomery [**582**, Sect. 3.2].

2. This will prove a lemma of van der Corput needed in the next problem. N and H are fixed with $H < N$. Given $\{u_n\}_{n=1}^{N}$, define u_n to be zero if $n \leq 0$ or $n \geq N + 1$.

(a) Prove that

$$H \sum_{n=1}^{N} u_n = \sum_{k=1}^{N+H-1} \sum_{m=0}^{H-1} u_{k-m} \qquad (2.8.118)$$

(b) Prove that

$$H^2 \left| \sum_{n=1}^{N} u_n \right|^2 \leq (N + H - 1) \sum_{k=1}^{N+H-1} \left| \sum_{m=0}^{H-1} u_{k-m} \right|^2 \qquad (2.8.119)$$

(c) Prove that

$$\text{RHS of } (2.8.119) = (N + H - 1)H \sum_{n=1}^{N} |u_n|^2$$
$$+ 2(N + H - 1) \sum_{k=1}^{N-1} (H - k)H \sum_{n=1}^{N-k} u_n \bar{u}_{n+k} \qquad (2.8.120)$$

That LHS of (2.8.119) $\leq$ RHS of (2.8.120) is the required lemma.

3. This will prove and use Theorem 2.8.19.

(a) For any sequence $\{x_n\}_{n=1}^\infty \in \mathbb{R}$ and ℓ and H, prove that

$$\limsup \left| \frac{1}{N} \sum_{n=1}^N e^{2\pi i \ell x_n} \right|^2$$

$$\leq \frac{1}{H} + \frac{2}{H} \sum_{k=1}^{H-1} (H-k) \limsup \frac{1}{N} \left| \sum_{n=1}^{N-k} e^{2\pi i \ell (x_n - x_{n+k})} \right| \tag{2.8.121}$$

(b) Prove Theorem 2.8.19.

(c) For any β irrational and $k = 1, 2, \dots$, prove inductively that $\{\beta n^k\}_{n=1}^\infty$ is uniformly distributed mod 1.

4. (a) For any $k = 1, 2, \dots$ and α irrational, prove that the generalized skew shift, T_k, (2.8.12) is ergodic.

(b) Prove it is uniquely ergodic.

(c) If $P_k(n) = \beta n^k + a_1 n^{k-1} + \cdots + a_n$ with β irrational, prove that $\{P_k(n)\}_{n=1}^\infty$ is uniformly distributed mod 1.

Remark. See also Problem 6.

5. (a) For any N and any C^1-function, g on $(0, \infty)$, prove that

$$\left| \sum_{j=0}^{N-1} g(j) - \int_0^N g(y)\, dy \right| \leq \int_0^N |g'(y)|\, dy \tag{2.8.122}$$

(b) For $\rho \in (0, 1)$, $\ell \neq 0$, $\beta \neq 0$ both real, prove that $\lim_{N \to \infty} \frac{1}{N} \int_0^N \exp(2\pi i \ell \beta x^\rho)\, dx = 0$. (*Hint*: Let $y = x/N$ and use stationary phase on $\int_0^1 f(y) \exp(2\pi i \ell \beta N^\rho y^\rho)\, dy$ for f with $0 \leq f \leq 1$ and $\text{supp}(1-f) \subset (0, \varepsilon)$.)

(c) For any $\rho \in (0, 1)$, β real and nonzero, and $\ell \in \mathbb{Z} \setminus \{0\}$, prove that $\frac{1}{N} \sum_{n=0}^{N-1} \exp(2\pi i \ell \beta n^\rho) \to 0$ and conclude $\{\beta n^\rho\}_{n=0}^\infty$ is uniformly distributed mod 1.

(d) If $\rho \in (0, \infty) \setminus \mathbb{Z}$, β real and nonzero, prove that $\{\beta n^\rho\}_{n=0}^\infty$ is uniformly distributed mod 1. (*Hint*: Use Theorem 2.8.19.)

Remark. A full stationary phase analysis in (b) shows that $\sum_{n=0}^{N-1} e^{i\alpha n^\rho} = O(N^\rho)$ for any real α and $\rho \in (0, 1)$.

6. (a) Fix p/q rational. Suppose that $\{y_n\}_{n=0}^\infty$ is a sequence so that for $b = 0, 1, \dots, q-1$, we have $\{y_{mq+b}\}_{m=0}^\infty$ is uniformly distributed mod 1. For any k, prove that $\frac{p}{q} n^k + y_n \equiv x_n$ is uniformly distributed mod 1.

(b) Suppose you know $\{P(n)\}_{n=1}^\infty$ is uniformly distributed mod 1 for any nontrivial polynomial whose highest-order coefficient is irrational. Prove the same is true for any polynomial, P, with $P(0) = 0$ and at least one irrational coefficient.

7. This problem will provide an alternate way to go from estimates like (2.8.48) and (2.8.58) to $|A||[0,1] \setminus A| = 0$ without using Lebesgue's differentiation theorem.

   (a) Let $\mathcal{I}$ be a family of open intervals, I, in $[0,1]$ so that any (a,b) is a countable union of mutually disjoint I's in $\mathcal{I}$ plus a countable set of points. Suppose μ is a measure with no pure points and $A \subset [0,1]$ measurable so that for all $I \in \mathcal{I}$,

   $$\mu(A \cap I) \geq c\,\mu(A)\mu(I) \qquad (2.8.123)$$

   for some $c > 0$. For any open set U, prove that

   $$\mu(A \cap U) \geq c\,\mu(A)\mu(U) \qquad (2.8.124)$$

   (b) Prove that (2.8.124) holds for any Borel set U. (*Hint*: By regularity, find U_n open so $\chi_{U_n} \to \chi_U$ in $L^1([0,1], d\mu)$.)

   (c) Prove that $\mu(A)\mu([0,1] \setminus A) = 0$ so that $\mu(A) = 0$ or $\mu([0,1] \setminus A) = 0$.

   (d) Show this applies when $\mathcal{I}$ is the set of all open dyadic intervals and also when $\mathcal{I}$ is the set of open intervals $I^{\mathrm{int}}_{a_1,\dots,a_n}$ of (2.8.52).

   Remark. This argument in this context is from Knopp [**476**].

8. (a) Let $a = \sum_{n=1}^{\infty} (-1)^{n+1}/n^2$, $b = \sum_{n=1}^{\infty} 1/n^2 = \pi^2/6$. Prove that $b - a = \frac{1}{2}b$ and conclude that $a = \pi^2/12$.

   (b) Prove that

   $$\int_0^1 \frac{\log y}{1+y}\, dy = -\int_0^{\infty} \frac{x e^{-x}\, dx}{1 + e^{-x}} \qquad (2.8.125)$$

   (c) Prove that the RHS of (2.8.125) is $-\sum_{n=1}^{\infty} (-1)^{n+1}/n^2$. Conclude that (2.8.63) holds.

9. (a) Prove that $\{x \mid a_1(x) = a_1, \dots, a_\ell(x) = a_\ell\}$ is the interval between p_ℓ/q_ℓ and $(p_\ell + p_{\ell-1})/(q_\ell + q_{\ell-1})$, where $p_\ell/q_\ell = [a_1, \dots, a_\ell]$ and $p_{\ell-1}/q_{\ell-1} = [a_1, \dots, a_{\ell-1}]$.

   (b) If μ is Gauss measure, conclude that $\mu(\{x \mid a_1(x) = a_1, \dots, a_\ell(x) = a_\ell\}) = $ RHS of (2.8.107).

   (c) Conclude (2.8.107).

10. Compute the a.e. value of $\lim_{n \to \infty} \frac{1}{n} \sum_{j=1}^{n} \frac{1}{a_j(x)}$.

11. Let $f(z) = (az + b)/(cz + d)$ with $\det\left(\begin{smallmatrix} a & b \\ c & d \end{smallmatrix}\right) = 1$. Prove that $f'(z) = 1/(cz + d)^2$ and that $|f'(z)| = \operatorname{Im} f(z)/\operatorname{Im} z$ if $z \in \mathbb{C}^+$ and $a, b, c, d \in \mathbb{R}$.

12. (a) Let Γ be a Fuchsian group on $\mathbb{D}$ so that $\gamma(0) \neq 0$ for any $\gamma \in \Gamma$, $\gamma \neq 1$. Define $D(\Gamma)$, the *Dirichlet domain* of Γ to be $\{z \in \mathbb{D} \mid \forall \gamma \in \Gamma, \gamma \neq 1, |\gamma(z)| \geq |z|\}$. Prove that each orbit, $\{\gamma(\omega)\}_{\gamma \in \Gamma}$, intersects D^{int} in at most one point and D in at least one point. Then show D^{int} plus a subset of ∂D consisting of finitely many arcs of geodesics is a fundamental domain, F, in that each orbit intersects F in exactly one point.

    (b) Let $\Xi \subset \mathbb{PSU}(1,1)$ be $\{g \mid f_g(0) \in F\}$. Prove that any coset Γk has exactly one point in Ξ and that Ξ is a Baire set (indeed, an open set plus a nice subset of its boundary).

13. Prove that the central region of Figure 2.8.2 has finite hyperbolic volume and conclude that $\Gamma(2)$ is a lattice.

14. This will present the simple-minded proof that any $g \in \mathbb{SL}(2, \mathbb{R})$ is a limit of finite products of $\left(\begin{smallmatrix} 1 & s \\ 0 & 1 \end{smallmatrix}\right)$ and $\left(\begin{smallmatrix} 1 & 0 \\ s & 1 \end{smallmatrix}\right)$ for possibly different s.

    (a) Prove that any g is a product uav where $u, v \in \{\left(\begin{smallmatrix} \cos\theta & \sin\theta \\ -\sin\theta & \cos\theta \end{smallmatrix}\right)\}$ and $a \in \{\left(\begin{smallmatrix} \cosh\beta & \sinh\beta \\ \sinh\beta & \cosh\beta \end{smallmatrix}\right)\}$.

    (b) Let A be a 2×2 matrix. If $A_n = 1 + \frac{A}{n} + o(\frac{1}{n})$, prove that $A_n^n \to \exp(A)$.

    (c) Prove the Lie product formula: for any 2×2 matrices, A, B,

    $$(e^{tA/n} e^{tB/n})^n \to \exp(tA + tB)$$

    (d) If $\sigma^2 = 1$ (respectively, -1), prove that $\exp(t\sigma) = (\cosh t)1 + (\sinh t)\sigma$ (respectively, $(\cos t)1 + (\sin t)\sigma$).

    (e) Show that $\left(\begin{smallmatrix} 1 & s \\ 0 & 1 \end{smallmatrix}\right) = \exp(s\left(\begin{smallmatrix} 1 & 1 \\ 0 & 1 \end{smallmatrix}\right))$ and $\left(\begin{smallmatrix} 1 & 0 \\ s & 1 \end{smallmatrix}\right) = \exp(s\left(\begin{smallmatrix} 1 & 0 \\ 1 & 1 \end{smallmatrix}\right))$.

    (f) If $\sigma_\pm = \left(\begin{smallmatrix} 0 & 1 \\ \pm 1 & 0 \end{smallmatrix}\right)$, prove that $\sigma_\pm^2 = \pm 1$.

    (g) Conclude any $g \in \mathbb{SL}(2, \mathbb{R})$ is a limit of products of $\left(\begin{smallmatrix} 1 & s \\ 0 & 1 \end{smallmatrix}\right)$ and $\left(\begin{smallmatrix} 1 & 0 \\ s & 1 \end{smallmatrix}\right)$ for possibly different s.

15. (Toral Automorphisms) Let A be an $n \times n$ matrix with integral coefficients and $\det(A) = 1$. Thinking of $\mathbb{T}^n = \mathbb{R}^n / \mathbb{Z}^n$, A induces a map $\tau_A \colon \mathbb{T}^n \to \mathbb{T}^n$ by $\tau_A[\theta] = [A\theta]$.

    (a) Prove that A is a bijection on $\mathbb{T}^n$ that preserves the natural Euclidean measure on $\mathbb{T}^n$.

    (b) For $k_1, \ldots, k_n \in \mathbb{Z}^n$, define $\varphi_{\vec{k}}(\vec{\theta}) = \exp(2\pi i (\vec{k} \cdot \vec{\theta}))$. Prove that $\varphi_{\vec{k}}(\tau_A^\ell(\vec{\theta})) = \varphi_{\vec{k}}(\vec{\theta})$ if and only if $(A^\ell)^t \vec{k} = \vec{k}$. Then show in that case that A has an eigenvalue which is an ℓ-th root of unity.

    (c) Prove that τ_A is ergodic if A has no eigenvalue which is a root of unity.

    Remark. Let case $A = \left(\begin{smallmatrix} 2 & 1 \\ 1 & 1 \end{smallmatrix}\right)$ have eigenvalues $\lambda > 1$ and $\lambda^{-1} < 1$ so τ_A is ergodic. It is called the *Arnold cat* map because Arnold often showed images of a picture of a cat's face under τ_A.

2.9. Bonus Section: Subadditive Ergodic Theorem and Lyapunov Behavior

Let (Ω, Σ, μ, T) be a measurable dynamical system and $f \colon \Omega \to \{z \in \mathbb{C} \mid r < |z| < R\}$ for $0 < r < R < \infty$. Then, by applying the Birkhoff theorem to $\log|f|$, we see for a.e. ω,

$$\lim_{n\to\infty} |f(T^{n-1}\omega)\ldots f(\omega)|^{1/n} \tag{2.9.1}$$

exists (and is constant if T is ergodic). But what happens, if instead, we look at $A \colon \Omega \to \mathbb{GL}(d, \mathbb{C})$, the $d \times d$ invertible matrices? $\|AB\|$ is not, in general, $\|A\|\|B\|$, so $\log|A(T^{n-1}\omega)\ldots A(\omega)|^{1/n}$ is no longer an ergodic average. In spite of this, we'll prove the theorem below:

Theorem 2.9.1 (Furstenberg–Kesten Theorem). *Let (Ω, Σ, μ, T) be a measurable dynamical system and $A \colon \Omega \to \mathbb{GL}(d, \mathbb{C})$, a function with $\log^+\|A(\cdot)\| \in L^1(\Omega, d\mu)$. Then for a.e. ω,*

$$\lim_{n\to\infty} \|A(T^{n-1}\omega)\ldots A(\omega)\|^{1/n} \tag{2.9.2}$$

exists and is an invariant function (so constant if T is ergodic). If $\log^+\|A(\cdot)^{-1}\| \in L^1(\Omega, d\mu)$, also, then the limit is nonzero.

Remarks. 1. Here $\log^+(y) = \max(0, \log y)$.

2. $\lim_{n\to\infty} \frac{1}{n} \log\|A(T^{n-1}\omega)\ldots A(\omega)\|$ is called the *Lyapunov exponent* or *principal Lyapunov exponent*.

We'll prove this from

Theorem 2.9.2 (Kingman Ergodic Theorem). *Let (Ω, Σ, μ, T) be a measurable dynamical system and $\{g_n\}_{n=1}^\infty$ a sequence of functions on Ω. Suppose for all n, m,*

$$g_{n+m}(\omega) \leq g_n(\omega) + g_m(T^n\omega) \tag{2.9.3}$$

and that $g_1^+ \equiv \max(g_n, 0) \in L^1$.

Then for a.e. ω,

$$\lim_{n\to\infty} \frac{1}{n} g_n(\omega) \equiv G(\omega) \tag{2.9.4}$$

exists and is invariant. Indeed, if all $g_n \in L^1(\Omega, d\mu)$, then

$$G(\omega) = \inf \frac{1}{n} \mathbb{E}(g_n \mid \mathcal{I})(\omega) \tag{2.9.5}$$

Remarks. 1. There is a subtle difference between our intentions here and in the Birkhoff theorem. In that earlier theorem, when we say the limit exists, we mean in $(-\infty, \infty)$, that is, the limit is a.e. finite. Here, without additional assumptions, the limit can be $-\infty$, that is, it lies in $[-\infty, \infty)$. By (2.9.3), if

$$G_n(\omega) = \frac{1}{n} \mathbb{E}(g_n \mid \mathcal{I})(\omega) \qquad (2.9.6)$$

then

$$(n+m)G_{n+m}(\omega) \le nG_n(\omega) + mG_m(\omega) \qquad (2.9.7)$$

so, by Proposition 2.9.3 below, we have pointwise that

$$G(\omega) = \lim G_n(\omega) = \inf G_n(\omega) \qquad (2.9.8)$$

If

$$\sup \frac{1}{n} \|g_n\|_1 < \infty \qquad (2.9.9)$$

then $G \in L^1$ and so G is a.e. finite.

2. $g_1^+ \in L^1$ and (2.9.3) imply $g_n^+ \in L^1$. If $f^+ \in L^1$, one can define $\mathbb{E}(f \mid \mathcal{I})$, although it may be $-\infty$ on a positive measure set, and so, in fact, (2.9.5) is always true.

3. This result is also known as the *subadditive ergodic theorem*.

After proving this, we'll turn to analyzing rates of exponential decay or growth for random matrix products applied to vectors. As background to why subadditivity should be relevant and why (2.9.5) holds, we recall

Proposition 2.9.3. *If $\{a_n\}_{n=1}^{\infty}$ is a subadditive sequence of reals, that is,*

$$a_{n+m} \le a_n + a_m \qquad (2.9.10)$$

Then

$$\lim_{n \to \infty} \frac{a_n}{n} = \inf_m \frac{a_m}{m} \qquad (2.9.11)$$

Proof. Fix m. Let $n = mr + q$, $q \in (1, \ldots, m)$. Then

$$\limsup_{n \to \infty} \frac{a_n}{n} \le \limsup_{n \to \infty} \frac{ra_m}{n} + \limsup \frac{1}{n} \max_{1 \le j \le m} a_j = \frac{a_m}{m} \qquad (2.9.12)$$

Thus,

$$\limsup_{n \to \infty} \frac{a_n}{n} \le \inf_m \frac{a_m}{m} \le \liminf \frac{a_n}{n} \le \limsup \frac{a_n}{n} \qquad (2.9.13)$$

so the limit exists and (2.9.11) holds. $\qquad \square$

The strategy for proving the Kingman theorem will be different from the maximal inequality plus density approach that we've used so far because there is no obvious dense set to play the role that $\{c\mathbb{1}+g-g\circ T\}$ played in L^1. As a warmup, we provide yet another proof of the Birkhoff theorem that introduces the strategy we'll use for the Kingman theorem. Interestingly enough, while we won't use a maximal *inequality*, we'll use the maximal theorem, Theorem 2.6.14. As in that theorem, we define $f^\sharp$, g_n, h_n, and E_n given by (2.6.54), (2.6.56), (2.6.57), and (2.6.58), respectively.

Alternate Proof of Theorem 2.6.14. As in the first proof, if we show for each n that

$$q \equiv f\chi_{E_n}, \qquad \int q(x)\,dx \geq 0 \tag{2.9.14}$$

then a simple application of dominated convergence proves (2.6.55).

Fix $m > n$ (we'll eventually take $m \to \infty$). For each ω, consider the points $\omega, T\omega, \ldots, T^{m-1}\omega$ which we call steps $1, 2, \ldots, m$. Successively color each one green, red, or yellow. Start with ω. If $\omega \notin E_n$, color 1 red. If $\omega \in E_n$, for some first $\ell \leq n$, $g_\ell(\omega) > 0$ where g_ℓ is given by (2.6.56). Color $1, \ldots, \ell$ green in that case.

Color the remaining sites inductively as follows: Suppose j is the first as yet uncolored site. If $j \geq m - n + 2$, there are at most $n - 1$ sites, including j, uncolored. Color them all yellow. If $j \leq m - n + 1$, then color it red if $T^{j-1}\omega \notin E_n$. If $T^{j-1}\omega \in E_n$, we know for $k = 0, \ldots, n-1$, we have $j - 1 + k \leq m - 1$, that is, there are at least n sites left. Pick ℓ so $g_\ell(T^{j-1}\omega) > 0$, and color the next ℓ sites green.

When we are done, we have some number of red sites, some green in groups, and at most n yellow sites. At the red sites, $q(T^{j-1}\omega) = 0$ since $T^{j-1}\omega \notin E_n$. At the yellow sites, $q(T^{j-1}\omega) \geq -f_-(\omega)$, where $f_-(\omega) = \max(0, -f(\omega))$ is the negative part of f. The green sites are in groups $j_r, \ldots, j_r + \ell_r - 1$ with

$$\sum_{s=j_r}^{j_r+\ell_r-1} f(T^{s-1}\omega) > 0 \tag{2.9.15}$$

If $f(\omega) > 0$, $g_1(\omega) > 0$, so $h_n(\omega) > 0$, so $\omega \in E_n$. Equivalently, $\omega \notin E_n \Rightarrow f(\omega) \leq 0$, that is, for all ω,

$$q(\omega) \geq f(\omega) \tag{2.9.16}$$

Thus, (2.9.15) implies the sum is positive if f is replaced by q. Thus, if there are $y(\omega)$ yellow sites,

$$\sum_{j=1}^{m} q(T^{j-1}\omega) \geq \sum_{\text{yellow } j\text{'s}} q(T^{j-1}\omega)$$

$$\geq - \sum_{j=m-y(\omega)+1} f_-(T^{j-1}\omega)$$

$$\geq - \sum_{j=m-n+1}^{m} f_-(T^{j-1}\omega) \tag{2.9.17}$$

Integrating over ω and using the invariance of the measure

$$m \int q(\omega)\,d\mu(\omega) \geq -n \int f_-(\omega)\,d\omega \tag{2.9.18}$$

Divide by m and take $m \to \infty$, noting $n/m \to 0$, to conclude (2.9.14). $\square$

Next, we want to prove the Birkhoff theorem from Theorem 2.6.14 without going through a maximal inequality. For any function f on Ω, which is finite, define

$$\bar{A}(\omega; f) = \limsup \mathrm{Av}_n(\omega; f), \qquad \underline{A}(\omega; f) = \liminf \mathrm{Av}_n(\omega; f) \qquad (2.9.19)$$

Clearly,

$$\underline{A}(\,\cdot\,; f) = -\bar{A}(\,\cdot\,; -f), \qquad |\bar{A}(\omega; f)| \leq A(\omega; |f|) \qquad (2.9.20)$$

and since $|(\frac{n-1}{n})\mathrm{Av}_{n-1}(\omega; f \circ T) - \mathrm{Av}_n(\omega; f)| \leq \frac{1}{n} f(\omega)$, one easily sees that for a.e. ω,

$$\bar{A}(T\omega; f) = \bar{A}(\omega; f), \qquad \underline{A}(T\omega; f) = \underline{A}(\omega; f) \qquad (2.9.21)$$

Lemma 2.9.4. *Let* $f \in L^1$ *and suppose* $h \in L^1$, $h \circ T = h$, *and for all* ω, $h(\omega) \leq \bar{A}(\omega; f)$. *Then*

$$\int (f - h)(\omega)\, d\mu(\omega) \geq 0 \qquad (2.9.22)$$

Proof. Let $g = f - h + \varepsilon$ with $\varepsilon > 0$. Since h is invariant,

$$\bar{A}(\omega; g) = \bar{A}(\omega; f) - h(\omega) + \varepsilon \geq \varepsilon$$

It follows that for each ω and some n, $\mathrm{Av}_n(\omega; g) > 0$, that is, $g^\sharp(\omega) > 0$ for all ω, so $\{\omega \mid g^\sharp(\omega) > 0\} = \Omega$ and (2.6.55) says $\int_\Omega g(\omega)\, d\mu(\omega) \geq 0$, so for all ε, the integral in (2.9.22) is bigger than $-\varepsilon$. By taking $\varepsilon \downarrow 0$, we have (2.9.22). $\qquad \square$

Alternate Proof of the Birkhoff Ergodic Theorem (Theorem 2.6.9). By a simple argument (Problem 1), if the limit exists pointwise, a.e., and has integral $\int f(\omega)\, d\mu(\omega)$, it must be $\mathbb{E}(f \mid \mathcal{I})$, so we need only prove that for all $f \in L^1$ and a.e. ω,

$$\bar{A}(\omega; f) \leq \underline{A}(\omega; f), \qquad \int \bar{A}(\omega; f)\, d\mu(\omega) = \int f(\omega)\, d\mu(\omega) \qquad (2.9.23)$$

Suppose $g \in L^1$, $g \geq 0$. Fix $K > 0$. Let $h_K(\omega) = \min(\bar{A}(\omega, g); K)$, which is invariant, in L^1 and obeys $h_K \leq \bar{A}(\,\cdot\,; g)$. Thus, by the lemma,

$$0 \leq \int h_K(\omega)\, d\mu(\omega) \leq \int g(\omega)\, d\mu(\omega) \qquad (2.9.24)$$

Using the monotone convergence theorem, by taking $K \to \infty$, we see $\bar{A}(\,\cdot\,; g) \in L^1$.

Now let $f \in L^1$. By the inequality in (2.9.20), we see $\bar{A}(\omega; f) \in L^1$, so we can apply (2.9.22) to $h = \bar{A}(\,\cdot\,; f)$ to see

$$\int \bar{A}(\omega; f)\, d\mu(\omega) \leq \int f(\omega)\, d\mu(\omega) \qquad (2.9.25)$$

Since $\underline{A}(\,\cdot\,;f) = -\bar{A}(\,\cdot\,;-f)$, this inequality for $-f$ implies

$$\int f(\omega)\,d\mu(\omega) \leq \int \underline{A}(\omega;f)\,d\mu \qquad (2.9.26)$$

In particular, $\int[\bar{A}(\omega;f) - \underline{A}(\omega;f)]\,d\mu(\omega) \leq 0$, so since $\underline{A} \leq \bar{A}$, we conclude the first half of (2.9.23). (2.9.25)/(2.9.26) then imply the second half. □

To make the proof of Theorem 2.9.2 look like the proof above, we'll prove the theorem for the $-g$'s, that is, we'll prove a superadditive ergodic theorem. Moreover, we note that if

$$\tilde{g}_n = g_n - n\mathrm{Av}_n(\,\cdot\,;g_1^+) \qquad (2.9.27)$$

then $\tilde{g}_n$ is subadditive in the sense of (2.9.3) and $\tilde{g}_1 \leq 0$ (so $\tilde{g}_n \leq 0$ and, of course, $\tilde{g}_1^+ = 0 \in L^1$). Moreover, since the Birkhoff theorem implies $\mathrm{Av}_n(\,\cdot\,;g_1^+)$ has a limit a.e. g_n/n has a limit if and only if $\tilde{g}_n/n$ does. Thus, Theorem 2.9.2 is implied by:

Theorem 2.9.5. *Let* (Ω, Σ, μ, T) *be a measurable dynamical system and* $\{g_n\}_{n=1}^\infty$ *a sequence of* L^1-*functions on* Ω *with*

$$g_{n+m}(\omega) \geq g_n(\omega) + g_m(T^n\omega) \qquad (2.9.28)$$

for all n, m *and a.e.* $\omega \in \Omega$. *Suppose* $g_1 \geq 0$, *so by* (2.9.28),

$$g_n(\omega) \geq 0 \qquad (2.9.29)$$

Then for a.e. ω,

$$\lim_{n\to\infty} \frac{1}{n} g_n(\omega) \equiv G(\omega) \qquad (2.9.30)$$

exists and is invariant. Indeed,

$$G(\omega) = \sup \frac{1}{n} \mathbb{E}(g_n \mid \mathcal{I})(\omega) \qquad (2.9.31)$$

Lemma 2.9.6. *Let* (Ω, Σ, μ, T) *be a measurable dynamical system. Let* $\{B_n\}_{n=1}^\infty \subset \Sigma$ *with* $B_{n+1} \subset B_n$ *and* $\mu(\cap_{n=1}^\infty B_n) = 0$. *Then* $\mathbb{E}(\chi_{B_n} \mid \mathcal{I})(\omega) \to 0$ *for a.e.* ω.

Proof. Since $\mathbb{E}(\cdot \mid \mathcal{I})$ preserves order, $f_n \equiv \mathbb{E}(\chi_{B_n} \mid \mathcal{I})$ is monotone decreasing to a limit f_∞. By the monotone convergence theorem, $\|f_\infty\|_1 = \lim\|\mathbb{E}(\chi_{B_n} \mid \mathcal{I})\|_1 = \lim\|\chi_{B_n}\|_1 = 0$, so $f_n \downarrow 0$ a.e. □

Proof of Theorem 2.9.5. We'll prove the limit (2.9.30) exists and leave (2.9.31) to Problem 2. Define

$$\overline{G}(\omega) = \limsup \frac{1}{n} g_n(\omega) \qquad (2.9.32)$$

and for $K = 1, 2, \ldots,$

$$\overline{G}_K(\omega) = \min(\overline{G}(\omega), K) \tag{2.9.33}$$

Fix ε and L and define (what will determine our red sites)

$$R_{K,L,\varepsilon} = \left\{ \omega \,\middle|\, \max_{\ell=1,\ldots,L} \frac{g_\ell(\omega)}{\ell} \leq G_K(\omega) - \varepsilon \right\} \tag{2.9.34}$$

For K, ε fixed, clearly $R_{K,L+1,\varepsilon} \subset R_{K,L,\varepsilon}$, and

$$\bigcap_L R_{K,L,\varepsilon} = \left\{ \omega \,\middle|\, \sup_\ell \frac{g_\ell(\omega)}{\ell} \leq G_K(\omega) - \varepsilon \right\} = \emptyset \tag{2.9.35}$$

since $\limsup g_n/n \leq \sup g_n/n$.

With L, K, ε fixed, for any ω and $n > L$, color $k \in \{1, \ldots, n\}$ inductively as follows. If k is the smallest as yet uncolored site, color k yellow if $k \in \{n - L + 1, \ldots, n\}$ and color k red if $T^{k-1}\omega \in R_{K,L,\varepsilon}$. If k has not been colored yellow or red, $T^{k-1}\omega \notin R_{K,L,\varepsilon}$, so there is a smallest ℓ, which is in $\{1, \ldots, L\}$, so that

$$g_\ell(T^{k-1}\omega) \geq \ell(G_K(\omega) - \varepsilon) \tag{2.9.36}$$

Color $k, k+1, \ldots, k+\ell-1$ all green.

Let $r_1, \ldots, r_m$ be the red sites, $c_1, \ldots, c_p$ the beginning of the green chains, of lengths $\ell_1, \ldots, \ell_p$, and $y_1, \ldots, y_q$ the yellow sites. By counting sites,

$$m + q + \sum_{k=1}^p \ell_k = n \tag{2.9.37}$$

and, of course,

$$q \leq L \tag{2.9.38}$$

By superadditivity and $g_1 \geq 0$,

$$g_n(\omega) \geq \sum_{k=1}^p g_{\ell_k}(T^{c_k-1}\omega)$$

$$\geq \left(\sum_{k=1}^p \ell_k \right)(G_K(\omega) - \varepsilon) \qquad \text{(by (2.9.36))}$$

$$= (n - m - q)(G_K(\omega) - \varepsilon) \qquad \text{(by (2.9.37))}$$

so, since L is fixed, (2.9.38) implies

$$\liminf_{n \to \infty} \frac{g_n(\omega)}{n} \geq \left(1 - \lim_{n \to \infty} \left(\frac{m}{n} \right) \right)(G_K(\omega) - \varepsilon) \tag{2.9.39}$$

But $m = \#\{k \in \{1, \ldots, n\} \mid T^{k-1}\omega \in R_{K,L,\varepsilon}\}$, so by the Birkhoff ergodic theorem, for a.e. ω,

$$\lim_{n \to \infty} \frac{m}{n} = \mathbb{E}(\chi_{R_{K,L,\varepsilon}} \mid \mathcal{I})(\omega) \tag{2.9.40}$$

By (2.9.35) and the lemma, this number goes to zero as $L \to \infty$. Thus, since (2.9.39) holds for all L, we conclude

$$\liminf \frac{g_n(\omega)}{n} \geq (G_K(\omega) - \varepsilon) \tag{2.9.41}$$

Taking $\varepsilon = 1/t$, $t = 1, 2, \ldots$, and $K = 1, 2, \ldots$, we see for a.e. ω, $\liminf g_n(\omega)/n \geq \limsup g_n(\omega)/n$. Thus, the limit exists a.e. $\qquad\square$

Proof of Theorem 2.9.1. Let

$$g_n(\omega) = \log \| A(T^{n-1}\omega) \ldots A(\omega) \| \tag{2.9.42}$$

(2.9.3) follows from $\|AB\| \leq \|A\| \|B\|$ and $g_1^+ \in L^1$ by explicit hypothesis, so the limit exists by Proposition 2.9.3.

Since

$$1 \leq \| A(T^{n-1}\omega) \ldots A(\omega) \| \, \| A(T^{n-1}\omega)^{-1} \| \ldots \| A(\omega)^{-1} \|$$

we see that

$$g_n^-(\omega) \leq \sum_{j=0}^{n-1} \log^+ \| A(T^j\omega)^{-1} \| \tag{2.9.43}$$

so $\log^+ \| A(\,\cdot\,)^{-1} \| \in L^1$ implies (2.9.9), so $|G| < \infty$ a.e., that is, e^G is finite and nonzero a.e. $\qquad\square$

Before turning to a more detailed analysis of matrix products, here is one more example of the use of the Kingman theorem.

Example 2.9.7 (A Percolation Model). Let (Ω, Σ, μ, T) be a measurable dynamical system. Suppose for each unordered pair $e = \{\alpha, \beta\} \subset \mathbb{Z}^\nu$ with $|\alpha - \beta| = 1$ (i.e., each edge of $\mathbb{Z}^\nu$ as a lattice), we are given a nonnegative random variable $u_e(\omega)$ with values in $(0,1)$ so that

$$u_{\{(\alpha_1, \ldots, \alpha_\nu), (\beta_1, \ldots, \beta_\nu)\}}(T\omega) = u_{(\alpha_1+1, \alpha_2, \ldots, \alpha_\nu), (\beta_1+1, \beta_2, \ldots, \beta_\nu)}(\omega) \tag{2.9.44}$$

(i.e., T translates the edges). For $\alpha \in \mathbb{Z}^\nu$, define

$$p_\alpha(\omega) = \inf \left\{ \sum_{e_j, j=1}^{k} u_{e_j}(\omega) \,\middle|\, e_k \ldots e_1 \text{ is a path from } \vec{0} \text{ to } \alpha \right\} \tag{2.9.45}$$

This is a model of percolation from 0 to α. p_α represents the time to get from 0 to α.

If $\nu = 1$,

$$p_n(\omega) = \sum_{j=0}^{n-1} u_{\{j, j+1\}}(\omega) = \sum_{j=0}^{n-1} u_{\{0,1\}}(T^j\omega) \tag{2.9.46}$$

and the Birkhoff theorem assures the a.e. existence of $\lim n^{-1} p_n(\omega)$.

For general ν, it is easy to see (Problem 3) that

$$p_{(n+m,0,\ldots,0)}(\omega) \le p_{(n,0,\ldots,p)}(\omega) + p_{(m,0,\ldots,0)}(T^n\omega) \qquad (2.9.47)$$

so the Kingman theorem applies and $\lim n^{-1}p_{(n,0,\ldots,0)}(\omega)$ exists for a.e. ω. □

Our final topic is the multiplicative ergodic theorem which will say that under suitable L^1-hypotheses, in the context of Theorem 2.9.1, we have for each $u \in \mathbb{C}^d$, $u \ne 0$,

$$\lim_{n\to\infty} \|A(T^{n-1}\omega)\ldots A(\omega)u\|^{1/n} \qquad (2.9.48)$$

exists. For most u, this will be the limit of the norms but, in general, there will be ω-dependent lower-dimensional spaces on which the limit is smaller. The start of the argument is some linear algebra—we recall some facts about $d \times d$ complex matrices that can be found, for example, in Sections 1.3, 2.4, and 3.5 of Part 4:

(1) For any invertible matrix, A, we can uniquely write (polar decomposition)

$$A = U|A| \qquad (2.9.49)$$

where U is unitary and $|A|$ is self-adjoint and positive.

(2) The eigenvalues of $|A|$, ordered by $\mu_1(A) \ge \cdots \ge \mu_d(A) > 0$, are called *singular values* of A. One picks mutually orthogonal one-dimensional projections, $Q_1(A), \ldots, Q_d(A)$, so that $Q_j(A)\varphi = \varphi \Rightarrow |A|\varphi = \mu_j(A)\varphi$ ($\Rightarrow \|A\varphi\| = \mu_j(A)\|\varphi\|$). If the μ's are unequal, the Q_j's are unique; if some μ's are equal, there are choices to make. In particular, for any φ,

$$\mu_d(A)\|\varphi\| \le \|A\varphi\| \le \mu_1(A)\|\varphi\| \qquad (2.9.50)$$

We also have

$$\|A\varphi\|^2 = \sum_{j=1}^{d}\|AQ_j(A)\varphi\|^2 = \sum_{j=1}^{d}\mu_j(A)^2\|Q_j(A)\varphi\|^2 \qquad (2.9.51)$$

(3) For $k \le d$, one can define $\wedge^k(\mathbb{C}^d)$ an $\binom{d}{k}$-dimensional space, an alternating product $\varphi_1 \wedge \cdots \wedge \varphi_k \in \wedge^k(\mathbb{C}^d)$ for any $\varphi_1, \ldots, \varphi_k$, and $\wedge^k(A)\colon \wedge^k(\mathbb{C}^d) \to \wedge^k(\mathbb{C}^d)$ so that

$$\wedge^k(A)(\varphi_1 \wedge \cdots \wedge \varphi_k) = A\varphi_1 \wedge \cdots \wedge A\varphi_k \qquad (2.9.52)$$

and

$$\|\varphi_1 \wedge \cdots \wedge \varphi_k\| \le \prod_{j=1}^{k}\|\varphi_j\| \qquad (2.9.53)$$

Moreover,

$$\wedge^k(AB) = \wedge^k(A)\wedge^k(B) \qquad (2.9.54)$$

and, by (2.9.50) for $\varphi \in \wedge^k(\mathbb{C}^d)$,

$$\left(\prod_{j=1}^{k} \mu_{d-j+1}(A)\right)\|\varphi\| \leq \|\wedge^k(A)\varphi\| \leq \left(\prod_{j=1}^{k} \mu_j(A)\right)\|\varphi\| \qquad (2.9.55)$$

The following is the key perturbative lemma:

Lemma 2.9.8. *If B and D are invertible $d \times d$ matrices, then for any j and k,*

$$\|Q_j(B)Q_k(D)\| \leq \frac{\|BD^{-1}\|\mu_k(D)}{\mu_j(B)} \qquad (2.9.56)$$

$$\|Q_j(B)Q_k(D)\| \leq \frac{\|DB^{-1}\|\mu_j(B)}{\mu_k(D)} \qquad (2.9.57)$$

Proof. Since $[Q_j(B)Q_k(D)]^* = Q_k(D)Q_j(B)$, (2.9.56) and symmetry imply (2.9.57), so we'll prove (2.9.56). Suppose

$$Q_k(D)\varphi = \varphi \qquad (2.9.58)$$

Then, by (2.9.51),

$$\mu_j(B)\|Q_j(B)\varphi\| = \|BQ_j(B)\varphi\| \leq \|B\varphi\| \leq \|BD^{-1}\| \, \|DQ_k(D)\varphi\|$$
$$= \|BD^{-1}\|\mu_k(D) \, \|\varphi\| \qquad (2.9.59)$$

so for any φ,

$$\|Q_j(B)Q_k(D)\varphi\| \leq \|BD^{-1}\|\mu_k(D)\mu_j(B)^{-1}\|Q_k(D)\varphi\| \qquad (2.9.60)$$

proving (2.9.56). $\qquad \square$

Suppose $A_1, A_2, \ldots$ is a sequence of $d \times d$ invertible matrices. We say they have *Lyapunov behavior* if and only if for $j = 1, \ldots, d$,

$$\lim_{m \to \infty} \frac{1}{m} \log \mu_j(A_m \ldots A_1) = \gamma_j \qquad (2.9.61)$$

exists. The γ's are called the *Lyapunov exponents*. Since $\|\wedge^j(A_m \ldots A_1)\| = \prod_{k=1}^{j} \mu_k(A_m \ldots A_1)$, we see that the limits in (2.9.61) exist if and only if, for each j,

$$\lim_{m \to \infty} \frac{1}{m} \log\|\wedge^j(A_m \ldots A_1)\| = \gamma_1 + \cdots + \gamma_j \qquad (2.9.62)$$

Theorem 2.9.9 (Ruelle–Oseledec Theorem). *Let $A_1, A_2, \ldots$ be a sequence of invertible $d \times d$ matrices with Lyapunov behavior, and suppose*

$$\lim_{m \to \infty} \frac{1}{m} \log\|A_m\| = \lim_{m \to \infty} \frac{1}{m} \log\|A_m^{-1}\| = 0 \qquad (2.9.63)$$

Then for any $\varphi \in \mathbb{C}^d$, $\varphi \neq 0$,

$$\lim_{m \to \infty} \|A_m \ldots A_1\varphi\|^{1/m} \qquad (2.9.64)$$

exists and is finite. Indeed, there are subspaces, $V_0, \ldots, V_d$ with $V_{j-1} \subset V_j$, $\dim(V_j) = j$, $V_0 = \{0\}$, $V_d = \mathbb{C}^d$ so that

$$\varphi \in V_j \setminus V_{j-1} \Rightarrow \lim \|A_m \ldots A_1 \varphi\|^{1/m} = e^{\gamma_{d-j+1}} \qquad (2.9.65)$$

Remarks. 1. We emphasize there is no randomness in this theorem—it is for a deterministic sequence.

2. Obviously, if $\gamma_{d-j} > \gamma_{d-j+1}$, V_j is uniquely determined by (2.9.65), and if $\gamma_{d-j} = \gamma_{d-j+1}$, there is an ambiguity in choice of V_j.

3. If $d = 2$, which is a commonly used special case, the argument can be streamlined; see Problem 4.

Proof. We'll consider the case $\gamma_1 > \gamma_2 > \cdots > \gamma_d$. If some $\gamma_{j+1} = \gamma_j$, the bookkeeping is a little bit more complicated, but the proof is not harder; see Problem 5.

Let

$$Q_j^{(n)} = Q_j(A_n \ldots A_1), \qquad \mu_j^{(n)} = \mu_j(A_n \ldots A_1) \qquad (2.9.66)$$

Pick ε so that

$$\min(\gamma_j - \gamma_{j+1}) > 4d\varepsilon \qquad (2.9.67)$$

and then pick $C \in (1, \infty)$ so that for $j = 1, \ldots, d$,

$$C^{-1} \exp(n(\gamma_j - \varepsilon)) \leq \mu_j^{(n)} \leq C \exp(n(\gamma_j + \varepsilon)) \qquad (2.9.68)$$

$$\|A_n\| \leq Ce^{\varepsilon n}, \qquad \|A_n^{-1}\| \leq Ce^{\varepsilon n} \qquad (2.9.69)$$

By these estimates and (2.9.56)/(2.9.57) with $B = A_{n+1} \ldots A_1$, $D = A_n \ldots A_1$, we see for a constant C_1 and

$$\|Q_j^{(n)} Q_k^{(n+1)}\| \leq C_1 \exp[(-|\gamma_j - \gamma_k| + 3\varepsilon)n] \qquad (2.9.70)$$

Below we'll prove that for all $\ell > 0$, $j \neq k$, we have

$$\|Q_j^{(n)} Q_k^{(n+\ell)}\| \leq K \exp[(-|\gamma_j - \gamma_k| + 3d\varepsilon)n] \qquad (2.9.71)$$

for a suitable constant, K.

If P and R are two projections, then $P - R = P - PR - (R - PR) = P(1 - R) - (1 - P)R$. So, since $\|A\| = \|A^*\|$,

$$\|P - R\| \leq \|P(1 - R)\| + \|R(1 - P)\| \qquad (2.9.72)$$

Writing $1 - Q_j^{(n+1)} = \sum_{k \neq j} Q_k^{(n+1)}$, and similarly for (n), we get

$$\|Q_j^{(n+1)} - Q_j^{(n)}\| \leq \sum_{k \neq j} \|Q_k^{(n)} Q_j^{(n+1)}\| + \|Q_j^{(n)} Q_k^{(n+1)}\| \qquad (2.9.73)$$

which is $O(e^{-qn})$ for some $q > 0$ by (2.9.71) and (2.9.67). It follows that

$$Q_j^{(\infty)} = \lim_{n \to \infty} Q_j^{(n)} \qquad (2.9.74)$$

exists. By (2.9.71) for $j \neq k$,

$$\|Q_j^{(n)} Q_k^{(\infty)}\| \leq K \exp[(-|\gamma_j - \gamma_k| + 3d\varepsilon)n] \qquad (2.9.75)$$

Thus, by (2.9.68), we have for $\varphi \in \mathrm{Ran}(Q_k^{(\infty)})$ that

$$\|A_n \ldots A_1 \varphi\| \leq \sum_{j=1}^{d} C e^{n(\gamma_j + \varepsilon)} \|Q_j^{(n)} Q_k^{(\infty)}\| \, \|\varphi\| \qquad (2.9.76)$$

$$\leq KC \sum_{j \neq k} e^{n(\gamma_j + 3d\varepsilon + C - |\gamma_j - \gamma_k|)} \|\varphi\| + C e^{n(\gamma_k + \varepsilon)} \|\varphi\| \quad (2.9.77)$$

by (2.9.75) for $j \neq k$ and $\|Q_j^{(n)} Q_j^{(\infty)}\| \leq 1$ for $j = k$. Since $\gamma_j - |\gamma_j - \gamma_k| \leq \gamma_k$, we see

$$\limsup \|A_n \ldots A_1 \varphi\|^{1/n} \leq e^{\gamma_k} \qquad (2.9.78)$$

since ε can be taken to 0.

On the other hand, by (2.9.55) applied to $\psi = \varphi_d \wedge \cdots \wedge \varphi_{d-k+1}$ with $\varphi_\ell \in \mathrm{Ran}(Q_\ell^{(\infty)})$ with $\|\varphi_\ell\| = 1$ so $\|\psi\| = 1$,

$$\liminf \prod_{j=1}^{k} \|A_n \ldots A_1 \varphi_{d-j+1}\|^{1/n} \geq e^{\sum_{j=1}^{k} \gamma_{d-j+1}} \qquad (2.9.79)$$

This and (2.9.78) implies

$$\varphi_j \in \mathrm{Ran}(Q_j^{(\infty)}), \quad \varphi_j \neq 0 \Rightarrow \lim \|A_n \ldots A_1 \varphi_j\| = e^{\gamma_j}$$

Take V_k to be the span of $\{\mathrm{Ran}(Q_{d-j+1}^{(\infty)})\}_{j=1}^{k}$ and we get (2.9.65).

Thus, we are reduced to proving (2.9.71) from (2.9.70). Define

$$P_j^{(n)} = \sum_{\ell=1}^{j} Q_j^{(n)}, \qquad M_j^{(n)} = \sum_{\ell=j}^{d} Q_j^{(n)} \qquad (2.9.80)$$

Since $Q_j^{(n)} P_j^{(n)} = Q_j^{(n)} M_j^{(n)} = Q_j^{(n)}$, (2.9.71) is implied by

$$\|P_j^{(n)} M_{j+q}^{(n+\ell)}\| \leq K \exp[-(|\gamma_{j+q} - \gamma_j| + 3dq)n] \qquad (2.9.81)$$

for $q = 1, \ldots, d-1$ and the result with P and M reversed. We'll prove (2.9.81); the P-M reversed proof is identical.

We start with $q = 1$ by writing

$$P_j^{(n)} M_{j+1}^{(n+\ell)} = P_j^{(n)} (P_j^{(n+1)} + M_{j+1}^{(n+1)}) M_{j+1}^{(n+\ell)}$$

$$= P_j^{(n)} P_j^{(n+1)} (P_j^{(n+2)} + M_{j+1}^{(n+2)}) M_{j+1}^{(n+\ell)} + P_j^{(n)} M_{j+1}^{(n+1)} M_{j+1}^{(n+\ell)}$$

$$= \sum_{p=1}^{\ell-1} P_j^{(n)} \ldots P_j^{(n+p-1)} M_{j+1}^{(n+p)} M_{j+1}^{(n+\ell)} \qquad (2.9.82)$$

so, by (2.9.70),

$$\|P_j^{(n)} M_{j+1}^{(n+\ell)}\| \leq \sum_{p=1}^{\ell-1} \|P_j^{(n+p-1)} M_{j+1}^{(n+p)}\| \tag{2.9.83}$$

$$\leq C_1 d^2 \sum_{p=0}^{\infty} \exp[(-|\gamma_{j+1} - \gamma_j| + 3\varepsilon)(n + p)] \tag{2.9.84}$$

$$\leq C_2 \exp[(-|\gamma_{j+1} - \gamma_j| + 3\varepsilon)n] \tag{}$$

with $C_2 = C_1 d^2 (1 - \exp(-|\gamma_{j+1} - \gamma_j| + 3\varepsilon))^{-1}$.

Now suppose we have (2.9.81) for $q = 1, \ldots, q_0 - 1$ and all j, n, ℓ with a constant K replaced by C_{q_0}. Use an expansion like (2.9.82) with $P_j + M_{j+1}$ replaced by $P_j + \sum_{r=1}^{q_0-1} Q_{j+r} M_{j+q_0}$, expanding at each step only the $P \ldots PM$ terms. The analog of (2.9.83) is then

$$\|P_j^{(n)} M_{j+q_0}^{(n+\ell)}\|$$
$$\leq \sum_{p=1}^{\ell-1} \left(\|P_j^{(n+p-1)} M_{j+q_0}^{(n+p)}\| + \sum_{r=1}^{q_0-1} \|P_j Q_{j+r}^{(n+p-1)}\| \, \|Q_{j+r}^{(n+p-1)} M_{j+q_0}^{(n+\ell)}\| \right)$$
$$\tag{2.9.85}$$

The first term is controlled with (2.9.70) and the second by the induction hypothesis. This proves the result for $\ell = q_0$. $\qquad\square$

We use this to prove an ergodic result.

Theorem 2.9.10 (Multiplicative Ergodic Theorem). *Let* (Ω, Σ, μ, T) *be a measurable dynamical system and* $A \colon \Omega \to \mathbb{GL}(d, \mathbb{C})$ *a measurable map of* Ω *into the invertible* $d \times d$ *matrices. Suppose that* $\log\|A(\omega)\|$, $\log\|A(\omega)^{-1}\|$ *lie in* $L^1(\Omega, d\mu)$. *Then for a.e.* ω, $A(\omega), A(T\omega), \ldots, A(T^n\omega), \ldots$ *have Lyapunov behavior and there exist subspaces* $0 = V_0(\omega) \subset V_1(\omega) \subset V_2(\omega) \subset \cdots \subset V_d(\omega) = \mathbb{C}^d$ *so that* $\dim(V_j(\omega)) = j$, *and if* $\varphi = V_j(\omega) \setminus V_{j-1}(\omega)$ *for* $j = 1, 2, \ldots, d$, *then*

$$\lim_{n\to\infty} \|A(T^{n-1}\omega) \ldots A(\omega)\varphi\|^{1/n} = e^{\gamma_{d-j+1}} \tag{2.9.86}$$

Remarks. 1. This is sometimes called *Oseledec's theorem*.

2. As the proof shows, γ_j is ω-dependent but invariant, so constant, if T is ergodic. The $V_j(\omega)$ are definitely ω-dependent but measurable (Problem 6).

Proof. Applying the Furstenberg–Kesten theorem to $\wedge^{\ell}(A(\,\cdot\,))$, we see $\mu_j(A(T^{n-1}\omega) \ldots A(\omega))^{1/n}$ has a limit, so we have Lyapunov behavior.

By hypothesis and the Birkhoff theorem, $\frac{1}{n}\sum_{j=0}^{n-1} \log\|A(T^j\omega)\|$ has a limit, and the same for $\|A(T^j\omega)^{-1}\|$, so (2.9.63) holds since $\frac{1}{n}\sum_{j=1}^{n} \alpha_j \equiv$

$A_n \to A_\infty$, finite $\Rightarrow \frac{\alpha_n}{n} = A_n - (\frac{n-1}{n})A_{n-1} \to 0$. Thus, the Ruelle–Osceledec theorem applies and we get the existence of $V_j(\omega)$. $\qquad\square$

Notes and Historical Remarks. The Furstenberg–Kesten theorem is from 1960 [**298**]. In 1965, Hammersley–Welsh [**350**], in the context of Example 2.9.7, introduced the general notion of subadditive processes and Kingman [**470**] proved his ergodic theorem motivated by this.

Our proof of the maximal ergodic theorem using the three kinds of sites is taken from Keane–Petersen [**457**], whose argument we also follow in going from it to the Birkhoff theorem without a maximal inequality.

Our proof of the subadditive ergodic theorem follows Steele [**774**]. Interestingly enough, while the proof is similar to the proof Keane–Petersen found fifteen years later, Keane–Petersen make no mention of Steele. The reason is that they have a common influence in the work of Kamae [**444**] and Katznelson–Weiss [**454**].

Steele's proof, like many others, relies on the Birkhoff theorem; see Avila–Bochi [**22**] for a direct proof which therefore gets the Birkhoff theorem also.

Liggett [**536**] has proven an extension of the Kingman theorem, stated in terms of random variables in a triangular array $\{g_{n,m}(\omega)\}_{0 \le n < m}$—for our setting $g_{n,n+k}(\omega) = g_n(T^{k-1}\omega)$—with an invariant transformation replaced by invariance of some joint distributions. His hypotheses are weaker (but not for the $g_n(T^{k-1}\omega)$ case), but conclusions are the same. His result has some applications to certain probabilistic models.

While there is no maximal inequality approach to the subadditive theorem, the similarity of the Keane–Petersen and Steele proofs suggests there should be an approach to the Kingman theorem through a suitable positive integral theorem for maximal functions (i.e., maximal theorems like Theorem 2.6.14). There is—going back to Derriennic [**217**]. See the book of Krengel [**491**] for more on this approach.

The multiplicative ergodic theorem is due to Oseledec [**624**] in 1968. Ten years later, Ruelle [**707**] (see also Raghunathan [**668**]) realized that part of the argument was not random at all and singled out the result we have called the Ruelle–Oseledec theorem. Our proof follows Ruelle.

Instead of $B_n(\omega) = A(T^{n-1}\omega)\dots A(\omega)$, one can think of a family $\{B_n(\omega)\}_{n=1}^\infty$ obeying

$$B_{n+m}(\omega) = B_n(T^m\omega)B_m(\omega)$$

If $\varphi(\omega) = A(\omega)$, $\tilde{T}(\omega, C) = (T\omega, \varphi(\omega)C)$ on $\Omega \times \mathbb{GL}(n, \mathbb{C})$ is precisely a group extension in the language of Section 2.8.1, so this section is linked to that subsection. In particular, consistent with that language, $\{B_n\}$ is called a *cocycle*.

The Notes to Section 3.7 discuss an application of Kingman's theorem to the theory of ergodic Jacobi matrices.

Before leaving the subject, we mention there are results guaranteeing unequal Lyapunov exponents for iid matrices when the support of the distribution of the individual matrices is large. For example, if the distribution is one on $\mathbb{SL}(d, \mathbb{C})$, whose support generates the entire group, then these exponents are unequal. These results go back to Furstenberg [**296**]. For a discussion of the $d = 2$ case with applications to random orthogonal polynomials, see [**749**, Sect. 10.6].

Problems

1. If it is known for every $f \in L^1$ that $L(f)(\omega) = \lim_{n \to \infty} \frac{1}{n} \sum_{j=0}^{n-1} f(T^j \omega)$ exists and that $\int L(f)(\omega) \, d\mu(\omega) = \int f(\omega) \, d\mu(\omega)$, prove that $L(f) = \mathbb{E}(f \mid \mathcal{I})$. (*Hint*: First show that $L(f)$ is invariant, then that $L(gf) = gL(f)$ for any invariant g in L^∞.)

2. Knowing the limit exists in (2.9.30) for any superadditive nonnegative sequence, prove it is given by (2.9.31). (*Hint*: Follow Problem 1.)

3. Verify (2.9.47).

4. Suppose $d = 2$ in Theorem 2.9.9.

   (a) First reduce to the case $A_n \in \mathbb{SL}(2, \mathbb{C})$ for all n.

   (b) In that case where $\gamma_2 = -\gamma_1 \equiv \gamma$, prove if $P_n = Q_2(A_n)$, then $\|P_{n+1} - P_n\| \leq C_\varepsilon e^{-(2\gamma - \varepsilon)n}$ so $P_\infty = \lim P_n$ exists.

   (c) Prove that $\limsup \|A_n \ldots A_1 P_\infty\|^{1/n} \leq e^{-\gamma}$.

   (d) Complete the proof of Theorem 2.9.9.

   (*Note*: See Section 10.5 of [**749**].)

5. Give the proof of Theorem 2.9.9 if some γ's are equal. (*Hint*: Show it suffices to control $\|P_j^{(n)} M_{j+q}^{(n+\ell)}\|$ for the case $\gamma_j > \gamma_{j+1}$ and $\gamma_{j+q-1} > \gamma_{j+q}$.)

6. (a) Let $P(\omega), R(\omega)$ be measurable families of orthogonal projections in $\mathbb{C}^d$ with $\dim \operatorname{Ran}(P(\omega)) = \ell$, $\dim \operatorname{Ran}(R(\omega)) = \ell - j$, $\operatorname{Ran}(R(\omega)) \subset \operatorname{Ran}(P(\omega))$. Prove there is a measurable family, $Q(\omega)$, of orthogonal projections with $\operatorname{Ran}(R(\omega)) \subset \operatorname{Ran}(Q(\omega)) \subset \operatorname{Ran}(P(\omega))$ and $\dim(Q(\omega)) = \ell - 1$.

   (b) Prove that the spaces $V_j(\omega)$ in Theorem 2.9.10 can be chosen so that if $P_j(\omega)$ is the orthogonal projection onto $V_j(\omega)$, then $P_j(\omega)$ is measurable. (*Hint*: Use (a) if some γ_j's are equal.)

2.10. Martingale Inequalities and Convergence

We begin with an example that illustrates the kind of convergence problems we are interested in.

Example 2.10.1 (Random Series). Let $\{\sigma_n\}_{n=1}^{\infty}$ be random signs, that is, we take $\Omega = \{-1, 1\}^{\mathbb{Z}_+}$ with the product measure $d\mu$ assigning weight $\frac{1}{2}$ to both $+1$ and -1 (so the $\sigma_n(\omega) \equiv \omega_n$ are iidrv). For $\gamma \in (0, \infty)$, consider the sum $\sum_{n=1}^{\infty} \sigma_n n^{-\gamma}$. If $\gamma > 1$, this converges for all choices of σ_n. If $\sigma_n = (-1)^n$, by the alternating series theorem, we get convergence for all $\gamma > 0$. How about if σ_n is a typical random set of signs? We'll prove below that one has convergence (to a finite limit) for a.e. ω if $\gamma > \frac{1}{2}$, and in Problem 1 that one gets divergence (with all points in $\mathbb{R} \cup \{\pm\infty\}$ as limit points) if $0 < \gamma \leq \frac{1}{2}$.

To see what this has to do with the theme of this chapter, let $\{a_n\}_{n=1}^{\infty}$ be a general ℓ^2-sequence. Since $\{\sigma_n\}_{n=1}^{\infty}$ are an orthonormal family in $L^2(\Omega, d\mu)$, $\sum_{n=1}^{\infty} a_n \sigma_n$ defines a function $A(\omega) \in L^2$, that is, $\sum_{n=1}^{N} a_n \sigma_n \to A$ in L^2. The question of convergence of the series for individual $\{\sigma_n(\omega)\}_{n=1}^{\infty}$ is precisely the question of pointwise convergence a.e. that we have studied elsewhere in this chapter. The key will be a maximal inequality that also implies the Hardy–Littlewood maximal inequality.

In fact, we'll obtain a still more general result. If $\{g_n\}_{n=1}^{\infty}$ are mutually independent random variables and $\mathbb{E}(g_n) = 0$, $\sum_{n=1}^{\infty} \mathbb{E}(|g_n|^2) < \infty$, we'll prove that $\sum_{n=1}^{N} g_n(\omega)$ converges to a finite limit for a.e. ω. $\qquad\square$

The framework we'll use is a general one, although we'll restrict to the discrete case and probability measures (as the Notes explain, there are continuous time results and infinite measure versions). We recall (see Theorem 2.6.6) that if (Ω, Σ, μ) is a probability measure space and $\Sigma' \subset \Sigma$ a subalgebra, then for any $f \in L^1(\Omega, \Sigma, \mu)$, there is a unique L^1-function, called the *conditional expectation*, $\mathbb{E}(f \mid \Sigma')$, which is Σ'-measurable, and so that for all $h \in L^{\infty}(\Omega, \Sigma', d\mu)$, one has that

$$\int f(\omega) h(\omega)\, d\mu(\omega) = \int \mathbb{E}(f \mid \Sigma')(\omega) h(\omega)\, d\mu(\omega) \qquad (2.10.1)$$

Moreover, $\mathbb{E}(\,\cdot \mid \Sigma')$ is positivity-preserving and a contraction on each L^p. If h is bounded and Σ'-measurable, then

$$\mathbb{E}(fh \mid \Sigma') = h\mathbb{E}(f \mid \Sigma') \qquad (2.10.2)$$

Definition. A (*measure-theoretic*) *filtration* of a measure space, (Ω, Σ), is a sequence $\{\Sigma_n\}_{n=1}^{\infty}$ of subalgebras of Σ so that for all n, $\Sigma_n \subset \Sigma_{n+1}$. We set Σ_{∞} to the σ-algebra generated by $\bigcup_{n=1}^{\infty} \Sigma_n$.

Example 2.10.2 (Filtrations). (1) If $\{f_n\}_{n=1}^{\infty}$ is a family of measurable functions, let Σ_n be the σ-algebra generated by

$\{f_j^{-1}((a,b))\}_{a,b\in\mathbb{R},\,j=1,\ldots,n}$, the smallest σ-algebra in which $f_1,\ldots,f_n$ are all measurable. Σ_n is a filtration called the *intrinsic filtration* of the f's.

(2) Let $\Omega = \times_{n=1}^\infty X_n$ where each X_n is compact. Let Σ_n be the σ-algebra generated by $\{A_1 \times \cdots \times A_n \mid A_j \subset X_j,\ A_j \text{ Baire}\}$. This is a filtration that shows the relevance to probability and to gambling models.

(3) In $\mathbb{R}^\nu$, let Σ_n be the σ-algebra generated by products of intervals $(\frac{j}{2^n}, \frac{j+1}{2^n}]$, $j \in \mathbb{Z}$. For $\nu = 1$ and taking $[0,1]$ instead of $\mathbb{R}$, we get what we'll call the *restricted dyadic filtration*.

As these examples show, going from Σ_n to Σ_{n+1} adds information to the system.   □

Definition. A sequence of functions, $\{f_n\}_{n=1}^\infty$ is said to be *adapted* to a filtration $\{\Sigma_n\}_{n=1}^\infty$ if and only if each f_n is Σ_n-measurable. It is said to be *predictable* if f_1 is constant and each f_n, $n \geq 2$, is Σ_{n-1}-measurable. A function, N, with values in $\mathbb{Z}_+$ is called a *stopping time* if

$$\{\omega \mid N(\omega) = n\} \in \Sigma_n \tag{2.10.3}$$

The name, "predictable," comes from the gambling (or investing strategy) picture. f_n describes the result after n steps (e.g., winnings after n spins of the roulette wheel). In forming f_n, which is a result of some random event plus a strategy based on your history in the past, h_n, based on your history, is Σ_{n-1}-measurable. Stopping times are so named because the model is to stop gambling at time n based on the situation after n steps.

Here is the key object of this section:

Definition. Let $\{\Sigma_n\}_{n=1}^\infty$ be a filtration. A sequence, $\{f_n\}_{n=1}^\infty$, of $L^1(\Omega, d\mu)$-functions is called a *martingale* (more properly, a Σ_n-martingale) if it is adapted to Σ_n and if for all n,

$$\mathbb{E}(f_{n+1} \mid \Sigma_n) = f_n \tag{2.10.4}$$

If all $f_n \in L^p$ for some $p \geq 1$, we say $\{f_n\}$ is an L^p-martingale. $\{f_n\}_{n=1}^\infty$ is called a *submartingale* if

$$\mathbb{E}(f_{n+1} \mid \Sigma_n) \geq f_n \tag{2.10.5}$$

and *supermartingale* if

$$\mathbb{E}(f_{n+1} \mid \Sigma_n) \leq f_n \tag{2.10.6}$$

Given a sequence of functions without explicit Σ_n's, we say it is an *intrinsic martingale* (or just a *martingale*) if it is a martingale with the intrinsic filtration.

It is easy to see (Problem 3) that if $\{f_n\}_{n=1}^\infty$ is a martingale, for all $n < m$,

$$\mathbb{E}(f_m \mid \Sigma_n) = f_n \tag{2.10.7}$$

and similarly for sub- and supermartingales.

Martingales are models of results of fair games, supermartingales where the house has an advantage, and submartingales of the best bull market you've ever seen. We postpone exploring examples until we present two tools that help for (sub)martingales.

Theorem 2.10.3 (Jensen's Inequality for Conditional Expectation). *Let* $\Sigma' \subset \Sigma$, $f \in L^1(\Omega, \Sigma, d\mu)$ *and* φ: *the convex hull of* $\mathrm{Ran}(f) \to \mathbb{R}$ *convex. If* $\varphi(f) \in L^1(\Omega, \Sigma, d\mu)$, *we have*

$$\varphi(\mathbb{E}(f \mid \Sigma')) \leq \mathbb{E}(\varphi(f) \mid \Sigma') \qquad (2.10.8)$$

In particular, if $\{f_n\}_{n=1}^{\infty}$ *is a martingale (respectively, submartingale) and* φ *is convex (respectively, convex and increasing) so that* $\varphi(f_n) \in L^1$ *for all* n, *then* $\varphi(f_n)$ *is a submartingale.*

Proof. We just follow the proof of the ordinary Jensen inequality in Theorem 5.3.14 of Part 1. For any convex function, φ, and all x, $D^+\varphi(x) = \lim_{\varepsilon \downarrow 0}(\varphi(x + \varepsilon) - \varphi(x))/\varepsilon$ exists, and for all x and y,

$$\varphi(y) - \varphi(x) \geq (D^+\varphi)(x)(y - x) \qquad (2.10.9)$$

In particular, if $g(\omega) = \mathbb{E}(f \mid \Sigma)(\omega)$, we have

$$\varphi(f(\omega)) - \varphi(g(\omega)) \geq (D^+\varphi)(g(\omega))(f(\omega) - g(\omega)) \qquad (2.10.10)$$

Since $(D^+\varphi)(g(\omega))$ is Σ-measurable, by (2.10.2),

$$\mathbb{E}((D^+\varphi \circ g)(f - g) \mid \Sigma) = (D^+\varphi \circ g)\mathbb{E}(f - g \mid \Sigma) = 0$$

so, since $\mathbb{E}(\,\cdot\mid \Sigma)$ is positivity-preserving,

$$\mathbb{E}(\varphi \circ f \mid \Sigma) \geq \mathbb{E}(\varphi \circ g \mid \Sigma) = \varphi \circ g$$

which is (2.10.8).

If $\{f_n\}$ is a martingale, by (2.10.8),

$$\mathbb{E}(\varphi \circ f_{n+1} \mid \Sigma_n) \geq \varphi \circ \mathbb{E}(f_{n+1} \mid \Sigma_{n+1}) = \varphi \circ f_n \qquad (2.10.11)$$

If $\{f_n\}$ is a submartingale and φ is monotone, we get $\geq$ in the last step in (2.10.11). $\qquad\square$

Proposition 2.10.4. (a) *If* $\{f_n\}_{n=1}^{\infty}$ *is a martingale,* $\{|f_n|\}_{n=1}^{\infty}$ *is a submartingale.*

(b) *If* $\{f_n\}_{n=1}^{\infty}$ *is an* L^p-*martingale for some* $p > 1$, *then* $|f_n|^p$ *is a submartingale. In particular, if* f_n *is an* L^2-*martingale,* $f_n^2 - \mathbb{E}(f_n^2)$ *is a submartingale.*

(c) *If* $\{f_n\}$ *is a submartingale (respectively, supermartingale),* $\{h_n\}_{n=1}^{\infty}$ *is predictive and nonnegative, and*

$$g_n = f_1 + \sum_{j=2}^{n} h_j(f_j - f_{j-1}) \qquad (2.10.12)$$

then g_n is a submartingale (respectively, supermartingale). If $\{f_n\}_{n=1}^{\infty}$
is a martingale, this is true without positivity of $\{h_n\}_{n=1}^{\infty}$.

(d) *If $\{f_n\}_{n=1}^{\infty}$ is a martingale (respectively, submartingale, supermartin-*
gale) and N a stopping time, then $g_n(\omega) = f_{\min(n,N(\omega))}(\omega)$ is a mar-
tingale (respectively, submartingale, supermartingale).

Remark. In terms of gambling strategy, (d) says if you have a fair game
and change the strategy to stop at time N depending on the history of what
has happened before, until that time, you still only have a fair game.

Proof. (a) and (b) follow immediately from the fact that $\varphi(x) = |x|^p$ is
convex for all $p \geq 1$. Since $\mathbb{E}(f_n^2)$ is a constant, the final assertion in (b) is
immediate.

(c) $\mathbb{E}(g_n \mid \Sigma_{n-1}) = g_{n-1} + \mathbb{E}(h_n(f_n - f_{n-1}) \mid \Sigma_{n-1})$

$$= g_{n-1} + h_n(\mathbb{E}(f_n \mid \Sigma_{n-1}) - f_{n-1}) \qquad (2.10.13)$$

from which all results are evident.

(d) We'll do the martingale case for definiteness. Let $h_n = \chi_{\{\omega \mid N(\omega) \geq n\}}$.
Then $h_n = 1 - \chi_{\{\omega \mid N(\omega) \leq n-1\}}$ is Σ_{n-1}-measurable. So, by (c),

$$g_n(\omega) = f_1 + \sum_{j=2}^{n} h_j(f_j - f_{j-1}) \qquad (2.10.14)$$

is a martingale. If $N(\omega) \geq n$, $g_n = f_n$, and if $N(\omega) = j \leq n$, $g_n = f_1 + \sum_{j=2}^{j}(f_j - f_{j-1}) = f_j$, so $g_n = f_{\min(n,N(\omega))}$. $\qquad \square$

Example 2.10.5 (Examples of Martingales). (1) (Sums of independent
random variables) Let X_j be independent random variables, each L^1.
Let X_0 be a constant function and let $f_n = \sum_{j=0}^{n} X_j$. If $\mathbb{E}(X_j) = 0$
(resp. > 0, resp. < 0) for each $j = 1, 2, \ldots$, then f_n is a martingale
(respectively, submartingale, respectively, supermartingale). The example, $\sum_{n=1}^{N} a_n \sigma_n$ of Example 2.10.1, is a martingale. More generally,
if $\mathbb{E}(X_j) \neq 0$, $f_n = \sum_{j=0}^{n}[X_j - \mathbb{E}(X_j \mid \Sigma_{j-1})]$ is a martingale.

(2) ("The Martingale") This is a gambling strategy going back at least to
the eighteenth century. You play a game where you bet any amount and
either double your bet or lose it depending on a fair coin. Bet one coin.
If you win, stop. If not, double your bet. If you win, stop. Continue
indefinitely. If f_n is your net winnings/losses after n steps, f_n is a
martingale. $\mathbb{E}(f_n) = 0$ for all n, but it is not hard to see (Problem 4)
that for a.e. ω, $f_n(\omega) \to 1$ as $n \to \infty$. If you have infinite money and
the house has no betting limits, you can have a strategy that "always"
wins.

(3) ("The Las Vegas Tourist") Some visitors to Las Vegas who go for en-
tertainment, not gambling, strive to bet no more than a certain total

amount and keep betting until they have none of the set-aside money left. Put simply, suppose the bet is one dollar in a win/lose one dollar game. You stop if your stake at step n is 0. Let f_n be $\sum_{j=0}^n X_j$ of (1), where $X_0 = 1$, $X_j = \pm 1$, $j \geq 1$, with equal probability (so f_n is random walk starting at 1). Let $N(\omega) = n \Leftrightarrow f_j(\omega) > 0$, $j = 1, 2, \ldots, n - 1$, $f_n(\omega) = 0$, that is, N is the first time you hit 0. Then $g_n = f_{\min(N,n)}$ is a model of the Las Vegas tourist. It is a martingale by (d) of Proposition 2.10.4. We'll see later that it is interesting from a theoretical point of view.

(4) (Products; de Moivre's Martingale) Let $\{X_n\}_{n=1}^\infty$ be iidrv with $\mathbb{E}(X_n) = 1$. Then (Problem 5(a)), $f_n = \prod_{j=1}^n X_j$ is a martingale. In particular, if $\{\sigma_n\}_{n=1}^\infty$ are iidrv with $\sigma_n = +1$ with probability p and $\sigma_n = -1$ with probability $q = 1 - p$, define $X_n = \sum_{j=1}^n \sigma_j$. Then (Problem 5) $f_n = (\frac{q}{p})^{X_n}$ is a martingale, called *de Moivre's martingale*.

(5) (Dyadic Hardy–Littlewood) Let Σ_n be the restricted dyadic partition on $[0, 1]$. Let ν be a probability measure on $[0, 1]$ with $\nu(\{1\}) = 0$ and define

$$f_n(x) = \sum_{j=0}^{2^n - 1} 2^{-n} \nu\left(\left(\frac{j}{2^n}, \frac{j+1}{2^n}\right]\right) \chi_{(\frac{j}{2^n}, \frac{j+1}{2^n}]}(x) \qquad (2.10.15)$$

It is easy to see that f_n is a Σ_n-martingale. We give it the name we do because, as we'll see, Doob's inequality and martingale convergence for this are closely related to the Hardy–Littlewood maximal theorem and to Lebesgue's theorem. □

With these lengthy preliminaries, that made the basic definitions and illustrated them, done, we can describe the main results that we'll prove:

(1) A maximal inequality on $\sup_n |f_n(\omega)|$ for martingales

(2) A proof that L^p-martingales (with bounded L^p-norms) for $p > 1$ converge in L^p and also pointwise a.e. $p = 1$ is more subtle—one has pointwise a.e. convergence, but perhaps not convergence in L^1. Problem 10 will lead the reader through the usual proof of L^1-convergence; instead, we'll provide an approach that goes through some interesting intermediate steps.

(3) A decomposition theorem that any submartingale is the sum of a martingale and an increasing family of functions

(4) A convergence theorem for nonnegative L^2-submartingales

(5) A convergence theorem for any nonnegative martingale

(6) A convergence theorem for L^1-martingales.

We begin with the maximal theorem. Given a submartingale, define

$$f_n^*(\omega) = \sup_{j \leq n} |f_j(\omega)|, \qquad f^*(\omega) = \sup_j |f_j(\omega)| \qquad (2.10.16)$$

Theorem 2.10.6 (Doob's Inequality). (a) *For any nonnegative submartingale, $\{f_j\}_{j=1}^{\infty}$, $\lambda > 0$, and $n = 1, 2, \ldots$,*

$$\mu(\{\omega \mid f_n^*(\omega) > \lambda\}) \leq \lambda^{-1}\mathbb{E}(f_n \chi_{\{\omega \mid f_n^*(\omega) > \lambda\}}) \leq \lambda^{-1}\mathbb{E}(f_n) \qquad (2.10.17)$$

and if $\sup\|f_n\|_1 < \infty$,

$$\mu(\{\omega \mid f^*(\omega) > \lambda\}) \leq \lambda^{-1}\sup_n\|f_n\|_1 \qquad (2.10.18)$$

(b) *For any martingale,* (2.10.17), *with* f_n *replaced by* $|f_n|$, *and* (2.10.18) *hold.*

Remark. Recall (see Theorem 4.6.10 of Part 1) that Markov's inequality says that for any $f_n \geq 0$,

$$\mu(\{\omega \mid f_n(\omega) > \lambda\}) \leq \lambda^{-1}\mathbb{E}(f_n \chi_{f_n(\omega) > \lambda}) \leq \lambda^{-1}\mathbb{E}(f_n) \qquad (2.10.19)$$

It is striking that for submartingales, we have essentially the same upper bound on the potentially much larger $\sup_{0 \leq j \leq n} f_j(\omega)$. There is also the related Lévy inequality (see Theorem 7.2.9 of Part 1).

Proof. (a) Let

$$\tau(\omega) = \inf\{j \mid f_j(\omega) > \lambda\} \qquad (2.10.20)$$

Then $\tau(\omega) = n \Leftrightarrow f_n(\omega) > \lambda, f_1(\omega) \leq \lambda, \ldots, f_{n-1}(\omega) \leq \lambda$, so τ is a stopping time. Moreover, $f_n^*(\omega) > \lambda \Leftrightarrow \tau(\omega) \leq n$. Thus,

$$\mathbb{E}(f_n \chi_{\{\omega \mid f_n^*(\omega) > \lambda\}}) = \sum_{j=1}^{n} \mathbb{E}(f_n \chi_{\{\omega \mid \tau(\omega) = j\}})$$

$$= \sum_{j=1}^{n} \mathbb{E}(\mathbb{E}(f_n \mid \Sigma_j)\chi_{\{\omega \mid \tau(\omega) = j\}})$$

$$\geq \sum_{j=1}^{n} \mathbb{E}(f_j \chi_{\{\omega \mid \tau(\omega) = j\}}) \qquad (2.10.21)$$

$$\geq \lambda\mathbb{E}(\chi_{\{\omega \mid \tau(\omega) \leq n\}})$$

$$= \lambda\mu(\{\omega \mid f_n^*(\omega) > \lambda\})$$

which is (2.10.17). (2.10.21) uses the submartingale property.

(2.10.18) follows by noting $\mu(\{\omega \mid f^*(\omega) > \lambda\}) = \lim \mu(\{\omega \mid f_n^*(\omega) > \lambda\})$ by the monotone convergence theorem.

(b) is immediate from (a), given that if f_n is a martingale, $|f_n|$ is a submartingale. $\qquad\qquad\square$

Corollary 2.10.7 (Kolmogorov's Inequality). *Let $\{g_n\}_{n=1}^{\infty}$ be a sequence of independent random variables on (Ω, μ, Σ), each in L^2, with all $\mathbb{E}(g_j) = 0$.*

Let $f_n = \sum_{j=1}^{n} g_j$. Then

$$\mu\big(\sup_{j=1,\dots,n} |f_j(\omega)| > \lambda \big) \le \lambda^{-2} \int |f_n(\omega)|^2 \, d\mu(\omega) \qquad (2.10.22)$$

Proof. As noted in Proposition 2.10.4(b) and Example 2.10.5(1), $|f_n|^2$ is a submartingale, so this is (2.10.17) if we note that $(|f.|^2)_n^* = (f_n^*)^2$. $\qquad\square$

Remark. One can go directly from this to a.e. convergence of $\sum_{j=1}^{n} g_j$ when $\sum_{j=1}^{\infty} \|g_j\|^2 < \infty$ (Problem 6).

To get our first a.e. convergence result, we need the following:

Definition. Let $\{f_n\}_{n=1}^{\infty}$ be a martingale adapted to a filtration $\{\Sigma_n\}_{n=1}^{\infty}$. We say $f_\infty \in L^1$ *covers* the martingale if f_∞ is Σ_∞-measurable and

$$f_n = \mathbb{E}(f_\infty \mid \Sigma_n) \qquad (2.10.23)$$

for all n.

Proposition 2.10.8. *If $\{f_n\}_{n=1}^{\infty}$ is an L^p-martingale for some $p \in (1, \infty)$ and $\sup_n \|f_n\|_p < \infty$, then there is $f_\infty \in L^p$ covering $\{f_n\}_{n=1}^{\infty}$.*

Proof. Since $p > 1$, $\{g \in L^p \mid \|g\|_p \le S \equiv \sup_n \|f_n\|_p\}$ is compact in the weak topology (because $p > 1$, the weak and weak-* topology agree by Theorem 4.9.1 of Part 1, so Theorem 5.8.2 of Part 1 applies). So there exists $n_1 < n_2 < \dots$ with $n_j \to \infty$ and $f_\infty \in L^p(\Omega, \Sigma_\infty, d\mu)$, so that for all $h \in L^q(\Omega, \Sigma_\infty, d\mu)$ ($q = p/(p-1)$), $\int h f_{n_j} \, d\mu \to \int h f_\infty \, d\mu$. In particular, if $g \in L^\infty(\Omega, \Sigma_{n_j}, d\mu)$, we have that (by the martingale property)

$$\int g f_{n_j} \, d\mu = \int g f_{n_{j+1}} \, d\mu = \dots = \int g f_\infty \, d\mu \qquad (2.10.24)$$

This implies

$$f_{n_j} = \mathbb{E}(f_\infty \mid \Sigma_{n_j})$$

It follows that for any $k < n_j$,

$$f_k = \mathbb{E}(f_{n_j} \mid \Sigma_k) = \mathbb{E}(f_\infty \mid \Sigma_k)$$

Since $n_j \to \infty$, (2.10.23) holds for all n. $\qquad\square$

Theorem 2.10.9 (L^p-Martingale Convergence). *Let $\{f_j\}_{j=1}^{\infty}$ be a martingale covered by some $f_\infty \in L^1(\Omega, \Sigma_\infty, d\mu)$. For some $p \in [1, \infty)$, suppose*

$$\sup_n \|f_n\|_p < \infty \qquad (2.10.25)$$

Then $f_\infty \in L^p$, $f_n \to f_\infty$ in L^p and also pointwise a.e. In particular, if (2.10.25) holds for $p \in (1, \infty)$, (without a priori supposing $\{f_n\}_{n=1}^{\infty}$ is covered), then f_n converges a.e.

Remark. If $f_\infty \in L^1$ covers $\{f_n\}_{n=1}^\infty$, we have (2.10.25) automatically for $p = 1$ (since $\mathbb{E}(\cdot \mid \Sigma)$ is an L^p-contraction).

Proof. By Proposition 2.10.8, if (2.10.25) holds, $f_\infty \in L^p$. For each $f \in L^p(\Omega, \Sigma_\infty, d\mu)$, define T_n by

$$T_n f = \mathbb{E}(f \mid \Sigma_n) \tag{2.10.26}$$

The T_n's are uniformly bounded; indeed, $\|T_n f\|_p \leq \|f\|_p$. Moreover, Doob's inequality implies

$$\mu(\{\lambda \mid \sup_n |(T_n f)(\omega)| > \lambda\}) \leq \lambda^{-1}\|f\|_p \tag{2.10.27}$$

so, by Theorem 2.1.1, it suffices to prove $T_n f \to f$ in L^p and pointwise a.e. for a dense set of f's.

If $f \in L^p(\Omega, \Sigma_k, d\mu)$ for some k, $T_n f = f$ for all $n \geq k$, so the convergence results are trivial. Since $\bigcup_k \Sigma_k$ generates Σ_∞, $\bigcup_k L^p(\Omega, \Sigma_k, d\mu)$ is dense in $L^\infty(\Omega, \Sigma_\infty, d\mu)$. $\qquad\square$

As an immediate corollary, we resolve Example 2.10.1.

Theorem 2.10.10 (Kolmogorov's Random L^2 Series Theorem). *Let $\{f_n\}_{n=1}^\infty$ be a sequence of independent, L^2, random variables with*

$$\sum_{n=1}^\infty \|f_n\|_2^2 < \infty \tag{2.10.28}$$

and $\mathbb{E}(f_n) = 0$. Then for almost all ω, $\sum_{n=1}^\infty f_n(\omega)$ converges to a finite limit.

Remark. In particular, if $\{\sigma_n\}_{n=1}^\infty$ are random signs and $\gamma > \frac{1}{2}$, $\sum_{n=1}^N \sigma_n n^{-\gamma}$ has a limit for a.e. ω.

Proof. $g_n = \sum_{j=1}^n f_j$ is a martingale with $\|g_n\|_2^2 = \sum_{j=0}^n \|f_j\|^2$. So, by (2.10.28), $\sup_n \|g_n\|_2 < \infty$. We get convergence from Theorem 2.10.9. Since $f_\infty \in L^2$, $|f_\infty(\omega)| < \infty$ for a.e. ω. $\qquad\square$

Theorem 2.10.11. (a) (Levy 0-1 Law) *Let $\{\Sigma_n\}_{n=1}^\infty$ be a filtration and Σ_∞ the σ-algebra generated by $\bigcup_{n=1}^\infty \Sigma_n$. If $A \in \Sigma_\infty$, then for a.e. ω, $\mathbb{E}(\chi_A \mid \Sigma_n)(\omega)$ converges and the limit for a.e. ω is either 0 or 1.*

(b) (Kolmogorov 0-1 Law) *Let $\{(\Omega_j, \mathcal{F}_j, \mu_j)\}_{j=1}^\infty$ be a family of probability measure spaces and $(\Omega = \times_{j=1}^\infty \Omega_j, \mathcal{F}, \mu = \otimes_{j=1}^\infty \mu_j)$ the infinite product. Let $\tilde{\Sigma}_j$ be the σ-algebra generated by $\bigcup_{k=j}^\infty \mathcal{F}_j$ and $\tilde{\Sigma}_\infty = \bigcap_{j=1}^\infty \tilde{\Sigma}_j$, the tail σ-algebra. If $A \in \tilde{\Sigma}_\infty$, $\mu(A)$ is 0 or 1.*

(c) *Let $\{f_j\}_{j=1}^\infty$ be a family of independent random variables. Then either $\sum_{j=1}^n f_j(\omega)$ converges for a.e. ω or it fails to converge for a.e. ω.*

Remarks. 1. A typical event which lies in the *tail algebra* is that if f_j is $\mathcal{F}_j$-measurable, then the set of ω where $\sum_{j=1}^n f_j(\omega)$ converges is tail algebra-measurable since it doesn't depend on finitely many values.

2. As K. L. Chung [**160**] says about Levy's law: "The reader is urged to ponder over the meaning of this result and judge for himself whether it is obvious or incredible." Given the effortless proof of the Kolmogorov 0-1 law, I vote for the latter.

Proof. (a) Let $f_n = \mathbb{E}(\chi_A \mid \Sigma_n)$. Then $\{f_n\}_{n=1}^\infty$ is a martingale in all L^p (including L^∞) so, by Theorem 2.10.9, $f_n(\omega)$ converges a.e. to $\chi_A(\omega)$.

(b) Let A be $\widetilde{\Sigma}_\infty$-measurable. Then $\mathbb{E}(\chi_A \mid \Sigma_n)$ is a constant since A is independent of $\bigcup_{j=1}^n \mathcal{F}_j$. Thus, $\mathbb{E}(\chi_A \mid \Sigma_n) = \mathbb{P}(A)$, the probability of A, for a.e. ω. So, the Levy theorem implies $\mathbb{P}(A)$ is 0 or 1.

(c) As noted in Remark 1, $\{\omega \mid \sum_{j=1}^n f_j(\omega) \text{ converges}\}$ is a tail algebra event, and so obeys a 0-1 law. $\qquad\square$

As a preliminary to the sequence of results that will culminate in the L^1-convergence theorem, we note the reason for the subtlety:

Example 2.10.12 (Example 2.10.5(3), revisited). Recall in this example that f_n was random walk on $\mathbb{Z}$ starting at $n = 1$, $N(\omega) = \min\{j \mid X_j(\omega) = 0\}$, and $g_n = f_{\min(N,n)}$. Since g_n is a martingale, $\mathbb{E}(g_n) = \mathbb{E}(g_0) = 1$. Since $g_n \geq 0$, $\|g_n\|_1 = 1$ for all n and, in particular, $\sup_n \|g_n\|_1$. On the other hand, it is not hard to see (Problem 7 or Problem 9) that $\lim_{n\to\infty} g_n(\omega) = 0$. Thus, we have pointwise convergence a.e. but not L^1-convergence. In this example, there is no g_∞ covering $\{g_n\}_{n=1}^\infty$. $\qquad\square$

We next turn to convergence results for submartingales—interesting for its own sake but also needed in our proof of the L^1-martingale theorem. The key is the following:

Theorem 2.10.13 (Doob Decomposition Theorem). *Any submartingale,* $\{f_n\}_{n=1}^\infty$, *can be written*

$$f_n = m_n + i_n \tag{2.10.29}$$

where m_n is a martingale and i_n is a predictable sequence with

$$0 \leq i_n \leq i_{n+1} \tag{2.10.30}$$

If $f_n \geq 0$, then

$$\|i_n\|_1 \leq \|f_n\|_1 \tag{2.10.31}$$

and if also $f_n \in L^2$, then

$$\|m_n\|_2 \leq \|f_n\|_2 \tag{2.10.32}$$

Remark. We can pick $i_1 = 0$ and then (Problem 14) m and i are unique.

Proof. For $j = 2, \ldots,$ define

$$q_j = \mathbb{E}(f_j \mid \Sigma_{j-1}) - f_{j-1} \tag{2.10.33}$$

which is predictable, L^1, and nonnegative. Thus, if we define

$$i_n = \sum_{j=2}^{n} q_j \tag{2.10.34}$$

we have that i_n is predictable, L^1, and (2.10.30) holds.

Define

$$m_n = f_n - i_n \tag{2.10.35}$$

Note that m_n is L^1 and

$$m_{n+1} = f_{n+1} - f_n + m_n - q_{n+1} = m_n + f_{n+1} - \mathbb{E}(f_{n+1} \mid \Sigma_n) \tag{2.10.36}$$

Thus,

$$\mathbb{E}(m_{n+1} \mid \Sigma_n) = m_n \tag{2.10.37}$$

so m_n is a martingale.

Since $\mathbb{E}(q_j) = \mathbb{E}(f_j) - \mathbb{E}(f_{j-1})$, we see that $\mathbb{E}(i_n) = \mathbb{E}(f_n) - \mathbb{E}(f_1)$. If $f_j \geq 0$, we have

$$\|i_n\|_1 = \|f_n\|_1 - \|f_1\|_1 \leq \|f_n\|_1$$

proving (2.10.31).

To get the L^2-bound, we look at (2.10.36). Noting that

$$\langle m_n, f_{n+1} - \mathbb{E}(f_{n+1} \mid \Sigma_n) \rangle = \mathbb{E}(m_n f_{n+1}) - \mathbb{E}(\mathbb{E}(m_n f_{n+1} \mid \Sigma_n)) = 0$$

we see that

$$\begin{aligned}
\|m_{n+1}\|^2 &= \|m_n\|^2 + \|f_{n+1} - \mathbb{E}(f_{n+1} \mid \Sigma_n)\|^2 \\
&= \|m_n\|^2 + \|f_{n+1}\|^2 - \|E(f_{n+1} \mid \Sigma_n)\|^2
\end{aligned} \tag{2.10.38}$$

since $\langle f_{n+1}, \mathbb{E}(f_{n+1} \mid \Sigma_n) \rangle = \|\mathbb{E}(f_{n+1} \mid \Sigma_n)\|^2$ because $h \mapsto \mathbb{E}(h \mid \Sigma_n)$ is an orthogonal projection in L^2.

Since f_n is a positive submartingale, we have that $0 \leq f_n \leq \mathbb{E}(f_{n+1} \mid \Sigma_n)$, so $-\|\mathbb{E}(f_{n+1} \mid \Sigma_n)\| \leq -\|f_n\|^2$ and (2.10.38) implies that

$$\|f_{n+1}\|^2 - \|m_{n+1}\|^2 \geq \|f_n\|^2 - \|m_n\|^2 \tag{2.10.39}$$

We have $i_1 = 0$, so $\|f\|_2^2 = \|m_1\|^2$ and thus, (2.10.32) holds inductively. $\square$

Theorem 2.10.14 (L^2 Nonnegative Submartingale Convergence Theorem). *If $\{f_n\}_{n=1}^{\infty}$ is a nonnegative submartingale with $\sup_n \|f_n\|_2 < \infty$, then $f_n(\omega)$ converges pointwise a.e. to a finite limit, $f_\infty(\omega)$, with $\|f_n - f_\infty\|_2 \to 0$.*

Proof. By the Doob decomposition, $f_n = m_n + i_n$ with m_n an L^2-martingale with

$$\sup_n \|m_n\|_2 < \infty \qquad (2.10.40)$$

by (2.10.27) and the hypothesis. By Theorem 2.10.9, $m_n \to m_\infty$ in L^2 and pointwise a.e.

By (2.10.39) and $\|i_n\|_2 \leq \|m_n\|_2 + \|f_n\|_2$, i_n is monotone increasing, so by the monotone convergence theorem, $i_n \to i_\infty$ in L^2 and pointwise a.e. $\qquad \square$

Theorem 2.10.15. *If $\{f_n\}_{n=1}^\infty$ is a nonnegative martingale, for a.e. ω, it converges to a finite limit or to $+\infty$. If $\sup_n \|f_n\|_1 < \infty$, the limit is finite almost everywhere.*

Proof. $\varphi(x) = e^{-x}$ is convex, so by Theorem 2.10.3, $g_n = e^{-f_n}$ is a submartingale. Clearly, $|g_n| \leq 1$, so in L^2, so $g_n \to g_\infty$ pointwise a.e. if $g_\infty(\omega) = 0$, $f_n(\omega) \to \infty$, and otherwise to a finite limit.

If $f_\infty = \lim f_n$ (could be infinite), by Fatou's lemma, $\int f_\infty(\omega) \, d\mu(\omega) \leq \sup \int f_n(\omega) \, d\mu$, so if $\sup_n \|f_n\|_1 < \infty$, $f_\infty \in L^1$, so a.e. finite. $\qquad \square$

Lemma 2.10.16. *Any martingale, $\{f_n\}_{n=1}^\infty$, with $\sup_n \|f_n\|_1 < \infty$ can be rewritten as $f_n = f_n^+ - f_n^-$, where $\{f_n^\pm\}_{n=1}^\infty$ are nonnegative martingales with $\sup_n \|f_n^\pm\|_1 < \infty$.*

Proof. Since $|f_j|$ is a submartingale, we have

$$\mathbb{E}(|f_{j+1}| \mid \Sigma_j) \geq |f_j| \qquad (2.10.41)$$

Thus, for any $j \geq n$,

$$h_j^{(n)} \equiv \mathbb{E}(|f_{j+1}| - |f_j| \mid \Sigma_n) \geq 0 \qquad (2.10.42)$$

On the other hand, for any $k > n$,

$$\sum_{j=n}^k h_j^{(n)} = \mathbb{E}(|f_{k+1}| - |f_n| \mid \Sigma_n) \leq \mathbb{E}(|f_{k+1}| \mid \Sigma_n) \qquad (2.10.43)$$

so

$$\sum_{j=n}^\infty \|h_j^{(n)}\|_1 \leq \sup_k \|f_{k+1}\|_1 < \infty \qquad (2.10.44)$$

Thus, since L^1 is complete, we can define

$$f_n^+ \equiv |f_n| + \sum_{j=n}^\infty h_j^{(n)} \qquad (2.10.45)$$

We have, of course, that $f_n^+ > 0$ and $f_n^+ \geq |f_n|$ so $f_n^- = f_n^+ - f_n \geq 0$. Moreover, by (2.10.44),

$$\sup_n \|f_n^+\|_1 \leq 2 \sup_n \|f_n\|_1 \qquad (2.10.46)$$

and similarly, from $f_n^+ - f_n = |f_n| - f_n + \sum_{j=n}^{\infty} h_j^{(n)}$,

$$\sup_n \|f_n^-\|_1 \leq 2 \sup_n \|f_n\|_1 \qquad (2.10.47)$$

Thus, it suffices to show that f_n^+ is a martingale. By (2.10.43), if $|f_n| + \sum_{j=n}^{k} h_j^{(n)} \equiv g_n^{(k)}$, we have

$$g_n^{(k)} = \mathbb{E}(|f_{k+1}| \mid \Sigma_n) \qquad (2.10.48)$$

since $\mathbb{E}(|f_n| \mid \Sigma_n) = |f_n|$. It follows that

$$\mathbb{E}(g_{n+1}^{(k)} \mid \Sigma_n) = g_n^{(k)} \qquad (2.10.49)$$

As $k \to \infty$, $g_n^{(k)}$ converges to f_n^+ in L^1, so (2.10.49) implies f_n^+ is a martingale. $\qquad \square$

Theorem 2.10.17 (L^1-Martingale Convergence Theorem). *Let $\{f_n\}_{n=1}^{\infty}$ be a martingale with $\sup_n \|f_n\|_1 < \infty$. Then $f_n(\omega)$ converges to a finite limit for a.e. ω.*

Proof. Immediate from Theorem 2.10.15 and Lemma 2.10.16. $\qquad \square$

Example 2.10.18 (Example 2.10.5(5) revisited). The Doob inequality implies for $f \in L^1([0,1])$ and $\mathcal{D}$ the set of all dyadic intervals that one has

$$\{x \mid \sup_{x \in I \in \mathcal{D}} |I|^{-1} \int_I |f(y)| \, dy > \lambda\} \leq \lambda^{-1} \|f\|_1 \qquad (2.10.50)$$

This is clearly a relative of the Hardy–Littlewood maximal inequality and, indeed (see below), one can derive the Hardy–Littlewood inequality from it and its translates.

If $f \in L^1$, it covers the martingale f_n of (2.10.15) and so Theorem 2.10.9 is "essentially" the Lebesgue theorem on differentiability. Theorem 2.10.17 is even applicable and illustrates the loss of the covering f for a general L^1-martingale. If ν is a general measure, the f_n's of (2.10.15) converge to the Radon–Nikodym derivative of the a.c. part of ν. To the extent that ν is singular, the limit does not cover $\{f_n\}_{n=1}^{\infty}$. $\qquad \square$

Theorem 2.10.19. *For any $f \in L^1(\mathbb{R})$, we have, with M_{HL} given by (2.3.1) with $\nu = 1$, that for all $\lambda > 0$,*

$$|\{x \mid M_{\mathrm{HL}} f(x) > \lambda\}| \leq 10 \lambda^{-1} \|f\|_1 \qquad (2.10.51)$$

Remarks. 1. No attempt has been made to optimize the constant. The point is that Doob's inequality implies a Hardy–Littlewood inequality, and so also a maximal ergodic inequality with some constant!

2. The same argument works in $\mathbb{R}^{\nu}$ for hypercubes with suitable constant (and so for some constant for $M_{\mathrm{HL}} f$).

Proof. Let $\mathcal{D}_{\pm} = \{I \pm \frac{1}{3} \mid I \in \mathcal{D}\}$. Given $f \in L([0,1])$, if we define a martingale as in (2.10.15), that is, for $x \in [0,1]$ (*Note*: the sum has a single term),

$$f_n^{\pm}(x) = \sum_{\substack{x \in I \subset \mathcal{D}_{\pm} \\ |I| = 2^{-n}}} |I|^{-1} \int_I |f(y)| \, dy \qquad (2.10.52)$$

then $\{f_n^{\pm}\}_{n=1}^{\infty}$ is a martingale and, for $f \in L^1([0,1], dx)$, Doob's inequality implies

$$f^{*,\pm}(x) \equiv \sup_{x \in I \subset \mathcal{D}^{\pm}} |I|^{-1} \int_I |f(y)| \, dy \qquad (2.10.53)$$

$$|\{x \in [0,1] \mid f^{*,\pm}(x) > \lambda\}| \leq \lambda^{-1} \|f\|_1 \qquad (2.10.54)$$

$\frac{1}{3}$ has a base 2 expansion .010101 and $\frac{2}{3}$ a base 2 expansion .101.... It follows that for any n and j, $(\frac{j}{2^n} + \frac{1}{3}\frac{1}{2^n}, \frac{j+1}{2^n} + \frac{1}{3}\frac{1}{2^n}]$ and $(\frac{j}{2^n} - \frac{1}{3}\frac{1}{2^n}, \frac{j+1}{2^n} - \frac{1}{3}\frac{1}{2^n}]$ are in $\mathcal{D}^+$ and $\mathcal{D}^-$ or vice-versa. Thus, for any x, $J_n(x) \equiv (x - \frac{1}{3}\frac{1}{2^n}, x + \frac{1}{3}\frac{1}{2^n}]$ lies in a union of the intervals of size 2^{-n} containing x, one each in $\mathcal{D}$, $\mathcal{D}_+$, and $\mathcal{D}_-$.

Since $|J_n(x)| = \frac{2}{3} 2^{-n}$, we see

$$|J_n(x)|^{-1} \int_{|J_n(x)|} |f(y)| \, dy \leq \frac{3}{2} \left[f^*(x) + f^{*,+}(x) + f^{*,-}(x) \right] \qquad (2.10.55)$$

Using (2.2.33), we get

$$f^{\sharp}(x) \equiv \sup_n |J_n(x)|^{-1} \int_{|J_n(x)|} |f(y)| \, dy \qquad (2.10.56)$$

$$|\{x \mid f^{\sharp}(x) > \lambda\}| \leq \frac{9}{2\lambda} \|f\|_1 \qquad (2.10.57)$$

If I is an interval with x at center and $|I| \leq \frac{2}{3}$, there is an n with $J_{n+1}(x) \subset I \subseteq J_n(x)$ so, since $|J_{n+1}| = \frac{1}{2}|J_n|$, we see for $f \in L^1((0,1), dx)$,

$$f^{\flat}(x) \equiv \sup_{|a| \leq \frac{1}{3}} (2a)^{-1} \int_{|y-x|<a} |f(y)| \, dy \qquad (2.10.58)$$

$$\{x \in [0,1] \mid f^{\flat}(x) > \lambda\} \leq 9\lambda^{-1} \|f\|_1 \qquad (2.10.59)$$

We can then restrict to $f \in L^1((\frac{1}{3}, \frac{2}{3}), dx)$, translated to $(-\frac{1}{6}, \frac{1}{6})$, and scale up to a factor of 6 to see a bound for $f \in [-1,1]$ of the form (2.10.59) for $x \in [-1,1]$ and $|a| \leq 2$. But if $|a| > 2$, $\int_{|y-x|\leq 2}|f(y)| \, dy = \int_{|y-x|\leq a}|f(y)| \, dy$ and $(2|a|)^{-1} > 2^{-1}$, so we get

$$f \in L^1([-1,1], dx) \Rightarrow |\{x \in [-1,1] \mid (M_{\mathrm{HL}}f)(x) > \lambda\}| \leq 9\lambda^{-1}\|f\|_1 \qquad (2.10.60)$$

If $|x| > 1$, $\sup(2a)^{-1} \int_{x-a}^{x+a} |f(y)| \, dy \leq [2(|x| - 1)]^{-1} \|f\|_1$, so

$$f \in L^1([-1, 1], dx) \Rightarrow |\{x \notin [-1, 1] \mid (M_{\mathrm{HL}} f)(x) \geq \lambda\}| \leq \lambda^{-1} \|f\|_1$$

and thus,

$$f \in L^1([-1, 1], dx) \Rightarrow |\{x \mid M_{\mathrm{HL}} f(x) > \lambda\}| \leq 10\lambda^{-1} \|f\|_1 \qquad (2.10.61)$$

By scaling, we get the same for $f \in L^1([-R, R])$ and then for all $f \in L^1$ by taking $R \to \infty$. $\square$

Notes and Historical Remarks.

> I went to the casino of Venice, taking all the gold I could get, and by means of what in gambling is called the martingale I won three or four times a day during the rest of the carnival.
>
> —G. Casanova, as translated in Mansuy [**557**][4]

Around 1930, Bachelier, Bernstein, Kolmogorov, and Lévy had results that we now recognize as special cases of martingale convergence theorems. But the fundamental notions were presented in important publications of Jean Ville (1910–89) in 1939 [**833**] and Joseph Doob (1910–2004) in 1940 [**224**] and firmly established in Doob's 1953 book [**225**].

While he didn't give a formal definition, Ville used the term "martingale" and proved inequalities and convergence theorems of the genre of Theorems 2.10.6 and 2.10.17. Doob phrased everything in a measure-theoretic context and, while giving Ville credit, was so clear that his name became associated to the theorems. In understanding the impact, it is useful to know that there was controversy at the time concerning the proper mathematical framework for probability theory. The approach now universally accepted of measure theory pioneered by Kolmogorov (see the Notes to Section 7.1 of Part 1) had challengers—in particular, an idea of von Mises where the fundamental axiomatic objects were random sequences. Ville's work was done in this context; Doob's work, especially his book, in a measure-theoretic context.

For more on the history of the martingale concept, see [**71, 177, 557**]; in particular, Mansuy [**557**] studies the history of the term, probably through the Provençal expression "jouga a la martegalo," meaning to play in an absurd and incomprehensible way, a term applied to the gambling strategy described in Example 2.10.5(3) and used already by name in the mid-eighteenth century as the Casanova (yes, that Casanova!) quote shows. Martingale may be from the name of a town Martigues. As Mansuy documents, martingale had at least two other traditional meanings: it was a kind of horse harness and, apparently, an eighteenth-century slang expression for

---

[4]G. Casanova, *Histoire de ma vie*, Vol. 4, Chap. VII, 1754,

prostitute. (And I avoided the temptation to invent a famous mathematician named Martin Gale.)

Our approach to the L^1-martingale convergence theorem from the L^2-submartingale result is from Isaac [**420**] (see also Báez-Duarte [**28**] and Garsia [**308**]). The idea behind Doob's inequality is in Problem 10. A third proof I learned from Garling [**304**] is in Problems 12 and 13.

Martingale notions can be extended to $f_t(x)$'s depending on $t \in (0, \infty)$. A key example is Brownian motion, and one place continuous-time martingales play a critical role is in the theory of Ito stochastic integrals. They also are important in the mathematical theory of finance. A pioneer in the continuous theory was P.-A. Meyer in a paper [**565**] and two books [**567, 208**]. In particular, he proved a continuum analog of the Doob decomposition theorem (which Doob first presented in his 1953 book [**225**])—the full version is sometimes called the Doob–Meyer decomposition. For two books on continuous martingales, see [**537, 446**].

There is an extension of the notion of martingale to measures, μ, on locally compact, σ-compact spaces, Ω, with $\mu(\Omega) = \infty$, for example, $d^\nu x$ on $\mathbb{R}^\nu$. A martingale is a sequence of functions, $\{f_n\}_{n=1}^\infty$ in $L^1(\Omega, d\mu)$, so for all $B \in \Sigma_n$ with $\mu(B) < \infty$, one has $\int_B f_{n+1}(x)\,d\mu(x) = \int_B f_n(x)\,d\mu(x)$. An example is a full dyadic partition on $\mathbb{R}^\nu$. The Doob inequality and L^p, $1 \le p < \infty$, convergence theorems still hold. See, for example, Garling [**304**].

As we saw in this section, Doob's inequality implies the Hardy–Littlewood inequality and Calderón's idea to get the maximal ergodic inequality from the Hardy–Littlewood inequality was explained in Theorem 2.6.15. Other approaches relating Doob's inequality to the maximal ergodic inequality an be found in [**417, 428, 609, 671, 701**].

Theorem 2.10.10 can be generalized to a comprehensive form:

Theorem 2.10.20 (Kolmogorov Three-Series Theorem). *Let $\{f_n\}_{n=1}^\infty$ be a sequence of independent random variables and let*

$$g_n(\omega) = \begin{cases} f_n(\omega) & \text{if } |f_n(\omega)| \le 1 \\ 0 & \text{if } |f_n(\omega)| > 1 \end{cases}$$

Then $\sum_{n=1}^\infty f_n(\omega)$ converges for a.e. ω if and only if

(i) $\sum_{n=1}^\infty \mathbb{P}(|f_n| \ge 1) < \infty$.
(ii) $\sum_{n=1}^N \mathbb{E}(g_n)$ *has a finite limit.*
(iii) $\sum_{n=1}^\infty \mathbb{E}(|g_n - \mathbb{E}(g_n)|^2) < \infty$.

If any of these conditions fails, $\sum_{n=1}^N f_n(\omega)$ has no unique finite limit for a.e. ω.

For a proof, see Durrett [**235**]. This theorem and, so also, Theorem 2.10.10 are from Kolmogorov [**481**].

A different generalization, clearly related to martingale-convergence theorems, is the following result of Lévy [**526**] (it is sometimes called the Lévy convergence theorem, but that name is more often applied to Theorem 7.3.2 of Part 1).

Theorem 2.10.21. *Let $\{f_n\}_{n=1}^{\infty}$ be independent random variables. Then $S_N \equiv \sum_{n=1}^{N} f_n(\omega)$ converges for a.e. ω if and only if it converges in probability (i.e., there is S_∞ with $\mu(\omega \mid |S_n - S_\infty| \geq \varepsilon) \to 0$ for each $\varepsilon > 0$).*

For a proof, see [**291**, Thm. 12.22]. Since L^2-convergence implies convergence in probability, this implies Theorem 2.10.10.

Martingale inequalities have deep consequences in harmonic analysis. The key player in this is Donald Burkholder (1927–2013); see, for example, [**124**] which characterizes $H^p(\mathbb{C}_+)$, $0 < p < 1$. His central contributions in this area are summarized in [**35**]; [**202**] has a collection of his papers and [**123**] an exposition of some of his work.

Problems

1. This problem needs the central limit theorem in the form proven in Theorem 7.3.5 of Part 1.

   (a) Let $\{a_n\}_{n=1}^{\infty}$ be a sequence going to zero and $\{\sigma_n\}_{n=1}^{\infty}$ random signs. Prove that

   $$\lim_{k\to\infty} \lim_{N\to\infty} \mathbb{P}\left(\left(\sum_{n=1}^{N} a_n^2 \right)^{-1/2} \left| \sum_{n=1}^{N} a_n \sigma_n \right| \leq \frac{1}{k} \right) = 0 \qquad (2.10.62)$$

   (b) If $\sum_{n=1}^{\infty} a_n^2 = \infty$, prove that $\mathbb{P}\left(\sum_{n=1}^{N} a_n \sigma_n \text{ converges to a finite limit} \right) = 0$.

   (c) Prove that when $\sum_{n=1}^{\infty} a_n^2 = \infty$ and $a_n \to 0$ that for a.e. σ, $\limsup_{n\to\infty} \sum_{n=1}^{N} a_n \sigma_n = \infty$ and $\liminf_{n\to\infty} \sum_{n=1}^{N} a_n \sigma_n = -\infty$. (*Hint*: use the Kolmogorov 0-1 law.)

   (d) If $\sum_{n=1}^{\infty} a_n^2 = \infty$ and $a_n \to 0$, prove for a.e. ω, the set of limit points of $\sum_{n=1}^{N} a_n \sigma_n$ is $\mathbb{R} \cup \{\infty\} \cup \{-\infty\}$.

2. Let $\sum_{n=1}^{\infty} |a_n|^2 < \infty$, $\sum_{n=1}^{\infty} |a_n| = \infty$, $a_n \geq 0$. Suppose $\{\sigma_n\}_{n=1}^{\infty}$ are iidrv with $\sigma_n = \pm 1$ and $\mathbb{E}(\sigma_n) > 0$. For a.e. σ, prove that $\sum_{n=1}^{N} a_n \sigma_n \to \infty$.

3. (a) Let $\mathcal{H}$ be a Hilbert space, $V \subset W \subset \mathcal{H}$ closed subspaces. If P (respectively, Q) is the orthogonal projection onto V (respectively, W), prove that $PQ = QP = P$.

(b) If $\Sigma_1 \subset \Sigma_2 \subset \Sigma$ are σ-algebras, prove that

$$\mathbb{E}(\mathbb{E}(f \mid \Sigma_2) \mid \Sigma_1) = \mathbb{E}(f \mid \Sigma_1)$$

4. (a) For the martingale of Example 2.10.5(2), prove that with probability 1, $f_n(\omega) \to 1$.

   (b) If you start out with \$1000 and play this strategy with a \$1 initial bet, what is the chance you'll go bust (i.e., leave the table having lost; and how much will you lose)?

5. (a) If X_n are iidrv with $\mathbb{E}(X_n) = 1$, prove that $f_n = \prod_{j=1}^{n} X_j$ is a martingale.

   (b) If $\sigma = 1$ (respectively, -1) with probability p (respectively, $q = 1-p$), prove that $X = (\frac{q}{p})^{\sigma}$ has $\mathbb{E}(X) = 1$.

6. Go directly from Corollary 2.10.7 to Theorem 2.10.10.

7. This problem requires you to know about the central limit theorem (see Theorem 7.3.4 of Part 1).

   (a) Let $\{x_n\}_{n=1}^{\infty}$ be iidrv with $x_n = \pm 1$, each with probability $\frac{1}{2}$. Let $f_n = \sum_{j=1}^{n} x_n$ (random walk). Using the central limit theorem, prove that with probability 1, $\limsup |f_n| = \infty$.

   (b) Prove with probability 1 that $\limsup f_n = \infty$, $\liminf f_n = -\infty$.

   (c) Prove with probability 1 that $f_n = -1$ infinitely often.

   (d) Show that the g_n of Example 2.10.12 has $\lim_{n\to\infty} g_n(\omega) = 0$ for a.e. ω.

   Remark. There are purely combinatorial approaches to the fact that random walk in one dimension is recurrent; see, for example, Durrett [**235**, Sect. 3.2].

8. This problem and the next will provide another proof of a recurrence result that relies on ergodic theory and stopping times instead of the central limit theorem. Let (Ω, Σ, μ, T) be an ergodic dynamical system and f an integer-valued $L^1(\Omega, d\mu)$-function with $\mathbb{E}(f) = 0$. Let $S_n(\omega) = \sum_{j=0}^{n-1} f(T^j\omega)$. Let $A = \{\omega \mid S_n(\omega) \neq 0, n = 1, 2, \dots\}$. This problem will show $\mu(A) = 0$, a kind of recurrence result.

   (a) If $T^{j_1}\omega, \dots, T^{j_\ell}\omega \in A$ for $j_1 < \cdots < j_\ell$, prove that $S_{j_1}(\omega), S_{j_2}(\omega), \dots, S_{j_\ell}(\omega)$ are all distinct.

   (b) Let $D_n(\omega)$ be the number of distinct values of $\{S_j(\omega)\}_{j=1}^{n}$. Prove that $D_n(\omega) \geq \sum_{j=0}^{n-1} \chi_A(T^{j-1}\omega)$ and conclude that $\mu(A) \leq \liminf \frac{1}{n} D_n(\omega)$ for a.e. ω.

(c) For any M, prove that

$$\limsup_{n\to\infty} \frac{\max_{1\le m\le n}|S_m|}{n} = \limsup_{n\to\infty} \frac{\max_{M\le m\le n}|S_m|}{n} \le \sup_{m\ge M} \frac{|S_m|}{m}$$

Then show that for a.e. ω,

$$\lim_{n\to\infty} \frac{\max_{1\le m\le n}|S_m(\omega)|}{n} = 0$$

(*Hint*: Birkhoff theorem.)

(d) Prove that $D_n \le 1 + 2(\max_{1\le m\le n}|S_m|)/n$ and conclude that $\mu(A) = 0$, that is, every ergodic random walk on $\mathbb{Z}$ returns to its starting point.

Remark. This argument (indeed, the stronger $\mu(A) = \lim D_n(\omega)/n$ for a.e. ω) appeared first in Spitzer [**771**] where he says it is joint unpublished work with Kesten and Whitman.

9. Let $\{f_n\}_{n=1}^{\infty}$ be iidrv with $f_n = \pm 1$, each with probability $\frac{1}{2}$. Let $S_n = \sum_{j=1}^{n} f_j$. In Problem 7, you proved with probability 1, for some first $n(\omega) > 0$, that $S_{n(\omega)}(\omega) = 0$.

   (a) Let $\tau_1(\omega) = \min\{j \mid S_j(\omega) = 0\}$. Prove that the joint distribution of $\{S_{\tau_1(\omega)+j}(\omega)\}_{j=1}^{\infty}$ and $\{S_j(\omega)\}$ are identical and conclude that $S_j(\omega) = 0$ for infinitely many j.

   (b) Let $\tau_\ell(\omega) = \min\{j > \tau_{\ell-1}(\omega) \mid S_j(\omega) = 0\}$. Prove that $\{S_{\tau_\ell(\omega)+1}(\omega)\}_{\ell=1}^{\infty}$ and $\{f_\ell(\omega)\}_{\ell=1}^{\infty}$ are identically distributed.

   (c) For a.e. ω, prove that $S_j(\omega) = -1$ for some j. Conclude that the g_n of Example 2.10.12 has $\lim_{n\to\infty} g_n(\omega) = 0$.

10. This problem will lead you through a version of the original and more common proof of the L^1-convergence theorem for (sub)martinagles. Let $\{f_n\}_{n=1}^{\infty}$ be a submartingale. Fix $a < b$ real. Define functions $\{\sigma_j, \tau_j\}_{j=1}^{\infty}$ with values in $\mathbb{Z}_+ \cup \{0\}$ by

$$\sigma_1(\omega) = \inf\{j \mid f_j(\omega) \le a\} \tag{2.10.63}$$

$$\tau_k(\omega) = \inf\{j > \sigma_k(\omega) \mid f_j(\omega) \ge b\} \tag{2.10.64}$$

$$\sigma_{k+1}(\omega) = \inf\{j > \tau_k(\omega) \mid f_j(\omega) \le a\}; \quad k \ge 1 \tag{2.10.65}$$

with σ or τ equal to ∞, if in the set whose inf we are finding is empty. These times track *upcrossings* across the interval and the inequality, (2.10.74), is called an *upcrossing inequality*. The intuition behind the proof below is that one can take a strategy of picking up $(b-a)$ through every upcrossing, so if the total winnings are limited by an L^1-condition, there is a limit on the number of upcrossings. But if $\limsup f_n(\omega) > \liminf f_n(\omega)$, picking a and b between them, you have to have infinitely many upcrossings. We need several additional functions:

$$g_n(\omega) = a + \max(0, f_n(\omega) - a) \qquad (2.10.66)$$

$$N_n(\omega) = \sup\{k \mid \tau_k \leq n\} \qquad (2.10.67)$$

$$N_\infty(\omega) = \lim N_n(\omega) \qquad \text{(exists since } N_n \leq N_{n+1}) \qquad (2.10.68)$$

$$h_n(\omega) = \begin{cases} 1 & \text{if } \sigma_j < n \leq \tau_k \text{ for some } k \\ 0 & \text{if } n \leq \sigma_1 \text{ or } \tau_k < n \leq \sigma_{k+1} \end{cases} \qquad (2.10.69)$$

$$k_n = 1 - h_n \qquad (2.10.70)$$

$$u_n(\omega) = \sum_{j=1}^n h_j(g_j - g_{j-1}) \qquad (2.10.71)$$

$$v_n(\omega) = g_1 + \sum_{j=2}^n k_j(g_j - g_{j-1}) \qquad (2.10.72)$$

(a) Prove that σ_j, τ_j are stopping times and that $\{g_n\}_{n=1}^\infty$ is a submartingale.

(b) Prove that h_n, k_n are predictive and that u_n, v_n are submartingales.

(c) Prove that

$$u_n \geq (b-a)N_n \qquad (2.10.73)$$

(d) Prove that $\mathbb{E}(v_n) \geq 0$, $u_n + v_n = g_n$, and then that

$$\mathbb{E}(N_n) \leq (b-a)^{-1}\mathbb{E}(g_n) \qquad (2.10.74)$$

(e) If

$$\sup_n \mathbb{E}(\max(f_n, 0)) < \infty \qquad (2.10.75)$$

prove that

$$\mathbb{E}(N_\infty) < \infty \qquad (2.10.76)$$

and conclude that $N_\infty(\omega) < \infty$ for a.e. ω.

(f) If $\limsup f_n(\omega) > b$ and $\liminf f_n(\omega) < a$, prove that $N_\infty(\omega) = \infty$.

(g) If (2.10.75) holds, prove that $\lim_{n\to\infty} f_n(\omega)$ exists for a.e. ω.

Remark. (2.10.74) is sometimes called *Doob's upcrossing inequality*. It appeared in [**224**]. We discussed the same kind of idea as used in the proofs of the ergodic theorem in the Notes to Section 2.6.

11. Let $\{f_n\}_{n=1}^\infty$ be a martingale and N a stopping time. Let $g_n = f_{\min(n,N(\cdot))}$. Prove that

$$\sup_n \|g_n\|_1 \leq \sup_n \|f_n\|_1 \qquad (2.10.77)$$

(*Hint*: For $k < n$, prove that $\mathbb{E}(\chi_{T(\cdot)=k}|f_k|) \leq \mathbb{E}(\chi_{T(\cdot)=k}|f_n|)$.)

12. Let $\{f_n\}_{n=1}^{\infty}$ be a martingale so that for some $q \in L^1$ and all n, $|f_n(\omega)| \le q(\omega)$ for all ω. Prove there is an $f_\infty \in L^1$ covering $\{f_n\}_{n=1}^{\infty}$ so that $f_n(\omega)$ converges a.e. (*Hint*: Write $f_n = h_n q$ with $h_n \in L^\infty$ and let h_∞ be a weak-$*$ (i.e., $\sigma(L^\infty, L^1)$) limit point of the h_n. Use $f_\infty = h_\infty q$.)

13. This will sketch yet another proof of the L^1-martingale convergence theorem (following Garling [**304**]). It will use Problems 11 and 12. Let $\{f_n\}_{n=1}^{\infty}$ be a martingale with $s = \sup_n \|f_n\|_1 < \infty$.

    (a) Fix $M \in (0, \infty)$ and let $N(\omega) = \min(n \mid |f_n(\omega)| > M)$ ($N = \infty$ if $\sup |f_n(\omega)| \le M$). Let $g_n = f_{\min(n, N(\cdot))}$ so (2.10.77) holds. Let

    $$q(\omega) = \begin{cases} |f_{N(\omega)}(\omega)| & \text{if } N(\omega) < \infty \\ M & \text{if } N(\omega) = \infty \end{cases}$$

    Prove that $q \in L^1$ and that for all n, $|g_n(\omega)| \le q(\omega)$. Conclude (by Problem 12 and Theorem 2.10.9) that for a.e. ω, $g_n(\omega)$ converges.

    (b) Use Doob's inequality to prove that $\mu(\{x \mid N(\omega) < \infty\}) \le s/M$.

    (c) Prove that $f_n(\omega)$ converges for a.e. ω with $N(\omega) = \infty$.

    (d) Prove that $f_n(\omega)$ converges for a.e. ω.

14. If f_n is a submartingale with $f_n = i_n + m_n = \tilde{i}_n + \tilde{m}_n$ with $i_n \le i_{n+1}$, $i_1 = 0$, $\{m_n\}_{n=1}^{\infty}$ a martingale and the same for $\tilde{i}_n, \tilde{m}_n$, prove that $i_n = \tilde{i}_n$, $m_n = \tilde{m}_n$.

15. This problem and the next will prove the strong law of large numbers from Theorem 2.10.10. In this problem, you'll suppose independence and

    $$\sum_{n=1}^{\infty} n^{-2} \mathbb{V}\mathrm{ar}(f_n) < \infty, \qquad \mathbb{V}\mathrm{ar}(g) = \|g - \mathbb{E}(g)\|_2^2 \tag{2.10.78}$$

    which holds if the $\{f_n\}_{n=0}^{\infty}$ are iidrv with $\mathbb{E}(|f_n|^2) < \infty$, and in the next problem that the $\{f_n\}_{n=0}^{\infty}$ are iidrv with $\mathbb{E}(|f_n|) < \infty$ (the optimal iidrv strong law).

    (a) Let $\{x_k\}_{k=1}^{\infty}$ be a sequence of reals and define

    $$u_n = \sum_{k=1}^{n} \frac{x_k}{k}, \qquad s_n = \sum_{k=1}^{n} x_k \tag{2.10.79}$$

    Prove inductively that

    $$s_n = n u_n - \sum_{k=1}^{n-1} u_k \tag{2.10.80}$$

    (b) If $u_n \to u_\infty$, a finite real, prove that $s_n/n \to 0$. (*Note*: This is a special case of what is known as *Kronecker's lemma*.)

(c) If $\{f_n\}_{n=1}^{\infty}$ is a sequence of independent L^2 random variables so that (2.10.78) holds, show that for a.e. ω,

$$\sum_{n=1}^{\infty} \frac{1}{n}(f_n(\omega) - \mathbb{E}(f_n))$$

converges to a finite limit and conclude that for a.e. ω,

$$\frac{1}{N}\sum_{n=1}^{N}(f_n(\omega) - \mathbb{E}(f_n)) \to 0$$

(d) If $\{f_n\}_{n=1}^{\infty}$ are iidrv with $\mathbb{E}(|f_n|^2) < \infty$, prove that for a.e. ω,

$$\frac{1}{N}\sum_{n=1}^{N} f_n(\omega) \to \mathbb{E}(f_1) \tag{2.10.81}$$

Note. The result in (c) was proven by Kolmogorov [**482**].

16. This problem will prove that if $\{f_n\}_{n=0}^{\infty}$ are iidrv with $\mathbb{E}(|f_1|) < \infty$, then (2.10.81) holds. This is called Kolmogorov's strong law of large numbers after [**483**], whose proof is followed below. ((b)–(d) is sometimes called Kolmogorov truncation.) Let

$$g_n = f_n \chi_{\{\omega \mid |f_n| \leq n\}}$$

(a) Prove that $\sum_{n=1}^{\infty} \mathbb{E}(|f_1| \geq n) \leq \mathbb{E}(|f_1| + 1) < \infty$.

(b) For a.e. ω, prove that eventually $f_n(\omega) = g_n(\omega)$ and conclude that $\frac{1}{N}\sum_{n=1}^{N}|g_n(\omega) - f_n(\omega)| \to 0$ for a.e. ω.

(c) Prove that $\mathbb{E}(g_n) \to \mathbb{E}(f_1)$ as $n \to \infty$.

(d) Prove that $\sum_{n=1}^{\infty} \mathbb{E}(g_n^2) n^{-2} = \mathbb{E}(|f_1|^2 G \circ f_1)$, where $G(y) = \sum_{n \geq |y|, n \geq 1} n^{-2}$. Prove that $|G(y)| \leq C(1 + |y|)^{-1}$, and then that $\sum_{n=1}^{\infty} \mathbb{E}(g_n^2) n^{-2} < \infty$.

(e) Using Problem 15, prove that (2.10.81) holds.

17. This problem will provide another proof that for every $\{a_n\}_{n=1}^{\infty}$ in ℓ^2, $\sum_{n=1}^{\infty} a_n \sigma_n$, where σ_n are random signs, converges for a.e. ω. It illustrates the close connection between martingales and the Hardy–Littlewood inequality. Define φ_n on $\mathbb{R}$ by $\varphi_0(t) = 1$ on $[0, 1)$, -1 on $[1, 2)$, and to have period 2. In $L^2([0, 1], dx)$, prove that $\{\varphi_n(\cdot)\}_{n=1}^{\infty}$, where $\varphi_n(t) = \varphi_0(2^n t)$ is an orthonormal set (is it an orthonormal basis?). Fix $\{a_n\}_{n=1}^{\infty}$ in ℓ^2.

(a) Let $s_n = \sum_{j=1}^{n} a_j \varphi_j$ in L^2. Show that $\|s_n - f\|_2 \to 0$ for some $f \in L^2$.

(b) For $t \in [0, 1]$, $a \equiv k2^{-n} < t < (k+1)2^{-n} \equiv b$, prove that

$$s_n(t) = \frac{1}{b - a}\int_a^b f(x)\,dx$$

(c) For Lebesgue a.e. t, prove that $s_n(t) \to f(t)$. (*Hint*: Lebesgue's differentiation theorem.)

(d) For a.e. ω, prove that $\sum_{j=1}^{n} a_j \sigma_j(\omega)$ has a limit.

2.11. The Christ–Kiselev Maximal Inequality and Pointwise Convergence of Fourier Transforms

In this section, we'll obtain some results on the convergence of Fourier integrals, for example, if

$$(\mathcal{F}_R f)(k) = (2\pi)^{-1/2} \int_{-R}^{R} f(x)e^{-ikx}\,dx \qquad (2.11.1)$$

for f a function on $\mathbb{R}$ with $\chi_{(-R,R)}f \in L^1$ for all R so the integral is defined. In Section 6.6 of Part 1, we saw if $f \in L^p(\mathbb{R}, dx)$, $1 \le p \le 2$, then $\mathcal{F}_R f \to \widehat{f}$ in $L^q(\mathbb{R}, dk)$-norm, $q = (1 - p^{-1})^{-1}$, but we are interested in pointwise convergence for a.e. k. We expect the key is

$$(\mathcal{F}^* f)(k) = \sup_R |\mathcal{F}_R f(k)| \qquad (2.11.2)$$

It is a celebrated result of Carleson (see the Notes) that $\mathcal{F}^*$ is weak-type $(2,2)$, that is (*Carleson's inequality*),

$$|\{k \mid (\mathcal{F}^* f)(k) > \lambda\}| \le c\lambda^{-2}\|f\|_2^2 \qquad (2.11.3)$$

which implies for $f \in L^2(\mathbb{R}, dx)$ and a.e. k,

$$(\mathcal{F}_R f)(k) \to \widehat{f}(k) \qquad (2.11.4)$$

The proof of (2.11.3) is beyond the scope of this book (but see the Notes), but we'll prove the easier result that there is C_p for $1 \le p < 2$ so that with $q = (1 - p^{-1})^{-1}$,

$$\|\mathcal{F}^* f\|_q \le C_p\|f\|_p \qquad (2.11.5)$$

which implies (2.11.4) for $f \in L^p$.

(2.11.5) will follow from a general maximal inequality for maps from L^p to L^q with $p < q$. Here is the setup:

Definition. Let (M, Σ, μ) be a σ-finite measure space with $L^2(M, d\mu)$ separable. A *continuous filtration* is a family of sets $\{A_\alpha\}_{\alpha \in \mathbb{R}}$ so that

(i) $\alpha > \beta \Rightarrow A_\beta \subset A_\alpha$ and $\mu(A_\alpha \setminus A_\beta) < \infty$

(ii) For all α, we have

$$\lim_{\varepsilon \downarrow 0} \mu(A_{\alpha+\varepsilon} \setminus A_\alpha) = \lim_{\varepsilon \downarrow 0} \mu(A_\alpha \setminus A_{\alpha-\varepsilon}) = 0 \qquad (2.11.6)$$

(iii) $\bigcap_\alpha A_\alpha = \emptyset, \quad \bigcup_\alpha A_\alpha = M \qquad (2.11.7)$

The two canonical examples are $A_\alpha = (-\infty, \alpha) \subset \mathbb{R} = M$ and μ a Baire measure with no pure points, and $A_\alpha = \{x \mid |x| < \alpha\} \subset \mathbb{R}^\nu = M$ and $\mu(\{x \mid |x| = \alpha\}) = 0$ for all α. Let χ_α be the characteristic function of A_α. If $f \in L^p(M, d\mu)$, (2.11.6)/(2.11.7) imply that $\alpha \mapsto \|f\chi_\alpha\|_p^p \equiv G(\alpha)$ is continuous with $\lim_{\alpha \downarrow -\infty} G(\alpha) = 0$, $\lim_{\alpha \to \infty} G(\alpha) = \|f\|_p^p$. Let $T \colon L^p(M, d\mu) \to L^q(N, d\nu)$ be bounded. The *Christ–Kiselev maximal function* is defined by

$$(T^* f)(x) = \sup_\alpha |T(\chi_\alpha f)(x)| \qquad (2.11.8)$$

Theorem 2.11.1 (Christ–Kiselev Maximal Inequality). *Let $1 \le p < q < \infty$. Let $\beta = p^{-1} - q^{-1} \in (0, 1)$. Let $T \colon L^p(M, d\mu) \to L^q(N, d\nu)$ be bounded with norm t. Let $\{A_\alpha\}_{\alpha \in \mathbb{R}}$ be a continuous filtration on M and $T^* f$ given by (2.11.8). Then for all $f \in L^p(M, d\mu)$,*

$$\|T^* f\|_q \le 2^{-\beta}(1 - 2^{-\beta})^{-1} t \|f\|_p \qquad (2.11.9)$$

Remark. $2^{-\beta}(1 - 2^{-\beta})^{-1} = \sum_{n=1}^\infty 2^{-n\beta}$ is monotone decreasing in β, so this constant is larger than its value at $\beta = 1$, that is, it is bigger than 1.

Proof. Without loss, suppose $\|f\|_p = 1$. For $\alpha \in \mathbb{R}$, define

$$G(\alpha) = \|f\chi_\alpha\|_p^p \qquad (2.11.10)$$

$G(\alpha)$ is monotone in α and continuous and runs from 0 to 1, so for $m = 1, 2, \ldots$, $j = 1, \ldots, 2^{m-1}$, define α_j^m by

$$G(\alpha_j^m) = \frac{j}{2^m} \qquad (2.11.11)$$

with α_j^m the smallest such solution if G has a flat piece.

Define

$$f_j^m = f(\chi_{\alpha_j^m} - \chi_{\alpha_{j-1}^m}) \qquad (2.11.12)$$

for $m = 1, 2, \ldots$, $j = 1, \ldots, 2^m$, where $\chi_{\alpha_{2^m}^m} \equiv 1$, $\chi_{\alpha_0^m} \equiv 0$. Thus,

$$\|f_j^m\|_p^p = 2^{-m} \qquad (2.11.13)$$

If $\alpha \in [0, 1]$ is real, write $G(\alpha) = \sum_{m=1}^\infty k_m/2^m$ with $k_m \in \{0, 1\}$ as a base two expansion. Define $j_m(\alpha) = \sum_{\ell=1}^m k_\ell 2^{m-\ell}$, so by the continuity of G, as functions in L^p (i.e., up to sets of measure 0),

$$f\chi_\alpha = \sum_{\substack{m=1 \\ k_m=1}} f_{j_m(\alpha)}^m \qquad (2.11.14)$$

Thus,

$$(T^* f)(x) \le \sum_{m=1}^\infty \Big[\sup_{1 \le j \le 2^m} |(T(f_j^m))(x)| \Big] \qquad (2.11.15)$$

$$\le \sum_{m=1}^\infty h_m(x) \qquad (2.11.16)$$

where

$$h_m(x) = \left(\sum_{j=1}^{2^m} |(T(f_j^{(m)}))(x)|^q \right)^{1/q} \tag{2.11.17}$$

This follows from $(\max_{\ell=1,\dots,L} |a_\ell|)^q \leq \sum_{\ell=1}^{L} |a_\ell|^q$.

By (2.11.13),

$$\|h_m\|_q^q = \int \sum_{j=1}^{2^m} |Tf_j^m(x)|^q \, d\nu(x) \leq 2^m t^q (\|f_j^m\|_p^p)^{q/p} = 2^{m(1-q/p)} t^q \tag{2.11.18}$$

so

$$\|h_m\|_q \leq 2^{-m\beta} t \tag{2.11.19}$$

Thus, by (2.11.16),

$$\|T^* f\|_q \leq \sum_{m=1}^{\infty} \|h_m\|_q \leq t \sum_{m=1}^{\infty} 2^{-m\beta} = t 2^{-\beta} (1 - 2^{-\beta}) \tag{2.11.20}$$

$\square$

As an immediate corollary, we have

Theorem 2.11.2. *Let $p \in [1, 2)$. Let $q = (1 - p^{-1})^{-1}$ and $\beta = \frac{1}{p} - \frac{1}{q} = \frac{2}{p} - 1$. Then with $\mathcal{F}^*$ given by (2.11.1)/(2.11.2), we have*

$$\|\mathcal{F}^* f\|_q \leq (2\pi)^{-\nu(p^{-1}-1/2)} 2^{-\beta} (1 - 2^{-\beta})^{-1} \|f\|_p \tag{2.11.21}$$

for all $f \in L^p(\mathbb{R}^\nu, d^\nu x)$. In particular, if $f \in L^p(\mathbb{R}^\nu, d^\nu x)$, $1 \leq p < 2$, we have (2.11.3) for a.e. k.

Proof. (2.11.21) is immediate from Theorem 2.11.1 and the Hausdorff–Young inequality (see Theorem 6.6.2 of Part 1). Since

$$\mu(\{k \mid (\mathcal{F}^* f)(k) > \lambda\}) \leq \lambda^{-q} \|\mathcal{F}^* f\|_q^q \tag{2.11.22}$$

we have a maximal inequality of the type needed to apply Theorem 2.1.1.

If $f \in L^1 \cap L^p$, then $\mathcal{F}_R f \to \hat{f}$ in $\|\cdot\|_\infty$, and so pointwise. This set is dense in L^p, so by Theorem 2.1.1, we get a.e. pointwise convergence for all $f \in L^p$. $\square$

To apply these ideas to Fourier series, we want to replace $L^p(M, d\mu)$ by $\ell^p(\mathbb{Z}) = \{\{a_n\}_{n=-\infty}^{\infty} \mid \|a\|_p = (\sum_{n=-\infty}^{\infty} |a_n|^p)^{1/p} < \infty\}$. One can make an argument mimicking the proof of Theorem 2.11.1 with an interpolation, but it is slicker to embed $\ell^p(\mathbb{Z})$ into $L^p(\mathbb{R}, dx)$. Thus, define $S \colon \ell^p \to L^p(\mathbb{R}, dx)$ and $\widetilde{S} \colon L^p(\mathbb{R}, dx) \to \ell^p$ by

$$(Sa)(x) = a_n \text{ if } n \leq x < n+1, \qquad (\widetilde{S}f)_n = \int_n^{n+1} f(x) \, dx \tag{2.11.23}$$

Clearly,

$$\|Sa\|_p = \|a\|_p, \qquad \|\widetilde{S}f\|_p \le \|f\|_p, \qquad \widetilde{S}S = \mathbb{1} \qquad (2.11.24)$$

since, by Hölder's inequality,

$$\left| \int_n^{n+1} f(x)\,dx \right|^p \le \int_n^{n+1} |f(x)|^p\,dx \qquad (2.11.25)$$

Theorem 2.11.3 (Discrete Christ–Kiselev Maximal Inequality). *Let* $1 \le p < q < \infty$. *Let* $\beta = p^{-1} - q^{-1} \in (0,1)$. *Let* $W\colon \ell^p(\mathbb{Z}) \to L^q(N, d\nu)$ *be a bounded map with norm* w *and* χ_m *the characteristic function of* $(-\infty, m]$ *for* $m \in \mathbb{Z}$. *Define*

$$(W^*a)(x) = \sup_m |[W(\chi_m a)](x)| \qquad (2.11.26)$$

Then for all $a \in \ell^p$,

$$\|W^*a\|_q \le 2^{-\beta}(1 - 2^{-\beta})^{-1}w\|a\|_p \qquad (2.11.27)$$

Proof. Let $T\colon L^p(\mathbb{R}, dx) \to L^q(N, d\nu)$ by $T = W\widetilde{S}$. Then since $\widetilde{S}$ is an ℓ^p-L^p-contraction, $\|T\| \le w$. Thus, by Theorem 2.11.1,

$$\|T^*f\|_q \le 2^{-\beta}(1 - 2^{-\beta})w\|f\|_p \qquad (2.11.28)$$

Since $S\chi_m = \chi_m S$ and $\widetilde{S}S = \mathbb{1}$, we have $Wa = TSa$ and

$$(W^*a)(x) \le (T^*(Sa))(x) \qquad (2.11.29)$$

It follows that (2.11.28) implies (2.11.27). $\qquad\square$

We can use this for a result on convergence of Fourier series. The result holds for any O.N. basis in an L^2 space. Perhaps surprisingly, boundedness of the functions $e^{inx}/\sqrt{2\pi}$ plays no role! We recall a simple fact:

Proposition 2.11.4. *For* $p > q$,

$$\|a\|_p \le \|a\|_q \qquad (2.11.30)$$

where $\|a\|_r^r = \sum_{n \in \mathbb{Z}} |a_n|^r$.

Proof. If $(|a|^\alpha)_n = |a_n|^\alpha$, we have $\||a|^\alpha\|_r = \|a\|_{\alpha r}^\alpha$, so $\|a\|_q = \||a|^q\|_1^{1/q}$. Thus, if $\|b\|_{p/q} \ge \|b\|_1$, we have $\|a\|_q = \||a|^q\|_1^{1/q} \le \||a|^q\|_{p/q}^{1/q} = \|a\|_p$.

We have thus shown that it suffices to consider the case $q = 1$. In that case,

$$|a_j|^p \le |a_j| \left(\sum_n |a_j| \right)^{p-1}$$

Summing over j, $\|a\|_p^p \le \|a\|_1^p$, proving (2.11.30). $\qquad\square$

Theorem 2.11.5 (Menshov's Theorem). *Let $\{\varphi_n\}_{n=1}^{\infty}$ be an orthonormal basis in some $L^2(M, d\mu)$. Given a sequence $\{c_n\}_{n=1}^{\infty}$, define*

$$c^*(x) = \sup_{n=1,2,\dots} \left| \sum_{j=1}^{n} c_j \varphi_j(x) \right| \tag{2.11.31}$$

Then for any $p < 2$, with $\beta = (\frac{1}{p} - \frac{1}{2})$, we have

$$\|c^*\|_{L^2(M, d\mu)} \le 2^{-\beta}(1 - 2^{-\beta})^{-1} \|c\|_{\ell^p} \tag{2.11.32}$$

Moreover, for any $c \in \ell^p$, $1 \le p < 2$, we have pointwise a.e. that

$$\sum_{n=1}^{n} c_n \varphi_n(x) \to \left(\sum_{n=1}^{\infty} c_n \varphi_n \right)(x) \tag{2.11.33}$$

(interpreting the sum on the right as an L^2-sum).

Proof. For any $\{c_n\}_{n=-\infty}^{\infty}$ in ℓ^p, $1 \le p < 2$, by the above proposition, $c \in \ell^2$, so $\sum_{n=1}^{\infty} c_n \varphi_n(x) \equiv W(c)$ defines an element of L^2 and

$$\|W(c)\|_{L^2} \le \|c\|_{\ell^2} \le \|c\|_{\ell^p}$$

Applying Theorem 2.11.3, we get (2.11.32). For a finite sequence, (2.11.33) is trivial, and such sequences are dense in ℓ^p, so the maximal inequality implies (2.11.33) for all $c_n \in \ell^p$. $\qquad\square$

Remark. For the special case of $\varphi_n(x) = e^{inx}/\sqrt{2\pi}$, Carleson's work implies this result is true for $c \in \ell^2$, and so for any element of $L^2(\partial\mathbb{D}, \frac{d\theta}{2\pi})$.

Notes and Historical Remarks. The maximal inequalities and method of proof are from Christ–Kiselev [**157**] (with earlier related work [**156, 471**]). Cho et al. [**151**] has extensions to Lorentz and Orlicz spaces.

　　Both applications in this section long predate the work of Christ–Kiselev. Theorem 2.11.5 is from Menshov in 1927 [**564**] and Theorem 2.11.2 is from Zygmund in 1936 [**876**].

　　Theorem 2.11.1 is false in the generality given for $p = q$. For example, if $1 < p < \infty$, the Hilbert transform is bounded from L^p to L^p (see Section 5.7), but if f is continuous and everywhere nonvanishing, $(T^*f)(x) = \infty$ for all x (because for such f's, $T(\chi_\alpha f)(\alpha) = \infty$).

　　Carleson's inequality is due to Carleson [**143**], with L^p, $1 < p < 2$ analogs due to Hunt [**412**]. The original proof was quite involved; Fefferman [**266**] and, most especially, Lacey–Thiele [**501**] have substantial improvements in the proof.

　　Carleson's result (and it is an analog of this that Hunt proves) shows that if $f \in L^2(\partial\mathbb{D}, \frac{d\theta}{2\pi})$, then the Fourier partial sums of $f^\sharp$ converge. This is false for suitable f in L^1 (see the Notes to Section 3.5 of Part 1).

Chapter 3

Harmonic and Subharmonic Functions

Big Notions and Theorems: Harmonic Function, Laplace–Beltrami Operator, MVP, Liouville's Theorem for Harmonic Functions, Maximum Principle, Dirichlet Problem, Free Green's Function, Classical Green's Function, Harmonic Measure, Symmetry of Classical Green's Function, Poisson Kernel, Poisson Representation, Harmonic Functions Are Real Analytic, Identity Principle, MVP Implies Harmonicity, Montel and Vitali for Harmonic Functions, Removable Singularities Theorem, Harnack's Inequality, Harnack's Principle, Kelvin Transform, Subharmonic Functions, SMP, f Analytic $\Rightarrow |f|^p$, $\log_+(|f|)$ Subharmonic, Potential, Antipotential, Subharmonic $\Leftrightarrow \Delta T \geq 0$, Riesz Decomposition, Eremenko–Sodin Proof of Picard, Lebesgue Spine, Perron Method, Perron Solution, Perron Solution Is Harmonic, Barrier, Weak Barrier $\Rightarrow$ Strong Barrier, Regular Point, Exterior Ball Condition, Poincaré's Criterion, Exterior Cone Condition, Zaremba's Criterion, Euler's Formula, Spherical Harmonics, Dimension of Degree d Spherical Harmonics, Eigenvalues of Laplace–Beltrami, Partial Wave Expansion, Zonal Harmonics, Fourier Transforms of Spherical Harmonics, Bessel Transform, Robin Constant, Equilibrium Measure, Equilibrium Potential, Polar Set, Continuity Principle, Lusin's Theorem for Potentials, Frostman's Theorem, Green's Function with a Pole at Infinity, Kellogg–Evans Theorem, Infinity Sets for Potentials, Polar Removable Singularities Theorem, Extended Maximum Principle, Extended Dirichlet Problem, Riemann Maps and Exterior Green's Function, Polar Sets in $\mathbb{C}$ Are Totally Disconnected, $A(\mathfrak{e}) \leq C(\mathfrak{e})$ with Equality if $\mathfrak{e}$ Is Connected, Equilibrium Measures on $\mathbb{C}$ as Harmonic Measures, Principle of Descent, Upper Envelope Theorem, Single and Double Layer, Neumann Boundary Conditions, Balayage, Fine Topology, Martin Boundary, Bernstein–Walsh Lemma, Regular Measure, Christoffel–Darboux Formula, Thouless Formula, DOS (aka Density of States), Regularity $\Rightarrow$ DOS = Equilibrium Measure, Widom's Theorem, Regularity and Root Asymptotics, Clock Space of Zeros, CD Universality, Denisov–Rakhmanov–Remling Theorem, Ergodic Jacobi Matrices, Anderson Model, Almost Matthieu Equation, Harmonic and Subharmonic Functions on Riemann Surfaces, Finite Bordered Riemann Subsurfaces, Harmonic Functions with Polar Singularities, Green's Functions on Riemann Surfaces, Bipolar Green's Function, Variational Principle for Green's Functions, Monotonicity of Green's Function in Region, Hyperbolic, Parabolic, and Elliptic Riemann Surfaces, Minimal Superharmonic Majorants, Hyperbolic Equivalent to Green's Function Existing at One (or at All) Points, Symmetry of the Green's Function, Existence of Bipolar Green's Function, Existence of Meromorphic Functions on any Riemann Surface

This chapter will discuss harmonic functions and their close relatives, subharmonic functions. Harmonic functions are the poor man's analytic function—or, more accurately, the real analyst's analytic function. We'll see many aspects reminiscent of analytic functions, with analogs of unique continuation along curves, Morera's theorem, Liouville's theorem, and Cauchy estimates.

In Section 5.4 of Part 2A, we already discussed harmonic functions on a region, Ω, in $\mathbb{C} \cong \mathbb{R}^2$, where we showed three real-valued classes of functions, u, on Ω were the same: (i) locally, $u = \operatorname{Re} f$ with f analytic, (ii) u is C^2 with $\Delta u = 0$, and (iii) the mean value property: that u is continuous and

$$\int_0^{2\pi} u(x_0 + re^{i\theta}) \frac{d\theta}{2\pi} = u(x_0) \tag{3.0.1}$$

for each x_0 and all small r (how small can be x_0-dependent). Since complex analytic functions don't make sense on $\mathbb{R}^n$, we'll drop (i), but a major result in Section 3.1 will be the equivalence of a condition that u is C^2 on Ω with $\Delta u = 0$ and of an analog of (3.0.1) that u is continuous and for $0 < r < R(x_0)$, one has

$$\fint_{S^{\nu-1}} u(x_0 + r\omega)\, d\omega = u(x_0) \tag{3.0.2}$$

where $\omega \in S^{\nu-1} \equiv \{\omega \in \mathbb{R}^\nu \mid |\omega| = 1\}$ with $d\omega$ the unique rotational invariant Borel probability measure on $S^{\nu-1}$.

As in the $\mathbb{R}^2$ case, a major role is played by the Dirichlet problem: given $f \in C(\partial\Omega)$, to find u harmonic in Ω, continuous on $\overline{\Omega}$ with $u \upharpoonright \partial\Omega = f$ (for Ω unbounded we'll need more). Especially useful will be the solution when $\Omega = B_\rho(x_0)$ is a ball, in which case the solution is $(0 < r < 1)$

$$u(x_0 + r\rho\omega) = \int P_r(\omega, \eta) f(\eta)\, d(\eta) \tag{3.0.3}$$

For $\mathbb{R}^2$, the Poisson kernel was $P_r(\omega, \eta) = (1 + r^2)/(1 + r^2 - 2r\cos(\omega - \theta))$ (when we view S^1 as $[0, 2\pi)$). For $\mathbb{R}^\nu$, we'll find in Section 3.1 that

$$P_r(\omega, \eta) = \frac{1 - r^2}{(1 + r^2 - 2r\omega \cdot \eta)^{\nu/2}} \tag{3.0.4}$$

where ω, η are thought of as unit vectors in $\mathbb{R}^\nu$, so points in $S^{\nu-1}$.

The second important class of functions we'll study are subharmonic functions, u. Continuity is replaced by upper semicontinuity and the mean value property, (3.0.2), by

$$\int_{S^{\nu-1}} u(x_0 + r\omega)\, d\omega \geq u(x_0) \tag{3.0.5}$$

Thus, u is harmonic $\Leftrightarrow$ u and $-u$ are both subharmonic. The analog to $\Delta u = 0$ will be our proof in Section 3.2 that a distribution, T, is equal a.e. to a subharmonic function if and only if the distributional derivative $\Delta T \geq 0$ (i.e., is a measure) and that any subharmonic function is locally L^1, so it defines a distribution with $\Delta u \geq 0$.

One reason we'll care about subharmonic functions is that if f is analytic on $\mathbb{D}$, $|f|^p$ is subharmonic for any $0 < p < \infty$ and that this implies

$$r \mapsto \int_0^{2\pi} |f(re^{i\theta})|^p \frac{d\theta}{2\pi} \tag{3.0.6}$$

is monotone and continuous in r.

A second reason concerns the use of subharmonic functions in proving the Dirichlet problem is solvable for certain regions, Ω. In Section 3.4, we'll construct harmonic functions as the sups of certain subharmonic functions. A key to this proof will be the flexibility of subharmonic functions. The max of two harmonic functions is not harmonic in general, but the max of two subharmonic functions is subharmonic. More importantly, we'll see a way of modifying subharmonic functions to keep them unchanged near the boundary but harmonic in disks.

A natural basis for $L^2(\partial\mathbb{D}, \frac{d\theta}{2\pi})$ is $\{e^{in\theta}\}_{n=-\infty}^{\infty}$. Notice that ($u$ complex-valued is called harmonic if $\operatorname{Re} u$ and $\operatorname{Im} u$ are) for $n = 0, 1, 2, \ldots$, $r^n e^{\pm in\theta} = z^n$ or $\bar{z}^n$ are harmonic, so $\{e^{in\theta}\}_{n=-\infty}^{\infty}$ are restrictions to $\partial\mathbb{D}$ of homogeneous harmonic polynomials in x and y. In the same way, the restrictions to $S^{\nu-1}$ of homogeneous, harmonic polynomials will be a nice basis of $L^2(S^{\nu-1}, d\omega)$—indeed, eigenfunctions of the Laplace–Beltrami operators. Section 3.5 studies these "spherical harmonics."

A common theme in this chapter will be the study of Dirichlet Green's functions, that is, for $y \in \Omega$, distributions $G_\Omega(\cdot, y)$ obeying $\Delta G_\Omega(\cdot, y) = -\delta(\cdot - y)$ which are, of course, harmonic functions on $\Omega \setminus \{y\}$, so we have that $\lim_{x \to \partial\Omega} G_\Omega(x, y) = 0$. If such a G_Ω exists, it can be proven (see Section 3.6) that if we extend $G_\Omega(\,\cdot\,,y)$ to $\mathbb{R}^\nu \setminus \{y\}$ by setting it to 0 on $\mathbb{R}^\nu \setminus \Omega$, then the extended function is subharmonic. Since it is harmonic on $\mathbb{R}^\nu \setminus \overline{\Omega}$ and on $\Omega \setminus \{y\}$, ΔG_Ω will be a measure $d\mu_y$ supported on $\partial\Omega$. Physically, if $\nu \geq 3$, it is the shielding charge density one gets if one puts a point charge at y and surrounds it with a grounded perfect conductor on $\partial\Omega$. This will not only link the theory of this chapter to physical potential theory but also gives us new tools in studying the Dirichlet problem. Because this is connected to the physics, it is often called potential theory. It is the subject of Sections 3.6–3.8.

Section 3.8 uses the tools of the earlier part of the chapter to conclude the details of the uniformization theorem of Section 8.7 of Part 2A.

In most of this chapter, we normalize Green's functions so that $\Delta G_\Omega = -\delta(\,\cdot\, - y)$ but, following common practice, in Sections 3.7 and 3.8 where $\nu = 2$, we drop the $(2\pi)^{-1}$ in G_0 so that $\Delta G_\Omega = -2\pi\delta(\,\cdot\, - y)$.

Notes and Historical Remarks. Given the mathematical and historical context of the subject, this chapter could just as well be called "Potential Theory." Indeed, many books on the subject have that name as all or part of their title. Our choice is an indication of the parts relevant to the applications we'll need. One surprise is how central this subject has been to the history of analysis.

The roots of the subject go back to Newton's work on gravity. If $\rho(x)$ is a density function of compact support for matter, its potential is (up to a constant)

$$U_\rho(x) = -\int \frac{\rho(y)}{|x-y|}\, d^3y \qquad (3.0.7)$$

(in that the force is given by $-\nabla U_\rho$). This function is harmonic on $\mathbb{R}^3 \setminus \text{supp}(\rho)$ and subharmonic on all of $\mathbb{R}^3$ (if $\rho(x) \geq 0$).

After input from Euler, the subject came into focus with the work of Lagrange, Legendre, Laplace (consider that $\Delta u = 0$ is *Laplace's equation* and that Δ is *Laplacian*), and Poisson. It was Lagrange who realized that gravitational fields were expressible in terms of potentials—the name "potential" is from Green (who called them "potential functions") and Gauss (who dropped "function").

Because of the physics relation, ideas from that subject lurk in the background throughout the theory. Central, of course, is the energy minimization problem that we studied already in Section 5.9 of Part 1. And ideas from physics play an important role.

There are three extensions of this area that we will not discuss much. First, there is a connection with probability theory. To understand the origins of this, let $\mathbb{E}_{\mathbf{x}}(\cdot)$ be expectation with respect to Brownian motion starting at $\mathbf{x}$, that is, $\mathbf{w}(t) = \mathbf{x} + \boldsymbol{b}(t)$, where $\boldsymbol{b}$ is ν-dimensional Brownian motion as described in Section 4.16 of Part 1. Let $\Omega \subset \mathbb{R}^\nu$ be a region with smooth boundary. For $\mathbf{x} \in \Omega$, since $\limsup_{t\to\infty} |\mathbf{w}(t)| = \infty$, there is a first time at which $\mathbf{w}(t) \in \partial\Omega$, called the *exit time* or *stopping time*, τ. Then if f is continuous on $\partial\Omega$, the function

$$u(\mathbf{x}) = \mathbb{E}_{\mathbf{x}}(f(\mathbf{w}(\tau))) \qquad (3.0.8)$$

obeys

$$\Delta u = 0, \qquad \lim_{\substack{\mathbf{x}\to\mathbf{x}_0\in\partial\Omega \\ \mathbf{x}\in\Omega}} u(\mathbf{x}) = f(\mathbf{x}_0) \qquad (3.0.9)$$

Given the connection of Brownian motion to $e^{t\Delta}$ (see (4.16.39) of Part 1), (3.0.8) is not surprising. Studying (3.0.9) by other methods will be the subject of Section 3.4.

Pioneers in the connection between probability and potential, especially in the context of general Markov processes, are Doob (e.g., [**227**]) and Hunt (e.g., [**409**]). Books on the subject that played important historical roles are Blumenthal–Getoor [**85**], Doob [**228**], Meyer [**567**], and Port–Stone [**664**].

A second extension which got a boost from the probabilistic connections is *abstract* or *axiomatic potential theory* which tries to formalize the maximum principle, Harnack's inequality, harmonic-measure notions that will be so important in our discussions. The pioneer is Brelot [**117**], with further developments by Bauer [**45, 46**] and Meyer [**566**]. See also the books of Bauer [**47**], Brelot [**118**], and Constantinescu–Cornea [**166**].

A different approach emphasizing the role of positivity-preserving semigroups is due to Beurling–Deny [**67**] and their theory of Dirichlet spaces; see also Fukushima [**293**].

A third extension treats Laplace's equation as the paradigmal second-order elliptic equation and discusses other equations of that type. Typical book presentations of this type are Gilbarg–Trudinger [**312**] and Han–Lin [**351**].

For monographs on harmonic/subharmonic functions and/or potential theory, see [**8, 13, 25, 83, 370, 376, 460, 507, 552, 577, 670, 818, 847**].

The books by Axler et al. [**25**] and Ransford [**670**] are introductory and pedagogically strong.

3.1. Harmonic Functions

In this section, we'll present a first strong definition of the most basic object, and then we'll have a much weaker definition which we'll show is equivalent. A key intermediate step will be the Poisson representation, a sort of harmonic function analog of the Cauchy integral formula which, like that formula, will have a cornucopia of consequences.

As in $\mathbb{C}$, a *region*, Ω, in $\mathbb{R}^\nu$ is a connected open set.

Definition. A real-valued function, u, on a region, Ω, is called *harmonic* if u is C^2 and

$$\Delta u = 0 \tag{3.1.1}$$

We are heading towards a proof that harmonic functions obey the mean value property. On $\mathbb{R}^\nu$, we pick coordinates by first picking $\omega_1, \ldots, \omega_{n-1}$ on $S^{\nu-1}$ and then letting

$$r = \left(\sum_{J=1}^{\nu} x_j^2 \right)^{1/2}, \qquad \omega_\ell = \omega_\ell \left(\frac{\mathbf{x}}{r} \right)$$

It can then be seen (Problem 1) that for $\nu \geq 2$,

$$\Delta = \frac{1}{r^{\nu-1}} \frac{\partial}{\partial r} r^{\nu-1} \frac{\partial}{\partial r} + \frac{1}{r^2} B \tag{3.1.2}$$

where B is a second-order differential operator in the ω's (called the *Laplace–Beltrami* operator) that obeys

$$B\mathbb{1} = 0, \qquad \langle g, Bf \rangle = \langle Bg, f \rangle \tag{3.1.3}$$

where $\langle \, \cdot \, , \, \cdot \, \rangle$ is the $L^2(S^{\nu-1}, d\omega)$ inner product. Here $d\omega$ is the normalized rotation invariant measure on $S^{\nu-1}$, so that on $\mathbb{R}^\nu$ in spherical coordinates,

$$d^\nu x = r^{\nu-1} \sigma_\nu \, dr d\omega \tag{3.1.4}$$

with σ_ν the area of the unit sphere in $\mathbb{R}^\nu$, that is (see Problem 5 of Section 4.11 of Part 1),

$$\sigma_\nu = \frac{2\pi^{\nu/2}}{\Gamma(\nu/2)} \tag{3.1.5}$$

Now suppose u is harmonic on Ω with $B_\rho(0)$ the ball of radius ρ about 0 in Ω. For $r \in [0, \rho)$, define

$$A(r) = \int_{S^{\nu-1}} u(r\omega) \, d\omega \tag{3.1.6}$$

$$= \langle 1, u(r \cdot) \rangle \tag{3.1.7}$$

By (3.1.3), we have

$$\langle 1, (Bu)(r \cdot) \rangle = \langle B1, u(r \cdot) \rangle = 0$$

so for $0 < r < \rho$,

$$0 = \langle 1, \Delta u \rangle = \left\langle 1, \frac{1}{r^{\nu-1}} \frac{\partial}{\partial r} r^{\nu-1} \frac{\partial}{\partial r} u \right\rangle = \frac{1}{r^{\nu-1}} \frac{d}{dr} r^{\nu-1} \frac{d}{dr} A(r) \tag{3.1.8}$$

or $(r^{\nu-1}A')' = 0$, that is,

$$\Delta u = 0 \Rightarrow A'(r) = \frac{C}{r^{\nu-1}} \qquad\qquad (3.1.9)$$

$$\Rightarrow A(r) = D - (\nu-2)^{-1}\frac{C}{r^{\nu-2}} \qquad (\nu \geq 3) \qquad (3.1.10)$$

$$= D + C\log r \qquad\qquad (\nu = 2) \qquad (3.1.11)$$

Since $\lim_{r\downarrow 0} A(r) = u(0)$, we see first that $C = 0$, and then that $A(r) = \lim_{r\downarrow 0} u(0)$ for all r, that is, we have proven:

Theorem 3.1.1 (*Mean Value Property* $\equiv$ MVP). *Let u be a harmonic function on a region Ω, $x_0 \in \Omega$, with $B_\rho(x_0) \subset \Omega$. Then for all $r \in [0, \rho)$,*

$$u(x_0) = \int_{S^{\nu-1}} u(x_0 + r\omega)\,d\omega \qquad (3.1.12)$$

Below (see Theorem 3.1.20), we'll prove a strong converse that if for each x_0, (3.1.2) holds for sufficiently small r, then u is harmonic.

To see the power of the MVP, we note the following analogs of the Liouville theorem:

Theorem 3.1.2. *Let u be a harmonic function on all of $\mathbb{R}^\nu$.*

(a) *If $u \geq 0$, u is constant.*
(b) *If u is bounded, u is constant.*

Remark. (a) is called *Bochner's theorem* or *Picard's theorem*.

Proof. (a) By integrating the MVP over radii, we get a version over balls ($\tau_\nu = \nu^{-1}\sigma_\nu$ is the volume of $B_1(x_0)$),

$$u(x_0) = (\tau_\nu R^\nu)^{-1}\int_{B_R(x_0)} u(y)\,d^\nu y \qquad (3.1.13)$$

Noting that there is a cancellation between integrals on $B_R(0) \cap B_R(x_0)$, we see for each fixed x_0,

$$|u(x_0) - u(0)| \leq (\tau_\nu R^\nu)^{-1}\int_{B_R(x_0)\triangle B_R(0)} |u(y)|\,d^\nu y \qquad (3.1.14)$$

If $R > |x_0|$, $B_R(0) \cap B_R(x_0) \supset B_{R-|x_0|}(0)$, while $B_R(0) \cup B_R(x_0) \subset B_{R+|x_0|}(0)$. Thus, since $u(y) \geq 0$,

$$|u(x_0) - u(0)| \leq (\tau_\nu R^\nu)^{-1}\int_{B_{R+|x_0|}(0)\backslash B_{R-|x_0|}(0)} u(y)\,d^\nu y \qquad (3.1.15)$$

$$= (R^\nu)^{-1}[(R + |x_0|)^\nu - (R - |x_0|)^\nu]u(0) \qquad (3.1.16)$$

by (3.1.13). This goes to zero as $R \to \infty$, so $u(x_0) = u(0)$, that is, u is constant.

(b) Let $M = \sup_x |u(x)|$. Then $M + u(y) \geq 0$ and is harmonic, so constant by (a). $\square$

The MVP shows that if $j \in C_0^\infty$ is spherically symmetric and has compact support so that $x_0 + \mathrm{supp}(j) \subset \Omega$ and

$$\int j(y)\, d^\nu y = 1 \tag{3.1.17}$$

then

$$u(x_0) = (j * u)(x_0) \tag{3.1.18}$$

Thus, if $\mathrm{dist}(x_0, \Omega) = r > 0$ and j is spherically symmetric and supported in $B_{r/2}(0)$ and obeys (3.1.17), then (3.1.18) holds with x_0 replaced by any $x \in B_{r/2}(x_0)$. It follows that u is C^∞ on $B_{r/2}(x_0)$, that is, if u is harmonic, it is C^∞. We'll even prove below, using different methods, that it is real analytic.

An immediate corollary of the MVP is

Corollary 3.1.3 (Maximum Principle: First Form). *Let u be a harmonic function on a region, Ω. If for some $x_0 \in \Omega$,*

$$u(x_0) = \sup_{y \in \Omega} u(y) \tag{3.1.19}$$

then u is constant.

Remark. Applying this to $-u$, we see there is also a minimum principle.

Proof. (3.1.19) implies $\alpha = \sup_{y \in \Omega} u(y) < \infty$. Let $C = \{x \in \Omega \mid u(x) = \alpha\}$. By (3.1.19), C is nonempty and it is obviously closed. If $x \in C$, for some ρ, $B_\rho(x) \subset \Omega$. By Theorem 3.1.1, for all $r < \rho$,

$$\int u(x + r\omega)\, d\omega = \alpha \tag{3.1.20}$$

Since u is continuous and $u(x + r\omega) \leq \alpha$ for all ω, (3.1.20) implies $u(x + r\omega) = \alpha$ for all ω and all $r \in [0, \rho)$, that is, C is open. Since Ω is connected, $C = \Omega$. $\square$

Corollary 3.1.4 (Maximum Principle: Second Form). *Let Ω be a bounded open set of $\mathbb{R}^\nu$. Let u be continuous on $\overline{\Omega}$ and harmonic on Ω. Then there exists $x \in \partial\Omega$ with*

$$u(x) = \sup_{y \in \overline{\Omega}} u(y) \tag{3.1.21}$$

Remark. It is easy to see (Problem 2) that if Ω is unbounded, either (3.1.21) holds for some $x \in \partial\Omega$ or there exists $x_n \to \infty$ so $u(x_n) \to \sup u(y)$.

Proof. Since $\overline{\Omega}$ is compact, u takes its sup at some point $y \in \overline{\Omega}$. If $y \in \Omega$, then by Corollary 3.1.3, u is constant, and so constant on $\overline{\Omega}$. Either way, u takes its maximum at a point in $\partial\Omega$. $\square$

Definition. Let Ω be a bounded region in $\mathbb{R}^\nu$. Let $f \in C(\partial\Omega)$. We say u solves the *Dirichlet problem* for f if and only if u is harmonic in Ω, continuous in $\overline{\Omega}$, and

$$u \restriction \partial\Omega = f \tag{3.1.22}$$

Corollary 3.1.5. *There is at most one solution to the Dirichlet problem for any $f \in C(\partial\Omega)$.*

Proof. If u_1 and u_2 are solutions and $v = u_1 - u_2$, then $v \in C(\overline{\Omega})$, $\Delta v = 0$ on Ω, and $v = 0$ on $\partial\Omega$. By Corollary 3.1.4, v takes its maximum on $\partial\Omega$ so $v \leq 0$ on Ω. Applying the same argument to $-v$, we see that $v \geq 0$. Thus, $v = 0$, so $u_1 = u_2$. $\qquad\square$

We next want to find a formula for u in terms of its boundary values. For now, we'll suppose Ω has a smooth boundary and that u is C^2 up to the boundary. For if u, v are two functions C^2 up to the boundary, then using

$$\text{div}(u \operatorname{grad} v - v \operatorname{grad} u) = u \triangle v - v \triangle u \tag{3.1.23}$$

we get for any u, v that

$$\int_\Omega (u \triangle v - v \triangle u)\, dx = \int_{\partial\Omega} \vec{n} \cdot (u \operatorname{grad} v - v \operatorname{grad} u)\, d\sigma \tag{3.1.24}$$

where $d\sigma$ is surface measure and $\vec{n}$ is an outward pointing normal (see Theorem 1.4.2).

The *free Green's function* in $\mathbb{R}^\nu$ is defined by

$$G_0(x, y) = [(\nu - 2)\sigma_\nu]^{-1} |x - y|^{-(\nu-2)}, \qquad \nu \geq 3 \tag{3.1.25}$$

$$= -\frac{1}{2\pi} \log(|x - y|), \qquad \nu = 2 \tag{3.1.26}$$

We set $G_0(x) = G_0(x, 0)$. By (3.1.2) on $\mathbb{R}^\nu \setminus \{0\}$, $\Delta G_0(x) = 0$ and the constant in front is chosen so that for any r,

$$\int_{S_r^{\nu-1}} -\frac{\partial}{\partial n} G_0(x)\, d\sigma = 1 \tag{3.1.27}$$

where $d\sigma = r^{\nu-1} \sigma_\nu\, d\omega$ is surface measure on $S_r^{\nu-1} = \{x \mid |x| = r\}$. This last shows that

$$\Delta G_0 = -\delta \tag{3.1.28}$$

as a distribution. This is discussed in Section 6.8 of Part 1 where the formula for G_0 is derived from computing the Fourier transform of k^{-2}. The form of Δ in spherical coordinates, (3.1.2), shows if u is spherically symmetric and harmonic in $\mathbb{R}^\nu \setminus \{0\}$, then $u(x) = c_1 + c_2 G_0(x)$, so if $\nu \geq 2$, G_0 is determined by spherical symmetry, (3.1.27), and $u(x) \to 0$ as $x \to \infty$. In $\nu = 2$, c_1 is a free constant that sets the scale. Our choice is determined by $G_0(x) = 0$ if $|x| = 1$.

Definition. Let Ω be a bounded region in $\mathbb{R}^\nu$. We say $G_\Omega(x, y)$ defined for $(x, y) \in (\overline{\Omega} \times \Omega) \setminus \Delta$ (where $\Delta = \{(x, y) \mid x = y\}$) is a *classical Green's function* for Ω if and only if

(i) G_Ω is C^2 in x on $\Omega \setminus \{y\}$, and for such x,

$$\Delta_x G_\Omega(\,\cdot\,, y) = 0 \qquad (3.1.29)$$

(ii) $G_\Omega(x, y) - G_0(x, y)$ has a harmonic extension in x to a neighborhood of $x = y$.

(iii) $G_\Omega(\,\cdot\,, y)$ is continuous for $x \in \overline{\Omega} \setminus \{y\}$ and $G_\Omega(x, y) = 0$ for all $(x, y) \in \partial\Omega \times \Omega$.

Remarks. 1. This is sometimes called the *Dirichlet Green's function*.

2. We'll see soon (see Theorem 3.1.26) that (ii) can be replaced by $G_\Omega(x, y) - G_0(x, y)$ is bounded for x near y.

3. We discuss classical Green's functions for unbounded regions in Section 3.6.

4. If it exists, G_Ω is unique since the difference of two such G_Ω's is harmonic on Ω and vanishing on $\partial\Omega$.

We are heading towards showing that solvability of the Dirichlet problem implies existence of a Green's function. As a preliminary:

Theorem 3.1.6. *Let Ω be a bounded open region in $\mathbb{R}^\nu$. Suppose for every $f \in C(\partial\Omega)$, the Dirichlet problem has a solution $u_f(x)$. Then for each $x \in \Omega$, there is a probability measure, $d\mu_x$, on $\partial\Omega$ so that*

$$u_f(x) = \int f(\omega) \, d\mu_x(\omega) \qquad (3.1.30)$$

Moreover, as $x \to x_\infty \in \partial\Omega$, $d\mu_x \to \delta_{x_\infty}$, the point mass at x_∞, weakly, that is, in $\sigma(\mathcal{M}(X), C(X))$-topology.

Remark. $d\mu_x$ is called *harmonic measure*. Notice that $d\mu_x$ is harmonic in x in that it is continuous in x and obeys the MVP.

Proof. Define

$$L_x(f) = u_f(x) \qquad (3.1.31)$$

Uniqueness implies L_x is linear in f. If $f \leq 0$, $u_f \leq 0$ by the maximum principle, so $u_{-f} \geq 0$, that is, $f \geq 0 \Rightarrow L_x(f) \geq 0$. Since $u_\mathbb{1}(x) = 1$, $L_x(\mathbb{1}) = 1$. Thus, by the Riesz–Markov theorem (Theorem 4.5.4 of Part 1), there is a probability measure μ_x with (3.1.30). The weak continuity is obvious. $\qquad \square$

Notice that if $E \subset \partial\Omega$ is a Borel set, $\tilde{u}_E(x) = \mu_x(E)$ defines a nonnegative function obeying the MVP, so a harmonic function (by Theorem 3.1.20).

By the maximum principle for $-\tilde{u}_E$, $\tilde{u}_E(x_0) = 0$ for one x_0 implies $\tilde{u}_E \equiv 0$. We have proven (modulo Theorem 3.1.20) that

Proposition 3.1.7. *Suppose Ω in $\mathbb{R}^\nu$ is a bounded region for which the Dirichlet problem can be solved. Then for all $x, y \in \Omega$, $d\mu_x$ and $d\mu_y$ are mutually absolutely continuous.*

If $\Omega = B_1(0)$, the unit ball, then by taking limits in the MVP, if the Dirichlet problem for f is solvable, then

$$u_f(0) = \int f(\omega)\, d\omega \tag{3.1.32}$$

where $d\omega$ is normalized surface measure. Thus, each $d\mu_x$ is a.c. wrt $d\omega$—something true for many other Ω's. In that case,

$$d\mu_x = P_\Omega(x, \omega)\, d\omega \tag{3.1.33}$$

for a function P_Ω, called the *Poisson kernel* for Ω.

If Ω is a bounded region and $\partial\Omega$ is a connected smooth $\nu-1$-dimensional manifold, the Dirichlet problem is solvable and (3.1.33) holds (see Corollary 3.4.10).

One of our intermediate goals will be to show the Dirichlet problem is solvable and to compute P_Ω for the ball (in fact, we'll guess P_Ω and use it to prove that the Dirichlet problem is solvable!).

Here is a major result on the existence of classical Green's functions:

Theorem 3.1.8. *Let $\Omega \subset \mathbb{R}^\nu$ be a bounded region for which the Dirichlet problem can be solved. Then a classical Green's function exists and obeys*

(a) $G_\Omega(x, y) > 0$, *all* $x \in \Omega \setminus \{y\}$ (3.1.34)

(b) $G_\Omega(x, y) = G_\Omega(y, x)$, *all* $x, y \in \Omega$, $x \neq y$ (3.1.35)

Remarks. 1. We'll prove a converse in Corollary 3.4.8.

2. The proof uses $G_0(x - y) \geq 0$ on $\Omega \times \Omega \setminus \{(x, x)\}$ which is true if $\nu \geq 3$ but may fail if $\nu = 2$. If $\nu = 2$, use $G_0(x - y) \geq -\frac{1}{2\pi} \log(\operatorname{diam}\Omega)$ instead.

3. If $\partial\Omega$ is smooth, there is a proof of (3.1.31) that uses $u(z) = G(z, x)\,\Delta_z G(z, y) - G(z, y)\,\Delta_z G(z, x) = G(y, x)\delta(z - y) - G(x, y)\delta(z - x)$ and (3.1.24). For $\nu = 2$, we'll present this in detail as a warmup to the result on Riemann surfaces; see Proposition 3.8.19.

Proof. For $y \in \Omega$, let $\delta G(x, y)$ solve the Dirichlet problem for $f_y(\omega) = -G_0(\omega - y)$, that is

$$\Delta_x(\delta G(\,\cdot\,, y)) = 0, \qquad \lim_{x \to \omega \in \partial\Omega} \delta G(x, y) = -G_0(\omega - y) \tag{3.1.36}$$

Then

$$G_\Omega(x, y) = G_0(x - y) + \delta G(x, y) \tag{3.1.37}$$

is the Dirichlet Green's function since (3.1.36) implies $\lim_{x \to \omega \in \partial\Omega} G_\Omega(x, y) = 0$, and clearly, $G_\Omega - G_0 = \delta G$ is harmonic on Ω.

Fix $y \in \Omega$. $G_\Omega(\cdot, y)$ is harmonic on $\widetilde{\Omega} = \Omega \setminus \{y\}$. If G_Ω is ever negative, the maximum principle for $-G_\Omega(\cdot, y)$ implies there is no $x_0 \in \Omega$ with $G_\Omega(x_0, y) = \min_{x \in \Omega} G_\Omega(x, y) \equiv \alpha$ and that there is $x_n \in \Omega \setminus \{y\}$ so $G_\Omega(x_n, y) \to \alpha$ and x_n has no limit point in $\Omega \setminus \{y\}$. Since $\overline{\Omega}$ is compact and $\lim_{x \to y} G_\Omega(x, y) = \infty$, we can find, by passing to a subsequence, $x_\infty \in \partial\Omega$, so $x_n \to x_\infty$. But $G_\Omega(x_n, y) \to 0$. So $\alpha < 0$ cannot hold, that is, we have proven (3.1.34) with ≥ 0. The maximum principle for $-G$ implies that $>$ holds.

Fix $y \in \Omega$ and $\varepsilon > 0$. Let

$$h_\varepsilon(x) = G_\Omega(y, x) - G_\Omega(x, y) + \varepsilon G_0(x - y) \qquad (3.1.38)$$

By (3.1.37),

$$G_\Omega(x, y) = G_0(x - y) - \int G_0(\omega - y) \, d\mu_x(\omega) \qquad (3.1.39)$$

which implies $G_\Omega(x, y)$ is harmonic in y on Ω. (If the reader worries about taking Δ_y inside the integral, use the fact proven in Theorem 3.1.20 that the MVP implies harmonicity since the integral obeys the MVP in y by Fubini's theorem.) Thus, h_ε is harmonic in x on $\Omega \setminus \{y\}$.

By (3.1.39), $G_\Omega(y, x) - G_\Omega(x, y)$ is bounded for x near y, so

$$\lim_{x \to y} h_\varepsilon(x) = \infty \qquad (3.1.40)$$

Since $G_\Omega(x_n, y) \to 0$ as $x_n \to x_\infty \in \partial\Omega$, and G_0 and $G(y, x)$ are nonnegative, we see that

$$x_n \to x_\infty \in \Omega \Rightarrow \liminf h_\varepsilon(x_n) \geq 0 \qquad (3.1.41)$$

The maximum principle for $-h_\varepsilon$ on $\Omega \setminus \{y\}$ implies $h_\varepsilon \geq 0$. Taking ε to zero, we find for all $x \in \Omega$,

$$G_\Omega(y, x) \geq G_\Omega(x, y) \qquad (3.1.42)$$

Since this holds for all y, symmetry implies (3.1.35). $\qquad \square$

We next want to obtain an explicit formula for the Poisson kernel of regions with smooth boundary. We say "obtain" rather than prove because we'll do a semiformal calculation and not give a formal proof. We'll use it for motivation but always separately verify the formula it provides for the Poisson kernel.

Suppose Ω has a smooth boundary and a Dirichlet Green's function, G_Ω, which is C^1 up to $\partial\Omega$. Suppose u is a function harmonic on Ω and C^1 up to $\partial\Omega$ with $u \upharpoonright \partial\Omega = f$. Fix $y \in \Omega$. Let Ω_ε be Ω with the ball $B_\varepsilon(y)$ removed (where $\varepsilon < \text{dist}(y, \mathbb{R}^\nu \setminus \Omega)$; see Figure 3.1.1). In (3.1.24), use Ω_ε in place of Ω and take $v(x) = G_\Omega(x, y)$. Since u and v are harmonic on Ω_ε, the left side

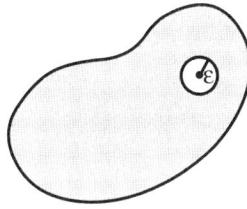

Figure 3.1.1. The domain Ω_ε.

of (3.1.24) is zero. Letting $\frac{\partial}{\partial n} = \vec{n} \cdot \text{grad}$ for the outward pointing normal for Ω and for $B_\varepsilon(y)$ (which is inward pointing for $\Omega \setminus B_\varepsilon$!), we have (using $G \equiv G_\Omega$)

$$\int_{\partial\Omega} f(\omega) \frac{\partial G}{\partial n}(\omega, y)\, d\sigma(\omega) = \int_{\partial B_\varepsilon(y)} \left[f(\omega) \frac{\partial G}{\partial n}(\omega, y) - G(\omega, y) \frac{\partial f}{\partial n} \right] d\sigma \tag{3.1.43}$$

where we used $G(\omega, y) = 0$ and don't have a minus sign in the right side because $\frac{\partial}{\partial n}$ is into $\Omega \setminus B_\varepsilon(y)$.

On the right side of (3.1.43),

$$d\sigma = \sigma_\nu \varepsilon^{\nu-1}\, d\omega \tag{3.1.44}$$

Since $\delta G = G - G_0$ is harmonic at y, the $f\frac{\partial(\delta G)}{\partial n}$ and $\delta G \frac{\partial f}{\partial n}$ terms are bounded, so because of the $\varepsilon^{\nu-1}$ in (3.1.44), they are $O(\varepsilon^{\nu-1})$. Since $G_0(\omega, y) = O(\varepsilon^{2-\nu})$ or $O(\log(\frac{1}{\varepsilon}))$, $\varepsilon^{\nu-1} G_0(\omega, y) \to 0$. Since $f(\omega) = f(y) + o(\varepsilon)$ and (3.1.27) holds, we see that the right side of (3.1.43) $= -f(y) + O(\varepsilon)$ (or $O(\varepsilon \log(\frac{1}{\varepsilon}))$) and so, since the left side of (3.1.43) is constant, it is $-f(y)$. That is, we have proven

$$f(y) = \int_{\partial\Omega} f(\omega) \left(-\sigma(\partial\Omega) \frac{\partial G}{\partial n}(\omega, y) \right) d\omega \tag{3.1.45}$$

where $\sigma(\partial\Omega)$ is the surface area of $\partial\Omega$ and $d\omega = \frac{d\sigma}{\sigma(\partial\Omega)}$ is normalized surface measure. This suggests

$$P_\Omega(y, \omega) = -\sigma(\partial\Omega) \frac{\partial G}{\partial n}(\omega, y) \tag{3.1.46}$$

which we also write as

$$d\mu_y(\omega) = -\frac{\partial G}{\partial n}(\omega, y)\, d\sigma(\omega) \tag{3.1.47}$$

since we'll formally use this form in some cases where $\partial\Omega$ is unbounded and $\sigma(\partial\Omega) = \infty$. Note that since $G_\Omega > 0$ in Ω,

$$-\frac{\partial G}{\partial n}(\omega, y) = \lim_{\varepsilon \downarrow 0} \frac{G(\omega - \varepsilon\hat{n}, y)}{\varepsilon} \geq 0 \tag{3.1.48}$$

as is required by $d\mu_y \in \mathcal{M}_{+,1}(\partial\Omega)$.

We now turn to some examples of Dirichlet Green's functions and Poisson kernels.

Example 3.1.9 (Half-Spaces). Let $\Omega = \mathbb{R}_+^{\nu-1} = \{x \in \mathbb{R}^\nu \mid x_\nu > 0\}$. Let

$$I(x_1, \ldots, x_\nu) = (x_1, \ldots, x_{\nu-1}, -x_\nu)$$

be reflection in the plane $\pi = \{(x_1, \ldots, x_{\nu-1}, 0)\}$. Then for all x and all $y \in \pi$, $y \cdot Ix = y \cdot x$ and $|x| = |Ix|$. Thus,

$$|x - y| = |Ix - y| \qquad (3.1.49)$$

for y in π, so $G_0(x - y) = G_0(Ix - y)$. Since $G_0(x - Iy)$ is harmonic on all of Ω if $x, y \in \Omega$, we see that

$$G_\Omega(x, y) = G_0(x - y) - G_0(x - Iy) \qquad (3.1.50)$$

has $G_\Omega(x, y) = 0$ if $x \in \pi$ by (3.1.49). This doesn't quite fit into our framework above since Ω is not bounded. A direct calculation shows for $y \in \Omega$, $x \in \pi$,

$$P_\nu^H(y, x) = -2 \frac{\partial G}{\partial x_\nu}(x, y) = \frac{2}{\sigma_\nu} \frac{y_\nu}{|x - y|^\nu} \qquad (3.1.51)$$

This is sometimes called the *Poisson kernel for the half-space*. As we'll see in Problem 13, if $f \in C_0(\pi)$, the continuous functions of compact support on π, then

$$u_f(y) = \int_{x \in \pi} P_\nu^H(y, x) f(x) \, d^{\nu-1}x \qquad (3.1.52)$$

is harmonic on $\mathbb{R}_+^{\nu-1}$ and is uniquely determined by

$$\lim_{y \to \infty} u_f(y) = 0, \qquad \lim_{y_\nu \downarrow 0} u_f(y_1, \ldots, y_{\nu-1}, y_\nu) = f(y_1, \ldots, y_{\nu-1}) \qquad (3.1.53)$$

Problem 14 discusses the Fourier transform of $P_\nu^H(y, \cdot)$.   □

Example 3.1.10 (Strips and Hypercubes). Let $\Omega_1 = \{x \in \mathbb{R}^\nu \mid 0 < x_\nu < 1\}$. The product of reflections in the two edges is translation by $(0, \ldots, 0, 2)$. We define

$$I_n(x_1, \ldots, x_\nu) = \begin{cases} (x_1, \ldots, x_{\nu-1}, n + 1 - x_\nu), & n \text{ odd} \\ (x_1, \ldots, x_{\nu-1}, n + x_\nu), & n \text{ even} \end{cases} \qquad (3.1.54)$$

Then

$$G_{\Omega_1}(x, y) = \sum_{n=-\infty}^{\infty} (-1)^n G(x, I_n(y)) \qquad (3.1.55)$$

By reflecting and translating in the ν-coordinate directions, one can write down $G_{\Omega_2}(x, y)$ for Ω_2 the hypercube $\{x \mid 0 < x_j < 1\}$ (Problem 3).   □

Example 3.1.11 (Balls). In $\mathbb{C}$, inversion in the unit circle took z to $1/\bar{z} = z/|z|^2$. This suggests the inversion for $S^{\nu-1}$ in $\mathbb{R}^\nu$ might take $r\omega (\omega \in S^{\nu-1})$ to $\omega/r = (r\omega)/r^2$, so we define for $x \in B_1(0)$,

$$Ix = \frac{x}{|x|^2} \tag{3.1.56}$$

(3.1.49) does not hold for $y \in S^{\nu-1}$, but notice that for such y, if $r = |x|$, then

$$|Ix - y|^2 = r^{-2} + 1 - 2x \cdot y r^{-2} = r^{-2}|x - y|^2 \tag{3.1.57}$$

Thus, to get cancellation on $S^{\nu-1}$, we need to not only invert x in $S^{\nu-1}$ but adjust the charge, that is, (3.1.57) implies that

$$G_\Omega(y, x) = \begin{cases} [(\nu - 2)\sigma_\nu]^{-1}|y - x|^{-(\nu-2)} - |x|^{-(\nu-2)}|y - Ix|^{-(\nu-2)}, & \nu \geq 3 \\ \frac{1}{2\pi}\left[\log(|y - x|^{-1}) - \log(|x||y - Ix|^{-1})\right], & \nu = 2 \end{cases} \tag{3.1.58}$$

obeys $G_\Omega(\omega, x) = 0$ if $\omega \in S^{\nu-1}$ and so is the Dirichlet Green's function for $\Omega = B_1(0)$, the unit ball. One sees the symmetry of $G_\Omega(y, x)$ explicitly by noting for any y,

$$|x|^2 |y - Ix|^2 = \left| |x|y - \frac{x}{|x|} \right|^2 = |x|^2|y|^2 - 2x \cdot y + 1 \tag{3.1.59}$$

is symmetric in x and y.

By direct calculation (Problem 4),

$$\frac{\partial}{\partial r} |r\omega - y|^{-(\nu-2)}\bigg|_{r=1} = -(\nu - 2)\frac{(1 - w \cdot y)}{|\omega - y|^\nu} \tag{3.1.60}$$

and

$$\frac{\partial}{\partial r} |r\omega - Iy|^{-(\nu-2)}\bigg|_{r=1} = -(\nu - 2)\frac{(|y|^2 - \omega \cdot y)}{|\omega - y|^\nu} \tag{3.1.61}$$

so (3.1.46) and (3.1.58) imply for $0 \leq r < 1$, $\omega, \eta \in S^{\nu-1}$,

$$P_B(r\eta, \omega) = \frac{1 - r^2}{|1 + r^2 - 2r\eta \cdot \omega|^{\nu/2}} \equiv P_r(\eta, \omega) \tag{3.1.62}$$

is the Poisson kernel of the ball. □

Remark. The fact that $|x|^{-(\nu-2)}$ is harmonic is a special case of the *Kelvin transform* (Problem 15). If $0 \in \Omega$ is a bounded region and u is harmonic in $\mathbb{R}^\nu \setminus \Omega$, then

$$Ku(x) = |x|^{-(\nu-2)}u(Ix) \tag{3.1.63}$$

is harmonic in $\Omega \setminus \{0\}$. If $\nu \geq 3$, it is harmonic at 0 if and only if $u(x) \to 0$ as $|x| \to \infty$. Conversely, if u is harmonic in Ω and Ku is given by (3.1.63), then Ku is harmonic in $\mathbb{R}^\nu \setminus \Omega$ with $u \to 0$ at ∞. Moreover, $K(Ku) = u$.

We'll only use the above to "guess" the Poisson kernel since we want to solve the Dirichlet problem. We have

Proposition 3.1.12. $P_r(\eta, \omega)$, *given by (3.1.62), obeys*

(a) $P_{|x|}(x/|x|, \omega)$ *is harmonic in* x.

(b) $P_r(\eta, \omega) > 0$ *for all* $r \in [0, 1)$ *and* $\omega, \eta \in S^{\nu-1}$.

(c) $\displaystyle\int P_r(\eta, \omega)\, d\eta = 1$ (3.1.64)

   for all $r \in [0, 1)$ *and* ω. *Here* $d\eta$ *is normalized measure on the sphere.*

(d) *For any* $\delta > 0$, $\sup_{|\eta - \omega| > \delta} P_r(\eta, \omega) \to 0$ *as* $r \uparrow 1$.

Proof. (a) $G_B(x, y)$ is harmonic in $x \in B \setminus \{y\}$, so since G_B is C^∞ in $\{(x, y) \mid x \neq y,\, x \neq Iy\}$, $\Delta_x \frac{\partial}{\partial r} G_B(x, r\omega) = \frac{\partial}{\partial r} \Delta_x G_B(x, r\omega) = 0$, that is, the function in question is harmonic.

(b) is obvious.

(c) Since $P_{r=0} \equiv 1$, this is just the MVP for the harmonic function in (a).

(d) Obvious since $1 - r^2 \to 0$, and as $r \uparrow 1$, $|r\eta - \omega|$ stays away from zero if $|\eta - \omega| > \delta$. $\square$

In the language of Section 3.5 of Part 1, $P_r(\eta, \omega)$ is an approximate identity. With this in hand:

Theorem 3.1.13 (Dirichlet Problem for the Ball). *For every* $f \in C(S^{\nu-1})$, *there is a unique* $u \in C(\overline{B_1(0)})$ *so that* u *is harmonic on* $B_1(0)$ *and* $u \upharpoonright S^{\nu-1} = f$. *Moreover,*

$$u(r\eta) = \int P_r(\eta, \omega) f(\omega)\, d(\omega)$$ (3.1.65)

Proof. We proved uniqueness in Corollary 3.1.5. To prove existence and (3.1.65), define u on B_1 by (3.1.65). Since $P_{|x|}(x/|x|, \omega)$ is harmonic, so is u. (Using uniform convergence of the integrals in $r\eta$ and of the difference quotients in ω so long as r is bounded away from 1, one can justify exchanging the integrals and derivatives.) Since P_r is an approximate identity by the arguments used to prove Theorem 3.5.11 of Part 1, uniformly in η, $\lim_{r\uparrow 1} u(r\eta) = f(\eta)$. This plus continuity of f shows

$$\lim_{\substack{x \in B_1 \\ x \to \omega \in S^{\nu-1}}} u(x) = f(\omega)$$

so u has a continuous extension to $\overline{B}_1$ with $u \upharpoonright S^{\nu-1} = f$. $\square$

This has lots and lots of immediate applications.

Theorem 3.1.14 (Poisson Representation). *Let u be a harmonic function on a region, Ω, in $\mathbb{R}^\nu$. Let $x_0 \in \Omega$ and $\delta < \operatorname{dist}(x_0, \mathbb{R}^\nu \setminus \Omega)$. Then for $x \in B_\delta(x_0)$,*

$$u(x) = \int P_{|x-x_0|/\delta}\left(\frac{x - x_0}{|x - x_0|}, \omega\right) u(x_0 + \delta\omega)\, d\omega \tag{3.1.66}$$

Proof. u is harmonic in $B_\delta(x_0)$, continuous on $\overline{B_\delta(x_0)}$, and so solves the Dirichlet problem for the data $u(x_0 + \delta\omega)$ on $\partial B_\delta(x_0)$. By the explicit form of the solution in (3.1.65) and uniqueness of solution, we have (3.1.66). $\quad\square$

Corollary 3.1.15. *Any harmonic function is C^∞, indeed, real analytic.*

Proof. A straightforward analysis (Problem 5) shows that if we define for $x \in B_1(0)$,

$$\widetilde{P}_\omega(x) = P_{|x|}\left(\frac{x}{|x|}, \omega\right) \tag{3.1.67}$$

then for each ω, $\widetilde{P}_\omega$ is real analytic near $x = 0$ with bounds

$$\left|\frac{1}{\alpha!} D^\alpha \widetilde{P}_\omega\Big|_{x=0}\right| \leq C_\nu^{|\alpha|} \tag{3.1.68}$$

for a ν-dependent constant C_ν independent of ω. Since (Problem 6)

$$\#(\alpha \mid |\alpha| = n) = \binom{n + \nu - 1}{\nu - 1} \leq 2^{n+\nu-1} \tag{3.1.69}$$

if $|x| < (2C_\nu)^{-1}$, then

$$\widetilde{P}_\omega(x) = \text{polynomials of degree } n \text{ in } x_1, \ldots, x_\nu + E_\omega^{(n)}(x) \tag{3.1.70}$$

$$|E_\omega^{(n)}| \leq 2^{\nu-1}(2C_\nu|x|)^{n+1}(1 - 2C_\nu|x|)^{-1} \tag{3.1.71}$$

Therefore, (3.1.66) proves u is C^∞ at x_0 with a Taylor series converging in the ball of radius $(2C_\nu)^{-1}\operatorname{dist}(x_0, \mathbb{R}^\nu \setminus \Omega)$. $\quad\square$

Example 3.1.16. Let $u(x, y)$ be the real function on $\mathbb{D}$

$$u(x, y) = \operatorname{Re}(1 - (x + iy))^{-1} \tag{3.1.72}$$

It is easy to see (Problem 7) that

$$u(x, y) = \sum_{m=0}^{\infty} \sum_{j=0}^{[m/2]} \binom{m}{2j} x^{m-2j}(-1)^j y^{2j} \tag{3.1.73}$$

and that this series converges absolutely if and only if

$$|x| + |y| < 1 \tag{3.1.74}$$

Thus, u is harmonic in $\mathbb{D}$, but the largest disk in which the series converges has radius $\sqrt{\frac{1}{2}}$, that is, not all of $B_{\operatorname{dist}(0, \mathbb{C}\setminus\mathbb{D})}(0)$. This is a big difference from analytic functions! Also, we see, unlike power series in one complex variable, the domain of convergence of a multivariable power series is not always a ball. We will return to analytic expansions in Theorem 3.5.10 and show that if one first combines all terms of homogeneous order m, one gets convergence in the full ball.   □

Real analyticity implies that some of the standard facts about analytic functions carry over to the harmonic case.

Theorem 3.1.17 (Identity Principle). *Let u be harmonic in a region, Ω, in $\mathbb{R}^\nu$. If $u = 0$ in a neighborhood of a point, $x_0 \in \Omega$, then $u \equiv 0$. Therefore, if u and v are harmonic and $u = v$ in a neighborhood of a point, $x_0 \in \Omega$, then $u = v$ on Ω.*

Proof. Let $Q = \{x \in \Omega \mid u = 0 \text{ in a neighborhood of } x\}$. Q is clearly open and, by hypothesis, $x_0 \in Q$, so Q is nonempty. If $\{x_n\}_{n=1}^\infty \subset Q$ and $x_n \to x_\infty \in \Omega$, then since u is C^∞, $D^\alpha u(x_\infty) = \lim_{n\to\infty}(D^\alpha u)(x_n) = 0$, so the Taylor series at x_∞ is 0. By real analyticity, $u = 0$ near x_∞, that is, Q is closed. Thus, Q is all of Ω.

The $u = v$ result follows from the $u = 0$ result by noting $u = v \Leftrightarrow u - v = 0$ and that $u - v$ is harmonic.   □

In fact, we have the following stronger result:

Theorem 3.1.18. *Let u be real analytic in a region, Ω, in $\mathbb{R}^\nu$. If u is not identically zero, then*

$$B = \{x \mid u(x) = 0\} \Rightarrow |B| = 0 \tag{3.1.75}$$

(i.e., u vanishes on a set of measure zero).

Remarks. 1. In particular, this applies to a harmonic function.

2. The idea of the proof can be used to show that B has Hausdorff dimension at most $\nu - 1$.

Proof. It is enough to prove this if Ω is a hypercube. For if $|B| > 0$, there are hypercubes, $\Lambda \subset \Omega$, so that $|B \cap \Lambda| > 0$ and then the special case plus the identity principle (whose proof only depended on real analyticity) implies $u \equiv 0$.

We prove the hypercube result by induction in ν. For $\nu = 1$, zeros are discrete, so the set of zeros in (a, b) is countable, so of Lebesgue measure zero.

For general ν, assuming we have the result for $\nu = 1$, fix $(x_1^{(0)}, \ldots, x_\nu^{(0)}) \in \Omega$ so $u(x^{(0)}) \neq 0$. By the one-dimensional case, for a.e. x_ν in (a_ν, b_ν), $u(x_1^{(0)}, \ldots, x_{\nu-1}^{(0)}, x_\nu) \neq 0$ (this uses that slices of hypercubes are connected).

Let S be the set of x_ν for which this is nonzero. For each $x_\nu^{(1)} \in S$, by the induction hypothesis, $u(x_1, \ldots, x_{\nu-1}, x_\nu^{(1)}) \neq 0$. Thus, the zeros with $x_\nu \in S$ are a set of measure zero and $|\{x \mid x_\nu \notin S\}| = 0$ so the zeros lie in a set of measure zero. $\qquad \square$

We can use the solvability of the Dirichlet problem for balls to prove a strong inverse MVP.

Definition. We say that a continuous function on a region $\Omega \subset \mathbb{R}^\nu$ is *weakly harmonic* if and only if for each $x_0 \in \Omega$, there exists $r_n(x_0) \to 0$ so that for all n, $B_{r_n(x_0)}(x_0) \subset \Omega$ and

$$\int u(x_0 + r_n(x_0)\omega) \, d\omega = u(x_0) \tag{3.1.76}$$

Note that, a priori, $r_n(x_0)$ need not be bounded away from zero on compacts, that is, there can be $x_m \to x_0 \in \Omega$ with $\sup_n r_n(x_m) \to 0$ even though $r_n(x_0) > 0$. We begin with:

Proposition 3.1.19 (Maximum Principle for Weakly Harmonic Functions). *Let u be continuous in $\overline{B_1(0)}$ and weakly harmonic in $B_1(0)$. Then there is $x_0 \in \partial B_1(0)$, so that*

$$u(x_0) = \sup_{y \in \overline{B_1(0)}} u(y) \tag{3.1.77}$$

Proof. Let M be the sup on the right side of (3.1.77). Let

$$K = \{x \in \overline{B_1(0)} \mid u(x) = M\} \tag{3.1.78}$$

If $\partial B_1(0) \cap K \neq \emptyset$, there is $x_0 \in \partial B_1(0)$ so (3.1.77) and we are done.

If not, by continuity of u and compactness of $\overline{B_1(0)}$, K is nonempty and closed, so

$$r = \sup\{|x| \mid x \in K\} < 1 \tag{3.1.79}$$

Since K is compact, pick $x_1 \in K$ so $|x_1| = r$. Since $r_n(x_1) \to 0$ as $n \to \infty$, find $r_n(x_1) < 1 - r$. Since (3.1.76) holds and u is continuous with $u(x_1 + r_n(x_1)\omega) \leq M$, we have $u(x_1 + r_n(x_1)\omega) = M$ for all ω. Thus, $x_1(\mathbb{1} + \frac{r_1(x_1)}{|x_1|}) \in K$, so there is an $x \in K$ with $|x| > r$. This contradiction shows that (3.1.77) holds for some $x_0 \in \partial B_1(0)$. $\qquad \square$

Theorem 3.1.20. *Every weakly harmonic function is harmonic.*

Proof. We start with two preliminaries. Because $r_n(x_0) \to 0$ for all x_0, if $\Omega_1 \subset \Omega$ and u is weakly harmonic on Ω, $u \upharpoonright \Omega_1$ is weakly harmonic on Ω_1. Secondly, since harmonic functions obey the MVP for all balls in Ω, if u is weakly harmonic on Ω and v is harmonic, $\pm(u - v)$ is weakly harmonic.

For any $x_0 \in B$, pick δ so that $B_\delta(x_0) \subset \Omega$. Let v be the solution of the Dirichlet problem for $B_\delta(x_0)$ with $v \restriction \partial B_\delta(x_0) = u \restriction \partial B_\delta(x_0)$. Thus, by the above, $\pm(u - v)$ are weakly harmonic on $B_\delta(x_0)$ and continuous on $\overline{B_\delta(x_0)}$. By the last proposition and $\pm(u - v) \restriction \partial B_\delta(x_0) = 0$, we see that $\pm(u - v) \leq 0$ on $\overline{B_\delta(x_0)}$, that is, $u = v$ on $\overline{B_\delta(x_0)}$, so $\Delta u(x_0) = 0$. $\qquad\square$

This is an analog of Morrera's theorem, and, as for that analog, implies a reflection principle for harmonic functions (see Problem 9).

Theorem 3.1.21. *Let $\{u_n\}_{n=1}^{\infty}$ be a sequence of functions harmonic on a region, Ω, in $\mathbb{R}^\nu$. Suppose u_∞ is a function with $\sup_{x \in K} |u_n(x) - u_\infty(x)| \to 0$ for each compact set $K \subset \Omega$. Then u_∞ is harmonic.*

Proof. For any $x_0 \in \Omega$, pick $\delta < \mathrm{dist}(x, \mathbb{R}^\nu \setminus \Omega)$ and $K = \overline{B_\delta(x_0)}$. Taking limits in (3.1.66) for u_n yields (3.1.66) for u_∞, so u_∞ is harmonic near x_0. $\qquad\square$

Theorem 3.1.22 (Cauchy Estimate for Harmonic Functions). *If u is harmonic in a neighborhood of $\overline{B_r(x_0)}$, then*

$$\left| \left(\frac{1}{\alpha!} D^\alpha u \right)(x_0) \right| \leq \left(\frac{C_\nu^{|\alpha|}}{r^{|\alpha|}} \right) \sup_{|x-x_0|=r} |u(x_0)| \qquad (3.1.80)$$

Proof. For $x_0 = 0$ and $r = 1$, this is immediate from (3.1.68) and the Poisson representation. For general x_0 and r, it follows by applying the special case to $v(y) = u(x_0 + ry)$. $\qquad\square$

Theorem 3.1.23 (Montel's Theorem for Harmonic Functions). *If $\{u_n\}_{n=1}^{\infty}$ is a sequence of harmonic functions in a region, Ω, with*

$$\sup_{n, x \in K} |u_n(x)| < \infty \qquad (3.1.81)$$

for every compact $K \subset \Omega$, then there is a subsequence, $\{u_{n_j}\}_{j=1}^{\infty}$ converging uniformly on every compact $K \subset \Omega$.

Proof. As in the analytic case (see Section 6.2 of Part 2A), pick compact $\{K_m\}_{m=1}^{\infty}$ with $K_m \subset K_{m+1}^{\mathrm{int}}$ so $\Omega = \bigcup_{m=1}^{\infty} K_m$. With $r_m = \frac{1}{2} \mathrm{dist}(K_m, \mathbb{R}^\nu \setminus K_{m+1}) > 0$ since $K_m \subset K_{m+1}^{\mathrm{int}}$, we have

$$\sup_{x \in K_m} |u_n'(x)| \leq C_1 r_m^{-1} \sup_{x \in K_{m+1}} |u_n(x)| \qquad (3.1.82)$$

so the $\{u_n\}_{n=1}^{\infty}$ are uniformly Lipschitz, and thus uniformly equicontinuous on each K_m. The Ascoli–Arzelà theorem (Theorem 2.3.14 of Part 1) completes the proof. $\qquad\square$

Theorem 3.1.24 (Vitali's Convergence Theorem for Harmonic Functions). *Let $\{u_n\}$ be a sequence of functions harmonic on a region, Ω, in $\mathbb{R}^\nu$. Suppose (3.1.81) holds for every compact $K \subset \Omega$ and that for all x in an open* *ball, B, in* Ω, $u_n(x)$ *converges. Then* u_n *converges to a harmonic function* u_∞ *uniformly on each compact subset of* Ω.

Remark. A ball is overkill—the proof shows one needs convergence on a set of uniqueness, that is, a set S so u harmonic in Ω and $u \restriction S \Rightarrow u \equiv 0$. For example, the boundary of a ball suffices.

Proof. By real analyticity and a connectedness argument (Problem 16), if $u \restriction B = 0$ and u is harmonic on Ω, then $u \equiv 0$.

By the last theorem, for some subsequence $u_{n(j)} \to u_\infty$ uniformly on compacts and any subsequence has a subsubsequence converging to a $\tilde{u}_\infty$. If u_n doesn't converge to u_∞, this implies there is another limit of some subsequence, that is, $\tilde{u}_\infty \not\equiv u_\infty$. But $\tilde{u}_\infty - u_\infty \equiv 0$ on B. By the above, $\tilde{u}_\infty - u_\infty \equiv 0$ on Ω. $\square$

Theorem 3.1.25. *Let* T *be a distribution on* $C_0^\infty(\Omega)$ *with* Ω *a region in* $\mathbb{R}^\nu$. *Suppose* $\Delta T = 0$. *Then there is a harmonic function,* u, *on* Ω *so that*

$$T(f) = \int f(x)u(x)\,d^\nu x \tag{3.1.83}$$

Remarks. 1. See Chapters 6 and 9 of Part 1 and Section 1.2.1 of this volume for a discussion of distributions and distributional derivatives.

2. In colloquial language, $T(x) = u(x)$.

Proof. By thinking of v, a harmonic function on $B_2(0)$, restricted to $B_1(0)$, as a Poisson kernel applied to $v \restriction B_\rho(0)$, $\frac{3}{2} < \rho < 2$, and smoothing in ρ, it is not hard to see (Problem 10) that for $x \in B_1(0)$, one has

$$v(x) = \int Q(x,y)v(y)\,d^\nu y \tag{3.1.84}$$

where $Q \in C^\infty(B_1(0) \times \{y \mid \frac{3}{2} < |y| < 2\})$ and $\bigcup_{x \in B_1} \text{supp}(Q(x,\cdot))$ is compact in $\{y \mid \frac{3}{2} < |y| < 2\}$.

Given $x_0 \in \Omega$, find r so $\overline{B}_{2r}(x_0) \subset \Omega$, and by convoluting $S \equiv T \restriction B_{2r}(x_0)$ with an approximate identity, find C^∞-functions, v_n, on $B_{2r}(x_0)$ so $v_n \to T$ as distributions and $\Delta v_n = 0$. By scaling and translation, suppose $x_0 = 0$ and $r = 1$.

By (3.1.84) and noting that $\int Q(x,y)v_n(y)\,d^\nu y \to T(Q(x,\cdot))$, we see that v_n converges pointwise on $B_1(0)$ and is uniformly bounded, so there is a harmonic limit v on $B_1(0)$. Thus, T is given locally by a harmonic function, so globally. $\square$

Theorem 3.1.26 (Removable Singularity Theorem). *Let* Ω *be a region in* $\mathbb{R}^\nu$ *and* $x_0 \in \Omega$. *If* u *is harmonic in* $\Omega \setminus \{x_0\}$ *and* u *is bounded on some* $B_\delta(x_0) \setminus \{x_0\}$ *in* Ω, *then* u *can be defined at* x_0 *in such a way that* u *is harmonic in* Ω.

Remark. The proof shows that u bounded near x_0 can be replaced by $\lim_{x \to x_0, \, x \neq x_0} |u(x_0)|/G_0(x - x_0) = 0$.

Proof. If $\nu = 2$, decrease δ if necessary so that $\delta < \frac{1}{2}$. Let v solve the Dirichlet problem on $B_\delta(x_0)$ with

$$v \upharpoonright \partial B_\delta(x_0) = u \upharpoonright \partial B_\delta(x_0) \tag{3.1.85}$$

For $\varepsilon > 0$, define $w_\varepsilon^\pm$ on $B_\delta(x_0) \setminus \{x_0\}$ by

$$w_\varepsilon^\pm(x) = \varepsilon G_0(x - x_0) \pm (u - v) \tag{3.1.86}$$

Since $u - v = 0$ on $\partial B_\delta(x_0)$,

$$\min_{x \in \partial B_\delta(x_0)} w_\varepsilon^\pm(x) > 0 \tag{3.1.87}$$

Since $u - v$ is bounded near x_0, as $x \to x_0$, $w_\varepsilon^\pm(x) \to \infty$.

Thus, by the maximum principle,

$$x \in B_\delta(x_0) \setminus \{x_0\} \Rightarrow w_\varepsilon^\pm(x) > 0 \tag{3.1.88}$$

Taking ε to zero,

$$x \in B_\delta(x_0) \setminus \{x_0\} \Rightarrow \pm(u(x) - v(x)) \geq 0 \Rightarrow u(x) = v(x) \tag{3.1.89}$$

So we see that if we set $u(x_0) = v(x_0)$, we get the desired result. $\qquad \square$

There is also a boundary version of this result:

Theorem 3.1.27. *Let u be a bounded harmonic on Ω, a region in $\mathbb{R}^\nu$. Fix $x_0 \in \partial\Omega$. Suppose $x_n \in \Omega$ and if $x_n \to x_\infty \in \partial\Omega \setminus \{x_0\}$ or $|x_n| \to \infty$ (in case Ω is unbounded), we have $\limsup u(x_n) \leq 0$. Then $u(x) \leq 0$ for all $x \in \Omega$. If $u(x_n) \to 0$ for all such sequences, $u(x) \equiv 0$.*

Remarks. 1. This says that single points in $\partial\Omega$ are negligible. We explore what sets are negligible in Section 3.6 and prove a stronger version of this in Theorem 3.6.13.

2. Example 3.4.2 will have an example where $u(x)$ has a limit at all points in $\partial\Omega$ except for a single point.

Proof. Since u is bounded and $G_0 > 0$, for any $\varepsilon > 0$, $u_\varepsilon(x) \equiv u(x) - \varepsilon G_0(x - x_0)$ has nonpositive $\limsup$ at all points, including x_0. Thus, by the maximum principle, $u_\varepsilon(x) \leq 0$. Taking $\varepsilon \downarrow 0$, $u(x) \leq 0$. The first result for u and $-u$ implies the second result. $\qquad \square$

Our final topic in this section involves Harnack's inequality and principle.

Theorem 3.1.28 (Harnack's Inequality for $B_1(0)$). *Let u be continuous on $\overline{B_1(0)}$ in $\mathbb{R}^\nu$ and harmonic in $B_1(0)$. Suppose $u > 0$. Then for $|x| < 1$,*

$$\frac{1 - |x|^2}{(1 + |x|)^\nu} \leq \frac{u(x)}{u(0)} \leq \frac{1 - |x|^2}{(1 - |x|)^\nu} \tag{3.1.90}$$

Proof. For any $\omega \in S^{\nu-1}$,

$$1 - |x| \leq |x - \omega| \leq 1 + |x| \tag{3.1.91}$$

(equality holds if x is parallel or antiparallel to ω). Thus,

$$\frac{1 - |x|^2}{(1 + |x|)^\nu} \leq P_{|x|}\left(\frac{x}{|x|}, \omega\right) \leq \frac{1 - |x|^2}{(1 - |x|)^\nu} \tag{3.1.92}$$

Multiply by $\frac{u(\omega)}{u(0)}$ (since $u > 0$, inequalities are preserved for all w) and integrate using $\int \frac{u(\omega)}{u(0)}\, d\omega = 1$ by the MVP. This yields (3.1.90). $\quad\square$

Theorem 3.1.29 (Harnack's Inequality). *Let Ω be a region in $\mathbb{R}^\nu$ and $K \subset \Omega$ a fixed compact set. Then there is a constant $C > 0$ so that for all positive harmonic functions u on Ω, one has for all $x, y \in K$ that*

$$C^{-1} \leq \frac{u(x)}{u(y)} \leq C \tag{3.1.93}$$

Proof. We proceed in three steps. First, we prove a bound of the form (3.1.93) for each pair of points $x_0, y_0 \in \Omega$. Second, we prove for each pair of points $x_0, y_0 \in \Omega$, there is a δ so a bound of the form (3.1.93) holds (fixed C) for all x, y with $x \in B_\delta(x_0)$, $y \in B_\delta(y_0)$. Then we use compactness of $K \times K$ to prove the general result.

Given x_0, y_0, we let γ be a curve in Ω from x_0 to y_0 and let $\varepsilon = \mathrm{dist}(\mathrm{Ran}(\gamma), \mathbb{R}^\nu \setminus \Omega)$. Then we find $x_0, \dots, x_n$ on γ so $x_n = y_0$ and $|x_{j+1} - x_j| < \frac{\varepsilon}{2}$ for $j = 0, 1, \dots, n - 1$. By Theorem 3.1.28, with

$$C_0 = \max\left(\frac{\frac{3}{4}}{(\frac{1}{2})^\nu}, \frac{(\frac{3}{2})^\nu}{\frac{3}{4}}\right)$$

we have

$$C_0^{-1} \leq \frac{u(x_{j+1})}{u(x_j)} \leq C_0$$

so (3.1.93) holds with $C = (C_0)^n$.

For any x_0, y_0, let C_1 be the C for which we have just proven (3.1.93) for x_0, y_0. Let $\delta = \frac{1}{2}\mathrm{dist}(\{x_0, y_0\}, \mathbb{R}^\nu \setminus \Omega)$. Then as above, (3.1.93) holds for $x \in B_\delta(x_0)$, $y \in B_\delta(y_0)$, with $C = C_1(C_0)^2$.

A simple compactness argument (Problem 8) goes from pairs of open sets to $K \times K$. $\quad\square$

Theorem 3.1.30 (Harnack's Principle). *Let $\{u_n\}_{n=1}^{\infty}$ be an increasing sequence of harmonic functions on a region Ω in $\mathbb{R}^{\nu}$, that is, for all n and all $x \in \Omega$,*

$$u_n(x) \leq u_{n+1}(x) \tag{3.1.94}$$

Then either

(i) *$u_n \to \infty$ uniformly on each compact $K \subset \Omega$, or*
(ii) *$u_\infty(x) = \sup_n u_n(x)$ is finite for all $x \in \Omega$ and $u_n \to u_\infty$ uniformly on compact subsets of Ω.*

Proof. By replacing u_n by $u_n - u_1 + 1$, we can suppose each $u_n > 0$. By Harnack's inequality for a pair of points $x_0, y_0 \in \Omega$, we see $u_n(x_0) \to \infty$ if and only if $u_n(y_0) \to \infty$. If $u_n(x_0) \to \infty$ at one point in a compact K, by Harnack's inequality, the convergence is uniform.

On the other hand, if $u_n(x_0)$ is bounded at a single point, by Harnack's inequality on $\{x_0\} \cup K$, u_n is bounded on each compact K. Since we have pointwise convergence at every point, Vitali's theorem implies uniform convergence on compacts. $\qquad\square$

Theorem 3.1.31. *Let $\{u_n\}_{n=1}^{\infty}$ be a sequence of positive harmonic functions on a region, Ω, in $\mathbb{R}^{\nu}$. Then either $u_n \to \infty$ uniformly on compact subsets or there is a subsequence $\{u_{n_j}\}_{j=1}^{\infty}$ and u_∞ harmonic on Ω so that u_{n_j} converges to u_∞ uniformly on compacts.*

Remark. This is reminiscent of the fact that analytic functions, F, on $\Omega \subset \mathbb{C}$, with $\operatorname{Re} F > 0$, are a normal family. Indeed, if $\Omega = \mathbb{D} \subset \mathbb{C}$, $u_n = \operatorname{Re} f_n$, and we can use the normal function result to obtain this theorem. So this is not merely an analogous result but an extension.

Proof. Pick $x_0 \in \Omega$. If $\lim_{n\to\infty} u_n(x_0) = \infty$ then, by Harnack's inequality, $u_n \to \infty$ uniformly on compacts. If not, then pick $n_1(j)$ so $\sup_j u_{n_1(j)}(x_0) < \infty$. By Harnack's inequality, $u_{n_1(j)}$ is bounded on compact subsets. So, by Theorem 3.1.23, a u_∞ and $u_{n_j} \to u_\infty$ on compacts exist. $\qquad\square$

Notes and Historical Remarks. The roots of "harmonic" function go back to the Greek word harmonia meaning in accord or to fit together, a term used in music especially to refer to harmonic notes. It was realized that harmonic notes were ones with frequencies with small rational multiples, typically small integer multiples and they are periodic. The functions sine and cosine were called harmonics. In 1867 Thomson (later Lord Kelvin) and Tait [**805**] introduced functions on the sphere $S^{\nu-1}$ in $\mathbb{R}^{\nu}$ that were restrictions of homogeneous polynomials obeying $\Delta u = 0$. When $\nu = 2$, these are the *circular harmonics*, $\sin(n\theta), \cos(n\theta)$. For $\nu \geq 3$, they called them *spherical harmonics*. Eventually the term harmonic was transferred to all solutions of $\Delta u = 0$.

As we've seen, Laplace, Legendre, and Poisson considered harmonic functions around 1800. The pioneers in the early nineteenth century were Gauss, Green, and Thomson, and slightly later, Dirichlet. Because much of the terminology comes from Riemann's inaugural dissertation [**679**] and Riemann learned the subject from Dirichlet, we have the names Dirichlet problem, Dirichlet boundary conditions, etc.

That solutions of $\Delta u = 0$ obey the MVP is a result of Gauss [**309**] in 1840 and is sometimes called Gauss' theorem. The converse, that the MVP for all small balls implies $\Delta u = 0$ is a result of Koebe [**477**] in 1906. Remarkably, it suffices for a function u on $\mathbb{R}^\nu$ to obey the MVP for each x but only two radii, r_1 and r_2, so long as r_1/r_2 is not an exceptional set determined by the Bessel function, $J_{\frac{1}{2}\nu-1}$. Netuka–Veselý [**604**] has a summary of various results of this genre.

For Riemann, harmonic functions and analytic functions were closely intertwined (see the Ahlfors quote at the start of Section 3.8) so the maximum principle and removable singularity theorems for both analytic and harmonic functions (at least for $\nu = 2$) go back to Riemann's dissertation [**679**].

That any positive harmonic function on all of $\mathbb{R}^\nu$ is constant is due to Bôcher [**86**] in 1902 although the theorem is often named after Picard's rediscovery [**645**] over twenty years later—there is often reference to the Liouville–Picard theorem. Our elegant proof involving the overlay of $B_R(0)$ and $B_R(x)$ is due to Nelson [**601**]. This unusual paper is only nine lines long (that really is lines, not pages) and is straight prose (or perhaps we should say, straight poetry) without a formula or the words definition or theorem.

The name Green's function is after his 1828 mémoire [**331**]. The term is generally used for fundamental solutions, especially of second-order elliptic equations, used in many contexts, so much so that there can be confusion. For example, in the theory of OPRL going beyond what we discuss in these volumes (see, for example, Simon [**757**]), one discusses Jacobi matrices, J. Since these are second-order difference equations, the mathematical physics community calls the matrix elements of its resolvent $\langle \delta_n, (J-z)^{-1}\delta_n \rangle$, the Green's function. But as we see in Section 3.7, there is the potential-theoretic Green's function (close to the one in this section) and these two "Green's functions" are very different objects!

The Poisson kernel and formula are named after Poisson [**658, 659**] although he only considered balls in $\mathbb{R}^3$. The notion of harmonic measure, at least on $\mathbb{C}$, goes back to Nevanlinna [**608**].

Harnack's inequality and Harnack's principle go back to his 1887 book [**364**]. Before the formulation of Montel's theorem twenty years later, many papers in function theory used Harnack-principle notions to get convergent subsequences.

We'll see later that if $\mathfrak{e} \subset \mathbb{C}$ is connected and simply connected, the exterior Green's function is connected to the Riemann map, φ, of $(\mathbb{C} \cup \{\infty\}) \setminus \mathfrak{e}$ to $\mathbb{D}$; see Theorem 3.6.24.

Problems

1. Define spherical coordinates inductively on $\mathbb{R}^\nu$ by supposing $(\rho, \theta_1, \ldots, \theta_{\nu-2})$ are coordinates for $\mathbb{R}^{\nu-1}$, and for $x = (y, x_\nu)$ with $y \in \mathbb{R}^{\nu-1}$ using

$$r = (\rho(y)^2 + x_\nu^2)^{1/2}, \qquad \cos(\theta_{\nu-1}(x)) = \frac{x_\nu}{r}$$

and using $r, \theta_1, \ldots, \theta_{\nu-1}$. For $\nu = 2$, we use $\cos(\theta_1) = x_2/r$, $r = (x_1^2 + x_2^2)^{1/2}$.

(a) Prove that

$$\frac{\partial}{\partial x_j} = \frac{x_j}{r}\frac{\partial}{\partial r} + \frac{1}{r}A_j \tag{3.1.95}$$

where A_j is a function of θ_j and $\frac{\partial}{\partial \theta_j}$.

(b) For $j \neq k$, define

$$J_{jk} = x_j\frac{\partial}{\partial x_k} - x_k\frac{\partial}{\partial x_j} \tag{3.1.96}$$

Prove that J_{jk} is a function of θ_j and $\frac{\partial}{\partial \theta_j}$, affine in $\frac{\partial}{\partial \theta_j}$.

(c) Prove that

$$J^2 \equiv \sum_{j<k}(J_{jk})^2 = \sum_{j=1}^{\nu}(r^2 - x_j^2)\frac{\partial^2}{\partial x_j^2}$$
$$- \sum_{j \neq k}x_j x_k\frac{\partial^2}{\partial x_k \partial x_k} - 2(\nu-1)r\frac{\partial}{\partial r} \tag{3.1.97}$$

(d) Prove that

$$r\frac{\partial}{\partial r} = \sum_{j=1}^{\nu}x_j\frac{\partial}{\partial x_j} \tag{3.1.98}$$

and that

$$r^2\frac{\partial^2}{\partial r^2} = \sum_{j=1}^{\nu}x_j^2\frac{\partial^2}{\partial x_j^2} + (\nu-1)r\frac{\partial}{\partial r} + \sum_{j \neq k}x_j x_k\frac{\partial^2}{\partial x_j \partial x_k} \tag{3.1.99}$$

(e) Conclude that

$$\Delta = \frac{\partial^2}{\partial r^2} + \frac{\nu - 1}{r} \frac{\partial}{\partial r} + \frac{1}{r^2} J^2 \tag{3.1.100}$$

proving (3.1.2) with $B = J^2$.

Remark. $-iJ$ is the quantum angular momentum and is self-adjoint. $B = J^2$ is negative definite (as is Δ).

2. Let $\Omega \subset \mathbb{R}^\nu$ be a region which is not bounded. Let u be harmonic in Ω. Prove that there exists $x_n \in \Omega$ with either $x_n \to x_\infty \in \partial\Omega$ or $|x_n| \to \infty$ so that $u(x_n) \to \sup_{y \in \Omega} u(y)$.

3. Develop a formula for G_Ω for Ω a hypercube in terms of reflections and translations.

4. Verify (3.1.60) and (3.1.61).

5. Let $\widetilde{P}_\omega(x)$ be given by (3.1.67).

   (a) Verify that for each ω in $S^{\nu-1}$, $\widetilde{P}_\omega(x_1, \dots, x_\nu)$ has an analytic continuation to $\mathcal{R} \equiv \{(x_1, \dots, x_\nu) \in \mathbb{C}^\nu \mid |x_j| \le \rho_\nu\}$ for some ρ_ν with uniform in ω bounds on $\widetilde{P}_\omega(x)$ in $\mathcal{R}$.

   (b) Verify (3.1.68). (*Hint*: Cauchy estimate.)

6. Verify the equality in (3.1.69).

7. Verify the details of Example 3.1.16.

8. Provide the final step in the proof of Theorem 3.1.29—the step going from pairs of balls to $K \times K$.

9. (*Reflection principle* for harmonic functions) Let $I \colon \mathbb{R}^\nu \to \mathbb{R}^\nu$ by $I(x_1, \dots, x_{\nu-1}, x_\nu) = (x_1, \dots, x_{\nu-1}, -x_\nu)$. Let $\mathbb{R}^\nu_\pm = \{x \mid \pm x_\nu > 0\}$ and $\pi_\nu = \{x \mid x_\nu = 0\}$. Suppose u is continuous on $\Omega \cap [\mathbb{R}_+ \cup \pi_\nu]$ and harmonic on $\Omega \cap \mathbb{R}_+$ with $u \restriction \Omega \cap \pi_\nu \equiv 0$. Prove that u has a harmonic extension to all of Ω. (*Hint*: See Theorem 5.5.3 in Part 2A.)

10. Provide the details of the construction of $Q(x, y)$ so (3.1.84) holds.

11. Let f be continuous on the unit sphere, $S^{\nu-1}$, in $\mathbb{R}^\nu$, $\nu \ge 3$. For $|x| > 1$, define

$$u_f(x) = \int \frac{|x|^2 - 1}{|x - \omega|} f(\omega) \, d\omega \tag{3.1.101}$$

Prove that u_f is harmonic on $\{x \mid |x| > 1\}$ and obeys $\lim_{|x| \to \infty} u_f(x) = 0$, $\lim_{x \to \omega} u_f(x) = f(\omega)$, and is the unique such function. (*Hint*: A useful preliminary will be to consider $u_{f \equiv 1}$ and show it is $|x|^{-(\nu-2)}$ by considering its asymptotics as $|x| \to \infty$ and using the form of spherically symmetric harmonic functions; see also Problem 15.)

12. This problem will prove that for all $y \in (0, \infty)$,

$$\int_{\mathbb{R}^{\nu-1}} \frac{y}{(y^2 + x^2)^{\nu/2}} \, d^{\nu-1}x = \frac{\sigma_\nu}{2} \tag{3.1.102}$$

(a) Prove the integral, call it $I(y)$, on the left of y is independent of y and conclude that

$$I(1) = \int_0^\infty e^{-y} I(y) \, dy \tag{3.1.103}$$

(b) Prove that

$$I(1) = \int_{\mathbb{R}_+^\nu} e^{-x_\nu} \frac{x_\nu}{|x|^\nu} \, d^\nu x \tag{3.1.104}$$

(c) Switch to polar coordinates with $\omega_\nu = x_\nu/|x|$ on $S^{\nu-1}$ to see that

$$I(1) = \int_{S_+^{\nu-1}} d\sigma \int_0^\infty e^{-r\omega_\nu} \omega_\nu \, dr \tag{3.1.105}$$

(d) Conclude

$$I(1) = \int_{S_+^{\nu-1}} d\sigma = \frac{\sigma_\nu}{2} \tag{3.1.106}$$

13. If $f \in C_0(\pi)$, verify u_f given by (3.1.52) is harmonic on $\mathbb{R}_+^{\nu-1}$ and is the unique such function obeying (3.1.53). (*Hint*: In confirming that as $y \downarrow 0$, $P_H(y, \cdot)$ is an approximate identity, you'll want to use Problem 12 which says $\int_{\mathbb{R}^{\nu-1}} P_H(y, x) \, d^{\nu-1}x = 1$.)

14. This problem will find the Fourier transform of the function $P_\nu^H(y, x)$ (of (3.1.51)) in the x variable.

(a) Let $f \in C_0^\infty(\pi)$ and suppose $u_f(x)$, $x \in \mathbb{R}_+^\nu$, solves the Dirichlet problem. Write

$$v_{f,y}(x) = u_f(y, x) \tag{3.1.107}$$

as a function on $\mathbb{R}^{\nu-1}$ and the $\nu - 1$-dimensional Fourier transform as $\hat{v}_{y,y}(k)$. Argue that the formulas

$$\frac{d^2}{dy^2} \hat{v}_{f,y}(k) = k^2 \hat{v}_{f,y}(k), \qquad \lim_{y \downarrow 0} \hat{v}_{f,y}(k) = \hat{f}(k) \tag{3.1.108}$$

should hold and conclude that one expects

$$\hat{v}_{f,y}(k) = e^{-|k|y} \hat{f}(k) \tag{3.1.109}$$

(b) Part (a) suggests one expects u_f to obey

$$u_f(y, x) = (2\pi)^{-(\nu-1)/2} \int e^{-|k|y} e^{ik\cdot x} \widehat{f}(k) \, d^{\nu-1}k \qquad (3.1.110)$$

Verify directly that the integral is harmonic in $\mathbb{R}_+^\nu$ and obeys (3.1.53).

(c) By letting f approach a delta function and using uniqueness of solutions, prove that

$$P_\nu^H(y, x) = (2\pi)^{-(\nu-1)} \int e^{-|k|y} e^{ik\cdot x} \, d^{\nu-1}k \qquad (3.1.111)$$

(d) By the Fourier inversion formula, show that

$$\int e^{-ik\cdot x} P_\nu^H(y, x) \, d^{\nu-1}k = e^{-|k|y} \qquad (3.1.112)$$

15. (Kelvin Transform) (a) Given u on an open set $\Omega \subset \mathbb{R}^\nu \setminus \{0\}$, define the Kelvin transform, Ku, by

$$(Ku)(x) = |x|^{2-\nu} u\left(\frac{x}{|x|^2}\right) \qquad (3.1.113)$$

Prove that if u is C^2, then so is Ku and

$$[\Delta(Ku)](x) = |x|^{-4} [K(\Delta u)](x) \qquad (3.1.114)$$

Conclude that if u is harmonic on Ω, Ku is harmonic on $\widetilde{\Omega} = \{x/|x|^2 \mid x \in \Omega\}$.

(b) Prove that $K(Ku) = u$.

(c) If $\nu \geq 3$ and u is harmonic on a region, Ω, with $0 \in \Omega$, prove that Ku defined on $\widetilde{\Omega} \setminus \{0\}$ obeys $\lim_{|x|\to\infty} (Ku)(x) = 0$.

(d) Let $\nu \geq 3$. If Ω is as in (c) and u is harmonic on $\widetilde{\Omega} \setminus \{0\}$ with $u(x) \to 0$ as $|x| \to \infty$, prove that $|x|^{\nu-2}(Ku)(x) \to 0$ as $|x| \to 0$ and conclude that 0 is a removable singularity for Ku. (*Hint*: See the remark after Theorem 3.1.26.)

16. Provide the connectedness argument needed in Theorem 3.1.24 to prove that if u is harmonic and vanishes on a ball, then $u \equiv 0$.

17. Let x_0 be a Lebesgue point of a set, B, of Lebesgue positive measure in $\mathbb{R}^\nu$. This problem will show how to pick $\{r_j, R_j\}_{j=1}^\infty$, so $R_n \leq 2^{-n}$, $R_j > r_j > R_{j+1}$, and (3.1.76) holds. Let

$$F(r, R) = \frac{|\{y \mid r < |y - x_0| < R, \, y \notin B\}|}{|\{y \mid r < |y - x_0| < R\}|}$$

(a) Assuming $\{r_j, R_j\}_{j=1}^{J-1}$ have been picked, prove you can pick R_J so that $R_J \leq 2^{-J}$ and $F(0, R_J) \leq 2^{-J-1}$.

(b) Then prove you can pick $r_J < R_J$ so $F(r_J, R_J) \leq 2^{-J}$.

The *Neumann problem* for a region Ω with smooth boundary, given $f \in C^1(\partial\Omega)$, is to find u harmonic in Ω so that u and ∇u have continuous extensions to $\overline{\Omega}$ with $n \cdot \nabla u = f$ where n is an inward-pointing norm. The next two problems concern the Neumann problem when $\Omega \subset \mathbb{C}$ is the Jordan region associated to a C^2 Jordan curve.

18. (a) Prove if the Neumann problem has a solution, then $\oint_{\partial\Omega} f(z)\, dz = 0$.

    (b) Prove that if there is a solution, it is not unique.

19. For the case $\Omega = \mathbb{D}$, show that if $f \in C^1(\partial\Omega)$ and $\int_0^{2\pi} f(e^{i\theta})\frac{d\theta}{2\pi} = 0$, then the Neumann problem has a solution given by explicit Fourier series and prove the solution is unique up to a constant.

3.2. Subharmonic Functions

Definition. A *subharmonic function* on a region, Ω, in $\mathbb{R}^\nu$ is a function $u \colon \Omega \to \mathbb{R} \cup \{-\infty\}$ so that

   (i) u is upper semicontinuous (henceforth usc).
   (ii) For each $x_0 \in \Omega$, there exists strictly positive sequence $\{r_n(x_0)\}_{n=1}^\infty \to 0$ so that $\overline{B_{r_n(x_0)}(x_0)} \subset \Omega$ and

$$u(x_0) \leq \int_{S^{\nu-1}} u(x_0 + r_n(x_0)\omega)\, d\omega \qquad (3.2.1)$$

   where $d\omega$ is normalized rotation invariant measure on $S^{\nu-1}$.
   (iii) For some x_0, $u(x_0) > -\infty$.

Remarks. 1. (iii) eliminates the case $u \equiv -\infty$ which most authors don't consider subharmonic, but some authors (e.g., Hayman–Kennedy [**370**]) do allow $u \equiv -\infty$.

2. usc functions are measurable and are bounded above on compact sets, so the integral in (3.2.1) is either convergent or diverges to $-\infty$. Indeed, we'll prove below (Corollary 3.2.13) that the integral is always finite, if $r_n > 0$.

3. We'll eventually show (see Corollary 3.2.12) that (3.2.1) holds for all r with $\overline{B_r(x_0)} \subset \Omega$.

4. (3.2.1) is called the *submean property* (SMP).

u is called *superharmonic* if $-u$ is subharmonic. Once we know (3.2.1) for all balls, we see u is harmonic if and only if it is both subharmonic and superharmonic.

Recall that f is usc if and only if for all a, $f^{-1}([-\infty, a))$ is open (equivalently, $f^{-1}([a, \infty))$ is closed). This implies that

$$x_n \to x_\infty \Rightarrow \limsup f(x_n) \le f(x_\infty) \tag{3.2.2}$$

and in a metric space, and thus for $\Omega \subset \mathbb{R}^\nu$, (3.2.2) implies usc. We'll say more about usc functions shortly.

In one dimension, u harmonic means $u'' = 0$, so u is affine. The submean property in one dimension is midpoint convexity (see Section 5.3 in Part 1), while u usc $\Rightarrow u$ measurable, so by Problem 2 of Section 5.3 in Part 1, in one dimension, subharmonic functions are exactly the convex functions. This shows that one reason for interest in subharmonic functions is as a higher-dimensional analog of one-dimensional convex functions.

We can now explain why usc is a good companion to the SMP. If f is usc, for any x_0 and $\delta > 0$, $\{x \mid f(x) < f(x_0) + \delta\}$ is open and contains x_0, so some $B_{\rho(x_0)}(x_0)$. Thus, if

$$\mathrm{Av}_r(f)(x_0) = \int f(x_0 + r\omega) \, d\omega \tag{3.2.3}$$

we have

$$0 < r < \rho(x_0) \Rightarrow \mathrm{Av}_r(f)(x_0) < f(x_0) + \delta \tag{3.2.4}$$

Thus,

$$\text{usc} \Rightarrow \limsup_{n \to \infty} \mathrm{Av}_{r_n(x_0)}(f)(x_0) \le f(x_0) \tag{3.2.5}$$

while

$$\text{SMP} \Rightarrow \liminf_{n \to \infty} \mathrm{Av}_{r_n(x_0)}(f)(x_0) \ge f(x_0) \tag{3.2.6}$$

so the two conditions together imply

$$u \text{ subharmonic} \Rightarrow u(x_0) = \lim_{n \to \infty} \mathrm{Av}_{r_n(x_0)}(f)(x_0) \tag{3.2.7}$$

Therefore, if u is subharmonic, changing just the value of $u(x_0)$ destroys subharmonicity, a statement false of SMP alone.

We next recall some useful properties of usc functions beyond the definition and equivalent (on a metric space) definition (3.2.2):

(1) If f is usc on a compact Hausdorff space, X, then f is bounded above and there is $x_0 \in X$ with

$$f(x_0) = \sup_{y \in X} f(y) \tag{3.2.8}$$

(see Theorem 2.3.13 of Part 1). For if x_n is such that $f(x_n) \to M = \sup_{y \in X} f(y)$, then we can find $x_\infty \in X$ and $x_{n_j} \to x_\infty$ so $f(x_\infty) \ge M$, which shows $M < \infty$ and $f(x_\infty) = M$.

(2) If $f_1, \ldots, f_k$ are usc, then so is $g = \max(f_1, \ldots, f_k)$ since $\{x \mid g(x) \ge a\} = \cup_{j=1}^k \{x \mid f_j(n) \ge a\}$ and finite unions of closed sets are closed.

It is easy to see the SMP for f_j implies it for g. Thus,

Proposition 3.2.1. *A max of a finite number of subharmonic functions is subharmonic.*

Remarks. 1. This is false for infinitely many f_j's since a countable union of closed sets may not be closed.

2. This proposition requires the result on the SMP holding for all r's, since otherwise the f_j might have SMP for different circles.

3. This shows it is easy to construct lots of subharmonic functions and, in some sense, there are many more subharmonic than harmonic functions, just as there are many more convex than affine functions.

(3) If $f_1 \geq \cdots \geq f_n \geq f_{n+1} \geq \ldots$ are a decreasing family of usc functions, then $g = \inf f_n = \lim f_n$ is usc since $\{x \mid g(x) \geq a\} = \cap_{n=1}^{\infty}\{x \mid f_n(x) \geq a\}$. By the monotone convergence theorem, if each f_n obeys SMP, then SMP is true for g (again using the fact that SMP holds for all balls in Ω). Thus,

Proposition 3.2.2. *A decreasing limit of subharmonic functions, f_n, on Ω is either subharmonic or is identically $-\infty$.*

(4) The last point implies that any decreasing limit of continuous functions is usc. Theorem 2.3.15 of Part 1 proves the converse: For any usc function, f, there are continuous functions, $\{f_n\}_{n=1}^{\infty}$, so $f_n \geq f_{n+1}$ and for all x, $f(x) = \inf f_n(x)$.

(5) On a metric space, (3.2.2) is equivalent to f usc, so if F is monotone increasing, continuous, and f usc, so is $F(f(\,\cdot\,))$.

Before turning to the general theory, let's discuss some examples. We begin with one way of constructing examples from other examples.

Theorem 3.2.3. *Let $F(x)$ be a real-valued convex, monotone increasing function on $(-\infty, \infty)$ and define $F(-\infty) = \lim_{t \to -\infty} F(t)$. Let u be a subharmonic function on a region, Ω, in $\mathbb{R}^\nu$. Then $F(u(x))$ is subharmonic also.*

Proof. By Jensen's inequality (see Theorem 5.3.14 of Part 1), the monotonicity of F and the SMP for u

$$F(u(x_0)) \leq F\left(\int u(x_0 + r_n(x_0)\omega)\,d\omega\right) \leq \int F(u(x_0 + r_n(x_0)\omega))\,d\omega \quad (3.2.9)$$

so $F(u(\,\cdot\,))$ has SMP. By point (5) above, $F(u(\cdot))$ is usc. $F((-\infty, \infty)) \subset \mathbb{R}$ and $u \not\equiv -\infty$ implies $F(u(\cdot)) \not\equiv -\infty$. $\square$

Theorem 3.2.4. *Let $G_\Omega(x,y)$ be the classical Green's function for a bounded region, or $G_0(x-y)$. Then for each y,*

$$u(x) = \begin{cases} -G_\Omega(x,y), & x \neq y \\ -\infty, & x = y \end{cases} \tag{3.2.10}$$

is subharmonic on Ω.

Proof. u is continuous away from y and $G_\Omega(x,y) \to \infty$ as $x \to y$, so u is continuous in the extended sense as a function with values in $[-\infty, \infty)$. If $x_0 \neq y$, then u is harmonic on $B_r(x_0)$ and continuous on $\overline{B_r(x_0)}$ if $r < |x_0 - y|$, so u has the MVP, and so SMP for such r's. If $x_0 = y_0$, $u(x_0) = -\infty$, so the SMP holds for all $r < \mathrm{dist}(y, \mathbb{R}^\nu \setminus \Omega)$. $\square$

In the other direction, we have: (Note the difference in sign from the last theorem. G_Ω, as extended below, is subharmonic near $\partial\Omega$ and superharmonic near y.)

Theorem 3.2.5. *Let $G_\Omega(x,y)$ be the classical Green's function for a region, Ω. Define $\widetilde{G}_\Omega(x,y)$ for $x \in \mathbb{R}^\nu \setminus \{y\}$, $y \in \Omega$, by*

$$\widetilde{G}_\Omega(x,y) = \begin{cases} G_\Omega(x,y), & x \in \overline{\Omega} \setminus \{y\} \\ 0, & x \notin \Omega \end{cases} \tag{3.2.11}$$

Then $\widetilde{G}_\Omega(\,\cdot\,, y)$ is subharmonic on $\mathbb{R}^\nu \setminus \{y\}$.

Proof. $\widetilde{G}$ is continuous, so usc. If $x_0 \notin \partial\Omega$, $\widetilde{G}_\Omega(x,y)$ is harmonic near x_0 and so obeys the MVP and so SMP at x_0. If $x_0 \in \partial\Omega$, $\widetilde{G}_\Omega(x_0, y) = 0$. So since $\widetilde{G}_\Omega(x,y) \geq 0$, the SMP holds at x_0. $\square$

The following is one reason that subharmonic functions are important.

Theorem 3.2.6. *Let $f(z)$ be a not identically zero analytic function on a region, Ω, in $\mathbb{C}$. Then*

$$u(z) = \log|f(z)| \tag{3.2.12}$$

is subharmonic.

Proof. If $g(z)$ is analytic and nonvanishing in a neighborhood of $\overline{\mathbb{D}_r(z_0)}$, then $\log|g(z_0)| = \mathrm{Re}\log(g(z))$ is harmonic. So if $f(z_0) \neq 0$, then $\log|f(z)|$ is harmonic near z_0.

If $f(z_0) = 0$, $f(z) = (z - z_0)^n g(z)$ for z near z_0 with $g(z)$ nonvanishing at z_0. Thus, by Theorem 3.2.4 (recall in $\mathbb{R}^2$, $-G_0(x) = \log|x|$),

$$\log|f(z)| = n\log|z - z_0| + \log|g(z)| \tag{3.2.13}$$

is subharmonic near z_0. $\square$

Example 3.2.7. $F(y) = e^{py}$ and $F_0(x) = x\chi_{[0,\infty)}(x)$ are convex and monotone on $[-\infty, \infty)$. Thus, if f is analytic on $\Omega \subset \mathbb{C}$,

$$|f(z)|^p = \exp(p\log|f(z)|), \qquad \log_+(|f(z)|) = F_0(\log|f(z)|) \qquad (3.2.14)$$

are subharmonic for all $p \in (0, \infty)$. $\square$

Example 3.2.8 (Antipotentials). A *potential* is a function on $\mathbb{R}^\nu$ ($\nu \geq 2$) of the form

$$\Phi_\mu(x) = \int G_0(x - y) \, d\mu(y) \qquad (3.2.15)$$

where μ is a positive measure of compact support. The integral is either convergent or equals $+\infty$ which we allow. For $N = 1, 2, \dots$, let

$$\Phi_\mu^{(N)}(x) = \int \min(N, G_0(x - y)) \, d\mu(y) \qquad (3.2.16)$$

By the continuity of $\min(N, G_0(x-y))$ in x and the dominated convergence theorem, $\Phi_\mu^N(x)$ is continuous. Since

$$\Phi_\mu(x) = \lim_N \Phi_\mu^{(N)}(x) = \sup_N \Phi_\mu^{(N)}(x) \qquad (3.2.17)$$

(this holds by the monotone convergence theorem), Φ_μ is lsc. Thus, $\Psi_\mu(x) \equiv -\Phi_\mu(x)$ is usc. Since $-G_0$ has the submean property, by Fubini's theorem, so does $\Psi_\mu(x)$. Thus, $\Psi_\mu(x)$, which we call an *antipotential*, is subharmonic.
                                                                                         $\square$

Remarks. 1. In fact (Problem 1), $-\Phi_\mu^{(N)}(x)$ is also subharmonic.

2. Φ_μ is the electrostatic potential of a positive particle in a density, μ, of positive charges. Ψ_μ is the electrostatic potential of a negative charge in such field and also of a gravitation attraction. Either could be called "a potential" and, indeed, the literature is split on whether the "potential generated by μ" is Φ_μ or $\Psi_\mu = -\Phi_\mu$. I use Φ_μ, although for this section, Ψ_μ is more natural. For lack of a better name for Ψ_μ, I have used the nonstandard "antipotential." Φ_μ will be a major player in Section 3.6.

Example 3.2.9 (Discontinuous Subharmonic Functions). From the examples so far, the reader might think subharmonic functions are continuous in extended sense (i.e., allowing the value $-\infty$). Here is an example to show this is not so. In $\mathbb{R}^\nu$, let $x_n = (\frac{1}{n}, 0, \dots)$ and let $\lambda_n > 0$ be such that $\sum_{n=1}^\infty \lambda_n < \infty$. Let Ψ be the antipotential of $\mu = \sum_{n=1}^\infty \lambda_n \delta_{x_n}$ or

$$\Psi(x) = -\sum_{n=1}^\infty \lambda_n G_0(x - x_n) \qquad (3.2.18)$$

Then $\Psi(x_n) = -\infty$ but

$$\Psi(0) = -\sum_{n=1}^\infty \lambda_n G_0(x_n) \qquad (3.2.19)$$

may be finite. For example, in ν dimensions, take $\lambda_n = n^{-\nu}$ so the sum is $-C \sum_{n=1}^{\infty} n^{-2}$ (or $-C \sum_{n=1}^{\infty} n^{-2} \log(n^{-1})$ in $\nu = 2$). Lest you think this is associated to Ψ taking the value $-\infty$, note that the max of two subharmonic functions is subharmonic. So if $-\Psi(0) \equiv \alpha < \infty$,

$$\widetilde{\Psi}(x) = \max(-2\alpha, \Psi(x)) \tag{3.2.20}$$

has $\widetilde{\Psi}(0) = -\alpha$, but $\widetilde{\Psi}(x_n) = -2\alpha$ with $x_n \to 0$ and $\widetilde{\Psi}$ bounded.

A closely related example is to let μ be the measure in $\mathbb{R}^{\nu}$ ($\nu \geq 3$)

$$\mu = (\nu - 2)\sigma_\nu x_1^{\nu-2} \chi_{[0,1]}(x_1) \delta((x_2, \ldots, x_\nu)) \, d^\nu x \tag{3.2.21}$$

that is, supported on $\{x \mid x_2 = \cdots = x_\nu, \, 0 < x_1 < 1\}$ with weight $x_1^{\nu-2} \, dx_1$ on that line. Then

$$\Psi_\mu((x_1, 0, \ldots, 0)) = - \int_0^1 (x_1 - y)^{-(\nu-2)} y^{\nu-2} \, dy$$

so $\Psi_\mu = -\infty$ for $0 < x_1 \leq 1$ while $\Psi_\mu = -1$ for $x_1 = 0$. This is discontinuous at $x = 0$ and so is $\max(-2, \Psi_\mu(x))$ which is also bounded. This example, due to Lebesgue [515], will reappear in Section 3.4; see Example 3.4.2. $\square$

As in the harmonic case, a key is the maximum principle:

Theorem 3.2.10 (Maximum Principle). *Let u be subharmonic on a region, $\Omega \subset \mathbb{R}^{\nu}$. Then there exists $\{x_n\}_{n=1}^{\infty}$ in Ω with either $|x_n| \to \infty$ or $x_n \to x_\infty \in \partial\Omega$ so that*

$$u(x_n) \to \sup_{y \in \Omega} u(y) \tag{3.2.22}$$

Remarks. 1. Once we have the SMP for all circles (see Corollary 3.2.12), we can prove (Problem 2) the version that says if the maximum is taken at an interior point, then u is constant.

2. Unlike for harmonic functions, there is no minimum principle for subharmonic functions.

Proof. Let

$$K = \big\{x \in \Omega \mid u(x) = \sup_{y \in \Omega} u(y) \equiv \alpha\big\} \tag{3.2.23}$$

If $K = \emptyset$ (in particular, if α is ∞), there is $x_n \in \Omega$ so that $u(x_n) \to \alpha$. By passing to a subsequence, we can suppose $|x_n| \to \infty$ or $x_n \to x_\infty \in \overline{\Omega}$. If $x_\infty \in \Omega$, $u(x_\infty) \geq \alpha$ by usc, so $x_\infty \in K$ and so K is nonempty. Thus, $x_\infty \in \partial\Omega$ and we have the desired sequence.

If $K \neq \emptyset$ and if K is unbounded, we can find $\{x_n\}_{n=1}^{\infty}$ with $|x_n| > n$ and $u(x_n) = \alpha$, providing the needed sequence.

If K is bounded, it is closed in Ω. If $\operatorname{dist}(K, \partial\Omega) = 0$, we can find $x_n \in K$ with $x_n \to x_\infty \in \partial\Omega$, providing the required sequence.

Finally, if K is bounded but $\text{dist}(K, \partial\Omega) = \delta > 0$, then K is compact and there exists $x \in K$, $y \in \partial\Omega$, so $\text{dist}(x, y) = \delta$. Pick r so $B_r(x) \subset \Omega$ and so that the SMP holds for B_r. Since $\alpha \equiv \int u(x + r\omega) \, d\omega$ and $u \le \alpha$, we see $u(x + r\omega) = \alpha$ for $d\omega$-a.e. ω, and so by usc for all ω. In particular, the point, ω, on $\partial B_r(x)$ on the line segment from x to y is in K. But $\text{dist}(\omega, y) = \delta - r$, so $\text{dist}(K, \partial\Omega) \le \delta - r$. This contraction shows this case cannot occur. $\square$

This has a corollary with many consequences:

Theorem 3.2.11. *Let u be subharmonic on a region, Ω in $\mathbb{R}^\nu$. Suppose $B_r(x_0) \subset \Omega$. Then for all $0 \le \rho < r$, we have*

$$u(x_0 + \rho\tau) \le \int P_{\rho/r}(\tau, \omega) u(x_0 + r\omega) \, d\omega \qquad (3.2.24)$$

where P is the Poisson kernel for $B_1(0)$.

Proof. Suppose $f \in C(S^{\nu-1})$ with

$$f(\omega) \ge u(x_0 + r\omega) \qquad (3.2.25)$$

for all ω. Define on $B_r(x_0)$

$$v(x_0 + \rho\tau) = \int P_{\rho/r}(\tau, \omega) f(\omega) \, d\omega \qquad (3.2.26)$$

that is, v is harmonic on $B_r(x_0)$ and solves the Dirichlet problem for f.

Then $q \equiv u - v$ is subharmonic on $B_r(x_0)$ so that $x_n \to x_\infty$, with $x_\infty = x_0 + r\omega_0 \in \partial B_r(x_0)$, implies

$$\limsup q(x_n) = \limsup u(x_n) - \lim v(x_n) \le u(x_\infty) - f(\omega_0) \le 0 \quad (3.2.27)$$

by (3.2.25).

By the maximum principle, $q \le 0$ so

$$u \le v \qquad (3.2.28)$$

Since $u(x_0 + r\omega)$ is usc, there are continuous $f_n \in C(S^{\nu-1})$ so $f_n \ge f_{n+1}$ with $f_n(\omega) \searrow u(x_0 + r\omega)$. Using the monotone convergence theorem for v_n (corresponding to f_n), we obtain (3.2.24). $\square$

Corollary 3.2.12. *Let u be a subharmonic function on a region, Ω, in $\mathbb{R}^\nu$. Then for every $x_0 \in \Omega$ and $r \in (0, \text{dist}(x_0, \Omega))$, we have*

$$u(x_0) \le \int_{S^{\nu-1}} u(x_0 + r\omega) \, d\omega \qquad (3.2.29)$$

$$u(x_0) \le \frac{1}{r^\nu \tau_\nu} \int u(x) \, d^\nu x \qquad (3.2.30)$$

where $\tau_\nu = \sigma_\nu / \nu$ is the volume of the unit ball in $\mathbb{R}^\nu$. Moreover,

$$r \mapsto \int_{S^{\nu-1}} u(x_0 + r\omega) \, d\omega \qquad (3.2.31)$$

is monotone increasing in r.

Remark. For now, the integrals could be $-\infty$.

Proof. (3.2.29) is (3.2.24) for $\rho = 0$. (3.2.30) is obtained from (3.2.29) by integrating in r' from 0 to r (using $d^\nu x = r^{\nu-1}\sigma_\nu \, dr d\omega$).

If $0 < \rho < r$, by (3.2.24),

$$\int u(x_0 + \rho\tau) \, d\tau \le \int \left[u(x_0 + r\omega) \int P_{\rho/r}(\tau, \omega) \, d\tau \right] d\omega$$
$$= \int u(x_0 + r\omega) \, d\omega \qquad (3.2.32)$$

on account of (3.1.64). This proves the monotonicity. $\qquad\square$

Corollary 3.2.13. *If u is subharmonic on a region, Ω, then u is locally L^1 (i.e., for any compact $K \subset \Omega$, $\int_K |u(x)| \, d^\nu x < \infty$) and for all x_0 and $0 < r < \text{dist}(x_0, \partial\Omega)$,*

$$\int u(x_0 + r\omega) \, d\omega > -\infty \qquad (3.2.33)$$

Proof. By a covering argument, to show u is locally L^1, it suffices to prove that for every x_0, there is an r with $\overline{B_r(x_0)} \subset \Omega$ and

$$\fint_{B_r(x_0)} u(y) \, d^\nu y > -\infty \qquad (3.2.34)$$

Let Q be the set of such x_0. If $u(x_0) \ne -\infty$, (3.2.34) holds for any $r < \text{dist}(x_0, \partial\Omega)$ by (3.2.30). Since $u \not\equiv -\infty$, this means Q is nonempty. Clearly, Q is open since $B_{r-|x-x_0|}(x) \subset B_r(x_0)$, so $x_0 \in Q$ with (3.2.34) $\Rightarrow B_r(x_0) \subset Q$.

Suppose $\{x_n\}_{n=1}^\infty \subset Q$ with $x_n \to x_\infty \in \Omega$. Let $\rho = \text{dist}(x_\infty, \partial\Omega)$ (or 1 if $\Omega = \mathbb{R}^\nu$). Pick x_n so $\text{dist}(x_n, x_\infty) < \rho/5$. Then there is $\tilde{x}$ in $B_{\rho/5}(x_n)$ with $u(\tilde{x}) > -\infty$. Thus, $B_{3\rho/5}(\tilde{x}) \subset \Omega$. So by (3.2.30), for all $\varepsilon > 0$, (3.2.2) holds for $B_r(x_0)$ replaced by $B_{3\rho/5-\varepsilon}(\tilde{x})$. Then if $r_0 < \rho/5 \le 3\rho/5 - |\tilde{x} - x_\infty|$, (3.2.34) holds for $B_{r_0}(x_0)$, that is, $x_\infty \in Q$. Therefore, Q is closed, open, and nonempty, so all of Ω.

By the monotonicity of $r \to \int u(x_0 + r\omega) \, d\omega$, we have

$$\int u(x_0 + r\omega) \, d\omega \ge \frac{1}{\tau_\nu r^\nu} \int_{B_r(x_0)} u(y) \, d^\nu y \qquad (3.2.35)$$

so u locally L^1 implies (3.2.33). $\qquad\square$

Now that we know u is locally L^1, it defines an ordinary distribution via ($f \in C_0^\infty(\Omega)$)

$$f \mapsto \int u(x) f(x) \, d^\nu x \qquad (3.2.36)$$

We are heading towards a major goal that $\Delta u \ge 0$ and conversely, any distribution T with $\Delta T \ge 0$ is a.e. a subharmonic function. We'll get this by showing that if u is subharmonic, then it is, in a sense, a decreasing limit of C^∞ subharmonic functions.

Lemma 3.2.14. *Let $j(x) \in C_0^\infty(B_1(0))$, $j \geq 0$, $\int j(y)\, d^\nu y = 1$, with j a function of $|x|$ only and $0 \leq |x| \leq |y| \Rightarrow j(y) \leq j(x)$. Let*

$$j_\lambda(x) = \lambda^{-\nu} j(\lambda^{-1} x) \tag{3.2.37}$$

Suppose u is subharmonic on $\Omega \subset \mathbb{R}^\nu$ and $B_r(x_0) \subset \Omega$. For $0 < \lambda < r$, define for $x \in B_{r-\lambda}(x_0)$,

$$(j_\lambda * u)(x) = \int j_\lambda(x-y) u(y)\, d^\nu y \tag{3.2.38}$$

Then $j_\lambda * u$ is subharmonic and C^∞ in x and $0 < \mu < \lambda \Rightarrow j_\mu * u \leq j_\lambda * u$, where both are defined.

Proof. Since j is C^∞, so is $j_\lambda * u$. Each $u(x-y)$ obeys SMP in x, so we see $j_\lambda * u$ obeys SMP. Write $j(x) = J(|x|)$ and change variables to find

$$(j_\lambda * u)(x) = \int j(w) u(x - \lambda w)\, d^\nu w$$

$$= \int_0^1 J(r) \left(\int_{S^{\nu-1}} u(x + \lambda r\omega)\, d\omega \right) \sigma_\nu r^{\nu-1}\, dr \tag{3.2.39}$$

which is monotone in λ by Corollary 3.2.12. $\qquad\square$

Given Ω, pick K_j compact so $K_j \subset K_{j+1}^{\mathrm{int}}$ and $\bigcup_{j=1}^\infty K_j = \Omega$. Then pick $r_1 > r_2 > \ldots$ so that $\mathrm{dist}(K_j, \partial\Omega) > r_j$. Let j_λ be as in the last lemma. Define u_j on K_j by

$$u_j = j_{r_j} * u \tag{3.2.40}$$

By the lemma, $u_j \geq u_{j+1}$ and, by (3.2.7), $\lim u_j(x) = u(x)$ for each x. We summarize:

Theorem 3.2.15. *Any subharmonic function is a decreasing limit pointwise and a distributional limit of a decreasing sequence of C^∞ subharmonic functions in the sense that there are C^∞-functions $u_n(x)$ defined and subharmonic on K_n^{int}, so $u_{n+1} \leq u_n$ on K_n^{int}, and for all x,*

$$u(x) = \lim u_n(x) \tag{3.2.41}$$

Corollary 3.2.16. *If u is subharmonic, then, as a distribution, $\Delta u \geq 0$.*

Proof. By Theorem 3.2.15, it suffices to prove this in case u is C^∞ and subharmonic. For any C^∞-function using

$$u(x) = u(x_0) + (x - x_0) \cdot \nabla u(x)$$

$$+ \sum_{j,k=1}^\nu (x-x_0)_j (x-x_0)_k \frac{\partial^2}{\partial x_i \partial x_j} u(x_0) + O(|x-x_0|^3) \tag{3.2.42}$$

we prove that

$$\int u(x_0 + r\omega)\,d\omega = u(x_0) + \frac{1}{\nu} r^2 \Delta u(x_0) + O(r^3) \tag{3.2.43}$$

so SMP $\Rightarrow (\Delta u)(x_0) \geq 0$. $\square$

Remark. We used $\int \omega_i\,d\omega = 0$, $\int \omega_i\omega_j\,d\omega = \frac{1}{\nu}\delta_{ij}$ (using $\omega_i \to -\omega_i$ symmetry and $\int(\sum_{i=1}^{\nu}\omega_i^2)\,d\omega = 1$).

Corollary 3.2.17. *Let u be subharmonic on a region, $\Omega \subset \mathbb{R}^\nu$. For $x_0 \in \Omega$ and $y = \log r$ ($\nu = 2$) or $-r^{2-\nu}$ ($\nu \geq 3$) for $y \in (-\infty, \rho)$ (ρ such that $r(\rho) = \mathrm{dist}(x_0, \partial\Omega)$), define*

$$f(y) = \int u(x_0 + r(y)\omega)\,d\omega \tag{3.2.44}$$

(i.e., $r(y) = e^y$ if $\nu = 2$ and $(-y)^{1/(2-\nu)}$ if $\nu \geq 3$). Then f is convex in y, and so continuous in r.

Proof. By taking limits and using Theorem 3.2.15, we can suppose u is C^∞. If

$$g(r) = \int u(x_0 + r\omega)\,d\omega \tag{3.2.45}$$

by $\Delta u \geq 0$, this implies (via (3.1.2)) that $r^{\nu-1}(r^{\nu-1}g')' \geq 0$. Since $r^{\nu-1}\frac{d}{dr} = \frac{d}{dy}$, this implies $f''(y) \geq 0$, so f is convex. $\square$

Finally, there is a converse to Corollary 3.2.16. We begin with

Proposition 3.2.18. *Let μ be a measure of compact support $K \subset \Omega$ a region in $\mathbb{R}^\nu$. Let Ψ_μ be its antipotential. As a distribution on Ω,*

$$\Delta\Psi_\mu = \mu \tag{3.2.46}$$

Proof. As noted in (3.1.28), $\Delta(-G_0) = \delta$ in the sense that for any $f \in C_0^\infty(\Omega)$,

$$\int (\Delta f)(y)[-G_0(x-y)]\,d^\nu y = f(x) \tag{3.2.47}$$

Integrating $d\mu(x)$ shows

$$\int (\Delta f)(y)\Psi_\mu(y)\,d^\nu y = \int f(x)\,d\mu(x) \tag{3.2.48}$$

which is (3.2.46). $\square$

Theorem 3.2.19. *If Ω is a region in $\mathbb{R}^\nu$ and T a distribution on Ω with $\Delta T \geq 0$, then for any compact $K \subset \Omega$, there is a measure μ on K with $\mu(\partial K) = 0$ and harmonic function, f, on K^{int} so that as distributions on K^{int},*

$$T \upharpoonright K^{\mathrm{int}} = \Psi_\mu + f \tag{3.2.49}$$

In particular T is subharmonic. μ is uniquely determined by (3.2.49) *and* $\text{supp}(\mu) \subset K$, $\mu(\partial K) = 0$.

Remarks. 1. By $T \upharpoonright K^{\text{int}}$, we mean the distribution on $C_0^\infty(K^{\text{int}})$ given by $g \mapsto T(g)$.

2. By Corollary 3.2.16, this says that any subharmonic function has the form (3.2.49) pointwise on K^{int}. This is called the *Riesz decomposition theorem.*

Proof. Let $\Delta T = \nu$ which is a σ-finite measure on Ω so it has a finite restriction to K. Let μ be the measure

$$\mu(A) = \nu(K^{\text{int}} \cap A) \tag{3.2.50}$$

which is a finite measure of compact support. Since

$$\Delta(T \upharpoonright K^{\text{int}}) = \mu \tag{3.2.51}$$

we have, by Proposition 3.2.18, that

$$\Delta(T \upharpoonright K^{\text{int}} - \Psi_\mu) = 0 \tag{3.2.52}$$

so the difference is a harmonic function by Theorem 3.1.25.

The uniqueness follows from the fact that on K^{int}, $\mu = \Delta T$. $\qquad\square$

Remark 2 is so important we state it as a theorem:

Theorem 3.2.20 (Riesz Decomposition Theorem). *Let u be subharmonic on a region, Ω, in $\mathbb{R}^\nu$. Let $S \subset \Omega$ be a region with $\overline{S} \subset \Omega$. Let*

$$\mu = \Delta u \upharpoonright \overline{S} \tag{3.2.53}$$

Then with Ψ_μ the antipotential for μ, we have

$$u = \Psi_\mu + f \tag{3.2.54}$$

where f is harmonic on S.

Notes and Historical Remarks. The formal definition of subharmonic including the fact that usc is the correct continuity condition and the actual name appeared in a 1926 paper of F. Riesz [**684**]. The paper had an important second part [**686**] published four years later although clearly planned since Riesz included "first part" in the title of the first paper.

Of course, given work on potential theory, subharmonicity had been widely used prior to Riesz' paper but Riesz realized both that the submean property and usc were the critical criteria (for example, Perron [**639**] considered continuous function less than their Poisson modification in the language of Section 3.4). And at that time, papers were more likely to use "solution of $\Delta u = 0$" than harmonic. Riesz specifically mentioned Poincaré [**656**] and

Hartogs [**368**] as prior motivating works and indicated he wanted to generalize Hardy's results [**352**] on convexity and monotonicity of averages of $\log|f|$ for analytic functions, f.

Riesz' paper required the subharmonic property for all small r. Very soon after Riesz' papers, Littlewood wrote a paper [**541**] that proved Corollary 3.2.12, i.e., it sufficed to have SMP for a sequence of r_n's going to zero.

The Riesz decomposition theorem is from Riesz [**686**]. Riesz himself called a totally different result of his the decomposition theorem (*Zerlegungssatz*), namely the idea that a decomposition of the spectrum of an operator into disjoint closed subsets induces a decomposition of the underlying space on which the operator acts (Theorem 2.3.3 of Part 4).

Riesz emphasized that subharmonicity was a higher-dimensional analog of one-dimensional convexity, a theme explored, in particular, by Hörmander [**399**].

Problems

1. Prove that the function $\Phi_\mu^{(N)}$ of (3.2.16) is superharmonic.

2. Given Corollary 3.2.12, prove that if u is subharmonic on a region Ω, and $\sup_{w \in \Omega} u(w) = u(z_0)$ for $z_0 \in \Omega$, then u is constant.

3.3. Bonus Section: The Eremenko–Sodin Proof of Picard's Theorem

In this section, we'll use subharmonic function theory to prove the little Picard theorem, namely,

Theorem 3.3.1. *Let f be an entire, nonconstant analytic function. Then f takes every value with at most one exception.*

As usual, we need only show that if f is never 0 or 1, then there is a contradiction. Let

$$U_1(z) = \log|f(z)|, \quad U_2(z) = \log\left|\frac{f(z)-1}{\sqrt{2}}\right|, \quad U_3(z) = 0 \qquad (3.3.1)$$

$$V(z) = \log\sqrt{|f(z)|^2 + 1} \qquad (3.3.2)$$

Since $f, f-1$ are nonvanishing, the U_j are harmonic. Since $\Delta = 4\bar\partial\partial$, a simple calculation shows that

$$\Delta V(z) = \frac{2|f'(z)|^2}{(|f(z)|^2 + 1)^2} \qquad (3.3.3)$$

so V is subharmonic.

We want to first control $V - U_j \vee U_k$ where, as usual, for real-valued functions, $(f \vee g)(x) = \max(f(x), g(x))$. For use in the Problems, we'll do this in greater generality.

Lemma 3.3.2. *Let $a, b \in \mathbb{C}^2$ with $\|a\| = \|b\| = 1$. Let P_a, P_b be the orthogonal projections onto the real subspace of $\mathbb{C}$ containing a and b, respectively. Then for all $\alpha \in \mathbb{C}^2$,*

(a) $(\alpha, P_a \alpha) + (\alpha, P_b \alpha) \geq (1 - |(a,b)|)\|\alpha\|^2$ $\qquad\qquad$ (3.3.4)

(b) $\displaystyle\max_{x=a,b}(\alpha, P_x \alpha) \leq \|\alpha\|^2 \leq \frac{2}{1 - |(a,b)|} \max_{x=a,b}(\alpha, P_x \alpha)$ $\qquad$ (3.3.5)

Remark. Since explicit constants aren't important, one could instead just use the equivalence of all norms on a finite-dimensional space (see Theorem 5.1.3 of Part 1).

Proof. Pick an orthonormal basis in $\mathbb{C}^2$, e_1, e_2 so $e_1 = a$. Then change b by $e^{i\theta}$ so $b_1 \geq 0$, that is, so $a \cdot b \geq 0$, and pick e_2 so $b_2 > 0$. Thus, if $|a \cdot b| = \cos\theta$, we have $a = \binom{1}{0}$, $b = \binom{\cos\theta}{\sin\theta}$. Thus,

$$A \equiv P_a + P_b = \begin{pmatrix} 1 + \cos^2\theta & \cos\theta\sin\theta \\ \cos\theta\sin\theta & \sin^2\theta \end{pmatrix} \qquad (3.3.6)$$

Thus, $\mathrm{Tr}(A) = 2$, $\det(A) = \sin^2\theta$, so the eigenvalues are $1 \pm \cos\theta$ and $A \geq (1 - \cos\theta)\mathbb{1}$, which is (3.3.4).

The first inequality in (3.3.5) is trivial and the second follows from (3.3.4) since

$$(\alpha, P_a\alpha) + (\alpha, P_b\alpha) \leq 2\max_x(\alpha, P_x\alpha) \qquad (3.3.7)$$

$\qquad\qquad\qquad\qquad\qquad\qquad\qquad\qquad\qquad\qquad\qquad\qquad\qquad\qquad$ $\square$

Proposition 3.3.3. *We have*

(a) $0 \leq V - U_1 \vee U_2 \leq \frac{1}{2}\log 2 + \frac{1}{2}\log\left(\dfrac{\sqrt{2}}{\sqrt{2}-1}\right)$ $\qquad\qquad$ (3.3.8)

(b) $0 \leq V - U_2 \vee U_3 \leq \frac{1}{2}\log 2 + \frac{1}{2}\log\left(\dfrac{\sqrt{2}}{\sqrt{2}-1}\right)$ $\qquad\qquad$ (3.3.9)

(c) $0 \leq V - U_1 \vee U_3 \leq \frac{1}{2}\log 2$ $\qquad\qquad\qquad\qquad\qquad\qquad$ (3.3.10)

Proof. (3.3.5) says

$$0 \leq \tfrac{1}{2}\log\|\alpha\|^2 - \max_{x=a,b}[\tfrac{1}{2}\log(\alpha, P_x\alpha)] \leq \tfrac{1}{2}\log 2 + \tfrac{1}{2}\log\left(\frac{1}{1 - |(a,b)|}\right)$$
$$(3.3.11)$$

If $\alpha = (f(z), 1)$, $a = \binom{1}{0}$, $b = \binom{1/\sqrt{2}}{-1/\sqrt{2}}$, $c = \binom{0}{1}$, then $U_1 = \frac{1}{2}\log(\alpha, P_a\alpha)$, $U_2 = \frac{1}{2}\log(\alpha, P_b\alpha)$, $U_3 = \frac{1}{2}\log(\alpha, P_c\alpha)$, and (a)–(c) follow from (3.3.11). $\quad\square$

Here is where we'll get the contradiction:

Proposition 3.3.4. *Let u_1, u_2, u_3 be harmonic in a region, Ω, and*

$$v = u_1 \vee u_2 = u_2 \vee u_3 = u_1 \vee u_3 \qquad (3.3.12)$$

Then v is harmonic.

Proof. It cannot happen that $u_1 > u_2 \vee u_3$, since then $u_1 \vee u_2 > u_2 \vee u_3$. That is, at each z, at least two u's are equal. Thus, if $B_{jk} = \{z \mid u_j(z) = u_k(z)\}$, then

$$\Omega = B_{12} \cup B_{23} \cup B_{13} \qquad (3.3.13)$$

so at least one of the three sets must have positive measure, say B_{12}. By Theorem 3.1.18, $u_1 = u_2$ on all of Ω, so $v = u_1 \vee u_2 = u_1$ is harmonic. $\qquad \square$

The basic idea will be to look at V and $\{U_j\}_{j=1}^{3}$ on suitable large disks, scaled in distance and function size, that is, $v_n(\zeta) = V(z_n + r_n \zeta)/M_n$ for $r_n \to \infty$, $z_n \to \infty$, and $M_n \to \infty$. Because of the $M_n \to \infty$, the bounds in Proposition 3.3.3 become equalities and Proposition 3.3.4 will get us the required contradiction. For this to work, we need to be sure the limiting v isn't harmonic.

There are two potential difficulties that could make v harmonic in the limit. First, V could have a harmonic piece that overwhelms the subharmonic piece, for example, $V(z) = |z| \cos \theta + \log \sqrt{1 + |z|^2}$. Any scaled limit will lose the $\log \sqrt{1 + |z|^2}$ term. We'll get around this by restricting V to a large disk using the Riesz decomposition and subtracting the harmonic part; that is, we'll have v_n's that are antipotentials of normalized measures, ν_n, in $\overline{\mathbb{D}}$.

We'll be able to pick a subsequence so $\nu_n \to \nu_\infty$ weakly in $C(\overline{\mathbb{D}})$ and we'll find v_n has a limit, v_∞, which is the antipotential of ν_∞. V_∞ will be harmonic on $\mathbb{D}$ if and only if ν_∞ is concentrated on $\partial \mathbb{D}$, so we'll need to pick our disks to avoid that. This is the purpose of Proposition 3.3.7 below. For this to work at all, we need to know:

Lemma 3.3.5. *If V is given by (3.3.2) and f is never zero and is non-constant, then*

$$\int_{\mathbb{C}} (\Delta V)(z) \, d^2 z = \infty \qquad (3.3.14)$$

Remark. We'll prove this from f everywhere nonvanishing. Problem 1 has a proof which shows that, for any analytic f, if (3.3.14) fails, then f is a polynomial.

Proof. By Green's formula applied to V and 1,

$$\int_{|z| \leq R} (\Delta V)(z) \, d^2 z = R \left[\frac{\partial}{\partial r} \int_0^{2\pi} V(re^{i\theta}) \, d\theta \right]\bigg|_{r=R} \qquad (3.3.15)$$

If the integral in (3.3.14) is $\beta < \infty$ (recall $\Delta V \geq 0$), then

$$\int_0^{2\pi} V(re^{i\theta}) \frac{d\theta}{2\pi} \leq C + \frac{\beta}{2\pi} \log r \qquad (3.3.16)$$

By $(U_1)_+ \leq V$, we see that

$$\int_0^{2\pi} (U_1)_+ \frac{d\theta}{2\pi} \leq C + \frac{\beta}{2\pi} \log r \qquad (3.3.17)$$

Since $\int_0^{2\pi} U_1 \frac{d\theta}{2\pi} = U_1(0)$, we see that

$$\int_0^{2\pi} |U_1(re^{i\theta})| \frac{d\theta}{2\pi} \leq 2C + \frac{2\beta}{2\pi} \log r + U_1(0) \qquad (3.3.18)$$

By the remark in Theorem 3.1.22, all derivatives $D^\alpha U_1(0) = 0$ for $|\alpha| > 0$, that is, U_1 is a constant, which implies $|f|$ is constant, which implies that f is constant. $\qquad \square$

Next, we need a simple piece of geometry:

Lemma 3.3.6. *There is an integer, N_0, so that $A \equiv \{z \mid \frac{1}{2} \leq |z| \leq 1\}$ can be covered by N_0 disks of radius $\frac{1}{4}$.*

Remark. A very crude estimate shows that $N_0 = 90$ will certainly do.

Proof. A is compact and clearly covered by $\{\mathbb{D}_{1/4}(x) \mid x \in A\}$ and so by a finite subfamily. $\qquad \square$

Proposition 3.3.7 (Rickman's Lemma). *Let μ be a positive Borel measure on $\mathbb{D}$ with $\mu(\mathbb{C}) = \infty$. Let N_0 be the number in Lemma 3.3.6. Then there exists $z_n \in \mathbb{C}$ and $r_n \in (0, \infty)$ with $r_n \to \infty$ so that*

(i) $\mu(\mathbb{D}_{r_n}(z_n)) \to \infty$ $\qquad\qquad\qquad\qquad\qquad\qquad\qquad$ (3.3.19)

(ii) $\mu(\mathbb{D}_{\frac{1}{2}r_n}(z_n)) \geq (2N_0 + 1)^{-1} \mu(\mathbb{D}_{r_n}(z_n))$ $\qquad\qquad\quad$ (3.3.20)

Proof. For $|z| < n$, define

$$\delta_n(z) = n - |z| \qquad (3.3.21)$$

Pick $z_n \in \mathbb{D}_n(0)$ so that

$$m_n \equiv \mu(\mathbb{D}_{\frac{1}{4}\delta(z_n)}(z_n)) \geq \frac{1}{2} \sup_{|z| \leq n} \mu(\mathbb{D}_{\frac{1}{4}\delta(z)}(z)) \qquad (3.3.22)$$

and let $r_n = \frac{1}{2}\delta(z_n)$.

By Lemma 3.3.6 and scaling, $\{z \mid \frac{1}{4}\delta(z_n) \leq |z - z_n| \leq \frac{1}{2}\delta(z_n)\}$ is covered by N_0 disks $D_{\frac{1}{8}\delta(z_n)}(z_j^{(n)})$, $j = 1, 2, \ldots, N$. Since $z_j^{(n)} \in \mathbb{D}_{\frac{1}{2}\delta(z_n)}(z_n)$, we see $\delta(z_j^{(n)}) \geq \frac{1}{2}\delta(z_n)$. Thus,

$$\mu(\mathbb{D}_{\frac{1}{8}\delta(z_j)}(z_j^{(n)})) \leq \mu(\mathbb{D}_{\frac{1}{4}\delta(z_j^{(n)})}(z_j^{(n)})) \leq 2m_n \qquad (3.3.23)$$

by (3.3.22). It follows that

$$\mu(\mathbb{D}_{r_n}(z_n)) \leq (2N_0 + 1)m_n \tag{3.3.24}$$

which is (3.3.20).

Taking $z = 0$ in (3.3.21), we see that

$$\mu(\mathbb{D}_{r_n}(z_n)) \geq \mu(\mathbb{D}_{\frac{1}{2}r_n}(z_n)) \geq \tfrac{1}{2}\,\mu(\mathbb{D}_{\frac{1}{4}n}(0))$$

goes to infinity as $n \to \infty$, proving (3.3.19). $\qquad\square$

Proof of Theorem 3.3.1. Define $\tilde{v}_n$ on $\mathbb{D}$ by

$$\tilde{v}_n(\zeta) = V(z_n + r_n\zeta) \tag{3.3.25}$$

where $\{z_n, r_n\}_{n=1}^{\infty}$ are the sequences chosen in Proposition 3.3.7 with $d\mu_0(x) = (\Delta v)(x)\,d^2x$. Since $\Delta\tilde{v}_n(\zeta) = r_n^2(\Delta V)(z_n + r_n\zeta)$ and $r_n^2\,d^2\zeta = d^2(r_n\zeta)$, we see that

$$\int_{\mathbb{D}} \Delta\tilde{v}_n(\zeta)\,d^2\zeta = \mu_0(\mathbb{D}_{r_n}(z_n)) \tag{3.3.26}$$

Let

$$M_n = \mu_0(\mathbb{D}_{r_n}(z_n)), \qquad d\mu_n(x) = M_n^{-1}(\Delta\tilde{v}_n)(\zeta)\,d^2\zeta \tag{3.3.27}$$

as a measure on $\overline{\mathbb{D}}$ with $\mu_n(\overline{\mathbb{D}}) = 0$. Then, by the Riesz decomposition,

$$\tilde{v}_n(\zeta) = M_n v_n(\zeta) + h_n(\zeta) \tag{3.3.28}$$

where v_n is the antipotential of μ_n and h_n is harmonic on $\mathbb{D}$.

Define

$$u_j^{(n)}(\zeta) = M_n^{-1}[U_j(z_n + r_n\zeta) - h_n(\zeta)] \tag{3.3.29}$$

Then each $u_j^{(n)}$ is harmonic. Since $M_n \to \infty$ and we have Proposition 3.3.3, we see that for any $j \neq k$,

$$|v_n - u_j^{(n)} \vee u_k^{(n)}| \to 0 \tag{3.3.30}$$

By passing to a subsequence, we can suppose $\mu_n \to \mu_\infty$ weakly as measures on $\overline{\mathbb{D}}$. By (3.3.20),

$$\mu_\infty(\overline{\mathbb{D}_{1/2}(0)}) \geq (2N + 1)^{-1} \tag{3.3.31}$$

Thus, if v_∞ is the antipotential of μ_∞, v_∞ is not harmonic.

Since $\mu_n(\mathbb{D}) = 1$,

$$v_n(\zeta) = \int \log|\zeta - w|\,d\mu_n(w) \tag{3.3.32}$$

and $\log|\zeta - w| \geq \log 2$ for $\zeta, w \in \mathbb{D}$, we see

$$\log 2 - u_j^{(n)}(\zeta) \geq 0 \tag{3.3.33}$$

because $u_j^{(n)} \leq v_n \leq \log 2$. By Theorem 3.1.31, we can pass to subsequences so each $u_j^{(n)}$ either converges uniformly to $-\infty$ on compacts of $\mathbb{D}$ or to $u_j^{(\infty)}$, a harmonic function.

Since (3.3.32) provides a bound on $\sup_n \int |v_n(\zeta)| \, d\zeta$, we cannot have two $u_j^{(n)}$ going to $-\infty$, since then v_n would. But if $u_j^{(n)} \to -\infty$ and $u_k^{(n)} \to u_k^{(\infty)}$, we would have $v_\infty = u_k^{(\infty)}$, which is impossible since v_∞ is not harmonic.

Thus, each $u_j^{(n)} \to u_j^{(\infty)}$, a harmonic function uniformly (on compact subsets of $\mathbb{D}$) and thus, $v_n = u_1^{(n)} \vee u_2^{(n)}$ also converges uniformly. By (3.3.30), $v_\infty, u_\infty^{(n)}$ obey Proposition 3.3.4. This implies v_∞ is harmonic. But we know Δv_∞ is not identically zero on $\overline{\mathbb{D}}_{1/2}(0)$. Thus, we have the required contradiction. $\qquad\qquad\square$

Notes and and Historical Remarks. As the section title says, our proof here is due to Eremenko–Sodin [**249, 250**], based in part on ideas they introduced earlier [**248**]. Their papers concern Nevanlinna theory and the arguments are more involved. Later, Eremenko [**246**] specialized to get Cartan's extension of Picard's theorem to projective varieties (see Problem 4 where we look at the special case where these varieties are hyperplanes). We have specialized further, as suggested to me by Sodin.

These results were followed up by Eremenko–Lewis [**247**] and Lewis [**527**]. In particular, Lewis has a variant that just uses harmonic functions and Harnack's inequality. It requires somewhat less machinery, but we prefer the version we present here.

Rickman's lemma is from Rickman [**678**] who used it in his extension of Picard's theorem to quasiregular mappings.

Problems

1. Let f be an entire function which is not constant.

   (a) Prove that

   $$\sup_{0 \leq \theta \leq 2\pi} \log|f(re^{i\theta})| \leq 3 \int_0^{2\pi} \log|f(2re^{i\theta})| \, \frac{d\theta}{2\pi} \qquad (3.3.34)$$

   (b) If

   $$\int_0^{2\pi} \log|f(re^{i\theta})| \, \frac{d\theta}{2\pi} \leq c_1 \log r + c_2 \qquad (3.3.35)$$

   prove that f is a polynomial.

   (c) If $v(re^{i\theta}) = \Delta \left[\log(1 + |f(\cdot)|^2)\right]|_{\cdot = re^{i\theta}}$, prove that if

   $$\int v(z) \, d^2z < \infty \qquad (3.3.36)$$

then (3.3.35) holds, and conclude that if (3.3.36) holds, then f is a polynomial.

(d) If f is a polynomial, prove that (3.3.36) holds.

2. (a) Let $a^{(1)}, \ldots, a^{(n+1)} \in \mathbb{C}^{n+1} \setminus \{0\}$. For $b \in \mathbb{C}^{n+1} \setminus \{0\}$, let H_b be the hyperplane in $\mathbb{CP}(n)$

$$H = \left\{ [(z_1, \ldots, z_{n+1})] \mid \sum b_j z_j = 0 \right\} \tag{3.3.37}$$

Prove that

$$\bigcap_{j=1}^{n+1} H_{a^{(j)}} = \emptyset \Leftrightarrow \{a^{(j)}\}_{j=1}^{n+1} \text{ are linearly independent} \tag{3.3.38}$$

(b) Let $P_{a^{(j)}}$ be the projection onto the multiples of $a^{(j)}$. Prove that $\sum_{j=1}^{n+1} P_{a^{(j)}} \geq c\mathbb{1}$ for some $c > 0$ if and only if $\{a^{(j)}\}_{j=1}^{n+1}$ are linearly independent.

3. Fix n and let $\mathcal{I}$ be the family of all subsets, I, of $\{1, \ldots, 2n+1\}$ with $\#(I) = n+1$.

(a) Suppose for each $I \in \mathcal{I}$, we have some $x_I \in I$. Prove that $\{x_I\}_{I \in \mathcal{I}}$ has at least $n+1$ distinct elements.

(b) Let $\{u_j\}_{j=1}^{2n+1}$ be a set of $(2n+1)$ real-valued functions on a set, X. Suppose for all $I \in \mathcal{I}$ (same v),

$$v = \bigwedge_{j \in I} u_j \tag{3.3.39}$$

Let

$$S_I = \{x \in X \mid \forall j \in I, \, v = u_j\} \tag{3.3.40}$$

Prove that $\bigcup_{I \in \mathcal{I}} S_I = X$. (*Hint*: Use (a).)

(c) Let $\{u_j\}_{j=1}^{2n+1}$ be harmonic functions on $\mathbb{D}$ so that for a fixed (subharmonic) v, we have (3.3.39). Prove that v is harmonic.

4. Prove that if $f \equiv (f_1, \ldots, f_{n+1})$ are $n+1$ entire analytic functions with no common zero, if $a^{(1)}, \ldots, a^{(2n+1)} \in \mathbb{C}^{n+1} \setminus \{0\}$ are $2n+1$ vectors so that for any $I \subset \mathcal{I}$ (the $\mathcal{I}$ of Problem 3), $\bigcap_{j \in I} H_{a^{(j)}} = \emptyset$, and if $f(z) \notin H_{a^{(j)}}$ for all j, then for an entire g, f_j/g is constant for each j. (*Hint*: Let $u^{(j)} = \log(|\sum_{k=1}^{n} a_k^{(j)} f_k|)$, $v = \log(\sqrt{|f_1|^2 + \cdots + |f_{n+1}|^2}$ and mimic the proof of the section using Problems 2 and 3.)

3.4. Perron's Method, Barriers, and Solution of the Dirichlet Problem

In this section, we'll present an approach to the Dirichlet problem for bounded regions, Ω, and, in particular, prove that the problem is solvable if Ω is convex or if Ω has a C^1 boundary. The basic idea, based on earlier work of Poincaré and Lebesgue, is due to Perron, whose name is applied to several notions below. The modern version that we follow is due to Brelot, with important intermediate steps by Wiener. So the method is sometimes called Brelot–Perron, Brelot–Perron–Wiener, or even BPW. To set the mood, we begin with some examples where the Dirichlet problem is *not* solvable.

Example 3.4.1 (Punctured Ball). Let $\Omega = B_1(0) \setminus \{0\}$, the unit ball with the center removed. Dirichlet data consists of a function, g, on $S^{\nu-1}$ and a value, α, at 0. If u is harmonic on Ω with boundary value, g, on $S^{\nu-1}$ and if $\lim_{|x|\to 0} u(x) = \alpha$, then u is bounded near 0, so 0 is a removable singularity. Thus, u is harmonic at 0 and so $\lim_{|x|\to 0} u(x) = \int g(\omega)\, d\omega$, that is, the Dirichlet problem for (g, α) is solvable if and only if

$$\alpha = \int g(\omega)\, d\omega \tag{3.4.1}$$

The problem is not solvable for *all* data. In particular (Problem 1), there is no classical Green's function for Ω. □

Example 3.4.2 (Lebesgue Spine). Let $\nu \equiv 3$. Let μ be the measure (3.2.21) of Example 3.2.9, that is, μ lives on $\{x \mid 0 < x_1 < 1, x_2 = x_3 = 0\}$ with density $4\pi x_1\, dx_1$. We saw the antipotential, Ψ_μ, has $\Psi_\mu((x_1, 0, 0)) = -\infty$ (respectively, -1) if $x_1 \in (0, 1]$ (respectively, $x_1 = 0$). As $x \to 0$ along the negative x_1-axis, $\Psi_\mu \to -1$. Indeed, this is true coming in along any ray, except the positive x_1-axis. A detailed analysis (Problem 2) shows that

$$\Psi_\mu(x) = -1 + x_1 \log(x_2^2 + x_3^2) + g(x) \tag{3.4.2}$$

with $\lim_{x\to 0} g(x) = 0$.

Let $\partial\Omega = \{x \mid \Psi_\mu(x) = -2\} \cup \{0\}$, a surface shown in Figure 3.4.1. The cusp is exponential, that is, by (3.4.2) asymptotic to $x_2^2 + x_3^2 = \exp(-1/x_1)$. Let Ω be the *exterior* of $\partial\Omega$. The function $\Psi_\mu(x)$ is harmonic on Ω and converges to -2 on $\partial\Omega \setminus \{0\}$, but does not have a continuous boundary value at 0 (every value between -2 and -1 is a limit point). $\Psi_\mu \to 0$ at ∞.

Clearly, for all $x \in \Omega$, $\Psi_\mu(x) \in [-2, 0]$ so $\Psi_\mu(x)$ is bounded there. Suppose $u(x)$ is harmonic in Ω and obeys

$$u(x_n) \to 0 \text{ as } x_n \to \infty, \qquad u(x_n) \to 2 \text{ as } x_n \to x_\infty \in \partial\Omega \tag{3.4.3}$$

Then by the maximum principle, u is bounded, and clearly, $u - \Psi_\mu$ is bounded and goes to zero at $\{\infty\} \cup (\partial\Omega \setminus \{0\})$. So, by Theorem 3.1.27, $u = \Psi_\mu$, which

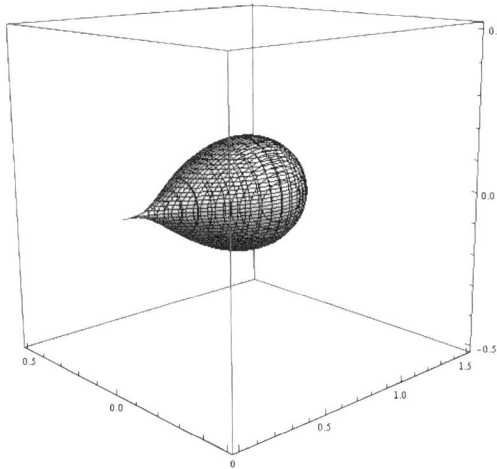

Figure 3.4.1. The Lebesgue spine.

shows u does not go to -2 at 0, that is, no harmonic u obeying (3.4.3) exists and the exterior Dirichlet problem is not solvable for $f \equiv -2$ on $\partial\Omega$.

By translating and scaling so 0 is inside $\partial\Omega$ and $\partial\Omega$ is inside $B_1(0)$ and using a Kelvin transform (see Problem 15 of Section 3.1), one gets a bounded region with exponential cusp pointing inwards for which the Dirichlet problem is not always solvable.   □

We now turn to Perron's method. We've previously only defined $\limsup$ for sequences. We say

$$\limsup_{\substack{x \in \Omega \\ x \to y}} f(x) = a \tag{3.4.4}$$

if for any sequence, $\{x_n\} \in \Omega$, with $x_n \to y$, we have $\limsup_n f(x_n) \leq a$, and for some sequence, it equals a. Equivalently, (3.4.4) holds if and only if

$$\lim_{\varepsilon \downarrow 0}[\sup\{f(x) \mid x \in \Omega, \, |x - y| \leq \varepsilon\}] = a \tag{3.4.5}$$

Since the sup is decreasing, the limit exists.

Definition. Let Ω be a bounded region. Let f be a bounded function on $\partial\Omega$. We say $u \in \mathcal{S}_f$, the *Perron trials* for f, if and only if

(i) u is subharmonic on Ω.
(ii) For all $x_0 \in \partial\Omega$,

$$\limsup_{\substack{x \to x_0 \\ x \in \Omega}} u(x) \leq f(x_0) \tag{3.4.6}$$

The *Perron solution*, $H_\Omega f$, is defined on Ω by

$$(H_\Omega f)(x) = \sup\{u(x) \mid u \in \mathcal{S}_f\} \tag{3.4.7}$$

We are heading towards a proof that $H_\Omega f$ is always harmonic on Ω and that, in many cases, it solves the Dirichlet problem for f when f is continuous on $\partial\Omega$. We note several simple facts:

(1) If $\alpha \equiv \inf_y f(y)$, then $u(x) \equiv \alpha$ is in $\mathcal{S}_f$, so $\mathcal{S}_f$ is nonempty and $(H_\Omega f)(x) \geq \alpha$.

(2) By the maximum principle, $u(x) \leq \sup_y f(y)$ for any $u \in \mathcal{S}_f$. We have thus proven that

$$\inf_{y\in\partial\Omega} f(y) \leq (H_\Omega f)(x) \leq \sup_{y\in\partial\Omega} f(y) \tag{3.4.8}$$

(3) Since $\mathcal{S}_f + \mathcal{S}_g \subset \mathcal{S}_{f+g}$, we have

$$H_\Omega(f+g) \geq H_\Omega f + H_\Omega g \tag{3.4.9}$$

In particular, since $H_\Omega(0) = 0$ (by (3.4.8)), we have

$$H_\Omega f \leq -H_\Omega(-f) \tag{3.4.10}$$

(4) We have that

$$v, w \in \mathcal{S}_f \Rightarrow \max(v,w) \in \mathcal{S}_f \tag{3.4.11}$$

because the max of subharmonic functions is subharmonic and

$$\limsup_{x\to\eta}[\max(v(x), w(x))] = \max\left(\limsup_{x\to\eta} v(x), \limsup_{x\to\eta} w(x)\right) \tag{3.4.12}$$

Example 3.4.1 (revisited). Suppose we know that $f \in C(\partial\mathbb{D})$ and that $H_\mathbb{D}(f)$ solves the Dirichlet problem for f (we'll prove this in Theorems 3.4.7 and 3.4.9). Let $\Omega = \mathbb{D}\setminus\{0\}$ and suppose g is defined on $\partial\Omega$ with $g \upharpoonright \partial\mathbb{D} = f$. Let $u = H_\mathbb{D}(f)$. Then $u(x) - \varepsilon G_0(x) \in \mathcal{S}_g^{(\Omega)}$ so $u - \varepsilon G_0 \leq H_\Omega(g)$. On the other hand, if $v \in \mathcal{S}_g^{(\Omega)}$, then $v - \varepsilon G_0 \in \mathcal{S}_f^{\mathbb{D}}$ so $v \leq u + \varepsilon G_0$, and thus, $H_\Omega(g) \leq u + \varepsilon G_0$. This shows that $H_\Omega(g) = u$ independently of $g(0)$ and illustrates a case where the Dirichlet problem is not solvable. $\quad\square$

The key to proving $H_\Omega f$ is harmonic is

Definition. Let $v \in \mathcal{S}_f$. Given $x_0 \in \Omega$ and $\rho < \text{dist}(x_0, \mathbb{R}^\nu \setminus \Omega)$, let $w \in C(B_\rho(x_0))$ be given by applying the Poisson kernel to $v(x_0 + \rho\omega)$, that is, if that v is continuous, w solves the Dirichlet problem for $v \upharpoonright \partial B(x_0)$. We define the *Perron modification*, $u \equiv \mathcal{P}_{x_0,\rho}(v)$, by

$$u(x) = \begin{cases} w(x), & x \in B_\rho(x_0) \\ v(x), & x \in \Omega \setminus B_\rho(x_0) \end{cases} \tag{3.4.13}$$

Lemma 3.4.3. *If $v \in \mathcal{S}_f$, then $u \equiv \mathcal{P}_{x_0,\rho}(v) \in \mathcal{S}_f$ for any $x_0 \in \Omega$ and $\rho < \text{dist}(x_0, \mathbb{R}^\nu \setminus \Omega)$. Moreover, $u \geq v$ on Ω.*

Proof. Since $u = v$ near $\partial\Omega$, u obeys the boundary value condition for $\mathcal{S}_f$ and we need only show that u is subharmonic.

Clearly, u is usc in $\Omega \setminus \partial B_\rho(x_0)$ and has SMP for any sphere disjoint from $\partial B_\rho(x_0)$. So we need only look at $x_\infty \in \partial B_\rho(x_0)$.

If $x_n \to x_\infty$ with $x_n \in \Omega \setminus \overline{B_\rho(x_0)}$, then $\limsup u(x_n) = \limsup v(x_n) \leq v(x_\infty) = u(x_\infty)$. If $g_n \in C(\partial B_\rho(x_0))$ are continuous functions with $g_n \geq g_{n+1}$ and $g_n \downarrow v$ and if u_n solves the Dirichlet problem for g_n on $\overline{B_\rho(x_0)}$, then $u_n \downarrow u$ on $\overline{B_\rho(x_0)}$, so $u \upharpoonright \overline{B_\rho(x_0)}$ is usc. It follows that u is usc on Ω.

By Theorem 3.2.11, $u \geq v$ on $B_\rho(x_0)$, and so on all of Ω. Thus, for $x_\infty \in \partial B_\rho(x_0)$ and r small,

$$\int u(x_\infty + r\omega)\, d\omega \geq \int v(x_\infty + r\omega)\, d\omega \geq v(x_\infty)$$

by the SMP for v. Thus, u has SMP on $\partial B_\rho(x_0)$. $\qquad\square$

Theorem 3.4.4. *$H_\Omega(f)$ is harmonic on Ω for any bounded f on $\partial\Omega$.*

Proof. Since harmonicity is a local property, it suffices to show that $w \equiv H_\Omega(f)$ is harmonic on each $B_\rho(x_0) \subset \Omega$. Given such an x_0 and ρ, find $v_n \in \mathcal{S}_f$ so

$$v_n(x_0) \leq v_{n+1}(x_0), \qquad \lim_{n\to\infty} v_n(x_0) = w(x_0) \tag{3.4.14}$$

Let $u_n = \mathcal{P}_{x_0,\rho}(v_n)$. Since

$$v_n(x) \leq u_n(x) \leq u_{n+1}(x) \leq w(x) \tag{3.4.15}$$

we see that

$$u_\infty = \lim_{n\to\infty} u_n \tag{3.4.16}$$

obeys

$$u_\infty(x) \leq w(x), \qquad u_\infty(x_0) = w(x_0) \tag{3.4.17}$$

By Harnack's principle (Theorem 3.1.30), u_∞ is harmonic on $B_\rho(x_0)$.

Let $t \in \mathcal{S}_f$ and define

$$t_n = \max(t, u_n) \tag{3.4.18}$$

By (3.4.11), $t_n \in \mathcal{S}_f$. Let s_n be the Perron modification $\mathcal{P}_{x_0,\rho}(t_n)$. As above, s_n converges to a harmonic function, s_∞ with $t \leq s_\infty$ (since $t \leq t_n \leq s_n \leq s_\infty$). Since $u_n \leq t_n \leq s_n$, we have $s_\infty \geq u_\infty$ so $s_\infty - u_\infty$ is a nonnegative harmonic function on $B_\rho(x_0)$. But, since $s_n \in \mathcal{S}_f$, $s_n(x_0) \leq w(x_0)$ so $u_\infty(x_0) = w_0(x_0) \leq s_\infty(x_0) \leq w(x_0)$. Thus $s_\infty - u_\infty$ vanishes at x_0.

By the maximum principle, $s_\infty = u_\infty$ on $B_\rho(x_0)$ so $t \leq u_\infty$ on $B_\rho(x_0)$. Since t was an arbitrary element of $\mathcal{S}_f$, $w \leq u_\infty$ on $B_\rho(x_0)$. Since $u_n \in \mathcal{S}_f$, $u_\infty \leq w$. Thus, $w = u_\infty$ is harmonic on $B_\rho(x_0)$. $\qquad\square$

The key to showing $H_\Omega f$ has f as boundary value (when suitable conditions hold) is the following:

Definition. Let Ω be a bounded region in $\mathbb{R}^\nu$ and $x_0 \in \partial\Omega$. A function, u, is called a *weak barrier* at x_0 if u is defined on $N \cap \Omega$ for N some open neighborhood of x_0, and obeys

  (i) u is subharmonic on $N \cap \Omega$

  (ii) $u < 0$ on $N \cap \Omega$

  (iii) $\lim\limits_{\substack{x \to x_0 \\ x \in N \cap \Omega}} u(x) = 0$ (3.4.19)

If u is defined on all of Ω and is bounded and for each $\delta > 0$

$$M(\delta) \equiv \sup\{u(x) \mid x \in \Omega, \, |x_0 - x| \geq \delta\} < 0 \qquad (3.4.20)$$

then u is called a *strong barrier*.

Definition. Let Ω be a region in $\mathbb{R}^\nu$. $x_0 \in \partial\Omega$ is called a *regular point* if there is a weak barrier at x_0 and a *singular point* if not.

If $G_\Omega(x, y)$ is a classical Green's function, $G_\Omega(\,\cdot\,, y)$ restricted to a neighborhood of $x_0 \in \partial\Omega$ is a weak barrier, so if a classical Green's function exists, all points in $\partial\Omega$ are regular, indeed $G_\Omega(x, y)$ only needs to exist for a single y.

The point of discussing both weak and strong barriers is that the existence of weak barriers is local, but one has:

Proposition 3.4.5. *Let Ω be a region in $\mathbb{R}^\nu$ and $x_0 \in \partial\Omega$. If x_0 is regular, there exists a strong barrier at x_0.*

Proof. Let $d_0(x) = \|x - x_0\|$ on $\partial\Omega$ and let

$$h(x) \equiv H_{d_0}(x) \qquad (3.4.21)$$

the harmonic function obtained by the Perron method. We'll prove that

$$\lim_{\substack{x \to x_0 \\ x \in \Omega}} h(x) = 0 \qquad (3.4.22)$$

in which case $-h(x)$ is a strong barrier. Since $d(x) = \|x - x_0\|$ on Ω lies in $\mathcal{S}_{d_0}$ (since $\|x - x_0\| = \sup_{y, \|y\|=1} y \cdot (x - x_0)$ clearly has SMP and is continuous, so subharmonic), $-h(x) \leq -\|x - x_0\|$ implying $M(\delta) \leq -\delta$.

Let w be a weak barrier at x_0, $x_n \to x_0$ with $x_n \in \Omega$. Let $\rho_n = \|x_n - x_0\|$. Let $B = \{x \mid \|x - x_0\| = \rho_n\}$, so $\partial B \cap \Omega \neq \emptyset$ since it contains x_n. Let σ be the normalized surface area measure on $\partial B \cap \Omega$ (since $\partial B \cap \Omega \neq \emptyset$ and $\partial B \cap \Omega$ is open on ∂B, $\sigma(\partial B \cap \Omega) > 0$). By regularity of the Borel measure σ, choose $F \subset \partial B \cap \Omega$ closed so that

$$\sigma\big((\partial B \cap \Omega) \setminus F\big) < \frac{\rho_n}{M} \qquad (3.4.23)$$

where $M = \sup_{x \in \partial\Omega} \|x - x_0\| = \sup_{x \in \partial\Omega} d_0(x)$ so

$$\|h\|_\infty \leq M \tag{3.4.24}$$

Let f be defined on ∂B by

$$f(y) = \begin{cases} M & \text{if } y \in (\partial B \cap \Omega) \setminus F \\ 0, & \text{otherwise} \end{cases} \tag{3.4.25}$$

and $P(f)$, the Poisson integral of f.

Let

$$k = \inf_{y \in F} \{-w(y)\} > 0 \tag{3.4.26}$$

k is strictly positive because $-w$ is pointwise strictly positive on F (since $F \subset \Omega$), F is closed and $-w$ is lsc so there is $y_0 \in F$ with $-w(y_0) = k$.

Let u be an element of $\mathcal{S}_{d_0}$ and define q on $B \cap \Omega$ by

$$q = u - \rho_n + \frac{M}{k} w - P(f) \tag{3.4.27}$$

Our goal is to prove that $q \leq 0$ on $B \cap \Omega$. Since q is subharmonic, we only need to prove boundary values are negative. Since there are three kinds of boundary points, the argument is a little tedious!

Let $z_\ell \to y \in F$ with $z_\ell \in B \cap \Omega$. Since

$$u(y) \leq h(y) \leq M \tag{3.4.28}$$

and $\frac{M}{k} w \leq \frac{M}{k}(-k) = -M$ and $f \geq 0 \Rightarrow -P(f) \leq 0$ on $B \cap \Omega$, we conclude for $y \in F$ that

$$\limsup_\ell q(z_\ell) \leq 0 \tag{3.4.29}$$

If $y \in \partial B \cap \Omega \setminus F$ which is an open subset of ∂B, we know that $P(f)(z_\ell) \to M$, so by (3.4.28), $\limsup_\ell [u(z_\ell) - P(f)(z_\ell)] \leq 0$, so since $w \leq 0$ and $-\rho_n \leq 0$, (3.4.29) holds.

Finally, if $y \in \partial\Omega \cap B$, u in $\mathcal{S}_{d_0}$ implies that $\lim_\ell \sup u(z_\ell) \leq d_0(y) \leq \rho_n$. Since $w \leq 0$ and $P(f) \geq 0$, we conclude that (3.4.29) holds.

Thus, (3.4.29) holds on all of $\partial(B \cap \Omega)$. By the maximum principle for subharmonic functions (Theorem 3.2.11) and the fact that q is subharmonic, we conclude that on $B \cap \Omega$:

$$u \leq \rho_n - \frac{M}{k} w + P(f) \tag{3.4.30}$$

Since $u \in \mathcal{S}_{d_0}$ is arbitrary, we see that

$$h(x_n) \leq \rho_n - \frac{M}{k} w(x_n) + \rho_n \tag{3.4.31}$$

where we used $P(f)(x_n) = \int_B f(x_n + \rho_n w) dw \leq M \frac{\rho_n}{M}$ by (3.4.23) and (3.4.25). Since w is a weak barrier, $w(x_n) \to 0$ and $\rho_n \to 0$, so (3.4.22) holds. $\qquad\square$

Here is the main result of the Perron method:

Theorem 3.4.6. *Let Ω be a bounded region and $x_0 \in \partial\Omega$ a regular point. Let f be a bounded function on $\partial\Omega$ and*

$$f_+ = \limsup_{\substack{x \in \partial\Omega \\ x \to x_0}} f(x), \qquad f_- = \liminf_{\substack{x \in \partial\Omega \\ x \to x_0}} f(x) \qquad (3.4.32)$$

Then

$$f_- \leq \liminf_{\substack{x \in \Omega \\ x \to x_0}} (H_\Omega f)(x) \leq \limsup_{\substack{x \in \Omega \\ x \to x_0}} (H_\Omega(f)) \leq f_+ \qquad (3.4.33)$$

In particular, if f (on $\partial\Omega$) is continuous at x_0, then $H_\Omega f$ extended to $\overline{\Omega}$ by setting it equal to f on $\partial\Omega$ is continuous at x_0.

Remark. The Ω bounded assumption is mainly for simplicity; see Problem 3 for the unbounded case.

Proof. If we prove (3.4.33) with $f_\pm$ replaced by $f_\pm \pm \varepsilon$ for arbitrary $\varepsilon > 0$, we are done by taking $\varepsilon \downarrow 0$. So fix ε and choose $\delta > 0$ so that

$$y \in \partial\Omega, |y - x_0| \leq \delta \Rightarrow f(x) \in [f_- - \varepsilon, f_+ + \varepsilon] \qquad (3.4.34)$$

Let u be a strong barrier for x_0 and let

$$C_\pm(\varepsilon) = \frac{\|f\|_\infty \mp f_\pm}{|M(\delta)|} \qquad (3.4.35)$$

We claim that

$$g_- \equiv C_-(\varepsilon)u + f_- - \varepsilon \in \mathcal{S}_f \qquad (3.4.36)$$

For u subharmonic and $C_-(\varepsilon) \geq 0$ implies g_- is subharmonic. Since $u \leq 0$,

$$x_n \to y \in \partial\Omega \cap \overline{B_\delta(x_0)} \Rightarrow \limsup_{n \to \infty} g(x_n) \leq f_- - \varepsilon \leq f(y) \qquad (3.4.37)$$

On $\Omega \setminus \overline{B_\delta(x_0)}$,

$$C_-(\varepsilon)u \leq M(\delta)|M(\delta)|^{-1}(\|f\|_\infty + f_-) = -\|f\|_\infty - f_- \qquad (3.4.38)$$

so

$$x_n \to y \in \partial\Omega \setminus \overline{B_\delta(x_0)} \Rightarrow \lim_{n \to \infty} g(x_n) \leq -\|f\|_\infty - \varepsilon \leq f(y) \qquad (3.4.39)$$

This proves (3.4.36).

Clearly, (3.4.36) implies

$$C_-(\varepsilon)u(x) + f_- - \varepsilon \leq (H_\Omega f)(x) \qquad (3.4.40)$$

so, since $u(x) \to 0$ as $x \to x_0$,

$$\liminf_{x \to x_0} (H_\Omega f)(x) \geq f_- - \varepsilon \qquad (3.4.41)$$

For the other direction, let $v \in \mathcal{S}_f$. We claim that for all $x \in \Omega$,

$$g_+(x) \equiv v(x) + C_+(\varepsilon)u(x) \leq f_+ + \varepsilon \qquad (3.4.42)$$

By the maximum principle (since g_+ is subharmonic), it suffices to prove that if $x_n \to y \in \partial\Omega$, then the lim sup of $g_+(x_n)$ is at most $f_+ + \varepsilon$.

If $y \in \overline{B_\delta(x_0)}$, since u is negative and, by (3.4.34), $\limsup_n v(x_n) \le f_+ + \varepsilon$, the limit in (3.4.42) is less than $f_+ + \varepsilon$. If for large n, $x_n \in \Omega \setminus \overline{B_\delta(x_0)}$, then $u(x) \le M(\delta)$, so

$$\limsup g_+(x_n) \le f(y) + M(\delta)|M(\delta)|^{-1}(\|f\|_\infty - f_+) \qquad (3.4.43)$$
$$= f(y) - \|f\|_\infty + f_+ \le f_+ \qquad (3.4.44)$$

proving (3.4.42) and

$$v \le f_+ + \varepsilon - C_+(\varepsilon)u(x) \qquad (3.4.45)$$

Taking sup over v, we get

$$(H_\Omega f)(x) \le f_+ + \varepsilon - C_+(\varepsilon)u(x) \qquad (3.4.46)$$

Since $u(x) \to 0$ as $x \to x_0$, we get

$$\limsup_{\substack{x \to x_0 \\ x \in \Omega}} (H_\Omega f)(x) \le f_+ + \varepsilon$$

We have thus proven (3.4.33). If f is continuous at x_0, $f_+ = f_-$, so $\lim_{x \to x_0, x \in \Omega}(H_\Omega f)(x) = f(x_0)$. Of course, continuity of f implies the extension is continuous for sequences in $\overline{\Omega}$ converging to x_0. $\qquad \square$

While the first part of the following is a corollary, the result is so important that we call it a theorem:

Theorem 3.4.7. *Let Ω be a bounded domain in $\mathbb{R}^\nu$. If every point in $\partial\Omega$ is regular, the Dirichlet problem is solvable for Ω and the solution is given by $H_\Omega(f)$ (for $f \in C(\partial\Omega)$). Conversely, if the Dirichlet problem for Ω is solvable, every point in $\partial\Omega$ is regular; indeed, at any point there is a strong barrier which is harmonic on Ω. The solution of the Dirichlet problem is given by $H_\Omega(f)$.*

Proof. The direct result is immediate from the last theorem. If the Dirichlet problem is solvable, given $x_0 \in \partial\Omega$, let $f \in C(\Omega)$ be given by

$$f(x) = -|x - x_0| \qquad (3.4.47)$$

Let u solve the Dirichlet problem for f. Then u is harmonic, so subharmonic on Ω. Clearly, $\lim_{x \to x_0, x \in \Omega} u(x) = f(x_0) = 0$. For any δ, u is negative and nonvanishing on $\{x \in \overline{\Omega} \mid |x - x_0| \ge \delta\}$ (on the boundary by (3.4.47) and on the point in Ω by the maximum principle), so $M(\delta) = \sup\{u(x) \mid x \in \overline{\Omega}, |x - x_0| \ge \delta\} < 0$. $\qquad \square$

We saw that if a classical Green's function exists for Ω, then each point of $\partial\Omega$ is regular. Thus,

Corollary 3.4.8. *If a classical Green's function exists for Ω a bounded region of $\mathbb{R}^\nu$, then the Dirichlet problem is solvable for Ω.*

Remarks. 1. We proved the converse in Theorem 3.1.8.

2. As we noted, it suffices for there to be a classical Green's function, $G(\,\cdot\,, y_0)$ at a single y_0 (i.e., $f(x)$ continuous on $\overline{\Omega} \setminus \{y_0\}$ with $f \restriction \partial\Omega = 0$ and $f(x) - G_0(x, y_0)$ harmonic on $\Omega \setminus \{y_0\}$) and this then implies a jointly continuous classical Green's function at all points.

3. In Section 3.6 we'll use this to show that when $\nu = 2$, any bounded connected open Ω with $\mathbb{R}^2 \setminus \Omega$ connected has a solvable Green's function.

The question becomes when we can prove that a point is regular. We'll describe two criteria: the exterior ball condition and the exterior cone condition. While the first condition is stronger than the second, we include it for two reasons: its simplicity and wide applicability and the fact that we'll use it to prove the cone condition result.

Definition. Let Ω be a region in $\mathbb{R}^\nu$ and $x_0 \in \partial\Omega$. We say x_0 obeys an *exterior ball condition* if there is a closed ball, $\overline{B_\delta(y)}$, so that $\overline{B_\delta(y)} \cap \overline{\Omega} = \{x_0\}$; see Figure 3.4.2.

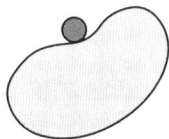

Figure 3.4.2. The exterior ball condition.

By Theorem 5.10.6 of Part 1, if $\overline{\Omega}$ is a compact convex set and $x_0 \in \partial\Omega$, there is a nonzero linear functional, ℓ, on $\mathbb{R}^\nu$ so $\overline{\Omega} \subset \{x \mid \ell(x) \leq \ell(x_0)\}$. If y_0 is perpendicular to $\{x \mid \ell(x) = 0\}$ with $\ell(y_0) = 1$, and $y = x_0 + \delta y_0$ for any $\delta > 0$, it is easy to see (Figure 3.4.3) that $\overline{B_\delta(y)} \cap \{x \mid \ell(x) \leq \ell(x_0)\} = \{x_0\}$, so the exterior ball condition holds at all points $x_0 \in \partial\Omega$. It is also easy

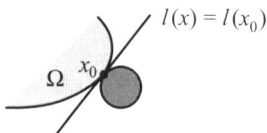

Figure 3.4.3. Convex sets obey an exterior ball condition.

to see (Problem 4) that if $\partial\Omega$ is a C^2-hypersurface, that is, there is a C^2 function defined on a neighborhood of $\overline{\Omega}$ so that

$$\Omega = \{x \mid F(x) < 0\}, \quad \partial\Omega = \{x \mid F(x) = 0\}, \quad x_0 \in \partial\Omega \Rightarrow \nabla F(x_0) \neq 0$$
$$(3.4.48)$$

then Ω obeys the exterior ball condition at all points.

Theorem 3.4.9 (Poincaré's Criterion). *Let Ω be a bounded open subset of $\mathbb{R}^\nu$ and $x_0 \in \partial\Omega$. If the exterior ball condition holds at x_0, then x_0 is a regular point.*

Proof. If $B_\delta(y_0) \cap \overline{\Omega} = \{x_0\}$, then

$$u(x) = G_0(x - y_0) - G_0(x_0 - y_0) \qquad (3.4.49)$$

is a strong barrier at x_0. $\qquad\square$

Corollary 3.4.10. *If Ω is a bounded, open convex set in $\mathbb{R}^\nu$ or is the interior of a bounded C^2-hypersurface, then the Dirichlet problem is solvable for Ω.*

The *standard cone* with opening angle $\alpha \in (0, \pi/2)$ is the set (see Figure 3.4.4)

$$C_\alpha = \{x \mid x_\nu \geq 0, \, \|(x_1, \ldots, x_{\nu-1}, 0)\| \leq x_\nu \tan\alpha\} \qquad (3.4.50)$$

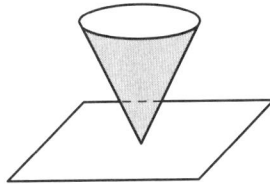

Figure 3.4.4. The standard cone.

Proposition 3.4.11. *Let $\Omega = B_1(0) \setminus C_\alpha$ for some $\alpha \in (0, \pi/2)$. Then 0 is a regular point of Ω.*

Remark. Every point in $\partial\Omega$ other than 0 obeys an exterior ball condition, so all of $\partial\Omega$ is regular points.

Proof. Define f on $\partial\Omega$ by $f(x) = -\|x\|$ and let $u = H_\Omega(f)$. We'll prove that u is a strong barrier at 0. Since $-1 \leq f \leq 0$, we have $-1 \leq u \leq 0$. Every point in $\partial\Omega \setminus \{0\}$ has an exterior ball, so by the Poincaré criterion,

$$\lim_{\substack{x \to x_0 \in \partial\Omega \setminus \{0\} \\ x \in \Omega}} u(x) = f(x_0)$$

so, in particular, by the maximum principle, $-1 < u < 0$ on Ω. It follows by compactness and $f(x) = -\|x\|$ that

$$-\delta \leq M(\delta) = \sup_{\substack{x \in \overline{\Omega} \\ \|x\| \geq \delta}} u(x) < 0 \tag{3.4.51}$$

We are thus left to prove

$$\liminf_{\substack{x \to 0 \\ x \in \Omega}} u(x) = 0 \tag{3.4.52}$$

Let $\Omega_{1/2} = \{\frac{1}{2}x \mid x \in \Omega\}$ and

$$c = -\inf_{x \in \partial B_{1/2} \cap \overline{\Omega}} u(x) \tag{3.4.53}$$

By the maximum principle for $-u$, $c < 1$, and since $u(x) \to -\frac{1}{2}$ at any point in $\partial \Omega_{1/2} \cap C_\alpha$, where $\|x\| = \frac{1}{2}$, we have

$$\frac{1}{2} \leq c < 1 \tag{3.4.54}$$

On $\Omega_{1/2}$, define

$$v(x) = cu(2x) - u(x) \tag{3.4.55}$$

We will prove on $\Omega_{1/2}$ that

$$v(x) \leq 0 \tag{3.4.56}$$

By Theorem 3.1.27, it suffices to prove the continuous extension of v to $\overline{\Omega}_{1/2} \setminus \{0\}$ has $v(x) \leq 0$ on $\partial \Omega_{1/2} \setminus \{0\}$.

On $\partial \Omega_{1/2} \cap (C_\alpha \setminus \{0\})$, we have $u(2x) = -2\|x\|$, $u(x) = -\|x\|$, so $v(x) = -\|x\|(2c - 1) \leq 0$ since $c \geq \frac{1}{2}$. On $\partial \Omega_{1/2} \cap \partial B_{1/2}(0)$, we have $u(2x) = -1$ and $u(x) \geq -c$ or $-u(x) \leq c$, so $v(x) \leq -c + c = 0$. Thus, (3.4.56) holds, or on $\Omega_{1/2}$,

$$u(x) \geq cu(2x) \tag{3.4.57}$$

Let $\beta = \liminf_{x \to 0} u(x) \leq 0$ and let $x_n \to 0$ so $u(x_n) \to \beta$. By passing to a subsequence, we can suppose $u(2x_n) \to \gamma$ and we have $\gamma \geq \beta$ since β is the $\liminf$. Thus, (3.4.57) implies that

$$\beta \geq c\gamma \Rightarrow (1 - c)\beta \geq 0 \Rightarrow \beta \geq 0 \Rightarrow \beta = 0 \tag{3.4.58}$$

since $c < 1$.   □

Definition. Let Ω be a region in $\mathbb{R}^\nu$ and $x_0 \in \partial\Omega$. We say x_0 obeys an *exterior cone condition* if there is $\alpha \in (0, \pi/2)$ and a rotation R and $\delta > 0$ so that (Figure 3.4.5)

$$\overline{\Omega} \cap (x_0 + R[C_\alpha]) \cap B_\delta(x_0) = \{x_0\} \tag{3.4.59}$$

Theorem 3.4.12 (Zaremba's Criterion). *If Ω is a bounded region and $x_0 \in \partial\Omega$ obeys an exterior cone condition, then x_0 is a regular point.*

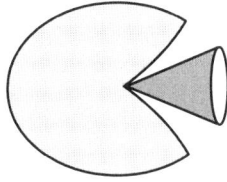

Figure 3.4.5. The exterior cone condition (Note: Not a pacman eating an ice cream cone).

Proof. By translating and rotating the strong barrier at 0 in the last example, we get a local barrier at x_0.   □

This allows the interior of C^1-surfaces and also certain kinds of cusps.

Notes and Historical Remarks. In 1911, the Polish mathematician, Zaremba [**871**] noted that the punctured ball showed that the Dirichlet problem is not always solvable. In this paper, Zaremba also showed that an exterior cone condition at all boundary points sufficed for solvability of the Dirichlet problem (essentially Theorem 3.4.12 without the language of barriers). In 1913, Lebesgue [**515**] found the example now called his spine.

Poincaré [**652**], in 1890, proved the result on solvability under a global exterior ball condition using a form of the notion of barrier; see also his book on potential theory [**654**]. Lebesgue [**516**] refined the notion of barrier and gave it the name "barrier."

Perron's construction of harmonic functions is from his 1923 paper [**639**] (using continuous rather than usc subharmonic functions). In several papers, Wiener [**853, 854, 855**] refined Perron's ideas, and in 1938, Brelot [**113**] brought the Perron theory into its standard modern form that we present here.

Proposition 3.4.5 is sometimes called Bouligand's lemma since he first proved it in 1926 [**101**]. The proof we use is that of Brelot [**113**].

Problems

1. Suppose the punctured disk, $\Omega \equiv \mathbb{D} \setminus \{0\}$, has a classical Green's function.

   (a) Prove that $G_\Omega(x, y)$ has a removable singularity at $x = 0$.

   (b) Prove that $\lim_{x \to 0, x \neq 0} G_\Omega(x, y) > 0$.

   (c) Conclude that G_Ω does not exist.

   Remark. While $\mathbb{D} \setminus \{0\}$ does not have a classical Green's function, it does have one in a natural sense, see Example 3.8.10.

2. Let Ψ_μ be the antipotential of Example 3.4.2. Prove that (with $\rho^2 = x_2^2 + x_3^2$)

$$-\Psi_\mu(x_1, x_2, x_3) = [(1 - x_1)^2 + \rho^2]^{1/2} - [x_1^2 + \rho^2]^{1/2}$$
$$+ x_1 \log|(1 - x_1) + [(1 - x_1)^2 + \rho^2]^{1/2}|$$
$$+ x_1 \log|x_1 + (x_1^2 + \rho^2)^{1/2}| - x_1 \log \rho^2$$

   and then that (3.4.2) holds.

3. Extend Theorem 3.4.6 to the exterior of a compact set with a condition that u's are required to go to zero at infinity.

4. Show that a C^2-hypersurface obeys an exterior ball condition.

3.5. Spherical Harmonics

A function $f : \mathbb{R}^\nu \setminus \{0\} \to \mathbb{R}$ (here and throughout this section, $\nu \geq 2$) is said to be homogeneous of degree λ ($\lambda \in \mathbb{R}$) if and only if for all $\mathbf{x} \in \mathbb{R}^\nu \setminus \{0\}$ and all $\rho \in (0, \infty)$,

$$f(\rho \mathbf{x}) = \rho^\lambda f(\mathbf{x}) \tag{3.5.1}$$

Proposition 3.5.1 (Euler's Formula). *Let f be a C^1-function on $\mathbb{R}^\nu \setminus \{0\}$. Then f is homogeneous of degree λ if and only if*

$$\mathbb{E}f \equiv \left[\sum_{j=1}^{\nu} x_j \frac{\partial}{\partial x_j} \right] f = \lambda f \tag{3.5.2}$$

Proof. If we change to spherical coordinates, $\omega_1, \ldots, \omega_{\nu-1}$, on $S^{\nu-1}$ (e.g., $\omega_j = x_j/r$, $j = 1, \ldots \nu - 1$; $r = (\sum_{j=1}^{\nu} x_j^2)^{1/2}$ will do) and r, it is easy to see (Problem 1) that

$$\sum_{j=1}^{\nu} x_j \frac{\partial}{\partial x_j} = r \frac{\partial}{\partial r} \tag{3.5.3}$$

Thus, if $g(r, \omega) = f(r\omega)$, $g : (0, \infty) \times S^{\nu-1} \to \mathbb{R}$, then (3.5.2) is equivalent to

$$\frac{\partial}{\partial r} g(r, \omega) = \frac{\lambda}{r} g(r, \omega) \tag{3.5.4}$$

which is solved by $g(r, \omega) = r^\lambda g(1, \omega)$ (and vice-versa). $\qquad\square$

Remark. It is interesting to check (3.5.3) directly for homogeneous polynomials (Problem 2).

Definition. A *spherical harmonic* of degree d in ν dimensions is the restriction to $S^{\nu-1}$ of a harmonic polynomial which is homogeneous (of necessity, of nonnegative integral degree).

These are important special functions. We'll see that they span $L^2(S^{\nu-1}, d\omega)$, and one can obtain an orthonormal basis of spherical harmonics. Each will be an eigenfunction of the Laplace–Beltrami operator. Before discussing them, we'll first see why we don't consider more general harmonic homogeneous functions.

Theorem 3.5.2. *Let u be a harmonic function on $\mathbb{R}^\nu \setminus \{0\}$ which is homogeneous of degree λ. Then*

(a) *If $\lambda \geq 0$, then λ is a nonnegative integer and u is a polynomial.*
(b) *If $-(\nu-2) < \lambda < 0$, then $u = 0$.*
(c) *If $\lambda \leq -(\nu-2)$, then λ is a nonpositive integer and there is a harmonic polynomial, v, of degree $-\lambda - (\nu-2)$ so that*

$$u(r\omega) = r^{(2-\nu)}v(r^{-1}\omega) \tag{3.5.5}$$

Remark. In particular, the restriction of an homogeneous, harmonic function to $S^{\nu-1}$ is a spherical harmonic. If $Y(\omega)$ is a spherical harmonic of degree d, then $r^d Y(\omega)$ and $r^{-d-(\nu-2)}Y(\omega)$ are the two homogeneous harmonic functions whose restriction to $S^{\nu-1}$ is Y.

Proof. (a) If $\lambda \geq 0$, u is bounded near $x = 0$, so 0 is a removable singularity and u is a harmonic function on $\mathbb{R}^\nu$. For any multi-index, α, $D^\alpha u$ is harmonic and homogeneous of degree $\lambda - |\alpha|$. Thus, for $\rho > 0$,

$$(D^\alpha u)(0) = \int (D^\alpha u)(\rho\omega)\, d\omega = \rho^{\lambda-|\alpha|} \int (D^\alpha u)(\omega)\, d\omega = \rho^{\lambda-|\alpha|}(D^\alpha u)(0) \tag{3.5.6}$$

Taking $\rho \to 0$ (if $|\alpha| < \lambda$) or $\rho \to \infty$ (if $|\alpha| > \lambda$), we see $(D^\alpha u)(0) = 0$ unless $|\alpha| = \lambda$. Thus, since u is given by its Taylor series, u is 0 unless λ is an integer, and if λ is an integer, u is a polynomial of total degree λ.

(b) By the remark after Theorem 3.1.26, if $u(x) = o(|x|^{-(\nu-2)})$, 0 is a removable singularity. Thus, u is continuous at 0, but if $u(x_0) \neq 0$, $u(\rho x_0) = \rho^\lambda u(x_0)$ diverges at $\rho = 0$. Thus, $u \equiv 0$.

(c) Since u is harmonic on $\mathbb{R}^\nu \setminus \{0\}$, its Kelvin transform (see Problem 15 of Section 3.1),

$$v(r\omega) \equiv (Ku)(r\omega) = r^{2-\nu}u(r^{-1}\omega) \tag{3.5.7}$$

is harmonic on $\mathbb{R}^\nu \setminus \{0\}$. It is obviously homogeneous of degree $-\lambda - (\nu-2) \geq 0$. Thus, by (a), v is a harmonic polynomial. Since $u = Kv$, we obtain (3.5.5). $\qquad \square$

We will see for each $d = 0, 1, 2, \ldots$ with $d \geq 0$, the set of spherical harmonics of degree d is a space of dimension, $D(\nu, d) \geq 1$. We'll pick an orthonormal basis $\{Y_{dm}\}_{m=1}^{D(\nu,d)}$ of this space. Traditionally, one allows complex-valued functions and takes the $Y_{d,m}$ complex-valued. We are heading towards a proof that $\{Y_{dm}\}_{m=1;d=0,1,2,\dots}^{D(\nu,d)}$ is a basis of $L^2(S^{\nu-1}, d\omega)$. We begin with a simple calculation. By Leibniz's rule,

$$\frac{\partial^2}{\partial x_i^2}[x_i^2 f] = x_i^2 \frac{\partial^2}{\partial x_i^2} f + 4x_i \frac{\partial f}{\partial x_i} + 2f \tag{3.5.8}$$

Thus, with $\mathbb{E}$ given by (3.5.2) and $x^2 = \sum_{i=1}^{\nu} x_i^2$, viewed as a multiplication operator, we have

$$[\Delta, x^2] = 4\mathbb{E} + 2\nu \tag{3.5.9}$$

Lemma 3.5.3. *Let P be a harmonic polynomial of degree d. Then for a constant $c(\nu, d, k) > 0$, we have*

$$\Delta(x^{2k+2}P) = c(\nu, d, k)x^{2k}P \tag{3.5.10}$$

Remark. The proof shows that

$$c(\nu, d, k) = 2(4k + 4d + 2\nu)(k+1) \tag{3.5.11}$$

Proof. We use induction on k. $k = 0$ follows from

$$\Delta(x^2 P) = x^2(\Delta P) + (4\mathbb{E} + 2\nu)P = (4d + 2\nu)P \tag{3.5.12}$$

by (3.5.2). If we know it for $x^{2k}P$ on the left, then

$$\Delta(x^{2k+2}P) = x^2 \Delta(x^{2k}P) + (4d + 8k + 2\nu)x^{2k}P$$
$$= [(4d + 8k + 2\nu) + c(\nu, d, k-1)]x^{2k}P \tag{3.5.13}$$

$\square$

Lemma 3.5.4. *If Y is a spherical harmonic of degree d and $Y^{\sharp}$ of degree $d^{\sharp} \neq d$, then*

$$\int \overline{Y(\omega)}\, Y^{\sharp}(\omega)\, d\omega = 0 \tag{3.5.14}$$

(where $d\omega$ is normalized rotation invariant measure on $S^{\nu-1}$ so $d^{\nu}x = \sigma_\nu r^{\nu-1}dr\, d\omega$ with σ_ν the surface area of $S^{\nu-1}$).

Proof. $\overline{Y}$ is harmonic, so $\overline{Y}\Delta Y^{\sharp} - Y^{\sharp}\Delta\overline{Y} = 0$ and Green's theorem implies

$$\int_{S^{\nu-1}} \left[\overline{Y}\frac{\partial}{\partial r}Y^{\sharp} - \left(\frac{\partial}{\partial r}\overline{Y}\right)Y^{\sharp}\right] d\omega = 0 \tag{3.5.15}$$

Since $r = 1$ on $S^{\nu-1}$,

$$\frac{\partial}{\partial r}Y^{\sharp} = r\frac{\partial}{\partial r}Y^{\sharp} = d^{\sharp}Y^{\sharp} \tag{3.5.16}$$

so the integral in (3.5.15) is $(d^{\sharp} - d)$ times the LHS of (3.5.14). Since $d^{\sharp} \neq d$, we get (3.5.14). $\square$

Let $\mathcal{P}(\nu, d)$ be polynomials in $x_1, \ldots, x_\nu$ of total degree at most d, $\mathcal{H}(\nu, d)$ the homogeneous polynomials of degree d, and $\mathcal{S}(\nu, d)$ the harmonic homogeneous polynomials of degree d, that is, the span of $\{r^d Y_{dm}\}_{m=1}^{D(\nu,d)}$. Then it is easy to see (Problems 3 and 4) that

$$\dim \mathcal{H}(\nu, d) = \dim \mathcal{P}(\nu - 1, d) = \binom{d + \nu - 1}{\nu - 1} \tag{3.5.17}$$

the binomial coefficient $\frac{(d+\nu-1)!}{(\nu-1)!d!}$.

Theorem 3.5.5. *We have*

$$\mathcal{H}(\nu, d) = \sum_{j=0}^{[\frac{d}{2}]} x^{2j} \mathcal{S}(\nu, d - 2j) \tag{3.5.18}$$

where, by $\sum$, we mean every element of $\mathcal{H}$ is a unique sum of elements of $\{x^{2j}\mathcal{S}(\nu, d - 2j)\}_{j=0}^{[\frac{d}{2}]}$.

Proof. We use induction on d in steps of 2. For $d = 0, 1$, (3.5.18) says $\mathcal{H}(\nu, d) = \mathcal{S}(\nu, d)$, which is obvious since $\deg u \leq 1 \Rightarrow \Delta u = 0$.

Suppose we know (3.5.18) for d. x^2 is a one–one map of $\mathcal{H}(\nu, d)$ to $\mathcal{H}(\nu, d+2)$, so $\{x^{2(j+1)}\mathcal{S}(\nu, d-2j)\}_{j=0}^{[\frac{d}{2}]}$ are linearly independent in $\mathcal{H}(\nu, d+2)$. By (3.5.12), Δ is a bijection of $x^{2(j+1)}\mathcal{S}(\nu, d - 2j)$ to $x^{2j}\mathcal{S}(\nu, d - 2j)$, so $\Delta \colon \mathcal{H}(\nu, d + 2) \to \mathcal{H}(\nu, d)$ is a surjection.

Since Δ is a bijection on $\mathcal{H}_0 \equiv \sum_{j=0}^{[\frac{d}{2}]} x^{2j+2}\mathcal{S}(\nu, d-2j)$, this sum is disjoint from $\mathcal{S}(\nu, d + 1)$, the kernel of Δ. Moreover, if $\varphi \in \mathcal{H}(\nu, d + 2)$, we can find $\psi \in \mathcal{H}_0$ so $\Delta\varphi = \Delta\psi$, that is, $\varphi = \psi + (\varphi - \psi)$, so $\varphi \in \mathcal{S}(\nu, d + 2) \dotplus \mathcal{H}_0$ ($\dotplus$ means independent direct sum), that is, we have proven (3.5.18) for $d + 2$. $\qquad\square$

Corollary 3.5.6. *$D(\nu, d)$, the dimension of $\mathcal{S}(\nu, d)$ is*

$$\binom{d + \nu - 1}{\nu - 1} - \binom{d + \nu - 3}{\nu - 1} \tag{3.5.19}$$

Remarks. 1. For $d = 0, 1$, we interpret the second $(\cdot)$ as 0, (i.e., $D(\nu, 0) = 1$, $D(\nu, 1) = \nu$.

2. For ν fixed, $D(\nu, d)$ is a polynomial in d of degree $(\nu - 2)$ (the $d^{\nu-1}$ terms cancel). In particular,

$$D(3, d) = 2d + 1 \tag{3.5.20}$$

$$D(4, d) = (d + 1)^2 \tag{3.5.21}$$

The first is familiar to students of physics.

3. In Problem 5, the reader will verify an alternate formula for $D(\nu, d)$, namely,

$$D(\nu, d) = \frac{d + \nu - 2}{\nu - 2} \binom{\nu - 3 + d}{\nu - 3} \tag{3.5.22}$$

Proof. Since Δ is a surjection of $\mathcal{H}(\nu, d)$ to $\mathcal{H}(\nu, d - 2)$, its kernel is the difference of the dimensions. Thus, (3.5.17) implies (3.5.19). $\qquad\square$

The next corollary is so important that we call it a theorem!

Theorem 3.5.7. *The sums of spherical harmonics are $\|\cdot\|_\infty$-dense in $C(\mathcal{S}^{\nu-1})$. The $\{Y_{dm}\}_{m=1;d=0,1,2,\ldots}^{D(\nu,d)}$ are an orthonormal basis for $L^2(S^{\nu-1}, d\omega)$. Moreover, if B is the Laplace–Beltrami operator on $S^{\nu-1}$, then*

$$-BY_{dm} = d(d + \nu - 2)Y_{dm} \tag{3.5.23}$$

Remark. If $\nu = 3$, we have the familiar $d(d + 1)$ of the physics student.

Proof. Every polynomial is a sum of homogeneous polynomials and, by (3.5.18), every homogeneous polynomial restricted to $S^{\nu-1}$ (where $x^2 = 1$) is a sum of spherical harmonics. Thus, the restriction of any polynomial to $S^{\nu-1}$ is a sum of spherical harmonics.

If $f \in S^{\nu-1}$ and $g \in C_0^\infty(1 - \delta, 1 + \delta)$ with $g(1) = 1$, then $h(r\omega) = g(r)h(\omega)$ is in $C([-1 - \delta, 1 + \delta]^\nu)$ and so, by the classical Weierstrass approximation theorem, can be approximated uniformly by polynomials in $(x_1, \ldots, x_\nu)$. Thus, the restriction of h to $S^{\nu-1}$, that is, f is a limit of restrictions to the sphere of polynomials and so of sums of spherical harmonics. This proves the first assertion of the theorem.

By Lemma 3.5.4, $\{Y_{dm}\}_{m-1;d=0,1,2,\ldots}^{D(\nu,d)}$ is an orthonormal set. The L^2-closure of their span contains all of $C(S^{\nu-1})$ (since $\|\cdot\|_\infty$ convergence implies $\|\cdot\|_2$ convergence), and so all of $L^2(S^{\nu-1}, d\omega)$, that is, they are an orthonormal basis.

By (3.1.2) and $\Delta(r^d Y) = 0$, we see that (3.5.23) holds. $\qquad\square$

This lets us reduce the study of $-\Delta$ on $\mathbb{R}^\nu$ in many cases to problems on $(0, \infty)$.

Theorem 3.5.8. *Let $f \in \mathcal{S}(\mathbb{R}^\nu)$. Then there exist functions $\{f_{dm}\}_{m=1;d-0,1,\ldots}^{D(\nu,d)}$ C^∞ on $(0, \infty)$ so that*

$$f(r\omega) = \sum_{d,m} r^{-(\nu-1)/2} f_{dm}(r) Y_{dm}(\omega) \tag{3.5.24}$$

converging in $L^2(S^{\nu-1}, d\omega)$ for each r. We have

$$\|f\|^2_{L^2(\mathbb{R}^\nu, d^\nu x)} = \sigma_\nu \sum_{d,m} \|f_{dm}\|^2_{L^2(\mathbb{R}, dr)} \qquad (3.5.25)$$

and

$$(\Delta f)_{dm} = \left[\frac{d^2}{dr^2} - \frac{(\nu-1)(\nu-3)}{4r^2} - \frac{d(d+\nu-2)}{r^2} \right] f_{dm} \qquad (3.5.26)$$

Remark. For $f \in \mathcal{S}$, one can prove uniform convergence in ω for r in closed subsets of $(0, \infty)$.

Proof. Define f_{dm} by

$$f_{dm}(r) = \int r^{(\nu-1)/2} f(r\omega) \overline{Y_{dm}(0)} \, d\omega \qquad (3.5.27)$$

so (3.5.24) and the L^2-convergence is immediate from orthogonality and completeness. This also implies

$$\int r^{\nu-1} |f(r\omega)|^2 \, d\omega = \sum_{d,m} |f_{dm}(r)|^2 \qquad (3.5.28)$$

Since $d^\nu r = \sigma_\nu r^{\nu-1} \, dr d\omega$, we get (3.5.25).

Given (3.1.2), we need to first check, as operators on functions (Problem 9)

$$r^{-(\nu-1)/2} \left[\frac{d}{dr} r^{\nu-1} \frac{d}{dr} r^{-(\nu-1)/2} \right] = \frac{d^2}{dr^2} - \frac{(\nu-1)(\nu-3)}{r^2} \qquad (3.5.29)$$

and then use (3.5.23) to get (3.5.26). $\qquad \square$

Remark. In the language of unbounded operators on $L^2(\mathbb{R}^\nu)$ (see Chapter 7 of Part 4), $-\Delta$ on $L^2(\mathbb{R}^\nu)$ is unitarily equivalent under the unitary given by the expansion (3.5.24) and its inverse given by

$$f_{dm}(r) = r^{(\nu-1)/2} \int_{\mathbb{S}^{\nu-1}} Y_{dm}(w) f(rw) dw \qquad (3.5.30)$$

to $\underset{d,m}{\oplus} h_{dm}$ where

$$h_{dm} = -\frac{d^2}{dr^2} + \frac{d(d+\nu-2)}{r^2} + \frac{(\nu-1)(\nu-3)}{4r^2} \qquad (3.5.31)$$

In the language of Section 7.4 of Part 4, h_{dm} is limit point at ∞ and is limit point at 0, except in the cases $\nu = 2, 3, d = 0$ when it is limit circle. In these two cases, a boundary condition is needed at $r = 0$. In $\nu = 3, d = 0$, it is $u(0) = 0$.

Those who have studied the $\nu = 3$ case know that there is in each $\mathcal{S}(3,d)$ a unique element (up to a constant) invariant under rotations about the z-axis which is given by $P_d(\cos\theta)$ in the standard spherical coordinates where $z = r\cos\theta$. Here P_d is a Legendre polynomial. We will generalize this to ν-dimensions and show P_d is, up to constant, a reproducing kernel!

Definition. The *zonal harmonic* of degree d with pole at ω_0 in $S^{\nu-1}$ is the function (Y_{dm} is an orthonormal basis for $S(\nu,d)$)

$$Z_d(\omega,\omega_0) = \sum_{m=1}^{D(\nu,d)} \overline{Y_{dm}(\omega_0)}\, Y_{dm}(\omega) \tag{3.5.32}$$

Theorem 3.5.9. (a) *For any $Y \in S(\nu,d)$, we have*

$$\int Z_d(\omega,\omega_0)Y(\omega)\,d\omega = Y(\omega_0) \tag{3.5.33}$$

and Z_d is the unique element of $S(\nu,d)$ with this property.

(b) *Z_d is real-valued, independent of choice of orthonormal basis, and only a function of $\omega \cdot \omega_0$ and*

$$\omega \in S^{\nu-1} \Rightarrow |Z_d(\omega,\omega_0)| \leq Z_d(\omega_0,\omega_0) \tag{3.5.34}$$

(c) $Z_d(\omega_0,\omega_0) = \|Z_d(\,\cdot\,,\omega_0)\|^2 = D(\nu,d)$ ⠀⠀⠀⠀⠀⠀ (3.5.35)

(d) *If $\omega\cdot\omega_0 = \cos\theta$, Z_d is a polynomial $P_d^{(\nu)}$ in $\cos\theta$ of degree d that obeys, for $d \neq d^\sharp$,*

$$\int_{-1}^{1} \overline{P_{d^\sharp}^{(\nu)}(x)}\, P_d^{(\nu)}(x)(1-x^2)^{(\nu-3)/2}\,dx = 0 \tag{3.5.36}$$

The $P_d^{(\nu)}(x)$ are orthogonal polynomials of the measure $(1-x^2)^{(\nu-3)/2}\,dx$.

(e) *$Z_d(\omega,\omega_0)$ is, up to a constant multiple, the only spherical harmonic of degree d invariant under rotations about the ω_0-axis. Any L^2-function invariant under rotations about the ω_0-axis lies in the L^2-span of $\{Z_d(\,\cdot\,,\omega_0)\}_{d=0}^{\infty}$.*

Proof. (a), (b) We write $Y(\omega) = \sum_{m=1}^{D(\nu,d)} a_m Y_{dm}(\omega)$. Then

$$\int \overline{Z_d(\omega,\omega_0)}\, Y(\omega)\,d\omega = \sum_{m=1}^{D(\nu,d)} a_m Y_{dm}(\omega_0) = Y(\omega_0) \tag{3.5.37}$$

initially as functions on L^2, but then, since all harmonic functions are continuous, as a pointwise statement.

If now $\widetilde{Y}_{dm}$ is another orthonormal basis, (3.5.37) says

$$\langle \widetilde{Y}_{dm}, Z_d(\,\cdot\,,\omega_0)\rangle = \overline{\widetilde{Y}_{dm}(\omega_0)} \tag{3.5.38}$$

so

$$Z_d(\omega, \omega_0) = \sum_{d=1}^{D(\nu,d)} \overline{\widetilde{Y}_{dm}(\omega_0)}\, \widetilde{Y}_{dm}(\omega) \tag{3.5.39}$$

This only depended on (3.5.37) and shows Z_d is the unique function obeying (3.5.37) and (3.5.32) is independent of basis.

Since $Y \in S(\nu, d) \Rightarrow \operatorname{Re} Y \in S(\nu, d)$, we know there is a choice of $\{Y_{dm}\}$ that are all real-valued, so Z is real-valued and (3.5.37) implies (3.5.32).

Rotations take homogeneous harmonic functions to themselves, so $f \mapsto f \circ R$ (R a rotation) is a unitary map of $L^2(S^{\nu-1}, d\nu)$ mapping $S(\omega, d)$ to itself. In particular, if $\{Y_{dm}\}_{m=1}^{D(\nu,d)}$ is an orthonormal basis, so is $\{Y_{dm} \circ R\}_{m=1}^{D(\nu,d)}$. Basis independence of Z_d implies

$$Z_d(R\omega, R\omega_0) = Z_d(\omega, \omega_0) \tag{3.5.40}$$

It is easy to see (Problem 6) if $\widetilde{\omega} \cdot \widetilde{\omega}_0 = \omega \cdot \omega_0$ for four points on $S^{\nu-1}$, there is a rotation with $R\omega = \widetilde{\omega}$, $R\omega_0 = \widetilde{\omega}_0$. Thus, Z_d is only a function of $\omega \cdot \omega_0$. In particular, for all $\omega, \omega_0 \in S^{\nu-1}$,

$$Z_d(\omega, \omega) = Z_d(\omega_0, \omega_0) \tag{3.5.41}$$

Thus, by the Cauchy inequality for sums,

$$|Z_d(\omega, \omega_0)| \le Z_d(\omega, \omega)^{1/2} Z_d(\omega_0, \omega_0)^{1/2} = Z(\omega_0, \omega_0)$$

proving (3.5.34).

(c) By the orthonormality of $\{Y_{dm}(\omega)\}_{m=1}^{D(\nu,d)}$,

$$\|Z_d(\,\cdot\,, \omega_0)\|^2 = \sum_{m=1}^{D(\nu,d)} |Y_{dm}(\omega_0)|^2 = Z_d(\omega_0, \omega_0)$$

Since $\int |Y_{dm}(\omega)|^2 \, d\omega = 1$,

$$\int Z_d(\omega, \omega) \, d\omega = D(\nu, d) \tag{3.5.42}$$

By (3.5.41), the integral is $Z_d(\omega_0, \omega_0)$.

(d) Pick $\omega_0 = (1, 0, 0, \ldots, 0)$. $Z_d(\omega, \omega_0)$ is a harmonic homogeneous polynomial in $\omega_1, \ldots, \omega_\nu$, the components of ω. It is invariant on $S^{\nu-1}$ under rotations about ω_0, and so, by homogeneity, as a polynomial on $\mathbb{R}^\nu$, it is invariant under such rotations. Thus, it is a function of ω_1 and $\omega_2^2 + \cdots + \omega_\nu^2$. Restricted to $S^{\nu-1}$, $\omega_2^2 + \cdots + \omega_\nu^2 = 1 - \omega_1^2$, so as a function on $S^{\nu-1}$, $Z_d(\omega, \omega_0)$ is a polynomial in $\omega_1 = \omega \cdot \omega_0$ only.

If $\omega_0 = (1, 0, \ldots, 0)$ is fixed and $\omega_0 \cdot \omega = \cos\theta$, the surfaces of constant $\omega_0 \cdot \omega$ are $\nu - 2$-dimensional spheres of radius $\sin\theta$ (see Figure 3.5.1). Thus,

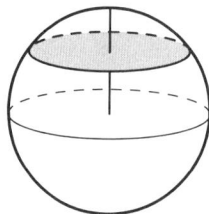

Figure 3.5.1. Slices of spheres.

for any function f of $\omega_0 \cdot \omega$,

$$\int f(\omega \cdot \omega_0)\, d\omega = C_\nu \int_0^{2\pi} f(\cos\theta)[\sin\theta]^{\nu-2}\, d\theta$$

$$= C_\nu \int_{-1}^1 f(x)(1-x^2)^{(\nu-2)/2}\, dx \qquad (3.5.43)$$

In particular, if $d \neq d^\sharp$,

$$\int \overline{Z_{d^\sharp}(\omega,\omega_0)}\, Z_d(\omega,\omega_0)\, d\omega = 0$$

by orthonormality of spherical harmonics of differing degrees. Thus, (3.5.43) for $f = \overline{Z}_{d^\sharp} Z_d$ implies (3.5.36).

(e) (3.5.36) and $\deg P_d(x) \leq d$ implies $\deg P_d(x) = d$ and if $c_d = \|P_d\|_{L^2}^{-1}$, $\{c_d P_d\}_{d=0}^\infty$ is an orthonormal basis for $L^2([-1,1],(1-x^2)^{(\nu-3)/d}\,dx)$. Undoing (3.5.43), we see it is an orthonormal basis. For ω_0 fixed, $\{P_d(\omega \cdot \omega_0)\}_{d=0}^\infty$ is an orthonormal basis for those L^2-functions on $S^{\nu-1}$ invariant under rotations about the ω_0-axis. This proves the second statement in (e).

In particular, if Y is a rotationally invariant function,

$$Y = \sum_{d=0}^\infty c_d^2 \langle Z_d, Y \rangle Z_d \qquad (3.5.44)$$

If Y is also a degree d_0 spherical harmonic, $\langle Z_d, Y \rangle = 0$ for $d \neq d_0$ so Y is a multiple of Z_d, finishing the proof. $\qquad\square$

The weight $\chi_{[-1,1]}(x)(1-x^2)^{(\nu-3)/2}$ is a Jacobi weight (i.e., of the form $\chi_{[-1,1]}(x)(1-x)^\alpha(1+x)^\beta$) in the language of Example 14.4.17 of Part 2B. Here

$$\alpha = \beta = \frac{\nu-3}{2} \qquad (3.5.45)$$

Up to a normalization constant, $\nu = 2$ (i.e., $\alpha = \beta = -\frac{1}{2}$) are Chebyshev polynomials (of the first kind); indeed, $Y_\pm^d = e^{\pm id\theta}$ and $Z_d(\cos\theta) = 2\cos(d\theta) = 2T_d(\cos\theta)$. When $\nu = 3$, $\alpha = \beta = 0$, that is, we have the Legendre polynomials up to normalization. Indeed, $P_d(1) = 1$ and $Z_d(\cos\theta = 1) = 2d + 1$, so

$$(\nu = 3) \qquad Z_d(\cos\theta) = (2d+1)P_d(\cos\theta) \tag{3.5.46}$$

For general ν, $\alpha = \beta$ and we have Gegenbauer, also called *hyperspherical polynomials*.

Next, we turn to spherical harmonic expansions. Several preliminary bounds will be useful. By (3.5.32)/(3.5.35), $\sum_m |Y_{dm}(\omega)|^2 = D(\nu, d)$, so

$$\sup_\omega |Y_{dm}(\omega)| \leq D(\nu, d)^{1/2} \tag{3.5.47}$$

As we've seen for ν fixed, $D(\nu, d)$ is a polynomial of degree $\nu - 2$ in d, so

$$D(\nu, d) \leq C_\nu (d+1)^{\nu-2} \tag{3.5.48}$$

A general fact is:

Theorem 3.5.10. *Let f be harmonic on $\{x \in \mathbb{R}^\nu \mid |x| < 1\}$. Then for $0 < r < 1$,*

$$f(r\omega) = \sum_{d=0}^\infty r^d \sum_{m=1}^{D(\nu,d)} a_{dm} Y_{dm}(\omega) \tag{3.5.49}$$

If f is invariant under rotations about the ω_0-axis (so f is a function of r and $\omega \cdot \omega_0$), then

$$f(r\omega) = \sum_{d=0}^\infty r^d b_{dm} Z_d(\omega, \omega_0) \tag{3.5.50}$$

These series converge uniformly in r, ω on each set of the form $(r, \omega) \in [0, 1 - \delta] \times S^{\nu-1}$ with $\delta > 0$.

Proof. Fix d, m. Let v be the harmonic function

$$v(r\omega) = r^d \overline{Y_{dm}(\omega)} \tag{3.5.51}$$

and define

$$a_{dm}(r) = \int f(r\omega) \overline{Y_{dm}(\omega)} \, d\omega \tag{3.5.52}$$

Since $v\Delta f - f\Delta v = 0$, Green's theorem implies on $B_r(0)$ for any $r \in (0, 1)$ that

$$\int v(r\omega) \frac{\partial}{\partial r} f(r\omega) \, d\omega = \int f(r\omega) \frac{\partial}{\partial r} v(r\omega) \, d\omega \tag{3.5.53}$$

which in terms of (3.5.52) says

$$r^d \frac{d}{dr} a_{dm}(r) = d r^{d-1} a_{dm}(r) \tag{3.5.54}$$

If $a_{dm}(r_0) \neq 0$ for some $r_0 \in (0, 1)$, we can integrate from r_0 to r to get

$$a_{dm}(r) = \left(\frac{r}{r_0}\right)^d a_{dm}(r_0) \tag{3.5.55}$$

which clearly also holds if $a_{dm}(r) = 0$ for all r. Thus, (3.5.55) always holds, which yields (3.5.49) initially converging in ω in $L^2(S^{\nu-1}, d\omega)$ sense.

For each $\rho_0 \in (0, 1)$, (3.5.52) and $\int |Y_{dm}|^2 d\omega = 1$ implies $|a_{dm}(\rho_0)| \leq \sup |f(\rho_0 \omega)| \leq C_{\rho_0}$, so

$$|a_{dm}| \leq C_\rho \rho_0^{-d} \tag{3.5.56}$$

Thus, by (3.5.47),

$$\sum_{m=1}^{D(\nu,d)} |a_{dm}| |Y_{dm}(\omega)| \leq D(\nu, d)^{3/2} C_\rho \rho_0^{-d}$$

Since $D(\nu, d)$ is polynomial bounded, for $r < \rho_0$,

$$\sum_{d=0}^{\infty} D(\nu, d)^{3/2} \left(\frac{r}{\rho_0} \right)^d < \infty \tag{3.5.57}$$

uniformly for $r < \rho_0(1 - \frac{\varepsilon}{2})$. Thus, (3.5.49) has the claimed uniform convergence which must be the L^2-limit.

Suppose f is invariant under rotations about the ω_0-axis. A simple induction in d, looking at $\lim_{r \downarrow 0} r^{-k}[f(r\omega) - \sum_{d=0}^{k=1} r^d \dots]$, proves each function $\sum_{m=1}^{D} a_{dm} Y_{dm}(\omega)$ is invariant under such rotations, so $b_{dm} Z_d(\omega, \omega_0)$. $\qquad \square$

Example 3.5.11 (Poisson Kernel). Here is how we can compute $a_{dm}(r)$ directly without a priori showing it is $a_{dm} r^d$. Since $r^d Y_{dm}(\omega)$ is harmonic,

$$\int P_r(w, \omega') Y_{dm}(\omega') \, d\omega' = r^d Y_{dm}(\omega) \tag{3.5.58}$$

Thus, (3.5.49) becomes

$$P_r(\omega, \omega') = \sum_{d=0}^{\infty} r^d \sum_{m=1}^{D(\nu,d)} \overline{Y_{dm}(\omega)} \, Y_{dm}(\omega') \tag{3.5.59}$$

$$= \sum_{d=0}^{\infty} r^d Z_d(\omega, \omega') \tag{3.5.60}$$

We note that without knowing a Poisson kernel exists, we can use (3.5.59) to *define* $P_r(\omega, \omega')$ and note that (3.5.58) holds. Thus, $\int P_r(\omega, \omega') f(\omega') \, d\omega'$ is a harmonic function for any finite linear combination of Y_{dm}'s. Since they are dense in $\| \cdot \|_\infty$ in $C(S^{\nu-1})$ and $P_r(\omega, \omega')$ is uniformly bounded in $r \in [0, 1 - \delta]$, $\omega, \omega' \in S^{\nu-1}$, we conclude that $P_r(\omega, \omega')$ defined by (3.5.59) is a Poisson kernel and the Dirichlet problem is solvable for the ball. $\qquad \square$

Example 3.5.12 (Multipole Expansion). Fix $x \in \mathbb{R}^\nu$ (with $\nu \geq 3$), $x \neq 0$, and define f on $\{y \mid |y| < |x|\}$ by

$$f(y) = |x - y|^{-(\nu-2)} \tag{3.5.61}$$

which we know is harmonic in y. Moreover, it is clearly invariant under rotations of y about the axis through 0 and x. Thus,

$$|x - y|^{-(\nu-2)} = \sum_{d=0}^{\infty} \frac{|y|^d}{|x|^{d+\nu-2}} \, \alpha_{\nu,d} Z_d\left(\frac{y}{|y|}, \frac{x}{|x|}\right) \qquad (3.5.62)$$

where the Z_d term at $y = 0$ is interpreted as 1 if $d = 0$ and is irrelevant if $d \geq 1$ since it is multiplied by $|y|^d$.

To evaluate $\alpha_{\nu,d}$, suppose $|x| = 1$, $|y| = r < 1$, and $t = y \cdot x / |y|$. In terms of $P_d^{(\nu)}$ obeying $Z_d(\omega, \omega_0) = P_d^{(\nu)}(\omega \cdot \omega_0)$, we have

$$(1 + r^2 - 2rt)^{(\nu-2)/2} = \sum_{d=0}^{\infty} \alpha_{\nu,d} r^d P_d^{(\nu)}(t) \qquad (3.5.63)$$

By (3.5.35), $P_d^{(\nu)}(1) = D(\nu, d)$, so (3.5.63) becomes

$$(1 - r)^{-(\nu-2)} = \sum_{d=0}^{\infty} \alpha_{\nu,d} D(\nu, d) r^d \qquad (3.5.64)$$

On the other hand, by the binomial theorem (or computing derivatives in Taylor's formula),

$$(1 - r)^{-\nu/2} = \sum_{d=0}^{\infty} \frac{(\nu - 2) \ldots (\nu - 3 + d)}{d!} r^d \qquad (3.5.65)$$

$$= \sum_{d=0}^{\infty} \binom{\nu - 3 + d}{\nu - 3} r^d \qquad (3.5.66)$$

Thus, using (3.5.22),

$$\alpha_{\nu,d} = D(\nu, d)^{-1} \binom{\nu - 3 + d}{\nu - 3} = \frac{\nu - 2}{2d + (\nu - 2)} \qquad (3.5.67)$$

We thus have the *multipole expansion* for the Coulomb potential

$$|y| < |x| \Rightarrow |x - y|^{-(\nu-2)} = \sum_{d=0}^{\infty} \frac{\nu - 2}{2d + (\nu - 2)} \frac{|y|^d}{|x|^{d+\nu-2}} Z_d\left(\frac{y}{|y|}, \frac{x}{|x|}\right) \qquad (3.5.68)$$

In particular, if $\nu = 3$, then $P_d^{(3)}(x) = (2d + 1) P_d(x)$, with P_d the Legendre polynomials, and we have that

$$(\nu = 3), \quad |y| < |x|, \quad |x - y|^{-1} = \sum_{d=0}^{\infty} \frac{|y|^d}{|x|^{d+1}} P_d\left(\frac{x \cdot y}{|x||y|}\right) \qquad (3.5.69)$$

$\square$

As a final topic, we want to explore Fourier transforms of spherical harmonics and their connection to Bessel functions. The Bessel function of order α obeys

$$J_\alpha(z) = \frac{1}{\sqrt{\pi}\,\Gamma(\alpha + \frac{1}{2})} \left(\frac{z}{2}\right)^\alpha \int_{-1}^1 e^{izt}(1 - t^2)^{\alpha - \frac{1}{2}}\, dt \qquad (3.5.70)$$

((14.5.22) of Part 2B). For this section, we'll only be interested in α's with $2\alpha \in \mathbb{Z}$, and we could take (3.5.70) as a definition of J_α, although that would miss important properties like the ODE it obeys and that $z^{-\alpha}J_\alpha(z)$ is an entire function.

We begin with

Theorem 3.5.13. *For each fixed $\omega' \in S^{\nu-1}$, $t \in \mathbb{R}$,*

$$\frac{\sigma_\nu}{(2\pi)^{\nu/2}} \int_{S^{\nu-1}} e^{-it\omega \cdot \omega'}\, d\omega = \frac{J_{(\nu-2)/2}(t)}{t^{(\nu-2)/2}} \qquad (3.5.71)$$

Remarks. 1. Here σ_ν is area of $S^{\nu-1}$, that is (see Problems 5 and 6 of Section 4.11 of Part 1),

$$\sigma_\nu = \frac{2(\sqrt{\pi})^\nu}{\Gamma(\frac{\nu}{2})} \qquad (3.5.72)$$

2. The left side of (3.5.71) is the Fourier transform of the normalized measure on the unit sphere.

Proof. Without loss (since rotation invariance says the integral on the left side is ω'-independent), we can take $\omega' = (0, \ldots, 0, -1)$. Using $x_\nu = \cos\theta$ and $\omega_{\nu-1}$ for coordinates on $S^{\nu-2}$, we see that

$$d\omega_\nu(\omega_{\nu-1}, \theta) = c_\nu[\sin\theta]^{\nu-2}\, d\theta d\omega_{\nu-1}$$

since the slices at fixed θ are $\nu - 2$-dimensional spheres of radius $\sin\theta$. Here c_ν is picked so that $d\omega_\nu$ is a probability measure if $d\omega_{\nu-1}$ is. To compute c_ν, we note

$$\int_0^\pi (\sin\theta)^{\nu-2}\, d\theta = \int_{-1}^1 (1 - x^2)^{(\nu-3)/2}\, dx \qquad (3.5.73)$$

$$= \frac{\sqrt{\pi}\,\Gamma(\frac{\nu-1}{2})}{\Gamma(\frac{\nu}{2})} \qquad (3.5.74)$$

by (4.11.64)/(4.11.65) of Part 1. Thus,

$$c_\nu = \frac{\Gamma(\frac{\nu}{2})}{\sqrt{\pi}\,\Gamma(\frac{\nu-1}{2})} \qquad (3.5.75)$$

and

$$\int e^{-it\omega \cdot \omega'}\, d\omega = c_\nu \int_{-1}^1 (1 - x^2)^{(\nu-3)/2} e^{ixt}\, dx \qquad (3.5.76)$$

$$= c_\nu \sqrt{\pi}\, \Gamma\left(\frac{\nu-1}{2}\right) 2^{(\nu-2)/2} \frac{J_{(\nu-2)/2}(t)}{t^{(\nu-2)/2}} \qquad (3.5.77)$$

by (3.5.70).

The constant in front of $J_{(\nu-2)/2}(t)$ is

$$2^{(\nu-2)/2}\Gamma\left(\frac{\nu}{2}\right) = (2\pi)^{\nu/2}\frac{\Gamma(\frac{\nu}{2})}{2\pi^{\nu/2}} = \frac{(2\pi)^{\nu/2}}{\sigma_\nu} \qquad (3.5.78)$$

proving (3.5.71). $\square$

Definition. The *Bessel transform*, aka Hankel transform, is defined for functions $g \in \mathcal{S}(\mathbb{R})$ with $g(-x) = g(x)$ by

$$\tilde{g}_\nu(k) = k^{-(\nu-2)/2}\int_0^\infty J_{(\nu-2)/2}(kt)t^{\nu/2}g(t)\,dt \qquad (3.5.79)$$

Corollary 3.5.14. *Let $f \in \mathcal{S}(\mathbb{R}^\nu)$ so that for some even $g \in \mathcal{S}(\mathbb{R})$,*

$$f(\mathbf{x}) = g(|\mathbf{x}|) \qquad (3.5.80)$$

Then

$$\hat{f}(\mathbf{k}) = \tilde{g}_\nu(|\mathbf{k}|) \qquad (3.5.81)$$

Remark. Once one has this for $\mathcal{S}$, it can be extended to $\mathcal{S}'$, L^p, etc.

Proof. $d^\nu x = \sigma_\nu(r^{\nu-1}dr)\,d\omega_\nu$ (normalized $d\omega_\nu$), so

$$\hat{f}(k) = (2\pi)^{-\nu/2}\sigma_\nu \int_0^\infty g(r)r^{n-1}\left[\int e^{-i|k|r\omega\cdot\omega'}\,d\omega_\nu\right]dr \qquad (3.5.82)$$

$$= \int_0^\infty r^{\nu-1}(kr)^{-(\frac{\nu}{2}-1)}g(r)J_{(\nu-2)/2}(kr)\,dr \qquad (3.5.83)$$

is (3.5.81), given (3.5.79). $\square$

We are heading towards analogous formulae when $f(x) = g(|x|)$ is replaced by $f(x) = g(|x|)P_d(x)$, where P_d is a homogeneous harmonic polynomial of degree d.

Theorem 3.5.15. *Let $P_d(x)$ be a homogeneous harmonic polynomial of degree d. Let*

$$f(x) = e^{-\frac{1}{2}x^2}P_d(x) \qquad (3.5.84)$$

Then

$$\hat{f}(k) = i^{-d}f(k) \qquad (3.5.85)$$

Remark. As we discussed in Section 6.4 of Part 1, iterating $\hat{\ }$ four times gives the identity and thus, $L^2(\mathbb{R}^\nu)$ is spanned by eigenfunctions of $\hat{\ }$, where the eigenvalues are among $\pm 1, \pm i$.

Proof. Since P_d is harmonic, for any $y \in \mathbb{R}^\nu$ and $r \in (0, \infty)$,

$$\int P_d(r\omega + y)\,d\omega = P_d(y) \tag{3.5.86}$$

Thus, integrating dr, for any radially symmetric $f \in \mathcal{S}(\mathbb{R}^\nu)$ (i.e., of the form (3.5.80)),

$$\int f(x) P_d(x + y)\,d^\nu x = P_d(y) \int f(x)\,d^\nu x \tag{3.5.87}$$

In particular, taking $f(x) = (2\pi)^{-\nu/2} e^{-\frac{1}{2}x^2}$ (so $\int f(x)\,d^\nu x = 1$),

$$(2\pi)^{-\nu/2} \int e^{-\frac{1}{2}x^2} P_d(x + y)\,d^\nu x = P_d(y) \tag{3.5.88}$$

or

$$(2\pi)^{-\nu/2} \int e^{-\frac{1}{2}(x-y)^2} P_d(x)\,d^\nu x = P_d(y) \tag{3.5.89}$$

Both sides of (3.5.89) are easily seen to be entire analytic functions of $(y_1, y_2, \ldots, y_\nu)$, so (3.5.89) is valid for complex y. In particular, taking $y = -ik$ with $k \in \mathbb{R}^\nu$, we get

$$(2\pi)^{-\nu/2} \int e^{-\frac{1}{2}x^2} e^{+\frac{1}{2}k^2} e^{-ik\cdot x} P_d(x)\,d^\nu x = P_d(-ik) = (-i)^d P_d(k) \tag{3.5.90}$$

which is (3.5.85). $\qquad\qquad\qquad\qquad\qquad\qquad\qquad\qquad\qquad\qquad\square$

We are heading towards a proof that if $Y_d \in \mathcal{S}(\nu, d)$ is a ν-variable spherical harmonic of degree d, then

$$\frac{\sigma_\nu}{(2\pi)^{\nu/2}} \int_{S^{\nu-1}} e^{-it\omega\cdot\omega'} Y_d(\omega)\,d\omega = i^{-d} \frac{J_{d+\frac{1}{2}(\nu-2)}(t)}{t^{(\nu-2)/2}} Y_d(\omega') \tag{3.5.91}$$

This implies that if $g \in \mathcal{S}(\mathbb{R})$ is even and

$$f(x) = g(|x|)|x|^d Y_d\left(\frac{x}{|x|}\right) \tag{3.5.92}$$

then

$$\widehat{f}(k) = i^{-d} \widetilde{g}_{\nu+2d}(|k|)|k|^d Y_d\left(\frac{k}{|k|}\right) \tag{3.5.93}$$

by mimicking the proof of Corollary 3.5.14 (Problem 7).

Indeed, we'll prove (3.5.93) and use it to deduce (3.5.91) by letting g approach an approximate identity at $k = 1$. What we'll do is note that Theorem 3.5.15 is (3.5.93), first for $g(x) = e^{-x^2/2}$ and, by scaling, for $e^{-x^2/2a}$. The following identity result is thus critical:

Lemma 3.5.16. *Let $\mathcal{H}_\alpha = \{\varphi$ measurable on $[0, \infty)\big| \int_0^\infty r^{\alpha-1}|\varphi(r)|^2\,dr < \infty\}$ with*

$$\|\varphi\|_\alpha = \left(\int_0^\infty r^{\alpha-1}|\varphi(r)|^2\,dr\right)^{1/2} \tag{3.5.94}$$

Then the finite linear combinations of $\{e^{-x^2/2a}\}_{a\in(0,\infty)}$ are dense in $\mathcal{H}_\alpha$.

Proof. Since $x \mapsto e^{-x^2}$ maps $[0, \infty]$ bijectively to $[0, 1]$, polynomials in e^{-x^2} are $\|\cdot\|_\infty$-dense in the continuous functions on $[0, \infty]$ (i.e., continuous functions on $[0, \infty)$ with a limit at infinity). Thus, if f is a continuous function on $[0, \infty)$ with compact support, there is a polynomial, P_n, in e^{-x^2} so $\sup_x |P_n(e^{-x^2}) - e^{x^2} f(x)| < \frac{1}{n}$. Thus, $\sup_x |e^{x^2} [e^{-x^2} P_n(e^{-x^2}) - f(x)]| < \frac{1}{n}$, so for all α, $\|e^{-x^2} P_n(e^{-x^2}) - f\|_\alpha \to 0$. Thus, linear combinations of $\{e^{-mx^2}\}_{m=1}^\infty$ are dense in $\mathcal{H}_\alpha$, that is, we only need $a = \{\frac{1}{2m}\}_{m=1}^\infty$! $\square$

Define $T_\nu : \mathcal{H}_\nu \to \mathcal{H}_\nu$ for $\nu = 1, 2, 3, \ldots$ by

$$(T_\nu \varphi)(k) = \tilde{\varphi}_\nu(k) \tag{3.5.95}$$

By (3.5.81), the fact that $\hat{}$ is an isometry on $L^2(\mathbb{R}^\nu)$ and that $\int |g(|x|)|^2 \, d^\nu x = \sigma_\nu \|g\|_\nu^2$, we conclude T_ν is an isometry of $\mathcal{H}_\nu \to \mathcal{H}_\nu$. By the Fourier inversion formula, it is onto.

Lemma 3.5.17.

(a) $T_\nu(e^{-x^2/2a}) = a^{\nu/2} e^{-ak^2/2}$ $\tag{3.5.96}$

(b) *If P_d is a homogeneous harmonic polynomial on $\mathbb{R}^\nu$, then*

$$\widehat{e^{-|x|^2/2a} P_d(x)}\,(k) = i^{-d} a^{d + \frac{\nu}{2}} e^{-ak^2/2} P_d(a^{1/2} k) \tag{3.5.97}$$

Proof. Define S_a on $\mathcal{S}(\mathbb{R}^\nu)$ by

$$(S_a g)(x) = g\left(\frac{x}{a^{1/2}}\right) \tag{3.5.98}$$

Then

$$\widehat{S_a g}\,(k) = (2\pi)^{-\nu/2} \int e^{-ik \cdot x} g\left(\frac{x}{a^{1/2}}\right) d^\nu x = a^{\nu/2} \hat{g}(a^{1/2} k) = a^{\nu/2} S_{1/a} \hat{g} \tag{3.5.99}$$

(3.5.96) is true for $a = 1$ (i.e., $e^{-\frac{1}{2}|x|^2}$ is its own Fourier transform), so for all a by (3.5.99).

Since $e^{-|x|^2/2a} P_d(x) = a^{d/2} S_a(e^{-x^2/2} P_d(x))$, (3.5.92) says

$$\text{LHS of (3.5.97)} = i^{-d} a^{d/2} a^{\nu/2} e^{-a|k|^2/2} P_d(a^{1/2} k)$$

$$= \text{RHS of (3.5.97)} \qquad \square$$

Theorem 3.5.18. *Fix P_d a homogeneous harmonic polynomial of degree d on $\mathbb{R}^\nu$.*

(a) *Let*

$$f(x) = g(|x|) P_d(x) \tag{3.5.100}$$

for $g \in \mathcal{H}_{\nu+2d}$.

Then

$$\widehat{f}(k) = i^{-d}(T_{\nu+2d}f)(k)P_d(k) \qquad (3.5.101)$$

that is, (3.5.93) holds.

(b) *(3.5.91) holds.*

Proof. (a) Because of continuity of the Fourier and Bessel transforms, it suffices to prove (3.5.101) for an L^2-dense set. By Lemma 3.5.16, it suffices to prove it for $f(x) = e^{-x^2/2a}$. This is immediate from (3.5.96) and (3.5.97). $\square$

Combining this with Theorem 6.8.1 of Part 1, one gets (Problem 8)

Theorem 3.5.19. *Let $0 < \gamma < \nu$ and on $\mathbb{R}^\nu$,*

$$f(x) = |x|^{-\gamma} Y_d\left(\frac{x}{|x|}\right) \qquad (3.5.102)$$

for a spherical harmonic Y_d of degree d. Then

$$\widehat{f}(k) = i^{-d} 2^{\frac{\nu}{2}-\gamma} \frac{\Gamma(\frac{1}{2}(\nu-\gamma)+\frac{d}{2})}{\Gamma(\frac{\gamma}{2}+\frac{d}{2})} |k|^{-(\nu-\gamma)} Y_d\left(\frac{k}{|k|}\right) \qquad (3.5.103)$$

Example 3.5.20 (Spherical Harmonic Expansion of Plane Waves)**.** We want to look at $e^{i\mathbf{k}\cdot\mathbf{x}}$ for

$$\mathbf{k} = k\boldsymbol{\omega}, \qquad \mathbf{x} = x\boldsymbol{\omega}' \qquad (3.5.104)$$

and expand in functions of ω and ω'. $e^{it\boldsymbol{\omega}\cdot\boldsymbol{\omega}}$ is invariant under rotations about the ω'-axis, so

$$e^{it\omega\cdot\omega'} = \sum_{d=0}^{\infty} a_d(t) Z_d(\omega \cdot \omega') \qquad (3.5.105)$$

By the orthogonality of the Z_d's,

$$a_d(t) = \|Z_d\|^{-2} \int e^{it\omega\cdot\omega'} Z_d(\omega \cdot \omega') \, d\omega \qquad (3.5.106)$$

$$= \|Z_d\|^{-2} Z_d(\omega' \cdot \omega') \frac{(2\pi)^{\nu/2}}{\sigma_\nu} i^d \frac{J_{d+\frac{1}{2}(\nu-2)}(t)}{t^{(\nu-2/2)}} \qquad (3.5.107)$$

Noting $\|Z_d\|^2 = Z_d(\omega', \omega')$ (3.5.35), we get

$$e^{i\mathbf{k}\cdot\mathbf{r}} = \sum_{d=0}^{\infty} \frac{(2\pi)^{\nu/2}}{\sigma_\nu} i^d \frac{J_{d+\frac{1}{2}(\nu-2)}(kr)}{(kr)^{(\nu-2)/2}} Z_d(\boldsymbol{\omega} \cdot \boldsymbol{\omega}') \qquad (3.5.108)$$

$$= \sum_{d=0}^{\infty} \frac{(2\pi)^{\nu/2}}{\sigma_\nu} i^d \frac{J_{d+\frac{1}{2}(\nu-2)}(kr)}{(kr)^{(\nu-2)/2}} \sum_{m=0}^{D(\nu,d)} \overline{Y_{dm}(\omega)} Y_{dm}(\omega') \qquad (3.5.109)$$

called the *plane wave expansion*. If $\nu = 3$, $\frac{(2\pi)^{\nu/2}}{\sigma_\nu} = \frac{(2\pi)^3}{4\pi} = (\frac{\pi}{2})^{1/2}$. With this in mind, one defines the *spherical Bessel function, j_ℓ,* by

$$j_\ell(t) = \sqrt{\frac{\pi}{2}} \frac{J_{\ell+\frac{1}{2}}(t)}{t^{1/2}} \tag{3.5.110}$$

so that

$$(\nu = 3) \qquad e^{ik \cdot r} = \sum_{\ell=0}^{\infty} i^\ell (2\ell + 1) j_\ell(kr) P_d(\omega \cdot \omega') \tag{3.5.111}$$

$$= \sum_{d=0}^{\infty} i^d j_d(kr) \sum_{m=0}^{d} \overline{Y_{dm}(\omega)} \, Y_{dm}(\omega') \tag{3.5.112}$$

(the second differs from the form in (14.5.34) of Part 2B because there, the Y_{dm} are normalized for $4\pi \, d\omega$ measure).

If $\nu = 2$, $\frac{(2\pi)^{\nu/2}}{\sigma_\nu} = 1$ and (3.5.109) becomes

$$e^{ikr\cos\theta} = \sum_{d=-\infty}^{\infty} i^{|d|} J_{|d|}(kr) e^{id\theta} \tag{3.5.113}$$

$\square$

Notes and and Historical Remarks. Our normalization convention for spherical harmonics is different from the standard one. Everyone agrees they should have $L^2(S^{\nu-1}, d\omega)$-norm 1. However, we define $d\omega$ to be the normalized rotation-invariant measure, but most references use the measure induced by Lebesgue measure on $\mathbb{R}^\nu$, that is, so $d\omega$ has total mass σ_ν. The difference is factors of $\sqrt{\sigma_\nu}$. Therefore, for us, $\sqrt{3/2}\,(x_1 + ix_2)$ is a dimension-3, degree-1 spherical harmonic, while for others it is $\sqrt{3/8\pi}\,(x_1 + ix_2)$.

Spherical harmonics were first introduced (for dimension 3) by Legendre [**518**] and Laplace [**508**]. (Legendre was only published in 1785, but was actually first and known to Laplace.) They were studying Newton's potential $|x - y|^{-1}$. Legendre invented his polynomials as occurring in the multipole expansion ($|y| < |x|$)

$$\frac{1}{|x - y|} = \sum_{n=0}^{\infty} \frac{|y|^n}{|x|^{n+1}} P_n(\cos\theta) \tag{3.5.114}$$

where $x \cdot y = |x||y|\cos\theta$. Laplace separated variables in trying to solve $\Delta u = 0$. In spherical coordinates where

$$x_1 = r\sin\theta\cos\varphi, \qquad x_2 = r\sin\theta\sin\varphi, \qquad x_3 = r\cos\theta \tag{3.5.115}$$

separating variables, that is, looking for solutions of the form $f_{\ell m}(r)g_{\ell m}(\theta)\psi_{\ell m}(\varphi)$ leads to $f_{\ell m}(r) = r^\ell$, $\psi_{\ell m}(\varphi) = e^{\pm im\varphi}$ (Laplace, of course, used $\cos(m\varphi), \sin(m\varphi))$, $g_{\ell m}$ obeys a complicated ODE whose solutions are now called associated Legendre polynomials.

In 1867, William Thomson and Peter Tait, in their *Treatise on Natural Philosophy* [**805**], emphasized the homogeneous harmonic polynomial aspect. They introduced the name "spherical harmonic" because the two-dimensional analogs were $\cos(m\theta), \sin(m\theta)$, which are the standard "harmonic function" in the sense of musical harmonics. In the end, their name spherical harmonic not only stuck but led to the more general term "harmonic function" for solutions of $\Delta u = 0$. While on the subject of names, some authors use "spherical harmonic" for the harmonic polynomials on $\mathbb{R}^\nu$ and "surface harmonic" for their restrictions to $S^{\nu-1}$.

William Thomson (1824–1907) was a British scientist. Shortly after his birth, William's father was made professor of mathematics at Glasgow where William, himself, served as Professor of Natural Philosophy for over fifty years starting when he was only twenty-two! Eventually (in 1892), he was made Baron Kelvin of Largs and much of his work, for example, the temperature scale (based on his notion of absolute zero) is known under this name. By the end of his career, he was certainly the most famous British physicist since Newton. His interests were wide, ranging from mathematics (where we've seen his work on harmonic functions and as a popularizer of Green's work) to physics (where as the reconciler of the work of Carnot and Joule, he founded the science of thermodynamics) to engineering (where he was instrumental in the success of the transatlantic telegraph cable). Indeed, it was the work on the telegraph cable that gained him fame, fortune, and his being ennobled. As one of his biographers (Watson [**843**]; for other biographies see [**538, 763, 804**]) remarked, Kelvin was always right in the first part of his career and seemed to often be wrong in the latter part where he championed incorrect theories of the age of the Earth and Sun and opposed Darwin's theory, the theory of atoms, and radioactivity as not physical phenomena. His contribution to mathematics would have made him a notable figure but, like Newton, he illustrates how broad and profound a mathematical figure can be.

We have suppressed the connections between spherical harmonics and representations of $\mathbb{SO}(\nu)$, the group of rotations in ν-dimensions. As we noted, the space $\mathcal{S}(\nu, d)$ is left invariant by $\mathbb{SO}(\nu)$, so as a finite-dimensional space, it supports a representation which is unitary under the $L^2(S^\nu, d\omega)$ inner product. It can also be shown to be irreducible (see Section 6.8 of Part 4 for representation theory terminology). For $\nu = 3$, every irreducible representation of $\mathbb{SO}(3)$ occurs this way, but that is not true if $\nu \geq 4$ (see, e.g., Simon [**746**] for a discussion of representations of $\mathbb{SO}(\nu)$.)

The representation of $\mathbb{SO}(\nu)$ induces a representation of so(ν), its Lie algebra by first-order differential operators. If $\{L_\beta\}_{\beta=1}^{\dim \text{so}(\nu)}$ is an orthonormal basis, $L^2 = \sum_\beta L_\beta^2$ is a multiple of the Laplace–Beltrami operator. One labels the Y_{dm} (i.e., determines m) by picking a Weyl algebra (maximal abelian subalgebra) and choosing the Y_{dm} to be eigenvectors of all the generators. In $\nu = 3$ with spherical coordinates, the Weyl algebra is one-dimensional and its generator is conventionally taken to be $\frac{1}{i}\frac{\partial}{\partial \varphi}$, that is $Y_{dm}(\theta, \varphi) = f_{dm}(\theta)e^{im\theta}$.

One consequence of (3.5.114) is that (with our normalization of $Y_{\ell m}$), if $d\rho$ is a charge density supported in $\{y \mid |y| < R\}$ and $|x| > R$, then

$$\int \frac{d\rho(y)}{|x-y|} = \sum_{\ell=0}^{\infty} \frac{\alpha_{\ell m}}{|x|^{\ell+1}} Y_{\ell m} \qquad (3.5.116)$$

where

$$\alpha_{\ell m} = \int |y|^\ell Y_{\ell m}(\omega) \, d\rho(|y|\omega) \qquad (3.5.117)$$

If $\alpha_{00} \neq 0$ (i.e., the total charge $\int d\rho(y) \neq 0$), the leading term for large $|x|$ is $\alpha_{00}|x|^{-1}$. But if not, the $\ell = 1$ term involving $\int y_j \, d\rho(y)$, the *dipole moment*, is the leading term. If the dipole moment is 0, the quadruple matters, etc. Hence the name "multipole" expansion.

Our approach to the Fourier transforms of spherical harmonics times the probability measure on the sphere, including the Gaussian calculation and the $\nu + 2d$ shift, is from Stein–Weiss [**785**]. Earlier approaches use properties of special functions like addition formula for Bessel functions.

Problems

1. If f is C^∞ near $\mathbf{x} \neq 0$ in $\mathbb{R}^\nu$, prove that the component of ∇f parallel to $\mathbf{x}$ is $\frac{\partial f}{\partial r}$ and conclude that (3.5.3) holds.

2. Check (3.5.3) directly when applied to a monomial $f(x) = x_1^{n_1} \ldots x_\nu^{n_\nu}$, using that $f(\lambda x) = \lambda^{n_1 + \cdots + n_\nu} f(x)$.

3. Set up a one–one correspondence between monomials of degree at most d in $\nu - 1$ variables and monomials of degree exactly d in ν variables and conclude that

$$\dim \mathcal{H}(\nu, d) = \dim \mathcal{P}(\nu - 1, d) \qquad (3.5.118)$$

4. Prove that the number of monomials of exact degree d in ν variables is $\binom{d+\nu-1}{\nu-1}$. (*Hint*: If $\nu - 1$ points are picked from $\{1, 2, \ldots, d + \nu - 1\}$, show they determine $n_1, n_2, \ldots, n_\nu$ in $\{0, 1, 2, \ldots\}$ so that $\sum_{j=1}^{\nu} n_j = d$.)

5. Let Δ_α be Laplacian in $x_1, \dots, x_\alpha$. If p is a polynomial in $(x_1, \dots, x_\nu)$ homogeneous of degree d, write

$$p(x_1, \dots, x_\nu) = \sum_{j=0}^{d} x_\nu^j q_j(x_1, \dots, x_{\nu-1})$$

(a) Prove that $\Delta_\nu p = 0$ if and only if for $j = 0, \dots, d-2$, we have

$$\Delta_{\nu-1} q_j = -(j+2)(j+1)q_{j+2} \qquad (3.5.119)$$

(b) Prove that q_0, q_1 are arbitrary homogeneous polynomials of degree d and $d-1$, respectively, and they determine $q_2, \dots, q_d$.

(c) Conclude $\dim \mathcal{S}(\nu, d) = \dim \mathcal{H}(\nu-1, d) + \dim \mathcal{H}(\nu-1, d-1)$.

(d) Prove (3.5.22).

6. For $\omega_1, \omega_2, \omega_3, \omega_4$ in $S^{\nu-1}$, prove there is a rotation $R \in \mathbb{SO}(\nu)$ so that $R\omega_1 = \omega_3$, $R\omega_2 = \omega_4$ if and only if $\omega_1 \cdot \omega_2 = \omega_3 \cdot \omega_4$.

7. Show directly that (3.5.91) implies (3.5.93).

8. Prove Theorem 3.5.19.

9. (a) Prove that

$$r^\alpha \frac{d}{dr} r^{-\alpha} = \frac{d}{dr} - \frac{\alpha}{r} \qquad (3.5.120)$$

(b) Verify (3.5.29).

3.6. Potential Theory

Let μ be a finite Borel measure on $\mathbb{R}^\nu$ ($\nu \geq 2$) of compact support. The *potential* generated by μ is the function

$$\Phi_\mu(x) = \int G_0(x-y)\, d\mu(y) \qquad (3.6.1)$$

where G_0 is the free Green's function given by (3.1.25)/(3.1.26). We note that if $\nu \geq 3$, $G_0 \geq 0$ so the integral is well-defined although it may be infinite for some x. Since μ has compact support, if $\nu = 2$, G_0 is bounded below so the integral still exists. Thus, Φ_μ is well-defined although it may sometimes be $+\infty$. We saw (see Example 3.2.8) that Φ_μ is superharmonic and, in particular, lsc. When $\nu = 2$, G_0 has a factor of $(2\pi)^{-1}$. Often, especially in applications to complex analysis, this factor of $(2\pi)^{-1}$ is dropped. Indeed, we'll do so in Sections 3.7 and 3.8 below. So the reader will need some care in comparing those sections to this one.

From one point of view, "potential theory" is the study of potentials but, as we've noted, some use it as a synonym for the theory of harmonic and subharmonic functions, so much so that one mathematician I know said he felt that the standard reference on potential theory is a book entitled *Subharmonic Functions* [**370**].

For me, the essence of "potential theory," what makes it a subset of the theory of subharmonic functions, is the central role of the Coulomb energy

$$\mathcal{E}(\mu) = \int G_0(x-y) \, d\mu(x) \, d\mu(y) \tag{3.6.2}$$

(which differs by a multiplicative constant from the definition we used in (5.9.3) of Part 1). That said, there are books with the name "potential theory" in their title which never define $\mathcal{E}(\mu)$ or which discuss it as an afterthought. In our discussion here, $\mathcal{E}(\mu)$ will play a central role.

In this regard, we recall Theorem 1.2.1 (which recapped results from Part 1). If $\mathfrak{e} \subset \mathbb{R}^\nu$ is compact, we say $\mathfrak{e}$ has zero capacity if $\mathcal{E}(\mu) = \infty$ for all nonzero measures, μ, on $\mathfrak{e}$. If $\mathfrak{e}$ does not have zero capacity, we define the *Robin constant*, $R(\mathfrak{e})$, by

$$R(\mathfrak{e}) = \inf_{\mu \in \mathcal{M}_{+,1}(\mathfrak{e})} \mathcal{E}(\mu) \tag{3.6.3}$$

and *capacity* of $\mathfrak{e}$ by $G_0(C(\mathfrak{e})) = R(\mathfrak{e})$. Theorem 1.2.1 asserts that if $R(\mathfrak{e}) < \infty$, then there is a unique probability measure, $\rho_\mathfrak{e}$, on $\mathfrak{e}$, called the *equilibrium measure* for $\mathfrak{e}$ with

$$\mathcal{E}(\rho_\mathfrak{e}) = R(\mathfrak{e}) \tag{3.6.4}$$

Our initial focus will be on $\rho_\mathfrak{e}$ and its potential, $\Phi_{\rho_\mathfrak{e}} \equiv \Phi_\mathfrak{e}$. We will link it to a Green's function, $G_\mathfrak{e}$, and show that this Green's function can be used to construct barriers. A major result of that part of this section will be the Kellogg–Evans theorem that for any bounded open Ω in $\mathbb{R}^\nu$, the set of singular points in $\partial\Omega$ is small in the sense that any measure, μ, with support in the singular points has $\mathcal{E}(\mu) = \infty$. We'll then use the Kellogg–Evans theorem to study an extended Dirichlet problem for general bounded domains, Ω. Next, we'll explore what the Riemann mapping theorem says when $\nu = 2$ and finally prove a useful technical result about convergence of potentials that we'll need in the next section. Theorem 3.6.24 will relate $G_\mathfrak{e}$ to the Riemannian map of $(\mathbb{C} \setminus \mathfrak{e}) \cup \{\infty\}$ to $\mathbb{D}$ in case $\nu = 2$ and $\mathfrak{e}$ is compact and connected and simply connected.

We want to note some links of Φ_μ and $\mathcal{E}(\mu)$. Clearly, by Fubini's theorem,

$$\mathcal{E}(\mu) = \int \Phi_\mu(x) \, d\mu(x) \tag{3.6.5}$$

Fubini's theorem also implies that

$$\int \Phi_\mu(x) \, d\eta(x) = \int \Phi_\eta(x) \, d\mu(x) \tag{3.6.6}$$

for any two measures μ and η (since both sides equal $\int G_0(x-y) \, d\mu(x) \, d\eta(y)$, which some call their *mutual energy*).

Definition. A *polar set*, is a Borel set, S, in $\mathbb{R}^\nu$ so that $\mathcal{E}(\mu) = \infty$ for any μ with support in S (equivalently if $C(\mathfrak{e}) = 0$ for all compact sets $\mathfrak{e} \subset S$). If some property, $P(x)$, holds except for $x \in S$, a polar set, we say P holds *quasi-everywhere* (q.e.).

It is easy to see that a countable union of polar sets is polar.

Proposition 3.6.1. *Let S be a polar set and μ a Baire measure on $\mathbb{R}^\nu$ with compact support so that $\mathcal{E}(\mu) < \infty$. Then,*

$$\mu(S) = 0 \qquad\qquad (3.6.7)$$

In particular, the Lebesgue measure, $|S|$, of S is zero.

Remarks. 1. That $|S| = 0$ is one sense in which polar sets are small. In particular, something that holds q.e. holds Lebesgue a.e.

2. Much more is true than $|S| = 0$. In fact (Problem 1), S has Hausdorff dimension at most $\nu - 2$. It can be of that dimension, e.g., $\{(x, y, z) \mid y = z = 1, x \in [-1, 1]\}$ is a polar set in $\mathbb{R}^3$ (Problem 2). This is significant because we'll later talk about polar subsets of $\partial\Omega$ with Ω a bounded open set and $\partial\Omega$ has dimension at least $\nu - 1$.

Proof. It suffices to suppose that S is compact by inner regularity of μ. If $\nu = 2$, we can then scale so that $S \cup \operatorname{supp}(\mu) \subset \{|x| < \frac{1}{2}\}$. Thus, if $x, y \in \operatorname{supp}(\mu)$, $G_0(x - y) \geq 0$, i.e., if μ_S is the measure $\mu_S(B) = \mu(S \cap B)$, then

$$\mathcal{E}(\mu_S) \leq \mathcal{E}(\mu) < \infty \qquad\qquad (3.6.8)$$

since $G_0(x - y) \geq 0$ on $\operatorname{supp}(\mu) \times \operatorname{supp}(\mu)$. If $\mu(S) \neq 0$, μ_S is not the zero measure. Since $\operatorname{supp}(\mu_S) = S$ (since S is compact), we have a contradicition to the assumption that S is polar.

Since Lebesgue measure, μ_R, on $B_R = \{x \mid |x| \leq R\}$ is a finite measure of compact support with $\mathcal{E}(\mu_R) < \infty$, $\mu_R(S) = 0$ so $|S| = \lim_{R \to \infty} \mu_R(S) = 0$. $\qquad\square$

We need the following technical result below and again in the next section.

Theorem 3.6.2 (Continuity Principle). *Let μ be a measure on $\mathbb{R}^\nu$ of compact support $\mathfrak{e}$. Then Φ_μ is continuous (and finite) at $x_0 \in \operatorname{supp}(\mu)$ if and only if $\Phi_\mu \upharpoonright \mathfrak{e}$ is continuous (and finite) at x_0.*

Proof. Clearly, continuity of Φ_μ implies continuity of $\Phi_\mu \upharpoonright \mathfrak{e}$ so we need only prove the converse. Let $\Phi_\mu \upharpoonright \mathfrak{e}$ be continuous. Φ_μ is always continuous on $\mathbb{R}^\nu \setminus \mathfrak{e}$ so pick $x_0 \in \mathfrak{e}$. Since $\Phi_\mu(x_0) < \infty$, given ε, there is $\delta < \frac{1}{2}$, so that

$$\int_{|x-x_0|<\delta} G_0(x - x_0) \, d\mu(x) < \varepsilon \qquad\qquad (3.6.9)$$

Let $\mu_1 = \mu \restriction \{x \mid |x - x_0| \geq \delta\}$, $\mu_2 = \mu - \mu_1$. Since Φ_{μ_1} is trivially continuous at x_0, so is $\Phi_{\mu_2} \restriction \mathfrak{e} = \Phi_\mu \restriction \mathfrak{e} - \Phi_{\mu_1} \restriction \mathfrak{e}$. Thus, there is $\eta \in (0, \frac{1}{2})$ so that

$$|y - x_0| < \eta, \quad y \in \mathfrak{e} \Rightarrow \Phi_{\mu_2}(y) < \varepsilon \qquad (3.6.10)$$

If x is arbitrary, pick $y(x) \in \mathfrak{e}$ so that

$$|x - y(x)| = \inf_{w \in \mathfrak{e}} |x - w| \qquad (3.6.11)$$

Then, for any $w \in \mathfrak{e}$,

$$|y(x) - w| \leq |x - y(x)| + |x - w| \leq 2|x - w| \qquad (3.6.12)$$

so

$$G_0(x - w) \leq \begin{cases} 2^{\nu-2} G_0(y(x) - w), & \nu \geq 3 \\ G_0(y(x) - w) + \log(2), & \nu = 2 \end{cases} \qquad (3.6.13)$$

We also note that if $\nu = 2$, since $\delta < \frac{1}{2}$, (3.6.9) implies that

$$\mu(\{x \mid |x - x_0| < \delta\}) \leq \varepsilon (\log 2)^{-1}, \quad \nu = 2 \qquad (3.6.14)$$

If $|x - x_0| < \frac{1}{2}\eta$, then by (3.6.12), $|y(x) - x_0| < \eta$ so if $\nu \geq 3$

$$|x - x_0| < \frac{1}{2}\eta \Rightarrow \Phi_{\mu_2}(x) \leq 2^{\nu-2}\Phi_{\mu_2}(y(x)) \leq 2^{\nu-2}\varepsilon \qquad (3.6.15)$$

and if $\nu = 2$,

$$|x - x_0| < \frac{1}{2}\eta \Rightarrow \Phi_{\mu_2}(x) \leq \Phi_{\mu_2}(y(x)) + (\log 2)\,\varepsilon\,(\log 2)^{-1} \leq 2\varepsilon \qquad (3.6.16)$$

We thus have that if $|x - x_0| < \frac{1}{2}\eta$, then

$$|\Phi_\mu(x) - \Phi_\mu(x_0)| \leq |\Phi_{\mu_1}(x) - \Phi_{\mu_1}(x_0)| + \Phi_{\mu_2}(x) + \Phi_{\mu_2}(x_0)$$
$$\leq |\Phi_{\mu_1}(x) - \Phi_{\mu_1}(x_0)| + (2^{\nu-1} + 1)\varepsilon \qquad (3.6.17)$$

Since Φ_{μ_1} is continuous at x_0,

$$\limsup_{x \to x_0} |\Phi_\mu(x) - \Phi_\mu(x_0)| \leq (2^{\nu-1} + 1)\varepsilon \qquad (3.6.18)$$

Since ε is arbitrary, Φ_μ is continuous at x_0. $\qquad\qquad\square$

Corollary 3.6.3. *Let μ be a measure on $\mathbb{R}^\nu$ of compact support, $\mathfrak{e}$, with $\mathcal{E}(\mu) < \infty$, and let $\varepsilon > 0$. Then, there is K of compact support so that with $\mu_K(B) = \mu(K \cap B)$, we have that*

(a) Φ_{μ_K} *is continuous on $\mathbb{R}^\nu$*

(b) $\mu(\mathfrak{e} \setminus K) < \varepsilon$ $\qquad\qquad\qquad\qquad\qquad\qquad\qquad\qquad\qquad$ (3.6.19)

Proof. $\mathcal{E}(\mu) = \int_{\mathfrak{e}} \Phi_\mu(x)\, d\mu(x)$ so $\Phi_\mu(x) \in L^1(\mathfrak{e}, d\mu)$. Thus, by Lusin's theorem (Theorem 4.4.8 of Part 1), there exists K compact so that (3.6.19) holds and $\Phi_\mu \restriction K$ is continuous. Let $\tilde{\mu} = \mu - \mu_K$ which is also a positive measure. $\Phi_{\mu_K} \restriction K$ and $\Phi_{\tilde{\mu}} \restriction K$ are both lsc as potentials. Thus, $\Phi_{\mu_K} \restriction K = \Phi_\mu \restriction K - \Phi_{\tilde{\mu}} \restriction K$ is usc so $\Phi_{\mu_K} \restriction K$ is continuous. By the theorem, Φ_{μ_K} is globally continuous. $\qquad\square$

Remark. Picking $\varepsilon = \frac{1}{2}\mu(\mathfrak{e})$, we get $\rho = \mu_K$ with $\rho \neq 0$, $\mathrm{supp}(\rho) \subset \mathrm{supp}(\mu) = \mathfrak{e}$ and Φ_ρ continuous, i.e.,

Corollary 3.6.4. *If $\mathfrak{e}$ is a compact nonpolar set, there exists a probability measure, ρ on $\mathfrak{e}$, so that Φ_ρ is continuous on $\mathbb{R}^\nu$ and $\mathcal{E}(\rho) < \infty$.*

Corollary 3.6.5. *For any potential, Φ_μ, with $\mathrm{supp}(\mu) = \mathfrak{e}$, we have that*

$$\sup_{x \in \mathbb{R}^\nu} \Phi_\mu(x) = \sup_{x \in \mathfrak{e}} \Phi_\mu(x) \qquad (3.6.20)$$

Proof. If $\mathcal{E}(\mu) = \infty$, $\int_{\mathfrak{e}} \Phi_\mu(x)\, d\mu(x) = \infty$ so both sups in (3.6.20) are infinite. Thus, we can suppose that $\mathcal{E}(\mu) < \infty$.

Fix $\varepsilon > 0$. By the last corollary, pick K obeying (a) and (b). If $x \notin \mathfrak{e}$ and $\tilde{\mu} = \mu - \mu_K$,

$$\Phi_\mu(x) = \Phi_{\mu_K}(x) + \Phi_{\tilde{\mu}}(x)$$
$$\leq \Phi_{\mu_K}(x) + G_0(\mathrm{dist}(x, \mathfrak{e}))\,\varepsilon \qquad (3.6.21)$$

Since Φ_{μ_K} is continuous, the maximum principle for harmonic functions (and $\Phi_{\mu_K} \to 0$ at ∞ if $\nu \geq 3$ and to $-\infty$ if $\nu = 2$) implies

$$\Phi_\mu(x) \leq \sup_{y \in \mathfrak{e}} \Phi_{\mu_K}(y) + \varepsilon\, G_0(\mathrm{dist}(x, \mathfrak{e})) \qquad (3.6.22)$$
$$\leq \sup_{y \in \mathfrak{e}} \Phi_\mu(y) + \varepsilon\, G_0(\mathrm{dist}(x, \mathfrak{e})) \qquad (3.6.23)$$

If $\nu \geq 3$, we used $\Phi_{\mu_K} \leq \Phi_\mu$ and if $\nu = 2$, we scale so diam $\mathfrak{e} < 1$ so we still have $\Phi_{\mu_K} \leq \Phi_\mu$. Since ε is arbitrary, we have that (3.6.20) holds. $\qquad\square$

Here is a key fact about equilibrium measures:

Theorem 3.6.6 (Frostman's Theorem). *Let $\mathfrak{e}$ be a compact subset of $\mathbb{R}^\nu$ with $R(\mathfrak{e}) < \infty$, $\rho_{\mathfrak{e}}$ its equilibrium measure, and $\Phi_{\mathfrak{e}}$, the potential of $\rho_{\mathfrak{e}}$. Then,*

(a) *On $\mathfrak{e}$, $\Phi_{\mathfrak{e}}(x) = R(\mathfrak{e})$ for q.e. x, in fact, $\{x \in \mathfrak{e} \mid \Phi_{\mathfrak{e}}(x) \neq R(\mathfrak{e})\}$ is a polar F_σ set.*

(b) *On $\mathfrak{e}^{\mathrm{int}}$, $\Phi_{\mathfrak{e}}(x) = R(\mathfrak{e})$.*

(c) *$\rho_{\mathfrak{e}}$ is supported on $\partial\mathfrak{e}$, the topological boundary of $\mathfrak{e}$.*

(d) *$\Phi_{\mathfrak{e}}(x) \leq R(\mathfrak{e})$ on all of $\mathbb{R}^\nu$.* $\qquad (3.6.24)$

(e) If $\nu \geq 3$, $\Phi_{\mathfrak{e}}(x) \geq 0$ and if $\mathfrak{e} \subset \{x \mid |x| \leq R_0\}$ for $\nu = 2$,

$$\Phi_{\mathfrak{e}}(x) \geq -\frac{1}{2\pi} \log(|x| + R_0). \tag{3.6.25}$$

(f) For q.e. $x_0 \in \mathfrak{e}$, $\lim_{x \to x_0} \Phi_{\mathfrak{e}}(x) = \Phi_{\mathfrak{e}}(x_0)$. $\tag{3.6.26}$

Proof. We first claim that

$$\mathfrak{e}_n = \{x \in \mathfrak{e} \mid \Phi_{\mathfrak{e}}(x) \leq R(\mathfrak{e}) - \frac{1}{n}\} \tag{3.6.27}$$

is a closed polar set. If not, let ν be a probability measure supported on $K \subset \mathfrak{e}_n$ with $\mathcal{E}(\nu) < \infty$. We want to show how to decrease the energy by borrowing a little of the charge in $\rho_{\mathfrak{e}}$ and placing it in a small multiple of ν.

Since

$$\int \Phi_{\mathfrak{e}}(x) \, d\rho_{\mathfrak{e}}(x) = R(\mathfrak{e}) \tag{3.6.28}$$

we can find $S \subset \mathfrak{e}$, closed (by regularity of $\rho_{\mathfrak{e}}$) so $\rho_{\mathfrak{e}}(S) > 0$ and $\Phi_{\mathfrak{e}}(x) \geq R(\mathfrak{e})$ on S. Let $\rho_{\mathfrak{e},S}(B) = \rho_{\mathfrak{e}}(S)^{-1} \rho_{\mathfrak{e}}(S \cap B)$ and for $\lambda \in (0, \rho_{\mathfrak{e}}(S))$

$$\eta_\lambda = \rho_{\mathfrak{e}} + \lambda(\nu - \rho_{\mathfrak{e},S}) \tag{3.6.29}$$

η_λ is a positive measure since $\lambda < \rho_{\mathfrak{e}}(S)$. Since $\mathcal{E}(\rho_{\mathfrak{e},S}) < \infty$ and $\mathcal{E}(\nu) < \infty$ we know integrals like $\int \Phi_\nu(x) \, d\rho_{\mathfrak{e},S}(x)$ are finite (if $\nu \geq 3$, $G_0 \geq 0$ and such manipulations are straightforward; if $\nu = 2$, one uses the trick of scaling everything inside $\{x \mid |x| < \frac{1}{2}\}$ (Problem 3)). Thus,

$$\mathcal{E}(\eta_\lambda) = \mathcal{E}(\rho_{\mathfrak{e}}) + 2\lambda \int \Phi_{\mathfrak{e}}(x)(d\nu(x) - d\rho_{\mathfrak{e},S}(x))$$

$$+ \lambda^2 \left(\mathcal{E}(\nu) + \mathcal{E}(\rho_{\mathfrak{e},S}) - 2 \int \Phi_\nu(x) \, d\rho_{\mathfrak{e},S} \right) \tag{3.6.30}$$

By construction, $\Phi_{\mathfrak{e}}(x) \leq R(\mathfrak{e}) - \frac{1}{n}$ on $\mathrm{supp}(\nu)$ and $\Phi_{\mathfrak{e}}(x) \geq R(\mathfrak{e})$ on $\mathrm{supp} \, d\rho_{\mathfrak{e},S}$. Thus,

$$\mathcal{E}(\eta_\lambda) \leq \mathcal{E}(\rho_{\mathfrak{e}}) - \lambda \frac{1}{n} + A\lambda^2 \tag{3.6.31}$$

for $A < \infty$. It follows for λ small, $\mathcal{E}(\eta_\lambda) < \mathcal{E}(\rho_{\mathfrak{e}})$ violating the definition of $\rho_{\mathfrak{e}}$. This contradiction shows that $\mathfrak{e}_n$ is a polar set.

Secondly, we claim that

$$\{x \in \mathrm{supp}(d\rho_{\mathfrak{e}}) \mid \Phi_{\mathfrak{e}}(x) > R(\mathfrak{e})\} \tag{3.6.32}$$

is empty. For let $x_0 \in \mathrm{supp}(d\rho_{\mathfrak{e}})$ with $\Phi_{\mathfrak{e}}(x_0) > R(\mathfrak{e}) + \frac{1}{n}$. By the first claim, $\{x \in \mathfrak{e} \mid \Phi_{\mathfrak{e}}(x) < R(\mathfrak{e})\}$ is a polar set, so by Proposition 3.6.1, $\rho_{\mathfrak{e}}(\{x \mid \Phi_{\mathfrak{e}}(x) < R(\mathfrak{e})\}) = 0$. Since

$$\int \Phi_{\mathfrak{e}}(x) \, d\rho_{\mathfrak{e}}(x) = R(\mathfrak{e}) \tag{3.6.33}$$

we conclude that

$$\rho_{\mathfrak{e}}(\{x \mid \Phi_{\mathfrak{e}}(x) > R(\mathfrak{e})\}) = 0 \tag{3.6.34}$$

Since $\Phi_{\mathfrak{e}}$ is lsc, $\{y \mid \Phi_{\mathfrak{e}}(y) > R(\mathfrak{e}) + \frac{1}{n}\}$ is open so there is $r > 0$ with $\Phi_{\mathfrak{e}}(y) > R(\mathfrak{e}) + \frac{1}{n}$ on $B_r^{x_0} = \{y \mid |y - x_0| < r\}$. Since $x_0 \in \text{supp}(d\rho_{\mathfrak{e}})$, $\rho_{\mathfrak{e}}(B_r^{x_0}) > 0$ violating (3.6.34). Thus, as claimed, the set in (3.6.32) is empty. With these two claims in hand, we turn to proving statements (a)–(e) of the theorem:

(d) Let $K = \text{supp}(d\rho_{\mathfrak{e}})$. By the fact that (3.6.32) is empty,

$$\sup_{x \in K} \Phi_{\mathfrak{e}}(x) \le R(\mathfrak{e}) \tag{3.6.35}$$

(of course, once we prove (a), we know that the sup is $R(\mathfrak{e})$). By (3.6.20), we get (3.6.24).

(a) By (d), $\Phi_{\mathfrak{e}}(x) \le R(\mathfrak{e})$ for every $x \in \mathfrak{e}$ and by the fact that each $\mathfrak{e}_n$ given by (3.6.27) is a polar set which is closed (since $\Phi_{\mathfrak{e}}$ is usc), $\bigcup_{n=1}^\infty \mathfrak{e}_n$ is a polar F_σ and is the set where $\Phi_{\mathfrak{e}}(x) \ne R(\mathfrak{e})$.

(b) Let $x_0 \in \mathfrak{e}^{\text{int}}$. If δ_0 is such that $\{x \mid |x - x_0| < \delta_0\} \subset \mathfrak{e}$, then by (a) and Proposition 3.6.1, for Lebesgue a.e. x, in that ball, $\Phi_{\mathfrak{e}}(x) = R(\mathfrak{e})$. So the average over any ball of radius $\delta < \delta_0$ is $R(\mathfrak{e})$. Since $\Phi_{\mathfrak{e}}$ is superharmonic, this average converges to $\Phi_{\mathfrak{e}}(x_0)$, i.e., $\Phi_{\mathfrak{e}}(x_0) = R(\mathfrak{e})$.

(c) As distributions, by (3.2.46)

$$\rho_{\mathfrak{e}} = -\Delta\Phi_{\mathfrak{e}} \tag{3.6.36}$$

By (b), $\rho_{\mathfrak{e}}(\mathfrak{e}^{\text{int}}) = 0$.

(e) For $y \in \mathfrak{e}$, $|x - y| \le |x| + R_0$ so, by monotonicity of $(2\pi)^{-1}\log(|x|^{-1})$, $\frac{1}{2\pi}\log(|x-y|^{-1}) \ge -\frac{1}{2\pi}\log(|x| + R_0)$ which implies (3.6.25) since $d\rho_{\mathfrak{e}}$ is a probability measure.

(f) If $\Phi_{\mathfrak{e}}(x_0) = R(\mathfrak{e})$, by lsc $\liminf_{x \to x_0} \Phi_{\mathfrak{e}}(x) \ge R(\mathfrak{e})$. By (3.6.24), $\limsup_{x \to x_0} \Phi_{\mathfrak{e}} \le R(\mathfrak{e})$. Thus, (3.6.26) holds. $\square$

The following is an aside in the sense that we won't use it later (except that on the unbounded component $\Phi_{\mathfrak{e}} < R(\mathfrak{e})$ which is easier than the rest) but it illuminates what is going on. Given $\mathfrak{e} \subset \mathbb{R}^\nu$ compact,

$$\mathbb{R}^\nu \setminus \mathfrak{e} = \bigcup_{j=0}^J U_j \tag{3.6.37}$$

where the U_j are disjoint open connected sets ($J = 0, \ldots$ or ∞), the components of $\mathbb{R}^\nu \setminus \mathfrak{e}$. We can label them so that U_0 is the only unbounded set among the U_j's. ∂U_0 is called the *outer boundary* of $\mathfrak{e}$ and denoted $O\partial(\mathfrak{e})$.

Theorem 3.6.7. *Let $\mathfrak{e}$ be a compact subset of $\mathbb{R}^\nu$. Let $\widetilde{\mathfrak{e}}$ be $\mathfrak{e}$ "with the holes filled in," i.e., $\mathfrak{e} \cup \bigcup_{j=1}^\infty U_j = \mathbb{R}^\nu \setminus U_0$. Then:*

(a) $\rho_{\mathfrak{e}} = \rho_{\widetilde{\mathfrak{e}}}$, $R(\mathfrak{e}) = R(\widetilde{\mathfrak{e}})$. $\tag{3.6.38}$

(b) $\rho_{\mathfrak{e}}$ is supported on $O\partial(\mathfrak{e})$.

(c) $\Phi_{\mathfrak{e}}(x) = R(\mathfrak{e})$ on each bounded component, $U_j, j \geq 1$.   (3.6.39)

(d) $\Phi_{\mathfrak{e}}(x) < R(\mathfrak{e})$ on U_0 the unbounded component.   (3.6.40)

Proof. (a) Let $\Phi = \Phi_{\mathfrak{e}}$, $\widetilde{\Phi} = \Phi_{\widetilde{\mathfrak{e}}}$. Since $\mathfrak{e} \subset \widetilde{\mathfrak{e}}$, $\widetilde{\Phi}(x) = R(\widetilde{\mathfrak{e}})$ for q.e. $x \in \widetilde{\mathfrak{e}}$ and so $\rho_{\mathfrak{e}}$-a.e. $x \in \mathfrak{e}$ (since polar sets give zero weight to $\rho_{\mathfrak{e}}$). It follows that

$$R(\widetilde{\mathfrak{e}}) = \int \widetilde{\Phi}(x)\,d\rho_{\mathfrak{e}}(x) \tag{3.6.41}$$

Similarly, since $\operatorname{supp}(\rho_{\widetilde{\mathfrak{e}}}) \subset \partial\widetilde{\mathfrak{e}} = O\partial\mathfrak{e} \subset \mathfrak{e}$, we have that $\Phi(x) = R(\mathfrak{e})$ for $\rho_{\widetilde{\mathfrak{e}}}$-a.e. x. Thus,

$$R(\mathfrak{e}) = \int \Phi(x)d\rho_{\widetilde{\mathfrak{e}}}(x) \tag{3.6.42}$$

By (3.5.6), $R(\mathfrak{e}) = R(\widetilde{\mathfrak{e}})$!

   Since $\rho_{\widetilde{\mathfrak{e}}}$ is supported on $\partial\widetilde{\mathfrak{e}} \subset \mathfrak{e}$ and $\mathcal{E}(\rho_{\widetilde{\mathfrak{e}}}) = R(\widetilde{\mathfrak{e}}) = R(\mathfrak{e})$, by uniqueness of minimizers, we conclude that $\rho_{\widetilde{\mathfrak{e}}} = \rho_{\mathfrak{e}}$.

(b) Since $\rho_{\widetilde{\mathfrak{e}}}$ is supported by $\partial\widetilde{\mathfrak{e}} = O\partial\mathfrak{e}$, and $\rho_{\mathfrak{e}} = \rho_{\widetilde{\mathfrak{e}}}$, we conclude that $\rho_{\mathfrak{e}}$ is supported by $O\partial\mathfrak{e}$.

(c) $U_j \subset \widetilde{\mathfrak{e}}^{\,\text{int}}$ so $\rho_{\widetilde{\mathfrak{e}}} = \rho_{\mathfrak{e}}$ is equal to $R(\widetilde{\mathfrak{e}}) = R(\mathfrak{e})$ everywhere on U_j.

(d) $\Phi_{\mathfrak{e}}(x)$ is harmonic on U_0 and bounded above by $R(\mathfrak{e})$. If it were equal to $R(\mathfrak{e})$, by the maximum principle, $\Phi_{\mathfrak{e}}$ would equal $R(\mathfrak{e}) > 0$ on all of U_0. But $\Phi_{\mathfrak{e}}(x) \to 0$ as $|x| \to \infty$ ($\nu \geq 3$) or $\Phi_{\mathfrak{e}}(x) \to -\infty$ as $|x| \to \infty$ ($\nu = 2$). $\quad\square$

We define the *Green's function* of a nonpolar compact set, $\mathfrak{e}$, by

$$G_{\mathfrak{e}}(x) = R(\mathfrak{e}) - \Phi_{\mathfrak{e}}(x) \tag{3.6.43}$$

(also called the *Green's function with a pole at infinity*). We have

Theorem 3.6.8. $G_{\mathfrak{e}}(x)$ *is a nonnegative function on* $\mathbb{R}^\nu$ *which is everywhere subharmonic and harmonic on* $\mathbb{R}^\nu \setminus \mathfrak{e}$. *Moreover,*

(a) $G_{\mathfrak{e}}(x) > 0$ *if* x *is in the unbounded component of* $\mathbb{R}^\nu \setminus \mathfrak{e}$.
(b) *For q.e. $x_0 \in \mathfrak{e}$, we have*

$$\lim_{x \to x_0} G_{\mathfrak{e}}(x) = 0 \tag{3.6.44}$$

(c) *We have the following asymptotics at $x = \infty$:*

$$G_{\mathfrak{e}}(x) = -G_0(x) + R(\mathfrak{e}) + O(|x|^{-(\nu-1)}) \tag{3.6.45}$$

(d) *If $\nu \geq 3$, $G_{\mathfrak{e}}(x)$ is bounded (below by 0 and above by $R(\mathfrak{e})$). If $\nu = 2$, $G_{\mathfrak{e}}(x)$ is bounded on the set $\{x \mid |x| \leq R\}$ (above by $R(\mathfrak{e})$ and below by $-\frac{1}{2\pi}\log(R + R_0)$).*

Remarks. 1. We'll later prove (Theorem 3.6.17) that these conditions uniquely determine $G_{\mathfrak{e}}$ (indeed, we need less than the full asymptotics).

2. The asymptotics in (3.6.45) is very different if $\nu = 2$ or $\nu \geq 3$. If $\nu = 2$, $-G_0 \to \infty$ as $|x| \to \infty$ and $G_0(x)$ is the dominant term. If $\nu \geq 3$, $G_0(x) = O(|x|^{-(\nu-2)})$ and $R(\mathfrak{e})$ is the dominant term.

Proof. All but (c) are direct consequences of the properties of $\Phi_{\mathfrak{e}}$ proven in Theorems 3.6.6 and 3.6.7. To prove (3.6.45), we note that uniformly for y in $\mathfrak{e}$, as $|x| \to \infty$,

$$G_0(x - y) = G_0(x) + O(|x|^{-(\nu-1)}) \tag{3.6.46}$$

by an easy calculation (Problem 4) so

$$\Phi_{\mathfrak{e}}(x) = -G_0(x) + O(|x|^{-(\nu-1)}) \tag{3.6.47}$$

$\square$

The point, of course, is that strictly positive harmonic functions going to 0 at a boundary point are the hallmark of barriers. Of course, these functions are exterior to $\mathfrak{e}$, not interior (if, say $\mathfrak{e} = \overline{\Omega}$ with Ω a bounded region) but the Kelvin transform

$$(Ku)(x) = |x|^{-(\nu-2)} u(x |x|^{-2}) \tag{3.6.48}$$

will invert interior and exterior! Here is the result:

Theorem 3.6.9 (Kellogg–Evans Theorem). *Let $\Omega \subset \mathbb{R}^\nu$, $\nu \geq 2$ be a bounded domain. Then the set of singular points in $\partial\Omega$ is an F_σ polar set.*

Remark. For now, we'll only prove that there is a F_σ polar set, $S \subset \partial\Omega$ so that all $x \in \partial\Omega \setminus S$ are the regular points, so the singular points are a subset of a polar set. Later, (Corollary 3.6.16), we'll prove that points in S are all singular proving that the singular points are Borel; indeed an F_σ set.

Proof. Let $\mathbb{R}^\nu \setminus \Omega = \bigcup_{j=0}^J U_j$ be the decomposition (3.6.37) where U_j are the components of the complement and U_0 is the unbounded component. By translating, we can suppose that $0 \in \Omega$. Let $\mathfrak{e} = \{x^{-1} \mid x \in \overline{U}_0\} \cup \{0\}$. Then $G_{\mathfrak{e}}$ is harmonic and positive on $\mathbb{R}^\nu \setminus \mathfrak{e}$ (since by construction $\mathbb{R}^\nu \setminus \mathfrak{e}$ has a single component.

The Kelvin transform $Q = K G_{\mathfrak{e}}$ is a positive harmonic function on $(\mathbb{R}^\nu \setminus U_0) \setminus \{0\} = [\Omega \setminus \{0\}] \cup \bigcup_{j=1}^j U_j$. $G_{\mathfrak{e}} \to 0$ as $x \to x_0 \in \partial\mathfrak{e} \setminus T$ for a polar set T. Q goes to zero as $x \to x_0 \in \partial U_0 = O\partial\Omega$ except for $x_0 \in S_0 = \{x^{-1} \mid x \in T\}$ which it is easy to see is a polar set (Problem 5). Q is a weak barrier at any point $x_0 \in \partial U_0$ where $Q(x) \to 0$ as $x \to x_0$. Thus, $\partial U_0 \setminus S_0$ are all regular points by definition.

For later purposes, we note that there is a constant c so that $h = Q - c\,G_0$ is harmonic also at $x = 0$ and h is bounded on Ω. This follows from the properties of $G_{\mathfrak{e}}$.

For $j \geq 1$, let $\mathfrak{e}_j = \overline{U}_j$. Then $G_{\mathfrak{e}_j}$ goes to 0 as $x \in \mathbb{R}^\nu \setminus \mathfrak{e}_j \to x_0 \in \partial\mathfrak{e}_j \setminus S_j$ for all polar sets $S \subset \partial\mathfrak{e}_j$. At any such x_0, $G_{\mathfrak{e}_j}$ is a weak barrier for x_0 and Ω, so if

$$S = S_0 \cup \bigcup_{j=1}^{j} S_j$$

then all $x \in \partial\Omega \setminus S$ are regular. For $\nu \geq 3$, we note that on Ω, $G_{\mathfrak{e}_j}$ is bounded. $\square$

By the Perron theory, for any $f \in C(\partial\Omega)$, there is a harmonic function on Ω, $u = H_\Omega f$ so for q.e. $x_0 \in \partial\Omega$, $(H_\Omega f)(x_0) \to f(x_0)$ as $x \to x_0$ with $x \in \Omega$. It is natural to think of this as the solution of a kind of Dirichlet problem for f, but we are missing the uniqueness part of the picture and it is that to which we turn. We'll prove that a bounded harmonic function h on Ω so that for q.e. $x_0 \in \partial\Omega$, $h(x) \to 0$ as $x \in \Omega \to x_0$ has $h \equiv 0$. This will imply uniqueness. The key will be an extended maximum principle for subharmonic functions where we only know something about boundary values off a polar set and for that, the key will be a removable singularities theorem. Look at the proof which shows how to remove a single point in Theorem 3.1.26—it relied on the existence of a superharmonic function which is infinite at the point, namely $G_0(x - x_0)$. Thus, we begin by looking for potentials (the prototypical superharmonic function) that are infinite exactly on some polar set.

Proposition 3.6.10. *For any measure μ of compact support, $\{x \mid \Phi_\mu(x) = \infty\}$ is a polar G_δ.*

Proof. Since Φ_μ is lsc, $\{x \mid \Phi_\mu(x) > n\}$ is open so

$$Q = \{x \mid \Phi_\mu(x) = \infty\} = \bigcap_{n=1}^{\infty} \{x \mid \Phi_\mu(x) > n\} \qquad (3.6.49)$$

is a G_δ.

Suppose that Q is not polar. Then there is a measure ν_1 supported on a compact $K_1 \subset Q$ with $\mathcal{E}(\nu_1) < \infty$ and then by Corollary 3.6.3 a measure ν on $K \subset K_1 \subset Q$ with

$$\sup_{x \in \mathbb{R}^\nu} \Phi_\nu(x) = M < \infty \qquad (3.6.50)$$

Since $\Phi_\mu = \infty$ on $K \subset Q$ and ν is supported on K

$$\int \Phi_\mu(x)\,d\nu(x) = \infty \qquad (3.6.51)$$

But by (3.6.50)

$$\int \Phi_\nu(x)d\mu \leq M < \infty \qquad (3.6.52)$$

By (3.6.6), these last two facts are inconsistent. Thus, Q is polar. ☐

It is a fact (see the Notes) that the converse of this theorem is true, i.e., if Q is a polar G_δ, there is a measure μ so that $\{x \mid \Phi_\mu(x) = \infty\}$ is Q but we'll settle for the weaker

Proposition 3.6.11. *Let Q be a compact polar set. Then, there is a measure μ of compact support so that*

$$Q = \{x \mid \Phi_\mu(x) = \infty\} \qquad (3.6.53)$$

Proof. Let $K_n = \{x \mid \operatorname{dist}(x, Q) \leq \frac{1}{n}\}$. Since Q is compact,

$$\bigcap_n K_n = Q \qquad (3.6.54)$$

Since $K_n^{\mathrm{int}} \neq \emptyset$, $R(K_n) < \infty$. Let ν_n be the equilibrium measure for K_n so each $\nu_n \in \mathcal{M}_{+,1}(K_1)$ and thus ν_n has a weak limit point ν_∞. Since $\operatorname{supp}(\nu_n) \subset K_m$ if $m \leq n$, by (3.6.54), $\operatorname{supp}(\nu_\infty) = Q$. Since Q is polar, $\mathcal{E}(\nu_\infty) = \infty$. Since $\mathcal{E}$ is lsc (see Lemma 5.9.2 of Part 1), $\mathcal{E}(\nu_n) \to \infty$, so, by passing to a subsequence which we still label as ν_n and K_n, we still have (3.6.54) and

$$R(K_n) = \mathcal{E}(\nu_n) \geq 2^n \qquad (3.6.55)$$

Let

$$\mu = \sum_n 2^{-n} \nu_n \qquad (3.6.56)$$

We claim that μ obeys (3.6.53).

By (c) of Theorem 3.6.6, $\Phi_{\nu_n}(x) = R(K_n)$ for $x \in Q \subset K_n^{\mathrm{int}}$. Thus, if $x \in Q$,

$$\Phi_\mu(x) \geq \sum_n 2^{-n} R(K_n) \geq \sum_n 2^{-n} 2^n = \infty \qquad (3.6.57)$$

By (3.6.54), if $x \notin Q$, there is an n_0 with $x \notin K_{n_0}$. If $m \geq n_0$ and $K_m \subset K_{n_0}$, so $\operatorname{dist}(x, K_m) \geq \operatorname{dist}(x, K_{n_0}) \equiv d_{n_0}(x)$, so since $G_0(x)$ is decreasing in $|x|$,

$$\Phi_{\nu_m}(x) \leq G_0(d_{n_0}(x)) \qquad (3.6.58)$$

Thus, $\sum_{m=n_0}^\infty 2^{-m} \Phi_{\nu_m}(x) \leq 2^{1-n_0} G_0(d_{n_0}(x))$. Since $\Phi_{\nu_m}(x) \leq R(K_m) \leq R(K_{n_0})$ if $m < n_0$

$$\Phi_\mu(x) \leq R(K_{n_0}) + 2^{1-m} G_0(d_{n_0}(x)) < \infty \qquad (3.6.59)$$

Therefore, (3.6.53) holds. ☐

Theorem 3.6.12. (Removable Singularities Theorem for Subharmonic Functions). *Let Ω be an open domain in $\mathbb{R}^\nu$, K a closed polar set in $\mathbb{R}^\nu$ and u a function subharmonic on $\Omega \setminus K$ with*

$$\sup_{x \in \Omega \setminus K} u(x) < \infty \qquad (3.6.60)$$

Then u has an extension to all of Ω which is subharmonic and this extension is unique.

Remark. Later in this section and also in Section 6.8 of Part 2A, we'll discuss the related but distinct question of extending analytic functions. Since analytic functions can be expressed in terms of harmonic functions, any set removable for bounded harmonic functions will be removable for analytic functions but there are sets removable for bounded analytic functions but not for bounded harmonic functions.

Proof. By (3.2.1), if u is subharmonic on Ω for all $x_0 \in \Omega$,

$$u(x_0) = \lim_{r \downarrow 0} |\{x \mid |x - x_0| \leq r\}|^{-1} \int_{|x-x_0| \leq r} u(x) \, d^\nu x \qquad (3.6.61)$$

Since K is polar, $|K| = 0$ by Proposition 3.6.1 so changing u on K doesn't change the integral on the right of (3.6.61) so doesn't change u at any point. This proves uniqueness.

Let $\Omega_n = \Omega \cap \{x \mid |x| < n\}$, $K_n = K \cap \{x \mid |x| \leq n\}$. If we can extend u as a function on $\Omega_n \setminus K_n$ to Ω_n, it is easy to see by uniqueness, we can extend from Ω to $\Omega \setminus K$ so henceforth we can suppose that K is compact.

Define u on K by

$$x \in K \Rightarrow u(x) = \limsup_{y \to x; \, y \in \Omega \setminus K} u(y) \qquad (3.6.62)$$

It is easy to see (Problem 6) that u defined this way is usc so we need only prove the submean property.

Fix $x_0 \in K$ and δ_m so that $B_m = \{x \mid |x - x_0| \leq \delta_m\} \subset \Omega$ and the spherical measure of $K \cap \partial B_m$ is zero. Since $|K| = 0$, we can find $\delta_m \downarrow 0$ with this property. By the construction in the proof of Theorem 3.2.11, we can find continuous functions v_n on B_m so v_n is harmonic on B_m^{int}, and

$$\forall \omega : v_n(x_0 + \delta_m \omega) \geq u(x_0 + \delta \omega) \qquad (3.6.63)$$

$$\lim_{n \to \infty} \int v_n(x_0 + \delta_m \omega) \, d\omega = \int u(x_0 + \delta_m \omega) \, d\omega \qquad (3.6.64)$$

Let μ be a measure with $\{x \mid \Phi_\mu(x) = \infty\} = K$ and consider the function on B_m^{int}

$$u - v_n - \varepsilon \, \Phi_\mu(x) \equiv f(x) \qquad (3.6.65)$$

It is subharmonic on $B_m^{\text{int}} \setminus K$. As x_ℓ in $B_m \setminus K \to x_\infty \in K$, u is bounded above, v_n is bounded and by lsc of Φ_μ, $\Phi_\mu(x_\ell) \to \Phi_\mu(x_\infty) = \infty$. Thus, $f(x_\ell) \to -\infty$ as $x_\ell \to K$ so f is usc on B_m^{int}. It is obviously submean at $x \in K$, so f is subharmonic on all of B_m.

If $x_\ell \to x_\infty \in \partial B_m$, $\limsup u(x_\ell) \leq u(x_\infty) \leq v_n(x_\infty)$ since u is usc and (3.6.63) holds. Similarly, $\limsup -\Phi_\mu(x_\ell) \leq -\Phi_\mu(x_\infty)$. Since v_n is continuous, we conclude that

$$\limsup_{\substack{x \to K \cup \partial B_m \\ x \in B_m \setminus K}} f(x) \leq -\varepsilon \inf_{x \in \partial B_m} \Phi_\mu(x) \tag{3.6.66}$$

By the maximum principle for subharmonic functions (Theorem 3.2.10)

$$\sup_{x \in B_m^{\text{int}}} u(x) - v_n(x) - \varepsilon \Phi_\mu(x) \leq -\varepsilon \inf_{x \in \partial B_m} \Phi_\mu(x) \tag{3.6.67}$$

Since Φ_μ is lsc, it takes its minimum on ∂B_m and so the inf is finite. Therefore, for $x \in B_m \setminus K$, $u(x) \leq v_m(x)$. Since v is continuous, the definition (3.6.62) implies that $u(x) \leq v_m(x)$ also for $x \in K$, i.e.,

$$u(x) \leq \lim v_m(x) = \int u(x_0 + \delta_m \omega)\, d\omega \tag{3.6.68}$$

by (3.6.64), i.e., u has SMP. $\square$

Theorem 3.6.13 (Extended Maximum Principle). *Let Ω be a bounded domain in $\mathbb{R}^\nu$. Let u be a subharmonic function on Ω which is bounded above. Suppose there is a polar set, $S \subset \partial \Omega$ so that $x_n \in \Omega \to x_\infty \in \partial \Omega \setminus S \Rightarrow \limsup_{n \to \infty} u(x_n) \leq 0$. Then, $u(x) \leq 0$ on Ω.*

Proof. Fix $\varepsilon > 0$. Let

$$S_\varepsilon = \{x \in \partial \Omega \mid \lim_{\substack{y \to x \\ y \in \Omega}} u(y) \geq \varepsilon\} \tag{3.6.69}$$

A simple argument (Problem 7) implies that S_ε is closed. Define v on $\mathbb{R}^\nu \setminus S_\varepsilon$ by

$$v(x) = \begin{cases} \max(u, \varepsilon) & \text{on } \Omega \\ \varepsilon & \text{on } \mathbb{R}^\nu \setminus (\Omega \cup S_\varepsilon) \end{cases}$$

v is usc on Ω and is continuous on $\mathbb{R}^\nu \setminus \Omega$ and on $\partial \Omega \setminus S_\varepsilon$ by definition of S_ε. Thus, v is usc. In Ω, v obeys SMP since u does and for $x \in \mathbb{R}^\nu \setminus \Omega \cup S_\varepsilon$, since globally $v \geq \varepsilon$, and $v(x_0) = \varepsilon$, SMP is trivial. Thus, v is subharmonic on $\mathbb{R}^\nu \setminus S_\varepsilon$ and so, by Theorem 3.6.13 (and u bounded above) u has a subharmonic extension to all of $\mathbb{R}^\nu$.

For any $x \in \mathbb{R}^\nu$, for R large, $x + R\omega \in \mathbb{R}^\nu \setminus \Omega$ for all $\omega \in S^\nu$ since Ω is bounded. Thus, by SMP, $v(x) \leq \varepsilon$, and so $v(x) = \varepsilon$, i.e., $u \leq \varepsilon$ on Ω. Since ε is arbitrary, $u \leq 0$ on Ω. $\square$

This implies the uniqueness of the solution of the generalized Dirichlet problem.

Theorem 3.6.14. *Let Ω be a bounded domain in $\mathbb{R}^\nu$. Then, for each continuous f on $\partial\Omega$, there exists a unique function u_f on Ω obeying u_f is bounded, harmonic in Ω, and for q.e. $x_0 \in \partial\Omega$, $u_f(x) \to f(x_0)$ as x goes to x_0.*

Proof. Existence follows from the Perron method and the Kellogg–Evans theorem. If u, v are two such functions, $w_\pm = \pm(u - v)$ are both bounded above and subharmonic on Ω with $\lim_{x\to x_0} w_\pm(x) = 0$ for q.e. $x_0 \in \partial\Omega$ (the union of the exceptional polar sets u and v). By Theorem 3.6.13, $w_\pm \equiv 0$ in Ω so $u = v$. □

As an immediate consequence we get

Theorem 3.6.15. *The map $f \mapsto H_\Omega(f)$ is linear in x. For each $x \in \Omega$, there is a probability measure μ_x on $\partial\Omega$ so that*

$$H_\Omega(f)(x) = \int_{\partial\Omega} f(y)\, d\mu_x(y) \tag{3.6.70}$$

$x \mapsto d\mu_x$ is weakly continuous in x for $x \in \Omega$ and at any regular point, $x_0 \in \partial\Omega$, $d\mu_x \to \delta_{x_0}$ as $x \to x_0$.

Remarks. 1. $d\mu_x$ is called the *harmonic measure*.

2. As in the classical case (see Proposition 3.1.7), Harnack's inequality implies that the $d\mu_x$'s are all mutually absolutely continuous.

Proof. The linearity is an immediate consequence of uniqueness. Clearly, for f and x, $f \mapsto (H_\Omega f)(x)$ is a positive linear functional on $C(\partial\Omega)$ with $(H_\Omega 1)(x) \equiv 1$, so the probability measure exists. The regularity properties of individual $(H_\Omega f)(x)$ in x imply the regularity of μ_x in x. □

Corollary 3.6.16. *The union of the sets $S_0, S_1, \ldots$ of the proof of Theorem 3.6.9 is precisely the set of singular points.*

Proof. We've already proven that any point not in $\bigcup_{j=0}^J S_j$ is regular. So we only need to show that a regular point, x_0, in ∂U_j is not in S_j, i.e., has the corresponding $G_{\mathfrak{e}_j}(x_0) = 0$.

Consider, first, $j \neq 0$; for notational simplicity, take $j = 1$. Let f be the function on $\partial\Omega$ which is given by $G_{\mathfrak{e}_1} \upharpoonright \partial\Omega$ on $\bigcup_{j\neq 1} \partial U_j$ and 0 on ∂U_1. By Theorem 3.6.8, $G_{\mathfrak{e}_1}$ is bounded on Ω and continuous on $\bigcup_{j\neq 1} \partial U_j$ and by (3.6.44), $\lim_{x\to x_0, x\in\Omega} G_{\mathfrak{e}_1}(x) = f(x_0)$ for q.e. $x_0 \in \partial\Omega$. Thus, by uniqueness of the generalized Dirichlet problem, $G_{\mathfrak{e}_1} \upharpoonright \Omega = H_\Omega(f)$. At any regular point, $x_0 \in \partial U_1$, $H_\Omega(f) \to 0$ so $x_0 \notin S_1$ by definition of S_1.

S_0 was defined to be the complement of the points, x_0, on ∂U_0 where $Q(x) \to 0$ as $x \to x_0$ where Q was a function of the form $h + cG_0$ with $c > 0$ and h a bounded harmonic function on $(\Omega \cup \bigcup_{j=1}^J U_j)$. Let $f = -cG_0 \upharpoonright \partial\Omega$. Then, since $h(x) \to -c\,G_0(x)$ as $x \to x_0$, we see that by the uniqueness result, $h = H_\Omega f$. As above, if $x_0 \in \partial U_0$ is regular, $Q(x) \to 0$ as $x \to x_0$, so $x_0 \notin S$. $\square$

Definition. Let Ω be a bounded open subset of $\mathbb{R}^\nu$. The Green's function, $G_\Omega(x, y)$, is defined on $\Omega \times \Omega$ by

$$G_\Omega(x, y) = G_0(x - y) + H_\Omega(-G_0(\cdot - y) \upharpoonright \partial\Omega)(x) \qquad (3.6.71)$$

(with $G_\Omega(x, x) = \infty$).

Theorem 3.6.17. *For each $y \in \Omega$, $G_\Omega(x, y)$ is the unique function obeying*

(i) $G_\Omega(\cdot, y)$ *is superharmonic on Ω and harmonic on $\Omega \setminus \{y\}$.*
(ii) $G_\Omega(x, y) \to 0$ *as $x \in \Omega \to x_0 \in \partial\Omega$ for q.e. $x_0 \in \partial\Omega$.*
(iii) $G_\Omega(\cdot, y) - G_0(\cdot - y)$ *is bounded on Ω.*

   Moreover,

(a) $G_\Omega > 0$ *and jointly continuous on $\Omega \times \Omega$.*   (3.6.72)

(b) $G_\Omega(x, y) = G_\Omega(y, x)$.   (3.6.73)

(c) *If $s(x)$ is a nonnegative function with $s - G_0(\cdot - y)$ superharmonic, then $s(x) \geq G_\Omega(x, y)$.*

Remark. Some authors define G_Ω by property (c) rather than (3.6.71); see the discussion in Section 3.8.

Proof. (i)–(iii) are immediate from what we have proven about H_Ω and uniqueness. Positivity, i.e., (a), follows from the extended maximum principle applied to $-G_\Omega(\cdot, y)$. Joint continuity follows from a simple argument (Problem 9). (b) is proven by the same argument that worked for (3.1.35) for classical Green's functions. It only needed harmonic measure. (c) follows from Theorem 3.6.13 and the fact that $u \equiv -(s - G_0)$ is subharmonic with $\limsup_{x \to x_\infty} u(x) \leq 0$ for q.e. $x_0 \in \partial\Omega$. $\square$

Next we want to turn to the exterior problem where we'll focus on $\nu = 2$ because the striking result that the equilibrium measure is the same as the harmonic measure at infinity is specific to $\nu = 2$ as the next example shows. Problem 10 will say something about the exterior problem when $\nu \geq 3$.

Example 3.6.18. Let $\nu \geq 3$. Let $\Omega = \mathbb{R}^\nu \setminus \{x \mid \|x\| \leq 1\}$. Then both $u(x) = 1$ and $u(x) = |x|^{-(\nu-2)}$ are harmonic on Ω so $1 - |x|^{-(\nu-2)}$ is harmonic on Ω but goes to 0 on $\partial\Omega$. Thus, the solutions of the exterior Dirichlet problem are not unique. In Problem 10, the reader will show there is existence and uniqueness if one demands that $u(x) \to 0$ at ∞. Thus, there is a natural harmonic measure but it is not a probability measure, and the natural "harmonic measure at ∞" is the zero measure. We'll see shortly that when $\nu = 2$, the exterior harmonic measure is a probability measure.   □

We start with the exterior problem for a nonpolar compact subset, $\mathfrak{e}$, in $\mathbb{C}$.

Theorem 3.6.19. *Let $\mathfrak{e}$ be a nonpolar, compact subset of $\mathbb{C}$ so that $\mathbb{C} \setminus \mathfrak{e}$ is connected. For each $f \in C(\mathfrak{e})$, there is a unique function, u, on $\mathbb{C} \setminus \mathfrak{e}$ so that*

(a) *u is bounded and harmonic.*
(b) *For q.e. $z_0 \in \mathfrak{e}$, $\lim_{z \to z_0, z \notin \mathfrak{e}} u(z) = f(z_0)$.*

   Moreover,
(c) *$u(\infty)$ can be defined so that*

$$\tilde{u}(w) = u(w^{-1}) \tag{3.6.74}$$

   is harmonic at $w = 0$.
(d) *For all $z \in (\mathbb{C} \setminus \mathfrak{e}) \cup \{\infty\}$, there is a probability measure $\mu_z^{(\mathfrak{e})}$ on $\mathfrak{e}$, so that for all $f \in C(\mathfrak{e})$,*

$$u_f(z) = \int f(\zeta) \, d\mu_z^{(\mathfrak{e})}(\zeta) \tag{3.6.75}$$

   $z \mapsto \mu_z^{(\mathfrak{e})}$ is continuous and obeys the mean value property.

Proof. (a), (b), (c) By Perron's method, we can find $u_f(z) = H_{\mathbb{C} \setminus \mathfrak{e}}(f)(z)$ harmonic on $\mathbb{C} \setminus \mathfrak{e}$ and bounded so $w = 0$ is a removable singularity in $u(w^{-1})$. By Frostman's theorem, $G_{\mathfrak{e}}$ converges to zero at q.e. $z_0 \in \mathfrak{e}$ and so is a barrier for such z_0. Thus, by Theorem 3.4.6, (b) holds and we have existence. If v is a second solution, $u = v$ by (b) and the extended maximum principle extended to this unbounded case (Theorem 3.6.13) proving uniqueness.

(d) Since $u_f \equiv 1$ if $f = 1$ and $f > 0 \Rightarrow u_f > 0$, $\mu_z^{(\mathfrak{e})}$ exists by the Riesz–Markov theorem. The mean value property and continuity are immediate from those properties of each u_f.   □

As a consequence, we have

Theorem 3.6.20. *Let $\mathfrak{e} \subset \mathbb{C}$ be compact. Then there exists nonconstant bounded harmonic functions on $(\mathbb{C} \cup \{\infty\}) \setminus \mathfrak{e}$ if and only if $\mathfrak{e}$ is nonpolar.*

Proof. If $\mathfrak{e}$ is nonpolar, then $\mathfrak{e}$ has more than one point, so there are nonconstant continuous functions $f \in C(\mathfrak{e})$. If u is the function of Theorem 3.6.19, then we have a bounded harmonic function which cannot be constant by (b) of that theorem.

Conversely, if $\mathfrak{e}$ is polar and u is a bounded harmonic function on $(\mathbb{C} \cup \{\infty\}) \backslash \mathfrak{e}$, by Theorem 3.6.12, u has an extension to all of $\mathbb{C}$ which is harmonic. Thus, by Theorem 3.1.2, u is constant. $\qquad\square$

We also have the following characterization of closed polar sets in $\mathbb{R}^\nu$ including $\nu = 2$ (and so $\mathbb{C}$).

Theorem 3.6.21. *If $\mathfrak{e} \subset \Omega$ is a compact polar subset of Ω, an open domain of $\mathbb{R}^\nu$, then any u bounded and harmonic in $\Omega \backslash \mathfrak{e}$ has an harmonic extension to all of Ω. Conversely, if $\mathfrak{e}$ is nonpolar, there exists u bounded and harmonic on $\mathbb{R}^\nu \backslash \mathfrak{e}$ with a unique continuous extension to $\mathbb{R}^\nu$ which is not harmonic on all of $\mathbb{R}^\nu$.*

Proof. The first statement follows from Theorem 3.6.12 applied to u and $-u$. If $\nu \geq 3$, apply Corollary 3.6.4 and take $u = \Phi_\rho$. If $\nu = 2$, find ρ as in that corollary. Φ_ρ is not bounded at infinity so pick $f \in C(\mathfrak{e})$ so $0 \leq f \leq 1$ and $\alpha = \int f d\rho \neq 0$, and f is not a.e. constant wrt ρ. Then $\Phi_{f\rho}$ and $\Phi_{(1-f)\rho}$ are both lsc and their sum is continuous, so they are both continuous. Let $d\eta = \alpha^{-1}(f d\rho)$ so $d\eta \neq d\rho$ but $\int d\rho = 1 = \int d\eta$. Take $u = \Phi_\rho - \Phi_\eta$ which is bounded and continuous at infinity. $\qquad\square$

The Riemann mapping theorem says a lot about regular points when $\Omega \subset \mathbb{C}$. We'll suppose for the next few results that the reader is familiar with the discussion in Sections 8.1 and 8.2 (and soon also 8.8) of Part 2A.

Theorem 3.6.22. *Let Ω in $\mathbb{C}$ be a bounded region which is simply connected, i.e., $\widehat{\mathbb{C}} \backslash \Omega$ is connected. For $z_0 \in \Omega$, let $f(z, z_0)$ be the Riemann map $z \in \Omega \to f(z, z_0) \in \mathbb{D}$ determined by $f(z_0, z_0) = 0$ and $f'(z, z_0) \mid_{z=z_0} > 0$. Then*

(a) $G_0(z, z_0)$ *defined by*

$$|f(z, z_0)| = e^{-2\pi G_0(z,z_0)}, \quad G_0(z, z_0) \in \mathbb{R} \qquad (3.6.76)$$

is a classical Green's function.

(b) *The Dirichlet problem is solvable for Ω.*

Remarks. 1. $f'(z, z_0) \mid_{z=z_0} > 0$ is never used because once $f(z_0, z_0) = 0$, f is determined up to multiplication by $e^{i\theta}$ which doesn't affect $|f(z, z_0)|$.

2. As discussed in the Notes to Section 8.1 of Part 2A, one can go backwards and use a classical Green's function to prove the Riemann mapping theorem.

3. On $\mathbb{C}$, it is common to define potentials and $G_\mathfrak{e}$ to have $\log |z|^{-1}$ rather than $(2\pi)^{-1} \log |z|^{-1}$ singularities. If we do that, (3.6.76) doesn't have the 2π.

Proof. (a) $G_0(z, z_0) \equiv -\log|f(z, z_0)|$ is harmonic and nonnegative for $z \in \Omega \setminus \{z_0\}$ and since $f(z, z_0)$ has a simple zero at z_0, $G_0(z, z_0) - \log|z - z_0|^{-1}$ is harmonic at z_0. By Theorem 8.2.1 of Part 2A, for z_0 fixed, $G(z, z_0) \to 0$ as $z \to \partial\Omega$.

(b) As noted in Corollary 3.4.8, for any $z_0 \in \Omega$, $G(z, z_0)$ restricted to a neighborhood of $w_0 \in \partial\Omega$ is a barrier at w_0, so by Theorem 3.4.7, the Dirichlet problem is solvable. $\square$

Let Ω in $\mathbb{C}$ be bounded and open (not necessarily connected but with finitely many components) and suppose that its outer boundary $O\partial(\Omega)$ is a finite distance from $\partial\Omega \setminus O\partial(\Omega)$. We can apply Theorem 3.6.22 to each bounded component, $\tilde{\Omega}$, of $\mathbb{C} \setminus O\partial(\Omega)$ to get a classical Green's function for $\tilde{\Omega}$ which provides barriers for each $w \in O\partial(\Omega)$. Similarly, if $K \subset \partial\Omega$ is a component a finite distance from $\partial\Omega \setminus K$ and K is not a single point, we can apply the Riemann mapping theorem to the unbounded component of $\mathbb{C} \setminus K$ and obtain barriers at any point in K. The result is

Corollary 3.6.23. *Let Ω be a bounded open domain in $\mathbb{C}$ which is n-connected so that each component of $\mathbb{C} \setminus \Omega$ has more than a single point. Then the Dirichlet problem is solvable for Ω.*

We can use this idea of Riemann maps on the unbounded component of $\widehat{\mathbb{C}} \setminus \mathfrak{e}$ to see that

Theorem 3.6.24. *Let $\mathfrak{e}$ be a compact connected subset of $\mathbb{C}$ with more than one point. Then:*

(a) *$\mathfrak{e}$ is nonpolar.*

(b) *If f is the Riemann map of the component of $\widehat{\mathbb{C}} \setminus \mathfrak{e}$ containing ∞ with $f(\infty) = 0$, $f'(\infty) > 0$, then*
$$|f(z)| = e^{-2\pi G_{\mathfrak{e}}(z)} \tag{3.6.77}$$

(c) *$\mathfrak{e}$ is regular for potential theory in that $G_{\mathfrak{e}}$ is globally continuous and vanishes on all of $\mathfrak{e}$.*

(d) *Near $z = \infty$,*
$$f(z) = \frac{C(\mathfrak{e})}{z} + O(|z|^{-2}) \tag{3.6.78}$$
where $C(\mathfrak{e})$ is the capacity of $\mathfrak{e}$.

(e) *The Riemann map, f, is the Ahlfors function for $\widehat{\mathbb{C}} \setminus \mathfrak{e}$.*

(f) *$A(\mathfrak{e}) = C(\mathfrak{e})$.*

Remarks. 1. The Ahlfors function and the analytic capacity, $A(\mathfrak{e})$, are the subject of Section 8.8 of Part 2A.

2. In (3.6.78), $C(\mathfrak{e})$ is defined as $\exp(-2\pi R(\mathfrak{e}))$ with the definition of the Robin constant via $(2\pi)^{-1}\log(|x|^{-1})$.

Proof. (a) Re f is a nonconstant bounded harmonic function so $\mathfrak{e}$ is nonpolar by Theorem 3.6.20.

(b), (c) The argument is the same as in the proof of Theorem 3.6.22.

(d) With the definition of $C(\mathfrak{e})$ in Remark 2, (3.6.45) says for $z = \infty$,

$$-2\pi G_{\mathfrak{e}}(z) = \log(|z|^{-1}) + \log(C(\mathfrak{e})) + O(|z|^{-1}) \tag{3.6.79}$$

which implies that $|f(z)| = C(\mathfrak{e})|z|^{-1} + O(|z|^{-2})$ which given the normalization of f via $f'(\infty) > 0$ implies (3.6.78).

(e) As discussed in the Notes to Section 8.1 and in Section 8.8 of Part 2A, the Riemann map is the Ahlfors map.

(f) Immediate from the definition of $A(\mathfrak{e})$ and part (e). $\qquad\square$

Corollary 3.6.25. *Any polar subset of $\mathbb{C}$ is totally disconnected.*

Remark. This is not true in $\mathbb{R}^\nu$, $\nu \geq 3$ since for such ν, $\mathfrak{e} = \{x \mid -1 \leq x_1 \leq 1, x_2 = x_3 = \cdots = x_\nu = 0\}$ is a polar set as we've seen.

Proof. If $\mathfrak{e}_1 \subset \mathfrak{e}$ is a connected component and $\mathfrak{e}_1$ has more than one point, $C(\mathfrak{e}_1) > 0$ by the theorem so $C(\mathfrak{e}) \geq C(\mathfrak{e}_1) > 0$. $\qquad\square$

As a final element of this aside on Riemann mappings:

Theorem 3.6.26. *For any compact subset $\mathfrak{e} \subset \mathbb{C}$:*

(a) *If $\mathfrak{e}$ is nonpolar and f is analytic on $\widehat{\mathbb{C}} \setminus \mathfrak{e}$ with values in $\mathbb{D}$ and $f(\infty) = 0$, then*

$$|f(z)| \leq e^{-2\pi G_{\mathfrak{e}}(z)} \tag{3.6.80}$$

(b) $A(\mathfrak{e}) \leq C(\mathfrak{e})$ $\qquad\qquad\qquad\qquad\qquad\qquad\qquad\qquad$ (3.6.81)

Remarks. 1. By Theorem 3.6.20, if $\mathfrak{e}$ is polar, any analytic f on $\widehat{\mathbb{C}} \setminus \mathfrak{e}$ with $f(\infty) = 0$ has $f \equiv 0$ so $A(\mathfrak{e}) = C(\mathfrak{e}) = 0$.

2. $C(\mathfrak{e})$ in (3.6.81) has the even factor of 2π as in the last theorem.

Proof. (a) Let

$$u(z) = -2\pi G_{\mathfrak{e}}(z) + \log|f(z)| \tag{3.6.82}$$

u is subharmonic on $\widehat{\mathbb{C}} \setminus \mathfrak{e}$ since on $\mathbb{C} \setminus \mathfrak{e}$, $G_{\mathfrak{e}}$ is harmonic and $\log|f(z)|$ is subharmonic. At ∞, both have a $\log|z|^{-1}$ singularity which cancel, so ∞ is a removable singularity and $u(x)$ is harmonic there. For a.e. $w \in \mathfrak{e}$, if $z_n \to w$, $z_n \in \widehat{\mathbb{C}} \setminus \mathfrak{e}$, then $\lim G_{\mathfrak{e}}(z_n) = 0$ while, since $|f(z_n)| \leq 1$, $\limsup \log|f(z_n)| \leq 0$. Thus, by Theorem 3.6.13, $u(z) \leq 0$ on $\widehat{\mathbb{C}} \setminus \mathfrak{e}$, i.e., (3.6.80) holds.

(b) As in the last theorem, $e^{-2\pi G_{\mathfrak{e}}(z)} = \frac{C(\mathfrak{e})}{|z|} + O(|z|^{-2})$ near ∞, so by definition of $A(\mathfrak{e})$, we have (3.6.81). $\qquad\square$

Theorem 3.6.27. *Let $\mathfrak{e}$ be a nonpolar, compact subset of $\mathbb{C}$ so that $\mathbb{C} \setminus \mathfrak{e}$ is connected. For each $a \in \mathbb{C} \setminus \mathfrak{e}$, there is a unique function, $g(z;a)$, for $z \in [\mathbb{C} \setminus \mathfrak{e} \cup \{a\}] \cup \{\infty\}$ harmonic on that set so that*

(a) *For q.e. $z_0 \in \mathfrak{e}$, $\lim_{z \to z_0,\, z \notin \mathfrak{e}} g(z;a) = 0$.*

(b) *$g(z:a) - G_0(z - a)$ is bounded near $z = a$ (and so harmonic there).*

Moreover,

(c) *If $g(z;\infty) \equiv G_{\mathfrak{e}}(z)$, we have that for all $a \in \mathbb{C} \setminus \mathfrak{e}$*

$$g(z;a) = (2\pi)^{-1} \int \log|\zeta - a|\, d\mu_z^{(\mathfrak{e})}(\zeta)$$
$$-(2\pi)^{-1} \log|z - a| + g(z,\infty) \tag{3.6.83}$$

(d) *$g(z;a) = g(a;z)$* (3.6.84)

Remarks. 1. We'll define $g(z;a)$ via conformal mapping and existence of $G_{\tilde{\mathfrak{e}}}$ for suitable $\tilde{\mathfrak{e}}$. Then, we'll derive (c) but one could use (3.6.83) as a definition.

2. Our proof of (d) will rely on results from Section 3.8.

3. $g(z;a)$ is called the *Green's function with a pole at a* (in line with the discussion in Section 3.8).

Proof. (a), (b) Let $T(z) = (z-a)^{-1}$. Then $T^{-1}(w) = w^{-1} + a$. Let $\tilde{\mathfrak{e}} = T[\mathfrak{e}]$. Since T maps a to infinity and $\log(|T(z)|) = -\log|z - a|$, we see that

$$g(z;a) = G_{\tilde{\mathfrak{e}}}(T(z)) \tag{3.6.85}$$

g has the required properties. Uniqueness follows from the extended maximum principle as in the last theorem.

(c) The right side of (3.6.83) is harmonic on $\mathbb{C} \setminus (\mathfrak{e} \cup \{a\})$. Since the integral converges to $\log|z_0 - a|$ as $z \to z_0 \in \mathfrak{e}$ for q.e. z_0 and $g(z;\infty)$ vanishes q.e. on $\mathfrak{e}$, we see the right side obeys (a). Since

$$g(z;\infty) = (2\pi)^{-1} \log|z| + R(e) + O\left(\frac{1}{|z|}\right) \tag{3.6.86}$$

and $\log|z| - \log|z - a| = O(|z|^{-1})$, the right side is bounded as $z \to \infty$; indeed, by (3.6.85),

$$\lim_{z \to \infty} \text{RHS of (3.6.83)} = (2\pi)^{-1} \int \log|\zeta - a|\, d\mu_\infty^{(\mathfrak{e})} + R(e) \tag{3.6.87}$$

Thus, by the uniqueness we get (3.6.83).

(d) $g(z;a)$ is a Green's function in the sense of Section 3.8, i.e., viewing $\mathbb{C} \cup \{\infty\} \setminus \mathfrak{e}$ is a Riemann surface. Thus (3.6.84) follows from Proposition 3.8.19. $\qquad\square$

Corollary 3.6.28. *Let $\mathfrak{e}$ be a nonpolar, compact subset of $\mathbb{C}$ so that $\mathbb{C} \setminus \mathfrak{e}$ is connected. Then the equilibrium measure, $\rho_{\mathfrak{e}}$, is the same as harmonic measure at infinity, $d\mu_{\infty}^{\mathfrak{e}}$.*

Proof. (3.6.83) and (3.6.87) say that

$$\lim_{z \to \infty} g(z; a) = R(\mathfrak{e}) - \Phi_{\mu_{\infty}^{(\mathfrak{e})}}(a) \tag{3.6.88}$$

Since $g(z; a) = g(a; z)$ and g is jointly continuous on the Riemann surface, we see that this is the same as

$$g(a; \infty) = R(\mathfrak{e}) - \Phi_{\mu_{\infty}^{(\mathfrak{e})}}(a) \tag{3.6.89}$$

But

$$g(a; \infty) = R(\mathfrak{e}) - \Phi_{\rho_{\mathfrak{e}}} \tag{3.6.90}$$

so

$$\Phi_{\rho_{\mathfrak{e}}} = \Phi_{\mu_{\infty}^{(\mathfrak{e})}} \tag{3.6.91}$$

For any measure ν of compact support

$$\nu = -\Delta \Phi_{\nu} \tag{3.6.92}$$

so (3.6.91) $\Rightarrow \rho_{\mathfrak{e}} = \mu_{\infty}^{(\mathfrak{e})}$. $\qquad\qquad\square$

Finally, we study when $\nu_n \to \nu$ weakly implies $\Phi_{\nu_n}(x) \to \Phi_{\nu}(x)$.

Theorem 3.6.29 (Principle of Descent). *Let $K \subset \mathbb{R}^{\nu}$ be compact. Let $\{\nu_n\}_{n=1}^{\infty}, \nu$ be measures on $\mathbb{R}^{\nu}$ with $\operatorname{supp} \nu_n$, $\operatorname{supp} \nu \subset K$ and suppose that $\nu_n \to \nu$ weakly. Then, for all $x \in \mathbb{R}^{\nu}$,*

$$\lim_{n \to \infty} \inf \Phi_{\nu_n}(x) \geq \Phi_{\nu}(x) \tag{3.6.93}$$

with equality if $x \notin \bigcap_{n=1}^{\infty} \overline{\left(\bigcup_{m=n}^{\infty} \operatorname{supp}(\nu_m) \right)} \equiv \mathfrak{e}$

Proof. If $x \notin \mathfrak{e}$, eventually $x \notin \overline{\bigcup_{m=n}^{\infty} \operatorname{supp}(\nu_m)} \equiv \mathfrak{e}_m$, so $G_0(x - \cdot)$ is continuous on $\mathfrak{e}_m$, so we have equality in (3.6.93)

If $\Phi_{\mu}^{(N)}$ is given by (3.2.16), then since $\min(N, G_0(x))$ is continuous for any $N < \infty$ and any x, as $n \to \infty$

$$\Phi_{\nu_n}^{(N)}(x) \to \Phi_{\nu}^{(N)}(x) \tag{3.6.94}$$

Since $\Phi_{\nu_n} \geq \Phi_{\nu_n}^{(N)}$, we see that

$$\lim_{n \to \infty} \inf \Phi_{\nu_n}(x) \geq \Phi_{\nu}^{(N)}(x) \tag{3.6.95}$$

Taking $N \to \infty$ yields (3.6.93). $\qquad\qquad\square$

Example 3.6.30. Let ν be such that $\Phi_\nu(x)$ is discontinuous (see Example 3.2.9); say $\Phi_\nu(x_n) \to \alpha > \Phi_\nu(x)$. Let ν_n be the translate of ν by $x_n - x$ so $\Phi_{\nu_n}(x) = \Phi_\nu(x_n)$. Since $x_n - x \to 0$, $\nu_n \to \nu$ weakly. Clearly equality does not hold in (3.6.93). See Problem 8 for another example. Thus, we can't hope for (3.6.93) for all x but the next theorem shows it happens for most x $\qquad\square$

Theorem 3.6.31 (Upper Envelope Theorem). *Under the hypotheses of Theorem 3.6.29, equality holds in (3.6.93) for q.e. x.*

Proof. Let $S = \{x \mid \liminf \Phi_{\nu_n}(x) > \Phi_\nu(x)\}$ which is a Borel set. If $R(S) > 0$, then, by Corollary 3.6.3, there is a measure μ supported on a compact $\mathfrak{e} \subset S$ with $\Phi_\mu(x)$ continuous and, in particular, bounded on K.

Thus,

$$\int \Phi_\mu(x)\, d\nu(x) = \lim_{n\to\infty} \int \Phi_\mu(x)\, d\nu_n(x) \qquad (3.6.96)$$

$$= \lim_{n\to\infty} \int \Phi_{\nu_n}(x)\, d\mu(x) \qquad (3.6.97)$$

$$\geq \int \liminf \Phi_{\nu_n}(x)\, d\mu(x) \qquad (3.6.98)$$

$$> \int \Phi_\nu(x)\, d\mu(x) \qquad (3.6.99)$$

$$= \int \Phi_\mu(x)\, d\nu(x) \qquad (3.6.100)$$

which shows such a μ cannot exist, so $R(S) = 0$. In the above, (3.6.96) follows from the continuity of Φ_μ and weak convergence, (3.6.97) and (3.6.100) are (3.6.6), (3.6.98) is Fatou's lemma and the fact that $\Phi_\eta(x) > 0$ in dimension 3 or more and in dimension 2, that $\Phi_\eta(x) \geq -\eta(\mathbb{R}^\nu)\log(2R_0)$ if $\operatorname{supp}\eta \subset K$ and R_0 is the radius of a ball containing K. Finally, (3.6.99) follows from $\operatorname{supp}\mu \subset S$ and the definition of S. $\qquad\square$

Notes and Historical Remarks. Initially, because of its relevance to physics, and later, additionally, because of its importance to complex analysis (in part because of Riemann's use of it in his proof of his mapping theorem), potential theory has been a central object of mathematics for almost 300 years, and its greatest practitioners are almost a who's-who of analysts: Lagrange, Legendre, Laplace, Poisson, Gauss, Green, and Thomson (aka Kelvin) in the period around 1800; Dirichlet, Riemann, Schwarz, and Neumann in the period centered sixty years after; Poincaré, Hilbert, Zaremba, and Lebesgue in the period around 1905; Kellogg, Frostman, Evans, Perron, Wiener, F. Riesz, and Brelot in the period 1925–1940; and, at mid-century, a quartet of French mathematicians: Brelot, Cartan, Choquet, and Deny.

Since then, the field has blossomed into many branches: general elliptic theory, probability and potential theory, and some axiomatic approaches going back to Brelot and to Deny.

As we've noted, to some extent, the theory of subharmonic functions is the same as that of potentials, and the books listed at the end of the introduction to this chapter often include "potential" in their title and certainly in their content.

In 1886, Robin [695] studied the problem of minimizing $\mathcal{E}$ as the way of understanding how charge redistributed in a perfect conductor. The Robin constant is named after this work. The minimization problem is sometimes called *Robin's problem* and $\Phi_{\mathfrak{e}}$ the *Robin potential*.

The continuity principle (Theorem 3.6.2) is due to Vasilesco [830] and Evans [254] in 1935. The potential theory version of Lusin's theorem (Corollary 3.6.3) which is proven with Lusin's theorem was noted by Choquet [152] twenty years later. Frostman's theorem (which Tsuji [818] called the fundamental theorem of potential theory) is due to Frostman [292].

The Kellogg–Evans theorem, using rather different methods from what we do, was proven in two dimensions by Kellogg [460] and for three dimensions by Evans [255]. Evans' proof works in all dimensions $\nu \geq 3$.

Proposition 3.6.10 is a result of Cartan [146]. The converse, i.e., that any polar G_δ is the set where some potential is infinite is a result of Deny [215]. The argument we use to prove the weaker converse, Proposition 3.6.27 is from Ransford's book [670] whose approach to the removable singularities theorem and the extended maximum principle we partially follow (his proof relies on a Liouville-type theorem only true in two dimensions).

Brelot is responsible for the removable singularity theorem, Theorem 3.6.12 [115] and for the upper envelope theorem, Theorem 3.6.31 [112]. There were many intermediate theorems between the removability of an isolated point singularity for bounded harmonic functions (for $\nu = 2$, essentially a result of Riemann) and Brelot's result.

Harmonic measure has been extensively studied, so much so that there are books on the subject [306]. F. and M. Riesz [690] proved for the interior of a Jordan curve, that the harmonic measure is absolutely continuous with respect to arc length measure on the boundary, and Dahlberg [182] has a kind of higher-dimensional analog—if a bounded region, Ω, in $\mathbb{R}^\nu$ has a Lipschitz boundary, harmonic measure is mutually absolutely continuous wrt $(\nu - 1)$-dimensional Hausdorff measure on $\partial\Omega$. Makarov [553] proved that for any simply connected bounded domain $\Omega \subset \mathbb{R}^2$, $d\nu_x(E) = 0$ if E has dimension $s < 1$ and that $d\nu_x$ is mutually singular relative to t-dimensional Hausdorff dimension for any $t > 1$. Makarov [554] has also shown that harmonic measure on the classical Cantor set, which has uniform dimension $\log 2 / \log 3$ lives on a set of dimension strictly smaller than this dimension.

This section is the last of several that discuss the solution of the Dirichlet problem so we mention the variety of approaches that have been used to conquer this central problem of classical analysis and classical mathematical physics: (1) Minimization, aka, the Dirichlet principle; (2) Layer and double-layer potentials; (3) Schwarz alternation; (4) Balayage; (5) Perron's method.

We discussed the Dirichlet principle in the Notes to Section 8.1 of Part 2A. There are also closely related variational approaches to the definition of capacity—a good reference for discussions of this subject is Lieb–Loss [**533**].

The method of single and double-layer potentials attempts to write a solution, u, of the Dirichlet problem on Ω, a bounded set with smooth boundary, either as a *single layer*

$$u(x) = \int G_0(x - y) \, d\mu_s(y) \qquad (3.6.101)$$

or *double layer* (also called dipolar layer)

$$u(x) = \int \frac{\partial}{\partial n_y} G_0(x - y) \, d\mu_d(y) \qquad (3.6.102)$$

Carl Neumann (1832–1925) was a pioneer in studying the double-layer method [**605**] deducing an integral equation for h in $d\mu_d(y) = h(y) \, d\sigma(y)$ which he tried to solve via a geometric series (after this work, often called the *Neumann series* in a much more general context). The solution of double-layer problems via this integral equation was a triumph of Fredholm theory which we discuss in Section 3.3 of Part 4. Neumann also considered what we now call *Neumann boundary conditions* and the *Neumann problem* (i.e., given g, finding u with $\Delta u = 0$ and $\frac{\partial u}{\partial n} \upharpoonright \partial \Omega = g$).

The Schwarz alternation method attempts to solve the Dirichlet problem for Ω by approximating Ω from within by a union of balls. The Dirichlet problem for such a union is solved by a general method of solving for $\Omega_1 \cup \Omega_2$ given the solutions for Ω_1 and Ω_2 by iterating alternating between the regions—hence, the name of the technique.

Balayage, a French term meaning "sweeping out," is a method due to H. Poincaré (1854–1912) [**652, 656**](Section 12.2 of Part 2B has a capsule biography). Given a measure ν on Ω, its *balayage* is a measure μ on $\partial \Omega$ so that on $\mathbb{R}^\nu \setminus \Omega$, $\Phi_\mu = \Phi_\nu$. Its relevance to the Dirichlet problem can be seen by the fact that if $x \in \Omega$, its harmonic measure, $d\mu_x$, is the balayage of δ_x, the point mass at x.

Potential theory has been generalized in a variety of ways. The theory of Riesz potentials (named after M. Riesz [**693**] but studied earlier by Frostman [**292**]) replaces G_0 by $|x - y|^{-\alpha}$ for some $\alpha \in (0, \nu)$. Landkof's book [**507**] studies them in detail. Bessel potentials replace G_0, the fundamental solution for $-\Delta$, by the fundamental solution for $-\Delta + 1$, which is a Bessel function. Early works on this subject are Aronszajn–Smith [**16**] and Calderón [**127**].

There is some work on extending the theory of potentials to fundamental solutions of general second-order elliptic operators, see Gilbarg–Trudinger [**312**].

There are two approaches to axiomatizing potential theory. One approach (in several variants) has the cone of subharmonic functions as its basic object and the other a quadratic form on L^2 space, the analog of the energy, $\int (\nabla u)^2 \, d^\nu x$, fundamental to the Dirichlet principle. Brelot [**117, 119**] and Bauer [**47**] are the masters of the first approach which is related to variants that axiomatize the relation of harmonic functions to Brownian motion—a subject known as "probabilistic potential theory" whose pioneers are Doob [**226**] and Hunt [**409**] (see also the book of Meyer [**567**]).

The quadratic-form axiomatic approach is due to Beurling–Deny [**67, 68**] following up on earlier work of Deny [**216, 147**]. The subject is expounded in the book of Fukushima [**293**].

Some other ideas that have been important in the theory that we should mention in passing are the general theory of capacities (in which a major result is the *Choquet capacity theorem* that for analytic sets (including all Borel sets), the capacity defined by approximation from within by compacts is the same as the capacity defined by approximating from without by open sets), the fine topology, and the Martin boundary.

The fine topology is intimately related to the notion of thin set. Brelot [**114, 116**] called a set, A, *thin* at x (in $\overline{A}$) if there existed a subharmonic function, u defined in a neighborhood of x so that $u(x) > \lim_{y \to x, y \in A} \sup u(y)$). (Recall if A is open, one has equality in the lim sup for all $x \in A$). Cartan, in a letter to Brelot, defined the *fine topology* as the weakest topology in which all subharmonic functions are continuous and pointed out that A is thin at $x \notin A$ if and only if $\mathbb{R}^\nu \setminus A$ is a neighborhood of x in the fine topology. Both notions have been useful.

The *Martin boundary* (which goes back to Martin [**560**]) rectifies the fact that if the Dirichlet problem isn't solvable, the Green's function (defined via Theorem 3.6.17) doesn't have limits at singular points on $\partial\Omega$. One replaces irregular points by sets of possible limits. In the end, one finds a compactification $\widehat{\Omega}$ of any bounded open $\Omega \subset \mathbb{R}^\nu$, so that for each $y \in \Omega$, $G(\,\cdot\,, y)$ has a continuous extension to $\widehat{\Omega}$. The Martin boundary is $\widehat{\Omega} \setminus \Omega$.

This is intimately related to the theory of compactification (discussed in Section 6.5 of Part 4) and the Shilov and Choquet boundaries (discussed in Section 6.10 of Part 4).

Problems

1. In this problem, the reader will show that a compact polar set, Q, in $\mathbb{R}^\nu$ has Hausdorff dimension at most $\nu - 2$, i.e., for any $\alpha > (\nu - 2)$ and any $\varepsilon > 0$, there is an open cover of Q by balls, $B_{x_j}^{\rho_j} = \{x \mid |x - x_j| < \rho_j\}$ so that $\sum \rho_j^\alpha \le \varepsilon$. Let μ be the measure of compact support with $\mu(\mathbb{R}^\nu) = 1$, so that Q obeys (3.6.53).

   (a) Fix $x \in Q$. Pick β with $\nu - 2 < \beta < \alpha$ and suppose that $r_n = 2^{-n}$ and for some c, we have that for all large n

   $$\mu(B_x^{r_n}) \le c(r_n)^\beta \tag{3.6.103}$$

   Prove that then $\Phi_\mu(x) < \infty$. (*Hint:* $\Phi_\mu(x) \le \sum_{n=N_0}^{\infty} G_0(r_{n+1}) \mu(B_x^{r_n})$.)

   (b) Conclude that for any δ and $x \in Q$, there is an r_x so that

   $$\mu(B_x^{r_x}) \ge \delta^{-1}(r_x)^\alpha \tag{3.6.104}$$

   (c) Use Vitali's covering theorem to prove that there is $x_1, \ldots, x_\ell \in Q$ so $Q \subset \bigcup_{j=1}^{\ell} B_{x_j}^{3r_{x_j}}$ and that $\{B_{x_j}^{r_{x_j}}\}_{j=1}^{\ell}$ are disjoint.

   (d) Let $\rho_j = 3r_{x_j}$ and prove that

   $$\sum \rho_j^\alpha \le 3^\alpha \delta \tag{3.6.105}$$

   (e) Conclude that Q has dimension at most $\nu - 2$.

2. (a) Let I be the line segment in $\mathbb{R}^3$,

   $$I = \{(x, y, z) \mid |x| \le 1, y = z = 0\} \tag{3.6.106}$$

   Let μ be uniform measure (i.e., $\frac{1}{2}dx$) on I. Prove that

   $$I = \{(x, y, z) \in \mathbb{R}^3 \mid \Phi_\mu(x) = \infty\} \tag{3.6.107}$$

   (b) Conclude that I is a polar set. (*Hint:* Proposition 3.6.10.)

   (c) In $\mathbb{R}^\nu$, prove that $S = \{x \mid x_{\nu-1} = x_\nu = 0, |x| \le 1\}$ is a polar set of dimension exactly $\nu - 2$.

3. (a) If $\nu \ge 3$ and $\mathcal{E}(\rho) < \infty$, $\mathcal{E}(\eta) < \infty$ for two measures of compact support, prove that

   $$\int |\Phi_\eta(x)| \, d\rho(x) < \infty \tag{3.6.108}$$

   (*Hint:* See (1.2.50).)

   (b) If $\nu = 2$ and ρ, η are supported in $\{x \mid |x| < \frac{1}{2}\}$ with $\mathcal{E}(\rho) < \infty$, $\mathcal{E}(\eta) < \infty$, prove that $\Phi_\eta(x) > 0$ on $\operatorname{supp}\rho$ and (3.6.108) holds.

(c) Use scaling to show that if $\nu = 2$ for any two measures of compact support and finite energy, $\Phi_\eta(x)$ is bounded below on supp ρ and (3.6.108) holds for $|\Phi_\eta(x)|$.

4. (a) Verify (3.6.46) for $\nu = 2$.

   (b) Verify (3.6.46) for $\nu \geq 3$.

5. Let $\mathfrak{e}$ be a compact set and F a C^1-function in a neighborhood of $\mathfrak{e}$ with $c^{-1}|x - y| \leq |F(x) - F(y)| \leq c|x - y|$ for some $c \in (1, \infty)$ and all x, y in $\mathfrak{e}$. Let S be a polar set. Prove that $F[S]$ is a polar set.

6. Let Ω be an open domain in $\mathbb{R}^\nu$ and $K \subset \Omega$ a compact set nowhere dense in Ω. Let u be a usc function on $\Omega \setminus K$ which is bounded above. Extend u to Ω by using (3.6.60). Prove that u is usc on Ω. (*Hint*: If $x_n \to x$ all in K, find $y_n \in \Omega \setminus K$ so that $|y_n - x_n| \leq \frac{1}{n}$ and $u(y_n) \geq u(x_n) - \frac{1}{n}$.)

7. Let u be usc on Ω, a domain in $\mathbb{R}^\nu$ so that u is bounded above. Prove that S_ε given by (3.6.69) is closed. (*Hint*: See Problem 6.)

8. Let μ_n be the probability measure that gives weight $\frac{1}{2^n}$ to each point of the form $\frac{j}{2^n}, j = 1, 2, \ldots, 2^n$ in $[0, 1]$.

   (a) Prove that $\mu_n \to \mu_\infty =$ Lebesgue measure on $[0, 1]$.

   (b) If $x \in [0, 1]$ has $2^m x \in \mathbb{Z}_+$ for some m, prove that $\Phi_n(x) = \infty$ for all large n and $\Phi_\infty(x) < \infty$ so that $\liminf \Phi_{\mu_n}(x) > \Phi_{\mu_\infty}(x)$.

9. Prove that G_Ω given by (3.6.71) is jointly continuous in x, y on $\Omega \times \Omega \setminus \{(x, x) \mid x \in \Omega\}$. (*Hint*: If $f_n \to f$ uniformly in $C(\partial\Omega)$, prove that $H_\Omega f_n \to H_\Omega f$ uniformly on Ω.)

10. Fix $\nu \geq 3$. Let $\mathfrak{e} \subset \mathbb{R}^\nu$ be a compact set so that $\mathbb{R}^\nu \setminus \mathfrak{e}$ is connected and $\mathfrak{e}^{\mathrm{int}} \neq \emptyset$. Suppose also $0 \in \mathfrak{e}^{\mathrm{int}}$ (otherwise you could just use a shifted Kelvin transform).

    (a) By using the Kelvin transform, solve the exterior Dirichlet problem, i.e., prove for every $f \in C(\partial\mathfrak{e})$, there is a unique u harmonic on $\mathbb{R}^\nu \setminus \mathfrak{e}$ with $\lim_{x \to \infty} u(x) = 0$ and so that for q.e. $x_0 \in \partial\mathfrak{e}$, $\lim_{x \to \infty, x \notin \mathfrak{e}} u(x) = f(x_0)$.

    (b) Show that there are nonzero bounded harmonic functions on $\mathbb{R}^\nu \setminus \mathfrak{e}$ so that for q.e. $x_0 \in \partial\mathfrak{e}$, $\lim_{x \to x_0, x \notin \mathfrak{e}} u(x) = 0$.

11. Let $\mathfrak{e}_1 \subset \mathfrak{e}_2 \subset \mathbb{C}$ be compact sets. Prove that for all z, $G_{\mathfrak{e}_1}(z) \geq G_{\mathfrak{e}_2}(z)$ with strict inequality for $z \notin \mathfrak{e}_2$ if $\mathfrak{e}_2 \setminus \mathfrak{e}_1$ is not polar.

3.7. Bonus Section: Polynomials and Potential Theory

In this section, we want to discuss one application of potential theory dear to my heart, namely the connection of the subject to growth of polynomials, especially orthogonal polynomials. We illustrate some of the special features of 2D potential theory using $\mathbb{R}^2 \simeq \mathbb{C}$.

In this section, we modify our definition of potential and Green's function by a factor of 2π so

$$\Phi_\mu(z) = \int \log|z-w|^{-1}\, d\mu(w) \tag{3.7.1}$$

and energies are given by

$$\mathcal{E}(\mu) = \int \log|z-w|^{-1}\, d\mu(z)\, d\mu(w) \tag{3.7.2}$$

If $\mathfrak{e}$ is compact, we still define its Robin constant as $\inf_{\mu \in \mathcal{M}_{+,1}(\mathfrak{e})} \mathcal{E}(\mu)$ which is 2π times the old Robin constant and $G_\mathfrak{e}(z)$ is 2π times the old one, i.e.,

$$G_\mathfrak{e}(z) = R(\mathfrak{e}) - \Phi_{\rho_\mathfrak{e}}(z) \tag{3.7.3}$$

We also recall that the capacity of $\mathfrak{e}$ is given by (again with 2π's dropped)

$$C(\mathfrak{e}) = \exp(-R(\mathfrak{e})) \tag{3.7.4}$$

If $q_n(z)$ is a polynomial on $\mathbb{C}$ of degree n, then $\log|q_n(z)|$ is subharmonic (by Theorem 3.2.6) and diverges as $n\log|z|$ near infinity. That suggests that it is $-n\Phi_\mu(z)$ for a suitable μ, and we'll see what μ is below (when q_n is monic).

We begin with a general bound expressed in terms of Green's functions. Recall that if $\mathfrak{e}$ is a compact subset of $\mathbb{C}$, there is a function, $G_\mathfrak{e}(z)$, its Green's function, which is harmonic on U_0, the unbounded component of $\mathbb{C} \setminus \mathfrak{e}$, strictly positive there, vanishing on $\mathbb{C} \setminus \overline{U_0}$, and which obeys

(i) $\lim_{z \to z_0} G_\mathfrak{e}(z) = 0$ for q.e. $z_0 \in \partial U_0$. $\tag{3.7.5}$

(ii) $G_\mathfrak{e}(z) = \log|z| + R(\mathfrak{e}) + O(|z|^{-1})$ as $|z| \to \infty$. $\tag{3.7.6}$

Theorem 3.7.1 (Bernstein–Walsh Lemma). *Let $\mathfrak{e}$ be a compact nonpolar subset of $\mathbb{C}$. Let*

$$\|f\|_\mathfrak{e} = \sup_{z \in \mathfrak{e}} |f(z)| \tag{3.7.7}$$

Then, for any polynomial, q_n, of degree n, we have for all z that

$$|q_n(z)| \le \|q_n\|_\mathfrak{e}\, \exp(nG_\mathfrak{e}(z)) \tag{3.7.8}$$

Proof. Since $G_\mathfrak{e} \ge 0$, (3.7.8) is obvious on $\mathfrak{e}$. If U_j is a bounded component of $\mathbb{C} \setminus \mathfrak{e}$, $\|q_n\|_{U_j} = \|q_n\|_{\partial U_j}$ by the maximum principle for analytic functions and $\|q_n\|_{\partial U_j} \le \|q_n\|_\mathfrak{e}$ since $\partial U_j \subset \mathfrak{e}$, so again (3.7.8) is obvious.

Thus, we need only prove (3.7.8) on U_0. Fix $\varepsilon > 0$ and let

$$f_\varepsilon(z) = -(n+\varepsilon)G_\mathfrak{e}(z) + \log|q_n(z)| - \log\|q_n\|_\mathfrak{e} \tag{3.7.9}$$

By (3.7.6) and $\log |q_n(z)| = n \log |z| + O(1)$ as $|z| \to \infty$, we see that $\lim_{|z| \to \infty} f_\varepsilon(z) = -\infty$, so we can pick R large so that $\mathfrak{e} \subset B^R \equiv \{z \mid |z| < R\}$ and $f_\varepsilon(z) < 0$ on $\mathbb{C} \setminus B^R$.

$f_\varepsilon(z)$ is subharmonic on $\Omega \equiv B^R \cap U_0$ and has nonnegative limits as $z \to \partial B^R$ with $z \in \Omega$. As $z \to z_0 \in U_0$, $\log |q(z)| - \log \|q(z)\|_{\mathfrak{e}}$ has a negative limit and $G_{\mathfrak{e}}(z) \to 0$ for q.e. z_0. Thus, by the extended maximum principle, Theorem 3.6.13, $f_\varepsilon(z) \leq 0$ on $U_0 \cap B^R$ and so on U_0, i.e., (3.7.8) holds with n replaced by $n + \varepsilon$. Since ε is arbitrary, we can take $\varepsilon \downarrow 0$. $\qquad \square$

We can now identify the potential associated to $\log |q_n(z)|$.

Proposition 3.7.2. *Let q_n be a degree n polynomial whose leading coefficient is A^{-1} and zeros (counting multiplicity) $\{z_j\}_{j=1}^n$. Let $d\nu^{(q)}$ be the normalized zero counting measure, i.e.,*

$$d\nu^{(q)} = \frac{1}{n} \sum_{j=1}^{n} \delta_{z_j} \qquad (3.7.10)$$

Then,

$$\log |q_n(z)|^{1/n} = -\frac{1}{n} \log |A| - \Phi_{\nu^{(q)}}(z) \qquad (3.7.11)$$

where $\Phi_{\nu^{(q)}}$ is the potential of $\nu^{(q)}$.

Proof. By the fundamental theorem of algebra

$$q_n(z) = A^{-1} \prod_{j=1}^{n} (z - z_j) \qquad (3.7.12)$$

Thus,

$$\log |q_n(z)|^{1/n} = -\frac{1}{n} \log |A| + \frac{1}{n} \sum_{j=1}^{n} \log |z - z_j| \qquad (3.7.13)$$

which is (3.7.11) given (3.7.10). $\qquad \square$

The remainder of this section will apply the theory to orthogonal polynomials on the real line (OPRL). We will study the basics in Chapter 4 of Part 4, especially Sections 4.1 and 4.3 of Part 4. We start out with a probability measure, $d\mu$, on $\mathbb{R}$ which is nontrivial (i.e., not supported on a set with only finitely many points) and compact support. We'll let $\mathfrak{e}$ be the essential support of $d\mu$, i.e., $\text{supp}(d\mu)$ with isolated points removed, i.e.,

$$\mathfrak{e} = \{x \in \mathbb{R} \mid \mu((x - \varepsilon, x + \varepsilon) \setminus \{x\}) > 0\} \qquad (3.7.14)$$

Because μ is nontrivial, $\{x^n\}_{n=0}^{\infty}$ are independent in $L^2(\mathbb{R}, d\mu)$ so we can use Gram-Schmidt to form the monic orthogonal polynomials, $\{P_n\}_{n=0}^{\infty}$ (i.e., $P_n(x) = x^n+$ lower-order and $\int P_n(x) P_\ell(x) \, d\mu(x) = 0$ if $n \neq \ell$) and orthonormal polynomials, $p_n = P_n / \|P_n\|_{L^2(\mathbb{R}, d\mu)}$. We need the following facts:

Fact 1 (Theorem 4.1.1 of Part 4). There are associated to μ Jacobi parameters, $\{a_n\}_{n=1}^\infty$, $\{b_n\}_{n=1}^\infty$ with $a_n > 0$ and $b_n \in \mathbb{R}$, so that the p_n obey the recursion relations, $n = 0, 1, \ldots$

$$zp_n(z) = a_{n+1}p_{n+1}(z) + b_{n+1}p_n(z) + a_n p_{n-1}(z); \quad p_0(z) = 1, \ p_{-1}(z) = 0 \tag{3.7.15}$$

(since $p_{-1} = 0$, a_0 doesn't enter).

Fact 2 (Proposition 4.1.2 of Part 4). $\sup_n(|a_n| + |b_n|) < \infty$.

Fact 3 (Theorem 4.1.1 of Part 4). $\|P_n\|_{L^2} = a_1 \cdots a_n$. $\tag{3.7.16}$

Fact 4 (Proposition 4.1.4 of Part 4). All the roots of p_n are real and simple and the roots of p_n and p_{n+1} strictly interlace.

Fact 5 (Problem 4 of Section 4.1 of Part 4). If $\mathrm{supp}(\mu) \subset [\alpha, \beta]$, all the zeros of p_n lie in $[\alpha, \beta]$.

Fact 6 (Problem 4 of Section 4.1 of Part 4). If $(\gamma, \lambda) \cap \mathrm{supp}(\mu) = \emptyset$, each p_n has at most one zero in (γ, λ).

Fact 7 (Corollary 4.3.11 of Part 4). If $C(\mathfrak{e})$ is the capacity of $\mathfrak{e}$, the essential support of μ, then

$$\limsup_{n \to \infty} \|P_n\|_{L^2(\mathbb{R}, d\mu)}^{1/n} \leq C(\mathfrak{e}) \tag{3.7.17}$$

We'll find another proof of this below, but it is useful to know it now as a guidepost to the definition below.

Definition. Let μ be a nontrivial probability measure on $\mathbb{R}$ with compact support and with essential support $\mathfrak{e}$. We say μ is *regular* if equality holds in (3.7.17) and the limit exists, i.e.,

$$\lim_{n \to \infty} (a_1 \cdots a_n)^{1/n} = C(\mathfrak{e}) \tag{3.7.18}$$

Our primary goals next are to prove if μ has a large enough absolutely continuous piece, then μ is regular and to show regularity implies the existence of a limit for the zero counting measure, indeed to the equilibrium measure of $\mathfrak{e}$ and root asymptotics for $|p_n|$ q.e. on $\mathfrak{e}$ and on all of $\mathbb{C} \setminus \mathrm{cvh}(\mathrm{supp}(\mu))$ where $\mathrm{cvh}(\mathrm{supp}(\mu))$ is the convex hull of $\mathrm{supp}(\mu)$, i.e., the smallest interval $[\alpha, \beta] \subset \mathbb{R}$ with $\mathrm{supp}(\mu) \subset [\alpha, \beta]$. We'll also say some things about not regular measures. A key a priori result we'll need will prove growth of $|p_n(z)|$ for $z \in \mathbb{C} \setminus [\alpha, \beta]$. We are heading towards

Theorem 3.7.3. *Let μ be a nontrivial probability measure on $\mathbb{R}$ with* $\mathrm{cvh}(\mathrm{supp}(\mu)) = [\alpha, \beta]$. *Let*

$$D = \sup_n |a_n|, \quad d(z) = \mathrm{dist}(z, [\alpha, \beta]) \tag{3.7.19}$$

Then, for any $z \in \mathbb{C}$ and any n:

$$|p_n(z, d\mu)|^2 \geq \left(\frac{d}{D}\right)^2 \left(1 + \left(\frac{d}{D}\right)^2\right)^{n-1} \tag{3.7.20}$$

In particular, if $z \notin [\alpha, \beta]$, we have

$$\liminf |p_n(z, d\mu)|^{1/n} \geq \left(1 + \left(\frac{d}{D}\right)^2\right)^{1/2} > 1 \tag{3.7.21}$$

As a preliminary, we need a classical equality for OPRL, the *Christoffel–Darboux formula* (aka *CD formula*).

Proposition 3.7.4 (Christoffel–Darboux Formula). *For any OPRL and any $x, z \in \mathbb{C}$*

$$K_n(x, z) \equiv \sum_{j=0}^{n} p_j(x) \, p_j(z) = \frac{a_{n+1}[p_{n+1}(x)p_n(z) - p_{n+1}(z)p_n(x)]}{x - z} \tag{3.7.22}$$

Remark. The left side of (3.7.22) is called the *CD kernel*, or *Christoffel–Darboux kernel*.

Proof. Let $C_n(x, z)$ be the numerator of the right side of (3.7.22). Take (3.7.15) for x and multiply by $p_n(z)$ and subtract (3.7.15) for z multiplied by $p_n(x)$ to get

$$(x - z)p_n(x) \, p_n(z) = C_n(x, z) - C_{n-1}(x, z) \tag{3.7.23}$$

Since $C_{-1}(x, z) = 0$, if we sum (3.7.23) from $j = 0$ to n, we get

$$(x - z) \sum_{j=0}^{n} p_j(x) \, p_j(z) = C_n(x, z)$$

which is (3.7.22)   □

Proof of Theorem 3.7.3. Fix $z \in \mathbb{C}$ and $n \in \mathbb{N}$ and consider

$$\varphi_n(x) = \sum_{j=0}^{n} p_j(z) \, p_j(x) \tag{3.7.24}$$

viewed as a function in $L^2(\mathbb{R}, d\mu)$. Since $\{p_j\}_{j=0}^{n}$ are orthonormal

$$\|\varphi_n\|_{L^2(\mathbb{R}, d\mu)}^2 = \sum_{j=0}^{n} |p_j(z)|^2 \tag{3.7.25}$$

On the other hand, the CD formula says that as functions in L^2,

$$(x - z)\varphi_n = -a_{n+1}p_{n+1}(z)p_n + a_{n+1}p_n(z)p_{n+1} \tag{3.7.26}$$

Since p_{n+1} is orthogonal to $\{p_j\}_{j=0}^{n}$, we see that

$$\langle \varphi_n, (x - z)\varphi_n \rangle = -a_{n+1}p_{n+1}(z)p_n(z) \tag{3.7.27}$$

so for any $\omega \in \mathbb{C}$ with $|\omega| = 1$, we have that

$$|\langle \varphi_n, \omega \cdot (x - z)\varphi_n \rangle| \leq a_{n+1}|p_{n+1}(z)p_n(z)| \tag{3.7.28}$$

For any convex set, C, in an inner-product space, V, if there is x_0 in C with $|z - x_0| = \text{dist}(z, C)$, then with $\omega = \frac{(x_0 - z)}{|z - x_0|}$, we have for all $x \in C$ that

$$\text{Re}[\omega \cdot (x - z)] \geq |z - x_0| \tag{3.7.29}$$

(Problem 1). Thus, there is ω in (3.7.28) so that

$$\text{Re}[\langle \varphi_n, \omega \cdot (x - z)\varphi_n \rangle] \geq d \, \|\varphi_n\|^2 \tag{3.7.30}$$

Therefore, by (3.7.25)

$$|p_{n+1}(z)p_n(z)| \geq \frac{d}{D} \sum_{j=0}^{n} |p_j(z)|^2 \tag{3.7.31}$$

Since $\left(\sqrt{\frac{D}{d}}\,|p_n(z)| - \sqrt{\frac{d}{D}}\,|p_{n+1}(z)|\right)^2 \geq 0$, this implies that

$$2\frac{d}{D} \sum_{j=0}^{n} |p_j(z)|^2 \leq \frac{d}{D} \, |p_n(z)|^2 + \frac{D}{d} \, |p_{n+1}(z)|^2 \tag{3.7.32}$$

so, we see that

$$|p_{n+1}(z)|^2 \geq \left(\frac{d}{D}\right)^2 \sum_{j=0}^{n} |p_j(z)|^2 \tag{3.7.33}$$

which in turn implies that

$$\sum_{j=0}^{n+1} |p_j(z)|^2 \geq \left(1 + \left(\frac{d}{D}\right)^2\right) \sum_{j=0}^{n} |p_j(z)|^2 \tag{3.7.34}$$

Therefore, since $p_0(z) = 1$, by induction in n

$$\sum_{j=0}^{n} |p_j(z)|^2 \geq \left(1 + \left(\frac{d}{D}\right)^2\right)^n \tag{3.7.35}$$

so (3.7.33) implies (3.7.20) which yields (3.7.21) if $d \neq 0$. $\qquad\square$

Theorem 3.7.5 (Thouless Formula). *Let $\{a_n, b_n\}_{n=1}^{\infty}$ be the Jacobi parameters of a measure $d\mu$ with $\text{supp}(d\mu) \subset [\alpha, \beta] \subset \mathbb{R}$. Let $\{p_n\}_{n=0}^{\infty}$ be the orthonormal polynomials. Let $n_j \to \infty$ be such that*

(i) $\displaystyle \lim_{j \to \infty} (a_1 \cdots a_{n_j})^{1/n_j} \to A_\infty \in (0, \infty)$ $\tag{3.7.36}$

(ii) $d\nu_{n_j} \to d\nu_\infty$ *weakly as* $j \to \infty$ $\tag{3.7.37}$

Then, for all $z \in \mathbb{C} \setminus [\alpha, \beta]$

$$\lim_{j \to \infty} |p_{n_j}(z)|^{1/n_j} = A_\infty^{-1} \, e^{-\Phi_{\nu_\infty}(z)} \tag{3.7.38}$$

Moreover, for $z \in [\alpha, \beta]$, we have

$$\overline{\lim_{j \to \infty}} |p_{n_j}(z)|^{1/n_j} \leq A_\infty^{-1} e^{-\Phi_{\nu_\infty}(z)} \tag{3.7.39}$$

and (3.7.38) holds for q.e. $z \in [\alpha, \beta]$.

Proof. By Proposition 3.7.2, if $A_n = (a_1 \cdots a_n)^{1/n}$

$$|p_{n_j}(z)|^{1/n_j} = A_{n_j}^{-1} e^{-\Phi_{\nu_{n_j}}(z)} \tag{3.7.40}$$

By Fact 5 about OPRL, supp $d\nu_{n_j} \subset [\alpha, \beta]$. Thus the principle of descent (Theorem 3.6.29) implies (3.7.39) and implies (3.7.38) if $z \notin [\alpha, \beta]$. That (3.7.38) holds for q.e. $z \in [\alpha, \beta]$ follows from the lower envelope theorem (Theorem 3.6.31). □

Corollary 3.7.6. *For any OPRL on a bounded set, if* $\liminf (a_1, \ldots a_n)^{1/n}$ *> 0 and if $d\nu_\infty$ is a limit point of the zero counting measures $d\nu_{q_n}$ (given by (3.7.10)), then ν_∞ has no pure points.*

Remarks. 1. This is an illustration of the usefulness of potential theory ideas.

2. This says $E \mapsto \nu_\infty((-\infty, E]) \equiv N(E)$ is a continuous function. In fact (Problem 2), one can prove that it is log-Hölder continuous.

3. ν_∞ is called a *density of zeros* or density of states. If ν_n converges to ν_∞, we call it the *density of states* and we say the *density of states exists* (aka the DOS exists).

Proof. Pass to a further subsequence so that (3.7.36) holds. Since $\liminf (a_1 \cdots a_n) > 0$, $A_\infty \in (0, \infty)$. By Theorem 3.7.3 and the Thouless formula, for all $z \in \mathbb{C} \setminus [\alpha, \beta]$

$$1 \leq A_\infty^{-1} e^{-\Phi_{\nu_\infty}(z)} \tag{3.7.41}$$

so for such z,

$$\Phi_{\nu_\infty}(z) \leq -\log(A_\infty) \tag{3.7.42}$$

Since Φ_{ν_∞} is superharmonic, for $x \in [\alpha, \beta]$

$$\Phi_{\nu_\infty}(x) = \lim_{r \downarrow 0} (\pi r^2)^{-1} \int_{|x-z| \leq r} \Phi_{\nu_\infty}(z) \, d^2z \tag{3.7.43}$$

so (3.7.42) holds for all $z \in \mathbb{C}$. If x_0 is a pure point of ν_∞, $\Phi_{\nu_\infty}(x_0) = \infty$, so (3.7.42) implies there are no pure points. □

Theorem 3.7.7. *Suppose that the hypotheses of Theorem 3.7.5 hold. Then*

(a) $\mathcal{E}(\nu_\infty) \leq -\log(A_\infty)$. \hfill (3.7.44)

(b) $A_\infty \leq C(\mathfrak{e})$. \hfill (3.7.45)

(c) *For any OPRL*

$$\limsup(a_1 \cdots a_n)^{1/n} \leq C(\mathfrak{e}) \qquad (3.7.46)$$

(d) *One has equality in (3.7.45) if and only if $\nu_\infty = \rho_\mathfrak{e}$, the equilibrium measure for $\mathfrak{e}$.*

(e) *If μ is regular, then the density of states exists and is the equilibrium measure for $\mathfrak{e}$.*

(f) *If μ is regular, then for any $z \notin [\alpha, \beta]$*

$$\lim_{n \to \infty} |p_n(z)|^{1/n} = \mathfrak{e}^{G_\mathfrak{e}(z)} \qquad (3.7.47)$$

and this holds for q.e. $z \in [\alpha, \beta]$.

Proof. (a) Since (3.7.42) holds on all of $\mathbb{R}$, we have

$$\mathcal{E}(\nu_\infty) = \int \Phi_{\nu_\infty}(x) \, d\nu_\infty(x) \leq -\log(A_\infty) \qquad (3.7.48)$$

(b) By Fact 6 about OPRL, $\mathrm{supp}(\nu_\infty) \subset \mathfrak{e}$. Thus, by (3.7.44)

$$R(\mathfrak{e}) \leq \mathcal{E}(\nu_\infty) \leq -\log(A_\infty) \qquad (3.7.49)$$

so

$$A_\infty \leq \exp(-R(\mathfrak{e})) = C(\mathfrak{e})x$$

(c) Find n_j so $(a_1 \cdots a_{n_j})^{1/n_j} \to \limsup (a_1 \cdots a_n)^{1/n}$ and by passing to a subsequence so $d\nu_{n_j}$ converges. Then (3.7.45) implies (3.7.46).

(d) By (3.7.49), one has equality in (3.7.45) if and only if $\mathcal{E}(\nu_\infty) = R(\mathfrak{e})$, i.e., by uniqueness of minimizer, only if $\nu_\infty = \rho_\mathfrak{e}$.

(e) If μ is regular, any limit point of ν_n is $\rho_\mathfrak{e}$ so the limit exists and is $\rho_\mathfrak{e}$.

(f) Given the relation of $G_\mathfrak{e}$ and $\Phi_{\rho_\mathfrak{e}}$, this is just a restatement of the Thouless formula and the fact that $A_\infty = C(\mathfrak{e})$ and $\nu_\infty = \rho_\mathfrak{e}$. $\qquad \square$

One might wonder whether, if the DOS exists and equals ν_∞, then μ must be regular. Not quite, but something is true. We need a preliminary that we'll use several times.

Lemma 3.7.8. *Let μ be a nontrivial probability measure of compact support on $\mathbb{R}$. Then for μ-a.e. x, we have*

$$\limsup |p_n(x)|^{1/n} \leq 1 \qquad (3.7.50)$$

Proof. Since $\int |p_n(x)|^2 \, d\mu(x) = 1$, we have

$$\int \sum_{n=0}^\infty (1 + n^2)^{-1} |p_n(x)|^2 \, d\mu(x) = \sum_{n=0}^\infty (1 + n^2)^{-1} < \infty \qquad (3.7.51)$$

Since the sum is in L^1, it is μ-a.e. x finite, i.e., for μ-a.e. x, there is $C(x)$ so that

$$|p_n(x)| \le C(x)\,(1 + n^2)^{1/2} \tag{3.7.52}$$

Since $\lim_{n \to \infty} [C(x)\,(1 + n^2)^{1/2}]^{1/n} = 1$, this implies (3.7.50) $\square$

Theorem 3.7.9. *Let μ be a nontrivial probability measure with compact support. Suppose that the DOS exists and equals $\rho_{\mathfrak{e}}$, the equilibrium measure for $\mathfrak{e} = \text{ess supp}(d\mu)$. Then either μ is regular, or there is a polar set $S \subset \mathbb{R}$ with $\mu(\mathbb{R} \setminus S) = 0$.*

Remark. Remarkably, as we'll explain in the Notes, there exist examples where the DOS is $\rho_{\mathfrak{e}}$ but which are not regular.

Proof. If μ is not regular, there is n_j so (3.7.36) holds with $A_\infty < C(\mathfrak{e})$. Thus, by the Thouless formula for q.e. x in $\mathfrak{e}$, we have that

$$\lim_{j \to \infty} |p_{n_j}(x)|^{1/n_j} = \frac{C(\mathfrak{e})}{A_\infty} e^{-G_{\mathfrak{e}}(x)} = \frac{C(\mathfrak{e})}{A_\infty} \tag{3.7.53}$$

since $G_{\mathfrak{e}}(x) = 0$ q.e. on $\mathfrak{e}$ (by Theorem 3.6.6).

Since $\frac{C(\mathfrak{e})}{A_\infty} > 1$, this is inconsistent with (3.7.50). That implies $\mu(\mathbb{R} \setminus (S_1 \cup S_2 \cup S_3)) = 0$ where $S_1 = \{x \in \mathfrak{e} \mid G_{\mathfrak{e}}(x) > 0\}$, $S_2 = \{x \mid (3.7.38) \text{ fails}\}$, $S_3 =$ countable set of eigenvalues. Since $S = S_1 \cup S_2 \cup S_3$ is a union of polar sets, we have the desired result. $\square$

Of course, we want effective criteria for μ to be regular. Here is one.

Theorem 3.7.10 (Widom's Theorem). *Let μ be a nonnegative measure on $\mathbb{R}$ with compact support and let $\mathfrak{e} = \text{ess supp}(\mu)$. If $d\rho_{\mathfrak{e}}$ is absolutely continuous with respect to $d\mu$, then μ is regular.*

Remarks. 1. It is not hard to see (Problem 3) that if $d\mu = g(x)d\rho_{\mathfrak{e}}(x) + d\tilde{\mu}_s$ where $d\tilde{\mu}_s$ is $\rho_{\mathfrak{e}}$-singular, then $d\rho_{\mathfrak{e}}$ is $d\mu$-a.c. if and only if $g(x) > 0$ for $\rho_{\mathfrak{e}}$-a.e. x.

2. There is a much stronger result due to Stahl–Totik if $\mathfrak{e} = [-1, 1]$. It is discussed in the Notes.

Proof. Let A be a limit point of $\|p_n\|^{1/n}$ and pass to a subsequence n_j so $\|p_{n_j}\|^{1/n_j} \to A$ and $d\nu_{n_j} \to d\nu_\infty$ weakly. By Corollary 3.7.6 and the Thouless formula for μ-a.e. x, we have that

$$\limsup_{j \to \infty} \exp(-\Phi_{\nu_{n_j}}(x)) \le A \tag{3.7.54}$$

Let S_1 be the set where this fails so $\mu(S_1) = 0$.

By the upper envelope theorem for q.e. x

$$\limsup_{j \to \infty} \exp(-\Phi_{\nu_{n_j}}(x)) = \exp(-\Phi_{\nu_\infty}(x)) \tag{3.7.55}$$

Let S_2 be the set where this fails. Clearly $\rho_\mathfrak{e}(S_2) = 0$ since S_2 is a polar set and $\mathcal{E}(\rho_\mathfrak{e}) < \infty$. Since $\rho_\mathfrak{e}$ is μ-a.c., $\rho_\mathfrak{e}(S_1) = 0$, i.e., $\rho_\mathfrak{e}(S_1 \cup S_2) > 0$ so for $\rho_\mathfrak{e}$-a.e. x, we have

$$\log(A^{-1}) \le \Phi_{\nu_\infty}(x) \tag{3.7.56}$$

Thus

$$\log(A^{-1}) \le \int \Phi_{\nu_\infty}(x)\, d\rho_\mathfrak{e}(x) \tag{3.7.57}$$

$$= \int \Phi_\mathfrak{e}(x)\, d\nu_\infty(x) \tag{3.7.58}$$

$$\le \log(C(\mathfrak{e})^{-1}) \tag{3.7.59}$$

where (3.7.58) is (3.6.6) and (3.7.59) comes from $\Phi_{\rho_\mathfrak{e}}(x) \le R(\mathfrak{e})$, i.e., (3.6.24).

Therefore $C(\mathfrak{e}) \le A$. Since, as we've seen, $A \le C(\mathfrak{e})$, we conclude that $A = C(\mathfrak{e})$, i.e., every limit point of $(a_1 \cdots a_n)^{1/n}$ is $C(\mathfrak{e})$, so μ is regular. $\square$

To apply this, it is useful to know a little more about regularity properties of $\rho_\mathfrak{e}$. We need a preliminary:

Lemma 3.7.11. *Let f be a Herglotz function of $\mathbb{C}_+$. Let $(\alpha, \beta) \subset \mathbb{R}$. Then*

(a) *If $\operatorname{Im} f$ is bounded on $\{z \mid \operatorname{Re} z \in (\alpha, \beta),\ \operatorname{Im} z \in (0, 1)\}$ and for Lebesgue a.e. x in (α, β),*

$$\lim_{\varepsilon \downarrow 0} \operatorname{Im} f(x + i\varepsilon) = 0 \tag{3.7.60}$$

then f has an analytic continuation into a neighborhood of (α, β) and (3.7.60) holds for all $x \in (\alpha, \beta)$.

(b) *If f has a continuous extension to $\mathbb{C}_+ \cup (\alpha, \beta)$ and*

$$\operatorname{Re} f(x) = 0 \text{ for all } x \in (\alpha, \beta) \tag{3.7.61}$$

then $\operatorname{Im} f(x) > 0$ for all $x \in (\alpha, \beta)$.

Proof. (a) We use a Herglotz representation (see (5.9.14))

$$f(z) = az + b + \int \left(\frac{1}{y - z} - \frac{1}{1 + y^2} \right) d\rho(y) \tag{3.7.62}$$

where $a > 0$, $b \in \mathbb{R}$ so

$$\operatorname{Im} f(x + i\varepsilon) = a\varepsilon + \int \frac{\varepsilon}{(x - y)^2 + \varepsilon^2}\, d\rho(y) \tag{3.7.63}$$

Write $d\rho = w(x)dx + d\rho_s$. By Theorem 2.5.4, since $\operatorname{Im} f$ is bounded near (α, β), $\rho_s(\alpha, \beta) = 0$. By (3.7.60) and (2.5.20), $w(x) = 0$ for a.e. x and thus

$\rho(\alpha, \beta) = 0$. By (3.7.62), f has an analytic continuation through (α, β) and (3.7.60) holds for all $x \in (\alpha, \beta)$.

(b) Since f is Herglotz, we can define $\log f$ analytic in $\mathbb{C}_+$, so $0 < \text{Im}(\log f) < \pi$ on $\mathbb{C}_+$. By (3.7.61), $\lim_{\varepsilon \downarrow 0} (\log f(x + i\varepsilon)) = \frac{\pi}{2}$ on (α, β). Applying the strong reflection principle (Theorem 5.5.3 of Part 2A) to $-i \log f - \frac{\pi}{2}$, we conclude that $\log f$ has an analytic continuation through (α, β). In particular, f is never zero there. Since $\text{Re} f = 0$ on (α, β), and $\text{Im} f \geq 0$, we conclude that $\text{Im} f > 0$ on (α, β). $\qquad \square$

Theorem 3.7.12. *Let $\mathfrak{e}$ be a compact, nonpolar subset of $\mathbb{R}$. Suppose $(\alpha, \beta) \subset \mathfrak{e}$. Then $\rho_{\mathfrak{e}} \upharpoonright (\alpha, \beta)$ is absolutely continuous wrt Lebesgue measure with a Radon–Nikodym derivative which is strictly positive on (α, β). Moreover, $G_{\mathfrak{e}}(x) = 0$ for all $x \in (\alpha, \beta)$.*

Remarks. 1. If $[\alpha, \beta]$ is an isolated component of $\mathfrak{e}$, one can say a lot more about the form of the Radon–Nikodym derivative; see Problem 6.

2. The proof shows an explicit formula for $\frac{d\rho(\mathfrak{e})}{dx}(x_0)$ for all $x_0 \in (\alpha, \beta)$, namely

$$\frac{d\rho(\mathfrak{e})}{dx}(x_0) = \frac{1}{\pi} \frac{\partial G_{\mathfrak{e}}}{\partial y}(x_0 + i0) \equiv \frac{1}{\pi} \lim_{\varepsilon \downarrow 0} \frac{G_{\mathfrak{e}}(x_0 + iy)}{y} \tag{3.7.64}$$

Proof. Define $f(z)$ to be analytic on $\mathbb{C}_+$ with

$$\text{Im} f(z) = G_{\mathfrak{e}}(z) \tag{3.7.65}$$

and $\text{Re} f(i) = 0$. Since $G_{\mathfrak{e}}$ is harmonic on $\mathbb{C}_+$, this is possible. Since $G_{\mathfrak{e}} \geq 0$ on $\mathbb{C}_+$ and $\lim_{z \to x_0} G_{\mathfrak{e}}(z) = 0$ for q.e. and so a.e. $x_0 \in \mathfrak{e}$ and since $G_{\mathfrak{e}}$ is bounded on compact subspaces of $\mathbb{C}_+$, f obeys the hypotheses of Lemma 3.7.11(a). It follows that f has an analytic continuation through (α, β).

For $z \in \mathbb{C}_+$

$$h(z) \equiv \int \frac{d\rho_{\mathfrak{e}}(x)}{x - z} = \frac{d}{dz} f(z) \tag{3.7.66}$$

so $h(z)$ has an analytic continuation through (α, β). By Theorem 2.5.4, we conclude that $\rho_{\mathfrak{e}} \upharpoonright (\alpha, \beta)$ is absolutely continuous with a weight $\frac{1}{\pi} \text{Im} h(x+i0)$ which is analytic.

Since $G_{\mathfrak{e}}(\bar{z}) = G_{\mathfrak{e}}(z)$, we have that for any $x \in (\alpha, \beta)$

$$G_{\mathfrak{e}}(x) = \frac{2}{\pi r^2} \int_{|z-x| \leq r, \, \text{Im} z > o} G_{\mathfrak{e}}(z) \, d^2 z \tag{3.7.67}$$

and thus since f is continuous up to (α, β) with $\text{Im} f(x) = 0$ for all $x \in (\alpha, \beta)$, we conclude that (3.7.65) holds also on (α, β) and $G_{\mathfrak{e}}(x) = 0$ there.

Since $\text{Re} f = 0$ on (α, β), $\text{Re} \frac{d}{dz} f = 0$ on (α, β), h obeys the hypotheses of Lemma 3.7.11(b) and thus the weight is strictly positive on (α, β). $\qquad \square$

Corollary 3.7.13. *Let* $\mathfrak{e} = \bigcup_{j=1}^{\ell+1} [\alpha_j, \beta_j]$ *with* $\alpha_1 < \beta_1 < \alpha_2 < \ldots < \beta_{\ell+1}$ *in* $\mathbb{R}$ *and* $\ell \in \mathbb{N}$. *Let*

$$d\mu(x) = w(x)dx + d\mu_s \qquad (3.7.68)$$

where $d\mu_s$ *is singular with respect to Lebesgue measure and* $\mu_s \upharpoonright \mathbb{R} \setminus \mathfrak{e}$ *is pure point. If* $w(x) > 0$ *for a.e.* x, *then* μ *is regular.*

Remarks. 1. One can prove, by very different methods, a considerably stronger result than $(a_1 \cdots a_n)^{1/n} \to C(\mathfrak{e})$; see the Notes.

2. A set like this $\mathfrak{e}$ is sometimes called a *finite gap set*.

Proof. On $\mathfrak{e}^{\mathrm{int}}$, $d\rho_{\mathfrak{e}}(x) = f(x)dx$. Since $\bigcup_{j=1}^{\ell+1} \{\alpha_j, \beta_j\}$ is a finite set and $\rho_{\mathfrak{e}}$ gives zero weight to polar sets on all of $\mathbb{R}$, $d\rho_{\mathfrak{e}}(x) = f(x)dx$ with $f > 0$ on $\mathfrak{e}^{\mathrm{int}}$. Let S be a set with $|S| = 0$ and $\mu_s(\mathbb{R} \setminus S) = 0$. Define

$$g(x) = \begin{cases} 0, & x \in S \text{ or in } \mathbb{R} \setminus \mathfrak{e}^{\mathrm{int}} \\ \frac{f(x)}{w(x)}, & x \in \mathfrak{e}^{\mathrm{int}} \end{cases} \qquad (3.7.69)$$

Then $d\rho_{\mathfrak{e}} = g(x)d\mu$ is a.e wrt $d\mu$, so μ is regular. $\qquad\square$

Our next result about regularity concerns some equivalent statements.

Theorem 3.7.14. *Let* μ *be a probability measure supported on a compact set, with* $\mathfrak{e} = \mathrm{ess\ supp}(\mu)$ *a nonpolar set. Then the following are equivalent:*

(1) *μ is regular.*
(2) *For all z in $\mathbb{C}$*

$$\limsup |p_n(z)|^{1/n} \leq e^{G_{\mathfrak{e}}(z)} \qquad (3.7.70)$$

(3) *For q.e. $x \in \mathfrak{e}$*

$$\limsup |p_n(x)|^{1/n} \leq 1 \qquad (3.7.71)$$

Moreover, if that holds and $\mathfrak{e} \subset [\alpha, \beta]$ we have that

$$\lim_{n \to \infty} |p_n(z)|^{1/n} = e^{G_{\mathfrak{e}}(z)} \qquad (3.7.72)$$

everywhere in $\mathbb{C} \setminus [\alpha, \beta]$ and q.e. on $\mathfrak{e}$ (where $G_{\mathfrak{e}}(x) = 0$ q.e.).

Proof. We've already proven that regularity implies the "moreover" claim so we only need to prove that $(1) \Rightarrow (2) \Rightarrow (3) \Rightarrow (1)$. We've already proven $(1) \Rightarrow (2)$ and $(2) \Rightarrow (3)$ is trivial so we need only prove $(3) \Rightarrow (1)$.

Let n_j be a sequence so $(a_1 \cdots a_{n_j})^{1/n_j} \to A = \liminf (a_1 \cdots a_n)^{1/n}$ and $d\nu_{n_j} \to d\nu_{\infty}$ for some ν_{∞}. We first claim that $A > 0$, for if $A = 0$, by the same argument that led to the Thouless formula and (3.7.71), we see that $\Phi_{\nu_{\infty}}(x) = \infty$ for q.e. $x \in \mathfrak{e}$ and this is inconsistent with $\mathfrak{e}$ nonpolar and Proposition 3.6.10.

If A is finite, we use the same argument as in the proof of Theorem 3.7.10. By the Thouless formula again (and the lower envelope theorem) and (3.7.71), for q.e. $x \in \mathfrak{e}$ we have

$$\Phi_{\nu_\infty}(x) \geq \log\left(A^{-1}\right) \qquad (3.7.73)$$

Since $\rho_\mathfrak{e}$ is supported on $\mathfrak{e}$ gives zero weight to polar sets, we see that

$$\log\left(A^{-1}\right) \leq \int \Phi_{\nu_\infty}(x)\,d\rho_\mathfrak{e}(x)$$

$$= \int \Phi_{\rho_\mathfrak{e}}(x)d\nu_\infty(x)$$

$$\leq \log\left(C(\mathfrak{e})^{-1}\right)$$

by the same arguments that went from (3.7.56) to (3.7.59). Thus $A \geq C(\mathfrak{e})$, so, since $A \leq C(\mathfrak{e})$, we have regularity. $\qquad\square$

Lastly, we want to say something about some not regular measures. Suppose μ is a nontrivial probability measure supported on $[\alpha, \beta]$ with $\mathfrak{e} = \operatorname{ess\,supp}(\mu)$, a nonpolar set. Suppose $A_\infty = \lim_{n\to\infty}(a_1 \cdots a_n)^{1/n}$ exists and the DOS, $d\nu_\infty$ exists. We define the *Lyapunov exponent*, $\gamma(z)$, on $\mathbb{C}$ by

$$\gamma(z) = -A_\infty - \Phi_{\nu_\infty}(z) \qquad (3.7.74)$$

(WARNING: This is different from the more usual definition of Lyapunov exponent although it is q.e to one version of the usual definition and equal to another version; see the Notes.)

Thus, the Thouless formula says that for q.e. $z \in \mathbb{C}$ and all $z \in \mathbb{C}\setminus[\alpha,\beta]$, we have that

$$\lim_{n\to\infty} \frac{1}{n}\log(|p_n(z)|) = \gamma(z) \qquad (3.7.75)$$

Theorem 3.7.15. *Let μ be a measure as above where the DOS and A_∞ exist. Let $T \subset \mathfrak{e}$ be a Borel set so that $\gamma(x) > 0$ for $x \in T$. Then $\mu \restriction T$ is supported by a polar set, i.e., there exists a polar set, $S \subset T$, so that $\mu(T \setminus S) = 0$.*

Remarks. 1. Of course $\mu(T)$ could be zero. The most interesting case is where $T = \mathfrak{e}$ so $\mu(T)$ must be nonzero.

2. At first sight, it may appear unlikely that this can happen, but it does in two of the most heavily studied examples in mathematical physics; see the Notes.

3. Of course, this implies S has Hausdorff dimension zero.

Proof. This is an immediate consequence of Lemma 3.7.8 and (3.7.75)! For let $S = \{x \in T \mid (3.7.50) \text{ holds}\}$. By Lemma 3.7.8, $\mu(T \setminus S) = 0$. But if (3.7.50) holds and $x_0 \in \mathfrak{e}$, then (3.7.75) must fail since $\gamma(\mathbf{x}_0) > 0$. Thus S is contained in the polar set where (3.7.75) fails. $\qquad\square$

Notes and Historical Remarks. The Bernstein–Walsh inequality is named after Walsh's 1935 book [**841**] with an early related result in Bernstein [**61**]. In essence, it appears already in Szegő's seminal 1924 paper [**794**] which proved the result we discuss in Section 4.3 of Part 4 on root asymptotics for the $\|\cdot\|_\infty$ norm of Chebyshev polynomials (Theorem 4.3.10, aka the Faber–Fekete–Szegő theorem, of Part 4).

Szegő understood the connection between potentials generated by zero counting measures and root asymptotics so, while the Thouless formula was not explicitly in their work, it would not have surprised them. Earlier Faber [**256**] had equivalent formulae in some cases, but phrased it in terms of conformal maps rather than potential theory. Fifty years later, the formula was explicitly written down in the physics literature, usually named after Thouless [**806**], although Herbert and Jones [**378**] had essentially the same result a few years earlier. This work was done in the context of Lyapunov exponents for ergodic discrete Schrödinger operators, a subject we discuss later in these Notes. A first careful mathematical proof that worried about $x \in \text{supp}(d\mu)$ appeared in Avron–Simon [**24**]. The ergodic-theoretic definition of $\gamma(z)$ for *all* $z \in \mathbb{C}$ is due to Craig–Simon [**176**].

In 1940, Erdős–Turán [**245**] studied measures on $[-1, 1]$ where the measure is $d\mu = w(x)\,dx$ and assuming inf $w > 0$ proved the density of zeros (even more) was given by what we know is the equilibrium measure for $[-1, 1]$ (although they didn't note that) so obtaining an early result combining the notion of regularity and Widom's theorem. In a series of five papers [**819, 820, 821, 822, 823**], Ullman developed the theory of regular measures when $\mathfrak{e} = [-1.1]$. In a remarkable book, Stahl–Totik [**772**] not only initiated the general theory for any compact nonpolar set in $\mathbb{C}$ (not merely $\mathbb{R}$) but brought the subject to maturity. Simon [**751**] reviewed some of their results and discussed the applications to ergodic Jacobi matrices (see below).

The CD kernel and formula, which date to the second half of the nineteenth century, are named after Christoffel [**159**] and Darboux [**186**] and are extensively used in the OPRL literature (there is also an OPUC analog; see [**748**]). One unexpected connection to the subject of this section is that the weak limits of the zero counting measuring, $d\nu_n$ are exactly the same as the weak limits of the measures $d\mu_n(x) = \frac{1}{n+1}K_n(x,x)d\mu(x)$. This is a theorem of Simon [**754**] and can be thought of as a boundary condition independent of the DOS (see Avron–Simon [**24**]). Simon [**752**] has a review of results on the CD kernel; see also the discussion of regularity below.

Lemma 3.7.8 for the theory of Schrödinger operators (in the form that in ν-dimensions there is a spectral-theorem type eigenfunction expansion with eigenfunctions obeying $|\varphi(x)| \leq C(1 + |x|)^{\nu/2+\varepsilon}$) goes back at least to Berezanskiĭ [57] (see also Simon [744] and Last–Simon [511]).

Widom's theorem (Theorem 3.7.10) is from Widom [852]; the proof we give follows van Assche [826] (in a form discussed by Simon [751]).

Much more than regularity is true if $w(x) > 0$ for a.e. x and there is additional information on $\mathfrak{e}$. This is either via refined information on the zeros or via more information on the Jacobi parameters.

The extra zero information involves the fine structure of the zeros which says that asymptotically they are locally equal spaced, explicitly for a.e. x_0 in supp(μ) for all $A > 0$ and n large, the zeros of $p_n(x)$ in $\left(x_0 - \frac{A}{n}, x_0 + \frac{A}{n}\right)$ are spaced up to errors of order $o(\frac{1}{n})$ the same distance, namely $\frac{1}{\rho_\mathfrak{e}(x_0)n}$ apart from the next one. This was dubbed *clock spacing* in [750] since for OPUC when $d\rho_\mathfrak{e} = \frac{d\theta}{2\pi}$, the zeros look like numerals on a clock.

Erdős–Turán [245] proved clock spacing for their examples on $[-1, 1]$ and Last–Simon [512] proved it for many sets of $\{a_n, b_n\}_{n=0}^\infty$ with $|a_n - 1| + |b_n| \to 0$. In two beautiful papers, Lubinsky [549, 550] showed that the key was nonzero weight in μ (Lubinsky needed the weight strictly bounded away from zero, later refinements only needed $\log w(x)$ to be locally L^1). He focused on something that had been discussed earlier in the random matrix literature called *CD kernel universality* but only proven there in a slightly different context and for real analytic weights. It says that the CD kernel, K_n, of (3.7.22) obeys for $x_0 \in \mathfrak{e}$, $A, B > 0$

$$\lim_{n\to\infty} \frac{K_n(x_0 - \frac{A}{n}, x_0 + \frac{B}{n})}{K_n(x_0, x_0)} \to \operatorname{sinc}(\pi\rho_\mathfrak{e}(x_0)(A + B)) \qquad (3.7.76)$$

where $\operatorname{sinc}(y) = \sin(y)/y$.

Lubinsky found two different approaches to proving universality. The first relied on comparison theorem for the CD kernel and the second on product formulae for entire functions. Both require only local conditions on w plus a single global condition—regularity for $\mathfrak{e}$. Levin–Lubinsky [523] showed that universality implies clock spacing of the zeros (a result obtained earlier by Freund [288]).

Using the first method which Lubinsky only applied to $\mathfrak{e} = [-1, 1]$, Simon [753] and Totik [810] proved (3.7.76) for $\mathfrak{e}$'s with $\overline{(\mathfrak{e}^{\mathrm{int}})} = \mathfrak{e}$ and Findley [275] obtained results only requiring that $\log w$ be locally L^1. Using the second method, Avila–Last–Simon [23] even obtained (3.7.76) for some cases where $\mathfrak{e}$ is a (positive Lebesgue measure) Cantor set.

The second result stronger than regularity that holds when $\mathfrak{e}$ is special, most notably an interval. The idea goes back to Rakhmanov [669] who looked at OPUC (OPUC and Verblunsky coefficients are discussed in Section 4.4 of Part 4) with $d\mu = w(\theta)\frac{d\theta}{2\pi} + d\mu_s$ and showed if $w(\theta) > 0$ for a.e. θ in $[0, 2\pi)$, then the Verblunsky coefficients, $\{\alpha_n\}_{n=0}^{\infty}$, obey $\alpha_n \to 0$. In 2004, Denisov [212] proved the long open conjecture for OPRL that if ess supp($d\mu$) $= [-2, 2]$ and $d\mu = w(x)\,dx + d\mu_s$ with $w(x) > 0$ for a.e. $x \in [-2, 2]$, then $a_n \to 1$, $b_n \to 0$ which, of course, implies that $(a_1 \cdots a_n)^{1/n} \to 1 = C([-2, 2])$ showing this is stronger than regularity.

If $\mathfrak{e}$ has a finite number of gaps and there is $\{a_n^{(0)}, b_n^{(0)}\}$ periodic so the corresponding $d\mu_0$ has $\mathfrak{e}$ as spectrum, then Damanik–Killip–Simon [184] proved an analog of the Denisov–Rakhmanov theorem for $\mathfrak{e}$ in terms of approach to an isospectral torus. Remling [677] then proved an ultimate theorem of this type demonstrating this approach to the isospectral torus for any finite gap set, and even more generally, that all the right limits are reflectionless. For a discussion of the meaning of *isospectral torus*, *right limits*, and *reflectionless Jacobi matrices*, see Simon[757].

While it is not hard to see that regularity on $[-2, 2]$ does not imply $a_n \to 1$, $b_n \to 0$, Simon [755] has shown that regularity does imply that

$$\frac{1}{n}\sum_{j=1}^{n}(|a_j - 1| + |b_j|) \to 0 \qquad (3.7.77)$$

(this paper also has examples of regularity without $a_n \to 1$). At first sight, it seems surprising that $(a_1 \cdots a_n)^{1/n} \to 1$ alone implies anything about the b_j's! But there is a second condition, namely ess supp(μ) $= [-2, 2]$ and that together with $(a_1 \cdots a_n)^{1/n} \to 1$ is quite strong.

Stahl–Totik [772] have proven the following; (Simon[751] has a short version of the proof).

Theorem 3.7.16. *Let μ be a measure with ess supp(μ) $= [\alpha, \beta]$ an interval. Suppose that for any $\eta > 0$*

$$\lim_{m \to \infty} |\{x \in [\alpha, \beta] \mid \mu([x - \frac{1}{m}, x + \frac{1}{m}]) \le e^{-m\eta}\}| = 0 \qquad (3.7.78)$$

Then μ is regular.

(3.7.78) is an expression of the notion that μ is spread out uniformly locally. It implies, for example, (Problem 4) that if ν gives weight $\frac{1}{n^2}2^{-n}$ to all $x \in [\alpha, \beta]$ of the form $\frac{k}{2^n}$ with k odd, then $\mu = \nu/\nu([\alpha, \beta])$ is regular. On the other hand, it is easy to show (Problem 5) that if $\{x_m\}_{m=1}^{\infty}$ is an arbitrary dense subset of $[\alpha, \beta]$, then for small enough y, $\mu = (1 - y)\sum_{m=1}^{\infty} y^{m-1}\delta_{x_m}$ is a not regular measure which shows the exponential in (3.7.78) is optimal. Using Theorem 3.7.16, one can prove the result of Wyneken [864], that given any ν with ess supp(ν) $= [\alpha, \beta]$, there is a regular measure μ equivalent (i.e., each a.c. wrt the other) to ν (this consequence can be found in Simon [751]).

Theorem 3.7.15 is due to Simon [**751**]. Earlier Pastur [**631**] and Ishii [**421**] using more elementray methods had proven that the measure $\mu \restriction T$ was singular wrt Lebesgue measure. Jitomirskaya–Last had conjectured that the measure was supported on a set of zero Hausdorff dimension which is a consequence of Simon's result.

One might think that a positive Lyapunov exponent on $\operatorname{supp}(\mu)$ is an unusual if not impossible occurrence, but it holds in cases of the most heavily studied models in mathematical physics: the almost Matthieu and Anderson models.

To set the stage for presenting these models, we discuss ergodic Jacobi matrices. [**145, 181, 632, 749**] is a set of books on this subject (the last on ergodic OPUC rather than OPRL) and Jitomirskaya [**431**] a review article.

Let $(\Omega, \Sigma, \mu, \tau)$ be a measurable dynamical system with τ ergodic for μ. Let $A, B : \Omega \to \mathbb{R}$ be bounded measurable functions with

$$\varepsilon \leq A(\omega) \leq \varepsilon^{-1} \tag{3.7.79}$$

for some $\varepsilon > 0$. For each $\omega \in \Omega$ define

$$a_n(\omega) = A(\tau^{n-1}\omega), \quad b_n(\omega) = B(\tau^{n-1}\omega) \tag{3.7.80}$$

and let $d\mu_\omega$ be the measure on $\mathbb{R}$ with Jacobi parameters $\{a_n(\omega), b_n(\omega)\}_{n=1}^\infty$.

Define

$$C_\omega(z) = \frac{1}{A(\omega)} \begin{pmatrix} z - B(\omega) & -1 \\ A(\omega)^2 & 0 \end{pmatrix} \tag{3.7.81}$$

and

$$T_n(z; \omega) = C_{T^{n-1}\omega}(z) C_{T^{n-2}\omega}(z) \cdots C_\omega(z) \tag{3.7.82}$$

so that if $p_n(x; \{a_j, b_j\}_{j=0}^\infty)$ are the orthonormal OPRL for a given set of Jacobi parameters

$$T_n(z; \omega) \begin{pmatrix} 1 \\ 0 \end{pmatrix} = \begin{pmatrix} p_n(z; \{a_j(\omega), b_j(\omega)\} \\ a_n(\omega) p_{n-1}(z; \{a_j(\omega), b_j(\omega)\}) \end{pmatrix} \tag{3.7.83}$$

and

$$T_n(z; \omega) \begin{pmatrix} 0 \\ -1 \end{pmatrix} = \begin{pmatrix} q_n(z) \\ a_n q_{n-1}(z) \end{pmatrix} \tag{3.7.84}$$

$q_n(z)$ is a polynomial of degree $n - 1$ called the *second kind* polynomials. In fact, by construction

$$q_n(z) = a_1^{-1} p_{n-1}(z; \{a_{j+1}, b_{j+1}\}_{j=1}^\infty) \tag{3.7.85}$$

Of course

$$T_n(z) = \begin{pmatrix} p_n(z) & -q_n(z) \\ a_n p_{n-1}(z) & -a_n q_{n-1}(z) \end{pmatrix} \tag{3.7.86}$$

Second kind polynomials are an important element of spectral analysis for both the ergodic and general case, see [**757**]. They appear again in our study of the moment problem in Section 7.7 of Part 4.

By the Birkhoff ergodic theorem (Theorem 2.6.9), one can prove for a.e. ω, $d\nu_{n,\omega}$ has a limit $d\nu_\infty$ which is independent of ω and that $\frac{1}{n}\sum_{j=1}^{n}\log a_j(\omega) = \log A_\infty$ exists and is independent of ω. By the Furstenberg–Kesten theorem (Theorem 2.9.1) for each $z \in \mathbb{C}$, for a.e. ω

$$\gamma(z) = \lim_{n\to\infty}\frac{1}{n}\log(\|T_n(z;\omega)\|) \tag{3.7.87}$$

exists. It is called the *Lyapunov exponent* and is the more usual definition of this term.

Since the A_∞ and DOS exist for a.e. ω, for Ω_0, the set of $\omega \in \Omega$ for which it does exist, the Thouless formula implies that if $z \notin \mathbb{R}$, $\lim|p_n(z)|^{1/n}$ is given by that formula. It can be proven that the $n-1$ zeros of q_n interlace the n zeros of p_n so the DOS exists for q and is the same. Thus for $z \notin \mathbb{R}$, $\lim|p_n(z)|^{1/n} = \lim|q_n(z)|^{1/n}$ and is thus $\gamma(z)$. That is, γ defined by (3.7.87) is given by the Thouless formula if $z \notin \mathbb{R}$.

Craig–Simon [**176**] proved that $\gamma(z)$ given by (3.7.87) is suharmonic (see Problem 7), so the Thouless formula holds for *all* $z \in \mathbb{C}$. Examples of ergodic Jacobi matrices include random and almost periodic models.

The basic random model is the *Anderson model*—Anderson received the Nobel prize for his prediction of dense point spectrum—i.e., localized electron states in solids. In this model, $a_n \equiv 1$ and b_n are iidrv with uniform distribution $[-\lambda, \lambda]$. T_n is a random product where the two Lyapunov exponents (2.9.62) are negatives of each other since $\det(T_n) = 1$. Thus Furstenberg's theorem mentioned in the Notes to Section 2.9 implies $\gamma(z) > 0$ for all z. Theorem 3.7.15 and further analysis show that the spectrum is pure point for a.e. ω, i.e., μ lies on a countable set. The density of states looks very different from equilibrium measures (since $C[-2-\lambda, 2+\lambda] > 1$, those measures are not regular)—the density vanishes rapidly at the endpoints—known as *Lifschitz tails*.

The second model has been dubbed the *almost Matthieu operator* and is almost periodic. $a_n \equiv 1$ and

$$b_n = \lambda\cos(\pi\alpha n + \theta) \tag{3.7.88}$$

where $\lambda > 0$ and α are parameters with α irrational and θ labels points in $\Omega = \partial\mathbb{D}$ with $\tau(e^{i\varphi}) = e^{i(\varphi+\pi\alpha)}$ and $B(e^{i\theta}) = \lambda\cos(\pi\alpha + \theta)$. Since α is irrational, τ is ergodic.

A result found by the physicist Aubry [20] and first proven by Avron–Simon [24] called *Aubry duality* says that for $0 < \lambda < 2$

$$\gamma_{4/\lambda,\alpha}(z) = \log\left(\frac{2}{\lambda}\right) + \gamma_{\lambda,\alpha}\left(\frac{2z}{\lambda}\right) \tag{3.7.89}$$

Thus if $\lambda > 2$, since $\gamma \geq 0$, we see that

$$\lambda > 2 \Rightarrow \gamma_{\lambda,\alpha} \geq \log\left(\frac{\lambda}{2}\right) > 0 \tag{3.7.90}$$

so, Theorem 3.7.5 is applicable and $\mathrm{supp}(\mu_\theta)$ is a polar set for a.e. θ (actually it can be proved for every θ). If the irrational α has good Diophantine properties, it is a result of Jitomirskaya [430], that for $\lambda > 2$ and Lebesgue a.e. θ, μ_θ is pure point but it is a result of Jitomirsakya–Simon [432], based in part on ideas of Gordon [327], that for Baire generic θ, μ_θ is purely singular continuous (but still supported by a polar set). It is a result of Avron–Simon [24] that for $\lambda > 2$ when α is a Liouville number, μ_θ has no pure points, so is purely singular continuous and therefore supported on a polar set.

Aubry duality plus the Thouless formula implies $d\nu_{\infty,4/\lambda}$ is a scaled $d\nu_{\infty,\lambda}$ and thus since supports are also scaled, it is also the equilibrium measure for $\mathrm{ess\,supp}(d\mu_{4/\lambda,\theta})$. But the A_∞ is 1 while, if $\lambda > 2$, the capacity of $\mathfrak{e}$ is $\log(\frac{\lambda}{2}) > 1$, so we have explicit examples where the DOS is the equilibrium measure but the underlying μ is not regular!

There are also results (of Simon [745]) using ideas of Kotani [487] that in this general ergodic case, if $\gamma(x) = 0$ on a set of positive Lebesgue measure, μ has an absolutely continuous piece with a.e. nonzero weight on the set where $\gamma(x) = 0$.

We note that while our discussion involves ergodic Jacobi matrices, much of the literature deals with whole line operators.

Problems

1. Let C be a convex subset in an inner product V. Let $z \notin C$ and suppose $x_0 \in C$ with

$$|z - x_0| \leq |z - x|, \quad \forall\, x \in C \tag{3.7.91}$$

   Prove there is $\omega \in V$ with $|\omega| = 1$ so that

$$\mathrm{Re}(\langle \omega, (x - z)\rangle) \geq |z - x_0|$$

   for all $x \in C$. (*Hint*: Separating hyperplane theorems.)

2. Let ν be a positive measure of compact support on $\mathbb{R}$ so that $\inf_{x \in \mathbb{R}} \int \log|x - y|\, d\nu(y) > -\infty$. Prove that for some C and all $x < y$

in $\mathbb{R}$ with $|x - y| < \frac{1}{2}$, we have

$$\nu([x, y]) \leq C \left(\log|x - y|^{-1} \right)^{-1} \tag{3.7.92}$$

Remark. This is a result of Craig–Simon [**176**].

3. Let μ, ρ be two finite measures on $\mathbb{R}$ with

$$d\mu = g(x)\, d\rho(x) + d\tilde{\mu}_s \tag{3.7.93}$$

with $d\tilde{\mu}_s$ singular wrt $d\rho$. Prove that ρ is a.c. wrt μ if and only if $g(x) > 0$ for $d\rho$-a.e. x.

4. On $[0, 1]$, let ν be the measure giving weight $\frac{1}{n^2} \frac{1}{2^n}$ to each $\frac{k}{2^n}$, $0 < k < 2^n$ with k odd. Prove that $\nu([0, 1]) < \infty$ and that $\mu = \nu/\nu[0, 1]$ is regular by the Stahl–Totik theorem.

5. Let $\{x_n\}_{n=0}^\infty$ be a dense subset of $[0, 1]$. Let $0 < \lambda < 1$. Let μ be the measure $(1 - \lambda) \sum_{n=0}^\infty \lambda^n \delta_{x_n}$.

   (a) Let $Q = \prod_{j=0}^{n-1} (x - x_j)$. Prove that $\|Q\|^2_{L^2(d\mu)} \leq \lambda^n$.

   (b) Prove that $(a_1 \cdots a_n)^{1/n} \leq \lambda^{1/2}$ (*Hint*: See Section 4.3 of Part 4 for minimum properties of $\|P_n\|$.)

   (c) If $\lambda < \frac{1}{16}$, prove that μ is not regular.

   (d) Prove a result when $\mu(\{x_n\}) = (1 - \lambda)\lambda^n$ is replaced by $\mu(\{x_n\}) \leq C\lambda^n$.

6. (a) Suppose that $[c, d] \subset \mathfrak{e}$ and $(c - \varepsilon, d + \varepsilon) \cap \mathfrak{e} = [c, d]$. Prove that

$$\frac{d\rho_\mathfrak{e}}{dx} \upharpoonright [c, d] = g(x) \left[(d - x)(x - c) \right]^{-1/2} \tag{3.7.94}$$

where g is real analytic on $(c - \varepsilon, d + \varepsilon)$.

   (b) Let $\mathfrak{e} = [\alpha_1, \beta_1] \cup [\alpha_2, \beta_2] \ldots \cup [\alpha_{\ell+1}, \beta_{\ell+1}]$ with $\alpha_1 < \beta_1 < \alpha_2 < \ldots < \beta_{\ell+1}$. Prove that for some $x_j \in [\beta_j, \alpha_{j+1}]$, $j = 1, \ldots, \ell$

$$\frac{d\rho_\mathfrak{e}}{dx} = \frac{1}{\pi} \prod_{j=1}^\ell |x - x_j| \prod_{j=1}^{\ell+1} |x - \alpha_j|^{-1/2} |x - \beta_j|^{-1/2} \chi_\mathfrak{e}(x) \tag{3.7.95}$$

(*Hint*: Use a Herglotz representation for $\log\left[\int \frac{d\rho_\mathfrak{e}(x)}{x - z} \right]$. *Note*: This idea is from Craig [**175**].)

7. In the context of ergodic Jacobi matrices, let

$$\gamma_n(z) = \frac{1}{n} \mathbb{E}\left(\log \|T_n(z; \omega)\| \right) \tag{3.7.96}$$

Prove that $\gamma_n(z)$ is subharmonic and $\gamma(z) = \inf_n \gamma_n(z)$ and conclude that $\gamma(z)$ is subharmonic. (*Note*: This argument is from Craig–Simon [**176**].)

3.8. Harmonic Function Theory of Riemann Surfaces

> Next to the geometric interpretation, the leading mathematical idea in
> Riemann's paper is the importance attached to Laplace's equation. He
> virtually puts equality signs between two-dimensional potential theory and
> complex function theory. Riemann's aim was to make complex function
> theory a powerful tool in real analysis.
>
> *—Lars Alhfors* [7]

A *Riemann surface*, S, is a connected Hausdorff topological space with a
countable cover, $\{U_j\}_{j=1}^N$, by open sets and maps, $f_j : U_j \to \mathbb{C}$, so that f_j is a
homeomorphism of f_j and $f_j[U_j]$ and for all i, j, $g_{ij} : f_j[U_i \cap U_j] \to f_i[U_i \cap U_j]$
by $g_{ij}(z) = f_i \circ f_j^{-1}(z)$ is an analytic bijection. We discussed the basics of
such objects in Sections 7.1 and 8.7 of Part 2A and, in particular, defined
analytic (respectively, meromorphic) functions, $f \colon V \subset S$ to $\mathbb{C}$ (V open
in S), with each $f \circ f_j : f_j^{-1}[V \cap U_j]$ to $\mathbb{C}$ analytic (respectively, meromorphic).
One can also define analytic maps between Riemann surfaces in which case
meromorphic functions on S are just analytic maps of S to $\widehat{\mathbb{C}}$, the Riemann
sphere.

In this section, we'll discuss the theory of harmonic and subharmonic
functions on a general Riemann surface. We do this not merely for its
intrinsic interest, but for its importance to the study of Riemann surfaces:
In Section 8.7 of Part 2A, we proved the uniformization theorem—that the
only simply connected Riemann surfaces are, up to analytic bijection, $\mathbb{C}$,
$\mathbb{D}$, and $\widehat{\mathbb{C}}$, assuming four basic facts for harmonic functions on Riemann
surfaces that we'll prove in this section. We'll also prove that any Riemann
surface has nontrivial meromorphic functions.

We begin with the basic definitions and simplest facts dragged over from
open sets in $\mathbb{C}$. A *coordinate patch* is an open set $V \subset S$ and an analytic
bijection, f, from V to $f[V] \subset \mathbb{C}$. If D is a closed disk in $f[V]$, $f^{-1}[D^{\text{int}}] = U$
is called a *coordinate disk*. Since the Dirichlet problem is solvable for D^{int}
in $f[V]$, given a continuous function, g, from a neighborhood, N of $f^{-1}[\overline{D}]$
to $\mathbb{R}$, we can define the *Perron modification*, $\mathcal{P}_U g$, of g by

$$\mathcal{P}_U\, g(z) = \begin{cases} g(z), & z \in N \setminus f^{-1}[D] \\ h(z), & z \in f^{-1}[D] \end{cases} \tag{3.8.1}$$

where $h \circ f^{-1}$ is harmonic on D with boundary values on ∂D given by
$g \circ f^{-1}$. While this definition would seem to depend on the coordinate map,
f, as well as D, it does not since the solution of the Dirichlet problem is
unique and harmonic functions are taken to other harmonic functions under
conformal coordinate change (see below).

Definition. Let $V \subset S$ be an open subset of a Riemann surface S. A function $u : V \to \mathbb{R}$ is called *harmonic* if for every $z_0 \in V$, there is a coordinate disk, U, containing z_0, so that $\overline{U} \subset V$ and a function f defined and analytic in a neighborhood of U so that on U,

$$u = \operatorname{Re} f \tag{3.8.2}$$

u is called *subharmonic* if it is continuous, and as a map of V to $[-\infty, \infty)$ for every $z_0 \in V$, there is a coordinate disk U with $\overline{U} \subset V$, so that on all of V

$$u \leq \mathcal{P}_U\, u \tag{3.8.3}$$

Remarks. 1. It is easy to see that (Problem 1) if $\Delta, \widetilde{\Delta}$ are Laplacian in two local coordinate systems near z_0 then for a function $F(z)$, $\widetilde{\Delta} = F(z)\Delta$ so $\widetilde{\Delta}u = 0 \Leftrightarrow \Delta u = 0$. Thus, we can alternatively define harmonic in terms of $\Delta u = 0$ in any local coordinate system.

2. One can develop the notion of subharmonic on a Riemann surface requiring only that u is usc (as we do in $\mathbb{C}$) but it is technically simpler to require continuity only in the extended sense to $[-\infty, \infty)$ and suffices for the applications we are interested in.

3. As with Remark 1, subharmonicity can be expressed locally by requiring the distributional Laplacian be positive.

4. Once (3.8.3) holds for one U for each z_0, it holds for all U by the same arguments that works on $\mathbb{R}^\nu$.

5. We call $u : V \to [-\infty, \infty)$ *superharmonic* if $-u$ is subharmonic.

Fact 1 (Maximum Principle). If u is subharmonic on V, open in S, and $u(z_0) = sup_{z \in V}\, u(z)$ for $z_0 \in V$, then u is constant.

This is because the set of z_0 where this holds is closed by continuity and open by the local submean property. Once one has Fact 1, one concludes that there exists $\{z_n\}_{n=1}^\infty \subset V$ with no limit point in V so $u(z_n) \to \sup_z\, u(z)$.

Fact 2 (Harnack's Inequality). For any $K \subset U \subset S$ with K compact and U open and connected, there is a constant C so that for all $x, y \in K$, u harmonic on U and $u > 0$ on U, we have

$$C^{-1} \leq \frac{u(x)}{u(y)} \leq C \tag{3.8.4}$$

The proof is identical to Theorem 3.1.29 given that any $x, y \in U$ are connected by a curve.

Fact 3 (Montel and Vitali Theorems). If u_n is a family of functions all harmonic on an open, connected set $U \subset S$, and for each compact $K \subset U$, $\sup_{n, z \in K} |u_n(z)| < \infty$, then there is a subsequence uniformly convergent on each compact $K \subset U$. If u_n converges pointwise on some open ball to a finite limit, then u_n converges uniformly on all compacts in U. The proofs are unchanged from Theorem 3.1.23 and 3.1.24.

Fact 4 (Removable Singularities Theorem). Since harmonicity is local and we have a removable singularities theorem for disks in $\mathbb{C}$, we see that (by Theorem 3.1.26) if $U \subset S$ is open, $q \in U$ and u is harmonic on $U \setminus \{q\}$ and bounded in a neighborhood of q, then one can define $u(q)$ so that u is harmonic in all of U.

Fact 5 (Perron Families and Perron's Principle). A *Perron family* on an open connected set $U \subset S$ is a nonempty family, $\mathcal{F}$, of subharmonic functions on U with two properties:

(i) $u, v \in \mathcal{F} \Rightarrow \max(u, v) \in \mathcal{F}$;
(ii) If D is a coordinate disk with $\overline{D} \subset U$ and $u \in \mathcal{F}$, then $\mathcal{P}_D u \in \mathcal{F}$.

By a simple analysis using Harnack's Principle, one can prove that (Problem 2)

Theorem 3.8.1 (Perron's Principle). *Let $\mathcal{F}$ be a nonempty Perron family of functions on U, an open connected subset of S. Let*

$$w(z) = \sup_{u \in \mathcal{F}} u(z) \qquad (3.8.5)$$

Then either w is a harmonic function on U or $w(z) = +\infty$ for all $z \in U$.

If $w < \infty$ (respectively, $w \equiv \infty$), we say $\mathcal{F}$ has a finite (respectively, infinite) majorant; w is called the *majorant* of $\mathcal{F}$.

Fact 6 (Barriers and the Dirichlet Problem). The argument of regularity at points with barriers and external cone conditions implies the following

Theorem 3.8.2. *Let S be a Riemann surface. Let $R \subset S$ be an open subset with the following properties:*

(i) $\overline{R}$ *is compact.*
(ii) ∂R *has finitely many components, each a closed curve.*
(iii) $\partial(S \setminus R) = \partial R$.
(iv) *Each component of ∂R is a finite number of real analytic curves with noncuspal intersection points.*

Then the Dirichlet problem is solvable for R in the sense that for any $f \in C(\partial R)$, there is a unique continuous function u on $\overline{R}$ with u harmonic on R and $u \restriction \partial R = f$.

Proof. By (i), if $x_n \in R$ has no limit point in R, it must have a limit point in ∂R. Therefore, any v continuous on $\overline{R}$ and harmonic on R must take its maximum value on ∂R, so, as usual we have uniqueness.

Given f, we let $\mathcal{F}$ be the set of u's subharmonic on R with the following for all $z_0 \in \partial R$

$$\limsup_{z \in R, z \to z_0} u(z) \leq f(z_0) \tag{3.8.6}$$

As in Section 3.4, this is a Perron family. By the maximum principle and (3.8.6) for all $u \in \mathcal{F}$, $\sup_{z \in R} u(z) \leq \sup_{z_0 \in \partial R} f(z_0) < \infty$, so the Perron majorant, $\sup_{u \in \mathcal{F}} u(t)$ is a bounded harmonic function.

(iii) and (iv) plus Theorem 3.4.12 imply the existence of a barrier at any $z_0 \in \partial R$, so by Proposition 3.4.5 and Theorem 3.4.6, $u(x) \to f(x_0)$ as $x \to x_0 \in \partial\Omega$. This proves existence. $\qquad\square$

This completes our introductory remarks. We begin the further discussion by showing general Riemann surfaces can be approximated by very nice ones.

Definition. A *finite bordered Riemann subsurface* is a connected open subset, R, of a Riemann surface S obeying (i)–(iii) of Theorem 3.8.2 plus

(iv) Each component of ∂R is a single real analytic curve.

Theorem 3.8.3. (a) *For any Riemann surface, S, there is an increasing family $R_1 \subset R_2 \subset \ldots$ of finite bordered Riemann subsurfaces so that $S = \bigcup_{j=1}^{\infty} R_j$.*

(b) *For S that is not compact, each R_j has $\partial R_j \neq \emptyset$*

(c) *For S compact, for any $S \setminus \{z_0\}$, there is an increasing union of finite bordered subsurfaces as in* (a) *with $\bigcup_{j=1}^{\infty} R_j = S \setminus \{z_0\}$ and $\partial R_j \neq \emptyset$.*

Remarks. 1. In case (c), one can pick a coordinate patch with $f(z_0) = 0$ and $f[S] \supset \mathbb{D}$ and let $R_n = f \setminus \{z \mid |f(z)| \leq \frac{1}{n}\}$.

2. We'll see later why $\partial R \neq \emptyset$ is important (see Theorem 3.8.17).

3. In the proof, ∂ and $\bar{\partial}$ refer to the operations on functions described in Section 2.1 of Part 2A (see (3.8.49) below).

Proof. Since $\partial R_j = \emptyset \Rightarrow R_j$ is open and closed $\Rightarrow R_j = \overline{R_j} = S$, we see (a) $\Rightarrow$ (b). Since $S \setminus \{z_0\}$ is not compact, (b) $\Rightarrow$ (c). Thus we need only prove (a). Since S has a countable cover by open sets homeomorphic to open subsets of $\mathbb{C}$, we can find compacts $K_1 \subset K_2 \subset \ldots$ with $\cup K_j = S$. We will prove that any connected compact, K, is contained in a finite bordered Riemann subsurface with $\partial R \neq \emptyset$ in which case we get R_{j+1} by asking it contain $\widetilde{K}_{j+1} = K_{j+1} \cup \overline{R_j}$.

If S is compact, take $K_j \equiv S = R_j$, so we can suppose S is not compact. Since K is compact and any $z \in S$ lies in a coordinate disk, we can cover S by a finite union, U, of coordinate disks, each of which has nonempty boundary.

By adjusting the radii slightly, we can suppose U obeys the hypotheses, (i)–(iv), of Theorem 3.8.2.

Pick $z_0 \in U \setminus K$ and a coordinate disk D containing z_0 with $\overline{D} \subset U \setminus K$. Then $V = U \setminus \overline{D}$ also obeys the hypotheses of Theorem 3.8.3, so we can solve the Dirichlet problem for V for $f \equiv 1$ on ∂D and $f \equiv 0$ on ∂U. If u is the solution, $u > 0$ on V by the maximum principle for $-u$, so by compactness of K, $\varepsilon_0 = \inf_{z \in K} u(z) > 0$.

Locally $g = \partial u$ defined on local coordinates is analytic since $\overline{\partial}(\partial u) = 0$ and not identically zero, since u is not constant. Thus, in V, u has isolated critical points so the critical values are discrete in $(0, 1)$. We can thus find $\varepsilon \in (0, \varepsilon_0)$ not a critical value of u, i.e., $B = \{x \mid u(x) = \varepsilon\}$ has no critical points of u and so is a finite union of closed analytic curves.

Since K is connected, it lies in a single connected component of $\widetilde{R} = \{x \mid u(x) > \varepsilon\}$. This component is the required R. $\square$

We now turn to the two main heros of this section.

Definition. Let $U \subset S$ be an open subset of a Riemann surface and $p_1, \ldots, p_\ell \in U$ and $\eta_1, \ldots \eta_\ell \in \mathbb{Z} \setminus \{0\}$. A function, u on $U \setminus \{p_1, \ldots, p_\ell\}$ is called harmonic on U except for *polar singularities* of *charge* η_j at p_j if u is harmonic at $U \setminus \{p_1, \ldots, p_j\}$ and, for each j in a local coordinate system z_j near p_j with $z_j(p_j) = 0$, we have that for z_j near p_j, $u(p) + \eta_j \log\left(|z_j(p_j)|\right)$ has a removable singularity at p_j.

Remark. If f is analytic near $z = 0$ with $f(0) = 0$ and $f'(0) \neq 0$, $\log|f(z)| - \log|z|$ is harmonic near zero. This shows that this definition is independent of choice of coordinate system near p_j.

Definition. Let p_0, p_1 be distinct points on a Riemann surface, S. A *bipolar Green's function*, $G(z; p_0, p_1)$ is a function harmonic on S except for polar singularities at p_0, p_1 with respective charges 1 and -1 and so that for any neighborhoods N_0 and N_1 of p_0 and p_1, G is bounded on $S \setminus (N_0 \cup N_1)$.

Definition. Let $p_0 \in S$ be a Riemann surface. A *Green's function* for S with singularity at p_0 is a function u with the following properties:

(i) u is harmonic on S except for a polar singularity with charge 1 at p_0.
(ii) $u > 0$ on S.
(iii) If $\widetilde{u}$ is any other function obeying (i) and (ii), then

$$\widetilde{u} \geq u \tag{3.8.7}$$

We then write $u(z) = G(z; p_0)$. We'll use $G_S(z; p)$ if we need to be explicit about S.

Clearly, if a Green's function exists, it is unique. Note that these Green's functions look like $\log\left(|z|^{-1}\right)$ in local coordinates near p. This differs by a factor of 2π from the normalization in Sections 3.1, 3.2, and 3.6 where the singularity is $(2\pi)^{-1}\log(|z|^{-1})$ but agrees with the normalization in Section 3.7. Since $\lim_{z\to p} G(z;p) = \infty$ (because $-\log|z| \to \infty$ as $|z| \to 0$), if we set $G(p;p) = \infty$, it is a superharmonic function on all of S. To see the power of the minimization property, (iii), we have

Proposition 3.8.4. *If a Green's function $G(\,\cdot\,;p)$ exists, we have that*

$$\inf_{z\in S} G(z;p) = 0 \tag{3.8.8}$$

Proof. Since $G > 0$, $\alpha = \inf_{z\in S} G(z;p) \geq 0$. If $\alpha > 0$, $G(z;p) - \alpha$ obeys (i), (ii) so $G(z;p) - \alpha \geq G(z;p) \Rightarrow \alpha \leq 0$. Thus, $\alpha = 0$. $\square$

Remark. It is tempting to think that as in the case of classical Green's functions, $\lim_{z\to\infty} G(z;p) = 0$ in the sense that

$$\{z_n\} \in S, \quad z_n \text{ has no limit point in } S \Rightarrow \lim_{n\to\infty} G(z_n;p) = 0 \tag{3.8.9}$$

but in general this is false (see Example 3.8.10). However, we'll prove it is true for finite bordered Riemann surfaces (see Theorem 3.8.17).

In Section 8.7 of Part 2A, we proved the uniformization theorem assuming four facts we intend to prove below:

Fact 1. A Green's function exists at one point p_0 in S, a simply connected Riemann surface, if and only if it exists at all points (we used a somewhat different definition of Green's function there—so we'll need to prove the equivalence).

Fact 2. If S has a bounded nonconstant analytic function, then a Green's function exists and if S is simply connected and a Green's function exists, then S has nonconstant bounded analytic functions.

Fact 3. If a Green's function exists, then

$$G(z;p) = G(p;z) \tag{3.8.10}$$

Fact 4. For any S and any $p_0 \neq p_1$ in S, a bipolar Green's function exists.

Our definition of Green's function in Section 8.7 of Part 2A was slightly different—the definition there only works for simply connected S—below (see Theorem 3.8.16), we also prove it is equivalent to the one here for such S.

In our analysis, two tools will be used multiple times: first that the maximum principle implies that if g is superharmonic on U and u subharmonic on U and for any sequence $\{z_n\}$ without a limit point in U, we have that $\liminf (g(z_n) - u(z_n)) \geq 0$, then $u(z) \leq g(z)$ for all z and second, that if

$g(z) = f(z) + \log(|z|)$ is subharmonic on a coordinate disk $D = \{z \mid |z| \leq \rho\}$ and if $\sigma < \rho$, then

$$\sup_{|z|=\sigma} |f(z)| \leq \sup_{|z|=\rho} |f(z)| + \log\left(\frac{\rho}{\sigma}\right) \tag{3.8.11}$$

(since $\sup_{|z|=\sigma} |g(z)| \leq \sup_{|z|=\rho} |g(z)|$) which is useful because typically $\sup_{|z|=\sigma} |f(z)| \geq \sup_{|z|=\rho} |f(z)|$ if, say $f \to \infty$ as $|z| \to 0$.

Example 3.8.5 (Simply Connected S). G, if it exists, is superharmonic but compact Riemann surfaces have no nonconstant superharmonic functions by the following argument: If u is such a function, it takes its minimum value at some point q, but then the maximum principle for $-u$ implies u is constant. Thus, compact Riemann surfaces have no Green's functions and, in particular, $\widehat{\mathbb{C}}$ does not.

For $S = \mathbb{D}$,

$$G_{\mathbb{D}}(z; 0) = -\log(|z|) \tag{3.8.12}$$

is a Green's function as it clearly obeys (i), (ii). If u also obeys (i), (ii), $h \equiv u + \log(|z|)$ is harmonic on $\mathbb{D}$ and since $u > 0$ on $\mathbb{D}$ and $\log|z| \to 0$ on $\partial\mathbb{D}$, $\liminf_{z \to z_0 \in \partial\mathbb{D}} h(z) \geq 0$. Thus by the maximum principle for $-h$, $h \geq 0$ on $\mathbb{D}$, i.e., $u \geq G_{\mathbb{D}}(z; 0)$ given by (3.8.12) so this choice obeys (iii).

If $w \in \mathbb{D}$, by a conformal mapping argument

$$G_{\mathbb{D}}(z; w) = -\log\left(\left|\frac{z - w}{1 - \bar{w}z}\right|\right) \tag{3.8.13}$$

is a Green's function, i.e., for $\mathbb{D}$ a Green's function exists for all $p \in \mathbb{D}$.

Suppose $u(z) \equiv G(z; 0)$ exists for $S = \mathbb{C}$. Then, for any θ, $u(e^{i\theta}z)$ also obeys (i), (ii) so, by (iii), $u(e^{i\theta}z) \geq u(z)$ for all z and θ. Replacing z by $e^{-i\theta}z$, we see that $u(e^{i\theta}z) = u(z)$, i.e., $u(z)$ is only a function of $|z|$. By writing $\Delta u = 0$ in polar coordinates, the only rotationally invariant harmonic functions on $\mathbb{C} \setminus \{0\}$ are of the form $\alpha \log(|z|) + \beta$ for $\alpha, \beta \in \mathbb{C}$. (ii) implies $\alpha = -1$, i.e., for some β, $u(z) = -\log(z) + \beta$. But $\lim_{|z|\to\infty}[-\log(|z|) + \beta] = -\infty$, so u cannot be positive. Thus (using translation invariance), for $S = \mathbb{C}$, $G(z; p)$ exists for no p. $\qquad\square$

Example 3.8.6. Let $\Omega \subset \mathbb{C}$ be a bounded region. In Theorem 3.6.17 we discussed $G_{\Omega}(x; y)$ and proved many of its properties. In particular, (iii) of that theorem implies that G_{Ω} is a Green's function for Ω viewed as a Riemann surface. $\qquad\square$

Theorem 3.8.7 (Variational Principle for the Green's Function). *Let S be a Riemann surface and $q \in S$. Let $\mathcal{F}_q$ be the set of functions u on $S \setminus \{q\}$ obeying*

(i) *u is nonnegative and vanishes outside a compact set $K \subset S$.*

(ii) u is subharmonic on S.

(iii) For some coordinate disk D with center q, $u(p) + \log(|z(q)|)$ has a subharmonic extension to D.

Then $\mathcal{F}_q$ is a Perron family and its majorant is finite if and only if a Green's function exists for q and, in that case,

$$G(z;q) = \sup_{u \in \mathcal{F}_q} u(z) \qquad (3.8.14)$$

Proof. It is easy to prove that $\mathcal{F}_q$ is a Perron family. Since $u = 0 \in \mathcal{F}_q$, it is nonempty and $G \geq 0$. Suppose $h(z)$ is a function obeying conditions (i) and (ii) of the definition of G. Then, near q,

$$u - h = [u + \log(|z(\cdot)|)] - [h + \log(|z(\cdot)|)] \qquad (3.8.15)$$

is subharmonic so $u - h$ is subharmonic on all of S. Since $h \geq 0$ and u vanishes off a compact set, by the maximum principle, $u \leq h$.

Since this is true of all u, we conclude if G exists then $\widetilde{G} \equiv \sup_{u \in \mathcal{F}_q} u(z)$ is the finite majorant of a Perron family and $\widetilde{G} \leq h$ for any h obeying (i), (ii) of the definition of G. Thus, if we prove that when the Perron has a finite majorant, $\widetilde{G}$ obeys (i), (ii), we see that $\widetilde{G}$ is the Green's function.

Since $u \equiv 0 \in \mathcal{F}$, $\widetilde{G} \geq 0$ is immediate and as the majorant of a Perron family, it is harmonic on $S \setminus \{q\}$. If we prove that $\widetilde{G}(p) + \log(|z(p)|)$ is bounded near $z(p) = 0$ (i.e., near q), it is a removable singularity and so $\widetilde{G}$ obeys condition (i). Since then $\widetilde{G}(p) \to \infty$ as $p \to q$, $\widetilde{G}$ is not constant and so $\widetilde{G} \geq 0 \Rightarrow \widetilde{G} > 0$ by the maximum principle for $-\widetilde{G}$ (which is subharmonic).

Pick ε, so $D_0 = \{z \mid |z(p)| \leq \varepsilon\}$ is a coordinate disk about q. Let

$$u(p) = \begin{cases} \log(\varepsilon/|z(p)|), & |z(p)| \leq \varepsilon \\ 0, & p \notin D_0 \end{cases} \qquad (3.8.16)$$

$u \geq 0$ and is subharmonic on $S \setminus \{q\}$ (since it is harmonic away from ∂D_0 and on ∂D_0, $u(p) = 0$ which is less than any average of u) and $u(p) + \log(|z(p)|)$ is harmonic at q. Thus $u \in \mathcal{F}_q$. And therefore, since $u \leq \widetilde{G}$,

$$z(p) \in D_0 \Rightarrow \widetilde{G}(p) + \log(|z(p)|) \geq \log(\varepsilon) \qquad (3.8.17)$$

so $\widetilde{G}(p) + \log(|z(p)|)$ is bounded from below.

Since $\widetilde{G}$ is harmonic on $S \setminus \{q\}$, $s = \sup_{\{p \mid |z(p)| = \varepsilon\}} \widetilde{G}(z) < \infty$. If $u \in \mathcal{F}_q$, since $u + \log(|z(\cdot)|)$ is subharmonic if $p \in D_0$

$$u(p) + \log(|z(p)|) \leq \sup_{\{p \mid |z(p)| = \varepsilon\}} [u(p) + \log(\varepsilon)]$$

$$\leq s + \log(\varepsilon) \qquad (3.8.18)$$

Taking the sup over u, we see that

$$z(p) \in D_0 \Rightarrow \widetilde{G}(p) + \log(|z(p)|) \leq s + \log(\varepsilon) \qquad (3.8.19)$$

Thus, as desired, $\widetilde{G}(p) + \log(|z(p)|)$ has a removable singularity at q. □

Corollary 3.8.8. *Let U be an open connected subset of a Riemann surface S. If $G_S(z;q)$ exists for some $q \in U$, then $G_U(z;q)$ exists and*

$$G_U(z;q) \leq G_S(z;q) \qquad (3.8.20)$$

Proof. Clearly $\mathcal{F}_q(U) \subset \mathcal{F}_q(S)$, so, by (3.8.14), $\mathcal{F}_q(U)$ has a finite majorant and (3.8.20) holds. □

Corollary 3.8.9. *Let S be a Riemann surface and $U_1 \subset U_2 \subset \ldots$ connected open subsets of S with $S = \bigcup_{j=1}^{\infty} U_j$. If $q \in U_\ell$ and $G_S(z;q)$ exists, then*

$$G_S(z;q) = \sup_{j \geq \ell} G_{U_j}(z;q) = \lim_{j \to \infty} G_{U_j}(z;q) \qquad (3.8.21)$$

Proof. By the last corollary, each $G_{U_\ell}(\,\cdot\,;q)$ exists and

$$G_{U_\ell}(z;q) \leq G_{U_{\ell+1}}(z;q) \leq G_S(z;q) \qquad (3.8.22)$$

so

$$\lim_{\ell \to \infty} G_{U_\ell}(z;q) = \sup_\ell G_{U_\ell}(z;q) \leq G_S(z;q) \qquad (3.8.23)$$

Let $u \in \mathcal{F}_q(S)$. Thus, $\text{supp}(u) \equiv K$ is compact and so $K \subset U_\ell$ for some ℓ and thus $u \in \mathcal{F}_q(U_\ell)$. It follows that

$$u \leq G_{U_\ell}(z;q) \leq \sup_\ell G_{U_\ell}(z;q) \qquad (3.8.24)$$

Since u is arbitrary

$$G_S(z;q) \leq \sup_\ell G_{U_\ell}(z;q) \qquad (3.8.25)$$

□

Example 3.8.10. Let S be any Riemann surface so that $G_S(z;q)$ exists for some $q \in S$ and obeys (3.8.9). Let $p \neq q$ and $U = S \setminus \{p\}$. Then, by Corollary 3.8.8, $G_U(z;q)$ exists. Since $0 \leq G_U(z;q) \leq G_S(z;q)$, $G_U(z;q)$ is bounded near p, so p is a removable singularity and G_U is harmonic on $S \setminus \{q\}$. Moreover, $G_S - G_U$ has q as a removable singularity also. Thus, it is a positive harmonic function which by (3.8.9) goes to zero as z_n leaves S so by the maximum principle is 0. It follows that

$$G_U = G_S \qquad (3.8.26)$$

In particular, if $z_n \to p$, $G_U(z;q) \to G_S(p;q) > 0$ and thus (3.8.9) fails for U! In particular, if $S = \mathbb{D} \setminus \{0\}$, (3.8.9) fails for S. We note by an approximation argument (Problem 3) that (3.8.26) holds even if S doesn't obey (3.8.9). □

To prove that Green's functions exist at one point if and only if they exist at all points, we'll need another pair of notions.

Definition. A Riemann surface, S, is called *hyperbolic* if there exist non-constant bounded subharmonic functions on all of S. It is called *elliptic* if it is compact and *parabolic* if it is not compact and not hyperbolic.

Remarks. 1. As we've noted, there are no nonconstant subharmonic functions on compact Riemann surfaces so elliptic is a separate class from hyperbolic.

2. While parabolic is defined by negatives, it has some special properties. By the exterior Dirichlet problem for a compact set $K \subset S$ we mean, given $f \in C(\partial K)$, a continuous function u on $S \backslash K$ which is bounded and harmonic on $S \backslash \overline{K}$ and $u \upharpoonright \partial K = f$. In Problem 4, the reader will show that for any coordinate disk, D, the Dirichlet problem has no unique solution if S is hyperbolic but has unique solutions if S is parabolic!

Consider now a coordinate disk, D, in S, a noncompact Riemann surface. Let $\mathfrak{J}_D$ be the set of functions, u, on $S \backslash \overline{D}$ obeying

(i) $0 \leq u \leq 1$.
(ii) $u = 0$ outside a compact subset, K with $\overline{D} \subset K^{\text{int}}$.

$\mathfrak{J}_D$ is a Perron family whose majorant, ω_D, is clearly finite since $0 \leq u \leq 1$. ω_D, extended to S by setting $\omega_D(z) = 1$ if $z \in D$, is called the *minimal superharmonic majorant* of D since (Problem 5) it is the smallest superharmonic function, $g \geq \chi_D$, the characteristic function of D.

Proposition 3.8.11. ω_D *is harmonic on* $S \backslash D$, $0 \leq \omega \leq 1$ *and*

$$\lim_n \omega_D(z_n) = 1 \qquad (3.8.27)$$

if $z_n \to z_\infty \in \partial D$.

Remark. One can also use barriers to prove (3.8.27).

Proof. By the Perron construction, ω is harmonic. Suppose in a local coordinate system $D = \{p \mid |z(p)| < \sigma\}$. Pick $\rho > \sigma$, so that $\widetilde{D} = \{p \mid |z(p)| \leq \rho\}$ is a bijection of $\widetilde{D}$ and $\mathbb{D}$, i.e., z is defined in a neighborhood of $\widetilde{D}$. Define

$$u(p) = \begin{cases} \log(\rho/z(p)) \left[\log(\rho/\sigma)\right]^{-1}, & z \in \widetilde{D} \\ 0, & z \notin \widetilde{D} \end{cases} \qquad (3.8.28)$$

Clearly u is continuous, of compact support with $0 \leq u \leq 1$. u is harmonic on $(\widetilde{D})^{\text{int}} \cup S \backslash \widetilde{D}$ and has the submean property at points in $\partial \widetilde{D}$, since $u \geq 0$ and $u \upharpoonright \partial \widetilde{D} = 0$.

Thus $\omega \geq u$ and since $u \to 1$ as $z_n \to z_\infty \in \partial D$, so does ω. $\qquad \square$

By the maximum principle either $0 < \omega < 1$ on $S \setminus D$ or ω is constant and in that case, by the proposition, $\omega = 1$. We thus have that each disk is either of

Type 1. $0 < \omega_D < 1$ on $S \setminus D$ or
Type 2. $\omega_D \equiv 1$.

Theorem 3.8.12. *Let S be a Riemann surface. Then the following are equivalent:*

(1) *S is hyperbolic.*
(2) *For one $p_0 \in S$, $G(z; p_0)$ exists.*
(3) *For every $p \in S$, $G(z; p)$ exists.*
(4) *One coordinate disk with some center p_0 is of type 1.*
(5) *Every coordinate disk is of type 1.*

In particular, a Green's function exists at one point if and only if it exists at all points.

Proof. Letting $(2)_{p_0}$ indicate G exists for a specific p_0, we'll prove $(1) \Rightarrow (5) \Rightarrow (4)_{p_0} \Rightarrow (2)_{p_0} \Rightarrow (1)$

Since p_0 is arbitrary $((5) \Rightarrow (4)_{p_0}$ is trivial!), we get also $(5) \Rightarrow (3) \Rightarrow (2)_{p_0} \Rightarrow (1)$.

$(1) \Rightarrow (5)$. Let g be a bounded nonconstant subharmonic function. Then $h \equiv \|g\|_\infty - g$ is a superharmonic, nonnegative, nonconstant function. By the maximum principle for g, h doesn't take its infimum so $h > 0$ and on the compact set $\overline{D}$, there is ω_1 where

$$h(\omega_1) = \inf_{z \in \overline{D}} h(z) \tag{3.8.29}$$

By replacing h by $h/h(\omega_1)$, we can suppose

$$\inf_{z \in \overline{D}} h(z) = 1 \tag{3.8.30}$$

By the maximum principle again, $h \upharpoonright \overline{D}$ must take its minimum value at some point $\omega_2 \in \partial D$. If $h(z) \geq 1$ for all $z \in S \setminus D$, then ω_2 is a global minimum which violates the maximum principle since g is nonconstant. Thus there is $\omega_3 \in S \setminus D$ with $h(\omega_3) < 1$.

Let $u \in \mathfrak{I}_D$ and let $K = \text{supp}(u) \cup \overline{D}$ which is compact. $f = u - h$ is subharmonic on $K^{\text{int}} \setminus \overline{D}$. On ∂K, f is continuous and negative since $u \upharpoonright \partial K = 0$ and $h \geq 0$. As $z_n \to z_\infty \in \partial D$, $h(z_n) \to h(z_\infty) \geq 1$ and $\limsup u(z_n) \leq 1$, so $\limsup f(z_n) \leq 0$. Thus, by the maximum principle $u \leq h$. Since u is arbitrary, $\omega_D \leq h$. Thus, $\omega_D(\omega_3) < 1$ and D must be of type 1.

$(4)_{p_0} \Rightarrow (2)_{p_0}$. Let the disk be $D = \{p \mid |z(p)| < \sigma\}$ with $z(p_0) = 0$. Pick $\rho > \sigma$ with z a bijection on $\overline{B} = \{p \mid |z(p)| \leq \rho\}$. For any function f, define $f_\lambda = \sup_{\{p \mid |z(p)| = \lambda\}} |f(p)|$ when $0 < \lambda < \rho$. We claim that for all $p \in S \setminus D$ and all $u \in \mathcal{F}_{p_0}$, we have that

$$u(p) \leq u_\sigma \omega(p) \tag{3.8.31}$$

and that

$$u_\sigma \leq u_\rho + \log (\rho/\sigma) \tag{3.8.32}$$

Assuming we have this, (3.8.31) implies

$$u_\rho \leq u_\sigma \omega_\rho \tag{3.8.33}$$

and thus

$$(1 - \omega_\rho)u_\sigma \leq \log(\rho/\sigma)$$

Since $0 < \omega < 1$ on ∂B and ω is continuous, $\omega_\rho < 1$ and

$$u_\sigma \leq (1 - \omega_\rho)^{-1} \log(\rho/\sigma) \tag{3.8.34}$$

This holds for all u and shows the Perron family $\mathcal{F}_{p_0}$ has a finite majorant and thus, by Theorem 3.8.7, $G(z; p_0)$ exists. Thus we only need to prove (3.8.31) and (3.8.32).

Since $u \in \mathcal{F}_{p_0}$, $u(p) + \log(|z(p)|)$ is subharmonic in a neighborhood of $\overline{B}$. Therefore, $u(p) + \log (|z(p)|) \leq u_\rho + \log(\rho)$ for $p \in \overline{B}$ and, in particular

$$u_\sigma + \log(\sigma) \leq u_\rho + \log(\rho) \tag{3.8.35}$$

which is (3.8.32).

Consider the function $g \equiv u - u_\sigma \omega$ on $S \setminus D$, subharmonic on $S \setminus \overline{D}$ and continuous. If $\{z_n\}$ has no limit point in $S \setminus D$, either there is a subsequence with $z_{n(j)}$ outside $K = \text{supp}(u)$ or with $z_{n(j)} \to z_\infty \in \partial D$. If $z_{n(j)}$ is outside K, $u(z_{n(j)})$ is zero while $u_\sigma \omega > 0$ so $g(z_{n(j)}) \leq 0$. If $z_{n(j)} \to z_\infty \in \partial D$, since $\omega \restriction \partial D = 1$ and $u(z_\infty) \leq u_\sigma$, $\lim g(z_{n(j)}) \leq 0$. Thus, by the maximum principle, (3.8.31) holds.

$(2)_{p_0} \Rightarrow (1)$. Since $G(p; p_0) + \log(|z(p)|)$ is bounded as $p \to p_0$ and $\log(|z(p_0)|) \to -\infty$, we see $G(p; p_0) \to \infty$ as $p \to p_0$. Since $\inf_p G(p; p_0) = 0$, we conclude G takes every value. Thus

$$f(p) = \min(1, G(p; p_0)) \tag{3.8.36}$$

is nonconstant, bounded, and superharmonic, $-f(p)$ is nonconstant, bounded, and subharmonic so S is hyperbolic. $\qquad \square$

The proof of (3.8.31) works also if $u(p) = G(p; p_0)$ and any suitable harmonic ω to show that

Corollary 3.8.13. *If S is a Riemann surface, D a coordinate disk centered at $p_0 \in S$ and ω is a function harmonic in $S \setminus \overline{D}$ and continuous on $S \setminus D$, with $\omega \restriction \partial D \equiv 1$ and $\omega \not\equiv 1$, then S is hyperbolic and for all $p \in S \setminus D$*

$$G(p; p_0) \leq (\sup_{w \in \partial D} G(w; p_0))\omega(p) \tag{3.8.37}$$

From Theorem 3.8.12, we have thus proven Fact 1 (modulo a different meaning of Green's function). Here is half of Fact 2:

Proposition 3.8.14. *If S is a Riemann surface and f is a nonconstant bounded analytic on S, then S is hyperbolic (and so Green's functions exist).*

Proof. $\operatorname{Re} f$ is a bounded harmonic function on S nonconstant by the Cauchy–Riemann equations and f nonconstant. Thus S is hyperbolic. □

In Section 8.7 of Part 2A, we proved if S is a simply connected Riemann surface and u is harmonic on $S \setminus \{p_1, \ldots, p_\ell\}$ with polar singularities at the p_j of chosen η_j, then there is a meromorphic function, f, on S which is analytic and nonvanishing on $S \setminus \{p_1, \ldots p_\ell\}$ with zeros of order η_j at those p_j with $\eta_j > 0$ and poles of order $-\eta_j$ at those p_j with $\eta_j < 0$. Conversely, if f is such a function, $u = \log|f|$ is harmonic with polar singularities and charges as indicated.

In the proof of Theorem 3.8.7, the fact that h was harmonic on $S \setminus \{q\}$ wasn't needed, only that $h > 0$ and superharmonic, thus we have

Proposition 3.8.15. *Let S be a hyperbolic Riemann surface. Let h be a positive function which is superharmonic on $S \setminus \{q\}$ and so that $h(p) + \log(|z(p)|)$ is superharmonic near q in a local coordinate system with $z(q) = 0$. Then*

$$h(p) \geq G(p; q) \tag{3.8.38}$$

We thus have:

Theorem 3.8.16. *Let S be a simply connected hyperbolic Riemann surface. Then, for each $q \in S$, there is an analytic function $\varphi : S \to \mathbb{C}$ so that*

(a) $|\varphi(p)| = \exp(-G(p; q)), \quad p \in S \setminus \{q\}.$ $\tag{3.8.39}$
(b) $\varphi(p) = 0 \Leftrightarrow p = q$ *and this zero is simple.*
(c) $|\varphi(p)| \leq 1$ *for all p.*
(d) *If ψ is any analytic function on S with $|\psi(p)| \leq 1$ for all p and $\psi(q) = 0$, then $|\psi(p)| \leq |\varphi(p)|$ for all p.*

Proof. This is immediate from Proposition 3.8.15 by taking $h(p) = -\log|\psi(p)|$. □

Proposition 3.8.14 and Theorem 3.8.16 imply in the case of that theorem that $G(z; q)$ is a Green's function in the sense used in Section 8.7 of Part 2A and that Fact 2 holds.

Before leaving direct consequences of Theorem 3.8.12, we note

Theorem 3.8.17. *Let S be a Riemann surface and $R \subset S$ a finite bordered Riemann subsurface with $\partial R \neq \emptyset$ (i.e., $R = S$ is not compact). Then R is hyperbolic and for any $q \in R$, and $p_n \to p_\infty \in \partial R$, we have*

$$G_R(p_n; q) \to 0 \tag{3.8.40}$$

Remarks. 1. Thus $G_R(\,\cdot\,; q)$ set to 0 on ∂R is continuous on $\overline{R}$.

2. Since $\overline{R}$ is compact, if $\{p_n\}_{n=1}^\infty$ has no limit point in R, it must have one in ∂R.

Proof. Let D be a coordinate disk centered at q with $\overline{D} \subset R$. Then $R \setminus \overline{D}$ is a finite bordered Riemann surface with $\partial(R \setminus \overline{D}) = \partial R \cup \partial D$ (disconnected union). By Theorem 3.8.2, we can find ω harmonic on $R \setminus \overline{D}$, continuous on $\overline{R} \setminus D$ and $\omega \upharpoonright \partial R \equiv 0$, $\omega \upharpoonright \partial D \equiv 1$. Since $\partial R \neq \emptyset$, $\omega \not\equiv 1$. By Corollary 3.8.13, R is hyperbolic and on $R \setminus D$, $G(p; q) \leq C\omega(p)$ so (3.8.40) holds. $\qquad\square$

Next, we turn to symmetry of G. The strategy will be to prove it for finite bordered Riemann subsurfaces and then use Theorem 3.8.3 and Corollary 3.8.9 to go to general hyperbolic Riemann surfaces. As a warmup, we discuss $\Omega \subset \mathbb{C}$ a bounded region with a real analytic boundary. Fix $x, y \in \Omega$, $x \neq y$ and small balls B_1 about x and B_2 about y with $B_1 \cap B_2 = \emptyset$ and $\overline{B_1} \cup \overline{B_2} \subset \Omega$. Let

$$u(z) = G_\Omega(z; x); \quad v(z) = G_\Omega(z; y) \tag{3.8.41}$$

Since

$$\vec{\nabla} \cdot \left(u \vec{\nabla} v - v \vec{\nabla} u \right) = u \Delta v - v \Delta u = 0 \tag{3.8.42}$$

on $\Omega \setminus (\overline{B_1} \cup \overline{B_2})$ and $u \vec{\nabla} v - v \vec{\nabla} u = 0$ on $\partial\Omega$ (by Theorem 3.8.17), we see, using Stokes' theorem that

$$\int_{\partial B_1 \cup \partial B_2} \left[u \frac{\partial v}{\partial n} - v \frac{\partial u}{\partial n} \right] d\sigma = 0 \tag{3.8.43}$$

Since u only blows up logarithmically as B_1 shrinks, the first term only contributes on ∂B_2 as both balls shrink and similarly, for the second term, i.e.,

$$\int_{\partial B_2} u \frac{\partial v}{\partial n} d\sigma = \int_{\partial B_1} v \frac{\partial u}{\partial n} d\sigma \tag{3.8.44}$$

The normalization of this section has $\Delta G_\Omega(z; x) = -2\pi\delta(x - z)$, so (3.8.44) implies

$$-2\pi u(y) = -2\pi v(x) \tag{3.8.45}$$

i.e.,

$$G_\Omega(y; x) = G_\Omega(x; y) \tag{3.8.46}$$

On a general Riemann surface, one may not be able to use a single coordinate system, so one needs some bookkeeping to replace $u\Delta v - v\Delta u$ and that's what differential forms are for. While neither Δu nor $d^2 z$ is invariant under local change of coordinates, for conformal changes $w = F(z)$, $d^2 w = |F'(z)|^2 d^2 z$ (see (2.2.22) of Part 2A) and $\Delta_w u = |F'(z)|^{-2}\Delta_z w$ so $\Delta u \, d^2 z$ is invariant, a good sign given the two-form we want is $(u\Delta v - v\Delta u)d^2 z$. These calculations use the Cauchy–Riemann equation for F.

Recall (see Sections 2.1 and 8.7 of Part 2A) the Wirtinger calculus. Given local coordinates $x + iy$, define the one-forms

$$dz = dx + i\,dy, \quad d\bar{z} = dx - i\,dy \tag{3.8.47}$$

so with $d^2 z = dx \wedge dy$, we have

$$dz \wedge d\bar{z} = -2i \, d^2 z \tag{3.8.48}$$

One also defines

$$\partial = \frac{1}{2}\left(\frac{\partial}{\partial x} - i\frac{\partial}{\partial y}\right), \quad \bar{\partial} = \frac{1}{2}\left(\frac{\partial}{\partial x} + i\frac{\partial}{\partial y}\right) \tag{3.8.49}$$

so the Cauchy–Riemann equations are $\bar{\partial} f = 0$ and for analytic f's

$$\bar{\partial} f = 0 \Rightarrow \partial f = f' \tag{3.8.50}$$

Basically, ∂ and $\bar{\partial}$ are picked so that under the pairing of one-forms and tangent vectors, $(\partial, \bar{\partial})$ is the dual basis to $(dz, d\bar{z})$. i.e.,

$$\langle dz, \partial \rangle = \langle d\bar{z}, \bar{\partial} \rangle = 1, \quad \langle dz, \bar{\partial} \rangle = \langle d\bar{z}, \partial \rangle = 0 \tag{3.8.51}$$

and one finds

$$du \equiv \frac{\partial u}{\partial x}dx + \frac{\partial u}{\partial y}dy = (\partial u)dz + (\bar{\partial} u)d\bar{z} \tag{3.8.52}$$

du is a one-form, i.e., invariant under any $\mathbb{R}^2$ change of coordinates, but since if $w = F(z)$ and F is analytic

$$dw = F'(z)dz, \quad \partial_z u = F'(z)\partial_w u \tag{3.8.53}$$

we see that for analytic coordinate changes $(\partial u)dz$ and $(\bar{\partial} u)d\bar{z}$ are separate coordinate-invariant objects.

In local (x, y)-coordinates, one defines

$$*(a\,dx + b\,dy) = (b\,dx - a\,dy) \tag{3.8.54}$$

Using the change of coordinates $(a, b) \to (c, e)$ induced by $a\,dx + b\,dy = c\,dz + e\,d\bar{z}$, one sees (Problem 6) that

$$*(c\,dz + e\,d\bar{z}) = i(c\,dz - e\,d\bar{z}) \tag{3.8.55}$$

Therefore, $*(du)$ is invariant under conformal coordinate changes.

One also finds that for a one-form

$$d(cdz + ed\bar{z}) = dc \wedge dz + de \wedge d\bar{z}$$

$$= \bar{\partial}c(d\bar{z} \wedge dz) + \partial e(dz \wedge d\bar{z}) \qquad (3.8.56)$$

$$= -z(\partial e - \bar{\partial}c)\, d^2z \qquad (3.8.57)$$

on account of (3.8.52). Since $\Delta u = 4\partial\bar{\partial}u = 4\bar{\partial}\partial u$, we see that

$$d(* \, du) = (\Delta u)\, d^2z \qquad (3.8.58)$$

If $w_1 = adx + bdy$, $w_2 = pdx + qdy$, then, by a simple calculation

$$w_1 \wedge (*w_2) = -(ap + bq)\, dx \wedge dy \qquad (3.8.59)$$

is symmetric in (a, b) and (p, q) so

$$w_1 \wedge (*w_2) = w_2 \wedge (*w_1) \qquad (3.8.60)$$

In particular, $du \wedge * \, dv = dv \wedge * \, du$ which implies that

$$d(u(*dv) - v(*du)) = (u\Delta v - v\Delta u)\, d^2z \qquad (3.8.61)$$

where we used (3.8.60) with $w_1 = du$, $w_2 = dv$ and (3.8.58).

In particular, if $u(p) = G_R(p; x)$, $v(p) = G_R(p; y)$ where R is a bordered Riemann surface $x, y \in R$, and

$$w = u(*dv) - v(*du) \qquad (3.8.62)$$

then $dw = 0$ on $R \backslash (\overline{B}_1 \cup \overline{B}_2)$ where B_1(respectively, B_2) is a small coordinate disk containing x (respectively, y) with $\overline{B}_1 \cup \overline{B}_2 \subset R$. u vanishes on ∂R, so by the Schwarz reflection principle, u has a harmonic extension to a neighborhood of ∂R and we can view $\overline{R} \backslash B_1 \cup B_2$ as contained in a slightly bigger open set in S.

Thus, by Stokes' Theorem (Theorem 6.5.2 of Part 2A) and $u = v = 0$ on ∂R, we obtain

$$\int_{\partial B_1 \cup \partial B_2} [u * dv - v * du] = 0 \qquad (3.8.63)$$

As in the case $\Omega \subset \mathbb{C}$, we can shrink the disks to see that $2\pi(u(y) - v(x)) = 0$. Indeed, in local coordinates with $z(y) = 0$, $*dv = i[(\partial v)\, dz - \bar{\partial}v d\bar{z}] = 2i\, \mathrm{Im}((\partial v)dz)$ and $\partial v = \frac{1}{2z} +$ bounded so the boundary integral $u * dv$ in the calculation can be viewed as a use of residues! We have thus proven

Proposition 3.8.18. *Let R be a finite bordered Riemann subsurface of a Riemann surface S with $\partial R \neq \emptyset$. For every $x \neq y \in R$, we have that*

$$G_R(x; y) = G_R(y; x) \qquad (3.8.64)$$

We are now ready to prove Fact 3:

Proposition 3.8.19. *Let S be a hyperbolic Riemann surface with Green's function $G_S(x;y)$. Then for all $x, y \in S$, we have that*

$$G_S(x;y) = G_S(y;x) \qquad (3.8.65)$$

Proof. By Theorem 3.8.3, we can find finite bordered Riemann surfaces $R_1 \subset R_2 \subset \dots$ with $\partial R_j \neq \emptyset$ and $S = \bigcup_{j=1}^{\infty} R_j$ and by Corollary 3.8.9

$$G_S(x;y) = \lim_{j \to \infty} G_{R_j}(x;y) \qquad (3.8.66)$$

This and (3.8.64) imply (3.8.65).   □

To complete the tools needed to prove uniformization, we need to construct bipolar Green's functions. If S has a Green's function, $G_S(z;p)$, then $G(z; p_0, p_1) = G_S(z; p_0) - G_S(z; p_1)$ is a bipolar Green's function. The idea will be to approximate S or $S \setminus \{z_0\}$ by a sequence of bordered finite Riemann subsurfaces and take limits. The following elegant a priori bound will be the key: since we'll use Harnack's inequality, we'll want a definition: Given a Riemann surface T and $K \subset T$ compact, $C(T; K)$ is the optimal constant so that for all $x, y \in K$ and positive harmonic functions, u, on T, we have that

$$C^{-1} \leq \frac{u(x)}{u(y)} \leq C \qquad (3.8.67)$$

Proposition 3.8.20. *Let R_0 be a bordered finite Riemann subsurface of a Riemann surface S with $\partial R_0 \neq \emptyset$. Let $p_0, p_1 \in R_0$ and let B_0, B_1 be coordinate disks about p_0 and p_1 of radii ρ_0, ρ_1 lying in R_0. Let Σ_0, Σ_1 be the disks of radii $\sigma_i = \frac{1}{2}\rho_j$. Then for every bordered finite Riemann subsurface, R_1, with $R_0 \subset R_1$ and $\partial R_1 \neq \emptyset$, we have*

$$|G_{R_1}(q; p_0) - G_{R_1}(q; p_1)| \leq (2 \log 2)\, C\left(R_0 \setminus (\overline{\Sigma_0} \cup \overline{\Sigma_1}); \partial B_0 \cup \partial B_1\right) \quad (3.8.68)$$

for all $q \in R_1 \setminus (B_0 \cup B_1)$.

Remark. The point, of course, is that the right side of (3.8.68) only depends on R_0, not R_1.

Proof. Let $G^{(i)}(q) \equiv G_{R_1}(q; p_i)$ for $i = 0, 1$. Let

$$M_i = \sup_{q \in \partial \Sigma_i} G^{(i)}(q) \qquad (3.8.69)$$

Since $R_1 \setminus \overline{\Sigma_0}$ has compact closure and $G^{(0)} \upharpoonright \partial R_1 = 0$, we have that

$$0 \leq G^{(0)}(q) \leq M_0, \quad q \in R_1 \setminus \Sigma_0 \qquad (3.8.70)$$

$G^{(0)}(q) + \log(|z(q)|)$ is harmonic in B_0. Thus

$$M_0 \leq \log(2) + \sup_{\partial B_0} G^{(0)}(q) \tag{3.8.71}$$

so there is a point $q_0 \in \partial B_0$ so that

$$M_0 \leq \log(2) + G^{(0)}(q_0) \tag{3.8.72}$$

Let $u(q) = M_0 - G^{(0)}(q)$. By (3.8.70), u is a positive harmonic function on $R_0 \setminus \overline{\Sigma}_0 \cup \overline{\Sigma}_1$ and by (3.8.72), $u(q_0) \leq \log(2)$. It follows by Harnack's inequality that (with $C \equiv C(R_0 \setminus \overline{\Sigma}_0 \cup \overline{\Sigma}_1; \partial B_0 \cup \partial B_1)$)

$$\sup_{q \in \partial B_0 \cup \partial B_1} u(q) \leq C \log(2) \tag{3.8.73}$$

Therefore

$$q \in \partial B_0 \cup \overline{B_1} \Rightarrow M_0 - C \log(2) \leq G^{(0)}(q) \leq M_0 \tag{3.8.74}$$

Note that we have replaced ∂B_1 on the left by $\overline{B_1}$ which we can do since $G^{(0)}(q)$ is harmonic in B_1. In particular,

$$G^{(0)}(p_1) = G(p_1; p_0) \in [M_0 - C \log(2), M_0] \tag{3.8.75}$$

Similarly

$$q \in \overline{B_0} \cup \partial B_1 \Rightarrow M_1 - C \log(2) \leq G^{(1)}(q) \leq M_1 \tag{3.8.76}$$

$$G(p_0; p_1) \in [M_1 - C \log(2), M_1] \tag{3.8.77}$$

Since $G(p_0; p_1) = G(p_1; p_0)$, we have

$$M_1 - C \log(2) \leq M_0, \quad M_0 - C \log(2) \leq M_1 \Rightarrow |M_0 - M_1| \leq C \log(2) \tag{3.8.78}$$

By this and (3.8.74)/(3.8.76)

$$|G^{(0)}(q) - G^{(1)}(q)| \leq |M_1 - M_0| + C \log(2) \leq 2C \log(2)$$

on $\partial B_1 \cup \partial B_2$. Since the difference of Green's functions is harmonic on $R_1 \setminus (\overline{B_1} \cup \overline{B_2})$ and goes to zero on ∂R_1, we get (3.8.68). $\qquad \square$

This lets us prove Fact 4:

Theorem 3.8.21. *For any Riemann surface, S, and any $p_0 \neq p_1 \in S$, there is a bipolar Green's function.*

Proof. Suppose first that S is not compact. By Theorem 3.8.3, we can write $S = \bigcup_{j=1}^{\infty} \tilde{R}_j$ where $\tilde{R}_j$ is a finite bordered Riemann subsurface, $\tilde{R}_j \subset \tilde{R}_{j+1}$ and $\partial \tilde{R}_j \neq \emptyset$. By Theorem 3.8.17, each $\tilde{R}_j$ has a Green's function. Given $p_0 \neq p_1$, pick $\tilde{R}_\ell$ with $p_0, p_1 \in \tilde{R}_\ell \equiv R_0$. Let B_0, B_1 be as in Proposition 3.8.20 and $R_1 = \tilde{R}_j$, some $j \geq \ell$.

By that proposition

$$|G_{\widetilde{R}_j}(q;p_0) - G_{\widetilde{R}_j}(q;p_1)| \leq C \tag{3.8.79}$$

all $q \in \widetilde{R}_j \setminus (B_0 \cup B_1)$. By Montel's theorem, a subsequence of the $G_{\widetilde{R}_j}$ converges as $j \to \infty$ to a function, h, harmonic on $S \setminus (B_0 \cup B_1)$ obeying

$$|h(q)| \leq C \tag{3.8.80}$$

By increasing the radii of B_j we get slightly larger coordinate disks, D_j, with $\overline{B}_j \subset D_j$ and $\partial \widetilde{R}_j \subset S \setminus (B_0 \cup B_1)$. Since each $G_{\widetilde{R}_j}(q;p_0) - G_{\widetilde{R}_j}(q;p_j) + \log(z_0(q))$ is harmonic in a neighborhood of D_0 and convergent on ∂D_0, we get uniform convergence to a function harmonic on D_0, so $h(q)$ has a pole of order 1 at p_0. Similarly, it has pole of order -1 at p_1, i.e., h is the required bipolar Green's function.

If S is a compact Riemann surface, pick $p_2 \neq p_0, p_1$ and let $\widetilde{S} = S \setminus \{p_2\}$ which is noncompact. Hence $G_{\widetilde{S}}(q;p_0,p_1)$ exists and by (3.8.80) (or by the definition of bipolar Green's function) this function is bounded in a punctured neighborhood of p_2. Thus, p_2 is a removable singularity and $G_{\widetilde{S}}(q;p_0,p_1)$ can be extended to a $G_S(q;p_0,p_1)$. □

Finally, we'll use bipolar Green's functions to construct meromorphic functions.

Theorem 3.8.22. *Any Riemann surface has (nonconstant) meromorphic functions.*

Proof. Here's a not quite right proof which we'll fix. Since $\Delta = 4\overline{\partial}\partial$, if u is harmonic, $\overline{\partial}(\partial u) = 0$ so ∂u is analytic. Note that in a local coordinate system, $\partial \log|z| = \frac{1}{2}\partial \log \bar{z}z = \frac{1}{2}\partial(\log z) + \frac{1}{2}\partial \log \bar{z} = \frac{1}{2}z^{-1}$ since $\log \bar{z}$ is anti-analytic. Thus, if u is a bipolar Green's function, ∂G is is meromorphic.

The problem with this argument is that ∂ is coordinate-dependent! But as noted in (3.8.47), ∂u and ∂v transform in the same way, so if u and v are bipolar Green's functions, say at p_0, p_1 and p_0, p_2 for three distinct points p_0, p_1, p_2, then $\frac{\partial u}{\partial v}$ is locally meromorphic and coordinate independent. (*Warning*: $\frac{\partial u}{\partial v}$ is *not* the derivative of u with respect to v but a ratio $(u_x - iu_y)/(v_x - iv_y)$). This gives us a meromorphic function with a pole at p_1 and a zero at p_2 but poles also at critical points of v and zeros at critical points of u. □

Notes and Historical Remarks. Potential-theoretic ideas were central to the original proofs of Koebe [478] and Poincaré [657] and for the approach of Hilbert [384] which influenced much of the later work. Our presentation here follows in places that of Farkas–Kra [260] and Gamelin [302].

Problems

1. (a) Let $w = f(z)$ be a local coordinate change. Prove that

$$\partial_z = (\partial_z f)\partial_w, \quad \overline{\partial}_z = (\overline{\partial}_z \overline{f})\overline{\partial}_w \tag{3.8.81}$$

(b) Prove that

$$\Delta_z = |f'(z)|^2 \Delta_w \tag{3.8.82}$$

2. Let $\mathcal{F}$ be a Perron family. In this problem, the reader will prove Perron's principle, Theorem 3.8.1.

(a) Fix a coordinate disk, D and $\{z_j\}_{j=1}^{\infty}$ a dense subset of D. Define $w(z)$ by (3.8.5). If $w(z)$ is not identically infinite on D, suppose $w(z_0) < \infty$. Prove there exists $\{u_j\}_{j=1}^{\infty}$ subharmonic on D so that $u_1 \leq u_2 \leq \ldots$ and $w(z_j) = \sup_\ell u_\ell(z_j)$.

(b) Using Perron modifications, show that one can choose the u_j's in (a) to be harmonic.

(c) Let $u(z) = \sup_\ell u_\ell(z)$. Prove that $u(z) < \infty$ for all $z \in D$ (*Hint*: Harnack's principle, Theorem 3.1.30).

(d) If $v \in \mathcal{F}$, prove that $v(z_j) \leq u(z_j)$ and conclude that $v(z) \leq u(z)$ for all $z \in D$ so that $w = u$ is finite on D.

(e) Complete the proof by covering a curve from z_0 to a given point $\tilde{z} \in S$ by coordinate disks.

3. Extend the analysis of Example 3.8.10 to any S for which $G_S(z; q)$ exists but for which (3.8.9) might not hold. (*Hint*: Use Theorems 3.8.3 and (3.8.19).)

4. (a) Let D be a coordinate disk for S, a hyperbolic Riemann surface. Prove that there are nonzero bounded harmonic functions, u, on $S \setminus \overline{D}$, continuous on $S \setminus D$, so that $u \upharpoonright \partial D = 0$ so the exterior Dirichlet problem does not have unique solutions.

(b) Conversely, suppose there is a nonzero bounded harmonic function, u on $S \setminus \overline{D}$, continuous on $S \setminus D$, so that $u \upharpoonright \partial D = 0$. By replacing u by $-u$ if need be, suppose $\sup u(z) = \alpha > 0$. Let

$$v(x) = \begin{cases} \max(u(x), 0)), & x \in S \setminus D \\ 0, & x \in D \end{cases} \tag{3.8.83}$$

Prove that v is subharmonic and nonconstant and conclude that S is hyperbolic.

(c) Prove the exterior Dirichlet problem for $D \subset S$ has at most one bounded solution if S is parabolic.

5. (a) If D is a coordinate disk in S a noncompact Riemann surface, prove that ω_D, extended to S by setting $\omega_D(z) = 1$ if $z \in D$, is superharmonic.

   (b) Let v be superharmonic on S with $v \geq \chi_D$. Prove that for every $u \in \mathfrak{I}_D$ we have $u \leq v$ and conclude that $\omega_D \leq v$.

6. Verify (3.8.55)

7. In this problem, the reader will prove that if S is a hyperbolic Riemann surface, then $G_S(x; y)$ is jointly continuous on $S \times S \setminus \{(x, x) \mid x \in S\}$. So suppose $x_n \to x_\infty$, $y_n \to y_\infty$, $x_\infty \neq y_\infty$. Pick disjoint compact neighborhoods, U of x_∞ and V of y_∞ and by throwing away the first few (x_n, y_n), if necessary, suppose for all n, we have that $x_n \in U$, $y_n \in V$.

   (a) Prove that $G(x_\infty; y_n) \to G(x_\infty; y)$ and, in particular, $\{G(x_\infty; y_n)\}$ is bounded.

   (b) Prove that $G(\cdot, y_n)$ are uniformly bounded on U. (*Hint*: Harnack's inequality.)

   (c) Prove that uniformly for $x \in U$, $G(x; y_n) \to G(x; y_\infty)$.

   (d) Prove that $G(x_n; y_n) \to G(x_\infty; y_\infty)$.

Chapter 4

Bonus Chapter:
Phase Space Analysis

Like the vague sighings of a wind at even
That wakes the wavelets of the slumbering sea
And dies on the creation of its breath

—The Fairy from *Queen Mab, A Philosophical Poem*
Percy Bysshe Shelley, 1813 [**735**]

Big Notions and Theorems: Uncertainty Principle, Hardy's Inequality, Hardy's Uncertainty Principle, Benedicks Theorem, Amrein–Berthier Theorem, Landau-Pollack Uncertainty Principle, Shannon Entropy, Hirschmann Uncertainty Principle, Singular Support, Wavefront Set, Partial Differential Operator, Elliptic Regularity, Pseudodifferential Operator, Pseudolocality of ΨDO, ΨDO Calculus, Fourier Integral Operator, Weyl Calculus, Wigner Distribution Function, Gaussian Coherent States, General Coherent States, Upper and Lower Symbols, Berezin–Lieb Inequalities, Admissible Vectors, Square Integrable Representation, Grossmann–Morlet–Paul Theorem, Bloch Coherent States, Continuous Wavelets, Gabor Lattices, von Neumann Lattice, Complete, Incomplete, Overcomplete, Completeness of the von Neumann Lattice, Bessel Sequence, Frame, Riesz Basis, Balian–Low Theorem, Haar Wavelets, MRA (aka Multiresolution Analysis), Scaling Function, Mother Wavelet, Structure Constants, Scaling Filter, Meyer Wavelets, Vanishing Moments of Smooth Decaying Mother Wavelets, Daubechies Wavelets, Fejér-Riesz Theorem, Wavelet Kernel, L^p and C_∞ Wavelet Expansions

This chapter could be called "More Fourier Transforms" and could have occurred as part of Chapter 6 of Part 1. The common theme in this chapter is the simultaneous consideration of f and $\widehat{f}$, put differently, of a function simultaneous in x- and k-space. The first section, which deals with a rigidity of phase space in that it cannot be that both f and $\widehat{f}$ are well localized, sets the stage for the rest of the chapter. One way of avoiding the difficulties of the uncertainty principle is to consider cones in k-space, the subject of Section 4.2.

Analysis on phase space is sometimes called microlocal analysis, especially the study of two kinds of operators defined by phase space functions—pseudodifferential operators, aka ΨDO, and Fourier Integral Operators. The basic theory of the former is the subject of Section 4.3 and further developments are a part of Chapter 6, especially Sections 6.3, 6.4, and 6.5.

The final three sections discuss expansions of functions in what are called coherent states, especially the states associated to continuous groups. The general theory is in Section 4.4, especially continuous (i.e., integral) expansions. Sections 4.5 and 4.6 discuss discrete expansions—for the Heisenberg group in Section 4.5 and for the affine group (wavelets) in Section 4.6. These discrete expansions are widely used in a vast array of practical applications.

4.1. The Uncertainty Principle

This is the first section on the use of phase space ideas, that is, the simultaneous consideration of a function and its Fourier transform, an idea that will reoccur in Part 4. In this section, we'll discuss the notion that it cannot be that both f and $\widehat{f}$ are small off of small sets. We begin with the classical Heisenberg uncertainty principle and then turn to various refinements.

Given a tempered distribution f on $\mathbb{R}$, we define

$$
\text{Var}_f(X) = \begin{cases} \infty & \text{if } xf \notin L^2 \text{ or } f \notin L^2 \\ \dfrac{\int x^2 |f(x)|^2 dx}{\int |f(x)|^2 dx} - \left(\dfrac{\int |xf(x)|^2 dx}{\int |f(x)|^2 dx} \right)^2 & \text{if } f, xf \in L^2 \end{cases}
$$

$$(4.1.1)$$

and

$$
\text{Var}_f(P) = \text{Var}_{\widehat{f}}(X) \tag{4.1.2}
$$

"Var" is short for variation and the name comes from the notion that in quantum mechanics $d\mu_x = |f(x)|^2 d^\nu x / \|f\|_2^2$ is the probability distribution for x (see Section 7.1 of Part 1) and if $\mathbb{E}(\cdot) = \int \cdot \, d\mu_x(y)$ then

$$
\text{Var}(h) = \mathbb{E}((h - \mathbb{E}(h))^2) = \mathbb{E}(h^2) - \mathbb{E}(h)^2 \tag{4.1.3}
$$

where

$$
\mathbb{E}(X) = \int y \, d\mu_x(y) \tag{4.1.4}
$$

is defined if $\mathbb{E}(X^2) < \infty$.

According to quantum mechanics, $d\mu_x$ is the distribution of position and $d\mu_k$, the same object for $\widehat{f}$, is the distribution of momentum. While quantum mechanics is only motivational and therefore not needed for statements, it is helpful to understand some of the background.

Theorem 4.1.1 (Heisenberg Uncertainty Principle). *For any f, we have*

$$\mathrm{Var}_f(X)\mathrm{Var}_f(P) \geq 1/4 \qquad (4.1.5)$$

We only have equality if for some $x_0, p_0 \in \mathbb{R}$, $a \in (0, \infty)$, we have that $f/\|f\|_2 = e^{i\eta}\varphi_{x_0,p_0,a}$ where $\eta \in \mathbb{R}$ and

$$\varphi_{x_0,p_0,a}(x) = (\pi a)^{-1/4}\exp(ip_0x - [(x - x_0)^2/2a])$$

Remark. The proof shows that in $\mathbb{R}^\nu$, we have the same result for $\mathrm{Var}_f(X_j)\mathrm{Var}_f(P_j)$ (same j) for each component. This leads to the bound

$$\left[\sum_j \mathrm{Var}_f(X_j)\right]\left[\sum_j \mathrm{Var}_f(P_j)\right] \geq \nu/4 \qquad (4.1.6)$$

As a preliminary, which will be useful later, we want to define some groups of unitaries on $L^2(\mathbb{R}^\nu, d^\nu x)$ (which define abstract groups of interest). The *Heisenberg group* is parameterized by $x_0, p_0 \in \mathbb{R}^\nu$ and $\omega \in \partial\mathbb{D}$ ($= \mathbb{T}^1$, the one-dimensional torus) via

$$(U(x_0, p_0, \omega)f)(x) = \omega e^{ip_0x}f(x - x_0) \qquad (4.1.7)$$

which obeys the group relation

$$U(x_0, p_0, \omega_0)U(x_1, p_1, \omega_1) = U(x_0 + x_1, p_0 + p_1, \omega_0\omega_1 e^{-ip_1x_0}) \qquad (4.1.8)$$

We'll also sometime need the *affine Heisenberg–Weyl group* which has an additional parameter $a \in (0, \infty)$ so

$$V(x_0, p_0, a, \omega) = a^{-\nu/2}\omega e^{ip_0x}f\left(\frac{x - x_0}{a}\right) \qquad (4.1.9)$$

We leave the group relation to the reader—see Problem 1

These groups enter in considerations of Theorem 4.1.1 in two ways. First,

$$\varphi_{x_0,p_0,a} = V(x_0, p_0, a^{1/2}, 1)\varphi_{x_0=0,p_0=0,a=1} \qquad (4.1.10)$$

Second, if X is multiplication by x and P by k, after taking Fourier transforms, then

$$V(x_0, p_0, a, \omega)\,X\,V(x_0, p_0, a, \omega)^{-1} = a^{-1}(X - x_0) \qquad (4.1.11)$$

$$V(x_0, p_0, a, \omega)\,P\,V(x_0, p_0, a, \omega)^{-1} = a(P - p_0) \qquad (4.1.12)$$

Proof of Theorem 4.1.1. If $\mathrm{Var}_f(X)$ or $\mathrm{Var}_f(P)$ is ∞, there is nothing to prove (since $\mathrm{Var}_f(X) > 0$) so suppose both are finite. By (4.1.11)/(4.1.12), we can suppose by looking at $V\left(\mathbb{E}_f(X), \mathbb{E}_f(X), \sqrt{2\mathrm{Var}_f(X)}, 1\right)f/\|f\|$ that

$$\mathbb{E}_f(X) = \mathbb{E}_f(P) = 0, \ \mathrm{Var}_f(X) = \frac{1}{2}, \ \|f\| = 1 \qquad (4.1.13)$$

and by (4.1.10) that when (4.1.13) holds, (4.1.5) has equality if any only if

$$f = e^{i\eta}\varphi_{x_0=0,p_0=0,a=1} \tag{4.1.14}$$

By a limiting argument (Problem 2), we can suppose that $f \in \mathcal{S}$. We note the Heisenberg commutation relation that is for $f \in \mathcal{S}$

$$(xf)' - xf' = f \tag{4.1.15}$$

$$\left[\frac{d}{dx}, x\right] = \mathbb{1} \tag{4.1.16}$$

Then, if $\|f\|^2 = 1$, by an integration by parts,

$$1 = \left\langle f, \left[\frac{d}{dx}, x\right] f \right\rangle = -2\,\mathrm{Re} < \frac{d}{dx}f, xf >$$

$$\leq 2\mathrm{Var}_f(X)^{1/2}\mathrm{Var}_f(P)^{1/2} \tag{4.1.17}$$

by the Schwarz inequality. This is (4.1.5).

To have equality in (4.1.5), we must have $\langle \frac{d}{dx}f, xf \rangle$ negative and $\frac{df}{dx} = \lambda(xf)$ for some λ in $\mathbb{C}$ (because of the condition for equality in the Schwarz inequality). For $\langle \frac{d}{dx}f, xf \rangle$ to be negative, we must have $\lambda < 0$. Since $\mathrm{Var}_f(X) = \frac{1}{2}$, we must also have $\mathrm{Var}_f(P) = \frac{1}{2}$ so $|\lambda| = 1$ and f must obey

$$\frac{df}{dx} = -xf \tag{4.1.18}$$

which occurs precisely when $f(x) = \gamma e^{-\frac{1}{2}x^2}$ where $\gamma \in \mathbb{C}$. $\|f\| = 1$ implies (4.1.14) holds. $\square$

While (4.1.5) is the original uncertainty principle of quantum mechanics, mathematicians have come to use the term to refer to the notion that functions cannot be well localized in both x-space (i.e., f) and k-space (i.e., $\hat{f}$) and the rest of the section will be devoted to several other results that extend, illuminate and, for some, address shortcomings of the original Heisenberg uncertainty principle. These results will fall under four main banners.

First, we'll explore the notion that not only can't $\mathbb{E}(P^2)$ and $\mathbb{E}(X^2)$ be too small, how small they can be increases if we want it for multiple f's. To state a version, we note since (4.1.5) implies $\mathbb{E}(p^2)^{1/2}\mathbb{E}(x^2)^{1/2} \geq \frac{1}{2}$ and $\alpha^2 + \beta^2 \geq 2\alpha\beta$, (4.1.5) implies for any f

$$\mathbb{E}(p^2 + x^2) \geq 1. \tag{4.1.19}$$

We'll actually see below that (4.1.19) in turn implies (4.1.5). We'll prove the following:

Theorem 4.1.2. *Let $X \subset \mathcal{S}(\mathbb{R})$ be a vector space of dimension n. Then*

$$\sup_{\|f\|=1, f\in X} \langle f, (-\frac{d^2}{dx^2} + x^2)f \rangle \geq 2n - 1 \tag{4.1.20}$$

If one has equality in (4.1.20), then X is the space of the first n elements of the Hermite basis.

Remarks. 1. The left side of (4.1.20) is $\mathbb{E}_f(p^2 + x^2)$.

2. One can replace $X \subset \mathcal{S}(\mathbb{R})$ by $f \in X \Rightarrow f, xf, f' \in L^2(\mathbb{R})$.

3. There are ν-dimensional analogs (see Problem 3).

    The second result concerns the following limitation of the Heisenberg uncertainty principle. The physics texts like to assert that the uncertainty principle is responsible for the stability of the hydrogen atom, that is, that while classically $p^2 - |x|^{-1}$ on $\mathbb{R}^3 \times \mathbb{R}^3$ is not bounded from below,

$$\inf_{\|f\|=1, f \in \mathcal{S}(\mathbb{R}^3)} \langle f, (-\Delta - \frac{1}{|x|})f \rangle > -\infty \tag{4.1.21}$$

The argument goes that for the expression in (4.1.21) to be very negative, f must be concentrated near $x = 0$, so, by the uncertainty principle, p must be large. Quantitatively, if $\langle f, |x|^{-1}f \rangle = O(r^{-1})$, then f is concentrated mainly in a ball of radius r, $\langle f, x^2 f \rangle = O(r^2)$ so $\langle f, p^2 f \rangle \geq O(r^{-2})$. While $-O(r^{-1})$ diverges as $r \downarrow 0$, $O(r^{-2}) - O(r^{-1})$ does not! At a heuristic level, this analysis is correct, but the Heisenberg version of the uncertainty principle is not strong enough to implement the argument. If f has two pieces, one near $x = 0$ and one near very large $|\vec{x}|$, $\langle f, |x|^{-1}f \rangle$ and $\langle f, x^2 f \rangle$ can both be very large! What will be sufficient though is

Theorem 4.1.3 (Hardy's Inequality). *Let $f \in \mathcal{S}(\mathbb{R}^\nu)$, vanishing near $x = 0$. Then*

$$\frac{(\nu - 2)^2}{4} \int \frac{|f(x)|^2}{|x|^2} d^\nu x \leq \int |\vec{\nabla} f(x)|^2 d^\nu x \tag{4.1.22}$$

If $\nu \geq 3$, one can drop the condition that f vanish near $x = 0$.

Remarks. 1. There are other results called Hardy's inequality; see the Notes.

2. This is intimately connected to Sobolev inequalities, a subject discussed in Section 6.3.

3. It is easy to see (Problem 4) that if $\nu = 3$, (4.1.22) implies

$$\langle f, (-\Delta - \frac{1}{|x|})f \rangle \geq -\|f\|^2 \tag{4.1.23}$$

showing the connection to the stability of hydrogen.

4. It is not hard (Problem 5) to see that, if $\nu \geq 3$, (4.1.22) holds for all $f \in L^2$ with $|x|^{-1/2}f \in L^2$ and $\vec{\nabla} f \in L^2$ (distributional derivatives). For $\nu = 1$, if $f, f' \in L^2$, then f is a continuous function and (4.1.22) holds for all $f \in L^2$ with $\nabla f \in L^2$ and $f(0) = 0$. See Problem 6 for a result when $\nu = 2$.

Our third extension involves the issue of rapid decay in both x- and k-space.

Theorem 4.1.4 (Hardy's Uncertainty Principle—Weak Form). *Suppose $f \in L^2$ with*

$$|f(x)| \leq Ce^{-\frac{a}{2}|x^2|}, \quad |\hat{f}(k)| \leq Ce^{-\frac{b}{2}|x|^2} \qquad (4.1.24)$$

for some C and $a > 0, b > 0$. If

$$ab > 1 \qquad (4.1.25)$$

then $f = 0$.

Remark. We'll prove this as a special case of a result that implies if $ab = 1$, then f is a multiple of $e^{-\frac{a}{2}|x|^2}$

Our fourth and final result will prove a stronger version of

Theorem 4.1.5 (Benedicks–Amrein–Berthier). *Let $A, B \subset \mathbb{R}^\nu$ both be Borel sets of the finite Lebesque measure. If $f \in L^2$ with $\mathrm{supp}\, f \subset A$ and $\mathrm{supp}\, \hat{f} \subset B$, then $f = 0$.*

Remarks. 1. If A is bounded, f is real analytic (see Section 11.1 of Part 2A) so $f = 0$ if $\mathrm{supp}\, \hat{f}$ has a positive measure set of zeroes (i.e., $|\mathbb{R}^\nu \backslash B| \neq 0$ suffices in this case). Thus, in one sense, this seems a weak result but it is not evident how to effectively use the finite measure hypothesis.

2. $\mathrm{supp}\, f \subset A$ can be replaced by $f(x) = 0$ for all $x \notin A$ (and $\mathrm{supp}\, \hat{f} \subset B$ by a similar condition), i.e., one need not take the closure.

Having presented most of the theorems of the rest of the section, we turn to proofs.

Proof of Theorem 4.1.2. Since $\dim(X) = n$, we can find $f \in X$ with $\|f\| = 1$ and $f \perp \{\varphi_j\}_{j=0}^{n-2}$, the first $n-1$ Hermite basis kernels. Thus, (by (6.4.1), (6.4.20), and Theorem 6.4.3 of Part 1),

$$\langle f, (p^2 + x^2)f \rangle \geq \sum_{j=n-1}^{\infty} (2j+1)|\langle \varphi_j, f \rangle|^2 \qquad (4.1.26)$$

$$\geq 2n - 1 \qquad (4.1.27)$$

proving (4.1.20). The only way to get equality in (4.1.27) is for $f = \varphi_{n-1}$, i.e., if the only $f \in X \cap \{\varphi_1, \ldots, \varphi_{n-2}\}^\perp$ is φ_{n-1} so X is spanned by $\{\varphi_j\}_{j=0}^{n-1}$ □

Before leaving this subject, we note that the volume in "classical phase space" of $\{\langle x, p \rangle \mid x^2 + p^2 \leq 2n - 1\}$ is

$$\pi(2n - 1) = 2\pi n \left[1 + \mathrm{O}\left(\frac{1}{n}\right)\right] \qquad (4.1.28)$$

The mantra associated with this is that "each quantum state takes 2π volume in phase space and this is asymptotically exact in the large number of states' limit." In $\mathbb{R}^\nu$, 2π becomes $(2\pi)^\nu$. This is a theme we'll explore at the end of Section 7.5 of Part 4.

Proof of Theorem 4.1.3. We begin with a formal calculation. Let

$$\partial_j \equiv \frac{\partial}{\partial x_j} \qquad (4.1.29)$$

as an operator and note that applied to the function $\frac{x_j}{r^2}$,

$$\frac{\partial}{\partial x_j}\left[\frac{x_j}{r^2}\right] = \frac{1}{r^2} - \frac{2x_j^2}{r^4} \qquad (4.1.30)$$

If we use $\frac{x_j}{r^2}$ for the multiplication operator, then

$$\left[\partial_j, \frac{x_j}{r^2}\right] = \frac{1}{r^2} - \frac{2x_j^2}{r^4} \qquad (4.1.31)$$

Therefore, for $\alpha \in \mathbb{R}$

$$\sum_{j=1}^{\nu}\left[-\partial_j + \frac{\alpha x_j}{r^2}\right]\left[\partial_j + \frac{\alpha x_j}{r^2}\right] = -\Delta + \frac{\alpha^2}{r^2} - \frac{\nu\alpha}{r^2} + \frac{2\alpha}{r^2} \qquad (4.1.32)$$

If $f \in \mathcal{S}(\mathbb{R}^\nu)$ vanishes near $r = 0$, we can apply (4.1.32) to f and take the inner product with f. Noting that, by an integration by parts,

$$\langle f, (-\partial_j + \frac{\alpha x_j}{r^2})(\partial_j + \frac{\alpha x_j}{r^2})f\rangle = \|(\partial_j + \frac{\alpha x_j}{r^2})f\|^2 \geq 0 \qquad (4.1.33)$$

we conclude that for all α

$$\left[(\nu - 2)\alpha - \alpha^2\right]\langle f, \frac{1}{r^2}f\rangle \leq \langle f, (-\Delta)f\rangle \qquad (4.1.34)$$

$(\nu - 2)\alpha - \alpha^2$ is maximized at $\alpha = \frac{\nu-2}{2}$ where it is $\frac{(\nu-2)^2}{4}$. Another integration by parts leads to (4.1.22) □

Remark. Note that (4.1.33) implies that to get equality in (4.1.22), one needs $\nabla_j f = -\alpha x_j r^{-1} f$ for all j which implies $f(x_j) = C|x|^\beta$ for suitable β. This is never in L^2 so the inequality in (4.1.22) is always strict.

We now turn to variants of the Hardy variational principle where we'll need some results on analytic functions, namely the following which is a special case of Corollary 5.1.3 of Part 2A.

If f is analytic in $\Omega = \{z \mid 0 < \text{Arg}(z) < \alpha\}$ for some $\alpha < \pi/2$ and if f is continuous on $\overline{\Omega}$ and for some $A, B, C < \infty$,

(i) $|f(x)| \leq Ae^{B|z|^2}$ for all $z \in \overline{\Omega}$ (4.1.35)

(ii) $|f(x)| \leq C$ for all $z \in \partial\Omega$ (4.1.36)

Then, (4.1.36) holds on all of $\overline{\Omega}$.

This is a result of Phragmén–Lindelöf type. It is proved as follows: by rotating, suppose, instead, that $\Omega = \{z \mid |\mathrm{Arg}\, z| < \frac{\alpha}{2}\}$. Let $g(z) = e^{-\varepsilon z^\beta} f(z)$ where $\pi/\alpha > \beta > 2$. Since $\beta < \pi/\alpha, \mathrm{Re}\, z^\beta > 0$ on $\overline{\Omega} \setminus \{0\}, |g(z)| \leq |f(z)|$ on $\overline{\Omega}$ and $e^{B|z|^2}|e^{-\varepsilon z^\beta}| \to 0$ in $\overline{\Omega}$, so one can apply the maximum principle on a sector union a large circle, to see that $|g(z)| \leq C$ on $\overline{\Omega}$. Thus, for all ε, $|f(z)| \leq C|e^{\varepsilon z^\beta}|$. Thus taking $\varepsilon \to 0$, one gets $|f(z)| \leq C$.

We will need the following consequence of the Phragmén–Lindelöf result:

Proposition 4.1.6. *Suppose f is an entire function which obeys*

$$|f(z)| \leq Ce^{D|\mathrm{Re}\, z|^2}(1 + |z|)^N, \quad |f(z)| \leq Ce^{D|\mathrm{Im}\, z|^2}(1 + |z|)^N \qquad (4.1.37)$$

for some $C, D > 0$. Then f is a polynomial in z of degree at most N.

Proof. On the first quadrant,

$$g(z) = (z + (1 + i))^{-N} f(z)$$

obeys (4.1.37) for $N = 0$, so it suffices to prove (4.1.37) for $N = 0$ on that quadrant implies f is bounded there (for then the original f is bounded in each quadrant by $C(1 + |z|)^N$, so a polynomial (by Theorem 3.1.9 of Part 2A)). Suppose f is analytic in $\Omega = \{z \mid 0 < \mathrm{Arg}\, z < \frac{\pi}{2}\}$, continuous on $\overline{\Omega}$, and obeys

$$|f(z)| \leq Ce^{D(\mathrm{Re}\, z)^2}, \quad |f(z)| \leq Ce^{D(\mathrm{Im}\, z)^2} \qquad (4.1.38)$$

Fix $\varepsilon > 0$ and let Ω_ε be the sector bounded by $\mathbb{R}_+$ and $\{z \mid \mathrm{Re}\, z = \varepsilon\, \mathrm{Im}\, z, \mathrm{Re}\, z > 0\}$, so Ω_ε has opening angle (strictly) less than $\pi/2$. On $\mathbb{R}_+$, $\mathrm{Im}\, z = 0$, so

$$|f(z)| \leq C, \quad z \in \mathbb{R}_+ \qquad (4.1.39)$$

by the second equation in (4.1.38). On the other hand, by the first equation,

$$|f(z)| \leq Ce^{D\varepsilon(\mathrm{Re}\, z)(\mathrm{Im}\, z)}, \quad z \in \partial\Omega_\varepsilon \setminus \mathbb{R}_+ \qquad (4.1.40)$$

Let

$$g_\varepsilon(z) = f(z)e^{i\frac{\varepsilon}{2}Dz^2} \qquad (4.1.41)$$

so

$$|g_\varepsilon(z)| \leq |f(z)|e^{-\varepsilon D\, \mathrm{Re}\, z\, \mathrm{Im}\, z} \qquad (4.1.42)$$

Thus $|g_\varepsilon(z)| \leq C$ on $\partial\Omega_\varepsilon$. Clearly $|f(z)| \leq Ce^{D|z|^2}$ so g_ε obeys the same bound. Thus, the Phragmén–Lindelöf result applies and we see that $|g_\varepsilon(z)| \leq C$. Thus, for each z in Ω_ε, $|f(z)| \leq Ce^{\frac{\varepsilon}{2}D|z|^2}$. Taking $\varepsilon \to 0$ yields the desired result. $\qquad \square$

Here is a strong version of Hardy's uncertainty principle.

Theorem 4.1.7 (Hardy Uncertainty Principle—Strong Form). *Let τ be a tempered distribution so that both $e^{\frac{1}{2}x^2}\tau$ and $e^{\frac{1}{2}p^2}\hat{\tau}$ are in $\mathcal{S}'(\mathbb{R}^\nu)$. Then, τ is $e^{-\frac{1}{2}x^2}$ times a polynomial in x.*

Remark. To say $e^{\frac{1}{2}x^2}\tau$ is in $\mathcal{S}'$, means there is $\sigma \in \mathcal{S}'$ so for $f \in \mathcal{S}$,

$$\tau(f) = \sigma(e^{-\frac{1}{2}x^2}f) \tag{4.1.43}$$

Proof. We'll use the Segal–Bargmann transform, $\mathbb{B}$, of Section 6.4 of Part 1. Suppose first $e^{\frac{1}{2}x^2}\tau$ is a finite complex measure, $d\mu$. Then, by (6.4.96) of Part 1,

$$|(\mathbb{B}\tau)(z)| = \left| \int \exp\left(-\left(x - \frac{z}{\sqrt{2}}\right)^2\right) d\mu(x) \right| \tag{4.1.44}$$

$$\leq \int \exp\left(\frac{|\operatorname{Im} z|^2}{2}\right) \exp\left(-\left(x - \frac{\operatorname{Re} z}{\sqrt{2}}\right)^2\right) d\mu \tag{4.1.45}$$

$$\leq \|\mu\| \exp\left(\frac{|\operatorname{Im} z|^2}{2}\right) \tag{4.1.46}$$

In general, there are multi-indices α_j, β_j and finite measures μ_j so

$$e^{\frac{1}{2}x^2}\tau = \sum_{j=1}^{\ell} D^{\alpha_j}\left(x^{\beta_j}\mu_j\right) \tag{4.1.47}$$

Each term in (4.1.44) then leads to a polynomial in x and z (since derivatives of the exponential are polynomials) so a polynomial in z and $x - \frac{z}{\sqrt{2}}$. Since any polynomial in $(x - \frac{z}{\sqrt{2}})$ times $\exp(-(x - \frac{z}{\sqrt{2}})^2)$ is bounded and μ_j is a finite measure, we see (4.1.46) just has an extra $|Q(z)|$ for some polynomial, Q, i.e.,

$$|(\mathbb{B}\tau)(z)| \leq C(1+|z|)^N \exp\left(\frac{|\operatorname{Im} z|^2}{2}\right) \tag{4.1.48}$$

By the assumption on $\hat{\tau}$ and $(\mathbb{B}\tau)(-iz) = (\mathbb{B}\hat{\tau})(z)$ (see (6.4.90) of Part 1), we also have

$$|(\mathbb{B}\tau)(z)| \leq C(1+|z|)^N \exp\left(\frac{|\operatorname{Re} z|^2}{2}\right) \tag{4.1.49}$$

Proposition 4.1.6 applies, so $\mathbb{B}\tau$ is a polynomial, and by Theorem 6.4.12 of Part 1, τ is a finite linear combination of Hermite basis functions, i.e., a polynomial times $e^{-\frac{1}{2}x^2}$. $\qquad\square$

Remark. The observant reader may have noted that we proved Proposition 4.1.6 for $z \in \mathbb{C}$ but applied it to $z \in \mathbb{C}^\nu$. A simple induction (Problem 7) lets one apply the one-variable result iteratively to prove that f is a polynomial in $z_1, \ldots, z_\nu$.

Proof of Theorem 4.1.4. By scaling, we can suppose that $b = 1$. Then $a > 1$, so (4.1.24) also holds for $a = 1$, i.e., f is a polynomial multiplied by $e^{-\frac{1}{2}x^2}$. This does not decay as $e^{-\frac{a}{2}x^2}$ for $a > 1$ unless $f = 0$. $\qquad\square$

Finally, we turn to our fourth theorem and prove a stronger version of Theorem 4.1.5. A *Benedicks set* is a set $A \subset \mathbb{R}^\nu$ so that for a.e. $x \in [0,1]^\nu \equiv \Delta$, we have that $(x + \mathbb{Z}^\nu) \cap A$ is finite. Since

$$|A| = \int_\Delta [\sharp(x + \mathbb{Z}^\nu) \cap A] d^\nu x \qquad (4.1.50)$$

we see that $|A| < \infty$ implies that A is a Benedicks set so the next theorem implies Theorem 4.1.5. But not all Benedicks sets are finite measure, for example, in $\mathbb{R}^2$, $\{(x_1, x_2) \mid |x_1 x_2| \leq 1\}$ is an infinite area Benedicks set. The following includes Theorem 4.1.5.

Theorem 4.1.8 (Benedicks Theorem). *Let A and B be Benedicks sets. Let $f \in L^1(\mathbb{R}^\nu)$. If $f(2\pi x) = 0$ for $x \notin A$ and $\hat{f}(k) = 0$ for $k \notin B$, then $f = 0$.*

Remark. The result does not hold if f is replaced by a tempered distribution. The Poisson summation formula in the form (6.6.52) of Part 1 says that

$$\tau = \sum_{n=-\infty}^{\infty} \delta(x - 2\pi n) \Rightarrow \hat{\tau} = (2\pi)^{-\nu/2} \sum_{m=-\infty}^{\infty} \delta(k - m) \qquad (4.1.51)$$

so there are A, B of Lebesque measure zero so that the hypothesis holds but $\tau \neq 0$.

Proof. For f in $L^1(\mathbb{R}^\nu)$, and a.e. $x \in 2\pi\Delta$, we have for all $q \in \Delta$, that (note $\hat{f} \in C(\mathbb{R}^\nu)$ so $\hat{f}(q + m)$ is well-defined for all q, m)

$$e^{-iq \cdot x} \sum_{n \in \mathbb{Z}^\nu} e^{-2\pi i n \cdot q} f(x + 2\pi n) = C\text{-}\lim_N \sum \frac{1}{\sqrt{2\pi}} \hat{f}(q + m) e^{im \cdot x} \qquad (4.1.52)$$

where C-lim means the Cesàro limit $\frac{1}{N} \sum_{m=1}^{N} \sum_{|q_j| \leq m} \cdots$. This is a higher-dimensional analog of (6.6.54) of Part 1 and the reader will prove it in Problem 8,

Under the hypothesis for a.e. $q \in \Delta$, $\{m \mid \hat{f}(q + m) \neq 0\}$ is finite, so in (4.1.52) the C-lim is a finite sum.

Suppose first that

$$|\{x \in 2\pi\Delta \mid f(x + 2\pi n) = 0 \text{ for all } n\}| > 0 \qquad (4.1.53)$$

Fix $q \in \Delta$. Then, $\sum_{m \in \mathbb{Z}^\nu} \hat{f}(q+m) e^{im \cdot x}$ is a trigonometric polynomial whose zeros have positive measure and so by Proposition 4.1.9 below, $\hat{f}(q+m) = 0$ for each q and m; i.e., $\hat{f} = 0$ so $f = 0$.

More generally, since $\sharp(x + \mathbb{Z}^\nu) \cap A$ is finite for a.e. x, we can find M so that

$$|\{x \in 2\pi\Delta \mid \sharp[(x + 2\pi\mathbb{Z}^\nu) \cap 2\pi A] < (2\pi M)^\nu\}| > 0 \qquad (4.1.54)$$

Let

$$g(x) = f(xM) \qquad (4.1.55)$$

Then, $g \upharpoonright 2\pi\Delta$ is related to $f \upharpoonright 2\pi\Delta M$ which is $M^\nu(2\pi)^\nu$ cubes and orbits $\{x + 2\pi n\}$ for g are $\{Mx + 2\pi Mn\}$ for f. If x lies in the set in (4.1.54), one of the orbits $\{y + 2\pi MN\}$ where $My \in \{x + 2\pi n\}$ must be empty since there are M^ν such orbits and fewer points on all of the orbits. It follows that g obeys (4.1.53), so g and thus f is zero. $\qquad \square$

Proposition 4.1.9. (a) *If $P(x_1, \ldots, x_\nu)$ is a nonzero polynomial in ν variables, $\{x \mid P(x) = 0\}$ has Lebesque measure 0.*

(b) *If $Q(e^{i\theta_1}, \ldots e^{i\theta_\nu})$ is a nonzero polynomial in $e^{\pm i\theta_j}$, then $\{(e^{i\theta_1}, \ldots e^{i\theta_\nu}) \mid Q(e^{i\theta_1}, \ldots, e^{i\theta_\nu}) = 0\}$ has $d\theta_1 \ldots d\theta_\nu/(2\pi)^\nu$ measure zero.*

Remarks. 1. We only need (b) here but since we needed (a) in Section 6.9 of Part 1 and the proofs are the same, we state both now. The result extends easily to real analytic functions.

2. The zeroes of P are a variety and usually have codimension 1. One can prove the Hausdorff dimension of the zero set is at most $\nu - 1$.

Proof. We use a simple induction on ν. For $\nu = 1$, P has finitely many zeros in $\mathbb{C}$ and so in $\mathbb{R}$ (and $e^{ik\theta}Q(e^{i\theta})$ is a polynomial in $z = e^{i\theta}$ for suitable k) and thus the set is of measure zero.

Since P is not identically zero, we write

$$P(x_1, \ldots x_\nu) = \sum_{j=0}^{m} x_\nu^j R_j(x_1, \ldots x_{\nu-1}) \qquad (4.1.56)$$

Some R_{j_0} is not identically zero. If $R_{j_0}(x_1^{(0)} \ldots, x_{\nu-1}^{(0)}) \neq 0$, then $P(x_1^{(0)}, \ldots x_{\nu-1}^{(0)}, x)$ is a nonzero polynomial in x which has finitely many zeros so, by the induction hypothesis, for a.e. $x^{(0)} \in \mathbb{R}^{\nu-1}$, $\{(x^{(0)}, x) \mid P(x^{(0)}, x) = 0\}$ has Lebesque measure zero. Then, by Fubini's theorem, P is nonzero a.e. The proof in part (b) is identical. $\qquad \square$

Our next result is an improvement of this if $|A|, |B| < \infty$.

Theorem 4.1.10 (Amrein–Berthier Theorem). *Let $A, B \subset \mathbb{R}^\nu$ have finite Lebesque measure. Then there is $C > 0$, so that for all $f \in L^2(\mathbb{R}^\nu)$*

$$\|f\|_2 \leq C \left[\|\chi_{A^c} f\|_2 + \|\chi_{B^c} \hat{f}\|_2 \right] \qquad (4.1.57)$$

Remarks. 1. χ_D is the characteristic function of D. Henceforth $\| \cdot \|$ will be $\| \cdot \|_2$.

2. If f vanishes off A and $\hat{f}$ off B, then the right side of (4.1.54) is zero, so this theorem implies the special case of Theorem 4.1.8 when $|A| < \infty$, $|B| < \infty$.

3. We'll use a functional analytic argument. There is also an operator-theoretic argument. In terms of the P, Q of this proof, one shows (4.1.57) is implied by $\|PQ\| < 1$ (Problem 9). What makes $|A| < \infty, |B| < \infty$ special is that PQ is compact in that case (see Problem 10).

Proof. Let $Pf = \chi_A f$, $Qf = (\chi_B \hat{f})^{\vee}$. These are projections. Below we will prove, $\exists C_2$ so that

$$Qf = f \Rightarrow \|f\| \leq C_2 \|(1-P)f\| \qquad (4.1.58)$$

Knowing this, we see that

$$\|f\| \leq \|Qf\| + \|(1-Q)f\| \qquad (4.1.59)$$
$$\leq C_2 \|(1-P)Qf\| + \|(1-Q)f\| \qquad (4.1.60)$$
$$\leq C_2 \|(1-P)f\| + C_2 \|(1-P)(1-Q)f\| + \|(1-Q)f\| \qquad (4.1.61)$$
$$\leq C_2 \|(1-P)f\| + (C_2+1)\|(1-Q)f\| \qquad (4.1.62)$$

which proves (4.1.57) with $C = C_2 + 1$. In the above, (4.1.59) and (4.1.61) come from the triangle inequality using $f = Qf + (1-Q)f$ and $Qf = f - (1-Q)f$. (4.1.62) uses (4.1.58) and $Q(Qf) = Qf$ and (4.1.62) uses $\|1 - P\| = 1$.

If (4.1.58) fails, there exits $\{f_n\}_{n=1}^{\infty}$ with $\|f_n\| = 1, Qf_n = f_n$ and $\|(1-P)f_n\| \leq n^{-1}$. By passing to a subsequence, we can suppose that $f_n \to f_\infty$ weakly, (i.e., for all $g \in L^2, \langle g, f_n \rangle \to \langle g, f \rangle$) for some f_∞ with $\|f_\infty\| \leq 1$. Since $Qf_n = f_n$, we have $Qf_\infty = f_\infty$. Since $\|(1-P)f_n\| \to 0$, $Pf_\infty = f_\infty$.

For $x \in A$,

$$f_n(x) = (2\pi)^{-\nu/2} \langle e^{ix\cdot} \chi_B, \hat{f}_n \rangle \qquad (4.1.63)$$

Thus, since $|B| < \infty$, $e^{ix\cdot}\chi_B \in L^2$ and so

$$f_n(x) \to f_\infty(x) \qquad (4.1.64)$$

by the weak convergence. (4.1.63) also implies, by the Schwarz inequality, that

$$|f_n(x)| \leq (2\pi)^{-\nu/2} |B|^{1/2} \qquad (4.1.65)$$

Since $\chi_A \in L^2$ also, by the dominated convergence theorem $\|f_n \chi_A - f_\infty \chi_A\| \to 0$, so $\|f_n - f_\infty\| \to 0$ since $\|(1 - \chi_A)f_n\| \to 0$.

It follows that $\|f_\infty\| = 1$. But we saw above that $Pf_\infty = f_\infty = Qf_\infty$ which implies, by Theorem 4.1.8, that $f_\infty = 0$. This contradiction proves the theorem. $\square$

We turn to one final topic. There is a quantitative version of the relation on $\|PQ\|$ and how big $\|Pf\|$ and $\|Qf\|$ can be that provides a quantitative expression about simultaneous localization.

Theorem 4.1.11 (Landau–Pollack Uncertainty Principle). *Let P, Q be two orthogonal projections on a Hilbert space. Then for any φ with $\|\varphi\| = 1$, we have that*

$$\arccos(\|P\varphi\|) + \arccos(\|Q\varphi\|) \geq \arccos(\|PQ\|) \tag{4.1.66}$$

Remarks. 1. $\theta = \arccos(x)$ is defined for $x \in [-1, 1]$ by $\cos(\theta) = x$ with $\theta \in [0, \pi]$.

2. $\arccos(1) = 0$, so if $\|PQ\|$ is small, it cannot be that both $\|P\varphi\|$ and $\|Q\varphi\|$ are near 1. Indeed (Problem 11), if $\|PQ\| = 0$, then (4.1.66) is equivalent to

$$\|P\varphi\|^2 + \|Q\varphi\|^2 \leq \|\varphi\|^2 \tag{4.1.67}$$

(which is just Bessel's inequality, since $P\varphi \perp Q\varphi$ in that case).

3. Further analysis (Problem 12) shows that if $\operatorname{Ran} P + \operatorname{Ran} Q$ is not dense, and $P, Q \neq 0$, then for any $\alpha, \beta \in (0, 1)$ with $\arccos(\alpha) + \arccos(\beta) > (\|PQ\|)$, there is φ with $\|P\varphi\| = \alpha, \|Q\varphi\| = \beta$. If PQ is compact, we can replace $> \arccos(\|PQ\|)$ by $\geq \arccos(\|PQ\|)$. In that sense, this inequality is optimal.

4. When P is $\chi_{[-a,a]}(X)$ and Q is $\chi_{[-b,b]}(P)$, one can show that $\|PQ\|$ is an "explicit" function of a, b expressible in terms of prolate spheroidal functions; see the Notes.

5. As we'll see, if we let θ_{ij} be the angle between vectors, v_i, v_j in a real Hilbert space, we'll prove when v_1, v_2, v_3 are all on one side of a hyperplane, then

$$\theta_{23} + \theta_{13} \geq \theta_{12} \tag{4.1.68}$$

For all $\theta_{ij} \in (0, \pi/2)$, the weaker result

$$2\sin\left(\tfrac{1}{2}\theta_{12}\right) \leq 2\sin\left(\tfrac{1}{2}\theta_{13}\right) + 2\sin\left(\tfrac{1}{2}\theta_{23}\right) \tag{4.1.69}$$

is a special case of the triangle inequality for a distance between subspaces (see the Notes and Problem 13). This confirms the idea that (4.1.68), and so (4.1.69), is some kind of triangle inequality.

6. The naive geometry for $\mathbb{R}^2$ is that φ lies between $P\varphi$ and $Q\varphi$ and that $\langle \varphi, (PQ)\varphi \rangle \geq 0$ but as Figure 4.1.1 shows, it can happen that φ does not lie between $P\varphi$ and $Q\varphi$ or that $\langle \varphi, (PQ)\varphi \rangle \leq 0$.

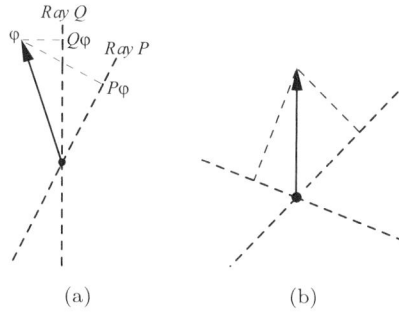

Figure 4.1.1. Examples where (a) φ does not lie between $P\varphi$ and $Q\varphi$ and (b) where $\langle \varphi, PQ\varphi \rangle \leq 0$.

7. The set of $a = \|P\varphi\|$ and $b = \|Q\varphi\|$ that obey (4.1.66) is the intersection of $[0, 1] \times [0, 1]$ and an ellipse (see Problem 23).

Proof. Let v_1, v_2, v_3 be three nonzero vectors in $\mathbb{R}^2$ all lying in a half-space (i.e., so for some $w \neq 0$, $\langle w, v_j \rangle \geq 0$ for all j).

Define θ_{ij} by

$$\cos(\theta_{ij}) = \frac{\langle v_i, v_j \rangle}{\|v_i\| \|v_j\|}, \quad \theta_{ij} \in [0, \pi] \tag{4.1.70}$$

Then one of the v's lies between the others so one of the θ_{ij}'s is the sum of the other two. It follows that either equality holds in (4.1.68) (if v_3 lies between v_1 and v_2) or $\max(\theta_{23}, \theta_{13}) \geq \theta_{12}$, and so (4.1.68) holds. In all cases (4.1.68) holds.

Next let v_1, v_2, v_3 lie in $\mathbb{R}^3$ with

$$\langle v_2, v_3 \rangle \geq 0, \quad \langle v_1, v_3 \rangle \geq 0$$

Let R be the orthogonal projection onto the span of v_1 and v_2 and let $v_4 = Rv_3$. Then

$$\cos(\theta_{14}) = \frac{\langle v_1, Rv_3 \rangle}{\|v_1\| \|Rv_3\|} = \frac{\langle Rv_1, v_3 \rangle}{\|v_1\| \|Rv_3\|} \geq \frac{\langle v_1, v_3 \rangle}{\|v_1\| \|v_3\|} = \cos(\theta_{13}) \tag{4.1.71}$$

so

$$\theta_{13} \geq \theta_{14}, \quad \theta_{23} \geq \theta_{24} \tag{4.1.72}$$

(In (4.1.71), we used $\langle v_1, v_3 \rangle \geq 0$ and $\|Rv_3\| \leq \|v_3\|$). By the two-dimensional case, (and noting that v_1, v_2, v_4 lie in the half-space $\{u \mid Ru = u, \langle u, v_4 \rangle \geq 0\}$, we have that

$$\theta_{12} \leq \theta_{14} + \theta_{24} \leq \theta_{13} + \theta_{24} \tag{4.1.73}$$

proving (4.1.68).

Now let $\mathcal{H}$ be a complex or real Hilbert space. On $\mathcal{H}$ define

$$\langle v, w \rangle_r = \mathrm{Re}\langle v, w \rangle \qquad (4.1.74)$$

Then, (4.1.68) holds, where θ is defined by (4.1.70) with $\langle \, , \, \rangle_r$ replacing $\langle v_i, v_j \rangle$ so long as

$$\mathrm{Re}\langle v_1, v_3 \rangle \geq 0, \quad \mathrm{Re}\langle v_1, v_2 \rangle \geq 0 \qquad (4.1.75)$$

Given f, with $\|f\| = 1$, (4.1.68) is obvious if $Pf = 0$ and $Qf = 0$ (since $\arccos(\theta) = 1$), so suppose $Pf = v_1 \neq 0$, $Qf = v_2 \neq 0$ and let $f = v_3$. Then

$$\|Pf\| = \frac{\|Pf\|^2}{\|f\|\|Pf\|} = \frac{\langle f, Pf \rangle}{\|f\|\|Pf\|} \qquad (4.1.76)$$

so

$$\theta_{13} = \arccos(\|Pf\|), \quad \theta_{23} = \arccos(\|Qf\|)$$

and

$$\cos(\theta_{12}) = \mathrm{Re}\, \frac{\langle Pf, Qf \rangle}{\|Pf\|\|Qf\|} = \mathrm{Re}\, \frac{\langle Pf, (PQ)Qf \rangle}{\|Pf\|\|Qf\|} \qquad (4.1.77)$$

$$\leq \|PQ\|$$

so $\theta_{12} \geq \arccos(\|PQ\|)$ and (4.1.68) implies (4.1.66) $\qquad \square$

Notes and Historical Remarks. The uncertainty principle in the context of harmonic analysis has been popular enough that there is a book by Havin and Joricke [**369**] on the subject and review articles by Benedetto [**52**] and Folland–Sitaran [**277**].

While it is the harmonic analysis context that interests us, I'd be remiss if I didn't mention the historical quantum physics context. In trying to understand the physical basis of his matrix mechanics, Heisenberg [**375**], in 1927, realized it was impossible to accurately measure both the position and momentum of an electron, even in principle. For example, to make an accurate measurement of position with light, the wave length of the light would have to be the order of a desired uncertainty in the position and a photon (the minimal amount of light) of very small wave length has very large energy which will knock the electron and produce an uncertainty in its velocity. The German word Heisenberg used was closer to "indeterminacy" but the English translation of his quantum theory book several years later used "uncertainty," and that name stuck. Uncertainty is closely related to but distinct from quantum weirdness and to measurement theory—the discussion of those issues is way beyond the scope of this book.

Using the notation,

$$\sigma_X = \mathrm{Var}_f(X)^{1/2}, \quad \sigma_P = \hbar\mathrm{Var}_f(P)^{1/2} \qquad (4.1.78)$$

(restoring Planck's constant, h; note $\hbar = h/2\pi$), Heisenberg stated his uncertainty principle as

$$(\Delta x)(\Delta p) \geq h \qquad (4.1.79)$$

without precise definition of what was Δx and Δp. The form (4.1.5) and proof by commutation were found independently in 1927–28 by Kennard [462] and Weyl [851] (who said that he learned the proof from Pauli).

Robertson [694] stated a version of the uncertainty principle for general operators, A, B; if $\|f\| = 1$ and

$$\langle C \rangle = \langle f, Cf \rangle, \quad \sigma_C = [\langle C^2 \rangle - \langle C \rangle^2]^{1/2} \qquad (4.1.80)$$

Robertson says (Problem 14)

$$\sigma_A \sigma_B \geq \tfrac{1}{2}|\langle [A, B] \rangle| \qquad (4.1.81)$$

Schrödinger [719] proved the stronger ($\{A, B\} = AB + BA$)

$$\sigma_A^2 \sigma_B^2 \geq |\tfrac{1}{2}\langle \{A, B\} \rangle - \langle A \rangle \langle B \rangle|^2 + |\tfrac{1}{2}\langle [A, B] \rangle|^2 \qquad (4.1.82)$$

(4.1.81) (or (4.1.82)) is sometimes called the *Schrödinger–Robertson uncertainty relations.*

The aspect of the uncertainty principle that a function, f, cannot be concentrated while at the same time, $\widehat{f}$ is concentrated was rediscovered in the situation where the variables, rather than position and (quantum) momentum, are time and frequency, a critical element of both optics and communication theory. This was presented in seminal papers of Gabor [299] in 1946 and Shannon [734] in 1948. Even earlier, Wiener, in his autobiography [857] says that he lectured on this in Göttingen in 1925. But there was no paper or precise statement of what Wiener said. Moreover, there is no reason to think that Heisenberg, a student in physics in Göttingen at the time, went to the mathematics lectures. In addition, the Fourier transform view is relevant to Schrödinger's version of quantum mechanics, not the wave mechanics view that Heisenberg was still using in 1927.

One improvement of the uncertainty principle that we won't prove, involves the *Shannon entropy, $E(\rho)$,* defined by

$$d\rho(x) = \rho(x)dx \Rightarrow E(\rho) = -\int \rho(x) \log \rho(x)dx \qquad (4.1.83)$$

The *Hirschmann uncertainty principle* says that for $f \in L^2(\mathbb{R}^\nu)$

$$E(|f|^2) + E(|\widehat{f}|^2) \geq \nu \log(\pi e) \qquad (4.1.84)$$

Notice that, if $\rho(x) = c^{-1}\chi_A(x)$ where $|A| = c$, then $E(\rho) = \log c$, so if c is very small, E is very negative and if c is large, E is large. Thus, if ρ is concentrated, E is very negative. Therefore, (4.1.84) says it cannot be that both $|f|^2$ and $|\hat{f}|^2$ are concentrated! *Shannon's inequality* (see Problem 15) says that

$$e^{2E(\rho)} \leq (2\pi e) \det[V(\rho)] \tag{4.1.85}$$

where

$$V_{ij}(\rho) = \int x_i x_j \rho(x)dx - \left(\int x_i \rho(x)dx\right)\left(\int x_j \rho(x)dx\right) \tag{4.1.86}$$

Thus, when $\nu = 1$, (4.1.84) implies that

$$(2\pi e)^2 V(|f|^2) V(|\hat{f}|^2) \geq (e^{\log(\pi e)})^2 \tag{4.1.87}$$

i.e.,

$$V(|f|^2)V(|\hat{f}|^2) \geq \tfrac{1}{4} \tag{4.1.88}$$

Therefore, (4.1.84) is a strengthening of the Heisenberg uncertainty principle, (4.1.5).

(4.1.84) was conjectured by Hirschman [385] who proved the weaker result when $\log(e\pi)$ is replaced by 0, proving this by the method of Problem 16, i.e., differentiating the Hausdorff–Young inequality but with classical constants. The full result was obtained independently by Beckner [50] and Bialynicki-Birula–Mycielski [69]. The proof (see Problem 16) used Beckner's optimal constant in the Hausdorff–Young inequality ([69] relied on Beckner's announcement which had optimal Hausdorff-Young but not its application to prove Hirschmann's conjecture).

Hardy's inequality, Theorem 4.1.3, while related to, should not be confused with the Hardy–Littlewood–Sobolev inequalities which we discuss in Section 6.2. Rather, it is named after Hardy's paper of 1920 [354] where he proved (but not with the optimal constants below) for $1 < p < \infty$,

$$\sum_{n=1}^{\infty} \left|\frac{a_1 + \ldots + a_n}{n}\right|^p \leq \left(\frac{p}{p-1}\right)^p \sum_{n=1}^{\infty} |a_n|^p \tag{4.1.89}$$

$$\int_0^\infty \left|\frac{1}{x}\int_0^x f(t)dt\right|^p dx \leq \left(\frac{p}{p-1}\right)^p \int_0^\infty |f(x)|^p dx \tag{4.1.90}$$

This may not look like (4.1.22), but if $p = 2$ and $g(x) = \int_0^x f(t)dt$, then (4.1.90) says that

$$\frac{1}{4}\int_0^\infty \left|\frac{g(x)}{x}\right|^2 dx \leq \int |g'(x)|^2 dx \tag{4.1.91}$$

which is the one-dimensional case of (4.1.22). An extended form of (4.1.90) will play a role in our discussion of real interpolation in Section 6.1. The optimal constants in (4.1.88)/(4.1.90) are due to Landau and Hardy; see the Notes to Section 6.1.

There have been many extensions of Hardy-type inequalities; for example if Ω is an open convex set in $\mathbb{R}^\nu$, $d(x) = \text{dist}(x, \mathbb{R}^\nu \backslash \Omega)$, and $f \in C_0^\infty(\Omega)$, then

$$\int |f(x)|^2 d(x)^{-2} d^\nu x \le 4 \int |(\nabla f)(x)|^2 d^\nu x \qquad (4.1.92)$$

as discussed in the review article of Davies [**197**]. For review of this type of inequality and thus inequalities motivated by Hardy's inequality see the review articles [**197, 120**] and books [**494, 619**].

The idea that quantum states take up $(2\pi)^\nu$ volume in phase space (in units with $\hbar = 1$, so 2π is really $2\pi\hbar = h$) will recur in Section 7.5 of Part 4. We'll see it is asymptotically correct in a regime where the number of states is large.

There is a refinement of this notion that doesn't allow sets that are too thin. For example, in two dimensions, $-\Delta + x_1^2 x_2^2$ has discrete spectrum (see Example 3.16.6 of Part 4) even though $|\{(p_1, p_2, x_1, x_2) \mid p_1^2 + p_2^2 + x_1^2 x_2^2 \le C\}|$ is infinite for all C—the tails of $x_1 x_2$ all are thin. This extra idea with application to PDEs is due to Fefferman–Phong [**269**] (see also Fefferman [**268**]).

What we have called the Heisenberg group is usually called the *Weyl group* by physicists. It is ironic that physicists name the group after a mathematician while mathematicians name it after a physicist! The Lie algebra of the Heisenberg group has $2\nu + 1$ generators with $[p_j', x_k'] = -i\delta_{jk}\mathbb{1}$, $[p_j, p_k] = [x_j, x_k] = 0$, $[p_j, \mathbb{1}] = [x_j, \mathbb{1}] = 0$, exactly (in $\hbar = 1$ units) the Heisenberg commutation relation if $\mathbb{1}$ is the identity; hence the name we used. But it was Weyl in [**851**] who realized that there was an underlying Lie group with the group relations (4.1.8)—so the name Weyl group is more appropriate. It is not used by mathematicians because there is a totally different "Weyl group," a finite group of reflections that arose in Weyl's theory of representations of the compact Lie groups (see, for example, Simon [**746**])

Some authors exploit the Heisenberg group as a fundamental tool in harmonic analysis—see Howe [**402**] for a review and reference. One of the fundamental theorems in the theory of the Heisenberg group is the Stone–von Neumann uniqueness theorem, named after Stone [**787**] and von Neumann [**836**], that there is a unique irreducible representation of the Heisenberg group with

$$U(\vec{0}, \vec{0}, \omega_0) = \omega_0 \mathbb{1} \qquad (4.1.93)$$

namely the standard one (4.1.7) (see Simon [**746**] for more on group representations). The reader will prove this in Problem 17. We note that some authors use $\mathbb{R}^\nu \times \mathbb{R}^\nu \times \mathbb{R}$ rather than $\mathbb{R}^\nu \times \mathbb{R}^\nu \times \partial\mathbb{D}$ for the Heisenberg group; if $\alpha \in \mathbb{R}$ replaces $\omega \in \partial\mathbb{D}$, the multiplier $\omega_0 \omega e^{-ip_1 \cdot x_0}$ is replaced by

$\alpha_0 + \alpha_1 - p_1 \cdot x_0$. $\alpha \mapsto \omega = e^{i\alpha}$ is a natural map of this group many-to-one onto the smaller one.

Theorem 4.1.4 is due to Hardy [**356**]. The usual proof applies Proposition 4.1.6 to $e^{\frac{1}{2}k^2} \widehat{f}(k)$ (see Problem 18). The idea of getting a distribution version by using the Segal-Bargmann transform is due to Bonami–Demange [**94**]. These authors also obtain results where multiplication by $e^{\frac{1}{2}|x|^2}$ is replaced by exponentials of nonpositive definite forms. Morgan [**585**] has results where $e^{-a|x|^2}$, $e^{-b|k|^2}$ are replaced by $e^{-a|x|^p}$, $e^{-b|x|^q}$ for dual indices (see [**94**]).

Theorem 4.1.5 was first proved by Benedicks in a 1976 preprint which has only been published in 1985 [**53**]. His proof uses the Poisson representation formula much as we did in the text. In 1977, Amrein–Berthier [**11**] proved (4.1.57) using very different methods. In the language of our proof, they noted PQ was compact (in the language of Section 3.8 of Part 4 it is a Hilbert-Schmidt operator) and used that to see that $\|PQ\| < 1$ (see Problems 9 and 19). The extension from $|A| < \infty$, $|B| < \infty$ to what we call Benedicks sets is due to Bonami–Demange [**94**]. Nazarov [**599**] proved there is a universal constant D_ν, only depending on dimension so the constant C (4.1.57) can be taken as $\exp(D_\nu |A||B|)$. There is a conjecture (see Izu–Lakey [**422**]) that for fixed $|A||B|$, C is minimized by balls.

(4.1.68) is an expression of a triangle inequality for a distance between one-dimensional subspaces in a real Hilbert space. It is related to, but distinct from, a general notion of distance between general subspaces of a Banach space, X. Given subspaces $M, N \subset X$, one defines

$$\widehat{d}(M, N) = \sup_{u \in M, \|u\|=1} \left(\inf_{v \in N, \|v\|=1} \|u - v\| \right) \qquad (4.1.94)$$

$$d(M, N) = \max(\widehat{d}(M, N), \widehat{d}(N, M)) \qquad (4.1.95)$$

This, as a distance between M and N, goes back to Hausdorff (and is the general Hausdorff distance between the unit spheres in M and N). It is discussed in Section IV.2.1 of Kato [**448**] and obeys the triangle inequality. In the case of one-dimensional M and N, if $\theta \in [0, \frac{\pi}{2}]$ is the angle between M and N, $d(M, N) = 2\sin\left(\frac{\theta}{2}\right)$ (see Problem 20) and (4.1.69) is the triangle inequality. This is strictly weaker than (4.1.68).

In [**504**], Landau–Pollak asked and answered the question about what joint values are possible for $(\|Pf\|, \|Qf\|)$ if P is the projection on f's in $L^2(\mathbb{R}, dx)$ with support in $[-a, a]$ and Q for f's with support of $\widehat{f}$ in $[-b, b]$. Their method works for any pair of projections (i.e., (4.1.66) holds). Their proof is algebraic (see also the write-up in Dym–McKean [**237**]) rather than

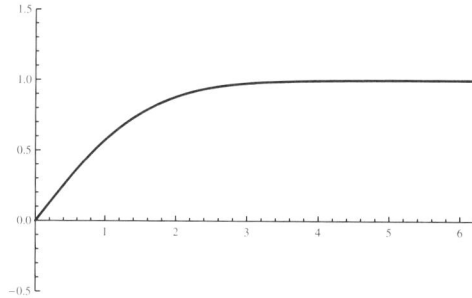

Figure 4.1.2. $\|PQ\|$ as a function of ab.

the simple geometric proof we present, fleshing out an argument in Folland–Sitaram [**277**]. [**504**] is one of five related papers on the subject by Landau, Pollak, Slepian and Widom [**761, 505, 760, 506**].

Landau-Pollak [**504**] also found $\|PQ\|$ for the above x-space interval of $[-a, a]$ and k-space interval of $[-b, b]$. They note that PQP is an integral operator on $L^2([-a, a], dx)$ with integral kernel

$$K(x, y) = \frac{\sin(b(x - y))}{\pi(x - y)} \tag{4.1.96}$$

They are able to obtain the eigenvalues and eigenfunctions (and so norm) of PQP in the form of solutions of the differential equation in t ($' = d/dt$)

$$((a^2 - t^2)f')' - b^2 t^2 f = \lambda f \tag{4.1.97}$$

Normalized to $a = 1$, these are prolate spheroidal functions. $\|PQ\|$ is only a function of ab and is graphed in Figure 4.1.2.

There has been further research on the eigenvalues of the above PQP. By scaling, these are only a function $c = 4ab$, so let $T_c = P_{a=1} Q_c P_{a=1}$. Then $\operatorname{Tr}(T_c) = \frac{c}{2\pi}$ by a simple calculation (Problem 21). This is consistent with the notion of a state per 2π volume in phase space, since $(-a, a) \times (-b, b)$ has volume c.

Landau-Widom [**506**] proves that if $N(T_c \geq \alpha)$ is the number of eigenvalues of T_c in $(\alpha, 1]$, then as $c \to \infty$ (the semiclassical limit)

$$N(T_c \geq \alpha) = \frac{c}{2\pi} + \frac{1}{\pi^2} \log\left(\frac{1 - \alpha}{\alpha}\right) \log(c) + o(\log c) \tag{4.1.98}$$

This result of Landau–Widom implies T_c has $\frac{c}{2\pi} + O(\log c)$ eigenvalues very near 1 and, for α near 0, only $O(\log c)$ eigenvalues in $(1 - \alpha, \alpha)$, see Figure 4.1.3 and see Problem 22. Landau [**503**] also found that

$$\left| N(T_c \geq \frac{1}{2}) - \frac{c}{2\pi} \right| \leq 1 \tag{4.1.99}$$

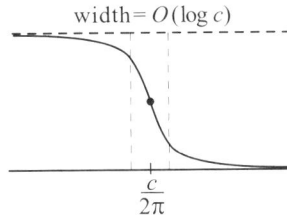

Figure 4.1.3. $N(T_c \geq 0)$ has a narrow transition region.

There has been application of this class of ideas to data recovery in a celebrated paper of Candès, Romberg, and Tao [**137**]. For a review of these ideas connected to PQ as above, see Izu–Lakey [**422**].

Finally, we note that there has been work on the uncertainty principle for finite groups. Donoho–Stark [**223**] proved that for $\mathbb{Z}_n$ and $x \neq 0$

$$(\#\{j \mid x_j \neq 0\})(\#\{j \mid (\mathcal{F}_n x)_j \neq 0\}) \geq n \tag{4.1.100}$$

where, as in the Notes to Section 3.5 of Part 1, $\mathcal{F}_n$ is the DFT:

$$(\mathcal{F}_n x)_j = \frac{1}{n} \sum_{l=0}^{n-1} x_l e^{2\pi i j l/n} \tag{4.1.101}$$

Since $\min_u (u + cu^{-1}) = 2c^{1/2}$, (4.1.100) implies

$$\#\{j \mid x_j \neq 0\} + \#\{j \mid (\mathcal{F}_n x)_j \neq 0\} \geq 2\sqrt{n} \tag{4.1.102}$$

They also proved that one has equality in (4.1.100) if and only if x is a multiple of the characteristic function of a coset of some subgroup of $\mathbb{Z}_n$. This suggests the result might depend on number-theoretic properties of n. Indeed, Tao [**798**] has shown that when $n = p$ is prime,

$$\#\{j \mid x_j \neq 0\} + \#\{j \mid (\mathcal{F}_p x)_j \neq 0\} \geq p + 1 \tag{4.1.103}$$

a substantial improvement on (4.1.102) in that case.

Problems

1. Compute the group law on $\mathbb{R}^\nu \times \mathbb{R}^\nu \times [0, \infty) \times \partial\mathbb{D}$ associated to the operator, $V(x_0, p_0, a, \omega)$ of (4.1.9).

2. Given f with f, xf, $k\widehat{f} \in L^2(\mathbb{R})$, prove there exists $f_n \in \mathcal{S}(\mathbb{R})$, so $\|f - f_n\|_2 + \|x(f - f_n)\|_2 + \|k(\widehat{f} - \widehat{f}_n)\|_2 \to 0$ and conclude that (4.1.5) for $f \in \mathcal{S}(\mathbb{R})$ implies it for all $f \in L^2(\mathbb{R})$.

3. (a) Prove that the eigenvalues of $-\Delta + x^2$ on $\mathbb{R}^\nu$ are $2m + \nu$, $m = 0, 1, 2, \ldots$.

   (b) Find the multiplicity of $2m + \nu$ as a function of m and ν.

   (c) State and prove an analog of Theorem 4.1.2 for $X \subset \mathcal{S}(\mathbb{R}^\nu)$.

4. Prove (4.1.23) from (4.1.22) for $\nu = 3$.

5. (a) If h$\in L^2(\mathbb{R}^\nu)$ with $\vec{\nabla} h \in L^2(\mathbb{R}^\nu)$ (distributional derivative) and if $h(x) = 0$ for $|x|$ small, prove (4.1.22) for $f \in \mathcal{S}(\mathbb{R}^\nu)$ vanishing near zero implies it for h.

   (b) Let $\nu \geq 3$. Let $f \in L^2(\mathbb{R}^\nu)$, $\vec{\nabla} f \in L^2(\mathbb{R}^\nu)$. Let g be a C^∞-function of $\mathbb{R}^\nu$ which is 1 for $|x| > 1$, and 0 if $|x| < 1/2$. Let $f_n(x) = g(nx)f(x)$. Prove that $\|f_n - f\|_2 + \|x^{-1/2}(f_n - f)\|_2 + \|\nabla f_n - \nabla f\|_2 \to 0$.

   (c) For $\nu \geq 3$, prove (4.1.22) for all f with $f, \nabla f \in L^2(\mathbb{R}^\nu)$.

   (d) Prove (4.1.22) when $\nu = 1$ for all f with $f, f' \in L^2(\mathbb{R})$ and $f(0) = 0$.

6. (4.1.22) provides no information if $\nu = 2$. Here, we'll find a replacement if $\nu = 2$ that has additional logarithms for functions supported inside a sphere.

   (a) Let $u \in C_0^\infty(\mathbb{R}^2)$ and f a smooth function on $\mathbb{R}$ with $f(-r) = f(r)$. By looking at

   $$\int \sum_{j=1}^{2} \left| \frac{\partial u}{\partial x_j} + x_j f(r) u \right|^2 d^2x \qquad (4.1.104)$$

   prove that if u is supported on $\mathbb{R}^2 \setminus \{0\}$, then

   $$\int |\nabla u|^2 dx^2 \geq \int |u(x)|^2 (2f(r) + rf'(r) - r^2 f(r)^2) d^2x \qquad (4.1.105)$$

   (b) Pick $f(r) = g(r)/r^2$ to find

   $$\int |\nabla u|^2 dx^2 \geq \int |u(x)|^2 (r^{-1} g'(r) - g(r)^2 r^{-2}) d^2x \qquad (4.1.106)$$

   (c) Picking $g(r) = -\alpha \left[\log(r/R_0) \right]^{-1}$ and optimizing α, prove for all $u \in C_0^\infty(\{x \mid |x| > R_0\})$, we have

   $$\int |\nabla u|^2 d^2x \geq \frac{1}{4} \int |u(x)|^2 x^{-2} \left[\log(x/R_0)^2 \right]^{-1} d^2x \qquad (4.1.107)$$

   Remark. It is known (see [**742**]) that if f is continuous, $f \geq 0$ and $f \not\equiv 0$, then inf spec$(-\Delta - f) < 0$, so one can't hope to get an estimate without a restriction like supp$(u) \subset \{x \mid |x| > R_0\}$. For more on eigenvalues when $\nu = 2$, see Laptev–Solomyak [**509**] and references there.

7. The proof we gave for Theorem 4.1.7 works when $\nu = 1$. Provide the induction argument to make it work for general ν.

8. Prove (4.1.52) by finding an explicit multi-dimensional Fejér kernel as a product of one-dimensional kernels.

9. (a) If P, Q are projections and $Qf = f$, show that $\|(1 - P)f\| \geq (1 - \|PQ\|)\|f\|$ and show that $\|PQ\| < 1 \Rightarrow$ (4.1.58).

(b) If $\|PQ\| = 1$, show for any $\varepsilon > 0$, there is f with $\|f\| = 1$ and $Qf = f$ and $\|(1 - P)f\| < \varepsilon$. Conclude that $(4.1.58) \Rightarrow \|PQ\| < 1$.

10. This problem assumes the reader knows that if A is a compact operator on a Hilbert space, $\mathcal{H}$, then there exists $\varphi \in \mathcal{H}$, $\varphi \neq 0$, with $\|A\varphi\| = \|A\|\|\varphi\|$ (see Section 3.5, especially (3.5.8) of Part 4).

   (a) If P and Q are orthogonal projections and $\|PQ\varphi\| = \|\varphi\|$, prove that $P\varphi = Q\varphi = \varphi$.

   (b) If PQ is compact and $\|PQ\| = 1$, prove there is $\varphi \in \operatorname{Ran} P \cap \operatorname{Ran} Q$, $\|\varphi\| = 1$.

   (c) If PQ is compact and $\operatorname{Ran} P \cap \operatorname{Ran} Q = \{0\}$, prove that $\|PQ\| < 1$.

11. Let P, Q be orthogonal projections.

   (a) If $PQ = 0$ prove that (4.1.67) holds.

   (b) Let $\theta_1, \theta_2 \in [0, \pi/2]$. Prove that $\cos^2(\theta_1) + \cos^2(\theta_2) \leq 1 \Leftrightarrow \theta_1 + \theta_2 \geq \frac{\pi}{2}$.

   (c) When $PQ = 0$, show that (4.1.61) is equivalent to (4.1.67).

12. (a) In $\mathbb{R}^2$, let P and Q be the projections onto two one-dimensional subspaces of angle θ (in $[0, \pi/2]$) between the subspaces. Prove that $\|PQ\| = \cos(\theta)$.

   (b) In the context of (a), for any $\alpha \in [0, \theta]$, prove that there is a unit vector f in $\mathbb{R}^2$ so that $\|Pf\| = \cos\alpha$, $\|Qf\| = \cos(\theta - \alpha)$ (so one has equality in (4.1.66)).

   (c) Prove that in the general case, if there exists $g \perp \operatorname{Ran} P + \operatorname{Ran} Q$, $g \neq 0$, then for any α, β in $[0, \pi/2]$ with $\alpha + \beta > \arccos(\|PQ\|)$, there is a φ with $\|P\varphi\| = \cos(\alpha)$ and $\|Q\varphi\| = \cos(\beta)$. If there is an η with $\|PQ\eta\| = \|PQ\|$ (and in particular if PQ is compact), prove the same is true if $\alpha + \beta = \arccos(\|PQ\|)$.

13. Let P, Q, R be projections onto three one-dimensional subspaces.

   (a) If d is given by (4.1.94)/(4.1.95) and $\theta = \arccos(\|PQ\|)$, prove that

   $$d(P, Q) = 2\sin(\tfrac{1}{2}\theta) \tag{4.1.108}$$

   (b) If $\theta, \varphi, \theta + \varphi \in (0, \pi/2]$, prove that

   $$\sin(\theta + \varphi) > \sin(\theta) + \sin(\varphi) \tag{4.1.109}$$

   and conclude that (4.1.68) is a strict improvement on (4.1.69) .

14. Prove (4.1.81).

15. This problem will prove Shannon's inequality, (4.1.85).

   (a) By using affine changes of variable, show it suffices to prove the result in the case where $V_{ij}(\rho) = \delta_{ij}$, $\int x_i \rho(x) dx = 0$ for $i = 1, \dots, \nu$.

(b) In that case, let $\phi(k) = (2\pi)^{\nu/2} e^{|k|^2/2} \rho(k)$, $d\gamma(x) = (2\pi)^{-\nu/2}$ $e^{-|x|^2/2} dx$. Prove that

$$\int \phi(x) \log[\phi(x)] d\gamma \geq 0 \qquad (4.1.110)$$

(*Hint*: Show that $g(y) = y \log y$ is convex and use Jensen's inequality.)

(c) Deduce that $\frac{\nu}{2} \log(2\pi e) - E(\rho) \geq 0$ which is (4.1.85).

Remark. This is an unpublished proof of Beckner which appears in Folland-Sitaram [**277**].

16. (a) By differentiating the Hausdorff–Young inequality at $p = 2$, prove that $E(|f|^2) + E(|\hat{f}|^2) \geq 0$.

(b) Using the optimal constant of the Hausdorff–Young inequality, (6.6.87) of Part 1, prove (4.1.84).

17. This problem will lead the reader through a proof of the Stone–von Neumann uniqueness theorem (essentially a variant of von Neumann's original proof). It will be convenient to use a slightly different parameterization than (4.1.7), namely

$$\tilde{U}(x, p, \omega) = e^{ix \cdot \frac{p}{2}} U(x, p, \omega) \qquad (4.1.111)$$

so that (4.1.8) becomes

$$\tilde{U}(x_0, p_0, \omega_0) \tilde{U}(x_1, p_1, \omega_1) = \tilde{U}(x_0 + x_1, p_0 + p_1, \omega e^{-\frac{i}{2}(p_1 \cdot x_0 - p_0 \cdot x_1)}) \quad (4.1.112)$$

In terms of the generators P and X of x-space translations and multiplication, (i.e., $P = -id/dx$, $X =$ multiplication by x)

$$U(x, p, \omega) = \omega \exp(ipX) \exp(ixP) \qquad (4.1.113)$$

while

$$\tilde{U}(x, p, \omega) = \omega \exp[i(pX + xP)] \qquad (4.1.114)$$

We'll also use $\tilde{U}(x, p) = \tilde{U}(x, p, \omega \equiv 1)$. We'll suppose some familiarity with the language of group representations, see Section 6.8 of Part 4.

(a) Given a realization of (4.1.112) with

$$U(0, 0, \omega) = \omega \mathbb{1} \qquad (4.1.115)$$

define

$$\Phi \equiv (2\pi)^{\nu/4} \int e^{-(x^2 + p^2)/2} \tilde{U}(x, p) dx dp \qquad (4.1.116)$$

Prove that

$$\Phi \tilde{U}(x, p) \Phi = e^{-(x^2 + p^2)/4} \Phi \qquad (4.1.117)$$

based only on (4.1.112).

(b) Let π be a continuous representation of the Heisenberg group with $\pi(\tilde{U}(0,0,\omega)) = \omega\mathbb{1}$ and let $\pi(x,p) = \pi(\tilde{U}(x,p))$ so

$$\pi(x_0,p_0)\pi(x_1,p_1) = e^{\frac{i}{2}(p_0\cdot x_1 - p_1\cdot x_0)}\pi(x_0 + x_1, p_0 + p_1) \qquad (4.1.118)$$

$$\pi(\Phi)\pi(x,p)\pi(\Phi) = e^{-(x^2+p^2)/4}\pi(\Phi) \qquad (4.1.119)$$

Prove that

$$\pi(a,b)\pi(\Phi)\pi(a,b)^{-1} = (2\pi)^{\nu/4}\int e^{-(x^2+p^2)/2}e^{i(bx-ap)}\pi(x,p)\,dx\,dp \quad (4.1.120)$$

and use this to prove $\pi(\Phi)$ is not the zero operator. (*Hint:* Fix $\varphi \in \mathcal{H}$, note that $\langle \varphi, \pi(x,p)\varphi\rangle e^{-(x^2+p^2)/2}$ is in $L^2(\mathbb{R}^{2\nu}) \cap C(\mathbb{R}^{2\nu})$, and not a.e. 0 and use the fact that the Fourier transform is injective.)

(c) Prove that $\pi(\Phi)$ is a nonzero projection (if $\mathcal{H} \neq \{0\}$).

(d) For any other representation, π_1, and any unit vectors $\eta \in \operatorname{Ran} \pi(\Phi)$, $\eta_1 \in \operatorname{Ran} \pi_1(\Phi)$, prove that

$$\langle \eta_1, \pi_1(x,p)\eta_1\rangle = \langle \eta, \pi(x,p)\eta\rangle \qquad (4.1.121)$$

(e) Given π and $\eta \in \operatorname{Ran} \pi(\Phi)$, let $\mathcal{H}_\eta^{(\pi)}$ be the closure of the span of $\{\pi(x,p)\eta \mid x,p, \in \mathbb{R}\}$. Prove that each $\pi(x,p)$ leaves $\mathcal{H}_\eta^{(\pi)}$ invariant and that $\pi \upharpoonright \mathcal{H}_\eta^{(\pi)}$ is unitarily equivalent to $\pi_1 \upharpoonright \mathcal{H}_{\eta_1}^{(\pi_1)}$.

(f) If, $\eta, \tilde{\eta} \in \operatorname{Ran} \pi(\Phi)$ and $\langle \eta, \tilde{\eta}\rangle = 0$, prove that $\mathcal{H}_\eta^{(\pi)} \perp \mathcal{H}_{\tilde{\eta}}^{(\pi)}$.

(g) Let $\{\eta_j\}_{j=1}^J$ be an orthonormal basis for $\operatorname{Ker} \pi(\Phi)$. Prove that the orthonormal complement of $\bigoplus_{j=1}^J \mathcal{H}_{\eta_j}^{(\pi)}$ is $\{0\}$. (*Hint:* If π_1 is the restriction of π to this orthonormal complement, prove that $\pi_1(\Phi) = 0$).

(h) Prove that the representation (4.1.7) is irreducible and conclude that any representation is unitarily equivalent to a direct sum of copies of (4.1.7)

18. (a) If f obeys (4.1.24) with $a = b = 1$, prove that $g(z) = e^{\frac{1}{2}z^2}\widehat{f}(z)$ can be analytically continued from $z \in \mathbb{R}$ to all of $\mathbb{C}$ and by Proposition 4.1.6 must be identically 1.

19. (a) Let P, Q be orthogonal projections. Suppose $\|PQ\varphi\| = \|\varphi\|$. Prove that $\varphi \in \operatorname{Ran} P \cap \operatorname{Ran} Q$.

(b) If PQ is compact, prove that

$$\|PQ\| < 1 \Leftrightarrow \operatorname{Ran} P \cap \operatorname{Ran} Q = \{0\}.$$

20. Let M, N be one-dimensional spaces with θ defined by $\cos\theta = |\langle u, v\rangle|/\|u\|\|v\|$ for any nonzero $u \in M$, $v \in N$. Let d be given by (4.1.95). Prove that $d(M, N) = 2\sin\left(\frac{\theta}{2}\right)$.

21. This problem and the next assume that the reader is familiar with trace class and Hilbert-Schmidt operators (see Chapter 3 of Part 4).
    (a) Let $f, g, h \in L^\infty(R^\nu)$ with $g \in L^2(\mathbb{R})$ and $f, h \in L^4(\mathbb{R}^\nu)$. Let $A = f(X)g(P)h(X)$. Prove that

    $$\mathrm{Tr}(A) = (2\pi)^{-\nu}\left(\int g(k)d^\nu k\right)\left(\int f(x)h(x)d^\nu x\right).$$

    (*Hint:* See (3.8.31) of Part 4.)

    (b) With T_c as given in the paragraph after (4.1.97), prove that $\mathrm{Tr}(T_c) = c/2\pi$.

22. (a) For any projections P and Q, prove that
    $$PQP - (PQP)^2 = PQ(1 - P)QP \qquad (4.1.122)$$

    (b) On $L^2(\mathbb{R})$, let P_a be the projection onto all functions of x supported in $(-a, a)$ and Q_b onto all functions with $\hat{f}$ supported in $(-b, b)$. Prove that as $a \to \infty, b = 1$, the Hilbert-Schmidt norm of $PQ(1 - P)$ obeys $\|P_aQ_{b=1}(1 - P)\|^2 \leq C\log(a)$.

    (c) With T_c as given in Problem 21, prove that as $c \to \infty$,
    $$\mathrm{Tr}(T_c(1 - T_c)) \leq c\log(c) \qquad (4.1.123)$$

    (d) Prove that if $N(\beta \geq T_c \geq \alpha)$ is the number of eigenvalues of T_c in (α, β), then for any fixed $\alpha \in (0, 1/2)$,
    $$N(1 - \alpha \geq T_c \geq \alpha) = O(\log(c))$$

23. The boundary of the set of $\alpha = \|P\varphi\|$, $\beta = \|Q\varphi\|$ with $\arccos(\alpha) + \arccos(\beta) \geq \arccos(\|PQ\|)$ is clearly given by $\alpha = 0, 0 \leq \beta \leq 1 - \|PQ\|$, $\beta = 0, 0 \leq \alpha \leq 1 - \|PQ\|$, and the curve, $0 \leq \alpha, \beta$ and
    $$\cos(\arccos(\alpha) + \arccos(\beta)) = \|PQ\| \qquad (4.1.124)$$

    (a) Prove (4.1.124) is equivalent to
    $$\alpha\beta - \sqrt{1 - \alpha^2}\sqrt{1 - \beta^2} = \|PQ\| \qquad (4.1.125)$$

    (b) Prove that (4.1.125) is equivalent to
    $$\alpha^2 + \beta^2 - 2\alpha\beta\|PQ\| = 1 - \|PQ\|^2 \qquad (4.1.126)$$

    (c) Prove that (4.1.126) is an ellipse, so the set of all (α, β) is the intersection of an ellipse with $[0, 1] \times [0, 1]$.

    Remark. This is an observation of Lenard [**520**].

4.2. The Wavefront Sets and Products of Distributions

In this section, we'll refine singularities of distributions and use this refined analysis to discuss products of two distributions. As we'll see shortly, two tempered distributions can have a product which is not tempered so we'll need to discuss ordinary distributions. In Chapter 9 of Part 1, we discussed ordinary distributions and, in particular, defined a topology on $C_0^\infty(\mathbb{R}^\nu)$ in which (ordinary) distributions are exactly the continuous linear functions.

For this section, it will suffice to have the definition of ordinary distributions. An *ordinary distribution* on $\mathbb{R}^\nu$, τ, is a linear map, $\tau : C_0^\infty(\mathbb{R}^\nu) \to \mathbb{C}$ so that for each $R > 0$, there is $N_R \in \mathbb{N}$ and $C_R \in (0, \infty)$ so that

$$|\tau(f)| \leq C_R \sum_{|\alpha| \leq N_R} \|D^\alpha f\|_\infty \qquad (4.2.1)$$

for all f with

$$\text{supp}(f) \subset \{x \mid |x| \leq R\} \qquad (4.2.2)$$

One reason that we mainly discussed tempered distributions in Chapter 6 of Part 1 is that ordinary distributions don't have natural Fourier transforms—they'll enter in this section because ordinary distributions have localizations that are tempered.

Definition. Let τ be an ordinary distribution on $\mathbb{R}^\nu$ and $g \in C_0^\infty(\mathbb{R})$. For $f \in \mathcal{S}$, define

$$(g\tau)(f) = \tau(gf) \qquad (4.2.3)$$

which is a tempered distribution. A *localization* of τ at $x_0 \in \mathbb{R}^\nu$ is a tempered distribution of the form $g\tau$ where $g \in C_0^\infty(\mathbb{R}^\nu)$ with $g \equiv 1$ in a neighborhood of x_0.

Definition. Two ordinary distributions, τ and σ, are said to be *equal near* $x_0 \in \mathbb{R}^\nu$, if there is $g \in C_0^\infty(\mathbb{R}^\nu)$, $g \equiv 1$ in a neighborhood of x_0, so that $g\tau = g\sigma$.

It is easy to see that this is an equivalence relation.

Definition. Let τ be an ordinary distribution. We say $x_0 \in \mathbb{R}^\nu$ is a *regular point* for τ, if and only if there is a localization of τ at x_0 which lies in C_0^∞, i.e., if and only if there is $F \in C_0^\infty(\mathbb{R}^\nu)$ and $\varepsilon > 0$ so $f \in C_0^\infty(\mathbb{R}^\nu)$ with $\text{supp}(f) \subset \{x \mid |x - x_0| < \varepsilon\} \Rightarrow \tau(f) = \int F(x)f(x)\,d^\nu x$. If x_0 is not a regular point for τ, we call it a *singular point*. The set of singular points for τ is called the *singular support* of τ, written $s\text{-supp}(\tau)$.

It is obvious that the set of regular points is open so $s\text{-supp}(\tau)$ is closed. It is a subset of $\text{supp}(\tau)$ (see Problem 6.2.14 of Section 6.2 of Part 1).

We want to consider how one might define the product of two distributions, τ and σ. A little thought suggests that $\delta(x)^2$ shouldn't be meaningful (since for functions f, $f\delta = [f(0)]\,\delta$). Something like $\mathcal{P}(\frac{1}{x})$ which depends on cancellations shouldn't have a natural square either (there are regularizations of $1/x^2$ but they aren't unique). The following shows under some circumstances that there is a natural meaning to $\tau\,\sigma$:

Theorem 4.2.1. *Let* τ, σ *be two ordinary distributions on* $\mathbb{R}^\nu$ *so that*

$$s\text{-}\mathrm{supp}(\tau) \cap s\text{-}\mathrm{supp}(\sigma) = \emptyset \tag{4.2.4}$$

Then, there exists a unique distribution, ω, *so that*

$$x_0 \notin s\text{-sing}(\tau),\ g \in C_0^\infty(\mathbb{R}^\nu)\ \text{so that}\ \tau = g\ \text{near}\ x_0 \Rightarrow \omega = g\sigma\ \text{near}\ x_0$$
$$x_0 \notin s\text{-sing}(\sigma),\ g \in C_0^\infty(\mathbb{R}^\nu)\ \text{so that}\ \sigma = g\ \text{near}\ x_0 \Rightarrow \omega = g\tau\ \text{near}\ x_0$$

Proof. Uniqueness is an immediate consequence of the fact (Problem 1) that if τ, σ are two distributions equal near each point in $\mathbb{R}^\nu$, they are equal. Existence is an easy application of partitions of unity. Noting that this is a special case of Theorem 4.2.6 below, we'll leave the details to the reader (Problem 2). $\qquad\square$

Example 4.2.2. Let $\tau_\pm = \exp(\pm ie^x)$ as a C^∞-function on $\mathbb{R}$. Since $\tau_\pm \in L^\infty$, they are tempered distributions. Clearly $s\text{-sing}(\tau_\pm) = \emptyset$. Let $\sigma = \tau'_+ = ie^x\tau_+$. It is a tempered distribution as the derivative of one and clearly $s\text{-sing}(\sigma) = \emptyset$. By the theorem, $\sigma\tau_-$ is a well-defined product. Indeed, as an ordinary distribution

$$\sigma\tau_- = ie^x \tag{4.2.5}$$

which is *not* tempered. This example shows why we need to consider ordinary, not only tempered, distributions in this section. $\qquad\square$

Example 4.2.3. Let τ be a distribution on $\mathbb{R}$ of (6.2.69) of Part 1.

$$\tau \equiv \mathcal{P}\left(\frac{1}{x}\right) - i\pi\delta(x) = \lim_{\varepsilon\downarrow 0} \frac{1}{x + i\varepsilon} \tag{4.2.6}$$

One can compute the Fourier transform of τ explicitly:

$$\widehat{\tau}(k) = -(2\pi)^{-1/2}(2\pi i)\theta(k) \tag{4.2.7}$$

where θ is the characteristic function of $(0, \infty)$. To see this, let

$$\tau_\varepsilon(x) = \frac{1}{x + i\varepsilon} \tag{4.2.8}$$

and compute that

$$\widehat{\tau}_\varepsilon(k) = -(2\pi)^{-1/2}(2\pi i)\,e^{-\varepsilon k}\,\theta(k) \tag{4.2.9}$$

and use (4.2.6). The latter can be seen either by a direct calculation of $\widehat{\tau}_\varepsilon$ using contour integration (Problem 3) or by computing the inverse Fourier transform of the right side of (4.2.9) and then using the Fourier inversion formula (Problem 4),

For the current discussion, the point of (4.2.7) is that

$$(2\pi)^{-1/2}(\widehat{\tau} * \widehat{\tau})(k) = -\int (2\pi)^{-3/2}(2\pi)^2 \theta(k - \ell)\,\theta(\ell)\,dk$$

$$= -(2\pi)(2\pi)^{-1/2}\,k\theta(k) \qquad (4.2.10)$$

$$= (ik)\,\widehat{\tau}(k) \qquad (4.2.11)$$

where (4.2.10) comes from doing the integral (noting it is over ℓ with $0 < \ell < k$). Since (6.2.37) of Part 1 says

$$\widehat{\tau^2} = (2\pi)^{-1/2}\,\widehat{\tau} * \widehat{\tau} \qquad (4.2.12)$$

we see that τ^2 has a natural meaning and expect that

$$\tau^2 = -\tau' \qquad (4.2.13)$$

Remarkably, even though each of the individual products δ^2, $\mathcal{P}^2$, $\mathcal{P}\delta$ seem unreasonable, τ^2 does not. Indeed, each convolution $\widehat{\delta} * \widehat{\delta}$, $\widehat{\mathcal{P}} * \widehat{\mathcal{P}}$, $\widehat{\delta} * \widehat{\mathcal{P}}$ leads to not absolutely convergent integrals. $\qquad\square$

This example suggests we want to refine the notion of singular support: if $x_0 \in s\text{-sing}(\tau)$, then all localizations $g\tau$ fail to be C^∞ and thus $\widehat{g\tau}$ fails to decay polynomially fast. But it may be the failure only involves some directions in k-space.

Definition. Let τ be a distribution on $\mathbb{R}^\nu$. A point $(x_0, k_0) \in \mathbb{R}^\nu \times (\mathbb{R}^\nu \setminus \{0\})$ is called a *regular directed point* for τ if and only if there exist neighborhoods N of x_0 and M of k_0 and a C_0^∞-function g with $g \equiv 1$ in N so that for all $m \in \mathbb{N}$, $\lambda \in (0, \infty)$ and $k \in M$, we have that

$$|\widehat{g\tau}(\lambda k)| \le C_m (1 + |\lambda|)^{-m} \qquad (4.2.14)$$

The complement in $\mathbb{R}^\nu \times (\mathbb{R}^\nu \setminus \{0\})$ of the regular directed points is called the *wavefront set*, WF(τ), of τ.

Note that since $g \in C_0^\infty$, $\widehat{g\tau}(k) = (2\pi)^{-\nu/2}\tau(ge^{-ik\cdot})$ defines a polynomially bounded function all of whose derivatives are polynomially bounded. We'll need the following lemma about such functions.

Lemma 4.2.4. *Let M be an open set in $\mathbb{R}^\nu$ disjoint from 0. Let S be a polynomially bounded function all of whose derivatives are polynomially bounded so that for all m, there is C_m so that for all $k \in M$*

$$|S(\lambda k)| \le C_m (1 + |\lambda|)^{-m} \qquad (4.2.15)$$

for all $\lambda > 0$. Let $h \in \mathcal{S}(\mathbb{R}^\nu)$ and $\widetilde{M} \subset M$ a compact subset. Then, $h * S$ obeys an estimate like (4.2.15) for all $k \in \widetilde{M}$.

Remark. Since $h \in \mathcal{S}$, it is easy to see that the integral defining $h * S(k)$ converges for all $k \in \mathbb{R}^\nu$.

Proof. We describe the intuition leaving the details to Problem 5. For $k \in \widetilde{M}$, we have

$$(S * h)(\lambda k) = \int S(\ell) \, h(\lambda k - \ell) \, dl \qquad (4.2.16)$$

For those $\ell_0 \notin \bigcup_{x>0} \lambda M$ as $\lambda \to \infty$, $(\lambda k - \ell_0) \to \infty$ and the decay of h forces the decay of the contribution to the integral for ℓ near ℓ_0. If $\ell_0 \in \bigcup_{\lambda>0} \lambda M$, $|\lambda k - \ell_0|$ may not go to zero but the decay of S forces decay of the integral. $\square$

The following describes the main properties of $\mathrm{WF}(\tau)$.

Theorem 4.2.5. *Let τ be a distribution on $\mathbb{R}^\nu$. Then*

(a) $\mathrm{WF}(\tau)$ *is a closed subset of* $\mathbb{R}^\nu \times (\mathbb{R}^\nu \setminus \{0\})$.
(b) *For each $x \in \mathbb{R}^\nu$, $\mathrm{WF}_x(\tau) \equiv \{k \mid (x, k) \in \mathrm{WF}(\tau)\}$ is a homogeneous set, i.e., $k \in \mathrm{WF}_x(\tau)$, $\lambda > 0 \Rightarrow \lambda k \in \mathrm{WF}_x(\tau)$.*
(c) $\mathrm{WF}(\tau + \sigma) \subset \mathrm{WF}(\tau) \cup \mathrm{WF}(\sigma)$
(d) $\{x \mid \mathrm{WF}_x(\tau) \neq \emptyset\}$ *is* $s\text{-}\mathrm{supp}(\tau)$
(e) *If $\tau \in \mathcal{S}'$ and $\hat{\tau}$ has support in a closed cone C, then for each x, $\mathrm{WF}_x(\tau) \subset C$.*
(f) *If M is a C^∞-diffeomorphism of $\mathbb{R}^\nu$ to $\mathbb{R}^\nu$ (i.e., M is a bijection and M and M^{-1} are C^∞) and $\tau \circ M$ is the distribution*

$$(\tau \circ M)(f) = \tau(g^{-1}[f \circ M^{-1}]) \qquad (4.2.17)$$

$g = \det(\partial M_i / \partial x_j)$ *and if $M_* : \mathbb{R}^\nu \times \mathbb{R}^\nu \setminus \{0\} \to \mathbb{R}^\nu \times \mathbb{R}^\nu \setminus \{0\}$ by*

$$M_*(x, k) = (M(x), dM_x^*(k))$$

where dM_x is the differential of M at x and dM_x^* its adjoint, then

$$\mathrm{WF}(\tau \circ M) = M_*[\mathrm{WF}(\tau)] \qquad (4.2.18)$$

Proof. (a)–(c) and (e) are all easy and we'll leave (f) to Problem 6. (d) comes from Lemma 4.2.4 as follows: Suppose that $\mathrm{WF}_x(\tau) = \emptyset$. For each $k \in S^{\nu-1} \subset \mathbb{R}^\nu$, there is a neighborhood N_k of x and neighborhood M_k of k and $g_k \in C_0^\infty(\mathbb{R}^\nu)$ so that $g_k \equiv 1$ in N_k and for each m there is C_m with $|\widehat{g_k \tau}(\lambda \ell)| \leq C_m(1 + \lambda|\ell|)^{-m}$ for $\lambda \in (0, \infty)$.

By a simple covering argument (Problem 7), we can find $k_1, \ldots, k_n$ and $\widetilde{M}_{k_j}$ an open neighborhood of k_j with $\overline{\widetilde{M}_{k_j}} \subset M_{k_j}$ and $\bigcup_{j=1}^n \widetilde{M}_{k_j} \supset S^{\nu-1}$. Let $g = g_{k_1} \cdots g_{k_m}$. By the lemma applied to $[g_{k_1} \cdots g_{k_{j-1}} g_{k_{j+1}} \cdots g_{k_m}](g_{k_j} \tau)$, we see

$$|\widehat{g\tau}(\lambda \ell)| \leq \widetilde{C}_m(1 + \lambda|\ell|)^{-m} \qquad (4.2.19)$$

for $\ell \in \widetilde{M}_{k_j}$. Thus, $g\tau$ decays faster than any power uniformly in all directions $\ell \in S^{\nu-1}$.

Pick any $f \in C_0^\infty(\mathbb{R}^\nu)$, $f \equiv 1$ in a neighborhood N of x. Then, $fg \equiv 1$ in $N \cap \bigcap_{j=1}^n N_{k_j}$ and $\widehat{(fg)\tau} = (2\pi)^{-\nu/2}\widehat{f} * \widehat{g\tau}$ has all derivatives decaying since we can put derivatives on $\widehat{f}$ and use the lemma. We have thus proven that $x \notin s\text{-}\mathrm{supp}(\tau)$.

The converse is easy. If $x \notin s\text{-}\mathrm{supp}(\tau)$, by definition no $\langle x, k\rangle \in \mathrm{WF}(\tau)$. $\qquad\square$

Here is the major result on products of distributions in terms of wavefront sets.

Theorem 4.2.6. *Let τ, σ be two distributions on $\mathbb{R}^\nu$. Define*

$$\mathrm{WF}(\tau) \oplus \mathrm{WF}(\sigma) = \{(x, k_1 + k_2) \mid (x, k) \in \mathrm{WF}(\tau), (x, k_2) \in \mathrm{WF}(\sigma)\}$$
$$(4.2.20)$$

Suppose this set contains no element of the form $(x, 0)$. Then there exists a distribution, $\kappa = \tau\sigma$ which is the product of τ and σ in the sense that for all $x \in \mathbb{R}^\nu$, there is $f \equiv 1$ near x with $f \in C_0^\infty(\mathbb{R}^\nu)$ and

$$\widehat{f^2\kappa}(k) = (2\pi)^{-\nu/2}\int (\widehat{f\tau})(\ell)(\widehat{f\sigma})(k - \ell)d^\nu\ell \qquad (4.2.21)$$

where for this f, the integral on the right converges absolutely for each $k \in \mathbb{R}^\nu$. Moreover, κ is unique and

$$\mathrm{WF}(\tau\sigma) \subset \mathrm{WF}(\tau) \cup \mathrm{WF}(\sigma) \cup [\mathrm{WF}(\tau) \oplus \mathrm{WF}(\sigma)] \qquad (4.2.22)$$

Before proving this theorem, we return to Example 4.2.3.

Example 4.2.3 (revisited). It follows from (4.2.7) and the form of τ that $\mathrm{WF}(\tau) = \{(0, k) \mid k > 0\}$. Thus $\mathrm{WF}(\tau) \oplus \mathrm{WF}(\tau) = \mathrm{WF}(\tau)$ doesn't contain any $(x, 0)$ so τ^2 exists. On the other hand, $\mathrm{WF}(\delta) = \{(0, k) \mid k \neq 0\}$ and $\mathrm{WF}(\delta) \oplus \mathrm{WF}(\delta)$ contains $(0, 0)$. $\qquad\square$

Proof of Theorem 4.2.6. We again sketch the key notions leaving the details to the reader (Problem 8). By localizing, we can suppose that there are cones $\Gamma_1 = \mathrm{WF}_{x_0}(\tau)$, $\Gamma_2 = \mathrm{WF}_{x_0}(\sigma)$ with $0 \notin \Gamma_1 + \Gamma_2$ and so that $\widehat{\tau}(k)$ (respectively, $\widehat{\sigma}(k)$) decays faster than any polynomial in k if $k \notin \Gamma_1$ (respectively, $k \notin \Gamma_2$) and so that for all k, $|\widehat{\tau}(k)| + |\widehat{\sigma}(k)| \leq C(1 + |k|)^n$ for some n. (In actuality, because we want uniform decay outside Γ_1, we need to slightly fatten Γ_1 and Γ_2 to cones $\widetilde{\Gamma}_1, \widetilde{\Gamma}_2$ that still obey $0 \notin \widetilde{\Gamma}_1 + \widetilde{\Gamma}_2$ and have $\Gamma_i \subset (\widetilde{\Gamma}_i)^{\mathrm{int}}$.)

We need to show that the integral

$$I(k) = \int_{\ell_2 = k - \ell_1} \widehat{\tau}(\ell_1) \widehat{\sigma}(\ell_2) \, d^\nu \ell_1 \qquad (4.2.23)$$

converges absolutely and determine its decay. We break this up into four integrals over $\ell_1 \in \Gamma_1, \ell_2 \in \Gamma_2$; $\ell_1 \in \Gamma_1, \ell_2 \notin \Gamma_2$; $\ell_1 \notin \Gamma_1, \ell_2 \in \Gamma_2$; $\ell_1 \notin \Gamma_1, \ell_2 \notin \Gamma_2$. In all but the first integral, either $\widehat{\tau}$ or $\widehat{\sigma}$ has much faster decay than the possible growth of the other factor, so the most dangerous integral is the one over $\ell_1 \in \Gamma_1, \ell_2 \in \Gamma_2$.

Here is the key observation: since $\ell_1 + \ell_2 \neq 0$, $(\ell_1 \cdot \ell_2)/\|\ell_1\| \|\ell_2\|$ is bounded away from -1, i.e., for such $\ell_1 \ell_2$ we have a fixed $\theta > 0$ so that

$$\|\ell_1 + \ell_2\|^2 \geq \theta(\|\ell_1\|^2 + \|\ell_2\|^2) \qquad (4.2.24)$$

That implies that for k fixed, the integration region $\{\ell_1, \ell_2) \mid \ell_2 = k - \ell_1, \ell_1 \in \Gamma_1, \ell_2 \in \Gamma_2\}$ is a ball of radius $\theta^{-1/2} \|k\|$ so that integral converges and is polynomially bounded in k and unless $k \in \Gamma_1 + \Gamma_2$ is 0. Thus, that piece of the integral defines a tempered distribution with local wavefront set in $\Gamma_1 + \Gamma_2$. $\qquad\qquad\qquad\qquad\qquad\qquad\qquad\qquad\qquad\qquad\square$

Notes and Historical Remarks. The name and definition of wavefront set are due to L. Hörmander [**396**]. A similar notion appeared slightly earlier in Sato [**718**]. Hörmander's definition was in terms of pseudodifferential operators but it is a theorem in his paper that his definition is equivalent to the one we use. Hörmander's paper also includes the application to products of distributions.

A second paper of Duistermaat and Hörmander [**232**] applies the theory to PDE's. Included in these applications are the refined version of elliptic regularity (as described in the Notes to the next section) and a refined version of propagation of singularities that says if a distribution τ solves $P(x, D)(\tau) = 0$, and $\langle x_0, k_0 \rangle \in \mathrm{WF}(\tau)$, then the entire bicharacteristic strip containing $\langle x_0, k_0 \rangle$ lies in $\mathrm{WF}(\tau)$.

The transformation rules in (f) of Theorem 4.2.5 imply $\mathrm{WF}(\tau)$ for distributions on manifolds are naturally subsets of the cotangent bundle.

Problems

1. (a) Let τ, σ be two distributions equal near each point of $\mathbb{R}^\nu$. Prove that $\tau = \sigma$. (*Hint*: Given $f \in C_0^\infty(\mathbb{R}^\nu)$ prove that $\tau(f) = \sigma(f)$ by using a partition of unity subordinate to the open cover $\{U_x\}$ where $g_x \tau \equiv g_x \sigma$ with $g_x \equiv 1$ on U_x.)

   (b) Use (a) to conclude uniqueness in Theorem 4.2.1.

2. Prove existence in Theorem 4.2.1 by using partitions of unity.

3. Verify (4.2.9) by contour integration.

4. If $\sigma_\varepsilon(k)$ is the right side of (4.2.9), compute $\check{\sigma}_\varepsilon$ and conclude that $\sigma_\varepsilon = \widehat{\tau}_\varepsilon$.

5. Fill in the details of the proof of Lemma 4.2.4.

6. Prove (f) of Theorem 4.2.5.

7. Provide the details of the covering argument used in the middle of the proof of (d) of Theorem 4.2.5.

8. This problem will lead you through the details of the proof of Theorem 4.2.6. So fix x_0, $f \in C_0^\infty(\mathbb{R}^\nu)$ with $f \equiv 1$ near x_0 so that $f\tau$ and $f\sigma$ have Fourier transforms each bounded by $C(1+|k|)^n$ and with fast decay outside $\Gamma_1 = \mathrm{WF}_{x_0}(\tau)$ and $\Gamma_2 = \mathrm{WF}_x(\sigma)$, respectively. For simplicity of notation, absorb f into τ and σ, i.e., just use τ and σ. We are interested in

$$I(k) = \int_{\ell_2 = k - \ell_1} \widehat{\tau}(\ell_1)\widehat{\sigma}(\ell_2)\, d^\nu \ell_1 \tag{4.2.25}$$

We want to prove the integral converges absolutely and defines a function of k which is polynomially bounded and with rapid decay outside $\Gamma_1 \cup \Gamma_2 \cup (\Gamma_1 + \Gamma_2)$.

(a) Show there exist cones $\widetilde{\Gamma}_1$, $\widetilde{\Gamma}_2$ so $\overline{\Gamma}_j \setminus \{0\} \subset (\widetilde{\Gamma}_j)^{\mathrm{int}}$ and $\widetilde{\Gamma}_1 + \widetilde{\Gamma}_2 \not\ni 0$.

(b) Show $I = \sum_{j=1}^5 I_j$ where I_1 is over $|\ell_1| < 1$ or $|\ell_2| < 1$, $I_2 \cdots I_5$ over subsets of $|\ell_j| > \frac{1}{2}$ and

$$I_2 : \ell_1 \in \widetilde{\Gamma}_1, \quad \ell_2 \in \widetilde{\Gamma}_2$$
$$I_3 : \ell_1 \in \widetilde{\Gamma}_1, \quad \ell_2 \notin \widetilde{\Gamma}_2$$
$$I_4 : \ell_1 \notin \widetilde{\Gamma}_1, \quad \ell_2 \in \widetilde{\Gamma}_1$$
$$I_5 : \ell_1 \notin \widetilde{\Gamma}_1, \quad \ell_2 \notin \widetilde{\Gamma}_2$$

(c) Show the integral for I_1 is over a bounded region (so convergent) and is polynomially bounded in k with rapid decay $k \notin \Gamma_1 \cup \Gamma_2$.

(d) Using the argument in the text, show that the integral defining I_2 is over a bounded (but k-dependent) region is polynomially bounded in k and has rapid decay outside $\widetilde{\Gamma}_1 + \widetilde{\Gamma}_2$.

(e) Show I_3 yields a convergent integral that is polynomially bounded and decays rapidly for $k \notin \widetilde{\Gamma}_1$. Similarly control I_4.

(f) Show that the I_5 integral is convergent and decays as $|k| \to \infty$. (*Hint:* $k = \ell_1 + \ell_2$ implies either $|\ell_1| \geq \frac{|k|}{2}$ or $|\ell_2| \geq \frac{|k|}{2}$.)

4.3. Microlocal Analysis: A First Glimpse

> The notion of pseudodifferential operator has grown in recent years out of attempts to describe an algebra of linear operators large enough to contain all differential operators and, whenever, possible, their inverses. Of course, it is important here that the new operators be close enough to differential operators so that it is convenient to compute with them.
>
> —*L. Nirenberg* [**615**]

A *partial differential operator* (PDO) is a map

$$f \mapsto \sum_{|\alpha| \leq m} a_\alpha(x)(D^\alpha f)(x) \equiv P(x, D)f \qquad (4.3.1)$$

where, at least in this section, we suppose the a_α are real-valued C^∞-functions so the PDO maps $C_0^\infty(\mathbb{R}^\nu)$ to itself. The m in (4.3.1) is called the *order* of the PDO (if some $a_\alpha(x)$ with $|\alpha| = m$ is nonzero). To avoid some minor technicalities, we'll suppose each $D^\beta a_\alpha$ is bounded and only look at tempered distributions. But the analysis is local and easily extends to ordinary distributions.

A PDO is called *elliptic* if for all $\mathbf{x}$ in $\mathbb{R}^\nu$ and all $\mathbf{k}$ in $\mathbb{R}^\nu \setminus \{0\}$

$$\sum_{|\alpha| = m} a_\alpha(x)k^\alpha \neq 0 \qquad (4.3.2)$$

This, of course, implies that for some $c, d > 0$,

$$\left| \sum_{|\alpha| \leq m} a_\alpha(x)(ik)^\alpha \right| \geq c|k|^m - d \qquad (4.3.3)$$

for all $\mathbf{k}$. In general, c and d can be $\mathbf{x}$-dependent but, for simplicity, we'll suppose that (4.3.3) holds uniformly in $\mathbf{x}$. Again, the analysis is local and Theorem 4.3.1 below holds without this uniformity assumption.

A major theme of this section will be a proof of

Theorem 4.3.1 (Elliptic Regularity for PDO). *Let $P(x, D)$ be an elliptic PDO with (4.3.3) for all $(x, k) \in \mathbb{R}^{2\nu}$ and let τ be a tempered distribution so that $P(x, D)\tau$ is a C^∞-function on some open $\Omega \subset \mathbb{R}^\nu$. Then, τ is a C^∞-function on Ω.*

Our point is not so much the theorem but the tools needed to prove it—a kind of analysis on phase space—part of what is called *microlocal analysis*. This theory has two sets of main tools: pseudodifferential operators and Fourier integral operators. Here, we'll only need and only develop some of the theory of pseudodifferential operators.

Two devices will dominate the technical side of this section: one is integration by parts combined with repeated use of

$$iw_j \, e^{ik \cdot w} = \frac{\partial}{\partial k_j} \, e^{ik \cdot w} \tag{4.3.4}$$

and the other is a souped-up version of the construction of Theorem 15.1.2 of Part 2B where we proved that any $\{a_n\}_{n=0}^\infty$ was the asymptotic series at 0 of a function $f(z)$ analytic in $\mathbb{C} \setminus (-\infty, 0]$. Basically, given functions $\{a_n(k)\}_{n=0}^\infty$ which are $O(|k|^{-n})$ as $|k| \to \infty$ in $\mathbb{R}^\nu$, we can find $f(k)$ so that $f(k) - \sum_{j=0}^N a_j(k) = O(|k|^{-N-1})$ as $|k| \to \infty$.

Notice that (4.3.1) can be rewritten when $f \in \mathcal{S}(\mathbb{R}^\nu)$ as

$$P(x, D)f(x) = (2\pi)^{-\nu/2} \int a(x, k) e^{ix \cdot k} \widehat{f}(k) d^\nu k \tag{4.3.5}$$

where

$$a(x, k) = \sum_{|\alpha| \le n} a_\alpha(x)(ik)^\alpha \tag{4.3.6}$$

The key will be to allow more general a's in (4.3.5) than polynomials in k.

Definition. A *symbol* of type $S_{\rho\delta}^m(\mathbb{R}^\nu)$ where $m \in \mathbb{R}$, $\rho, \delta \in [0, 1]$ is a C^∞-function, $a(x, k)$ on $\mathbb{R}^\nu \times \mathbb{R}^\nu$ so that for any multi-indices, α, β on $\mathbb{R}^\nu$, there is $C_{\alpha\beta}$ with

$$|(D_x^\beta D_k^\alpha a)(x, k)| \le C_{\alpha\beta} \langle k \rangle^{m - \rho|\alpha| + \delta|\beta|} \tag{4.3.7}$$

where $\langle k \rangle = (1 + |k|^2)^{1/2}$. $S_{10}^m \equiv S^m$ are called *classical symbols*.

Given a symbol, $a \in S_{\rho\delta}^m$ and $f \in \mathcal{S}(\mathbb{R}^\nu)$, we define

$$(\mathrm{Op}(a)f)(x) = (2\pi)^{-\nu/2} \int a(x, k) e^{ik \cdot x} \widehat{f}(k) d^\nu k \tag{4.3.8}$$

It can be written

$$(\mathrm{Op}(a)f)(x) = (2\pi)^{-\nu} \int a(x, k) e^{i(x-y) \cdot k} f(y) d^\nu k \, d^\nu y \tag{4.3.9}$$

Recall the action of the Heisenberg group (when $\omega = 1$), i.e., (4.1.7):

$$(U(x_0, p_0)f)(x) = e^{ip_0 \cdot x} f(x - x_0) \tag{4.3.10}$$

so, by (4.1.8)

$$(U(x_0, p_0)^{-1} f)(x) = e^{-ix_0 \cdot p_0} (U(-x_0, -p_0)f)(x) \tag{4.3.11}$$

A straightforward calculation (Problem 1) then shows that

$$U(x_0, p_0)\mathrm{Op}(a)U(x_0, p_0)^{-1} = \mathrm{Op}(\tau_{x_0, p_0} a) \tag{4.3.12}$$

where

$$(\tau_{x_0, p_0} a)(x, k) = a(x - x_0, k - p_0) \tag{4.3.13}$$

As we'll explain in the Notes, there are a distinct but closely related family of objects called the *Weyl calculus*:

$$(\text{Weyl}(a)f)(x) = (2\pi)^{-\nu} \int a\left(\frac{1}{2}(x+y), k\right) e^{i(x-y)\cdot k} f(y) d^\nu y \, d^\nu k \quad (4.3.14)$$

We also define for A a function, called an *amplitude*, on $\mathbb{R}^\nu \times \mathbb{R}^\nu \times \mathbb{R}^\nu$

$$(\text{Op}^\sharp(A)f)(x) = (2\pi)^{-\nu} \int e^{ik\cdot(x-y)} A(x, y; k) f(y) d^\nu y \, d^\nu k \quad (4.3.15)$$

where for now the integral is formal. If a is a symbol, we define two amplitudes

$$a^{(x)}(x, y; k) = a(x, k); \quad a_-^{(y)}(x, y; k) = a(y, -k) \quad (4.3.16)$$

so that

$$\text{Op}^\sharp(a^{(x)}) = \text{Op}(a) \quad (4.3.17)$$

We also define

$$\text{Op}^t(a) = \text{Op}^\sharp(a_-^{(y)}) \quad (4.3.18)$$

This is defined so that with

$$j(g)[h] = \int g(x)h(x) d^\nu x \quad (4.3.19)$$

(i.e., the real inner product used for $\mathcal{S}, \mathcal{S}'$ duality), we have that for all $f, g \in \mathcal{S}$ (again formally)

$$j(g)[\text{Op}(a)f] = j(\text{Op}^t(a)g)[f] \quad (4.3.20)$$

Here are the basic facts defining $\text{Op}(a)$ and $\text{Op}^t(a)$ as actual operators.

Theorem 4.3.2. *Let* $a \in S_{\rho\delta}^m$, $m \in \mathbb{R}$, $0 \le \rho \le 1, 0 \le \delta \le 1$. *Then*

(a) *For every* $f \in \mathcal{S}(\mathbb{R}^\nu)$, *the integral in* (4.3.8) *converges absolutely for all* x *and defines a function in* $\mathcal{S}(\mathbb{R}^\nu)$.

(b) *Let* $\delta < 1$. *For all* $f \in \mathcal{S}(\mathbb{R}^\nu)$, *the integral defining* $\text{Op}^t(a)$ *converges as an iterated integral, first in* y *and then in* k *and defines a function on* $\mathcal{S}(\mathbb{R}^\nu)$.

(c) *If* $j : \mathcal{S}(\mathbb{R}^\nu) \to \mathcal{S}'(\mathbb{R}^\nu)$ *is given by* (4.3.19), *then, for* $f, F \in \mathcal{S}(\mathbb{R}^\nu)$,

$$j(\text{Op}(a)F)(f) = j(F)(\text{Op}^t(a)f) \quad (4.3.21)$$

(d) *If* $\delta < 1$, $\text{Op}(a)$ *has a* $\sigma(\mathcal{S}', \mathcal{S})$ *continuous extension to a map of* $\mathcal{S}'$ *to* $\mathcal{S}'$ *given by*

$$(\text{Op}(a)\tau)(f) = \tau(\text{Op}^t(a)f) \quad (4.3.22)$$

(e) *If* $\delta < 1$, *for some* $q \in \mathbb{Z}_+$, $\tau \in \mathcal{S}'(\mathbb{R}^\nu)$ *has* $\hat\tau = (1 - \Delta_k)^q \sigma$ *with*

$$|\sigma(k)| \le C\langle k\rangle^L \quad (4.3.23)$$

and for $\ell \in \mathbb{Z}_+$, *we have that*

$$m < -(L + \ell + \nu) \quad (4.3.24)$$

then $\eta = \mathrm{Op}(a)(\tau)$ is a C^ℓ function of x with all $D^\alpha \eta$, $|\alpha| \leq \ell$ polynomially bounded in x.

Remark. All results about mapping $\mathcal{S}$ to $\mathcal{S}$ have associated bounds in terms of the $C_{\alpha\beta}$ of (4.3.7)

Proof. (a) Since $|\widehat{f}(k)| \leq C\langle k \rangle^{-\nu-m-1}$ and $\langle k \rangle^{-m} a$ is bounded (by (4.3.7) with $\alpha = \beta = 0$), the integral converges. Since (on account of $\delta \leq 1$)

$$|D_x^\beta(a(x,k)e^{ix\cdot k})| \leq C\langle k \rangle^{|\beta|} \tag{4.3.25}$$

the same argument shows that $\mathrm{Op}(a)f$ is a C^k-function.

Next, notice that

$$(ix)^\alpha e^{ix\cdot k} = D_k^\alpha(e^{ix\cdot k}) \tag{4.3.26}$$

Using this in (4.3.8) and then integrating by parts proves each $x^\alpha D_x^\beta(\mathrm{Op}(a)f)$ is bounded so $\mathrm{Op}(a)f \in \mathcal{S}$.

(b) We are interested in the integral

$$(\mathrm{Op}^{\mathrm{t}}(a)g)(x) = (2\pi)^{-\nu} \int \left(\int a(y,k)\, e^{-ik\cdot(x-y)} g(y)\, d^\nu y \right) d^\nu k \tag{4.3.27}$$

Since a is bounded in y and $g \in \mathcal{S}(\mathbb{R})$, the y integral converges. We claim with the $e^{-ik\cdot x}$ taken away, the integral is a function of k lying in $\mathcal{S}$ so the $d^\nu k$ is just the Fourier transform of a function in $\mathcal{S}$, so in $\mathcal{S}$, proving the result we want.

To see the y integral decays in k, write

$$e^{ik\cdot y} = \left[(1 - \Delta_y)^\ell\, e^{ik\cdot y} \right] \left[1 + k^2 \right]^{-\ell} \tag{4.3.28}$$

and integrate by parts. Since $\Delta_y a$ adds $|k|^{2\delta}$ factors, we see the integral is $O(\langle k \rangle^n \langle k \rangle^{-2\ell(1-\delta)})$ giving any polynomial decay desired. Adding D_β^k terms only adds at most $\langle y \rangle^\beta$ which is no problem since $g \in \mathcal{S}$ and we see that the intermediate integral is in $\mathcal{S}$.

(c) (4.3.21) is just (4.3.20)

(d) (4.3.21) and (6.2.55) of Part 1 implies (4.3.22) for $\tau = j(F)$ and then the general theory of extensions gets the claimed result.

(e) By approximating σ with functions, f_n, in $\mathcal{S}$ so that $\widehat{f}_n(k) \to \sigma(k)$, and $|f_n(k)| \leq \widetilde{C}\langle k \rangle^L$ and integrating $(1 - \Delta_k)^q$ by parts (using $|(1 - \Delta_k)^q[e^{ik\cdot x}a(x,k)]| \leq (1 + |x|)^r$ for suitable r), one see $\mathrm{Op}(a)(\tau)$ is a limit of C^ℓ functions with uniform bounds on D^α, $|\alpha| \leq \ell$ derivatives. One gets the polynomially bounded result by looking at the estimates.

$\square$

Operators $\mathrm{Op}(a)$ on $\mathcal{S}$ or $\mathcal{S}'$ with $a \in S^m_{\rho\delta}$ are called *pseudodifferential operators* (ΨDO). The set of all such $\mathrm{Op}(a)$ is denoted $\Psi^m_{\rho\delta}$ and $\Psi_{\rho\delta} = \bigcup_{m=-\infty}^{\infty} \Psi^m_{\rho\delta}$. We let $\Psi^{-\infty}$ stand for the set of $\mathrm{Op}(a)$ where $a \in \bigcap_m S^m_{\rho\delta}$. It is independent of $\rho\delta$ since

$$a \in \bigcap_m S^m_{\rho\delta} \Leftrightarrow \forall \alpha, \beta, \ell \; |D^\beta_x D^\alpha_k a(x,k)| \leq C_{\ell\alpha\beta}\langle k\rangle^{-\ell} \qquad (4.3.29)$$

Notice that this says if $a \in \bigcap_m S^m_{\rho\delta}$, then $\mathrm{Op}(a)\tau$ is a C^∞-function with polynomially bounded derivatives. Let C^∞_{poly} be the set of such functions. Then,

$$B \in \Psi^{-\infty} \Rightarrow \forall \tau \in \mathcal{S}', \; B\tau \in C^\infty_{\mathrm{poly}} \qquad (4.3.30)$$

An operator B from $\mathcal{S}'$ to $\mathcal{S}'$ is called *local* if

$$\mathrm{supp}(B\tau) \subset \mathrm{supp}(\tau) \qquad (4.3.31)$$

PDO's are local but ΨDO's, in general, are not: for example, convolution with any $f \in \mathcal{S}(\mathbb{R}^\nu)$ is a ΨDO (Problem 2) and is certainly not local.

In the previous section we defined the singular support, $s\text{-sing}(\tau)$, of a distribution τ as the set of points where τ is not locally given by a C^∞-function. A map, B, is called *pseudolocal* if and only if

$$s\text{-sing}(B\tau) = s\text{-sing}(\tau) \qquad (4.3.32)$$

We are heading towards the first critical fact for ΨDO's.

Theorem 4.3.3. Ψ*DO's are pseudolocal, i.e., for any* $a \in S^m_{\rho\delta}$ *(*$m \in \mathbb{R}$, $0 < \rho \leq 1$, $0 \leq \delta < 1$*) and* $\tau \in \mathcal{S}'(\mathbb{R}^\nu)$ *we have*

$$s\text{-}\mathrm{supp}(\mathrm{Op}(a)\tau) \subset s\text{-}\mathrm{supp}(\tau) \qquad (4.3.33)$$

Before turning to the proof of this theorem, we want to note that Theorem 4.3.1 can be phrased as

$$s\text{-}\mathrm{supp}(\tau) \subset s\text{-}\mathrm{supp}(P(x,D)\tau) \qquad (4.3.34)$$

Thus, (4.3.33) would imply Theorem 4.3.1 if we could show that $P(x,D)$ had an inverse which is a ΨDO. We'll prove something slightly weaker, but one can see why (4.3.33) is an important part of the proof of Theorem 4.3.1. The key to (4.3.33) is that ΨDO's have kernels that are C^∞ with rapid decay off diagonal.

Theorem 4.3.4. *Let* $U, V \subset \mathbb{R}^\nu$ *be open and bounded with* $\overline{U} \subset V$ *and let* $B = \mathrm{Op}(a)$ *for* $a \in S^m_{\rho\delta}$ *(*$m \in \mathbb{R}$, $0 < \rho \leq 1$, $0 \leq \delta < 1$*). Then there is a function* $K(x,y)$, $x \in U$, $y \in \mathbb{R}^\nu \setminus V$ *obeying for all* $\ell \in \mathbb{Z}_+$,

$$|(D^\alpha_x D^\beta_y K)(x,y)| \leq C_{\alpha,\beta,\ell}(1 + |y|)^{-\ell} \qquad (4.3.35)$$

(*i.e.*, K *is* C^∞ *in* x *and* y *with all derivatives decaying in* y) *so that if* $f \in \mathcal{S}(\mathbb{R}^\nu)$ *with* supp $f \subset \mathbb{R}^\nu \setminus V$ *and* $x \in U$, *then*

$$(Bf)(x) = \int K(x,y)f(y)d^\nu y \tag{4.3.36}$$

Remark. By the Schwartz kernel theorem (Theorem 6.4.10 of Part 1), there is a distribution K on $\mathbb{R}^\nu \times \mathbb{R}^\nu$ so that for $f, g \in \mathcal{S}(\mathbb{R}^\nu)$

$$\int g(x)(Bf)(x)d^\nu x = \int K(x,y)g(x)f(y)d^\nu x\, d^\nu y \tag{4.3.37}$$

This theorem implies $s\text{-supp}(K) \subset \Delta \equiv \{(x,x) \mid x \in \mathbb{R}^\nu\}$ and basically says precisely that (together with a decay estimate).

Proof. Formally B has a kernel given by

$$K(x,y) = (2\pi)^{-\nu} \int a(x,k)e^{ik(x-y)}d^\nu k \tag{4.3.38}$$

obtained by writing out $\hat{f}$ in (4.3.8) and formally interchanging the k and x integrals. Of course, the formality of this formula can be seen in that there is no reason for the integral to converge.

That said, we will continue to proceed formally. Multiplying K by $(x-y)^\gamma$ and using

$$i^\gamma(x-y)^\gamma e^{ik\cdot(x-y)} = D_k^\gamma e^{ik\cdot(x-y)} \tag{4.3.39}$$

and integrating by parts we get formally that

$$(-i)^\gamma(x-y)^\gamma K(x,y) = (2\pi)^{-\nu} \int e^{ik\cdot(x-y)} D_k^\gamma(a(x,k))d^\nu k \tag{4.3.40}$$

So long as

$$m - \rho|\gamma| < -\nu \tag{4.3.41}$$

the integral converges absolutely and, since $\rho > 0$, this happens for γ large.

For $x \in U$, $y \in \mathbb{R}^\nu \setminus V$, $|x-y|$ is bounded away from zero, so (4.3.40) can be used to define K in that region. By taking $|\gamma|$ large and varying the indices, one sees this function is $O(\langle y \rangle^{-\ell})$ for all k as $|y| \to \infty$. That means

$$\int K(x,y)f(y)d^\nu y$$

converges absolutely and one can go backwards to show it gives $(Bf)(x)$ when supp $f \subset \mathbb{R}^\nu \setminus V, x \in U$ (Problem 3).

Similarly $D_x^\alpha D_y^\beta K(x,y)$ is given by a sum of integrals like (4.3.40) with $\alpha = \alpha_1 + \alpha_2$ and $D_x^{\alpha_1}a(x,k)$ and $(ik)^{\alpha_1+\beta}(-1)^{|\beta|}$. Again writing (4.3.40), making sure that

$$m - \rho|\gamma| < -\nu - |\alpha| - |\beta| \tag{4.3.42}$$

one gets bounds of the integral and allows a proof of the fact that K is C^∞ with decay in $|y|$. We leave the details to the reader in Problem 4. □

Proof of Theorem 4.3.3. We must show that if τ is C^∞ near $x_0 \in \mathbb{R}^\nu$, then $\mathrm{Op}(a)\tau$ is C^∞ near x_0 also.

By Theorem 6.4.9 of Part 1,

$$\tau = (-\Delta + 1 + x^2)^M F \tag{4.3.43}$$

where F is a bounded continuous function and $M \in \mathbb{Z}_+$. Pick $U \subset U_1$, both open with $x_0 \in U$ and τ given by a C^∞-function on U_1. It is not hard to see (Problem 5) that F is C^∞ on U_1.

Pick $\varphi \in C_0^\infty$ with $\varphi = 1$ on U and $\varphi = 0$ on $\mathbb{R}^\nu \setminus U_1$ and write

$$\tau = \tau_1 + \tau_2$$

where τ_1 (respectively, τ_2) comes from putting φF (respectively, $(1 - \varphi)F$) in (4.3.43).

Since F is C^∞ on U_1, so is τ_1. Thus $\tau_1 \in \mathcal{S}$ so $\mathrm{Op}(a)\tau_1 \in \mathcal{S}$. Thus, $\mathrm{Op}(a)\tau_1$ is C^∞ on U.

By integrating by parts (and taking limits from $f \in \mathcal{S}$ to $(1 - \varphi)F$—see Problem 6) one has for x in a neighborhood of x_0 that

$$(\mathrm{Op}(a)\tau_2)(x) = \int \left[(-\Delta_y + 1 + y^2)^M K(x, y)\right] F(y) d^\nu y$$

is C^∞ also. Thus, $\mathrm{Op}(a)\tau$ is C^∞ near x_0. $\square$

We need one more tool before we put everything together. To invert $P(x, D)$, we'll need to invert $(1 - r)$ where $r \in S_{\rho\delta}^k$ (with $k < 0$). It is obvious we'll need to use a geometric series and so "convergence" of that series will be needed. We'll also see that series arise when looking at products of $\mathrm{Op}(a_1)$ and $\mathrm{Op}(a_2)$. Thus, we turn to such infinite series.

Definition. A *formal symbol*, $\tilde{a}$, in $FS_{\rho\delta}^m$ is a sequence $m = m_1 > m_2 > m_3 > \dots$ with $m_j \to -\infty$ and symbols $\{a_j\}_{j=1}^\infty$ with $a_j \in S_{\rho\delta}^{m_j}$.

Notice that in trying to form $a_1 + a_2 + \dots$ things get better as $|k| \to \infty$ since once $m_j < 0$, $a_j \to 0$ at infinity. This will allow us to always get convergence so long as we don't care about $\Psi^{-\infty}$ pieces.

Theorem 4.3.5. *Let* $m \in \mathbb{R}$, $\rho, \delta \in [0, 1]$ *with* $\delta < 1$. *Let* $\tilde{a}$ *be a formal symbol in* $FS_{\rho\delta}^m$. *Then there exists a symbol* $a \in S_{\rho\delta}^m$ *so that for each* J

$$a - \sum_{j=1}^{J} a_j \in S_{\rho\delta}^{m_{J+1}} \tag{4.3.44}$$

If $a, a^\sharp$ *are two symbols in* $S_{\rho\delta}^m$ *obeying* (4.3.44), *then*

$$\mathrm{Op}(a) - \mathrm{Op}(a^\sharp) \in \Psi^{-\infty} \tag{4.3.45}$$

Remarks. 1. We use $\widetilde{\mathrm{Op}}(\tilde{a})$ for the unique operator mod $\Psi^{-\infty}$ constructed via (4.3.44).

2. As we noted above, this can be viewed as a sophisticated variant of Theorem 15.1.2 of Part 2B.

Proof. By Theorem 4.3.2 (e), if $a, a^{\sharp}$ obey (4.3.44), then $a - a^{\sharp} \in S_{\rho\delta}^{m_J}$ for all m_J and so in $\bigcap_m S_{\rho\delta}^m$ proving (4.3.45). Thus, we need only prove existence of a's obeying (4.3.44).

Pick $\varphi \in C^{\infty}(\mathbb{R}^{\nu})$ only a function of $|k|$ so that

$$\varphi(k) = \begin{cases} 0 & \text{if } |k| \leq \frac{1}{2} \\ 1 & \text{if } |k| \geq 1 \end{cases} \tag{4.3.46}$$

We will show that one can pick $t_j \to \infty$ so fast that

$$a(x, k) = \sum_{j=1}^{\infty} \varphi\left(\frac{k}{t_j}\right) a_j(x, k) \tag{4.3.47}$$

converges in (x, k) (uniformly for x in any given compact $K \subset \mathbb{R}^{\nu}$, so that $a \in S_{\rho\delta}^m$ and obeys (4.3.44).

We begin noting, if $t \geq 1$ for any α,

$$\left| D_k^{\alpha} \left[\varphi\left(\frac{k}{t}\right) \right] \right| = t^{-|\alpha|} \left| (D_k^{\alpha}\varphi)\left(\frac{k}{t}\right) \right|$$

$$= |k|^{-\alpha} \left(\frac{|k|}{t}\right)^{\alpha} \left| (D_k^{\alpha}\varphi)\left(\frac{k}{t}\right) \right|$$

$$\leq 2^{\alpha} \langle k \rangle^{-\alpha} \sup_y y^{\alpha} |(D_y^{\alpha}\varphi)(y)| \tag{4.3.48}$$

using that since $t > 1$ and $\operatorname{supp}\varphi \subset [\frac{1}{2}, \infty)$, we have $|k| > \frac{1}{2}\langle k \rangle$ on $\operatorname{supp}(\varphi(\cdot/t))$.

(4.3.48) implies that if

$$a_j^{(t)}(x, k) = \varphi(k/t) a_j(x, k) \tag{4.3.49}$$

then, uniformly in t,

$$\sup_k \langle k \rangle^{-m_j + \rho|\alpha| - \delta|\beta|} |D_x^{\beta} D_k^{\alpha} a_j^{(t)}(x, k)| \leq C_{j,\alpha,\beta} \tag{4.3.50}$$

In particular, each $a_j^{(t)}$ is a symbol in $S_{\rho\delta}^{m_j}$.

Moreover, by (4.3.50), since $m_{j-1} > m_j$ and $a_j(x, k)$ is supported in $\{k \mid |k| > \frac{t}{2}\}$, we can pick t_j so that

$$\sup_{|\alpha|+|\beta| \leq j} \sup_{x,k} \langle k \rangle^{-m_{j-1} + \rho|\alpha| - \delta|\beta|} |D_x^{\beta} D_k^{\alpha} a_j^{(t_j)}(x, k)| \leq 2^{-j} \tag{4.3.51}$$

This in turn implies $\sum_{j=1}^{\infty} D_x^{\beta} D_k^{\alpha} a^{(t_j)}(x,k)$ converges uniformly on compacts to $D_x^{\beta} D_k^{\alpha} a$ for a C^{∞}-function a and that for each α, β

$$\sup_k \langle k \rangle^{-m_J + \rho|\alpha| - \delta|\beta|} |D_x^{\beta} D_k^{\alpha} (a - \sum_{j=1}^{J} a_j^{(t_j)})(x,k)| < \infty \qquad (4.3.52)$$

so that $a - \sum_{j=1}^{J} a^{(t_j)}$ lies in $S_{\rho\delta}^{m_J}$. Since

$$a - \sum_{j=1}^{J} a^{(t_j)} = a - \sum_{j=1}^{J+1} a^{(t_j)} + a_{J+1}^{(t_{J+1})}$$

we see it is actually in $S_{\rho\delta}^{m_{J+1}}$ as required. $\square$

The last tool we need is that $\Psi_{\rho\delta}$ is a graded algebra whose leading order (but not lower) is given by multiplication of symbols. One approach to this will involve kernels with three variables related to (4.3.9). As a preliminary, it will be helpful to have a characterization of $\Psi^{-\infty}$ in terms of integral kernels:

Theorem 4.3.6. *An operator B from $\mathcal{S}$ to C_{poly}^{∞} lies in $\Psi^{-\infty}$ if and only if*

$$(Bf)(x) = \int K(x,y)f(y)d^{\nu}y \qquad (4.3.53)$$

where K is a C^{∞}-function of x,y obeying (for all multi-indices, α, β, and all $N \in \mathbb{Z}_+$)

$$|D_x^{\alpha} D_y^{\beta} K(x,y)| \le C_{\alpha,\beta,N}(1 + |x-y|^2)^{-N} \qquad (4.3.54)$$

Proof. Let $B = \text{Op}(a)$ with $a \in \bigcap_m S_{\rho\delta}^m$. By (4.3.9), B has an integral kernel

$$K(x,y) = (2\pi)^{-\nu} \int a(x,k)e^{i(x-y)\cdot k}d^{\nu}k \qquad (4.3.55)$$

By hypothesis, a has rapid decay uniformly in k, so K is continuous in x,y and bounded. By (4.3.40), K has rapid decay in $x-y$, so (4.3.54) holds for $\alpha = \beta = 0$. One can repeat this for derivatives in x,y and so see that K obeys (4.3.54).

Conversely, if K obeys (4.3.54), let $\widetilde{K}(u,v)$ be defined by

$$\widetilde{K}(u,v) = K(u, u-v), \quad K(x,y) = \widetilde{K}(x, x-y) \qquad (4.3.56)$$

$\widetilde{K}$ has rapid decay in v, i.e., lies in $\mathcal{S}(\mathbb{R}^{\nu})$ in v, uniformly in u. Thus

$$a(x,k) = \int \widetilde{K}(x,v)e^{-ik\cdot v}d^{\nu}v$$

lies in $\bigcap_m S_{\rho\delta}^m$ and, by Fourier inversion

$$\widetilde{K}(x,v) = (2\pi)^{-\nu} \int a(x,k)e^{ik\cdot v}d^{\nu}v \qquad (4.3.57)$$

so by (4.3.56)

$$K(x,y) = (2\pi)^{-\nu} \int a(x,k)e^{ik\cdot(x-y)}d^\nu v \tag{4.3.58}$$

i.e., $B = \mathrm{Op}(a)$ $\square$

An *amplitude* in $\mathcal{A}_{\rho\delta}^m$ is a function $A(x,y;k)$ on $\mathbb{R}^{3\nu}$ obeying

$$|(D_k^\alpha D_x^\beta D_y^\gamma A)(x,y;k)| \le C_{\alpha,\beta,\gamma}\langle k\rangle^{m-\rho|\alpha|+\delta(|\beta|+|\gamma|)} \tag{4.3.59}$$

Given $A \in \mathcal{A}$, we define $\mathrm{Op}^\sharp(A)$ by

$$(\mathrm{Op}^\sharp(A)f)(x) = (2\pi)^{-\nu} \int e^{ik\cdot(x-y)} A(x,y;k)f(y)d^\nu y\, d^\nu k \tag{4.3.60}$$

Unlike symbols where the integral defining $\mathrm{Op}(a)$ is convergent (for $f \in \mathcal{S}(\mathbb{R}^\nu)$), we define $\mathrm{Op}^\sharp(A)$, at least a priori (we'll see soon that $\mathrm{Op}^\sharp(A)$ is an $\mathrm{Op}(a) \mod \Psi^{-\infty}$) as a map of $\mathcal{S}$ to $\mathcal{S}'$, i.e., give meaning to $j(g)[\mathrm{Op}^\sharp(A)f]$. Then, for fixed x,y, $A(x,y;k)$ is a distribution in k so

$$K(x,y) = (2\pi)^{-\nu} \int A(x,y;k)e^{ik\cdot(x-y)}d^\nu k \tag{4.3.61}$$

is a distribution and then

$$j(g)[\mathrm{Op}^\sharp(A)f] = \int g(x)K(x,y)f(y)d^\nu x\, d^\nu y \tag{4.3.62}$$

interpreted as the distribution K applied to $g \otimes f$. Of course, if $A(x,y;k) = a(x,k)$ is independent of y, $\mathrm{Op}^\sharp(A) = \mathrm{Op}(a)$.

Proposition 4.3.7. *If $A \in \mathcal{A}_{\rho\delta}^m$ is supported in $\{(x,y) \mid |x-y| \ge \varepsilon\}$ for some $\varepsilon > 0$, then $\mathrm{Op}^\sharp(A)$ lies in $\Psi^{-\infty}$.*

Proof. This is just a repeat of Theorem 4.3.4!

$$D_x^\beta D_y^\gamma \left[(x-y)^{2L}K(x,y)\right] = (2\pi)^{-\nu} \int [D_x^\beta D_y^\gamma e^{ik\cdot(x-y)}(-\Delta_k)^L A(x,y;k)]d^\nu k \tag{4.3.63}$$

by using $(x-y)_j e^{ik\cdot(x-y)} = -i\partial_j e^{ik\cdot(x-y)}$ and integrating by parts. Picking L so $2\rho L - (|\beta| + |\gamma|) > \nu$, we see convergent integrals and so $K(x,y)$ is a C^∞-function in x,y with derivatives of rapid decay in $x-y$, so $\mathrm{Op}^\sharp(A)$ lies in $\Psi^{-\infty}$ by Theorem 4.3.6. $\square$

Because of this, when convenient, we'll suppose that A is supported on $\{(x,y) \mid |x-y| \le 1\}$, since we can decompose any A into a sum $A_1 + A_2$ with A_1 suppported on this set and A_2 on the set $\{(x,y) \mid |x-y| \ge \frac{1}{2}\}$.

Theorem 4.3.8. *Let $\delta < \rho$. Let $A \in \mathcal{A}_{\rho\delta}^m$. Let $\tilde{a}$ be the formal symbol*

$$\tilde{a}(x,k) = \sum_\alpha \frac{(-i)^{|\alpha|}}{\alpha!}(D_k^\alpha D_y^\alpha A)(x,y;k)\,|_{x=y} \tag{4.3.64}$$

Then $\widetilde{\mathrm{Op}}(\tilde{a}) = \mathrm{Op}^\sharp(A) \mod \Psi^{-\infty}$.

Remarks. 1. $D_k^\alpha D_y^\alpha A \mid_{x=y} \in S_{\rho\delta}^{m-|\alpha|(\rho-\delta)}$ so $\delta < \rho$ implies this is really a series with orders going to $-\infty$. In the classical case, $\delta = 0$, $\rho = 1$, so $\delta < \rho$ holds.

2. Thus by Theorem 4.3.5, each $\mathrm{Op}^\sharp(A)$ lies in Ψ^m.

3. The leading term is just $A(x, x; k)$ i.e.,

$$\tilde{a}(x, k) - A(x, x, k) \in S_{\rho\delta}^{m-(\rho-\delta)} \tag{4.3.65}$$

Proof. As we've seen, formal integrals like (4.3.60) make sense if we multiply by $g(x) \in \mathcal{S}(\mathbb{R}^\nu)$ and integrate. We'll write and manipulate formal integrals in this proof without explicitly showing they make sense when integrated. We'll also use formulae like

$$\int e^{ip\cdot z} d^\nu p = (2\pi)^\nu \delta(z) \tag{4.3.66}$$

(as expression of the Fourier inversion formula, see (6.2.57) of Part 1)

$$\int e^{ip\cdot z}(ip)^\alpha d^\nu p = (2\pi)^\nu D^\alpha \delta(z) \tag{4.3.67}$$

Justifying these things is straightforward (Problem 7).

In (4.3.62) write $f(y) = (2\pi)^{-\nu/2} \int \widehat{f}(q) e^{iq\cdot y} d^\nu q$ to see that

$$(\mathrm{Op}^\sharp(A)f)(x) = (2\pi)^{-\nu} \int a^\flat(x, q) e^{iq\cdot x} \widehat{f}(q) d^\nu q \tag{4.3.68}$$

$$a^\flat(x, q) = (2\pi)^{-\nu} \int e^{i(k-q)\cdot(x-y)} A(x, y; k) d^\nu y \, d^\nu k \tag{4.3.69}$$

We don't yet know that $a^\flat$ is a symbol. The point of the proof will be to show that (mod $\Psi^{-\infty}$), and that it is given by the formal symbol (4.3.64). Change integration variables in (4.3.69) via $z = y - x$, $p = k - q$ to see that

$$a^\flat(x, k) = (2\pi)^{-\nu} \int B(x; p, p + k) d^\nu p \tag{4.3.70}$$

$$B(x; p, k) = \int e^{-ip\cdot z} A(x, x + z; k) d^\nu z \tag{4.3.71}$$

To see where we are heading, note that, by (4.3.66),

$$(2\pi)^{-\nu} \int B(x; p, k) d^\nu p = A(x, x; k) \tag{4.3.72}$$

which is a symbol and the leading term in (4.3.64). So the key will be to expanding $B(x; p, p + k)$ in the third variable about $p = 0$.

We need to start with an estimate on B. We claim for any multi-indexes, α, β, γ, and any $\mu \in \mathbb{N}$,

$$|(D_x^\alpha D_p^\beta D_k^\gamma B)(x; p, k)| \leq C_{\alpha,\beta,\gamma} \langle k \rangle^{m+\delta(|\alpha|+2\mu)-\rho|\gamma|} \langle p \rangle^{-2\mu} \tag{4.3.73}$$

To see this note that $D_x^\alpha D_k^\gamma$ moves through the integral to hit a and get $(D_x^{\alpha_1} D_y^{\alpha_2} D_k^\gamma A)\,|_{y=x+z}$ $(\alpha = \alpha_1 + \alpha_2)$. The D_p^β gives an $(-iz)^\beta$ and we use $|z| \leq 1$. Finally multiplying by $\langle p \rangle^{2\mu}$ becomes $\langle 1 - \Delta_z \rangle^\mu$ which after integrating by parts is $(1 - \Delta_y)A\,|_{y=x+z}$ which is $O(\langle k \rangle^{2\delta\mu})$.

For each N, write $B(x; p, p+k)$ as a Taylor series about $p = 0$ in the last variable (but not the middle)

$$|B(x; p, p+k) - \sum_{|\alpha| < N} \frac{1}{\alpha!} D_k^\alpha B(x; p, k) p^\alpha|$$
$$\leq C_{\nu,N} |p|^N \langle p \rangle^{-2\mu} \sup_{0 \leq t < 1} [\langle k + tp \rangle^{m+2\delta\mu-\rho N}] \tag{4.3.74}$$

where we use (4.3.73) on $(D_k^\gamma B)(x; p, k+tp)$ in Taylor series with remainder.

If $|p| \leq \frac{1}{2}|k|$, we take $2\mu = N$ in (4.3.74) (and $\sup_{0 \leq t \leq 1} \langle k + tp \rangle^{-1} \leq 2\langle k \rangle^{-1}$) to bound the right of (4.3.74) by $C\langle k \rangle^{m-(\rho-\delta)N}$ if N is so large that $N(\rho - 1) > m$. If $|p| \geq \frac{1}{2}|k|$ so $\langle p \rangle^{-1} \leq 2\langle k \rangle^{-1}$, we take μ large and get a bound by $\langle k \rangle^{-\ell} \langle p \rangle^{-\ell}$ for any ℓ.

We now integrate dp. By the rapid p and k decays, the integral over $|p| > \frac{1}{2}|k|$ is $O(\langle k \rangle^{-\ell})$ for all ℓ. The integral over $|p| < \frac{1}{2}|k|$ has $|k|^\nu$ times the $O(\langle k \rangle^{m-(\rho-\delta)N})$ bound. We also use (4.3.67) to see that

$$(2\pi)^{-\nu} \int D_k^\alpha B(x; p, k) p^\alpha d^\nu p = (-i)^{|\alpha|} D_k^\alpha D_y^\alpha A(x, y; k)\,|_{x=y} \tag{4.3.75}$$

The result is

$$|a^\flat(x, k) - \sum_{|\alpha| < N} \frac{(-i)^{|\alpha|}}{\alpha!} D_k^\alpha D_y^\alpha A(x, y; k)|_{x=y} \leq C\langle k \rangle^{m+\nu-(\rho-\delta)N} \tag{4.3.76}$$

By taking N larger, we get the asymptotic series estimates for $\tilde{a}$ in (4.3.64). We can do the same estimate for $D_x^\beta D_k^\gamma$ in (4.3.75) and so obtain (4.3.64). $\quad\square$

With this result in hand, we get a number of powerful, immediate consequences. They are corollaries but we call them theorems because of their intrinsic importance.

Theorem 4.3.9. *Let $a \in S_{\rho\delta}^m$ with $0 \leq \delta < \rho \leq 1$. Then*

$$\mathrm{Op}^\mathrm{t}(a) = \mathrm{Op}(a^\mathrm{t}) \pmod{\Phi^{-\infty}} \tag{4.3.77}$$

where

$$a^\mathrm{t}(x, k) = \sum_\alpha \frac{(i)^{|\alpha|}}{\alpha!} (D_k^\alpha D_x^\alpha a)(x, -k) \tag{4.3.78}$$

Moreover, if

$$a^*(x, k) = \sum_\alpha \frac{(-i)^{|\alpha|}}{\alpha!} (D_k^\alpha D_x^\alpha \bar{a})(x, k) \tag{4.3.79}$$

then

$$\mathrm{Op}(a^*) = \mathrm{Op}(a)^* \quad (\mathrm{mod} \; \Phi^{-\infty}) \tag{4.3.80}$$

where $\mathrm{Op}(a)^*$ means the Hilbert space adjoint.

Remarks. 1. $\bar{a}$ in (4.3.79) means the complex conjugate of the function a.

2. In particular, if $A \in \Psi^m_{\rho\delta}$ and $0 \le \delta < \rho \le 1$, then $A^* \in \Psi^m_{\rho\delta}$.

Proof. By (4.3.18), $\mathrm{Op}^{\mathrm{t}}(a) = \mathrm{Op}^{\sharp}(a^{(y)}_{-})$ with $a^{(y)}_{-}$ given by (4.3.16), so (4.3.78) is just (4.3.64).

  Since $\langle g, f \rangle = j(\bar{g})(f)$, we see that

$$\mathrm{Op}(a)^* g = \overline{(\mathrm{Op}^{\mathrm{t}}(a)\bar{g})} \tag{4.3.81}$$

which yields (4.3.79) $\qquad\qquad\qquad\qquad\qquad\qquad\qquad\qquad\qquad\qquad\qquad \square$

  Note that (4.3.77) implies that

$$a \in S^m_{\rho\delta} \Rightarrow a^* - \bar{a} \in S^{m-(\rho-\delta)}_{\rho\delta} \tag{4.3.82}$$

Theorem 4.3.10. *Let $a_j \in S^{m_j}_{\rho\delta}$, $j = 1, 2$ with $0 \le \delta < \rho \le 1$. Then*

$$a_1 \circ a_2 \equiv \sum_{\alpha} \frac{(-i)^{|\alpha|}}{\alpha!} D^{\alpha}_k (a_1(x,k)(a^t_2)(y,-k)) \,|_{x=y} \tag{4.3.83}$$

is a symbol in $S^{m_1+m_2}_{\rho\delta}$ with

$$\mathrm{Op}(a_1)\,\mathrm{Op}(a_2) = \mathrm{Op}(a_1 \circ a_2) \quad \mathrm{mod} \; \Psi^{-\infty} \tag{4.3.84}$$

We have that

$$a_1 \circ a_2 - a_1 a_2 \in S^{m_1+m_2-(\rho-\delta)}_{\rho\delta} \tag{4.3.85}$$

so that

$$[\mathrm{Op}(a_1), \mathrm{Op}(a_2)] \in \Psi^{m_1+m_2-(\rho-\delta)}_{\rho\delta} \tag{4.3.86}$$

Remark. Further calculation (see the Notes) shows that

$$a_1 \circ a_2(x,k) = \sum_{\alpha} \frac{(-i)^{|\alpha|}}{\alpha!} (D^{\alpha}_k a_1(x,k))(D^{\alpha}_x a_2(x,k)) \tag{4.3.87}$$

Proof. Let $a^{(y)}$ be the amplitude $A(x,y;k) = a(y,k)$. As we saw in the proof of Theorem 4.3.2(b), $(\mathrm{Op}^{\sharp}(a^{(y)}g))(x)$ is a Fourier transform so

$$\widehat{\mathrm{Op}^{\sharp}(a^{(y)})}g(k) = (2\pi)^{-\nu/2} \int a(y,k) e^{-ik\cdot y} g(y)\, d^{\nu}y \tag{4.3.88}$$

Thus

$$(\mathrm{Op}(b)\,\mathrm{Op}^{\sharp}(a^{(y)})g)(x) = (2\pi)^{-\nu/2} \int b(x,k)e^{ik\cdot x}\,\overline{(\mathrm{Op}^{*}(a^{(y)})g)}(k)\,d^{\nu}k$$

(4.3.89)

$$= (2\pi)^{-\nu} \int b(x,k)a(y,k)e^{ik\cdot(x-y)}g(y)\,d^{\nu}y\,d^{\nu}k$$

(4.3.90)

$$= \left[\mathrm{Op}^{\sharp}(b(x,k)a(y,k))g\right](x)$$

(4.3.91)

Up to $\Phi^{-\infty}$ errors, $\mathrm{Op}(a_2^{\mathrm{t}})$ is the transpose of $\mathrm{Op}(a_2)$. Since $(\mathrm{Op}(a_2)^{\mathrm{t}})^{\mathrm{t}} = \mathrm{Op}(a_2)$, we see

$$\mathrm{Op}(a_2) = \mathrm{Op}(a_2^{\mathrm{t}})^{\mathrm{t}} = \mathrm{Op}^{\sharp}((a_2^{\mathrm{t}})_{-}^{(y)})$$

(4.3.92)

This plus (4.3.91) says that

$$\mathrm{Op}(a_1)\,\mathrm{Op}(a_2) = \mathrm{Op}^{\sharp}(a_1(x,k)a_2^{\mathrm{t}}(y,-k))$$

(4.3.93)

This plus (4.3.64) is (4.3.82).

The last two relations follow from this. $\square$

We can prove Theorem 4.3.1 by proving a stronger result. A symbol $S_{\rho\delta}^{m}$ is called an *elliptic symbol of order m* if and only if for some $c > 0$ and all x, k we have

$$|a(x,k)| \geq c\langle k\rangle^{m} - d$$

(4.3.94)

If $P(x,D)$ is an elliptic PDO of degree m, $a(x,k) = \sum_{|\alpha|\leq m} a_{\alpha}(ik)^{\alpha}$ is an elliptic symbol, so the following includes Theorem 4.3.1:

Theorem 4.3.11 (Elliptic Regularity for Elliptic ΨDO). *Let $0 \leq \delta < \rho \leq 1$. Let a be an elliptic symbol of order m with $m > 0$. Then for any tempered distribution, τ,*

$$s\text{-}\mathrm{supp}(\mathrm{Op}(a)\tau) = s\text{-}\mathrm{supp}(\tau)$$

(4.3.95)

Remark. Below we construct a paramatrix, q, with $\mathrm{Op}(q)\,\mathrm{Op}(a) = \mathbb{1}$ mod $\Psi^{-\infty}$. It can be shown by the methods below (Problem 8) that $\mathrm{Op}(a)\,\mathrm{Op}(q) = \mathbb{1}$ mod $\Psi^{-\infty}$ also.

Proof. Theorem 4.3.3 implies $s\text{-}\mathrm{supp}(\mathrm{Op}(a)\tau) \subset s\text{-}\mathrm{supp}(\tau)$. Suppose we construct a *paramatrix*, i.e., a ΨDO, $\mathrm{Op}(q)$ so that

$$\mathrm{Op}(q)\,\mathrm{Op}(a) = \mathbb{1} \quad \mathrm{mod}\ \Psi^{-\infty}$$

(4.3.96)

Then

$$\mathrm{Op}(q)[\mathrm{Op}(a)\tau] = \tau + C\tau$$

(4.3.97)

with $C \in \Psi^{-\infty}$ so $C\tau$ is C^∞. Thus

$$s\text{-}\operatorname{supp}(\tau) \subset s\text{-}\operatorname{supp}[\operatorname{Op}(q)(\operatorname{Op}(a)\tau)]$$
$$\subset s\text{-}\operatorname{supp}[\operatorname{Op}(a)\tau]$$

by Theorem 4.3.3 again. This proves (4.3.95). Thus, we are reduced to (4.3.96).

By (4.3.94), there is R so

$$|a(x,k)| \geq \tfrac{1}{2}\langle k \rangle^m \tag{4.3.98}$$

if $|k| > R$. Let φ be a function in $C_0^\infty(\mathbb{R}^\nu)$ so $\varphi(k) \equiv 1$ if $|x| \leq R$ and $\equiv 0$ if $|x| \geq 2R$. Let

$$q_0(x,k) = [1 - \varphi(k)]a(x,k)^{-1} \tag{4.3.99}$$

Then, $q_0 \in S_{\rho\delta}^{-m}$ as is easy to see since for $|k| \geq 2R$, $q_0 = a^{-1}$.

By Theorem 4.3.8, $q_0 \circ a = 1 - r$ where $r \in S_{\rho\delta}^{-(\rho-\delta)}$. Let

$$q = \sum_{j=0}^{\infty} r^j \circ q_0 \tag{4.3.100}$$

as a formal symbol. Here $r^j \equiv r \circ r \circ \ldots \circ r \in S_{\rho\delta}^{-j(\rho-\delta)}$ so $r^j \circ q \in S_{\rho\delta}^{-m-j(\rho-\delta)}$ and (4.3.100) is a formal symbol.

Since operator multiplication is associative, symbol multiplication is also (mod $\Psi^{-\infty}$). Thus,

$$q \circ a = \sum_{j=0}^{N} r^j \circ q_0 \circ a + S_{\rho\delta}^{-(N+1)(\rho-\delta)}$$
$$= \sum_{j=0}^{N} r^j - (1-r) + S_{\rho\delta}^{-(N+1)(\rho-\delta)}$$
$$= 1 + S_{\rho\delta}^{-(N+1)(\rho-\delta)}$$

Since N is arbitrary, $q \circ a = 1 \pmod{\bigcap_m S_{\rho\delta}^m}$, i.e., (4.3.96) holds. $\qquad\square$

Notes and Historical Remarks. The twenty-five years following 1950 revolutionized the theory of PDEs in two waves of innovation—the first concerned the language of distribution theory which allowed precise study of weak solutions including the Malgrange–Ehrenpreis theorem. The second involves phase space methods, aka microlocal analysis in the form of pseudo-differential operators and, more generally, Fourier integral operators. Most of the PDE books quoted in the Notes to Section 6.9 of Part 1 discuss these issues—in particular, Volumes III and IV of Hömander [397] are entitled, respectively, *Pseudodifferential Operators* and *Fourier Integral Operators*. In some ways, the biggest difference between [397] and the earlier [393] is the availability of these methods. For further books on these subjects see [**1, 169, 231, 240, 252, 497, 648, 715, 736, 803, 811, 812**]. In particular, the terse, condensed Chapter 0 of Taylor [**803**] considerably influenced my treatment here.

We return to ΨDO's in Sections 6.3–6.5 when we discuss mappings between the L^2-Sobolev spaces, H^s.

The seminal paper introducing ΨDO's as an algebra of operators defined by (4.3.8) is by Kohn–Nirenberg [**479**] (and independently in the briefer pair by Unterberger–Bokobza [**824, 825**]). Kohn–Nirenberg say they are using the name pseudodifferential at the same time as Friedrichs–Lax [**290**] so one should perhaps attribute the name to the Courant Institute! Kohn–Nirenberg approach many of the central questions of the subject including commutation, formal series, and pseudolocality.

The idea of expanding the notion of PDO's was certainly in the air at the time. One factor was the then recent work of Calderón–Zygmund and the explosion of interest in singular integrable operators (see Section 6.4) and the desire to absorb them and PDO's into a common framework. This is seen by the independent work of Kohn-Nirenberg and Unterberger–Bokobza as well as almost immediate following work by Hörmander [**392**], Palais [**626**], and Seeley [**721**]. In fact, Kohn–Nirenberg came to the general theory because they needed to estimate the commutators of some explicit singular integral operators.

Kohn–Nirenberg used a very restrictive class of symbols much smaller than $S^m_{\rho\delta}$. The class $S^m_{\rho\delta}$ we use is the most commonly used one. It goes back to Hörmander [**394**]. As you might guess, there is considerable literature on allowing symbols which aren't C^∞, especially in the x-variable; see, for example, Chapter 1 of Taylor [**803**].

A Fourier integral operator (FIO) is an operator of the form

$$(Bf)(x) = \int e^{i\Psi(x,y;k)} A(x,y;k) u(y) dy\, dk \qquad (4.3.101)$$

Here Ψ is C^∞ on $\mathbb{R}^\nu \times \mathbb{R}^\nu \times (\mathbb{R}^\nu \setminus \{0\})$, and obeys

(i) Ψ is homogeneous of degree 1 in k, i.e., for $\lambda > 0$

$$\Psi(x,y;\lambda k) = \lambda\Psi(x,y;k) \qquad (4.3.102)$$

(ii) Ψ is real-valued and nondegenerate in that $\nabla_{(x,y;k)}\Psi \neq 0$ for all $(x,y;k) \in \mathbb{R}^{2\nu} \times (\mathbb{R}^\nu \setminus \{0\})$.

Because of the nondegeneracy, one can use integration by parts to define $\int g(x)(Bf)(x)d^\nu x$ for all $g \in \mathcal{S}$ and so get B as a map of $\mathcal{S}(\mathbb{R}^\nu)$ to $\mathcal{S}'(\mathbb{R}^\nu)$.

FIO's were introduced by Hörmander [**396, 232**]. Notice that $\Psi(x,y;k) = k \cdot (x - y)$ is a phase function so any ΨDO is a FIO. Before

Hörmander's work, there were important examples of non-ΨDO maps that are FIO's in Maslov [**561**], Eskin [**251**], and Egorov [**239**].

Lars Hörmander (1931–2012) dominated the theory of PDEs for many years starting with his 1955 papers. He was both a major (arguably, the major) figure in the use of distributions and in microlocal analysis. He began his graduate study in Lund under M. Riesz and, after Riesz retired, with Lars Gårding (1919–2014) who steered him towards PDEs. He was a frequent visitor to the US and had regular positions at Stanford and the Institute for Advanced Study before returning to Lund in 1968. His students include A. Melin and J. Sjöstrand.

For a proof of (4.3.87), see the book of Shubin [**736**].

In terms of the operators $U(x_0, p_0)$ of (4.3.10), one can show that ($\hat{a}$ is $\mathbb{R}^{2\nu}$ Fourier transform)

$$(\operatorname{Op}(a)f)(x) = (2\pi)^{-\nu} \int \hat{a}(q,p)(U(p,-q)f)(x)d^\nu q\, d^\nu p \qquad (4.3.103)$$

(see Problem 9).

The funny sign on q in $U(p,-q)$ is due to the fact that if $P_j = \frac{1}{i}\frac{\partial}{\partial x_j}$, then by Taylor's theorem $(e^{iq\cdot P}f)(x) = f(x+q)$, so

$$U(p,-q) = e^{ip\cdot Q}\, e^{iq\cdot P} \qquad (4.3.104)$$

(where $(Q_j f)(x) = x_j f(x_j)$). Notice, since in Fourier space, $Q_j = -\frac{1}{i}\frac{\partial}{\partial p_j}$, $e^{ip\cdot Q}$ translates k to $k-p$, i.e., $e^{ipQ_j}P_j e^{-ipQ_j} = P_j - p$ while $e^{iqP_j}Q_j e^{-iqP_j} = Q_j + q$. We defined U by (4.3.10) to translate (P,Q) by (p_0, x_0), but in terms of generators $U(p,-q)$ is more natural. Even more natural is

$$W(p,q) = e^{i(p\cdot Q + q\cdot P)} \qquad (4.3.105)$$

which acts by

$$(W(p,q)f)(x) = e^{ipq/2}e^{iqx}f(x+p) \qquad (4.3.106)$$

The *Weyl calculus* (named after his book [**851**]) is then defined by

$$(\operatorname{Weyl}(a)f)(x) = (2\pi)^{-\nu}\int \hat{a}(q,p)W(p,q)d^\nu q\, d^\nu p \qquad (4.3.107)$$

A calculation much like that for (4.3.103) (see Problem 10) leads to (4.3.14). Grossmann–Loupias–Stein [**341**] and Howe [**401, 403**] have used this formalization and group-theoretic calculations to study ΨDO's.

The Weyl calculus is more symmetrical than the ΨDO theory. For example, since $W(p,q)^* = W(-p,-q)$ and $\hat{a}(-p,-q) = \overline{\hat{a}(p,q)}$, we see if $a(x,k)$ is real, then $\operatorname{Weyl}(a)$ is Hermitian symmetric (i.e., $\langle f, \operatorname{Weyl}(a)g\rangle = \langle \operatorname{Weyl}(a)f, g\rangle$ while $\operatorname{Op}(a)^* - \operatorname{Op}(a)$ is only a lower-order operator.

Notice if F, G are functions on $\mathbb{R}^\nu$ and

$$(F\otimes 1)(x,k) = F(x), \quad (1\otimes G)(x,k) = G(k) \qquad (4.3.108)$$

then

$$\text{Weyl}(F \otimes 1) = \text{Op}(F \otimes 1) = F(X), \quad \text{Weyl}(1 \otimes G) = \text{Op}(1 \otimes G) = G(P) \tag{4.3.109}$$

where

$$(G(P)f) = (G(k)\hat{f})^{\vee} \tag{4.3.110}$$

Related to the Weyl calculus is the *Wigner distribution*, which can be defined as a kind of dual to the Weyl calculus. Given $f \in L^2(\mathbb{R}^\nu, d^\nu x)$ with $\|f\|_2 = 1$, we define $P_f(x, k)$ by requiring that

$$\langle f, \text{Weyl}(a)f \rangle = \int P_f(x, k)a(x, k)d^\nu x \, d^\nu k \tag{4.3.111}$$

Notice that (4.3.109) implies that

$$\int P_f(x, k)d^\nu k = |f(x)|^2, \quad \int P_f(x, k)d^\nu k = |\hat{f}(k)|^2 \tag{4.3.112}$$

This is also true for the object with $\text{Op}(a)$ instead of $\text{Weyl}(a)$ but we'll see in a moment that P_f is real while the object with $\text{Op}(a)$ might not be.

By (4.3.107) and $\int f(x)\hat{g}(x)d^\nu x = \int \hat{f}(k)g(k)d^\nu k$, we see that (noting $\hat{a}(q, p)$ not $\hat{a}(p, q)$)

$$P_f(x, k) = (2\pi)^{-\nu}\langle f, \widehat{W(p, q)}f \rangle(x, k) \tag{4.3.113}$$

$$= (2\pi)^{-2\nu}\int e^{-i(xq+kp)}\langle f, W(p, q)f \rangle d^\nu p \, d^\nu q \tag{4.3.114}$$

Since

$$\langle f, W(p, q)f \rangle = \int e^{ipq/2} \, e^{iqy}\overline{f(y)}f(y + p)d^\nu y \tag{4.3.115}$$

we see that

$$P_f(x, k) = (2\pi)^{-2\nu}\int e^{-iq(\frac{p}{2}+y-x)} \, e^{-ikp}\overline{f(y)}f(y + p)d^\nu y \, d^\nu p \, d^\nu q \tag{4.3.116}$$

$$= (2\pi)^{-\nu}\int e^{-ikp}\overline{f(x - \frac{p}{2})}f(x + \frac{p}{2})d^\nu p \tag{4.3.117}$$

since

$$(2\pi)^{-\nu}\int e^{iq(\frac{p}{2}+y)-x}d^\nu q = \delta(y - (x - \frac{p}{2}))$$

(4.3.117) proves again that $\int P_f(x, k)d^\nu k = |f(x)|^2$.

By looking at the relation of Fourier transform and $W(p, q)$, one sees (Problem 11) that

$$P_{\hat{f}}(x, k) = P_f(-k, x) \tag{4.3.118}$$

so $\int P_{\hat{f}}(x, k)d^\nu k = |\hat{f}(x)|^2 \Rightarrow \int P_f(x, k)d^\nu x = |\hat{f}(k)|^2$, proving that again.

(4.3.117) implies directly that

$$\overline{P_f(x, k)} = P_f(x, k) \tag{4.3.119}$$

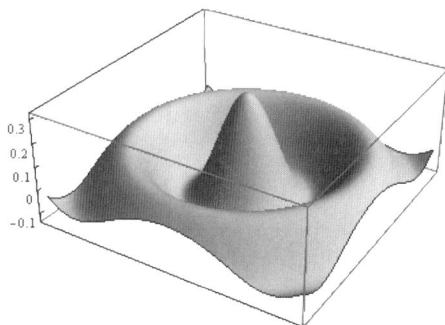

Figure 4.3.1. The Wigner distribution function for the sixth harmonic oscillator wave function.

i.e., $P_f(x, k)$ is real and that

$$|P_f(x, k)| \le (2\pi)^{-\nu} \qquad (4.3.120)$$

It is very tempting to think of $P_f(x, k)d^\nu x\, d^\nu k$ as a joint probability distribution of position and momentum since the marginals are $|f(x)|^2 d^\nu x$ and $|\hat{f}(k)|^2 d^\nu k$. (4.3.120) is then an expression of the uncertainty principle! Alas, while P_f is real, it is not, in general, positive as simple examples show (Problem 12). Figure 4.3.1 shows the Wigner distribution function of the sixth harmonic oscillator wave function and illustrates this negativity problem.

The Wigner distribution is named after his 1932 paper [**858**]. He used (4.3.117) as a definition. It is sometimes called the *Wigner–Ville distribution* since Ville [**834**] rediscovered it in 1948 in work in communications theory. The idea of defining it as a dual of Weyl(a), i.e., via (4.3.111) is due to Moyal [**590**] who also rediscovered. it.

The condition $s-\text{sing}(Pu) = s-\text{sing}(u)$ is so important that it deserves a name. An operator $p : \mathcal{S}' \to \mathcal{S}'$ with that property is called *hypoelliptic*. Theorem 4.3.1 says elliptic operators are hypoelliptic. Hörmander [**395**] has proven a remarkable theorem. Let $A_0, \ldots, A_\ell$ be first-order PDOs, i.e., $A_j = \sum_{k=1}^\nu a_{jk}(x)D_k + a_{j0}(x)$. We call $\sum_{k=1}^\nu a_{jk}(x_0)e_k$ (e_k is the basis element of $\mathbb{C}^\nu$ with a 1 in the kth slot) the leading symbol at x_0. Notice any commutator of A_j's is a first-order PDO. We say $\{A_0, \ldots, A_\ell\}$ obeys *Hörmander's condition* if at each x, the leading symbols of A_0, $\{A_j\}_{j=1}^\ell$, $\{[A_{j_1}, A_{j_2}]\}_{1 \le j_1 < j_2 \le \ell} \ldots \{[A_{j_1} \ldots A_{j_m}]\}_{j_1 \ldots j_m}$ span $\mathbb{C}^\nu$. The Hörmander operator is $H = A_0 + \sum_{j=1}^\nu A_j^2$ when the A's obey Hörmander's condition. An example of such an operator which is not elliptic (look at $x_j = 0$) is $\left(x_1 \frac{\partial}{\partial x_2}\right)^2 + \left(\frac{\partial}{\partial x_2}\right)^2$. Hörmander's theorem says such operators, H, are hypoelliptic.

The wavefront set of the last section is a significant part of the toolbag of microlocal analysis. Theorem 4.3.3 can be refined to $\mathrm{WF}(\mathrm{Op}(a)u) \subset \mathrm{WF}(u)$; indeed, something stronger is true:

$$\mathrm{WF}(\mathrm{Op}(a)u) \subset \mathrm{WF}(u) \cap \mathrm{ES}(a) \tag{4.3.121}$$

$\mathrm{ES}(a)$, for *essential support*, is the subset of $\mathbb{R}^\nu \times (\mathbb{R}^\nu \setminus \{0\})$ which is the complement of those (x_0, k_0) for which there is a neighborhood, N, with

$$|(D_x^\beta D_k^\alpha a)(x, \lambda k)| \leq C_{\alpha,\beta,\ell,N} \langle \lambda \rangle^{-\ell} \tag{4.3.122}$$

for all $\alpha, \beta, \ell \in \mathbb{Z}_+$ and $\lambda \in (0, \infty)$. This is discussed in Section 0.10 of Taylor [**803**] as well as the consequence that for elliptic operators (where $\mathrm{ES}(a) = \emptyset$) $\mathrm{WF}(\mathrm{Op}(a)u) = \mathrm{WF}(u)$ (indeed this follows from (4.3.121) and the existence of parametrics). Taylor (and many other references) also phrase propagation of singularities for smooth hyperbolic systems in terms of wavefront sets.

We mention that Section 6.5 continues the discussion of ΨDO's looking at their action on L^2- and $H^{s,2}$-Sobolev spaces. In particular, we prove the result of Calderón–Vaillancourt that if $a \in S_{0,0}^0$, then $\mathrm{Op}(a)$ is bounded from L^2 to L^2.

One important aspect of ΨDO's that we will not discuss are the trace class/Hilbert–Schmidt class properties of ΨDO's (see, for example Section 2.7 of Shubin [**736**]).

Problems

1. Verify (4.3.12)/(4.3.13) by first showing that (LHS of (4.3.12) $f)(x)$ is given by

$$(2\pi)^{-\nu} e^{ip_0 \cdot x} \int e^{i(x-x_0)\cdot k} a(x-x_0, k) e^{-ix_0 \cdot p} e^{-iy \cdot p_0} e^{-ik \cdot y} f(y+x_0) dy \, dk \tag{4.3.123}$$

   and changing variables $y + x_0 \to y$, $k \to k - p_0$ to get

$$(2\pi)^{-\nu} \int e^{ik \cdot (x-y)} a(x-x_0, k-p_0) f(y) dy \, dk \tag{4.3.124}$$

2. Let $a(x, k) = \hat{g}(k)$ where $g \in \mathcal{S}(\mathbb{R}^\nu)$

   (a) Prove that $a \in \bigcap_m S_{\rho\delta}^m$.

   (b) Prove that $\mathrm{Op}(a)$ is a convolution with $(2\pi)^{-\nu} g$ so such convolutions are in $\Psi_{\rho\delta}^m$ for all m, ρ, δ.

3. Justify that (4.3.36) holds for $x \in U$, $\mathrm{supp}\, f \subset \mathbb{R}^\nu \setminus V$ and K given by (4.3.40) for $|\gamma|$ large.

4. Provide the details that for $x \in U$, $y \in \mathbb{R}^\nu \setminus V$, the kernel K of Theorem 4.3.4 is C^∞ with decay in y.

5. (a) Using Mehler's formula (see Problem 4 of Section 4.11 of Part 1) find a formula for $e^{-t(-\Delta + x^2 + 1)}$ and show for any M that $(-\Delta + 1 + x^2)^{-M}$ has an integral kernel that is C^∞ off diagonal with rapid decay.

   (b) Prove that if F is bounded continuous, so that τ given by (4.3.43) is C^∞ on an open set, U, then F is C^∞ on U.

6. Justify the limit argument in the proof of Theorem 4.3.3.

7. Justify the formal steps in the proof of Theorem 4.3.8.

8. (a) Use the method of proof of Theorem 4.3.11, to find $\tilde{q} \in S_{\rho\delta}^{-m}$ so that $\mathrm{Op}(a)\,\mathrm{Op}(\tilde{q}) = 1 \mod \Psi^{-\infty}$.

   (b) Prove $\mathrm{Op}(q) - \mathrm{Op}(\tilde{q}) \in \Psi^{-\infty}$. (*Hint*: Look at $\mathrm{Op}(q)\,\mathrm{Op}(a)\,\mathrm{Op}(\tilde{q})$.)

   (c) Conclude that $\mathrm{Op}(a)\,\mathrm{Op}(q) = \mathbb{1} \mod \Psi^{-\infty}$.

9. This problem will verify (4.3.103).

   (a) By writing out $\hat{a}$, prove that

   $$\text{RHS of (4.3.103)} = (2\pi)^{-2\nu} \int a(y,k) e^{iq(x-y)} e^{-ikp} f(x+p) d^\nu q\, d^\nu p\, d^\nu k$$
   $$(4.3.125)$$

   (b) Use $\int e^{iq\cdot(x-y)} d^\nu q = (2\pi)^\nu \delta(x-y)$ followed by doing the y integral to see

   $$\text{RHS of (4.3.125)} = (2\pi)^{-\nu} \int a(x,k) e^{-ikp} f(x+p) d^\nu p\, d^\nu k \qquad (4.3.126)$$

   followed by $p = y - x$, to see that (4.3.103) holds.

10. This will verify that (4.3.14) and (4.3.107) are consistent.

    (a) For the RHS of (4.3.107) show you get an integral like (4.3.125) but now $e^{iq(x-y)}$ is $\exp[iq(x + \frac{p}{2} - y)]$.

    (b) In place of (4.3.126) show that you get

    $$(\mathrm{Weyl}(a)f)(x) = (2\pi)^{-\nu} \int a\left(x + \frac{p}{2}, k\right) e^{-ik\cdot p} f(x+p) d^\nu p\, d^\nu k$$

    and that with $p = y - x$, this leads to (4.3.12).

11. This problem will prove (4.3.118). For simplicity, we restrict to $\nu = 1$. For this problem only, define $\mathcal{F}F = \hat{f}$ (so we have $\hat{}$ as an operator). Let $Q,\, P : \mathcal{S}(\mathbb{R}) \mapsto \mathcal{S}(\mathbb{R})$ by

    $$(Qf)(x) = xf(x), \quad (Pf)(x) = i^{-1} f'(x) \qquad (4.3.127)$$

    so (4.3.105) holds. Prove that

    $$\mathcal{F}P = Q\mathcal{F}, \quad \mathcal{F}Q = -P\mathcal{F} \qquad (4.3.128)$$

and that

$$\mathcal{F}W(p,q)\mathcal{F}^{-1} = W(q,-p) \tag{4.3.129}$$

12. This problem will compute $P_f(x,k)$ for f the characteristic function of $[-\frac{1}{2},\frac{1}{2}]$.

(a) Prove that $|x - \frac{p}{2}| \leq \frac{1}{2}$, $|x + \frac{p}{2}| \leq \frac{1}{2}$ if and only if $|x| \leq \frac{1}{2}$ and $|p| \leq 1 - 2|x|$.

(b) Prove that for $f(x) = \chi_{[-\frac{1}{2},\frac{1}{2}]}$ we have that

$$P_f(x,k) = \chi_{[-\frac{1}{2},\frac{1}{2}]}(x)\frac{\sin(k(1-2|x|))}{\pi k} \tag{4.3.130}$$

(c) Check that $\int P_f(x,k)dk$ is $|f(x)|^2$ in this case.

(d) Is $P_f(x,k)$ always positive?

4.4. Coherent States

Coherent states (including Glauber states) are elegant and very helpful for understanding what goes on in oscillating systems, including harmonic oscillators as well as many kinds of waves. They show us the correspondence between quantum oscillators and classical oscillators. Among other things, they make it clear that even in fully quantum mechanical systems, not everything is quantized, not all waves are quantized, not all states are discrete... If some of this comes as a surprise to you, you are not alone. A lot of smart people have gotten this wrong over the years. Indeed, the guy who figured this out the first time got the Nobel prize
—*John Denker* [**214**][1]

Note: In this section we'll make contact with the theory of reproducing kernel Hilbert space, especially L^2-reproducing kernels. The reader should do, or, at least look over, Problems 4–11 of Section 3.3 of Part 1, Problem 7 of Section 3.4 of Part 1, and Problem 17 of Section 6.6 of Part 1.

The next three sections deal with attempting to expand functions in terms of some standard functions that are localized in both x- and k-space (to the extent allowed by the uncertainty principle). Here, we'll focus on continuous expansions and, in the next two sections, on discrete sums.

Given the uncertainty principle limits, it is natural to use for $\mathbf{p}, \mathbf{q} \in \mathbb{R}^\nu$:

$$\varphi(\mathbf{x}; \mathbf{p}, \mathbf{q}) = e^{i\mathbf{p}\cdot\mathbf{x}}\varphi_0(\mathbf{x} - \mathbf{q}) \tag{4.4.1}$$

$$\varphi_0(\mathbf{x}) = \pi^{-\nu/4}\exp(-x^2/2) \tag{4.4.2}$$

[1]The reference is to Roy Glauber's 2005 Nobel Prize "for his contribution to the quantum theory of optical coherence."

which we saw are all the minimal uncertainty states with $\langle x^2 \rangle = \langle k^2 \rangle$ (see Theorem 4.1.1). These φ are called the *Gaussian coherent states, classical coherent states*, or *canonical coherent states*. Define for $f \in L^2(\mathbb{R}^\nu, d^\nu x)$

$$\mathcal{G}(f; p, q) = \langle \varphi(\,\cdot\,; p, q), f \rangle \qquad (4.4.3)$$

Here is the key starting point:

Theorem 4.4.1. *For every* $f \in L^2(\mathbb{R}^\nu, d^\nu x)$, *we have that* $\mathcal{G}(f; p, q) \in L^2(\mathbb{R}^{2\nu}, d^\nu p\, d^\nu q)$ *and*

$$\int |\mathcal{G}(f; p, q)|^2 \frac{d^\nu p\, d^\nu q}{(2\pi)^\nu} = \|f\|^2 \qquad (4.4.4)$$

Remarks. 1. At first sight, this looks like a Plancherel theorem and $\mathcal{G}$ like a Fourier transform. There are some very significant differences: For Fourier transforms, we take "inner products" with a nonnormalized plane wave while $\varphi \in L^2(\mathbb{R}^\nu)$. (4.4.4) only says $\mathcal{G}$ is an isometry; unlike a Fourier transform, it is *not* onto all of L^2.

2. We'll, in essence, have three proofs of (4.4.4)—the one here using Bargmann transforms, one using the fact that by a direct calculation

$$\int |\langle \varphi(\,\cdot\,; p, q), \varphi(\,\cdot\,; 0, 0) \rangle|^2 d^\nu p\, d^\nu q < \infty \qquad (4.4.5)$$

and the last is a special case of $e^{ip\cdot x} g(x - y)$ for general g proven using the Plancherel theorem.

Proof. Recall the Segal–Bargmann transform ((6.4.82) and Theorem 6.4.12 of Part 1)

$$(\mathbb{B}f)(z) = \pi^{-1/4} \int \exp\left(-\frac{1}{2}(z^2 + x^2) + \sqrt{2}\, z \cdot x \right) f(x) dx \qquad (4.4.6)$$

which is a unitary map of $L^2(\mathbb{R}^\nu, d^\nu x)$ to $L^2(\mathbb{C}^\nu, \pi^{-\nu} \exp(-|z|^2) d^{2\nu} z)$. A direct calculation (Problem 1) shows that

$$\mathcal{G}(f; p, q) = \exp\left(-\frac{1}{4}(p^2 + q^2) \right) (\mathbb{B}f)\left(\frac{q + ip}{\sqrt{2}} \right) \qquad (4.4.7)$$

$$= \exp\left(-\frac{1}{2}|z|^2 \right) (\mathbb{B}f)(z) \quad \Big|_{z = \frac{q+ip}{\sqrt{2}}} \qquad (4.4.8)$$

Given that, by a change of variables, when $z = (q + ip)/\sqrt{2}$

$$\frac{d^\nu p\, d^\nu q}{(2\pi)^\nu} = \frac{d^{2\nu} z}{\pi^\nu} \qquad (4.4.9)$$

(4.4.4) is just

$$\int |(\mathbb{B}f)(z)|^2 e^{-|z|^2} \frac{d^{2\nu} z}{\pi^\nu} = \|f\|^2 \qquad (4.4.10)$$

$\square$

An equivalent way to write (4.4.4) (or more properly the bilinear extension obtained by polarization) is to define

$$P(p,q) = \langle \varphi(\cdot\,; p,q), \cdot \rangle \varphi(\cdot\,; p,q) \qquad (4.4.11)$$

the projection onto multiples of $\varphi(\cdot, p, q)$ and note that

$$\int P(p,q) \frac{d^\nu p\, d^\nu q}{(2\pi)^\nu} = \mathbb{1} \qquad (4.4.12)$$

We proved this as a weak integral but it is easy to see it can be taken as a strongly convergent Riemann integral (the integrals over finite regions are norm-convergent Riemann integrals but the limit as $R \to \infty$ for $\int_{|p| \le R, |q| \le R} \cdots$ is only strongly convergent).

This example motivates the following

Definition. A set of *coherent states* is a triple $(\mathcal{H}, X, \mu)$ of a Hilbert space, $\mathcal{H}$, a locally compact, σ-compact topological space, X, and a Baire measure, μ, on X and a map, $x \mapsto \varphi(x)$ of $X \to \mathcal{H}$, of X into the nonzero vectors in $\mathcal{H}$ so that

(i) $x \mapsto \varphi(x)$ is continuous (in the norm topology on $\mathcal{H}$).
(ii) For any $\psi \in \mathcal{H}$

$$\int |\langle \varphi(x), \psi \rangle|^2 d\mu(x) = \|\psi\|^2 \qquad (4.4.13)$$

If $\|\varphi(x)\| = 1$, we say we have *normalized coherent states*.

By letting $\widetilde{\varphi}(x) = \varphi(x)/\|\varphi(x)\|$ and $d\widetilde{\mu} = \|\varphi(x)\|^2 d\mu(x)$, we get a normalized family, so we will henceforth assume, unless noted otherwise, that our family is normalized. There are interesting examples (see below), when X is an open subset of $\mathbb{C}^n$ and $x \mapsto \varphi(x)$ is antianalytic (so $x \mapsto \langle \varphi(x), f \rangle$ is analytic). Passing to $\widetilde{\varphi}$ normally destroys this analyticity, so there is sometimes a reason not to normalize.

Given a set of (not necessarily normalized) coherent states, if $K \subset X$ is compact, $x \mapsto \varphi(x)$ is uniformly continuous and bounded on K so uniformly in $\{\psi \mid \|\psi\|_{\mathcal{H}} < 1\}$, the functions

$$h_\psi(x) = \langle \varphi(x), \psi \rangle \qquad (4.4.14)$$

are uniformly bounded on K and uniformly equicontinuous there. Thus

$$Y = \{h_\psi(x) \mid \psi \in \mathcal{H}\} \qquad (4.4.15)$$

is a L^2-reproducing kernel Hilbert space in the sense of Problem 6 of Section 3.3 of Part 1. Conversely, given such an L^2-reproducing kernel Hilbert space, the vectors $\varphi(x) = k_x$ constructed in that problem are a family of coherent states. We thus see that coherent states are a redefintion of L^2-reproducing Hilbert space. There is a change of emphasis however—in the reproducing kernel Hilbert space view, the fundamental object is the family of functions, Y, and in the coherent states view, it is the $\varphi(x)$. Of course, (4.4.14) links the two views.

Example 4.4.2 (Bergman Coherent States). Let $\mathfrak{A}_2(\mathbb{D})$ be functions analytic on $\mathbb{D}$ and in $L^2(\mathbb{D}, d^2z)$. Then, a use of Cauchy estimates shows that $\{f \in \mathfrak{A}_2(\mathbb{D}) \mid \|f\|_2 \leq 1\}$ is uniformly bounded and uniformly equicontinuous on each compact $K \subset \mathbb{D}$. It follows that $\mathfrak{A}_2(\mathbb{D})$ is an L^2-reproducing Hilbert space and so there is a coherent state representation. One can compute the coherent states in closed form (for $\mathfrak{A}_2(\mathbb{D})$ but not for general $\mathfrak{A}_2(\Omega)$). This was done in Example 12.5.5 of Part 2B. One finds the unnormalized coherent states are

$$\varphi_w(z) = \pi^{-1}(1 - \bar{w}z)^{-2} \tag{4.4.16}$$

The φ_w are not normalized, instead

$$\langle \varphi_w, \varphi_w \rangle = \varphi_w(w) = \pi^{-1}(1 - \bar{w}w)^{-2}$$

so the normalized coherent states are

$$\tilde{\varphi}_w(z) = \pi^{-1/2}(1 - |w|^2)/(1 - \bar{w}z)^2 \tag{4.4.17}$$

(which is no longer anti-analytic in w) and the normalized measure is

$$d\tilde{\mu}(z) = \pi^{-1}(1 - |z|^2)^{-2}d^2z \tag{4.4.18}$$

Notice that $\int_{\mathbb{D}} d\tilde{\mu}(z) = \infty$. Section 12.5 of Part 2B has more on this example. ☐

Example 4.4.3 (Paley–Wiener Coherent States). Let PW_π be the subset of $L^2(\mathbb{R}, dx)$ with $\text{supp}(\hat{f}) \subset [-\pi, \pi]$. This Paley–Wiener space was studied in the second proof of Theorem 6.6.16 and Problem 20 of Section 6.6 of Part 1. If

$$\varphi_y(x) = \text{sinc}(x - y) \tag{4.4.19}$$

$(\text{sinc}(z) = \sin(\pi z)/\pi z)$, then

$$\hat{\varphi}_y(k) = (2\pi)^{-1/2}e^{-iyk}\chi_{[-\pi,\pi]}(k) \tag{4.4.20}$$

so $\varphi_y(x) \in PW_\pi$. If $f \in PW_\pi, \chi_{[-\pi,\pi]}(k)\hat{f}(k) = \hat{f}(k)$ so by the Fourier inversion formula

$$f(y) = \langle \overline{\hat{\varphi}}_y, \hat{f} \rangle = \langle \varphi_y, f \rangle \tag{4.4.21}$$

Thus,

$$\|f\|^2 = \int |f(y)|^2 dy = \int |\langle \varphi_y, f \rangle|^2 dy \tag{4.4.22}$$

and so $\{\varphi_y\}_{y \in \mathbb{R}}$ is a family of normalized coherent states. ☐

It is useful sometimes to consider a slightly weaker notion—a family of *coherent projections* is a triple $(\mathcal{H}, X, \mu)$ of a Hilbert space, a locally compact σ-compact topological space, X, and Baire measure, μ, on X and a map,

$x \mapsto P(x)$, of X to the rank one orthogonal projections on $\mathcal{H}$ so that

(i) $x \mapsto P(x)$ is norm-continuous.

(ii) For any $\psi \in \mathcal{H}$

$$\int \langle \psi, P(x)\psi \rangle d\mu(x) = 1 \tag{4.4.23}$$

If $\varphi(x)$ is a family of normalized coherent states, then

$$P(x) = \langle \varphi(x), \cdot \rangle \varphi(x) \tag{4.4.24}$$

is a family of coherent projections. In general (Problem 2), one can pick $\varphi(x)$ so (4.4.24) holds for piecewise continuous φ, but there might be obstructions to a globally continuous choice. The following elementary result is of interest even for normalized coherent states:

Proposition 4.4.4. *For a family of coherent projections*

$$\int d\mu(x) = \dim(\mathcal{H}) \tag{4.4.25}$$

Remark. If $\dim(\mathcal{H}) = \infty$, this means $d\mu$ is not a finite measure

Proof. For any finite orthogonal set, $\{\psi_n\}_{n=1}^N$, if $P(x) = \langle \varphi(x), \cdot \rangle \varphi(x)$, then

$$\sum_{n=1}^N \langle \psi_n, P(x)\psi_n \rangle = \sum_{n=1}^N |\langle \varphi(x), \psi_n \rangle|^2 \leq 1 \tag{4.4.26}$$

by Bessel's inequality, with equality if $\dim(\mathcal{H}) = N < \infty$. Thus, integrating, and using $\int \langle \psi_n, P(x)\psi_n \rangle d\mu(x) = \|\psi_n\|^2 = 1$, we see that

$$N \leq \int d\mu(x) \tag{4.4.27}$$

with equality if $\dim(\mathcal{H}) = N < \infty$. Since N can be taken arbitrarily if $\dim(\mathcal{H}) = \infty$, we have (4.4.25) in general. $\qquad \square$

Before turning to more examples, we make a brief aside on representing operators, $A \in \mathcal{L}(\mathcal{H})$, by functions on X.

Definition. If $(\mathcal{H}, X, \mu), \{P(x)\}_{x \in X}$ is a family of coherent projections and $A \in \mathcal{L}(\mathcal{H})$, we define the *lower symbol*, $L(A)$, to be the continuous function on X:

$$L(A)(x) = \mathrm{Tr}(AP(x)) \tag{4.4.28}$$

If there is an L^∞-function, f, on X so that

$$A = \int f(x)P(x)d\mu(x) \tag{4.4.29}$$

we say f is an *upper symbol* for A, written $U(A) = f$.

Remarks. 1. To define (4.4.28), one doesn't need to know about Tr on infinite-dimensional $\mathcal{H}$ (see Section 3.6 of Part 4); if $P(x) = \langle \varphi(x), \cdot \rangle \varphi(x)$, then $\mathrm{Tr}(AP(x)) = \langle \varphi(x), A\varphi(x) \rangle$.

2. By the Schwarz inequality

$$\int |\langle \psi, P(x)\eta \rangle| d\mu \leq \int \|P(x)\psi\| \, \|P(x)\eta\| d\mu(x)$$

$$\leq \left(\int \|P(x)\psi\|^2 d\mu(x) \right)^{1/2} \left(\int \|P(x)\eta\|^2 d\mu(x) \right)^{1/2}$$

$$= \|\psi\| \, \|\eta\|$$

3. We do not claim that a given A must have an upper symbol. Moreover, upper symbols need not be unique. Indeed, if $d = \dim(\mathcal{H}) < \infty$, $\dim(\mathcal{L}(\mathcal{H})) = d^2$ while $\dim(L^\infty) = \infty$, so many f's must go to zero in (4.4.29), i.e., there are an infinite-dimensional family of f's if there is one obeying (4.4.29).

4. In the physics literature, lower symbols are sometimes called *Husimi symbols* and upper symbols, *Glauber–Sudarshan symbols*.

5. (4.4.29) can be defined as a weak integral.

Symbols are important partially because of the following *Berezin–Lieb inequalities*, which the reader will prove in Problem 3 and which, as we'll explain in the Notes, are useful in establishing the classical limit in quantum statistical mechanics:

$$\int \exp(-L(A)(x)) d\mu(x) \leq \mathrm{Tr}(e^{-A}) \leq \int \exp(-U(A)(x)) d\mu(x) \qquad (4.4.30)$$

Turning from the aside on symbols, we focus on a general group-theoretic construction of coherent states. We'll need some facts about groups and their representations. G will be a locally compact, σ-compact group. In Section 4.19 of Part 1, we proved that every such group has a left invariant measure (called *Haar measure*), i.e., for all $f \in L^1(G, d\mu)$ and all $y \in G$:

$$\int f(yx) d\mu(x) = \int f(x) d\mu(x) \qquad (4.4.31)$$

Recall that the measure μ_R defined by $d\mu_R(x) = d\mu(x^{-1})$, i.e.,

$$\int f(x) d\mu_R(x) = \int f(x^{-1}) d\mu(x) \qquad (4.4.32)$$

is right invariant. G is called unimodular if $d\mu_R = d\mu$. By definition, $d\mu(yx) = d\mu(x)$. Since such a measure is unique up to a constant, $d\mu(xy)$

is a multiple of $d\mu(x)$ so the modular function is defined by (see (4.19.5) of Part 1)

$$\int f(xy^{-1})d\mu(x) = \Delta(y)\int f(x)d\mu(x) \qquad (4.4.33)$$

A representation of G is a Hilbert space, $\mathcal{H}$, and a continuous map $U : G \to \mathcal{L}(\mathcal{H})$ (continuous in the strong operator topology on U) so that each $U(x)$ is unitary and $U(x)U(y) = U(xy)$, $U(x^{-1}) = U(x)^*$. U is called *irreducible* if its only invariant subspaces (i.e., $V \subset \mathcal{H}$ so $\varphi \in V \Rightarrow \forall x \in G$, $U(x)\varphi \in V$) are $\{0\}$ and $\mathcal{H}$. We discuss group representation in Section 6.8 of Part 4 and, in particular, prove Schur's Lemma: U is irreducible $\Leftrightarrow$ $\{U(x)\}'_{x \in G} = \{c\mathbb{1} \mid c \in \mathbb{C}\}$ where $\{B\}'_{B \in \mathbb{B}} = \{C \in \mathcal{L}(\mathcal{H}) \mid \forall B \in \mathbb{B}, CB = BC\}$.

Suppose $\varphi_0 \in \mathcal{H}$, $\{U(x)\}_{x \in G}$ is an irreducible representation. Define $\varphi(x) = U(x)\varphi_0$ and consider

$$A = \int \langle \varphi(x), \cdot \rangle \varphi(x)d\mu(x) \qquad (4.4.34)$$

where, for now, we ignore the issue of convergence of the integral and proceed formally. For any $y \in G$,

$$\begin{aligned}
U(y)AU(y^{-1}) &= \int \langle \varphi(x), U(y)^* \cdot \rangle U(y)\varphi(x)d\mu(x) \\
&= \int \langle U(y)\varphi(x), \cdot \rangle U(y)\varphi(x)d\mu(x) \\
&= \int \langle \varphi(yx), \cdot \rangle \varphi(yx)d\mu(x) \\
&= A \qquad (4.4.35)
\end{aligned}$$

Formally applying Schur's lemma, $A = c\mathbb{1}$, so if $d\tilde{\mu} = c^{-1}d\mu$, then $\langle \mathcal{H}, G, d\tilde{\mu} \rangle$ and $x \mapsto \varphi(x)$ are a set of coherent states. There are two problems with this argument. The integral in (4.4.34) may not make sense and, even if it does, it may not define a bounded operator.

It seems unlikely that the integral defining A will make sense unless the integral for $\langle \varphi_0, A\varphi_0 \rangle$ does. This leads to the definition.

Definition. A vector $\varphi_0 \in \mathcal{H}$, the space of an irreducible representation, U, of G is called *admissible* if and only if

$$\int |\langle \varphi_0, U(x)\varphi_0 \rangle|^2 d\mu(x) < \infty \qquad (4.4.36)$$

An irreducible representation with a nonzero admissible vector is called a *square integrable representation*.

If φ_0 is admissible, so is $U(y)\varphi_0$ for any y (by translation invariance of $d\mu$—see Problem 4). Thus, so are vectors in the finite linear span admissible.

Since the closure of this span is invariant, we see that in any square integrable representation, the set of admissible vectors is dense. The following is the basic result for group-theoretic coherent vectors; unfortunately the proof relies on some amount of unbounded operator theory (we'll give quotations to Part 4)—the reader can either accept the theorem or learn the needed unbounded operator theory from Part 4.

Theorem 4.4.5 (Grossmann–Morlet–Paul Theorem). *Let U be square integrable representation of a group, G, with Haar measure, μ.*

(a) *There is a strictly positive self-adjoint operator, C, so that $Q(C)$, its form domain, is exactly the set of admissible vectors and so that if φ_0 is admissible, then for all $\psi \in \mathcal{H}$*

$$\int |\langle \psi, U(x)\varphi_0 \rangle|^2 d\mu(x) = \langle C^{1/2}\varphi_0, C^{1/2}\varphi_0 \rangle \, \|\psi\|^2 \qquad (4.4.37)$$

(b) *If G is unimodular, all vectors are admissible and $C = c\mathbb{1}$ for a constant C.*

(c) *If φ_0 is admissible with $\|\varphi_0\| = 1$ and $\tilde{\mu} = \|C\varphi_0\|^{-2}\mu$, then $(\mathcal{H}, G, \tilde{\mu})$ with $\varphi(x) = U(x)\varphi_0$ are a family of coherent states.*

Remarks. 1. If G is compact, so $f = \dim(\mathcal{H}) < \infty$ (see Theorem 6.8.2 of Part 4), then by Proposition 4.4.4, $\tilde{\mu} = f\mu$ (where $\mu(G) = 1$).

2. By polarization, once we have (4.4.37), we get for φ_0, φ_1 admissible and $\psi_0, \psi_1 \in \mathcal{H}$,

$$\int \langle U(x)\varphi_1, \psi_0 \rangle \, \langle \psi_1, U(x)\varphi_0 \rangle d\mu(x) = \langle \varphi_1, C\varphi_0 \rangle \, \langle \psi_1, \psi_0 \rangle \qquad (4.4.38)$$

Especially in the unimodular case, this can be viewed as an extension of Schur's famous orthogonality relations—see the Notes.

Proof. (a) For φ_0 admissible define

$$\mathcal{D}_{\varphi_0} = \{\psi \in \mathcal{H} \mid \langle \psi, U(x)\varphi_0 \rangle \in L^2(G, d\mu)\} \qquad (4.4.39)$$

This is clearly a subspace and contains φ_0 since φ_0 is admissible. Moreover, if $\psi \in \mathcal{D}_{\varphi_0}$ and $y \in G$

$$\langle U(y)\psi, U(x)\varphi_0 \rangle = \langle \psi, U(y^{-1}x)\varphi_0 \rangle \qquad (4.4.40)$$

so by the left invariance of $d\mu$, $U(y)\psi \in \mathcal{D}_{\varphi_0}$ also, and if

$$q_{\varphi_0}(\psi) = \int |\langle \psi, U(x)\varphi_0 \rangle|^2 d\mu(x) \qquad (4.4.41)$$

then

$$q_{\varphi_0}(U(y)\psi) = q_{\varphi_0}(\psi) \qquad (4.4.42)$$

Since $\mathcal{D}_{\varphi_0}$ is invariant under $\{U(y)\}_{y\in G}$, its closure is also, so, by irreducibility, $\mathcal{D}_{\varphi_0}$ is dense. Moreover, since $|\langle\psi, U(x)\varphi_0\rangle|^2$ obeys a parallelogram law, so does q_{φ_0}, i.e., q_{φ_0} is a positive quadratic form in the sense of Section 7.5 of Part 4. We claim that it is a closed form, i.e., if $\{\psi_n\}_{n=1}^{\infty}\subset\mathcal{D}_{\varphi_0}$, $\psi_n\to\psi_\infty$ in $\mathcal{H}$ and $\lim_{N\to\infty}\sup_{n,m\geq N}q_{\varphi_0}(\psi_n-\psi_m)=0$, then $\psi_\infty\in\mathcal{D}_{\varphi_0}$ and $q_{\varphi_0}(\psi_n-\psi_\infty)\to 0$.

Given such a sequence, let $f_n(x)=\langle\psi_n, U(x)\varphi_0\rangle\in L^2(G, d\mu)$. By hypothesis, $f_n(x)$ is Cauchy in $L^2(G, d\mu)$, so there is f_∞ in $L^2(G, d\mu)$ so that $\|f_n-f_\infty\|_{L^2}\to 0$. By passing to a subsequence, we can suppose for a.e. x, $f_n(x)\to f_\infty(x)$. Since $\psi_n\to\psi_\infty$ in $\mathcal{H}$, $f_n(x)\to\langle\psi_\infty, U(x)\varphi_0\rangle$. Thus $f_\infty(x)=\langle\psi_\infty, U(x)\varphi_0\rangle$, so $\psi_\infty\in\mathcal{D}_{\varphi_0}$ and by the L^2-convergence, $q_{\varphi_0}(\psi_n-\psi_\infty)\to 0$.

We have thus proven that q is closed. By Theorem 7.5.5 of Part 4, there is a positive self-adjoint operator, M, so that $Q(M)=\mathcal{D}_{\varphi_0}$ and $q_{\varphi_0}(\psi)=\langle\psi, M\psi\rangle$. By (4.4.43), each $U(y)$ leaves $Q(M)$ invariant and $\langle U(y)\psi, MU(y)\psi\rangle=\langle\psi, M\psi\rangle$. Thus each $U(y)$ defines a unitary operator of $\mathcal{H}_{+1}=Q(M)$ with norm $\|\psi\|_{+1}=(\langle\psi, M\psi\rangle+\|\psi\|^2)^{1/2}$. In terms of the scale, $\mathcal{H}_{+1}\subset\mathcal{H}\subset\mathcal{H}_{-1}$, discussed in Section 7.5 of Part 4, we see that, by duality, each $U(y)$ is unitary on $\mathcal{H}_{-1}$. This means that, for all $y\in G$, $U(y)(M+1)^{-1}U(y)^{-1}=(M+1)^{-1}$. Therefore, by Schur's lemma, $(M+1)^{-1}$ is a nonzero constant. It follows that $Q(M)=\mathcal{H}$ and $q_{\varphi_0}(\psi)=c(\varphi_0)\|\psi\|^2$.

As above, $c(\varphi_0)$, obeys a parallelogram law, and $c(\varphi_0)$ is a closed form. Thus, for some positive operator, C, $C(\varphi_0)=\langle\varphi_0, C\varphi_0\rangle$. This proves (4.4.37).

(b) Fix an admissible φ_0 with $\|\varphi_0\|=1$. Let $f(x)=\langle\varphi_0, U(x)\varphi_0\rangle$. Then

$$c(\varphi_0)=\int|f(x)|^2 d\mu(x) \tag{4.4.43}$$

and

$$c(U(y)\varphi_0)=\int|f(y^{-1}xy)|^2 d\mu(x) \tag{4.4.44}$$

$$=\int|f(xy)|^2 d\mu(x) \tag{4.4.45}$$

$$=\Delta(y)^{-1}c(\varphi_0) \tag{4.4.46}$$

So, in the unimodular case, $c(U(y)\varphi_0)=c(\varphi_0)$ and the same analysis we did to conclude that $q_{\varphi_0}(f)$ is $\mathrm{const}\|f\|^2$ implies $c(\varphi_0)=\mathrm{const}\|\varphi_0\|^2$, i.e., every state is admissible and

$$\int|\langle\psi, U(x),\varphi_0\rangle|^2 d\mu(x)=c\|\psi\|^2\|\varphi_0\|^2 \tag{4.4.47}$$

(c) is immediate from (4.4.37). $\qquad\square$

Example 4.4.6 (Bloch Coherent States). Let $\mathbb{SU}(2)$ be the group of 2×2 unitary matrices of determinant 1 (thought of as a two-fold cover of $\mathbb{SO}(3)$, the 3×3 rotation matrices; see the Notes). The irreducible representations, U_ℓ, of $\mathbb{SU}(2)$ are all finite dimensional. $\ell = 0, \frac{1}{2}, 1, \ldots$ and U_ℓ is $2\ell + 1$-dimensional, i.e., acts on $\mathbb{C}^{2\ell+1}$. L_z is the infinitesimal generator of rotation about the z-axis; see the Notes. In U_ℓ, L_z has eigenvalues $\ell, \ell - 1, \ldots, -\ell$, each with multiplicity 1. The coherent states built from $\varphi_0^{(\ell)}$, the $L_z = \ell$ states are called the *Bloch coherent states*. There is a subgroup, H, of $\mathbb{SU}(2)$ so $x \in H \Rightarrow \varphi_0$ is an eigenvector of $U^{(\ell)}(x)$. One often uses projections labelled by $\mathbb{SU}(2)/H$ which is S^2, the two-dimensional sphere. This is the space for describing the classical limit, $\ell \to \infty$; see the Notes. Note that $\mathbb{SU}(2)$ is compact, so unimodular. $\square$

Example 4.4.7 (Classical Coherent States, Revisited). Let $g \in L^2(\mathbb{R}^\nu, d^\nu x)$. For $(\mathbf{p}, \mathbf{q}) \in \mathbb{R}^{2\nu}$, let $g^{(\mathbf{p},\mathbf{q})}(x) = e^{i\mathbf{p}\cdot\mathbf{x}}g(\mathbf{x} - \mathbf{q})$. The classical coherent vectors were the special case where g is a Gaussian. The classical Gaussian result can be viewed as coherent states for the Heisenberg group (where the phase drops out of $|\langle g^{(p,q)}, f \rangle|^2$ and so we can integrate over $\mathbb{R}^{2\nu}$ rather than $\mathbb{R}^{2\nu} \times \partial\mathbb{D}$). The Heisenberg group is unimodular, and the Gaussian calculation shows the standard representation is square integrable. By Theorem 4.4.5, all states are admissible and the constant is the same for all g, i.e., for any $g, f \in L^2(\mathbb{R}^\nu)$, we have that

$$\int |\langle g^{(p,q)}, f \rangle|^2 \frac{d^\nu p \, d^\nu q}{(2\pi)^\nu} = \|f\|^2 \|g\|^2 \tag{4.4.48}$$

Our proof so far of this is rather involved. We use the Segal-Bargmann transform to handle the case where g is Gaussian (the only case where this transform gives information) and then a Schur lemma argument to go to general f.

But there is a direct simple calculation! Note that

$$(2\pi)^{-\nu/2} \int \overline{g^{(p,q)}(x)} f(x) d^\nu x = \widehat{(g^{(0,q)}f)}(p) \tag{4.4.49}$$

Thus by the Plancherel theorem

$$(2\pi)^{-\nu} \int |\langle g^{(p,q)}, f \rangle|^2 d^\nu p = \int |g^{(0,q)}f(x)|^2 d^\nu x \tag{4.4.50}$$

Therefore

$$\text{LHS of (4.4.48)} = \int |g(x-q)|^2 \, |f(x)|^2 \, d^\nu x \, d^\nu q \tag{4.4.51}$$

$$= \|g\|_2^2 \, \|f\|_2^2 \tag{4.4.52}$$

by first fixing x and doing the q integral.

In typical application, g is of compact support or at least of rapid decay. Thus, $\langle g^{(p,q)}, f \rangle$ is a Fourier transform of f cut off by multiplying by (a translation of) g. It is called the *windowed Fourier transform.*. $\qquad\square$

Example 4.4.8 (Continuous Wavelets). *Wavelets* are coherent states for the affine group. We'll look at $\mathbb{R}$ rather than $\mathbb{R}^\nu$, in part because it is simpler, but also because a main application is to signal processing where the variable is time. The group is parameterized by $b \in \mathbb{R}$ and $a \in \mathbb{R} \setminus \{0\}$. We are interested in the representation on $L^2(\mathbb{R}, dx)$

$$(U(a,b)\varphi) = |a|^{-1/2}\varphi((x-b)/a) \qquad (4.4.53)$$

The group operation is

$$(a,b)(\tilde{a}, \tilde{b}) = (a\tilde{a}, \tilde{b} + \tilde{a}b) \qquad (4.4.54)$$

As we saw in Problem 5 of Section 4.19 of Part 1,this group is not unimodular; left invariant Haar measure is given by

$$d\mu(a,b) = (da\, db)/|a|^2 \qquad (4.4.55)$$

We note that in that problem and in many situations, one takes $a \in (0, \infty)$ rather than $a \in \mathbb{R}\setminus\{0\}$. One reason that we take the larger group (the more typical group is a subgroup of index 2) is because the representation (4.4.53) is irreducible (see Problem 5) when we include $a = -1$. Without $a < 0$, the set of $\{\varphi \mid \hat{\varphi}(k) = 0 \text{ for } k < 0\}$ is a nontrivial invariant subspace. This said, one can prove completeness of the coherent states generated by applying to the smaller group φ if $\overline{\varphi(x)} = \varphi(x)$ (Problem 6).

Given $g, f \in L^2(\mathbb{R})$, let $g^{(a,b)}(x) = a^{-1/2}g\left(\frac{x-b}{a}\right)$ and note that, by the Plancherel theorem and

$$\hat{g}^{(a,b)}(k) = a^{1/2}e^{-ikb}\hat{g}(ak) \qquad (4.4.56)$$

we see that

$$\langle g^{(a,b)}, f \rangle = a^{1/2} \int e^{ikb}\overline{\hat{g}(ak)}\,\hat{f}(k)dk \qquad (4.4.57)$$

$$= (2\pi a)^{1/2}\check{h}(b) \qquad (4.4.58)$$

where $h(k) = \overline{\hat{g}(ak)}\hat{f}(k)$.

Thus, by the Plancherel theorem again

$$\int |\langle g^{(a,b)}, f \rangle|^2 db = (2\pi a)\int |\hat{g}(ak)|^2\,|\hat{f}(k)|^2 dk \qquad (4.4.59)$$

Since $a\, da/a^2 = da/a$ is scale invariant, we see that

$$\int |\langle g^{(a,b)}, f \rangle|^2 \frac{da\, db}{a^2} = 2\pi \int |\hat{g}(a)|^2\,|\hat{f}(k)|^2 dk\frac{da}{a} \qquad (4.4.60)$$

$$= C(g)\|f\|$$

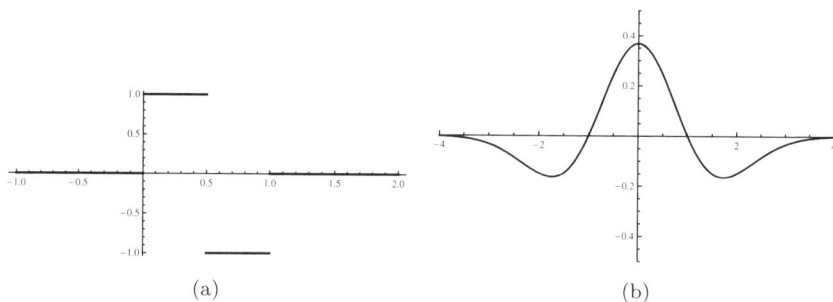

Figure 4.4.1. (a) The Haar and (b) Mexican hat wavelets.

where

$$C(g) = 2\pi \int |\hat{g}(a)|^2 \frac{da}{a} \tag{4.4.61}$$

We have thus proven the result below.   □

Theorem 4.4.9. *The representation* (4.4.53) *of the one-dimensional affine group is square integrable.* $\varphi \in L^2(\mathbb{R}, dx)$ *is admissible if and only if* $C(\varphi)$ *given by* (4.4.61) *is finite.*

The function φ generating the coherent states is called the *mother wavelet*. One popular example of a continuous wavelet are *the Haar wavelets*, those with the *Haar function* as the mother wavelet, i.e., where (Figure 4.4.1 (a))

$$\varphi(t) = \begin{cases} 1, & 0 \le t < \frac{1}{2} \\ -1, & \frac{1}{2} \le t \le 1 \\ 0, & t < 0 \text{ or } t \ge 1 \end{cases} \tag{4.4.62}$$

Since $\hat{\varphi}(0) = 0$, and $\varphi \in L^2$, we have that $C(\varphi) < \infty$. Another way is to arrange $\varphi = \eta'$ since then $\hat{\varphi}(k) = (ik)\hat{\eta}(k)$ will also vanish at $k = 0$ if η is smooth. Another popular choice is the *Mexican hat wavelet* (aka, *Ricker wavelet*), where the mother wavelet is (Figure 4.4.1 (b))

$$\varphi_\sigma(t) = \frac{2}{(\sqrt{3\sigma})\pi^{1/4}} \left(1 - \frac{t^2}{\sigma^2} \right) e^{-t^2/2\sigma^2} \tag{4.4.63}$$

(normalized so that $\|\varphi_\sigma\|^2 = 1$).

In this section, we discussed continuous expansions $f = \int \langle \varphi(x), f \rangle \, \varphi(x_0) \, d\mu(x)$. The set $\{\varphi(x)\}_{x \in X}$ is overcomplete except in the trivial case. In the analytic case, like the classical coherent states, for any ε, $\{\varphi(x)\}_{x \le \varepsilon}$ is complete in that by analyticity, if $\langle \varphi(x), \psi \rangle = 0$ for $|x| \le \varepsilon$, then $\langle \varphi(x), \psi \rangle = 0$ for all x so $\psi = 0$, i.e., the span of those φ is all of $\mathcal{H}$. This suggests that a discrete set of $\varphi(x)$ should be sufficient. In the next section, we'll consider

Gabor lattices, i.e., $\{e^{2\pi ibm}g(x-na)\}$ for a,b fixed and $m,n \in \mathbb{Z}$ and in the section after discrete wavelets, i.e., $\varphi(2^{-n}(x-m))$.

Notes and Historical Remarks. The modern theory of coherent states dates from foundational papers in the physics literature from the early 1960's. For the classical case, this literature defined for $z \in \mathbb{C}$:

$$\psi(z) = e^{-\frac{1}{2}|z|^2} \sum_{n=0}^{\infty} (n!)^{-1/2} z^n \varphi_n \qquad (4.4.64)$$

where φ_n is the eigenfunctions of $H_0 = -\frac{1}{2}\frac{d^2}{dx^2} + \frac{1}{2}x^2 - \frac{1}{2}$ with $H_0\varphi_n = n\varphi_n$, and noted that

$$\int \langle \psi(z), \cdot \rangle \psi(z) \frac{d^2 z}{\pi} = \mathbb{1} \qquad (4.4.65)$$

They also noted that the $\psi(z)$ are eigenfunctions of A:

$$A\psi(z) = z\psi(z) \qquad (4.4.66)$$

These states and representations appeared first in a 1960 paper of Klauder [**472**] with follow-up papers on an abstract theory of coherent states [**473, 474**]. In 1963, Glauber [**316, 317**], in fundamental work in quantum optics for which he received the Nobel prize, restated this representation and gave it the name "coherent states" since this described the coherent phase collections of photons needed to describe laser light. Shortly thereafter, Sudarshan [**792**] discussed representations of operators by integrals of coherent projections, i.e., what we have called upper symbols.

Earlier and/or contemporaneously not unrelated ideas appeared in the mathematics and physics literature. As we discussed in the Notes to Section 3.3 of Part 1, examples of reproducing kernel Hilbert spaces go back to 1907, and were extensively studied in the 1920's and a general theory of them was developed in the 1940's, but the specific class of those which we call L^2-reproducing kernel Hilbert spaces was not emphasized and the key objects were not the same as in the theory of coherent states.

As we discussed in Section 6.2 of Part 1, in the 1962–1963 period, Bargmann [**37**] and Segal [**722**] were discussing the Fock spaces of analytic functions which includes the $e^{|z|^2/2}\psi(z)$ of (4.4.59) and realizes (4.4.65). Bargmann was on Klauder's Ph.D. exam committee and this thesis contained the material of [**472**], so Bargmann may have been motivated by Klauder.

There is also a thread involving square integrable representations, certain infinite-dimensional representations on unimodular groups where one has

$$\int \overline{\langle \psi, U(a)\varphi \rangle} \, \langle \tilde{\psi}, U(a)\tilde{\varphi} \rangle \, da = c\langle \tilde{\psi}, \psi \rangle \, \langle \varphi, \tilde{\varphi} \rangle \qquad (4.4.67)$$

which, as we've seen, essentially define coherent states. In the mathematical literature, this goes back at least to Bargmann in 1947 [**36**] who found this for certain representations of $\mathbb{SL}(2, \mathbb{R})$ (which is a two-fold cover of the three-dimensional Lorentz group which Bargmann was explicitly studying) and more generally to Godement [**323**]. In the physics literature, Moyal found in 1949 [**590**] that (4.4.67) holds for the standard representation of the Heisenberg group (he was not explicitly looking at representations but at the action of x-space and k-space translations in quantum mechanics). Moyal's work was partial motivation for Klauder.

Closely related to this work on coherent states is the work of Gabor [**299**], done in the analysis of signals, but it was only in the 1980's that the relations between "Gabor analysis" in signal theory and coherent states in physics was appreciated.

In 1946, Bloch [**84**] wrote down a family of states of a quantum spin which was discussed in Example 4.4.6 and which are $\mathbb{SU}(2)$-coherent states. He never noted an analog of (4.4.65) so was far from the full theory of coherent states, but interestingly enough, he called the states he wrote down coherent mixtures.

The physicist's name Husimi for lower symbols and Glauber-Sudarshan for upper symbols come from Husimi [**413**], Glauber [**318**], and Sudarshan [**792**]. The names upper and lower symbol that we use are due to Berezin [**58**]. The Berezin-Lieb inequalities are from this paper and Lieb [**529**]. The simple proof of them that we give in Problem 3 are from Simon [**743**] although he says that the proofs are due to Lieb.

Lieb invented his inequalities to study the infinite spin limit of the pressure of quantum spin systems. When the Hamiltonians are quadratic, the fact that one gets upper and lower bounds allows one to control the infinite volume limit. The manifold for the Bloch coherent states is S^2 which is where the classical (infinite spin) limit lives.

For general compact Lie groups, the classical limits are coadjoint orbits and the sequences of spins are associated with multiples of maximal weights. Coherent states for this situation are discussed by Barut–Giradello [**41**], Gilmore [**313**], Perelomov [**638**], and Simon [**743**]. In particular, Simon discusses the classical limits of the pressure by extending Lieb's ideas.

Theorem 4.4.5 is due to Grossmann, Morlet, and Paul [**343**]. They use operator methods rather than form methods but the key idea of finding a closed form of an operator by taking all vectors for which something is L^2 is from their paper. Prior to their work, much of the physics literature's either appealed to Schur's lemma without worrying about the fact that operators were only formally defined and not a priori bounded or did calculations in explicit groups. In the mathematical literature, there was the work of

Bargmann and Godement on the unimodular case quoted above and work on certain square integrable representations of nonunimodular groups. [**230**, **139**, **642**].

For finite and compact groups, there are the celebrated *orthogonality relations* of Schur (see Theorem 6.8.13 of Part 4): If U and V are two irreducible representations on $\mathcal{H}$ and $\mathcal{K}$, respectively, and $\varphi, \psi \in \mathcal{H}, \eta, \kappa \in \mathcal{K}$, then

$$\int \overline{\langle \varphi, U(x)\psi \rangle} \, \langle \eta, V(x)\kappa \rangle d\mu(x) = \begin{cases} 0 & \text{if} \quad U \ncong V \\ d(U)\langle \varphi, \psi \rangle \, \langle \eta, \kappa \rangle & \text{if} \quad U \cong V \end{cases} \quad (4.4.68)$$

where $U \ncong V$ means not unitarily equivalent while $d(U) = \dim(\mathcal{H})$ which is finite in the finite or compact case. Here $d\mu$ is a normalized (i.e., $\mu(G) = 1$) Haar measure. (4.4.37) can be viewed as a generalization of this to the general nonunimodular square integrable representation (especially clear in the unimodular case).

As noted already, Bloch coherent states go back to Bloch [**84**]. For discussion of $\mathbb{SU}(2)$ as a cover of $\mathbb{SO}(3)$ and its Lie algebra and the representation thereof, see Simon [**746**] and Problem 19 of Section 7.3 of Part 2A.

The basic calculation of (4.4.60)/(4.4.61) for coherent states in the natural representation of the affine group on $L^2(\mathbb{R})$ goes back at least to Aslaksen–Klauder [**18**] in 1969. A few years earlier, Calderón [**129**] had proven in passing that in $\mathbb{R}^\nu$ if $\int \frac{|\hat{\varphi}(k)|^2}{|k|^\nu} d^\nu k < \infty$ then for an explicit φ-dependent constant, C_φ, with $\varphi_t(x) = t^{-\nu}\varphi(x/t)$, one has ($\|\cdot\| = L^2(\mathbb{R}^\nu)$-norm)

$$\int_0^\infty \|f * \varphi_t\|^2 \frac{dt}{t} = C_\varphi \|f\|^2 \quad (4.4.69)$$

a result (called the *Calderón reproducing formula*) equivalent when $\nu = 1$ to the coherent state resolution of the identity for the affine group. In 1984, Grossmann–Morlet [**342**] rediscovered the affine group coherent states in the context of signal analysis (specifically of seismic waves), and it is this use that helped jump start wavelet theory and their paper is most often quoted in the wavelet theory literature.

Problems

1. By using the integral kernel of the Segal—Bargmann transform, verify (4.4.7)

2. If $x \mapsto P(x)$ is a norm continuous family of rank one projections, prove for any $x_0 \in X$, there is an open neighborhood, N, of x_0 and $\{\varphi(x) \mid x \in N\}$ so that φ is continuous on N, $\|\varphi(x)\| = 1$ and $P(x) = \langle \varphi(x), \cdot \rangle \varphi(x)$. (*Hint*: Pick $\varphi(x_0)$, and then for x near x_0 so $\langle \varphi(x_0), \varphi(x) \rangle > 0$.)

3. This problem will prove the Berezin–Lieb inequalities, (4.4.30). It will use Jensen's inequality, (5.3.66) of Part 1, which says that for any bounded function, $f(x)$, and probability measure, $d\mu(x)$, we have that

$$\exp\left(\int f(x)d\mu(x)\right) \le \int e^{f(x)}d\mu(x) \tag{4.4.70}$$

We'll prove them only for the case that A is an Hermitian operator on a finite-dimensional space. Then there is an orthonormal family, $\{\varphi_j\}_{j=1}^{N}$ and real $\{\mu_j\}_{j=1}^{N}$ so that

$$A\varphi_j = \mu_j\varphi_j \tag{4.4.71}$$

We also suppose $P(x)$ is a family of rank one projections and $d\mu$ a measure on X so that

$$\int P(x)d\mu(x) = \mathbb{1} \tag{4.4.72}$$

(a) Let $c_j(x) = \langle \varphi_j, P(x)\varphi_j \rangle$. Prove that for all x

$$\sum_{j=1}^{n} c_j(x) = 1 \tag{4.4.73}$$

(b) If $L(A)(x) = \text{Tr}(AP(x))$ prove that

$$L(A)(x) = \sum_{j=1}^{N} \mu_j c_j(x) \tag{4.4.74}$$

and that

$$\text{Tr}(P(x)e^A) = \sum_{j=1}^{N} c_j(x)e^{\mu_j} \tag{4.4.75}$$

(c) Use Jensen's inequality for the measure on $\{1,\ldots,N\}$ with weights $c_j(x)$ to prove that for each x, we have that

$$\exp(L(A)(x)) \le \text{Tr}(P(x)e^A) \tag{4.4.76}$$

and conclude the first Berezin–Lieb inequality

$$\int e^{L(A)(x)}d\mu(x) \le \text{Tr}(e^A) \tag{4.4.77}$$

(d) Suppose that

$$A = \int U(A)(x)P(x)d\mu(x) \tag{4.4.78}$$

For each $j = 1, \ldots, N$, prove that

$$\langle \varphi_j, e^A \varphi_j \rangle = \exp \left(\int U(A)(x) c_j(x) d\mu(x) \right)$$

$$\leq \int e^{U(A)(x)} c_j(x) d\mu(x) \qquad (4.4.79)$$

and conclude the second Berezin–Lieb inequality

$$\mathrm{Tr}(e^A) \leq \int e^{U(A)(x)} d\mu(x) \qquad (4.4.80)$$

4. Given a representation, $U(x)$ of a locally compact, σ-compact group, G, on a Hilbert space, $\mathcal{H}$, define for $x \in G$ and $\psi \in \mathcal{H}$

$$f_\psi(x) = \langle \psi, U(x)\psi \rangle \qquad (4.4.81)$$

(a) For any $x, y \in G$ and $\varphi_0 \in \mathcal{H}$, show that

$$f_{U(y)\varphi_0}(x) = f_{\varphi_0}(y^{-1}xy) \qquad (4.4.82)$$

(b) If μ is (left-invariant) Haar measure and $\Delta(y)$ is the modular function, prove that

$$\int |f_{U(y)\varphi_0}(x)|^2 d\mu(x) = \Delta(y^{-1}) \int |f_{\varphi_0}(x)|^2 d\mu(x) \qquad (4.4.83)$$

and conclude that φ_0 is admissible if and only if $U(y)\varphi_0$ is admissible.

5. (a) If $\mathcal{H} \subset L^2(\mathbb{R}, dx)$ is invariant under all maps $U(1, b)$ (U given by (4.4.53)), prove that there is a Borel set, $K \subset \mathbb{R}$ so that $\mathcal{H} = \{f \mid f(k) = 0 \text{ for all } k \notin K\}$.

(b) If $\mathcal{H} \subset L^2(\mathbb{R}, dx)$ is invariant under all maps $U(a, b)$ with $a > 0, b \in \mathbb{R}$, prove that $\mathcal{H}$ is one of four subspaces.

(c) If $\mathcal{H} \subset L^2(\mathbb{R}, dx)$ is invariant under all maps $U(a, b)$ with $a \neq 0, b \in \mathbb{R}$, prove $\mathcal{H} = \{0\}$ or L^2, i.e., U is irreducible for this group.

6. Let g obey (4.4.61) and look at

$$A = \int_{b \in \mathbb{R}, a > 0} \langle g^{(a,b)}, \cdot \rangle g^{(a,b)} \frac{da \, db}{a^2}$$

(a) Prove that A is a bounded operator that commutes with all $(U(a, b))_{b \in \mathbb{R}, a > 0}$.

(b) Prove that $A = \alpha P_+ + \beta P_-$ for some $\alpha, \beta \in \mathbb{R}$ and $P_\pm$ the projection onto all $f \in L^2$ with $\hat{f}(\pm k) = 0$ for $k \leq 0$.

(c) If $\overline{g(x)} = g(x)$, prove that $\alpha = \beta$.

4.5. Gabor Lattices

> Let us take for instance a short segment of Mozart's *Magic Flute*. If
> we represent this piece of music as a function of time, we may be able
> to perceive the transition from one note to the next, but we get little
> insight about which notes are in play. On the other hand the Fourier
> representation may give us a clear indication about the prevailing notes in
> terms of the corresponding frequencies, but information about the moment
> of emission and duration of the notes is masked in the phases. Although
> both representations are mathematically correct, but one does not have to
> be a member of the Vienna Philharmonic Orchestra to find neither of them
> very satisfying. According to our hearing sensations we would intuitively
> prefer a representation which is local both in time and frequency, like mu-
> sic notation, which tells the musician which note to play at a given moment.
>
> —*Hans G. Feichtinger and Thomas Strohmer*
> in Introduction to [**272**]

Fix $g \in L^2(\mathbb{R})$ (there are results for $\mathbb{R}^\nu$, but for simplicity, we focus on
$\nu = 1$). Fix $a, b > 0$. The *Gabor lattice* generated by g is the family of
functions

$$g_{(n,m)}(x) = e^{2\pi i m b x} g(x - na) \tag{4.5.1}$$

The case where g is given by φ_0 in (4.4.2) is called the *classical Gabor
lattice.*

In this section, we want to consider expansion in $\{g_{(n,m)}\}_{n,m \in \mathbb{Z}}$. If
$ab > 1$, we call the lattice, *subcritical*, if $ab < 1$, *supercritical*, and if $ab = 1$
critical. The case $ab = 1$ is also called a *von Neumann lattice*. Notice the
volume of a fundamental cell for the lattice is $2\pi ab$, so the critical case is
consistent with the notion "one state per 2π volume in phase space."

The simplest property that you can ask for a family of vectors concerns
completeness.

Definition. A family of vectors $\{\psi_k\}_{k=1}^\infty$ in a Hilbert space, $\mathcal{H}$, is called
complete if their span is dense in $\mathcal{H}$, equivalently if

$$\langle \varphi, \psi_k \rangle = 0, \quad k = 1, 2, \ldots \Rightarrow \varphi = 0 \tag{4.5.2}$$

If $\{\psi_k\}_{k=1}^\infty$ is not complete, we call it *incomplete*. If some proper subset
of $\{\psi_k\}_{k=1}^\infty$ is complete, we say $\{\psi_k\}_{k=1}^\infty$ is *overcomplete*. If the set remains
complete when any finite subset is removed, we call it *strongly overcomplete*.

Here is the basic result for the classical Gabor lattice.

Theorem 4.5.1. *The classical Gabor lattice is*

(a) *strongly overcomplete if $ab < 1$;*
(b) *incomplete if $ab > 1$;*
(c) *complete if $ab = 1$. The set remains complete if a single vector is
    removed and is incomplete if two vectors are removed.*

Completeness of the classical Gabor lattice involves ψ with $\mathcal{G}(\psi; p, q)$ vanishing at suitable $\{p_n, q_n\}$. By (4.4.7), it is thus constructed with possible zeros of the Segal–Bargmann transform of ψ. It is not surprising then that the key to the proof of Theorem 4.5.1 is the following

Proposition 4.5.2. *Fix $\alpha, \beta > 0$. There exists an entire analytic function g so that*

(a) *g vanishes exactly at the points*

$$\mathcal{L}_{\alpha, \beta} = \{\alpha n + i\beta m \mid n, m \in \mathbb{Z}\} \tag{4.5.3}$$

(b) *$\exp(-\frac{\pi}{2\alpha\beta}|z|^2)\, |g(z)|$ is doubly periodic, i.e., invariant under $z \to z + \alpha$ and $z \to z + i\beta$.*

Remarks. 1. The proof of this result will rely on analytic function theory presented in Part 2A. As one might expect, given the double periodicity, the theory of elliptic functions is critical.

2. The $\exp(\frac{\pi}{2\alpha\beta}|z|^2)$ growth for λ on $|g(z)|$ is the minimum forced by the zeros and Jensen's formula (see Section 9.8 of Part 2A). Cover the plane with rectangles of sides α and β centered at the points in $\mathcal{L}_{\alpha, \beta}$. A large disk $\mathbb{D}_r$ (r large) has area πr^2. Since each rectangle has area α, β, we expect (and can prove), that the number, $n(r)$, of points in $\mathbb{D}_r \cap \mathcal{L}_{\alpha, \beta}$ is close to $\pi r^2/\alpha\beta$ for large r. Thus for large r_0,

$$\int_0^{r_0} \frac{n(r)}{r}\, dr \cong \frac{\pi}{2\alpha\beta} r^2 \tag{4.5.4}$$

Thus, by Problem 3 of Section 9.8 of Part 2A, if g has zeros on $\mathcal{L}_{\alpha, \beta}$, then $\int_0^{2\pi} \log(|g(re^{i\theta})|)\, \frac{d\theta}{2\pi}$ must grow as $\frac{\pi}{2\alpha\beta} r^2$. This is the sense in which the growth can't be any slower.

Proof. As noted, we'll need some facts about elliptic functions, specifically, the Weierstrass σ-function (one could instead use Jacobi theta functions). Let

$$\sigma(z) = z \prod_{w \in \mathcal{L}_{\alpha, \beta} \setminus \{0\}} E_2(z/w), \quad E_2(z) = (1-z)e^{(z+\frac{1}{2}z^2)} \tag{4.5.5}$$

We need the following properties of σ (proven in Section 10.4 of Part 2A):

(1) There are ω_1, ω_2 so that

$$\sigma(z + \alpha) = -e^{\omega_1(z + \alpha/2)}\sigma(z), \quad \sigma(z + i\beta) = -e^{\omega_2(z + i\beta/2)}\sigma(z) \tag{4.5.6}$$

(2) ω_1 is real and $\omega_2 = iw_2$ with w_2 real.

(3) $\omega_1(i\beta) - \omega_2\alpha = 2\pi i$, equivalently

$$\beta\omega_1 - w_2\alpha = 2\pi \tag{4.5.7}$$

As an explanation of where these come from, one defines $\zeta = \sigma'/\sigma$, $\mathcal{P} = -\zeta'$. $\mathcal{P}$, the Weierstrass $\mathcal{P}$-function, is doubly periodic (and meromorphic). This implies ζ is doubly periodic up to a constant, i.e.,

$$\zeta(z + \alpha) = \zeta(z) + \omega_1, \quad \zeta(z + i\beta) = \zeta(z) + \omega_2 \tag{4.5.8}$$

(4.5.6) is an integrated form of this.

The Legendre relation (4.5.7) is just an expression that $\oint \zeta(z)dz = 2\pi i$ if one integrates around the boundary of the rectangle $\{x + iy \mid |x| \leq \frac{\alpha}{2}, |y| \leq \frac{\beta}{2}\}$. This is because ζ has a single simple pole in this rectangle.

The reality of ω_1 and $-i\omega_2$ follow from reality of $\mathcal{P}$ on $\mathbb{R}$ and $i\mathbb{R}$ (by conjugation symmetry of $\mathcal{L}$), $\zeta' = -\mathcal{P}$ and (4.5.8).

(4.5.6) implies that

$$
\begin{aligned}
|\sigma(x + iy + \alpha)| &= e^{\omega_1(x+\alpha/2)} |\sigma(x + iy)| \\
|\sigma(x + iy + i\beta)| &= e^{-w_2(y+\beta/2)} |\sigma(x + iy)|
\end{aligned}
\tag{4.5.9}
$$

Next notice that

$$A((x + \alpha)^2 - x^2) = 2A\alpha\left(x + \frac{\alpha}{2}\right) \tag{4.5.10}$$

Thus, we take

$$A = \frac{\omega_1}{2\alpha}, \quad B = -\frac{w_2}{2\beta} \tag{4.5.11}$$

and see that

$$\exp(-Ax^2 - By^2) |\sigma(x + iy)| \tag{4.5.12}$$

is doubly periodic.

If

$$C = \tfrac{1}{2}(A + B), \quad D = \tfrac{1}{2}(A - B) \tag{4.5.13}$$

then

$$Ax^2 + By^2 = C|z|^2 + D\operatorname{Re} z^2 \tag{4.5.14}$$

Therefore, if we define

$$g(z) = e^{-Dz^2}\sigma(z) \tag{4.5.15}$$

then

$$e^{-C|z|^2}|g(z)| \tag{4.5.16}$$

is doubly periodic.

Finally, note that

$$
\begin{aligned}
\tfrac{1}{2}(A + B) &= \frac{1}{2}\left[\frac{\omega_1}{2\alpha} - \frac{w_2}{2\beta}\right] = \frac{\beta\omega_1 - \alpha w_2}{4\alpha\beta} \\
&= \frac{\pi}{2\alpha\beta}
\end{aligned}
\tag{4.5.17}
$$

by (4.5.7). Thus, (4.5.16) is (b) of Proposition 4.5.2. $\square$

We also need the elementary

Lemma 4.5.3. *Let f be an entire analytic function. Then,*

$$\int |f(z)|^2 (1+|z|^2)^{-1} d^2 z < \infty \qquad (4.5.18)$$

implies that f is 0.

Proof. By the Cauchy integral formula, for any $r > 0$,

$$f(0) = \int_0^{2\pi} f(re^{i\theta}) \frac{d\theta}{2\pi} \qquad (4.5.19)$$

so by the Schwarz inequality (and $\int_0^{2\pi} 1 \frac{d\theta}{2\pi} = 1$)

$$|f(0)|^2 \le \int_0^{2\pi} |f(re^{i\theta})|^2 \frac{d\theta}{2\pi} \qquad (4.5.20)$$

Thus,

$$2\pi |f(0)|^2 \int_0^R (1+r^2)^{-1} r \, dr \le \int_{|z| \le R} (1+|z|^2)^{-1} |f(z)|^2 d^2 z \qquad (4.5.21)$$

As $R \to \infty$, the right side is finite by (4.5.18). The left side is infinite unless $f(0) = 0$, so we conclude $f(0) = 0$.

Since (4.5.18) implies

$$\int |f(z)|^2 (1+|z-z_0|^2)^{-1} d^2 z < \infty$$

for all z_0, by applying the above to $f(z + z_0)$, we conclude $f(z_0) = 0$ for all z_0. $\qquad \square$

Proof of Theorem 4.5.1. By (4.4.8), we take

$$\alpha = \frac{a}{\sqrt{2}}, \quad \beta = \frac{2\pi b}{\sqrt{2}} \qquad (4.5.22)$$

so

$$\frac{\pi}{2\alpha\beta} = \frac{1}{2ab} \qquad (4.5.23)$$

If $ab > 1$, let g be the function of Proposition 4.5.2 with α, β given by (4.5.22). Then $e^{-z^2/2ab}|g|$ is doubly periodic and continuous, so bounded. Thus, since $ab > 1$, $1 - 1/ab > 0$ so

$$e^{-z^2/2}|g| = e^{-(1-1/ab)z^2/2}[e^{-z^2/2ab}|g|] \qquad (4.5.24)$$

is in $L^2(\mathbb{C}, d^2 z)$ which implies that $g \in \mathcal{F}$, the Fock space. Let $\psi = \mathbb{B}^{-1} g$. By (4.4.8), $\mathcal{G}(\psi; 2\pi nb, ma) = 0$ for all n, m. But $\psi \ne 0$, so $\psi \perp \{e^{2\pi inbx} \varphi_0(x - ma)\}$ proving incompleteness.

If $ab = 1$, and $z_0, z_1 \in \mathcal{L}_{\alpha,\beta}$, then $e^{-|z|^2/2}(z - z_0)^{-1}(z - z_1)^{-1} g$ is entire and in $L^2(\mathbb{C}, d^2 z)$ since $g(z_0) = g(z_1) = 0$, If $\psi = \mathbb{B}^{-1}((z - z_1)^{-1}(z - z_1)g)$

then $\psi \neq 0$ and $\psi \perp \{e^{2\pi i n b x}\varphi_0(x-ma)\}_{na+2\pi nb \neq z_0, z_1}$ proving that the von Neumann lattice with two vectors removed is not complete.

Suppose next that $ab \leq 1$ and $\psi \perp \{e^{2\pi i n b x}\varphi_0(x-ma)\}$, $m, n \in \mathbb{Z}$. Let $f = \mathbb{B}\psi$. Then $g(z) = 0 \Rightarrow f(z) = 0$ so $h(z) = f(z)/g(z)$ is an entire function.

Pick δ, so that $\{z \mid \delta < |z| < 2\delta\} \subset \{x + iy \mid |x| \leq \frac{a}{2}, |y| \leq \pi b\}$. As with (4.5.20),

$$|h(0)| = \frac{1}{3\pi\delta^2}\int_{\delta<|z|<2\delta}|h(z)|^2 d^2z \qquad (4.5.25)$$

Using the Poisson kernel (see Section 5.3 of Part 2A), we find C so that

$$\int_{|w|<\frac{\delta}{2}}|h(w)|^2 d^2w \leq C\int_{\delta<|z|<2\delta}|h(z)|^2 d^2z \qquad (4.5.26)$$

We conclude that for any entire function, h and $N \in \mathbb{Z}_+$

$$\int_{\substack{z=x+iy,|x|\leq(N+1/2)\alpha \\ |y|\leq(N+1/2)\beta}}|h(z)|^2 d^2z \leq C_1 \int_{\substack{z=x+iy,|x|\leq(N+1/2)\alpha \\ \text{dist}(z,\mathcal{L}_{\alpha\beta})\geq\delta/2,|y|\leq(N+1/2)\beta}}|h(z)|^2 d^2z \qquad (4.5.27)$$

We can write

$$|h(z)|^2 = \frac{e^{-|z|^2}|f(z)|^2}{e^{-|z|^2}|g(z)|^2} \qquad (4.5.28)$$

and note that $\exp(-|z|^2) \geq \exp(-|z|^2/ab)$ (since $(ab)^{-1} \geq 1$). If $B = \min_{\text{dist}(z,\mathcal{L}_{\alpha\beta})\geq\delta/2}[e^{-|z|^2/2ab}|g(z)|] > 0$, we conclude that for $\text{dist}(z, \mathcal{L}_{\alpha\beta}) \geq \delta/2$,

$$|h(z)|^2 \leq B^{-2}e^{-|z|^2}|f(z)|^2 \qquad (4.5.29)$$

so by (4.5.27), $h \in L^2(\mathbb{C}, d^2z)$. By Lemma 4.5.3, $h = 0$ so $f = 0$, i.e., we have completeness.

When $ab < 1$, we can replace $g(z)$ by $g(z)/\prod_{j=1}^k(z-z_j)$ for any distinct $z_1, z_2, \ldots, z_k \in \mathcal{L}_{a,b}$ since $\prod_{j=1}^k(z-z_j)^2 \exp(-|z|^2(\frac{1}{ab}-1))$ is bounded.

When $ab = 1$, for any $z_j \in \mathcal{L}_{a,b}$, we can replace $g(z)$ by $g(z)/(z-z_j)$ and conclude that $|h(z)|^2(1+|z|^2)^{-1/2} \in L^2$ and still see that $h = 0$. Thus, we get strong overcompleteness if $ab < 1$ and completeness with a single lattice point removed if $ab = 1$. $\qquad \square$

We want to look beyond what we've just discussed in two ways. First, we want to consider conditions on families of vectors distinct from but related to completeness and, second, we want to consider coherent states for the Heisenberg group for more general generating vectors than the Gaussian, φ_0, i.e., Gabor lattices that are not classical. We'll only do the latter if $ab = 1$; in fact, we'll take $a = b = 1$ which is no loss since we can scale

$x \mapsto \lambda x$, $p \mapsto \lambda^{-1}p$ (which just changes g to another g_λ—for the classical case where we want $g = \varphi_0$, we can't scale and reduce general a, b to $a = b$).

Definition. Let $\{\psi_k\}_{k=1}^{\infty}$ be a family of nonzero vectors in a Hilbert space, $\mathcal{H}$, with $\sup_k |\psi_k| < \infty$. We say that ψ

(a) is a *Bessel sequence* if there is $B > 0$ so that for every $g \in \mathcal{H}$

$$\sum_{k=1}^{\infty} |\langle \psi_k, g \rangle|^2 \leq B \|g\|^2 \tag{4.5.30}$$

(b) a *frame* if there exists $0 < A < B$ so that for every $g \in \mathcal{H}$

$$A\|g\|^2 \leq \sum_{k=1}^{\infty} |\langle \psi_k, g \rangle|^2 \leq B\|g\|^2 \tag{4.5.31}$$

(c) a *Riesz basis* if there is an orthonormal basis $\{\psi_k\}_{k=1}^{\infty}$ and an invertible map, T on $\mathcal{H}$ so that

$$\psi_k = T\varphi_k \tag{4.5.32}$$

Every frame is a Bessel sequence. Moreover, every Riesz basis is a frame since

$$\sum_{k=1}^{\infty} |\langle \psi_k, g \rangle|^2 = \sum_{k=1}^{\infty} |\langle \varphi_k, T^*g \rangle|$$

$$= \|T^*g\|^2 \tag{4.5.33}$$

Thus, (4.5.31) holds with $B = \|T\|^2$ and $A = \|T^{-1}\|^{-2}$ since

$$\|g\|^2 = \|(T^*)^{-1}T^*g\|^2$$

$$\leq \|(T^*)^{-1}\|^2 \|T^*g\|^2$$

A frame is clearly complete since $\langle \varphi_k, g \rangle = 0$ for all k implies $\|g\|^2 \leq A^{-1} \sum |\langle \varphi_k, g \rangle|^2 = 0$. If $\varphi_{2k} = \varphi_{2k-1}$ where $\{\varphi_k\}_{k=1}^{\infty}$ is an orthonormal basis, then we have a frame which remains complete with many possible removables, i.e., it is overcomplete. As we'll see shortly, a frame can even be strongly overcomplete. Riesz bases are clearly not overcomplete.

If $\{\psi_k\}$ is a Riesz basis, then

$$T^*g = \sum_{n=1}^{\infty} \langle \varphi_n, T^*g \rangle \varphi_n$$

$$= \sum_{n=1}^{\infty} \langle \psi_n, g \rangle \varphi_n \tag{4.5.34}$$

Thus, if $\eta_n = (T^*)^{-1}\varphi_n$, we have

$$g = \sum_{n=1}^{\infty} \langle \psi_n, g \rangle \eta_n \tag{4.5.35}$$

Interestingly enough, even for frames, there is an expansion like (4.5.35) (Problem 1) but unlike the Riesz basis case, the η's are not unique (Problem 2).

Before leaving the classical Gabor lattice case, we want to note several facts about their properties with regard to these refined conditions.

Proposition 4.5.4. *For any positive a, b with φ_0 given by (4.4.4), $\{e^{2\pi imbx}\varphi_0(x - na)\}_{m,n\in\mathbb{Z}}$ is a Bessel sequence.*

Remark. The basic idea is the one we've used several times above that for analytic functions, h, $|h(0)|^2$ is bounded by a multiple of $\int_{|z|<\delta}|h(z)|^2 d^2z$. What makes it slightly subtle is that we want to multiply by $e^{-|z|^2/2}$ which is not the absolute value of an analytic function. We cannot ignore this factor since for $|z_0|$ large and $\delta > 0$ fixed, $e^{-|z|^2/2}$ varies considerably over $\{z \mid |z - z_0| \leq \delta\}$. We'll exploit the fact that

$$e^{-|z_0+w|^2/2} = e^{-|z_0|^2/2}e^{-\operatorname{Re}(\bar{z}_0 w)}e^{-1/2|w|^2} \qquad (4.5.36)$$

and $e^{-2\operatorname{Re}(\bar{z}_0 w)}$ is the absolute value of an analytic function and, over $\{w \mid |w| \leq \delta\}$, $e^{-1/2|w|^2}$ is bounded below.

Proof. We need to prove that for any $f \in \mathcal{F}$, the Fock space, we have that

$$\sum_{z\in\mathcal{L}_{\alpha,\beta}} e^{-|z|^2}|f(z)|^2 \leq B \int e^{-|z|^2}|f(z)|^2 d^2z \qquad (4.5.37)$$

For $z_0 \in \mathcal{L}_{\alpha,\beta}$, let $h(z) = e^{-\frac{1}{2}|z_0|^2}e^{-\bar{z}_0(z-z_0)}f(z)$. If δ is less than half the minimum distance between distinct points of $\mathcal{L}_{\alpha,\beta}$, then, for h entire

$$|h(z_0)|^2 \leq \frac{1}{\pi\delta^2}\int_{|z-z_0|\leq\delta}|h(z)|^2 d^2z \qquad (4.5.38)$$

Noting that, if $|z - z_0| < \delta$, then

$$e^{-|z|^2}|f(z)|^2 = |h(z)|^2 e^{-|z-z_0|^2}$$

$$\geq |h(z)|^2 e^{-\delta^2} \qquad (4.5.39)$$

and we conclude that (4.5.37) holds with $B = (\pi\delta^2)^{-1}e^{\delta^2}$. $\qquad\square$

As for other properties:

(a) If $ab > 1$, the lattice is not complete and so certainly not a frame.
(b) If $ab < 1$, the lattice is overcomplete and so not a Riesz basis. It is a frame—the proof, which we say more about in the Notes, uses the function g of Proposition 4.5.2.
(c) If $ab = 1$, we'll see below that the (von Neumann) lattice is not a frame.

We'll now focus on $a = b = 1$ but general g in $L^2(\mathbb{R})$ in (4.5.1). As a warmup, we consider the following example.

Example 4.5.5. Let $g = \chi_{[0,1]}$, the characteristic function of $[0,1]$. Then $\{g_{(n,m)}\}$ is an orthonormal family (!) and it is not hard to see that it is complete (Problem 4). Similarly, if $g(x) = \sin(\pi x)/\pi x$ (essentially the Fourier transform of χ !), $\{g_{(n,m)}\}$ is an orthonormal basis. Thus, Gabor lattices can even be an orthonormal basis. Notice that χ is not smooth, not even continuous, while the sinc function is smooth but has very slow decay. This is not coincidental as we'll see. $\qquad \square$

The key to a Gabor lattice when $a = b = 1$ will be the Zak transform, discussed in Problem 13 of Section 6.2 of Part 1. There is a good reason for this. In general, the Heisenberg group is not abelian—translations in x and k only commute up to a phase. But for the critical lattice, $\tilde{U}(1,0)$ and $\tilde{U}(0, 2\pi)$ commute. Thus, it is natural to look at some kind of joint spectral representation. As we'll see, that is precisely what the Zak transform is. Here is the definition and basic properties of the Zak transform.

(1) For $f \in L^2(\mathbb{R}, dx)$, the Zak transform is defined as a function on $\mathbb{R}^2$ by

$$(Zf)(x,k) = \sum_{j\in\mathbb{Z}} f(x+j)e^{-2\pi ijk} \qquad (4.5.40)$$

One can show (Problem 13(a) of Section 6.2 of Part 1) that for a.e. $x \in \mathbb{R}$, the function $a_j = f(x+j)$ is in ℓ^2 and (4.5.40) is interpreted as an $L^2(dk)$-convergent sum. Alternatively, for $f \in \mathcal{S}(\mathbb{R})$, (4.5.40) converges pointwise and one can extend to L^2 once one has (4.5.41) and (4.5.42).

(2) Using the Parseval relation, one proves (Problem 13(b) of Section 6.2 of Part 1) that

$$\int_0^1 dx \int_0^1 dk |(Zf)(x,k)|^2 = \int_{-\infty}^\infty |f(x)|^2 dx \qquad (4.5.41)$$

(3) For all $\ell \in \mathbb{Z}$, we have (Problem 13(c) of Section 6.2 of Part 1)

$$(Zf)(x+\ell,k) = e^{2\pi i\ell k}(Zf)(x,k), \quad (Zf)(x,k+\ell) = (Zf)(x,k) \quad (4.5.42)$$

(4) If $(Uf)(x) = f(x-1)$, $(Vf)(x) = e^{2\pi ix}f(x)$, then (Problem 5)

$$(ZUf)(x,k) = e^{-2\pi ik}(Zf)(x,k), \quad (ZVf)(x,k) = e^{2\pi ix}(Zf)(x,k) \quad (4.5.43)$$

which shows how Z realizes a joint spectral representation for U, V which obeys $UV = VU$.

Thus,

$$(Zg_{(n,m)})(x,k) = e^{-2\pi ink}e^{2\pi imx}(Zg)(x,k) \qquad (4.5.44)$$

Let $\Delta \equiv [0,1]\times[0,1]$. Since Z is unitary from $L^2(\mathbb{R})$ to $L^2(\Delta)$ (i.e., (4.5.41)), we have that

$$\langle g_{(n,m)}, f\rangle_{L^2(\mathbb{R})} = \langle Zg, E_{n,m}(x,k)Zf\rangle_{L^2([0,1]\times[0,1])} \qquad (4.5.45)$$

398 4. Phase Space Analysis

where

$$E_{n,m}(x,k) = \exp(2\pi i(nk - mx)) \tag{4.5.46}$$

so $\langle g_{(n,m)}, f \rangle$ are the Fourier series coefficients of $(\overline{Zg})(Zf)$. By the Parseval relation for this series, we see that

$$\sum_{n,m} |\langle g_{(n,m)}, f \rangle|^2 = \int |(Zg)(x,k)|^2 |Zf(x,k)|^2 dx\, dk \tag{4.5.47}$$

where both sides can be infinite (at the same time).

Before finding the consequences of this relation, we note an interesting property of Zf in cases where Zf is continuous on $\mathbb{R}^2$.

Proposition 4.5.6. *Let Zf be continuous on $\mathbb{R}^2$. Then Zf has a zero on Δ.*

Proof. Since $(Zf)(x, k+1) = (Zf)(x,k)$ if $x \in [0,1]$, $\gamma_x(t) = (Zf)(x,t)$ defines a continuous curve in $\mathbb{C}$ which is continuous in x. Suppose Zf is everywhere nonvanishing. Then the curves lie in $\mathbb{C} \setminus \{0\}$ so

$$\tilde{\gamma}_x(t) = \gamma_x(t)/|\gamma_x(t)| \tag{4.5.48}$$

is a curve on $\partial\mathbb{D}$ and has a winding number. Since $(Zf)(x+1,t) = e^{2\pi i t}(Zf)(x,t)$, the winding number of $\gamma_{t=1}$ is one more than the winding number of $\gamma_{t=0}$. This violates continuity of γ_t in t, so we conclude that ZF must have a zero. $\qquad\square$

Corollary 4.5.7. *Suppose $g \in L^2(\mathbb{R})$ obeys*

$$|g(x)| \le C(1 + |x|)^{-\alpha}, \quad |\hat{g}(k)| \le C(1 + |k|)^{-\alpha} \tag{4.5.49}$$

for some $\alpha > 1$. Then Zg is continuous on $\mathbb{R}^2$ and so has a zero in Δ.

Proof. Since $\hat{g} \in L^1$, g is continuous by the Riemann–Lebesgue lemma. By the bound on g, the sum in (4.5.40) is locally uniformly convergent. Since each term is continuous, so is the limit. $\qquad\square$

Theorem 4.5.8. *Let $g \in L^2(\mathbb{R})$ have $\|g\|_{L^2} = 1$. Let $a = b = 1$. Then $\{g_{(n,m)}\}_{n,m\in\mathbb{Z}}$ is*

(a) *a Bessel sequence if and only if Zg is a bounded function;*
(b) *a frame if and only if for some $A, B > 0$, $A \le |Zg(x,k)|^2 \le B$ for a.e. x, k;*
(c) *an orthonormal basis if and only if $|Zg(x,k)| = 1$ for a.e. x, k;*
(d) *a Riesz basis if and only if it is a frame (see (b));*
(e) *complete if and only if Zg is a.e. nonvanishing;*
(f) *If Zg is continuous on Δ with a single zero at $(x_0, y_0) = p_0$ which is simple in the sense that for $0 < c_1 < c_2 < \infty$ and some $\varepsilon > 0$, we have*

$$c_1|(x,y) - p_0| \le |(Zg)(x,y)| \le c_2|(x,y) - p_0| \tag{4.5.50}$$

then $\{g_{(n,m)}\}_{n,m\in\mathbb{Z}}$ remains complete if any single $g_{(n,m)}$ is removed but becomes incomplete if any two of them are removed.

Remarks. 1. If Zg is C^1 and its derivative as a map of $\mathbb{R}^2$ to $\mathbb{C} = \mathbb{R}^2$ is invertible at p_0, then (4.5.50) holds.

2. There are results if Zg has exactly k simple zeros (Problems 10, 11): if any $k+1$ functions are removed, one loses completeness and there are some sets of k $g_{(n,m)}$ that can be removed while keeping completeness.

Proof. (a) Given (6.4.52) of Part 1 and the fact that $\{Zg \cap \Delta \mid g \in L^2(\mathbb{R})\}$ is $L^2(\Delta)$, $\{g_{(n,m)}\}$ is a Bessel sequence if and only if for all $h \in L^2(\Delta)$

$$\int |Wg)(x,k)|^2 |h(x,k)|^2 dx\,dk \leq B \int |h(x,k)|^2 dx\,dk \qquad (4.5.51)$$

and this holds if and only if $|Wg| \leq \sqrt{B}$ for a.e. (x,k).

(b) The proof is the same as for (a).

(c) We claim first that $\{g_{(n,m)}\}$ is an orthonormal basis if and only if for all φ

$$\sum_{n,m} |\langle \varphi, g_{(n,m)}\rangle|^2 = \|\varphi\|^2 \qquad (4.5.52)$$

If $\{g_{(n,m)}\}$ is an orthonormal basis, (4.5.52) is the Parseval relation. Conversely, if (4.5.52) holds, taking $g_{(n_0,m_0)}$ for φ and using $\|g_{(n_0,m_0)}\| = 1$, we see that (4.5.52) implies $\langle g_{(n_0,m_0)}, g_{(n,m)}\rangle = 0$ for all $(n,m) \neq (n_0, m_0)$.

(d) By (b), $\sqrt{A} \leq |Zg| \leq \sqrt{B}$. Let $\tilde{T}$ be multiplication by Zg on $L^2(\Delta)$. Then, $\|\tilde{T}\| \leq \sqrt{B}$ while $\|(\tilde{T})^{-1}\| \leq \sqrt{A^{-1}}$. Moreover, by (4.5.44), $g_{(n,m)}$ is $\tilde{T}e_{nm}$ where e_{nm} is an orthonormal basis. Thus, if $T = W^{-1}\tilde{T}W$, $g_{(n,m)}$ is an image of an orthonormal basis (essentially the first one in Example 4.5.5) under T.

(e) Since Z is unitary and $Zg_{(n,m)} = \overline{E}_{n,m}Zg$, we see $\{g_{(n,m)}\}$ is complete if and only if for all $f \in L^2(\Delta)$ we have

$$\forall_{n,m} \int \overline{Zg(x,k)} f(x,k) E_{n,m}(x,k) dx\,dk = 0 \Rightarrow f = 0 \qquad (4.5.53)$$

The integral in (4.5.53) is a Fourier coefficient and an L^1-function vanishes a.e. if and only if all its Fourier coefficients are zero (for the converse, use the fact that the Cesàro sum of Fourier series converge in L^1). Thus, (4.5.53) is equivalent to

$$(\overline{Zg})f = 0 \Rightarrow f = 0 \qquad (4.5.54)$$

This is clearly true if and only if for a.e. (x,k), $(Zg)(x,k) \neq 0$.

(f) Since Z is an isometry, we need only study whether there is $f \in L^2(\Delta)$ with $(\overline{Zg})f$ orthogonal to certain subsets of $\{E_{n,m}\}$. If one is missing,

$(\overline{Zg})f \in L^1$ has only one nonzero Fourier coefficient, so by Fejér's theorem, $(\overline{Zg})f = aE_{n_0,m_0}$ with $a \in \mathbb{C}$. But then, for a.e. $(x,k) \in \Delta$,

$$f = aE_{n_0,m_0}/(\overline{Zg}) \tag{4.5.55}$$

Since $E_{n_0,m_0}(p_0) \neq 0$, (4.5.50) implies this $f \notin L^2$ unless $a = 0$, i.e., $(\overline{Zg})f = 0$ so $f = 0$. Thus, $\{(Zg)E_{n,m}\}_{(n,m)\neq(n_0,m_0)}$ is complete.

For any two $(n_0,m_0) \neq (n_1,m_1)$, let

$$f = (E_{n_0,m_0} - aE_{n_1,m_1})/\overline{Zg} \tag{4.5.56}$$

$$a = E_{n_0,m_0}(p_0)\overline{E_{n_1,m_1}(p_0)} \tag{4.5.57}$$

Since the numerator vanishes clearly at p_0 (since the numerator is C^1) by (4.5.50), f is bounded near p_0 and continuous elsewhere, so bounded. Thus $\{(Zg)E_{n,m}\}_{(n,m)\neq(n_0,m_0),(n_1,m_1)}$ is not complete $\qquad \square$

Example 4.5.5 (revisited). If $g = \chi_{(0,1)}$ and $x \in (0,1]$, then the sum in (4.5.40) only has $j = 0$, so

$$Z\chi_{(0,1]}(x,k) = 1 \text{ if } (x,k) = \Delta \tag{4.5.58}$$

(this seems to violate (4.5.41) but Zg is discontinuous and (4.5.42) holds for a.e. (x,k). Thus $|Z\chi| = 1$ for a.e. (x,k) and, so, $\{g_{(n,m)}\}$ is an orthonormal basis. A similar analysis works for $g(x) = \sin(\pi x)/\pi x$ (Problem 4).

Example 4.5.9. We can revisit the classical Gabor lattices (aka von Neumann lattices) in light of Theorem 4.5.8. The Zak transform of the canonical Gaussian is a nonvanishing factor times a Jacobi theta function (Problem 6). This vanishes at only one point in Δ (mod 1), so, by (e) of the theorem, we recover the fact that the von Neumann lattice is complete. By (f), we recover the result about removing one or two functions. Since Zg has a zero and is continuous, it is not a.e. bounded away from zero so we see that the von Neumann lattice is not a frame. $\qquad \square$

Combining Corollary 4.5.7, Theorem 4.5.8, and the fact that a continuous function with a zero is not a.e. bounded away from zero, we get

Corollary 4.5.10. *If $\{g_{(n,m)}\}$ is a frame, then for any $\alpha > 1$, it cannot be that (4.5.49) holds.*

There is a related result called the (extended) Balian–Low theorem that if $\{g_{(n,m)}\}$ is a frame, then either $\int x^2|g(x)|^2dx = \infty$ or $\int k^2|\hat{g}(x)|^2dk = \infty$. In some ways, (4.5.49) is a weaker hypothesis than the Balian–Low hypothesis in that if (4.5.49) holds, one needs $\alpha > \frac{3}{2}$ for the Heisenberg-type integrals to be finite (put differently, there are examples where the corollary says the $\{g_{(n,m)}\}$ are not a frame even though $\int x^2|g(x)|^2dx = \int k^2|\hat{g}(x)|^2dk = \infty$). But the Balian–Low hypotheses are not strictly stronger, and the corollary does not imply their theorem. The proof of this result (in the extended form) is rather subtle (as seen by the fact that both of the original proofs were incorrect!). We discuss this proof and its history in the Notes. In Problem 7, the reader will prove their original (unextended) result, that if $\{g_{(n,m)}\}$ is an orthonormal basis, then one of the variance integrals is infinite.

Finally, we should mention that there is a general result that if for some $g \in L^2$, (4.5.1) is complete, then $ab \leq 1$; see the Notes.

Notes and Historical Remarks. There is a huge literature in the signal processing community on the questions of this section starting with Gabor's 1946 suggestion [**299**] that one could use the classical Gabor lattice to expand general signals (L^2-functions). For books on the subject, see [**158, 373, 728, 185**]; Heil [**372**] has a historical review article on some aspects of the subject.

The classical critical case where $ab = 1$ is called the von Neumann lattice because von Neumann considered it in his book on foundations of quantum mechanics [**839**]. He stated that it was easy to see the family was complete but provided no proof! Theorem 4.5.1 was proven independently by Bargmann et al. [**38**] and Perelomov [**637**]. Perelomov had the full theorem with a version of the proof we give. Bargmann et al. had the results for the full lattices but not the results on one or two missing states in the critical case. They used a construction similar to the one we used but only for the case $ab > 1$—unlike Perelomov, they only proved upper bounds which suffice to prove incompleteness. They handled $ab < 1$ by a Jensen formula argument related to Remark 2 after Proposition 4.5.2. Their proof of the critical case is a variant of the Zak transform proof of Bacry–Grossmann–Zak [**27**]—see Theorem 4.5.8.

The term "frame" was introduced by Duffin–Schaeffer [**229**] who, in particular, discussed expansions like those in Problem 1. They were studying expansions of functions on $[0, 2\pi]$ in $\{e^{\pm i\mu_n x}\}_{n=0}^{\infty}$ where the μ_n might not be integers but were close to them. (Young [**866**] is a comprehensive look at such expansions.) There were sporadic uses of the idea before then, but, in the West, none that were systematic. Before them, N. K. Bari [**39, 40**] systematically discussed related ideas but her work seems to be forgotten in the West except for the use of the names "Riesz basis" and "Bessel sequence" which she introduced (and whose origin has been forgotten in the frames community: several experts I consulted had no idea where that name came from).

As we mentioned, it is a theorem that a classical Gabor lattice with $ab < 1$ is a frame. This was conjectured by Daubechies–Grossmann [**190**] who proved it for rational (a, b) with $ab < 1$. The full result is in Lyubarskii [**551**] and Seip–Wallstón [**727, 729**]. The idea of their proof is to use the analytic function g of Proposition 4.5.2. One proves that for functions f in the Fock space, with $z_{nm} = n\alpha + im\beta$ (and $\alpha\beta < \pi$), one has that

$$f(z) = \sum_{m,n} \frac{f(z_{nm})}{g'(z_{nm})} \frac{g(z)}{z - z_m} \tag{4.5.59}$$

This is shown by looking at $\frac{f}{g} - \sum_{m,n} \frac{f(z_{nm})}{g'(z_{nm})} \frac{1}{z-z_m}$ and proving this is an entire analytic function (which is obvious) going to zero at infinity. Once one has this expansion, one uses it to show that

$$\int |f(z)|^2 e^{-|z|^2} \frac{d^2z}{\pi} \le A^{-1} \sum_{n,m} |f(z_{nm})|^2 e^{-|z_{nm}|^2} \tag{4.5.60}$$

which implies the lower frame bound (by taking f a Segal–Bargmann transform).

As we've seen, the classical coherent states when $ab = 1$ are not a frame so one cannot expect good L^2-convergent expansions of general $f \in L^2(\mathbb{R})$ as $\sum_{n,m \in \mathbb{Z}} \alpha_{nm} g_{nm}(x)$. But Janssen [**425**] has studied distributionally convergent expansions and proven that for the classical case, there are usually (i.e., $\sigma(\mathcal{S}', \mathcal{S})$) convergent expansions of L^2-functions.

The Zak transform is named after a construction of Zak [**868, 869, 870**]. As a tool to prove the Plancherel theorem, as we do in Problem 13 of Section 6.2 of Part 1, it appeared almost twenty years earlier in Gel'fand [**310**] as an aside not used except for that. Gel'fand, in turn, noted this should be viewed as a specialization of a general proof that Weil [**845**] gave of the Plancherel theorem for generally compact abelian groups. For this and related ideas in Berezin [**59**], what we (and most others) call the Zak transform is sometimes called the *Berezin–Weil–Zak tranform*.

The idea of using the Zak transform to study completeness of $\{g_{(n,n)}\}$ is from Bacry–Grossmann–Zak [**27**]. The results on completeness if one but not two states are removed when Zg has a simple zero (i.e., (f) of Theorem 4.5.8) is from Boon–Zak [**96**] and Janssen [**426**]. The results for multiple zeros (see Problems 7, 10) are from Boon–Zak–Zucker [**97**].

The original Balian–Low theorem from their papers [**31, 548**] asserted that if

$$\int x^2 |g(x)|^2 dx < \infty, \quad \int k^2 |\hat{g}(k)|^2 dk < \infty \tag{4.5.61}$$

then $\{g_{(n,m)}\}_{n,m \in \mathbb{Z}}$ could not be an orthonormal basis. Problem 7 has an elegant proof of this result by Battle [**42**]. Balian and Low each had the same error in their papers (Battle's proof is correct): namely they assumed that (4.5.61) implies Zg is continuous on $\mathbb{R}^2$. Essentially, they assumed a function f, on $\mathbb{R}^2$ in L^2 with L^2-gradient is continuous, and this is false although it is "almost true." This error, a patch to it and an extension to $\{g_{(n,m)}\}_{n,m \in \mathbb{Z}}$

not even being a frame was noted by Coifman and Semmes—their result was presented in a review article of Daubechies [**188**]. Daubechies–Janssen [**191**] then showed how to modify Battle's argument to get the frame result.

As noted, if $ab > 1$, no $\{g_{(n,m)}\}_{n,m\in\mathbb{Z}}$ can be complete. This is a subtle result proved by Baggett [**29**] using representation theory of discrete Heisenberg groups. Heil [**372**] has a review article on the subject.

Gröchenig [**333**] has extended some of the theory of Gabor frames to more general locally compact abelian groups than $\mathbb{R}^\nu$.

Problems.

1. (a) Let $\{\varphi_n\}_{n=1}^\infty$ be a Bessel seuqence. For $\psi \in \mathcal{H}$, define

$$(T\psi)_n = \langle \varphi_n, \psi \rangle \qquad (4.5.62)$$

Prove that T maps $\mathcal{H}$ to ℓ^2 and is a bounded map.

(b) Prove that $T^*a = \sum_{n=1}^\infty a_n\varphi_n$ (including that this sum converges).

(c) Prove that $\{\varphi_n\}_{n=1}^\infty$ is a frame if and only if T is invertible.

(d) If $\{\varphi_n\}_{n=1}^\infty$ is a frame, let

$$\eta_n = (T^*T)^{-1}\varphi_n \qquad (4.5.63)$$

Prove that for any $\psi \in \mathcal{H}$,

$$\psi = \sum_{n=1}^\infty \langle \varphi_n, \psi \rangle \eta_n \qquad (4.5.64)$$

Remarks. 1. These results are from Duffin–Schaeffer [**229**].

2. The η_n are called the *canonical dual frames*.

2. (a) Let ψ_n be an orthonormal basis for $\mathcal{H}$ and

$$\varphi_{2n} = \varphi_{2n-1} = \psi_n$$

Prove that $\{\psi_n\}_{n=1}^\infty$ is a frame. Find the η_n of (4.5.63). Find the other sets of η_n for which (4.5.64) holds.

(b) Let $\{\varphi_n\}_{n=1}^\infty$ be a frame. Prove that the η_n of (4.5.64) are unique if and only if the $\{\varphi_n\}_{n=1}^\infty$ are a Riesz basis.

3. In $\mathbb{C}^2$, let

$$\varphi_1 = (\frac{\sqrt{3}}{2}, -\frac{1}{2}), \quad \varphi_2 = (0, 1), \quad \varphi_3 = (-\frac{\sqrt{3}}{2}, -\frac{1}{2}) \qquad (4.5.65)$$

This is sometimes called the Mercedes frame or the peace frame (See Figure 4.5.1). Prove that this is a frame, compute the canonical dual frame and all other η's obeying (4.5.64).

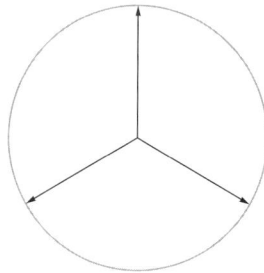

Figure 4.5.1. The Mercedes frame.

4. (a) Let $g = \chi_{[0,1]} \in L^2(\mathbb{R}, dx)$. Prove that $\{g_{(n,m)}\}_{n,m \in \mathbb{Z}}$ is an orthonormal basis.

(b) Prove that if $g(x) = \sin(\pi x)/\pi x$, then $\{g_{(n,m)}\}_{n,m \in \mathbb{Z}}$ is an orthonormal basis.

5. Verify (4.5.43).

6. This problem will rely on on the Jacobi Θ function which can be defined by, for $z \in \mathbb{C} \setminus \{0\}$, $q \in \mathbb{D}$

$$\Theta(z|q) = \sum_{n=-\infty}^{\infty} q^{n^2} z^n \tag{4.5.66}$$

and which obeys (see Section 10.5 of Part 2A)

$$\Theta(z|q) = 0 \Leftrightarrow z = q^{2n-1}, \quad n \in \mathbb{Z} \tag{4.5.67}$$

(a) Let φ_0 be given by (4.4.2). Prove that

$$(Z\varphi_0)(x,k) = \pi^{-1/4} e^{-\frac{1}{2}x^2} \Theta(e^{-2\pi i k} e^{-x} | e^{-\frac{1}{2}}) \tag{4.5.68}$$

(b) Prove that the zeros of $Z\varphi_0$ on $\mathbb{R}^2$ are points of the form $(n + \frac{1}{2}, m)$; $n, m \in \mathbb{Z}$, i.e., one zero per unit cell.

7. This problem will lead the reader through a proof of the original theorem of Balian–Low that if $g \in L^2$ and

$$\int x^2 |g(x)|^2 dx < \infty, \quad \int k^2 |\hat{g}(k)|^2 dk < \infty \tag{4.5.69}$$

then, $\{g_{(n,m)}\}$ is not an orthonormal basis. Define

$$(Xf)(x) = xf(x), \quad (Pf)(x) = (k\hat{f})^{\vee} \tag{4.5.70}$$

if $xf \in L^2$, respectively, $k\hat{f} \in L^2$. Let

$$(T_k f)(x) = f(x - k), \quad (M_n f)(x) = e^{2\pi i n x} f(x) \tag{4.5.71}$$

(a) If $g, xg \in L^2$ and $(M_n T_k g)_{n,k \in \mathbb{Z}}$ are orthonormal, prove that

$$\langle M_n T_k g, Xg \rangle = \langle M_n T_k Xg, g \rangle \tag{4.5.72}$$

(b) Prove that $[M_n, T_k] = 0$

(c) Conclude that

$$\langle M_n T_k g, Xg \rangle = \langle Xg, M_{-n} T_{-k} g \rangle \tag{4.5.73}$$

(d) Prove that if also $k\hat{g} \in L^2$, then

$$\langle M_n T_k g, Pg \rangle = \langle Pg, M_{-n} T_{-k} g \rangle \tag{4.5.74}$$

(e) Conclude that if $\{g_{nm}\}$ is an orthonormal basis and $xg, k\hat{g} \in L^2$, then

$$\langle Xg, Pg \rangle = \langle Pg, Xg \rangle \tag{4.5.75}$$

(*Hint*: Expand in the g_{nm} basis.)

(f) By integrating by parts, prove that

$$\langle Pg, Xg \rangle - \langle Xg, Pg \rangle = -i\|g\|^2 \tag{4.5.76}$$

and conclude that $\{g_{nm}\}$ cannot be an orthonormal basis.

Remark. The proof is due to Battle[**42**].

8. Suppose $\{\varphi_n\}_{n=1}^\infty$ is an orthonormal basis for a Hilbert space, $\mathcal{H}$, and that $\{\psi_n\}_{n=1}^\infty$ is a family of vectors obeying

$$A \equiv \sum_{n=1}^\infty \|\varphi_n - \psi_n\|^2 < \infty \tag{4.5.77}$$

and

$$\sum_{n=1}^\infty |a_n|^2 < \infty \, (a_n \in \mathbb{C}) + \sum_{n=1}^\infty a_n \psi_n = 0 \Rightarrow a_n = 0 \tag{4.5.78}$$

This problem will help you show that $\{\psi_n\}_{n=1}^\infty$ is a Riesz basis.

(a) Define $S : \mathcal{H} \to \mathcal{H}$ by

$$S\eta = \sum_{n=1}^\infty \langle \varphi_n, \eta \rangle \, (\varphi_n - \psi_n) \tag{4.5.79}$$

Prove that for any unit vector, κ,

$$|\langle \kappa, S\eta \rangle|^2 \le A \, \|\eta\|^2 \tag{4.5.80}$$

so that S is a bounded operator.

(b) Prove that

$$\sum_{n=1}^\infty \|S \varphi_n\|^2 < \infty \tag{4.5.81}$$

and conclude that S is Hilbert–Schmidt and, in particular, compact.

(c) Let $T = 1-S$. Prove that $T\varphi_n = \psi_n$ and, by (4.5.78), that $\ker(T) = 0$.

(d) Prove that T is invertible and so conclude that $\{\psi_n\}_{n=1}^{\infty}$ is a Riesz basis.

(e) Conversely, if T is invertible and $\{\varphi_n\}_{n=1}^{\infty}$ an orthonormal basis and if $\psi_n = T\varphi_n$ (so $\{\psi_n\}_{n=1}^{\infty}$ is a Riesz basis) if $1 - T$ is Hilbert–Schmidt, prove that $\{\psi_n\}_{n=1}^{\infty}$ obeys (4.5.77) and (4.5.78).

Remarks. 1. This problem needs some basic facts about compact and Hilbert–Schmidt operators as discussed in Chapter 3 of Part 4.

2. This result is due to Bari [**40**] so a Riesz basis that obeys $1 - T$ Hilbert–Schmidt is called a *Bari basis*.

3. Since orthonormal sets $\{\psi_n\}_{n=1}^{\infty}$ obey (4.5.78), this result shows that if $\{\varphi_n\}_{n=1}^{\infty}$ is an orthonormal basis and $\{\psi_n\}_{n=1}^{\infty}$ an orthonormal set and (4.5.77) holds, then $\{\psi_n\}_{n=1}^{\infty}$ is complete. This was proven by George Birkhoff [**73**] in 1917; see also Birkhoff–Rota [**78**] written by George's son. The latter paper was unaware of Bari's work. These two papers were interested in the question to prove completeness of Sturm–Liouville orthonormal systems using their known asymptotics and the completeness of eigenfunctions of $-u'' = \lambda u$ on $[a, b]$ with suitable boundary conditions.

9. Let $\{\varphi_n\}_{n=1}^{\infty}$ be an orthonormal basis for a Hilbert space, $\mathcal{H}$, and $\{\psi_n\}_{n=1}^{\infty}$ another set of vectors that obeys

$$\left| \sum a_n(\varphi_n - \psi_n) \right| \le \lambda \sqrt{\sum_n |a_n|^2} \qquad (4.5.82)$$

(a) Let S be given by (4.5.79). Prove that

$$\|S\| \le \lambda \qquad (4.5.83)$$

(b) If $|\lambda| < 1$, so $T = 1 - S$ is invertible, prove that $\{\psi_n\}_{n=1}^{\infty}$ is a Riesz basis.

Remarks. 1. This is a result of Paley–Wiener [**627**], a pioneering paper in expansions of $L^2(-\pi, \pi)$-functions in $\{e^{i\lambda_n x}\}_{n=-\infty}^{\infty}$.

2. This result plus a calculation (see the book of Young [**866**]) yields the Kadec $\frac{1}{4}$ theorem [**437**] that if $\sup_n |\lambda_n - n| < \frac{1}{4}$ for a sequence of real numbers, $\{\lambda_n\}_{n=-\infty}^{\infty}$, then $\{e^{i\lambda_n x}\}_{n=-\infty}^{\infty}$ is a Riesz basis for $L^2(-\pi, \pi)$. It is known that $\frac{1}{4}$ is the optimal constant.

3. Ortega-Cerdà–Seip [**623**] have determined the exact set of real sequences $\{\lambda_n\}_{n=-\infty}^{\infty}$ for which $\{e^{i\lambda_n x}\}_{n=-\infty}^{\infty}$ is a frame for $L^2(-\pi, \pi)$.

10. Suppose that Zg has k simple zeros at points $p_1, \ldots, p_k$ in Δ.

(a) Given any $k + 1$ pairs $\{(n_i, m_i)\}_{i=1}^{k+1}$, prove that there exists $(\alpha_1, \ldots, \alpha_{k+1}) \in \mathbb{C}^{k+1}$ so that

$$\sum_{j=1}^{k-1} \alpha_j E_{n_j, m_j}(p_\ell) = 0;\ \ell = 1, \ldots, k \qquad (4.5.84)$$

(b) Prove that $\{E_{n,m}(Zg)\}_{n,m \in \mathbb{Z}}$ with $\{(n_i, m_i)\}_{i=1}^{k+1}$ removed is not a complete set for $L^2(\Delta)$.

11. Suppose that Zg has k simple zeros at points $p_1, \ldots, p_k$ in Δ.

(a) For any $p \in \Delta$, $p \neq (0,0)$ mod 1, prove that

$$\frac{1}{N^2} \#((n, m) \mid |n| \leq N, |m| \leq N, E_{(n,m)}(p) = 1) \to 0 \qquad (4.5.85)$$

as $N \to \infty$.

(b) Prove that there is $(n_0, m_0) \in \mathbb{Z}^2$ so that for all $i \neq j$ in $\{1, \ldots, k\}$, we have that

$$E_{(n_0, m_0)}(p_i) \neq E_{(n_0, m_0)}(p_j) \qquad (4.5.86)$$

(c) Prove that for $\alpha \in \mathbb{C}^k$,

$$\sum_{j=1}^{k} \alpha_j E_{((j-1)n_0, (j-1)m_0)}(p_\ell) = 0,\ \ell = 1, \ldots, k$$

implies $\alpha_j \equiv 0$. (*Hint*: How many zeros can a polynomial of degree $k - 1$ have?)

(d) Prove that $\{E_{(n,m)}(Zg)\}$ with $\{E_{((j-1)n, (j-1)m)}(Zg)\}_{j=1}^{k}$ is complete.

4.6. Wavelets

This sparse coding makes wavelets excellent tools in data compression. For example, the FBI has standardized the use of wavelets in digital fingerprint image compression. The compression ratios are on the order of 20:1, and the difference between the original image and the decompressed one can be told only by an expert. ... Coifman and his Yale team used wavelets to clean noisy sound recordings, including old recordings of Brahms playing his First Hungarian Dance on the piano.

—Brani Vidakovic and *Peter Mueller* in [**832**]

In the last section, we discussed Heisenberg group discrete coherent states $\{e^{2\pi i m x} g(x - n)\}_{m,n \in \mathbb{Z}}$ and saw if they are an orthonormal basis then g cannot be both very smooth and have fast decay (indeed either $\int |g'(x)|^2 dx = \infty$ or $\int x^2 |g(x)|^2 dx = \infty$). In this section, we'll instead consider discrete wavelets, i.e., coherent states of the affine group. Explicitly, we'll consider $\{2^{\ell/2} \psi(2^\ell x - k)\}_{\ell, k \in \mathbb{Z}}$ where $\psi(x)$ is a function called the *mother wavelet*.

We construct ψ's in $\mathcal{S}$ where the set is an orthonormal basis (Meyer wavelets) and, more significantly, where ψ has compact support and a fixed but large number of derivatives (Daubechies wavelets). We'll also show that ψ cannot be both C^∞ and have exponential decay.

At the risk of mixing metaphors in a subject with mother wavelets, we begin with the granddaddy of all wavelets, the Haar basis.

Example 4.6.1 (Haar wavelets). In $L^2([0,1], dx)$ consider the wavelet spaces $W_0 \subset W_1 \subset \ldots \subset W_\ell \subset \ldots$ where W_ℓ is the 2^ℓ-dimensional space of functions constant on each interval $J_{\ell,k} \equiv [\frac{k}{2^\ell}, \frac{k+1}{2^\ell})$, $k = 0, \ldots, 2^\ell - 1$. Let P_ℓ be the orthogonal projection onto W_ℓ, i.e.,

$$(P_\ell f)(x) = \sum_{k=0}^{2^\ell - 1} \chi_{[\frac{k}{2^\ell}, \frac{k+1}{2^\ell})}(x) \frac{1}{2^{-\ell}} \int_{k/2^\ell}^{(k+1)/2^\ell} f(y) dy \qquad (4.6.1)$$

If f is continuous, it is easy to see that $\|f - P_\ell f\|_\infty \to 0$ so L^2-$\lim P_\ell f = f$ (since $\|P_\ell f\|_\infty \le \|f\|_\infty$ and we can use the dominated convergence theorem or one can prove uniform convergence directly since

$$\|P_\ell f - f\|_\infty \le \sup_{|x-y| \le 2^{-\ell}} |f(x) - f(y)| \qquad (4.6.2)$$

). Thus, by density

$$s\text{-}\lim_{\ell \to \infty} P_\ell = \mathbb{1}, \quad \text{i.e.,} \quad \overline{\bigcup_{\ell=0}^\infty W_\ell} = L^2([0,1], dx) \qquad (4.6.3)$$

The spaces W_ℓ have obvious and natural orthonormal bases: If $\varphi(x)$ on $\mathbb{R}$ is $\chi_{[0,1)}$, then $S_\ell \equiv \{2^{\ell/2}\varphi(2^\ell x - k)\}_{k=0}^{2^\ell-1}$, i.e., L^2-normalized characteristic functions of $J_{\ell,k}$ are an orthnormal basis. Here ℓ is fixed and k varied. Clearly $\bigcup_{\ell=0}^\infty S_\ell$ is an overcomplete set normalized but not orthonormal.

It is natural to get an orthonormal basis by using Gram–Schmidt. If

$$\varphi_{\ell,k}(x) = \{2^{\ell/2}\varphi(2^\ell x - k)\} \qquad (4.6.4)$$

then $\varphi_{1,0}$ is not orthogonal to $\varphi_{0,0}$, so we orthonormalize and get that the second function is

$$\psi(x) = \chi_{[0,\frac{1}{2})}(x) - \chi_{[\frac{1}{2},1)}(x) \qquad (4.6.5)$$

Of course $\varphi_{1,1}(x) = 2^{-1/2}[\varphi_{0,0} - \psi(x)]$, so we don't get a third function. It is easy to see that up to $W_{\ell-1}$, we have $2^{\ell-1}$ functions in our Gram–Schmidt basis. Of the 2^ℓ $\varphi_{\ell,k} \in S_\ell$, we only use $2^{\ell-1}$ of the form $2^{-1/2}(\varphi_{\ell,2k-1} - \varphi_{\ell,2k})$, that is, we have constructed an orthonormal basis for $L^2([0,1], dx)$, $\{\varphi_{0,0}\} \cup \{\psi_{\ell,k}\}_{\ell=0,1,\ldots,\,k=0,\ldots,2^{\ell-1}}$ where

$$\psi_{\ell,k}(x) = 2^{\ell/2}\psi(2^\ell x - k) \qquad (4.6.6)$$

This is the *Haar basis* for $L^2([0,1], dx)$.

To extend the idea to $\mathbb{R}$, we let V_ℓ be the functions constant on each interval $[\frac{k}{2^\ell}, \frac{k+1}{2^\ell}]$ where now, first we replace $0 \leq k \leq 2^\ell - 1$ by $k \in \mathbb{Z}$ and $\ell = 0, 1, \ldots$ by $\ell \in \mathbb{Z}$. Thus,

$$\ldots \subset V_{-1} \subset V_0 \subset V_1 \subset \ldots \tag{4.6.7}$$

is now two-sided. P_ℓ, the projection onto V_ℓ, is still given by (4.6.1) but now the sum goes $k = -\infty$ to $k = \infty$. If f is continuous of compact support, then $\|P_\ell f - f\|_\infty \to 0$ with supports inside a fixed compact set (for $\ell \to \infty$) so $\|P_\ell f - f\|_2 \to 0$ and (4.6.3) still holds.

In addition,

$$s\text{-}\lim_{\ell \to -\infty} P_\ell = 0, \text{ i.e., } \bigcap_{\ell=-\infty}^{\infty} V_\ell = \{0\} \tag{4.6.8}$$

since $\varphi = P_\ell \varphi \Rightarrow \|\varphi\|_\infty \leq 2^{\ell/2} \|\varphi\|_2$, for then $\|\varphi\|_2^2 \geq 2^{-\ell} |\varphi(x)|^2$ by looking at the $J_{\ell,k}$ interval with x. Since $2^{\ell/2} \to 0$ as $\ell \to -\infty$, we get (4.6.8).

We can define $\varphi_{\ell,k}$ by (4.6.4), but now $k \in \mathbb{Z}$. For ℓ fixed, $\{\varphi_{\ell,k}\}_{k=-\infty}^{\infty}$ is an orthonormal basis of V_ℓ. Our $L^2([0,1], dx)$-result shows that for any fixed ℓ_0

$$\{\varphi_{\ell_0,k}\}_{k=-\infty}^{\infty} \bigcup_{\ell=\ell_0}^{\infty} \{\psi_{\ell,k}\}_{k\in\mathbb{Z}}$$

is an orthonormal basis for $L^2(\mathbb{R}, dx)$. If we take $\ell_0 \to -\infty$ and exploit (4.6.8) we see that

$$\{\psi_{\ell,k}\}_{l,k\in\mathbb{Z}} \tag{4.6.9}$$

is an orthonormal basis for $L^2(\mathbb{R}, dx)$ called the *Haar basis* for $L^2(\mathbb{R}, dx)$.

Remark. Closely related to the Haar basis are the *Rademacher functions* on $[0,1]$

$$r_n(x) = 2^{-n/2} \sum_{k=0}^{2^n-1} \psi_{n,k}(x) \tag{4.6.10}$$

which is an orthonormal but not complete family. Another definition is

$$r_n(x) = \text{sgn}(\sin(2^n \pi x)) \tag{4.6.11}$$

Notice that any nontrivial product of different r_j's has integral zero so since $r_j^2 = 1$, the $\{r_j\}_{j=1}^{\infty}$ are a sequence of iidrv and a realization of a coin toss or random walk. They thus prodive a link between probability and analysis on $[0,1]$. They are also useful in the theory of Banach spaces, X, where one considers $\int_0^1 \|\sum_{j=1}^n r_j(x) y_j\| dx$ for $\{y_j\}_{j=1}^n \subset X$. The Rademacher functions were introduced by him in [**667**].

The Haar basis has a strong convergence property. □

Theorem 4.6.2. *Let P_ℓ be the projection onto $W_\ell \subset L^2([0,1], dx)$ (respectively, $V_\ell \subset L^2(\mathbb{R}, dx)$). Then as $\ell \to \infty$:*

(a) *If f is a bounded continuous function on $[0,1]$ (respectively, uniformly continuous on $\mathbb{R}$) then*

$$f = L^\infty\text{-}\lim P_\ell f \tag{4.6.12}$$

(b) *If $f \in L^p([0,1], dx)$, $1 \le p < \infty$ (respectively, $L^p(\mathbb{R}, dx)$)*

$$f = L^p\text{-}\lim P_\ell f \tag{4.6.13}$$

Remark. The point of course is that this implies in both cases that P_ℓ has an expansion in terms of the orthonormal basis. Therefore,

$$f = \sum \langle \psi_{\ell,k}, f \rangle \psi_{\ell,k} \tag{4.6.14}$$

is a norm convergent sum in $\|\cdot\|_\infty$ or $\|\cdot\|_p$. In the $[0,1]$ case, the sum means $\sum_{\ell \le k, k=0,\ldots 2^\ell - 1}$ which is a finite sum with limits. In the $\mathbb{R}$ case, it may mean $\lim_{L \to \infty} \lim_{K \to \infty} \sum_{\ell \le L} \sum_{|k| \le K}$.

Proof. For f continuous of compact support, we have shown the required convergence in $L^1 \cap L^\infty$ above so if $c_p = \sup_\ell \|P_\ell\|_{L^p, L^p} < \infty$, we get the result by a density argument for $1 \le p < \infty$. From the formula (4.6.1) and its $\mathbb{R}$-analog, $\|P_\ell f\|_\infty \le \|f\|_\infty$ and $\|P_\ell f\|_1 \le \|f\|_1$. Thus, by interpolation, $\|P_\ell f\|_p \le \|f\|_p$ for all $1 \le p \le \infty$.

From (4.6.1), we also see (4.6.2) on $\mathbb{R}$, so convergence on the bounded uniformly continuous functions follows. □

Remark. In terms of the Haar basis on $L^2(\mathbb{R})$, $P_\ell f$ is still an infinite sum even if f has compact support. In the latter case, if $\ell > k$, $(P_\ell - P_k)f$, i.e., the projection on $V_\ell \cap V_k^\perp$ is a finite sum and it is natural to be interested in L^p-convergence of

$$\lim_{\substack{\ell \to \infty \\ k \to -\infty}} (P_\ell - P_k)f$$

to f. Of course, one has L^2-convergence. Given that on L^p we have $P_\ell f \to f$, one wants to show $P_k f \to 0$ in L^p. It is not hard to see (Problem 1) that if $f \in L^\infty$ and has compact support, then $\|P_k f\|_\infty \to 0$, so for $f \in C_\infty(\mathbb{R})$, the continuous functions vanishing at ∞, the limit above in $\|\cdot\|_\infty$ is f. But there is a problem when $p = 1$: if $f \ge 0$, $\|P_j f\|_1 = \|f\|_1$ for all j (by a simple calculation), so $\|P_j f\|_p \to 0$ fails for $p = 1$! It is true though for $1 < p < \infty$. Rather than prove that here, we'll prove it as a special case of a much more general result at the end of this section. Thus, $\lim_{\ell \to \infty} P_\ell f = f$ is often called *convergence at small scales* (since $(1 - P_\ell)f$ is a projection onto structures at $2^{-\ell}$ or smaller) while $\lim_{j \to -\infty} P_j f = 0$ is called *convergence at large scales*. What we've seen is that there is no L^1-convergence at large scales.

Motivated by the Haar basis, we introduce a basic definition:

Definition. A *multiresolution analysis* (MRA) is a sequence $\{V_n\}_{n=-\infty}^{\infty} \subset L^2(\mathbb{R})$ and a distinguished real-valued function, φ, called the *scaling function* so that

(a) (4.6.7) holds, i.e., $V_{-1} \subset V_0 \subset V_1 \ldots$.

(b) $\displaystyle\bigcup_{\ell=-\infty}^{\infty} V_\ell$ is dense in $L^2(\mathbb{R})$; $\displaystyle\bigcap_{l=-\infty}^{\infty} V_\ell = \{0\}$.   (4.6.15)

(c) $f \in V_\ell \Leftrightarrow f(2^{-\ell}\cdot) \in V_0$.   (4.6.16)

(d) $\{\varphi(\cdot - k)\}_{k \in \mathbb{Z}}$ is an orthonormal basis for V_0.

We begin by analyzing when $\{\varphi(\cdot - k)\}_{k \in \mathbb{Z}}$ is an orthonormal set with the elegant:

Theorem 4.6.3. *Let $\varphi \in L^2(\mathbb{R})$. Then $\{\varphi(\cdot - n)\}_{n \in \mathbb{Z}}$ is an orthonormal set if and only if for a.e. p*

$$\sum_{m=-\infty}^{\infty} |\widehat{\varphi}(p + 2\pi m)|^2 = (2\pi)^{-1} \qquad (4.6.17)$$

Remarks. 1. Much of the literature uses the convention $\int e^{-2\pi i p x} \varphi(x) dx$ for Fourier transform. If one does, both 2π's in (4.6.17) become 1.

2. This proof uses the interplay of continuous and discrete Fourier analysis that will reoccur several times below.

Proof. Let $\varphi_n(x) = \varphi(x - n)$ so that

$$\widehat{\varphi}_n(p) = e^{-ipn}\widehat{\varphi}(p) \qquad (4.6.18)$$

Thus,

$$\{\varphi_n\}_{n=-\infty}^{\infty} \text{ is orthonormal} \Leftrightarrow \int |\widehat{\varphi}(p)|^2 e^{-ipn} dp = \delta_{n0} \qquad (4.6.19)$$

Define

$$F(p) = \text{LHS of (4.6.17)} \qquad (4.6.20)$$

which is periodic in p with period 2π. Since $\widehat{\varphi} \in L^2(\mathbb{R})$, $F(p) \in L^1([0, 2\pi])$ and the sum is a.e. finite. Since e^{-ipn} is also periodic with that period, (4.6.19) becomes

$$\{\varphi_n\}_{n=-\infty}^{\infty} \text{ is orthonormal} \Leftrightarrow \int_0^{2\pi} F(p) e^{-ipn} dp = \delta_{n0} \qquad (4.6.21)$$

$F(p) = (2\pi)^{-1}$ obeys the RHS of (4.6.21) and, by uniqueness of L^1-functions with given Fourier coefficients, it is the unique such function. $\square$

Another key observation is the following. If (d) and (c) hold, then $\{2^{1/2}\varphi(2\cdot-n)\}_{n\in\mathbb{Z}}$ is an orthonormal basis for V_1. Since $\varphi\in V_0\subset V_1$, we can expand $\varphi(x)$ in this basis (recall that we assume that φ is real)

$$\varphi(x)=\sum_{n\in\mathbb{Z}}a_n\varphi(2x-n),\quad a_n=2\int\varphi(x)\varphi(2x-n)dx \qquad (4.6.22)$$

φ obeying (4.6.22) is called a *refinable function* and this equation is called the *refinement equation*. The $\{a_n\}_{n\in\mathbb{Z}}$ are called the *structure constants* of the MRA. They obey (since $2^{1/2}\varphi(2\cdot-n)$ are orthonormal)

$$\sum_{n\in\mathbb{Z}}|a_n|^2=2<\infty \qquad (4.6.23)$$

Notice also, since

$$\varphi(x-k)=\sum_{n\in\mathbb{Z}}a_n\varphi(2x-2k-n) \qquad (4.6.24)$$

$$=\sum_{n\in\mathbb{Z}}a_{n+2k}\varphi(2-n) \qquad (4.6.25)$$

we have that

$$\sum_{n\in\mathbb{Z}}a_n\,a_{n+2k}=2\delta_{k0} \qquad (4.6.26)$$

There is a partial converse to this observation:

Proposition 4.6.4. *Let $\varphi\in L^2(\mathbb{R})$ obey $\{\varphi(\cdot-n)\}_{n\in\mathbb{Z}}$ is orthonormal and so that we have (4.6.22) where the sum holds in L^2. Let V_ℓ be the L^2-closure of the span of $\{\varphi(2^\ell\cdot-n)\}_{n\in\mathbb{Z}}$. Then, $\{V_\ell\}$ obeys (4.6.7) and (4.6.17).*

Proof. (4.6.17) is immediate. By (4.6.22), $V_0\subset V_1$. This implies $V_\ell\subset V_{\ell+1}$ for any ℓ by using $f\mapsto f(2^{-\ell}\cdot)$ scaling. Thus, we have (4.6.7)   □

That leaves open the question of when $s\text{-}\lim_{\ell\to-\infty}P_\ell=0$ and $s\text{-}\lim_{\ell\to\infty}P_\ell=\mathbb{1}$, a question we now discuss. The first is automatic.

Proposition 4.6.5. *Under the hypothesis of Proposition 4.6.4, if $P_\ell=$ projection onto V_ℓ, we have*

$$s\text{-}\lim_{\ell\to-\infty}P_\ell=0 \qquad (4.6.27)$$

Proof. We begin by noting that if we define the unitary operator

$$(U_\ell f)(x)=2^{-\ell/2}f(2^{-\ell}x) \qquad (4.6.28)$$

then $f\in V_\ell\Leftrightarrow U_\ell f\in V_0$ implies that

$$P_\ell=U_\ell^{-1}P_0U_\ell \qquad (4.6.29)$$

so that

$$\|P_\ell f\|^2=\|P_0U_\ell f\|^2 \qquad (4.6.30)$$

Clearly supp $f \subset [-R, R] \Rightarrow$ supp $U_\ell f \subset [-2^\ell R, 2^\ell R]$ which implies

$$|\langle \varphi_n, U_\ell f \rangle|^2 \leq \|f\|^2 \int_{-2^\ell R}^{2^\ell R} |\varphi_n(x)|^2 \, dx$$

$$= \|f\|^2 \int_{n-2^\ell R}^{n+2^\ell R} |\varphi(x)|^2 \, dx \qquad (4.6.31)$$

so that

$$\|P_\ell f\|^2 = \sum_{n=-\infty}^{\infty} |\langle \varphi_n, U_\ell f \rangle|^2$$

$$\leq \|f\|^2 \sum_{n=-\infty}^{\infty} \int_{n-2^\ell R}^{n+2^\ell R} |\varphi(x)|^2 \, dx \qquad (4.6.32)$$

If ℓ is so negative that $2^\ell R < \frac{1}{2}$, then the intervals are disjoint and if $S_\ell = \bigcup_{n=-\infty}^{\infty} [n - 2^\ell R, n + 2^\ell R]$ then S_ℓ is decreasing as $\ell \downarrow -\infty$. Thus, (4.6.32) implies

$$\|P_\ell f\|^2 \leq \|f\|^2 \int_{S_\ell} |\varphi(x)|^2 \, dx \to 0$$

because $S \mapsto \mu(S) \equiv \int_S |\varphi(x)|^2 \, dx$ is a finite absolutely continuous measure and $\bigcap_\ell S_\ell = \mathbb{Z}$ has measure zero. Such f are dense as R increases, so we have (4.6.27). $\qquad\square$

To handle s-$\lim_{\ell \to \infty} P_\ell$, we need:

Lemma 4.6.6. *Suppose that $\{\varphi_n(\,\cdot - n)\}_{n \in \mathbb{Z}}$ is an orthonormal set and let P_0 be the projection onto their closed span. Then for $g \in L^2$*

$$\text{supp} \, \widehat{g} \subset [-\pi, \pi] \Rightarrow \|P_0 g\|^2 = 2\pi \int_{-\pi}^{\pi} |\widehat{\varphi}(k) \, \widehat{g}(k)|^2 \, dk \qquad (4.6.33)$$

Proof.

$$\langle \varphi_n, g \rangle = \langle \widehat{\varphi_n}, \widehat{g} \rangle = \int e^{ikn} \overline{\widehat{\varphi}}(k) \, \widehat{g}(k) \, dk \qquad (4.6.34)$$

by (4.6.17). Since supp $\widehat{g} \subset [-\pi, \pi]$, $h \equiv \overline{\widehat{\varphi}} \, \widehat{g} \in L^1([-\pi, \pi])$ and (4.6.34) says that

$$\langle \varphi_n, g \rangle = 2\pi h^\sharp_{-n} \qquad (4.6.35)$$

in terms of Fourier series. Thus,

$$\|P_0 g\|^2 = \sum_n (2\pi)^2 |h^\sharp_n|^2$$

$$= (2\pi)^2 \int_{-\pi}^{\pi} |\overline{\widehat{\varphi}(k)} \, \widehat{g}(k)|^2 \, \frac{dk}{2\pi}$$

which is (4.6.33). $\qquad\square$

Proposition 4.6.7. *Suppose $\varphi \in L^2(\mathbb{R})$ with $\{\varphi(\cdot - n)\}_{n\in\mathbb{Z}}$ of the form that (4.6.17) holds and so that $\widehat{\varphi}(k)$ is continuous at $k = 0$. Then*

$$\underset{\ell\to\infty}{s\text{-}\lim}\, P_\ell = \mathbb{1} \Leftrightarrow \pm\widehat{\varphi}(0) = (2\pi)^{-1/2} \qquad (4.6.36)$$

In particular, if $\varphi \in L^1(\mathbb{R}) \cap L^2(\mathbb{R})$

$$\underset{\ell\to\infty}{s\text{-}\lim}\, P_\ell = \mathbb{1} \Leftrightarrow \pm\int \varphi(x)\, dx = 1 \qquad (4.6.37)$$

Proof. Suppose $\operatorname{supp}\widehat{g} \subset [-K, K]$ is C^∞. Then, $\operatorname{supp}\widehat{U_\ell g} \subset [-2^{-\ell}K, 2^{-\ell}K]$ and $\widehat{U_\ell g}(k) = 2^{\ell/2}\widehat{g}(2^\ell k)$. Once ℓ is so large that $2^{-\ell}K < \pi$, we can use the lemma to see that

$$\|P_\ell g\|^2 = \|P_0 U_\ell g\|^2 = 2\pi \int_{-\pi}^{\pi} |\widehat{\varphi}(k)|^2\, 2^\ell |\widehat{g}(2^\ell k)|^2 dk$$

So, since $\widehat{\varphi}$ is continuous and $2^\ell |\widehat{g}(2^\ell \cdot)|^2 \|g\|_2^{-2}$ is an approximate identity, we see that

$$\lim_{\ell\to\infty} \|P_\ell g\|^2 = 2\pi|\widehat{\varphi}(0)|^2 \|g\|^2 \qquad (4.6.38)$$

If $s\text{-}\lim_{\ell\to\infty} P_\ell = \mathbb{1}$, clearly this implies $2\pi|\widehat{\varphi}(0)|^2 = 1$. Conversely, if $P = s\text{-}\lim P_\ell$ (which exists since $V_\ell \subset V_{\ell+1}$) is not all of $\mathcal{H}$, we can find g with $\widehat{g} \in C_0^\infty$ and $\|Pg\| < \|g\|$ which shows $2\pi|\widehat{\varphi}(0)|^2 < 1$. Thus, we have proven that

$$\underset{\ell\to\infty}{s\text{-}\lim}\, P_\ell = \mathbb{1} \Leftrightarrow 2\pi|\widehat{\varphi}(0)|^2 = \mathbb{1} \qquad (4.6.39)$$

Since φ is real, $\widehat{\varphi}(0)$ is real so $\widehat{\varphi}(0) = \pm 1/\sqrt{2\pi}$. $\qquad \square$

Remarks. 1. The proof shows by taking $\widehat{g}$ a multiple of a characteristic function on $[-R, R]$, that even if $\widehat{\varphi}$ is not continuous, $\lim_{n\to\infty} 2^{n+1} \int_{-2^{-n}}^{2^{-n}} |\widehat{\varphi}(k)|^2\, dk$ exists and is $(2\pi)^{-1}$ if and only if $s\text{-}\lim_{\ell\to\infty} P_\ell = \mathbb{1}$.

2. If φ obeys the hypothesis, so does $-\varphi$. To normalize the sign, we henceforth suppose that $\operatorname{Re}\widehat{\varphi}(k) > 0$ is true for all small k.

We summarize what we've proven about MRA's.

Theorem 4.6.8. *Let $\varphi \in L^2(\mathbb{R})$, real-valued, $\varphi_{\ell,k}(x) \equiv \varphi(2^\ell x - k)$, and V_ℓ is the closed span of $\{\varphi_{\ell,k}\}_{k\in\mathbb{Z}}$ Suppose $\widehat{\varphi}$ is continuous at 0 (in particular, this is true if $\varphi \in L^1$). Then, $\{V_\ell\}_{\ell\in\mathbb{Z}}$, φ are a MRA if and only if*

(a) $\{\varphi_{\ell=0,k}\}_{k\in\mathbb{Z}}$ *are orthonormal. Equivalently*

$$\sum_{m=-\infty}^{\infty} |\widehat{\varphi}(p + 2\pi m)|^2 = (2\pi)^{-1} \qquad (4.6.40)$$

(b) $\varphi \in V_1$, *i.e., there is $\{a_n\}_{n\in\mathbb{Z}} \in \ell^2$ so (4.6.22) holds.*

(c) $|\widehat{\varphi}(0)| = (2\pi)^{-1/2}$ (4.6.41)

(d) $m \neq 0$ in $\mathbb{Z} \Rightarrow \widehat{\varphi}(2\pi m) = 0$ (4.6.42)

Remark. (4.6.42) follows from (4.6.40) and (4.6.41)

We want to link this up to mother wavelets—that is, find ψ so that $\{\psi(\cdot - n)\}_{n \in \mathbb{Z}}$ are orthonormal and span $V_1 \cap V_0^{\perp}$ which will imply that $\{\psi(2^{\ell} \cdot -n)\}_{\ell, n \in \mathbb{Z}}$ are an orthonormal basis. Remarkably, one can write down such a ψ explicitly in terms of the structure constants. We'll just pull it out of a hat but Problem 3 will show how to "derive" it and also alternate choices.

Theorem 4.6.9. *Let $\{V_{\ell}\}_{\ell \in \mathbb{Z}}$, φ be a MRA. Define*

$$\psi(x) = \sum_{n \in \mathbb{Z}} (-1)^n a_{1-n} \varphi(2x - n) \qquad (4.6.43)$$

where $\{a_n\}_{n \in \mathbb{Z}}$ are the structure constants for the MRA. Then,

(a) *For all $j, m \in \mathbb{Z}$,*

$$\int \psi(x-j) \varphi(x-m)\, dx = 0, \quad \int |\psi(x)|^2\, dx = 1 \qquad (4.6.44)$$

(b) *$\{\psi(\cdot - j)\}_{j \in \mathbb{Z}} \cup \{\varphi(\cdot - m)\}_{m \in \mathbb{Z}}$ are an orthonormal basis for V_1.*

(c) *$\{\psi(2^{\ell} \cdot -n)\}_{n, \ell \in \mathbb{Z}}$ are an orthonormal basis for $L^2(\mathbb{R})$, i.e., ψ is a mother wavelet.*

Proof. (a) The second equality in (4.6.44) is immediate from (4.6.23) and $\|\varphi(2 \cdot -n)\|^2 = \frac{1}{2}$. It suffices by $x \to x - j$ invariance, to prove the first if $j = 0$ which given that $\{\varphi(2 \cdot -n)\}$ is orthogonal is equivalent to

$$S_k \equiv \sum_n (-1)^n a_{1-n}\, a_{n+2k} = 0 \qquad (4.6.45)$$

for each k. In S_k, change the summation index to m defined by

$$n = -m - 2k + 1 \qquad (4.6.46)$$

so $(-1)^n = -(-1)^m$. Since $1 - n = m + 2k$ and $n + 2k = 1 - m$, we see that

$$S_k = \sum_m -(-1)^m a_{m+2k}\, a_{1-m} = -S_k \qquad (4.6.47)$$

i.e., $2S_k = 0 \Rightarrow S_k = 0$.

(b) Define $\psi_j = \psi(\cdot - j)$, $\varphi_j = \varphi(\cdot - j)$, $\varphi_j^{(1)} = \varphi(2 \cdot -j)$. By $\langle \varphi_j^{(1)}, \varphi_m^{(1)} \rangle = \frac{1}{2}\delta_{jm}$, we see that

$$\langle \psi_j, \varphi_n^{(1)} \rangle = \tfrac{1}{2}(-1)^n a_{1-n+2j}, \quad \langle \varphi_j, \varphi_n^{(1)} \rangle = \tfrac{1}{2} a_{n+2k} \qquad (4.6.48)$$

If $\tilde{j} = j - n$, then

$$\sum_j |a_{1-n+2j}|^2 = \sum_{\tilde{j}} |a_{n+2\tilde{j}+1}|^2 \tag{4.6.49}$$

Thus, for n fixed,

$$\sum_j |\langle \psi_j, \varphi_n^{(1)} \rangle|^2 + \sum_m |\langle \varphi_m, \varphi_n^{(1)} \rangle|^2 = \frac{1}{4} \left(\sum_j |a_{n+2\tilde{j}+1}|^2 + \sum_m |a_{n+2m}|^2 \right)$$

$$= \frac{1}{4}(2) = \frac{1}{2} = \|\varphi_n^{(1)}\|^2 \tag{4.6.50}$$

It follows that each $\varphi_n^{(1)}$ is in the span of the ψ's and φ's. Since $\{\varphi_n^{(1)}\}$ are a basis for V_1, we have the result.

(c) $\bigcap_j V_j = \{0\}$ and $\overline{\bigcup_j V_j} = L^2$ implies

$$L^2 = \bigoplus_{\ell=-\infty}^{\infty} (V_{\ell+1} \cap V_\ell^\perp) \tag{4.6.51}$$

Since (a), (b) imply $\{\psi(\cdot - j)\}_j$ span $V_1 \cap V_0^\perp$, we see $\{\psi(2^\ell \cdot -j)\}_j$ span $V_\ell \cap V_{\ell-1}^\perp$ $\square$

As a final general result before turning to explicit examples, we want to translate (4.6.22) to Fourier transform language. If $\varphi_n^{(1)} = \varphi(2 \cdot -n)$, then

$$\widehat{\varphi_n^{(1)}}(k) = \frac{1}{2} e^{-\frac{1}{2}ink} \widehat{\varphi}\left(\frac{k}{2}\right) \tag{4.6.52}$$

If we define

$$m_0(k) = \frac{1}{2} \sum_{n \in \mathbb{Z}} a_n e^{-ink} \tag{4.6.53}$$

then (4.6.22) is equivalent to

$$\widehat{\varphi}(k) = m_0\left(\frac{k}{2}\right) \widehat{\varphi}\left(\frac{k}{2}\right) \tag{4.6.54}$$

Conversely, if $m_0 \in L^2_{\text{loc}}(\mathbb{R})$ with

$$m_0(k + 2\pi) = m_0(k) \tag{4.6.55}$$

then the $a_n \in \ell^2$ defined by (4.6.53) obey (4.6.22). The function $m_0(k)$ is called the *scaling filter* or *low-pass filter*. We want to note one property of this function.

Proposition 4.6.10. *If $m_0(k)$ is the scaling filter for a MRA, then*

$$|m_0(k)|^2 + |m_0(k + \pi)|^2 = 1 \tag{4.6.56}$$

Proof. (4.6.54) plus the 2π periodicity of m_0 implies for any integer j

$$|\widehat{\varphi}(k + (2j)2\pi)|^2 = |m_0\left(\frac{k}{2}\right)|^2 \, |\widehat{\varphi}\left(\frac{k}{2} + 2\pi j\right)|^2 \qquad (4.6.57)$$

$$|\widehat{\varphi}(k + (2j+1)2\pi)|^2 = |m_0\left(\frac{k}{2} + \pi\right)|^2 \, |\widehat{\varphi}\left(\frac{k}{2} + \pi + 2\pi j\right)|^2 \qquad (4.6.58)$$

Summing over j and using (4.6.17) implies

$$(2\pi)^{-1} = \left[\, |m(k)|^2 + |m(k+\pi)|^2 \,\right](2\pi)^{-1} \qquad (4.6.59)$$

which yields (4.6.56) $\square$

Example 4.6.11 (Meyer Wavelets). This will produce wavelets with $\psi \in \mathcal{S}(\mathbb{R})$. Pick $g \in \mathcal{S}(\mathbb{R})$ with

(i) $0 \le g(k) \le 1$ for all x.
(ii) $g(k) = g(-k)$.
(iii) $g(k) = 1$, $\quad -\frac{2\pi}{3} < |k| < \frac{2\pi}{3}$.
(iv) $\operatorname{supp} g \subset \left[-\frac{4\pi}{3}, \frac{4\pi}{3}\right]$.
(v) $g(k)^2 + g(2\pi - k)^2 = 1$ for $0 < k < 2\pi$. $\qquad (4.6.60)$

It is easy to see (Problem 2) that such a g exists.

Let $\varphi = (2\pi)^{-1/2}\widecheck{g}$ so that $\widehat{\varphi} = (2\pi)^{-1/2}g$. $\widehat{\varphi}$ has compact support so φ is real analytic. We first claim that φ obeys (4.6.17). For $\operatorname{supp} g$ has size $\frac{2}{3}(4\pi)$, so at most two points from $\{2\pi n\}_{n \in \mathbb{Z}}$ have $g(k + 2\pi n) \ne 0$ and then (4.6.60) yields (4.6.17) at that k. By construction, $\widehat{\varphi}(0) = (2\pi)^{-1/2}$.

Finally define $m_0(k)$ on $[-\pi, \pi]$ by

$$m_0(k) = g(2k), \quad |k| \le \pi \qquad (4.6.61)$$

and extend to be 2π periodic. We claim that (4.6.54) holds. For if $|k| \le \frac{2\pi}{3}$ then $\widehat{\varphi}(\frac{k}{2}) = (2\pi)^{-1/2}$, so $\widehat{\varphi} = (2\pi)^{-1/2}g(k) = m_0(\frac{k}{2})(2\pi)^{-1/2} = m_0(\frac{k}{2})\widehat{\varphi}(\frac{k}{2})$. If $\frac{2\pi}{3} \le |k| \le \frac{4\pi}{3}$, $m_0(\frac{k}{2}) = (2\pi)^{1/2}\widehat{\varphi}(k) = g(k) = 0$, so (4.6.54) holds. If $|k| > \frac{4\pi}{3}$, $\widehat{\varphi}(k) = \widehat{\varphi}(\frac{k}{2}) = 0$ so (4.6.54) holds. Thus, by Theorem 4.6.8, φ generates a MRA. Since m_0 is C^∞ on $[-\pi, \pi]$, a_n decays faster than

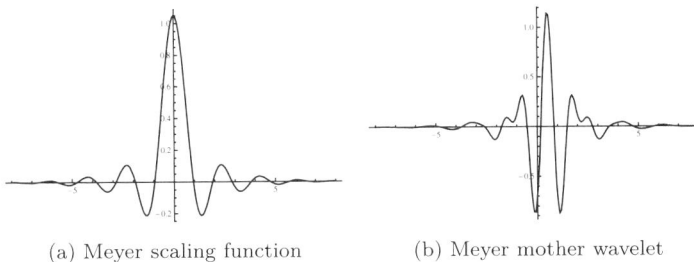

(a) Meyer scaling function   (b) Meyer mother wavelet

Figure 4.6.1. Meyer wavelets.

any power, so ψ is also in $\mathcal{S}(\mathbb{R})$. Figure 4.6.1 shows φ and ψ for a Meyer wavelet. $\square$

We summarize:

Theorem 4.6.12. *There exist wavelets, ψ, in $\mathcal{S}(\mathbb{R})$.*

The next result will imply that if a wavelet, f, decays quickly enough and is smooth enough, then f has vanishing moments. We are interested in the case $f = g$ below but it helps to separate the two parts. As usual, if $h \in L^2$, we define for $\ell, k \in \mathbb{Z}$

$$h_{\ell,k}(x) = 2^{\ell/2} h(2^\ell x - k) \tag{4.6.62}$$

Theorem 4.6.13. *Fix $L \in \mathbb{Z}_+$. Let $f, g \in L^2(\mathbb{R})$ and obey*

$$\int f_{\ell,k}(x)\, g_{p,q}(x)\, dx = 0 \tag{4.6.63}$$

for all $(\ell, k) \neq (p, q)$ in $\mathbb{Z}^2$. Suppose

(i) $|f(x)| \leq C(1 + |x|)^{-\alpha}$ *for $\alpha > L + 1$*
(ii) g *is $C^{(L)}$ with $\|g^{(j)}\|_\infty < \infty$ for $j = 0, 1, \ldots, L$*

Then

$$\int x^j f(x)\, dx = 0, \quad j = 0, 1, 2, \ldots, L \tag{4.6.64}$$

Proof. The idea will be for ℓ large, f is very concentrated near a single point, so we can pick the point with g nonzero to conclude successive moments of f are nonzero. We'll prove it for $j = 0, 1$ and leave the induction to the reader.

Let g be continuous. Since the dyadic rationals are dense, pick J and K so $g(2^{-J}K) \neq 0$. Then, by (4.6.63) for all $j > J$ and $j > 0$ (for $k = 2^{j-J}K$),

$$\int 2^j f(2^j(x - 2^{-J}K))\, g(x)\, dx = 0 \tag{4.6.65}$$

Since g is bounded and f decays faster than $(1 + |x|)^{-1-\varepsilon}$, it is easy to prove that as $j \to \infty$ the integral converges to

$$g(2^{-J}K) \int f(x)\, dx \tag{4.6.66}$$

so by (4.6.65), $\int f(x)\, dx = 0$.

Now suppose that g is C^1 and $L \geq 1$. Now pick J, K so that $g'(2^{-J}K) \neq 0$. Since $\int f(x)\, dx = 0$ as proven above,

$$\int 2^{2j} f(2^j(x - 2^{-J}K))\, [\,g(x) - g(2^{-J}K)\,]\, dx = 0 \tag{4.6.67}$$

Since g is C^1, we can write

$$g(x) - g(2^{-J}K) = (x - 2^{-J}K)h(x) \tag{4.6.68}$$

where h is bounded and continuous and $h(2^{-J}K) = g'(2^{-J}K) \neq 0$. Let $\tilde{f}(x) = x f(x)$. Then (4.6.67) says

$$\int 2^j \, \tilde{f}(2^j(x - 2^{-J}K))h(x) \, dx = 0 \qquad (4.6.69)$$

As above, given the bound on f which implies a bound on $\tilde{f}$, we conclude $\int \tilde{f}(x) \, dx = \int x f(x) \, dx = 0$.

An obvious iteration implies the general result. $\qquad\qquad\qquad \square$

Corollary 4.6.14. *Let f be a mother wavelet which lies in $\mathcal{S}(\mathbb{R})$. Then,*

$$\int x^\ell f(x) \, dx = 0 \qquad (4.6.70)$$

for all $\ell = 0, 1, 2, \ldots$. There is no such f which also obeys

$$|f(x)| \leq A e^{-B|x|} \qquad (4.6.71)$$

In particular, there are no compact support, C^∞, mother wavelets.

Remark. The issue of f's which obey (4.6.70) is closely connected to nonuniqueness of the moment problem as discussed in Section 5.6 of Part 1 and Section 7.7 of Part 4; see, in particular, (5.6.23) of Part 1.

$$f(x) = \chi_{[0,\infty)}(x) \, x^{-\log(x)} \sin(2\pi \log x) \qquad (4.6.72)$$

obeys (4.6.70) and is $\mathcal{S}(\mathbb{R})$ although it is not claimed to be a mother wavelet.

Proof. That (4.6.70) holds is a consequence of the last theorem. If (4.6.71) also holds, then

$$\hat{f}(k) = (2\pi)^{-1/2} \int e^{-ikx} f(x) \, dx \qquad (4.6.73)$$

has an analytic contribution to $|\text{Im } k| < B^{-1}$. (4.6.70) says that this analytic function has zero Taylor series at $k = 0$, so, for k real, $\hat{f}(k) = 0$, so $f = 0$. $\quad \square$

While there are no C^∞-wavelets of compact support, we are heading towards proving that for any ℓ, there are C^ℓ wavelets of compact support (as $\ell \to \infty$, the size of the support diverges). Next we'll prove

Theorem 4.6.15 (Daubechies' Theorem). *For any $\ell = 0, 1, \ldots$, there exist mother wavelets, ψ, of compact support so that ψ is C^ℓ.*

Figure 4.6.2 shows the Daubechies mother wavelet with $N = 4$ (in terms of the parameter N of (4.6.112) below).

Of course, the point is not merely existence but a constructive proof that provides the ψ for explicit calculations. We'll construct φ which is C^ℓ of compact support with only finitely many nonzero a_n—if we do that, by

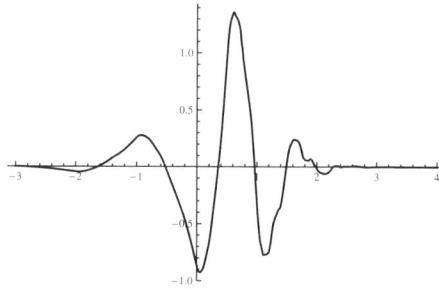

Figure 4.6.2. The $N = 4$ Daubechies wavelet.

(4.6.36), ψ also is C^ℓ of compact support. To get φ which is C^ℓ, it suffices to find φ so that for some $C, \varepsilon > 0$

$$|\widehat{\varphi}(k)| \le C(1 + |k|)^{-1-\ell-\varepsilon} \tag{4.6.74}$$

We'll use the scaling filter, m_0, to construct these wavelets because of two basic facts: φ has compact support if and only if m_0 is a trigonometric polynomial (i.e., a finite sum of e^{-ink}) and, in that case, one can recover φ (and so ψ) from m_0. In this next result, we assume that we know tht we have a φ and a MRA. Later, we'll see what we need to be sure that the φ constructed from m yields a MRA.

Theorem 4.6.16. *Let φ obey the hypotheses of Theorem 4.6.8. Then the following are equivalent:*

(1) φ has compact support.
(2) $\{n \mid a_n \ne 0\}$ is finite.
(3) m_0 is a trigonometric polynomial.

If these equivalent conditions hold, then ψ also has compact support and

$$\widehat{\varphi}(k) = (2\pi)^{-1/2} \prod_{j=1}^{\infty} m_0(k/2^j) \tag{4.6.75}$$

where the product is uniformly convergent on compact subsets of $\mathbb{R}$.

Proof. (2) and (3) are clearly equivalent. If (1) holds, and $\operatorname{supp} \varphi \subset [-R, R]$ and the integral in (4.6.22) which define a_n is not zero, then there is an x with $|2x - n| \le R$ and $|x| \le \ell$ so $|n| \le |2x - n| + 2|x| \le 3R$, i.e.,

$$|n| > 3R \Rightarrow a_n = 0 \tag{4.6.76}$$

so (2) holds.

Conversely, if (3) holds, m_0 is C^1 with bounded derivatives. Moreover, $\widehat{\varphi}(0) = (2\pi)^{-1/2}$ and (4.6.54) imply $m_0(1) = 1$, so for real k,

$$|m_0(k) - 1| \le C|k| \tag{4.6.77}$$

and thus

$$\sum_{n=1}^{\infty} |m_0(k/2^n) - 1| \leq C \sum_{n=1}^{\infty} |k|/2^n \qquad (4.6.78)$$

proving the uniform convergence (on each $\{k \mid |k| \leq K\}$) of the product in (4.6.75).

By (4.6.54) and iteration,

$$\widehat{\varphi}(k) = \prod_{j=1}^{n} m_0(k/2^j)\widehat{\varphi}(k/2^n) \qquad (4.6.79)$$

By hypothesis, $\widehat{\varphi}$ is continuous at 0, so using $\widehat{\varphi}(0) = (2\pi)^{-1/2}$, we conclude that $\widehat{\varphi}$ obeys (4.6.75).

Here is one proof that φ has compact support. Problem 4 has a second proof using Payley–Wiener ideas. By hypothesis, for some M,

$$m_0(k) = \sum_{|n| \leq M} a_n e^{ikn} \qquad (4.6.80)$$

Thus, if we define m_N by

$$m_N(k) = \prod_{j=1}^{N} m_0(k/2^j)$$
$$= \sum_{\ell} a_\ell^{(N)} e^{ik\ell} \qquad (4.6.81)$$

where the ℓ's in the sum have the form

$$\ell = \frac{n_1}{2} + \frac{n_2}{4} + \ldots + \frac{n_N}{2^N}, \quad |n_j| \leq M \qquad (4.6.82)$$

so all ℓ obey

$$|\ell| \leq M \qquad (4.6.83)$$

It follows that

$$\check{m}_N(x) = (2\pi)^{1/2} \sum_{\ell} a_\ell^{(N)} \delta(x - \ell) \qquad (4.6.84)$$

is supported in x in $[-M, M]$.

By (4.6.56), $|m_0(k)| \leq 1$. Thus, $|m_N(k)| \leq 1$, so the pointwise limit $\widehat{\varphi}(k)(2\pi)^{1/2}$ obeys

$$\lim_{N \to \infty} (2\pi)^{-1/2} \int g(k) m_N(k)\, dk = \int \widehat{\varphi}(k) g(k)\, dk$$

for all $g \in \mathcal{S}(\mathbb{R})$. Thus φ is the weak limit in $\mathcal{S}'(\mathbb{R})$ of $\check{m}_N$ and so φ is also supported in $[-M, M]$.

This also shows that $\widehat{\varphi}$ obeys (4.6.22). If φ has compact support and only finitely many a_n's are nonzero, (4.6.43) implies that ψ has compact support. $\qquad \square$

This leads to the question of when a given m_0 leads to a φ defined by (4.6.75) with φ a scaling function. Here is the result:

Theorem 4.6.17. *Let m_0 be a 2π periodic continuous function obeying*

(i) $m_0(0) = 1, \quad m_0(-k) = \overline{m_0(k)}$.

(ii) $|m_0(k)|^2 + |m_0(k+\pi)|^2 = 1$. $\qquad\qquad\qquad\qquad$ (4.6.85)

(iii) $m_0(k) \neq 0$ for $|k| \leq \frac{\pi}{2}$.

(iv) *For some $C, |m_0(k) - 1| \leq C|k|$ for all k.*

Define φ by (4.6.75). Then φ generates a MRA with m_0 as its scaling filter.

Remarks. 1. The new element is the requirement that m_0 be nonvanishing on $[-\frac{\pi}{2}, \frac{\pi}{2}]$. After the proof, we give an example that shows that without some condition on the zeros, the conclusion fails.

2. One doesn't require as much regularity as we state. For example, Hölder continuity near 0 suffices but we want to apply this to an m_0 which is C^∞, so we don't try to optimize.

Proof. As noted earlier, the product defining $\widehat{\varphi}$ for m_0 converges uniformly on each $[-K, K]$. Since $\widehat{\varphi}(k) = m_0(\frac{k}{2})\widehat{\varphi}(\frac{k}{2})$ by construction, (4.6.22) holds as does $\widehat{\varphi}(0) = (2\pi)^{-1/2}$. To apply Theorem 4.6.8, all we need is to prove the orthonormality of $\{\varphi(\cdot - n)\}_{n\in\mathbb{Z}}$.

We define φ_N, for $N = 1, 2, \ldots$, φ_N by

$$\widehat{\varphi}_N(k) = \begin{cases} (2\pi)^{-1/2} \prod_{j=1}^N m_0(\frac{k}{2^j}) & \text{if } |k| \leq 2^N\pi \\ 0 & \text{if } |k| > 2^N\pi \end{cases} \qquad (4.6.86)$$

We will prove that $\varphi \in L^2$ and that $\|\varphi_N - \varphi\|_2 \to 0$ and also that $\{\varphi_N(\cdot - n)\}_{n\in\mathbb{Z}}$ is orthonormal. Since then

$$\langle \varphi(\cdot - n), \varphi(\cdot - m)\rangle = \lim_{N\to\infty} \langle \varphi_N(\cdot - n), \varphi_N(\cdot - m)\rangle$$
$$= \delta_{nm} \qquad (4.6.87)$$

This will imply that $\{\varphi(\cdot - n)\}_{n\in\mathbb{Z}}$ is orthonormal.

We will prove orthonormality of $\{\varphi_N(\cdot - n)\}$ by using Theorem 4.6.3 and induction in N. For $N = 1$, $\widehat{\varphi}_1$ is supported on $[-2\pi, 2\pi]$. If k is not an integral multiple of 2π, $\{k + 2\pi m\}_{m\in\mathbb{Z}}$ intersects $[-2\pi, 2\pi]$ in exactly two points, say (by 2π periodicity of m_0), k and $k + 2\pi$, so

$$|\widehat{\varphi}_1(k)|^2 + |\widehat{\varphi}_1(k+2\pi)|^2 = ((|m_0(k/2)|^2 + m_0|(k/2+\pi)|^2))(2\pi)^{-1} = (2\pi)^{-1}$$

so $\widehat{\varphi}_1$ obeys (4.6.17)

Suppose $\widehat{\varphi}_{N-1}$ obeys (4.6.17). Note that

$$\widehat{\varphi}_N(k) = m_0(k/2)\widehat{\varphi}_{N-1}(k/2) \qquad (4.6.88)$$

and compute

$$\sum_{m=-\infty}^{\infty} |\widehat{\varphi}_N(k+2\pi m)|^2 = \sum_{m=-\infty}^{\infty} |\widehat{\varphi}_N(k+(2\pi)2m)|^2 + |\widehat{\varphi}_N(k+(2\pi)(2m+1))|^2$$

$$= \sum_{m=-\infty}^{\infty} |m_0(k/2)|^2 |\widehat{\varphi}_{N-1}(k/2+2\pi m)|^2$$
$$+ |m_0(k/2+\pi)|^2 |\widehat{\varphi}_{N-1}(k/2+\pi+2\pi m)|^2 \tag{4.6.89}$$

$$= (|m_0(k/2)|^2 + |m_0(k/2+\pi)|^2)(2\pi)^{-1} \tag{4.6.90}$$

$$= (2\pi)^{-1} \tag{4.6.91}$$

(4.6.89) uses (4.6.88), (4.6.90) follows from the induction hypothesis, and (4.6.91) uses (4.6.84).

We've thus proven $\{\varphi_N(\cdot-m)\}_{m\in\mathbb{Z}}$ is an orthonormal set. In particular,

$$\int |\widehat{\varphi}_N(k)|^2 \, dk = 1 \tag{4.6.92}$$

But as we've seen, $\varphi_N \to \varphi$ pointwise, so by Fatou's lemma

$$\int |\widehat{\varphi}(k)|^2 \, dk \le 1 \tag{4.6.93}$$

i.e., $\widehat{\varphi} \in L^2(\mathbb{R})$. We claim that for a constant C and all k and N,

$$|\widehat{\varphi}_N(k)| \le C|\widehat{\varphi}(k)| \tag{4.6.94}$$

If we prove this, then dominated convergence and the pointwise convergence show that $\|\varphi_N - \varphi\|_2 = \|\widehat{\varphi}_N - \widehat{\varphi}\|_2 \to 0$ and the proof is complete.

As we noted on $[-\pi, \pi]$, the product defining $\widehat{\varphi}$ converges uniformly so $\widehat{\varphi}$ is continuous there. Moreover, since $|k| \le \pi \Rightarrow |\frac{k}{2^j}| \le \frac{\pi}{2}$ ($j = 1, 2, \ldots$) no factor is zero. So $\widehat{\varphi}$ is continuous and nonvanishing on $[-\pi, \pi]$, i.e.,

$$Q \equiv \inf_{|k|\le\pi} |\widehat{\varphi}(k)| > 0 \tag{4.6.95}$$

If $|k| > 2^N \pi$, $\widehat{\varphi}_N(k) = 0$ and (4.6.94) is trivial. Note that

$$|k| \le 2^N \pi \Rightarrow (2\pi)^{1/2}\widehat{\varphi}_N(k) = \frac{\widehat{\varphi}(k)}{\widehat{\varphi}(2^{-N}k)} \tag{4.6.96}$$

and that $2^{-N}k \in [-\pi, \pi]$ so

$$|\widehat{\varphi}_N(k)| \le (2\pi)^{-1/2}Q^{-1}|\widehat{\varphi}(k)| \tag{4.6.97}$$

proving (4.6.94) with $C = (2\pi)^{-1/2}Q^{-1}$. $\square$

Example 4.6.18. Let

$$\varphi(x) = \tfrac{1}{3}\chi_{[0,3]} \qquad (4.6.98)$$

Then, because $e^{2\alpha} - 1 = (e^\alpha - 1)(e^\alpha + 1)$, we have

$$\widehat{\varphi}(k) = (2\pi)^{-1/2}\left(\frac{e^{3ik} - 1}{3ik}\right) = m_0(k/2)\,\widehat{\varphi}(k/2) \qquad (4.6.99)$$

$$m_0(k) = \tfrac{1}{2}\left(e^{3ik} + 1\right) \qquad (4.6.100)$$

Notice that $|m_0(k)| = |\cos(\tfrac{3k}{2})|$, so $|m_0(k+\pi)| = |\sin(\tfrac{3k}{2})|$ and (4.6.84) holds. m_0 obeys (i), (ii), and (iv) of the theorem. The translates of φ are neither normalized nor orthonormal so the conclusion of the theorem fails. Of course, $\widehat{\varphi}(\tfrac{\pi}{3}) = 0$ and $\tfrac{\pi}{3} < \tfrac{\pi}{2}$, so (iii) fails. This shows some hypothesis like (iii) is needed. $\qquad\square$

Example 4.6.1 (revisited). This will illustrate some of the choices we'll eventually make. It will help, in general, to focus on

$$M_0(k) \equiv |m_0(k)|^2 \qquad (4.6.101)$$

The condition

$$M_0(k) + M_0(k+\pi) = 1 \qquad (4.6.102)$$

and $\cos^2(\theta) + \sin^2(\theta) = 1$ suggests $M_0(k)$ is related to $\cos^2(k)$. Since $\sin(k) = \cos(k + \tfrac{\pi}{2})$ though, we need to have

$$M_0(k) = \cos^2(k/2) \qquad (4.6.103)$$

M_0 has to be a trigonometric polynomial, while $\cos(\tfrac{k}{2})$ is not (if $z = e^{ik}$, $\cos(\tfrac{k}{2}) = \tfrac{1}{2}(z^{1/2} + z^{-1/2})$), $\cos^2(\tfrac{k}{2}) = \tfrac{1}{2}(1 + \cos(k))$ is so we seem to be on the right track. We can't take $m_0(k) = \cos(\tfrac{k}{2})$ since that is neither a trigonometric polynomial nor 2π periodic but notice if

$$m_0(k) = \tfrac{1}{2}(1 + e^{ik}) \qquad (4.6.104)$$

then m_0 is 2π periodic and a trigonometric polynomial and for k real:

$$\overline{m_0(k)}\,m_0(k) = \cos^2(k/2) = \tfrac{1}{4}(2 + e^{ik} + e^{-ik}) \qquad (4.6.105)$$

While we were led to this by considerations of simplicity and (4.6.102), this is in fact the m_0 for the Haar wavelet! For $\varphi(x) = \chi_{[0,1]}$ implies

$$(2\pi)^{1/2}\widehat{\varphi}(k) = (e^{ik} - 1)/ik = e^{ik/2}\sin(k/2)/(k/2) \qquad (4.6.106)$$

We have

$$a_0 = a_1 = 1, \quad a_n = 0 \quad (n \neq 0, 1) \qquad (4.6.107)$$

Thus

$$m_0(k) = \tfrac{1}{2}(1 + e^{ik}) \qquad (4.6.108)$$

as can also be seen from $m_0(k/2) = \widehat{\varphi}(k)/\widehat{\varphi}(k/2)$ and (4.6.106).

Notice also in this case, (4.6.43) provides the standard Haar wavelet. Finally, (4.6.75) is equivalent to

$$\frac{\sin(k)}{k} = \prod_{j=1}^{\infty} \cos\left(\frac{k}{2^j}\right) \tag{4.6.109}$$

which was proven using a product formula in Problem 16 of Section 9.2. of Part 2A. This, in turn, implies

$$\prod_{j=1}^{J} \cos\left(\frac{k}{2^j}\right) = \frac{\sin(k)}{2^J \sin\left(\frac{k}{2^J}\right)} \tag{4.6.110}$$

something we'll need later. $\qquad\square$

(4.6.109) has a moral. $\prod_{j=1}^{\infty} m\left(\frac{k}{2^j}\right)$ is a product of functions, each of which is periodic (with growing periods). One might have guessed that such a product has no decay since each individual factor has no decay. The explicit formula, (4.6.109) shows such a product does have decay. If $|m_0(k)|^2 \leq [\cos(\frac{k}{2})]^N$ for N large, we'd get $|\widehat{\varphi}(k)| \leq C|k|^{-N}$ near infinity and so get $\widehat{\varphi}$ smooth. The argument will need to be more subtle, but this notion is part of it—we'll want $m_0(k)$ to have a large factor of $\cos(\frac{k}{2})$ in front.

We now turn to the initial steps of Daubechies' construction. Fix $N \in \{1, 2, \ldots\}$. We'll need to take N large to get smoother and smoother wavelets. We note that

$$1 = \left(\cos^2\left(\frac{k}{2}\right) + \sin^2\left(\frac{k}{2}\right)\right)^{2N+1} =$$
$$\sum_{j=0}^{2N+1} \binom{2N+1}{j} \left[\sin^2\left(\frac{k}{2}\right)\right]^j \left[\cos^2\left(\frac{k}{2}\right)\right]^{2N+1-j} \tag{4.6.111}$$

Thus, if we define

$$M_0(k) \equiv \sum_{j=0}^{N} \binom{2N+1}{j} \left[\sin^2\left(\frac{k}{2}\right)\right]^j \left[\cos^2\left(\frac{k}{2}\right)\right]^{2N+1-j} \tag{4.6.112}$$

and note that $k \to k + \pi$ interchanges $\sin^2(\frac{k}{2})$ and $\cos^2(\frac{k}{2})$ (and so j and $2N+1-j$), we get by (4.6.111) that

$$M_0(k) + M_0(k + \pi) = 1 \tag{4.6.113}$$

We also have that

$$M_0(k) = \left[\cos^2\left(\frac{k}{2}\right)\right]^{N+1} P_N\left(\sin^2\left(\frac{k}{2}\right)\right) \tag{4.6.114}$$

$$P_N(y) = \sum_{j=0}^{N} \binom{2N+1}{j} y^j (1-y)^{N-j} \qquad (4.6.115)$$

which gives us the desired high number of $\cos(\frac{k}{2})$ factors.

One immediate consequence is that

$$M_0(k) \geq 0, \quad M_0(k) = 0 \text{ only if } k = \pi + 2\pi j;\ j \in \mathbb{Z} \qquad (4.6.116)$$

for all terms in (4.6.112) are nonnegative and the first is nonzero if $\cos(\frac{k}{2}) \neq 0$. If $\cos(\frac{k}{2}) = 0$, $M_0(k) = 0$ by (4.6.114).

Of course, we need m_0, not just M_0, but here the following rescues us:

Theorem 4.6.19 (Fejér-Riesz Theorem). *Let $g(z)$ be a finite Laurent series in z so that*

$$g(e^{i\theta}) \geq 0 \qquad (4.6.117)$$

for all $e^{i\theta} \in \mathbb{D}$. Then, there is a polynomial $P(z)$ so that

$$g(z) = P(z)\overline{P\left(\frac{1}{\bar{z}}\right)} \qquad (4.6.118)$$

Moreover, P can be chosen so that

$$P(1) = g(1)^{1/2} \qquad (4.6.119)$$

and so that

$$g(\bar{z}) = \overline{g(z)} \Rightarrow P(\bar{z}) = \overline{P(z)} \qquad (4.6.120)$$

Remark. We will also use (and reprove) this theorem in our discussion of the spectral theorem for unitary operators in Section 5.5 of Part 4.

Proof. Since $g(z)$ is real on $\partial\mathbb{D}$, $\overline{g((\frac{1}{\bar{z}}))} = g(z)$ on $\partial\mathbb{D}$ and so, by analyticity, on $\mathbb{C}$. In particular, if $g(z) = 0$, then $g(\frac{1}{\bar{z}}) = 0$. Moreover, $g(e^{i\theta}) \geq 0$ implies all zeros on $\partial\mathbb{D}$ are of even order. Hence, we can order the zeros (counting multiplicity) $z_1, \ldots, z_{2N}$ so $z_j = (\bar{z}_{2N+1-j})^{-1}$ and $z_1, \ldots z_N \in \mathbb{D}$ and $z_{N+1}, \ldots z_{2N} \in \mathbb{C} \setminus \mathbb{D}$. Let

$$P(z) = c\prod_{j=1}^{N}(z - z_j) \qquad (4.6.121)$$

where c will be chosen shortly. Then,

$$g(z)/P(z)\overline{P\left(\frac{1}{\bar{z}}\right)} = h(z) \qquad (4.6.122)$$

has no singularities and no zeros on $\mathbb{C} \setminus \{0\}$ and is a Laurent polynomial. It follows that $h(z) = dz^k$. By hypothesis, $h(e^{i\theta}) \geq 0$. It follows that $k = 0$, $d > 0$, and we can choose c so that $P(1) \geq 0$ and $d = 1$. We thus have proven (4.6.118) and (4.6.119).

If $g(\bar{z}) = \overline{g(z)}$, the zeros of g are complex conjugates to each other and so those of P are also. It follows that we can take $c > 0$ in (4.6.121) and $P(\bar{z}) = \overline{P(z)}$. $\qquad\square$

Applying this to the $M_0(z)$ constructed above, we see there is a polynomial in e^{ik}, $m_0(k)$, (i.e., $a_n \neq 0 \Rightarrow n \geq 0$) so that

$$m_0(-k) = \overline{m_0(k)} \qquad (4.6.123)$$

Thus the $\hat{\varphi}$ constructed via (4.6.75) is real-valued since m obeys all the hypotheses of Theorem 4.6.17 and we have constructed a MRA for each N. All that remains is to prove that

$$|\widehat{\varphi(k)}| \leq C(1 + |k|)^{-\alpha_N} \qquad (4.6.124)$$

with $\alpha_N \to \infty$ as $N \to \infty$.

Lemma 4.6.20. *Let* P_N *be given by* (4.6.115) *Then,*

$$P_N(y) = \sum_{k=0}^{N} \binom{N+k}{k} y^k \qquad (4.6.125)$$

Proof. Let $R_N(y)$ be the right side of (4.6.125). By computing derivatives (or using the binomial theorem), for $|y| < 1$

$$(1-y)^{-N-1} = \sum_{k=0}^{\infty} \binom{N+k}{k} y^k = R_N(y) + O(y^{N+1}) \qquad (4.6.126)$$

Thus,

$$(1-y)^{N+1} R_N(y) = 1 + O(y^{N+1}) \qquad (4.6.127)$$

On the other hand, by the binomial theorem (how we constructed P_N after all),

$$1 = (y + (1-y))^{2N+1} = \sum_{k=0}^{2N+1} \binom{2N+1}{k} y^k (1-y)^{2N+1-k} \qquad (4.6.128)$$

$$= (1-y)^{N+1} P_N(y) + O(y^{N+1}) \qquad (4.6.129)$$

It follows that

$$(1-y)^{N+1} (P_N(y) - R_N(y)) = O(y^{N+1}) \qquad (4.6.130)$$

so

$$P_N(y) - R_N(y) = O(y^{N+1}) \qquad (4.6.131)$$

Since $P_N - R_N$ is a polynomial of degree at most N, we conclude that $P_N = R_N$. $\qquad\square$

Corollary 4.6.21. *We have*

$$\sup_{0 \le y \le 1} P_N(y) = P_N(1) = \binom{2N+1}{N} \qquad (4.6.132)$$

$$\le \frac{2^{2N+1}}{\sqrt{\pi N}} \qquad (4.6.133)$$

Proof. Since, as a polynomial in y, all coefficients are positive, $P_N(y)$ is monotone in y on $[0, 1]$ so the first equality in (4.6.132) holds. By (4.6.115), $P_N(1) = \binom{2N+1}{N}$, since only the $j = N$ term is nonzero.

For the inequality, define

$$b_N = \sqrt{\pi N}\, 2^{-(2N+1)} \binom{2N+1}{N} \qquad (4.6.134)$$

By a simple calculation on how $\binom{k+2}{\ell+1}$ relates to $\binom{k}{\ell}$, we see that

$$\frac{b_{N+1}}{b_N} = \frac{(2N+3)}{(2N+4)} \sqrt{\frac{N+1}{N}} \qquad (4.6.135)$$

Since $(2N+3)^2(N+1) - (2N+4)^2 N = 9 + 5N$, we see that

$$b_1 \le b_2 \le \ldots \qquad (4.6.136)$$

By Stirling's formula (see, for example, Theorem 7.2.12 of Part 1),

$$\lim_{N \to \infty} b_N = 1 \qquad (4.6.137)$$

The last two equations prove (4.6.133). $\qquad\qquad\qquad\qquad\qquad\square$

The following which proves (4.6.124) completes the proof of Daubechies' theorem, Theorem 4.6.15.

Proposition 4.6.22. *For the $m(k)$ given by the above construction for N fixed, there is a constant C_N so that*

$$|\widehat{\varphi}(k)| \le C_N(1 + |k|)^{-(\beta_N+1)/2} \qquad (4.6.138)$$

$$\beta_N = \log_2(\sqrt{\pi N}) \qquad (4.6.139)$$

Remark. The key fact is that $\beta_N \to \infty$ as $N \to \infty$ so we can get any prescribed power decay if we take N large enough.

Proof. We use the formulae

$$|\varphi(k)| = (2\pi)^{-1/2} \prod_{j=1}^{\infty} m(k/2^j) \qquad (4.6.140)$$

$$|m(k)| \le 1, \quad |m(k)| \le K_N^{1/2}|\cos(k/2)|^{N+1}, \quad K_N = 2^{2N+1}/\sqrt{\pi N} \qquad (4.6.141)$$

For 2^j large compared to k, $k/2^j$ is small and we get no decay from $\cos(\frac{k}{2})$ to compensate for $K_N \geq 1$. Thus for

$$2^{J-1} \leq |k| \leq 2^J \qquad (4.6.142)$$

we use $|m(k)| \leq 1$ for all j in (4.6.130) with $j \geq J+1$. Therefore, when (4.6.132) holds

$$|\varphi(k)| \leq \left[\prod_{j=1}^{J} \cos\left(\frac{k}{2^{j+1}}\right) \right]^{N+1} K_N^{J/2} \qquad (4.6.143)$$

$$= \left[\frac{|\sin(k/2)|}{|2^J \sin(k/2^{J+1})|} \right]^{N+1} K_N^{J/2} \qquad (4.6.144)$$

$$\leq \left[\frac{\pi}{2^{J+1}|k|/2^{J+1}} \right]^{N+1} \left[\frac{2^{(2N+1)}}{\sqrt{\pi N}} \right]^{J/2} \qquad (4.6.145)$$

$$\leq \left[\frac{\pi}{|k|} \right]^{N+1} [2|k|]^{N+1/2} \left(\sqrt{\pi N} \right)^{-J/2} \qquad (4.6.146)$$

$$= 2^{N+1/2} \pi^{N+1} |k|^{-1/2} 2^{-J\beta_N/2} \qquad (4.6.147)$$

$$\leq 2^{N+1/2} \pi^{N+1} |k|^{-(\beta_N+1)/2} \qquad (4.6.148)$$

which implies (4.6.138) given that $|\varphi(k)| \leq (2\pi)^{-1/2}$ for $|k| \leq 1$.

In the above, we get (4.6.144) using (4.6.110). (4.6.145) then comes from $\sin(\theta) \geq \frac{2\theta}{\pi}$ if $|\theta| \leq \frac{\pi}{2}$ (since $\frac{k}{2^{J+1}} \leq \frac{1}{2} \leq \frac{\pi}{2}$) and from (4.6.133). (4.6.146) uses $2^J \leq 2|k|$ and (4.6.147) the definition of β_N. Finally, (4.6.148) uses $2^{-J} \leq |k|^{-1}$. $\qquad \square$

As a final topic, we want to discuss convergence of wavelet expansions. If φ is the scaling function and ψ the mother wavelet and $\{V_\ell\}_{\ell \in \mathbb{Z}}$ the MRA, we can write the orthogonal projection P_ℓ on V_ℓ as

$$(P_\ell f)(x) = \sum_{j=-\infty}^{\infty} \varphi(2^\ell x - j) \int 2^\ell \varphi(2^\ell y - j) f(y) \, dy \qquad (4.6.149)$$

We are interested in the *wavelet kernel*

$$\Phi(x, y) = \sum_{j=-\infty}^{\infty} \varphi(x - j)\varphi(y - j) \qquad (4.6.150)$$

Modulo issues of convergence and interchange of sum and integral (which we discuss below), we have

$$(P_\ell f)(x) = 2^\ell \int \Phi(2^\ell x, 2^\ell y) \, f(y) \, dy \qquad (4.6.151)$$

Under a simple hypothesis on the scaling function we can justify these for-mulae:

Theorem 4.6.23. *Suppose there exists a nonnegative bounded function,* $K(x)$ *on* $\mathbb{R}$, *so that*

$$|x| \geq |y| \Rightarrow K(x) \geq K(y) \tag{4.6.152}$$

and so that

$$|\varphi(x)| \leq K(x) \tag{4.6.153}$$

and

$$\int_{-\infty}^{\infty} K(x)\,dx < \infty \tag{4.6.154}$$

Then the series in (4.6.150) *converges absolutely and*

$$|\Phi(x,y)| \leq C_1 K\left(\frac{1}{2}|x-y|\right) \tag{4.6.155}$$

for a constant C_1. *Moreover, for any* $f \in L^p(\mathbb{R}, dx)$, *the integral in* (4.6.151) *converges for a.e.* x *and for a constant* C_2

$$\|P_\ell f\|_p \leq C_2\|f\|_p \tag{4.6.156}$$

for all ℓ *and all* p *with* $1 \leq p \leq \infty$.

Remark. If $|\varphi(x)| \leq C(1+|x|)^{-\alpha}$ for $\alpha > 1$, we can take $K(x) = C(1+|x|)^{-\alpha}$. In particular, bounded φ's which are of compact support (e.g., Daubechies wavelets) obey the hypothesis.

Proof. We look at

$$Q(x) = \sum_j K(x-j) \tag{4.6.157}$$

where a priori Q might be infinite. Q is periodic with period 1 so in evaluating

$$C_1 = 2 \sup_x Q(x) \tag{4.6.158}$$

we can look at $0 \leq x \leq 1$. In that case, for $j = 1, 2, \ldots$, $K(x-j) \leq K(j-1)$ (since $|j-1| \leq |j-x|$) and similarly $K(x-j) \leq K(-j)$ for $j = 0, -1, -2, \ldots$ since $-j \leq |x-j|$. Thus,

$$C_1 \leq 2\big[K(0) + K(1) + K(2) + \ldots\big]$$

$$\leq 2\Big[K(0) + \sum_{j=1}^{\infty} \int_{j-1}^{j} K(y)\,dy\Big]$$

$$\leq 2\Big[K(0) + \int_{0}^{\infty} K(y)\,dy\Big] < \infty$$

Now, define

$$R(x,y) = \sum_j K(x-j)K(y-j) \tag{4.6.159}$$

For any j, $|x-y| \leq |x-j| + |y-j|$ so either $|x-j| \geq \frac{1}{2}|x-y|$ or $|y-j| \geq \frac{1}{2}|x-y|$ or both. We can, therefore, overcount by summing over each set.

$$\sum_{j:|x-j|\geq\frac{1}{2}|x-y|} K(x-j)K(y-j) \leq \sum_{j:|x-j|\geq\frac{1}{2}|x-y|} K\left(\tfrac{1}{2}|x-y|\right)K(y-j)$$

$$\leq K\left(\tfrac{1}{2}|x-y|\right)Q(y)$$

$$\leq \tfrac{1}{2}C_1 K\left(\tfrac{1}{2}|x-y|\right) \tag{4.6.160}$$

so

$$R(x,y) \leq C_1 K\left(\tfrac{1}{2}|x-y|\right) \tag{4.6.161}$$

This proves the absolute convergence of the sum defining $\Phi(x,y)$ and the bound (4.6.155). If x is such that

$$\int K\left(2^{\ell-1}|x-y|\right)|f(y)|\,dy < \infty \tag{4.6.162}$$

then the integral in (4.6.151) converges (by (4.6.154) and the more general bound on $\sum_j |\varphi(x-j)|\,|\varphi(y-j)|$) and equals the sum.

By Lemma 2.4.2, if $(M_{\mathrm{HL}}f)(x) < \infty$ (which holds for a.e. x if $f \in L^p$), then (4.6.162) holds for all ℓ. Finally, by Young's inequality and

$$2^\ell \int K\left(2^{\ell-1}|u|\right)du = 2\int K(|u|)\,du \tag{4.6.163}$$

we see that

$$\|P_\ell f\|_p \leq 2\|K\|_1 \|f\|_p \tag{4.6.164}$$

proving (4.6.156) $\qquad\qquad\qquad\qquad\qquad\qquad\qquad\qquad\qquad\qquad\quad \square$

Theorem 4.6.24. *Let φ generate a MRA and obey the hypotheses of Theorem 4.6.23. Then,*

(a) *If f is a bounded uniformly continuous function on $\mathbb{R}$, we have*

$$\lim_{\ell\to\infty} \|P_\ell f - f\|_\infty = 0 \tag{4.6.165}$$

(b) *If $1 \leq p < \infty$ and $f \in L^p$, then*

$$\lim_{\ell\to\infty} \|P_\ell f - f\|_p = 0 \tag{4.6.166}$$

and for a.e. x

$$\lim_{l\to\infty} (P_\ell f)(x) = f(x) \tag{4.6.167}$$

(c) *If f is $C_\infty(\mathbb{R})$, the continuous functions vanishing at ∞, then*

$$\lim_{\ell\to-\infty} \|P_\ell f\|_\infty = 0 \tag{4.6.168}$$

(d) *If $f \in L^p$, $1 < p < \infty$, then*

$$\lim_{\ell\to-\infty} \|P_\ell f\|_p = 0 \tag{4.6.169}$$

Remarks. 1. Thus, if $f \in L^p$, $1 < p < \infty$ or $f \in C_\infty$, we have $\lim_{\substack{\ell \to \infty \\ k \to -\infty}} (P_\ell - P_k)f = f$ in L^p or $\|\cdot\|_\infty$.

2. We have seen that $\int \varphi(x)\,dx = (2\pi)^{1/2}\widehat{\varphi}(0) = 1$. By the Poisson summation formula (6.6.49) of Part 1, with $2\pi\alpha = 1$

$$\sum_{j=-\infty}^{\infty} \varphi(x-j) = \lim_{N \to \infty} \sum_{m=-N}^{N} \sqrt{2\pi}\,\widehat{\varphi}(2\pi m)e^{2\pi imx}$$
$$= 1$$

by (4.6.41), (4.6.42). Thus,

$$\int \Phi(x,y)\,dy = \int \Phi(y,x)\,dy = 1 \qquad (4.6.170)$$

which implies

$$P_\ell \mathbb{1} = \mathbb{1} \ (\text{so } \|P_\ell \mathbb{1}\|_\infty = 1), \quad \int (P_\ell f)(x)\,dx = \int f(x)\,dx \qquad (4.6.171)$$

so if $f \geq 0$, $\|P_\ell f\|_1 \geq \|f\|_1$ Thus, (4.6.169) fails for $p = 1$ and (4.6.168) for f only bounded uniformly continuous.

3. Formally, $P_\ell \mathbb{1} = \mathbb{1}$ and the fact that $\{\psi(2^\ell x - j)\}_{j \in \mathbb{Z}}$ lies in $\mathcal{H}_\ell^\perp \cap \mathcal{H}_{\ell+1}$ suggest $\langle \mathbb{1}, \psi(2^\ell \cdot -j) \rangle = 0$ (only formal since $\mathbb{1} \notin L^2$!). In fact, $\int \psi(y)\,dy = 0$ (see Problem 3), so in

$$f = \sum_{\ell,j} \langle \psi(2^\ell \cdot -j, f \rangle \psi(2^\ell \cdot -j)$$

again formally, the right side is zero if $f = \mathbb{1}$ dramatically illustrating the failure of the wavelet expansion if f is bounded and uniformly continuous but not in $C_\infty(\mathbb{R})$.

Proof. (a), (b) By (4.6.170)

$$(P_\ell f)(x) - f(x) = 2^\ell \int \Phi(2^\ell x, 2^\ell y)\,(f(y) - f(x))\,dy \qquad (4.6.172)$$

so for any p by Minkowski's inequality and (4.6.155)

$$\|P_\ell - f\|_p \leq \int 2^\ell K(2^{\ell-1}|u|)\,\|f_u - f\|_p\,du \qquad (4.6.173)$$

where $f_u(x) = f(x + u)$. Thus,

$$\|P_\ell f - f\|_p \leq \int K\left(\tfrac{1}{2}|u|\right)\|f_{2^{-\ell}u} - f\|_p\,du \qquad (4.6.174)$$

For f bounded and uniformly continuous, $\|f_u - f\|_\infty \to 0$ as $u \to 0$ and for $f \in L^p$, $\|f_u - f\|_p \to 0$ as $u \to 0$. By dominated convergence, we conclude that $\|P_\ell f - f\|_p \to 0$. Pointwise convergence is a simple maximal function argument given (4.6.172) and (4.6.155).

(c), (d) Let f be continuous and supported on $[-R, R]$. Then,

$$|(P_\ell f)(x)| \leq 2^\ell C_1 \int_{-R}^{R} |f(y)| \, K(2^{\ell+1}|x-y|) \, dy$$

$$\leq 2^{\ell+1} C_1 \|f\|_\infty K(0) R \qquad (4.6.175)$$

As $\ell \to -\infty$, we see $\|P_\ell f\|_\infty \to 0$. Since the continuous functions of compact support are dense in $C_\infty(\mathbb{R})$, we get (c).

This also shows for such f,

$$\int_{-R}^{R} |(P_\ell f)(x)|^p \, dx \to 0 \qquad (4.6.176)$$

If $|x| > R$,

$$|(P_\ell f)(x)| \leq 2^\ell C_1 \|f\|_1 K\left(\frac{1}{2} 2^\ell (|x| - R)\right) \qquad (4.6.177)$$

Thus,

$$\int_{|x| \geq R} |(P_\ell f)(x)|^p \, dx \leq 2^{p\ell} C_1^p \|f\|_1^p \int_{-\infty}^{\infty} |K(2^{\ell-1} y)|^p \, dy$$

$$= 2 \, 2^{\ell(p-1)} C_1^p \|f\|_1^p \|K\|_p^p$$

$$\to 0$$

as $\ell \to -\infty$ if $1 < p < \infty$. $\qquad \square$

Notes and Historical Remarks. For a broad array of textbook presentation of (discrete) wavelet theory, see [**82, 88, 189, 276, 381, 440, 569, 570, 571, 572, 573, 574, 595, 611, 647, 705, 832, 840**]. Daubechies [**189**] is a classic, the various books by Meyer (for which we list both the French originals and English translations) show the usefulness of these expansions in harmonic analysis (where wavelets can be viewed as a codification of the analysis at scales, an idea of Littlewood–Paley). Our presentation is heavily influenced by Daubechies [**189**] and Pinsky [**647**]. Recent literature has considered nonorthonormal wavelets and MRA. At least one expert informed me that my approach here is "old-fashioned."

Wavelets have been applied in a wide variety of disciplines from statistics to finance to physics to geophysics—[**44, 111, 311, 459, 556, 597, 635, 827, 831**] include applications and, sometimes, tutorial background. Because of these many applications, there is a huge literature and huge interest outside mathematics; for example, Daubechies [**189**] has over 20,000 citations in Google Scholar! Also, because of the applications, the subject is not taken as seriously as it might be by pure mathematicians despite its great mathematical beauty and applications in analysis.

[**374**] is a collection of historical papers on wavelets and Hubbard [**405**] is a semi-popular book on the history. Key figures in the development were Grossman and Mallet in the early days and Coiffman, Daubechies, and Meyer as the subject matured.

Haar bases are from Haar in 1910 [**346**]. He answered a question of Hilbert who wondered if $L^2([0,1])$ had an orthonormal basis for which the abstract Fourier series converged uniformly when f is continuous. Haar's solution is a bit of a cheat since his basis isn't continuous—at least for $L^2(\mathbb{R})$, there are other examples in this section that remedy this shortcoming. The notion of MRA is due to Mallat [**555**] based in part on ideas of Meyer. Meyer wavelets appeared in his Séminaire Bourbaki [**568**]. Even before Meyer, Strömberg [**789**] had constructed for each finite k, C^k wavelets with exponential decay. Battle [**43**] and LeMarié [**519**] constructed wavelets with the same features but different from Strömberg's. Daubechies wavelets are from the paper [**187**]. The actual formula for m and the idea of using the Fejér–Riesz theorem are from her paper but some other details were simplified in later work by her and others which we make use of. That smoothness of wavelets implies vanishing moments (the idea behind Theorem 4.6.13) is due to Battle [**43**].

The Fejér–Riesz theorem is named after their papers [**273, 681**], part of the explosion of papers around 1910 on Carathéodory functions. Note that the analog for $\mathbb{R}$ is more elementary and older than their work. It has been known since Euler that if P is a polynomial with real coefficients, equivalently $P(z) = \overline{P(\bar{z})}$, then it can be factored in factors of the form $(z - a_j)^2 + b_j^2 = \overline{(\bar{z} - a_j - ib_j)}(z - a_j - ib_j)$ and of the form $z - c_j$ ($a_j, b_j, c_j \in \mathbb{R}$). $P(z) \geq 0$ for z real implies $(z - c_j)$ must occur an even number of times. This implies, for suitable Q, we have $P(z) = \overline{Q(\bar{z})}Q(z)$.

Problems

1. Let $f \in L^\infty$ have support in $[-R, R]$. If $2^{-\ell} > R$, prove that for the Haar basis, $\|P_\ell f\|_\infty \leq 2^\ell \|f\|_1$ and conclude that $\lim_{\ell \to -\infty} \|P_\ell f\|_\infty = 0$.

2. Show that a $g \in C^\infty$ of the type needed in Example 4.6.11 exists. (*Hint*: Pick $h \in C_0^\infty$ with $h(x) = h(-x)$, $0 \leq h \leq 1$, $h \equiv 1$ on $\left[-\frac{2\pi}{3}, \frac{2\pi}{3}\right]$, $h = 0$ on $\left[\frac{4\pi}{3}, \infty\right]$ and let $g(x) = h(x)\left[h(x)^2 + h(2\pi - x)^2\right]^{-1/2}$ on $[0, \infty)$.)

3. Suppose $\psi(x)$ (not a priori the one in (4.6.43)) obeys

$$\psi(x) = \sum_{n \in \mathbb{Z}} b_n \varphi(2x - n) \tag{4.6.178}$$

and
$$m_1(k) = \tfrac{1}{2} \sum_{n \in \mathbb{Z}} b_n e^{-ink} \qquad (4.6.179)$$

(a) Prove that
$$\widehat{\psi}(k) = m_1(k/2)\,\widehat{\varphi}(k/2) \qquad (4.6.180)$$

(b) Prove that $\{\psi(\cdot - j)\}$ is orthonormal if and only if for a.e. k
$$|m_1(k)|^2 + |m_1(k + \pi)|^2 = 1 \qquad (4.6.181)$$
(*Hint*: Look at Proposition 4.6.10.)

(c) Prove that $\langle \psi(\cdot - j), \varphi(\cdot - k) \rangle = 0$ for all $j \neq k$ if and only if
$$m_0(k)\overline{m_1(k + \pi)} + m_0(k + \pi)\overline{m_1(k)} = 0 \qquad (4.6.182)$$

(d) Prove that if (4.6.181) and (4.6.182) hold, then there is a 2π periodic function $\alpha(k)$ so that $|\alpha(k)| \equiv 1$ and
$$(m_1(k), m_1(k + 1/2)) = \alpha(k)\big(\overline{m_0(k + \pi)}, -\overline{m_0(k)}\big) \qquad (4.6.183)$$
and then that
$$\alpha(k + \pi) = -\alpha(k) \qquad (4.6.184)$$

(e) Prove that $\alpha(k) = -e^{-ik}$ satisfies all the conditions and leads to the choice
$$b_n = (-1)^n \bar{a}_{1-n} \qquad (4.6.185)$$
used in (4.6.43).

(f) Find other choices of α that work.

(g) Prove that
$$|m_0(k)|^2 + |m_1(k)|^2 = 1 \qquad (4.6.186)$$
and deduce any solution has $\{\varphi(\cdot - j)\} \cup \{\psi(\cdot - k)\}$ spanning $\mathcal{H}_1$.

(h) Show that $m_0(0) = 1$ implies $m_1(0) = 0$ and conclude that $\int \psi(x)\,dx = 0$.

4. This will provide an alternate approach to the part of the proof of Theorem 4.6.16, that if $\widehat{\varphi}$ is given by (4.6.75), where m_0 is a trigonometric polynomial and φ generates a MRA, then φ has compact support.

(a) Prove that $\widehat{\varphi}(k)$ is an entire function and that there is A_0 so that
$$k \in \mathbb{C}, \ |k| \leq 1 \Rightarrow |\widehat{\varphi}(k)| \leq A_0 \qquad (4.6.187)$$

(b) If $a_n = 0$ for $|n| > M$, prove for some N
$$|m_0(k)| \leq 2^N e^{M|\operatorname{Im} k|} \qquad (4.6.188)$$

(c) If $2^\ell \leq |k| \leq 2^{\ell+1}$, $\ell = 0, 1, 2, \dots$ and $k \in \mathbb{C}$, prove that
$$|\widehat{\varphi}(k)| \leq 2^N A_0 (2^\ell)^N e^{M|\operatorname{Im} k|} \qquad (4.6.189)$$

(d) Prove that for all $k \in \mathbb{C}$

$$|\widehat{\varphi}(k)| \leq 2^N A_0 (1 + |k|)^N e^{M|\mathrm{Im}\, k|} \qquad (4.6.190)$$

(e) Conclude that φ is supported in $[-M, M]$ (*Hint*: See Theorem 11.1.4 of Part 2A.)

Chapter 5

H^p Spaces and Boundary Values of Analytic Functions on the Unit Disk

> The shortest path between two truths in the real domain passes through the complex domain.
>
> —*Jacques Hadamard* (1865-1963)
> as quoted by J. Kahane [**438**]

Big Notions and Theorems: Hardy Spaces, Nevanlinna Class, Hardy's Convexity Theorem, F. and R. Nevanlinna Theorem, H^2, Pointwise Boundary Values for H^p ($p > 0$), Norm Boundary Values for H^p ($p > 0$), Riesz Factorization, F. and M. Riesz Theorem, Predual of H^1, Carathéodory Functions, Herglotz Representation for Carathéodory and h^1, h^1-Functions Are Differences of Positive Harmonic Functions, Kolmogorov's Theorem, Inner-Outer Factorization, M. Riesz Conjugate Function Theorem, Hilbert Transform, Weak-L^1 Hilbert Transform Bounds, Homogeneous Space, L^p ($1 < p < \infty$) Convergence of Fourier Series, $\mathbb{C}_+$ Herglotz Representation, Boole's Inequality, Maximal Hilbert Transform Inequality, Beurling's Theorem, H^p Duality, BMO, Fefferman Duality Theorem, Fefferman–Stein DensityTheorem, VMO, HMO, Coifman Atomic Decomposition, John–Nirenberg Inequalities, Hankel Matrix, Nehari's Theorem, Hartman's Theorem, Cotlar's Ergodic Hilbert Transform

In this chapter, we ask the following question and its extensions: Suppose f is a bounded analytic function on $\mathbb{D}$. What can be said about its boundary values on $\partial\mathbb{D}$? In fact, we'll prove that for Lebesgue a.e. θ, $\lim_{r\uparrow 1} f(re^{i\theta})$ exists. The subject is a fascinating mix of ideas from complex analysis (like Blaschke products) and real analysis (like pointwise convergence results).

In fact, we'll look in two wider contexts. First, rather than look only at bounded functions (called $H^\infty(\partial\mathbb{D})$), we consider the *Hardy spaces*, H^p, of those functions, f, analytic on $\mathbb{D}$ with

$$\|f\|_p = \sup_r \left(\frac{1}{2\pi} \int_0^{2\pi} |f(re^{i\theta})|^p \, d\theta \right)^{1/p} < \infty \qquad (5.0.1)$$

where, for $p = \infty$, we take instead

$$\|f\|_\infty = \sup_{z \in \mathbb{D}} |f(z)| \qquad (5.0.2)$$

We'll study H^p for all $p \in (0, \infty]$. For $p \geq 1$, we'll see that $\|\cdot\|_p$ is a norm. For $p < 1$, as for L^p with $p < 1$ (see Proposition 5.7.4 of Part 1), $\|\cdot\|_p$ is not a norm, but $d_p(f, g) = \|f - g\|_p^p$ is a metric. We'll also see that H^p is complete in $\|\cdot\|_p$ for $p \geq 1$ and in the d_p metric if $p < 1$. Besides H^p, we study the *Nevanlinna class*, N, of those f's with (here $\log_+(y) = \max(0, \log(y))$)

$$\sup_{0<r<1} \int_0^{2\pi} \log_+ |f(re^{i\theta})| \frac{d\theta}{2\pi} = \|f\|_N \qquad (5.0.3)$$

Since, for $y > 0$, $\max(1, y) \leq e^y$, we see $\log_+(|y|) \leq |y|$ and so, replacing y by $|y|^p$,

$$\log_+(|y|) \leq \frac{1}{p}(|y|)^p \qquad (5.0.4)$$

Thus, for $0 < p < p' \leq \infty$,

$$N \supset H^p \supset H^{p'} \qquad (5.0.5)$$

It will also be natural to consider real-valued harmonic functions, u, on $\mathbb{D}$ with

$$\|u\|_p = \sup_{0<r<1} \left(\frac{1}{2\pi} \int |u(re^{i\theta})|^p \, d\theta \right)^{1/p} < \infty \qquad (5.0.6)$$

which we'll denote by h^p. We'll study this for $1 \leq p \leq \infty$. It is easy to see (Problem 1) that there is a unique function, v, on $\mathbb{D}$ so $v(0) = 0$ and $u + iv$ is analytic. v is called the *conjugate harmonic function* to u.

In Section 5.1, we show the sup in (5.0.1) and (5.0.6) is $\lim_{r\uparrow1}$ and establish that H^p are complete vector spaces. We'll also prove a useful theorem of F. and R. Nevanlinna that any $f \in N$ (and so, any $f \in H^p$) is a ratio of g/h with $g, h \in H^\infty$ and h nonvanishing on $\mathbb{D}$. Section 5.2 is devoted to the simplest of the H^p spaces—H^2, where $f \in H^2$ can be described in terms of its Taylor coefficients, so it will be easy to prove there is $f^* \in L^2(\partial\mathbb{D}, \frac{d\theta}{2\pi})$ so (with $f_r(e^{i\theta}) = f(re^{i\theta})$) that $\|f^* - f_r\|_2 \to 0$ and so that for a.e. θ, $f^*(re^{i\theta}) = \lim_{r\uparrow1} f_r(e^{i\theta})$.

Section 5.3 provides one route to go from H^2 to general H^p (the other route, discussed in Section 5.2, is to use the theorem of the Nevanlinnas). We prove that any $f \in H^p$ can be written $f = gB$, where B is a Blaschke product, g is nonvanishing on $\mathbb{D}$, and $g \in H^p$ with $\|f\|_p = \|g\|_p$. Since $g = h^{2/p}$ with $h \in H^2$, we'll be able to carry over the boundary value theory to general H^p.

Sections 5.4–5.6 study h^p and define a more sophisticated factorization than the one in Section 5.3 (inner-outer factorization). h^1 enters, even to study H^p, because if $f \in H^p$ and nonvanishing on $\mathbb{D}$, $\log|f| \in h^1$, but not, in general, in any h^p, $p > 1$. h^1 is different from H^p and h^p, $p > 1$, in the fine structure of their boundary values. Section 5.7 discusses M. Riesz's famous theorem that for $1 < p < \infty$, $u \mapsto v$, the map of u in h^p to its conjugate is bounded, and Section 5.8 uses what we've learned earlier to return to issues of convergence of Fourier series which we studied in Section 3.5 of Part 1; we'll prove if $f \in L^p(\partial\mathbb{D}, \frac{d\theta}{2\pi})$, $1 < p < \infty$, then the Fourier series converges in L^p. Section 5.9 turns to the analysis of $H^p(\mathbb{D})$ where the upper half-plane, $\mathbb{C}_+$, replaces $\mathbb{D}$. Section 5.11 discusses the important space BMO. The remaining sections are more specialized bonus sections.

Notes and Historical Remarks. Five standard references for H^p spaces are Duren [**234**], Garnett [**305**], Katznelson [**453**], Koosis [**486**], and Rudin [**706**].

Problems

1. If u is real-valued and harmonic on $\partial\mathbb{D}$, prove that there exists a unique harmonic function, v, on $\partial\mathbb{D}$ so that $u + iv$ is analytic and $v(0) = 0$. (*Hint*: Cauchy–Riemann equations.)

5.1. Basic Properties of H^p

In this section, we make the formal definitions of H^p, N, and h^p, establish that they are complete vector spaces, and prove the useful theorem of the Nevanlinnas that any f in N—and so, any f in H^p—is a quotient of H^∞-functions (with the denominator nowhere vanishing).

Given any continuous function, F, on $\mathbb{D}$, we define, for $0 < p < \infty$ and $0 < r < 1$,

$$M_r^{(p)}(F) = \left(\int_0^{2\pi} |F(re^{i\theta})|^p \, \frac{d\theta}{2\pi} \right)^{1/p} \tag{5.1.1}$$

We also define

$$M_r^{(\infty)}(F) = \sup_{0 \le \theta \le 2\pi} |F(re^{i\theta})| \tag{5.1.2}$$

$$N_r(F) = \int_0^{2\pi} \log_+(|F(re^{i\theta})|) \, \frac{d\theta}{2\pi} \tag{5.1.3}$$

Since the L^p-norm is a norm for $1 \le p \le \infty$, we have

$$M_r^{(p)}(F+G) \le M_r^{(p)}(F) + M_r^{(p)}(G), \qquad 1 \le p \le \infty \tag{5.1.4}$$

and by (5.7.12) of Part 1, if

$$d_r^{(p)}(F) = [M_r^{(p)}(F)]^p \tag{5.1.5}$$

then

$$d_r^{(p)}(F+G) \le d_r^{(p)}(F) + d_r^{(p)}(G), \qquad 0 < p \le 1 \tag{5.1.6}$$

Since (Problem 1) for $x, y \in [0, \infty)$,

$$\log_+(x+y) \le \log_+(x) + \log_+(y) + \log(2) \tag{5.1.7}$$

we have

$$N_r(F+G) \le N_r(F) + N_r(G) + \log(2) \tag{5.1.8}$$

By Hölder's inequality (writing $F = \mathbb{1}F$), we have

$$p > p' \Rightarrow M_r^{(p')}(F) \le M_r^{(p)}(F) \tag{5.1.9}$$

Since (see (5.0.4)) $\log_+(x) \le p^{-1}x^p$ (for $x > 0$), we have

$$N_r(F) \le p^{-1}[M_r^{(p)}(F)]^p \tag{5.1.10}$$

For any continuous F on $\mathbb{D}$, we define

$$\|F\|_p = \sup_{0<r<1} M_r^{(p)}(F), \qquad 0 < p \le \infty \tag{5.1.11}$$

$$\|F\|_N = \sup_{0<r<1} N_r(F) \tag{5.1.12}$$

$$d^{(p)}(F) = \sup_{0<r<1} d_r^{(p)}(F), \qquad 0 < p \le 1 \tag{5.1.13}$$

These numbers may be $+\infty$. Despite the notation, $\|\cdot\|_N$ and $\|\cdot\|_p$ for $p < 1$ are not norms!

Definition. For $0 < p \le \infty$, the *Hardy space*, H^p (or $H^p(\mathbb{D})$), is the set of f analytic on $\mathbb{D}$ with $\|f\|_p < \infty$. The *Nevanlinna space*, N, is the set of f analytic in $\mathbb{D}$ so $\|f\|_N < \infty$. Functions in N are also said to be functions with *bounded characteristic*. The space, h^p, for $1 \le p \le \infty$, is the set of real-valued functions, u, harmonic on $\mathbb{D}$, so that $\|u\|_p < \infty$.

If f is analytic in $\mathbb{D}$ with $f(0) \ne 0$, we know $\log|f(z)|$ is subharmonic (see Theorem 3.2.6), so

$$\int_0^{2\pi} [\log_+|f(re^{i\theta})| - \log_-|f(re^{i\theta})|] \frac{d\theta}{2\pi} \ge \log|f(0)| \tag{5.1.14}$$

Equivalently,

$$\int_0^{2\pi} \log_-|f(re^{i\theta})| \frac{d\theta}{2\pi} \le N_r(f) - \log|f(0)| \tag{5.1.15}$$

Thus,

$$f \in N + f(0) \neq 0 \Rightarrow \sup_{0 < r < 1} \int_0^{2\pi} \log_- |f(re^{i\theta})| \frac{d\theta}{2\pi} < \infty \qquad (5.1.16)$$

From (5.1.9) and (5.1.10), we immediately have:

Proposition 5.1.1. (a) *If $p > p'$, then $H^p \subset H^{p'}$ with*

$$\|f\|_{p'} \leq \|f\|_p \qquad (5.1.17)$$

(b) *For any $p \in (0, \infty]$, $H^p \subset N$ with*

$$\|f\|_N \leq p^{-1} \|f\|_p^p \qquad (5.1.18)$$

Theorem 5.1.2 (Hardy's Convexity Theorem). *Let f be analytic in $\mathbb{D}$. Then for $0 < p \leq \infty$,*

(a) $M_r^{(p)}(f)$ *is monotone in r.*
(b) $N_r(f)$ *is monotone in r.*
(c) $\log(M_r^{(p)})(f)$ *is convex in $\log(r)$.*
For $1 \leq p < \infty$, (a) and (c) hold for f real-valued and harmonic on $\mathbb{D}$.

Proof. For $p = \infty$, (a) is the maximum principle and (c) the Hadamard three-circle theorem (see Theorem 3.6.8 of Part 2A), so we can suppose $p < \infty$. By Example 3.2.7, $|f|^p$ and $\log_+(|f|)$ are subhamornic, so (a)–(c) follow from Corollary 3.2.12 and Corollary 3.2.17. For the harmonic case, we instead use Theorem 3.2.3 and the fact that u harmonic $\Rightarrow |u|$ is subharmonic. $\qquad \square$

The following corollary is so central, we call it a theorem.

Theorem 5.1.3. (a) *If F is analytic in $\mathbb{D}$ (or u is real-valued and harmonic in $\mathbb{D}$ and $1 \leq p \leq \infty$), we can replace the sup in (5.1.11)–(5.1.13) by $\lim_{r \uparrow 1}$.*

(b) *For $1 \leq p \leq \infty$, H^p is a vector space and $\|\cdot\|_p$ is a norm on H^p and h^p.*

(c) *For $0 \leq p \leq 1$, H^p is a vector space and $d^{(p)}$ is a metric on H^p.*

(d) *N is a vector space.*

Proof. (a) is immediate from the monotonicity result. Then taking $\lim_{r \uparrow 1}$ in (5.1.4), (5.1.6), and (5.1.8), we get

$$\|f + g\|_p \leq \|f\|_p + \|g\|_p, \qquad 1 \leq p \leq \infty \qquad (5.1.19)$$

$$d^{(p)}(f + g) \leq d^{(p)}(f) + d^{(p)}(g) \qquad (5.1.20)$$

$$\|f + g\|_N \leq \|f\|_N + \|g\|_N + \log(2) \qquad (5.1.21)$$

(with infinity allowed). (b)–(d) are immediate. $\qquad \square$

Theorem 5.1.4. (a) *For any f in H^p $(0 < p \leq \infty)$, we have for all $r < 1$ that*

$$\sup_{|z| \leq r} |f(z)|^p \leq \frac{1+r}{1-r} \|f\|_p^p \tag{5.1.22}$$

The same estimate applies for $f \in h^p$, $1 \leq p \leq \infty$.

(b) *For any $f \in N$,*

$$\sup_{|z| \leq r} |f(z)| \leq \exp\left(\frac{1+r}{1-r} \|f\|_N \right) \tag{5.1.23}$$

Proof. By Theorem 3.2.11, if w is nonnegative and subharmonic in $\mathbb{D}$,

$$\sup_{|z| \leq r} w(z) \leq \frac{1+r}{1-r} \lim_{\rho \uparrow 1} M_\rho^{(p=1)}(w) \tag{5.1.24}$$

(a) follows by applying this to the subharmonic function $|f(z)|^p$ and (b) by applying it to the subharmonic function $\log_+ |f(z)|$. $\qquad\square$

Theorem 5.1.5. (a) *For $1 \leq p \leq \infty$, H^p and h^p are Banach spaces in $\|\cdot\|_p$.*

(b) *For $0 < p \leq 1$, H^p is a Fréchet space in the d_p metric.*

(c) *N is complete in the sense that if $\lim_{N \to \infty} \sup_{m,n \geq N} \|f_m - f_n\|_N = 0$, then there is f_∞ in N so $\|f_n - f_\infty\|_N \to 0$.*

Proof. By Theorem 5.1.4, if f_n is Cauchy in any of these spaces, then $f_n \restriction \mathbb{D}_r(0)$ is Cauchy in $\|\cdot\|_\infty$, so there is f_∞ on $\mathbb{D}$ which is a uniform limit of the f_n on compact subsets of $\mathbb{D}$. Thus, f_∞ is analytic (or harmonic). Moreover, by looking at $M_r^{(p)}$ or N_r and then taking $\lim_{r \uparrow 1}$, we see that in any "norm,"

$$\|f_\infty - f_n\| \leq \sup_{m \geq n} \|f_m - f_m\| \tag{5.1.25}$$

implying convergence. $\qquad\square$

We are heading towards a proof of our final result of the section.

Theorem 5.1.6 (F. and R. Nevanlinna Theorem). *A function, f, on $\mathbb{D}$ lies in N if and only if there exists $\varphi, \psi \in H^\infty$, with ψ nonvanishing on $\mathbb{D}$, so that $f = \varphi/\psi$.*

We need to recall the Poisson–Jensen formula, which is Theorem 9.8.2 of Part 2A. It says that if f is analytic in a neighborhood of $\overline{\mathbb{D}_R(0)}$, if $f(0) \neq 0$, and if $\{z_j\}_{j=1}^N$ are the zeros of f in $\mathbb{D}_R(0)$, then for all $z \in \mathbb{D}_R(0)$, we have

$$f(z) = \frac{f(0)}{|f(0)|} \prod_{j=1}^N b\left(\frac{z}{R}, \frac{z_j}{R}\right) \exp\left(\int \frac{Re^{i\theta} + z}{Re^{i\theta} - z} \log|f(Re^{i\theta})| \frac{d\theta}{2\pi} \right) \tag{5.1.26}$$

Here b is a Blaschke factor defined (for $w \neq 0$) by

$$b(z, w) = \frac{|w|}{w}\left(\frac{w-z}{1-\bar{w}z}\right) \tag{5.1.27}$$

(The proof is easy. By scaling, one can take $R = 1$ and use the complex Poisson formula for $\log[f(z)/\prod_{j=1}^{N} b(z, z_j)]$.)

Proof. If $f = \varphi/\psi$ for $\varphi, \psi \in H^\infty$, then we can, by multiplying φ and ψ by a small constant, suppose $\|\varphi\|_\infty \leq 1$, $\|\psi\|_\infty \leq 1$. Thus,

$$\log_+|f(z)| \leq -\log|\psi(z)|$$

Since ψ is nonvanishing, $\log|\psi(z)|$ is harmonic, so

$$\int \log_+|f(re^{i\theta})|\,\frac{d\theta}{2\pi} \leq -\log|\psi(0)| \tag{5.1.28}$$

proving that $f \in N$.

Conversely, suppose $f \in N$. If $f(z) = z^n g(z)$ with $g(0) \neq 0$, then $g \in N$ also. So it suffices to prove $g = \varphi/\psi$. Equivalently, we can suppose $f(0) \neq 0$. In that case, for $r < 1$, we can apply the Poisson–Jensen formula to see that on $\mathbb{D}_r(0)$, $f(z) = \varphi_r(z)/\psi_r(z)$, where

$$\varphi_r(z) = \frac{f(0)}{|f(0)|}\prod_{j=1}^{N_r} b\left(\frac{z}{r}, \frac{z_j}{r}\right)\exp\left(-\int \frac{re^{i\theta}+z}{re^{i\theta}-z}\log_-|f(re^{i\theta})|\,\frac{d\theta}{2\pi}\right) \tag{5.1.29}$$

$$\psi_r(z) = \exp\left(-\int \frac{re^{i\theta}+z}{re^{i\theta}-z}\log_+|f(re^{i\theta})|\,\frac{d\theta}{2\pi}\right) \tag{5.1.30}$$

Clearly, since $\log_\pm \geq 0$, we have that

$$\|\varphi_r\|_\infty \leq 1, \qquad \|\psi_r\|_\infty \leq 1 \tag{5.1.31}$$

and

$$\psi_r(0) \geq \exp(-\|f\|_N) \tag{5.1.32}$$

Define on $\mathbb{D}$

$$\tilde{\varphi}_r(z) = \varphi_r(rz), \qquad \tilde{\psi}_r(z) = \psi_r(rz) \tag{5.1.33}$$

By (5.1.31) and compactness of the unit ball in H^∞ in the topology of uniform convergence on each $\mathbb{D}_r(0)$, $r < 1$ (i.e., the theory of normal families; see Section 6.2 of Part 2A), we can find $r_n \to 1$ and φ, ψ so $\tilde{\varphi}_{r_n}, \tilde{\psi}_{r_n}$ converge to φ, ψ uniformly on compact subsets.

By Hurwitz's theorem (Theorem 6.4.1 of Part 2A), either $\psi \equiv 0$ or ψ is nonvanishing. By (5.1.32) (and it is here that $f \in N$ enters), $|\psi(0)| \geq \exp(-\|f\|_N) \neq 0$, so ψ is everywhere nonvanishing. Taking limits in $f(r_n z) = \tilde{\varphi}_{r_n}(z)/\tilde{\psi}_{r_n}(z)$, we see $f = \varphi/\psi$. $\qquad \square$

Notes and Historical Remarks. Hardy's convexity theorem is from Hardy [**352**]. Its proof using subharmonic functions is due to Riesz [**682**] who gave the name H^p and "Hardy space" after Hardy's paper. The theorem of the Nevanlinnas is from F. and R. Nevanlinna [**606**]. They were brothers—the one of Nevanlinna theory (whose biographical sketch is in the Notes to Section 17.0 of Part 2B) is Rolf.

The complex analysis we use here (Poisson representation, normal families, Jensen's formula) is reviewed in Section 1.3.

Problems

1. Prove (5.1.7) by considering the cases $0 < x \le y \le 1$, $0 < x \le 1 < y$, and $1 \le x < y$.

5.2. H^2

In this section, we'll study the basic properties of H^2- and h^2-functions, most of which extend to H^p, $1 \le p < \infty$, and some to H^p, $0 < p \le \infty$, and/or N. We'll see:

(1) If $f \in H^2$ and $f_r \in L^2(\partial\mathbb{D}, \frac{d\theta}{2\pi})$ by

$$f_r(e^{i\theta}) = f(re^{i\theta}) \tag{5.2.1}$$

then there is $f^* \in L^2$ so that as $r \uparrow 1$,

$$\|f_r - f^*\|_2 \to 0 \tag{5.2.2}$$

(2) If P_r is the Poisson kernel,

$$P_r(\theta, \varphi) = \frac{1 - r^2}{1 - 2r\cos(\theta - \varphi) + r^2} \tag{5.2.3}$$

then $f_r = P_r * f^*$, that is,

$$f(re^{i\theta}) = \int P_r(\theta, \varphi) f^*(e^{i\varphi}) \frac{d\theta}{2\pi} \tag{5.2.4}$$

There is also a complex Poisson representation writing f in terms of $\operatorname{Re} f^*$.

(3) There is a radial maximal function

$$(M_{\text{Rad}}f)(e^{i\theta}) = \sup_{0 < r < 1} |f(re^{i\theta})| \tag{5.2.5}$$

with

$$\|M_{\text{Rad}}f\|_{L^2(\partial\mathbb{D}, \frac{d\theta}{2\pi})} \le C\|f\|_{H^2} \tag{5.2.6}$$

for a universal constant C. Indeed, there is a nontangential maximal inequality.

(4) Pointwise limits exist, that is, for Lebesgue a.e. θ,

$$\lim_{r\uparrow 1} f(re^{i\theta}) = f^*(e^{i\theta}) \qquad (5.2.7)$$

Indeed, nontangential limits exist.
(5) We have that

$$g \in L^2\left(\partial\mathbb{D}, \frac{d\theta}{2\pi}\right) \text{ is an } f^* \Leftrightarrow \forall n > 0, \int_0^{2\pi} e^{in\theta} g(e^{i\theta}) \frac{d\theta}{2\pi} = 0 \qquad (5.2.8)$$

(6) If $u \in h^2$, then its conjugate function, $v \equiv Ku$, obeys ($\|\cdot\|_2$ here is h_2-norm, i.e., $\sup\|u_r\|_{L^2(\partial\mathbb{D}, \frac{d\theta}{2\pi})}$)

$$\|Ku\|_2 \le \|u\|_2 \qquad (5.2.9)$$

One might guess that the key is the Hilbert space structure of H^2 but, in fact, it is the connection to Taylor coefficients of f via the Plancherel theorem for Fourier series. This is specific to H^2 only.

Theorem 5.2.1. *Let f be analytic on the disk with*

$$f(z) = \sum_{n=0}^{\infty} a_n z^n \qquad (5.2.10)$$

(a) $f \in H^2 \Leftrightarrow \sum_{n=0}^{\infty} |a_n|^2 < \infty$ *and* $\qquad (5.2.11)$

$$\|f\|_{H^2}^2 = \sum_{n=0}^{\infty} |a_n|^2 \qquad (5.2.12)$$

(b) *If f is in H^2, define $f^* \in L^2(\partial\mathbb{D}, \frac{d\theta}{2\pi})$ by*

$$f^*(e^{i\theta}) = \sum_{n=0}^{\infty} a_n e^{in\theta} \qquad (5.2.13)$$

(L^2 sum). Then (with f_r given by (5.2.1))

$$\|f_r - f^*\|_2 \to 0 \quad as \ r \uparrow 1 \qquad (5.2.14)$$

Moreover, if $f^{(r)}(z) = f(rz)$ for $z \in \mathbb{D}$ and $r \in (0,1)$ then

$$\lim_{r\uparrow 1} \|f^{(r)} - f\|_{H^2(\mathbb{D})} = 0 \qquad (5.2.15)$$

(c) *If P_r is the Poisson kernel, then (5.2.4) holds.*
(d) *If $C(z, e^{i\theta})$ is the complex Poisson kernel (aka Schwarz kernel), then $f = C * \operatorname{Re} f$, that is,*

$$f(z) = i\operatorname{Im} f(0) + \int_0^{2\pi} \frac{e^{i\theta} + z}{e^{i\theta} - z} \operatorname{Re} f^*(e^{i\theta}) \frac{d\theta}{2\pi} \qquad (5.2.16)$$

(e) *If $M_{\mathrm{Rad}}f$ is given by (5.2.5), then*

$$(M_{\mathrm{Rad}}f)(e^{i\theta}) \leq (M_{\mathrm{HL}}f)(e^{i\theta}) \tag{5.2.17}$$

where $M_{\mathrm{HL}}f$ is the Hardy–Littlewood maximal function. In particular, (5.2.6) holds. A similar result with an extra constant C holds for the nontangential maximal function of (2.4.42) replacing M_{Rad}.

(f) *For a.e. θ,*

$$\lim_{r\uparrow 1} f(re^{i\theta}) = f^*(e^{i\theta}) \tag{5.2.18}$$

Remarks. 1. Using Theorem 2.4.11, one can replace the radial maximal function of (c) and radial limits of (e) by nontangential maximal function and nontangential limit. One might hope that one can go beyond the non-tangential condition, but in Examples 5.3.2 and 5.5.6, we'll find $f \in H^\infty$ so that the radial limit of f^* has $|f^*(e^{i\theta})| = 1$, but so that for *every* $z_\infty \in \partial\mathbb{D}$, there exists $z_n \in \mathbb{D}$ so that $\lim_{n\to\infty} f(z_n) = 0$!

2. If $p = 2$, we see $\operatorname{Re} f^*$ and $\operatorname{Im} f(0)$ determine f (by (5.2.16)) and we'll see this is true if $p \geq 1$. However, there exists $f \in \bigcap_{p<1} H^p$ so f is noncon-stant but $\operatorname{Re} f^* \equiv 0$ (see Example 5.3.6). We'll also see there are $f \in H^\infty$, f nonzero on $\mathbb{D}$, with $|f^*| = 1$, but so that f is not constant (see Theorem 5.6.2). Thus $|f^*|$ does not determine f.

Proof. (a) For $r < 1$,

$$f_r(e^{i\theta}) = f(re^{i\theta}) = \sum_{n=0}^{\infty} a_n r^n e^{in\theta} \tag{5.2.19}$$

By the orthonormality of $e^{in\theta}$, we have

$$\|f_r\|_{L^2}^2 = \sum_{n=0}^{\infty} |a_n|^2 r^{2n} \tag{5.2.20}$$

(showing $\|f_r\|_{L^2}$ is monotone in r). By the monotone convergence theorem for sums,

$$\sup_{0<r<1} \|f_r\|_2^2 = \lim_{r\uparrow 1} \|f_r\|_2^2 = \sum_{n=0}^{\infty} |a_n|^2 \tag{5.2.21}$$

which implies (a).

(b) If $f \in H^2$, we have

$$\|f^* - f_r\|_2^2 = \sum_{n=0}^{\infty} |a_n(1-r^n)|^2 = \sum_{n=0}^{\infty} |a_n|^2 (1-r^n)^2 \tag{5.2.22}$$

By the dominated convergence theorem for sums, (5.2.14) holds. (5.2.15) is immediate from $\|f\|_{H^2(\mathbb{D})} = \|f^*\|_{L^2(\partial\mathbb{D})}$ and $(f - f^{(r)})^* = f^* - f_r$.

(c) Fix $\rho < 1$. Then by the Poisson formula for functions analytic in a neighborhood of $\overline{\mathbb{D}}$ (applied to $\tilde{f}_\rho(z) \equiv f(\rho z)$) for $r < 1$,

$$f(\rho r e^{i\theta}) = \int P_r(\theta, \varphi) f_\rho(e^{i\varphi}) \frac{d\varphi}{2\pi} \qquad (5.2.23)$$

For $\theta, r < 1$ fixed, as $\rho \uparrow 1$, $P_r(\theta, \cdot) \in L^2$, $f_\rho \in L^2$, and $f(\rho r e^{i\theta}) \to f(r e^{i\theta})$ in L^2, proving (5.2.4).

(d) Identical to (c) using the complex Poisson formula.

(e) This is immediate from (2.4.13) and $\|P_r\|_{L^1} = 1$. The nontangential result follows from Theorem 2.4.11.

(f) This is Theorem 2.4.6. $\qquad\square$

The following is immediate; recall $f_n^\sharp = (2\pi)^{-1} \int e^{-in\theta} f(e^{i\theta}) \, d\theta$.

Corollary 5.2.2. $\{f^* \mid f \in H^2\} = \{g \in L^2 \mid g_n^\sharp = 0 \text{ for } n < 0\}$.

Proof. If $g_n^\sharp = 0$ for $n < 0$, $\sum_{n=0}^\infty g_n^\sharp z^n \equiv f(z) \in H^2$ with $f^* = g$. By construction, if $f \in H^2$, f^* has $(f^*)_n^\sharp = 0$ for $n < 0$. $\qquad\square$

Next, we discuss h^2 and its conjugate function map.

Theorem 5.2.3. If $u \in h^2$, then for $a_0 \in \mathbb{R}$ and $a_n \in \mathbb{C}$, $n = 1, 2, \ldots$,

$$u(re^{i\theta}) = a_0 + \sum_{n=1}^\infty (a_n r^n e^{in\theta} + \bar{a}_n r^n e^{-in\theta}) \qquad (5.2.24)$$

(convergent uniformly on compacts of $\mathbb{R}$) with (if we set $a_{-n} = \bar{a}_n$ for $n > 0$)

$$\sum_{n=-\infty}^\infty |a_n|^2 < \infty \qquad (5.2.25)$$

and (with $a_n = \bar{a}_{-n}$ for $n < 0$)

$$\|u\|_{h^2}^2 = \sum_{n=-\infty}^\infty |a_n|^2 \qquad (5.2.26)$$

Remark. Below we'll provide an alternate proof that if $u \in h^2$, then there is $v \in h^2$ with $v(0) = 0$ and $f = u + iv$ in H^2. From that, one can easily deduce Theorem 5.2.3 from Theorem 5.2.1.

Proof. Let u be harmonic on $\mathbb{D}$ and, for $0 < r < 1$, let

$$u(re^{i\theta}) = \sum_{n=-\infty}^\infty a_n(r) e^{in\theta} \qquad (5.2.27)$$

be its Fourier expansion. Since (see (5.4.9) of Part 2A)

$$(\Delta u)(re^{i\theta}) = \frac{1}{r} \frac{\partial}{\partial r} \left(r \frac{\partial}{\partial r} \right) u + \frac{1}{r^2} \frac{\partial^2}{\partial \theta^2} \qquad (5.2.28)$$

$\Delta u = 0$ implies

$$\frac{1}{r}\frac{d}{dr}\left(r\frac{d}{dr}a_n(r)\right) = \frac{n^2}{r^2}a_n(r) \tag{5.2.29}$$

which implies, for $n > 0$, that

$$a_n(r) = a_n r^n + c_n r^{-n} \tag{5.2.30}$$

Since $a_n(r)$ is bounded as $r \downarrow 0$, $c_n = 0$. For $n = 0$,

$$a_0(r) = \frac{1}{2\pi}\int u(re^{i\theta})\,d\theta = u(0)$$

is constant. Thus, u obeys (5.2.13) initially in the L^2-sense but then, since $|a_n\rho^n| \leq \sup_{|z|\leq\rho}|u(z)|$, we see for $r < \rho < 1$, $|a_n r^n| \leq c_\rho(r/\rho)^n$, so the sum converges in $\|\cdot\|_\infty$.

By (5.2.24),

$$\|u(re^{i\theta})\|_{L^2} = \sum_{n=-\infty}^{\infty}|a_n|^2 r^{2|n|} \tag{5.2.31}$$

so (5.2.25) and (5.2.26) hold. $\square$

Define

$$\text{sgn}(n) = \begin{cases} +1, & n > 0 \\ -1, & n < 0 \\ 0, & n = 0 \end{cases} \tag{5.2.32}$$

Theorem 5.2.4. *Let $u \in h^2$ be given by (5.2.24). Let*

$$v(re^{i\theta}) = \sum_{n=-\infty}^{\infty}a_n(-i\,\text{sgn}(n))r^{|n|}e^{in\theta} \tag{5.2.33}$$

Then v is the unique real harmonic function on $\mathbb{D}$ with $v(0) = 0$ and $u + iv$ analytic on $\mathbb{D}$. If $v \equiv Ku$, we have $Ku \in h^2$ (and so $u + iKu \in H^2$) and

$$\|Ku\|_2 \leq \|u\|_2 \tag{5.2.34}$$

Indeed,

$$\|Ku\|_2^2 + |u(0)|^2 = \|u\|_2^2 \tag{5.2.35}$$

Proof. By the argument in the last theorem, v is harmonic in $\mathbb{D}$ and since $-ia_n\,\text{sgn}(n) = ia_{-n}$, v is real-valued. Moreover, we have $v(0) = 0$ and

$$u + iv = a_0 + \sum_{n=1}^{\infty}2a_n z^n \tag{5.2.36}$$

is analytic, so $v = Ku$ as claimed.

By (5.2.26), we have (5.2.35), which implies (5.2.34). $\square$

There is an alternate proof of (5.2.34) that avoids Taylor series. If f is analytic on $\mathbb{D}$ and $f = u + iv$, then f^2 is analytic, so

$$\int_0^{2\pi} f(re^{i\theta})^2 \, \frac{d\theta}{2\pi} = f(0)^2 \qquad (5.2.37)$$

Taking real parts, we get

$$\int_0^{2\pi} [u^2(re^{i\theta}) - v^2(re^{i\theta})] \, \frac{d\theta}{2\pi} = u(0)^2 - v(0)^2$$

If $v(0) = 0$, this is (5.2.35).

Thus $v \in h^2$, so $u + iv$ lies in H^2. Therefore, v has a boundary value on $\partial\mathbb{D}$, which is, by (5.2.33), the L^2-sum

$$v^*(e^{i\theta}) = \sum_{n=-\infty}^{\infty} a_n(-i\,\mathrm{sgn}(n))e^{in\theta} \qquad (5.2.38)$$

The map $u^* \to v^*$ of $L^2(\partial\mathbb{D}, \frac{d\theta}{2\pi})$ to itself is sometimes also called the conjugate map. We'll instead call it the *Hilbert transform* (more properly, *circular Hilbert transform*; see the discussion in Section 5.7) and write $\sim$ for this map, so that $v* = \widetilde{(u^*)}$.

Our final topic is one approach to pointwise boundary values (and nowhere vanishing) of functions in N (and so, any H^p). In Section 5.4 (see Theorem 5.4.5), we'll provide a second proof of these facts. While those proofs are more complicated, they provide more and, in particular, H^p analogs of $\|f_r - f^*\|_p \to 0$. As a preliminary (to avoid $0/0$ limits), we need:

Lemma 5.2.5. *Let $f \in N$. Suppose $f \not\equiv 0$ and that $f^*(e^{i\theta}) = \lim_{r\uparrow 1} f(re^{i\theta})$ exists (in $\mathbb{C}$) for a.e. θ. Then*

$$\int \log_-|f^*(e^{i\theta})| \, \frac{d\theta}{2\pi} < \infty \qquad (5.2.39)$$

and, in particular, for a.e. θ, $f^*(e^{i\theta}) \neq 0$.

Remarks. 1. We'll prove shortly the pointwise limit exists a.e. for any $f \in N$.

2. Fatou proved his lemma for the use here!

Proof. Since $f \not\equiv 0$, we can find K so $f(z)/z^K$ is nonvanishing at $z = 0$, so without loss, we can and will suppose that $f(0) \neq 0$.

By (5.1.16),

$$\sup_{0<r<1} \int_0^{2\pi} \log_-|f(re^{i\theta})| \, \frac{d\theta}{2\pi} < \infty \qquad (5.2.40)$$

By hypothesis and continuity of $\log_-(\,\cdot\,)$, we have for a.e. θ,

$$\log_-|f(re^{i\theta})| \to \log_-|f^*(e^{i\theta})| \tag{5.2.41}$$

with $\log_-(y) = \infty$ allowed if $y = 0$.

By Fatou's lemma,

$$\int_0^{2\pi} \log_-|f^*(e^{i\theta})| \frac{d\theta}{2\pi} \leq \liminf_{r\uparrow 1} \int_0^{2\pi} \log_-|f(re^{i\theta})| \frac{d\theta}{2\pi} < \infty$$

by (5.2.39). $\qquad\square$

Theorem 5.2.6 (Fatou's Theorem). *Let $f \in N$ (in particular, for $f \in H^p$ for any $p > 0$), $f \not\equiv 0$. Then there exists f^* on $\partial\mathbb{D}$ so that for a.e. θ, we have*

(a) $\lim_{r\uparrow 1} f(re^{i\theta}) = f^*(e^{i\theta})$

(b) $f^*(e^{i\theta}) \neq 0$

Proof. By the F. and R. Nevanlinna theorem (Theorem 5.1.6), $f = \varphi/\psi$ with $\varphi, \psi \in H^\infty \subset H^2$. Thus, φ, ψ have radial limits and, by the lemma, they are a.e. nonzero. It follows that f has a finite, nonzero limit for a.e. θ. $\qquad\square$

Corollary 5.2.7. *Let $f, g \in N$ (e.g., $f, g \in H^p$ for some p). If $f^*(e^{i\theta}) = g^*(e^{i\theta})$ for a.e. θ, then $f = g$.*

Proof. $f \mapsto f^*$ is easily seen to be linear, so $f^* - g^* = 0 \Rightarrow (f-g)^* = 0 \Rightarrow f - g = 0$ (by the theorem). $\qquad\square$

Notes and Historical Remarks. We will discuss the history of these results which first appeared for H^∞ or H^1 in the Notes to the next section.

5.3. First Factorization (Riesz) and H^p

In this section, we'll show if $f \in H^p$ (respectively, N), then $f = Bg$, where $g \in H^p$ (respectively, N) with B a Blaschke product of the zeros of f and g nonvanishing on $\mathbb{D}$. This will let us prove L^p and a.e. pointwise convergence of f_r to f^* if $f \in H^p$ (N will have to wait until the next section). As an application, we'll also prove a theorem of F. and M. Riesz for complex measures on $\partial\mathbb{D}$ with $\mu_n^\sharp = 0$ for all $n < 0$. As summarized in Section 1.3, if b is the Blaschke factor (5.1.27) if $w \neq 0$ and $b(z,0) = z$, then

(1) For any sequence $\{z_n\}_{n=1}^\infty$ in $\mathbb{D}$ and any $z \in \mathbb{D}$,

$$B(z) = \prod_{j=1}^\infty b(z, z_j) \tag{5.3.1}$$

converges. Moreover,

$$\sum_{n=1}^{\infty}(1-|z_n|)=\infty \Rightarrow B\equiv 0 \tag{5.3.2}$$

$$\sum_{n=1}^{\infty}(1-|z_n|)<\infty \Rightarrow \forall z\in\mathbb{D},$$

$$\sum_{n=1}^{\infty}(1-|b(z,z_j)|)<\infty \Rightarrow (B(z)=0 \Leftrightarrow z=z_j) \tag{5.3.3}$$

(Theorem 9.9.4 of Part 2A).

(2) $f\in N$, with $f\not\equiv 0$, and $\{z_n\}_{n=1}^{\infty}$ the zeros of f imply

$$\sum_{n=1}^{\infty}(1-|z_n|)<\infty \tag{5.3.4}$$

(Theorem 9.9.2 of Part 2A).

We'll begin with a result on B^* (since $\|B\|_\infty \le 1$, B lies in H^∞, so in H^2 and so it has boundary values).

Theorem 5.3.1. *Let $\{z_n\}_{n=1}^{\infty}$ obey (5.3.4) and let B be the Blaschke product, (5.3.1). Then*

$$\lim_{r\uparrow 1}\int \log|B(re^{i\theta})|\,\frac{d\theta}{2\pi}=\int \log|B^*(e^{i\theta})|\,\frac{d\theta}{2\pi}=0 \tag{5.3.5}$$

In particular, for a.e. θ,

$$|B^*(e^{i\theta})|=1 \tag{5.3.6}$$

Proof. The "in particular" follows from (5.3.5) and $|B(z)| \le 1 \Rightarrow \log|B^*(e^{i\theta})| = 0$ for a.e. $\theta \Rightarrow |B^*(e^{i\theta})| = 1$ for a.e. θ.

Write

$$B=B_N\widetilde{B}_N \tag{5.3.7}$$

where

$$B_N=\prod_{j=1}^{N}b(\,\cdot\,,z_j), \qquad \widetilde{B}_N=\prod_{j=N+1}^{\infty}b(\,\cdot\,,z_j) \tag{5.3.8}$$

Since $|b(z,z_j)|=1$ on $\partial\mathbb{D}$ and each $b(z,z_j)$ is analytic in a neighborhood of $\mathbb{D}$, $|B_N(e^{i\theta})|=1$ and, by continuity,

$$\lim_{r\uparrow 1}\int \log|B_N(re^{i\theta})|\,\frac{d\theta}{2\pi}=0 \tag{5.3.9}$$

Thus, using the existence of the limits by monotonicity,

$$\lim_{r\uparrow 1}\int \log|B(re^{i\theta})|\,\frac{d\theta}{2\pi}=\lim_{r\uparrow 1}\int \log|\widetilde{B}_N(re^{i\theta})|\,\frac{d\theta}{2\pi}\ge \sum_{j=N+1}^{\infty}\log|z_j| \tag{5.3.10}$$

The integral is monotone in r, and so bigger than its value at 0 and $b(0, z_j) = z_j$, showing the final inequality.

By (5.3.4), $\lim_{N \to \infty} \sum_{j=N+1}^{\infty} \log|z_j| = 0$ (Problem 1), so

$$\lim_{r \uparrow 1} \int \log|B(re^{i\theta})| \frac{d\theta}{2\pi} = 0 \tag{5.3.11}$$

Applying Fatou's lemma to the nonnegative functions $-\log|B(re^{i\theta})|$, we see that

$$\int \log|B^*(e^{i\theta})| \frac{d\theta}{2\pi} \geq 0 \tag{5.3.12}$$

so, since $|B^*(e^{i\theta})| \leq 1$, we see that (5.3.5) holds. $\square$

Example 5.3.2. Let $\{\theta_n\}_{n=1}^{\infty}$ be such that $\{e^{i\theta_n}\}_{n=1}^{\infty}$ is dense in $\partial\mathbb{D}$ (e.g., $\theta_n = 2\pi q_n$, where q_n is a counting of the rationals in $[0,1]$). Let $z_n = (1 - n^{-\alpha})e^{i\theta_n}$ with $\alpha > 1$. Then (5.3.4) holds, so $\lim_{r\uparrow1}|B(re^{i\theta})| = 1$ for a.e. θ. But by the density of $\{e^{i\theta_n}\}_{n=1}^{\infty}$ for any $z_\infty \in \partial\mathbb{D}$, there is a subsequence $z_{n_j} \to z_\infty$ and $\lim_{j\to\infty} B(z_{n_j}) = 0$. Thus, we cannot go from nontangential limits to general limits (so long as f is not continuous on $\overline{\mathbb{D}}$, this is true—but it is nice to see such a dramatic example). See also Example 5.5.6. $\square$

Theorem 5.3.3. *Let $f \in N$. Let B be the Blaschke product of the zeros of f and $g = f/B$ so*

$$f = gB \tag{5.3.13}$$

Then $g \in N$ and

$$\|g\|_N = \|f\|_N \tag{5.3.14}$$

Proof. Since $|f| \leq |g|$, we have

$$\|f\|_N \leq \|g\|_N \tag{5.3.15}$$

On the other hand, if $x, y \in (0, \infty)$, then

$$\log_+ xy \leq \log_+ x + \log_+ y \tag{5.3.16}$$

(for if $xy \leq 1$, the LHS is zero, and if $xy \geq 1$, the LHS is $\log x + \log y \leq \log_+ x + \log_+ y$). Thus, since $|B|^{-1} \geq 1$,

$$\log_+|g(z)| \leq \log_+|f(z)| + \log_-|B(z)| \tag{5.3.17}$$

By (5.3.5), we get

$$\|g\|_N \leq \|f\|_N \tag{5.3.18}$$

proving $g \in N$ and (5.3.14). $\square$

Theorem 5.3.4 (Riesz Factorization). *Let $0 < p \leq \infty$. Let $f \in H^p$. Let B be the Blaschke product of the zeros of f and $g = f/B$ so (5.3.13) holds. Then $g \in H^p$ and*

$$\|g\|_p = \|f\|_p \tag{5.3.19}$$

Proof. Let B_N be given by (5.3.8) and $g_N = f/B_N = g\widetilde{B}_N$. For each $z \in \mathbb{D}$, $|\widetilde{B}_N(z)| \uparrow 1$, so by the monotone convergence theorem, for all $p \in (0, \infty]$, $0 < r < 1$, we have

$$M_r^{(p)}(g) = \lim_{N\to\infty} M_r^{(p)}(g_N) \tag{5.3.20}$$

On the other hand, since $|B_N| = 1$ on $\partial\mathbb{D}$ and $|B_N|$ is continuous on $\overline{\mathbb{D}}$, we have

$$\liminf_{r\uparrow 1} {}_\theta |B_N(re^{i\theta})| = 1 \tag{5.3.21}$$

so, since $\|h\|_p = \lim_{r\uparrow 1} M_r^{(p)}(h)$, $g_N \in H^p$ and

$$\|g_N\|_p = \|f\|_p \tag{5.3.22}$$

Since $M_r^{(p)}(h) \le \|h\|_p$, we see for all $0 < r < 1$,

$$M_r^{(p)}(g_N) \le \|f\|_p \tag{5.3.23}$$

so taking $N \to \infty$ and using (5.3.20),

$$M_r^{(p)}(g) \le \|f\|_p \tag{5.3.24}$$

Taking sup over r, we find $g \in H^p$ and

$$\|g\|_p \le \|f\|_p \tag{5.3.25}$$

Since $|f(z)| \le |g(z)|$, it is trivial that $\|f\|_p \le \|g\|_p$. Thus, (5.3.19) holds. $\square$

This allows us to prove the main result of this section:

Theorem 5.3.5. *Let $f \in H^p(\mathbb{D})$, $0 < p \le \infty$. Then*

(a) *If $M_{\mathrm{Rad}}f$ is given by (5.2.5), we have that*

$$\left(\int |M_{\mathrm{Rad}}f(e^{i\theta})|^p \frac{d\theta}{2\pi}\right)^{1/p} \le C_p \|f\|_{H^p} \tag{5.3.26}$$

The same result (with a larger C_p) holds for $M_{\mathrm{NT}}f$ given by (2.4.42).

(b) *There exists f^* in $L^p(\partial\mathbb{D}, \frac{d\theta}{2\pi})$ so that*

$$\|f_r - f^*\|_p \to 0, \qquad \|f_r\|_p \to \|f^*\|_p \tag{5.3.27}$$

as $r \uparrow 1$. Moreover, if $f^{(r)}(z) = f(rz)$ for $z \in \mathbb{D}$ and $r \in (0,1)$, then

$$\lim_{r\uparrow 1} \|f^{(r)} - f\|_{H^p(\mathbb{D})} = 0 \tag{5.3.28}$$

(c) *For a.e. θ,*

$$f_r(e^{i\theta}) \to f^*(e^{i\theta}) \tag{5.3.29}$$

(d) *If $p \ge 1$, (5.2.4) holds.*

(e) *If $p \ge 1$, (5.2.16) holds.*

(f) *For $p \ge 1$, $\{f^* \mid f \in H^p\} = \{g \in L^p(\partial\mathbb{D}, \frac{d\theta}{2\pi}) \mid g_n^\sharp = 0 \text{ for } n < 0\}$.*

Remark. The limitation of the Poisson and complex Poisson formulae to $p \geq 1$ is not an artifact of our proof. As the example below shows, the integral in (5.2.4) may not converge. Worse still, it can happen that $f \in H^p$, for any $p < 1$, $f \not\equiv 0$ but that $\operatorname{Re} f(e^{i\theta}) = 0$ for a.e. θ!

Proof. (a) Write $f = gB$, the Riesz factorization. By multiplying by a nonzero constant, we can suppose $g(0) = 1$. Since g is nonvanishing on $\mathbb{D}$, $\log g = q$, the branch with $q(0) = 0$ is analytic, and

$$g = \exp(q) \tag{5.3.30}$$

Let

$$h(z) = \exp\left(\frac{pq(z)}{2}\right) \tag{5.3.31}$$

Then $|h(z)|^2 = |g(z)|^p$ so $h \in H^2$ and

$$(M_{\text{Rad}}f)(e^{i\theta}) \leq (M_{\text{Rad}}g)(e^{i\theta}) \leq [M_{\text{Rad}}h(e^{i\theta})]^{2/p} \tag{5.3.32}$$

Thus, (5.3.26) follows from (5.2.6).

(b), (c) Since $h \in H^2$, $\lim_{r\uparrow 1} h(re^{i\theta}) = h^*(e^{i\theta})$ exists and is nonzero for a.e. θ (if $f \not\equiv 0$). Since q is determined by (5.3.31), $\lim_{r\uparrow 1} q(re^{i\theta}) = q^*(e^{i\theta})$ exists a.e. and so, by (5.3.30), $g^*(e^{i\theta}) = \lim_{r\uparrow 1} g(re^{i\theta})$ exists a.e. Since $B(re^{i\theta}) \to B^*(e^{i\theta})$ for a.e. θ, we have $f^*(e^{i\theta}) = \lim_{r\uparrow 1} f(re^{i\theta})$ exists a.e. (5.3.28) follows from $\|f^*\|_{L^p(\partial\mathbb{D})} = \|f\|_{H^p(\mathbb{D})}$ and $(f^{(r)} - f)^* = f_r - f^*$.

Clearly, $|f^*(e^{i\theta})| \leq (M_{\text{Rad}}f)(e^{i\theta})$ so $f^* \in L^p$. Moreover, $|f_r(e^{i\theta}) - f^*(e^{i\theta})|^p \leq 2^p[(M_{\text{Rad}}f)(e^{i\theta})]^p$, so the first equation in (5.3.27) holds by the dominated convergence theorem. Similarly, the second equation holds.

(d) We'll prove (d); (e) is similar. We have

$$f(r\rho e^{i\theta}) = P_r * f_\rho(e^{i\theta}) \tag{5.3.33}$$

Since $f_\rho \to f^*$ in L^p and $p \geq 1$, $P_r * f_\rho \to P_r * f^*$.

(f) If $f \in H^p$ for $p \in [1, \infty)$, then since $f_r \to f^*$ in L^1, we have for $n < 0$,

$$(f^*)_n^\sharp = \lim_{r\uparrow 1}(f_r)_n^\sharp = \lim_{r\uparrow 1}\int_{0, z=re^{i\theta}}^{2\pi} z^{-n}f(z)\,\frac{dz}{2\pi i z} = 0$$

by the Cauchy integral formula since $-n \geq 1$.

Conversely, if $g \in L^p$ with $g_n^\sharp = 0$ for $n < 0$, one can prove (Problem 2) that $f(re^{i\theta}) = (P_r * g)(e^{i\theta})$ defines an analytic function with $f \in H^p$ and $f^* = g$. $\qquad\square$

Example 5.3.6. Let f be given by

$$f(z) = \frac{1+z}{1-z} \tag{5.3.34}$$

Then it is easy to see (Problem 3) that $\operatorname{Re} f \in h^1$, $f \notin H^1$ but $f \in H^p$ for all $p < 1$ (see also Theorem 5.4.6 below). Since $\inf_\varphi P_r(\theta, \varphi) > 0$ for all $r < 1$, the integral in (5.2.4) is not absolutely convergent. Clearly,

$$f^*(e^{i\theta}) = \frac{1 + e^{i\theta}}{1 - e^{i\theta}} = \frac{\cos(\theta/2)}{-i \sin(\theta/2)} = i \cot(\theta/2) \qquad (5.3.35)$$

so $\operatorname{Re} f^*(e^{i\theta}) = 0$ for a.e. θ. Thus, for functions in h^1, $\operatorname{Re} f^*$ may not determine f. The integral in (5.2.16) exists but is not $f(z)$!

Notice also that

$$\operatorname{Re} f(z) = \frac{1 - |z|^2}{|1 - z|^2} \geq 0 \qquad (5.3.36)$$

so

$$\int_0^{2\pi} |\operatorname{Re} f(re^{i\theta})| \frac{d\theta}{2\pi} = \operatorname{Re} f(0) = 1 \qquad (5.3.37)$$

so $\operatorname{Re} f \in h^1$, but since $f \notin H^1$, $Kf \notin h^1$. $\square$

The Riesz factorization has an interesting consequence that will be useful in Section 5.7.

Theorem 5.3.7. *Let* $0 < p < q$. *Suppose* $f \in H^p$ *and* $f^* \in L^q$. *Then* $f \in H^q$ *with*

$$\|f\|_{H^q} = \|f^*\|_{L^q} \qquad (5.3.38)$$

Proof. Suppose first $p = 2$. Since $f \in H^2$, we have $f_r = P_r * f^*$ by Theorem 5.2.1(c). Since $\|P_r\|_1 = 1$, $g \mapsto P_r * g$ is bounded with norm 1 on each L^q. Thus, $f^* \in L^q \Rightarrow \|f_r\|_q \leq \|f^*\|_q$, so $f \in H^q$, as usual, $\|f\|_{H^q} = \|f^*\|_{L^q}$.

Now let p be general. Suppose first that f is nonvanishing on $\mathbb{D}$. Let $\alpha = p/2$. By defining $h = \exp(\alpha \log f)$, we have $|h|^2 = |f|^p$ and $|f^q| = |h|^{2q/p}$. $f \in H^p \Rightarrow h \in H^2 \Rightarrow h \in H^{2q/p}$ (by the case $p = 2$ above) $\Rightarrow f \in H^q$ (since $|f|^q = |h|^{2q/p}$). Tracking norms shows that (5.3.35) holds.

For general f, write $f = Bg$ and use $|g^*| = |f^*|$ to see g and so f lie in H^q. $\square$

Here is an application of H^1 theory:

Theorem 5.3.8 (F. and M. Riesz Theorem). *Let* μ *be a complex Baire measure on* $\partial\mathbb{D}$ *so that* $\mu_n^\sharp = 0$ *for* $n < 0$. *Then* μ *is absolutely continuous with respect to* $\frac{d\theta}{2\pi}$.

Proof. Define $f(z)$ on $\mathbb{D}$ by

$$f(re^{i\theta}) = \int P_r(\theta, \varphi) \, d\mu(\varphi) \qquad (5.3.39)$$

Since

$$P_r(\theta, \varphi) = \text{Re}\left[\frac{1 + re^{i(\theta - \varphi)}}{1 - re^{i(\theta - \varphi)}}\right] \tag{5.3.40}$$

$$= \text{Re}\left[1 + 2\sum_{n=1}^{\infty} r^n e^{in(\theta - \varphi)}\right]$$

$$= \sum_{n=-\infty}^{\infty} r^{|n|} e^{in(\theta - \varphi)} \tag{5.3.41}$$

we have, since $\mu_n^\sharp = 0$ for $n < 0$,

$$f(z) = \sum_{n=0}^{\infty} \mu_n^\sharp z^n \tag{5.3.42}$$

Here we need the uniform in φ convergence of (5.3.41) for each $r < 1$. Since $|\mu_n^\sharp| \leq \|\mu\|$, (5.3.42) defines a function analytic in $\mathbb{D}$.

Moreover, by (5.3.39),

$$M_r^{(1)} f \leq \|\mu\| \tag{5.3.43}$$

so $f \in H^1$. Thus, by Theorem 5.3.5, there exists $f^* \in L^1$ so $f_r \to f^*$ in L^1. It follows, using (5.3.42), that for $n \geq 0$,

$$\mu_n^\sharp r^n \to \int e^{-in\theta} f^*(e^{i\theta}) \frac{d\theta}{2\pi} \tag{5.3.44}$$

and $\int e^{-in\theta} f^*(e^{i\theta}) \frac{d\theta}{2\pi} = 0$ for $n < 0$. Thus, μ and $f^*(e^{i\theta}) \frac{d\theta}{2\pi}$ have the same moments. By density of Laurent polynomials in $C(\partial\mathbb{D})$, we see

$$d\mu = f^*(e^{i\theta}) \frac{d\theta}{2\pi} \tag{5.3.45}$$

is a.c. wrt $\frac{d\theta}{2\pi}$.   □

Here is an interesting Corollary of the F. and M. Riesz Theorem. Recall that $L^1(\partial\mathbb{D}, \frac{d\theta}{2\pi})$ is not the dual of any Banach space (see Example 5.11.2 of Part 1). But $H^1(\mathbb{D})$ is.

Corollary 5.3.9. *In $C(\partial\mathbb{D})$, let $\mathfrak{A}_- = \{f \mid f_n^\sharp = 0 \text{ for } n = 1, 2, \ldots\}$. Then $(C(\partial\mathbb{D})/\mathfrak{A}_-)^*$ is isometrically isomorphic to $H^1(\mathbb{D})$ under $L : H^1(\mathbb{D}) \to (C(\partial\mathbb{D})/\mathfrak{A}_-)^*$ by*

$$L(f)(g) = \int f^*(e^{i\theta}) g(e^{i\theta}) \frac{d\theta}{2\pi} \tag{5.3.46}$$

Remark. The proof that L^1 is not a dual space relies on the fact that its unit ball $(L^1)_1$ has no extreme points. That H^1 is a dual space implies its unit ball has lots of extreme points. The reader will identify them (de Leeuw–Rudin theorem) in Problems 2 and 3 of Section 5.6.

Proof. By Theorem 5.1.8 of Part 1 and the fact that $C(\partial\mathbb{D})$ is the set of signed measures on $\partial\mathbb{D}$, we see that $(C(\partial\mathbb{D})/\mathfrak{A}_-)^*$ is isometrically isomorphic to

$$\mathfrak{A}_-^\perp = \{\mu \in M(\partial\mathbb{D}) \,\Big|\, \int g\,d\mu = 0 \text{ for all } g \in \mathfrak{A}_-\} \tag{5.3.47}$$

Since $e^{in\theta} \in \mathfrak{A}_-$, μ in $\mathfrak{A}_-^\perp \Rightarrow \mu_n^\sharp = 0$ for $n < 0$. But, by Fejér's theorem (see Theorem 3.5.6 of Part 1), any $g \in \mathfrak{A}_-$ is a limit of linear combinations of $\{e^{in\theta}\}_{n=1}^\infty$ so

$$\mu \in \mathfrak{A}_-^\perp \Leftrightarrow \mu_n^\sharp = 0 \text{ for } n < 0 \tag{5.3.48}$$

By the last theorem, any such μ has the form $f^* \, d\theta/2\pi$ with $f \in H^1$ and the converse is true by Theorem 5.3.5. Since, for $f \in L^1$

$$\left\| f\frac{d\theta}{2\pi} \right\|_{M(\partial\mathbb{D})} = \|f\|_{L^1} \tag{5.3.49}$$

we see that $\mathfrak{A}_-^\perp$ is isometrically isomorphic to H^1. $\qquad\square$

Notes and Historical Remarks. The Riesz factorization theorem was proven in 1923 by F. Riesz [**683**] and the F. and M. Riesz theorem is from [**690**] in 1916. Most of the results in Theorem 5.3.5 predate the Riesz factorization. In 1906, Fatou [**261**], in one of the first applications on Lebesgue's integration theory, proved $f \in H^\infty$ has a.e. boundary values. F. and M. Riesz [**690**] extended this to H^1. Using his factorization theorem, Riesz [**683**] proved boundary values for all H^p, $p > 0$. He also proved $\log|f^*(e^{i\theta})| \in L^1$ if $f \not\equiv 0$, proving f^* is nonzero for a.e. θ. In that paper and [**685**], Riesz also showed $\|f_r - f^*\|_p \to 0$, $\|f_r\|_p \to \|f^*\|_p$. A different approach to pointwise limits is due to F. and R. Nevanlinna [**606**], as we noted in Section 5.1.

Problems

1. If $\sum_{n=1}^\infty (1-|z_n|) < \infty$ for $\{z_n\}_{n=1}^\infty \subset \mathbb{D}\backslash\{0\}$, prove that $\sum_{n=1}^\infty \log|z_n| < \infty$ so $\lim_{N\to\infty} \sum_{n=N}^\infty \log|z_n| = 0$. (*Hint*: If $x \in [\frac{1}{2}, 1]$, prove that $|\log x| \le C(1 - |x|)$ for some constant C.)

2. (a) If $g \in L^1$ with $g_n^\sharp = 0$ for $n < 0$, prove that f defined by

$$f(re^{i\theta}) = \int P_r(\theta, \varphi) g(e^{i\varphi}) \frac{d\varphi}{2\pi} \tag{5.3.50}$$

is analytic in $\mathbb{D}$. (*Hint*: Use (5.3.41) to prove an analog of (5.3.42).)

(b) Prove that $M_r^{(p)}(f) \le \|g\|_p$ if $g \in L^p$ and $p \ge 1$. Conclude that $f \in H^p$.

(c) For a.e. θ, prove that $\lim_{r\uparrow 1} f(re^{i\theta}) = g(e^{i\theta})$ and conclude that $f^* = g$.

3. (a) For all $r \in (0,1)$ and $\theta \in [0, \frac{\pi}{2}]$, prove that $|1 - re^{i\theta}| \geq C|1 - e^{i\theta}|$ for a constant C.

(b) If f is given by (5.3.34), prove that

$$|f(re^{i\theta})| \leq K|1 - e^{i\theta}|^{-1} \qquad (5.3.51)$$

for a constant K.

(c) Prove that $f \in H^p$ for any $p < 1$.

(d) Prove $f^* \notin L^1$ and conclude that $f \notin H^1$.

(e) Prove that $\operatorname{Re} f \in h^1$.

The next three problems are relevant to H^p spaces since we'll prove an inequality of Hardy (which appeared first in Hardy–Littlewood [358]; note that Theorem 4.1.3 discussed and Theorem 6.1.3 will discuss a different, albeit related, result also usually called *Hardy's inequality*) that if $f(z) = \sum_{n=0}^{\infty} a_n z^n \in H^1$, then

$$\sum_{n=0}^{\infty} \frac{|a_n|}{n+1} \leq \pi \|f\|_{H^1} \qquad (5.3.52)$$

4. Let $\psi \in L^\infty(\partial\mathbb{D}, \frac{d\theta}{2\pi})$, $f, g \in H^2$.

(a) If f, g are polynomials, prove that

$$\int_0^{2\pi} f(e^{i\theta})g(e^{i\theta})\psi(e^{i\theta}) \frac{d\theta}{2\pi} = \sum_{n,m} f_n^\sharp g_n^\sharp \psi_{-n-m}^\sharp \qquad (5.3.53)$$

(b) If $\psi_k^\sharp \geq 0$ for $k \leq 0$, prove that for all $f, g \in H^2$,

$$\left| \sum_{n,m=0}^{\infty} \psi_{-n-m}^\sharp f_n^\sharp g_m^\sharp \right| \leq \|\psi\|_\infty \|f\|_2 \|g\|_2 \qquad (5.3.54)$$

where the sum is absolutely convergent.

(c) If $\psi(e^{i\theta}) = -ie^{+i\theta}(\pi - \theta)$ for $\theta = [0, 2\pi)$, prove that for $k \leq 0$, $\psi_k^\sharp = (-k+1)^{-1}$.

(d) Prove Hardy's inequality (see the Notes to Section 5.7)

$$\left| \sum_{n,m=0}^{\infty} \frac{f_n^\sharp g_m^\sharp}{n+m+1} \right| \leq \pi \|f\|_2 \|g\|_2 \qquad (5.3.55)$$

5. If $\psi \in L^\infty$ with $\psi_k^\sharp \leq 0$ for $k \leq 0$ and $f(z) = \sum_{n=0}^{\infty} a_n z^n \in H^1$, prove that

$$\sum_{n=0}^{\infty} \psi_{-n}|a_n| \leq \|\psi\|_\infty \|f\|_{H^1} \qquad (5.3.56)$$

(*Hint*: Use Riesz factorization to prove $f = gh$ with $g, h \in H^2$ and $\|g\|_{H^2} = \|h\|_{H^2} = (\|f\|_{H^1})^{1/2}$ and use Problem 4.)

6. Prove the inequality of Hardy (5.3.52) if $f(z) = \sum_{n=0}^{\infty} a_n z^n \in H^1$. (*Hint*: See Problems 4(c) and 5.)

5.4. Carathéodory Functions, h^1, and the Herglotz Representation

With the approach of the last section, we treated H^p but not N. The Riesz factorization implies it suffices to study $g \in N$ which is nowhere vanishing. Thus, $g = \exp(q)$ where, by the definition of N and (5.1.16), we know $\operatorname{Re} q \in h^1$. Thus, we want to know about boundary values of functions in h^1, where a new aspect will occur—something that will play an important role in the next few sections. We saw a hint of the new aspect in Example 5.3.6 where $u = \operatorname{Re} f$ lies in h^1, $u \not\equiv 0$, but $u^*(e^{i\theta}) = 0$ for all θ. Thus, the pointwise boundary value of u does not determine u. The reason, as we will see, is that in a real sense, the boundary value is a measure, not a function (and in Example 5.3.6, the measure is a point mass at $e^{i\theta} = 1$).

We'll start with positive harmonic rather than h^1-functions (we'll eventually see that any h^1-function is a difference of positive harmonic functions—we note that, by the MVP, $u \geq 0 \Rightarrow u \in h^1$ since then $\int_0^{2\pi} |u(re^{i\theta})| \frac{d\theta}{2\pi} = u(0)$).

Definition. A *Carathéodory function* is a function, f, analytic on $\mathbb{D}$ so that

  (i) $\operatorname{Re} f(z) \geq 0$ for all z,   (5.4.1)

  (ii) $f(0) = 1$.   (5.4.2)

Remarks. 1. (ii) is a normalization condition. It is easy to see that if f obeys (i) and $f \not\equiv 0$, then $f = \alpha \tilde{f} + i\beta$, where $\alpha > 0$ and $\beta \in \mathbb{R}$, and $\tilde{f}(0) = 1$.

2. By the open mapping theorem for analytic functions (and (ii)), we have that $\operatorname{Re} f(z) > 0$ for all $z \in \mathbb{D}$.

Theorem 5.4.1 (Herglotz Representation Theorem). *If F is a Carathéodory function, then there is a probability measure, μ, on $\partial\mathbb{D}$ so that*

$$F(z) = \int \frac{e^{i\theta} + z}{e^{i\theta} - z} \, d\mu(\theta) \tag{5.4.3}$$

Remarks. 1. Clearly, μ determines F via (5.4.3). But since linear combinations of $\{\frac{e^{i\theta}+z}{e^{i\theta}-z}\}_{z\in\mathbb{D}}$ and their conjugate are dense in $C(\partial\mathbb{D})$ (Problem 1), F determines μ.

2. For any probability measure, μ, F given by (5.4.3) has

$$\operatorname{Re} F(re^{i\varphi}) = \int P_r(\theta, \varphi) \, d\mu(\varphi) \tag{5.4.4}$$

so $\operatorname{Re} F > 0$ and $F(0) = \int d\mu(\theta) = 1$, so (5.4.3) defines a Carathéodory function. These two remarks say that (5.4.3) sets up a one-one correspondence between $\mu \in \mathcal{M}_{+,1}(\partial\mathbb{D})$ and Carathéodory functions.

Proof. By the complex Poisson representation, we have for any $r, \rho \in (0,1)$,

$$F(r\rho e^{i\varphi}) = \int \frac{e^{i\theta} + re^{i\varphi}}{e^{i\theta} - re^{i\varphi}} \operatorname{Re} F(\rho e^{i\theta}) \frac{d\theta}{2\pi} \tag{5.4.5}$$

Thus, we define

$$d\mu_\rho(\theta) = \operatorname{Re} F(\rho e^{i\theta}) \frac{d\theta}{2\pi} \tag{5.4.6}$$

Since $\operatorname{Re} F \geq 0$, $d\mu_\rho$ is a positive measure, and since $\operatorname{Re} F$ is harmonic, $d\mu_\rho$ is a probability measure.

Suppose $F(z) = \sum_{n=0}^\infty a_n z^n$. Then expanding

$$\frac{e^{i\theta} + re^{i\varphi}}{e^{i\theta} - re^{i\varphi}} = 1 + 2\sum_{n=1}^\infty r^n e^{in(\varphi-\theta)} \tag{5.4.7}$$

we see that

$$\int e^{-in\theta} \, d\mu_\rho(\theta) = \begin{cases} 1, & n = 0 \\ \frac{1}{2} a_n \rho^n, & n > 0 \\ \frac{1}{2} \bar{a}_{-n} \rho^n, & n < 0 \end{cases} \tag{5.4.8}$$

(where we used that $\overline{\int e^{-in\theta} \, d\mu_\rho(\theta)} = \int e^{in\theta} \, d\mu_\rho(\theta)$). Thus, $d\mu_\rho$ is a family of measures with

$$\lim_{\rho\uparrow 1} \int e^{-in\theta} \, d\mu(\rho)(\theta) = \begin{cases} 1, & n = 0 \\ \frac{1}{2} a_n, & n > 0 \\ \frac{1}{2} \bar{a}_{-n}, & n < 0 \end{cases} \tag{5.4.9}$$

Since these are probability measures and finite sums of $\{e^{in\theta}\}_{n=-\infty}^\infty$ are dense in $C(\partial\mathbb{D})$, we see that for any continuous f, $\lim_{\rho\uparrow 1} \int f(\theta) \, d\mu_\rho(\theta)$ exists. The limit is clearly a normalized positive functional, so there is a probability measure $d\mu$ so $\mu_\rho \to \mu$ in the weak (i.e., $\sigma(M(\partial\mathbb{D}), C(\partial\mathbb{D}))$) topology.

Since $\frac{e^{i\theta}+re^{i\varphi}}{e^{i\theta}-re^{i\varphi}}$ lies in $C(\partial\mathbb{D})$ for r, φ fixed, (5.4.5) implies that F obeys (5.4.3). $\quad\square$

The same proof works for h^1!

Theorem 5.4.2. *Let $u \in h^1$. Then there is a signed measure, μ, on $\partial\mathbb{D}$ so that*

$$u(re^{i\varphi}) = \int P_r(\theta, \varphi) \, d\mu(\theta) \tag{5.4.10}$$

Proof. Let

$$d\mu_\rho(\theta) = u(\rho e^{i\theta}) \frac{d\theta}{2\pi} \tag{5.4.11}$$

Then the Poisson representation for an harmonic function, continuous on $\overline{\mathbb{D}}$ implies that

$$u(r\rho e^{i\varphi}) = \int P_r(\theta, \varphi) \, d\mu_\rho(\theta) \tag{5.4.12}$$

Moreover,

$$\|\mu_\rho\| = \|u(\rho e^{i\theta})\|_1 \tag{5.4.13}$$

so $u \in h^1 \Rightarrow \sup_\rho \|\mu_\rho\| < \infty$.

There is an analytic function, $F(z)$, on $\mathbb{D}$ so $\operatorname{Re} F = u$ and $F(0) = u(0)$ (i.e., $\operatorname{Im} F(0) = 0$). If $F(z) = \sum_{n=0}^\infty a_n z^n$, we see that μ_ρ still obeys (5.4.8), and so (5.4.9) holds. Since $\sup \|\mu_\rho\| < \infty$, a density argument shows $\mu_\rho \to \mu$ in the weak topology for a signed measure μ with $\|\mu\| \leq \sup_\rho \|\mu_\rho\|$. Taking limits in (5.4.12) yields (5.4.10). $\qquad\square$

Corollary 5.4.3. *A function, u, lies in h^1 if and only if it is a difference of two positive harmonic functions.*

Proof. As we saw above, any positive harmonic function is in h^1, so a difference is. Conversely, if u is in h^1, it has a representation (5.4.10). If $\mu = \mu_+ - \mu_-$ for $\mu_\pm \geq 0$, $u = u_+ - u_-$ where $u_\pm = P * \mu_\pm$. $\qquad\square$

The next corollary is so important, we call it a theorem.

Theorem 5.4.4. *If $u \in h^1$, there is u^* in $L^1(\partial\mathbb{D}, \frac{d\theta}{2\pi})$ so that for a.e. θ,*

$$\lim_{r\uparrow 1} u(re^{i\theta}) = u^*(e^{i\theta}) \tag{5.4.14}$$

If v is the harmonic conjugate for u, then there is v^*, a real-valued measurable function in $\partial\mathbb{D}$, so that for a.e. θ,

$$\lim_{r\uparrow 1} v(re^{i\theta}) = v^*(e^{i\theta}) \tag{5.4.15}$$

Remarks. 1. It is not in general true that $\|u_r - u^*\|_1 \to 0$ nor that $\|u^*\|_1 = \lim_{r\uparrow 1} \|u_r\|_1$. What is true is that $\|u^*\|_1 \leq \lim_{r\uparrow 1} \|u_r\|_1$.

2. We'll find another proof of this below; see the remark after Theorem 5.4.6.

Proof. By the last corollary, it suffices to prove this result if $u \geq 0$. In that case, let v be the conjugate harmonic function, so $F = u + iv$ has $\operatorname{Re} F \geq 0$. Let

$$g(z) = \exp(-F(z)) \tag{5.4.16}$$

Since $\operatorname{Re} F \geq 0$, $|g(z)| \leq 1$, that is, $g \in H^\infty$ so there is g^* which is a.e. nonzero, so $g(re^{i\theta}) \to g^*(e^{i\theta})$. Since $u = -\log|g|$ and g^* is a.e. not 0, $u + iv$ has a finite limit for a.e. θ.

Since (5.4.14) holds, by Fatou's lemma,

$$\|u^*\|_1 \leq \sup_r \|u_r\|_{L^1} = \|u\|_{h^1} \tag{5.4.17}$$

Thus define $v^*(e^{i\theta}) = \mathrm{Im}(\log g^*(e^{i\theta}))$. $\qquad\square$

This allows us a second proof of the existence of boundary values for functions in N:

Theorem 5.4.5. *Let $f \in N$. Then there is f^*, measurable on $\partial\mathbb{D}$, so that for a.e. θ,*

$$\lim_{r\uparrow 1} f(re^{i\theta}) = f^*(e^{i\theta}) \tag{5.4.18}$$

Moreover, if $f \not\equiv 0$, f^* is a.e. nonzero. Indeed,

$$\int |\log|f^*(e^{i\theta})|| \,\frac{d\theta}{2\pi} < \infty \tag{5.4.19}$$

Remark. We describe this as a different proof of Theorem 5.2.6, but it can be viewed as a different proof of Theorem 5.1.6. That $\log|f(z)/B(z)|$ is a difference of nonnegative harmonic functions implies that theorem. In fact, one can see a close connection between our proofs of Theorem 5.1.6 and of Corollary 5.4.3.

Proof. By the Riesz factorization,

$$f(z) = B(z)\exp(h(z)) \tag{5.4.20}$$

with $\mathrm{Re}\,h(z) = \log|f(z)/B(z)| \in h^1$. By Theorem 5.4.4, $h(z)$ has boundary values and $\mathrm{Re}\,h \in L^1$. $\qquad\square$

We have seen in Example 5.3.6 that a Carathéodory function, F, can have $\mathrm{Re}\,F \in h^1$ but $F \notin H^1$. The following lovely theorem shows, though, that it is almost in H^1:

Theorem 5.4.6 (Kolmogorov's Theorem). *Let F be a Carathéodory function. Then $F \in H^p(\mathbb{D})$ for all $p < 1$. Indeed,*

$$\|F\|_p \leq \left[\cos\left(\frac{p\pi}{2}\right)\right]^{-1/p} \tag{5.4.21}$$

Remarks. 1. As $p \uparrow 1$, $\cos(p\pi/2) \to 0$, so the bound in (5.4.21) diverges, as it must since there are F's not in H^1; see Example 5.3.6.

2. Since any $u \in h^1$ is $\alpha F_1 - \beta F_2$ with F_j Carathéodory and $\alpha + \beta = \|u\|_1$, this implies that if $u \in h^1$ and v is its harmonic conjugate, then

$$\|v\|_p \leq \left[\cos\left(\frac{p\pi}{2}\right)\right]^{-1/p}\|u\|_1 \tag{5.4.22}$$

3. This theorem (without the explicit constant) also follows from the weak-L^1 inequality, (5.7.61).

Proof. Fix $p < 1$. Let $G = F^p$ which is analytic (since $\operatorname{Re} F > 0$) with

$$|\arg G| \leq \frac{p\pi}{2} \tag{5.4.23}$$

Since G is analytic for any $r < 1$,

$$\int_0^{2\pi} \operatorname{Re}[G(re^{i\theta})] \frac{d\theta}{2\pi} = \operatorname{Re} G(0) = \operatorname{Re} F(0)^p = 1 \tag{5.4.24}$$

By (5.4.23) and monotonicity of $\cos\eta$ for $\eta \in [0, \frac{\pi}{2}]$,

$$\operatorname{Re}[G(re^{i\theta})] \geq |G(re^{i\theta})| \cos\left(\frac{p\pi}{2}\right) \tag{5.4.25}$$

and $|G| = |F|^p$, so (5.4.23) implies

$$\cos\left(\frac{p\pi}{2}\right) \int_0^{2\pi} |F(re^{i\theta})|^p \frac{d\theta}{2\pi} \leq 1 \tag{5.4.26}$$

which implies (5.4.21), $\qquad\square$

Remark. Since $u + iv \in H^p$, $p < 1$, has boundary values, this provides another proof that for any $u \in h^1$, its conjugate has boundary values.

Example 5.4.7. This example provides an analytic function $f(z)$ so $u = \operatorname{Re} f$ lies in $\bigcap_{p<1} h^p$ (with h^p for $p < 1$ defined in the obvious way!) but so that for a.e. θ, $\lim_{r\uparrow 1} f(re^{i\theta})$ fails to exist. Thus, $f \notin N$ and, in particular, in no H^p. Let $\varepsilon_n = \pm 1$. Define

$$f(z; \{\varepsilon_n\}_{n=1}^\infty) = \sum_{n=1}^\infty \varepsilon_n \frac{z^{2^n}}{1 - z^{2^{n+1}}} \tag{5.4.27}$$

Then one can show (references in the Notes):

(1) For every choice of $\{\varepsilon_n\}_{n=1}^\infty$, $\operatorname{Re} f \in \bigcap_{p<1} h^p$.
(2) If one puts the measure $\gamma(\varepsilon_n = \pm 1) = \frac{1}{2}$ and takes a product measure (i.e., in the language of Section 7.1 of Part 1, the ε_n are iid Bernoulli $p = \frac{1}{2}$ variables), then for a.e. choice of $\{\varepsilon_n\}_{n=1}^\infty$, f does not have radial boundary values for a.e. θ. $\qquad\square$

Notes and Historical Remarks. The Herglotz representation theorem for Carathéodory functions is due to Herglotz [**380**] and Riesz [**680**] in 1911. That any $u \in h^1$ is a difference of positive harmonic functions is a result of Plessner [**650**] in 1923. Kolmogorov's theorem was proven by him in 1925 in [**480**]. Zygmund [**875, 880**] proved a converse: If $u + iv \in H^1$ has $\inf_{z\in\mathbb{D}} u(z) > 0$, then $u \in h\log^+ h$.

Example 5.4.7 is due to Paley–Zygmund [**628**], based on Hardy–Littlewood [**362**] who studied the case $\varepsilon_n \equiv 1$. Duren's book [**234**] has a simplified presentation of this example. The earliest example of this type is due to Littlewood [**540**]. There is a connection of this to results on random power series having natural boundaries, as discussed in the Notes to Section 2.3 of Part 2A.

The map $F \mapsto f(z) = z^{-1}\left[\frac{F(z)-1}{F(z)+1}\right]$ sets up a 1–1 correspondence between Carathéodory functions and maps, f, of $\mathbb{D}$ to $\mathbb{D}$, called Schur functions. They are also discussed in Section 7.5 of Part 2A and Section 5.5 of Part 4.

Problems

1. For $z \in \mathbb{D}$, let f_z be the function on $C(\partial\mathbb{D})$

$$f_z(e^{i\theta}) = (e^{i\theta} + z)(e^{i\theta} - z)^{-1}$$

Prove that linear combinations of $\{f_z\}_{z\in\mathbb{D}}$ and their conjugate are dense in $C(\partial\mathbb{D})$. (*Hint*: Compute derivatives of f_z in z at $z = 0$.)

5.5. Boundary Value Measures

We saw there is a one-one correspondence between Carathéodory functions, $F(z)$, and probability measures, μ, on $\partial\mathbb{D}$, via

$$F(z) = \int \frac{e^{i\theta} + z}{e^{i\theta} - z}\, d\mu(\theta) \tag{5.5.1}$$

This formula describes how to go from μ to F. We saw the converse was via

$$\lim_{r\uparrow 1} d\mu_r(\theta) = d\mu(\theta) \tag{5.5.2}$$

(in the weak topology, i.e., $\sigma\big(\mathcal{M}(\partial\mathbb{D}), C(\partial\mathbb{D})\big)$) where

$$d\mu_r(\theta) = \operatorname{Re} F(re^{i\theta})\, \frac{d\theta}{2\pi} \tag{5.5.3}$$

We explore this connection further in this section (for reasons connected with the next section) by showing:

(1) If $F^*(e^{i\theta}) = \lim_{r\uparrow 1} F(re^{i\theta})$, then

$$d\mu_{\mathrm{ac}}(\theta) = \operatorname{Re} F^*(e^{i\theta})\, \frac{d\theta}{2\pi} \tag{5.5.4}$$

so that $\operatorname{Re} F^*(e^{i\theta}) = 0$ for a.e. $\theta \Leftrightarrow d\mu_{\mathrm{ac}} = 0$.

(2) If $\operatorname{Re} F \in h^p$ for some $p > 1$, then $d\mu_{\mathrm{sing}}(\theta) = 0$.

Theorem 5.5.1. (a) $\mu_{\mathrm{ac}} = 0 \Rightarrow \operatorname{Re} F(re^{i\theta}) \to 0$ as $r \uparrow 1$ *for a.e.* θ.

(b) (5.5.4) *holds.*

Remark. This theorem is part of Theorem 2.5.5. We prove it again here.

Proof. Suppose $\operatorname{Re} F^*(e^{i\theta}) = g(e^{i\theta})$. Then for all $h \in C(\partial\mathbb{D})$ with $h \geq 0$, we have that

$$\int h(e^{i\theta}) g(e^{i\theta}) \frac{d\theta}{2\pi} = \int h(e^{i\theta}) \left[\lim_{r\uparrow 1} \operatorname{Re} F(re^{i\theta}) \right] \frac{d\theta}{2\pi}$$

$$\leq \liminf_{r\uparrow 1} \int h(e^{i\theta}) \operatorname{Re} f(re^{i\theta}) \frac{d\theta}{2\pi} \tag{5.5.5}$$

$$= \int h(e^{i\theta}) \, d\mu(\theta) \tag{5.5.6}$$

Here (5.5.5) is Fatou's lemma and $h \geq 0$, $\operatorname{Re} F \geq 0$ and (5.5.6) is (5.5.2).

Thus,

$$g(e^{i\theta}) \frac{d\theta}{2\pi} \leq d\mu \tag{5.5.7}$$

If $(d\mu)_{\mathrm{ac}}$ is zero, this implies $g = 0$ for a.e. θ.

If $d\mu = g(\theta) \frac{d\theta}{2\pi}$ with $g \in L^p(\partial\mathbb{D})$ (we need $p = 1$ here but will use $p > 1$ below), then F given by (5.5.1) has $h_r(e^{i\theta}) \equiv \operatorname{Re} F(re^{i\theta}) = P_r * g$ and, by the analysis in Example 2.4.5, $h_r \to g$ pointwise a.e. and $\max|h_r(e^{i\theta})| \in L^p$, so by the dominated convergence theorem, $\|h_r - g\|_p \to 0$.

If now $d\mu = d\mu_{\mathrm{ac}} + d\mu_{\mathrm{sing}}$, write $F = F_{\mathrm{ac}} + F_{\mathrm{sing}}$. By (a), $\operatorname{Re} F^*_{\mathrm{sing}}(e^{i\theta}) = 0$, and by the above argument,

$$d\mu_{\mathrm{ac}}(\theta) = \operatorname{Re} F^*_{\mathrm{ac}}(e^{i\theta}) \frac{d\theta}{2\pi} = \operatorname{Re} F^*(e^{i\theta}) \frac{d\theta}{2\pi}$$

since $\operatorname{Re} F^*_{\mathrm{sing}}$ is a.e. zero. $\qquad\square$

Theorem 5.5.2. *If $p > 1$, F is a Carathéodory function, with μ given by (5.5.3), and $\operatorname{Re} F \in h^p$, then*

(a) $d\mu_{\mathrm{sing}} = 0$.

(b) $d\mu(e^{i\theta}) = \operatorname{Re} F^*(e^{i\theta}) \dfrac{d\theta}{2\pi}$ \hfill (5.5.8)

    with $\operatorname{Re} F^* \in L^p(\partial\mathbb{D}, \frac{d\theta}{2\pi})$.

(c) $\|\operatorname{Re} F_r - \operatorname{Re} F^*\|_p \to 0$ *as* $r \uparrow 1$. \hfill (5.5.9)

Remark. Note that the proof of (a) only depends on $\operatorname{Re} F_r \frac{d\theta}{2\pi} \to d\mu$ weakly and the L^p-bound, and neither on the Poisson formula nor pointwise a.e. convergence.

Proof. (a) Let $d\mu_r = \operatorname{Re} F_r \frac{d\theta}{2\pi}$. By construction of μ, $\mu_r \to \mu$ weakly. Since for any open set, A, and Baire measure, ν, on X, we have

$$\nu(A) = \sup_{\substack{f \in C(X) \\ 0 \leq f \leq \chi_A}} \int f(x) \, d\nu(x) \tag{5.5.10}$$

We have for any ε the existence of $f \in C(X)$, $0 \leq f \leq \chi_A$, so that

$$\mu(A) \leq \mu(f) + \varepsilon \leq \lim_{r\uparrow 1}(\mu_r(f) + \varepsilon) \leq [\liminf \mu_r(A)] + \varepsilon \qquad (5.5.11)$$

and thus, taking ε to zero (this is Theorem 4.14.4(4) of Part 1),

$$\mu(A) \leq \liminf \mu_r(A) \qquad (5.5.12)$$

If $|\cdot|$ is Lebesgue measure, $\frac{d\theta}{2\pi}$, then

$$\mu_r(A) = \int_A \chi_A(e^{i\theta}) \operatorname{Re} F_r(e^{i\theta}) \frac{d\theta}{2\pi} \leq |A|^{1/q}\|\operatorname{Re} F_r\|_p \qquad (5.5.13)$$

by Hölder's inequality. Here q is the dual index to p.

Since $\operatorname{Re} F \in h^q$,

$$C = \sup_r\|\operatorname{Re} F_r\|_p < \infty \qquad (5.5.14)$$

so (5.5.13) and (5.5.12) imply that

$$\mu(A) \leq C|A|^{1/q} \qquad (5.5.15)$$

By outer regularity of μ and $|\cdot|$ (see Theorem 4.5.6 of Part 1), for any set B, we can find open $A_n \supset B$ so $\mu(A_n) \downarrow \mu(B)$, $|A_n| \downarrow |B|$. Thus, (5.5.15) implies

$$\mu(B) \leq C|B|^{1/q} \qquad (5.5.16)$$

In particular, $|B| = 0 \Rightarrow \mu(B) = 0$, that is, μ is a.e. wrt $\frac{d\theta}{2\pi}$.

(b) Given (a), (5.5.8) is just (5.5.4), which we proved in Theorem 5.5.1. Thus, we need to prove $\operatorname{Re} F^* \in L^p$.

For M fixed, define

$$h_n^{(M)}(e^{i\theta}) = \min\left(M, \operatorname{Re} F\left(1 - \frac{1}{n}\right)(e^{i\theta})\right) \qquad (5.5.17)$$

$$h^{(M)}(e^{i\theta}) = \min(M, F^*(e^{i\theta})) \qquad (5.5.18)$$

Since $h_n^{(M)}$ is bounded by M and converges pointwise to $h^{(M)}$, $\|h_n^{(M)} \to h^{(M)}\|_p \to 0$. Thus,

$$\|h^{(M)}\|_p \leq \sup_n\|h_n^{(M)}\|_p \leq \sup_n\|\operatorname{Re} F_{1-\frac{1}{n}}\|_p = \|\operatorname{Re} F\|_{h^p} \qquad (5.5.19)$$

Taking $M \to \infty$ using the monotone convergence theorem implies $F^* \in L^p$.

(c) This follows from the general p argument of the proof of (b) of the last theorem. $\qquad \square$

We want to extend this last result to h^1 and h^p. The following is useful:

Proposition 5.5.3. *Let $u \in h^1$ and define*

$$d\mu_r^\pm = u_\pm(re^{i\theta}) \frac{d\theta}{2\pi} \qquad (5.5.20)$$

where $u_\pm = \min(0, \pm u)$. Then there exist $r_n \to 1$ so $d\mu_{r_n}^\pm \to d\mu^\pm$ (weakly) and

$$u(re^{i\theta}) = \int P_r(\theta, \varphi)[d\mu^+(\varphi) - d\mu^-(\varphi)] \qquad (5.5.21)$$

Proof. Since $\|(u_\pm)_r\|_1 \le \|u\|_{h^1}$, the measures $d\mu_r^\pm$ are nonnegative Borel measures on the compact set $\partial\mathbb{D}$, with bounded norm, by the Banach–Alaoglu theorem, we can find the required subsequence. Since $u_{r\rho} = P_r * (d\mu_\rho^+ - d\mu_\rho^-)$, (5.5.21) follows from taking limits. $\quad\square$

Theorem 5.5.4. *Let $u \in h^1$ and let $u_\pm = \min(0, \pm u)$. Let μ be the measure guaranteed by the Herglotz theorem. Suppose that for some $p > 1$,*

$$\sup_{0 < r < 1} \|(u_+)_r\|_p < \infty \qquad (5.5.22)$$

Then (μ_{sing} is the singular part of μ)

$$d\mu_{\text{sing}} \le 0 \qquad (5.5.23)$$

Proof. As noted, the proof of Theorem 5.5.2(a) only uses the weak convergence and L^p-bound. Thus, letting $\mu_\pm$ be as in the last theorem, we have $(d\mu_+)_{\text{sing}} = 0$, so $(d\mu)_{\text{sing}} = -(d\mu_-)_{\text{sing}} \le 0$. $\quad\square$

As the penultimate result of this section:

Theorem 5.5.5. *Let $1 < p < \infty$ and $u \in h^p$. Then $u^* \in L^p$, $u_r = P_r * u^*$, $\|u_r\|_p \to \|u^*\|_p$, and $\|u_r - u^*\|_p \to 0$.*

Proof. By the last result and its application to $-u$, we have $d\mu_{\text{sing}} = 0$ so

$$d\mu(\theta) = u^*(e^{i\theta}) \frac{d\theta}{2\pi} \qquad (5.5.24)$$

Thus, $u_r = P_r * u^*$ by (5.4.10).

The same argument that proves Theorem 5.5.2(b) implies $u^* \in L^p$. Thus, $\sup_{0 < r < 1} |u(re^{i\theta})| = (M_r u)(e^{i\theta})$ lies in L^p (by (2.4.15)). The dominated convergence theorem then implies that $\|u_r - u^*\|_p \to 0$ and this implies $\|u_r\|_p \to \|u^*\|_p$. $\quad\square$

Example 5.5.6. This will construct $g \in L^\infty$, g nonvanishing on $\mathbb{D}$, so that $|g^*(e^{i\theta})| = 1$, but for any $z \in \partial\mathbb{D}$, there is $z_n \in \mathbb{D}$ so $z_n \to z$ and $g(z_n) \to 0$. Let w_n be a dense subset of $\partial\mathbb{D}$. Let

$$g(z) = \exp\left(-\sum_{n=1}^\infty n^{-2}\left[\frac{w_n + z}{w_n - z}\right]\right) \qquad (5.5.25)$$

By Theorem 5.5.1, $|g^*(e^{i\theta})| = 1$. On the other hand, if $w_n = e^{i\theta_n}$, it is easy
to see (Problem 1(a)) $\lim_{r\uparrow 1} g(re^{i\theta_n}) = 0$. From this and density of the w_n,
it is easy to find the required z_n (Problem 1(b)).   □

Notes and Historical Remarks.   The L^p-results of this section have
analogs for $\mathbb{R}$ which supplement Theorem 2.5.4.

Problems

1. Let g be given by (5.5.25).

   (a) Prove that if $w_n = e^{i\theta_n}$, then $\lim_{r\uparrow 1} g(re^{i\theta_n}) = 0$. (*Hint*: Show that
   $|g(re^{i\theta_n})| \le \exp(-n^{-2}[\frac{1+r}{1-r}])$.)

   (b) For any $z_0 \in \partial\mathbb{D}$, find z_j in $\mathbb{D}$, so $z_j \to z_0$ and $g(z_j) \to 0$. (*Hint*: Pick
   n_j so $w_{n_j} \to z_0$ and then r_j so that $|g(r_j w_{n_j})| \le j^{-2}$.)

5.6. Second Factorization (Inner and Outer Functions)

In Section 5.3, we factored $f \in H^p$ as gB, where B is a Blaschke product and
g is nonvanishing (and, by Riesz's theorem, in H^p). Here we'll find a further
natural factorization of g. We begin with a pair of illustrative examples.

Example 5.6.1. Let

$$h_\pm(z) = \exp\left(\mp\frac{1+z}{1-z}\right) \equiv \exp(\mp q(z)) \tag{5.6.1}$$

$q(z)$ has $\operatorname{Re} q > 0$, that is, it is a Carathéodory function. Indeed,

$$q(z) = \int \frac{e^{i\theta} + z}{e^{i\theta} - z} \, d\mu_q(z) \tag{5.6.2}$$

where $d\mu_q$ is a point mass at $\theta = 0$. It is the simplest example where the
associated measure is purely singular. As we've seen, $\operatorname{Re} h_\pm^*(e^{i\theta}) = 0$ for
a.e. θ. Thus,

$$|h_\pm^*(e^{i\theta})| = 1 \tag{5.6.3}$$

Since $\operatorname{Re} q > 0$, we have $h_+ \in H^\infty$. Indeed,

$$\|h_+\|_\infty = 1 \tag{5.6.4}$$

In a way, h_+ has a really strong zero of infinite order at $z = 1$ in the sense
that along the positive real axis,

$$|h_+(x)| \le C_n(1-x)^n \tag{5.6.5}$$

for all n. That, of course, means that $h_- = h_+^{-1}$ has a really bad infinity
at $z = 1$. In fact, as we'll see, h_- does not lie in any H^p, although since
$\operatorname{Re}(\frac{1-z}{1+z}) \in h^1$, h_- does lie in N. This is the first hint that N can have some
functions much more singular than those in H^p. Note that since $|h_-^*| = 1$
on $\partial\mathbb{D}$, in a spectacular way, the maximum principle fails for h_-!   □

Definition. An *inner function* is a function $f \in H^\infty$ so that for a.e. θ,

$$|f^*(e^{i\theta})| = 1 \tag{5.6.6}$$

If f is also nonvanishing on $\mathbb{D}$ and $f \not\equiv 1$, we say f is a *singular inner function*.

Here are the two main theorems of this section.

Theorem 5.6.2. *Every singular inner function has the form*

$$f(z) = \exp\left(-\int \frac{e^{i\theta} + z}{e^{i\theta} - z} \, d\mu_{\mathrm{s}}(\theta)\right) \tag{5.6.7}$$

where $d\mu_{\mathrm{s}}$ is a (positive) singular measure. Every f of the form (5.6.7) is a singular inner function. In particular, singular inner functions have $\|f\|_\infty = 1$.

Definition. An *outer function* is one of the form

$$f(z) = \exp\left(\int \frac{e^{i\theta} + z}{e^{i\theta} - z} \, h(e^{i\theta}) \, \frac{d\theta}{2\pi}\right) \tag{5.6.8}$$

for a real-valued function, $h \in L^1(\partial\mathbb{D}, \frac{d\theta}{2\pi})$. As usual, $|f^*(e^{i\theta})| = e^{h(e^{i\theta})}$, that is, $h = \log|f^*|$.

Theorem 5.6.3. *Every function $f \in H^p$ for some $p > 0$ has a factorization*

$$f(z) = e^{i\psi} B(z) S(z) O(z) \tag{5.6.9}$$

where B is a Blaschke product, S is a singular inner function (or $S \equiv 1$) with $S(0) > 0$, O is an outer function, and $\psi \in [0, 2\pi)$. This factorization is unique.

We will see that some functions in N might not have such a factorization. Given the work we've done in the last two sections, the proofs will be very easy. Recall (5.0.4), which we write as $\log_+ x \leq x^q/q$. Picking $q = p/k$ ($k = 1, 2, \dots$) and raising to the k-th power, we get

$$(\log_+ x)^k \leq \left(\frac{k}{p}\right)^k x^p \tag{5.6.10}$$

This implies:

Lemma 5.6.4. *Let $g \in H^p(\mathbb{D})$ be everywhere nonzero. Then*

$$\sup_{0<r<1} \int [\log_+(|g(re^{i\theta})|)]^k \, \frac{d\theta}{2\pi} < \infty$$

for all k.

The following is stronger than Theorem 5.6.2:

Theorem 5.6.5. *Let $f \in H^p(\mathbb{D})$ for some $p > 0$ with $f(z) \neq 0$ for all $z \in \mathbb{D}$, and so that $|f^*(e^{i\theta})| = 1$ for a.e. θ. Suppose that $f(0) > 0$. Then f has the form (5.6.7) for some singular positive measure $d\mu_s$. In particular, $f \in H^\infty$ with $\|f\|_\infty = 1$.*

Proof. Let $h(z) = \log(f(z))$ whose real part is in h^1 since $f \in H^p \subset N$. Thus, for a measure ν, by (5.4.10),

$$h(z) = \int \frac{e^{i\theta} + z}{e^{i\theta} - z} \, d\nu(\theta) \tag{5.6.11}$$

By (5.5.8),

$$d\nu_{\mathrm{ac}}(\theta) = \log|f^*(e^{i\theta})| \frac{d\theta}{2\pi} = 0 \tag{5.6.12}$$

by (5.6.4). By Theorem 5.5.4 and Lemma 5.6.4, $d\nu_{\mathrm{sing}} \leq 0$. Thus, (5.6.7) and $d\mu_{\mathrm{ac}} = 0$ hold and $d\mu_s = -d\nu = -d\nu_{\mathrm{sing}} \geq 0$. $\qquad\square$

Proof of Theorem 5.6.3. Write $f = Bg$, the Riesz factorization. Let $h = \log g$. By Theorem 5.4.2,

$$h(z) = i\psi + \int \frac{e^{i\theta} + z}{e^{i\theta} - z} \, d\nu(\theta) \tag{5.6.13}$$

By (5.5.8),

$$d\nu_{\mathrm{ac}} = \log(|f^*(e^{i\theta})|) \frac{d\theta}{2\pi} \tag{5.6.14}$$

and as in the last theorem, $d\nu_{\mathrm{sing}} \leq 0$. We conclude that

$$g = e^h = e^{i\psi} OS \tag{5.6.15}$$

where O comes from $d\nu_{\mathrm{ac}}$ and S from $d\nu_{\mathrm{sing}}$. Uniqueness is left to the reader (Problem 1). $\qquad\square$

We can try the same argument for $f \in N$. We can't conclude that $d\nu_{\mathrm{sing}} \leq 0$. We thus get that

$$f = e^{i\psi} BO\left(\frac{S_+}{S_-}\right) \tag{5.6.16}$$

where $S_\pm$ are both singular inner functions. Of course, this also follows from Theorem 5.6.3 and the F. and R. Nevanlinna theorem (if we note a quotient of outer functions is an outer function).

Notes and Historical Remarks. The inner-outer factorization is due to Smirnov [**762**] in 1929. It was rediscovered and popularized by Beurling [**66**] who coined the names inner and outer.

Problems

1. Prove uniqueness of the decomposition (5.6.7).

   Let $(H^1)_1 = \{f \in H^1 \mid \|f\|_1 \le 1\}$ be the unit ball in H^1. The next two problems will prove the de Leeuw–Rudin theorem which identifies the extreme points of $(H^1)_1$ as those f with $\|f\|_1 = 1$ and which are outer, i.e., $f = e^{i\psi}O(z)$, i.e., $B(z) = S(z) = 1$ in (5.6.9).

2. This problem will show if $f(z) = e^{i\psi}O(z)$, $\|f\|_1 = 1$, and $g \in H^1$ obeys $\|f \pm g\|_1 \le 1$, then $g = 0$.

   (a) Show that this fact implies that f is an extreme point of $(H^1)_1$.

   (b) Prove $\|f\|_1 = 1$ and $\|f \pm g\|_1 \le 1$ implies that

   $$\|f \pm g\|_1 = 1 \tag{5.6.17}$$

   (c) Let

   $$k(z) = g(z)/f(z) \tag{5.6.18}$$

   Prove that $\lim_{r \uparrow 1} k(re^{i\theta})$ exists and obeys

   $$\int |1 \pm k(e^{i\theta})| \, |f(e^{i\theta})| \frac{d\theta}{2\pi} = 1 \tag{5.6.19}$$

   (d) Prove for any $w \in \mathbb{C}$, we have $|1 + w| + |1 - w| \ge 2$ and

   $$|1 + w| + |1 - w| = 2 \Rightarrow -1 \le w \le 1 \tag{5.6.20}$$

   (e) Prove that $|f(e^{i\theta})| \ne 0$ for a.e. θ and then that

   $$|1 + k(e^{i\theta})| + |1 - k(e^{i\theta})| = 2 \tag{5.6.21}$$

   for a.e. θ. Conclude that for a.e. θ

   $$-1 < k(e^{i\theta}) < 1 \tag{5.6.22}$$

   (f) Prove that $|g(e^{i\theta})| \le |f(e^{i\theta})|$ and use this to conclude that $|O_g| \le |f|$ on $\mathbb{D}$ and then that $k \in H^\infty(\mathbb{D})$.

   (g) Conclude that k is a real constant in $[-1, 1]$ (*Hint*: Use (5.6.22).) and then that $k = 0$. (*Hint*: Use (5.6.17).)

3. This problem will prove that if $f \in H^1$ with $\|f\|_1 = 1$ and $f = O(z)I(z)$ with $O(z)$ outer, $I(z) = B(z)S(z)$ inner and $I(z) \not\equiv 1$, then there exists $g \in H^1$ with (5.6.17) and $g \ne 0$.

   (a) Show that this completes the proof of the *deLeeuw–Rudin theorem* that the extreme points of $(H^1)_1$ are exactly those $f \in H^1$ with $\|f\|_1 = 1$ and which are outer.

(b) Prove that there is an $\alpha \in [0, \pi]$, so that

$$f(\alpha) \equiv \int |f(e^{i\theta})| \operatorname{Re}(e^{i\alpha} I(e^{i\theta})) \frac{d\theta}{2\pi} = 0 \qquad (5.6.23)$$

(*Hint*: Show $f(\pi) = -f(0)$.)

(c) Define

$$u(z) = e^{i\alpha} I(z), \quad g(z) = \tfrac{1}{2} e^{-i\alpha} O(z)(1 + u(z)^2) \qquad (5.6.24)$$

Prove that $g \equiv 0 \Rightarrow I(z) = 1$.

(d) Prove that on $\partial \mathbb{D}$

$$g(e^{i\theta}) = f(e^{i\theta}) \operatorname{Re}(u(e^{i\theta})) \qquad (5.6.25)$$

(*Hint*: Prove that $1 + u(e^{i\theta})^2 = 2u(e^{i\theta}) \operatorname{Re} u(e^{i\theta})$ since $|u(e^{i\theta})|^2 = 1$.)

(e) Prove that $|f(e^{i\theta}) \pm g(e^{i\theta})| = |f(e^{i\theta})| (1 \pm \operatorname{Re}(u(e^{i\theta})))$ and conclude that $\|f \pm g\|_1 = 1$.

Remarks. 1. The de Leeuw–Rudin theorem is from their paper [**207**]. These problems follow their proof.

2. The fact that for $1 < p < \infty$, L^p is uniformly convex (see Section 5.3 of Part 1) implies that H^p is uniformly convex. It is easy to see that if X is uniformly convex, the extreme points of X_1 are all $x \in X$ with $\|x\|_1 = 1$.

3. It is an unpublished result of Arens, Buck, Carlson, and Hoffman, that the extreme points of H_∞ are exactly those f with $\|f\|_\infty \leq 1$ and $\int \log(1 - |f(e^{i\theta})|) \frac{d\theta}{2\pi} = -\infty$. This is proven in the paper of de Leeuw–Rudin [**207**] or in Simon [**748**, Theorem 2.7.9].

5.7. Conjugate Functions and M. Riesz's Theorem

We have already seen that conjugate harmonic functions play an important role. We study them further here, proving the celebrated result of M. Riesz below and then proving that the conjugate of a Hölder continuous function is Hölder continuous:

Theorem 5.7.1 (M. Riesz's Theorem). *There exist constants M_p for $p \in (1, \infty)$ so for any $1 < q < 2$, $\sup_{p \in [q, q']} M_p < \infty$ (where $q^{-1} + (q')^{-1} = 1$), and we have for all $u \in h^p$ that Ku, its conjugate function, lies in h^p and*

$$\|Ku\|_p \leq M_p \|u\|_p \qquad (5.7.1)$$

Example 5.7.2. This will show K is not bounded on h^∞. Let S be the strip $\{w \mid -\frac{\pi}{2} < \operatorname{Re} w < \frac{\pi}{2}\}$. By the Riemann mapping theorem, there is an analytic $f \colon \mathbb{D} \to S$ which is a bijection with $f(0) = 0$. This f clearly has $\|\operatorname{Re} f\|_\infty = \frac{\pi}{2}$ and $\|\operatorname{Im} f\|_\infty = \infty$. Of course, the map f can be written explicitly (Problem 1)

$$f(z) = i \log\left(\frac{1+z}{1-z}\right) \tag{5.7.2}$$

Note that since $\arg(1 \pm z)$ is discontinuous at ± 1, $\operatorname{Re} f$ is not continuous on $\partial\mathbb{D}$ in this example; see Example 5.7.19 for a continuous u with Ku not bounded. In Problem 2, the reader will show for $f \in L^\infty$, the boundary value, $\widetilde{f}$, of Ku (if u has boundary value f), can't grow worse than logarithmically so the behavior in this example is as bad as it can get. In Section 5.11, we will identify this set of $\widetilde{f}$'s (BMO) and prove a general exponential bound (John–Nirenberg inequality). $\qquad\square$

Example 5.7.3. Let $u \in h^1$. If $Ku \in h^1$, then $f \equiv u + iKu \in H^1$, and so the associated measure, μ, with $u = P * \mu$ has $d\mu(\theta) = f^*(e^{i\theta})\frac{d\theta}{2\pi}$. Thus, it is absolutely continuous. Thus, if $u = P * \mu$, where μ has a singular part, then $Ku \notin h^1$. An example, of course, is $u(z) = \operatorname{Re}(\frac{1+z}{1-z})$ (Example 5.3.6). It is interesting though that there are a.c. μ's so $K(P * \mu) \notin h^1$ (relevant to the Hilbert transform). In Problems 3–5, the reader will show that

$$f(z) = \sum_{n=2}^{\infty} z^n (\log|n|)^{-1} \tag{5.7.3}$$

has $\operatorname{Re} f = P * (\operatorname{Re} f^*) \in h^1$ and $\operatorname{Im} f \notin h^1$. $\qquad\square$

We'll approach Theorem 5.7.1 by showing for all $u \in h^1$ that Ku has a boundary value. We study the map $f \mapsto \widetilde{f}$ of the boundary value of u to that of Ku, called the *Hilbert transform*. Since we'll show for $p > 1$ that $Ku(re^{i\theta}) = (P_r * (\widetilde{u^*}))(e^{i\theta})$, it will suffice to prove $\sim$ is bounded from L^p to L^p. We already know $\sim$ is a contraction on L^2 (see Theorem 5.2.4). The new element will be a weak inequality ($|\cdot|$ is $\frac{d\theta}{2\pi}$ measure)

$$|\{\theta \mid \tilde{u}(e^{i\theta}) \geq \lambda\}| \leq \frac{4\|u\|_1}{\lambda} \tag{5.7.4}$$

Then Theorem 2.2.10 (i.e., (L^1, L^2)-interpolation) will prove that $f \mapsto \widetilde{f}$ is bounded from L^p to L^p for $1 < p \leq 2$. Duality (for $\sim$ is skew adjoint) will imply that $\sim$ is bounded from L^p to L^p for $2 \leq p < \infty$. The Notes and Problems will discuss some other approaches to the proof of M. Riesz's theorem.

We begin by defining a companion kernel to P_r and its relation to K. Recall the complex Poisson formula can be written for $f(z) = u(z) + iv(z)$ analytic in a neighborhood of $\bar{\mathbb{D}}$ by

$$f(re^{i\theta}) = (C_r * \operatorname{Re} f)(e^{i\theta}) \tag{5.7.5}$$

where

$$C_r(\theta, \varphi) = \frac{e^{i\varphi} + re^{i\theta}}{e^{i\varphi} - re^{i\theta}} = 1 + 2\sum_{n=1}^{\infty} r^n e^{in(\theta-\varphi)} \tag{5.7.6}$$

Of course, we have the familiar

$$P_r(\theta, \varphi) \equiv \operatorname{Re} C_r(\theta, \varphi) = \frac{1 - r^2}{1 + r^2 - 2r\cos(\theta - \varphi)} \tag{5.7.7}$$

$$= \sum_{n=-\infty}^{\infty} r^{|n|} e^{in(\theta-\varphi)} \tag{5.7.8}$$

Similarly, we define

$$Q_r(\theta, \varphi) \equiv \operatorname{Im} C_r(\theta, \varphi) = \frac{2r\sin(\theta - \varphi)}{1 + r^2 - 2r\cos(\theta - \varphi)} \tag{5.7.9}$$

$$= -\sum_{n=-\infty}^{\infty} i\operatorname{sgn}(n) r^{|n|} e^{in(\theta-\varphi)} \tag{5.7.10}$$

with $\operatorname{sgn}(n)$ given by (5.2.32).

Note that for $\theta \neq \varphi$, $\lim_{r\uparrow 1} P_r(\theta, \varphi) = 0$ so that for $\theta \neq \varphi$,

$$\lim_{r\uparrow 1} Q_r(\theta, \varphi) = -i \lim_{r\uparrow 1} C_r(\theta, \varphi) = \frac{\cos(\frac{1}{2}(\theta - \varphi))}{\sin(\frac{1}{2}(\theta - \varphi))}$$

$$= \cot(\tfrac{1}{2}(\theta - \varphi)) \equiv Q_1(\theta, \varphi) \tag{5.7.11}$$

We use Q_r, P_r, C_r; $0 \leq r < 1$; for the maps on $L^1(\partial\mathbb{D}, \frac{d\theta}{2\pi})$ with those integral kernels, for example,

$$(Q_r f)(e^{i\theta}) = \int Q_r(\theta, \varphi) f(e^{i\varphi}) \frac{d\varphi}{2\pi} \tag{5.7.12}$$

Below, unless noted otherwise, functions f on $\partial\mathbb{D}$ are real-valued. The following consequence of Kolmogorov's theorem will be useful:

Theorem 5.7.4. *For any $p > 1$, there is A_p so*

$$\sup_{0<r<1} \|Q_r f\|_1 \leq A_p \|f\|_p \tag{5.7.13}$$

In particular, if $f \in L^p$, $(P_r + iQ_r)f \in H^1$, and so has L^1 boundary values.

Remarks. 1. We'll eventually prove from M. Riesz's theorem that one can put $\|Q_r f\|_p$ on the left, so this is just a useful preliminary.

2. This is an interpolation result: $\{Q_r\}_{0<r<1}$ are bounded, uniformly in r, as maps of L^2 to L^2 and L^1 to $L^{1-\varepsilon}$. Complex interpolation relies on duality, and so is not well suited to L^p, $p < 1$, but real interpolation works. The proof is closely related to the one in Theorem 2.2.10.

Proof. Since $\|f\|_1 \leq \|f\|_2$ and, by (5.2.33),

$$\sup_r \|Q_r f\|_2 \leq \|f\|_2 \tag{5.7.14}$$

we need only consider p fixed with $1 < p < 2$. It clearly suffices to show for f with $\|f\|_p = 1$, a bound

$$\|Q_r f\|_1 \leq C \tag{5.7.15}$$

with C independent of f and r (but dependent on p). Since

$$\|g\|_1 = \int m_g(t)\, dt \tag{5.7.16}$$

(see (2.2.5)), with $|\cdot| = \frac{d\theta}{2\pi}$ measure and

$$m_g(t) = |\{\theta \mid |g(e^{i\theta})| > t\}| \tag{5.7.17}$$

it suffices to prove a bound

$$m_{Q_r f}(t) \leq C t^{-1-\delta} \tag{5.7.18}$$

for $C < \infty$, $\delta > 0$ only dependent on p (but not on r or f with $\|f\|_p = 1$). For small t, we can use $m_g(t) \leq 1$.

Let $f_{>a}$, $f_{\leq a}$ be given by (2.2.42)/(2.2.43). Since $Q_r f = Q_r f_{>a} + Q_r f_{\leq a}$ and $|z + w| > t \Rightarrow |z| > t/2$ or $|w| > t/2$, we have

$$m_{Q_r f}(t) \leq m_{Q_r f_{>t}}(t/2) + m_{Q_r f_{\leq t}}(t/2) \tag{5.7.19}$$

Since $1 < p < 2$ and $\|f\|_p = 1$, it is easy to see (Problem 6) that

$$\|f_{\leq t}\|_2^2 \leq t^{2-p}, \qquad \|f_{>t}\|_1 \leq t^{1-p} \tag{5.7.20}$$

By this and (5.7.14),

$$\|Q_r f_{\leq t}\|_2^2 \leq t^{2-p} \tag{5.7.21}$$

so

$$m_{Q_r f_{\leq t}}(t/2) \leq 4t^{-2}(t^{2-p}) = 4t^{-p} \tag{5.7.22}$$

Define $s \in (0,1)$ by

$$\varepsilon = \tfrac{1}{2}(p-1), \qquad s = 1 - \varepsilon \tag{5.7.23}$$

By Theorem 5.4.6, for a constant D (depending only on s, and so only on p) for all $r \in (0,1)$ and all g,

$$\|Q_r g\|_s \leq D\|g\|_1 \tag{5.7.24}$$

so

$$m_{Q_r f_{>t}}(t/2) \leq 2^s t^{-s} D^s \|f_{>t}\|_1^s \tag{5.7.25}$$

$$= 2^s D^s t^{-ps} \tag{5.7.26}$$

By (5.7.23),

$$ps = (1 + 2\varepsilon)(1 - \varepsilon) = 1 + \varepsilon(1 - 2\varepsilon) > 1 \tag{5.7.27}$$

since $p < 2 \Rightarrow \varepsilon < \frac{1}{2}$. Therefore, by (5.7.22), (5.7.26), and (5.7.27), we have (5.7.18) for $\delta > 0$.

Since $L^p \subset L^1$, it is immediate for (5.7.13) that $(P_r + iQ_r)f \in H^1$ if $f \in L^p$. That it has L^1 boundary values then follows from Theorem 5.3.5.   □

Given f in $L^1(\partial \mathbb{D}, \frac{d\theta}{2\pi})$, Cf defined by

$$(Cf)(re^{i\theta}) = ((P_r + iQ_r)f)(e^{i\theta}) \tag{5.7.28}$$

is analytic and in H^p, all $p < 1$, by Theorem 5.4.6. Thus, it has pointwise boundary values. We therefore define

Definition. If $f \in L^1(\partial \mathbb{D}, \frac{d\theta}{2\pi})$ is real-valued, we define its *Hilbert transform*, $\tilde{f}$, by

$$\tilde{f}(e^{i\theta}) = \lim_{r \uparrow 1}(Q_r f)(e^{i\theta}) \tag{5.7.29}$$

As remarked earlier, some authors called this the conjugate of f—since it is close to, but distinct from, the Hilbert transform on $\mathbb{R}$ (see the Notes), it might be clearer to use "circular Hilbert transform." This is one of three definitions of $\tilde{f}$ and one goal below will be to prove that these other definitions agree (at least when $\tilde{f} \in L^1$). The other definitions are:

(1) $\tilde{f}$ is the unique function with

$$(\tilde{f})_n^\sharp = -i\operatorname{sgn}(n)f_n^\sharp \tag{5.7.30}$$

(this requires $\tilde{f} \in L^1$ for $(\tilde{f})^\sharp$ to be defined).

(2) Q_1 given by (5.7.11) is not in L^1, so we can't define $Q_1 * f$ as an absolutely convergent integral, but we can as a principal part. $\tilde{f}$ obeys

$$\tilde{f}(e^{i\theta}) = \lim_{\varepsilon \downarrow 0} \int_{|\varphi - \theta| > \varepsilon} Q_1(\theta, \varphi)f(e^{i\varphi})\frac{d\varphi}{2\pi} \tag{5.7.31}$$

for a.e. θ.

We begin with (5.7.30) by proving a result of considerable interest:

Theorem 5.7.5. *If $f \in L^1$ and $\tilde{f} \in L^1$, then*

(a) $(P_r + iQ_r)f \in H^1$. $\hfill$ (5.7.32)

(b) $Q_r f = P_r \tilde{f}$. $\hfill$ (5.7.33)

(c) $\|Q_r f - \tilde{f}\|_1 \to 0$. $\hfill$ (5.7.34)

In particular, if $f \in L^p$ for some $p > 1$, (a)–(c) hold.

Proof. By Corollary 5.4.3 and Theorem 5.4.6,

$$F(re^{i\theta}) = ((P_r + iQ_r)f)(e^{i\theta}) \tag{5.7.35}$$

lies in H^q for all $q < 1$. Moreover,

$$F^*(e^{i\theta}) = f(e^{i\theta}) + i\tilde{f}(e^{i\theta}) \tag{5.7.36}$$

by definition of $\tilde{f}$ as pointwise boundary values. If $\tilde{f} \in L^1$, then $F^* \in L^1$, so $F \in H^1$ by Theorem 5.3.7.

Since $F \in H^1$, we have

$$F(re^{i\theta}) = (P_r * F^*)(e^{i\theta}) \tag{5.7.37}$$

proving (5.7.33). Since $\|F_r - F^*\|_1 \to 0$, we have (5.7.34).

The final "in particular" statement follows from the "in particular" clause of Theorem 5.7.4.   □

Corollary 5.7.6. *If $\tilde{f} \in L^1$ (in particular, if $f \in L^p$ for some $p > 1$), we have that (5.7.30) holds.*

Proof. By (5.7.10),

$$(Q_r f)^\sharp_n = -ir^{|n|}\operatorname{sgn}(n) f^\sharp_n \tag{5.7.38}$$

By (5.7.34),

$$(\tilde{f})^\sharp_n = \lim_{r\uparrow 1}(Q_r f)^\sharp_n \tag{5.7.39}$$

□

Theorem 5.7.7 (Conjugate Function Duality). *Let $1 < p < \infty$. Let q be the dual index. (Below, all norms are $L^p(\partial\mathbb{D}, \frac{d\theta}{2\pi})$.)*

(a) *If there is a constant A_p with*

$$\sup_{0<r<1} \|Q_r f\|_p \leq A_p\|f\|_p \tag{5.7.40}$$

for all $f \in L^p$, then

$$\sup_{0<r<1} \|Q_r f\|_q \leq A_p\|f\|_q \tag{5.7.41}$$

(b) *If $\tilde{f} \in L^p$ for all $f \in L^p$ with*

$$\|\tilde{f}\|_p \leq B_p\|f\|_p \tag{5.7.42}$$

then for all $f \in L^q$, $\tilde{f} \in L^q$ and

$$\|\tilde{f}\|_q \leq B_p\|f\|_p \tag{5.7.43}$$

Remarks. 1. We'll eventually see that (5.7.40)/(5.7.42) hold (with ($B_p = A_p$) for $1 < p < \infty$, but our proof will use this duality result.

2. The same proof for (a) works if $p = 1$. For $p = \infty$, there is a subtlety which could be overcome, but since (5.7.40) fails for $p = \infty$ and $p = 1$, we don't try!

Proof. (a) $Q_r f$ is real analytic in $e^{i\theta}$ for any $f \in L^1$, so Q_r is a bounded map of L^s to L^s for all $s \in [1, \infty)$. Since Q_r has a real antisymmetric kernel, we have

$$\int f(e^{i\theta})(Q_r g)(e^{i\theta}) \frac{d\theta}{2\pi} = -\int (Q_r f)(e^{i\theta}) g(e^{i\theta}) \frac{d\theta}{2\pi} \qquad (5.7.44)$$

It follows that $\|Q_r\|_{\mathcal{L}(L^p)} = \|Q_r\|_{\mathcal{L}(L^q)}$.

(b) Since $p < \infty$, $q > 1$, so if $f \in L^q$, then $f + i\tilde{f} \in H^1$ and $Q_r f \to \tilde{f}$ in L^1. Taking g a Laurent polynomial, we see for $f \in L^q$ and g such a polynomial, we have

$$\int f(e^{i\theta})\tilde{g}(e^{i\theta}) \frac{d\theta}{2\pi} = -\int \tilde{f}(e^{i\theta}) g(e^{i\theta}) \frac{d\theta}{2\pi} \qquad (5.7.45)$$

since $Q_r g \to \tilde{g}$ in $\| \cdot \|_\infty$ in that case.

Thus, for such g,

$$\left| \int \tilde{f}(e^{i\theta}) g(e^{i\theta}) \frac{d\theta}{2\pi} \right| \le \|\tilde{g}\|_p \|f\|_q \le B_p \|g\|_p \|f\|_q \qquad (5.7.46)$$

Since Laurent polynomials are dense in L^p, we have $\tilde{f} \in L^q$ and (5.7.43) holds. $\square$

To head towards (5.7.31), we give the integral on the right a symbol: for $f \in L^1$ and $0 < r < 1$, we define

$$(H_r f)(e^{i\theta}) = \int_{|\varphi - \theta| > 1-r} \cot(\tfrac{1}{2}(\theta - \varphi)) f(e^{i\varphi}) \frac{d\varphi}{2\pi} \qquad (5.7.47)$$

Theorem 5.7.8. *For all $r \in (0, 1)$ and $f \in L^1$, we have*

$$|((Q_r - H_r)f)(e^{i\theta})| \le \left(\pi + \frac{2}{\pi}\right)(M_{\mathrm{HL}}f)(e^{i\theta}) \qquad (5.7.48)$$

where $M_{\mathrm{HL}}f$ is the Hardy–Littlewood maximal function.

We need two lemmas:

Lemma 5.7.9. *For all $\theta, \varphi \in [0, 2\pi]$ and $r \in (0, 1)$, if $|\theta - \varphi| > 1 - r$, then*

$$|Q_1(\theta, \varphi) - Q_r(\theta, \varphi)| < \pi P_r(\theta, \varphi) \qquad (5.7.49)$$

Proof. Without loss, we can suppose $\varphi = 0$, in which case we write $Q_r(\theta, 0) \equiv Q_r(\theta)$. We have, by simple algebra,

$$Q_1(\theta) - Q_r(\theta) = \frac{(1-r)^2 \sin\theta}{(1-\cos\theta)(1+r^2 - 2r\cos\theta)} = \frac{1-r}{1+r} Q_1(\theta) P_r(\theta) \quad (5.7.50)$$

Let $\theta_r = (1 - r)$. Then if $|\theta| \ge \theta_r$, we have

$$|Q_1(\theta)| \le \frac{1}{|\sin(\theta/2)|} \le \frac{1}{\sin(\theta_r/2)} \le \frac{\pi}{\theta_r} = \frac{\pi}{1-r} \qquad (5.7.51)$$

since for $\eta \in [0, \frac{\pi}{2}]$, we have $\sin \eta \geq \frac{2\eta}{\pi}$ (see the proof of Lemma 3.5.13 of Part 1). Since $1 + r > 1$, we get (5.7.49) from (5.7.50) and (5.7.41). $\quad\square$

Lemma 5.7.10. *For all $\theta, \varphi \in [0, 2\pi]$ and $r \in (0, 1)$,*

$$|Q_r(\theta, \varphi)| < \frac{2}{1 - r} \tag{5.7.52}$$

Proof. By (5.7.6),

$$|C_r(\theta)| < \frac{2}{|1 - re^{i\theta}|} \tag{5.7.53}$$

On the circle $\partial \mathbb{D}_r$, the point closest to 1 is r, that is,

$$|1 - re^{i\theta}| \geq (1 - r) \tag{5.7.54}$$

Since $|Q_r| < |C_r|$, we get (5.7.52). $\quad\square$

Proof of Theorem 5.7.8. Since

$$|(Q_r - H_r)f(e^{i\theta})| \leq \int_{|\theta-\varphi|>1-r} |Q_1(\theta, \varphi) - Q_r(\theta, \varphi)||f(e^{i\varphi})| \frac{d\varphi}{2\pi}$$
$$\int_{|\theta-\varphi|<1-r} |Q_r(\theta, \varphi)f(e^{i\varphi})| \frac{d\varphi}{2\pi} \tag{5.7.55}$$

the lemmas imply

$$\text{LHS of } (5.7.48) \leq \pi \int P_r(\theta, \varphi)|f(e^{i\varphi})| \frac{d\varphi}{2\pi}$$
$$+ \frac{2}{2\pi} \frac{2}{2(1-r)} \int_{|\theta-\varphi|<1-r} |f(e^{i\varphi})| \frac{d\varphi}{2\pi} \tag{5.7.56}$$

The first integral is bounded by $(M_{\text{HL}}f)(e^{i\theta})$ by Theorem 2.4.3 and the second integral divided by $2(1 - r)$ is bounded by M_{HL} by its definition as a sup. $\quad\square$

Proposition 5.7.11. *Let f be a C^1-function on $\partial \mathbb{D}$. Then as $r \uparrow 1$, $(Q_r f)(e^{i\theta})$ and $(H_r f)(e^{i\theta})$ converge uniformly to the same limit (which is, of course, $\tilde{f}(e^{i\theta})$).*

Proof. Since Q_r and Q_1 are odd functions of $\theta - \varphi$, their integrals are zero, so in the formulae for Q_r and H_r, we can replace $f(e^{i\varphi})$ by $f(e^{i\varphi}) - f(e^{i\theta})$ and use

$$|f(e^{i\theta}) - f(e^{i\varphi})| \leq |\theta - \varphi| \, \|f'\|_\infty \tag{5.7.57}$$

Since $|\theta - \varphi| Q_1(\theta, \varphi)$ is integrable, we get $(H_r f)(e^{i\varphi})$ converges uniformly as $r \uparrow 1$.

Using the same replacement of $f(e^{i\varphi})$ by $f(e^{i\varphi}) - f(e^{i\theta})$ and (5.7.57), we can replace (5.7.56) by

$$|(Q_r - H_r)f(e^{i\theta})| \leq \pi\|f'\|_\infty \int |\varphi| P_r(\varphi, 0) \frac{d\varphi}{2\pi} + \frac{\|f'\|_\infty}{\pi(1-r)} \int_{|\varphi|<1-r} |\varphi|\, d\varphi \tag{5.7.58}$$

which goes to zero as $r \uparrow 1$, that is, $\|(H_r - Q_r)f\|_\infty \to 0$ as $r \uparrow 1$. Thus, since H_r has a limit, so does Q_r and it is the same limit.

Since $(Q_r f)(e^{i\theta}) \to \tilde{f}(e^{i\theta})$, this common limit is $\tilde{f}(e^{i\theta})$. □

Theorem 5.7.12. *For any $f \in L^1(\partial\mathbb{D}, \frac{d\theta}{2\pi})$ and a.e. θ,*

$$\lim_{r\uparrow 1}(H_r f)(e^{i\theta}) = \tilde{f}(e^{i\theta}) \tag{5.7.59}$$

Proof. By the fact that C^1 is dense and (5.7.48) which says

$$\sup_{0<r<1} |(Q_r - H_r)f(e^{i\theta})| \leq \left(\pi + \frac{2}{\pi}\right)(M_{\text{HL}}f)(e^{i\theta}) \tag{5.7.60}$$

we get (by the usual maximal function argument) that for a.e. θ, $(Q_r - H_r)f(e^{i\theta}) \to 0$. Since $(Q_r f)(e^{i\theta}) \to \tilde{f}(e^{i\theta})$ for a.e. θ (by definition of $\tilde{f}$), we have (5.7.59). □

Remark. Below (see Theorem 5.7.18), we'll also prove L^p-convergence if $f \in L^p$ for $p > 1$.

We are now prepared to begin the steps that will lead to a proof of Theorem 5.7.1. The key will be a weak type inequality. For $f \in L^1(\partial\mathbb{D}, \frac{d\theta}{2\pi})$, we'll prove (with $|\cdot| = \frac{d\theta}{2\pi}$ measure)

$$|\{\theta \mid |\tilde{f}(e^{i\theta})| > t\}| \leq \frac{4\|f\|_1}{t} \tag{5.7.61}$$

We first claim it suffices to prove that for each r,

$$|\{\theta \mid |(Q_r f)(e^{i\theta})| > t\}| \leq \frac{4\|f\|_1}{t} \tag{5.7.62}$$

Lemma 5.7.13. *For any $r_n \uparrow 1$, we have*

$$|\{\theta \mid |\tilde{f}(e^{i\theta})| > t\}| \leq \liminf_n |\{\theta \mid |(Q_{r_n} f)(e^{i\theta})| > t\}| \tag{5.7.63}$$

In particular, (5.7.62) implies (5.7.61).

Proof. This is immediate from the pointwise a.e. convergence of $Q_r f$ to $\tilde{f}$ and Theorems 2.2.11 and 2.2.13. □

The proof of (5.7.62) will use a cleverly chosen harmonic function on $\mathbb{H}_+ = \{x + iy \mid x > 0\}$,

$$H(x,y) = \frac{x^2 + y^2 + x}{(x+1)^2 + y^2} = \left. \operatorname{Re}\left(\frac{z}{z+1}\right)\right|_{z=x+iy} \qquad (5.7.64)$$

Lemma 5.7.14. *Let H be given by (5.7.64) on $\mathbb{H}_+$. Then*

(a) H *is harmonic.*

(b) $H > 0$ *on* $\mathbb{H}_+$.

(c) $H(x,0) \leq x$. $\qquad (5.7.65)$

(d) H *maps* $\{x + iy \in \mathbb{H}_+ \mid |y| \geq 1\}$ *to* $[\frac{1}{2}, \infty)$.

(e) $H \leq 1$. $\qquad (5.7.66)$

Proof. (a) is immediate from (5.7.64) and analyticity of

$$F(z) = \frac{z}{z+1} \qquad (5.7.67)$$

on $\mathbb{H}_+$.

(b) is obvious.

(c) $F'(z) = 1/(1+z)^2$ so on $\mathbb{H}_+$, $|F'(z)| \leq 1$, so $\frac{\partial}{\partial x}H(x,0) \leq 1$. Since $H(0,0) = 0$, we see (5.7.65) holds.

(d) F maps -1 to ∞ and 1 to $\frac{1}{2}$ and has $\overline{F(z)} = F(\bar{z})$. Thus, the fractional linear transformation, F, maps the circle $\{z \mid |z| = 1\}$ to a circle or line containing ∞, so to a line (see Theorem 7.3.8 of Part 2A). The line has to contain $w = \frac{1}{2}$ and be invariant under conjugation. It follows that f maps $\{z \mid |z| = 1\}$ to $\{w \mid \operatorname{Re} w = \frac{1}{2}\}$. Since $F(0) = 0$ is to the left of the line, F maps $\{z \mid |z| \geq 1\}$ to $\{w \mid \operatorname{Re} w \geq \frac{1}{2}\}$, that is,

$$x^2 + y^2 \geq 1 \Rightarrow H(x,y) \geq \tfrac{1}{2} \qquad (5.7.68)$$

which implies (d).

(e) Obvious from (5.7.64) since $(x+1)^2 + y^2 = (x^2 + y^2 + x) + (x+1) > (x^2 + y^2 + x)$ on $\mathbb{H}_+$. $\qquad \square$

Theorem 5.7.15. *Let $f \in L^1(\partial \mathbb{D}, \frac{d\theta}{2\pi})$ be real-valued.* (5.7.62) *and so* (5.7.61) *holds.*

Proof. Suppose first that $f > 0$. Define G on $\mathbb{D}$ by

$$G(re^{i\theta}) = H((P_r f)(e^{i\theta}), \quad (Q_r f)(e^{i\theta})) = \operatorname{Re} F((P_r + iQ_r)f(e^{i\theta})) \quad (5.7.69)$$

is the real part of an analytic function, so harmonic. Note $f > 0 \Rightarrow P_r f > 0 \Rightarrow P_r f + iQ_r f \in \mathbb{H}_+$.

Since G is harmonic,

$$\int G(re^{i\theta}) \frac{d\theta}{2\pi} = G(0) = H(\|f\|_1, 0) \leq \|f\|_1 \qquad (5.7.70)$$

where we used $(P_{r=0}f) = \int f(e^{i\theta}) \frac{d\theta}{2\pi} = \|f\|_1$ (since $f \geq 0$) and (5.7.65).

Since $H > 0$, $G > 0$, so

$$\int G(re^{i\theta}) \frac{d\theta}{2\pi} \geq \int_{\{\theta \mid |(Q_r f)(e^{i\theta})| \geq 1\}} G(re^{i\theta}) \frac{d\theta}{2\pi} \geq \tfrac{1}{2} |\{\theta \mid |Q_r f(e^{i\theta})| > 1\}| \qquad (5.7.71)$$

where we used (d) of the last lemma to see $|(Q_r f)(e^{i\theta})| \geq 1 \Rightarrow G(re^{i\theta}) \geq \tfrac{1}{2}$. (5.7.22) and (5.7.21) imply

$$f \geq 0 \Rightarrow |\{\theta \mid |(Q_r f)(e^{i\theta})| > 1\}| \leq 2\|f\|_1 \qquad (5.7.72)$$

Any f can be written $f = f_+ - f_-$ with $f_{\pm} = \max(\pm f, 0)$ of disjoint support, so

$$\|f\|_1 = \|f_+\|_1 + \|f_-\|_1 \qquad (5.7.73)$$

Moreover, if $|Q_r f_+(e^{i\theta})| \leq 1$ and $|Q_r f_-(e^{i\theta})| \leq 1$, then $|Q_r f(e^{i\theta})| \leq 2$, so (5.7.72) implies

$$f \text{ real-valued} \Rightarrow |\{\theta \mid |(Q_r f)(e^{i\theta})| > 2\}| \leq 2\|f_+\|_1 + 2\|f_-\|_1 = 2\|f\| \qquad (5.7.74)$$

Replacing f by $\frac{2}{\lambda} f$ yields (5.7.62). $\qquad \square$

The following includes Theorem 5.7.1:

Theorem 5.7.16. $f \mapsto \tilde{f}$ maps $L^p(\partial\mathbb{D}, \frac{d\theta}{2\pi})$ to itself for any $p \in (1, \infty)$ with

$$\|\tilde{f}\|_p \leq M_p \|f\| \qquad (5.7.75)$$

with M_p uniformly bounded for $p \in [s, s']$ for any $s \in (1, 2)$. Moreover, (5.7.1) holds.

Proof. The last theorem says $f \mapsto \tilde{f}$ is bounded from L^1 to weak-L^1, while (5.7.75) implies

$$\|\tilde{f}\|_2 = \|K(P_r f)\|_{h^2} \leq \|P_r f\|_{h_2} = \|f\|_2 \qquad (5.7.76)$$

so $f \mapsto \tilde{f}$ is bounded from L^2 to itself (this also follows from (5.7.30)). By interpolation (Theorem 2.2.10), we get boundedness on L^p for $p \in (1, 2]$ with constants bounded on $(s, 2)$. This plus duality (Theorem 5.7.7) implies the full result (5.7.75).

Now let $u \in h^p$ and $f = u^*$. By Theorem 5.5.5, $u^* \in L^p$ and $u_r = P_r * f$, so

$$\|f\|_p = \|u\|_{h^p} \qquad (5.7.77)$$

By (5.7.33) and Theorem 5.5.5,

$$\sup_r \|Q_r f\|_p = \|\widetilde{f}\|_p \tag{5.7.78}$$

Since $Ku(re^{i\theta}) = (Q_r f)(e^{i\theta})$, (5.7.78) implies

$$\|Ku\|_{h^p} = \|\widetilde{f}\|_p \tag{5.7.79}$$

Thus, (5.7.1) is equivalent to (5.7.75). □

We can now control various maximal functions. For $f \in L^1$, define

$$(M_{\mathrm{Conj}}f)(e^{i\theta}) = \sup_r |(Q_r f)(e^{i\theta})| \tag{5.7.80}$$

$$(M_{\mathrm{HT}}f)(e^{i\theta}) = \sup_r |(H_r f)(e^{i\theta})| \tag{5.7.81}$$

$$(M_\Delta f)(e^{i\theta}) = \sup_r |(H_r f)(e^{i\theta}) - (Q_r f)(e^{i\theta})| \tag{5.7.82}$$

where, of course,

$$M_{\mathrm{HT}}f(e^{i\theta}) \le (M_{\mathrm{Conj}}f)(e^{i\theta}) + (M_\Delta f)(e^{i\theta}) \tag{5.7.83}$$

Theorem 5.7.17. *For any $f \in L^p$, $p > 1$,*

$$(M_{\mathrm{Conj}}f)(e^{i\theta}) \le (M_{\mathrm{HL}}\widetilde{f})(e^{i\theta}) \tag{5.7.84}$$

$$(M_{\mathrm{HT}}f)(e^{i\theta}) \le (M_{\mathrm{HL}}\widetilde{f})(e^{i\theta}) + \left(\pi + \frac{2}{\pi}\right)(M_{\mathrm{HL}}f)(e^{i\theta}) \tag{5.7.85}$$

In particular $M_{\mathrm{Conj}}f$ and $M_{\mathrm{HT}}f$ lie in L^p.

Proof. (5.7.84) follows from (5.7.33) and $\sup_r P_r g \le M_{\mathrm{HL}}g$. (5.7.85) then follows from (5.7.33) and (5.7.36). □

There is a weak-L^1 bound on $M_{\mathrm{HT}}f$ for $f \in L^1$ proven by different means; see Problem 6 of Section 5.9.

Theorem 5.7.18. *Let $p \in (1, \infty)$ and $f \in L^p$. Then*

(a) $\|Q_r f - \widetilde{f}\|_p \to 0$ *as $r \uparrow 1$.* $\tag{5.7.86}$

(b) $\|H_r f - \widetilde{f}\|_p \to 0$ *as $r \uparrow 1$.* $\tag{5.7.87}$

Proof. (a) follows from (5.7.33) and (b) from (5.7.85) and the fact that (5.7.87) holds in $f \in C^1$. □

Finally, we want to see what $f \mapsto \widetilde{f}$ (and $u \mapsto Ku$) does to (Hölder) continuity. We begin with two examples that delimit what one can hope for.

Example 5.7.19 (f continuous, but not $\widetilde{f}$). Let f be defined on $\partial\mathbb{D}$ but

$$f(e^{i\theta}) = \begin{cases} [\log(\theta^{-1})]^{-1}, & 0 < \theta < \frac{1}{2} \\ 2(1-\theta)(\log 2)^{-1}, & \frac{1}{2} \le \theta \le 1 \\ 0, & 1 \le \theta \le 2\pi \end{cases} \qquad (5.7.88)$$

Then f is piecewise C^1 on $(0, 2\pi)$ and since $\lim_{\theta \downarrow 0} \log(\theta^{-1}) = \infty$, f is continuous on $\partial\mathbb{D}$. Since $\int_0^{\frac{1}{2}} |\theta^{-1}| f(e^{i\theta}) \frac{d\theta}{2\pi} = \infty$ with a positive integrand, it is easy to see that $\widetilde{f}$ is continuous on $(0, 2\pi)$ but $\lim_{\theta \downarrow 0} \widetilde{f}(e^{i\theta}) = \infty$, so $\widetilde{f}$ is not continuous. If u is a harmonic function with boundary value f, then $u \in C(\partial\mathbb{D})$ but Ku is unbounded showing again that K is not bounded on h^∞. $\qquad\square$

Example 5.7.20 (f Lipschitz continuous, but not $\widetilde{f}$). Let

$$F(z) = (z - 1) \log(1 - z) \qquad (5.7.89)$$

for $z \in \mathbb{D}$ (the branch of log with $\log(1 - z) \mid_{z=0}= 0$). Let $f(e^{i\theta}) = \lim_{r \uparrow 1} \operatorname{Re} F(re^{i\theta})$. Then (Problem 9) it is easy to see f obeys $|f(\theta) - f(\psi)| \le C|\theta - \psi|$ but $|\widetilde{f}(e^{i\theta}) - \widetilde{f}(1)|/|\theta| \to \infty$ as $\theta \downarrow 0$. Thus, f is Lipschitz (aka Hölder continuous of order 1) but $\widetilde{f}$ is not. $\qquad\square$

These examples are at the borderline of Hölder continuous functions of order, α, (i.e., $|f(e^{i\theta}) - f(e^{i\psi})| \le C|\theta - \psi|^\alpha$), $0 \le \alpha \le 1$. Away from the borders, the Hilbert transform is well-behaved!

Theorem 5.7.21 (Plemelj–Privalov Theorem). *Let $0 < \alpha < 1$. If f, defined on $\partial\mathbb{D}$, is uniformly Hölder continuous of order α in $(0, 1)$, so is $\widetilde{f}$.*

Remarks. 1. If

$$\|f\|_{\Lambda_\alpha} = \sup_{\theta, \psi, \theta \ne \psi} \frac{|f(e^{i\theta})| - f(e^{i\psi})}{|\theta - \psi|^\alpha} + \|f_\infty\| \qquad (5.7.90)$$

then the proof shows that

$$\|\widetilde{f}\|_{\Lambda_\alpha} \le C_\alpha \|f\|_{\Lambda_\alpha} \qquad (5.7.91)$$

2. From this it is easy to see (Problem 10) if $u \in h^\infty$ is (uniformly) Hölder continuous of order α, so is Ku.

We'll prove this as a special case of a more general case that is stated in terms of the *modulus of continuity* of a function f on $\partial\mathbb{D}$:

$$\omega(\delta, f) = \sup_{|\theta - \varphi| \le \delta} |f(e^{i\theta}) - f(e^{i\psi})| \qquad (5.7.92)$$

Below we'll prove

Theorem 5.7.22. *If*

$$\int_0^{\pi/4} \frac{\omega(\delta, f)}{\delta} \, d\delta < \infty \tag{5.7.93}$$

then for $0 < \delta < \frac{\pi}{4}$, we have that

$$\omega(\delta, \tilde{f}) \leq C \left[\int_0^\delta \frac{\omega(y, f)}{y} \, dy + \int_0^{\pi/4} \omega(y, f) \frac{\delta}{y(y + \delta)} \, dy \right] \tag{5.7.94}$$

Remarks. 1. Since $\frac{\delta}{y+\delta} < 1$, the integral converges and since its limit as $\delta \downarrow 0$ is 0, this bound shows that $\omega(\delta, \tilde{f}) \to 0$ so $\tilde{f}$ is continuous.

2. (5.7.93) is a Dini condition (see Theorem 3.5.4 of Part 1). In Problem 11, the reader will show that if $\int_0^{\pi/4} \frac{\omega(\delta,f)}{\delta \log(\delta^{-1})} < \infty$, then $\tilde{f}$ obeys a Dini condition.

Proof of Theorem 5.7.21 given Theorem 5.7.22. Suppose that f obeys

$$\omega(\delta, f) \leq c_1 \delta^\alpha \tag{5.7.95}$$

Then, since $\alpha > 0$

$$\int_0^\delta \frac{\omega(y, f)}{y} \, dy \leq c_1 \int_0^\delta \frac{y^\alpha}{y} \, dy = c_1 \delta^\alpha / \alpha \tag{5.7.96}$$

Putting $y = \omega \delta$ in the integral

$$\int_0^{\pi/4} \frac{\omega(y, f) \, \delta}{y(y + \delta)} \, dy \leq c_1 \delta^\alpha \int_0^{\pi/4\delta} \frac{\omega^\alpha}{\omega(\omega + 1)} \, d\omega$$

$$\leq c_1 \delta^\alpha \int_0^\infty \frac{\omega^\alpha}{\omega(\omega + 1)} \, d\omega \tag{5.7.97}$$

Since $\alpha < 1$, the integrand, call it I_α, is finite. Thus, by (5.7.94), if $\delta < \pi/4$,

$$\omega(\delta, \tilde{f}) \leq C \, c_1 (\alpha^{-1} + I_\alpha) \, \delta^\alpha \tag{5.7.98}$$

Because $\omega(2\delta, g) \leq 2\omega(\delta, g)$ (by the triangle inequality), we conclude $\tilde{f}$ is Hölder continuous of order α. $\square$

To prove Theorem 5.7.22, we need one more lemma bounding Q_1:

Lemma 5.7.23. *For $0 < \theta < \frac{\pi}{4}$ and $0 < \theta < \varphi < \pi + \theta$, we have that for some $c_1 < \infty$*

$$|Q_1(\varphi - \theta) - Q_1(\varphi)| \leq c_1 \frac{|\theta|}{|\varphi - \theta| \, |\varphi|} \tag{5.7.99}$$

Proof. Since $Q_1'(\eta) = -\frac{1}{2}\left[\sin(\frac{1}{2}\eta)\right]^{-2}$ and for $0 < \eta < \frac{\pi}{2}$ we have

$$\sin(\eta) \geq \frac{2\eta}{\pi} \tag{5.7.100}$$

and, for $0 < \eta < \frac{5\pi}{8}$, $\sin(\eta) \geq c_2\eta$, we see that for such $0 < \psi < \pi + \frac{\pi}{4}$

$$|Q'(\psi)| \leq \frac{c_2^2}{2\psi^2} \tag{5.7.101}$$

Thus,

$$|Q_1(\varphi - \theta) - Q_1(\varphi)| \leq \int_0^1 |[\frac{d}{dt}Q_1(\varphi - t\theta)]| \, dt$$

$$\leq c_1|\theta| \int_0^1 \frac{dt}{|\varphi - t\theta|^2}$$

$$= c_1 \left[\frac{1}{|\varphi - \theta|} - \frac{1}{|\varphi|}\right] = \text{RHS of (5.7.99)}$$

$\square$

Proof of Theorem 5.7.22. By rotational invariance, we need only bound $|\widetilde{f}(e^{i\theta}) - \widetilde{f}(1)|$ for $0 < \theta < \frac{\pi}{4}$. By (5.7.59) and $\int_{1-r}^{2\pi-(1-r)} Q_1(\theta)\, d\theta = 0$, we have that for Hölder continuous f that

$$\widetilde{f}(e^{i\theta}) = \int Q_1(\theta - \varphi)[f(e^{i\theta}) - f(e^{i\varphi})]\frac{d\varphi}{2\pi} \tag{5.7.102}$$

Thus, if we define

$$I(\varphi, \theta) = Q_1(\theta - \varphi)[f(e^{i\varphi}) - f(e^{i\theta})] - Q_1(-\varphi)[f(e^{i\theta}) - 1] \tag{5.7.103}$$

we have that

$$\widetilde{f}(e^{i\theta}) - \widetilde{f}(1) = J_1 + J_2 + J_3 \tag{5.7.104}$$

$$J_1 = \int_0^\theta I(\varphi, \theta)\frac{d\varphi}{2\pi} \tag{5.7.105}$$

$$J_2 = \int_\theta^{\pi+\theta/2} [I(\varphi, \theta) - Q_1(-\varphi)(f(1) - f(e^{i\theta}))]\frac{d\varphi}{2\pi} \tag{5.7.106}$$

$$J_3 = \int_{-\pi+\theta/2}^0 I(\varphi, \theta) + Q_1(\theta - \varphi)(f(e^{i\theta}) - f(1))\frac{d\varphi}{2\pi} \tag{5.7.107}$$

That is, we've broken the remainder, after removing $(0, \theta)$, exactly in half and put in counter terms, which cancel since $f(e^{i\theta}) - f(1)$ is φ-independent and

$$\int_{-\pi+\theta/2}^0 Q_1(\theta - \varphi)\frac{d\varphi}{2\pi} = \int_{-\pi-\frac{\theta}{2}}^{-\theta} Q_1(-\varphi)\frac{d\varphi}{2\pi}$$

$$= -\int_\theta^{\pi+\theta/2} Q_1(-\varphi)\frac{d\varphi}{2\pi} \tag{5.7.108}$$

since $Q_1(-\psi) = -Q_1(\psi)$.

Using (5.7.100), we have that

$$|Q_1(\theta)| \le \frac{\pi}{|\theta|} \qquad (5.7.109)$$

Thus, for $0 < \varphi < \theta$,

$$|I(\varphi, \theta)| \le \frac{\pi \omega(\varphi, f)}{|\varphi|} + \frac{\pi \omega(\theta - \varphi, f)}{|\theta - \varphi|} \qquad (5.7.110)$$

so

$$|J_1| \le 2\pi \int_0^\theta \omega(\delta, f) \frac{d\delta}{\delta} \qquad (5.7.111)$$

On the other hand, the integrand in (5.7.106) is $[f(e^{i\varphi}) - f(e^{i\theta})] [K(\theta - \varphi) - K(-\varphi)]$ so using (5.7.99)

$$|J_2| \le c_1 \int_\theta^{\pi + \theta/2} \omega(\varphi - \theta, f) \frac{|\theta|}{|\varphi| |\theta - \varphi|} \frac{d\varphi}{2\pi}$$

$$\le c_1 \int_0^\pi \omega(y, f) \frac{|\theta|}{y(y + \theta)} \frac{dy}{2\pi} \qquad (5.7.112)$$

Since $\omega(2y, f) \le 2\omega(y, f)$, and $[y(y + \theta)]^{-1}$ is monotone, we can bound the integral from 0 to π by a multiple of the integral for 0 to $\frac{\pi}{4}$.

The $|J_3|$ bound is the same, so (5.7.94) is proven.   □

Notes and Historical Remarks.

> Some months ago you wrote " ... I have proved that two conjugate ... L^p functions, $p > 1$." I want the proof. Both I and my pupil Titchmarsh have tried in vain to prove it.
> G. H. Hardy, letter to M. Riesz, 1923, as quoted in [781]

The Hilbert transform comes in three closely related variants: the circular Hilbert transform was discussed here, the "ordinary" Hilbert transform on functions on $\mathbb{R}$:

$$f_H(x) = \lim_{\varepsilon \downarrow 0} \frac{1}{\pi} \int_{\varepsilon^{-1} \ge |x-y| \ge \varepsilon} f(y)(x - y)^{-1} dx \qquad (5.7.113)$$

and the *discrete Hilbert transform*

$$\alpha_n^{(H)} = \pi^{-1} \sum_{m \ne n} \alpha_m (n - m)^{-1} \qquad (5.7.114)$$

Hilbert studied these (especially on L^2 or ℓ^2 spaces) in his papers on integral equations [383], especially the first 1906 paper. He looked at the circular transform in his work on the Riemann–Hilbert problem. The discrete transform appeared in unpublished lectures and the Hilbert inequality

$$\left| \sum_{n \ne m} \frac{\bar{\alpha}_n \alpha_m}{n - m} \right| \le \pi \sum_{n = -\infty}^\infty |\alpha_n|^2 \qquad (5.7.115)$$

only appeared in Weyl [**848**], who said it was a result Hilbert lectured on. They didn't have the best constant which was found by Schur [**720**].

That these three are analogs of each other can be seen as follows. Formula (9.2.39) of Part 2A can be written:

$$\frac{1}{2}\cot\left(\frac{x}{2}\right) = \frac{1}{x} + \sum_{n=1}^{\infty}\left(\frac{1}{x+2\pi n} + \frac{1}{x-2\pi n}\right) \tag{5.7.116}$$

This implies that if f is a function on $\partial\mathbb{D}$ and g is defined on $\mathbb{R}$ by

$$g(\theta) = f(e^{i\theta}) \tag{5.7.117}$$

then, if the circular Hilbert transform exists at $e^{i\theta_0}$, we have that

$$g_H(\theta_0) = \widetilde{f}(e^{i\theta_0}) \tag{5.7.118}$$

From this, scaling (to replace functions periodic with period 2π, to period $2\pi L$) and limits ($L \to \infty$), one gets an analog of M. Riesz's theorem for g_H, namely $g \mapsto g_H$ maps $L^p(\mathbb{R})$ to itself for $1 < p < \infty$. One can go in the opposite direction from Riesz inequalities on $g \mapsto g_H$ to those on $f \to \widetilde{f}$.

Convolution in a sum with $\frac{1}{\pi n}$ is clearly a discrete analog of convolution in an integral with the principal value of $\frac{1}{\pi x}$. There is an analog of M. Riesz's theorem sometimes written as

$$\left|\sum_{n,m\geq 1}\frac{\alpha_n\beta_m}{n+m}\right| \leq \frac{\pi}{\sin(\frac{\pi}{p})}\|\alpha\|_p\|\beta\|_q \tag{5.7.119}$$

where q is the conjugate index and $\|\alpha\|_p = (\sum_{n=1}^{\infty}|\alpha_n|^p)^{1/p}$. One can go from this to the continuous result by taking limits. This result is due to Hardy and M. Riesz and appeared first [**363**] with attribution to Hardy–Riesz.

Theorem 5.7.1 was proved by Marcel Riesz in 1928 [**692**]. Riesz's proof is sketched in Problem 7. Actually, three years earlier than Riesz, Kolmogorov [**480**] had proven that for $f \in L^1$, Kf is in weak-L^1 but it wasn't realized that this implied the L^p-result until later. P. Stein [**786**] found an elegant proof using superharmonic functions (see Problem 8). Section 6.4 will discuss higher-dimensional analogs of Theorem 5.7.1 and includes another proof of this theorem.

In 1972, Pichorides [**644**] obtained the optimal constants for M. Riesz's theorem

$$\begin{aligned}\|Ku\|_p &\leq \tan\left(\pi/2p\right)\|u\|_p, \quad 1 < p \leq 2 \\ &\leq \cot\left(\pi/2p\right)\|u\|_p, \quad 2 \leq p < \infty\end{aligned} \tag{5.7.120}$$

He also obtained the best constants for Zygmund's $L^1 \log L$-bound for $\|Ku\|_1$ and for Kolmogorov's $\|Ku\|_p$ bound, $0 < p < 1$, when $u \in L^1$, $u \geq 0$. Essén [**253**] recovered (5.7.120) using superharmonic functions.

For $u \in L^1$, it is easy to see (Problem 4) that not only is $\left|\{\theta \mid |\widetilde{f}(e^{i\theta})| > t\}\right|$ of $O\,(1/t)$, it is $o(1/t)$.

Spanne [**770**] and Stein [**776**] have identified $\{\widetilde{f} \mid f \in L^\infty(\partial\mathbb{D})\}$; see Section 5.11.

Rather than discuss the map $f \mapsto \widetilde{f}$ on real-valued functions, many authors look at the map

$$f \mapsto \tfrac{1}{2}(f + i\widetilde{f}) + \int_0^{2\pi} f(e^{i\theta})\frac{d\theta}{2\pi} \equiv f_+ \qquad (5.7.121)$$

and extend it to complex-valued f's by making it complex linear. Then, in terms of Fourier coefficients,

$$f_+ = \sum_{n=0}^{\infty} f_n^\sharp e^{in\theta} \qquad (5.7.122)$$

so $(f_+)_+ = f_+$, i.e., f_+ is a projection called the *Riesz projection* (after Marcel Riesz). Of course, M. Riesz's theorem implies $f \mapsto f_+$ is bounded on L^p, $1 < p < \infty$. Hollenbeck–Verbitsky [**386**] has determined the best constant, namely

$$\|f_+\|_p \le [\sin\,(\pi/p)]^{-1}\,\|f\|_p \qquad (5.7.123)$$

Cima–Matheson–Ross [**161**] is a book on topics close to the Riesz projection.

Marcel Riesz (1886–1969) was Hungarian-born, a student of Fejér, and younger brother of Frigyes Riesz. He got his Ph.D. in 1907 and, in 1908, at the invitation of Mittag-Leffler, he went to Sweden, spending his professional career there (between retirement in 1952 and his return to Lund in 1962, he was at various American universities). He was appointed to a position in Stockholm University in 1911 and moved to a Professorship in Lund in 1926 staying there until his retirement. His students in Stockholm included Cramér and Hille and in Lund, Frostman, Hörmander, and Gårding. Besides his work on L^p-bounds of conjugate functions and the related Riesz–Thorin theorem, M. Riesz is known as a pioneer in extensions of linear functionals (see Theorem 5.5.8 of Part 1) and compactness criteria in L^p (see Section 3.16 of Part 4).

The Plemelj–Privalov theorem is named after their papers [**649, 665**] where the result appeared; see also Zygmund [**877**].

Problems

1. (a) Prove that $\mathcal{F}(z) = \frac{1+z}{1-z}$ maps $\mathbb{D}$ onto $\mathbb{H}_+ = \{z \mid \mathrm{Re}\,z > 0\}$. (Prove that $\mathcal{F}(1) = \infty$, $\mathcal{F}(-1) = 0$, $\mathcal{F}(0) = 1$.)

   (b) Prove that f given by (5.7.2) maps $\mathbb{D}$ onto $\{w \mid |\mathrm{Re}\,w| < \frac{\pi}{2},\ \mathrm{Im}\,w \in \mathbb{R}\}$.

2. In this problem, the reader will show that if $f \in L^\infty(\partial\mathbb{D})$, $\|f\|_\infty = 1$, then for any $\alpha < \frac{\pi}{2}$,

$$\int \exp\left(\alpha\,|\widetilde{f}(e^{i\theta})|\right) \frac{d\theta}{2\pi} \leq 2\sec(\alpha) \tag{5.7.124}$$

so the singularities are at worse logarithmic as they are in Example 5.7.2.

(a) Let u be $P_r f$ and $Ku = Q_r f$ so $\widetilde{f}$ is the boundary value of Ku. Define $g = -i(u + i\tilde{u})$ on $\mathbb{D}$ so $\operatorname{Re} g(0) = 0$. Prove that $|u(z)| \leq 1$ and then that since $\alpha < \frac{\pi}{2}$ and cos is even and decreasing on $(0, \alpha)$ that

$$\operatorname{Re}\left(e^{\alpha g(z)}\right) \geq e^{\alpha \tilde{u}(z)} \cos(\alpha) \tag{5.7.125}$$

(b) Since $\operatorname{Re}(e^{\alpha g(z)})$ is harmonic and $\operatorname{Re}(e^{\alpha g(0)}) \leq 1$, prove that

$$\int e^{\alpha \tilde{u}(re^{i\theta})} \frac{d\theta}{2\pi} \leq \sec(\alpha) \tag{5.7.126}$$

(c) Using $e^{\alpha(x)} \leq e^{\alpha x} + e^{-\alpha x}$, prove (5.7.124).

Remark. We'll study this issue in detail in Section 5.11. This is related to the John–Nirenberg inequalities.

The next three problems will show that

$$f(z) \equiv \sum_{n=2}^{\infty} z^n / \log(n) \tag{5.7.127}$$

is the Cauchy integral of an L^1-function but that $\operatorname{Im} f^*(e^{i\theta}) \notin L^1$, so the conjugate function of $\operatorname{Re} f \in h^1$ but $\operatorname{Im} f \notin h^1$.

3. This problem will show that if $\{a_n\}_{n=-\infty}^{\infty} \subset \mathbb{R}$ obeys:
   (i) $a_{-n} = a_n$,
   (ii) $a_n \geq 0$,
   (iii) $a_n \to 0$ as $|n| \to \infty$,
   (iv) $2a_n \leq a_{n+1} + a_{n-1}$ for $n \geq 1$, $\tag{5.7.128}$

then there is $f \in L^1(\partial\mathbb{D}, dz/2\pi)$ so $f_n^\sharp = a_n$. Note that (5.7.128) is a convexity condition. This explains many of the results below.
   It will help to define

$$(\delta b)_n = b_{n+1} - b_n \tag{5.7.129}$$

so that (5.7.128) says that $\delta(\delta a)_n \geq 0$ for $n \geq 0$.

(a) Prove $\delta a_n \geq 0$ and $m > n \Rightarrow (\delta a)_m \geq (\delta a)_n$.

(b) Using $(\delta a)_{n+1} \leq (\delta a)_n$, show that

$$2[a_{2^n+1} - a_{2^n-1}] \leq 2^n(\delta a)_{2^n} \leq a_{2^{n+1}+1} - a_{2^n} \tag{5.7.130}$$

and then that
$$k(\delta a)_k \to 0 \qquad (5.7.131)$$

(c) Show that
$$\sum_{n=1}^{N} n(\delta^2 a)_{n-1} = a_0 - a_N - N(a_N - a_{N+1}) \qquad (5.7.132)$$

and conclude that
$$\sum_{n=1}^{\infty} n(\delta^2 a)_{n-1} = a_0 < \infty \qquad (5.7.133)$$

(d) Let (see also Section 3.5 of Part 1)
$$S_N = \sum_{n=-N}^{N} a_n e^{in\theta}, \quad D_N = \sum_{n=-N}^{N} e^{in\theta}, \quad F_N = \frac{1}{N} \sum_{j=0}^{N-1} D_j \qquad (5.7.134)$$

By two summation by parts prove that
$$S_N(e^{i\theta}) = \sum_{n=1}^{N} (\delta^2 a)_{n-1} n F_n(e^{i\theta}) - (\delta a)_N (N+1) F_{N+1}(e^{i\theta}) + a_{N+1} D_N(e^{i\theta})$$
$$(5.7.135)$$

(e) Let $f(e^{i\theta}) = \sum_{n=1}^{\infty} n(\delta^2 a)_{n-1} F_n(e^{i\theta})$. Prove that $f \in L^1(\partial\mathbb{D})$ and $f_n^\sharp = a_n$.

4. Suppose $f \in L^1(\partial\mathbb{D})$ with $b_n = f_n^\sharp$ obeying $b_n = -b_{-n} \geq 0$ for $n \geq 0$. We'll prove that
$$\sum_{n=1}^{\infty} \frac{b_n}{n} < \infty \qquad (5.7.136)$$

(a) Let $F(e^{i\theta}) = \int_0^\theta f(e^{i\varphi}) d\varphi$ $\;(|\theta| < \pi)$. Prove that $\frac{dF}{d\theta} = f$ and conclude that
$$F_n^\sharp = (in)^{-1} f_n^\sharp \qquad (5.7.137)$$

(b) Prove that F is continuous and use Fejér's theorem (Theorem 3.5.6 of Part 1) to see that
$$\lim_{N\to\infty} 2 \sum_{n=1}^{N} \left(1 - \frac{n}{N+1}\right) \frac{f_n^\sharp}{n} = i(F(0) - F_{n=0}^\sharp)$$

and conclude that (5.7.136) holds.

5. Consider the function $f(z)$ of (5.7.127).

(a) Prove that $\operatorname{Re} f$ has L^1 boundary values (see Problem 3) and conclude that f is the Cauchy integral of an L^1-function.

(b) Prove that $\operatorname{Im} f$ does not have an L^1 boundary value, i.e., $\operatorname{Re} f \in h^1$ but $\operatorname{Im} f^* \notin L^1$. (*Hint:* Use Problem 4.)

6. Verify that $1 < p < 2$ and $\|f\|_p = 1$ implies that (5.7.20) holds.

7. This problem provides an alternate proof which is essentially M. Riesz's original proof [**692**] of boundedness of the conjugate map on $L^p, 1 < p < \infty$.

   (a) If $f = u + iv$ is analytic, with $v(0) = 0$, compute first $\mathrm{Re}(f(z)^6)$ and then show that

   $$\int_0^{2\pi} v(re^{i\theta})^6 \frac{d\theta}{2\pi} \leq \int_0^{2\pi} u(re^{i\theta})^6 \frac{d\theta}{2\pi} + 15 \int_0^{2\pi} 4(re^{i\theta})^2 v(re^{i\theta})^4 \frac{d\theta}{2\pi} \quad (5.7.138)$$

   (b) By using Holder's inequality, find C so that

   $$15 \int u(re^{i\theta})^2 v(re^{i\theta})^4 \frac{d\theta}{2\pi} \leq \frac{1}{2} \int v(re^{i\theta})^6 \frac{d\theta}{2\pi} + C \int u(re^{i\theta})^6 \frac{d\theta}{2\pi} \quad (5.7.139)$$

   and conclude M. Riesz's theorem for $p = 6$.

   (c) Similarly obtain the result for $p = 10, 14, 18, \dots$ and, by the Riesz–Thorin theorem for all $p \in [2, \infty)$. Then use duality for $p \in (1, 2]$.

   Remark. Riesz developed the Riesz–Thorin theorem for this application.

8. This provides another proof of the Riesz conjugate function theorem, due to P. Stein [**786**]. It is undoubtedly the simplest proof. We prove it for $p \in (1, 2]$ and then use duality.

   (a) Prove that $H(x, y) = |x^2 + y^2|^{p/2} - p(p-1)^{-1}|x|^p$ is superharmonic. (*Hint*: Compute ΔH.)

   (b) If $f = u + iv$ with $v(0) = 0$ and f analytic, integrate $|f(re^{i\theta})|^p - p(p-1)^{-1}|u(re^{i\omega})|^p = H(u(re^\tau), v(re^\tau))$ and use $H(u(0), v(0)) = -(p-1)^{-1}|u(0)|^p < 0$ to see that

   $$\int |f(re^{i\theta})|^p \frac{d\theta}{2\pi} \leq p(p-1)^{-1} \int |u(re^{i\theta})|^p \frac{d\theta}{2\pi} \quad (5.7.140)$$

   to get the M. Riesz bound.

9. Let $f(e^{i\theta}) = \mathrm{Re}[(e^{i\theta}-1) \log(e^{i\theta}-1)]$ so $\tilde{f}(e^{i\theta}) = \mathrm{Im}[(e^{i\theta}-1) \log(e^{i\theta}-1)]$. Prove that $\theta \mapsto f(e^{i\theta})$ is Lipschitz, but that $\lim_{\theta \downarrow 0} |\tilde{f}(e^{i\theta}) - \tilde{f}(1)|/|\theta| = \infty$. (*Hint*: Show that $\mathrm{Im}\, F'(z)$ is bounded on $\partial\mathbb{D}$, but $\mathrm{Re}\, F'(z)$ is not when F is given by (5.7.89).)

10. Extend the Plemelj–Privalov result on $\tilde{f}$ as a function on $\partial\mathbb{D}$ to bounds on the conjugate harmonic function on $\mathbb{D}$, i.e., if $u \in h^\infty$ obeys $|u(z) - u(w)| \leq C|w - z|^\alpha$ for all $w, z \in \mathbb{D}$ and $\alpha \in (0, 1)$, then Ku obeys the same bound (with a different C).

11. Using Theorem 5.7.22. prove that if $\int_0^{\pi/4} \dfrac{\omega(\delta, f)}{\delta \, \log(\delta^{-1})} \, d\delta \, < \, \infty$ then $\int_0^{\pi/4} \dfrac{\omega(\delta, \widetilde{f})}{\delta} \, d\delta < \infty.$

5.8. Homogeneous Spaces and Convergence of Fourier Series

In Section 3.5 of Part 1, we defined the classical Fourier series for f, an $L^1(\partial \mathbb{D}, \frac{d\theta}{2\pi})$-function, by

$$f_n^\sharp = \int_0^{2\pi} e^{-in\theta} f(e^{i\theta}) \frac{d\theta}{2\pi} \tag{5.8.1}$$

and the partial Fourier series sums by

$$S_N(f)(e^{i\theta}) = \sum_{k=-N}^N f_k^\sharp e^{ik\theta} \tag{5.8.2}$$

In Problem 10 of Section 5.4 of Part 1, we proved there were functions $f_\infty \in C(\partial \mathbb{D})$ and $f_1 \in L^1(\partial \mathbb{D})$ so that $\|S_N f_\infty\|_\infty \to \infty$ and $\|S_n f_1\|_1 \to \infty$ as $N \to \infty$. In particular, for $p = 1, \infty$, $f \in L^p$ does not imply $\|S_n f - f\|_p \to 0$. Our goal in this section will be to prove this implication does however hold for $1 < p < \infty$. This result is here because our proof will rely on M. Riesz's theorem on L^p-boundedness of the circular Hilbert transform. We'll prove a theorem for a class of abstract spaces that includes L^p.

Definition. A *homogeneous space* is a subspace, X, of $L^1(\partial \mathbb{D}, \frac{d\theta}{2\pi})$ which is a Banach space in a norm $\| \cdot \|_X$ that obeys

(1) $\|f\|_1 \leq \|f\|_X$.
(2) If τ_θ is defined by

$$(\tau_\theta f)(e^{i\varphi}) = f(e^{i(\varphi - \theta)}) \tag{5.8.3}$$

then for all $f \in X$ and θ,

$$\|\tau_\theta f\|_X = \|f\|_X \tag{5.8.4}$$

(3) For each $f \in X$,

$$\lim_{\theta \to 0} \|\tau_\theta f - f\|_X = 0 \tag{5.8.5}$$

Remark. (2) and (3) imply that $\theta \to \tau_\theta f$ is a continuous map of $[0, 2\pi]$ to X.

Example 5.8.1. (a) $L^p(\partial \mathbb{D}, \frac{d\theta}{2\pi})$, $1 \leq p < \infty$, are all homogeneous spaces.

(b) $C(\partial \mathbb{D})$ and $C^k(\partial \mathbb{D})$ (the C^k-functions with $\|f\|^{(k)} = \sum_{j=0}^k \|D^j f\|_\infty$) are homogeneous spaces.

Let X be a homogeneous space and $g \in C(\partial\mathbb{D})$. Then $e^{i\theta} \mapsto g(e^{i\theta})\tau_\theta$ is a continuous function from $\partial\mathbb{D}$ to $\mathcal{L}(X)$, the bounded operators on X, so we can form the Riemann integral

$$\mathcal{O}(g) = \int_0^{2\pi} g(e^{i\theta})\tau_\theta \frac{d\theta}{2\pi} \tag{5.8.6}$$

Suppose that $f \in X$. Since f is L^1, we have that

$$(\mathcal{O}(g)f)(e^{i\varphi}) = (g * f)(e^{i\varphi}) \tag{5.8.7}$$

$$\equiv \int g(e^{i\theta})f(e^{i(\varphi-\theta)}) \frac{d\theta}{2\pi} \tag{5.8.8}$$

Since $X \subset L^1$, (5.8.7)/(5.8.8) hold a.e. pointwise for any $f \in X$. $\qquad\square$

Theorem 5.8.2. *Let X be a homogeneous space.*

(a) *For any $g \in C(\partial\mathbb{D})$, $\mathcal{O}(g)$ maps X to X and*

$$\|\mathcal{O}(g)\| \le \|g\|_1 \tag{5.8.9}$$

(b) *If g_n is an approximate identity (i.e., obeys Proposition 3.5.10 (i)–(iii) of Part 1), then for any $f \in X$, we have*

$$\|\mathcal{O}(g_n)f - f\|_X \to 0 \tag{5.8.10}$$

(c) *If*

$$P_n(f)(e^{i\theta}) \equiv f_n^\sharp e^{in\theta} \tag{5.8.11}$$

then $P_n(f) \in X$.

(d) *If, for $f \in X$,*

$$S_N(f) = \sum_{k=-N}^{N} P_k(f) \tag{5.8.12}$$

$$C_N(f) = \frac{1}{N} \sum_{n=0}^{N-1} S_n(f) \tag{5.8.13}$$

then $S_N(f), C_N(f)$ lie in X.

(e) $$\lim_{N\to\infty} C_N(f) = f \qquad (\text{in } \|\cdot\|_X) \tag{5.8.14}$$

In particular, any $f \in X$ is a $\|\cdot\|_X$ limit of trigonometric polynomials and $f, g \in X$ with $f_n^\sharp = g_n^\sharp$ for all n implies $f = g$.

Remark. (c) implies that for each n, either $e^{in\theta} \in X$ or else $f_n^\sharp = 0$ for all $f \in X$. Either possibility can occur (e.g., $X = \{f \in L^1 \mid f_n^\sharp = 0$ for all odd $n\}$ with $\|\cdot\|_1$ is a space where both possibilities occur).

Proof. (a) Since $\|\mathcal{O}(g)\| \le \int |g(e^{i\theta})| \|\tau_\theta\| \frac{d\theta}{2\pi}$ and $\|\tau_\theta\| = 1$ (by (5.8.4)), this is immediate.

(b) The proof is the same as for $C(X)$ convergence. Since $\int g_n(e^{i\theta})\frac{d\theta}{2\pi} = 1$, we have that

$$f = \int g_n(e^{i\theta}) f \frac{d\theta}{2\pi} \tag{5.8.15}$$

so for any δ,

$$\|\mathcal{O}(g_n)f - f\| = \left\| \int (g_n(e^{i\theta}))(\tau_\theta - \mathbb{1}) f \frac{d\theta}{2\pi} \right\|$$

$$\leq \int_{|\theta|<\delta} |g_n(e^{i\theta})| \|\tau_\theta f - f\| \frac{d\theta}{2\pi} + 2\|f\| \int_{|\theta|<\delta} |g_n(e^{i\theta})| \frac{d\theta}{2\pi} \tag{5.8.16}$$

By condition (iii) of Proposition 3.5.10 of Part 1, the last integral in (5.8.16) goes to zero as $n \to \infty$. Thus, since $\|g_n\|_{L^1} = 1$, we get for any δ,

$$\limsup_{n\to\infty} \|\mathcal{O}(g_n)f - f\| \leq \sup_{|\theta|<\delta} \|\tau_\theta f - f\| \tag{5.8.17}$$

By (5.8.5), the sup goes to zero as $\delta \downarrow 0$.

(c) If $g(e^{i\theta}) = e^{in\theta}$, then $\mathcal{O}(g) = P_n$.

(d) Follows from (c).

(e) Since the Fejér kernel is an approximate identity (see Proposition 3.5.10 of Part 1), this follows from (b). Since $C_N(f)$ is a trigonometric polynomial, any $f \in X$ is a limit of trigonometric polynomials. If $f_n^\sharp = g_n^\sharp$ for all N, then $C_N(f) = C_N(g)$ for all N. So taking $N \to \infty$, $f = g$. $\square$

The remaining results rely on the uniform boundedness principle and the closed graph theorem (see Theorems 5.4.9 and 5.4.17 of Part 1).

Theorem 5.8.3. *Let X be a homogeneous space. The following are equivalent:*

(1) *$S_N f \to f$ for all $f \in X$.*
(2) *$\sup_N \|S_N\|_{\mathcal{L}(X)} < \infty$.*

Proof. $\underline{(1) \Rightarrow (2)}$. By the uniform boundedness principle, since $S_N f \to f \Rightarrow \sup_N \|S_N f\| < \infty$

$\underline{(2) \Rightarrow (1)}$. Let s be the sup. Let p be a trigonometric polynomial. For N large, $S_N p = p$. Thus, for such N,

$$\|S_N f - f\| \leq \|S_N(f - p)\| + \|f - p\| \leq (1+s)\|f - p\| \tag{5.8.18}$$

that is,

$$\limsup \|S_N f - f\| \leq (1+s)\|f - p\| \tag{5.8.19}$$

Since every f is a limit of trigonometric polynomials, the lim sup is 0. $\square$

Our final set of results requires one more property of X:

Definition. A homogeneous space, X, is called *doubly homogeneous* if for all n and all $f \in X$, $e^{in\theta} f(\theta) \in X$ and

$$\|e^{in\theta} f\|_X = \|f\|_X \tag{5.8.20}$$

For example, L^p is doubly homogeneous but C^1 is not.

We want to discuss two not a priori everywhere defined operators on X:

$$D(\tilde{\ }) = \{f \in X \mid \exists g \in X \text{ so that } g_n^\sharp = -i \operatorname{sgn}(n) f_n^\sharp\}$$

$$D(P_+) = \{f \in X \mid \exists h \in X \text{ so that } h_n^\sharp = 0 \text{ for } n < 0, \ h_n^\sharp = f_n^\sharp \text{ for } n \geq 0\}$$

Of course, we set $g = \tilde{f}$, $h = P_+(f)$. Notice that

$$\tfrac{1}{2}(f + i\tilde{f} + P_0(f)) = P_+(f) \tag{5.8.21}$$

that is,

$$D(\tilde{\ }) = D(P_+) \tag{5.8.22}$$

and if f is in one, so both, then (5.8.21) holds. Here is the main abstract result of this section. It links the Hilbert transform to norm convergence of Fourier series.

Theorem 5.8.4. *Let X be a doubly homogeneous space. Then the following are equivalent:*

(1) $D(\tilde{\ }) = X$.

(2) $D(\tilde{\ }) = X$ and there is C_1 with

$$\|\tilde{f}\| \leq C_1 \|f\| \tag{5.8.23}$$

for all $f \in X$.

(3) *There is a constant, C_1, so that for all trigonometric polynomials,*

$$\|\tilde{p}\| \leq C_1 \|p\| \tag{5.8.24}$$

(4) $D(P_+) = X$.

(5) $D(P_+) = X$ and there is a constant, C_2, with

$$\|P_+(f)\| \leq C_2 \|f\| \tag{5.8.25}$$

(6) *There is a constant, C_2, so that for all trigonometric polynomials,*

$$\|P_+(p)\| \leq C_2 \|p\| \tag{5.8.26}$$

(7) $\displaystyle\sup_N \|S_N\|_{\mathcal{L}(X)} < \infty.$ $\tag{5.8.27}$

(8) *For all $f \in X$, $\|S_N f - f\| \to 0$.*

Proof. $(1) \Rightarrow (2)$ and $(4) \Rightarrow (5)$. Both $\sim$ and P_+ have closed graphs (if $f_k \to f$ and $\tilde{f}_k \to g$, then $(f_k)_n^\sharp \to f_n^\sharp$ and $(\tilde{f}_k)_n^\sharp \to g_n^\sharp$ because $\|\cdot\|_1 \leq \|\ \|_X$. Thus, relations among $(f_k)_n^\sharp$ and $(\tilde{f}_k)_n^\sharp$ are preserved (i.e., the graph is closed), so the result follows from the closed graph theorem.

$(2) \Rightarrow (3)$ and $(5) \Rightarrow (6)$ are trivial.

$(3) \Rightarrow (1)$ and $(6) \Rightarrow (3)$. The graphs are closed and (5.8.24) (respectively, (5.8.26)) implies if $p_k \to f$, then $\tilde{p}_k$ (respectively, $P_+(p_k)$) are Cauchy. Thus, density of polynomials implies $D(\sim) = X$ (respectively, $D(P_+) = X$).

$(1) \Leftrightarrow (4)$ follows from (5.8.22).

$(4) \Leftrightarrow (7)$ is Theorem 5.8.3.

$(5) \Rightarrow (8)$. We have that

$$S_N(f) = e^{-iN\theta} P_+(e^{iN\theta} f) - e^{i(N+1)\theta} P_+(e^{-i(N+1)\theta}) \tag{5.8.28}$$

so by (5.8.20),

$$\|S_N\|_{\mathcal{L}(X)} \leq 2\|P_+\|_{\mathcal{L}(X)} \tag{5.8.29}$$

$(8) \Rightarrow (6)$. Suppose $p(e^{i\theta}) = \sum_{-M}^{M} a_k e^{ik\theta}$. Then

$$P_+(p) = e^{iM\theta} S_M(e^{-iM\theta} p) \tag{5.8.30}$$

so by (5.8.20),

$$\|P_+(p)\| \leq \left[\sup_M \|S_M\|_{\mathcal{L}(X)}\right] \|p\| \tag{5.8.31}$$

$\square$

Corollary 5.8.5. *For $f \in L^p(\partial\mathbb{D}, \frac{d\theta}{2\pi})$, $1 < p < \infty$, we have*

$$\|S_N f - f\|_p \to 0 \tag{5.8.32}$$

Proof. This is immediate from M. Riesz's theorem and the last theorem.

$\square$

Remarkably, (5.8.32) is equivalent to M. Riesz's theorem!

Notes and Historical Remarks. That M. Riesz's theorem implies L^p-convergence of Fourier series is a result in M. Riesz's original 1928 paper [**692**] on conjugate functions.

These results extend to Fourier transforms on $L^p(\mathbb{R}, dx)$, $1 < p < \infty$. By a density argument, it is enough to prove a bound

$$\|S_R f\|_p \leq C_p \|f\|_p \tag{5.8.33}$$

where C_p is R-independent and where

$$(S_R f)(x) = (2\pi)^{-\nu/2} \int_{|k| \leq R} e^{ik \cdot x} \, \widehat{f}(k) \, d^\nu k \tag{5.8.34}$$

By scaling, it suffices to prove this for $R = 1$. When $\nu = 1$, this is a difference of two shifted Hilbert transforms, so bounded on L^p by M. Riesz' Theorem. Thus, $\|S_R f - f\|_p \to 0$.

It is easy to extend these results to $\mathbb{R}^\nu$ if we sum (or integrate for Fourier transform) over rectangles. It thus came as a big surprise when Fefferman [**264**] proved that if $\nu \geq 2$, S_R is not bounded on any L^p, $p \neq 2$ (he'd earlier proven this if $p \leq \frac{2\nu}{\nu+1}$ or $p \geq \frac{2\nu}{\nu-1}$ [**263**], but it was conjectured to hold for $p \in \left(\frac{2\nu}{\nu+1}, \frac{2\nu}{\nu-1}\right)$). His proof relies on the existence of Besicovitch–Kakeya sets—sets of zero Lebesgue measure that contain a unit line segment in every direction.

5.9. Boundary Values of Analytic Functions in the Upper Half-Plane

In this section, we want to discuss analogs of $H^p(\mathbb{D})$, the Herglotz representation and conjugate function/Hilbert transform when $\mathbb{D}$ is replaced by $\mathbb{C}_+$, the upper half-plane. Because $\mathbb{C}_+$ and its boundary $\mathbb{R}$ are unbounded, there are some technical differences (e.g., dx on $\mathbb{R}$ is not a finite measure; there exist analytic functions on $\mathbb{C}_+$ whose boundary value has vanishing real part—albeit only unbounded functions) but the criteria are almost identical.

On one level, conformal mapping allows one to carry over much of the $H^p(\mathbb{D})$ theory to $H^p(\mathbb{C}_+)$. The conformal maps

$$\varphi(z) = \frac{i-z}{i+z}, \quad \psi(w) \equiv \varphi^{-1}(w) = i\frac{(1-w)}{(1+w)} \tag{5.9.1}$$

bijectively map $\mathbb{C}_+$ to $\mathbb{D}$ and $\mathbb{D}$ to $\mathbb{C}_+$. For later purposes, we note (for $-\pi < \theta < \pi$)

$$\psi(e^{i\theta}) = \tan\left(\frac{\theta}{2}\right); \quad \varphi\left(\tan\left(\frac{\theta}{2}\right)\right) = e^{i\theta} \tag{5.9.2}$$

$$\psi(-1) = \infty, \quad \varphi(\infty) = -1 \tag{5.9.3}$$

Most of the results can be obtained from the use of the conformal maps but there are subtleties: in the Herglotz theorem, point masses at $z = -1$ are special since $\psi(-1) = \infty$ and the map $f \to f \circ \varphi$ will not be a bijection of H^p spaces (but $f \in H^p(\mathbb{C}_+)$ will imply that $f \circ \varphi \in H^p(\mathbb{D})$). We'll begin with this pair of results. While one could obtain M. Riesz type conjugate bounds by mimicking our proof in Section 5.7 (see Theorem 5.7.15), we'll instead derive them via a remarkable equality of Boole.

Recall that Carathéodory functions on $\mathbb{D}$ are analytic with $\operatorname{Re} F \geq 0$ and normalization condition $F(0) = 1$. They obey, for a positive measure, μ,

$$F(w) = \int \frac{e^{i\theta} + w}{e^{i\theta} - w} \, d\mu(\theta) \tag{5.9.4}$$

where $\mu(\partial\mathbb{D}) = 1$ by $F(0) = 1$. If we drop the normalization condition, μ is no longer normalized and we need an additional $+ib$ term if $\operatorname{Im} F(0) = b$.

A *Herglotz function* also called a *Pick function* or a *Nevanlinna function* is a function, $f(t)$, analytic on $\mathbb{C}_+$ with

$$\operatorname{Im} z > 0 \Rightarrow \operatorname{Im} f(z) > 0 \tag{5.9.5}$$

Clearly, if f is Herglotz, then F defined on $\mathbb{D}$ by

$$F(w) = -if(\psi(w)) \tag{5.9.6}$$

has $\operatorname{Re} F > 0$. By the above considerations, $f(z) = iF(\varphi(z))$ has the form

$$f(z) = b + i \int \frac{e^{i\theta} + \varphi(z)}{e^{i\theta} - \varphi(z)} \, d\mu(\theta) \tag{5.9.7}$$

It is natural to replace the θ variable by $x = \psi(e^{i\theta})$, or $e^{i\theta} = \varphi(x)$. ψ maps $\partial\mathbb{D}$ to $\mathbb{R} \cup \{\infty\}$, so we separate out a potential point mass at -1,

$$V(\{-1\}) = a \tag{5.9.8}$$

Note that

$$\frac{-1 + \varphi(t)}{-1 - \varphi(t)} = \frac{-2z}{-2i} = -iz \tag{5.9.9}$$

and for $x \in \mathbb{R}$

$$\frac{\varphi(x) + \varphi(z)}{\varphi(x) - \varphi(z)} = -i\left(\frac{1 + xz}{x - z}\right) \tag{5.9.10}$$

We thus get the first form of the Herglotz representation theorem: there is a positive finite measure ν on $\mathbb{R}$ (obtained from $\mu \upharpoonright \partial\mathbb{D} \setminus \{-1\}$ by using ψ to pull it to $\mathbb{R}$),

$$f(z) = az + b + \int \frac{1 + xz}{x - z} \, d\mu(x) \tag{5.9.11}$$

Noting that

$$\frac{1 + xz}{x - z} = \frac{1 + x^2}{x - z} - x = \left(\frac{1}{x - z} - \frac{x}{1 + x^2}\right)(1 + x^2) \tag{5.9.12}$$

we write $d\rho(x) = (1 + x^2)d\mu$ so

$$\int \frac{d\rho(x)}{1 + x^2} < \infty \tag{5.9.13}$$

and we get the second form of the Herglotz representation theorem:

$$f(z) = az + b + \int \left(\frac{1}{x - z} - \frac{x}{1 + x^2}\right) d\rho(x) \tag{5.9.14}$$

If

$$\int d\rho(x) < \infty \tag{5.9.15}$$

the subtraction in (5.9.14) is an absolute convergent integral and we can write

$$f(z) = az + c + \int \frac{d\rho(x)}{x - z} \qquad (5.9.16)$$

We summarize in

Theorem 5.9.1 (Herglotz Representation Theorem). *Every Herglotz function has a representation of the form (5.9.11) where $a > 0$, $b \in \mathbb{R}$, $b = \mathrm{Re}[f(i)]$, and μ is a finite positive measure on $\mathbb{R}$. Equivalently, it has the form (5.9.14) when ρ is a positive measure obeying (5.9.13). If ρ obeys (5.9.15), then (5.9.16) holds. Moreover,*

(a) $\mathrm{Im}\, f(x_0 + iy_0) = ay_0 + \int \dfrac{d\rho(x)}{(x - x_0)^2 + y_0^2}$. $\qquad (5.9.17)$

(b) *f has nontangential boundary values on $\mathbb{R}$.*

(c) $a = \lim\limits_{y_0 \to \infty} \mathrm{Im}\, f(iy_0)/y_0$. $\qquad (5.9.18)$

(d) *If*

$$d\rho = w(x)dx + d\rho_{\mathrm{sing}} \qquad (5.9.19)$$

then

$$w(x) = \tfrac{1}{\pi}\, \mathrm{Im}\, f(x + i0) \qquad (5.9.20)$$

(e) $d\rho(\{x_0\}) = \lim\limits_{\varepsilon \downarrow 0} \varepsilon\, \mathrm{Im}\, f(x + i\varepsilon)$. $\qquad (5.9.21)$

(f) *$d\rho_{\mathrm{sing}}$ is supported on $\{x \mid \mathrm{Im}\, f(x + i\varepsilon) \to \infty\}$.*

(g) *$\tfrac{1}{\pi} \mathrm{Im}\, f(x + i\varepsilon)\, dx \to d\rho$ in the weak-$*$ topology defined as a function $d\rho$ on the continuous functions of compact support.*

Remarks. 1. It is because of the more direct connection of $d\rho$ with boundary values that (5.9.14) is the more commonly used and stated form of the Herglotz theorem.

2. (5.9.17) is the Poisson kernel for the upper half-plane as we'd expect to relate harmonic functions like $\mathrm{Im}\, f$ and their boundary values.

Proof. The statement before "Moreover" has already been derived from the Herglotz representation on $\mathbb{D}$. (a) is a simple calculation and (d) follows from the fact that F given by (5.9.6) has $e^{-F(w)} \in H^\infty(\mathbb{D})$ and H^∞ has boundary values. The rest are a restatement of Theorem 2.5.4 $\qquad \square$

As we'll see, the conformal map φ does not set up a bijection of $H^p(\mathbb{D})$ and $H^p(\mathbb{C}_+)$ if $0 < p < \infty$ but for $p = \infty$, so we can immediately describe $H^\infty(\mathbb{C}_+)$.

Theorem 5.9.2. *Let u be bounded and harmonic on $\mathbb{C}_+$. Then for Lebesgue a.e. $x \in \mathbb{R}$*

$$u^*(x) = \lim_{\varepsilon \downarrow 0} u(x + i0) \qquad (5.9.22)$$

and ($L^\infty(\mathbb{R})$-norm on the left; $L^\infty(\mathbb{C}_+)$ on the right)

$$\|u^*\|_\infty = \|u\|_\infty \qquad (5.9.23)$$

and for all $z_0 = x_0 + iy_0 \in \mathbb{C}_+$

$$u(z_0) = \frac{1}{\pi} \int \frac{y_0}{(x - x_0)^2 + y_0^2} u^*(x)\, dx \qquad (5.9.24)$$

If f is analytic on C_+ with $\operatorname{Im} f \in L^\infty(\mathbb{C}_+)$, then for a.e. $x \in \mathbb{R}$,

$$f^*(x) = \lim_{\varepsilon \downarrow 0} f(x + i0) \qquad (5.9.25)$$

exists with $\|\operatorname{Im} f^*\|_\infty = \|\operatorname{Im} f\|_\infty$ and for a real number, α,

$$f(z) = \alpha + \frac{1}{\pi} \int \frac{\operatorname{Im} f^*(x)}{x - z}\, dx \qquad (5.9.26)$$

If $f^* \in L^1(\mathbb{R}, dx)$, then

$$\alpha = \lim_{y \to \infty} f(x_0 + iy) \qquad (5.9.27)$$

for any $x \in \mathbb{R}$.

Proof. $U(w) = u(\psi(w))$, (respectively, $F(w) = f(\psi(w))$) is in $h^\infty(\mathbb{D})$, (respectively, analytic on $\mathbb{D}$) with $\operatorname{Im} F \in h^\infty(\mathbb{D})$. In particular, $U \in h^2(\mathbb{D})$, so by Theorem 5.2.3, U and F have nontangential boundary values on $\partial \mathbb{D}$, so (5.9.22) and (5.9.25) hold and

$$F(w) = \alpha + i \int_0^{2\pi} \frac{e^{i\theta} + w}{e^{i\theta} - w} \operatorname{Im} F^*(e^{i\theta})\, d\theta \qquad (5.9.28)$$

The same manipulations that went from (5.9.7) to (5.9.13) lead to (5.9.24) and (5.9.26).

By (5.9.22), $\|u^*\|_\infty \leq \|u\|_\infty$ and by (5.9.24) and $\int \frac{y_0}{(x-x_0)^2+y_0^2}\, dx = \pi$, we have $\|u\|_\infty = \|u^*\|_\infty$ proving (5.9.23). If $\operatorname{Im} f^* \in L^1$, the dominated convergence theorem implies (5.9.27). $\qquad \square$

One defines $H^p(\mathbb{C}_+)$, $0 < p \leq \infty$, as those functions, f, analytic in $\mathbb{C}_+$ with (for $p = \infty$; $H^\infty(\mathbb{C}_+)$ is the bounded analytic functions)

$$\|f\|_p^p = \sup_{0 < y < \infty} \int |f(x + iy)|^p\, dx < \infty \qquad (5.9.29)$$

We'll hit some of the high points of the H^p theory leaving some details to the references in the Notes.

We begin as we did with consequences of subharmonicity. $|f(z)|^p$ is subharmonic, so for any $z_0 = x_0 + iy_0$ in $\mathbb{C}_+$,

$$|f(x_0 + iy_0)|^p \leq (\pi y_0^2)^{-1} \int_{|z-z_0|\leq y_0} |f(z_0)|^p \, d^2z$$

$$\leq (\pi y_0^2)^{-1} \int_0^{2y_0} dy \int_{-\infty}^{\infty} dx |f(x+iy)|^p \, dx \, dy$$

$$\leq \frac{2}{\pi y_0} \|f\|_p^p \tag{5.9.30}$$

Thus, we have proven

Theorem 5.9.3. *If* $f \in H^p(\mathbb{C}_+)$, $0 < p < \infty$, *then*

$$|f(z)| \leq \left(\frac{2}{\pi \operatorname{Im} z}\right)^{\frac{1}{p}} \|f\|_p \tag{5.9.31}$$

This has a corollary that shows one difference with $H^p(\mathbb{D})$. Let $h^p(\mathbb{C}_+)$ be the real-valued harmonic function obeying (5.9.29).

Corollary 5.9.4. *Let* $u \in h^p(\mathbb{C}_+)$, $0 < p < \infty$. *Then there is at most one real-valued function* v *on* $\mathbb{C}_+$ *so* $a + iv \in H^p(\mathbb{C}_+)$.

Remark. We'll prove below when $1 < p < \infty$, there is such a v.

Proof. If v_1, v_2 are two such v's, $f_j = u + iv_j$ is analytic so $f_1 - f_2$ is constant (since $(\operatorname{Re}(f_1 - f_2)) = 0$) and in H^p, so $0(\frac{1}{|\operatorname{Im} z|^{1/p}})$ at ∞, so 0. $\square$

Next, we analyze $p = 2$ as we did for $H^p(\mathbb{D})$. The key there was to use Taylor series, and there does not seem to be a good analog of Taylor series which relies on circles. But there is another way of thinking of the approach to $H^2(\mathbb{D})$. If f is analytic in $\mathbb{D} \setminus \{0\}$, we can define Fourier coefficients for $0 < \rho < 1$ by

$$a_n(\rho) = \frac{1}{2\pi} \int_0^{2\pi} f(\rho e^{i\theta}) e^{-in\theta} \, d\theta \tag{5.9.32}$$

Then the Cauchy integral theorem (essentially constancy of Laurent series coefficients) implies that for $0 < \rho < r < 1$

$$\rho^n a_n(\rho) = r^n a_n(r) \Rightarrow a_n(\rho) = \left(\frac{r}{\rho}\right) a_n(r) \tag{5.9.33}$$

If f is analytic at 0, (5.9.32) implies $a_n(r) = 0$ for $n < 0$.

The analog for $H^2(\mathbb{D})$ is to look at the Fourier transform instead of Fourier series. We've actually proven the analog of (5.9.33) in our study of the Paley–Wiener theorem (Theorem 11.2.1 of Part 2A).

Theorem 5.9.5. (a) *Let $g \in L^2([0, \infty), dx)$. Then for $z \in \mathbb{C}_+$, the integral*

$$f(z) = (2\pi)^{-1/2} \int e^{ikz} g(k) \, dk \qquad (5.9.34)$$

is absolutely convergent and defines an element of $H^2(\mathbb{C}_+)$ with $\|f\|_{H^2} = \|g\|_2$.

(b) *Conversely, if $f \in H^2(\mathbb{C}_+)$ and*

$$g_y(k) = (2\pi)^{-1/2} \int e^{-ikx} f(x + iy) \, dx \qquad (5.9.35)$$

(shorthand for the L^2 Fourier transform), then for $0 < y < w$

$$e^{ky} g_y(k) = e^{kw} g_w(k) \qquad (5.9.36)$$

Moreover, each $g_w(k)$ is supported on $[0, \infty)$ and there exits $g \in L^2([0, \infty), dx)$ so that

$$g_y(k) = e^{-ky} g(k) \qquad (5.9.37)$$

and f is given by (5.9.34).

For (c)–(g), $f \in H^2(\mathbb{C}_+)$ and f and g are related by (5.9.34).

(c) *If f^* is the inverse Fourier transform of g, then*

$$f(\cdot + iy) \to f^*(\cdot) \text{ in } L^2(\mathbb{R}, dx) \text{ as } y \downarrow 0 \qquad (5.9.38)$$

(d) *We have that, for $x_0 + iy_0 \in \mathbb{C}_+$,*

$$f(x_0 + iy_0) = \frac{1}{\pi} \int \frac{y_0}{(x - x_0)^2 + y_0^2} f^*(x) \, dx \qquad (5.9.39)$$

(e) *We have that for $z \in \mathbb{C}_+$ that*

$$f(z) = \frac{1}{\pi i} \int \frac{1}{x - z} \operatorname{Re} f^*(x) \, dx \qquad (5.9.40)$$

(f) $\sup\limits_{y} |f(x + iy)| \leq (M_{\mathrm{HL}} f^*)(x).$ \qquad (5.9.41)

(g) *For a.e. x,*

$$\lim_{y \downarrow 0} f(x + iy) = f^*(x) \qquad (5.9.42)$$

Remarks. 1. There are nontangential versions of (f) and (g); see Theorem 2.4.10.

2. One can also prove (d) by computing the Fourier transform of $e^{-y|x|}$ (see Problem 1)

Proof. (a) Let h be a C^∞-function supported in $[-1, \infty)$ with $h \equiv 1$ on $[0, \infty)$. Then it is easy to see (Problem 2(a)) that $F_z = e^{-i\bar{z} \cdot} h(\cdot)$ is antianalytic in z for $z \in \mathbb{C}_+$ as an L^2-valued function. It follows that $f(z) = \langle F_z, g \rangle$

is analytic and the integral is absolutely convergent. By the Plancherel theorem,

$$\int |f(z+iy)|^2 dx = \int e^{-2ky}|g(k)|^2 dk$$

and by the dominated convergence theorem that

$$\sup_y \int e^{-2ky}|g(k)|^2 dk = \|g\|_{L^2}^2$$

(b) We borrow the argument used in the proof of Theorem 11.1.1 of Part 2A. Fix y_0, let, for $\operatorname{Im} z > -y_0$,

$$G_z(x) = f(x+iy_0+z) \tag{5.9.43}$$

$G_z(\,\cdot\,)$ is easily seen to be an L^2-valued analytic function of z (Problem 2(b)). For $z \in \mathbb{R}$ ($\,\hat{}\,$ in x)

$$\widehat{G}_z(k) = e^{ikz} g_{y_0}(k) \tag{5.9.44}$$

By analyticity, we conclude that

$$\widehat{G}_{\delta y}(k) = e^{-k\delta y} g_{y_0}(k) \tag{5.9.45}$$

i.e.,

$$g_{y_1}(k) = e^{-k(y_1-y_0)} g_{y_0}(k) \tag{5.9.46}$$

For any $\alpha > 0$, we have

$$\sup_{y_1 \in (0,\infty)} \int_{-\infty}^{-\alpha} e^{-k(y_1-y_2)}|g_{y_0}(k)|^2 dk < \infty \tag{5.9.47}$$

(for this is the inner product of g_{y_1} and a fixed L^2-function. Thus,

$$\sup_{y_1 > y_0} e^{\alpha(y_1-y_0)} \int_{-\infty}^{-\alpha} |g_{y_0}(k)|^2 dk < \infty \tag{5.9.48}$$

which shows that $g_{y_0}(\,\cdot\,)$ is supported on $[0,\infty)$ for each y_0.

Define

$$g(k) = e^{ky_0} g_y(k) \tag{5.9.49}$$

which is y_0-independent by (5.9.46). Since $f \in H^2(\mathbb{C}_+)$ and $\,\hat{}\,$ is an L^2-isometry,

$$\sup_{y_0} \int e^{-2ky_0}|g(k)|^2 dk < \infty \tag{5.9.50}$$

It follows from monotone convergence, that $g \in L^2$ and then by dominated convergence that

$$\|g_{y_0} - g\|_2 \to 0 \text{ as } y_0 \downarrow 0 \tag{5.9.51}$$

By the Fourier inversion formaula, (5.9.35) and (5.9.49) imply that f is related to g by (5.9.34).

(c) Let $f^* = \breve{g}$. Then, (5.9.51) and the fact that the inverse Fourier transform is an isometry imply (5.9.38).

(d) Fix $y_1 \in (0, \infty)$. We claim that for $y_0 > 0$

$$f(x_0 + iy_0 + iy_1) = \frac{1}{\pi} \int \frac{y_0}{(x - x_0)^2 + y_0^2} f(x + iy_1) \, dx \qquad (5.9.52)$$

Both sides of (5.9.52) are (complex) harmonic functions of $x_0 + iy_0$ and since $f(\cdot + iy_1)$ is bounded (by (5.9.31)) and continuous, they have the same limits as $y_0 \downarrow 0$.

Moreover, both go to zero as y_0 goes to ∞: f by (5.9.31) and the right side since if $h_{y_0}(x) = \frac{1}{\pi} \frac{y_0}{x^2 + y_0^2}$, then $\|h_{y_0}\|_\infty = \frac{1}{\pi y_0}$, $\|h_{y_0}\|_1 = 1$ (by scaling) so $\|h_{y_0}\|_2 \to 0$ and thus the right side $q \to 0$ by $\|q(x_0 + iy_0)\| \leq \|h_{y_0}\|_2 \|f(\cdot + iy_1)\|_2$.

Using (c), we can take $y_1 \downarrow 0$ to zero in (5.9.52) and get (5.9.39).

(e) Both sides are analytic in $\mathbb{C}_+$ and they have the same real parts by (5.9.39). Thus, they differ by an imaginary constant. By (5.9.31), $|f(iy)| \to 0$ as $y \to \infty$ and since, by monotone convergence,

$$\int \frac{1}{x_0^2 + y_0^2} \, dx \to 0 \text{ as } y_0 \to \infty \qquad (5.9.53)$$

the right side evaluated at $x_0 = 0$, $y_0 \to \infty$ goes to zero. Thus, the constant is 0.

(f), (g) are immediate from (5.9.39). Lemma 2.4.2, and Theorem 2.4.7 $\square$

Theorem 5.9.6. *Let $u \in h^2(\mathbb{C}_+)$. Then u has a boundary value u^* in the sense that $u^* \in L^2(\mathbb{R})$ and $\|u(\cdot + iy_0) - u^*(\cdot)\|_2 \to 0$. If v and v^* are defined by*

$$\widehat{v^*}(k) = -i \operatorname{sgn}(k) \widehat{u^*}(k) \qquad (5.9.54)$$

$$v(x_0 + iy_0) = \frac{1}{\pi} \int \frac{y_0}{(x - x_0)^2 + y_0^2} v^*(x) \, dx \qquad (5.9.55)$$

then v is real-valued and $u + iv \in H^2(\mathbb{C}_+)$. Moreover, for a.e. $x_0 \in \mathbb{R}$

$$v^*(x_0) = \lim_{\varepsilon \downarrow 0} \frac{1}{\pi} \int \frac{(x - x_0)}{(x - x_0)^2 + \varepsilon^2} u^*(x) \, dx \qquad (5.9.56)$$

$$\|u + iv\|_{H^2} = 2\|u\|_{h^2} \qquad (5.9.57)$$

Remarks. 1. By Corollary 5.9.4, v is the unique such harmonic conjugate.

2. We'll show below (see Corollary 5.9.9) that for a.e. x_0

$$v^*(x_0) = \lim_{\varepsilon \downarrow 0} \frac{1}{\pi} \int_{|x - x_0| > \varepsilon} \frac{u^*(x)}{x - x_0} \, dx \qquad (5.9.58)$$

is the Hilbert transform. This shows that the Hilbert transform is an isometry on L^2 since $|\operatorname{sgn}(k)| = 1$ for a.e. k.

Proof. Suppose first that u is continuous on $\overline{\mathbb{C}_+}$ and its restriction to $\mathbb{R}$, u^*, is in L^2. Define v^* by (5.9.54) and inverse Fourier transform. A function w in $L^2(\mathbb{R})$ has

$$w \text{ is real-valued} \Leftrightarrow \widehat{w}(-k) = \overline{\widehat{w}(k)} \tag{5.9.59}$$

Since $-i\operatorname{sgn}(-k) = \overline{-i\operatorname{sgn}(k)}$, u^* real-valued implies that v^* is real-valued.

By construction, $\widehat{u^* + iv^*}$ vanishes on $(-\infty, 0]$ so, by Theorem 5.9.5, $u + iv \in H^2(\mathbb{C}_+)$. Moreover,

$$\|u + iv\|_{H^2}^2 = \|u^* + iv^*\|_{L^2}^2 = \|\widehat{u^*}(k) + \widehat{iv^*}(k)\|_{L^2}^2 \tag{5.9.60}$$

$$= 4 \int_0^\infty |\widehat{u^*}(k)|^2 \, dk \tag{5.9.61}$$

$$= 2 \int_{-\infty}^\infty |\widehat{u^*}(k)|^2 \, dk \tag{5.9.62}$$

by (5.9.59). This proves (5.9.57).

In the general case, for any $y_0 > 0$, we can apply the above argument to $u_{y_0}(z) = u(iy_0 + z)$ and so get $f_{y_0}(z) \equiv f(iy_0 + z) \in H^2(\mathbb{D})$ with $\|f_{y_0}\|_{H^2} = 2\|u_{y_0}\|_{h^2} \leq 2\|u\|_{h^2}$. The $f(iy_0 + z)$ fit into a function, f, that lies in $H^2(\mathbb{D})$ with $\operatorname{Re} f = u$. Since $f \in H^2$ has L^2 boundary values, so does u. (5.9.54) and (5.9.57) hold by taking limits of $u_{y_0} + iv_{y_0}$ results.

We get (5.9.55) and (5.2.29) by taking the imaginary parts of (5.9.39) and (5.9.41) ☐

To develop the general H^p theory, one needs four steps:

(1) (Analyze Blaschke Products) The image of a Blaschke factor under φ is clearly

$$b_{z_0}(z) = \begin{cases} e^{i\alpha} \frac{z - z_0}{z - \bar{z}_0}, & z_0 \neq i \\ \frac{z - i}{z + i}, & z_0 = i \end{cases} \tag{5.9.63}$$

where α is such that $b_{z_0}(i) > 0$. The analog of the Blaschke condition $\sum_j 1 - |w_j|^2 < \infty$ is $\sum_j 1 - |\frac{i - z_j}{i + z_j}|^2 < \infty$. Notice that

$$1 - \left|\frac{i - z}{i + z}\right| = \frac{4 \operatorname{Im} z}{|i + z|^2} \tag{5.9.64}$$

$\operatorname{Im} z > 0 \Rightarrow 1 + |z|^2 \leq |i + z|^2 \leq 2(1 + |z|^2)$ so

Proposition 5.9.7. *There exits an $H^\infty(\mathbb{C}_+)$-function vanishing exactly at $\{z_j\}_{j=1}^\infty$ (counting multiplicity) if and only if*

$$\sum_{j-1}^\infty \frac{\operatorname{Im}(z_j)}{1 + |z_j|^2} < \infty \tag{5.9.65}$$

and if that holds then $\prod_{j=1}^\infty b_{z_j}(z_0)$ is such a function.

We'll see shortly that $H^p(\mathbb{C}_+)$-functions, $0 < p \leq \infty$ are of the form $f(z) = g(\psi(z))$ for some $g \in H^p(\mathbb{D})$ so f has zeros obeying (5.9.65).

(2) (Noncontainment) For $\mathbb{D}$, $H^p(\mathbb{D}) \subset H^r(\mathbb{D})$ if $p > r$. This is false for $\mathbb{C}_+$. $f(z) = 1$ is in $H^\infty(\mathbb{C}_+)$, but no other H^p. $f(z) = (z+i)^{-1}$ is in H^p for $p > 1$ but not H^1. This implies a natural guess that $f \in H^p(\mathbb{C}_+) \Leftrightarrow f \circ \varphi \in H^p(\mathbb{D})$ is false, but fortunately the really useful direction is true!

(3) $f \in H^p(\mathbb{C}_+) \Rightarrow f \circ \varphi \in H^p(\mathbb{D})$ for $|f(z)|^p$ is subharmonic and goes to zero at ∞. If

$$g_h(x_0 + iy_0) = \frac{1}{\pi} \int_{-\infty}^{\infty} \frac{y_0 |f(x + iy_0 + ih)|^p}{(x - x_0)^2 + y_0^2} \, dy_0 \qquad (5.9.66)$$

then g is harmonic on $\mathbb{C}_+$ and goes to zero at ∞ so $g_h - |f(\cdot + ih)|^p$ is superharmonic and is zero on $\mathbb{R} \cup \{\infty\}$. It follows that it is everywhere nonnegative, i.e.,

$$|f(x + iy + ih)|^p \leq \frac{1}{\pi} \int \frac{y}{(x - t)^2 + y^2} \, d\mu_h(t) \qquad (5.9.67)$$

$$d\mu_h(t) = |f(t + ih)|^p dt \qquad (5.9.68)$$

Since $f \in H^p(\mathbb{C}_+)$, $d\mu_h$ as measures on $\mathbb{R} \cup \{\infty\}$ of size at most $\|f\|_p^p$ have a weak-$*$ limit point $d\mu$ of mass (on $\mathbb{R}$) at most $\|f\|_p^p$ so that

$$|f(z)|^p \leq \frac{1}{\pi} \int \frac{\mathrm{Im}\, z}{|z - t|^2} \, d\mu(t) \qquad (5.9.69)$$

Let $g(w)$ obey $g(\varphi(z)) = f(z)$. By a change of variables (Problem 3), (5.9.69) becomes

$$|g(re^{i\theta})|^p \leq \frac{1}{2\pi} \int_0^{2\pi} \frac{1 - r^2}{1 + r^2 - 2\cos(\theta - \tau)} \, d\nu(\tau) \qquad (5.9.70)$$

where if $\frac{i-t}{i+t} = e^{it}$ (i.e., $t = \tan(\frac{\tau}{2})$), then

$$d\nu(\tau) = \frac{2 d\mu(t)}{1 + t^2} \qquad (5.9.71)$$

$\int_{-\infty}^{\infty} d\mu \leq \|f\|_p^p \Rightarrow \int d\nu \leq 2\|f\|_p^p$ and so $g \in H^p(\mathbb{D})$ with $\|g\|_p^p \leq \pi^{-1}\|f\|_p^p$. Thus, the zeros of f obey a Blaschke condition. Notice that $(z+i)^{-p/2}f(z)$ also is of the form $h(\varphi(z))$ showing again that $f \mapsto g$ is not onto all of $H^p(\mathbb{D})$.

(4) Once one has the Blaschke condition, one can prove a Riesz factorization and an inner-outer factorization. We also see functions in $H^p(\mathbb{C}_+)$ have pointwise boundary values.

This completes what we want to say about $H^p(\mathbb{C}_+)$. We turn finally to the analog of the M. Riesz theorem. We could mimic the weak-L^1 proof of Section 5.7 but instead we'll present a different proof that relies on a remarkable *equality* of Boole. We consider Hilbert transforms of positive measures; that is we define for μ a finite measure on $\mathbb{R}$

$$H\mu(x) = \lim_{\varepsilon\downarrow 0} \frac{1}{\pi} \int_{-\infty}^{\infty} \mathrm{Re}\left(\frac{1}{t - x - i\varepsilon}\right) d\mu(t) \tag{5.9.72}$$

This limit exists for a.e. x because $f(z) = \int (t - z)^{-1} d\mu(t)$ is analytic in $\mathbb{C}_+$ with $\mathrm{Im}\, f > 0$ so $g(z) \equiv e^{if(z)} \in H_\infty(\mathbb{C}_+)$ so g has boundary values which are a.e. nonzero so $f = i\log g$ has a limit.

We begin with an analog of Theorem 5.7.8—the proof is esentially identical but the calculations are simpler because $(t - z)^{-1}$ is simpler than the Cauchy kernel, $(e^{i\theta} + z)(e^{i\theta} - z)^{-1}$. We define

$$(H_\varepsilon\mu)(x) = \frac{1}{\pi} \int_{|x-t|>\varepsilon} \frac{d\mu(t)}{t - x} \tag{5.9.73}$$

$$(Q_\varepsilon\mu)(x) = \frac{1}{\pi} \int_{-\infty}^{\infty} \mathrm{Re}((t - x - i\varepsilon)^{-1}) d\mu(t) \tag{5.9.74}$$

Theorem 5.9.8. *We have for any positive measure μ that*

$$|(H_\epsilon\mu)(x) - (Q_\varepsilon\mu)(x)| \le (1 + \tfrac{2}{\pi})(M_{\mathrm{HL}}\mu)(x) \tag{5.9.75}$$

In particular, for Lebesgue a.e. x

$$H\mu(x) = \lim_{\varepsilon\downarrow 0}(H_\varepsilon\mu)(x) \tag{5.9.76}$$

Proof. We write the integral in $H_\varepsilon\mu - Q_\varepsilon\mu$ as two terms, I_1 and I_2. I_1 is the integral of the Q_ε kernel from $x - \varepsilon$ to $x + \varepsilon$ and I_2 the integral of the difference of kernels on $\{t \mid |x - t| > \varepsilon\}$.

Since the kernel of Q_ε is bounded by $|t - x - i\varepsilon|^{-1} \le \varepsilon^{-1}$, we see that

$$|I_1(x)| \le \frac{1}{\pi\varepsilon}\mu([x - \varepsilon, x + \varepsilon]) \le \frac{2}{\pi}(M_{\mathrm{HL}}\mu)(x) \tag{5.9.77}$$

To estimate I_2 we note that the absolute value of the difference of the kernels is

$$\left|\frac{1}{t - x} - \frac{t - x}{(t - x)^2 + \varepsilon^2}\right| \le \frac{\varepsilon}{|t - x|}\frac{\varepsilon}{(t - x)^2 + \varepsilon^2} \tag{5.9.78}$$

If $|t - x| > \varepsilon$, then $\frac{\varepsilon}{|t-x|} < 1$, so

$$|I_2(x)| \le \frac{1}{\pi}\int \frac{\varepsilon}{(t - x)^2 + \varepsilon^2}\, d\mu(x) \le (M_{\mathrm{HL}}\mu)(x) \tag{5.9.79}$$

by Lemma 2.4.2.

(5.9.75) implies $\lim_{\varepsilon\downarrow 0} H_\varepsilon(x)$ exists and equals $\lim_{\varepsilon\downarrow 0} Q_\varepsilon(x)$ if $M_{\mathrm{HL}}\mu(x) < \infty$ and $\lim_{\varepsilon\downarrow 0} Q_\varepsilon(x) \equiv H\mu(x)$ exists. Since each happens for a.e. x, we get the result. $\qquad\square$

Corollary 5.9.9. *If $f \in L^p(\mathbb{R}, dx)$, $1 \leq p < \infty$ and $H_\varepsilon f$, $Q_\varepsilon f$ are defined by (5.9.73)/(5.9.74) with $d\mu$ replaced by $f dx$, then for a.e. x, the limits as $\varepsilon \downarrow 0$ of $H_\varepsilon f(x)$ and $Q_\varepsilon f(x)$ exist and are equal.*

Proof. Write $f = g+h$ where g is f restricted to $(x-1, x+1)$ and $h = f-g$. Since x^{-1} restricted to $\mathbb{R} \setminus (-1,1)$ is in all L^q, $1 < q \leq \infty$, it is easy to prove that $H_\varepsilon h(y)$ and $Q_\varepsilon h(y)$ have equal limits as $\varepsilon \downarrow 0$ uniformly for $y \in (x - \frac{1}{2}, x + \frac{1}{2})$. Since $g\, dx$ is a linear combination of positive measures, Theorem 5.9.8 completes the proof. $\square$

The center of our analysis of the Hilbert transforms of measures will be a remarkable equality of Boole.

Theorem 5.9.10 (Boole's Equality). *Let μ be a finite (positive) point measure with a finite number of pure points. Then for each $\lambda > 0$,*

$$|\{x \mid \pm(H\mu)(x) > \lambda\}| = \frac{\|\mu\|}{\pi\lambda} \tag{5.9.80}$$

Remarks. 1. We emphasize that (5.9.80) has $=$ not $\leq$.

2. (5.9.80) holds for any singular measure; see the Notes.

Proof. Write $\mu = \sum_{j=1}^n c_j \delta(x - x_j)$ with $x_1 < x_2 < \ldots < x_n$ and $c_j > 0$ so that

$$H\mu(x) = \frac{1}{\pi} \sum_{j=1}^n \frac{c_j}{(x_j - x)} \tag{5.9.81}$$

Then $\frac{dH\mu}{dx} > 0$ so $H\mu$ is strictly monotone and goes to infinity at each x_j. Each interval $(-\infty, x_1)$, (x_1, x_2) has exactly one point $y_1, \ldots, y_n$ where $H\mu(x) = \lambda$ (since $H\mu \to 0$ as $x \mapsto \pm\infty$ and to $-\infty$ as $x \downarrow x_j$). Thus,

$$\text{LHS of (5.9.80)} = \sum_{j=1}^n (x_j - y_j)$$

$$= \sum_{j=1}^n x_j - \sum_{j=1}^n y_j \tag{5.9.82}$$

Clearly

$$H\mu(x) = \lambda \Leftrightarrow Q_\lambda(x) \equiv \lambda \prod_{j=1}^n (x_j - x) - \frac{1}{\pi} \sum_{j=1}^n c_j \prod_{k \neq j} (x_k - x) = 0 \tag{5.9.83}$$

$Q_\lambda(x)$ is a polynomial of degree n, so it has n zeros, so $y_1, \ldots, y_n$ are all the zeros. Now

$$(-1)^n Q_\lambda(x) = \lambda \big[x^n - \big(\sum_{j=1}^n x_j - \frac{1}{\pi\lambda} \sum_{j=1}^n c_j\big) x^{n-1} + \ldots \big] \tag{5.9.84}$$

so

$$\sum_{j=1}^{n} y_j = \sum_{j=1}^{n} x_j - \frac{1}{\pi\lambda} \sum_{j=1}^{n} c_j \tag{5.9.85}$$

By (5.9.82) and (5.9.85)

$$\text{LHS of (5.9.80)} = \frac{1}{\pi\lambda} \sum_{j=1}^{n} c_j = \frac{\|\mu\|}{\pi\lambda} \tag{5.9.86}$$

proving (5.9.80) for the $+$ in $\pm$. For the minus, let $\tilde{\mu} = \sum_{j=1}^{n} c_j \delta(x + x_j)$ and let $(H\tilde{\mu})(-x) = -(H\mu)(x)$ and get the $-$ from $+$ and so $\pm$. $\square$

In going beyond the classic Boole inequality, we begin by extending it to purely singular measures which are distinguished by the fact that

$$F_\mu(z) = \int \frac{d\mu(t)}{t - z} \tag{5.9.87}$$

has $F_\mu(x + i0)$ real for a.e. x; see (2.5.20).

Theorem 5.9.11. (5.9.80) *holds for any finite (positive) measure,* μ, *on* $\mathbb{R}$, *with* μ *purely singular, i.e.,* $\mu(\mathbb{R} \setminus \mathfrak{e}) = 0$ *for some set,* $\mathfrak{e}$, *of Lebesgue measure* 0.

Proof. We exploit the fact that for any $L^1(\mathbb{R}, dx)$-function, $\psi(x)$, we have that, as $\operatorname{Im} z \to \infty$, $|\operatorname{Re} z|$ bounded

$$\frac{\operatorname{Im} z}{\pi} \int \frac{\psi(x)dx}{(x - \operatorname{Re} z)^2 + (\operatorname{Im} z)^2} = \frac{1}{\pi \operatorname{Im} z} \int \psi(x)dx + O\left(\frac{1}{(\operatorname{Im} z)^2}\right) \tag{5.9.88}$$

Define for $w \in \mathbb{C}_+$

$$\varphi_\lambda(w) = \pi - \operatorname{Im}(\log(w - \pi\lambda)) \tag{5.9.89}$$

This function is harmonic on $\mathbb{C}_+$ and for $u \in \mathbb{R}$

$$\varphi_\lambda(u + i0) = \begin{cases} \pi, & u > \pi\lambda \\ 0, & u < \pi\lambda \end{cases} \tag{5.9.90}$$

Thus, if, with F_μ given by (5.9.87)

$$\psi_\lambda(z) = \frac{1}{\pi}\varphi_\lambda(F_\mu(z)) \tag{5.9.91}$$

has

$$|\{x \mid (H\mu)(x) > \lambda\}| = \int \psi_\lambda(x + i0)dx \tag{5.9.92}$$

since $F_\mu(x + i0)$ is real for a.e. x and so equal to $\pi(H\mu)(x)$.

Thus, by (5.9.92) and (5.9.88),

$$|\{x \mid H\mu(x) > \lambda\}| = \lim_{y \to \infty} y\varphi_\lambda(F(iy)) \tag{5.9.93}$$

As $y \to \infty$, $F(iy) = \frac{i}{y}\|\mu\| + O(y^{-2})$ and for w near 0, $\log(w - \pi\lambda) = \log(-\pi\lambda) - \frac{w}{\pi\lambda} + O(w^2)$ so $\varphi_\lambda(w) = \frac{(\mathrm{Im}\, w)}{\pi\lambda} + O(w^2)$. Thus, as $y \to \infty$

$$\varphi_\lambda(F(iy)) = \frac{\|\mu\|}{\pi\lambda y} + O(y^{-2}) \qquad (5.9.94)$$

so (5.9.93) implies (5.9.80) $\qquad\qquad\qquad\qquad\qquad\qquad \square$

Next we turn to a weak-L^1 bound on F_μ and so on $H\mu$ for general μ. One can get this by carefully approximating by finite point measures and using Boole's equality (see the Notes) but instead we'll use the idea of the last proof. We are going to need a function like the φ_λ of (5.9.90) but one that tests whether $w \in \mathbb{C}_+$ has $|w| < \lambda$ or not. For $\lambda > 0$, we define on $\mathbb{C}_+$

$$\Phi_\lambda(w) = \pi - \mathrm{Im}\log\left(\frac{w - \lambda}{w + \lambda}\right) \qquad (5.9.95)$$

Lemma 5.9.12. $\Phi_\lambda(w) > 0$ *on* $\mathbb{C}_+$ *and on* $\{w \mid |w| \geq \lambda\}$, *we have* $\Phi_\lambda(w) \geq \frac{\pi}{2}$.

Proof. We have that

$$\frac{w - \lambda}{w + \lambda} = \frac{-\lambda^2 + |w|^2 + 2i\lambda\,\mathrm{Im}\,w}{|w + \lambda|^2} \qquad (5.9.96)$$

If $\mathrm{Im}\,w > 0$, this lies in $\mathbb{C}_+$ and has argument $\frac{\pi}{2}$ exactly if $|w| = \lambda$ and argument in $(0, \frac{\pi}{2})$ (respectively, $(\frac{\pi}{2}, \pi)$) if $|w| < \lambda$ (respectively, $|w| > \lambda$). Since $\mathrm{Im}\log(z) = \mathrm{Arg}(z)$ taking into account that the π had a minus sign, we get the claimed result.

While the algebraic argument is simple, it obscures the geometric reason that the result is true (which explains the choice of function). Consider the triangle with vertices $-\lambda$, λ, and w. $\mathrm{Im}\log(w - \lambda)$ is the angle between (λ, ∞) and the line from λ to w, i.e., in Figure 5.9.1, $\pi - \beta$. Similarly, $\mathrm{Im}\log(w + \lambda)$ is the angle α. Thus $\mathrm{Im}\log\left(\frac{w-\lambda}{w+\lambda}\right)$ is $\pi - (\alpha + \beta)$, i.e., the angle of the vertex at w. As is well-known, the semicircle, $|z| = \lambda$, $\mathrm{Im}\,z > 0$ is exactly the locus of points where the angle is $\frac{\pi}{2}$ (the proof is recalled in Figure 5.9.1—given the isosceles triangle, $2\alpha + 2\beta = \pi$, so $\pi - (\alpha + \beta) = \frac{\pi}{2}$). For points outside

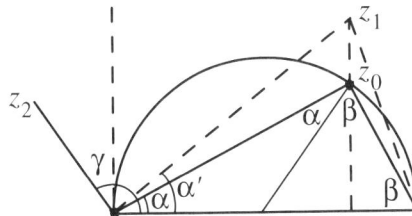

Figure 5.9.1. A geometric proof of Lemma 5.9.12.

of the semicircle, with $|\mathrm{Re}(w)| < \lambda$, like z_1 in the figure, $\alpha' > \alpha$, $\beta' > \beta$ so the angle at w is smaller than $\frac{\pi}{2}$. This is true for points with $\mathrm{Re}\, w > \lambda$, like z_2 in the figure since on the left the base angle is larger than $\frac{\pi}{2}$. Thus, $\pi - \Phi_\lambda(w)$ lies in $(0, \frac{\pi}{2})$ exactly if $|w| < \lambda$ proving the result. $\qquad\square$

Theorem 5.9.13. *For any finite (positive) measure, μ, on $\mathbb{R}$, we have that (with F_μ given by (5.9.87) and $H\mu$ by (5.9.72) so $(H\mu)(x) = \frac{1}{\pi} \mathrm{Im}\, F_\mu(x + i0))$:*

$$\left|\{x \mid |F_\mu(x + i0)| \geq \lambda\}\right| \leq \frac{4\|\mu\|}{\lambda} \qquad (5.9.97)$$

$$\left|\{x \mid |H\mu(x)| \geq \lambda\}\right| \leq \frac{4\|\mu\|}{\pi\lambda} \qquad (5.9.98)$$

Remark. It is known (see the Notes) that the optimal constant in (5.9.98) has $\frac{\|\mu\|}{\lambda}$ so this $\frac{4}{\pi}$ is close to optimal.

Proof. Since $|F_\mu(x + i0)| \geq \pi|(H\mu)(x)|$, (5.9.98) follows from (5.9.97). To prove that, write

$$\Phi_\lambda(z) = \frac{2}{\pi}\Phi_\lambda\big(F_\mu(z)\big) \qquad (5.9.99)$$

and note that, by the lemma,

$$\left|\{x \mid |F_\mu(x + i0)| > \lambda\}\right| \leq \left|\{x \mid \Psi_\lambda(x + i0) > 1\}\right|$$

$$\leq \int \Psi_\lambda(x + i0)dx \qquad (5.9.100)$$

$$= \lim_{y \to \infty} 2y\Phi_\lambda\big(F(iy)\big) \qquad (5.9.101)$$

If ε is small,

$$\Phi_\lambda(i\varepsilon) = \pi - \log\left[-\left(1 - \frac{2i\varepsilon}{\lambda}\right) + O(\varepsilon^2)\right] \qquad (5.9.102)$$

$$= \frac{2\varepsilon}{\lambda} + O(\varepsilon^2) \qquad (5.9.103)$$

Since $F(iy) = i\|\mu\|\, y^{-1} + O(y^2)$, this plus (5.9.101) implies (5.9.97). $\qquad\square$

Once one has this and the L^2-bounds, one can use interpolation and duality to get boundedness of H from L^p to L^p for $1 < p < \infty$.

This argument easily controls a maximal Hilbert transform as well. We instead suppose that

$$S \subset \{x \mid \sup_{\varepsilon > 0} |(H_\varepsilon \mu)|(x) > \lambda\} \qquad (5.9.104)$$

In (5.9.90), we only used that the limit exists to be sure that $\mu(\{x \pm \delta_x\}) = 0$. If we take $d\mu = f\, dx$, with $f \in L^1$, μ has no pure points so $\mu(\{x \pm \delta_x\}) = 0$ is automatic and the result of the proof goes through with no change. Noting that

$$\int_{a<|x-t|<b} \frac{d\mu(t)}{t-x} = (H_a\mu)(x) - (H_b\mu)(x) \qquad (5.9.105)$$

so

$$\sup_{a,b} \left| \int_{a<|x-t|<b} \frac{d\mu(t)}{t-x} \right| \leq 2\sup_{\varepsilon} |(H_\varepsilon\mu)(t)| \qquad (5.9.106)$$

and we have

Theorem 5.9.14. *For $f \in L^1(\mathbb{R})$ define the maximal Hilbert transform by*

$$(M_H f)(x) = \sup_{0<a<b} \left| \int_{a<|x-t|<b} \frac{f(t)dt}{t-x} \right| \qquad (5.9.107)$$

Then for a universal constant, C,

$$|\{x \mid M_H f(x) > \lambda\}| \leq C \|f\|_{L^1(\mathbb{R},dx)} \lambda^{-1} \qquad (5.9.108)$$

Remark. Since $\frac{1}{2}\cot(\frac{x}{2}) - \frac{1}{x}$ is bounded for $x \in [-\pi,\pi]$, one can deduce an L^1-maximal inequality for the circular Hilbert transform also (Problem 6).

Notes and Historical Remarks. Koosis [486] has a chapter on $H^p(\mathbb{C}_+)$ which impacted our presentation and has some details of the theory we only sketched. Duren [234] has a chapter on $H^p(\Omega)$ for fairly general Ω that goes beyond the $\mathbb{D}/\mathbb{C}_+$ theory and, in particular, the special role played by circles in $\mathbb{D}$ and boxes in $\mathbb{C}_+$.

While the name Herglotz representation is common for (5.9.14), Herglotz only studied $\mathbb{D}$ (see the discussion in the Notes to Section 5.4 for the reference to Herglotz's work and the simultaneous discovery of Riesz). Pick's and Nevanlinna's names are sometimes associated with what we call Herglotz functions because of their independent papers [646, 607]. They asked for which given $\{z_j\}_{j=1}^n$ in $\mathbb{D}$ and $\{w_j\}_{j=1}^n$ in $\mathbb{H}_+$, there exists a Carethéodory function f with $f(z_j) = w_j$. Pick only considered $\mathbb{D}$ but Nevanlinna considered also $\mathbb{C}_+$ using the conformal map φ. Both used the Herglotz representation for $\mathbb{D}$.

Boole's theorem is from his 1857 paper [95]. It was rediscovered several times in the middle of the last century, in particular by Loomis [544] and Stein–Weiss [784]. Loomis then used this result to prove the general result that $H\mu$ is in weak-L^1 for any measure, a result first proven by him. Stein–Weiss proved the related result that if E is any Borel set, then

$$|\{x \in \mathbb{R} \mid |H(\chi_E dx)| > \lambda\}| = \frac{2|E|}{\sinh(\pi\lambda)} \qquad (5.9.109)$$

which implies Boole's result (Problem 4). Our proof of Theorems 5.9.11 and 5.9.13 follows Aizenman–Warzel [10].

Using Brownian motion ideas, Davis [**200, 201**] found the optimal constant in (5.9.87) for signed measures, namely

$$|\{x \mid |(H\mu)(x)| > \lambda\}| \le \frac{\|\mu\|}{\lambda} \tag{5.9.110}$$

Boole's equality, (5.9.80), holds for any purely singular measure. This was proven by Hrušcev–Vinogradov [**404**]. A different approach using rank-one perturbation theory (see Section 5.8 of Part 4) was found independently by del Rio et al. [**210**] and Poltoratski [**662**].

If $f \in L^1$, one can write $f = g + h$ where $\|h\|_1 < \varepsilon$ and $\|g\|_2 < \infty$. Since

$$|\{x \mid |(Hg)(x)| > \lambda\}| \le \|g\|_2^2/\lambda^2 \tag{5.9.111}$$

we see that for $f \in L^1$

$$\lim_{|\lambda|\to\infty} |\lambda|\,|\{x \mid |Hf(x)| > \lambda\}| \tag{5.9.112}$$

exist. In turn, this implies (by the extended Boole equality) that

$$\lim_{\lambda\to\infty} \pi|\lambda||\{x \mid (\pm H\mu)(x) \ge \lambda\}| = \|\mu_s\| \tag{5.9.113}$$

In fact, Poltoratski [**662**] has shown that as $\lambda \to \infty$, $\frac{1}{2}\pi\lambda\chi_{\{x||H\mu(x)|>\lambda\}}(x)dx$ converges weakly to μ_s. Poltorastski et al. [**663**] have refined estimates that for certain sets $e \subset \mathbb{R}$,

$$\lim_{|\lambda|\to\infty} |\lambda|\,|\{x \mid |Hf(x)| > \lambda\} \cap e| = 0 \Rightarrow \mu_s(e) = 0 \tag{5.9.114}$$

There is extensive literature on H^p ($1 \le p \le \infty$) spaces on $\mathbb{R}^\nu_+$ starting with Fefferman–Stein [**271**]. These are real-valued functions in L^p, so that for $j = 1, \ldots, \nu$, the Riesz transforms $(k_j|k|^{-1}\widehat{f})^\vee \in L^p$. If $\nu = 1$, so the Riesz transform is the Hilbert transform, this is precisely the real parts of the boundary values of $H^p(\mathbb{C}_+)$-functions. For $1 < p < \infty$, $H^p(\mathbb{R}^\nu)$ is $L^p(\mathbb{R}^\nu)$ but, as for $\nu = 1$, H^1 is a strict subset of L^1. We'll say more about this in Section 5.11.

Problems

1. (a) In $\mathbb{R}$, compute the Fourier transform of $e^{-y|x|}$.

   (b) Use (a) plus (5.9.34) plus $g \in L^2([0,\infty))$ to prove (5.9.39).

2. (a) Let $f \in L^\infty(\mathbb{R})$ with $\text{supp}(f) \subset [a,\infty)$ for some $a \in \mathbb{R}$. Let

$$F_z(x) = e^{ixz}f(x) \tag{5.9.115}$$

Prove that for $z \in \mathbb{C}_+$, $F_z \in L^2$ and $z \mapsto F_z$ is an analytic L^2-valued function there.

(b)] If $f \in H^2(\mathbb{C}_+)$ prove that if

$$G_z(x) = f(x + z)$$

for $z \in \mathbb{C}_+$, then $z \mapsto G_z$ is an L^2-valued analytic function of z.

Remark. See Theorems 3.1.6 and 3.1.12 of Part 2A.

3. If φ is given by (5.9.1) and g, f are related by $g(\varphi(z)) = f(z)$, prove that (5.9.69) is equivalent to (5.9.70) when μ and ν are related by (5.9.71).

4. Prove that (5.9.109) for finite unions of intevals implies Boole's equality.

5. Prove that if $f_1, \ldots f_\ell$ are finitely many piecewise continuous functions on $\mathbb{R}$ (with one-sided limits at the points of discontinuity) and μ is a Baire measure on some $[-a, a]$ with no pure points at the points of discontinuity of the f_j, then there exist point measures ν_n on $[-a, a]$ so that for $j = 1, \ldots, \ell$, $\int f_j d\nu_n \to \int f_j \, d\mu$.

6. Let $f \in L^1(\partial \mathbb{D}, \frac{d\theta}{2\pi})$ and let $M_{\mathrm{HT}} f$ be given by (5.7.81) where H_r is the circular Fourier transform. Let g be defined on $[-6\pi, 6\pi]$ by (5.7.117) and let $M_H g$ be its maximal Hilbert transform given by (5.9.107). Prove that for $\theta \in [-\pi, \pi]$, $|(M_{\mathrm{HT}} f)(e^{i\theta})| \le (M_{\mathrm{HL}} g)(\theta) + (M_{\mathrm{HT}} g)(\theta)$ and conclude that $M_{\mathrm{HT}} f$ obeys a weak-L^1 inequality.

5.10. Beurling's Theorem

In this section, we'll discuss a problem in operator theory whose solution not only depends on H^p theory, but even the statement of the solution does. An important question in operator theory is, given a bounded operator, A, on a Banach space, X, to find all closed subspaces, $Y \subset X$, left invariant by A, that is, with $A[Y] \subset Y$. In Section 5.12 of Part 1,we showed that compact operators have invariant subspaces and in Section 3.4 of Part 4 we discussed the structure of invariant subspaces for compact operators on a Hilbert space. For self-adjoint operators, the spectral theorem provides a solution to this problem (see Chapter 5 of Part 4) but, in general, the structure of the set of invariant subspaces is a hard question, as we'll see by looking at one of the simplest nonnormal operators. On ℓ^2, let L be the left shift, that is,

$$L(a_1, a_2, \ldots) = (0, a_1, a_2, \ldots) \tag{5.10.1}$$

A simple model of L is the look at $H^2(\mathbb{D})$ which we saw (Theorem 5.2.1) is isometrically isomorphic to ℓ^2 under $I \colon \ell^2 \to H^2$ by

$$I(a_1, a_2, \ldots) = \sum_{n=0}^{\infty} a_{n+1} z^n \tag{5.10.2}$$

Clearly,

$$[ILI^{-1}]f = zf(z) \tag{5.10.3}$$

Thus, we are interested in knowing what subspaces of H^2 are left invariant by multiplication by z. This is answered by

Theorem 5.10.1 (Beurling's Theorem). *Let $Y \subset H^2(\mathbb{D})$ be a nonzero closed subspace invariant under multiplication by z. Then there is an inner function (i.e., $h \in H^\infty(\mathbb{D})$, with $|h(e^{i\theta})| = 1$ for a.e. θ), so*

$$Y = \{hf \mid f \in H^2(\mathbb{D})\} \tag{5.10.4}$$

h is unique if one requires for some $\ell \in \{0, 1, 2, \ldots\}$, $h(z) = z^\ell f(z)$ with $f(0) > 0$.

Remark. We note that for any h, Y given by (5.10.4) is invariant, so this classifies all the invariant subspaces.

By hypothesis, $zY \subset Y$ and by a simple argument (Problem 1), it is closed. Let

$$Z = \{g \in Y \mid g \perp zY\} \tag{5.10.5}$$

We'll eventually show Z is one-dimensional with Z spanned by an inner function, which will be the h of (5.10.4).

Lemma 5.10.2. (a) For $\ell \geq 1$,

$$z^\ell Z \perp Z \tag{5.10.6}$$

(b) For $\ell, k \geq 0$, $\ell \neq k$,

$$z^\ell Z \perp z^k Z \tag{5.10.7}$$

(c) $Y = \left[\bigoplus_{j=0}^{J} z^j Z \right] \oplus z^{J+1} Y.$ \hfill (5.10.8)

(d) $Y = \bigoplus_{j=0}^{\infty} z^j Z.$ \hfill (5.10.9)

(e) $h \in Z \Rightarrow |h(e^{i\theta})|$ is constant.

(f) Z is one-dimensional.

Proof. (a) is immediate since $z^\ell Z \subset zY$.

(b) Let $\ell < k$. Then z^ℓ is an isometry and $Z \perp z^{k-\ell} Z$, so $\langle z^\ell f, z^k y \rangle = \langle f, z^{k-\ell} y \rangle = 0$.

(c) By definition, $Y = Z \oplus zY$. Thus, $z^\ell Y = z^\ell Z \oplus z^{\ell+1} Y$, so (5.10.8) follows by a simple induction.

(d) This follows from (c) if we prove $\bigcap_{j=1}^{\infty} z^j Y = 0$. But if $f \in z^j Y$, f has a zero of order at least j at 0, so if $f \in \bigcap_{j=1}^{\infty} z^j Y$, f has a zero of infinite order, so is zero.

(e) By (5.10.6), if $h \in Z$, $h \perp z^j h$ for $j = 1, 2, \ldots$, that is,

$$\int e^{ij\theta} |h(e^{i\theta})|^2 \frac{d\theta}{2\pi} = 0 \qquad (5.10.10)$$

Taking complex conjugates, this also holds for $j = -1, -2, \ldots$. Thus, the L^1-function, $|h|^2$, has zero Fourier coefficients for all j but $j = 0$. By Fejér's theorem, $|h|^2$ is constant.

(f) By (d) and $Y \neq \{0\}$, $Z \neq \{0\}$. Let $h, q \in Z$ be nonzero. Then $2\operatorname{Re}(\alpha \bar{h} q) = |h + \alpha q|^2 - |h|^2 - |\alpha|^2 |q|^2$ is constant on $\partial \mathbb{D}$ for all complex α. It follows that $\bar{h} q$ is constant, that is, $q \in h(\bar{h}q)/|h|^2$ is a constant multiple of h on $\partial \mathbb{D}$ and so on $\mathbb{D}$. $\qquad \square$

Proof of Theorem 5.10.1. By (e) of the lemma, we can find $h \in Z$ with $|h(e^{i\theta})| = 1$. By (f) and (d) of the lemma, any $g \in Y$ has the form $\sum_{\nu=0}^{\infty} a_n z^n h$, where $\sum_{n=0}^{\infty} |a_n|^2 < \infty$, that is, $g = hf$ for $f \in H^2$.

If Y has the form (5.10.4) for h an inner function, then $\langle h, zhf \rangle = \int |h|^2 e^{i\theta} f(e^{i\theta}) \frac{d\theta}{2\pi} = \int e^{i\theta} f(e^{i\theta}) \frac{d\theta}{2\pi} = 0$, so $h \in Z$. Thus, by (f) of the lemma, h is unique up to a constant $e^{i\psi}$. The normalization condition fixes the constant. $\qquad \square$

Notes and Historical Remarks. Beurling's theorem is from his 1949 paper [66]. The slick proof we use is due to Halmos in 1961 [349].

Problems

1. If f_n are functions in $H^2(\mathbb{D})$, so $\|zf_n - g\|_2 \to 0$ for $g \in H^2(\mathbb{D})$, prove that there is $f_\infty \in H^2(\mathbb{D})$ so $g = zf_\infty$ and $\|f_n - f_\infty\|_2 \to 0$.

5.11. H^p-Duality and BMO

In this section, we answer three natural questions:

(1) (H^p-Duality) We saw (in Section 4.9 of Part 1) a natural duality of L^p and L^q if $1 \leq p < \infty$ and $q^{-1} = 1 - p^{-1}$. It is reasonable to expect that the dual of H^p is H^q. We'll prove this for $1 < p < \infty$ with two twists: the map of $(H^p)^*$ and H^q is antilinear, not linear, and it is not an isometry but only a topological isomorphism. For $p = 1$, the dual will be strictly larger than H^∞; it will be functions $f \in H^1(\mathbb{D})$ whose boundary values, f^*, are in the space, $\mathrm{BMO}(\partial \mathbb{D})$, of functions of bounded mean oscillation—this space, the hero of this section, will be defined below.

(2) (L^∞-extension of Riesz Boundedness) In Section 5.7, we saw that the conjugate map $f \mapsto \widetilde{f}$ takes L^p to L^p for $1 < p < \infty$ and L^1 to L^1_w but said nothing about L^∞. Of course, the L^p-result implies that $f \in L^\infty \Rightarrow \widetilde{f} \in \bigcap_{p<\infty} L^p$. Here we'll prove that $f \in L^\infty \Rightarrow \widetilde{f} \in$ BMO and that BMO is the smallest space containing L^∞ and $\widetilde{L^\infty} = \{\widetilde{f} \mid f \in L^\infty\}$.

(3) (BMO and Interpolation) Theorem 2.2.10 proved an interpolation result that the map of L^{p_j} to $L^{p_j}_w$ ($j = 0, 1$) maps L^p to L^p if $p_0 < p < p_1$. For $p_1 = \infty$, one interpreted L^∞_w as L^∞. But one can do better: for $p = \infty$, one only needs the map to take L^∞ to BMO. We will not prove this here (see the Notes), but it implies, for example, that one can prove the M. Riesz theorem by interpolating from L^1 to L^1_w and L^∞ to BMO without using duality.

We begin with the duality result for $1 < p < \infty$. One can't hope for a direct analog of the L^p-result where any $L \in (L^p)^*$ had the form $L = L_g$ for $g \in L^q$ where

$$L_g(f) = \int g(x)\, f(x)\, d\mu(x) \tag{5.11.1}$$

since for $f \in H^p$, $g \in H^q$:

$$\int g^*(e^{i\theta}) f^*(e^{i\theta}) \frac{d\theta}{2\pi} = f(0)\, g(0) \tag{5.11.2}$$

and there are linear functions on H^p other than $f \mapsto Cf(0)$. For H^2, which is a Hilbert space, we know that $(H^2)^* \simeq H^2$ anti-isomorphically under $g \mapsto \langle g, \cdot \rangle$. This suggests that we define for $f \in H^p(\mathbb{D})$, $g \in H^q(\mathbb{D})$

$$L_g(f) = \int \overline{g^*(e^{i\theta})}\, f^*(e^{i\theta}) \frac{d\theta}{2\pi} \tag{5.11.3}$$

Recall (see Theorem 5.8.4), that if P_+ is defined on functions on $\partial\mathbb{D}$ by

$$(P_+ f)^\sharp_n = \begin{cases} f^\sharp_n, & n \geq 0 \\ 0, & n < 0 \end{cases} \tag{5.11.4}$$

then for some $C_p < \infty$,

$$\|P_+ f\|_p \leq C_p \|f\|_p \tag{5.11.5}$$

$C_2 = 1$ but $C_p = C_q > 1$ if $p \neq 2$ and $C_p \to \infty$ as $p \to 1, \infty$.

Theorem 5.11.1 (Duality for H^p, $1 < p < \infty$). *Let $p \in (1, \infty)$ and $q = p/(p-1)$ the dual index. Then, there is a bijection of $H^p(\mathbb{D})^*$ and $H^q(\mathbb{D})$. Any $L \in H^p(\mathbb{D})^*$ is of the form L_g given by* (5.11.3) *with*

$$\|L_g\|_{(H^p)^*} \leq \|g\|_{H^q} \leq C_p \|L_g\|_{(H^p)^*} \tag{5.11.6}$$

$g \mapsto L_g$ *is an antilinear map.*

Remarks. 1. For $p = 1$, below, we'll identify $(H^1)^*$ as HMO, a space strictly larger than H^∞.

2. We recall that while $H^1 \neq (H^\infty)^*$, it is a dual space, see Corollary 5.3.9—in the language of this section, $H^1 = \mathfrak{A}(\overline{\mathbb{D}})^*$ where $\mathfrak{A}(\overline{\mathbb{D}})$ is functions analytic in $\mathbb{D}$, continuous in $\overline{\mathbb{D}}$.

Proof. If $L = L_g$, then $g_n^\sharp = \overline{L(e^{in\theta})}$, so, if there is a g, it is unique. By Hölder's inequality

$$\|L_g\|_{(H^p)^*} \leq \|g\|_{H^q} \tag{5.11.7}$$

For the converse, if $L \in (H^p)^*$, there is, by the Hahn-Banach theorem, an extension, $\tilde{L}$, of L to all of $L^p(\partial\mathbb{D}, d\theta/2\pi)$ with

$$\|\tilde{L}\|_{(L^p)^*} = \|L\|_{(H^p)^*} \tag{5.11.8}$$

By L^p-duality, there is an $h \in L^q$ with $\|h\|_{L^q} = \|\tilde{L}\|$ and

$$\tilde{L}(f) = \int \overline{h(e^{i\theta})}\, f(e^{i\theta})\, \frac{d\theta}{2\pi} \tag{5.11.9}$$

Since $1 < p < \infty$, $g^* \equiv P_+ h \in L^q$ and

$$g(z) = \sum_{n=0}^\infty h_n^\sharp z^n \tag{5.11.10}$$

lies in H^q with boundary value g^*. By construction, $L_g(e^{in\theta}) = \tilde{L}(e^{in\theta})$ for $n \geq 0$, so since $\{e^{in\theta}\}_{n\geq 0}$ is a total subset of H^p, $L = L_g$.

Finally

$$\|g\|_{H^q} = \|P_+ h\|_{L^q} \leq C_p \|h\|_q = C_p \|L\|_{(H^p)^*} \tag{5.11.11}$$

$\square$

To handle H^1, we need the space, BMO$(\partial\mathbb{D})$, of functions of bounded mean oscillation. This space is also defined on $\mathbb{R}^\nu$, and we'll develop part of the theory for $\mathbb{R}^\nu$. Q will stand for a standard hypercube. i.e.,

$$\underset{j=1}{\overset{\nu}{\times}} [a_j, b_j]$$

where $b_1 - a_1 = b_2 - a_2 = \ldots = b_\nu - a_\nu > 0$, in $\mathbb{R}^\nu$. On $\partial\mathbb{D}$, it will be an interval $\{e^{i\theta} \mid a \leq \theta \leq b\}$ where $|b - a| \leq 2\pi$. Given $f \in L^1_{\text{loc}}(\mathbb{R}^\nu)$ or $L^1(\partial\mathbb{D})$, and Q, we define its *average over Q*

$$f_Q = |Q|^{-1} \int_Q f(x)\, d^\nu x \tag{5.11.12}$$

(or $d\theta/2\pi$ on $\partial\mathbb{D}$) and its *mean oscillation over Q* by

$$|f - f_Q|_Q = \frac{1}{|Q|} \int_Q |f(x) - f_Q|\, d^\nu x \tag{5.11.13}$$

BMO($\mathbb{R}^\nu$), the functions of *bounded mean oscillation* are those $f \in L^1_{\text{loc}}$ with

$$\|f\|_{\text{BMO}} = \sup_Q |f - f_Q|_Q < \infty \tag{5.11.14}$$

$\|\cdot\|_{\text{BMO}}$ is not a norm on functions, since $f(x) \equiv \mathbb{1}$ has $\|f\|_{\text{BMO}} = 0$ but it is a norm on functions modulo constants, i.e., $\|f - g\|_{\text{BMO}} = 0 \Leftrightarrow \exists c \in \mathbb{C}$ so that $f = g + c\mathbb{1}$. One could imagine looking at $|Q|^{-1} \|f - f_Q\|_p$ rather than $p = 1$ but we'll see soon that the space is the same for $1 \leq p < \infty$.

We define the space HMO, the *Hardy space of bounded mean oscillation* as those $f \in H^1(\mathbb{D})$ so that the boundary value f^* of f is in BMO. We put the norm

$$\|f\|_{\text{HMO}} = |f(0)| + \|f^*\|_{\text{BMO}} \tag{5.11.15}$$

Because of the $f(0)$ term, this is a norm and not only a norm mod constants.

There are some equivalent norms on BMO derived from a comparison that holds cubewise. Fix $f \in L^1_{\text{loc}}$ and a cube Q. Define

$$O_Q(f) = |f - f_Q|_Q \tag{5.11.16}$$

$$A_Q(f) = \inf_{c \in \mathbb{C}} |f - c|_Q \tag{5.11.17}$$

$$D_Q(f) = |Q|^{-2} \int_Q \int_Q |f(x) - f(y)| \, d^\nu x \, d^\nu y \tag{5.11.18}$$

Proposition 5.11.2. *We have that*

$$A_Q(f) \leq O_Q(f) \leq D_Q(f) \leq 2A_Q(f) \tag{5.11.19}$$

Remark. In particular, $f \in BMO \Leftrightarrow \sup_Q A_Q(f) < \infty$.

Proof. Clearly $\inf_c |f - c|_Q \leq |f - f_Q|_Q$ since f_Q is a possible value of c; thus $A_Q(f) \leq O_Q(f)$. Note next that

$$f(x) - f_Q = \frac{1}{|Q|} \int_Q (f(x) - f(y)) \, d^\nu y \tag{5.11.20}$$

so

$$|f(x) - f_Q| \leq |Q|^{-1} \int_Q |f(x) - f(y)| \, d^\nu y \tag{5.11.21}$$

proving $O_Q(f) \leq D_Q(f)$. Finally, $|f(x) - f(y)| \leq |f(x) - c| + |f(y) - c|$ for any c from which we get

$$D_Q(f) \leq 2 |Q|^{-1} \int_Q |f(x) - c| \, d^\nu x \tag{5.11.22}$$

so taking infs, $D_Q(f) \leq 2A_Q(f)$. $\qquad \square$

Clearly $\|f\|_{\text{BMO}} \leq 2\|f\|_\infty$ so $L^\infty \subset \text{BMO}$.

Example 5.11.3. We'll show in this example that $\mathrm{BMO}(\mathbb{R})$ contains unbounded functions by proving that $f(x) = \log|x|$ is in BMO. The same argument works for $\log|\theta/2\pi|$, $|\theta| \leq \pi$ on $\partial\mathbb{D}$ and a similar argument works for $\log|x|$ on $\mathbb{R}^\nu$ (Problem 1). We'll see later (see Example 5.11.17) that $|\log|x||^\alpha$ is in BMO if $0 \leq \alpha \leq 1$ and not in BMO if $\alpha > 1$. Let $Q = [a, b]$.

Suppose first $0 < a < b$. Then, by (5.11.19)

$$|f - f_Q|_Q \leq \frac{2}{b-a} \int_a^b |\log x - \log b| \, dx$$

$$= \frac{2}{b-a} \int_a^b |\log(x/b)| \, dx$$

$$\leq \frac{2b}{b-a} \int_{a/b}^1 |\log(y)| \, dy$$

$$= \frac{2}{1-c} \int_c^1 |\log(y)| \, dy \tag{5.11.23}$$

$$\leq 2 \int_0^1 |\log(y)| \, dy \tag{5.11.24}$$

since the monotonicity of $|\log(y)|$ on $[0, 1]$ proves the average in (5.11.23) increases as c decreases. (In (5.11.23), we used $c = \frac{a}{b}$.)

If $-b \leq a < 0 < b$, let $y = x/b$ as above and $c = \log b$ to get

$$|f - c|_Q \leq \frac{b}{b-a} \left[\int_0^1 |\log(y)| \, dy + \int_0^{-a/b} |\log(y)| \, dy \right]$$

$$\leq \frac{2b}{b-a} \int_0^1 |\log(y)| \, dy \tag{5.11.25}$$

$$\leq 2 \int_0^1 |\log(y)| \, dy \tag{5.11.26}$$

where (5.11.26) has that $|\log(y)| > 0$ and $-a > 0$. Thus, by (5.11.19)

$$|f - f_Q|_Q \leq 2 \int_0^1 |\log(y)| \, dy \tag{5.11.27}$$

and $f \in \mathrm{BMO}$. $\qquad\square$

Example 5.11.4. Let f be the function on $\mathbb{R}$

$$f(x) = \begin{cases} |\log|x||, & x > 0 \\ 0, & x \leq 0 \end{cases} \tag{5.11.28}$$

Let $a \in (0, 1)$ and $Q = (-a, a)$. Then, using

$$\int_0^a \log x \, dx = a \log a - a \tag{5.11.29}$$

we see that

$$f_Q = -\tfrac{1}{2}\log a + \tfrac{1}{2} \tag{5.11.30}$$

On $(-a, 0)$, $|f - f_Q| = |f_Q|$ so

$$\frac{1}{2a}\int_{-a}^{0}|f - f_Q|\,dx = \frac{a|f_Q|}{2a} = \frac{1}{2}|f_Q| \to \infty \tag{5.11.31}$$

as $a \downarrow 0$ by (5.11.30). Thus, $f \notin \mathrm{BMO}$.

If $g(x) = \chi_{(0,\infty)}(x) \in L^\infty$, then $f = (\log|x|^{-1})g$. This shows products of BMO functions may not lie in BMO, so unlike L^∞, BMO is not an algebra. $\square$

$\log|x|$ is very suggestive. As noted in Example 5.7.2, $f(e^{i\theta}) = i\log(\frac{1+e^{i\theta}}{1-e^{i\theta}})$ has $f \in L^\infty$ but $\mathrm{Im}\,f$ looks like $-\log(1 - e^{i\theta})$ near $\theta = 0$ and $\log(1 + e^{i\theta})$ near $\theta = \pi$ so $\mathrm{Im}\,f \notin L^\infty$ but as above has $\mathrm{Im}\,f \in \mathrm{BMO}$. This suggests that $\widetilde{}$ might take L^∞ to BMO and that is our next goal. $\mathrm{BMO}(\mathbb{R}^\nu)$ like $L^\infty(\mathbb{R}^\nu)$ is scale-invariant so scaling which works for the Hilbert transform on $\mathbb{R}$ will make the argument work for $L^\infty(\mathbb{R})$ and $\mathrm{BMO}(\mathbb{R})$. Once we have that, (5.7.116)–(5.7.118) will get the result for $L^\infty(\partial\mathbb{D})$ to $\mathrm{BMO}(\partial\mathbb{D})$.

Theorem 5.11.5. *If $g \mapsto g_H$ is the $\mathbb{R}$-Hilbert transform and $f \mapsto \tilde{f}$ the circular Hilbert transform, we have*

$$g \in L^\infty(\mathbb{R}) \Rightarrow g_H \in \mathrm{BMO}(\mathbb{R}) \tag{5.11.32}$$

$$f \in L^\infty(\partial\mathbb{D}) \Rightarrow \tilde{f} \in \mathrm{BMO}(\partial\mathbb{D}) \tag{5.11.33}$$

Proof. As noted, since $f \in \mathrm{BMO}(\partial\mathbb{D}) \Leftrightarrow g(\theta) = f(e^{i\theta}) \in \mathrm{BMO}(\mathbb{R})$ (Problem 2), (5.11.32) implies (5.11.33) by (5.7.118). Thus, we need to prove for a universal constant, D, there is for each $g \in L^\infty$ and interval Q, a constant c_Q so that

$$|\tilde{g} - c_Q|_Q \leq D\,\|g\|_\infty \tag{5.11.34}$$

Since both sides of this inequality are covariant under scaling and translation (i.e., g to $g_{(a,b)}(x) = g(ax + b)$ and $Q \mapsto aQ + b$), it suffices to do this when $Q = (-1, 1)$ and $\|g\|_{L^\infty} = 1$. Moreover, we can suppose that g has compact support.

Let χ be the characteristic function of $(-2, 2)$ and write

$$g = g_1 + g_2; \quad g_1 = \chi g, \quad g_2 = (1 - \chi)g \tag{5.11.35}$$

Since $g_1 \in L^2$ with $\|g_1\|_2 \leq 2$, we have that

$$\frac{1}{|Q|}\int_{-1}^{1}|(g_1)_H(x)|\,dx \leq \frac{1}{2}\int_{-1}^{1}|(g_1)_H(x)|^2\,dx$$

$$\leq \frac{1}{2}\|g_1\|^2 \leq 2 \tag{5.11.36}$$

since $g \mapsto g_H$ is an isometry on $L^2(\mathbb{R})$.

Since g_2 has compact support avoiding $(-2,2)$ for $x \in (-1,1)$

$$(g_2)_H(x) = \int_{-\infty}^{\infty} g_2(y)((x-y)^{-1})\,dy \qquad (5.11.37)$$

Let

$$c = (g_2)_H(0) = -\int_{-\infty}^{\infty} y^{-1} g_2(y)\,dy \qquad (5.11.38)$$

Then, for $x \in (-1,1)$,

$$(g_2)_H(x) - c = \int_{-\infty}^{\infty} g_2(y)(xy^{-1}(x-y)^{-1})\,dy \qquad (5.11.39)$$

and so for such x, since $\|g_2\|_\infty \leq 1$

$$|(g_2)_H(x) - c| \leq 2 \int_2^{\infty} y^{-1}(y-1)^{-1}\,dy = \log(2) \qquad (5.11.40)$$

and thus

$$\frac{1}{|Q|} \int_{-1}^1 |(g_2)_H(x) - c|\,dx \leq \log(2) \qquad (5.11.41)$$

It follows that

$$\frac{1}{|Q|} \int_{-1}^1 |g_H(x) - c|\,dx \leq 2 + \log(2) \qquad (5.11.42)$$

We have thus proven (5.11.34) with $D = 2 + \log(2)$. $\qquad\square$

We say $f \in \mathrm{VMO}(\mathbb{R}^\nu)$, the functions of *vanishing mean oscillations*, if $f \in \mathrm{BMO}(\mathbb{R}^\nu)$ and

$$\lim_{a\downarrow 0} \sup_{Q,|Q|\leq a} |f - f_Q|_Q = 0 \qquad (5.11.43)$$

If $f \in \mathrm{BUC}(\mathbb{R}^\nu)$, the functions in BMO that are bounded and uniformly continuous, it is obvious that $f \in \mathrm{VMO}(\mathbb{R}^\nu)$ and one can prove that VMO is precisely the BMO closure of BUC in $\|\cdot\|_{\mathrm{BMO}}$. Moreover, $f \in \mathrm{VMO}$ if and only if $f \in \mathrm{BMO}$ and $\lim_{|y|\to 0}\|f(\,\cdot\, - y) - f\|_{\mathrm{BMO}} = 0$. If $f \in C_0^\infty(\mathbb{R}^\nu)$, $\tilde{f} \in \mathrm{BUC}(\mathbb{R}^\nu)$ by a simple argument, so that we see that $f \in \mathrm{BUC}(\mathbb{R}^\nu) \to \tilde{f} \in \mathrm{VMO}(\mathbb{R}^\nu)$ by a continuity argument. (All these facts are discussed in the references in the Notes).

Formally, if P_+ is given by (5.8.21), we see that if $L \in (H_1(\partial\mathbb{D}))^*$, there is a $g \in L^\infty$ so that $L = L_g = L_{P_+ g}$ (L_g given by (5.11.3)). $P_+ g$ may not be in L^∞ but it will lie in BMO and be the boundary value of a function in HMO. Thus, modulo some minor technicalities, we have half of (the other half will require some effort!)

Theorem 5.11.6 (Fefferman Duality Theorem). $H^1(\mathbb{D})^* = \mathrm{HMO}(\mathbb{D})$ *in the sense that for every $g \in \mathrm{HMO}(\mathbb{D})$ and $f \in H^1(\mathbb{D})$*

$$L_g(f) \equiv \lim_{r\uparrow 1} \int \overline{g^*(e^{i\theta})} f(re^{i\theta})\,d\theta/2\pi \qquad (5.11.44)$$

exists and

$$|L_g(f)| \le C\|g\|_{\mathrm{HMO}} \, \|f\|_{H^1} \qquad (5.11.45)$$

$g \mapsto L_g$ *is a complex antilinear bijection of* HMO *and* $(H_1(\mathbb{D}))^*$.

The half we've studied above says every $L \in H_1^*$ is an L_g. The harder half is that any $g \in$ HMO defines a bounded linear functional obeying (5.11.45). This half is essentially equivalent to

Theorem 5.11.7 (Fefferman–Stein Decomposition Theorem). *Any* $g \in$ BMO$(\partial\mathbb{D})$ *can be written*

$$g = f_1 + \widetilde{f_2} \qquad (5.11.46)$$

where $f_1, f_2 \in L^\infty$.

Remarks. 1. There exist many $f \in L^\infty$ with $\widetilde{f} \in L^\infty$ also. $f_1 + \widetilde{f_2} = (f_1 + f) + (f_2 - \widetilde{f})^{\sim}$, so the decomposition (5.11.46) is highly nonunique.

2. Similarly (see the Notes), $f \in$ VMO$(\partial\mathbb{D})$ if and only if $f = h + \widetilde{g}$ with $f, g \in C(\partial\mathbb{D})$.

We prove Theorem 5.11.6 and use it to prove Theorem 5.11.7 but one can go in the opposite direction (Problem 3). While one could avoid the need to exploit it, we'll use a result we only prove below (see Corollary 5.11.15): if $g \in$ BMO$(\partial\mathbb{D})$, then $g \in L^2(\partial\mathbb{D}, \frac{d\theta}{2\pi})$, with

$$\|g\|_{L^2} \le C(\|g\|_{\mathrm{BMO}} + \int |g(e^{i\theta})| \frac{d\theta}{2\pi}) \qquad (5.11.47)$$

for a universal constant C.

The key to our proof is a particular decomposition of functions in $H^1(\mathbb{D})$.

Definition. An *atom* of Re H^1 is a real-valued function, ϕ, in $L^1(\partial\mathbb{D}, \frac{d\theta}{2\pi})$ so that

  (i) ϕ is supported in an interval $I \subset \partial\mathbb{D}$
  (ii) $|\phi(e^{i\theta})| \le 1/|I|$ (where $|I| = \int_I \frac{d\theta}{2\pi}$)
  (iii) $\int \phi(e^{i\theta}) \frac{d\theta}{2\pi} = 0$

Since $\phi \in L^\infty \subset \bigcap_{p<\infty} L^p(\partial\mathbb{D})$, we have that $\widetilde{\phi} \in \bigcap_{p<\infty} L^p(\partial\mathbb{D})$, so $\phi + i\widetilde{\phi} \in H^1(\partial\mathbb{D}, \frac{d\theta}{2\pi})$, i.e., ϕ is the real part of an H^1-function. There are two preliminaries that help explain why atoms are so useful:

Proposition 5.11.8. *If* ϕ *is an atom,*

$$\|\widetilde{\phi}\|_1 \le D \qquad (5.11.48)$$

for a universal constant, D.

Remark. In our proof, $D = \frac{3\pi}{32} + \sqrt{3}$, but one can do much better.

Proof. Since $d\theta/2\pi$ is a probability measure and $\phi \mapsto \widetilde{\phi}$ is an isometry on $L^2(\partial\mathbb{D}, d\theta/2\pi)$, we have that

$$\|\widetilde{\phi}\|_1 \leq \|\widetilde{\phi}\|_2 = \|\phi\|_2 \leq |I|^{-1/2} \tag{5.11.49}$$

Thus, for I's with $|I| \geq \pi/2$, we have (5.11.48) with $D = \sqrt{\frac{2}{\pi}}$.

If $|I| \leq \frac{\pi}{2}$, we use the same argument we used to prove that $\widetilde{}$ mapped L^∞ to BMO. We can translate and suppose that

$$I = \{e^{i\theta} \mid |\theta| < \alpha\} \tag{5.11.50}$$

with $0 < \alpha < \frac{\pi}{4}$ and we define

$$J = \{e^{i\theta} \mid |\theta| < 3\alpha\} \subset \{e^{i\theta} \mid |\theta| < \pi\} \tag{5.11.51}$$

By (5.11.49) and the Schwarz inequality

$$\int_J |\widetilde{\phi}(e^{i\theta})| \frac{d\theta}{2\pi} \leq |J|^{1/2} \|\widetilde{\phi}\|_2 \leq \frac{(3|I|)^{1/2}}{|I|^{1/2}} \leq \sqrt{3} \tag{5.11.52}$$

For θ with $3\alpha \leq |\theta| \leq \pi$, we write

$$\widetilde{\phi}(e^{i\theta}) = \int_{|\eta| \leq \alpha} \left[\cot(\tfrac{1}{2}(\theta - \eta)) - \cot(\tfrac{1}{2}\theta) \right] \phi(e^{i\eta}) \frac{d\eta}{2\pi} \tag{5.11.53}$$

since ϕ has zero integral. If $g(\theta) = \cot(\frac{\theta}{2})$, then for $|\theta| \in (0, \pi]$

$$g'(\theta) = \frac{1}{2\sin^2(\frac{\theta}{2})} \leq \frac{\pi^2}{8\theta^2} \tag{5.11.54}$$

since $\sin(\psi) \geq \frac{2}{\pi}\psi$ for $\psi \in (0, \frac{\pi}{2})$ (see the argument after (3.5.55) in Part 1). If $|\eta| \leq \alpha$ and $|\theta| \geq 3\alpha$, then $|\theta - \eta| \geq \frac{2|\theta|}{3}$. This implies for such θ, η, one has that

$$|\cot(\tfrac{1}{2}(\theta - \eta)) - \cot(\tfrac{1}{2}\theta)| \leq \frac{9|\eta|\pi^2}{32|\theta|^2} \leq \frac{9\alpha\pi^2}{32|\theta|^2} \tag{5.11.55}$$

Thus, for $\theta \notin J$,

$$|\widetilde{\phi}(e^{i\theta})| \leq \frac{9\alpha\pi^2}{32|\theta|^2}$$

since $\int_I |\phi(e^{i\eta})| \frac{d\eta}{2\pi} \leq 1$. Therefore,

$$\int_{\partial\mathbb{D}\setminus J} |\widetilde{\phi}(e^{i\theta})| \frac{d\theta}{2\pi} \leq \frac{9\alpha\pi^2}{16} \int_{3\alpha}^{\pi} \frac{d\theta}{2\pi\theta^2}$$

$$\leq \frac{9\pi\alpha}{32} \int_{3\alpha}^{\infty} \frac{d\theta}{\theta^2} = \frac{3\pi}{32} \tag{5.11.56}$$

which proves (5.11.48). $\qquad\square$

Proposition 5.11.9. *Let ϕ be an atom and $g \in \mathrm{BMO}(\partial\mathbb{D})$. Then*

$$\left| \int \overline{g(e^{i\theta})}\phi(e^{i\theta})\frac{d\theta}{2\pi} \right| \leq \|g\|_{\mathrm{BMO}} \qquad (5.11.57)$$

Proof. Suppose I is the interval associated to ϕ. Since $\int_I \phi(e^{i\theta})\frac{d\theta}{2\pi} = 0$,

$$\text{L.H.S. of (5.11.57)} = \left| \int (\overline{g(e^{i\theta})} - \overline{g_I})\phi(e^{i\theta})\frac{d\theta}{2\pi} \right|$$

$$\leq \frac{1}{|I|}\int_I |g(e^{i\theta}) - g_I|\frac{d\theta}{2\pi} \leq \|g\|_{\mathrm{BMO}} \qquad (5.11.58)$$

$\square$

The key to proving Fefferman duality is the following whose proof we defer.

Theorem 5.11.10 (Coifman Atomic Decomposition). *Let $f \in H^1(\mathbb{D})$. Then there exist atoms $\{\phi_j\}_{j=1}^J$ (J finite or $J = \infty$) and constants a_j so that on $\partial\mathbb{D}$:*

$$\mathrm{Re}\, f^*(e^{i\theta}) = \mathrm{Re}\, f(0) + \sum_{j=1}^J a_j\phi_j(e^{i\theta}) \qquad (5.11.59)$$

and so that

$$\sum_{j=1}^J |a_j| < \infty$$

Indeed, there are universal constants, C_1, C_2, in $(0, \infty)$ so that

$$C_1\|f\|_{H^1} \leq |f(0)| + \sum_{j=1}^J |a_j| \leq C_2\|f\|_{H^1} \qquad (5.11.60)$$

We prove the following stronger result than the hard part of the Fefferman duality:

Theorem 5.11.11. *Let $g \in \mathrm{BMO}(\partial\mathbb{D})$. Then for all $f \in H^1(\mathbb{D})$*

$$L_g(f) = \lim_{r\uparrow 1} \int_0^{2\pi} \overline{g(e^{i\theta})}\, f(re^{i\theta})\frac{d\theta}{2\pi} \qquad (5.11.61)$$

exists and for a universal constant C_3

$$|L_g(f)| \leq C_3\|f\|_{H^1}(\|g\|_{\mathrm{BMO}} + \int |g(e^{i\theta})|\frac{d\theta}{2\pi}) \qquad (5.11.62)$$

Proof. By considering $\mathrm{Re}\, f$ and $\mathrm{Im}\, f$ separately (noting that $\mathrm{Im}\, f = \mathrm{Re}(-if)$ has an atomic decomposition also), we need only prove the result for

$$L_g^{(R)}(f) = \lim_{r\uparrow 1} \int_0^{2\pi} \overline{g(e^{i\theta})}(\mathrm{Re}\, f)(re^{i\theta})\frac{d\theta}{2\pi} \qquad (5.11.63)$$

For $f \in H^2$, we can define

$$L_g^{(R)}(f) = \int_0^{2\pi} \overline{g(e^{i\theta})} \operatorname{Re} f^*(e^{i\theta}) \frac{d\theta}{2\pi} \qquad (5.11.64)$$

since BMO $\subset L^2$. We'll show this obeys an estimate of the form (5.11.62) and then use that to define (5.11.63) in general.

If $f \in H^1$ and (5.11.59) is its atomic decomposition, let $f_N(e^{i\theta}) = f(0) + \sum_{j=1}^N a_j[\phi_j(e^{i\theta}) + i\tilde{\phi}_j(e^{i\theta})]$. By (5.11.60) and (5.11.48), $f_N \to f$ in H^1 and by (5.11.59) and (5.11.60) for $N \leq M$,

$$|L_g^{(R)}(f_N) - L_g^{(R)}(f_M)] \leq C \sum_{j=N}^\infty |a_j| \qquad (5.11.65)$$

$$|L_g^{(R)}(f_N)| \leq C\|f\|_{H^1}(\|g\|_{\mathrm{BMO}} + \|g\|_1) \qquad (5.11.66)$$

It follows that $L_g^{(R)}(f_N)$ has a limit that obeys

$$|L_g^{(R)}(f)| \leq C\|f\|_{H^1}(\|g\|_{\mathrm{BMO}} + \|g\|_1)$$

proving (5.11.62) for a functional that is defined by this limit.

We need only prove the function can be defined by (5.11.61). Given $f \in H^1$, define $f^{(r)}(z) = f(rz)$ for $0 < r < 1$. Since $f^{(r)} \in H^2$, $L_g(f^{(r)})$ is the integral in the limit in (5.11.61). Since $f^{(r)} \to f$ in H^1 (see 5.3.28) and L_g is continuous, $L_g(f) = \lim_{r \uparrow 1} L_g(f^{(r)})$. $\qquad \square$

As a first corollary, we prove the Fefferman–Stein decomposition theorem.

Proof of Theorem 5.11.7. Let $g \in$ BMO and suppose it real-valued (the full result for complex-valued functions follows by writing $g = g_r + ig_i$). As we sw above, L_g defined a bounded functional on $H^1(\mathbb{D})$, equivalently on $\{f \in L^1(\partial\mathbb{D}) \mid \tilde{f} \in L^1(\partial\mathbb{D})\}$. By the Hahn-Banach theorem, we can extend L_g to all of $L^1(\partial\mathbb{D})$ and so, by L^p-duality, find $h \in L^\infty(\partial\mathbb{D})$ so that

$$L_g(f) = \int \overline{h(e^{i\theta})} f^*(e^{i\theta}) \frac{d\theta}{2\pi} \qquad (5.11.67)$$

for all $f \in H^1$. Thus,

$$P_+ g = P_+ h = \frac{1}{2}(h + i\tilde{h} + h(0)) \qquad (5.11.68)$$

so $P_+ g \in L^\infty + \widetilde{L^\infty}$. Since $P_- g = \overline{P_+ g}$, we see that $P_- g \in L^\infty + \widetilde{L^\infty}$ so $g = P_+ g + P_- g - g(0) \in L^\infty + \widetilde{L^\infty}$. $\qquad \square$

Corollary 5.11.12. *If $g \in$ BMO, then $\tilde{g} \in$ BMO also. BMO is the smallest space containing L^∞ and left invariant by the circular Hilbert transform.*

Proof. Write $g = f_1 + \widetilde{f_2}$ with $f_j \in L^\infty$. Then, $\widetilde{g} = \widetilde{f_1} - (f_2 - f_2(0)) \in L^\infty$ also. Clearly if X contains L^∞ and is invariant under $\widetilde{}$, then $g = f_1 + \widetilde{f_2} \in X$ for all $f_j \in L^\infty$ so BMO $\subset X$. $\qquad\square$

Proof of Theorem 5.11.6. If $g_1, g_2 \in$ HMO and $L_{g_1} = L_{g_2}$ on H_1, then $(g_1)_n^\sharp = (g_2)_n^\sharp$ so g_1 and g_2 have the Taylor series and so are equal. This proves uniqueness.

Suppose $L \in (H^1)^*$. By the Hahn-Banach theorem, there is $h \in L^\infty$, so that $L = L_h$. Let $g^* = P_+ h$. Then, $L = L_{g^*}$. $P_+ h \in$ BMO $\subset L^2$ so $P_+ h$ is the boundary value of $g \in H_2$ with $g^* \in$ BMO so $g \in$ HMO.

By Theorem 5.11.11 every L_g, $g \in$ HMO lies in $(H^1)^*$. $\qquad\square$

We turn now to the proof of Coifman's atomic decomposition. Let $f \in H^1(\mathbb{D})$, $f \neq 0$, and let h be its nontangential maximal function, $M_{\mathrm{NT}} f^*$, given by (2.4.42). i.e.,

$$h(e^{i\theta}) = \sup_{z \in N(\theta) \cap \mathbb{D}} |f(z)| \tag{5.11.69}$$

It is easy to see that if, for $0 < \rho < 1$, h_ρ is the sup over $N(\theta) \cap \mathbb{D}_\rho(0)$, then (Problem 4) h_ρ is continuous, so (see Proposition 2.1.9 of Part 1), h is upper semicontinuous, i.e.,

$$O_\lambda = \{\theta \mid h(e^{i\theta}) > \lambda\} \tag{5.11.70}$$

is open. Notice also that $h(e^{i\theta}) \geq \inf_{|z| \leq 1/\sqrt{2}} |f(z)| > 0$ so

$$\inf h(e^{i\theta}) \equiv \lambda_0 \tag{5.11.71}$$

is strictly positive. If $\lambda > \lambda_O$, O_λ is a proper subset of $\partial\mathbb{D}$ and so a union of disjoint open intervals. Here is a key preliminary:

Lemma 5.11.13. *Let $\lambda > \lambda_0$. Let J be an open interval which is a connected component of O_λ. Then,*

$$\left| \int_J (\mathrm{Re}\, f^*)(e^{i\theta}) \frac{d\theta}{2\pi} \right| \leq 2\lambda |J| \tag{5.11.72}$$

Proof. Let $J = \{e^{i\theta} \mid \alpha < \theta < \beta\}$. Consider $\mathbb{D} \setminus [N(\alpha) \cup N(\beta)]$. There is a connected component of this set whose boundary, T, includes J. T is either a slice (i.e., J plus two lines) or a circular trapezoid (J and an arc of $\partial\mathbb{D}(\frac{1}{\sqrt{2}})$ plus two lines); see Figure 5.11.1. By the Cauchy integral formula, for $0 < r < 1$,

$$\int_T f(rz) \frac{dz}{2\pi i z} = 0 \tag{5.11.73}$$

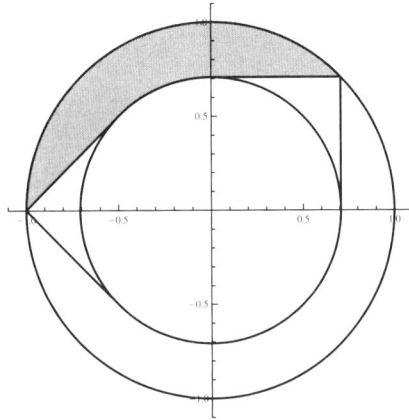

Figure 5.11.1. A circular trapezoid.

which by taking limits is true for $r = 1$ (since $f(re^{i\theta}) \to f^*(e^{i\theta})$ in $L^1(\partial\mathbb{D}, \frac{d\theta}{2\pi})$). Thus,

$$\left| \int_J \operatorname{Re} f^*(e^{i\theta}) \frac{d\theta}{2\pi} \right| \leq \left| \int_J f^*(e^{i\theta}) \frac{d\theta}{2\pi} \right|$$

$$\leq \int_{T \setminus J} |f(z)| \frac{d|z|}{2\pi|z|} \qquad (5.11.74)$$

Since $h(e^{i\alpha}) \leq \lambda$ and $h(e^{i\beta}) \leq \lambda$ (because $\alpha, \beta \notin O_\lambda$) and $T \setminus J \subset N(\alpha) \cup N(\beta)$, we see (since $|z| \geq \frac{1}{\sqrt{2}}$ on T)

$$\int_{T \setminus J} |f(z)| \frac{d|z|}{2\pi|z|} \leq \sqrt{2}\,\lambda \frac{|T \setminus J|}{2\pi} \qquad (5.11.75)$$

where $|T \setminus J|$ is Euclidean length (rather than $|J|$ which is Euclidean length over 2π). It is simple geometry (Problem 5) that

$$\frac{|T \setminus J|}{2\pi} \leq \sqrt{2}\,|J| \qquad (5.11.76)$$

The last three numbered formulas imply (5.11.72). $\qquad\qquad\square$

Proof of Theorem 5.11.10. The easy half of the theorem is that if $\operatorname{Re} f^*$ has the form (5.11.59), then

$$\|f\|_{H^1} \leq C_1^{-1}\Big(|f(0)| + \sum_{j=1}^{J} |a_j|\Big) \qquad (5.11.77)$$

For (5.11.59) implies that

$$f^*(e^{i\theta}) = f(0) + \sum_{j=1}^{J} a_j(\phi_j + i\widetilde{\phi}_j) \qquad (5.11.78)$$

Thus, by (5.11.48)

$$\|f^*\|_1 \le |f(0)| + (1+D)\sum_{j=1}^{J}|a_j| \qquad (5.11.79)$$

and (5.11.77) holds with $C_1 = (1+D)^{-1}$. The hard half is existence with the second inequality in (5.11.60). The proof will be constructive.

The proof of the hard half will be a variant of the construction of the Calderón–Zygmund decomposition which we turn to in Section 6.4. Pick n_0 so that $2^{n_0} \le \lambda_0 < 2^{n_0+1}$. Write, for $j = 1, 2, \ldots$,

$$O_{2^{n_0+j}} = \bigcup_k J_k^{(j)} \qquad (5.11.80)$$

and

$$\lambda_k^{(j)} = \int_{J_k^{(j)}} \operatorname{Re} f^*(e^{i\theta})\frac{d\theta}{2\pi} \qquad (5.11.81)$$

so by the lemma

$$|\lambda_k^{(j)}| \le 2^{n_0+j+1} \qquad (5.11.82)$$

Define functions for $j = 1, 2, \ldots$

$$g_j(e^{i\theta}) = \begin{cases} \operatorname{Re} f^*(e^{i\theta}), & e^{i\theta} \notin O_{2^{n_0+j}} \\ \lambda_k^{(j)}, & e^{i\theta} \in J_k^{(j)} \end{cases} \qquad (5.11.83)$$

$$b_j(e^{i\theta}) = \begin{cases} 0, & e^{i\theta} \in O_{2^{n_0+j}} \\ \operatorname{Re} f^*(e^{i\theta}) - \lambda_k^{(j)}, & e^{i\theta} \in J_k^{(j)} \end{cases} \qquad (5.11.84)$$

For $j = 0$, define

$$g_0(e^{i\theta}) = \operatorname{Re} f(0), \quad b_0(e^{i\theta}) = \operatorname{Re} f^*(e^{i\theta}) - \operatorname{Re} f(0) \qquad (5.11.85)$$

Then (with $J^{(0)} = \partial\mathbb{D}$)

(a) $g_j + b_j = \operatorname{Re} f^*$. $\qquad (5.11.86)$

(b) $|g_j(e^{i\theta})| \le 2^{n_0+j+1}$ for a.e. θ. $\qquad (5.11.87)$

(c) $\displaystyle\int_{J_k^{(j)}} b_j(e^{i\theta})\frac{d\theta}{2\pi} = 0$ $\qquad (5.11.88)$

(d) For a.e. $e^{i\theta} \in \partial\mathbb{D}$,

$$\lim_{j\to\infty} g_j(e^{i\theta}) = \operatorname{Re} f^*(e^{i\theta}) \qquad (5.11.89)$$

(a) is obvious, (c) follows for $j \ge 1$ by construction and for $j = 0$ from $\operatorname{Re} f(0) = \int \operatorname{Re} f^*(e^{i\theta})\frac{d\theta}{2\pi}$. For $j \ge 1$, (b) follows from (5.11.82) and that for a.e. $e^{i\theta} \in \partial\mathbb{D}$, $|\operatorname{Re} f^*(e^{i\theta})| \le h(e^{i\theta})$ by nontangential convergence, so for a.e. θ in $\mathcal{O}_{2^n}$, $|\operatorname{Re} f^*(e^{i\theta})| \le 2^{n_0+j}$. If $h(e^{i\theta}) < \infty$, then $g_j(e^{i\theta}) = \operatorname{Re} f^*(e^{i\theta})$ for j such that $2^{n_0+j} > h(e^{i\theta})$. This implies (d).

Because of (5.11.89) and (5.11.86)

$$\operatorname{Re} f^*(e^{i\theta}) = g_0 + \sum_{j=0}^{\infty} (g_{j+1} - g_j) \tag{5.11.90}$$

$$= \operatorname{Re} f(0) + \sum_{j=0}^{\infty} (b_j - b_{j+1}) \tag{5.11.91}$$

As j increases, $\mathcal{O}_{2^{n_0+j}}$ gets smaller, so for each k with $J_k^{(j)} \neq \emptyset$, there are a subset, $S(j,k)$, of the indices for $J_\ell^{(j+1)}$ so that

$$J_k^{(j)} \cap \mathcal{O}_{2^{n_0+j+1}} = \bigcup_{\ell \in S(j,k)} J_\ell^{(j+1)} \tag{5.11.92}$$

Define

$$\psi_k^{(j)} = (b_j - b_{J+1}) \upharpoonright J_k^{(j)} \tag{5.11.93}$$

(and set to 0 on $\partial \mathbb{D} \setminus J_k^{(j)}$). By (5.11.88),

$$\int_{J_k^{(j)}} \psi_k^{(j)} \frac{d\theta}{2\pi} = \int_{J_k^{(j)}} b_j \frac{d\theta}{2\pi} - \sum_{\ell \in S(j,k)} \int_{J_k^{(j)}} b_{j+1} \frac{d\theta}{2\pi} = 0$$

Moreover, since $|b_j - b_{j+1}| = |g_j + g_{j+1}| \leq |g_j| + |g_{j+1}|$ we see

$$|b_j - b_{j+1}| \leq 3 \, 2^{n_0+j+1} \tag{5.11.94}$$

so by

$$\phi_k^{(j)} = a_{j,k}^{-1} \psi_k^{(j)} \tag{5.11.95}$$

$$a_{j,k} = 3 \, 2^{n_0+j+1} |J_k^{(j)}| \tag{5.11.96}$$

is an atom and

$$\operatorname{Re} f^*(e^{i\theta}) = \operatorname{Re} f(0) + \sum_{j,k} a_{j,k} \phi_k^{(j)} \tag{5.11.97}$$

so we need only prove that

$$\sum_{j,k} |a_{j,k}| \leq C \|f\|_{H^1} \tag{5.11.98}$$

By (5.11.92), for each j

$$\sum_{k} |a_{j,k}| = 3 \, 2^{n_0+j+1} |\mathcal{O}_{2^{n_0+j}}|$$

$$= 6 \, 2^{n_0+j} \sum_{\ell=0}^{\infty} m_{j+\ell} \tag{5.11.99}$$

where

$$m_q = |\{x \mid 2^{n_0+q} < h(x) \leq 2^{n_0+q+1}\}| \tag{5.11.100}$$

Thus

$$\sum_{j,k}|a_{j,k}| = 6\sum_{j,\ell=0}^{\infty} 2^{n_0+1}m_{j+\ell} \tag{5.11.101}$$

$$= 6\sum_{k=0}^{\infty} 2^{n_0+k}m_k(1+\tfrac{1}{2}+\ldots+\tfrac{1}{2^\ell})$$

$$\leq 12\sum_{k=1}^{\infty} m_k 2^{n_0+k}$$

$$\leq 12\int h(e^{i\theta})\frac{d\theta}{2\pi} \tag{5.11.102}$$

$$\leq C\|f\|_{H^1} \tag{5.11.103}$$

by Theorem 5.3.5 for $p=1$. We get (5.11.102) by noting that $h(x)\geq 2^{n_0+k}$ on $\mathcal{O}_{2^{n_0+k}}$. □

Our final topic concerns the fact that BMO functions are almost L^∞ in the sense that $|\{x\in Q\mid Q^{-1}|f-f_Q|>\lambda\}|$ will decay exponentially in λ.

Theorem 5.11.14 (John–Nirenberg Inequalities). *There exist universal constants, A and B, depending on ν so that for functions in* BMO*($\mathbb{R}^\nu$) (or* BMO*($\partial\mathbb{D}$) where $\nu=1$), one has for all $f\in$* BMO *and all standard cubes, Q,*

$$\frac{1}{|Q|}\Big|\{x\in Q\mid |f-f_Q|>s\|f\|_{\mathrm{BMO}}\}\Big| \leq Ae^{-Bs} \tag{5.11.104}$$

Our proof will rely on the Calderón–Zygmund decomposition, so we defer it until Section 6.4 where we'll prove it with $A=e$, $B=(2^\nu e)^{-1}$ (certainly not optimal constants; see the Notes).

Corollary 5.11.15. *For any $1<p<\infty$ and $f\in$* BMO

$$\frac{1}{|Q|}\|f-f_Q\|_1 \leq \frac{1}{|Q|^{1/p}}\|f-f_Q\|_p \leq \left(pB^{-p}A\Gamma(p)\right)^{1/p}\|f\|_{\mathrm{BMO}} \tag{5.11.105}$$

In particular, $\sup_Q \frac{1}{|Q|^{1/p}}\|f-f_Q\|_p$ *are equivalent norms for all p and if* $f\in$ BMO*($\partial\mathbb{D}$), then $f\in\cap_{p=1}^{\infty}L^p(\partial\mathbb{D},\frac{d\theta}{2\pi})$.*

Proof. The first inequality in (5.11.105) follows from the fact that, on probability space, $\|g\|_1\leq\|g\|_p$ (by Hölder's inequality). For the second, note that if g obeys $|\{x\in X\mid g(x)\geq Cs\}|\leq Ae^{-Bs}$, then

$$\|g\|_p^p = \int p\lambda^{p-1}|\{x\mid g(x)\geq\lambda\}|\,d\lambda$$

$$\leq \int p\lambda^{p-1} A e^{-B\lambda/C}\, d\lambda$$

$$= \left(\frac{C}{B}\right)^p p\,\Gamma(p) A \qquad (5.11.106)$$

$\square$

Proposition 5.11.16. *Let $f \in$ BMO and $\alpha \in (0,1]$. Then, $|f|^\alpha \in$ BMO and*

$$\| |f|^\alpha \|_{\mathrm{BMO}} \leq 2\|f\|_{\mathrm{BMO}}^\alpha$$

Remark. In particular, $\alpha = 1$ says $f \in$ BMO $\Rightarrow |f| \in$ BMO. Since $f \vee g = \frac{1}{2}[f+g+|f-g|]$ and $f \wedge g = \frac{1}{2}[f+g-|f-g|]$, we see that the real functions in BMO are a lattice (but, as Example 5.11.4 shows, not a lattice with $f \in$ BMO and $|g| \leq |f| \Rightarrow |g| \in$ BMO).

Proof. We will exploit the facts we've already noted (see Theorem 5.2.6 of Part 1) that $\|s\|_{\ell^p} \leq \|s\|_{\ell^q}$ if $p \geq q$ and $\|f\|_{L^p} \leq \|f\|_{L^q}$ if $p \leq q$ and we have a probability measure. From the ℓ^p result, if $p = 1/\alpha$, we see that for $c, d \geq 0$,

$$(c^p + d^p)^{1/p} \leq c + d \qquad (5.11.107)$$

Setting $c = |x|^\alpha$, $d = |y|^\alpha$ for any complex x, y implies

$$|x+y|^\alpha \leq (|x|+|y|)^\alpha = (c^p + d^p)^{1/p} \leq c + d \leq |x|^\alpha + |y|^\alpha \qquad (5.11.108)$$

If $a = x + y$, $b = x$, we see for $a, b \in \mathbb{C}$:

$$|a|^\alpha \leq |b|^\alpha + |a-b|^\alpha \Rightarrow \big||a|^\alpha - |b|^\alpha\big| \leq |a-b|^\alpha \qquad (5.11.109)$$

using $a \leftrightarrow b$ symmetry.

Thus, integrating $\frac{dx}{|Q|}$ with $a = f(x)$, $b = f_Q$:

$$\big||f|^\alpha - |f_Q|^\alpha\big|_Q \leq \big||f - f_Q|^\alpha\big|_Q$$

$$\leq \big||f - f_Q|_Q\big|^\alpha \qquad (5.11.110)$$

$$\leq \|f\|_{\mathrm{BMO}}^\alpha \qquad (5.11.111)$$

where (5.11.110) uses $\int |g|\frac{dx}{Q} \leq \left(\int |g|^p \frac{dx}{|Q|}\right)^{1/p}$ with $g = |f - f_Q|^\alpha$.

Thus, by (5.11.19)

$$\big||f|^\alpha - (|f|^\alpha)_Q\big|_Q \leq 2\big||f|^\alpha - |f_Q|^\alpha\big|_Q \leq 2\|f\|_{\mathrm{BMO}}^\alpha \qquad (5.11.112)$$

$\square$

Example 5.11.17. Let $f_q(x) = |\log|x||^q$ for $|x| < 1$ and 0 for $|x| \geq 1$. Then, for all λ

$$\{x \in [-1,1] \mid f(x) > \lambda\} = 2e^{-\lambda^{1/q}} \qquad (5.11.113)$$

We conclude $q = 1$ (which we studied in Example 5.11.3) is borderline for the John–Nirenberg inequality. In particular, since (5.11.104) fails for f_q, $q > 1$, we see that $f_q \notin$ BMO if $q > 1$. This also shows that exp in (5.11.104) is optimal. Since $f_{q=1} \in$ BMO, Proposition 5.11.16 implies $f_q \in$ BMO, $0 < q \leq 1$. $\qquad\qquad\qquad\qquad\qquad\qquad\qquad\qquad\qquad\qquad\qquad\qquad$ $\square$

Notes and Historical Remarks. For book treatments of BMO see Garnett [**305**], Grafakos [**329**], and Koosis [**486**].

BMO was introduced by Fritz John in 1961 [**433**] in his work on elasticity. Shortly thereafter, he and Nirenberg [**434**] proved the John–Nirenberg inequality (the two papers appeared right after another). The proof we give (in Section 6.4 below) is essentially theirs and is no harder than any subsequent proofs.

The later proofs are more to improve the ν-dependence of the constant B in (5.11.104) than to simplify. One wants to keep B_ν away from zero. Our $B_\nu \sim 2^{-\nu}$ is very bad. John and Nirenberg remark without proof in their paper, that they can improve the constant to $O((\log \nu)/\nu)$ which is much better. Wik [**859**] has improved this to $O(\nu^{-1/2})$. Cwikel et al. [**180**] conjecture that there is a ν-independent B and make a plausible geometric conjecture that they prove would imply this. They and Bennett et al. (below) use decreasing rearrangements (see the discussion of this subject in the Notes to Section 6.1).

Fritz John (1910–1994) was born to a partially Jewish family in Berlin and studied in Göttingen from 1929–1933 where his advisor was Richard Courant. In his early career, he was most known for his work on the Radon transform and on elasticity, linear and nonlinear PDE's in the later half. He left Germany in 1933 because the political environment made it impossible to find employment there. After a year in Cambridge, England, he moved to the University of Kentucky and in 1946, with Courant's help, he moved to NYU where he spent the remainder of his career. His students include Clifford Gardner and Sergiu Klainerman.

The important role of BMO in general analysis was established by Fefferman's discovery of the dual of H^1. This was announced in [**265**] and proven in his joint paper with Stein [**271**] which also included their decomposition theorem that we saw is essentially equivalent to the duality result (see below for the context of their papers).

The fact that BMO is a kind of L_w^∞ for purposes of interpolation was developed by Bennett et al. [**55**]. Sarason [**717**] introduced VMO and used it as a tool in several papers. The decomposition theorem for VMO is due to him.

The atomic decomposition proof that we give is due to Coifman [**162**]— this decomposition first appeared in that paper although he remarks that Fefferman had noted that it was equivalent to his duality theorem. There are also (L^p, L^q) atomic decompositions more complicated than our L^∞ one; they are discussed, for example, in Grafakos [**329**].

We learned the proof we give of Proposition 5.11.16 from Grafakos.

The space we call HMO is often denoted BMOA (A for analytic)—HMO is new here.

Fefferman, Fefferman–Stein, and many later authors were motivated by more than the classic $H^p(\mathbb{D})$ Hardy spaces: $g \in L^p(\partial\mathbb{D})$ is the real part of an H^p-function if and only if $\tilde{g} \in L^p$ also. This is automatic for $1 < p < \infty$ but an extra condition for $p = 1$. Similarly, $g \in L^1(\mathbb{R})$ is in $\operatorname{Re} H^1(\mathbb{C}_+)$ if and only if its Hilbert transform, $g_H = (k|k|^{-1}\hat{g})^\vee$ is in L^1 also. This motivates the definition of the real-variable $H^p(\mathbb{R}^\nu)$ as those $f \in L^p(\mathbb{R}^\nu)$ so that for $j = 1, \ldots, \nu$, $(k_j|k|^{-1}\hat{g})^\vee$ is in L^p also. These are the "H^p spaces of several variables" that are the subject of the Fefferman–Stein paper. H^p, $1 < p < \infty$ is still L^p (these are the analogs of $\operatorname{Re} H^p(\mathbb{R})$) but not H^1 and Fefferman proved that BMO is the dual of $H^1(\mathbb{R}^\nu)$. A key paper in the study of these H^p spaces was also the one of Coifman-Weiss [**164**].

A *Hankel matrix* is a semi-infinite matrix $(a_{ij})_{0 \le i,j < \infty}$ depending on a sequence $\{\alpha_j\}_{j=0}^\infty$ so that

$$a_{ij} = \alpha_{i+j} \tag{5.11.114}$$

$$H(\alpha) = \begin{pmatrix} \alpha_0 & \alpha_1 & \alpha_2 & \cdots \\ \alpha_1 & \alpha_2 & \alpha_3 & \cdots \\ \alpha_2 & \alpha_3 & \alpha_4 & \cdots \\ \vdots & \vdots & \vdots & \ddots \end{pmatrix} \tag{5.11.115}$$

We think of it as acting on finite vectors $\{x_i\}_{i=1}^N$

$$(H(\alpha)x)_i = \sum_{j=0}^N \alpha_{i+j} x_j$$

It is a fundamental theorem of Nehari [**600**] that the following are equivalent (Problem 6):

(1) $\|H(\alpha)x\|_2 \le C \|x\|_2$ \hfill (5.11.116)

(2) There is $\varphi \in L^\infty(\partial\mathbb{D}, \frac{d\theta}{2\pi})$, so that for $n \ge 0$

$$\varphi_n^\sharp = \alpha_n \tag{5.11.117}$$

(3) $\sum_{j=0}^\infty \alpha_j z^j$ converges on $\mathbb{D}$ and defines a function in HMO. In that case

$$\|H(\alpha)\|_{\mathcal{L}(\ell^2)} = \inf\{\|\varphi\|_\infty \mid \varphi_n^\sharp = \alpha_n\} \tag{5.11.118}$$

It is a result of Hartman [**367**] that the following are equivalent (Problem 7)

(1) $H(\alpha)$ is a compact operator on $\ell^2(\mathbb{N})$

(2) There is $\varphi \in C(\partial\mathbb{D})$ so that $\varphi_n^\sharp = \alpha_n$

(3) $\sum_{j=0}^{\infty} \alpha_j z^j$ converges on $\mathbb{D}$ and defines a function in HMO whose boundary values lie in VMO.

Hartman and Nehari only proved the equivalence of (1) and (2) in their respective results, but once one has the decomposition theorems of Fefferman–Stein and Sarason, one gets the equivalence of (3) easily.

For a comprehensive book on Hankel operators, see Peller [**634**]. We've seen finite (and also semi-infinite) Hankel matrices in our discussion of the moment problem (see Sections 4.17 and 5.6 of Part 1 and Section 7.7 of Part 4).

Problems

1. Let $f(x) = \log|x|$ on $\mathbb{R}^\nu$. Prove that $f \in \mathrm{BMO}(\mathbb{R}^\nu)$. (*Hint*: First show it suffices to prove $\sup_B |f - f_B|_B < \infty$ for all balls $B = B_x^r = \{y \mid |y - x| \leq r\}$. Then, look at $g_B = C_B - f$ where $C_B = \sup_B f$. If $0 \notin B_x^r$, prove that $|g_B|_B \leq |g_{\tilde{B}}|_{\tilde{B}}$ where $\tilde{B} = B_x^{|x|}$ Then reduce to considering $B_0^{|x|}$.)

2. Let g be a periodic function on $\mathbb{R}$ with $g(x + 2\pi) = g(x)$. Let $f(e^{ix}) = g(x)$. Prove that $g \in \mathrm{BMO}(\mathbb{R}) \Leftrightarrow f \in \mathrm{BMO}(\partial\mathbb{D})$.

3. (a) If f is analytic on $\partial\mathbb{D}$ with $f(0) = 0$, show that

$$\left| \int f(re^{i\theta}) g(e^{i\theta}) \frac{d\theta}{2\pi} \right| = \left| \int f(re^{i\theta}) \tilde{g}(e^{i\theta}) \frac{d\theta}{2\pi} \right|$$

for any real-valued L^1-function g on $\partial\mathbb{D}$.

(b) Use (a) to show that if the Fefferman–Stein decomposition holds, then one has Theorem 5.11.1 and so the Fefferman duality theorem.

4. Let h be analytic on $\mathbb{D}$. Let $g_\rho(e^{i\theta})$ be defined by

$$g_\rho(e^{i\theta}) = \sup_{z \in N(\theta), |z| \leq \rho} |h(z)|$$

Prove that g_ρ is continuous in θ. (*Hint*: Prove that $|g_\rho(e^{i\theta}) - g_\rho(e^{i\psi})| \leq |e^{i\theta} - e^{i\psi}| \sup_{|z| \leq \rho} |h'(z)|$.)

5. Let $\ell(S)$ be the usual length for any arc in $\mathbb{C}$. In this problem, the reader will prove for T, the trapezoid of Lemma 5.11.13, that

$$\ell(T \setminus J) \leq \sqrt{2}\,\ell(J)$$

which is equivalent to (5.11.76) and that $\sqrt{2}$ is best possible.

(a) If H is the hypotenuse of an isoceles right triangle and T is the triangle, prove that
$$\ell(T \setminus H) = \sqrt{2}\,\ell(H)$$
and use this to conclude that $\ell(J)^{-1}\ell(T \setminus J) \to \sqrt{2}$ as $\ell(J) \to 0$.

(b) For an isosceles triangle with vertex angle (the one opposite the other side, S) larger than $90°$, prove that
$$\ell(T \setminus S) \le \sqrt{2}\,|S|$$

(c) Prove $\ell(T \setminus J) \le \sqrt{2}\,\ell(J)$ when J is so large that T has two arcs and two linear sides.

6. This problem will look at Hankel matrices of the form (5.11.115) and prove Nehari's theorem as described in the Notes.

(a) Let a, b be finite support sequences in $\ell^2(\mathbb{N})$ and define $f, g \in H^2(\mathbb{D})$ by $f(z) = \sum_{n=0}^{\infty} \bar{a}_n z^n$, $g(z) = \sum_{n=0}^{\infty} b_n z^n$. Prove that
$$\langle a, H(\alpha)b \rangle = \int \varphi(e^{i\theta})g(e^{-i\theta})\,f(e^{-i\theta})\,\frac{d\theta}{2\pi} \qquad (5.11.119)$$

if $\varphi_n^\sharp = a_n$.

(b) Conclude that if $\varphi \in L^\infty(\partial\mathbb{D}, \frac{d\theta}{2\pi})$, one has that
$$\|H(\alpha)\|_{\mathcal{L}(\ell^2)} \le \|\varphi\|_\infty \qquad (5.11.120)$$

(c) Suppose that $\sum_{n=0}^{\infty}|\alpha_n| < \infty$ so that $\mathcal{L} : H^1 \to \mathbb{C}$ by
$$\mathcal{L}(Q) = \sum_{m \ge 0} \alpha_m Q_m^\sharp \qquad (5.11.121)$$

is a bounded functional. Show that any $Q \in H^1$ can be decomposed as $Q = fg$ with $f, g \in H^2$ and $\|f\|_2 = \|g\|_2 = \|Q\|_1^{1/2}$ and conclude that, if also $H(\alpha)$ is a bounded map of $\ell^2(\mathbb{N})$ that
$$|\mathcal{L}(Q)| \le \|H(\alpha)\|_{\mathcal{L}(\ell^2)}\|Q\|_1 \qquad (5.11.122)$$

(d) Use the Hahn-Banach theorem to prove that there is $\varphi \in L^\infty(\partial\mathbb{D}, \frac{d\theta}{2\pi})$ with
$$\|\varphi\|_\infty \le \|H(\alpha)\|_{\mathcal{L}(\ell^2)}, \quad \varphi_m^\sharp = \alpha_m \qquad (5.11.123)$$

(e) Now suppose we only know that $H(\alpha)$ is a bounded map of $\ell^2(\mathbb{N})$ to itself without $\sum_{m=0}^{\infty}|\alpha_m| < \infty$. Let $\alpha_m^{(r)} = \alpha_m r^m$ for $0 < r < 1$. Prove that $\alpha \in \ell^\infty$, $\alpha^{(r)} \in \ell^1$, and $H(\alpha^{(r)})$ is a bounded map of ℓ^2 to itself with $\|H(\alpha^{(r)})\| \le \|H(\alpha)\|$. Conclude that
$$|\mathcal{L}_r(Q)| \le \|H(\alpha)\|_{\mathcal{L}(\ell^2)}\|Q\|_1 \qquad (5.11.124)$$

and then that $\mathcal{L}_r$ has a limit, $\mathcal{L}$, in H_1^* with (5.11.121) and $\mathcal{L}(e^{i\theta m}) = \alpha_m$.

(f) Conclude there is $\varphi \in L^\infty(\partial\mathbb{D}, \frac{d\theta}{2\pi})$ so (5.11.123) holds. Thus, we have the equivalence (a) and (b) in Nehari's theorem.

(g) Use the Fefferman–Stein decomposition to show that $\sum_{m=0}^\infty \alpha_m\, e^{im\theta} \in$ HMO if and only if there is $\varphi \in L^\infty(\partial\mathbb{D}, \frac{d\theta}{2\pi})$ with $\varphi_m^\sharp = \alpha_m$.

7. This problem will continue Problem 6 by proving Hartman's theorem on compact Hankel matrices.

(a) Given $\varphi \in L^\infty(\partial\mathbb{D}, \frac{d\theta}{2\pi})$, let $H^{(\varphi)}$ be the Hankel matrix $H(\alpha)$ with $\alpha_m = \varphi_m^\sharp$. If φ is a Laurent polynomial, prove that $H^{(\varphi)}$ is finite rank.

(b) If $\varphi \in C(\partial\mathbb{D})$, prove that $H^{(\varphi)}$ is a compact operator.

Note. The Hahn–Banach argument of the last problem says there is $\varphi \in L^\infty$ minimizing $\|\varphi\|_\infty$ among those φ's with $\varphi_m^\sharp = \alpha_m$. The next part will show that if there is a continuous φ obeying (5.11.117), there are other continuous $\widetilde{\varphi}$ obeying (5.11.117) with $\|\widetilde{\varphi}\|_\infty$ arbitrarily close to $\|H(\alpha)\|$ but it might not happen that there is one minimizing $\|\cdot\|_\infty$.

(c) Suppose φ is continuous. Prove that

$$\inf\{\|\psi\|_\infty \mid \psi_m^\sharp = \varphi_m^\sharp, m \geq 0, \psi \in L^\infty\} = \inf\{\|\psi\|_\infty \mid \psi_m^\sharp = \varphi_m^\sharp,$$
$$m \geq 0, \psi \in C(\partial\mathbb{D})\}$$
$$(5.11.125)$$

(*Hint:* For any $\psi \in L^\infty$ with $\psi_m^\sharp = \varphi_m^\sharp$, let $\psi^{(r)} = \varphi + (\psi - \varphi) * P_r$ where P_r is the Poisson kernel and note that $\|\psi^{(r)}\|_\infty \leq \|\varphi - \varphi * P_r\|_\infty + \|\psi\|_\infty$, $(\psi^{(r)})_n^\sharp = \varphi_n^\sharp$ and $\psi^{(r)}$ is continuous and finally that $\lim_{r\uparrow 1}\|\varphi - \varphi * P_r\|_\infty = 0$.)

(d) Prove that $\{H^{(\varphi)} \mid \varphi \in C(\partial\mathbb{D})\}$ is closed in the operator norm as a subset of $\mathcal{L}(\ell^2)$.

(e) Let R be right shift on $\ell^2(\mathbb{N})$, i.e., $Ra = (0, a_0, a_1, \ldots)$. Prove that for any rank-one operator and so any compact operator, K, on $\ell^2(\mathbb{N})$, we have that $\|KR^\ell\| \to 0$ as $\ell \to \infty$.

(f) Prove that $H(a)R^\ell = H(\{a_{n+\ell}\}_{n=0}^\infty)$ so that $H^{(\varphi)}R^\ell = H^{(\bar{z}^\ell\varphi)}$.

(g) Let $\widetilde{H}_-^\infty = \{\psi \in L^\infty(\partial\mathbb{D}, \frac{d\theta}{2\pi}) \mid \psi_n^\sharp = 0 \text{ for } n \geq 0\}$, i.e., $\psi \in \widetilde{H}_-^\infty \Leftrightarrow z^{-1}\bar{\psi} \in H^\infty$. Prove that if $H^{(\varphi)}$ is compact, there are $\psi_\ell \in \widetilde{H}_-^\infty$ so that $\|\bar{z}^\ell\varphi + \psi_\ell\|_\infty \to 0$ as $\ell \to \infty$.

(h) Prove there are polynomials P_ℓ of degree ℓ so that $\|H^{(\varphi)} - H^{(P_\ell)}\| \to 0$. (*Hint:* $\|\varphi + z^\ell\psi_\ell\|_\infty \to 0$ and $z^\ell\psi_\ell$ is a polynomial of degree ℓ mod $\widetilde{H}_-^\infty$.)

(i) Conclude that if $H^{(\varphi)}$ is compact, there is a continuous ψ with $H^{(\varphi)} = H^{(\psi)}$.

5.12. Cotlar's Theorem on Ergodic Hilbert Transforms

Our goal in this section is to prove the following which combines Hilbert transform ideas with ergodic theory—Section 2.6 has the basic language of ergodic theory.

Theorem 5.12.1 (Cotlar's Theorem—Continuum Case). *Let* $(\Omega, \Sigma, \mu, \{T_t\}_{t \in \mathbb{R}})$ *be a measure preserving flow. Then for any* $f \in L^1(\Omega, d\mu)$ *and a.e.* $\omega \in \Omega$, *we have that* $\lim_{\varepsilon \downarrow 0}(H_\varepsilon f)(\omega)$ *exists, where for* $\varepsilon \in (0,1)$

$$(H_\varepsilon f)(\omega) = \int_{\varepsilon \le |t| \le \varepsilon^{-1}} \frac{f(T_t x)}{t}\, dt \qquad (5.12.1)$$

We'll let the reader in Problems 1 and 2 prove the discrete analog.

Theorem 5.12.2 (Cotlar's Theorem—Discrete Case). *Let* (Ω, Σ, μ, T) *be an invertible measurable dynamical system. Then for any* $f \in L^1(\Omega, d\mu)$ *and a.e.* ω *in* Ω, *we have that* $\lim_{N \to \infty}(H_N^{(d)} f)(\omega)$ *exists, where for* $N \ge 1$

$$H_N^{(d)} f = \sum_{n=1}^{N} \frac{f(T^n \omega) - f(T^{-n} \omega)}{n} \qquad (5.12.2)$$

As usual, we first prove a maximal inequality and then convergence on a convenient dense subset. In Theorem 2.6.15, we saw (following Calderón) how to deduce a maximal ergodic inequality for the Hardy–Littlewood maximal inequality. Here we'll deduce a maximal ergodic Hilbert inequality from the one-dimensional maximal inequality of the maximal Hilbert transform, Theorem 5.9.14.

Theorem 5.12.3. *Define*

$$(M_{\mathrm{HE}} f)(\omega) = \sup_{0 < \varepsilon < 1} \left| \int_{\varepsilon < |t| < \varepsilon^{-1}} \frac{f(T_t \omega)}{t}\, dt \right| \qquad (5.12.3)$$

Then for a universal constant, C, *and all* $\lambda > 0$

$$\mu(\{\omega \mid (M_{\mathrm{HE}} f)(\omega) > \lambda\}) \le \frac{C\|f\|_1}{\lambda} \qquad (5.12.4)$$

Proof. As noted, we follow the strategy of Theorem 2.6.15. For $S > 1$, define $M_{\mathrm{HE},S}$ by (5.12.3) with $\sup_{0 < \varepsilon < 1}$ replaced by $\sup_{S^{-1} < \varepsilon < 1}$ so only t's with $|t| < S$ enter.

Given f in $L^1(\Omega, d\mu)$ and $\omega \in \Omega$ and $T > 0$, define

$$\varphi_{T,\omega}(s) = \begin{cases} f(T_S \omega), & -T < s < T \\ 0, & |s| \ge T \end{cases} \qquad (5.12.5)$$

As in the proof of Theorem 2.6.15, $\varphi \in L^1(\mathbb{R}, ds)$ for a.e. $\omega \in \Omega$. Fix $1 < S < T$. We note that if $|s| \leq T - S$, then

$$\left(M_{\mathrm{HE},S}f\right)(T_s\omega_0) = \sup_{S^{-1}<\varepsilon<1} \left| \int_{\varepsilon<|t|<\varepsilon^{-1}} \frac{f(T_{t+s}\omega_0)}{t} \, dt \right|$$

$$= \sup_{S^{-1}<\varepsilon<1} \left| \int_{\varepsilon<|t|<\varepsilon^{-1}} \varphi_{T,\omega_0} \frac{(s+t)}{t} \, dt \right| \qquad (5.12.6)$$

$$\leq M_{\mathrm{H}}\varphi_{T,\omega_0}(s) \qquad (5.12.7)$$

where M_{H} is given by (5.9.108). To get (5.12.6), we use $|s| < T - S$ and $|t| < \varepsilon^{-1} < S \Rightarrow |s + t| < T$.

Thus, by Theorem 5.9.14, with $|\cdot|$ = Lebesgue measure in s,

$$\left| \{ |s| \in [0, T - S] \mid \left(M_{\mathrm{HE},S}f\right)(T_s\omega_0) > \alpha \} \right| \leq C\alpha^{-1} \|\varphi_{T,\omega_0}\|_1 \qquad (5.12.8)$$

Now integrate over μ and use invariance of μ

$$\int \|\varphi_{T,\omega_0}\|_1 \, d\mu(\omega_0) = \int_{-T}^{T} \|f(T_s \cdot)\|_1 \, ds = 2T \|f\|_1 \qquad (5.12.9)$$

(where $\|\cdot\|_1$ on the left is $L^1(\mathbb{R}, ds)$ and $\|\cdot\|_1$ in the middle and right is $L^1(\Omega, d\mu)$). And on the left of (5.12.8) we get $2(T - S)\mu(\{\omega_0 \mid (M_{\mathrm{HE},S}f)(\omega_0) > \alpha\})$. Thus

$$\mu(\{\omega_0 \mid (M_{\mathrm{HE},S}f)(\omega_0) > \alpha\}) \leq \frac{2TC}{2(T - S)} \alpha^{-1} \|f\|_1$$

By taking $T \to \infty$ and then $S \to \infty$, we get (5.12.4) $\square$

Proof of Theorem 5.12.1. As usual, given the maximal inequality (5.12.4), we need only prove that the limit exists for a dense set of f's in $L^1(\Omega, d\mu)$. Let Y be the set of functions which are finite linear combinations of $\mathbb{1}$ and functions of the form

$$f(\omega) = \int \varphi(s)g(T_s\omega) \, ds \qquad (5.12.10)$$

where $\varphi \in C_0^\infty(\mathbb{R})$ with

$$\int \varphi(s) ds = 0 \qquad (5.12.11)$$

and $g \in L^\infty(\Omega, d\mu)$. Y is dense since taking φ to be a difference of shifted approximate identities, $\overline{Y}$ includes functions of the form $g(T_t\omega) - g(\omega)$ and the span of these and $\mathbb{1}$ is dense by the same argument as Proposition 2.6.13.

If $f \equiv 1$, $H_\varepsilon f = 0$, so the limit exists trivially. We are thus left with establishing the limit for f of the form (5.12.10) where $\varphi \in C_0^\infty$ obeys (5.12.11) and $g \in L^\infty$.

Define

$$h_\varepsilon(s) = \begin{cases} s^{-1}, & \varepsilon < |s| < \varepsilon^{-1} \\ 0, & |s| \geq \varepsilon \end{cases} \tag{5.12.12}$$

Then, for f of the form (5.12.22),

$$
\begin{aligned}
(H_\varepsilon f)(\omega) &= \int_{-\infty}^{\infty} \int_{\varepsilon < |t| < \varepsilon^{-1}} g(T_{t+s}\omega)\frac{1}{t}\varphi(s)\, dt\, ds \\
&= \int_{-\infty}^{\infty} g(T_u \omega)\left(\int_{\varepsilon < |u-s| < \varepsilon^{-1}} \varphi(s)\frac{1}{u-s}\, ds \right) d\mu \\
&= \int_{-\infty}^{\infty} g(T_u \omega)(h_\varepsilon * \varphi)(u)\, du \tag{5.12.13}
\end{aligned}
$$

Suppose we prove that

(a) $\lim_{\varepsilon \downarrow 0}(h_\varepsilon * \varphi)(u)$ exists for each u.

(b) For some $\Phi \in L^1(\mathbb{R})$, and all $\varepsilon < 1$,

$$|(h_\varepsilon * \varphi)(u)| \leq \Phi(u) \tag{5.12.14}$$

Then for a.e. (ω, u), the limit exists of the integrand in (5.12.13) and the integrand is bounded by $\|g\|_\infty \Phi(u)$ so the integral converges for a.e. ω completing the proof.

Note that since $\int_{\varepsilon < |s| < 1} s^{-1}\, ds = 0$,

$$(h_\varepsilon * \varphi)(u) = \int_{\varepsilon < |s| < 1}\left[\frac{\varphi(u-s)-\varphi(u)}{s}\right] ds + \int_{1 \leq |s| < \varepsilon^{-1}} \frac{\varphi(u-s)}{s}\, ds \tag{5.12.15}$$

$$\equiv k_{1,\varepsilon}(u) + k_{2,\varepsilon}(u) \tag{5.12.16}$$

Since $|\varphi(u-s)-\varphi(u)| \leq \|\varphi'\|_\infty s$, the limit $\lim_{\varepsilon \downarrow 0} k_{1,\varepsilon}(u)$ exists and if φ is supported in $[-R, R]$, we have

$$\sup_\varepsilon |k_{1,\varepsilon}(u)| \leq \begin{cases} 2\|\varphi'\|_\infty & \text{if } |u| \leq R \pm 1 \\ 0 & \text{if } |u| \geq R+1 \end{cases} \tag{5.12.17}$$

Since φ has compact support for each fixed u, the limit of $k_{2,\varepsilon}(u)$ exists proving (a).

For any u, clearly

$$|k_{2,\varepsilon}(u)| \leq \|\varphi\|_1 \tag{5.12.18}$$

If $|u| > R+1$, $\int_{1 \leq |s| \leq \varepsilon^{-1}} \varphi(u-s)\, ds = \int \varphi(u-s)\, ds = 0$ (by choice of φ) so for such u

$$|k_{2,\varepsilon}(u)| \leq \left| \int_{1 \leq s < \varepsilon^{-1}} |\varphi(u-s)|\left[\frac{1}{s} - \frac{1}{u}\right] ds \right| \tag{5.12.19}$$

$$\leq \int_{1 \leq s \leq \varepsilon^{-1}} |\varphi(u-s)|\,|u-s|\frac{1}{|s|}\frac{1}{|u|}\, ds \tag{5.12.20}$$

$$\leq \frac{1}{|u|} \frac{1}{|u| - R} \int |x| \, |\varphi(x)| \, dx \tag{5.12.21}$$

(5.12.16) and (5.12.21) provide an L^1 uniform upper bound on $|k_{2,\varepsilon}|$ and so prove (b) and complete the proof. $\qquad\square$

Notes and Historical Remarks. Cotlar's theorem is from [**173**]. The approach we use here is from Calderón [**131**].

Problems

1. This will prove a maximal inequality for the discrete Hilbert transform on sequences $a \in \ell_1(\mathbb{Z})$:

$$(\widetilde{M}_{\mathrm{H}} a)(n) = \sup_N (\widetilde{H}_N a)(n) \tag{5.12.22}$$

$$(\widetilde{H}_N a)(n) = \sum_{j=1}^{N} j^{-1} [a(n+j) - a(n-j)] \tag{5.12.23}$$

(a) Prove that for all $m \in \mathbb{Z}$, $|m| \geq 1$ and α with $|\alpha| \leq \frac{1}{2}$

$$\sup_{|\alpha| \leq \frac{1}{2}} \left| \int_m^{m+1} \frac{ds}{s+\alpha} - \frac{s}{m} \right| \leq \frac{C}{m^2} \tag{5.12.24}$$

(b) Given $a \in \ell_1$, let $\tilde{a}(x)$ be the function

$$\tilde{a}(x) = a_n \text{ if } n - \tfrac{1}{2} < x \leq n + \tfrac{1}{2} \tag{5.12.25}$$

Prove that with H_ε given by (5.12.1) for all $|\alpha| \leq \frac{1}{2}$

$$\left| (\widetilde{H}_N a)(n) - (H_{N^{-1}} \tilde{a})(n + \alpha) \right| \leq \left| (|g| * |a|)(n) \right| \tag{5.12.26}$$

for a sequence g in L^1.

(c) Using Theorem 5.9.14, prove there is a constant $\widetilde{C}$ with

$$\sharp \{ n \mid |\widetilde{M}_{\mathrm{H}} a(n)| > \alpha \} \leq \widetilde{C} \alpha^{-1} \|a\|_1 \tag{5.12.27}$$

2. Using Problem 1 and the proof of Theorem 5.12.1, prove Theorem 5.12.2.

Chapter 6

Bonus Chapter:
More Inequalities

In algebra, when one says $a = b$, it is a tautology and so uninteresting; while in analysis, when one says $a = b$, it is two deep inequalities.
—attributed to *S. Bochner*

If one only proves $a = b$ by showing $a \leq b$ and $b \leq a$, one has not understood the true reason that $a = b$.
—attributed to *E. Noether*

Big Notions and Theorems: Lorentz Space, (Generalized) Hardy's Inequality, Hunt–Marcinkiewicz Interpolation Theorem, HLS (aka Hardy–Littlewood–Sobolev) Inequalities, Stein–Weiss Inequality, Rellich's Inequality, Weak Young Inequality, Spherical Rearrangements, Brascamp–Lieb–Luttinger Inequality, Sobolev Space, $W^{\ell,p}(\mathbb{R}^\nu)$, $H^{s,p}(\mathbb{R}^\nu)$, Homogeneous Sobolev Estimates, Gagliardo–Nirenberg Inequalities, Nash Inequality, Inhomogeneous Sobolev Inequality, Sobolev Embedding Theorem, Rellich–Kondrachov Theorem, Poincaré Inequality, Fractional Sobolev Space, Besov Space, Calderón–Zygmund Operators, Calderón–Zygmund Decomposition, Riesz Potential, Hörmander–Mikhlin Multiplier Theorem, Littlewood–Paley Decomposition, T1 Theorem, Hörmander–Sobolev Smoothing Theorem, Calderón–Vaillancourt Theorem, Cotlar–Knapp–Stein Lemma, Hypercontractive Semigroup, Supercontractive Semigroup, Ultracontractive Semigroup, Varopoulos–Faber–Stroock Theorem, Markov Semigroup, Intrinsic Ultracontractivity, Classical Dirichlet Forms, Ornstein–Uhlenbeck Semigroup, Graph Laplacian, Bonami–Gross Semigroup, Bonami–Segal Lemma, Logarithmic Sobolev Inequalities, Gross' Theorem, Gross–Nelson Semigroups, Bonami–Gross Inequality, Nelson's Best Hypercontractive Estimate, Rosen's Lemma, Young's Inequality, Hypercontractivity and Semiboundedness, Hypercontractivity and Self-Adjointness, Lieb–Thirring Inequalities, Cwickel–Lieb–Rosenblum (CLR) Inequalities, Hardy–Lieb–Thirring Inequalities, Trace Inequality, Tomas–Stein Theorem, Knapp's Counterexample, Strichartz' Estimates, Bochner–Riesz Conjecture, Restriction Conjecture, Kakeya Sets, Kakeya Conjectures, Tauberian Theorems

Inequalities are the bread and butter of analysis. Hardly any major subject of these texts isn't suffused with inequalities: examples include the Bessel and Cauchy-Schwarz inequalities central to Hilbert space theory, Minkowski and Hölder inequalities of L^p-space theory, Jensen's and Markov's inequalities, Hausdorff–Young and Young inequalities related to Fourier analysis, Hardy's inequality and the related uncertainty principle, the Schwarz inequality and Cauchy estimates of complex analysis, Harnack's inequality in the theory of harmonic functions, maximal inequalities that have dominated this volume, and from Part 4 : the Schur–Lelesco–Weyl inequality and Kato's inequality. That said, there are some major inequalities that we haven't touched and that is the goal of this chapter, especially the new techniques needed to prove them.

The central theme will be various forms of "Sobolev inequalities"—how to go from L^p-properties of a function and some of its distributional derivatives to L^q-estimates on that function and sometimes its classical derivatives. If $f, \nabla f \in L^p$, then one expects that $(|k|\hat{f})^\vee$ is in L^p, so to see what this says about f, we need to know about $(|k|^{-1}\hat{f})^\vee$. This is, of course, convolution with the Fourier transform of $|k|^{-1}$ which we've seen is a constant times $|x - y|^{-(\nu-1)}$. Thus, the core inequality of this chapter is the Hardy–Littlewood–Sobolev (HLS) inequality: let p, q, α be such that $1 < p < \infty$, $1 < q < \infty$, $0 < \alpha < \nu$ and

$$\frac{1}{p} + \frac{1}{q} + \frac{\alpha}{\nu} = 2 \tag{6.0.1}$$

then, there is $C(p, q, \alpha, \nu)$ so that for all $f \in L^p(\mathbb{R}^\nu)$, $g \in L^q(\mathbb{R}^\nu)$

$$\int \frac{|f(x)|\,|g(y)|}{|x - y|^\alpha} d^\nu x\, d^\nu y \leq C\|f\|_p \|g\|_q \tag{6.0.2}$$

We'll prove (6.0.2) in Section 6.2 and apply it to the structure of Sobolev spaces in Section 6.3. In Section 6.4, we'll turn to a situation where the integral kernel, in essence has $\alpha = \nu$ so p and q are dual indexes. It is easy to see that (6.0.2) cannot hold in this case. But the Hilbert transform of Section 5.7 shows that

$$\lim_{\varepsilon \downarrow 0} \int_{|x-y|\geq \varepsilon} f(x)\frac{1}{x - y}g(y)dx\, dy \tag{6.0.3}$$

does have bounds (if $f \in L^p$, $g \in L^q$, $p^{-1} + q^{-1} = 1$, $1 < p < \infty$). The difference between $(x - y)^{-1}$ and $|x - y|^{-1}$ is the possibility of cancelation and Section 6.4 on Calderón–Zygmund theory will deal with kernels that have $|x|^{-\nu}$ decay but an extra angular dependence in x that averages to zero on the sphere. This will yield higher-dimensional analogs of M. Riesz's conjugate function (Hilbert transform) bounds. An example is the proof that $\partial_i\partial_j\Delta^{-2}$ is bounded on L^p, $1 < p < \infty$.

Section 6.5 will discuss some useful bounds on certain pseudodifferential operators. Section 6.6 is on hypercontractive operators. Section 6.7 involves eigenvalues of Schrödinger operators, Section 6.8 discusses restriction of Fourier transforms to submanifolds, and Section 6.9 is on results that let one undo averages. For example, we'll prove that if μ is a (positive) measure on $\mathbb{R}$ with

$$\int \frac{d\mu(x)}{x^2 + 1} < \infty \tag{6.0.4}$$

and $\alpha \in (0, 2)$, then

$$\lim_{R \to \infty} \frac{\mu([-R, R])}{R^\alpha} = a \Leftrightarrow \lim_{y \to \infty} \frac{1}{y^{\alpha-2}} \int \frac{d\mu(x)}{x^2 + y^2} = \frac{a(\alpha\pi/2)}{\sin(\pi\alpha/2)} \tag{6.0.5}$$

In this formula, $\Rightarrow$ is easy but $\Leftarrow$ is subtle in that it can fail if μ is only a signed measure.

Returning to the central result, (6.0.2), we note that it is reminiscent of the symmetric form of Young's inequality, (6.6.12) of Part 1. If $s = \frac{\nu}{\alpha}$, then (6.0.2) would exactly be Young's inequality if $|x|^{-\alpha}$ were in $L^s(\mathbb{R}^\nu)$. In fact, it is only in $L^s_w(\mathbb{R}^\nu)$, the weak-L^s space. Indeed, we'll prove (6.0.2) by proving a weak extension of Young's inequality. Young's inequality says $Tg = f * g$ maps L^q to L^r if $f \in L^p$, $1 \le p < \infty$, with $1 \le q \le \frac{p}{p-1}$ and $r = \left(\frac{1}{p} + \frac{1}{q} - 1\right)^{-1}$. We'll show this remains true if $f \in L^p_w$ so long as one avoids endpoints, i.e., we'll need $p > 1$, $1 < q < \frac{p}{p-1}$. In this regard, it is useful to understand that Young's inequality has such an extension but Hölder's does not:

Proposition 6.0.1. *Let f be a measurable function on X. Let μ be a σ-finite measure. Suppose for all $g \in L^4(X, d\mu)$, we have that $fg \in L^2(X, d\mu)$ with*

$$\|fg\|_2 \le D\|g\|_4 \tag{6.0.6}$$

where D is a g-independent constant. Then $f \in L^4(X, d\mu)$.

Remark. This is true for more general p, q than $p = q = 4$; we state for this case for notational simplicity.

Proof. Fix a and fix $C \subset X$ with $\mu(C) < \infty$

$$f_{a,C}(x) = \begin{cases} f(x), & x \in C, |f(x)| < a \\ 0, & \text{otherwise} \end{cases}$$

Let $g = f_{a,C}/\|f_{a,C}\|_4$ so $\|g\|_4 = 1$ and

$$\|fg\|_2^2 = \frac{\int_Y |f(x)|^4 d\mu(x)}{\left(\int_Y |f(x)|^4 d\mu(x)\right)^{1/2}} = \left(\int_Y |f(x)|^4 d\mu\right)^{1/2} \tag{6.0.7}$$

where $Y = \{x \mid x \in C, |f(x)| < a\}$. Thus, by (6.0.6)

$$\int_Y |f(x)|^4 d\mu \leq D^4 \qquad (6.0.8)$$

Take $a \to \infty$ and then C to be an increasing family of sets with finite measure whose union is X. Using the monotone convergence theorem shows that $f \in L^4$ with $\|f\|_4 \leq D$. $\qquad\qquad\qquad\qquad\qquad\qquad\qquad\qquad\qquad\qquad\qquad\quad\square$

The weak Young inequality will rely on the following

Theorem 6.0.2 (Original Marcinkiewicz Interpolation Theorem). *Let (M, Σ, μ) and (N, Ξ, ν) be two σ-finite measure spaces. Let $1 \leq p_0 \leq q_0 \leq \infty$, $1 \leq p_1 \leq q_1 \leq \infty$ and $T : L^{p_0}(M, d\mu) + L^{p_1}(M, d\mu)$ into the Ξ-measurable function on N so that*

$$\|Tf\|_{q_0, w}^* \leq C_0 \|f\|_{p_0}, \quad \|Tf\|_{q_1, w}^* \leq C_1 \|f\|_{p_1} \qquad (6.0.9)$$

Then for any $\theta \in (0, 1)$, there is a constant, C_θ depending only on θ, p_j, q_j, C_j so that

$$\|Tf\|_{q_\theta} \leq C\|f\|_{p_\theta} \qquad (6.0.10)$$

where

$$p_\theta^{-1} = \theta p_1^{-1} + (1 - \theta)p_0^{-1}, \quad q_\theta^{-1} = \theta q_1^{-1} + (1 - \theta)q_0^{-1} \qquad (6.0.11)$$

Theorem 6.0.3 (Hunt Interpolation Theorem). *Let (M, Σ, μ) and (N, Ξ, ν) be two σ-finite measure spaces. Let $p_0 < p_1$, $q_0 \neq q_1$, all in $[1, \infty]$ and suppose $T : L^{p_0}(M, d\mu) + L^{p_1}(M, d\mu)$ into the Ξ-measure function on N so that*

$$\|Tf\|_{q_j} \leq C_j \|f\|_{p_j} \ (j = 0, 1) \qquad (6.0.12)$$

Then for any $\theta \in (0, 1)$, there is a constant C depending only on θ, p_j, q_j, C_j so that

$$\|Tf\|_{q_\theta, w}^* \leq C\|f\|_{p_\theta, w}^* \qquad (6.0.13)$$

where p_θ, q_θ are given by (6.0.11).

One combines this to go from Young to weak Young. Suppose that

$$\frac{1}{p} + \frac{1}{q} - 1 = \frac{1}{r} \qquad (6.0.14)$$

Young says for $f \in L^p$, $T(g) = f * g$ maps L^q to L^r for $1 \leq q \leq \frac{p}{p-1}$ with r given by (6.0.14). Thus, by Hunt's theorem, T maps L_w^q to L_w^r for $1 < q < \frac{p}{p-1}$. Put differently, for $g \in L_w^q$, $S(f) = f * g$ maps L^p to L_w^r if $1 < p < \frac{q}{q-1}$ where $q > 1$. Since $q > 1$, $\frac{1}{p} > \frac{1}{r}$, i.e., $r > p$ and we can use the original Marcinkiewicz theorem to get rid of w, in L_w^r, i.e., S maps L^p to L^r so $f \in L^p$, $g \in L_w^q$ implies $f * g \in L^r$ where $1 < q < \frac{p}{p-1}$ and $1 < p < \frac{q}{q-1}$. This is weak Young.

If we try the same strategy with Hölder, instead of $r > p$, we find $r < p$. Thus Marcinkiewicz really is limited to $r \geq p$.

It is thus useful to understand why this distinction arises and this is clarified by Section 6.1. For each p, we'll introduce a one-parameter family of spaces $L^{p,r}$ where $r = p$ yields L^p and $r = \infty$ will yield L^p_w. Moreover, if $r_1 < r_2$, then $L^{p,r_1} \subset L^{p,r_2}$ and $\|f\|_{p,r_2} \leq \|f\|_{p,r_1}$. The Lorentz–Marcinkiewicz theorem will assert that if T_Ω maps $L^{p_0,1} + L^{p_1,1}$ to measurable functions and

$$\|Tf\|_{q_j,\infty} \leq C_j \|f\|_{p_j,1} \tag{6.0.15}$$

then for any r,

$$\|Tf\|_{q_\theta,r} \leq C \|f\|_{p_\theta,r} \tag{6.0.16}$$

In particular, taking $r = q_\theta$, we get

$$\|Tf\|_{q_\theta,q_\theta} \leq C \|f\|_{p_\theta,q_\theta} \tag{6.0.17}$$

When $p_\theta \leq q_\theta$, we have that $\|f\|_{p_\theta,q_\theta} \leq \|f\|_{p_\theta,p_\theta}$ and we get L^{p_θ} to L^{q_θ} boundedness. This "explains" the $p \leq q$ condition in the original theorem.

Notes and Historical Remarks. I heard the pair of dueling quotes that start the chapter when I was a student but haven't been able to find references so they may be apocryphal. Garling [**304**] is a lovely book on inequalities in linear analysis (that includes much more than we discuss but not all that we do).

6.1. Lorentz Spaces and Real Interpolation

As indicated in the introduction to the chapter, it will be useful to think of L^p and L^p_w as two spaces in a one-parameter family $L^{p,q}$. In this section, we'll define and briefly study these spaces and use them to prove Theorems 6.0.2 and 6.0.3 as special cases of a more general interpolation theorem. Along the way, we'll need to prove a family of inequalities on integrals due to Hardy.

Recall that given any measurable function, f, from a σ-finite measure space (M, Σ, μ) to $\mathbb{R}$ (or $\mathbb{C}$) which is a.e. finite and has $\mu(\{x \mid |f(x)| > t\}) < \infty$ for all $t > 0$, we defined f^* to be the unique monotone decreasing function from $[0, \infty)$ to $\mathbb{R}$ which is continuous from the right (i.e., $\lim_{t \downarrow t_0} f^*(t) = f^*(t_0)$) and so that $|f(x)|$ and $f^*(t)$ are equimeasurable, i.e., for all $\alpha > 0$,

$$m_f(\alpha) \equiv \mu(\{x \mid |f(x)| > \alpha\}) = |\{t \mid f^*(t) > \alpha\}| \tag{6.1.1}$$

It is constructed in Theorem 2.2.6 and given by

$$f^*(t) = \inf\{\alpha \mid m_f(\alpha) \leq t\} \tag{6.1.2}$$

The *Lorentz quasinorm* is defined, for $0 < p < \infty$, $0 < q < \infty$ by:

$$\|f\|_{p,q}^* = \left(\frac{q}{p} \int_0^\infty t^{q/p}[f^*(t)]^q \frac{dt}{t}\right)^{1/q} \tag{6.1.3}$$

The *Lorentz space*, $L^{p,q}(M, d\mu)$, is the set of functions for which $\|f\|_{p,q}^* < \infty$. For $q = \infty$,

$$\|f\|_{p,\infty}^* = \|f\|_{p,w}^* = \sup[t^{1/p} f^*(t)]$$

and $L^{p,\infty} = L_w^p$. It follows (Problem 1) from Problem 19 of Section 2.2 that $L^{p,q}$ is a vector space. $\|\cdot\|_{p,q}$ is not a norm but (Problem 1)

$$\|f + g\|_{p,q}^* \le 2^{1/p}(\|f\|_{p,q}^* + \|g\|_{p,q}^*) \tag{6.1.4}$$

As with L_w^p, for $1 < p < \infty$, if we define $\|f\|_{p,q}$ with f^* replaced by the Muirhead maximal function, (2.2.62), f^{**}, we get an equivalent norm when $p > 1$, indeed (Problem 2)

$$\|f\|_{p,q}^* \le \|f\|_{p,q} \le \frac{p}{p-1}\|f\|_{p,q}^* \tag{6.1.5}$$

If $q = p$,

$$\|f\|_{p,q=p}^* = \left(\int_0^\infty f^*(t)^p dt\right)^{1/p} = \|f\|_p \tag{6.1.6}$$

since f^* and f are equimeasurable, i.e.,

$$L^{p,p} = L^p \tag{6.1.7}$$

One should think of the q index as logarithmic in nature. If $f(t) = t^{-\alpha}(1 + \log|t|)^{-\beta}$ on $[1, \infty)$, then for $p = 1/\alpha$, $f \in L^{p,q}((1, \infty), dt)$ if and only if (Problem 3) $\beta q > 1$.

Notice that $f \in L^{p,q}$ if and only if f^* is in the L^q space with measure $\frac{q}{p} t^{\frac{q}{p}-1} dt$. It is also easy to see that $L^{p,q}(M, d\mu)$ with $\|\cdot\|_{p,q}$-norm is a Banach space for $p > 1$ (Problem 4). Here is a key fact:

Theorem 6.1.1. *If $0 < p < \infty$ and $1 \le q < r \le \infty$, then*

$$\|f\|_{p,r}^* \le \|f\|_{p,q}^* \tag{6.1.8}$$

and

$$L^{p,q}(M, d\mu) \subset L^{p,r}(M, d\mu) \tag{6.1.9}$$

Remarks. 1. It is for discrete cases like ℓ^p that spaces get larger as p increases while the opposite is true for finite measure spaces. Since $L^{p,q}$ is sort of like an L^q space, the ordering may seem surprising. The key is the monotonicity of f^*.

2. See Problems 5 and 6 for what happens when p varies.

Lemma 6.1.2. *Let g be a monotone decreasing function on $[0, \infty)$. For $1 \leq q < \infty$, define*

$$F(q) = \left(q \int_0^\infty t^q g(t)^q \frac{dt}{t} \right)^{1/q} \tag{6.1.10}$$

Then $F(q)$ is a monotone decreasing function of q.

Remarks. 1. The key to the proof will be Minkowski's inequality for integrals of functions, i.e.,

$$\left\| \int f_\alpha \, d\alpha \right\|_p \leq \int \|f_\alpha\|_p \, d\alpha \tag{6.1.11}$$

This will be a recurrent theme below.

2. The occurrence of $\frac{dt}{t}$ which is a Haar measure on the multiplicative group is consistent with the logarithmic scale nature of the Lorentz space.

Proof. Given q, define

$$h(t) = g(t^{1/q})^q \tag{6.1.12}$$

Using $F(q; g)$ for (6.1.10), i.e., making the g-dependence explicit, we have for any $r \geq q$ (using $s = t^{1/q}$ so $\frac{ds}{s} = \frac{1}{q}\frac{dt}{t}$) that

$$
\begin{aligned}
F(\frac{r}{q}; h) &= \left(\frac{r}{q} \int_0^\infty g(t^{1/q})^r t^{r/q} \frac{dt}{t} \right)^{q/r} \\
&= \left(r \int_0^\infty g(s)^r s^r \frac{ds}{s} \right)^{q/r} \\
&= F(r; g)^q
\end{aligned}
\tag{6.1.13}
$$

This allows us to reduce $1 \leq q < r < \infty \Rightarrow F(r) \leq F(q)$ to the case $q = 1$.

By a limiting argument, we can suppose g is continuous and strictly monotone in which case it has a monotone decreasing inverse, f. The wedding-cake representation, (2.2.17), says that

$$g = \int_0^\infty \chi_{[0, f(\alpha)]} \, d\alpha \tag{6.1.14}$$

Recognizing that $F(r)$ is the $L^r([0, \infty), \frac{dt}{t})$ norm of $k(t) = r^{1/r} t g(t)$, we set $k_\alpha(t) = r^{1/r} t \chi_{[0, f(\alpha)]}(t)$ and note that $k = \int k_\alpha \, d\alpha$ by (6.1.14), so by Minkowski's inequality:

$$F(r) = \|k\|_{L^r(\frac{dt}{t})} \leq \int \|k_\alpha\|_{L^r(\frac{dt}{t})} \, d\alpha \tag{6.1.15}$$

Now compute

$$\|k_\alpha\|_{L^r(\frac{dt}{t})} = \left(\int_0^{f(\alpha)} r t^{r-1} dt \right)^{1/r} = f(\alpha) \tag{6.1.16}$$

Thus,

$$\int \|k_\alpha\|_{L^r(\frac{dt}{t})} ds = \int f(\alpha)d\alpha = \|g\|_{L^1([0,\infty))} \tag{6.1.17}$$

$$= F(1) \tag{6.1.18}$$

where (6.1.17) comes from (6.1.14). Thus $F(r) \le F(1)$ proving monotonicity. $\qquad\square$

Proof of Theorem 6.1.1. Suppose first that $r = \infty$. For any t_0,

$$\|f\|_{p,q}^* \ge \left(\frac{q}{p}\int_0^{t_0} t^{p/q}[f^*(t)]^q \frac{dt}{t}\right)^{1/q} \tag{6.1.19}$$

$$\ge f^*(t_0)\left(\frac{q}{p}\int_0^{t_0}(t^{\frac{q}{p}-1})dt\right)^{1/q} \tag{6.1.20}$$

$$= t_0^{1/p} f^*(t_0) \tag{6.1.21}$$

where (6.1.20) used that f^* is monotone. Taking the sup over t_0, we get (6.1.19) in case $r = \infty$.

If $r < \infty$, we note that if p is given and

$$g(s) = f^*(s^p) \tag{6.1.22}$$

then using an $s = t^{1/p}$ change of variable

$$\|f\|_{p,q}^* = F(q;g) \tag{6.1.23}$$

where F is given by (6.1.10). Thus the lemma proves (6.1.8) when $q, r < \infty$. $\qquad\square$

To prove interpolation results, we'll need to control L^p-norms of indefinite integrals of f^*. The bounds we need are an extension of a classical result of Hardy. Both the original result and the generalization we need are consequences of Minkowski's inequality in the form (6.1.11). As a warmup, we start with the classical case:

Theorem 6.1.3 (Hardy's Inequality). *Let $1 < p < \infty$. Let f be an $L^p([0,\infty), dx)$-function on $[0,\infty)$. Then,*

$$\left(\int_0^\infty \left|\frac{1}{t}\int_0^t f(s)ds\right|^p dt\right)^{1/p} \le \frac{p}{p-1}\left(\int_0^\infty |f(s)|^p ds\right)^{1/p} \tag{6.1.24}$$

Remarks. 1. Since $f \in L^p$, $\int_0^t f(s)ds$ is well-defined (i.e., the integral is absolutely convergent).

2. If $g(t) = \int_0^t f(s)ds$, $p = 2$, this says $\int_0^\infty x^{-2}|g(x)|^2 dx \le 4\int|g'(x)|^2 dx$ which is the Hardy inequality of Theorem 4.1.3 when $\nu = 1$. Hence, the use of the same name!

3. This implies a Hardy inequality on sums; see Problem 7

Proof. Letting $s = tu$, we have that

$$a(t) \equiv \frac{1}{t}\int_0^t f(s)ds = \int_0^1 f(ut)du \tag{6.1.25}$$

so with $f_u(t) = f(ut)$, we see that $a(\,\cdot\,) = \int_0^1 f_u(\,\cdot\,)du$. Thus

$$\text{LHS of (6.1.24)} = \|\int_0^1 f_u du\|_p$$

$$\leq \int_0^1 \|f_u\|_p du \tag{6.1.26}$$

by (6.1.11). But using $s = ut$,

$$\|f_u\|_p = \left(\int_0^\infty |f(ut)|^p dt\right)^{1/p} = u^{-1/p}\|f\|_p \tag{6.1.27}$$

and $\int_0^1 u^{-1/p}du = (1-\frac{1}{p})^{-1} = \frac{p}{p-1}$. Thus RHS of (6.1.26) = RHS of (6.1.24).

$\square$

We'll need an enhanced version of this inequality with two parameters, $\alpha > 0$ and $\theta \in \mathbb{R}$. Let f be a nonnegative measurable function on $[0,\infty)$. Define

$$H_{\alpha,\theta}(f)(t) = t^{-\alpha}\int_0^t s^\theta f(s)\frac{ds}{s} \tag{6.1.28}$$

$$K_{\alpha,\theta}(f)(t) = t^\alpha \int_t^\infty s^\theta f(s)\frac{ds}{s} \tag{6.1.29}$$

We'll also use H_α (respectively, K_α) for $H_{\alpha,\theta=1}$ (respectively, $K_{\alpha,\theta=1}$).

Theorem 6.1.4 (Generalized Hardy Inequality). *For $1 \leq p < \infty$ and all f:*

$$\left(\int_0^\infty |H_{\alpha,\theta}(f)(t)|^p \frac{dt}{t}\right)^{1/p} \leq \frac{1}{\alpha}\left(\int_0^\infty |s^{\theta-\alpha}f(s)|^p \frac{ds}{s}\right)^{1/p} \tag{6.1.30}$$

$$\left(\int_0^\infty |K_{\alpha,\theta}(f)(t)|^p \frac{dt}{t}\right)^{1/p} \leq \frac{1}{\alpha}\left(\int_0^\infty |s^{\theta+\alpha}f(s)|^p \frac{ds}{s}\right)^{1/p} \tag{6.1.31}$$

Remarks. 1. We only supposed $f \geq 0$ for the integrals defining H and K to be well-defined. If f is such that the integrals make sense (e.g., f bounded with compact support in $(0,\infty)$), these inequalities hold.

2. While we'll separately prove the two results, if $g(s) = f(1/s)$, then

$$(K_{\alpha,\theta}f)(t) = (H_{\alpha,-\theta}g)(1/t) \tag{6.1.32}$$

(Problem 8 (a)) which can be used to deduce (6.1.31) from (6.1.30) (Problem 8 (b)).

3. It may not appear that this generalizes Hardy's inequality but if $\alpha = \frac{p-1}{p}$ and $\theta = 1$ using $1 - \alpha = 1/p$ and $\alpha p + 1 = p$, one sees that (6.1.24) is indeed a special case of (6.1.30).

Proof. We mimic the proof of Theorem 6.1.3. Start with $\theta = 1$. Then changing variables from s to tu

$$H_\alpha(f)(t) = \int_0^1 f_u(t)du, \quad f_u(t) = t^{1-\alpha}f(tu) \tag{6.1.33}$$

so, by Minkowski's inequality

$$\text{LHS of (6.1.30)} \le \int_0^1 \|f_u\|_{L^p((0,\infty),dt/t)}du \tag{6.1.34}$$

and using $ut = s$:

$$\|f_u\|_{L^p((0,\infty),dt/t)} = \left(\int_0^\infty |t^{1-\alpha}f(tu)|^p\frac{dt}{t}\right)^{1/p} \tag{6.1.35}$$

$$= [\alpha u^{\alpha-1}] \text{ RHS of (6.1.30)}$$

$$= \text{RHS of (6.1.30)} \tag{6.1.36}$$

since $\alpha > 0 \Rightarrow \int_0^1 u^{\alpha-1}du = \alpha^{-1}$, we get (6.1.30) when $\theta = 1$.

Similarly

$$(K_\alpha f)(t) = \int_1^\infty f_u(t)du, \quad f_u(t) = t^{1+\alpha}f(tu) \tag{6.1.37}$$

and the remaining steps are similar but note that we use $\alpha > 0 \Rightarrow \int_1^\infty u^{-1-\alpha}du = \alpha^{-1}$.

Noting that

$$H_{\alpha,\theta}(f) = H_{\alpha,\theta=1}(s^{\theta-1}f), \quad K_{\alpha,\theta}(f) = K_{\alpha,\theta=1}(s^{\theta-1}f) \tag{6.1.38}$$

we obtain the general θ result from the special case $\theta = 1$.   □

We need one last set of results. Recall $f_{>\alpha}$ and $f_{\le\alpha}$ defined by (2.2.42) and (2.2.43): $f_{>\alpha} = f\,\chi_{\{x|f(x)>\alpha\}}$; $f_{\le\alpha} = f\,\chi_{\{x|f(x)\le\alpha\}}$.

Proposition 6.1.5. *Fix w and let $\alpha = f^*(w)$.*

(a) $(f_{>\alpha})^* = (f^*)\chi_{[0,m_f(\alpha))}$ \hfill (6.1.39)

$\qquad\quad \le (f^*)\chi_{[0,w]}$ \hfill (6.1.40)

(b) $(f_{\le\alpha})^* \le f^*$ \hfill (6.1.41)

(c) $(f_{\le\alpha})^* \le \alpha$ \hfill (6.1.42)

Remark. (b), (c) hold for all α.

Proof. (a) By definition of m:

$$m_{f_{>\alpha}}(\beta) = \begin{cases} m_f(\beta) & \text{if } \beta \geq \alpha \\ m_f(\alpha) & \text{if } \beta \leq \alpha \end{cases} \qquad (6.1.43)$$

Thus, by (2.2.20), $f_{>\alpha}^*(t) = \inf\{\beta \mid m_{f_{>\alpha}}(\beta) \leq t\}$ so

$$f_{>\alpha}^* = \begin{cases} 0 & \text{if } t \geq m_f(\alpha) \\ f^*(t) & \text{if } t < m_f(\alpha) \end{cases} \qquad (6.1.44)$$

proving (6.1.39). By Problem 19(d) of Section 2.2, $m_f(f^*(w)) \leq w$, so (6.1.39) implies (6.1.40).

(b) An immediate consequence of $|f_{\leq \alpha}| \leq |f|$ and Problem 19(a) of Section 2.2.

(c) If $\beta > \alpha$, then $m_{f_{\leq\alpha}}(\beta) = m_{f_{\leq\alpha}}(\alpha)$. Thus, the inf in (2.2.20) is never larger than α which proves (6.1.42) $\qquad\square$

We now have all the tools in hand for a general interpolation theorem for Lorentz spaces:

Theorem 6.1.6 (Hunt–Marcinkiewicz Interpolation Theorem). *Let (M, Σ, μ) and (N, Ξ, ν) be two σ-finite measure spaces. Let $1 \leq p_0 < p_1 < \infty$ and $1 \leq q_0, q_1 \leq \infty$ with $q_0 \neq q_1$. Let T be a sublinear map of $L^{p_0,1}(M, d\mu) + L^{p_1,1}(M, d\mu)$ into the Ξ-measurable functions on N so that*

$$\|Tf\|_{q_j,\infty} \leq C_j\|f\|_{p_j,1}, \quad j = 0, 1 \qquad (6.1.45)$$

Then, for any $\theta \in (0,1)$, there is a constant C depending on the p_j, q_j, θ and the C_j so that for all $r \in [1, \infty]$, we have that $\left(\frac{1}{p_\theta} = \frac{\theta}{p_1} + \frac{(1-\theta)}{p_0}, \frac{1}{q_\theta} = \frac{\theta}{q_1} + \frac{(1-\theta)}{q_0}\right)$

$$\|Tf\|_{q_\theta,r} \leq C\|f\|_{p_\theta,r} \qquad (6.1.46)$$

Remarks. 1. One can, of course, get explicit bounds on C and as in the special case in Theorem 2.2.10, these constant diverge as θ^{-1/p_0} (respectively, $(1-\theta)^{-1/p_1}$) as $\theta \downarrow 0$ (respectively, $\theta \uparrow 1$).

2. This holds also for $p_1 = \infty$, if $L^{p_1,1}$ is interpreted as L^∞.

Proof. We'll suppose all indices are finite (leaving the case of infinite values to the reader) and we'll use C for a generic finite constant that can change from formula to formula! The idea will be to decompose f in an α-dependent way (as we did in the proof of Theorem 2.2.10, but with a more subtle split).

Fix θ and let $q = q_\theta, p = p_\theta$. Define

$$\gamma = \frac{q_0^{-1} - q_1^{-1}}{p_0^{-1} - p_1^{-1}} \in \mathbb{R} \setminus \{0\} \qquad (6.1.47)$$

Define

$$\widetilde{f}_{>\alpha}(x) = \begin{cases} f(x) & \text{if } |f(x)| > f^*(\alpha^\gamma) \\ 0, & \text{otherwise} \end{cases} \tag{6.1.48}$$

$$\widetilde{f}_{\leq\alpha} = f - \widetilde{f}_{>\alpha} \tag{6.1.49}$$

As with (2.2.45), the sublinearity and $(x + y \geq \alpha \Rightarrow x \geq \frac{\alpha}{2}$ or $y \geq \frac{\alpha}{2})$ implies that

$$(Tf)^*(\alpha) \leq (T\widetilde{f}_{\geq\alpha})^*\left(\frac{\alpha}{2}\right) + (T\widetilde{f}_{<\alpha})^*\left(\frac{\alpha}{2}\right) \tag{6.1.50}$$

Writing out the definition of $\|T(f)\|_{q,r}^*$ and using (6.1.50) we get that

$$\|T(f)^*\|_{q,r} \leq I_0 + I_1 \tag{6.1.51}$$

where

$$I_0 = \left(\frac{r}{q}\int_0^\infty \left[\alpha^{1/q}(T\widetilde{f}_{>\alpha})\left(\frac{\alpha}{2}\right)\right]^r \frac{d\alpha}{\alpha}\right)^{1/r},$$

$$I_1 = \left(\frac{r}{q}\int_0^\infty \left[\alpha^{1/q}(T\widetilde{f}_{\leq\alpha})\left(\frac{\alpha}{2}\right)\right]^r \frac{d\alpha}{\alpha}\right)^{1/r} \tag{6.1.52}$$

Since $p_0 < p_1$, we want to estimate the large values of f by p_0-norms, so we start by noting that

$$\|\widetilde{f}_{>\alpha}\|_{p_0,1}^* = \frac{1}{p_0}\int_0^\infty s^{1/p_0}(\widetilde{f}_{>\alpha})^*(s)\frac{ds}{s} \tag{6.1.53}$$

$$\leq \frac{1}{p_0}\int_0^{\alpha^\gamma} s^{1/p_0} f^*(s)\frac{ds}{s} \tag{6.1.54}$$

by (6.1.40). We plug this into the bound (6.1.45) for $j = 0$, into

$$(T\widetilde{f}_{>\alpha})^*(t) \leq C_0 t^{-1/q_0}\|\widetilde{f}_{>\alpha}\|_{p_0,1}^* \tag{6.1.55}$$

and see that

$$I_0^r \leq C\int_0^\infty \left(\alpha^{\frac{1}{q}-\frac{1}{q_0}}\int_0^{\alpha^\gamma} s^{1/p_0} f^*(s)\frac{ds}{s}\right)^r \frac{d\alpha}{\alpha} \tag{6.1.56}$$

$$\leq C\int_0^\infty \left(u^{\frac{1}{p}-\frac{1}{p_0}}\int_0^u s^{1/p_0} f^*(s)\frac{ds}{s}\right)^r \frac{du}{u} \tag{6.1.57}$$

where we used a $u = \alpha^\gamma$ change of variable and

$$\gamma^{-1}\left(\frac{1}{q} - \frac{1}{q_0}\right) = \frac{1}{p} - \frac{1}{p_0} \tag{6.1.58}$$

Recognizing the definition of $H_{\frac{1}{p_0}-\frac{1}{p},\frac{1}{p_0}}$ of (6.1.28), we see that (note $\alpha = \frac{1}{p_0} - \frac{1}{p} > 0$)

$$I_0^r \leq C \int_0^\infty ((H_{\frac{1}{p_0}-\frac{1}{p},\frac{1}{p_0}}f^*)(u))^r \frac{du}{u} \tag{6.1.59}$$

$$\leq C \int_0^\infty (s^{1/p}f^*(s))^r \frac{ds}{s} \tag{6.1.60}$$

$$= C(\|f\|_{p,r}^*)^r \tag{6.1.61}$$

where we used Hardy's inequality, (6.1.30), and $\theta - \alpha = \frac{1}{p_0} - (\frac{1}{p_0} - \frac{1}{p}) = \frac{1}{p}$.

I_1 is handled similarly. In place of (6.1.40), we use (6.1.41) (when $s \geq \alpha^\gamma$) and (6.1.42) (when $s < \alpha^r$) to see that

$$\|\tilde{f}_{\leq\alpha}\|_{p_1,1}^* \leq \alpha^{\gamma/p_1} f^*(\alpha^\gamma) + \frac{1}{p_1} \int_{\alpha^\gamma}^\infty s^{1/p_1} f^*(s) \frac{ds}{s} \tag{6.1.62}$$

and

$$(T\tilde{f}_{\leq\alpha})^*(t) \leq C_1 t^{-1/q_1} \|\tilde{f}_{\leq\alpha}\|_{p_1,1}^* \tag{6.1.63}$$

Plugging in (6.1.52), we see I_1 is an $L^r([0,\infty), \frac{d\alpha}{\alpha})$-norm of a sum which is bounded by the sum of norms so

$$I_1 \leq I_2 + I_3 \tag{6.1.64}$$

$$I_2 = C\left(\int_0^\infty [\alpha^{\frac{1}{q}-\frac{1}{q_1}+\frac{\gamma}{p_1}} f^*(\alpha^\gamma)] \frac{d\alpha}{\alpha}\right)^{1/r} \tag{6.1.65}$$

$$= C\left(\int_0^\infty (u^{1/p}f^*(u))^r \frac{du}{u}\right)^{1/r} \tag{6.1.66}$$

$$= C\|f\|_{p,r}^* \tag{6.1.67}$$

where (6.1.66) uses $\gamma^{-1}(\frac{1}{q} - \frac{1}{q_1} + \frac{\gamma}{p_1}) = \frac{1}{p} - \frac{1}{p_1} + \frac{1}{p_1} = \frac{1}{p}$.

Finally I_3^r looks just like the integral for I_0^r in (6.1.54) except that $\int_0^{\alpha^\gamma}$ is replaced by $\int_{\alpha^\gamma}^\infty$ and we have $\frac{1}{q} - \frac{1}{q_1}$ replacing $\frac{1}{q} - \frac{1}{q_0}$ and $\frac{1}{p_1}$ replacing $\frac{1}{p_0}$. We get a formula like (6.1.59) except that H is replaced by $K_{\frac{1}{p}-\frac{1}{p_1},\frac{1}{p_1}}$ (now $\frac{1}{p} - \frac{1}{p_1} > 0$) and $\theta + \alpha = \frac{1}{p} - \frac{1}{p_1} + \frac{1}{p_1} = \frac{1}{p}$. Putting together (6.1.51) with the estimates for I_0, I_2, and I_3 yields (6.1.47). $\qquad\square$

As corollaries, we get what we called the Hunt and original Marcinkiewicz interpolation theorems.

Proof of Theorem 6.0.3. This is just Theorem 6.1.6 with $r = \infty$, since $\|f\|_{p,\infty}^* \leq \|f\|_p \leq \|f\|_{p,1}^*$ says that an L^p to L^p bound implies an $L^{p,1}$ to $L^{p,\infty}$ bound. $\qquad\square$

Proof of Theorem 6.0.2. Since $\|f\|_p \le \|f\|_{p,1}^*$ implies the hypothesis of Theorem 6.1.6, so taking $r = p_\theta$, we see that

$$\|Tf\|_{q_\theta,p_\theta}^r \le C\|f\|_{p_\theta,q_\theta}^* \qquad (6.1.68)$$

The hypothesis $p_j \le q_j$ for $j = 0, 1$ implies $p_\theta \le q_\theta$ so, by Theorem 6.1.1, $\|\cdot\|_{q_\theta,p_\theta}^* \ge \|\cdot\|_{q_\theta,q_\theta}^*$. Since $\|\cdot\|_{s,s}^* = \|\cdot\|_s$, we get (6.0.10). $\square$

Notes and Historical Remarks. In contrast to the Riesz–Thorin theorem which uses analytic function methods and so is called the *complex interpolation method*, the generalization of the method of this section is the *real interpolation method*. Two classic monographs on the subject are Bergh–Löfström [**60**] and Bennett–Sharpley [**56**].

What we have called the original Marcinkiewicz interpolation theorem is due to Jósef Marcinkiewicz (1910–1940), a Polish mathematician. It was announced in 1939 [**559**]. Before he could publish the details, the Second World War broke out. Marcinkiewicz was a Polish nationalist and, despite the fact that his colleagues in England, where he was working, urged him to stay, he returned to Poland to take up his commission as an officer in the Polish army reserves. He was captured by the Russians and taken to a POW camp. With an eye to the aftermath of the war, the Russians systematically killed captured Polish officers and intelligentsia, including a notorious massacre in the Katyn Forest in March 1940 of over 20,000. It is believed that Mancinkiewicz was killed there or somewhat later in 1940. Before he left for the war, he had given some mathematical manuscripts to his parents for safekeeping, but his parents were arrested by the Russians and sent to a camp where they died of hunger. In 1956, Antoni Zygmund (1900–1992), who was Marcinkiewicz's advisor, published the details of his results [**878**].

Marcinkiewicz interpolation is typically worse than Riesz–Thorin in that there is a constant lost and that constant diverges. The tensor power trick (see Tao [**799**]) can sometimes overcome this. This is illustrated in Problem 9.

In the 1960's, there was a flurry of activity trying to exploit Marcinkiewicz's ideas that mostly focused on abstract spaces. There were important contributions of Krein–Semenov [**490**], Peetre–Lions [**539**], and Peetre [**633**]. In 1964, Hunt [**410**] found one of the most elegant and useful results for function spaces—what we called the Hunt–Mancinkiewicz interpolation theorem. Our proof is very close to his–see also Hunt [**411**] for more on his ideas on Lorentz spaces.

Lorentz spaces were introduced by G. C. Lorentz [**546**] in 1950. For more on them, see Lopez Safont [**545**].

For any function, f, one can define the sequence $\alpha_n(f) = 2^n m_f(2^n)^{1/p}$. The *dyadic Lorentz norm* is the ℓ^q-norm of this sequence. It can be proven that it is equivalent to the usual Lorentz norm and helps illustrate the logarithmic aspect of the Lorentz scale. For more on this and on Lorentz spaces and interpolation, see Tao [**800**].

There is a Hölder inequality for Lorentz spaces due to O'Neil [**618**] that says if $1 \leq p_1, p_2, r_1, r_2 \leq \infty$ and $\frac{1}{p_3} = \frac{1}{p_1} + \frac{1}{p_2}$ and $\frac{1}{r_3} = \frac{1}{r_1} + \frac{1}{r_2}$, then $f \in L^{p_1, r_1}, g \in L^{p_2, r_2} \Rightarrow fg \in L^{p_3, r_3}$ with a constant bound relating norms.

The history of Hardy inequalities is discussed in Kufner et al. [**493**]. Hardy started considering integral estimates on $\frac{1}{x} \int_0^x f(s)ds$ in a 1915 paper [**353**] and proved the continuous version of (6.1.24) in 1920 [**354**] but without the best constant. By looking at $f(x) = x^{-\frac{1}{p} - \varepsilon}$, he did note in that paper that the constant couldn't be better than $\frac{p}{p-1}$. In 1921, Landau [**502**] found the best constant for the discrete inequality and sent a letter to Hardy, which Hardy had published four years later. Hardy then found the optimal constant for (6.1.24) in 1925 [**355**]. This paper includes a remark of Landau that one can obtain the discrete from the continuous estimate (see Problem 7).

It is remarkable how much mathematics from that era was done by letters, often never published and often written to Hardy. Hardy sometimes then had letters with significant results published in the *Journal of the London Mathematical Society* (in addition to this letter of Landau, an important result of Riesz on convolution and rearrangement appeared in this way). The approach we use for the proof of Hardy inequalities from Minkowski's on integrals goes back to Ingham. It appears already in the 1934 book of Hardy, Littlewood, and Pòlya [**363**].

There is an enormous literature on Hardy inequalities and their extensions—see, for example, the review article of Oguntuase and Persson [**616**] and the two books of Kufner and collaborators [**619, 495**].

Lorentz spaces are only some of the zoo of function spaces that are useful in analysis. We'll see Sobolev spaces in Section 6.3. Among the more significant others are Besov spaces and Triebel-Lizorkin spaces, both with three indices! Those and other spaces are discussed in the books of Triebel [**815, 817**].

Problems

1. (a) Prove that (where $\|\cdot\|_{p,q}^* = \infty$ is allowed), then for any complex λ,

$$\|\lambda f\|_{p.q} = |\lambda| \, \|f\|_{p,q} \qquad (6.1.69)$$

(b) Prove that for any $p \in (1, \infty)$, $q \in [1, \infty]$

$$\|f + g\|_{p,q}^* \le 2^{1/p}(\|f\|_{p,q}^* + \|g\|_{p,q}^*) \tag{6.1.70}$$

(*Hint*: Use (2.2.90) with $t_1 = t_2 = \frac{t}{2}$.)

(c) Prove that $L^{p,q}$ is a vector space.

2. (a) Prove that

$$t^{1/p}f^{**}(t) = H_{\alpha=1-\frac{1}{p},\theta=1}(f^*) \tag{6.1.71}$$

(b) Use Hardy's inequality to prove that if $p > 1$,

$$\|f\|_{p,q} \le \frac{p}{p-1}\|f\|_{p,q}^* \tag{6.1.72}$$

(c) Use $f^* \le f^{**}$ to prove that

$$\|f\|_{p,q}^* \le \|f\|_{p,q} \tag{6.1.73}$$

3. For $\alpha > 1$, $\beta > 0$, let

$$f_{\alpha,\beta}(t) = t^{-\alpha}(1 + \log|t|)^{-\beta} \tag{6.1.74}$$

for $t \in [1, \infty]$. Prove that

$$f_{\alpha,\beta} \in L^{\frac{1}{\alpha},q}([1,\infty), dt)$$

if and only if $\beta > q^{-1}$. (*Hint*: Show that the $L^{p,q}$-properties of f don't change if it is extended to $[0, \infty)$ by setting it to 1 on $[0, 1]$.)

4. Let $p > 1$.

(a) Prove that $\|\cdot\|_{p,q}$ is a norm on $L^{p,q}$. (*Hint*: See (2.2.95).)

(b) Prove that $L^{p,q}$ is complete in $\|\cdot\|_{p,q}$.

5. Let $\ell^{p,q}$ be the $L^{p,q}$ space for $M = \mathbb{Z}_+$ with counting measure. Prove that if $1 \le p_1 < p_2 \le \infty$ and q_1, q_2 are arbitrary in $[1, \infty]$, then $\ell^{p_1,q_1} \subset \ell^{p_2,q_2}$.

6. Let $1 \le p_1 < p_2 \le \infty$ and q_1, q_2 are arbitrary in $[1, \infty]$. Prove that $L^{p_2,q_2}([0,1], dx)$ is contained in $L^{p_1,q_1}([0,1], dx)$.

7. Given a sequence, $\{a_n\}_{n=1}^{\infty}$, let f be the function with

$$f(x) = a_n \quad \text{if } n - 1 \le x < n \tag{6.1.75}$$

(a) Suppose first that $a_1 \ge a_2 \ge \cdots \ge 0$. Compute

$$g(x) = \frac{1}{x}\int_0^x f(t)dt$$

and show that it is decreasing. (*Hint*: Compute g'.)

(b) Conclude that in this case for $1 < p < \infty$,

$$\sum_{n=1}^{\infty} \left(\frac{1}{n} \sum_{k=1}^{n} a_k \right)^p \leq \left(\frac{p}{p-1} \right)^p \sum_{n=1}^{\infty} a_n^p \qquad (6.1.76)$$

(c) Prove that if a_k^* is the symmetric decreasing rearrangement of $|a_k|$, then

$$\sum_{n=1}^{\infty} |\frac{1}{n} \sum_{k=1}^{n} a_k|^p \leq \sum_{n=1}^{\infty} |\frac{1}{n} \sum_{k=1}^{n} a_k^*|^p \qquad (6.1.77)$$

(d) Conclude that (6.1.76) holds for all sequences. This is the *sharp discrete Hardy inequality*.

Remark. This method appeared in Hardy [**355**] who said that he got the idea from Landau.

8. (a) Verify (6.1.32).

(b) Use (6.1.32) to show (6.1.30) implies (6.1.31).

9. Fix p_0, p_1 and let $p_\theta = ((1-\theta)p_0^{-1} + \theta p_1^{-1})^{-1}$. The Marcinkiewicz theorem says that if S obeys $\|Sf\|_{p_0} \leq \|f\|_{p_0}$ and $\|Sf\|_{p_1} \leq \|f\|_{p_1}$, then $\|Sf\|_{p_\theta} = C(\theta)\|f\|_{p_\theta}$ where $C(\theta)$ doesn't depend on S (or the underlying measure space) but only on θ (and p_0, p_1).

On the other hand, the Riesz–Thorin theorem implies the result with $C(\theta) \equiv 1$. In this problem, the reader will go from the general $C(\theta)$ result to $C(\theta) = 1$ illustrating the "tensor power trick." The reader will need the Bonami–Segal lemma (Theorem 6.6.15) proven later in part (c).

(a) Prove that there is a map $\otimes^n S \colon L^{p_t}(X^n M, \otimes^n d\mu)$ to itself for $t = 0, 1$ so that $(\otimes^n S)(\otimes^n f) = \otimes^n (Sf)$ (where $(\otimes^n g)(x_1, \ldots, x_n) = g(x_1) \cdots g(x_n)$) and so that for $t = 0, 1$, $\|\otimes^n Sg\|_{p_t} \leq \|g\|_{p_t}$.

(b) Prove that $\|\otimes^n Sg\|_{p_\theta} \leq C(\theta) \|g\|_{p_\theta}$.

(c) Prove that $\|Sf\|_{p_\theta} \leq C(\theta)^{1/n} \|f\|_{p_\theta}$ for all n and conclude that $\|Sf\|_{p_\theta} \leq \|f\|_{p_\theta}$!

6.2. Hardy-Littlewood–Sobolev and Stein–Weiss Inequalities

Our main goal in this section will be to prove two inequalities (or rather more general inequalities that imply them). The first is the HLS inequality named after Hardy, Littlewood, and Sobolev.

Theorem 6.2.1 (HLS Inequalities). *Let $1 < p, q < \infty$ and $0 < \alpha < \nu$. Suppose that $\frac{1}{p} + \frac{1}{q} + \frac{\alpha}{\nu} = 2$. Then there is a constant $C(\alpha, \nu; p, q)$ so that*

$$\int \frac{|f(x)| |g(x)|}{|x-y|^\alpha} d^\nu x \, d^\nu y \leq C(\alpha, \nu; p, q) \|f\|_p \|g\|_q \qquad (6.2.1)$$

For the second inequality, recall that Hardy's inequality (see Theorem 4.1.3) says if $\nu \geq 3$ and $f, |x|^{-1}f, \nabla f \in L^2(\mathbb{R}^\nu)$ (distributional derivatives), then

$$\int \frac{|f(x)|^2}{|x|^2}d^\nu x \leq \frac{4}{(\nu-2)^2}\int |\vec{\nabla}f|^2 d^\nu x \qquad (6.2.2)$$

This can be rephrased as follows: Let $g \in \mathcal{S}(\mathbb{R}^\nu)$. Then $f = (|k|^{-1}\hat{g})^\vee$ clearly has $\nabla f \in L^2$. Moreover, since (see Theorem 6.8.1 of Part 1) f is a constant times the convolution of $|x-y|^{-(\nu-1)}$ and g, we see that f is continuous with at least $|x|^{-(\nu-1)}$ decay so $f, |x|^{-1}f \in L^2$ also. Thus (6.2.2) says that

$$\| |x|^{-1}|p|^{-1}g \|_2 \leq 2(\nu-2)^{-1}\|g\|_2 \qquad (6.2.3)$$

where

$$|p|^{-\alpha}g = (|k|^{-\alpha}\hat{g})^\vee \qquad (6.2.4)$$

and $\|\cdot\|_2$ is L^2-norm. This originally holds for $g \in \mathcal{S}(\mathbb{R}^\nu)$ but holds for $g \in L^2$ if we define $|x|^{-1}|p|^{-1}g$ as a limit. The second inequality generalizes this but replacing $^{-1}$ by $^{-\alpha}$ and L^2 by L^q.

Theorem 6.2.2 (Stein–Weiss Inequality). *Let $0 < \alpha < \nu$ and $1 < q < \frac{\nu}{\alpha}$ Then there exists a constant $\widetilde{C}(\alpha,\nu;q)$ so that*

$$\| |x|^{-\alpha}|p|^{-\alpha}g \|_q \leq \widetilde{C}(\alpha,\nu;q)\|g\|_q \qquad (6.2.5)$$

Remarks. 1. By duality, if $\frac{\nu}{\nu-\alpha} < r < \infty$, then

$$\| |p|^{-\alpha}|x|^{-\alpha}g \|_r \leq \widetilde{C}(\alpha,\nu;\frac{r}{r-1})\|g\|_r \qquad (6.2.6)$$

2. (6.2.5) holds for $g \in \mathcal{S}$ and then by density for all $g \in L^q$ if $|x|^{-\alpha}|p|^{-\alpha}$ is defined as a limit. Alternatively, (6.2.5) can be rewritten as

$$\int \frac{|h(x)|\,|g(y)|}{|x|^{-\alpha}|x-y|^{\nu-\alpha}}d^\nu x\, d^\nu y \leq C\|h\|_{\frac{q}{q-1}}\|g\|_q \qquad (6.2.7)$$

3. For $\nu \geq 5$, this implies Rellich's inequality (with the optimal constant below)

$$\left[\frac{\nu(\nu-4)}{4}\right]^2 \int \frac{|f(x)|^2}{x^4}d^\nu x \leq \int |\Delta f(x)|^2 d^\nu x \qquad (6.2.8)$$

(see Example 7.4.26 and Problem 10 of Section 7.4 in Part 4).

We'll obtain these two theorems as special cases of Young-type inequalities. Notice that (6.2.1) is reminiscent of Young's inequality in symmetric form (see (6.6.12) of Part 1). If $|\cdot|^{-\alpha}$ were in $L^{\nu/\alpha}(\mathbb{R}^\nu)$, (6.2.1) would exactly be a special case of Young's inequality. $|\cdot|^{-\alpha}$ is not in $L^{\nu/\alpha}$ but it is in the weak-$L^{\nu/\alpha}$ space $L_w^{\nu/\alpha} = L^{\nu/\alpha,\infty}$. Thus, Theorem 6.2.1 is implied by

Theorem 6.2.3 (Weak Young Inequality). *Let $1 < p, q, r < \infty$ so that*

$$\frac{1}{p} + \frac{1}{q} = 1 + \frac{1}{r} \tag{6.2.9}$$

Then, for all $f \in L^p_w(\mathbb{R}^\nu)$, $g \in L^q(\mathbb{R}^\nu)$, we have that $f * g \in L^r(\mathbb{R}^\nu)$ and

$$\|f * g\|_r \leq C(p, q)\|f\|^*_{p,w}\|g\|_q \tag{6.2.10}$$

for a constant $C(p, q)$ which is only dependent on p, q.

Remarks. 1. (6.2.10) is intended in the sense that the integral defining $f * g$ is a.e. absolutely convergent and then the integral obeys (6.2.10).

2. While our proof gives $C(p, q)$ independent of ν, the optimal constant is not claimed to be ν-independent and, indeed, is known not to be in some cases; see the Notes.

3. Warning: While we call this the weak Young inequality, it is stronger than (and, indeed, implies) the Young inequality.

4. This and Theorem 6.2.4 below only depend on Young's inequality and so hold in any locally compact abelian group.

Proof. We follow the idea sketched in the introduction to the chapter. Fix q and g. Then $f \mapsto f * g$ is a bounded map from L^p to L^r for $1 \leq p \leq \frac{q}{q-1}$ where r is given by (6.2.9). We can thus apply the special case of the Hunt–Marcinkiewicz theorem we called Hunt's interpolation theorem (Theorem 6.0.3) to conclude if $1 < p < \frac{q}{q-1}$ (note the strict inequality), then

$$\|f * g\|^*_{r,w} \leq C_1(p, q)\|f\|^*_{p,w}\|g\|_q \tag{6.2.11}$$

Now fix $f \in L^p_w$ and consider the map $g \mapsto f * g$. It is a bounded map of L^q to L^r_w if $1 < q < \frac{p-1}{p}$ and r is given by (6.2.9). In particular, since $p > 1, \frac{1}{q} > \frac{1}{r}$ or $r > q$. Thus, given any $q \in (1, \frac{p-1}{p})$, we can find $1 < q_0 < q < q_1 < \frac{p-1}{p}$ and define r_j by (6.2.9). Since $r_j \geq q_j$, we can apply the special case of the Hunt–Marcinkiewicz theorem which we called the original Marcinkiewicz theorem (Theorem 6.0.2). (6.2.9) is preserved by convex combinations in $\frac{1}{p}, \frac{1}{r}$, so we obtain (6.2.10) $\qquad \square$

Similarly, since $|p|^{-\alpha}$ is a convolution with a multiple of $|x|^{-(\nu-\alpha)}$, Theorem 6.2.2 is a special case of

Theorem 6.2.4 (Weak Stein–Weiss Estimate). *Let $1 < s < \infty$ and $1 < q < \infty$ so that*

$$\frac{1}{s} + \frac{1}{q} > 1 \tag{6.2.12}$$

Then for all $f \in L^s(\mathbb{R}^\nu)$, $g \in L^q(\mathbb{R}^\nu)_w$, $h \in L^{q/q-1}_w(\mathbb{R}^\nu)$, *we have that* $h(g*f) \in L^s(\mathbb{R}^\nu)$ *and*

$$\|h(g*f)\|_s \le C_2(q,s)\|h\|^*_{\frac{q}{q-1},w}\|g\|^*_{q,w}\|f\|_s \qquad (6.2.13)$$

Proof. Fix $f \in L^s$, $g \in L^{q/q-1}_w$. By Theorem 6.2.3 and Hölder's inequality, if $h \in L^t$, $0 \le t^{-1} \le 2 - s^{-1} - q^{-1}$, then

$$\|h(g*f)\|_r \le C(q,s)\|h\|_t\|g\|^*_{q,w}\|f\|_s \qquad (6.2.14)$$

where

$$r^{-1} = t^{-1} + s^{-1} + q^{-1} - 1 \qquad (6.2.15)$$

Thus, applying Hunt's interpolation theorem (Theorem 6.0.3), interpolating between $t = \infty$ and $t = (s^{-1} + q^{-1} - 1)^{-1}$, we see that for $0 < t^{-1} < 2 - s^{-1} - q^{-1}$, i.e., $t < \infty, r > 1$

$$\|h(g*f)\|^*_{r,w} \le C\|h\|^*_{t,w}\|g\|^*_{q,w}\|f\|_s \qquad (6.2.16)$$

In particular, if $t = q/q - 1 < \infty$, $r = s > 1$, so we find

$$\|h(g*f)\|_{s,w} \le C\|h\|^*_{q/q-1,w}\|g\|^*_{q,w}\|f\|_s \qquad (6.2.17)$$

Fixing q, then $g \in L^q_w$, $h \in L^{q/q-1}_w$, we have that $f \mapsto h(g*f)$ is bounded from L^s to L^s_w for all s in $1 < s < q/q - 1$. Given such an s, we can find s_0, s_1 so $1 < s_0 < s < s_1 < q/q - 1$ and apply the original Marcinkiewicz theorem to get (6.2.17) with $\|\cdot\|_s$ on the left (and a larger constant). $\square$

Notes and Historical Remarks. As we'll see in the next section, HLS inequalities are essentially equivalent to certain estimates involving derivatives and, as such, there are many precursors discussed in the Notes to that section. As an integral inequality, (6.2.1) goes back to Hardy–Littlewood [**359, 361**] in 1928 for $\nu = 1$ with an important later contribution by Riesz [**687**]. The ν-dimensional result is due to Sobolev [**767**] in 1938 who developed it in connection with the estimates of the next section.

Sobolev proved his ν-dimensional inequality from the one-dimensional inequality. The key observation is that if $x = (\tilde{x}, x_\nu)$ with $\tilde{x} \in \mathbb{R}^{\nu-1}$ and $x_\nu \ne 0$, then as a function of $\tilde{x}$, $|x|^{-s} = ||\tilde{x}|^2 + |x_\nu|^2|^{-s/2}$ is in $L^r(\mathbb{R}^{\nu-1})$ so long as $rs > \nu - 1$ and, by scaling,

$$\int |(\tilde{x}, x_\nu)|^{-sr} d^{\nu-1}\tilde{x} = C|x_\nu|^{\nu-1-sr} \qquad (6.2.18)$$

Given $\alpha < \nu$, one picks $r = \frac{\nu}{\alpha}$, $s = \alpha$ (so $rs = \nu > \nu - 1$) and uses Minkowski's inequality in x and Young's inequality in $\tilde{x}$ to reduce a ν-dimensional convolution of $|x|^{-\alpha}$ and f to a one-dimensional convolution of $|x_\nu|^{-\alpha/\nu}$ and partial integral of f. The Hardy–Littlewood one-dimensional inequality finishes the proof.

Theorem 6.2.2 is a special case of the following theorem of Stein–Weiss [**783**] which the reader will prove in Problem 2:

Theorem 6.2.5 (Generalized Stein–Weiss Inequality). *Let* $1 < r, q < \infty$, $0 \le \alpha, \beta, \gamma < \nu$ *so that*

$$
\begin{aligned}
r^{-1} + \nu^{-1}(\alpha + \beta + \gamma) + q^{-1} &= 2 \\
r^{-1} + q^{-1} &\ge 1 \\
r^{-1} + \nu^{-1}\alpha &< 1 \\
q^{-1} + r^{-1}\beta &< 1
\end{aligned}
\tag{6.2.19}
$$

Then there is a constant $C(\alpha, \beta, \gamma; p, q, \nu)$ *so that*

$$
\int \frac{|f(x)|\,|g(y)|}{|x|^\alpha |x-y|^\beta |y|^\gamma} d^\nu x\, d^\nu y \le C\|f\|_r \|g\|_q
\tag{6.2.20}
$$

Theorem 6.2.2 is the special case $\gamma = 0$, $\beta = \nu - \alpha$, $r = q/(q-1)$ (and Theorem 6.2.1 is $\alpha = \gamma = 0$).

The way we've presented the results of this section, Theorems 6.2.1/6.2.2 appear as special cases of Theorems 6.2.3/6.2.4 but given one beautiful additional result, the special cases actually imply the general result. This result is the Brascamp–Lieb–Luttinger (BLL) inequality of [**110**]. Given f, a measurable function on $\mathbb{R}^\nu$, one defines its spherical rearrangement, f^*, as the unique function on $\mathbb{R}^\nu$ which is a monotone decreasing, nonnegative function of $|x|$ alone, lower semicontinuous (i.e., $\lim_{r \downarrow r_0} f(r) = f(r_0)$) and so that f^* is equimeasurable with $|f|$. This is *not* our earlier f^* but is clearly very closely related (see Problem 3).

The BLL inequalities say:

Theorem 6.2.6 (Brascamp–Lieb–Luttinger Inequalities). *Let* $f_1, \ldots, f_\ell$ *be nonnegative measurable functions on* $\mathbb{R}^\nu$ *and* $\{a_{jm}\}_{1 \le j \le \ell; 1 \le m \le n}$ *an* $\ell \times n$ *matrix. Then,*

$$
\int_{(\mathbb{R}^\nu)^n} \prod_{j=1}^\ell f_j\Big(\sum_{m=1}^n a_{jm}\mathbf{x}_m\Big) \prod_{n=1}^n d^\nu x_m \le \int_{(\mathbb{R}^\nu)^n} \prod_{j=1}^\ell f_j^*\Big(\sum_{m=1}^n a_{jm}\mathbf{x}_m\Big) \prod_{m=1}^n d^\nu x_m
\tag{6.2.21}
$$

For a pedagogic presentation of the proof, see Simon [**756**]. The point is that if $f \in L_w^{\nu/\alpha}(\mathbb{R}^\nu)$, then ($\tau_\nu = $ volume of unit ball in $\mathbb{R}^\nu$)

$$
|f^*(x)| \le \|f\|_{\nu/\alpha,w}^* \tau_\nu^{-\alpha/\nu} |x|^{-\alpha}
\tag{6.2.22}
$$

Thus, Theorem 6.2.6 and Theorem 6.2.1 (respectively, Theorem 6.2.2) imply Theorem 6.2.3 (respectively, Theorem 6.2.4).

In an illustration of Arnold's principle, we note what have come to be called BLL inequalities appeared almost 20 years earlier in Rogers [**697, 698, 699**].

Especially if one uses the BLL inequalities, there are a variety of alternate proofs of the main results of this section, many of which use more "concrete" techniques than the soft interpolation theory central to our proof. Probably the simplest (given BLL) is to use the approach of Herbst (see below) to get Theorem 6.2.2, use BLL to get Theorem 6.2.4 and note that this implies Theorem 6.2.3 and so Theorem 6.2.4 since if $p^{-1} + q^{-1} = r^{-1}$, then $L^r = L^p L^q \subset L^p L^q_w$ (where $XY = \{fg \mid f \in X, g \in Y\}$). Other approaches to the HLS inequalities are discussed in the text of Lieb–Loss [**533**]. Sogge [**769**] has a proof using the Calderón–Zygmund decomposition of Section 6.4.

Lieb [**532**] (see also an alternate proof in Frank–Lieb [**282**]) has found the optimal constant $C(\alpha, \nu; p, q)$ in (6.2.1) when $p = q$ (so $p = 2\nu/(2\nu - \alpha)$); explicitly

$$C(\alpha, \nu; p, p) = \pi^{\alpha/2} \frac{\Gamma(\frac{\nu}{2} - \frac{\alpha}{2})}{\Gamma(\nu - \frac{\alpha}{2})} \left[\frac{\Gamma(\frac{\nu}{2})}{\Gamma(\nu)} \right]^{\frac{\alpha}{\nu} - 1} \tag{6.2.23}$$

The optimizer is $f(x) = g(x) = (1 + |x|^2)^{-\nu/p}$.

Herbst [**379**] has found the optimal constants in (6.2.5); explicitly

$$\widetilde{C}(\alpha, \nu; q) = 2^{-\alpha} \frac{\Gamma(\frac{1}{2}(\frac{\nu}{q} - \alpha))\Gamma(\frac{1}{2}\frac{\nu}{r})}{\Gamma(\frac{1}{2}(\frac{\nu}{r} + \alpha))\Gamma(\frac{1}{2}\frac{\nu}{q})} \tag{6.2.24}$$

where r is the dual index to q, i.e., $r = q/(q-1)$. This formula is valid when $\nu > q\alpha$. In particular, if $q = r = 2$, $\widetilde{C} = 2^{-\alpha}\Gamma(\frac{1}{2}(\frac{\nu}{2} - \alpha))/\Gamma(\frac{1}{2}(\frac{\nu}{2} + \alpha))$. For example, if $\alpha = 1$, $\widetilde{C} = \frac{2}{(\nu - 2)}$, so we recover Hardy's inequality (6.2.2). If $\alpha = 2$, $\widetilde{C} = \frac{4}{\nu(\nu - 4)}$ and we recover (6.2.8).

Herbst's proof relies on an elegant observation. If $f \geq 0$, $f \in L^1(\mathbb{R}^\nu, d^\nu x)$, then the norm of $g \mapsto f * g$ as a map of L^p to L^p is exactly $\|f\|_1$ (for all p). $|x|^{-\alpha}|p|^{-\alpha}$ is an operator with explicit positive integral kernel that commutes with dilation $x \mapsto \lambda x$. Thus, after a change of variable $|x| \mapsto u = \log|x|$, the operator should commute with translation and so be a convolution. This idea, implemented properly to account for angular variables, proves the boundedness and optimal constant. The use of dilations in the results like Theorem 6.2.2 goes back to Hardy–Littlewood–Pólya [**363**] and Stein–Weiss [**783**].

Problems

1. Prove the *weak Hausdorff–Young inequalities*: if $1 < p < 2$, there is $C(p, \nu)$ so that $\|\widehat{f}\|^*_{p/p-1,w} \leq C(p, \nu)\|f\|^*_{p,w}$.

2. (a) By mimicking the proof of Theorem 6.2.4, prove that if $1 < r, q, s, t < \infty$, $0 < \beta < \nu$ so that $r^{-1} + t^{-1} + s^{-1} + q^{-1} + \nu^{-1}\beta = 2$, $r^{-1} + t^{-1} < 1$, $q^{-1} + s^{-1} < 1$, then

$$\int \frac{|f(x)|\,|h(x)|\,|q(y)|\,|g(y)|}{|x-y|^\beta}\,d^\nu x\,d^\nu y \leq C\|f\|_r \|h\|_{t,w}^* \|q\|_{s,w}^* \|g\|_q \qquad (6.2.25)$$

(b) Deduce Theorem 6.2.5 from this.

3. Let $f^\sharp$ be the spherical rearrangement discussed in the Notes and f^* the function in Section 2.2. Prove that

$$f^*(t) = f^\sharp((t/\tau_\nu)^{1/2}) \qquad (6.2.26)$$

where τ_ν is the volume of the unit ball in $\mathbb{R}^\nu$.

6.3. Sobolev Spaces; Sobolev and Rellich–Kondrachov Embedding Theorems

In this section, we'll discuss a class of function spaces that are extremely useful in the theory of partial differential equations, especially the elliptic theory. Of great significance are results that suitable L^p-hypotheses on distributional derivatives imply classical continuity and differentiability. For example, in ν dimensions, if u is an L^2-function whose ℓ-th distributional Laplacian $\Delta^\ell u$ is also in L^2, then so long as $4\ell > \nu$, u is a continuous function and has classical continuous derivatives of order m, so long as $4\ell > \nu + 2m$.

At the end of the section, we'll present a simple application of this machinery to PDE's, including that if V is C^∞ and u is a distributional solution of $(-\Delta + V)u = 0$, then u is C^∞. In fact, this application will only use Sobolev spaces, $H^{\alpha,p}$ for $p = 2$ where the theory is much simpler for two reasons: First, this is the only value of p for which $\hat{}$ maps L^p to itself; for example, for general p in $(1, \infty)$, it is a deep fact (proven in the next section) that $f \mapsto (k_j(1+k^2)^{-1/2}\hat{f})^\vee$ maps L^p to itself, but since $\hat{}$ is an isometry on L^2, it is trivial for $p = 2$ (indeed, that will be an important input to the general p proof). Second, suitable multiplication operators map L^2 to L^q where $q < 2$, so we can use the Hausdorff–Young inequalities. This means the L^2 theory (which is all we need in our application) is easy to develop: the reader will do it in Problems 1–5.

There are a number of issues that arise in the theory that we intend to avoid in this introduction (but see the books listed in the Notes). We'll mainly discuss the spaces on all of $\mathbb{R}^\nu$ and say a little about the spaces of functions supported in an open set $\Omega \subset \mathbb{R}^\nu$. We'll mainly avoid the subject of restrictions of Sobolev space functions to Ω where the boundary properties of Ω are critical.

There is also the issue of fractional derivatives and the analog of Sobolev spaces for fractional exponent. Everyone agrees on the "right" definition of Sobolev spaces of integral exponent: $f \in W^{\ell,p}$ for $\ell \in \mathbb{Z}_+$ if the distributional derivatives, $D^\alpha f$, for all multi-indices α with $|\alpha| \leq \ell$, lie in L^p (with the obvious norms). What to use when ℓ is not an integer is not so natural; some of the possibilities are discussed in the Notes. A common choice called fractional Sobolev spaces, usually denoted $W^{s,p}$, is defined in terms of Hölder-like conditions, i.e., on the ratio $|f(x+y) - f(y)|/|y|^\theta$. We'll instead use the spaces, $H^{s,p}$ of functions with $((1+|k|^2)^{s/2}\widehat{f})^\vee \in L^p$ most often (for reasons that will be clear soon) called the *space of Bessel potentials* and sometimes *generalized Sobolev spaces*. We'll often drop "generalized." A critical result we'll prove below is that if $s = \ell \in \mathbb{Z}_+$, then $H^{s,\ell} = W^{s,\ell}$. It is also true that for $p = 2$, $W^{s,2} = H^{s,2}$ for all $s > 0$ (see Problem 5).

As a preliminary, we'll want to study the integral kernel of $f \mapsto ((k^2 + 1)^{-s/2}\widehat{f})^\vee$, which is clearly important in studying $H^{s,p}$. By (6.2.44) of Part 1, this is $(2\pi)^{-\nu/2}$ times the inverse Fourier transform of $(k^2 + 1)^{-s/2}$. Then, we define $J^{(s,\nu)}(x)$, the *Bessel kernel* by

$$((k^2 + 1)^{-s/2}\widehat{f})^\vee = J^{(s,\nu)} * f \tag{6.3.1}$$

To compute this Fourier transform, we use the trick of Section 6.8 of Part 1 and write $(k^2 + 1)^{-s/2}$ as a superposition of Gaussians whose Fourier transforms we know, i.e.,

$$(k^2 + 1)^{-s/2} = \Gamma\left(\frac{s}{2}\right)^{-1} \int_0^\infty t^{(\frac{s}{2}-1)} e^{-t(1+k^2)} dt \tag{6.3.2}$$

where, of course,

$$\Gamma(s) = \int_0^\infty t^{s-1} e^{-t} dt \tag{6.3.3}$$

For $s = 2$, we did this already in Section 6.9 of Part 1. We thus find that

$$J^{(s,\nu)}(x) = \Gamma\left(\frac{s}{2}\right)^{-1} (4\pi)^{-\nu/2} \int_0^\infty e^{-t} t^{\frac{1}{2}s-1-\frac{1}{2}\nu} \exp(-x^2/4t) dt \tag{6.3.4}$$

We'll discuss the features of $J^{(s,\nu)}$ we need from this integral representation but to write an explicit formula in terms of Bessel functions, we need to recall the integral representation (14.7.20) of Part 2B for K_β, the MacDonald function, aka modified Bessel function of the second kind:

$$K_\beta(z) = \int_0^\infty e^{-z\cosh(w)} \cosh(\beta w) dw \tag{6.3.5}$$

Proposition 6.3.1. *Let $s > 0$. Then*

(a) $J^{(s,\nu)}(x) = (2\pi)^{-\nu/2}\Gamma\left(\frac{s}{2}\right)^{-1} 2^{-(s-2)/2}|x|^{-(\nu-s)/2} K_{\frac{1}{2}|\nu-s|}(x) \tag{6.3.6}$

(b) $J^{(s,\nu)}(x) = (2\pi)^{-\nu/2}\Gamma\left(\dfrac{s}{2}\right)^{-1} 2^{-s/2}|x|^{-(\nu-s)}$

$$\int_0^\infty \exp\left(-\frac{1}{2}[x^2 y + y^{-1}]\right) y^{-\frac{1}{2}(s-\nu)-1} dy \qquad (6.3.7)$$

(c) $J^{(s,\nu)}(x) = (2\pi)^{-\nu/2}\Gamma\left(\dfrac{s}{2}\right)^{-1} 2^{-s/2}|x|^{-(\nu-s)/2}$

$$\int_0^\infty e^{-\frac{|x|}{2}(u+u^{-1})} u^{-\frac{1}{2}(\nu-s)-1} du \qquad (6.3.8)$$

(d) $J^{(s,\nu)}(x)$ is C^∞ on $\mathbb{R}^\nu \setminus \{0\}$.

(e) For any multi-index γ, in the region $|x| > 1$

$$|D^\gamma J^{(s,\nu)}(x)| \le C^{(1)}_{\gamma,s,\nu}|x|^{-[\nu-(s-1)]/2} e^{-|x|}, \quad \text{for } |x| > 1 \qquad (6.3.9)$$

(f) If $s > \nu$, $J^{(s,\nu)}$ is bounded.

(g) If $s < \nu$, then

$$|J^{(s,\nu)}(x)| \le C_{s,\nu}|x|^{-(\nu-s)} \qquad (6.3.10)$$

(h) If $s = 1$, then for $|x| > 0$,

$$\vec{\nabla} J^{(s=1,\nu)}(x) = \vec{x}\tilde{J}^{(\nu)}(x); \ |\tilde{J}^{(\nu)}(x)| \le |x|^{-\nu-1} \qquad (6.3.11)$$

(i) If $s > \nu + \ell$, $\ell = 1, 2, \ldots$, then $J^{(s,\nu)}$ is a C^ℓ function on $\mathbb{R}^\nu$.

(j) If $s > \nu$, $J^{(s,\nu)} \in \bigcap_{1 \le p \le \infty} L^p(\mathbb{R}^\nu)$

(k) If $s < \nu$, $J^{(s,\nu)} \in \bigcap_{1 \le p < \nu(\nu-s)^{-1}} L^p(\mathbb{R}^\nu)$

Remarks. 1. It is because of the construction (6.3.6) that this is called the Bessel kernel and that one speaks of Bessel potentials and Bessel potential spaces.

2. If $s = \nu$, then (6.3.10) is replaced by $C_{s,\nu} \log(2 + |x|)$ and (j) is replaced by $1 \le p < \infty$ (Problem 6).

3. There are also results on higher derivatives near $|x| = 0$ (Problem 7).

4. In (g), it is easy to show that $|x|^{(\nu-s)/2} J^{(s,\nu)}|x|$ has a nonzero limit as $|x| \downarrow 0$.

Proof. (a)–(c) (c) follows from (6.3.4) by the change of variable, $y = 2t/|x|^2$ and (b) then from $u = |x|y$. (a) follows from (b) by letting $u = e^w$. The resulting integral is of the form $\int_{-\infty}^\infty G(w)dw = \int_0^\infty (G(w) + G(-w))dw$ yielding (a), i.e., (6.3.5).

(d) is immediate from (6.3.8) noting that for $|x| \ne 0$, the integral is convergent at (6.3.8), indeed, by Theorem 3.1.6 of Part 2A, $J^{(s,\nu)}$ has an analytic continuation to $\operatorname{Re} x > 0$ (one can prove much more—using the properties of Bessel functions).

(e) The representation (6.3.8) and the fact that $\inf_{u\in\mathbb{R}}(u + u^{-1}) = 2$ imply the $e^{-|x|}$ bound that is critical. The extra $|x|^{-[\nu-(s-1)]/2}$ prefactor comes from Laplace's method (see Section 15.2 of Part 2B).

(f), (g) This is immediate using $e^{-t} \leq 1$ in (6.3.4), scaling out x^2 and noting that if $s < \nu$, $\int_0^\infty \exp(-1/t)\, t^{\frac{1}{2}s-\frac{1}{2}\nu-1}dt < \infty$.

(h) From (6.3.4), we see that for x

$$\vec{\nabla} J^{(s,\nu)}(x) = \vec{x}\,\widetilde{J}^{(\nu,s)}(x) \qquad (6.3.12)$$

$$\widetilde{J}^{(\nu,s)}(x) = \frac{1}{2}\Gamma\!\left(\frac{s}{2}\right)^{-1}(4\pi)^{-\nu/2}\int_0^\infty e^{-t}\, t^{\frac{1}{2}s-2-\frac{1}{2}\nu}\exp(-x^2/4t)dt \qquad (6.3.13)$$

We get an upper bound using $e^{-t} \leq 1$ and scaling out x, noting that $\int_0^\infty e^{-1/4t}t^{\frac{1}{2}s-2-\frac{1}{2}\nu} < \infty$, if $s = 1$.

(i) is immediate when $s > \nu + 2\ell$ by taking derivatives in (6.3.4) and noting that all the integrals converge if $x = 0$ and $s > \nu + 2\ell$. To get the $\nu + \ell$ result, one needs to use the x's in the derivatives of $\exp(-x^2/4t)$ and some $|x|^{-1}$ (see Problem 8).

(j), (k) are immediate given the bounds at 0 and at ∞. $\qquad\square$

The fact that, $J^{(s,\nu)}$ is L^1 in many cases, makes the integral operator it defines bounded on all the L^p's and more. $\vec{\nabla} J^{(s=1,\nu)}$ is non-L^1 making it a singular integral kernel. The $\vec{x}/|x|^{\nu+1}$ is reminiscent of the Hilbert transform when $\nu = 1$, so one could hope it still is bounded on L^p, $1 < p < \infty$. Indeed, this is what we'll prove in Section 6.4, i.e.,

Fact (see Corollary 6.4.7). In $\mathbb{R}^\nu$, for $j = 1, \ldots, \nu$, $f \mapsto [(k_j(k^2+1)^{-1/2})\widehat{f}]^{\vee}$ is bounded from L^p to itself if $1 < p < \infty$.

With these lengthy preliminaries out of the way, we can turn to the basic definitions:

Definition. Let $\ell \in \mathbb{N}$. We say $f \in W^{\ell,p}(\mathbb{R}^\nu)$, the *Sobolev space* of order ℓ and index p if and only if for all multi-indices, α, with $|\alpha| \leq \ell$, $D^\alpha f \in L^p(\mathbb{R}^\nu)$. We define the *Sobolev norm* to be

$$\|f\|_{W^{l,p}} = \sum_{|\alpha|\leq\ell} \|D^\alpha f\|_p \qquad (6.3.14)$$

Definition. Let $s \in \mathbb{R}$, $H^{s,p}(\mathbb{R}^\nu)$, the *generalized Sobolev space* of order s and index p is the set of tempered distributions, f, with

$$\|f\|_{H^{s,p}} = \left\|[(1 + k^2)^{s/2}\widehat{f}]^{\vee}\right\|_p < \infty \qquad (6.3.15)$$

Remarks. 1. It is straightforward to prove that these spaces are complete with $C_0^\infty(\mathbb{R}^\nu)$ dense (Problems 9 and 10) and that $H^{s,2}$ and $H^{-s,2}$ are naturally dual (Problem 11).

2. Since $\alpha = 0$ is included in (6.3.14), elements of $W^{\ell,p}$ are functions. If $s > 0, f \mapsto [(1+k^2)^{-s/2}\widehat{f}]^{\vee}$ is convolution with an L^1-function so

$$s > 0 \Rightarrow H^{s,p} \subset L^p \qquad (6.3.16)$$

More is true (Problem 12):

$$s_1 > s_2 \Rightarrow H^{s_1,p} \subset H^{s_2,p} \qquad (6.3.17)$$

3. We'll prove below that if $s = \ell \in \mathbb{N}$, $H^{\ell,p} = W^{\ell,p}$ justifying the name "generalized Sobolev space." We'll often drop "generalized" but we warn the reader that there are spaces denoted by $W^{s,p}$, $s \in \mathbb{R}$, called fractional Sobolev spaces, which (for $s \notin \mathbb{N}$) are distinct from $H^{s,p}$ (at least for general p).

Theorem 6.3.2. *For any ν and $\ell \in \mathbb{N}$,*

$$W^{\ell,p}(\mathbb{R}^\nu) = H^{\ell,p}(\mathbb{R}^\nu)$$

including an equivalence of norms.

Proof. Suppose first that $f \in H^{\ell,p}(\mathbb{R}^\nu)$. Then $g = ((1+k^2)^{\ell/2}\widehat{f})^{\vee} \in L^p$ and $\|g\|_p = \|f\|_{H^{p,p}(\mathbb{R}^\nu)}$. By the fact mentioned above, $h \mapsto (k_j(1+k^2)^{-1/2}\widehat{h})^{\vee} \equiv Q_j h$ is bounded on L^p.

Also, since $Q_0 h = ((1+k^2)^{-1/2}\widehat{h})^{\vee}$ is convolution with a function in L^1, Q_0 is also bounded on L^p. Using $\|f\|_{p,p}$ for the $\mathcal{L}(L^p)$-norm of an operator, noting that for $|\alpha| = j \leq \ell$

$$D^\alpha f = \Big(\prod_{i=1}^{\nu} Q_j^{\alpha_j} Q_0^{\ell-k}\Big)g \qquad (6.3.18)$$

we have that

$$\|D^\alpha f\|_p \leq \Big[\prod_{j=1}^{\nu}\|Q_j\|_{p,p}^{\alpha_j}\Big]\|Q_0\|_{p,p}^{\ell-k}\|g\| \qquad (6.3.19)$$

which proves $f \in W^{\ell,p}(\mathbb{R}^\nu)$ and

$$\|f\|_{W^{\ell,p}(\mathbb{R}^\nu)} \leq C\|f\|_{H^{\ell,p}(\mathbb{R}^\nu)} \qquad (6.3.20)$$

Conversely, suppose $f \in W^{\ell,p}$. For each α with $|\alpha| \leq \ell$, $D^\alpha f \in L^p$. As noted in the proof of the first half, for any β with $|\beta| \leq \ell$, $h \to ((1+k^2)^{-\ell/2}\widehat{D^\beta f})^{\vee}$ maps L^p to L^p. Thus, for any α, β with $|\alpha|, |\beta| \leq \ell$, $((1+k^2)^{-\ell/2}\widehat{D^{\alpha+\beta}f} \in L^p$, i.e., if $|\gamma| \leq 2\ell$, $((1+k^2)^{-\ell/2}\widehat{D^\alpha f})^{\vee} \in L^p$. Since $(1-\Delta)^\ell$

is a sum of D^γ's with $|\gamma| \leq 2\ell$, we conclude that $((1+k^2)^{-\ell/2}(1+k^2)^\ell \widehat{f})^\vee \in L^p$, i.e., $f \in H^{\ell,p}$. The proof shows that

$$\|f\|_{H^{\ell,p}(\mathbb{R}^\nu)} \leq C\|f\|_{W^{\ell,p}(\mathbb{R}^\nu)} \tag{6.3.21}$$

$$\square$$

Our next topic concerns Sobolev and Gagliardo–Nirenberg estimates which can be thought of as a priori estimates for $f \in C_0^\infty(\mathbb{R}^\nu)$ bounding some $\|\cdot\|_p$-norm of f or $D^\alpha f$ by $\|\cdot\|_q$-norms of f or $D^\beta f$. Since, by density of C_0^∞, they also imply say that some $H^{s,p}$ lies in some $H^{t,q}$, they are also called *Sobolev embedding theorems*. The density argument we have in mind is if X, Y are the Banach spaces of functions, each with $C_0^\infty(\mathbb{R}^\nu)$ dense and so that $X \subset \mathcal{S}'(\mathbb{R}^\nu)$ and convergence in X implies convergence in $\mathcal{S}'$ and suppose that for $f \in C_0^\infty$,

$$\|f\|_X \leq C\|f\|_Y \tag{6.3.22}$$

then $Y \subset X$ (as sets of distributions) and (6.3.22) holds for all $f \in Y$. For a given $f \in Y$, find $f_n \in C_0^\infty$ so $\|f - f_n\|_Y \to 0$. Then, by (6.3.4), f_n is Cauchy in X so $f_n \to g$ in X. Thus, $f = g$ as a distribution, so $Y \subset X$. By taking limits, (6.3.22) holds for all $f \in Y$. Below, we'll state estimates for $C_0^\infty(\mathbb{R}^\nu)$ without commenting explicitly that they imply embedding theorems. We begin with the simplest, but in some ways subtlest of these estimates:

Theorem 6.3.3 (Homogeneous Sobolev Estimates). *Let $0 < s < \nu$. Let $1 < q < p < \infty$ so that*

$$\frac{1}{p} = \frac{1}{q} - \frac{s}{\nu} \tag{6.3.23}$$

Let $|D|^s$ be the map on distributions, $|D|^s \tau = (|k|^s \widehat{\tau})^\vee$. Then, if f is such that $|D|^s f \in L^q$, we have that

$$\|f\|_p \leq C\| |D|^s f \|_q \tag{6.3.24}$$

Moreover, for any $f \in \mathcal{S}(\mathbb{R}^\nu)$,

$$\|f\|_p \leq C\|f\|_{s,q} \tag{6.3.25}$$

where $\|\cdot\|_{s,p}$ is the $H^{s,p}$ norm.

Remarks. 1. The part of our name "homogeneous" comes from the following. For $\lambda > 0$, let $f_\lambda(x) = f(\lambda x)$ so

$$\|f_\lambda\|_p = \left(\int |f(\lambda x)|^p d^\nu x \right)^{1/p}$$

$$= \lambda^{-\nu/p}\|f\|_p \tag{6.3.26}$$

while $|D|^s f_\lambda = \lambda^s (|D|^s f)_\lambda$ and so

$$\| |D|^s f_\lambda \|_q = \lambda^{s-\nu/q}\| |D|^s f \|_q \tag{6.3.27}$$

Both sides of (6.3.24) are homogeneous functions of λ of the same order precisely because (6.3.23) holds. (6.3.24) cannot hold for any other values of p (if s, q are fixed) because if the two sides were homogeneous of different order, the inequality would fail for either λ small (if $p^{-1} < q^{-1} - s\nu^{-1}$) or λ large (if $p^{-1} > q^{-1} - s\nu^{-1}$). On the other hand, we'll see inequalities like (6.3.25) hold for a range of p.

2. At first sight, (6.3.25) seems weaker than (6.3.24) since $\|f\|_{s,q}$ has a $\|f\|_q$ built into it. But scaling shows that if (6.3.25) holds, then for any $\varepsilon > 0$,

$$\|f\|_p \leq C\|((k^2 + \varepsilon)^{s/2}\widehat{f})^{\vee}\|_q \tag{6.3.28}$$

(C independent of ε), so taking $\varepsilon \downarrow 0$, we see (6.3.25) implies (6.3.24) if we use scaling.

3. (6.3.25) is somewhat cleaner since $(1 + k^2)^{s/2}$ maps $\mathcal{S}$ to $\mathcal{S}$ for any value of s and thus we can interpret (6.3.25) as an a priori estimate with no work. It would require some effort to prove that $|D|^s f \in L^q$ for all $f \in \mathcal{S}$. Most books only discuss $s \in \mathbb{Z}_+$ and use $\|D^s f\|_q$ meaning $\sum_{|\alpha| \leq s}\|D^\alpha f\|_q$ so this is not an issue.

4. There is a result also if $p = 1$, $s = 1$; see Problems 18 and 19.

Proof. If $g = |D|^s f$ and $h = [(1 + k^2)^{s/2}\widehat{f}]^{\vee}$, then $f = C_{s,\nu}^{(1)}|x|^{-(\nu-s)} * g$ and $f = J^{(s,\nu)} * h$. Since $J^{(s,\nu)}(x)$ is bounded by $C_{s,\nu}|x|^{-(\nu-s)}$ (see (6.3.10)), $|f| \leq C_{s,\nu}|x|^{-(\nu-s)} * |h|$. By the HLS inequality, we see that $\|f\|_p \leq C\|g\|_q$ and $\|f\|_p \leq C\|h\|_q$. $\qquad\square$

Theorem 6.3.4 (Gagliardo–Nirenberg Inequalities). *Let $\theta \in [0, 1]$, $0 < s < \nu$, $1 < q < \frac{\nu}{s}$, $1 \leq r \leq \infty$ and*

$$\frac{1}{p} = \theta\left(\frac{1}{q} - \frac{s}{\nu}\right) + \frac{(1 - \theta)}{r} \tag{6.3.29}$$

Then, for suitable constants,

$$\|u\|_p \leq C\||D|^s u\|_q^\theta \|u\|_r^{(1-\theta)} \tag{6.3.30}$$

$$\|u\|_p \leq C[\|u\|_{s,q}]^\theta \|u\|_r^{(1-\theta)} \tag{6.3.31}$$

Remarks. 1. This is usually restated in the form (6.3.30) for $s \in \mathbb{Z}_+$ with $\||D|^s u\|_q$ replaced by

$$\|D^s u\|_q = \sum_{|\alpha|=s}\|D^\alpha u\|_q \tag{6.3.32}$$

2. If $0 \leq t < s < \nu$ with $t, s \in \mathbb{Z}_+$, where $\theta \geq t/s$, and

$$\frac{1}{p} = \frac{t}{\nu} + \theta\left(\frac{1}{q} - \frac{s}{\nu}\right) + \frac{(1 - \theta)}{r} \tag{6.3.33}$$

one has also (using (6.3.32)) that

$$\|D^t u\|_p \leq C \|D^s u\|_q^\theta \|u\|_r^{(1-\theta)} \tag{6.3.34}$$

also called *Gagliardo–Nirenberg inequalities*. In Problem 13, the reader will prove an $H^{s,p}$ analog of this.

3. The special case $p = q = 2$, $r = 1$, $s = 1$ $(\nu \geq 2)$ so $\theta = \nu/(\nu + 2)$, i.e.

$$\|u\|_2 \leq C \| \, |Du| \, \|_2^{\nu/\nu+2} \|u\|_1^{2/\nu+2} \tag{6.3.35}$$

is called *Nash's inequality*. There is a simple direct proof; see Problem 14. There are cases where $q \geq \nu/s$ (but so that the right side of (6.3.29) is positive) where (6.3.30) still holds; see Theorem 6.7.3 in Section 6.7.

Proof. It follows from Hölder's inequality (see for example, Theorem 5.2.5 of Part 1) that if $1 \leq p, r, y \leq \infty$

$$\frac{1}{p} = \frac{\theta}{r} + \frac{(1 - \theta)}{y} \tag{6.3.36}$$

then

$$\|u\|_p \leq \|u\|_r^\theta \|u\|_y^{1-\theta} \tag{6.3.37}$$

Thus, given q, s, r obeying the conditions of the theorem, let

$$\frac{1}{y} = \frac{1}{q} - \frac{s}{\nu} \tag{6.3.38}$$

so $q < \frac{\nu}{s} \Rightarrow y < \infty$ and $y > q \Rightarrow y > 1$. The homogeneous Sobolev estimates imply

$$\|u\|_y \leq C\|D^s u\|_q, \quad \|u\|_y \leq C\|u\|_{s,q} \tag{6.3.39}$$

This plus (6.3.37) implies (6.3.30), (6.3.31). $\square$

Theorem 6.3.5 (Inhomogeneous Sobolev Estimates). *Let $0 < s < \nu$, $1 < q < \nu/s$. Let p be given by (6.3.23) and*

$$p \geq y \geq q \tag{6.3.40}$$

Then

$$H^{s,q} \subset L^y \tag{6.3.41}$$

and

$$\|u\|_y \leq C\Big(\| \, |D|^s u\|_q + \|u\|_q \Big) \tag{6.3.42}$$

$$\|u\|_y \leq C\|u\|_{s,q} \tag{6.3.43}$$

Remarks. 1. (6.3.41)/(6.3.43) is sometimes called (part of) the *Sobolev embedding theorem.*

2. We'll get (6.3.42) from a suitable Gagliardo–Nirenberg estimate with $y = q$ using

$$x^\theta y^{1-\theta} \leq \theta x + (1-\theta)y \qquad (6.3.44)$$

It would appear that since (6.3.44) is, in general, strict, that (6.3.42) is weaker than this Gagliardo–Nirenberg inequality, but using scaling (Problem 15), one can recover the value of θ and $\|u\|_r \leq \| |D|^s u\|_q^\theta \|u\|_q^{1-\theta}$ from (6.3.42)!

3. The second proof below, when $r < p$, relies on Young's convolution inequality. Unlike the weak Young equality needed for the homogeneous Sobolev inequality, this holds at the endpoints. Thus, this theorem holds when $q = 1$, $y < p$ (Problem 16).

First Proof. We want to take $r = q$ and $p = y$ (this is the p of (6.3.29) which was set to ν in the proof of (6.3.22)) in (6.3.29), so

$$\frac{1}{y} = \frac{1}{q} - \frac{\theta s}{\nu} \qquad (6.3.45)$$

Thus, we *define*

$$\theta = \frac{\nu}{s}\left(\frac{1}{q} - \frac{1}{y}\right) \qquad (6.3.46)$$

$p \geq y \geq q$ means $0 \leq \theta \leq 1$. Thus, by Theorem 6.3.4,

$$\|u\|_y \leq C\| |D|^s u\|_q^\theta \|u\|_q^{1-\theta}, \quad \|u\|_y \leq C\|u\|_{s,q}^\theta \|u\|_q^{(1-\theta)} \qquad (6.3.47)$$

Young's equality (usually written $wz \leq \frac{w^a}{a} + \frac{z^b}{b}$ where $a^{-1} + b^{-1} = 1$) implies $x^\theta y^{1-\theta} \leq \theta x + y^{1-\theta}$ (for take $w = x^\theta$, $z = y^{1-\theta}$, $a = \theta^{-1}$, $b = (1-\theta)^{-1}$), so the first inequality in (6.3.47) implies (6.3.42). Since $s > 0$ implies $\|u\|_q \leq \|u\|_{s,q}$ (because $J^{(\nu,s)} \in L^1$), the second inequality in (6.3.47) implies (6.3.43). $\qquad \square$

Second Proof of (6.3.43). Let $v = \left[(1+k^2)^{s/2}\,\widehat{u}\right]^{\vee}$ so $u = J^{(s,\nu)} * v$. By (6.3.9), (6.3.10),

$$J^{(s,\nu)} \in \left[\bigcap_{1 \leq p < \nu/\nu - s} L^p\right] \cap L_w^{\nu/\nu - s} \qquad (6.3.48)$$

so (6.3.43) follows from Young (if $q \geq y > p$) and weak Young (if $y = p$). $\qquad \square$

The same proof (as the second proof) proves not only if $H^{s,q} \subset L^y$ (under the conditions above) but for any s_1 (Problem 17),

$$H^{s+s_1,q} \subset H^{s_1,y} \qquad (6.3.49)$$

So far, we have only proven results if $s < \nu/q$. As s increases, the maximal r of (6.3.43) (i.e., p given by (6.3.23)) gets larger, i.e., closer to ∞. One might expect if $s > \nu/q$, u is actually a continuous function and for s even larger, a smoother and smoother function. That's the content of one of the major results of this section towards which we are heading. For $\ell \in \mathbb{N}$, define C_∞^ℓ to be those continuous functions, f, which have classical derivatives of order α, for all $|\alpha| \leq \ell$ and so that $D^\alpha f \to 0$ at ∞ for all such α.

Theorem 6.3.6 (Sobolev Embedding Theorem). *Let* $1 < q < \infty$. *Let* $\ell \in \mathbb{N}$. *Then*

$$s > \ell + \nu/q \Rightarrow H^{s,q} \subset C_\infty^\ell \qquad (6.3.50)$$

and

$$\sum_{|\alpha| \leq \ell} \|D^\alpha u\|_\infty \leq C\|u\|_{s,q} \qquad (6.3.51)$$

Remarks. 1. The Sobolev embedding theorem is usually the combination of this and (6.3.41)/(6.3.43).

2. There are also results for $q = 1$ (see Problem 18).

3. While there is no L^∞-result if $s = \ell + \nu/q$, there is a BMO result (as defined in Section 5.11)—see the Notes and Problem 4.

4. One can prove more than just continuity. If $s = \frac{\nu}{q} + \ell + \theta$ where $\ell \in \mathbb{N}$ and $\theta \in (0,1)$, then any $f \in H^{s,q}$ is in C_∞^ℓ and each $D^\alpha f$ with $|\alpha| = \ell$ is Hölder continuous of order θ.

5. For $q = 1$, this result is true; indeed, we only need $s \geq \ell + \nu$.

Lemma 6.3.7. *If* $s > \frac{\nu}{q}$, *then* $H^{s,q} \subset C_\infty$, *i.e., any* $u \in H^{s,q}$ *lies in* C_∞ *and*

$$\|u\|_\infty \leq C\|u\|_{s,q} \qquad (6.3.52)$$

Proof. Suppose $s < \nu$. Then $q > \nu/s$, so the dual index $p < \nu(\nu - s)^{-1}$ (for $q^{-1} < \frac{s}{\nu}$, so $p^{-1} > 1 - \frac{s}{\nu} = \frac{\nu-s}{\nu}$). It follows from Proposition 6.3.1(j), that $J^{(\nu,s)} \subset L^p$. If $s \geq \nu$, pick $s_0 \in \left(\frac{\nu}{q}, \nu\right)$ and use the fact that $s > s_0 \Rightarrow H^{s,q} \subset H^{s_0,q}$ (Problem 12) and $\|u\|_{s_0,q} \leq C\|u\|_{s,q}$ to see we need only prove the result when $s < \nu$. In that case, if $v = (1 + k^2)^{s/2}u$, then $u = J^{(s,\nu)} * v \in L^p * L^q \subset C_\infty$ by Theorem 6.6.4 of Part 1 and $\|u\|_\infty \leq C\|v\|_q$ by Hölder's inequality (continuity follows from the density of C_0^∞ in L^p and L^q and Hölder's inequality). $\qquad\square$

Lemma 6.3.8. *Let* $s \in \mathbb{R}$ *and* $1 < q < \infty$. *Then,* D_j *given by* $D_j u = \partial u/\partial x_j$ *maps* $H^{s,q}$ *to* $H^{s-1,q}$, *i.e.,* $u \in H^{s,q} \Rightarrow D_j u \in H^{s-1,q}$ *and*

$$\|D_j u\|_{s-1,q} \leq C\|u\|_{s,q} \qquad (6.3.53)$$

For any multi-index, α, $u \in H^{s,q} \Rightarrow D^\alpha u \in H^{s-|\alpha|,q}$,

$$\|D^\alpha u\|_{s-|\alpha|,q} \le C\|u\|_{s,q} \qquad (6.3.54)$$

Proof. (6.3.54) follows from (6.3.53) by iteration so we need only prove (6.3.53). By the fact listed after Proposition 6.3.1 (and proven in the next section), $\delta_j f = (k_j (1+k^2)^{-1/2} \widehat{f})^\vee$ maps L^q to L^q. Since δ_j clearly commutes with $(1+k^2)^{-(s-1)/2}$ for any s, we see δ_j maps $H^{s,q}$ to $H^{s,q}$. Since $u \mapsto [(1+k^2)^{1/2} \widehat{u}]^\vee$ maps $H^{s,q}$ to $H^{s-1,q}$, we see that $D_j f = [ik_j \widehat{f}]^\vee$ maps $H^{s,q}$ to $H^{s-1,q}$. $\qquad\square$

Lemma 6.3.9. *Let u be a distribution with $D^\alpha u \in C_\infty$ for all α with $|\alpha| \le \ell$. Then, $u \in C_\infty^\ell$.*

Proof. By a simple iteration, it follows that it suffices to prove this for $\ell = 1$. Let $D_j u = f_j$. Let $j_{(n)}$ be an approximate identity of compact support. By continuity of u and f_j, if $u^{(n)} = j_{(n)} * u$, $f_j^{(n)} = j_{(u)} * f_j$, then $u^{(n)} \to u$, $f_j^{(n)} \to f_j$ uniformly on compacts. Moreover, $\vec\nabla u^{(n)} = \vec{f}^{(n)}$ in the classical sense, so

$$u^{(n)}(x) - u^{(n)}(y) = \int_0^1 (x-y) \cdot \vec{f}^{(n)}(\theta x + (1-\theta)y)\, d\theta$$

so taking limits

$$u(x) - u(y) = \int_0^1 (x-y) \cdot \vec{f}(\theta x + (1-\theta)y)\, d\theta$$

This, plus the local uniform continuity of $\vec{f}$ proves that u is C^1 with $\vec\nabla u = \vec{f}$. $\qquad\square$

Proof of Theorem 6.3.6. Let $u \in H^{s,q}$. By Lemma 6.3.8, for all α with $|\alpha| \le \ell$, $D^\alpha u \in H^{s-|\ell|,q}$. Since $s - |\ell| > \nu/q$, by Lemma 6.3.7, each $D^\alpha u \in C_\infty$. Thus, by Lemma 6.3.9, $u \in C_\infty^\ell$. (6.3.46) follows from (6.3.54) and (6.3.52) and (6.3.51) $\qquad\square$

We turn now to open subsets, $\Omega \subset \mathbb{R}^\nu$, where there are two natural Sobolev spaces:

Definition. Let Ω be an open subset of $\mathbb{R}^\nu$. We denote by $W_0^{\ell,p}(\Omega)$ (respectively, $H_0^{s,p}(\Omega)$) the closures in $W^{\ell,p}(\mathbb{R}^\nu)$ (respectively, $H^{s,p}(\mathbb{R}^\nu)$) of the $C_0^\infty(\Omega)$-functions. We denote by $W^{\ell,p}(\Omega)$, the distributions, u, on Ω so that for all $|\alpha| \le \ell$, $D^\alpha u \in L^p(\Omega, d^\nu x)$.

Example 6.3.10. Sobolev embedding theorems say that $W^{1,2}(\mathbb{R}^2)$ is contained in all L^p, $2 \le p < \infty$. It can be shown that this is true for $W^{1,2}(B_1(0))$, with $B_1(0)$ the unit ball. But consider

$$\Omega_\alpha = \{(x,y) \in \mathbb{R}^2 \mid 0 < x < 1,\, 0 < y < x^\alpha\} \qquad (6.3.55)$$

which has a cusp at $(0,0)$ if $\alpha > 1$. Let $f_\beta(\mathbf{k}) = |k|^{-\beta}$ for $\beta > 0$. This fails to lie in $W^{1,2}(\mathbb{R}^2)$ but since on Ω_α, $x < |\mathbf{r}| < 2x$, one sees that

$$\int_\Omega |\mathbf{r}|^{-\beta} \leq \int_0^1 x^{-\beta} x^\alpha dx < \infty \text{ if } \alpha - \beta > -1 \qquad (6.3.56)$$

and the integral is ∞ if $\alpha - \beta \leq -1$.

$$f_\beta \in W^{1,2} \Leftrightarrow \alpha > 2\beta + 1 \qquad (6.3.57)$$

$$f_p \in L^p \Leftrightarrow \alpha > p\beta - 1 \qquad (6.3.58)$$

In particular, given any $p > 2$, we can pick $\beta > 1/(p-2)$ and then α so $p\beta - 1 > \alpha > 2\beta + 1$ so find $f \in W^{1,2}(\Omega_\alpha)$ but $f \notin L^p$. $\square$

This example shows that theorems about $W^{\ell,s}(\Omega)$ will require regularity conditions on $\partial\Omega$ and additional work. The most common conditions are one of C^k, Lipschitz, or internal cone conditions. This is beyond the scope of this introduction, so we'll say no more about it and focus on $W_0^{\ell,p}(\Omega)$. We'll be interested in compactness results. We'll require the following which is a special case of a compactness result of M. Riesz, Theorem 3.16.1 of Part 4.

Theorem 6.3.11. *Let $\Omega \subset \mathbb{R}^\nu$ be a bounded open set. A set $S \subset L^p(\Omega, d^\nu x)$ has compact closure in $L^p(\Omega, d^\nu x)$ if and only if*

(a) $\sup_{f \in S} \|f\|_p < \infty$

(b) *S is uniformly L^p-continuous, i.e., $\forall \varepsilon > 0$, $\exists \delta$ such that*

$$\sup_{\substack{f \in S \\ |y| < \delta}} \sup_{y \in \mathbb{R}^\nu} \int_{\{x | x, x-y \in \Omega\}} |f(x) - f(x-y)|^p d^\nu x \leq \varepsilon^p \qquad (6.3.59)$$

The idea of the proof is to find a uniformly equicontinuous family of functions near to all $f \in S$ and appeal to the Ascoli–Arzelà theorem. We do this by looking at $f * \eta$ with η in C_0^∞ and supported in $\{y \mid |y| < \delta\}$.

Theorem 6.3.12 (Rellich–Kondrachov Embedding Theorem). *Let $\Omega \subset \mathbb{R}^\nu$, $\nu \geq 2$, be bounded. Let $1 < q < \nu$ and $q \leq y < p$ with*

$$\frac{1}{p} = \frac{1}{q} - \frac{1}{\nu} \qquad (6.3.60)$$

Then, $\left(W_0^{1,q}(\Omega)\right)_1 = \{f \mid \|f\|_{1,q} \leq 1\}$ is a compact subset of $L^y(\Omega, d^\nu x)$.

Remarks. 1. As we'll see below, this fails for $y = p$.

2. This is often stated as saying the map, i.e., $W_0^{1,q}(\Omega, d^\nu x) \to L^y(\Omega, d^\nu x)$ by $f \mapsto f$ is a compact map in the sense of Chapter 3 of Part 4.

3. There is also a compact embedding theorem for $H^{s,2}$ for all $s > 0$, not just $s = 1$; see Problem 22.

Proof. Let $S = (W_0^{1,q}(\Omega))_1$. We'll first show that $\overline{S}$ is compact in L^y and then that S is closed in $\|\cdot\|_y$. Since $q \leq y < p$, there is $0 < \theta \leq 1$ so that

$$\frac{1}{y} = \frac{1-\theta}{p} + \frac{\theta}{q} \tag{6.3.61}$$

so by Hölder's inequality (see Theorem 5.2.5 in Part 1),

$$\|f\|_y \leq \|f\|_q^\theta \|f\|_p^{1-\theta} \tag{6.3.62}$$

Let now $f \in C_0^\infty(\Omega)$. Then,

$$f(x-y) - f(x) = -\int_0^1 y \cdot \nabla f(x - \theta y) d\theta \tag{6.3.63}$$

so that taking $L^q(\mathbb{R}^\nu)$-norms

$$\|f(\cdot - y) - f(\cdot)\|_q \leq |y| \|f\|_{1;q} \tag{6.3.64}$$

By the Sobolev embedding theorem,

$$\|f(\cdot - y) - f\|_p \leq C\|f(\cdot - y) - f\|_{1;q} \leq 2C\|f\|_{1;q} \tag{6.3.65}$$

It follows from (6.3.62) that

$$\|f(\cdot - x) - f\|_y \leq C_1 |x|^\theta \|f\|_{1;q} \tag{6.3.66}$$

This extends to the closure of $C_0^\infty(\Omega)$ in $W^{1,q}(\mathbb{R}^\nu)$ and implies that

$$\text{LHS of (6.3.59) (for } p = y) \leq (C_1 \delta^\theta)^y$$

so (6.3.59) holds (take $\delta = (\varepsilon/C_1)^{1/\theta}$). Thus $\overline{S}$ is compact in L^y by Theorem 6.3.11.

S is closed by the standard argument from the theory of compact operators: Since $1 < q < \infty$, $(L^q)_1$ is compact in the $\sigma(L^q, (L^q)^*)$-topology (since C^q is reflexive and one can use the Banach–Alaoglu theorem). It follows that $(W_0^{1,q}(\Omega))_1$ is closed in the weak-$\sigma(W_0^{1,q}, C_0^\infty(\Omega))$ topology (since $C_0^\infty(\Omega)$ is dense in $(L^q)^*$, this is the same as the weak topology). Let $f_n \in S$ and $f \to f_n$ in $\|\cdot\|_y$ norm, we can find $g \in S$ so that $f_{n_j} \xrightarrow{w} g \in \mathcal{S}$ for some subsequences. Thus, $\langle \varphi, f_{n_j} \rangle \to \langle \varphi, g \rangle$ for $\varphi \in C_0^\infty(\Omega)$. It follows that $\langle \varphi, f - g \rangle = 0$ for all $\varphi \in C_0^\infty(\Omega)$ so $f = g$, i.e., S is closed in $\|\cdot\|_y$ $\quad\square$

Example 6.3.13. This will prove that $(W_0^{q,1})_1$ is *never* compact in L^p for p given by (6.3.60). This will be a consequence of the homogeneous nature of the Sobolev estimate in that case—i.e., of scale invariance. Let $x_0 \in \Omega$ and pick a nonzero $\varphi \in C_0^\infty(\mathbb{R}^\nu)$ with $\operatorname{supp} \varphi \subset \{x \mid |x| < 1\}$. Let

$$\varphi_n(x) = n^{\nu/p} \varphi(n(x - x_0)) \tag{6.3.67}$$

Then, for n large, $\operatorname{supp}(\varphi_n) \subset \{x \mid |x - x_0| \leq \frac{1}{n}\} \subset \Omega$. By scaling, $\|\varphi\|_p = c_1$, $\|\nabla \varphi_n\|_q = c_2$ (since $\frac{\nu}{p} + 1 = \frac{\nu}{q}$), and $\|\varphi_n\|_q = c_3 n^{(q-p)\nu/p} \to 0$ (for $c_1, c_2, c_3 \in (0, \infty)$).

Thus, φ_n lies in a ball in $W_0^{q,1}$. If $(W_0^{q,1})_1$ were compact, some subsequence of φ_n would converge in L^p-norm to a φ_∞ with $\|\varphi_\infty\|_p = c_1$. Since $\varphi_n \to 0$ in $\sigma(L^p, L^{p'})$, no such φ_∞ exists. $\qquad\square$

While there are elementary proofs of Poincaré's inequality, we want to show it follows from Relllich's embedding theorem. As a preliminary we need:

Lemma 6.3.14. *Let $\Omega \subset \mathbb{R}^\nu$ be nonempty. Then, the constant (in Ω) functions do not lie in $W_0^{1,p}(\Omega)$ for any p, $1 < p < \infty$.*

Proof. Let $\varphi_n \in C_0^\infty(\Omega)$ so that φ_n converges to $f \equiv c\chi_\Omega$ in $W^{1,p}(\Omega)$. Then $\nabla\varphi_n$ is Cauchy and so has a limit g in L^p. Since $\nabla\varphi_n$ vanishes off Ω, g vanishes off Ω. For any $\psi \in C_0^\infty(\Omega)$,

$$\int \psi g \, d^\nu x = \lim \int \psi \nabla\varphi_n = \int (\nabla\psi)\varphi_n$$

$$= c \int \nabla\psi = 0 \qquad\qquad (6.3.68)$$

Thus $g = 0$ on Ω so $g \equiv 0$. It follows that f has zero derivatives as a distribution in $\mathcal{S}(\mathbb{R}^\nu)$, i.e., $f = c$ on $\mathbb{R}^\nu$ so $c = 0$. $\qquad\square$

Corollary 6.3.15 (Poincaré's Inequality). *For any bounded $\Omega \subset \mathbb{R}^\nu$ and $1 < q < \nu$, there is a constant C so that for all $f \in W_0^{1,q}(\Omega)$, we have*

$$\|f\|_q \leq C\|\nabla f\|_q \qquad\qquad (6.3.69)$$

Proof. If not, there exists for $n = 1, 2, \ldots$, $f_n \in W_0^{1,q}(\Omega)$ so that $\|f_n\|_q = 1$, and

$$\|f_n\|_q = 1 \geq n\|\nabla f_n\|_q \qquad\qquad (6.3.70)$$

By compactness of balls in $W_0^{1,q}(\Omega)$, there is a subsequence $f_{n(j)}$ so $f_{n(j)} \to f_\infty$ in $L^q(\Omega)$. In particular, $\|f_\infty\|_q = 1$. By (6.3.15), $\|\nabla f_n\|_q \leq n^{-1} \to 0$, so for any $\varphi \in C_0^\infty(\Omega)$, $\int f_\infty(\nabla\varphi)d^\nu x = 0$, i.e., $\nabla f_\infty = 0$ as a distribution. It follows that f_∞ is constant, so by the lemma, $f_\infty = 0$ is inconsistent with $\|f_\infty\|_q = 1$. This contradiction proves (6.3.69) holds for some C. $\qquad\square$

As a final topic, we want to use Sobolev space technology to prove:

Theorem 6.3.16. *Let φ be any ordinary distribution on $\mathbb{R}^\nu$ that obeys*

$$\Delta\varphi = f\varphi + g \cdot \vec{\nabla}\varphi \qquad\qquad (6.3.71)$$

where f and $\{g_j\}_{j=1}^\nu$ are C^∞-functions. Then φ is a C^∞-function.

Remarks. 1. We, of course, proved a stronger result than this in Section 6.3 but that was with a more involved machinery.

2. We emphasize that we only use $H^{s,2}$ (i.e., $p = 2$) where the theory is much simpler—see Problems 1 and 2.

3. This is a local theorem so it holds for distributions $\varphi \in \mathcal{D}(\Omega)'$.

For the rest of this section, we'll drop the ^{2} and just use H^s for $H^{s,2}$. Since there are no Hardy spaces in the neighborhood, we won't worry about the, in principle, ambiguous notation. We start with

Proposition 6.3.17. *Fix* $F \in C_0^\infty(\mathbb{R}^\nu)$. $f \mapsto Ff$ *is a bounded map of each* H^s *to itself.*

Proof. $\widehat{Ff}(k) = (2\pi)^{-\nu/2}(\widehat{F} * \widehat{f})(k)$. If $s \geq 0$, we write $\langle k \rangle = (1 + |k|^2)^{1/2}$ and note what is sometimes called *Peetre's inequality*: if $s \geq 0$, then

$$\langle k \rangle^s \leq 2^s \langle \ell \rangle^s \langle k - \ell \rangle^s \tag{6.3.72}$$

It is enough to prove this when $s = 1$. We note that

$$|\vec{\nabla}\langle k \rangle| = |\vec{k}|(1 + |k|^2)^{-1/2} \leq 1 \tag{6.3.73}$$

which implies

$$\langle k \rangle \leq \langle \ell \rangle + |k - \ell| \leq \langle \ell \rangle + \langle k - \ell \rangle \tag{6.3.74}$$

$$\leq 2\langle \ell \rangle \langle k - \ell \rangle \tag{6.3.75}$$

since $\langle \ell \rangle, \langle k - \ell \rangle \geq 1$.

Thus, if $g_\ell(k) = \langle k - \ell \rangle^s \widehat{f}(k - \ell)$, then

$$\langle k \rangle^s |\widehat{Ff}(k)| \leq (2\pi)^{-\nu/2} 2^s \int \langle \ell \rangle^s \widehat{F}(\ell) |g_\ell(k)| d^\nu \ell$$

so by the triangle inequality (for integrals of functions)

$$\|\langle \cdot \rangle^s \widehat{Ff}(\cdot)\|_2 \leq c \|g_0\|_2 \int \langle l \rangle^s |\widehat{F}(\ell)| d^\nu \ell \tag{6.3.76}$$

since $\|g_\ell(k)\|_2 = \|g_0(k)\|_2$ independently of ℓ. Since $\widehat{F}$ decays faster than any power, the integral in (6.3.76) is finite and

$$\|Ff\|_s \leq C\|f\|_s \tag{6.3.77}$$

Interchanging k and ℓ in (6.3.72) shows that for $s \geq 0$

$$\langle k \rangle^{-s} \leq 2^s \langle \ell \rangle^{-s} \langle k - \ell \rangle^s \tag{6.3.78}$$

so repeating the argument above (6.3.77) holds for $-s$, i.e., for all $s \in \mathbb{R}$ rather than just $s \geq 0$. $\qquad \square$

We say a distribution, τ, is in $H^s_{\mathrm{loc};R}$ if and only if for all $F \in C_0^\infty(\mathbb{R}^\nu)$ supported in $\{x \mid |x| < R\}$, $F\tau \in H^s$.

Lemma 6.3.18. *For any distribution* τ *and any* R, *there is an* s *(perhaps depending on* R*) so that* $\tau \in H^s_{\mathrm{loc};R}$.

Proof. By the definition of distribution (see Chapter 9 of Part 1), there is an N_R and C_R so that for $f \in C_0^\infty(\{x \mid |x| < R\})$,

$$|\tau(f)| \le C_R \sum_{\alpha \le N_R} |D^\alpha f|_\infty \tag{6.3.79}$$

Thus, if F is in $C_0^\infty(\mathbb{R}^\nu)$, for all $f \in \mathcal{S}(\mathbb{R}^\nu)$

$$|(F\tau)(f)| \le C_{R,1} \sum_{|\alpha| \le N_R} \|D^\alpha f\|_\infty$$

using Leibniz's rule. Thus $F\tau$ is tempered and using $f(x) = g(x)e^{ik \cdot x}$ where $g \in \mathcal{S}(\mathbb{R}^\nu)$ is 1 on $\{x \mid |x| < R\}$, we see that

$$|(\widehat{F\tau})(k)| \le C\langle k \rangle^{N_R} \tag{6.3.80}$$

which means $F\tau \in H^s$ where

$$s = \frac{-N_R}{2} - \nu - 1 \tag{6.3.81}$$

(since $\langle k \rangle^{-(\nu+1)} \in L^2(\mathbb{R}^\nu)$). $\qquad\square$

Proposition 6.3.19. *Let φ be a distribution.*

(a) $\varphi \in H^s \Rightarrow \nabla\varphi \in H^{s-1}$
(b) $\varphi \in H^s_{\text{loc};R} \Rightarrow \nabla\varphi \in H^{s-1}_{\text{loc};R}$
(c) $\varphi, \Delta\varphi \in H^s \Rightarrow \varphi \in H^{s+2}$
(d) $\varphi, \Delta\varphi \in H^s_{\text{loc};R} \Rightarrow \varphi \in H^{s+2}_{\text{loc};R}$

Proof. (a) Clearly $\langle k \rangle^{s-1} |\vec{k}\hat\varphi| \le \langle k \rangle^s |\hat\varphi|$ which implies the claim.

(b) If $F \in C_0^\infty(\{x \mid |x| < R\})$, then

$$F\nabla\varphi = \nabla(F\varphi) - (\nabla F)\varphi \tag{6.3.82}$$

If $\varphi \in H^s_{\text{loc};R}$, $(\nabla F)\varphi \in H^s \subset H^{s-1}$ and by (a), $\nabla(F\varphi) \in H^{s-1}$, so $F(\nabla\varphi) \in H^s$.

(c) Since $\langle k \rangle^{s+2} = (1+k^2)\langle k \rangle^s$, we see that

$$\|\varphi\|_{s+2}^2 = \|\Delta\varphi\|_s^2 + \|\varphi\|_s^2 \tag{6.3.83}$$

(d) Suppose $\varphi, \Delta\varphi \in H^s_{\text{loc};R}$ and F is a generic function in $C_0^\infty(\{x \mid |x| < R\})$. Note that

$$\Delta(F\varphi) = F\Delta\varphi + 2\nabla F \cdot \nabla\varphi + (\Delta F)\varphi \tag{6.3.84}$$

By (b) and the hypothesis, $\nabla\varphi \in H^{s-1}_{\text{loc};R}$, so by (6.3.84) $\Delta(F\varphi) \in H^{s-1}$ and so is $F\varphi$. Thus, by (c), $F\varphi \in H^{s+1}$. Since F is generic, $\varphi \in H^{s+1}_{\text{loc};R}$, so $\nabla\varphi \in H^s_{\text{loc};R}$ by (b). Thus, (6.3.84) implies $F\varphi, \Delta(F\varphi) \in H^s$ so $F\varphi \in H^{s+2}$, i.e., $\varphi \in H^{s+2}_{\text{loc};R}$. $\qquad\square$

Proof of Theorem 6.3.16. By Theorem 6.3.6, $\bigcap_{s=0}^{\infty} H^s$ is contained in the C^{∞}-functions, so it suffices to prove that $\varphi \in H^s_{\text{loc};R}$ for each s and R.

Fix R. By Lemma 6.3.18, $\varphi \in H^{s_0}_{\text{loc};R}$ for some s_0.

For a generic $F \in C_0^{\infty}(\{x \mid |x| < R\})$, by (6.3.71) we have

$$\Delta(F\varphi) = Ff\varphi + (Fg) \cdot \nabla\varphi + 2\nabla F \cdot \nabla g + \Delta F\varphi \qquad (6.3.85)$$

Each term is in H^{s_0} or H^{s_0-1} so $F\varphi \in H^{s_0+1}$ by Proposition 6.3.19.

Thus, we've improved from $H^{s_0}_{\text{loc};R}$ to $H^{s_0+1}_{\text{loc};R}$. Iterating we have $\varphi \in H^s_{\text{loc};R}$ for all s. $\qquad\square$

Notes and Historical Remarks. The fundamental theorem of calculus immediately implies that if $\varphi \in C_0^{\infty}((0,1))$, then

$$\|\varphi\|_{\infty} \leq \left\|\frac{d\varphi}{dx}\right\|_1$$

so it is not surprising that inequalities of Sobolev type have a long history predating Sobolev. Naumann [**598**] has a review article on this history. An important motivation for the more sophisticated aspects was the attempt to prove a version of the Dirichlet principle, i.e., Riemann's claim that one could solve the Dirichlet problem by minimizing $\int_{\Omega} |\nabla f(x)|^2 dx$ over all f with a given set of values on $\partial\Omega$—a claim that was refuted by Weierstrass (see the Notes to Section 8.1 of Part 2A.) For example, in 1933, Tonelli [**809**] proved if $\nabla u \in L^p(\mathbb{D}, d^2 x)$ for $p > 2$ for $\mathbb{D}$ the unit disc in $\mathbb{C}$, then u is uniformly continuous. In that year, Nikodym [**613**] considered the space of functions on a bounded domain $\Omega \subset \mathbb{R}^3$ with $\int_{\Omega} |\nabla u|^2 d^3 x < \infty$, $\int_{\Omega} |u|^2 d^3 x < \infty$ and proved it complete (this is, of course, $W^{1,2}$). In this early work, one difficulty was the meaning of ∇u since weak derivatives were only considered a few years later. For example, Nikodym assumed u was absolutely continuous along almost every ray and defined directional derivatives by Lebesgue's theory. Calkin [**135**] and Morrey [**587**] considered what we'd now call $W^{1,p}(\Omega)$.

These involve attempts to describe spaces of functions. There is even earlier work on inequalities between functions and derivatives. As early as 1890, Poincaré [**652**] proved for convex sets, Ω, of diameter $d(\Omega)$, in $\mathbb{R}^{\nu}$ that when $f \in C_0^{\infty}(\Omega)$

$$\int |f|^2 d^{\nu} x \leq \frac{C_{\nu}}{d^2} \int |\nabla f|^2 d^{\nu} x \qquad (6.3.86)$$

now called *Poincaré's inequality*. In fact, he showed if f is C^1 in a neighborhood of Ω, it holds if $\int |f|^2 d^{\nu} x$ is replaced by $\int (f - \frac{1}{|\Omega|} \int f d^{\nu} x)^2 d^{\nu} x$. In fact, we've seen (6.3.86) without convexity. Friedrichs [**289**] proved an L^p-analog of (6.3.86) in 1927. In 1908, Hadamard [347] proved and used (in his study of elastic plates) the inequalities

$$\max_{0 \leq x \leq a} |u(x)|^{2+q} \leq 2^{2+q} \int_0^a |u'(t)|^2 dt \int_0^a |u(t)|^q dt \tag{6.3.87}$$

which is an inequality of the Gagliardo–Nirenberg type.

But it was Sobolev, who, in a series of papers starting in 1935 and including [764, 765, 766], systematically looked at what are now called Sobolev spaces and used them to define weak solutions of PDEs, thereby developing a precursor of distribution theory. He codified his work in a 1950 book [768]. Nash's inequality is due to John Nash [596], one of his last papers before he was sidelined by paranoid schizophrenia.

Gagliardo [301] and Nirenberg [614] discussed the inequalities now named after them.

Best constants in homogeneous Sobolev inequalities for $W^{1,p}$ were found by Aubin [19] and Talenti [795] who proved for $1 < p < \nu$ and $q = \frac{\nu p}{\nu - p}$, one has on $\mathbb{R}^\nu$ that

$$\|u\|_q \leq \pi^{-1/2} \nu^{-1/p} \left[\frac{(p-1)}{(\nu - p)} \right]^{1 - \frac{1}{p}} \left\{ \frac{\Gamma(\nu)\Gamma(1 + \frac{\nu}{2})}{\Gamma(\frac{\nu}{p})\Gamma(1 + \nu - \frac{\nu}{p})} \right\}^{1/\nu} \tag{6.3.88}$$

where the optimal u (showing the constant is best possible) is

$$u(x) = [a + b|x|^{p/p-1}]^{1-\nu/p} \tag{6.3.89}$$

Del Pino–Dolbeault [209] has best constants in certain Gagliardo–Nirenberg inequalities (those involving $\|\nabla u\|_2$).

Sobolev embedding fails at the end point; i.e., if $q = \nu/s$, and $f \in H^{q,s}$, it is not true that $f \in L^\infty$ but it is sometimes in BMO; see Problem 4. Similarly, if $s = \ell + \nu/q$, $H^{s,q}$ may not lie in C_∞^ℓ but it is $C_\infty^{\ell-1}$ and the distribution's ℓ-th derivative is in BMO.

The Rellich–Kondrachov embedding theorem is named after Rellich [676] who proved the L^2-case and Kondrachov [484] who had the L^p-result.

Finally, we turn to issues of *fractional Sobolev spaces*. In the PDE literature, the name Sobolev space is often used for $H^{s,p}$ especially if $p = 2$ (we'll see in a moment when $p = 2$ the definitions are equivalent). The name "fractional Sobolev space" or "fractional-order Sobolev space" is used for a space defined when $0 < s < 1$ (if $s = n + \theta$ with $n \in \mathbb{Z}_+$, one demands $D^\alpha f \in L^p$ if $|\alpha| \leq p$ and $D^\alpha f \in W^{\theta,p}$ when $|\alpha| = n$) by requiring

$$\|u\|_{s,p}^{\sim} = \left(\int_{\mathbb{R}^\nu} \int_{\mathbb{R}^\nu} \frac{|u(x) - u(y)|^p}{|x - y|^{sp+\nu}} \, dx \, dy \right)^{1/p} < \infty \tag{6.3.90}$$

and $W^{s,p}$ as the set with

$$\|u\|_{s,p} = \|u\|_{\tilde{s,p}} + \|u\|_p < \infty \tag{6.3.91}$$

It is not hard to see that $W^{s,2} = H^{s,2}$ (Problem 5).

There is considerable literature on two three-parameter classes of spaces: Besov spaces, B^s_{pq} (after Besov [**65**]) and Triebel or Triebel–Lizorkin spaces, F^s_{pq} (after Lizorkin [**543**] and Triebel [**813, 814**]). The original definition of Besov space has (again, for $0 < s < 1$ for simplicity)

$$\left(\|f\|_p^q + \int_0^\infty \left| \frac{\|f(\,\cdot\, + h) - f(\,\cdot\,)\|_p}{h^s} \right|^q \frac{dh}{h} \right)^{1/q} < \infty \tag{6.3.92}$$

$H^{s,2} = B^s_{2,2}$ and more generally if $1 < p \le 2$,

$$B^s_{p,p} \subset H^{s,p} \subset B^s_{p,2}$$

and if $2 \le p < \infty$,

$$B^s_{p,2} \subset H^{s,p} \subset B^s_{p,p}$$

The above has a kind of Lorentz space flavor and indeed, Besov spaces can be defined via real interpolation as $W^{n,p}$ interpolation spaces.

Using Littlewood–Paley ideas (see the discussion in Section 6.4), one can define Besov spaces in terms of sums of $\widehat{f}$ restricted to regions $2^\ell < |k| < 2^{\ell+1}$ and this is the way Triebel–Lizorkin spaces are defined.

There is a huge literature on Sobolev spaces, much of it dealing with the classes, $H^{s,p}$, $W^{s,p}$, $B^s_{p,q}$ and $F^s_{p,q}$. For example, see [**3, 60, 56, 211, 365, 521, 562, 710, 802, 816**].

As a parting note, we mention Sobolev inequalities; their connection to CLR and Lieb–Thirring inequalities will be discussed in Section 6.7.

Problems

The first five problems concern the simplified theory available for $H^{s,p}$ if $p = 2$. They rely on the fact that both $\widehat{}$ and $\check{}$ are bounded when $p = 2$.

1. Let $\ell \in \mathbb{Z}_+$. Prove that $((1 + k^2)^{\ell/2} \widehat{f})^\vee \in L^2$ if and only if $D^\alpha f$ ($|\alpha| \le \ell$) are all in L^2. (*Hint*: Compute $\|(1 + k^2)^{\ell/2} \widehat{f}\|_2$.)

2. Let $s \le \nu/2$. This will prove if $f \in H^{s,2}$ and $\frac{1}{p} > \frac{1}{2} - \frac{s}{\nu}$, then $f \in L^p$ (with a bounded relation of norms).

   (a) Prove that $(1 + k^2)^{-s/2} \in L^r$ if and only if $r > \frac{\nu}{s}$.

   (b) If $g \in L^2$, prove that $(1 + k^2)^{-s/2} \widehat{g} \in L^q$ if $\frac{1}{q} < \frac{s}{\nu} - \frac{1}{2}$.

   (c) If $f \in H^{s,2}$, prove $f \in L^p$, if $\frac{1}{p} > \frac{1}{2} - \frac{s}{\nu}$. (*Hint*: Use Young's and the Hausdorff-Young inequality.)

3. This will prove the more subtle homogeneous Sobolev estimates when $s < \frac{\nu}{2}$, $p^{-1} = \frac{1}{2} - \frac{s}{\nu}$ and we want to show $L^p \subset H^{s,2}$. For $f \in L^2$, define for $\eta > 0$

$$f_{1,\eta} = (\chi_{|k| \leq \eta}(k)\, \widehat{f}\,)^\vee, \quad f_{2,\eta} = f - f_{1,\eta} \tag{6.3.93}$$

(a) Prove that

$$m_f(\lambda) \leq m_{f_{1,\eta}}\left(\frac{\lambda}{2}\right) + m_{f_{2,\eta}}\left(\frac{\lambda}{2}\right) \tag{6.3.94}$$

(b) Prove that

$$\|f_{1,\eta}\|_\infty \leq \frac{C}{\nu - 2s} \| |D|^s f\|_2 \, \eta^{\frac{\nu}{2} - s} \tag{6.3.95}$$

so that if

$$\eta(\lambda) = \frac{\lambda(\nu - 2s)}{4C\| |D|^s f\|_2} \tag{6.3.96}$$

we have that

$$m_{f_{1,\eta(\lambda)}}\left(\frac{\lambda}{2}\right) = 0 \tag{6.3.97}$$

(c) Prove that if $\eta(\lambda)$ is given by (6.3.96), then

$$\|f\|_p^p \leq 4p \int_0^\infty \lambda^{p-3} \|f_{2,\eta(\lambda)}\|_2^2 d\lambda \tag{6.3.98}$$

(*Hint*: Bound $f_{2,\eta(\lambda)}\left(\frac{\lambda}{2}\right)$ by $\|f_{2,\eta(\lambda)}\|^2$ and λ.)

(d) Complete the proof that

$$\|f\|_p^p \leq C_{p,\nu}\| |D|^s f\|_2^p$$

4. This problem will show that if $f \in L^{2,\frac{q}{2}}$ (where the p in (6.3.1) for $q = 2$ is ∞), then $f \in \mathrm{BMO}$ and also provide an example where f is not in L^∞. Let B be a ball and $f_{1,\eta}, f_{2,\eta}$ given by (6.3.93).

(a) Prove that

$$\int_B |f - f_B| \frac{dx}{|B|} \leq \|f_{1,\eta} - (f_{1,\eta})_B\|_{L^2(B, \frac{dx}{|B|})} + \frac{2}{|B|^{1/2}} \|f_{2,\eta}\|_2$$

(b) Let $\eta = R^{-1}$ where R is the radius of B. Prove that

$$\frac{2}{|B|^{1/2}} \|f_{2,R^{-1}}\|_2 \leq C_\nu \left(\int_{|k| \geq R^{-1}} |k|^\nu |\widehat{f}(k)|^2 \right)^{1/2} \tag{6.3.99}$$

(c) Prove that

$$\|f_{1,\eta} - (f_{1,\eta})_B\|_{L^2(B_0 \frac{dx}{|B|})} \leq R\|\lambda f_{1,\eta}\|_\infty$$

$$\leq R\eta |f|_{H^{\frac{\nu}{2},2}} \tag{6.3.100}$$

and conclude that $f \in \mathrm{BMO}$.

(d) Let $g \in C_0^\infty(\mathbb{R}^\nu)$, only a function of $|x|$ with $g \equiv 1$ near $x = 0$ and $\mathrm{supp}(g) \subset \{x \mid |x| < \frac{1}{2}\}$. Let

$$f(x) = \log(|x|^{-1})g(x) \tag{6.3.101}$$

Prove that $f \in H^{\frac{\nu}{2},2}$ but f is unbounded. (*Hint*: Compute $(f, D^\alpha f)$ for $|\alpha| \le \nu$.)

Remark. In Problems 3 and 4, we follow Chemin–Xu [150] although the idea may well predate them.

5. Let $s \in (0,1)$. This problem will prove the equivalence of the $H^{s,2}$ and the norm (6.3.91) when $p = 2$.

(a) Let $\mathbf{z}$ be an $\mathbb{R}^\nu$-valued variable and z_1 its first component. Prove that

$$C_{\nu,s} \equiv \int \frac{|e^{iz_1} - 1|^2}{|z|^{\nu+2s}} d^\nu z < \infty \tag{6.3.102}$$

(b) Prove that for any $\mathbf{k} \in \mathbb{R}^\nu$

$$\int \frac{|e^{i\mathbf{k}\cdot\mathbf{z}} - 1|^2}{|z|^{\nu+2s}} dz = C_{\nu,s}|k|^{2s} \tag{6.3.103}$$

(c) Prove that for any $f \in \mathcal{S}(\mathbb{R}^\nu)$, we have

$$\int_{\mathbb{R}^\nu} \int_{\mathbb{R}^\nu} \frac{|f(\mathbf{x}) - f(\mathbf{y})|^2}{|\mathbf{x} - \mathbf{y}|^{\nu+2s}} d^\nu x \, d^\nu y = C_{\nu,s} \int |k|^{2s} |\hat{f}(\mathbf{k})|^2 d^\nu k \tag{6.3.104}$$

(*Hint*: Let $\mathbf{y} = \mathbf{x} + \mathbf{z}$ and use the Plancherel theorem in the x-variable.)

(d) Conclude that $\|\cdot\|_{W^{s,2}}$ and $\|\cdot\|_{H^{s,2}}$ are equivalent norms.

6. Prove for $\nu \ge 2$ and $|x| < \frac{1}{2}$, there is $C_{\nu,\nu}$ so that

$$|J^{(\nu,\nu)}(x)| \le C_{\nu,\nu} \log(|x|^{-1}) \tag{6.3.105}$$

7. Prove that $|D^\alpha J^{(\nu=1,\nu)}| \le C_\alpha |x|^{-\nu-|\alpha|}$

8. Prove (h) of Proposition 6.3.1 when $\nu + 2\ell \ge s > \nu + \ell$.

9. Prove completeness of $H^{s,p}$ given by (6.3.15).

10. Prove C_0^∞ is dense in each $H^{s,p}$.

11. Let $L \in (H^{s,2})^*$.

(a) Prove there is $g \in H^{-s,2}$ so that for every $f \in H^{s,2}$, $\overline{\hat{g}}\hat{f} \in L^2(\mathbb{R}^\nu)$ and

$$L(f) = \int \overline{\hat{g}(k)} f(k) d^\nu k \tag{6.3.106}$$

(b) If $s < 0$ and $f \in \mathcal{S}(\mathbb{R}^\nu)$, prove that $g, f \in L^2(\mathbb{R}^\nu)$ and

$$L(f) = \int \overline{g(x)} f(x) d^\nu x \qquad (6.3.107)$$

12. (a) If $s > 0$, prove that $J^{(\nu,s)} *$ maps L^p to L^p.

   (b) If $s > 0$, prove that $J^{(\nu,s)} *$ maps $H^{t,p}$ to $H^{t,p}$ for all $t \in \mathbb{R}$, $p \in (1,\infty)$.

   (c) If $p \in (1,\infty)$ and $s_1 > s_2$, prove that $H^{s_1,p} \subset H^{s_2,p}$.

13. Let p, q, r, θ, s, t obey (6.3.34). Prove that $H^{s,q} \cap L^r \subset H^{t,p}$ and that for some C

$$\|f\|_{t,p} \le C \|f\|_{s,q}^\theta \|f\|_r^{1-\theta} \qquad (6.3.108)$$

14. This will provide a direct proof of Nash's inequality in $\mathbb{R}^\nu$, $\nu \ge 2$:

$$\|f\|_2 \le C \|\vec{\nabla} f\|_2^{\nu/\nu+2} \|f\|_1^{2/\nu+2} \qquad (6.3.109)$$

   (a) Prove that

$$\|f\|_2^2 \le \tau_\nu R^\nu \|\widehat{f}\|_\infty^2 + R^{-2} \|\widehat{\nabla f}\|_2^2$$

   (b) Prove (6.3.109) by using $\|\widehat{f}\|_\infty \le (2\pi)^{-\nu/2} \|f\|_1$ and optimizing over R.

15. Prove a Gagliardo–Nirenberg inequality from (6.3.42) by applying (6.3.42) to $u_\lambda(x) = u(\lambda x)$ and then optimizing over λ.

16. (a) Let $s < \nu$, $1 \le r < \nu/\nu - s$. Prove that $H^{s,1} \subset L^r$ by using Young's convolution inequality.

   (b) If $-\infty < t < s < t + \nu$, prove that $H^{s,1} \subset H^{t,r}$.

17. If s, q, and y obey the hypothesis of Lemma 6.3.9, prove that $H^{s+s_1,q} \subset H^{s_1,y}$ for any $s_1 \in \mathbb{R}$.

   The methods of this section relying on weak Young inequalities don't work if the initial space is an L^1 space. The next problems explore some results when $p = 1$. The first pair are classical and may be regarded as a general part of "Sobolev estimates."

18. Let π_j map $\mathbb{R}^\nu$ to $\mathbb{R}^{\nu-1}$ by dropping coordinate j. In this problem, the reader will inductively prove that if $u_1, \ldots, u_\nu \in L^{\nu-1}(\mathbb{R}^{\nu-1})$ and u on $\mathbb{R}^\nu$ is defined by

$$u(x) = \prod_{j=1}^\nu u_j\big(\pi_j(x)\big) \qquad (6.3.110)$$

then

$$\|u\|_{L^1(\mathbb{R}^\nu)} \le \prod_{j=1}^\nu \|u_j\|_{L^{\nu-1}(\mathbb{R}^{\nu-1})} \qquad (6.3.111)$$

(a) Prove the result for $\nu = 2$.

    Now suppose the result is known for $\nu = M$ and consider $\nu = M + 1$.

(b) Prove that for each fixed value of x_{M+1}:

$$\int_{\mathbb{R}^M} u(x)dx_1 \ldots dx_M \leq \|u_{M+1}\|_{L^M(\mathbb{R}^M)}$$

$$\left(\int_{\mathbb{R}^M} \prod_{j=1}^{M} |u_j(\pi_j(x))|^{M/M-1} dx_1 \ldots dx_M \right)^{\frac{M-1}{M}} \quad (6.3.112)$$

(*Hint*: Hölder's inequality.)

(c) By the induction hypothesis, prove that (for x_{M+1} fixed)

$$\int_{\mathbb{R}^M} \prod_{j=1}^{M} |u_j(\pi_j(x))|^{M/M-1} dx_1 \ldots dx_M \leq$$

$$\prod_{j=1}^{M} \left(\int_{\mathbb{R}^{M-1}} |u_j(\pi_j(x))|^M dx_1 \ldots \widehat{dx_j} \ldots dx_M \right)^{1/M-1} \quad (6.3.113)$$

where $\widehat{dx_j}$ means no dx_j is included.

(d) Prove that for x_{M+1} fixed,

$$\int_{\mathbb{R}^M} |u(x)| dx_1 \ldots dx_M \leq \|u_{M+1}\|_{L^M(\mathbb{R}^M)}$$

$$\prod_{j=1}^{M} \left(\int_{\mathbb{R}^{M-1}} |u_j(\pi_j(x))|^M dx_1 \ldots \widehat{dx_j} \ldots dx_M \right)^{1/M} \quad (6.3.114)$$

(e) Complete the proof by integrating dx_M (and using Hölder).

19. In this problem the reader will prove for $u \in \mathcal{S}(\mathbb{R}^\nu)$, $\nu \geq 2$, we have with $p = \frac{\nu}{\nu-1}$ (so $\frac{1}{p} + \frac{1}{\nu} = 1$) that

$$\|u\|_p \leq \|\vec{\nabla} u\|_1 \quad (6.3.115)$$

Throughout $u \in \mathcal{S}(\mathbb{R}^\nu)$

(a) Prove that with $\rho_1 : \mathbb{R}^{\nu-1} \times \mathbb{R} \to \mathbb{R}^\nu$ given by

$$[\rho_1(y,s)]_k = \begin{cases} y_k, & k < j \\ s, & k = j \\ y_{k+1}, & k > j \end{cases}$$

(i.e., inserting s as the j-th coordinate), one has that

$$|u(x)| \leq \int_{-\infty}^{\infty} \left| \frac{\partial u}{\partial x_j}(\rho_1(\pi_j(x), s)) \right| ds \quad (6.3.116)$$

(b) Let u_j be defined on $\mathbb{R}^{\nu-1}$ by

$$u_j(y) = \int_{-\infty}^{\infty} \left| \frac{\partial u}{\partial x_j}(\rho_j(y,s)) \right| ds \qquad (6.3.117)$$

Prove that

$$|u(x)|^{\frac{\nu}{\nu-1}} \le \prod_{j=1}^{\nu} |u_j(\pi_j(x))|^{\frac{1}{\nu-1}} \qquad (6.3.118)$$

(c) Prove that

$$\|u\|_p \le \prod_{j=1}^{\nu} \left\| \frac{\partial u}{\partial x_j} \right\|_1^{1/\nu} \qquad (6.3.119)$$

and then that (6.3.115) holds. (*Hint*: Problem 18.)

20. Let $1 < q < \nu$ and $\beta = \frac{q(\nu-1)}{\nu-q}$, $r = \frac{\nu q}{\nu-q}$. By applying (6.3.115) to $|f|^\beta$, prove that

$$\|f\|_r \le \beta \|\vec{\nabla} f\|_q \qquad (6.3.120)$$

which is the general homogeneous Sobolev inequality when $s = 1$.

21. Let $\ell \in \mathbb{N}$ and $s > \ell + \nu$, prove that $H^{s,1} \subset C_\infty^\ell$ (*Hint*: Use Problem 16 to go through $H^{s-\varepsilon,p_\varepsilon}$ with $p_\varepsilon > 1$.)

22. For Ω open and bounded in $\mathbb{R}^\nu$, let $H_0^{s,2}(\Omega)$ be the closure of $C_0^\infty(\Omega)$ in $H^{s,2}$.

    (a) If $s > t$, prove that the unit ball in $H_0^{s,2}(\Omega)$ is compact in $H^{t,2}(\Omega)$. (You'll need Theorem 3.1.17 of Part 4.)

    (b) If $s < \frac{\nu}{2}$ and $2 \le y < \frac{2\nu}{\nu-2s}$, prove that the unit ball in $H_0^{s,2}(\Omega)$ is compact in $H^{s,y}$.

6.4. The Calderón–Zygmund Method

In this section, we'll discuss a method for studying certain *singular integral operators* on $L^p(\mathbb{R}^\nu)$, $1 < p < \infty$. Our results will include extensions of the M. Riesz L^p-boundedness of the Hilbert transform to higher dimensions. Our operators will formally have the form

$$(Tf)(x) = \int K(x-y)f(y)d^\nu y \qquad (6.4.1)$$

The nonsingular integral operator theory works for $K \in L^1$ but our K's will not be L^1. That means that typically the integral will not converge absolutely. The Hilbert transform should come to mind. There are two basic assumptions we'll make on K (actually by integrating in from infinity (6.4.3) & $K \to 0$ at $\infty \Rightarrow$ (6.4.2) with $A = B\nu^{-1}$).

(1) $|K(x)| \leq A|x|^{-\nu}$. $\qquad\qquad$ (6.4.2)

(2) K is C^1 on $\mathbb{R}^\nu \setminus \{0\}$ and $|\vec{\nabla} K(x)| \leq B|x|^{-\nu-1}$. $\qquad$ (6.4.3)

We'll sometimes make a third assumption:

(3) For each $0 < r < \infty$,

$$\int_{S^{\nu-1}} K(r\omega)\, dS(\omega) = 0 \qquad\qquad (6.4.4)$$

where dS is the normalized rotation-invariant measure on the sphere $S^{\nu-1}$. We note that one of the basic results, Theorem 6.4.5 below does not require (6.4.4).

A C^1-function on $\mathbb{R}^\nu \setminus \{0\}$, K, obeying (6.4.4) with $K(\lambda x) = \lambda^{-\nu} K(x)$ for all $x \in \mathbb{R}^\nu \setminus \{0\}$, $\lambda \in (0, \infty)$ is called a *classical Calderón–Zygmund kernel* and the corresponding T is a *classical Calderón–Zygmund operator*.

When (6.4.4) holds, we'll see that one can define Tf for $f \in \mathcal{S}(\mathbb{R}^\nu)$ by a principle-part integral. But there will be other cases where T will have a different direct definition, for example $Tf = (F\hat{f})^{\vee}$ for some polynomially bounded function F in which case we can define $Tf \in L^2$ for $f \in \mathcal{S}(\mathbb{R}^\nu)$. The question then is what we mean by $K(x - y)$ being the integral kernel of T.

Definition. Let T be a linear map of $\mathcal{S}(\mathbb{R}^\nu)$ to $L^1_{\mathrm{loc}}(\mathbb{R}^\nu)$. We say a function $K(x, y)$ on $\mathbb{R}^\nu \times \mathbb{R}^\nu$ is an *off-diagonal kernel* for T if and only if

(i) For each compact set, Ω, and $\varepsilon > 0$,

$$\sup_{\substack{x \in \Omega \\ |y-x| > \varepsilon}} |K(x, y)| < \infty \qquad\qquad (6.4.5)$$

(ii) If $f \in \mathcal{S}(\mathbb{R}^\nu)$ and $\mathrm{supp}(f) \subset \Omega$ compact, then for $x \notin \Omega$

$$(Tf)(x) = \int K(x, y)\, f(y)\, d^\nu y \qquad\qquad (6.4.6)$$

We note that if T extends to a bounded map of L^r to L^r for $1 \leq r < \infty$ (i.e., for all $f \in \mathcal{S}$, $Tf \in L^p$ and for some C, $\|Tf\|_r \leq C\|f\|$), then (6.4.6) extends to $f \in L^r$ with $\mathrm{supp}\, f \subset \Omega$.

One can ask to what extent K determines T and the answer is up to a bounded multiplication operator, i.e., in Problem 2, the reader will show if T_1 and T_2 are bounded maps of L^r to L^r for some $1 \leq r < \infty$ and they have the same off-diagonal kernel, then for $H \in L^\infty$, $(T_2 f)(x) = (T_1 f)(x) + H(x) f(x)$ (and conversely T_1 and $T_1 + H$ have the same off-diagonal kernel).

While historically the kernels studied are of the form $K(x - y)$ as indicated, the ideas are also applicable to operators formally of the form

$$(Tf)(x) = \int K(x, y) f(y)\, d^\nu y \qquad\qquad (6.4.7)$$

which we'll refer to as the two-variable kernel. Conditions (1)–(3) are then replaced by

(1TV) $|K(x,y)| \leq A|x-y|^{-\nu-1}$. $\qquad\qquad$ (6.4.8)

(2TV) $|\vec{\nabla}_x K(x,y)| + |\vec{\nabla}_y K(x,y)| \leq B|x-y|^{-\nu-1}$. $\qquad$ (6.4.9)

(3TV) For each $0 < r < \infty$ and $x, y \in \mathbb{R}^\nu$,

$$\int_{S^{\nu-1}} K(x, x - r\omega)dS(\omega) = 0 = \int K(y + r\omega, x)dS(\omega) \qquad (6.4.10)$$

We'll have one major general result in this section: if (1)–(2) hold and T, suitably defined, maps L^{r_0} to itself for some r_0 in $(1, \infty)$, then T maps L^r to itself for all $r \in (1, \infty)$. In the next section (see Theorem 6.5.8), we'll complete the analysis when (1)–(3) hold by showing they imply T is bounded on L^2.

In many applications, one knows $\widehat{K}$ is an L^∞-function so boundedness on L^2 will be easy. We'll prove our main result by proving an L^1 to L^1_w bound and then use Marcinkiewicz interpolation (Theorem 2.2.10).

Two canonical examples are the singular Riesz-potential, $K(x) = x_j |x|^{-\nu-1}$ (this is a classical Calderón–Zygmund kernel) and $K(x) = \frac{\partial}{\partial x_j} J^{(1,\nu)}(x)$, the gradient of the Bessel potential of order 1 (here we'll easily get an L^{r_0} bound when $r_0 = 2$)—the L^p-boundedness in the later case was used in the last section (see, for example Theorem 6.3.2).

We begin by defining T which is formally (6.4.1). When (6.4.4) holds, the following works:

Proposition 6.4.1. *Let K obey (1)–(3) above. For $\varepsilon > 0$ and $f \in \mathcal{S}(\mathbb{R}^\nu)$, define*

$$(T_\varepsilon f)(x) = \int_{|x-y|>\varepsilon} K(x-y)f(y)d^\nu y \qquad (6.4.11)$$

Then, for any $R > 0$, the following pointwise limit exists:

$$(Tf)(x) = \lim_{\varepsilon \downarrow 0}(T_\varepsilon f)(x) = \int K(y)[f(x-y) - f(x)\chi_{|y|\leq R}(y)]d^\nu y \quad (6.4.12)$$

Remarks. 1. In fact, using the ideas of this section, one can prove when (1)–(3) and an L^{r_0} bounded condition holds, then for any f in L^p, $1 < p < \infty$, $T_\varepsilon f$ has an L^p-limit (Problem 1).

2. The proof shows the limit is uniform in x.

Proof. By (6.4.4),

$$\int K(y)\chi_{|y|\leq R}(y)d^\nu y = 0 \qquad (6.4.13)$$

Thus,

$$(T_\varepsilon f)(x) = \int_{|y|>\varepsilon} K(y)f(x-y)d^\nu y \qquad (6.4.14)$$

$$= \int_{|y|>\varepsilon} K(y)[(f(x-y) - f(x)\chi_{|y|\leq R}(y)]d^\nu y \qquad (6.4.15)$$

Since $f \in \mathcal{S}(\mathbb{R}^\nu)$, the integral over $|y| > R$ is uniformly integrable. For the integral over $|y| < R$, we note that

$$|f(x-y) - f(x)| \leq |y|\,\|\nabla f\|_\infty \qquad (6.4.16)$$

so by (6.4.2), the integral in (6.4.12) over $|y| < R$ is convergent so the limit as $\varepsilon \downarrow 0$ exists. $\qquad\square$

If (6.4.4) is not assumed, one can obviously define $\langle f, Tg\rangle$ if $f, g \in \mathcal{S}(\mathbb{R}^\nu)$ with $\text{dist}(\text{supp}(f), \text{supp}(g)) > 0$. It is easy to see (Problem 2), that if one proves that for such f, g, one has $\langle f, Tg\rangle = \langle f, Sg\rangle$ for S mapping $L^r(\mathbb{R}^\nu)$ to $L^r(\mathbb{R}^\nu)$, then S is uniquely determined up to an L^∞-multiplication operator, i.e., $\langle f, (S_1 - S_2)g\rangle = 0$ for all such f, g implies $S_1 - S_2$ is a multiplication operator. Below when we say T is bounded from L^r to L^r we mean in this sense.

The key to controlling Tf will be to decompose, for each $\alpha > 0$, any nonnegative f in L^1 into $b + g$ (for "bad" and "good") by singling out a carefully chosen disjoint family of bad dyadic cubes (a dyadic cube is a product of $(\frac{j}{2^n}, \frac{j+1}{2^n}]$, some $n \in \mathbb{Z}$ (not just $\mathbb{Z}_+$) and $j \in \mathbb{Z}$ with the same n for each term in the product). The bad cubes, $\{Q_k\}_{k=1}^\infty$, will have $f(x)$ large in the sense that

$$g_k = \frac{1}{|Q_k|}\int_{Q_k} f(x)d^\nu x \qquad (6.4.17)$$

is larger than α. The key will be that we can pick the Q_k, so that g_k isn't too large, indeed bounded by $2^\nu \alpha$. b will live on $\bigcup_k Q_k$ and $b = f - g_k$ on Q_k. By a variant of the Lebesgue differentiation theorem, $g(x) = f(x) \leq \alpha$ a.e. on $\mathbb{R} \setminus \bigcup_k Q_k$, so overall $\|g\|_\infty \leq 2^\nu \alpha$. By construction $\|g\|_1 = \|f\|_1 < \infty$, so $g \in L^r$. Because $\int_{Q_k} b(x)d^\nu x = 0$, we can write $\int_{y \in Q_k} K(x-y)b(y)d^\nu y = \int [K(x-y) - K(x-y_0)]b(y)d^\nu y$ to control Tb.

As indicated, we need a variant of the Lebesgue differentiation theorem. By standard cubes of side 2^{-n} we mean cubes of the form $\prod_{i=1}^\nu (\frac{j_i}{2^n}, \frac{j_i+1}{2^n}]$, $j_i \in \mathbb{Z}$. For each x, let $C_n(x)$ be that standard cube of side 2^{-n} with $x \in C_n(x)$. We have the following lemma:

Lemma 6.4.2. *For each $f \in L^1(\mathbb{R}^\nu)$, for a.e. x, we have*

$$f(x) = \lim_{n \to \infty} 2^{n\nu} \int_{C_n(x)} f(y) d^\nu y \tag{6.4.18}$$

Remark. In terms of dyadic filtration of Example 2.10.2(iii), this is just a martingale convergence theorem.

Proof. $y \in C_n(x) \Rightarrow |x - y| \le 2^{-n}\sqrt{\nu}$, so (6.4.18) holds if f is a continuous function of compact support. As usual, it suffices to show that

$$(M_C f)(x) = \sup_n 2^{n\nu} \int_{C^n(x)} |f(y)| d^\nu y \tag{6.4.19}$$

is bounded as a map of L^1 to L^1_w.

As noted above, $C_n(x)$ is contained in the ball of radius $2^{-n}\sqrt{\nu}$ about x. The volume of this ball is $2^{-n\nu}\nu^{\nu/2}\tau_\nu$ (where τ_ν is the volume of the unit ball). This implies that

$$(M_C f)(x) \le \nu^{\nu/2}\tau_\nu M_{\mathrm{HL}}f(x) \tag{6.4.20}$$

so the Hardy–Littlewood maximal inequality proves what we need here. $\square$

Theorem 6.4.3 (Calderón–Zygmund Decomposition Theorem). *Given a nonnegative function $f \in L^1(\mathbb{R}^\nu)$ and $\alpha \in (0, \infty)$, there is a disjoint family $\{Q_j\}_{j=1}^J$ (J finite or $J = \infty$) of standard dyadic cubes (perhaps of different sizes) and measurable functions, b and g (with $g \ge 0$), so that*

(a) $f = b + g$ (6.4.21)

(b) $\sum_j |Q_j| \le \alpha^{-1}\|f\|_1$ (6.4.22)

(c) *On $\mathbb{R}^\nu \setminus \bigcup_{j=1}^J Q_j$,*

$$f(x) = g(x) \in [0, \alpha], \quad x \notin \bigcup_{j=1}^\infty Q_j \tag{6.4.23}$$

(d) $g(x) = g_j$, *a constant, on Q_j where*

$$g_j = \frac{1}{|Q_j|} \int_{Q_j} f(x) d^\nu x \tag{6.4.24}$$

(e) $\|g\|_\infty \le 2^\nu \alpha$ (6.4.25)

(f) *If $b_j = b \upharpoonright Q_j$, then*

$$\int b_j(x) d^\nu x = 0 \tag{6.4.26}$$

(g) $\|g\|_1 = \|f\|_1, \quad \|b\|_1 \le 2\|f\|_1$ (6.4.27)

(h) $\|b_j\|_1 \le 2^{\nu+1}\alpha|Q_j|$ (6.4.28)

(i) *For any r in $[1, \infty)$, we have that*

$$\frac{\|g\|_r^r}{\alpha^r} \leq 2^{\nu(r-1)} \frac{\|f\|_1}{\alpha} \tag{6.4.29}$$

Remarks. 1. (6.4.23) implies that

$$\{x \mid f(x) > \alpha\} \subset \cup_{j=1}^{\infty} Q_j \tag{6.4.30}$$

2. This theorem is stated for functions on $\mathbb{R}^\nu$. That we need all of $\mathbb{R}^\nu$ is only used at the initial choice of n_0 with (6.4.31) below. We can look at functions f defined on a dyadic cube, C_0, and replace $\mathbb{R}^\nu$ by C_0 so long as $\alpha > |C_0|^{-1} \int_{C_0} f(y) d^\nu y$. We start with n_0 so $|C_0| = 2^{-n_0\nu}$ and (6.4.31) holds. One can easily modify to allow C_0 to be any size cube so long as the Q_j are then among the cube, obtained by subdividing C_0 successively into 2^ν pieces with $|Q_j| = 2^{-n_j\nu}|C_0|$.

Proof. Pick n_0 so negative that

$$2^{n_0\nu} \int f(y) d^\nu y < \alpha \tag{6.4.31}$$

Thus, if Q is any cube of side 2^{-n_0}

$$|Q|^{-1} \int_Q f(y) d^\nu y < \alpha \tag{6.4.32}$$

Look at all the standard cubes of side 2^{-n_0-1}. If one of these cubes has $|Q|^{-1} \int_Q f(y) d^\nu y \geq \alpha$, use it as a Q_j. Since

$$|Q_j| \leq \alpha^{-1} \int_{Q_j} f(y) d^\nu y \tag{6.4.33}$$

we have (6.4.22) for these cubes. Thus, there are only finitely many of these cubes of volume $2^{-\nu(n_0-1)}$ added to the collection.

Next, subdivide any cubes not included in the first set of cubes into standard cubes of side 2^{n_0-2}. Add to the collection of cubes, the cubes of this size (not contained in earlier cubes!) for which (6.4.33) holds. Iterate to get a disjoint collection $\{Q_j\}_{j=1}^J$ (J finite or infinite) so that (6.4.33) holds for each cube and so (6.4.22) holds.

Define g_j by (6.4.24). Here is the key observation. If Q_j has sides 2^{-k_j}, then $Q_j \subset \widetilde{Q}_j$ a standard cube of side 2^{-k_j+1}. Since we didn't stop with $\widetilde{Q}_j$, we know that

$$|\widetilde{Q}_j|^{-1} \int_{\widetilde{Q}_j} f(y) d^\nu y \leq \alpha \tag{6.4.34}$$

Since $|Q_j| = 2^{-\nu}|\widetilde{Q}_j|$ and $\int_{Q_j} f(y) d^\nu y \leq \int_{\widetilde{Q}_j} f(y) d^\nu y$, this implies that

$$g_j \leq 2^\nu \alpha \tag{6.4.35}$$

If $x \notin \bigcup_{j=1}^{J} Q_j$, then

$$\sup_n \frac{1}{|C_n(x)|} \int_{C_n(x)} f(x) d^\nu x \leq \alpha \qquad (6.4.36)$$

so, by Lemma 6.4.2, for a.e. x,

$$x \notin \bigcup_{j=1}^{J} Q_j \Rightarrow f(x) \leq \alpha \qquad (6.4.37)$$

We now define

$$g(x) = \begin{cases} g_j, & x \in Q_j \\ f(x,) & x \notin \bigcup_{j=1}^{J} Q_j \end{cases} \qquad (6.4.38)$$

$$b(x) = f(x) - g(x) \qquad (6.4.39)$$

(a), (c), and (d) hold by definition. We confirm the other conditions.

(b) follows from (6.4.33). Since $x \notin \cup Q_j$ implies $g(x) = f(x) \leq \alpha$, (6.4.35) implies (e). By (6.4.24),

$$\int_{Q_j} g(x) d^\nu x = \int_{Q_j} f(x) d^\nu x \qquad (6.4.40)$$

so (f) holds. Since $f = g$ on $\mathbb{R} \setminus \bigcup_{j=1}^{J} Q_j$ and (6.4.26) holds, we get $\|g\|_1 = \|f\|_1$ and then $\|b\|_1 \leq \|f\|_1 + \|g\|_1$ implies (g). Similarly $\|b_j\|_1 \leq 2g_j|Q_j|$ by (6.4.26), so (6.4.35) implies (h). We get (i) by

$$\frac{\|g\|_r^r}{\alpha^r} \leq \frac{\|g\|_\infty^{r-1} \|g\|_1}{\alpha^r} \leq \frac{(2^\nu \alpha)^{r-1} \|g\|_1}{\alpha^r}$$

$\square$

As an aside before discussing weak-L^1 bounds for Calderón–Zygmund operators, we provide the promised proof of the John–Nirenberg inequality for BMO functions. Recall that the BMO norm on locally L^1-functions on $\mathbb{R}^\nu$ (modulo constants) is defined by

$$\|f\|_{\text{BMO}} = \sup_C \frac{1}{|C|} \int_C |f(x) - f_C| d^\nu x \qquad (6.4.41)$$

$$f_C = \frac{1}{|C|} \int_C f(x) d^\nu x \qquad (6.4.42)$$

where C is a cube with edges parallel to the coordinates axes. (We'll call these standard cubes.) The inequality we seek to prove is

$$\frac{1}{|C|} \left| \{x \in C \mid |f(x) - f_C| > s \|f\|_{\text{BMO}} \} \right| \leq A e^{-Bs} \qquad (6.4.43)$$

where

$$A = e, \quad B = (2^\nu e)^{-1} \qquad (6.4.44)$$

for all $f \in \text{BMO}$ and standard cubes C.

Proof of Theorem 5.11.14. Define

$$F(s) = \sup_{\substack{f \in \mathrm{BMO} \\ C, \|f\|_{\mathrm{BMO}} = 1}} \frac{1}{|C|} \left| \left\{ x \in C \mid \frac{1}{|C|} |f(x) - f_C| > s \right\} \right| \qquad (6.4.45)$$

Certainly the sup is finite since $F(s) \leq 1$ is trivial (and with a little effort the a priori estimate $F(s) \leq s^{-1}$ can be proven since $|f - f_C| \in L^1(C, d^\nu x/|C|)$ with norm at most 1).

Fix now, f, with $\|f\|_{\mathrm{BMO}} = 1$ and standard cube C. By adding a constant, suppose $f_C = 0$. Pick $\alpha > 1$ (we'll eventually pick $\alpha = e$) and apply the Calderón–Zygmund decomposition to the function

$$h = \frac{1}{|C|} |f - f_C| = \frac{1}{|C|} |f| \qquad (6.4.46)$$

and α. Since $\|f\|_{\mathrm{BMO}} = 1$, $\|h\|_1 \leq 1$ and we have (6.4.30) and can use the construction on C rather than $\mathbb{R}^\nu$.

Pick $s > 0$. Since $s + 2^\nu \alpha > \alpha$, by (6.4.30),

$$\left\{ x \in C \mid \frac{1}{|C|} |f(x)| > s + 2^\nu \alpha \right\} \subset \bigcup_{j=1}^{\infty} Q_j \qquad (6.4.47)$$

By the construction of the Q_j, $h_{Q_j} \leq 2^\nu \alpha$. Thus, if $\frac{1}{|C|} |f(x)| > s + 2^\nu \alpha$, $h(x) - h_{Q_j} \geq s + 2^\nu \alpha - 2^\nu \alpha \geq s$. We conclude that

$$\left| \left\{ x \in C \mid \frac{1}{|C|} |f(x)| > s + 2^\nu \alpha \right\} \right| \leq \sum_j \left| \left\{ x \in Q_j \mid h(x) - h_{Q_j} \geq s \right\} \right| \qquad (6.4.48)$$

$$\leq \sum_j |Q_j| F(s) \leq \alpha^{-1} F(s) \qquad (6.4.49)$$

by (6.4.22) and $\|h\|_1 \leq 1$.

Since f is an arbitrary BMO function, we conclude

$$F(s + 2^\nu \alpha) \leq \alpha^{-1} F(s) \qquad (6.4.50)$$

Now take $\alpha = e$. We conclude that if $B = (2^\nu e)^{-1}$ and $F(s) \leq A e^{-Bs}$, then

$$F(s + 2^\nu e) \leq A e^{-1} e^{-Bs} = A e^{-B(s + 2^\nu e)} \qquad (6.4.51)$$

since $e^{-B(2^\nu e)} = e^{-1}$. If $0 \leq s \leq 2^\nu e$, the trivial bound $F(s) \leq 1$ implies $F(s) \leq A e^{-Bs}$ if $A = e^{B 2^\nu e}$. Thus, we get (6.4.43) by iteration. $\qquad \square$

We return to our main subject of proving a bound on singular integral operators. As preparation for proving an L^1 to weak-L^1 bound on kernels obeying (6.4.2)–(6.4.3), we make use of (6.4.3).

Lemma 6.4.4. *Let h be an L^1-function supported in a ball $B_R(x_0)$ about some point x_0. Suppose that*

$$\int h(y)d^\nu y = 0 \tag{6.4.52}$$

Then, with T given by (6.4.1) and K obeying (6.4.3),

$$\int_{|x-x_0|\geq 2R} |(Th)(x)|d^\nu x \leq 2^\nu B\sigma_\nu \|h\|_1 \tag{6.4.53}$$

with σ_ν the area of the unit sphere in $\mathbb{R}^\nu$.

Remarks. 1. The reader may be concerned that we've only defined Tf for $f \in \mathcal{S}(\mathbb{R}^\nu)$ and we have not assumed h is smooth. We note that since $|x - x_0| \geq 2R$ and $|y| \leq R$, K is bounded in (6.4.1) for all y in $\mathrm{supp}(f)$ so $h \in L^1$ implies Th defined by (6.4.1) is well-defined.

2. What is critical is that (6.4.53) is independent of R connected to the scale covariance of the estimate (6.4.3).

3. All that is needed for this lemma is that K be an off-diagonal kernel for T.

Proof. By translation invariance of T, we can suppose for notational simplicity that $x_0 = 0$. By (6.4.52) for any x

$$Th(x) = \int [K(x-y) - K(x)]h(y)d^\nu y \tag{6.4.54}$$

We note that for $|x| \geq 2R, |y| \leq R$

$$|K(x-y) - K(x)| = \left| \int_0^1 \frac{d}{d\theta}[K(x-\theta y)]d\theta \right|$$

$$\leq \int_0^1 |y \cdot \nabla K(x-\theta y)|d\theta$$

$$\leq RB(|x| - R)^{-\nu-1} \tag{6.4.55}$$

where (6.4.55) used (6.4.3), $|x - \theta y| \geq |x| - R$ and $|y| \leq R$.

Thus, for each y with $|y| \leq R$,

$$\int_{|x|\geq 2R} |K(x-y) - K(x)|d^\nu x \leq BR\sigma_\nu \int_{2R}^\infty (r-R)^{-\nu-1}r^{\nu-1}dr$$

$$\leq 2^{\nu+1}BR\sigma_\nu \int_{2R}^\infty r^{-2}dr \tag{6.4.56}$$

$$= 2^\nu \sigma_\nu B \tag{6.4.57}$$

(6.4.56) used $r - R \geq \frac{1}{2}r$ if $r \geq 2R$. (6.4.53) is immediate from this estimate and (6.4.54). $\qquad\square$

We are now ready for one of the main general results of the Calderón–Zygmund theory:

Theorem 6.4.5 (Calderón–Zygmund Weak-L^1 Estimate). *Suppose K is a C^1-function on $\mathbb{R}^\nu \setminus \{0\}$ obeying (6.4.2)–(6.4.3) and for some $r \in (1, \infty)$, we have an operator T so that for all $f \in \mathcal{S}(\mathbb{R}^\nu)$ (and so, by taking limits, all $f \in L^r(\mathbb{R}^\nu)$), that*

$$\|Tf\|_r \leq C\|f\|_r \tag{6.4.58}$$

Suppose that K is an off-diagonal kernel for T. Then, for a constant D, we have for all $f \in L^1$ that

$$|\{x \mid |Tf(x)| \geq \alpha\}| \leq D\|f\|_1 \alpha^{-1} \tag{6.4.59}$$

and for all $p \in (1, \infty)$, there is D_p with

$$\|Tf\|_p \leq D_p\|f\|_p \tag{6.4.60}$$

Remarks. 1. Our estimates below (certainly not optimal) show that

$$D \leq C^r 2^{r+1} 2^{\nu(r-1)} + 2\tau_\nu \nu^{\nu/2} + 2^{\nu+2} \sigma_\nu B \tag{6.4.61}$$

which doesn't scale properly under multiplying K by a constant γ! By using (6.4.61) for γK and optimizing γ, one finds

$$D \leq c_{\nu,r}(C + B) \tag{6.4.62}$$

for a constant $c_{\nu,r}$.

2. Note that we do not assume (6.4.4).

3. The interpolation estimates as obtained directly hold on $(1, r)$ and (r, ∞) and diverge as $p \to r$. However, once we have them, we can repeat the argument for $r_1 \in (r, \infty)$ and so get boundedness on $(1, r_1)$. This shows D_p in (6.4.39) is bounded on each interval $[p_0, p_0/(p_0 - 1)]$ for any $p_0 > 1$.

Proof. Without loss, we can suppose $f \geq 0$. Fix α and let $g, b, \{Q_j\}_{j=1}^J$ be the functions and cubes in the Calderón–Zygmund decomposition for that α. Let $\widetilde{Q}_j$ be the ball of radius $2 \cdot 2^{-n_j}(\frac{1}{2}\sqrt{\nu}) \equiv 2R_j$ with center the center of Q_j where Q_j has side 2^{-n_j}. Thus $\widetilde{Q}_j$ is a ball with twice the radius of the smallest ball containing Q_j. Clearly (with τ_ν the volume of the unit ball)

$$|\widetilde{Q}_j| = \tau_\nu \nu^{\nu/2} |Q_j| \tag{6.4.63}$$

As usual, write $Tf = Tb + Tg$ and note that

$$\{x \mid |Tf(x)| > \alpha\} \subset \{x \mid |Tb(x)| \geq \tfrac{\alpha}{2}\} + \{x \mid |Tg(x)| \geq \tfrac{\alpha}{2}\} \tag{6.4.64}$$

and thus

$$\{x \mid |Tf(x)| > \alpha\} \leq ① + ② + ③ \tag{6.4.65}$$

$$① = |\{x \mid |Tg(x)| \geq \tfrac{\alpha}{2}\}|,$$

$$② = \left|\bigcup_{j=1}^{J} \tilde{Q}_j\right|, \tag{6.4.66}$$

$$③ = \{x \mid |Tb(x)| > \tfrac{\alpha}{2}; x \notin \cup_{j=1}^{J}\tilde{Q}_j\}$$

Since $\|Tg\|_r^r \leq C^r\|g\|_r^r$, we have

$$① \leq 2^r C^r\|g\|_r^r\alpha^{-r} \leq 2^r 2^{\nu(r-1)}C^r\|f\|_1\alpha^{-1} \tag{6.4.67}$$

by (6.4.29). By (6.4.22) and (6.4.63)

$$② = |\cup_{j=1}^{J}\tilde{Q}_j| \leq \sum_{j=1}^{J}|\tilde{Q}_j| \leq \tau_\nu\nu^{\nu/2}\|f\|_1\alpha^{-1} \tag{6.4.68}$$

$$③ \leq \tfrac{2}{\alpha}\int_{\mathbb{R}\setminus\widetilde{\cup Q_k}}|(Tb)(x)|d^\nu x \tag{6.4.69}$$

$$\leq \tfrac{2}{\alpha}\sum_j \int_{\mathbb{R}\setminus\widetilde{\cup Q_k}}|(Tb_j)(x)|d^\nu x \tag{6.4.70}$$

$$\leq \tfrac{2}{\alpha}\sum_j \int_{\mathbb{R}\setminus\tilde{Q}_j}|(Tb_j)(x)|d^\nu x \tag{6.4.71}$$

since $\mathbb{R}\setminus\cup\tilde{Q}_k \subset \mathbb{R}\setminus\tilde{Q}_j$. If x_j is the center of Q_j, $Q_j \subset B_{R_j}(x_j)$ and $\mathbb{R}\setminus\tilde{Q}_j$ is the exterior of $B_{2R_j}(x_j)$. Thus, the lemma applies and, by (6.4.53)

$$③ \leq \tfrac{2}{\alpha}2^\nu\sigma_\nu B\sum_j\|b_j\|_1 = \tfrac{2}{\alpha}2^\nu\sigma_\nu B\|b\|_1 \leq \tfrac{4}{\alpha}2^\nu\sigma_\nu B\|f\|_1 \tag{6.4.72}$$

by (6.4.26). We have thus proven (6.4.59).

The Marcinkiewicz interpolation theorem (Theorem 2.2.10) implies (6.4.3) for $p \in (1, r)$. T^*, the (Banach space) adjoint of T also is an integral operator with kernel $K(y - x)$ and it is bounded from $L^{r'}$ to $L^{r'}$ where r' is the dual of r. Thus, by the above T^* is bounded from L^p to itself if $p \in (1, r')$ so, T is bounded for $p \in (r, \infty)$. Thus we have (6.4.60) for all p. $\qquad\square$

With essentially no change in the proof (Problem 3), one can replace conditions (1)–(2) by (1TV)–(2TV):

Theorem 6.4.6. *Suppose $K(x, y)$ is C^1 on $\{(x, y) \in \mathbb{R}^\nu \times \mathbb{R}^\nu \mid x \neq y\}$ and obeys (6.4.8)/(6.4.9). Then the operator T of (6.4.7) obeys the conclusions of Theorem 6.4.5, i.e., an L^r bound implies an L^1_w and L^p, $1 < p < \infty$, bound.*

$F(k)$ is called an L^p *Fourier multiplier* if $f \mapsto (F\hat{f})^\vee$ is bounded on L^p. All the multipliers we prove something about below are L^p-multipliers for

$1 < p < \infty$ but there exist some that are only for restricted values of p (necessarily an interval in $\frac{1}{p}$ by Riesz–Thorin and symmetric about $\frac{1}{2}$ by duality).

The multipliers we'll control (two with our bare hands and then a general theorem that actually includes the two special cases) are all smooth except at $k = 0$. An intruiging heavily studied example is the Bochner–Riesz multipliers $(1 - k^2)_+^\delta$, $\delta \in (0, \infty)$. These are singular on the sphere $|k| = 1$. It is known if $2\delta + 1 < \nu$, then this is not an L^p-multiplier if $\left|\frac{1}{p} - \frac{1}{2}\right| \geq \frac{2\delta + 1}{2\nu}$. It is known it is an L^p-multiplier if $\left|\frac{1}{p} - \frac{1}{2}\right| < \frac{\delta}{\nu - 1}$ and an open conjecture (called the *Bochner–Riesz conjecture*) that it is if $\left|\frac{1}{p} - \frac{1}{2}\right| < \frac{2\delta + 1}{2\nu}$. This is discussed further in the Notes.

Corollary 6.4.7. *$F(k) = k_j(1 + k^2)^{-1/2}$ on $\mathbb{R}^\nu$ is an an L^p Fourier multiplier for $1 < p < \infty$.*

Proof. This is a singular integral operator with kernel $\partial_j J^{(1,\nu)}$ obeying (6.4.2)–(6.4.4) by Proposition 6.3.1. Since $k_j(1 + k^2)^{-1/2}$ is a bounded function on $\mathbb{R}^\nu$

$$\|k_j(1 + k^2)^{-1/2}\widehat{f}\|_2 \leq \|\widehat{f}\|_2$$

so since $\check{}$ is unitary, $\partial_j J^{(1,\nu)}*$ is bounded from $L^2(\mathbb{R}^\nu)$ to itself. Theorem 6.4.5 thus implies $k_j(1 + k^2)^{-1/2}$ is an L^p Fourier multiplier. $\square$

Corollary 6.4.8. *The singular Riesz potential $K(x) = x_j|x|^{-\nu-1}$ defines an operator bounded on all L^p, $1 < p < \infty$.*

Remarks. 1. $\nu = 1$ corresponds to the Hilbert transform so this provides another proof of M. Riesz' theorem on conjugate harmonic functions.

2. One can also deduce this from Corollary 6.4.7 by scaling and limits.

Proof. K obviously obeys (6.4.2)–(6.4.4). Moreover, by Theorem 6.8.1 of Part 1, if $\nu \geq 2$, the Fourier transform of $|k|^{-1}$ is $c_\nu|x|^{-\nu+1} \equiv c_\nu F(x)$. $\nabla_j F(x) = -(\nu - 1)x_j|x|^{-\nu-1}$ so, $\widehat{K} = k_j|k|^{-1}$ is bounded on $\mathbb{R}^\nu$ so convolution with K is bounded on L^2 (for $\nu = 1$, one needs a separate calculation, see Theorem 5.2.4, that $\widehat{K}(k) = \text{sgn}(k)$). Thus, Theorem 6.4.5 implies convolution with K is bounded on all L^p, $1 < p < \infty$. $\square$

Theorem 6.4.9. *Let F be a $C^{\nu+2}$ function on $\mathbb{R}^\nu \setminus \{0\}$ that obeys*

$$|(D^\alpha F)(k)| \leq C|k|^{-|\alpha|}, \quad \text{all } \alpha \text{ with } 0 \leq |\alpha| \leq \nu + 2 \qquad (6.4.73)$$

Then F is an L^p-multiplier for all p in $(1, \infty)$.

Remarks. 1. This is a special case (but only in that it requires more smoothness) of the *Hörmander–Mikhlin multiplier theorem*.

2. In particular, if $a(x,k)$ is an x-independent symbol in $S^0_{1\delta}$, $\mathrm{Op}(a)$ is bounded on L^p, $1 < p < \infty$. We'll say more about $S^0_{1\delta}$ symbols in Section 6.5.

Proof. We'll prove $\check{F}$ is a convolution operator with a kernel obeying (6.4.2)–(6.4.3). By hypothesis ($\alpha = 0$), $F \in L^\infty$, so $f \mapsto (F\hat{f})^\vee$ is bounded on L^2. Thus, by Theorem 6.4.5, $f \mapsto (F\hat{f})^\vee$ is bounded on all L^p, $1 < p < \infty$.

We'll use a technique known as the *Littlewood–Paley decomposition* which uses the idea of breaking a function into pieces that live near $|k| \sim 2^n$, $n \in \mathbb{Z}$. Let ψ be a function in $C_0^\infty(\mathbb{R}^\nu)$ with $\psi(x)$ only a function of $|x|$, $\psi(x) = 1$ if $0 \le |x| \le 1$, $\psi(x) = 0$ if $|x| > 2$ and $\psi(x)$ monotone in $|x|$. For each $n \in \mathbb{Z}$, let

$$\varphi_n(k) = \psi(k/2^n) - \psi(k/2^{n-1}) \tag{6.4.74}$$

Then φ_n obeys

(i) $\displaystyle\sum_{n=-\infty}^{\infty} \varphi_n(k) = 1, \quad k \in \mathbb{R}^\nu \setminus \{0\}.$ $\tag{6.4.75}$

(ii) $\varphi_n(k) = \varphi_0(2^{-n}k).$ $\tag{6.4.76}$

(iii) $0 \le \varphi_n \le 1$, $\varphi_n \in C_0^\infty(\mathbb{R}^\nu)$, $\quad \mathrm{supp}\,\varphi_n \subset \{x \mid 2^{n-1} < |x| < 2^{n+1}\}.$ $\tag{6.4.77}$

Let $F_n(k) = \varphi_n(k)F(k)$. Notice that (6.4.76) implies $(D^\alpha\varphi_n)(k) = 2^{-|\alpha|n}\varphi_0(2^{-n}k)$ and $k/2^n$ lies between $\frac{1}{2}$ and 2 on $\mathrm{supp}\,\varphi$. It follows, by Leibniz's rule, that for all n and $0 \le |\alpha| \le \nu + 2$,

$$|(D^\alpha F_n)(x)| \le C|k|^{-\alpha} \tag{6.4.78}$$

with C independent of n.

Let

$$K_n(x) = (2\pi)^{-\nu/2}\check{F}_n(x) = (2\pi)^{-\nu}\int e^{ik\cdot x}F_n(k)\,d^\nu k \tag{6.4.79}$$

Since $F_n(k)$ is bounded by $\|F\|_\infty$ and supported in $\{k \mid 2^{n-1} < |k| < 2^{n+1}\}$

$$\int |F_n(k)|\,d^\nu k \le \tau_\nu 2^{(n+1)\nu}\|F\|_\infty \tag{6.4.80}$$

and, in particular, K_n is a continuous function. Since

$$\left(\frac{\vec{x}}{i|x|^2}\cdot\vec{\nabla}_k\right)^\ell e^{ik\cdot x} = e^{ik\cdot x} \tag{6.4.81}$$

We can integrate by parts noting that, by (6.4.78),

$$\int_{2^{n-1}<|k|<2^{n+1}} |k|^{\nu-\ell}d^\nu k \le C_\ell 2^{n(\nu-\ell)} \tag{6.4.82}$$

The result is for $\ell = 0, 1, \ldots, \nu + 2$

$$|x|^\ell|K_n(x)| \le C2^{n(\nu-\ell)} \tag{6.4.83}$$

Optimizing on ℓ for each $|x|$, i.e., using $\ell = 0$ or $(\nu + 2)$, we see

$$|K_n(x)| \leq C|x|^{-\nu} \begin{cases} (2^n|x|)^{\nu} & \text{if } 2^{n-1}|x| < 1 \\ (2^n|x|)^{-2} & \text{if } 2^n|x| \geq 1 \end{cases} \qquad (6.4.84)$$

Thus for $2^{-n_0} < |x| < 2^{-n_0+1}$, we get two geometric series with ratios 2^{-4} or $2^{-\nu}$ and initial term bounded by 2^2 or 2^{ν}. Thus

$$\sum_{n=1}^{\infty} |K_n(x)||x|^{\nu}$$

converges uniformly to a continuous function $|x|^{\nu} K(x)$.

For ∇K, the argument is similar; ∇K_n has an extra $\vec{k}$ so an extra factor of 2^n and (6.4.83) becomes

$$|x|^{\ell}|\nabla K_n(x)| \leq C 2^{n(\nu+1-\ell)} \qquad (6.4.85)$$

and (6.4.84) becomes

$$|(\nabla K_n)(x)| \leq C|x|^{-\nu-1} \begin{cases} (2^n|x|)^{\nu+1} & \text{if } 2^{n-1}|x| < 1 \\ (2^n|x|)^{-1} & \text{if } 2^n|x| \geq 1 \end{cases} \qquad (6.4.86)$$

so $\sum |x|^{\nu+1}|\nabla K_n|$ converges uniformly. $\qquad \square$

Notes and Historical Remarks. The construction of a Calderón–Zygmund decomposition and its use to prove Theorem 6.4.5 goes back to Calderón–Zygmund [**134**]. They also proved in case

$$K(\lambda x) = \lambda^{-\nu} K(x) \qquad (6.4.87)$$

and (6.4.4) holds that T mapped L^2 to L^2.

Our construction of the CZ decomposition has the flavor of martingales—a proof that explicitly uses martingale ideas including Doob's inequality can be found in the book of Garling [**304**]. Stein [**780**] has a proof exploiting balls and a variant of the Vitali covering theorem.

Antoni Zygmund (1900–1992) was born in Warsaw and after a period of evacuation to the Ukraine during the First World War, he received his Ph.D. under the influence of A. Rajchman. From the start, he was interested in the subject of trigonometric series, the title of his enormously influential book. In 1940, he fled from Vilna to the US, initially to Mount Holyoke and the University of Pennsylvania and from 1947 to the University of Chicago. His students include Calderón, E. Stein, P. Cohen, and Fabes. He is undoubtedly best known as the expert on Fourier analysis and for the work described in this section on singular integral operators.

Alberto Calderón (1920–1998) was born in Mendoza, Argentina. His father convinced him that it was impractical to hope to make a living as a mathematician so he studied engineering at the University of Buenos Aires. But he continued to study mathematics and had enough of a local reputation that when Zygmund came to visit Buenos Aires in 1948, Calderón was assigned to be his assistant. Zygmund arranged a Rockefeller Fellowship for Calderón to visit Chicago and Stone then convinced him to get a formal degree, which he did under the supervision of Zygmund.

After a few years of post-doctoral study, with the splash of his joint work with Zygmund, he returned to a professorship at Chicago where he spent most of the rest of his career. His reputation is based not only on the original Calderón–Zygmund work on singular integral operators, but also on applications of that theory to PDE's and on work in interpolation theory and in ergodic theory as we've seen in other parts of these volumes. Calderón's students include Christ, Kenig, Rivière, Seeley, and Torchinsky.

As you might expect, a method as clean as this has been extensively refined. The most common conditions drop a requirement that the integral kernel be a convolution (i.e., function of $x - y$) and define a *Calderón–Zygmund kernel* to be a continuous function $K(x, y)$ on $\mathbb{R}^{2\nu} \setminus \{(x, y) \mid x = y\}$ obeying, for some $C, \delta > 0$

(a) $|K(x, y)| \leq \dfrac{C}{|x - y|^\nu}$. $\qquad\qquad\qquad\qquad\qquad\qquad\qquad$ (6.4.88)

(b) For all x, x' with $|x - x'| \leq \frac{1}{2}|x - y|$ and $y \neq x, x'$

$$|K(x, y) - K(x', y)| + |K(y, x) - K(y, x')| \leq \frac{C|x - x'|^\delta}{(|x - y| + |x' - y|)^{\nu+\delta}} \quad (6.4.89)$$

One can then define for $f \in \mathcal{S}(\mathbb{R}^\nu)$ and $\varepsilon > 0$

$$(T_\varepsilon f)(x) = \int_{|x-y| \geq \varepsilon} K(x, y) f(y) d^\nu y \qquad\qquad (6.4.90)$$

T_ε^* is the object with $K(x, y)$ replaced by $K(y, x)$. The argument in Lemma 6.4.4 works if (6.4.3) is replaced by (6.4.60).

A remarkable theorem of David–Journé [**192**] then says that there is a C_2 independent of ε so that

$$\|T_\varepsilon(f)\|_2 \leq C_2 \|f\|_2, \quad \forall \varepsilon > 0, \forall f \in \mathcal{S}(\mathbb{R})^\nu \qquad (6.4.91)$$

if and only if for some constant D and all dyadic cubes, Q, and all $\varepsilon > 0$

$$\|T_\varepsilon \chi_Q\|_2 \leq D|Q|^{1/2}, \quad \|T_\varepsilon^* \chi_Q\|_2 \leq D|Q|^{1/2} \qquad (6.4.92)$$

(i.e., one only needs (6.4.91) for characteristic functions of dyadic cubes). Here we use the fact that (6.4.90) is a convergent integral if f is a bounded function of compact support. (6.4.92) is often easy to check in concrete cases. Once one has (6.4.91), one can get a limit as $\varepsilon \downarrow 0$ and weak-L^1 estimates and so L^p, $1 < p < \infty$ estimates with conditions like (6.4.4).

The form we stated here is the version of Chousionis and Tolsa [**153**]. The original David–Journé version was stated in terms of $T1, T^*1$ in BMO— so the theorem is usually called the *T1 Theorem*.

There are many books on (or with substantial parts on) singular integral operators including [**1, 155, 324, 325, 576, 593, 712, 777**]. Many of these books discuss more general results than we have here on the L^p Fourier multiplier problem, a subject with foundational contribution by Marcinkiewicz [**558**], Michlin [**575**], and Hörmander [**391**]. For a nice exposition of the multiplier problems, see Grafakos [**328**].

The Bochner–Riesz multipliers take their name from Bochner–Riesz means—for Fourier series on $(\partial\mathbb{D})^\nu$, this is defined by

$$(B_R^\alpha f)(\theta_1, \dots \theta_\nu) = \sum_{m \in \mathbb{Z}^\nu |m| \leq R} \left(1 - \frac{|m|^2}{R^2}\right)^\alpha f^\sharp(m_1, \dots, m_\nu) e^{2\pi i m \cdot \theta}$$

(6.4.93)

a kind of spherical analog of Cesàro means. It is named after the ($\nu = 1$) work of M. Riesz [**691**] and general ν work of Bochner [**87**]. The Bochner–Riesz conjecture is a cutting edge of technology for which the Kakeya problem seems to be central. In this regard, a paper of Fefferman [**264**] is seminal. In this paper, he proved $\chi_{|k| \leq 1}$ is not an L^p-multiplier for $p \neq 2$! See the Notes to Section 5.8. For much more on the subject of Bochner–Riesz means and multipliers, see the book of Grafakos [**329**] and discussion in the Notes to Section 6.8. We will say more about Bochner–Riesz means in Section 6.8.

The Littlewood–Paley idea of 2^n decompositions is often used in harmonic analysis. We'll see it again especially in Sections 6.5 and 6.8. It goes back to their series of papers [**542**].

Problems

1. Use the method of the proof of Theorem 6.4.5 to prove if the hypothesis of the theorem holds, then $\lim \|(T - T_\varepsilon)f\|_p = 0$ for any $f \in L^p$, $1 < p < \infty$.

2. Let T_1, T_2 be two bounded operators on $L^p(\mathbb{R}^\nu, d^\nu x)$, for some $1 \leq p < \infty$. Suppose they have the same off-diagonal kernel. Let $S = T_1 - T_2$.

   (a) Let E, F, be bounded Borel sets with $\text{dist}(E, F) > 0$. With χ_C the characteristic function of C, prove that

   $$\langle \chi_E, S\chi_F \rangle = 0$$

   (6.4.94)

   (b) Prove (6.4.94) if only $E \cap F = \emptyset$.

   (c) Prove that on bounded Borel sets, the function

   $$\mu(E) = \langle \chi_E, S\chi_E \rangle$$

   (6.4.95)

   is a σ-finite measure.

(d) Prove μ is a.c. wrt dx so $d\mu = h(x)\,d^\nu x$.

(e) For any f, g continuous of compact support, prove that

$$\langle f, Sg \rangle = \int f(x)\,h(x)g(x)\,d^\nu x$$

so S is multiplication by h.

(f) Prove that $h \in L^\infty$.

3. (a) If K obeys (6.4.9), prove Lemma 6.4.4 for the T of (6.4.7).

(b) Prove Theorem 6.4.6.

4. If (6.4.9) is replaced by

$$|K(x, y_1) - K(x, y_2)| + |K(y_1, x) - K(y_2, x)| \leq \frac{C|y_1 - y_2|^\delta}{|x - y_1|^{\nu+\delta}} \qquad (6.4.96)$$

for some $C > 0$, $\delta \in (0, 1]$ and all y_1, y_2 with $|y_1 - y_2| < \frac{1}{2}|x - y_1|$, prove that Theorem 6.4.6 still holds with a minor change in proof.

6.5. Pseudodifferential Operators on Sobolev Spaces and the Calderón–Vaillancourt Theorem

In Section 4.3, we presented the theory of pseudodifferential operators (ΨDO) in $\Psi_{\rho\delta}^m$ (defined below) with $0 \leq \delta < \rho \leq 1$. Here, we will study their action on the sets of L^2-Sobolev spaces, H^s ($= H^{s,2}$) (with something also about $H^{s,p}$). We'll prove for $\delta < \rho$, $A \in \Psi_{\rho\delta}^m$ maps H^s to H^{s+m} and then turn to the more subtle result of Calderón–Vaillancourt that $A \in \Psi_{\rho\rho}^m$ maps L^2 to L^2 (we'll prove this when $\rho = 0$).

A *symbol* $a \in S_{\rho\delta}^m$ ($m \in \mathbb{R}, 0 \leq \delta, \rho \leq 1$) is a function, $a(x, k)$ on $\mathbb{R}^{2\nu}$ with ($\langle x \rangle = (|x|^2 + 1)^{1/2}$)

$$|(D_x^\beta D_k^\alpha a)(x, k)| \leq C_{\alpha\beta}\langle k \rangle^{m-|\alpha|\rho+|\beta|\delta} \qquad (6.5.1)$$

The operators, $\mathrm{Op}(a)$, associated to a symbol a are defined by

$$(\mathrm{Op}(a)f)(x) = (2\pi)^{-\nu/2} \int a(x, k)e^{ik\cdot x}\widehat{f}(k)\,d^\nu k \qquad (6.5.2)$$

which we showed maps $\mathcal{S}(\mathbb{R}^\nu)$ to itself. We denote $\{\mathrm{Op}(a) \mid a \in S_{\rho\delta}^m\}$ by $\Psi_{\rho\delta}^m$ and $\{\mathrm{Op}(a) \mid a \in \bigcap_m S_{\rho\delta}^m\}$ by $\Psi_{\rho\delta}^{-\infty}$.

In Section 4.3, we proved for $0 \leq \delta < \rho \leq 1$

(1) (see Theorem 4.3.10) $A_j = \Psi_{\rho\delta}^{m_j}$, $j = 1, 2 \Rightarrow A_1 A_2 \in \Psi_{\rho\delta}^{m_1+m_2}$.

(2) (see Theorem 4.3.10) $A_j \in \Psi_{\rho\delta}^{m_j}$, $j = 1, 2 \Rightarrow [A_1, A_2] \in \Psi_{\rho\delta}^{m_1+m_2-(\rho-\delta)}$.

(3) (see Theorem 4.3.9) $A \in \Psi_{\rho,\delta}^m \Rightarrow A^* \in \Psi_{\rho\delta}^m$.

(4) (see the proof of Theorem 4.3.4) $A \in \Psi^m_{\rho\delta}$ had a distributional integral kernel, $K(x,y)$ which was a function for $x \neq y$ and obeyed for any $\delta, \ell > 0$:

$$|K(x,y)| \leq C_{\delta,\ell}|x-y|^{-\ell} \quad (\text{if } |x-y| \geq \delta) \tag{6.5.3}$$

Here, we'll use these facts as well as results on leading symbols (of $A_1 A_2$, and A^*). We are heading towards the first main result of this section.

Theorem 6.5.1. *Let $0 \leq \delta < \rho \leq 1$ and let $A \in \Psi^m_{\rho\delta}$ and $s \in \mathbb{R}$. Then A maps H^s to H^{s+m} and for some C_s:*

$$\|Af\|_{s+m} \leq C_s \|f\|_s \tag{6.5.4}$$

Remarks. 1. We prove (6.5.4) for $f \in \mathcal{S}(\mathbb{R}^\nu)$ but once one has it, one can use density of $\mathcal{S}$ in H^s to define Af for $f \in H^s$ and (6.5.4) holds for all $f \in H^s$.

2. Many CZOs are in some $\Psi^0_{\rho\delta}$ so this includes an L^2-bound for such maps. But general CZ kernels don't obey (6.5.3) for all ℓ so only some CZOs are ΨDOs.

Lemma 6.5.2. *Let $0 \leq \delta < \rho \leq 1$. If $m < -\nu$, then $A \in \Psi^m_{\rho\delta}$ is a bounded map of L^2 to itself.*

Proof. By (6.5.2), A has a distributional kernel

$$K(x,y) = (2\pi)^{-\nu} \int e^{ik\cdot(x-y)} a(x,k)\, d^\nu k \tag{6.5.5}$$

If $a \in S^m_{\rho\delta}$, with $m < -\nu$, then $\sup_x \int |a(x,k)|\, d^\nu k = C < \infty$ so

$$|K(x,y)| \leq (2\pi)^{-\nu} C \tag{6.5.6}$$

Combining this with (6.5.3), we see that

$$|K(x,y)| \leq g(x-y) \tag{6.5.7}$$

where $g(x) = C_1 \langle x \rangle^{-\nu-1}$. Since $g \in L^1$, Young's inequality says that

$$\| g * |f| \|_2 \leq C_2 \|f\|_2 \tag{6.5.8}$$

which implies that

$$\left\| \int K(\cdot, y) f(y)\, d^\nu y \right\|_2 \leq C_2 \|f\|_2 \tag{6.5.9}$$

i.e., $\|Af\|_2 \leq C_2 \|f\|_2$ $\square$

Lemma 6.5.3. *Let $0 \leq \delta < \rho \leq 1$. If $m < 0$, then $A \in \Psi^m_{\rho\delta}$ is a bounded map of L^2 to itself.*

Proof. Pick n so $2^{n+1}m < -\nu$. Since $A^* \in \Psi^m_{\rho\delta}$ (by (3) above), by repeated use of (1), $(A^*A)^{2^n} \in \Psi^{2^{n+1}m}_{\rho\delta}$ so by Lemma 6.5.2

$$\| (A^*A)^{2^n} f\|_2 \leq C\|f\|_2 \tag{6.5.10}$$

Thus,

$$\|Af\|^2 = \langle f, A^*Af \rangle \tag{6.5.11}$$
$$\leq \|f\| \, \|A^*Af\| \tag{6.5.12}$$
$$\leq \|f\|^{3/2} \, \|(A^*A)^2 f\|^{1/2} \tag{6.5.13}$$
$$\leq \cdots \leq \|f\|^{2-1/2^n} \|(A^*A)^{2^n} f\|^{1/2^n}$$
$$\leq C^{1/2^n} \|f\|_2^2 \tag{6.5.14}$$

$\square$

Lemma 6.5.4. *Let $0 \leq \delta < \rho \leq 1$. Then, any $A \in \Psi^0_{\rho\delta}$ is a bounded map of L^2 to itself.*

Proof. Let b be the symbol of A^*A. Then $b \in S^0_{\rho\delta}$ is a bounded function. Moreover, since A^*A is self-adjoint, its symbol is real up to lower order, i.e.,

$$\operatorname{Im} b \in S^{-(\rho-\delta)}_{\rho\delta} \tag{6.5.15}$$

Let $M = 2\|b\|_\infty$. Then,

$$\operatorname{Re}(M - b(x,k)) \geq \frac{1}{2}M \tag{6.5.16}$$

so (since $z \mapsto z^{1/2}$ is C^∞ with bounded derivatives on $\{z \mid \operatorname{Re} z > \frac{1}{2}M\}$)

$$c(x,k) = (M - \operatorname{Re} b(x,k))^{1/2} \tag{6.5.17}$$

is also a symbol in $S^0_{\rho\delta}$.

Let $C = \operatorname{Op}(c)$. The symbol of C^* is $\overline{C(x,k)}$ plus an element of $S^{-(\rho-\delta)}_{\rho\delta}$. The symbol of a product is the product of symbols plus a lower-order symbol. Then, $C^*C = \operatorname{Op}(q)$ where

$$q = M - \operatorname{Re} b + S^{-(\rho-\delta)}_{\rho\delta} \tag{6.5.18}$$
$$= M - b + r \tag{6.5.19}$$

with (using (6.5.15))

$$r \in S^{-(\rho-\delta)}_{\rho\delta} \tag{6.5.20}$$

Thus,

$$0 \leq \|Cf\|^2 = \langle f, (C^*C)f \rangle$$
$$= M\|f\|^2 - \langle f, A^*Af \rangle + \langle f, \operatorname{Op}(r)f \rangle \tag{6.5.21}$$

so

$$\|Af\|^2 \le M\|f\|^2 + \langle f, \mathrm{Op}(r)f \rangle$$
$$\le (M + \|\mathrm{Op}(r)\|)\,\|f\|^2 \tag{6.5.22}$$

Since $\|\mathrm{Op}(r)\|_{L^2 \to L^2} < \infty$ by (6.5.20) and Lemma 6.5.3, we have proven A is bounded from L^2 to L^2.   □

Proof of Theorem 6.5.1. Suppose first $A \in \Psi^0_{\rho\delta}$. Let $s \in \mathbb{R}$. Define Λ^s on $\mathcal{S}(\mathbb{R}^\nu)$ by

$$\Lambda^s f = (\langle k \rangle^s \widehat{f}\,)^\vee \tag{6.5.23}$$

Thus, Λ^s is the ΨDO with symbol $\langle k \rangle^s$ which lies in $\Psi^s_{1\,0}$ and so in all $\Psi^s_{\rho\delta}$.

By property (2) of ΨDO after (6.5.2) , $[\Lambda^s, A] \in \Psi^{s - \langle \rho - \delta \rangle}_{\rho\delta}$ so

$$\Lambda^s A \Lambda^{-s} = A + [\Lambda^s, A]\Lambda^{-s} \in \Psi^0_{\rho\delta} + \Psi^{-(\rho-\delta)}_{\rho\delta}$$

Thus, by Lemma 6.5.4,

$$\|\Lambda^s A \Lambda^{-s} f\|_{L^2} \le C_s \|f\|_{L^2} \tag{6.5.24}$$

Applying this to $f = \Lambda^s g$ and using $\|g\|_{H^s} = \|\Lambda^s g\|_{L^2}$ we see that

$$\|Ag\|_{H^s} \le C_s \|g\|_{H^s} \tag{6.5.25}$$

proving Theorem 6.5.1 for $m = 0$.

For general m, let $B = \Lambda^{-m} A \in \Psi^0_{\rho\delta}$. Thus,

$$\|Af\|_{m+s} = \|\Lambda^m B f\|_{m+s}$$
$$= \|Bf\|_s$$
$$\le C\|f\|_s \tag{6.5.26}$$

by the $m = 0$ case.   □

We can extend this to L^p-bounds: for simplicity, we handle the case $\rho = 1$, $\delta = 0$. In that case, $\mathrm{Op}(a)$ looks like $f \mapsto (F\widehat{f}\,)^\vee$ where F obeys (6.4.73). The Paley–Littlewood decomposition argument in Theorem 6.4.9 extends easily (Problem 1) to prove that $\mathrm{Op}(a), a \in S^0_{1\,0}$, has an integral kernel, K, obeying

$$|K(x,y)| \le C|x - y|^{-\nu} \tag{6.5.27}$$

$$|\nabla_y K(x,y)| \le C|x - y|^{-\nu-1} \tag{6.5.28}$$

$\mathrm{Op}(\nabla_x a)$ has integral $(\nabla_x + \nabla_y)K(x,y)$ (since $(\nabla_x + \nabla_y)e^{ik(x-y)} = 0$) so $(\nabla_x + \nabla_y)K$ obeys an estimate like (6.5.27). Using that plus (6.5.28) shows that

$$|\nabla_x K(x,y)| \le C|x - y|^{-\nu-1} \tag{6.5.29}$$

This plus Theorem 6.4.6 plus the L^2-bounded result we have just proven shows that $\mathrm{Op}(a)$ is bounded from L^p to L^p, $1 < p < \infty$. The same argument using Λ^s that worked for L^2 works for L^p and we have:

Theorem 6.5.5. *If $A \in \Psi^m_{\rho=1\,\delta=0}$ and $s \in \mathbb{R}$, then, for $1 < p < \infty$, A maps $H^{s,p}$ to $H^{s,p+m}$ and for some $C_{s,p}$,*

$$\|Af\|_{H^{s+m,p}} \leq C_{s,p}\|f\|_{H^{s,p}} \tag{6.5.30}$$

The second main result we'll prove in this section is

Theorem 6.5.6 (Calderón–Vaillancourt Theorem). *Let $a \in S^0_{00}$. Then $\mathrm{Op}(a)$ is a bounded map on $L^2(\mathbb{R}^\nu)$.*

While we'll only prove it for $\rho = 0$, the result is true for all $S^0_{\rho\rho}$, $\rho < 1$; see the reference in the Notes. The idea will be to write a as a sum $\{a_{\ell p}\}_{\ell,p \in \mathbb{Z}}$ with $a_{\ell p}$ localized near $x = \ell$, $k = p$. We'll put the sum together using the following elegant result:

Lemma 6.5.7 (Cotlar–Knopp–Stein Lemma aka CKS Lemma). *Let $\{A_j\}_{j=1}^\infty$ be a family of bounded operators on a Hilbert space $\mathcal{H}$. Suppose*

$$\sup_j \sum_k \|A_j^* A_k\|^{1/2} = M < \infty, \quad \sup_k \sum_j \|A_k A_j^*\|^{1/2} = Q < \infty \tag{6.5.31}$$

Define for $N = 1, 2, \dots$

$$S_N = \sum_{j \leq N} A_j \tag{6.5.32}$$

Then, for all N,

$$\|S_N\| \leq \sqrt{MQ} \tag{6.5.33}$$

Remarks. 1. In applications, it is easy to see that S_N has a weak limit (on a dense set), S_∞, which is clearly bounded. In fact, under the hypotheses, it is easy to see s-lim $S_N = S_\infty$ exists (see Problem 2).

2. Let $\mathrm{coker}(A) \equiv \mathrm{ker}(A)^\perp = \overline{\mathrm{Ran}(A^*)}$. If

$$\mathrm{Ran}(A_1) \perp \mathrm{Ran}(A_2) \text{ and } \mathrm{coker}(A_1) \perp \mathrm{coker}(A_2) \tag{6.5.34}$$

then it is easy to see that (Problem 3)

$$\|A_1 + A_2\| = \max(\|A_1\|, \|A_2\|) \tag{6.5.35}$$

But $\mathrm{Ran}(A_1) \perp \mathrm{Ran}(A_2) \Leftrightarrow A_1^* A_2 = 0$ so (6.5.34) $\Leftrightarrow A_1^* A_2 = A_1 A_2^* = 0$. The condition on the sums (6.5.31) by $A_j^* A_k$ and $A_k A_j^*$ go to zero rapidly as $|j - k| \to \infty$. In light of the above, it is often called an *almost orthogonality* condition.

Proof. Clearly (6.5.27) and $\|A^*A\| = \|A\|^2$ (see (3.7.4) of Part 1 and Problem 1 of Section 3.7 of Part 1) implies that

$$\sup_j \|A_j\| \leq M \qquad (6.5.36)$$

(take $k = j$ in the sum). Let

$$T_N = S_N^* S_N \qquad (6.5.37)$$

Then $\|B\|^2 = \|B^*B\|$ and $T_N = T_N^*$ imply that

$$\|T_N^{2^n}\| = \|S_N\|^{2^{n+1}} \qquad (6.5.38)$$

We have that

$$T_N^m = \sum_{j_1,k_1,j_2,\ldots,k_m=1,\ldots,N} A_{j_1}^* A_{k_1} A_{j_2}^* A_{k_2} \ldots A_{k_m} \qquad (6.5.39)$$

and the pair of estimates

$$\|A_{j_1}^* \ldots A_{k_m}\| \leq \|A_{j_1}^* A_{k_1}\| \ldots \|A_{j_m}^* A_{k_m}\| \qquad (6.5.40)$$

$$\|A_{j_1}^* \ldots A_{k_m}\| \leq \|A_{j_1}^*\| \|A_{k_1} A_{j_2}^*\| \ldots \|A_{k_{m-1}} A_{j_m}^*\| \|A_{k_m}\|$$

Then, using (6.5.36)

$$\|A_{j_1}^* \ldots A_{k_m}\| \leq M \|A_{j_1}^* A_{k_1}\|^{1/2} \|A_{k_1} A_{j_2}\|^{1/2} \ldots \|A_{j_m}^* A_{k_m}\|^{1/2} \qquad (6.5.41)$$

Plugging this into the norm estimate based on (6.5.39), summing first on k_m, then $j_m, \ldots$ yields

$$\|T_N^m\| \leq N M^{m+1} Q^{m-1} \qquad (6.5.42)$$

where the N come from the fact that after doing the sums over $k_m, j_m, \ldots k_1$, we still have to sum over j_1.

Here comes the trick. The N looks like a disaster since we want an N-independent bound but (6.5.38) says

$$\|S_N\|^{2^{n+1}} \leq N M^{2^n+1} Q^{2^n-1} \qquad (6.5.43)$$

We get to take roots and use $\lim_{n\to\infty} N^{1/2^n} = 1$ for all fixed N, the $n \to \infty$ limit of the 2^n-th root of (6.5.43) is (6.5.33). $\qquad \square$

As an aside to the direct theme of this section, we want to note this result can be used to provide the other half of the machinery to proving any kernel $K(x)$ C^1 on $\mathbb{R}^\nu \setminus \{0\}$ and obeying

$$|K(x)| \leq A|x|^{-\nu} \qquad (6.5.44)$$

$$|\vec{\nabla} K(x)| \leq B|x|^{-\nu-1} \qquad (6.5.45)$$

$$\int_{S^{\nu-1}} K(r\omega) dS(\omega) = 0 \qquad (6.5.46)$$

defines via (6.4.11) a bounded operator on $L^p(\mathbb{R}^\nu)$. By Theorem 6.4.5, we need only prove it bounded on $L^2(\mathbb{R}^\nu)$ and we'll use the CKS lemma to do that.

We'll see the Littlewood–Paley 2^n decomposition used in the last section and again below. We do it though in x-space. Thus,

(i) $\displaystyle\sum_{n=-\infty}^{\infty} \varphi_n(x) = 1$ on $\mathbb{R}^\nu \setminus \{0\}$. $\qquad$ (6.5.47)

(ii) $\varphi_n(x) = \varphi_0(2^{-n}x)$. $\qquad$ (6.5.48)

(iii) $\varphi_n \in C_0^\infty(\mathbb{R}^\nu \setminus \{0\})$, $\quad$ supp $\varphi_n \subset \{x \mid 2^{n-1} < |x| < 2^{n+1}\}$. $\qquad$ (6.5.49)

Theorem 6.5.8. *Let K obey* (6.5.44)–(6.5.46). *Then T defined by* (6.4.11) *is a bounded map of $L^2(\mathbb{R}^\nu)$ to itself.*

Remark. This argument works with minor changes (Problem 4) for the variable kernels.

Proof. Define

$$K_n(x) = \varphi_n(x)K(x) \qquad (6.5.50)$$

and

$$(C_n f)(x) = \int K_n(x-y)f(y)\, d^\nu y \qquad (6.5.51)$$

Since $|K_n|$ is bounded by $A\, 2^{-\nu(n-1)}$ on supp φ_n,

$$\int |K_n(x)|\, d^\nu x \le \tau_\nu \big(2^{(n+1)\nu} - 2^{(n-1)\nu}\big) A \big(2^{-\nu(n-1)}\big)$$

$$\le 3\tau_\nu A \qquad (6.5.52)$$

(τ_ν = volume of unit ball in $\mathbb{R}^\nu$) each K_n is an L^1-function, so by Young's inequality

$$\|C_n\|_{L^2 \to L^2} \le 3\tau_\nu A \qquad (6.5.53)$$

The key estimate will be on $\widetilde{K}_n * K_m$ where $\widetilde{K}_n(x) = \overline{K_n(-x)}$. Without loss, we can suppose that $m > n+4$, since for $|n-m| \le 3$, $\|C_n^* C_m\| \le (3\tau_\nu A)^2$ and $C_m^* C_n = (C_n^* C_m)^*$.

Since $(\nabla\varphi_m)(x) = 2^{-m}(\nabla\varphi_0)(2^{-m}x)$ and $|x| \le 2^{m+1}$ on supp(φ_m), we have $|(\nabla\varphi_m)(x)| \le 2\|\nabla\varphi_0\|_\infty |x|^{-1}$, so by Leibniz's rule,

$$|(\nabla K_m)(x)| \le \widetilde{B}|x|^{-\nu-1} \qquad (6.5.54)$$

Notice next that if $|y| < \frac{1}{2}|x|$, then $|x - ty| \ge \frac{1}{2}|x|$ for all $t \in [0,1]$, so writing $K_n(x-y) - K_n(x)$ as an integral of ∇K along $\{x - ty \mid 0 \le t \le 1\}$, we see that (if $|y| < \frac{1}{2}|x|$)

$$|K_m(x-y) - K_m(y)| \le \widetilde{B}\, |y|\, 2^{\nu+1}\, |x|^{-\nu-1} \qquad (6.5.55)$$

(since $(\nabla K_m)(x-ty) = 0$ if $x - ty \notin \text{supp } K_n$).

Since $\int \widetilde{K}_m(y)\,d^\nu y = 0$ on account of (6.5.45)

$$(\widetilde{K}_n * K_m)(x) = \int \left[K_m(x-y) - K_m(x) \right] \widetilde{K}_n(y)\,d^\nu y \qquad (6.5.56)$$

By hypothesis, for the integrand in (6.5.56) to be nonzero, we need $|y| \leq 2^{n+1} \leq 2^{m-3}$ and either $|x| > 2^{m-1}$ or $|x-y| > 2^{m-1}$. Either way, $|x| > 2^{m-2} \geq 2|y|$, so we can use (6.5.55). Thus,

$$\|\widetilde{K}_n * K_m\| \leq \int_{\substack{2^{m-2} \leq |x| \leq 2^{m+2} \\ 2^{n-1} \leq |y| \leq 2^{n+1}}} A\,|y|^{-\nu}\,|y|\,\widetilde{B}\,2^{n+1}\,|x|^{-\nu-1}\,d^\nu x\,d^\nu y$$

$$\leq C\,2^{-m}\,2^n \qquad (6.5.57)$$

for a constant C. Since $m > n$, and we have the symmetry, we see that

$$\|C_n^* C_m\| \leq C\,2^{-|m-n|} \qquad (6.5.58)$$

The argument for $C_n C_m^*$ is identical. The CKS lemma implies that $\sum_n C_n$ defines a bounded operator on L^2. $\qquad\square$

Proposition 6.5.9. *Let* $A \in L^2(M \times N, d\mu \otimes d\nu)$. *For* $f \in L^2(N, d\nu)$ *and* $x \in M$

$$(\mathrm{O}(A)f)(x) = \int A(x,y) f(y)\,d\nu(y) \qquad (6.5.59)$$

Then, the integral is absolutely convergent for a.e. $x \in M$, $\mathrm{O}(A)f \in L^2(M, d\mu)$ *and*

$$\|\mathrm{O}(A)f\|_2 \leq \|A\|_{L^2(M \times N, d\mu \otimes d\nu)}\|f\|_2 \qquad (6.5.60)$$

Remark. We'll repeat an argument that appears also in Example 3.1.15 of Part 4 because the argument is so simple.

Proof. For μ-a.e. x, $A(x, \cdot) \in L^2(N)$. For such x, the integral in (6.5.59) is absolutely convergent and, by the Schwarz inequality bounded by $\|A(x, \cdot)\|_{L^2(N, d\mu)}\|f\|_2$. Squaring and integrating proves $\mathrm{O}(A)f \in L^2$ and that (6.5.60) holds. $\qquad\square$

Proposition 6.5.10. *Let* $X_j, P_j : \mathcal{S}(\mathbb{R}^\nu) \to \mathcal{S}(\mathbb{R}^\nu)$, $j = 1, \ldots, \nu$ *by*

$$(X_j f)(x) = x_j f(x), \quad (P_j f)(x) = -i\frac{\partial f}{\partial x_j}(x) \qquad (6.5.61)$$

Then, for $x_0, k_0 \in \mathbb{R}^\nu$ *and* $a \in S^m_{\rho\delta}$ *as operators from* $\mathcal{S}(\mathbb{R}^\nu)$ *to* $\mathcal{S}'(\mathbb{R}^\nu)$.

(a) $(X_j - (x_0)_j)\,\mathrm{Op}(a) = \mathrm{Op}((x-x_0)_j a)$ $\qquad (6.5.62)$

(b) $[X_j, \mathrm{Op}(a)] = \mathrm{Op}\!\left(i\dfrac{\partial a}{\partial k_j}\right)$ $\qquad (6.5.63)$

(c) $\mathrm{Op}(a)(P_j - (k_0)_j) = \mathrm{Op}(-(k-k_0)_j a)$ $\qquad (6.5.64)$

(d) $[P_j, \mathrm{Op}(a)] = \mathrm{Op}\!\left(-i\dfrac{\partial a}{\partial x_j}\right)$ $\qquad (6.5.65)$

Remark. Notation like $b \equiv (x - x_0)_j a$ is shorthand for the relation

$$b(x, k) = (x_j - x_{0,j}) a(x, k) \tag{6.5.66}$$

Proof. We use the fact that $\mathrm{Op}(a)$ formally has integral kernel K given by (6.5.5). The formal nature is justified by looking at $\int f(x) K(x, y) g(y) \, d^\nu x \, d^\nu y$ for $f, g \in \mathcal{S}(\mathbb{R}^\nu)$.

(a) $(X_j - (x_0)_j) \mathrm{Op}(a)$ has an integral kernel given by (6.5.5) where $a(x, k)$ is replaced by $(x - x_0)_j a(x, k)$.

(b) $[X_j, \mathrm{Op}(a)]$ has integral kernel given by (6.5.5) with a replaced by $(x - y)_j a(x, k)$. We use $(x - y)_j e^{ik \cdot (x-y)} = -i \frac{\partial}{\partial k_j} e^{ik(x-y)}$ and integrate by parts to get (6.5.63).

(c) When looking at $\int K(x, y)(P_j g)(y) \, d^\nu y$, we can integrate by parts and use $-i \frac{\partial}{\partial y_j} e^{ik(x-y)} = -k_j e^{ik(x-y)}$ and so get (6.5.64).

(d) Integrating by parts as in (c), we see that $[P_j, \mathrm{Op}(a)]$ has integral kernel $-i \left[\frac{\partial}{\partial x_j} + \frac{\partial}{\partial y_j} \right] K$. Noting that $\left(\frac{\partial}{\partial x_j} + \frac{\partial}{\partial y_j} \right) \left(e^{ik(x-y)} \right) = 0$, we get (6.5.65). $\qquad \square$

Proposition 6.5.11. *Let $U \subset \mathbb{R}^{2\nu}$ be a bounded open set. For $(x_0, k_0) \subset \mathbb{R}^{2\nu}$, let $U_{(x_0, k_0)} = (x_0, k_0) + U$. Then, for any nonnegative even integers, n_1, m_1, n_2, m_2, there is C and ℓ so that for any $a \in S_{00}^0$ with*

$$\mathrm{supp}(a) \subset U_{(x_0, k_0)} \tag{6.5.67}$$

we have that $\langle (X - x_0) \rangle^{n_1} \langle (P - k_0) \rangle^{m_1} \mathrm{Op}(a) \langle (X - x_0) \rangle^{m_2} \langle (P - k_0) \rangle^{n_2}$ is a bounded operator on L^2 and

$$\| \langle (X - x_0) \rangle^{n_1} \langle (P - k_0) \rangle^{m_1} \mathrm{Op}(a) \langle (X - x_0) \rangle^{n_2} \langle (P - k_0) \rangle^{m_2} \|_{\mathcal{L}(L^2(\mathbb{R}^\nu))} \leq$$
$$C \sum_{|\alpha| + |\beta| \leq \ell} \| D_x^\alpha D_k^\beta a \|_\infty \tag{6.5.68}$$

Remarks. 1. $\langle (X - x_0) \rangle^{n_1}$ means the operator from $\mathcal{S}$ to $\mathcal{S}$ and $\mathcal{S}'$ to $\mathcal{S}'$ $\left(\sum_{j=1}^r (X_j - x_{0\,j})^2 + 1 \right)^{n_1/2}$ and similarly $\langle (P - k_0) \rangle^{m_1}$.

2. A priori, the operator maps $\mathcal{S}$ to $\mathcal{S}'$. This asserts if $f \in \mathcal{S}$, the operator takes f to L^2 and there is an operator bound as indicated.

3. We make no attempt here and in the proof of Theorem 6.5.6 to optimize the number of derivatives needed but it is not a lot; indeed, there are various results with fewer derivatives needed for Theorem 6.5.6; e.g., each α_j and β_j need only be 0 or 1—see the Notes.

Proof. By Proposition 6.5.10, the operator in question is a sum of ΨDO's with symbol of the form $(x - x_0)^\gamma (k - k_0)^\delta (D_x^\alpha D_k^\beta a)(x, k)$. These are in L^2 with bounds dominated by $\| D_x^\alpha D_k^\beta a \|_\infty$, so by Proposition 6.5.9, we get (6.5.68). $\qquad \square$

Proof of Theorem 6.5.6. We begin by noting that by Peetre's inequality, as operators on $L^2(\mathbb{R}^\nu)$

$$\|\langle X - x_0 \rangle^{-L} \langle X - x_1 \rangle^{-L}\| \leq C_\ell \langle x_0 - x_1 \rangle^{-L} \qquad (6.5.69)$$

Working in Fourier transform space, one gets

$$\|\langle P - k_0 \rangle^{-L} \langle P - k_1 \rangle^{-L}\| \leq C_\ell \langle k_0 - k_1 \rangle^{-L}$$

It follows, using Proposition 6.5.11, that if $a_0, a_1 \in S^0_{00}$ with $\operatorname{supp} a_j \subset U_{(x_j, k_j)}$ that

$$\|\operatorname{Op}(a_1)^* \operatorname{Op}(a_2)\| + \|\operatorname{Op}(a_1) \operatorname{Op}(a_2)^*\| \leq C_L (\langle k_1 - k_2 \rangle + \langle x_1 - x_2 \rangle)^{-L} \qquad (6.5.70)$$

where C_L depends on L^∞-norms of a L-dependent number of derivatives of a_1 and a_2. This holds if $\langle x_1 - x_2 \rangle \geq \langle k_1 - k_2 \rangle$ by writing

$$\operatorname{Op}(a_1)^* \operatorname{Op}(a_2) =$$
$$[\langle X - x_1 \rangle^L \operatorname{Op}(a_1)]^* [\langle X - x_1 \rangle^{-L} \langle X - x_2 \rangle^{-L}] [\langle X - x_2 \rangle^L \operatorname{Op}(a_2)] \qquad (6.5.71)$$

and then using Proposition 6.5.11 and (6.5.69). A similar argument works for $\langle k_1 - k_2 \rangle \geq \langle x_1 - x_2 \rangle$ and for $\operatorname{Op}(a_1) \operatorname{Op}(a_2)^*$.

Let U be the hypercube of side 3 centered at $(\mathbf{0}, \mathbf{0}) \subset \mathbb{R}^{2\nu}$. It is easy (Problem 5) to find $\varphi \in C_0^\infty(\mathbb{R}^{2\nu})$, $\operatorname{supp} \varphi \subset U$, $0 \leq \varphi \leq 1$ so that

$$\sum_{x_j, k_j \in \mathbb{Z}^\nu} \varphi(\,\cdot\, - x_j, \,\cdot\, - k_j) = 1 \qquad (6.5.72)$$

Letting $a_{j\ell} = a\varphi(\,\cdot\, - x_j, \,\cdot\, - k_\ell)$, we write $A_{jl} = \operatorname{Op}(a_{j\ell})$ and use (6.5.70) and the CKS lemma to get a bound on $\operatorname{Op}(a)$. $\qquad \square$

Notes and Historical Remarks. Our presentation of the $S^m_{\rho\delta}$ results for $0 \leq \delta < \rho \leq 1$ follows Hörmander [**398**]. The CKS lemma is named after Cotlar [**172**] and Knapp–Stein [**475**]. Cotlar required the A_j to be self-adjoint and to commute and required $\|A_j A_k\| \leq C\, 2^{-|j-\ell|}$. His paper had the application that appears as Theorem 6.5.8. Knapp–Stein had the full result and the proof we give, which is quite different from Cotlar's. Stein had found the result (and independently, Cotlar) between the two papers (in 1955 and 1971) but neither published it until Stein needed it in his joint work with Knapp. For this reason, some authors call this Cotlar's lemma or the Cotlar–Stein lemma. We follow the published record.

Mischa Cotlar (1912–2007) was born in the Ukraine. His father was a mill manager and was persecuted after the Revolution so much so that Mischa was barred from school. With his family, he migrated to Uruguay in 1928. Despite only one year of formal schooling, he taught himself mathematics and, with the help of a local expert and via correspondence with

Fréchet, he published his first paper in 1935. In those years he earned a living as a piano player, first in disreputable bars in Montevideo, and then in a classical piano trio. Also in 1935, he moved to Buenos Aires, Argentina where for a while he had a job as a research assistant until the authorities realized he had only one year of formal schooling!

After the Second World War, several American mathematicians, including Zygmund, Stone, and George Birkhoff, visited Argentina. Birkhoff was so impressed with Cotlar that he arranged a Guggenheim fellowship[1]. So, in 1951, Cotlar went to Yale to study ergodic theory with Kakutani. But when it came time to be awarded his Ph.D., Yale refused given his lack of a high-school or college diploma! Stone, relying on the University of Chicago's well-known flexibility, brought him to Chicago where, in 1953, he got a degree under Zygmund and Calderón (who himself was Argentinian and had been "discovered" by Zygmund during a trip to Buenos Aires). Cotlar spent the rest of his career in Buenos Aires and Caracas[2], most of it as a Professor there. We've seen him not only in this section, but also in Section 5.12. For a charming set of reminiscences of Cotlar, see Horváth [**400**] and Sadosky [**713**]. [**714**] has short pieces in his honor.

The Calderòn–Vaillancourt theorem is due to them [**132, 133**] and has spawned an enormous literature. Howe [**401, 403**] has a Weyl operator calculus version phrased in terms of the Heisenberg group. For additional literature (including the $S^0_{\rho\rho}$ result which appeared first in the second Calderón–Vaillancourt paper), see Beals [**48**], Coifman–Meyer [**163**], Cordes [**168**], Hwang [**414**], and Kato [**447**]. Several try to minimize the number of derivatives needed. Hwang's proof uses elementary tools (and, in particular, does not need the CKS lemma).

Problems

1. Extend the Paley–Littlewood argument used in Theorem 6.4.9 to prove (6.5.27)/(6.5.28) for $\mathrm{Op}(a)$ when $a \in S^0_{10}$.

2. Under the hypotheses of Lemma 6.5.7, prove that for all $\varphi \in \mathcal{H}$, $\sum_{j \leq N} A_j \varphi$ has a limit in $\mathcal{H}$.

3. When A_1, A_2 obey (6.5.34), prove that (6.5.35) holds.

4. Extend Theorem 6.5.8 to two variable kernels obeying (6.4.8)–(6.4.10).

5. Construct the function $\varphi \in C^\infty_0(\mathbb{R}^{2\nu})$ obeying (6.5.72).

---

[1] The reality was more complicated. George Birkhoff applied for the fellowship and was told it had been approved but he died before he could inform Cotlar. It was several years later after George's son, Garrett, also a Harvard mathematician, found the letter from the foundation in his father's papers, that Cotlar was actually able to use the fellowship.

[2] This bland statement masks the fact that Cotlar only left Argentina for Venezuela after a right-wing junta took governmental control and started persecuting University faculty, including some beatings. Cotlar returned to Argentina towards the end of his life after the fall of the junta.

6.6. Hypercontractivity and Logarithmic Sobolev Inequalities

Here are two motivating questions for consideration in this section, both connected to the moral of Sobolev inequalities that the smothing of resolvents of elliptic differential operators can often be captured by estimates saying they take L^p to L^q with $q > p$ (so the possible local singularities are less severe):

(1) (Hypercontractivity, i.e., smoothing of e^{-tA} for suitable A on the set of $L^p(M, d\mu)$, $1 \le p \le \infty$) An L^p-*contractive semigroup* is a semigroup of self-adjoint contractions on $L^2(M, d\mu)$ so that for all $p \in [1, \infty]$, $\|e^{-tA} f\|_p \le \|f\|_p$ if $f \in L^2 \cap L^p$. We'll explore when it maps L^2 to L^p for some $p > 2$, called a hypercontractive estimate. Of course, if $(A + 1)^{-1}$ maps L^2 to L^p, so does e^{-tA} since we can write $e^{-tA} = (A+1)^{-1}[(A+1)e^{-tA}]$ and note that $(A+1) e^{-tA}$ maps L^2 to L^2 by the spectral theorem. So these estimates are weaker than Sobolev estimates and can hold more generally as we'll see. This theory is in part marked by considerable use of interpolation—we'll mainly use Riesz-Thorin and Stein but a few times Marcinkiewicz.

(2) (Sobolev type estimates in infinite dimensions) At first glance, this seems unlikely. In $\mathbb{R}^\nu$, $(-\Delta + 1)^{-1/2}$ maps L^2 to L^{p_ν} where $\nu \ge 2$ and $p_\nu = 2\nu/(\nu - 2)$. As $\nu \to \infty$, $p_\nu \downarrow 2$. It is easy to understand why p_ν is ν-dependent: If $(f \otimes g)(x; y) = f(x) g(y)$, then $\|f \otimes g\|_p = \|f\|_p \|g\|_p$, i.e., L^p-norms are multiplicative. But $\nabla(f \otimes g) = (\nabla f) \otimes g + f \otimes \nabla g$ is additive. Taking into account that log turns products into sums, one has a glimmer that an estimate that product spaces might respect is

$$\int |f(x)|^2 \log|f(x)| \, dx \le c_1 \langle f, Af \rangle \tag{6.6.1}$$

when $\|f\|_2 = 1$. These are the logarithmic Sobolev estimates that are a hero of this section. A second difficulty in dealing with products is that to go to infinite dimensions, one doesn't want products of dx's on $\mathbb{R}$ but probability measures. Indeed, one of our most interesting examples will be the Gaussian products of Section 4.12 of Part 1. In fact, after an initial result on Sobolev estimates, we'll only look at $L^2(M, d\mu)$ with $\mu(M) = 1$.

A little more thought suggests that to get even more useful estimates from (6.6.1), one might want to exponentiate so (6.6.1) might link up to hypercontractivity. Indeed, since

$$\frac{d}{dt} \|f\|_{p+\alpha t}^{p+\alpha t} \Big|_{t=0} = \frac{d}{dt} \int e^{(p+\alpha t) \log|f(x)|} \, d\mu(x) \Big|_{t=0} \tag{6.6.2}$$

$$= \alpha \int |f(x)|^p \log|f(x)| \, d\mu(x) \tag{6.6.3}$$

one sees that log Sobolev estimates are the differential form of hypercontractive estimates and we'll be able to go between one and the other by differentiation and integration.

As the above discussion suggests, we need some facts from the theory of unbounded operators on L^2—all found in Part 4 (see especially, Sections 5.1, 7.1, 7.2, 7.5, and 7.6).

(1) A *self-adjoint semigroup* on Hilbert space is a family $\{P_t\}_{t\geq 0}$ of bounded self-adjoint operators on $\mathcal{H}$ with

$$P_t P_s = P_{t+s}, \ P_{t=0} = \mathbb{1}, \quad \forall\, \varphi \in \mathcal{H}, t \mapsto P_t\varphi \text{ is norm continuous on } [0,\infty)$$
$$(6.6.4)$$

There is then a (perhaps unbounded and only densely defined) self-adjoint operator, $A \geq \alpha\mathbb{1}$; $\alpha = \log\|e^{-A}\|$ so that

$$P_t = e^{-tA} \tag{6.6.5}$$

A is called the *generator* of P_t. Conversely, if A is an unbounded self-adjoint operator on $\mathcal{H}$, with $A \geq \alpha\mathbb{1}$ for some $\alpha \in \mathbb{R}$, then P_t defined by (6.6.5) is a self-adjoint semigroup.

(2) These facts follow from the spectral theorem: A (possibly unbounded) self-adjoint operator is unitarily equivalent to a direct sum of multiplications by x on $L^2(\mathbb{R}, d\nu_j)$, i.e., there are probability measures, $\{\nu_j\}_{j=1}^N$, $N = 1, 2, \ldots$ or $N = \infty$, and a unitary V of $\mathcal{H}$ to $\oplus_{j=1}^N L^2(\mathbb{R}, d\nu_j)$ so that

$$(VAV^{-1}f)_j(x) = xf_j(x) \tag{6.6.6}$$

Here the domain of A is all of $\mathcal{H}$ if its spectrum

$$\sigma(A) = \overline{\bigcup_j \text{supp}(d\nu)} \tag{6.6.7}$$

is bounded and otherwise

$$D(A) = \{\psi \in \mathcal{H} \mid \sum_{j=1}^N \int x^2\, |(V\psi)_j(x)|^2\, d\nu_j < \infty\} \tag{6.6.8}$$

The form domain if $A \geq \alpha\mathbb{1}$ for some $\alpha \in \mathbb{R}$ is

$$Q(A) = \{\psi \in \mathcal{H} \mid \sum_{j=1}^N \int |x|\, |(V\psi)_j(x)|^2\, d\nu_j < \infty \tag{6.6.9}$$

with

$$q_A(\psi) = \sum_{j=1}^N \int x\, |(V\psi)_j(x)|^2\, d\nu_j \tag{6.6.10}$$

The relation to (1) is that if $V^{-1}P_{t=1}V$ is multiplication by $y \in [0, \|P_{t=1}\|]$, then A is multiplication by $-\log(y) \in [-\log\|P_{t=1}\|, \infty)$. Conversely, e^{-tA}

is defined by the functional calculus, if VAV^{-1} is multiplication by x and F is bounded on $\sigma(A)$ then

$$VF(A)V^{-1} = \text{ multiplication by } F(x)$$

(3) One can define the domain and form domain of the generator, A, of P_t by

$$D(A) = \{\varphi \mid \overline{\lim}_{t\downarrow 0} t^{-1} \|(P_t - \mathbb{1})\varphi\| < \infty\}; \; A\varphi = \lim_{t\downarrow 0} t^{-1}(\mathbb{1} - P_t)\varphi \quad (6.6.11)$$

$$Q(A) = \{\varphi \mid \overline{\lim}_{t\downarrow 0} t^{-1} |\langle\varphi, (\mathbb{1} - P_t)\varphi\rangle| < \infty\}; \; q_A(\varphi) = \lim_{t\downarrow 0} t^{-1}\langle\varphi, (\mathbb{1} - P_t)\varphi\rangle$$
$$(6.6.12)$$

(4) In terms of the generator, A, there are Beurling–Deny criteria for when e^{-tA} is positivity preserving and when it is an L^p-contraction

$$\left[\varphi \geq 0 \Rightarrow e^{-tA}\varphi \geq 0\right] \Leftrightarrow \left[\varphi \in Q(A) \Rightarrow |\varphi| \in Q(A)\right] + \left[q_A(|\varphi|) \leq q_A(\varphi)\right]$$
$$(6.6.13)$$

and if these conditions hold, then $(\varphi \wedge 1 = \min(\varphi, \mathbb{1}))$

$$\left[\|e^{-tA}\varphi\|_p \leq \|\varphi\|_p; \; 1 \leq p \leq \infty, t \geq 0\right] \Leftrightarrow$$
$$\forall \varphi \geq 0 \left[\varphi \in Q(A) \Rightarrow \varphi \wedge 1 \in Q(A)\right] + \left[q_A(\varphi \wedge 1) \leq q_A(\varphi)\right] \quad (6.6.14)$$

We will only use these criteria once here, but some of the theorems in this section and the next are sometimes stated by others in terms of them rather than equivalent semigroup conditions.

(5) While not connected to semigroups specifically, we recall the following consequence of the Dunford–Pettis theorem (see Theorem 4.9.7 of Part 1): If $L^1(M, d\mu)$ is separable and $B : L^1(M, d\mu) \to L^\infty(M, d\mu)$, there is a bounded measurable function $b(x, y)$ so that

$$(Bf)(x) = \int b(x, y) f(y) d\mu(y) \quad (6.6.15)$$

and conversely, if $b \in L^\infty(M \times M)$, then B defined by (6.6.15) is bounded from L^1 to L^∞ and

$$\|Bf\|_\infty \leq \|b\|_\infty \|f\|_1 \quad (6.6.16)$$

In particular, if P_t is an L^p-continuous semigroup with $\|P_t f\|_\infty \leq C_t \|f\|_1$, then there is $p_t \in L^\infty(M \times M, d\mu \otimes d\mu)$ so that

$$(P_t f)(x) = \int p_t(x, y) f(y) d\mu(y) \quad (6.6.17)$$

(6) We'll occasionally need some simple facts about compact operators and the trace ideals, especially the trace class and Hilbert–Schmidt operators. These issues are discussed in Chapter 3 of Part 4.

Definition. An L^p-contractive semigroup, e^{-tA}, is called

(a) *hypercontractive* if there are $t_0 > 0$ and C so that for all $f \in L^2$, $e^{-t_0 A} f \in L^4$ and

$$\|e^{-t_0 A} f\|_4 \leq C \|f\|_2 \tag{6.6.18}$$

(b) *supercontractive* if for all $t > 0$, there is a C_t so that for all $f \in L^2$, $e^{-tA} f \in L^4$ and

$$\|e^{-tA} f\|_4 \leq C_t \|f\|_2 \tag{6.6.19}$$

(c) *ultracontractive* if for all $t > 0$, there is C_t so that for all $f \in L^2$, $e^{-tA} f \in L^\infty$ and

$$\|e^{-tA} f\|_\infty \leq C_t \|f\|_2 \tag{6.6.20}$$

After our detour to study a consequence of ultracontractivity with a suitable bound on C_t, we study consequences of this for L^p- to L^q-bounds for general $p < q$ and see there is nothing special about $(2,4)$ or $(2,\infty)$. Conceptually, it will be useful to know one fact relating the constants C_t and $C_t^{(1)}$ defined by

$$\|e^{-tA} f\|_2 \leq C_t \|f\|_1, \quad \|e^{-tA} f\|_\infty \leq C_t^{(1)} \|f\|_1 \tag{6.6.21}$$

namely

$$C_t = (C_{2t}^{(1)})^{1/2} \tag{6.6.22}$$

For by duality,

$$\|e^{-tA} f\|_2 \leq C_t \|f\|_1 \Rightarrow \|e^{-tA} f\|_\infty \leq C_t \|f\|_2 \Rightarrow$$
$$\|e^{-2tA} f\|_\infty \leq C_t^2 \|f\|_1 \Rightarrow C_{2t}^{(1)} \leq C_t^2$$

while

$$\|e^{-tA} f\|_2^2 = \langle f, e^{-2tA} f \rangle \leq \|f\|_1 \|e^{-2tA} f\|_\infty$$
$$\leq C_{2t}^{(1)} \|f\|_1^2 \Rightarrow C_t^2 \leq C_{2t}^{(1)}$$

Example 6.6.1 (Free Laplacian Semigroup). Let $A = -\Delta$ on $L^2(\mathbb{R}^\nu, d^\nu x)$. e^{-tA} is the heat kernel, i.e.,

$$(e^{t\Delta} f)(x) = \int p_t(x-y) f(y) \, d^\nu y, \quad p_t(x) = (4\pi t)^{-\nu/2} e^{-|x|^2/4t} \tag{6.6.23}$$

as discussed in Section 6.9 of Part 1. Notice that $p_t \in \bigcap_{1 \leq p \leq \infty} L^p(\mathbb{R}^\nu, d^\nu x)$ with

$$\|p_t\|_p = [p^{-1/p}(4\pi t)^{-(1-p^{-1})}]^{\frac{\nu}{2}} \tag{6.6.24}$$

with $p^{-1/p} = 1$ if $p = \infty$ (since $\lim_{p\to\infty} p^{1/p} = 1$).

We thus have by Young's inequality for $q \geq p$

$$\|e^{t\Delta} f\|_q \leq r^{-\nu/2r}[(4\pi t)^{-\nu/2}]^{\frac{1}{p}-\frac{1}{q}} \|f\|_p \tag{6.6.25}$$

where $r = [(1 - \frac{1}{p}) + \frac{1}{q}]^{-1}$. (The constant in (6.6.25) is not optimal.) In particular, $e^{t\Delta}$ is a contraction on each L^p and

$$\|e^{t\Delta}\|_{L^1 \to L^\infty} \leq D_\nu^{(1)} t^{-\nu/2}; \quad \|e^{t\Delta}\|_{L^2 \to L^\infty} \leq D_\nu t^{-\nu/4} \tag{6.6.26}$$

$\square$

Remarkably, the bounds in (6.6.26) are essentially equivalent to one of the Sobolev estimates based on abstract considerations.

Theorem 6.6.2 (Varopoulos–Fabes–Stroock Theorem). *Let e^{-tA} be a contraction semigroup on $L^p(M, d\mu)$. Let $\nu > 0$. Consider the three conditions:*

(1) *(ultracontractivity) For some $C > 0$*

$$\|e^{-tA} f\|_2 \leq C t^{-\nu/4} \|f\|_1 \tag{6.6.27}$$

(2) *(Sobolev estimates when $\nu > 2$) For some $C > 0$*

$$\|f\|_{2\nu/(\nu-2)}^2 \leq C q_A(f) \tag{6.6.28}$$

(3) *(Nash estimate) For some $C > 0$*

$$\|f\|_2^{2+4\nu^{-1}} \leq C q_A(f) \|f\|_1^{4/\nu} \tag{6.6.29}$$

Then for all $\nu > 0$, (1) $\Leftrightarrow$ (3) and if $\nu > 2$, (1) $\Leftrightarrow$ (2) $\Leftrightarrow$ (3).

Remarks. 1. The constants C in (6.6.27)–(6.6.29) are certainly different although the proof provides relations among them. Here and below, we use C for a constant which is independent of t (but may depend on ν) and which can change from equation to equation! We also emphasize ν need not be an integer in these abstract considerations which apply to $A = (-\Delta)^\gamma$ for suitable γ (Problem 1).

2. $(A + 1)^{-1/2} = \Gamma(\frac{1}{2})^{-1} \int e^{-tA} e^{-t} t^{-1/2} dt$ so given that (6.6.27) implies $\|e^{-tA} f\|_{2\nu/(\nu-2)} \leq C t^{-1/2} \|f\|_2$ (by Stein interpolation; see Theorem 6.6.4 below) and that (6.6.28) is equivalent to $\|A^{-1/2} f\|_{2\nu/(\nu-2)} \leq C \|f\|_2$, the integral for $(A + 1)^{-1/2}$ is logarithmically divergent at $t = 0$ suggesting a Marcinkiewicz interpolation might be relevant as we will see it is. But these considerations do provide a proof of inhomogeneous Sobolev estimates in this setting (Problem 2).

3. For $A = -\Delta$, this provides another proof of the Sobolev estimates of Theorem 6.3.3 for $q = 2$, $s = 1$, $\nu \geq 2$.

Proof. (2) $\Rightarrow$ (3) if $\nu > 2$. Let $\theta = \nu/(\nu + 2)$ so that

$$\frac{1}{2} = (1 - \theta) + \theta \left[\frac{\nu - 2}{2\nu} \right] \tag{6.6.30}$$

and by Theorem 5.2.5 of Part 1 $(1 - \theta = 2/(\nu + 2))$,

$$\|f\|_2 = \|f\|_1^{2(\nu+2)^{-1}} \|f\|_{2\nu(\nu-2)^{-1}}^{\nu(\nu+2)^{-1}} \tag{6.6.31}$$

$$\leq C \|f\|_1^{2(\nu+2)^{-1}} q_A(f)^{\nu[2(\nu+2)]^{-1}} \tag{6.6.32}$$

by (6.6.28). Raising to the $\frac{2(\nu+2)}{\nu} = 2 + \frac{4}{\nu}$ power yields (6.6.29).

$(3) \Rightarrow (1)$. It suffices to prove (6.6.27) when $f \in L^1 \cap L^2$. In that case, for $t \geq 0$ define

$$f_t = e^{-tA} f, \quad N(t) = \|f_t\|_2^{-4/\nu} \tag{6.6.33}$$

This is natural since (6.6.27) is equivalent to

$$N(t) \geq C\, t\, \|f\|_1^{-4/\nu} \tag{6.6.34}$$

where C is independent of t and f. $N(t)$ is continuous and nonnegative on $[0, \infty)$ and C^1 on $(0, \infty)$, so (6.6.34) is implied by

$$\frac{dN}{dt} \geq C \|f\|_1^{-4/\nu} \tag{6.6.35}$$

Writing

$$N(t) = \langle f, e^{-2tA} f \rangle^{-2/\nu} \tag{6.6.36}$$

we see that

$$\frac{dN}{dt} = \frac{4}{\nu}\, q_A(f_t)\, N(t)^{-1-\frac{2}{\nu}} \tag{6.6.37}$$

$$= \frac{4}{\nu}\, q_A(f_t)\, \|f_t\|_2^{-2-4/\nu} \tag{6.6.38}$$

$$\geq \frac{4}{\nu}\, C^{-1} \|f_t\|_1^{-4/\nu} \tag{6.6.39}$$

$$\geq \frac{4}{\nu}\, C^{-1} \|f\|_1^{-4/\nu} \tag{6.6.40}$$

where (6.6.39) is (6.6.29) and (6.6.40) comes from $\|f_t\|_1 \leq \|f\|_1$.

$(1) \Rightarrow (3)$. By (6.6.27)

$$C^2\, t^{-\nu/2}\, \|f\|_1^2 \geq \|e^{-tA} f\|_2^2 \tag{6.6.41}$$

$$= \langle f, f \rangle + \int_0^{2t} \frac{d}{ds} \langle e^{-sA} f, f \rangle\, ds \tag{6.6.42}$$

$$= \langle f, f \rangle - \int_0^{2t} \langle e^{-sA} A f, f \rangle\, ds$$

$$\geq \langle f, f \rangle - 2t\, q_A(f) \tag{6.6.43}$$

since going to a spectral representation and using $xe^{-sx} \leq x$ for $x \geq 0$ implies that

$$\langle e^{-sA} A f, f \rangle \leq q_A(f) \tag{6.6.44}$$

Thus, for all t,

$$\|f\|_2^2 \le C^2 \, t^{-\nu/2} \, \|f\|_1^2 + 2t \, q_A(f) \tag{6.6.45}$$

The optimum t is $D \, q_A(f)^{-2/(\nu-2)} \, \|f\|_1^{4/(\nu+2)}$ which leads to (6.6.29).

(1) $\Rightarrow$ (2) if $\nu > 2$. We start with

$$A^{-1/2} = \Gamma\!\left(\frac{1}{2}\right)^{-1} \int_0^\infty t^{-1/2} e^{-tA} \, dt \tag{6.6.46}$$

true applied to $f \in L^1 \cap L^\infty \subset L^2$ by the spectral theorem. We next note by Theorem 5.2.5 of Part 1 applied to $e^{-tA} f$ (or by Riesz–Thorin), we have that for $1 \le p \le \infty$,

$$\|e^{-tA} f\|_p \le C \, t^{-(1-\frac{1}{p})\frac{\nu}{2}} \, \|f\|_1 \tag{6.6.47}$$

(for this holds if $p = 1$ or $p = \infty$) by the equivalence in (6.6.21). Thus, by duality, for $2 \le q \le \infty$,

$$\|e^{-tA} f\|_\infty \le C \, t^{-\nu/2q} \, \|f\|_q \tag{6.6.48}$$

Now write

$$A^{-1/2} f = g_T + h_T \tag{6.6.49}$$

where g_T, h_T come from (6.6.46) with $\int_0^\infty$ replaced by $\int_0^T$ and $\int_T^\infty$. Suppose that $1 < q < \nu$ so that $\nu/2q > \frac{1}{2}$ and $t^{-1/2} \, t^{-\nu/2q}$ is integrable at infinity to see that

$$\|h_T\|_\infty \le C \, T^{1/2-\nu/2q} \, \|f\|_q \tag{6.6.50}$$

Given λ, pick T so that

$$\frac{\lambda}{2} = C \, T^{1/2-\nu/2q} \, \|f\|_q \tag{6.6.51}$$

and thus $|A^{-1/2} f| > \lambda \Rightarrow |g_T| > \frac{\lambda}{2}$ and

$$
\begin{aligned}
|\{x \mid |(A^{-1/2} f)(x)| > \lambda\}| &\le |\{x \mid |g_T(x)| > \tfrac{\lambda}{2}\}| \\
&\le \left(\tfrac{\lambda}{2}\right)^{-q} \|g_T\|_q^q \\
&\le \left(\tfrac{\lambda}{2}\right)^{-q} \left[\Gamma\!\left(\tfrac{1}{2}\right)^{-1} 2 T^{1/2} \|f\|_q^q\right] \tag{6.6.52}
\end{aligned}
$$

using $\|e^{-tA} f\|_q \le \|f\|_q$. Plugging in the relation of T to $\|f\|_q$ and λ yields

$$|\{x \mid |(A^{-1/2} f)(x)| > \lambda\}| \le C \lambda^{-r} \|f\|_q^r \tag{6.6.53}$$

$$\frac{1}{r} = \frac{1}{q} - \frac{1}{\nu} \tag{6.6.54}$$

Thus, $A^{-1/2}$ maps $L^q \to L_w^r$ for $1 < q < \nu$ where r is given by (6.6.54) and so $q < r$. By Marcinkiewicz interpolation (Theorem 6.0.2), it is bounded from L^q to L^r. In particular, taking $q = 2$ (in $(1, \nu)$ since $\nu > 2$), we get

$$\|A^{-1/2} f\|_{2\nu/(\nu-2)} \leq C \|f\|_2 \tag{6.6.55}$$

initially for f in $L^1 \cap L^\infty$ but then by a density argument for $f \in L^2$.

If $g \in Q(A)$, we set $f = A^{1/2} g$ and get (6.6.28) for g. $\square$

In the remainder of this section, we'll restrict to cases where $d\mu$ is a probability measure.

Here is one way $L^p(M, d\mu_0)$-contractive semigroups arise naturally. Suppose that $e^{-tA_0}\varphi \geq 0$ if $\varphi \geq 0$ and there is an L^2-function φ_0 with

$$A_0 \varphi_0 = \alpha \varphi_0, \quad \alpha = \inf \sigma(A_0) \tag{6.6.56}$$

By an infinite-dimensional analog of the Perron–Frobenius theorem (Theorem 7.5.3 of Part 1), $\varphi_0 \geq 0$ and in many cases (see the Notes) α is a simple eigenvalue and $\varphi_0 > 0$. In that case, one can define an operator on $L^2(M, \varphi_0^2 \, d\mu_0)$ by

$$A f = \varphi_0^{-1} (A_0 - \alpha_0)(f \varphi_0) \tag{6.6.57}$$

equivalently,

$$e^{-tA} f = \varphi_0^{-1} e^{-t(A_0 - \alpha_0)} (\varphi_0 f) \tag{6.6.58}$$

e^{-tA} is called the *intrinsic semigroup* associated to A_0. A, especially realized as a Dirichlet form (Theorem 6.6.8 below) is sometimes called a *ground state representation*.

On $L^2(M, \varphi_0^2 \, d\mu_0)$, e^{-tA} is unitarily equivalent to $e^{t\alpha_0} e^{-tA_0}$ on $L^2(M, d\mu_0)$ but there is no such connection with L^p's, $p \neq 2$. Indeed,

$$e^{-tA} \mathbb{1} = \mathbb{1} \tag{6.6.59}$$

This implies that e^{-tA} is an L^p-contractive semigroup by a general result.

Definition. A *Markov semigroup* is a self-adjoint semingroup, e^{-tA} which obeys

$$f \geq 0 \Rightarrow e^{-tA} f \geq 0 \tag{6.6.60}$$

$$e^{-tA} \mathbb{1} = \mathbb{1} \tag{6.6.61}$$

If

$$e^{-tA} \mathbb{1} \leq \mathbb{1} \tag{6.6.62}$$

the semigroup is called a *sub-Markovian semigroup*. If (6.6.60) holds, e^{-tA} is called a *positivity preserving semigroup*.

Remark. Some authors, e.g., Davies [**196**] call what we call sub-Markovian by the name Markov. But, for example, Guionnet–Zegarlarski [**345**] use our definition of Markov.

Proposition 6.6.3. *A positivity preserving semigroup is an L^p-contractive semigroup if and only if it is sub-Markovian. In particular, every Markov semigroup and so every intrinsic semigroup associated to a positivity preserving e^{-tA_0} with strictly positive φ_0 obeying $e^{-tA_0}\varphi_0 = \varphi_0$ is L^p-contractive.*

Proof. For any real θ, $\mathrm{Re}(|f| - e^{i\theta}f) \geq 0$ so noting that e^{-tA} positivity preserving implies it is reality preserving, we get

$$\mathrm{Re}\big(e^{-tA}(|f| - e^{i\theta}f)\big) \geq 0 \tag{6.6.63}$$

or

$$\mathrm{Re}[e^{-tA}(e^{i\theta}f)(x)] \leq e^{-tA}|f|(x) \tag{6.6.64}$$

for a.e. x and each real θ. For any complex number z

$$|z| = \sup_{\theta \text{ rational}} \mathrm{Re}(e^{i\theta}z) \tag{6.6.65}$$

so for a.e. x

$$|(e^{-tA}f)(x)| \leq e^{-tA}|f|(x) \tag{6.6.66}$$

We needed to use the θ rational trick because we only have (6.6.64) for a.e. x so we want to deal with a countable union of measure zero sets.

If $e^{-tA}\mathbb{1} \leq \mathbb{1}$, then since $0 \leq |f| \leq \|f\|_\infty \mathbb{1}$ and e^{-tA} preserves inequalities, we see that

$$0 \leq e^{-tA}|f| \leq \|f\|_\infty e^{-tA}\mathbb{1} \leq \|f\|_\infty \mathbb{1} \tag{6.6.67}$$

This plus (6.6.66) implies

$$\|e^{-tA}f\|_\infty \leq \|f\|_\infty \tag{6.6.68}$$

Duality implies an $L^1 \to L^1$ contraction and then by Riesz–Thorin interpolation, we get an L^p-contraction, i.e., we have proven sub-Markovian implies L^p-contraction.

Conversely, if e^{-tA} is an L^∞-contraction, $\|e^{-tA}\mathbb{1}\|_\infty \leq 1$, i.e., for all x

$$0 \leq (e^{-tA}\mathbb{1})(x) \leq 1 \tag{6.6.69}$$

i.e., e^{-tA} is sub-Markovian. $\qquad\square$

One of the themes below will be when the intrinsic semigroup associated to $-\Delta + V$ on $L^2(\mathbb{R}^\nu, d^\nu x)$, where $V(x) \to \infty$ as $|x| \to \infty$ is an intrinsically hypercontractive or ultracontractive semigroup, meaning the intrinsic semigroup is hypercontractive or ultrcontractive. We'll prove that $-\Delta + |x|^\alpha$ is intrinsically ultracontractive (respectively, hypercontractive) if $\alpha > 2$ (respectively, $\alpha = 2$). It is known (see the Notes) that for $\alpha < 2$, it is not intrinsically hypercontractive.

We begin our study of general hypercontractive semigroups by showing that there is nothing special about L^2 to L^4.

Theorem 6.6.4. *Let e^{-tA} be an L^p-contractive semigroup. Suppose for some p_0, q_0 with $1 < p_0 < q_0 < \infty$, there is T_{p_0,q_0} so that e^{-tA} is bounded from L^{p_0} to L^{q_0}. Then for all $1 < p < q < \infty$, there is $T_{p,q}$ so that $e^{-T_{p,q}A}$ maps L^p to L^q. Moreover, for any $1 < p < \infty$, and $0 < t < 1$, there is $\alpha > 0$ and $C_t \to 1$ as $t \downarrow 0$ so that*

$$\|e^{-tA}f\|_{pe^{\alpha t}} \le C_t \|f\|_p \tag{6.6.70}$$

Remark. The "p" in L^p-contractive is a general symbol as part of the name (short for *all L^p*). It has no relation to the p that appears later.

Proof. The proof is an orgy of interpolation theory! We first claim without loss we can suppose that $p_0 = 2$, $q_0 > 2$. For if $p_0 < 2$, interpolate between $e^{-TH} : L^{p_0} \to L^{q_0}$ and L^∞ to L^∞ to get $e^{-TH} : L^2 \to L^r$ where $r = 2q_0/p_0 > 2$ and if $p_0 > 2$ interpolate between $L^1 \to L^1$ and $L^{p_0} \to L^{q_0}$.

If $p_0 = 2$ and $T = T_{2,q_0}$, we let

$$F(z) = e^{-zTA}, \quad 0 \le \operatorname{Re} z \le 1 \tag{6.6.71}$$

If $\operatorname{Re} z = 0$, $F(z)$ is bounded with norm 1 from L^2 to L^2 and if $\operatorname{Re} z = 1$, $F(z) = e^{-TA}e^{-iyTA}$ maps L^2 to L^2 and then to L^{q_0} with norm C. It follows by the Stein interpolation theorem (see Problem 4 of Section 5.2 of Part 2A) that for $t \in [0,1]$

$$\|e^{-tTA}f\|_{q_t} \le C^t \|f\|_2 \tag{6.6.72}$$

where $q_t^{-1} = tq_0^{-1} + (1-t)\frac{1}{2}$. This implies (6.6.70) for $p = 2$. This handles L^2 to L^q for $2 \le q \le q_0$.

By interpolating between this and $\|e^{-tA}f\|_1 \le \|f\|_1$ and $\|e^{-tA}f\|_\infty \le \|f\|_\infty$, we get $q_t(p)$ running from p to $q_1(p) > p$ so that

$$\|e^{-tA}\|_{q_t(p)} \le C(p)^t \|f\|_p \tag{6.6.73}$$

In particular, for $N = 0, 1, \ldots$

$$\|e^{-TA}f\|_{q(\frac{q}{2})^N} \le C_N \|f\|_{q(\frac{q}{2})^{N-1}} \tag{6.6.74}$$

so

$$\|e^{-NTA}f\|_{q(\frac{q}{2})^N} \le C_1 C_2 \ldots C_N \|f\|_2 \tag{6.6.75}$$

which combining with the argument above proves for any $q > 2$, there is $T_{2,q}$ to $e^{-T_{2,q}A}$ mapping L^2 to L^q.

Interpolating from that to $L^1 \to L^1$ or $L^\infty \to L^\infty$ yields general L^p to L^q. (6.6.73) also implies (6.6.70) $\qquad\square$

We will use $\|B\|_{p,q}$ for the L^p- to L^q-norm of B.

The following says that being hypercontractive and not ultracontractive is a borderline unstable phenomenon in that if A^α generates an L^p-contractive semigroup for $\alpha \in [\alpha_0, \alpha_1]$, there is at most one $\tilde\alpha$ in the interval where $e^{-t\widetilde{A^\alpha}}$ is hypercontractive but not ultracontractive and in that case, e^{-tA^α} is ultracontractive if $\alpha > \widetilde{\alpha}$ and not even hypercontractive if $\alpha < \widetilde{\alpha}$.

Theorem 6.6.5. *Let e^{-tA} be an L^p-contractive semigroup. Suppose for some $c > 0$ and $\alpha \in (0,1)$, e^{-cA^α} is a bounded map of L^2 to L^4. Then, e^{-tA} is ultracontractive.*

Proof. By the spectral theorem,

$$\|e^{cA^\alpha} e^{-tA}\|_{2,2} = \sup_{s>0} e^{cs^\alpha - ts} \equiv e^{g(t)} \tag{6.6.76}$$

By scaling,

$$g(t) = c_1 \, t^{-\alpha/(1-\alpha)} \tag{6.6.77}$$

Clearly,

$$\|e^{-tA}\|_{2,4} \le \|e^{-cA^\alpha}\|_{2,4} \, \|e^{cA^\alpha - tA}\|_{2,2}$$

$$\le D e^{g(t)} \tag{6.6.78}$$

for $D = \|e^{-cA^\alpha}\|_{4,2} < \infty$. By interpolation from L^∞ to L^∞

$$\|e^{-tA}\|_{2^{n+1}, 2^{n+2}} \le \left(D e^{g(t)}\right)^{1/2^n} \tag{6.6.79}$$

Given t, let

$$t_n = 6t\pi^{-2}(n+1)^{-2} \tag{6.6.80}$$

so that

$$\sum_{n=0}^{\infty} t_n = t \tag{6.6.81}$$

and

$$\|e^{-tA}\|_{2,\infty} \le \prod_{n=0}^{\infty} \|e^{-t_n A}\|_{2^{n+1}, 2^{n+2}}$$

$$\le \exp\left(\sum_{n=0}^{\infty} g(t_n)/2^n\right) D^{1+1/2+1/4+\dots}$$

$$= D^2 \exp \sum_{n=0}^{\infty} \frac{c_1}{2^n} \left(\frac{\pi^2}{6}(n+1)^2\right)^{\frac{\alpha}{1-\alpha}} < \infty \tag{6.6.82}$$

$\square$

We next want to see what intrinsic ultracontractivity says about the relation of eigenfunctions and integral kernels. So we suppose that X is a locally compact space and e^{-tA_0} has a jointly continuous strictly positive integral kernel, $a_t(x,y)$, i.e., for

$$a_t \in C(X \times X), \quad a_t(x,y) > 0 \text{ all } t > 0, \, x, y \in X \tag{6.6.83}$$

$$(e^{-tA_0} f)(x) = \int a_t(x,y) \, f(y) \, d\mu_0(y) \tag{6.6.84}$$

We also suppose for all $t > 0$

$$\mathrm{Tr}(e^{-tA_0}) < \infty \tag{6.6.85}$$

which by Mercer's theorem (see Theorem 3.10.9 of Part 4), is equivalent to

$$\mathrm{Tr}(e^{-tA_0}) = \int a_t(x,x) \, d\mu_0(x) < \infty \tag{6.6.86}$$

It is known (see the Notes) that if $A_0 = -\Delta + V$ on $L^2(\mathbb{R}^\nu)$ with $V \geq 0$ and $V(x) \to \infty$ at ∞, then, if e^{-tA_0} is intrinsically hypercontractive, (6.6.85) holds so in this important special case, there is no loss in assuming it.

Moreover, if e^{-tA} is ultracontractive on a probability measure space, we know that (6.6.86) holds for e^{-tA} since $a_t(x,x)$ is bounded. So in the context of intrinsically defined semigroups, using that e^{-tA_0} on $L^2(X, d\mu_0)$ is unitarily equivalent to e^{-tA} on $L^2(X, d\mu)$, we have that (6.6.85) holds for e^{-tA_0} and e^{-tA_0} is intrinsically ultracontractive.

Theorem 6.6.6. *Let e^{-tA} be an ultracontractive semigroup on $L^2(X, d\mu)$ with $\mu(X) < \infty$. Then, e^{-tA} is trace class and, in particular, a compact operator (so A has a complete set of eigenvectors, φ_n with $A\varphi_n = \alpha_n \varphi_n$ and $\alpha_n \to \infty$).*

Proof. By the Dunford–Pettis theorem (Theorem 4.9.7 of Part 1), e^{-tA} has an integral kernel $a_t(x,y)$ in $L^\infty(X \times X)$, so since $\mu(X) < \infty$, e^{-tA} is Hilbert–Schmidt and $e^{-tA} = \left(e^{-tA/2}\right)^2$ is trace class. $\qquad\square$

Define

$$b_t(x) = a_t(x,x)^{1/2} \tag{6.6.87}$$

so the positive definiteness of $a_t(x,y)$ as a kernel implies

$$0 < a_t(x,y) \leq b_t(x) \, b_t(y) \tag{6.6.88}$$

Theorem 6.6.7. *Let e^{-tA_0} be an L^2-semigroup obeying (6.6.86) and $a_t(x,y)$ continuous and strictly positive. Let $A_0\varphi_0 = \alpha\varphi_0$ with $\|\varphi_0\| = 1$, $\varphi_0 > 0$, and $\alpha = \inf \sigma(A_0)$. Then the following are equivalent:*

(1) *e^{-tA_0} is intrinsically ultracontractive.*

(2) *For all $t > 0$, there is c_t so that for all $f \in L^2$*

$$|e^{-tA_0} f(x)| \leq c_t \, \|f\|_2 \, \varphi_0(x) \tag{6.6.89}$$

(3) *For all $t > 0$, there exists d_t so that*

$$b_t(x) \leq d_t \, \varphi_0(x) \tag{6.6.90}$$

(4) *For all $t > 0$, there exists f_t so that*

$$a_t(x,y) \leq f_t \, \varphi_0(x) \, \varphi_0(y) \tag{6.6.91}$$

(5) *For all $t > 0$, there is g_t so that*

$$a_t(x, y) \geq g_t \, b_t(x) \, b_t(y)$$

Proof. We'll show that $(1) \Leftrightarrow (2) \Leftrightarrow (3) \Leftrightarrow (4) \Rightarrow (5) \Rightarrow (3)$.

$\underline{(1) \Leftrightarrow (2)}$. $e^{-tA} = \varphi_0^{-1} \, e^{-tA_0} \, \varphi_0$, implies

$$\varphi_0 \, e^{-tA} g = e^{-tA_0} \, (\varphi_0 g) \tag{6.6.92}$$

and $\|\varphi_0 g\|_{L^2(X, d\mu_0)} = \|g\|_{L^2(X, d\mu)}$. Thus,

$$\|e^{-tA} g\|_\infty \leq c_t \, \|g\|_{L^2(X, d\mu)} \Leftrightarrow$$
$$\forall x \, |(e^{-tA_0} \varphi_0 g)(x)| \leq c_t \, \|\varphi_0 g\|_{L^2(X, d\mu)} \varphi_0(x)$$

which says (6.6.89) is equivalent to ultracontractivity of e^{-tA}.

$\underline{(1) \Rightarrow (3)}$. Let $\tilde{a}_t(x, y) = a_t(x, y) \, \varphi_0(x)^{-1} \, \varphi_0(y)^{-1}$ be the integral kernel of e^{-tA}. If e^{-tA} is ultracontractive, $e^{-tA} \colon L^1 \to L^\infty$, so $|\tilde{a}_t(x, y)| \leq \|e^{-tA_0}\|_{L^1 \to L^\infty} \equiv d_t^2$. Then,

$$\tilde{b}_t(x) \equiv b_t(x) \, \varphi_0(x)^{-1} = [a_t(x, x)\varphi_0(x)^{-2}]^{-1/2} = \tilde{a}_t(x, x)^{1/2} \leq d_t \tag{6.6.93}$$

$\underline{(3) \Rightarrow (1)}$. By (6.6.88), $|\tilde{a}_t(x, y)| \leq d_t^2$ so

$$\|e^{-tA} f\|_\infty \leq d_t^2 \, \|f\|_1 \tag{6.6.94}$$

proving ultracontractivity.

$\underline{(3) \Rightarrow (4)}$. Immediate from (6.6.88).

$\underline{(4) \Rightarrow (3)}$. Take $x = y$.

$\underline{(4) \Rightarrow (5)}$. Fix t. Pick K compact so that

$$\int_{X \setminus K} \varphi_0(x)^2 \, d\mu_0(x) \leq e^{-\alpha t} (2 f_t)^{-1} \tag{6.6.95}$$

Then,

$$e^{-\alpha t} \varphi_0(x) = \int_X a_t(x, y)\varphi_0(y) d\mu_0(y)$$
$$\leq f_t \int_{X \setminus K} \varphi_0(x)\varphi_0(y)^2 \, d\mu_0(y) + \int_K a_t(x, y)\varphi_0(y) \, d\mu(y) \tag{6.6.96}$$
$$\leq \tfrac{1}{2} e^{-\alpha t}\varphi_0(x) + \int_K a_t(x, y)\varphi_0(y) \, d\mu_0(y) \tag{6.6.97}$$

Thus,

$$\int_K a_t(x, y) \, \varphi_0(y) \, d\mu(y) \geq \tfrac{1}{2} e^{-\alpha t}\varphi_0(x) \tag{6.6.98}$$

Since a_t and φ_0 are continuous and strictly positive and K is compact,

$$\gamma \equiv \min_{z,x \in K} \left\{ \frac{a_t(z,w)}{\varphi_0(z)\varphi_0(w)} \right\} \tag{6.6.99}$$

is strictly positive.

Since $a_t \geq 0$,

$$a_{3t}(x,y) \geq \int_{w \in K, z \in K} a_t(x,z)\, a_t(z,w)\, a_t(w,y)\, d\mu_0(z)\, d\mu_0(w)$$

$$\geq \gamma \left(\tfrac{1}{2} e^{-\alpha t} \right)^2 \varphi_0(x)\, \varphi_0(y) \tag{6.6.100}$$

since $a_t(z,w) \geq \gamma \varphi_0(x)\varphi_0(y)$ and we can then use (6.6.98) and $a_t(w,y) = a_t(y,w)$. This proves (5) given that (4) $\Rightarrow$ (3).

$\underline{(5) \Rightarrow (3)}$. From the assumed (5),

$$\varphi_0(x) = e^{\alpha t} \int a_t(x,y)\varphi_0(y)\, d\mu_0(y)$$

$$\geq e^{\alpha t} g_t b_t(x) \int b_t(y)\varphi_0(y)\, d\mu_0(y) \tag{6.6.101}$$

$$\geq e^{\frac{1}{2}\alpha t} g_t\, b_t(x) \int \varphi_0(y)^2\, d\mu_0(y) = e^{\frac{1}{2}\alpha t} g_t\, b_t(x) \tag{6.6.102}$$

where we used

$$\varphi_0(x)^2 \leq e^{\alpha t} \sum_n e^{-\alpha_n t} |\varphi_n(x)|^2 \tag{6.6.103}$$

$$= b_t(x)^2 e^{\alpha t} \tag{6.6.104}$$

since

$$a_t(x,y) = \sum_n e^{-\alpha_n t} \varphi_n(x)\varphi_n(y) \tag{6.6.105}$$

is an absolutely convergent series (see the proof of Mercer's theorem, Theorem 3.11.9 of Part 4). $\qquad \square$

Our next topic is to understand two of the basic examples of the subject, so much so they mark critical points in the history and in the applications— optimal intrinsic hypercontractive estimates for $-\frac{d^2}{dx^2} + x^2$ and for the simplest 2×2 matrix self-adjoint semigroup.

Since intrinsic $-\frac{d^2}{dx^2} + x^2$ is a special case of an intrinsic Schrödinger operator, we pause to say more about A when $A_0 = -\Delta + V(x)$ on $L^2(\mathbb{R}^\nu)$. Let $\varphi_0(x) > 0$ for all x with $A_0\, \varphi_0 = \alpha\varphi_0$. Let f be a C_0^∞-function. Let M_f be multiplication by f, i.e., $M_f\, g = fg$. Formally,

$$\left[[(A_0 - \alpha), M_f], M_f \right] = -2|\nabla f|^2 \tag{6.6.106}$$

where $[B, C] = BC - CB$. If $V \geq 0$ and is in $L^2_{\text{loc}}(\mathbb{R})$, then Theorem 7.6.1 of Part 4 shows that M_f takes the domain of A_0 to itself and (6.6.106) is true when applied to vectors in $D(A_0)$. Note that

$$\langle \varphi_0, [[A_0 - \alpha, M_f], M_f] \varphi_0 \rangle = -2 \langle M_f \varphi_0, (A_0 - \alpha) M_f \varphi_0 \rangle \qquad (6.6.107)$$

since $(A_0 - \alpha) \varphi_0 = 0$.

Thus, for any $f \in C_0^\infty$, we have shown that

$$\langle f \varphi_0, (A_0 - \alpha) f \varphi_0 \rangle = \int |\nabla f(x)|^2 \varphi_0(x)^2 \, dx \qquad (6.6.108)$$

Taking into account how A is defined on $L^2(\mathbb{R}, d\mu)$, we have shown the following

Theorem 6.6.8. *Let $V \geq 0$ on $\mathbb{R}^\nu$ be in $L^2_{\text{loc}}(\mathbb{R}^\nu)$ and suppose $A_0 = -\Delta + V$ (defined as the closure of the operator on $C_0^\infty(\mathbb{R}^\nu)$). Suppose $\varphi_0 > 0$ obeys $\varphi_0 \in D(A_0)$, $A_0 \varphi_0 = \alpha \varphi_0$ and $\|\varphi_0\| = 1$. If A is the generator of the intrinsic semigroup associated to A and $d\mu(x) = \varphi_0(x)^2 dx$, then*

$$\langle f, Af \rangle = \int |\nabla f(x)|^2 \, d\mu(x) \qquad (6.6.109)$$

Remarks. 1. We have proven this for $f \in C_0^\infty(\mathbb{R}^\nu)$ but this is a form core (by Theorem 7.6.2 of Part 4), so it is easy to show (Problem 3) that $Q(A) = \{f \in L^2(\mathbb{R}^\nu, d\mu) \mid \nabla f \in L^2(\mathbb{R}^\nu, d\mu)\}$ and (6.6.109) holds for all $f \in Q(A)$ if $\langle f, Af \rangle$ is interpreted as $q_A(f)$.

2. Because of the similarity to Dirichlet energies (see the Notes to Section 8.1 of Part 2A), operators defined by (6.6.109) are called *classical Dirichlet forms*. They obey both Beurling–Deny criteria because e^{-tA} is a Markov semigroup. In their theory, Beurling–Deny call any operator obeying both of their criteria Dirichlet semigroups. Some authors follow this convention for that term and others use the name for what we call classical Dirichlet forms. We've added "classical" for that reason.

If φ_0 is C^1 and locally bounded away from zero (which can often be proven must hold) and $f \in C_0^\infty$ is real, then

$$\int (\vec{\nabla} f)^2 \varphi_0^2 \, d^\nu x = - \int f (\vec{\nabla} \cdot [(\vec{\nabla} f) \varphi_0^2]) \, d^\nu x \qquad (6.6.110)$$

$$= \int f(-\Delta f - 2\varphi_0^{-1} \vec{\nabla} \varphi_0 \cdot \vec{\nabla} f) \varphi_0^2 \, d^\nu x \qquad (6.6.111)$$

It follows that

Theorem 6.6.9. *Under the hypotheses of Theorem 6.6.8, if also φ_0 is C^1 and $f \in C_0^\infty(\mathbb{R}^\nu)$, then*

$$Af = -\Delta f - \vec{g} \cdot \vec{\nabla} f \qquad (6.6.112)$$

where

$$\vec{g}(x) = -2\varphi_0(x)^{-1}\vec{\nabla}\varphi_0(x) = -2\vec{\nabla}\log\varphi_0(x) \qquad (6.6.113)$$

Remark. If $\int|\vec{\nabla}\varphi_0|^2\,d^\nu x < \infty$, then $g \in L^2(\mathbb{R}, d\mu)$ and using that $D(A) \subset Q(A)$ means $f \in D(A)$ has $\vec{\nabla}f \in L^2(\mathbb{R}, d\mu)$, one can prove (Problem 4), that under this extra hypothesis on φ_0, $D(A) = \{f \mid f, \vec{\nabla}f, -\Delta f - \vec{g}\cdot\vec{\nabla}f \in L^2(\mathbb{R}, d\mu)\}$ and (6.6.112) holds for all $f \in D(A)$.

For later purposes, we want to note if A is a Dirichlet form, then for any $g \in L^1 \cap L^\infty \cap D(A)$ nonnegative, we have

$$Q_A(g^{p/2}) = \left(\tfrac{p}{2}\right)^2 (p-1)^{-1}\langle g^{p-1}, Ag\rangle \qquad (6.6.114)$$

which is immediate if we note that

$$\nabla g^{p/2} = \tfrac{p}{2}g^{(p/2-1)}\nabla g, \quad \nabla g^{p-1} = (p-1)g^{p-2}\nabla g \qquad (6.6.115)$$

so

$$Q_A(g^{p/2}) = \left(\tfrac{p}{2}\right)^2 \int g^{p-2}|\nabla g|^2 \qquad (6.6.116)$$

$$= \left(\tfrac{p}{2}\right)^2 (p-1)^{-1}\langle\nabla g^{p-1}, \nabla g\rangle \qquad (6.6.117)$$

$$= \left(\tfrac{p}{2}\right)(p-1)^{-1}\langle g^{p-1}, Ag\rangle \qquad (6.6.118)$$

since $g \in D(A)$. We will prove below (Theorem 6.6.13) that for general sub-Markovian semigroups while (6.6.114) might fail, we have that

$$Q_A(g^{p/2}) \leq \left(\tfrac{p}{2}\right)^2 (p-1)^{-1}\langle g^{p-1}, Ag\rangle \qquad (6.6.119)$$

If (6.6.119) holds, we say that A obeys a *sub-Dirichlet inequality*.

Example 6.6.10 (Ornstein–Uhlenbeck semigroup). This is also called the *Gauss semigroup* or *Hermite semigroup*. One takes $A_0 = -\Delta + \tfrac{1}{4}x^2$ in $\mathbb{R}^\nu$. By the analysis in Section 6.4 of Part 1

$$\varphi_0(x) = (2\pi)^{-\nu/4}\exp(-\tfrac{1}{4}x^2), \quad A_0\varphi_0 = \tfrac{\nu}{2}\varphi_0 \qquad (6.6.120)$$

A_0 has eigenvalues $n + \tfrac{\nu}{2}$ (if $\nu = 1$, they are simple for all n; for $\nu \geq 2$, there is a degeneracy growing as $O(n^\nu/\nu!)$). One has that

$$d\mu(x) = (2\pi)^{-\nu/2}\exp(-\tfrac{1}{2}x^2)\,d^\nu x \qquad (6.6.121)$$

By (6.6.112),(6.6.113)

$$A = -\Delta + \vec{x}\cdot\vec{\nabla} \qquad (6.6.122)$$

It is traditional to pick this normalization so that $\int x_j^2\,d\mu(x) = 1$, rather than the $\tfrac{1}{2}$ it would be if we took $-\Delta + x^2$. We know that eigenfunctions of A_0 are polynomials times φ_0, so eigenfunctions of A are polynomials and so not in L^∞. It follows that e^{-tA} cannot be ultracontractive since $\varphi_n = e^{t\alpha_n}e^{-tA}\varphi_n$ would then be bounded but they are hypercontractive. Below, we'll prove Nelson's best hypercontractive estimate: e^{-tA} is a contraction of L^p to L^q if

$$e^{-t} \leq \sqrt{\frac{p-1}{q-1}} \qquad (6.6.123)$$

and unbounded from L^p to L^q if (6.6.123) fails. $\square$

Next we look at some discrete examples. Hypercontractivity is interesting, even for suitable 2×2 matrices! Let G be a *finite simple graph* i.e., k points, $x_1, \ldots, x_k$ with an incidence relation, i.e., some set, S, of pairs $(i,j) \in \{1, \ldots, k\}^2$ so that $(i,i) \notin S$ and $(i,j) \in S \Rightarrow (j,i) \in S$. We also suppose that $\{1, \ldots, k\}$ is connected, i.e., for any i, m, there is $j_1, \ldots, j_\ell$ so $(i, j_1), (j_1, j_2), \ldots, (j_\ell, m)$ are all in S. We'll call G *connected* in that case.

Example 6.6.11 (Graph Laplacian). Let G be a finite simple graph. If $\sharp(G) = k$, we set μ_0 to be the normalized counting measure, i.e., $\mu_0(\{p\}) = \frac{1}{k}$ for all $p \in G$; $\ell^2(G, \mu_0)$ is the corresponding L^2 space. Define B on $\ell^2(G, \mu_0)$ by

$$(B\varphi)(j) = \frac{1}{2} \sum_{i \in G | (j,i) \in S} [\varphi(j) - \varphi(i)] \qquad (6.6.124)$$

Thus (Problem 5)

$$\langle \varphi, B\varphi \rangle = \frac{1}{2k} \sum_{(j,i) \in S} |\varphi(j) - \varphi(i)|^2 \qquad (6.6.125)$$

Notice that $B\mathbb{1} = 0$ and $B \geq 0$. Since G is connected, $\ker(B)$ is exactly the constant function.

It is a useful fact that graph Laplacians obey a sub-Dirichlet bound (see (6.6.119)) $\square$

Proposition 6.6.12. *Let B be a graph Laplacian. Then, for $1 < p < \infty$ and any nonnegative function φ, we have that*

$$\langle \varphi^{p/2}, B\, \varphi^{p/2} \rangle \leq (\tfrac{p}{2})^2\, (p-1)^{-1} \langle \varphi^{p-1}, B\, \varphi \rangle \qquad (6.6.126)$$

Proof. By (6.6.125) and polarization,

$$\langle \psi, B\varphi \rangle = \frac{1}{2k} \sum_{(p,q) \in S} \overline{(\psi(p) - \psi(q))}\, (\varphi(p) - \varphi(q)) \qquad (6.6.127)$$

Thus, (6.6.126) is implied by (using $\alpha = \varphi(i)^{p/2}$, $\beta = \varphi(j)^{p/2}$ and interchanging α and β if need be)

$$1 < p < \infty,\ 0 \leq \alpha \leq \beta < \infty \Rightarrow |\beta^{p/2} - \alpha^{p/2}|^2 \leq \frac{p^2}{4(p-1)}\, (\beta - \alpha)\, (\beta^{p-1} - \alpha^{p-1}) \qquad (6.6.128)$$

To see this, write

$$\left(\beta^{p/2} - \alpha^{p/2}\right)^2 = \left(\frac{p}{2}\int_\alpha^\beta s^{p/2-1}\,ds\right)^2 \tag{6.6.129}$$

$$\leq \frac{p^2}{4}(\beta - \alpha)\int_\alpha^\beta s^{p-2}\,ds \tag{6.6.130}$$

$$= \frac{p^2}{4(p-1)}(\beta - \alpha)\left(\beta^{p-1} - \alpha^{p-1}\right) \tag{6.6.131}$$

where (6.6.130) uses the Schwarz inequality. □

This same argument works in a more general context if one uses the details of the proof of the second Beurling–Deny criterion (Theorem 7.6.5 of Part 4).

Theorem 6.6.13. *Let $A \geq 0$ be a self-adjoint operator so that e^{-tA} is a sub-Markovian semigroup. Then (6.6.119) holds for $g \in L^1 \cap L^\infty \cap D(A)$, i.e., A obeys a sub-Dirichlet inequality.*

Proof. It suffices to prove that for any $q \in L^1 \cap L^\infty$, $g \geq 0$, we have that for $C = e^{-tA}$

$$\langle g^{p/2}, (\mathbb{1} - C) g^{p/2}\rangle \leq \left(\frac{p}{2}\right)^2 (p-1)^{-1}\langle g^{p-1}, (\mathbb{1} - C) g\rangle \tag{6.6.132}$$

since we can divide by t and take $t \downarrow 0$. We'll prove this for any positivity-preserving map C obeying

$$\|Cg\|_1 \leq \|g\|_1 \tag{6.6.133}$$

As in the proof of Theorem 7.6.5 in Part 4, by an approximation argument, it suffices to prove (6.6.132) if (M, μ) has M a finite set of points. Also as in the proof of that theorem in that case, there are $K_{ij} \geq 0$ and $m_j \geq 0$ so that

$$\langle x, (\mathbb{1} - C) y\rangle = \sum_{i,j=1}^n (\bar{x}_i - \bar{x}_j)(y_i - y_j)K_{ij} + \sum_{j=1}^n m_j \bar{x}_j y_j \tag{6.6.134}$$

By (6.6.128), for $x_\ell \geq 0$

$$|x_i^{p/2} - x_j^{p/2}|^2 \leq \frac{p^2}{4(p-1)}(x_i - x_j)(x_i^{p-1} - x_j^{p-1}) \tag{6.6.135}$$

Moreover,

$$p^2 = (p-2)^2 + 4p - 4 \geq 4(p-1) \Rightarrow \frac{p^2}{4(p-1)} \geq 1 \tag{6.6.136}$$

so

$$x_j^p \leq \frac{p^2}{4(p-1)}x_i x_i^{p-1} \tag{6.6.137}$$

Thus, by (6.6.134), we have (6.6.132) for the discrete case. □

Example 6.6.14 (Bonami–Gross Semigroup). Let $G = \{1, -1\}$ with $S = \{(1, -1), (-1, 1)\}$. Thus, in the natural orthonormal basis $\sqrt{2}\delta_1, \sqrt{2}\delta_{-1}$

$$B = \begin{pmatrix} \frac{1}{2} & -\frac{1}{2} \\ -\frac{1}{2} & \frac{1}{2} \end{pmatrix} \tag{6.6.138}$$

The semigroup generated by B, i.e., e^{-tB}, is the one-dimensional *Bonami–Gross semigroup*. The ν-dimensional Bonami–Gross semigroup acts on $G_\nu = \{1, -1\}^\nu$ with $B = B_1 + \cdots + B_\nu$ so $e^{-tB} = e^{-tB_1} \otimes \cdots \otimes e^{-tB_\nu}$ where B_j is the one-dimensional operator on the jth-coordinate. B is the Laplacian for G_ν with $((j_1, \ldots, j_\nu), (i_1, \ldots, i_\nu)) \in S \Leftrightarrow \sum_{\ell=1}^\nu |i_\ell - j_\ell| = 1$, i.e., G_ν are the vertices of a hypercube with incidence relation given by the edges. We'll prove that e^{-tB} is a contraction from L^p to L^q if and only if (6.6.123) holds. $\qquad\square$

Notice that the Bonami–Gross semigroup e^{-tB_ν} in ν dimensions is the ν-fold product of e^{-tB_1}. This is also true of Ornstein–Uhlenbeck semigroups. Thus, the following result reduces the results we seek to the ($\nu = 1$)-case:

Theorem 6.6.15 (Bonami–Segal Lemma). *Let $(M, d\mu)$ and $(N, d\nu)$ be probability measure spaces. Let C (respectively, D) be maps on $L^p(M, d\mu)$ (respectively, $L^p(N, d\nu)$) which are contractions on each L^p space and so that D is positivity preserving. For a given $1 \leq p \leq q < \infty$, suppose that C is a bounded map of $L^p(M, d\mu)$ to $L^q(N, d\nu)$ with norm $\|C\|_{p,q}$ and similarly for D. Then, there is a unique linear map $C \otimes D \colon L^p(M \times N, d\mu \otimes d\nu) \to L^q(M \times N, d\mu \otimes d\nu)$ so that*

$$\|C \otimes D\| = \|C\|_{p,q} \|D\|_{p,q}, \quad (C \otimes D)(f \otimes g) = Cf \otimes Dg \tag{6.6.139}$$

Proof. Suppose first that C and D have integral kernels (i.e., $(Cf)(m) = \int C(m, m')f(m')d\mu(m')$. Let $C \otimes D$ be the operator which has integral kernel $C(m, m') D(n, n')$ which obeys the second condition in (6.6.139). (By a density argument uniqueness is easy.) Let $F \in L^p(M \times N)$ and $G \in L^{q'}(M \times N)$ where q' is the dual index of q. Then,

$$\left| \int G(m, n)\left[(C \otimes D)F\right](m, n)d\mu(m)d\nu(n') \right| =$$

$$\left| \int G(m, n)C(m, m')D(n, n')F(m', n')d\mu(m)d\mu(n')d\nu(n)d\nu(n') \right| \tag{6.6.140}$$

Since C maps L^p to L^q for each fixed n, n', we have

$$\left| \int G(m, n)C(m, m')F(m', n')d\mu(m')d\mu(m) \right| \leq \|C\|_{p,q} \|G(\cdot, n)\|_{q'} \|F(\cdot, n')\|_{p'} \tag{6.6.141}$$

Thus,

$$\text{LHS of (6.6.140)} \leq \|C\|_{p,q} \int \|G(\cdot,n)\|_{q'} D(n,n') \|F(\cdot,n')\|_p \, d\nu(n)\, d\nu(n') \tag{6.6.142}$$

where we used positivity of D. Using the L^p- to L^q-boundedness of D we get

$$\text{LHS of (6.6.140)} \leq \|C\|_{p,q} \|D\|_{p,q} \|G\|_{q'} \|F\|_p \tag{6.6.143}$$

which is the other half of (6.6.139).

It is easy to show that any operator from L^p to L^q can be approximated by operators (with not larger norm) with integral kernels (See Problem 6) so $\|C \otimes D\| \leq \|C\| \|D\|$ is proven in general.

Since

$$\frac{\|(C \otimes D)f \otimes g\|}{\|f \otimes g\|} = \frac{\|Cf\|}{\|f\|} \frac{\|Dg\|}{\|g\|} \tag{6.6.144}$$

it follows that $\|C \otimes D\| \geq \|C\| \|D\|$.   □

The second general tool we need is the relation between hypercontractive estimates and their infinitesimal form. As a preliminary, we need the following which we'll only state for Markov semigroups. In that case, it is useful to define:

$$\mathcal{D} = \bigcup_{s>0} e^{-sA} L^\infty, \quad \mathcal{D}_+ = \bigcup_{s>0} e^{-sA}\{u \in L^\infty \mid u \geq 0\}$$

$$\mathcal{D}_{++} = \bigcup_{s>0} e^{-sA}\{u \in L^\infty \mid \exists \varepsilon > 0 \text{ so } \varepsilon \mathbb{1} \leq u\} \tag{6.6.145}$$

Of course, $\mathcal{D}_{++} \subset \mathcal{D}_+ \subset \mathcal{D} \subset D(A) \cup \bigcup_{1 \leq p \leq \infty} L^p$. $\mathcal{D}$ is L^p-dense in each L^p and $\mathcal{D}_{++}$ and $\mathcal{D}_+$ are L^p-dense in $\{u \in L^p \mid u \geq 0\}$. Moreover (Problem 7), $\mathcal{D}$ is an L^p-operator core, $1 < p < \infty$, for the L^p-generator of e^{-tA}.

Proposition 6.6.16. (a) *For $f \in \mathcal{D}_{++}$, $G(p) = \|f\|_p^p$ is C^1 in p on $[0,\infty)$ and*

$$G'(p) = \int f(x)^p \log f(x) d\mu(x) \tag{6.6.146}$$

(b) *For $f \in \mathcal{D}_{++}$, $Q(p) = \|f\|_p$ is C^1 and*

$$Q'(p) = (p\|f\|_p^{p-1})^{-1} \left[\int f(x)^p \log f(x) - \|f\|_p^p \log\|f\|_p \right] \tag{6.6.147}$$

(c) *For $f \in \mathcal{D}_{++}$ and p fixed, $F(s) = \|e^{-sA}f\|_p^p$ is C^1 on $[0,\infty)$ and (with $f_s = e^{-sA}f$)*

$$F'(s) = - \int f_s(x)^{p-1} (Af_s)(x) \, d\mu(x) \tag{6.6.148}$$

(d) *If $f \in \mathcal{D}_{++}$, and p is fixed and $R(s) = \|e^{-sA}\|_p$ then,*

$$R'(s) = -(p\|f\|_p^{p-1})^{-1} \int f_s(x)^{p-1} (Af_s)(x)d\mu(x) \qquad (6.6.149)$$

(e) *If $p(s) \in [1, \infty)$ is a C^1-function of s and $M(s) \in (-\infty, \infty)$ is a C^1-function of s, then for $f \in \mathcal{D}_{++}$*

$$H(s) = \log[e^{-M(s)}\|f_s\|_{p(s)}]; \quad f(s) = e^{-sA}f \qquad (6.6.150)$$

is C^1 and

$$\frac{dH}{ds} = -M'(s) + \|f_s\|_{p(s)}^{-p(s)} p(s)^{-1} p'(s)$$

$$\left\{ \int f_s(x)^{p(s)} \log f_s(x)d\mu(x) - \|f_s\|_{p(s)}^{p(s)} \log\|f_s\|_{p(s)} \qquad (6.6.151) \right.$$

$$\left. - \int f_s(x)^{p-1}(Af)(x)d\mu(x) \right\}$$

Proof. (a) Write

$$G(p) = \int e^{p \log f(x)} d\mu(x) \qquad (6.6.152)$$

Since $|e^x - e^y| \leq |x - y|\,(e^x + e^y)$ (by integrating the derivative), we see that the difference quotient $[e^{(p+\delta p) \log f(x)} - e^{p \log f(x)}]/\delta p$ is uniformly bounded so (6.6.152) follows from the dominated convergence theorem.

(b) $\log G(p) = p \log Q(p)$ so

$$G'(p)\,G(p)^{-1} = \log Q(p) + p\,Q'(p)Q(p)^{-1} \qquad (6.6.153)$$

so

$$Q'(p) = (Q/pG)[G' - Q \log Q] \qquad (6.6.154)$$

which is (6.6.147).

(c) Write with $f_s = e^{-sA}f$

$$s^{-1}[f_s^p - f^p] = \{[f_s^p - f^p]/[f_s - f]\}\{[f_s - f]/s\} \qquad (6.6.155)$$

Since $|a^p - b^p| \leq p|a - b|[a^{p-1} + b^{p-1}]$ (by integrating the derivatives), the first factor in (6.6.155) converges to pf^{p-1} in any L^q space, $q < \infty$, by the dominated convergence theorem. The second factor converges in L^2 to $-Af$. Taking $q = 2$, we get the result at $s = 0$. The proof for general s is identical since $F(s - s_0) = \|e^{-sA}f_{s_0}\|^p$.

(d) Since p is fixed and

$$R(s) = F(s)^{\frac{1}{p}} \qquad (6.6.156)$$

$$R'(s) = \tfrac{1}{p}F(s)^{-1+\frac{1}{p}} F'(s) \qquad (6.6.157)$$

which is (6.6.148).

(e) Follows from (a), (b), and the chain rule and $\|f\|_p^{-p+1}\|f\|_p^{-1} = \|f\|_p^{-p}$. $\square$

This suggests that we single out a specific inequality.

Definition. We say that A, a generator of a Markov semigroup, obeys an L^p-*logarithmic Sobolev inequality* if there are constants $c(p)$ and $\Gamma(p)$ so that for all $f \in \mathcal{D}_{++}$

$$\int f(x)^p \log f(x) \leq c(p)\langle f^{p-1}, Af\rangle + \Gamma(p)\,\|f\|_p^p + \|f\|_p^p \log\|f\|_p \quad (6.6.158)$$

$c(p)$ is called the *local constant* and $\Gamma(p)$, the *local norm*.

Here is the connection between log Sobolev inequalities and hypercontractivity.

Theorem 6.6.17 (Gross' Theorem). *Let e^{-tA} be a Markov semigroup. Suppose for $p \in [p_0, p_1)$, A obeys a p-log Sobolev inequality with local norm, $\Gamma(p)$, and local constant, $c(p)$, both continuous in p. Suppose*

$$t = \int_{p_0}^{p_1} \frac{c(p)}{p}\,dp, \quad M = \int_{p_0}^{p_1} \Gamma(p)\frac{dp}{p} \quad (6.6.159)$$

are both finite. Then for all $f \in L^{p_0} \cap L^{p_1}$, we have

$$\|e^{-tA}f\|_{p_1} \leq e^M \|f\|_{p_0} \quad (6.6.160)$$

Conversely, if $p_0(t), p(t), M(t)$ are C^1-functions of $t \in [0, \delta)$ for some $\delta > 0$ with $p_0(t) < p_1(t)$ for $t > 0$ and $p_0(0) = p_1(0) \equiv p$, $M(0) = 0$ so that

$$\|e^{-tA}\|_{p_1(t)} \leq e^{M(t)} \|f\|_{p_0(t)} \quad (6.6.161)$$

for all $f \in L^1 \cap L^\infty$, then A obeys a p-log Sobolev inequality with local constant and local norm given by

$$c(p) = \frac{p}{p_1'(0) - p_0'(0)}, \quad \Gamma(p) = M'(0)\,c(p) \quad (6.6.162)$$

Remarks. 1. Suppose one has

$$\|e^{-tA}\|_{q(t,p)} \leq e^{M_p(t)} \|f\|_p \quad (6.6.163)$$

for C^1-functions q and M_p (for each $p \in [p_0, p_1]$ where q and M are also C^1 in p). Then by the second half of the theorem, one gets a p-log Sobolev inequality for all $p \in [p_0, p_1)$ and so one can integrate it to get a bound in (6.6.163). It can happen (see Example 6.6.18 below) that in this round trip the t needed to go from some p to q gets larger and/or M could increase; that is, the global and infinitesimal inequalities are not equivalent. For the case of Gross–Nelson semigroups, though, they are.

2. We add the M, t finite condition because $p_1 = \infty$ is allowed; that is, we can use log Sobolev inequalities, as we'll see, to get ultracontractive estimates.

Proof. For $s \in (0, t)$, define $\widetilde{p}(s)$ by

$$s = \int_{p_0}^{\widetilde{p}(s)} \frac{c(p)}{p} \, dp, \quad \widetilde{c}(s) = c(\widetilde{p}(s)), \quad M(s) = \int_{p_0}^{\widetilde{p}(s)} \Gamma(p) \frac{dp}{p} \qquad (6.6.164)$$

Note that

$$\frac{d\widetilde{p}(s)}{ds} = \frac{\widetilde{p}(s)}{\widetilde{c}(s)}, \quad \frac{dM(s)}{ds} = \frac{\Gamma(\widetilde{p}(s))}{\widetilde{c}(s)} \qquad (6.6.165)$$

Let

$$H(s) = \log\left[e^{-M(s)} \|e^{-sA} f\|_{p(s)}\right] \qquad (6.6.166)$$

Then by (6.6.151) and $M'(s)\widetilde{p}(s)\widetilde{p}'(s)^{-1} = \Gamma(\widetilde{p}(s))$,

$$H'(s) = \left[\|f_s\|_{\widetilde{p}(s)}^{\widetilde{p}(s)} \widetilde{p}(s)\widetilde{c}(s)\right]^{-1} \widetilde{p}'(s)$$

$$\left\{\int f_s(x)^{\widetilde{p}(s)} \log f_s(x) d\mu(x) - \|f_s\|_{\widetilde{p}(s)}^{\widetilde{p}(s)} \log\|f\|_{p(s)} \qquad (6.6.167)\right.$$

$$\left. - \Gamma(p(s)) \|f_s\|_{\widetilde{p}(s)}^{\widetilde{p}(s)} - \int f_s(x)^{\widetilde{p}-1}(Af_s)(x)d\mu(x)\right\}$$

$$\leq 0$$

for all $f \in \mathcal{D}_{++}$, if and only if the $L^{\widetilde{p}(s)}$-logarithmic Sobolev inequality holds with local norm $\Gamma(p(s))$ and local constant $c(p(s))$.

Thus, if the log Sobolev inequality holds for all p in $[p_0, p_1)$, $H(t) \leq H(0) = \|f\|_{p_0}$, i.e., (6.6.160) holds.

Conversely taking the derivatives of (6.6.161) as in (6.6.167) yields the claimed log Sobolev inequality at p. □

Example 6.6.18 (General Hypercontractive Semigroup). As we saw (Theorem 6.6.4), for an L^p-contractive semigroup

$$\|e^{-TA}\|_{2,4} = C \qquad (6.6.168)$$

implies the explicit estimate of $T_{p,q}$ and $C_{p,q}$ for all $p \leq q$ in $(1, \infty)$ with

$$\|e^{-T_{p,q}A}\|_{p,q} = C_{p,q}$$

This, in turn (Problem 8), implies L^p-log Sobolev inequalities with

$$c(p) = \begin{cases} 2T, & 2 \leq p < \infty \\ 2T(p-1)^{-1}, & 1 < p \leq 2 \end{cases}, \quad \Gamma(p) = 4 \log C \qquad (6.6.169)$$

If one integrates back from these log Sobolev inequalities, one gets a time $\widetilde{T}_{2,4} = 2T \log 2$, i.e., something is lost by using just interpolation and differentiation. This can be understood as follows: the $C_{p,q}$, $T_{p,q}$ obtained by integrating log Sobolev inequalities ($T_{p,q}$ depends on two parameters and $c(p)$ only on one so we expect a relation)

$$p < r < q \Rightarrow T_{p,q} = T_{p,r} + T_{r,q}, \quad C_{p,q} = C_{p,r} C_{r,q} \tag{6.6.170}$$

If the initial $T_{p,q}$ or $C_{p,q}$ do not obey this, then one has to lose information on the round trip. □

Example 6.6.19 (Gross–Nelson Semigroups). We say e^{-tA} is a *Gross–Nelson semigroup* if and only if

$$e^{-t} \le \sqrt{\frac{p-1}{q-1}} \Rightarrow \|e^{-tA}f\|_q \le \|f\|_p \tag{6.6.171}$$

Thus, one has L^p-log Sobolev inequalities with local norm, 0, and local constant, $c(r)$, given by

$$\int_p^q \frac{c(r)}{r}\, dr = \frac{1}{2}[\log(q-1) - \log(p-1)] \tag{6.6.172}$$

i.e., taking derivatives at $q = p$ for p fixed

$$c(p) = \frac{p}{2(p-1)} \tag{6.6.173}$$

Integrating back up, shows L^p-log Sobolev estimates with (6.6.173) and $\Gamma(p) = 0$ implies (6.6.171) since $c(p)$ given by (6.6.173) obeys (6.6.172). Note that one loses nothing in going infinitesimal and integrating, because $T_{p,q}$ given by

$$e^{-T_{p,q}} = \sqrt{\frac{p-1}{q-1}} \tag{6.6.174}$$

obeys (6.6.170) (and $C_{p,q} \equiv 1$ obeys (6.6.170) also).

Recall that, by Theorem 6.6.13, any sub-Markovian semigroup obeys a sub-Dirichlet inequality, i.e., (6.6.119) holds. Suppose such a semigroup also obeys an L^2-log Sobolev inequality with $c(2) = 1$, $\Gamma(2) = 0$, i.e.,

$$\int |f|^2 \log|f| d\mu \le \langle |f|, A\,|f| \rangle + \|f\|_2^2 \log\|f\|_2 \tag{6.6.175}$$

Replace $|f|$ by $|f|^{p/2}$ and use the sub-Dirichlet inequality to see that

$$\frac{p}{2} \int |f|^p \log|f| \le \langle |f|^{p/2}, A\,|f|^{p/2} \rangle + \frac{p}{2}\|f\|_p^p \log\|f\|_p \tag{6.6.176}$$

$$\le \left(\frac{p}{2}\right)^2 \frac{1}{p-1} \langle |f|^{p-1}, A\,f \rangle + \frac{p}{2}\|f\|_p^p \log\|f\|_p \tag{6.6.177}$$

Dividing by $\left(\frac{p}{2}\right)$, we see that A obeys an L^p-log Sobolev inequality with $c(p)$ given by (6.6.173) and $\Gamma(p) = 0$. We have thus proven the theorem below. □

Theorem 6.6.20. *Let e^{-tA} be a sub-Markovian semigroup. Then the following are equivalent:*

(1) e^{-tA} *is a Gross–Nelson semigroup, i.e.,* (6.6.171) *holds.*

(2) *For all $p \in (1, \infty)$, an L^p-log Sobolev inequality holds with $c(p) = p/[2(p-1)]$, $\Gamma(p) = 0$, i.e., for all $f \in \mathcal{D}_{++}$*

$$\int \|f\|^2 \log|f| d\mu \leq \frac{p}{2(p-1)} \langle |f|^{p-1}, A\,|f| \rangle + \|f\|_p^p \log\|f\|_p \qquad (6.6.178)$$

(3) (6.6.175) *holds (i.e.,* (6.6.178) *for $p = 2$).*

(4) *For all p in $(2 - \varepsilon, 2)$ (and some $\varepsilon > 0$)*

$$e^{-t} \leq \sqrt{p-1} \Rightarrow \|e^{-tA}f\|_2 \leq \|f\|_p \qquad (6.6.179)$$

Proof. The only missing step is $(4) \Rightarrow (3)$ which follows by taking derivatives at $p = 2$. $\qquad \square$

Example 6.6.21. Let e^{-tA} be a sub-Markovian semigroup. Suppose it obeys an L^2-log Sobolev inequality with $c(2) = \varepsilon$ for any ε and corresponding $\Gamma(2) = b(\varepsilon)$, i.e.,

$$\int |f|^2 \log|f| \leq \varepsilon \langle f, Af \rangle + b(\varepsilon) \|f\|_2^2 + \|f\|_2^2 \log\|f\|_2 \qquad (6.6.180)$$

We claim if for all small ε,

$$b(\varepsilon) \leq \exp(\varepsilon^{-a}), \quad a < 1 \qquad (6.6.181)$$

then e^{-tA} is ultracontractive (and we'll see examples later where the bound holds, with $a = 1$ which are not ultracontractive). For since $p > 2$ implies $\frac{p}{2}\left(\frac{1}{p-1}\right) < 1$, putting $|f|^p$ for $|f|$ in (6.6.180) shows there are L^p-log Sobolev inequalities, for $p > 2$ with

$$c(p) = \varepsilon, \quad \Gamma(\varepsilon, p) = 2b(\varepsilon)/p \qquad (6.6.182)$$

Fix $t > 0$. Pick $\alpha > 1$ so $a\alpha < 1$ (a as in (6.6.181))

$$d(\alpha) \int_2^\infty \frac{dp}{p(\log p)^\alpha} = 1 \qquad (6.6.183)$$

If we use (6.6.159) with $p_0 = 2$, $p_1 = \infty$ and

$$c(p) = \frac{td(\alpha)}{(\log p)^\alpha} \qquad (6.6.184)$$

then

$$\int_{p_0}^{p_1} \frac{c(p)}{p} dp = 1, \quad M = \int_{p_0}^{p_1} \frac{2b(c(p))}{p^2} dp < \infty \qquad (6.6.185)$$

since $b(c(p)) \leq \exp(D(\log p)^{a\alpha}) = o(p^\delta)$ for all δ making the integral convergent. We summarize in the theorem below. $\qquad \square$

Theorem 6.6.22. *Let e^{-tA} be a sub-Markovian semigroup. Suppose (6.6.180) where*

$$b(\varepsilon) \leq C_1 \exp(c_2 \varepsilon^{-a}) \tag{6.6.186}$$

for some $a < 1$. Then e^{-tA} is ultracontractive.

With these tools in hand, we can turn to optimal hypercontractive estimates for the Ornstein–Uhlenbeck and Bonami–Gross semigroups:

Theorem 6.6.23 (Bonami–Gross Inequality). *The ν-dimensional Bonami–Gross semigroup e^{-tB} is a Gross–Nelson semigroup, indeed a stronger result is true; e^{-tB} is a contraction from L^p to L^q if and only if*

$$e^{-t} \leq \sqrt{\frac{p-1}{q-1}} \tag{6.6.187}$$

Proof. By the Bonami–Segal lemma (Theorem 6.6.15), it suffices to prove the result for $\nu = 1$. In fact, if for some t, p, q, $\|e^{-tB_1}\|_{p,q} > 1$, then $\lim_{\nu \to \infty} \|e^{-tB_\nu}\|_{p,q} = \infty$.

Since e^{-tB} is positivity preserving, we need only look at $e^{-tB} f$ for $f > 0$, i.e., $f = \left(\frac{1+x}{1-x}\right)$ for $|x| < 1$, $x \in \mathbb{R}$. If $\beta = e^{-t}$, then

$$e^{-tB} \left(\frac{1+x}{1-x}\right) = \left(\frac{1+\beta x}{1-\beta x}\right) \tag{6.6.188}$$

Thus, we want to explore for which $\beta < 1$, $q < p$, we have

$$\{\tfrac{1}{2}[(1+\beta x)^q + (1-\beta x)^q]\}^{\frac{1}{q}} \leq \{\tfrac{1}{2}[(1+x)^p + (1-x)^p]\}^{\frac{1}{p}} \tag{6.6.189}$$

for $-1 < x < 1$.

Consider first x small. Then for $y = x$ or $y = \beta x$

$$\left\{\tfrac{1}{2}[(1+y)^r + (1-y)^r]\right\}^{\frac{1}{r}} = \left\{1 + \frac{r(r-1)}{2}y^2 + O(y^4)\right\}^{\frac{1}{r}}$$

$$= 1 + \frac{r-1}{2}y^2 + O(y^4)$$

so (6.6.189) holds to leading order for small x if and only if

$$\left(\frac{q-1}{2}\right)\beta^2 \leq \left(\frac{p-1}{2}\right) \Leftrightarrow \beta \leq \sqrt{\frac{p-1}{q-1}} \tag{6.6.190}$$

In particular, this shows that if (6.6.187) fails, then $\|e^{-tB}\|_{p,q} > 1$.

Consider next $q = 2$, $p < 2$ and note that (6.6.189) is the same as

$$(1 + \beta^2 x^2)^{\frac{p}{2}} \leq \tfrac{1}{2}[(1+x)^p + (1-x)^p] \tag{6.6.191}$$

Since $p < 2$, we have that $\frac{p}{2} < 1$ so $y \mapsto (1+y)^{\frac{p}{2}}$ is concave and

$$(1+y)^{\frac{p}{2}} \leq (1 + \tfrac{p}{2}y) \tag{6.6.192}$$

(this is Bernoulli's inequality (see (5.3.79) of Part 1)). Thus, (6.6.191) is implied by

$$1 + \tfrac{p}{2}x^2\beta^2 \le \tfrac{1}{2}[(1+x)^p + (1-x)^p] \tag{6.6.193}$$

By the binomial theorem for $|x| < 1$

$$\frac{1}{2}[(1+x)^p + (1-x)^p] = \sum_{m=0}^{\infty} \binom{p}{2m} x^{2m} \ge 1 + \frac{p(p-1)}{2}x^2 \tag{6.6.194}$$

since $\binom{p}{2m} > 0$ on account of $1 < p < 2$ and $p(p-1)(p-2)\ldots(p-2m-1) \ge 0$ as a product of two positive factors and the $2m - 2$ negative ones. Thus, (6.6.193) (and so (6.6.191)) is implied by

$$\frac{p}{2}\beta^2 \le \frac{p(p-1)}{2} \Leftrightarrow \beta \le \sqrt{p-1} \tag{6.6.195}$$

which is (6.6.187). Therefore, (d) of Theorem 6.6.20 holds and so e^{-tB} is a Gross–Nelson semigroup. $\square$

Quite remarkably, we'll deduce the optimal hypercontractive estimates for the Ornstein–Uhlenbeck semigroup from that for B_ν taking $\nu \to \infty$ and using the central limit theorem to get the Gaussian measure. The key (beside the central limit theorem) will be a calculation of B_ν on a special class of functions on $\{-1,1\}^\nu$. Let g be a general function on $\{-1,1\}^\nu$ viewed as $g(y_1,\ldots,y_\nu)$ with each $y_j = \pm 1$. We claim that

$$(B_j g)(y) = \tfrac{1}{4}[g(y) - g(y-2\delta_j)](1+y_j) - \tfrac{1}{4}[g(y+2\delta_j) - g(y-2\delta_j)](1-y_j) \tag{6.6.196}$$

Here $(y - 2\delta_j)_k = y_k - 2\delta_{jk}$. Of course, if $y_j = -1$, $g(y-2\delta_j)$ is not defined but that doesn't matter since in that case $1 + y_j = 0$. Thus, the right side of (6.6.196) is defined. To check (6.6.196), it suffices to prove it when $\nu = 1$ because of the product/sum nature of B_ν. In case $\nu = 1$, (6.6.196) says that

$$(Bg)(\pm 1) = \pm\tfrac{1}{2}[g(1) - g(-1)] \tag{6.6.197}$$

which is correct for $g(1) = g(-1)$ and for $g(1) = -g(-1)$, a basis in $\mathbb{C}^2$.

Now on $\{-1,1\}^\nu$, define

$$x_\nu = (y_1 + \ldots + y_\nu)/\nu^{\frac{1}{2}}, \quad h_\nu = 2\nu^{-\frac{1}{2}} \tag{6.6.198}$$

Thus x_ν takes $\nu + 1$ values only separated by h_ν.

Let f be an arbitrary continuous function on $\mathbb{R}$. Noting that

$$x_\nu(y \pm 2\delta_j) = x_\nu(y) \pm h_\nu \tag{6.6.199}$$

we see that $\sum_{j=1}^{\nu} B_j f$ is also a function only of x_ν:

$$\sum_{j=1}^{\nu} B_j f(x_\nu) = \sum_{j=1}^{\nu} \tfrac{1}{4}(f(x_\nu) - f(x_\nu - h_\nu))(1 + y_j)$$

$$- \sum_{j=1}^{\nu} \tfrac{1}{4}(f(x_\nu + h_\nu) - f(x_\nu))(1 - y_j) \qquad (6.6.200)$$

$$= \tfrac{1}{4}[f(x_\nu) - f(x_\nu - h_\nu)][\nu + \nu^{\frac{1}{2}} x_\nu]$$

$$- \tfrac{1}{4}[f(x_\nu + h_\nu) - f(x_\nu)][\nu - \nu^{\frac{1}{2}} x_\nu] \qquad (6.6.201)$$

$$= (D_\nu f)(x_\nu) \qquad (6.6.202)$$

where

$$(D_\nu f)(x) = -h_\nu^{-2}\{f(x_\nu + h_\nu) + f(x_\nu - h_\nu) - 2f(x_\nu)\}$$
$$+ x_\nu (2h_\nu)^{-1}\{f(x_\nu + h_\nu) - f(x_\nu - h_\nu)\} \qquad (6.6.203)$$

If f is C^∞ of compact support and $x_\nu \to x$ and $\nu \to \infty$ (so $h_\nu \to 0$), this formally converges to $-f''(x) + x f'(x)$ which is the generator of the Ornstein–Uhlenbeck semigroup given in (6.6.122) on $\mathbb{R}^\nu$ when $\nu = 1$. This will allow us to prove

Theorem 6.6.24 (Nelson's Best Hypercontractive Estimate). *Let e^{-tA} be the Ornstein–Uhlenbeck semigroup for $\nu = 1$ (i.e., $A = -\frac{d^2}{dx^2} + x\frac{d}{dx}$ on $L^2(\mathbb{R}, (2\pi)^{-\frac{1}{2}} \exp(-\frac{1}{2}x^2))$, equivalently, A is the intrinsic semigroup to the harmonic oscillator normalized to have eigenvalues $0, 1, 2, \ldots$). Then e^{-tA} is a Gross–Nelson semigroup, i.e.,*

$$e^{-t} \le \sqrt{\frac{p-1}{q-1}} \Rightarrow \|e^{-tA}f\|_q \le \|f\|_p \qquad (6.6.204)$$

Moreover, if the left side of (6.6.204) fails, e^{-tA} is not bounded from L^p to L^q. In addition, the same is true for $\mathbb{R}^\nu$.

Remarks. 1. In place of limits of operators, one could look at the quadratic forms of B_ν and A.

2. One should be able to get the lack of boundedness by using the fact that if the right side of (6.6.204) fails, then $\lim_{\nu \to \infty} \|e^{-tB_\nu}\|_{p,q} = \infty$ but instead, we'll use explicit calculations.

3. Once one has (6.6.204) for $\mathbb{R}^\nu$ and all ν, one can take ν to infinity and get hypercontractive estimates for the Gaussian processes of Section 4.12 of Part 1—see the Notes.

Proof. The general $\mathbb{R}^\nu$ case follows from the $\mathbb{R}$ case and the Bonami–Segal lemma (together with an extension to handle unbounded operators; see Problem 9).

By Theorem 6.6.20, it suffices to prove an L^2-log Sobolev inequality, i.e.,

$$\int |f(x)|^2 \log|f(x)|\, d\mu(x) \leq \langle |f|, A\, |f| \rangle + \|f\|_2^2 \log\|f\|_2 \qquad (6.6.205)$$

By a simple limiting argument (Problem 10), it suffices to prove this for $f \geq 0$ and $f \in C_0^\infty(\mathbb{R})$.

By the Bonami–Gross inequality and Theorem 6.6.20, with x_ν given by (6.6.74), we have

$$\int f\big(x_\nu(y_1, \ldots y_\nu)\big)^2 \log f(x_\nu)\, d\eta_\nu(y) \leq \langle f, B_\nu f \rangle + \|f\|_{2,\nu}^2 \log\|f\|_{2,\nu} \quad (6.6.206)$$

where η_ν is a counting measure on $\{-1, 1\}^\nu$.

Since $\langle y_j^2 \rangle = 1$, $\langle y_j \rangle = 0$, by the central limit theorem (see Theorem 7.3.4 of Part 1), $d\eta_\nu(y)$ converges weakly (on continuous functions of compact support) to the Gaussian measure with $\langle x^2 \rangle = 1$, $\langle x \rangle = 0$, i.e., the measure $d\mu$ of (6.6.145). Thus (since $f^2 \log f$ is continuous), the first and third terms of (6.6.206) converge to those of (6.6.205).

Write

$$(B_\nu f)(x_\nu) = (Af)(x_\nu) + (B_\nu f - Af)(x) \qquad (6.6.207)$$

and note that, by the central limit theorem,

$$\int f(x_\nu)(Af)(x_\nu)d\eta_\nu \to \int f(x)(Af)(x)\, d\mu(x) \qquad (6.6.208)$$

The convergence of $|(B_\nu f)(x_\nu) - (Af)(x_\nu)|$ to zero is uniform in x, so its integral over $d\eta_\nu$ goes to zero. Thus, we've proven (6.6.205) and so (6.6.204).

For the case

$$\sqrt{\frac{p-1}{q-1}} < e^{-t} \qquad (6.6.209)$$

we use the functions $: e^{\alpha x} :$ of Problem 4 of Section 4.11 of Part 1. Let $\{\varphi_n\}_{n=0}^\infty$ be the normalized eigenfunctions of A with $A\varphi_n = n\varphi_n$. Then

$$: e^{\alpha x} : \equiv \sum_{n=0}^\infty \frac{\alpha^n}{n!} \varphi_n = e^{\alpha x - \frac{1}{2}\alpha^2} \qquad (6.6.210)$$

where the second equality is a calculation (see part (c) of the above referenced problem). By $A\varphi_n = n\varphi_n$, we see that

$$e^{-tA} : e^{\alpha x} : \ = \ : e^{\alpha e^{-t}x} : \qquad (6.6.211)$$

By the second equality in (6.6.210), we compute (Problem 11)

$$\|: e^{\alpha x} :\|_p = e^{\frac{1}{2}\alpha^2(p-1)} \qquad (6.6.212)$$

Thus,

$$\frac{\|e^{-tA} : e^{\alpha x} :\|_q}{\|: e^{\alpha x} :\|_p} = \exp(\tfrac{1}{2}\alpha^2[e^{-2t}(q-1)-(p-1)]) \qquad (6.6.213)$$

If (6.6.209) holds, the quantity in [ ] is strictly positive, so the LHS of (6.6.213) $\to \infty$ as $\alpha \to \infty$ implying e^{-tA} is not bounded from L^p to L^q. □

We turn next to intrinsic ultracontractivity for Schrödinger operators $H = -\Delta + V$ with $V(x) = |x|^a$ (one can handle more general V's–see the Notes). We saw for $a = 2$, one has intrinsic hypercontractivity but not supercontractivity. One can show (see the Notes) for $a < 2$, one doesn't even have intrinsic hypercontractivity. We are heading towards showing for $a > 2$, one has intrinsic ultracontractivity. We saw in Theorem 6.6.20 that the key is to prove that (6.6.180) holds with effective estimates on $b(\varepsilon)$ as $\varepsilon \downarrow 0$.

Since we'll deal with both $L^2(\mathbb{R}^\nu, d^\nu x)$- and $L^2(\mathbb{R}^\nu, \varphi_0^2 d^\nu x)$-norms, we use $\|\cdot\|_2$ for the former and $\|\cdot\|_{2,\varphi_0}$ for the latter. Clearly (6.6.180) is implied by

$$\|f\|_{2,\varphi_0} = 1 \Rightarrow \log|f| \le \varepsilon A + b(\varepsilon) \qquad (6.6.214)$$

as an $L^2(\mathbb{R}^\nu, \varphi_0^2 d^\nu x)$-operator inequality. By the relation of H and A, this is equivalent to

$$\|f\|_{2,\varphi_0} \ge 1 \Rightarrow \log|f| \le \varepsilon H + b(\varepsilon) - \alpha \varepsilon \qquad (6.6.215)$$

The key will be

Theorem 6.6.25 (Rosen's Lemma). *Let $H = -\Delta + V$ on $L^2(\mathbb{R}^\nu)$ with $V \ge 0$ have a strictly positive eigenfunction φ_0; $H\varphi_0 = \alpha\varphi_0$. Suppose for some $\delta > 0$*

$$-\log\varphi_0 \le \delta H + g(\delta) \qquad (6.6.216)$$

Then (6.6.215) holds with (for all ε with $\varepsilon/2$ an allowed δ)

$$b(\varepsilon) = a_\nu - \nu\log(\varepsilon) + g(\tfrac{\varepsilon}{2}) + \alpha\varepsilon + \frac{\varepsilon}{2} \qquad (6.6.217)$$

where a_ν is a ν-dependent constant independent of V.

Remark. The point is that this reduces a family of estimates to a single one.

Proof. By a Sobolev estimate with $s = \tfrac{1}{4}$ and $q = 2$ (see Theorem 6.3.3)

$$\|f\|_p \le C_\nu \|(-\Delta)^{\frac{1}{4}} f\|_2, \quad p = \tfrac{1}{2} - \tfrac{1}{4\nu} \qquad (6.6.218)$$

Thus, by Hölder's inequality, if $W \in L^{2\nu}(\mathbb{R}^\nu, d^\nu x)$,

$$\|W^{\frac{1}{2}}f\|_2 \le C_\nu \|W^{\frac{1}{2}}\|_{4\nu} \|(-\Delta)^{\frac{1}{4}}f\|_2 \qquad (6.6.219)$$

or

$$W \le C_\nu^2 \|W\|_{2\nu} (-\Delta)^{\frac{1}{2}} \le \frac{1}{2} C_\nu^2 \|W\|_{2\nu}(-\Delta + 1) \qquad (6.6.220)$$

Fix β to be picked later (depending on ε) and put ($f > 0$)

$$W(x) = \log(\beta \, f \varphi_0) \qquad (6.6.221)$$

For some constant D_ν, one has that for all $x \in [0, \infty)$

$$\log(x)^{2\nu} \le D_\nu x^2 \qquad (6.6.222)$$

Thus,

$$
\begin{aligned}
\|W\|_{2\nu} &\le \left(\int D_\nu \, \beta^2 f(x)^2 \varphi_0^2(x) \, d^\nu x \right)^{1/2\nu} \\
&= D_\nu^{1/2\nu} \beta^{1/\nu}
\end{aligned}
\qquad (6.6.223)
$$

by $\|f\|_{2,\varphi_0} = 1$.

By (6.6.220)

$$
\begin{aligned}
\log(\beta|f|\varphi_0) &\le \tfrac{1}{2}\beta^{1/\nu} \, e^{a_\nu/\nu}(-\Delta + 1) \\
&\le \tfrac{1}{2}\beta^{1/\nu} \, e^{a_\nu/\nu}(H + 1)
\end{aligned}
\qquad (6.6.224)
$$

for a suitable constant a_ν (built from C_ν and D_ν) since $V \ge 0 \Rightarrow -\Delta \le H$.

Given ε, pick β so

$$\frac{\varepsilon}{2} = \tfrac{1}{2}\beta^{1/\nu} \exp(a_\nu/\nu) \Rightarrow \log \beta = \nu \log \varepsilon - a_\nu \qquad (6.6.225)$$

Then, (6.6.224) becomes

$$
\begin{aligned}
\log|f| &\le -\nu \log \varepsilon + a_\nu - \log(\varphi_0) + \frac{\varepsilon}{2}(H + 1) \\
&\le \varepsilon H + g\!\left(\frac{\varepsilon}{2}\right) - \nu \log \varepsilon + a_\nu + \frac{\varepsilon}{2}
\end{aligned}
$$

which proves (6.6.215)/(6.6.217) $\square$

Corollary 6.6.26. *Under the setup of Theorem 6.6.25 if (6.6.216) holds for all $\delta > 0$, e^{-tA} is supercontractive. If it holds for all such δ and*

$$g(\delta) \le C_1 \exp(C_3 \delta^{-a})$$

for some $a < 1$, then e^{-tA} is ultracontractive.

Proof. Since e^{-tA} obeys a sub-Dirichlet inequality and a L^2-Sobolev inequality for any local constant $\varepsilon > 0$, we see that $e^{-tA} \colon L^2 \to L^4$ for any t with $e^{-t/\varepsilon} \le 1/\sqrt{3}$. Since ε is arbitrary, we get a supercontractive estimate.

The second result follows from Theorems 6.6.22 and 6.6.25. $\square$

Remark. While we don't have a result saying (6.6.216) for all δ is necessary for supercontractivity, the case $-\Delta + \frac{1}{4}x^2$ which is the paradigmal example of hypercontractivity without supercontractivity has (6.6.216) only for sufficiently large δ and $-\frac{d^2}{dx^2} + x^\alpha$, $\alpha < 2$ which is known not to be hypercontractive (see the Notes) has (6.6.216) for no δ (see the Remarks after the next theorem).

Theorem 6.6.27. (a) *For $a > 2$, $-\Delta + |x|^a$ on $L^2(\mathbb{R}^\nu, d^\nu x)$ is intrinsically ultracontractive.*

(b) *$-\frac{d^2}{dx^2} + x^2 \left[\log(|x| + 2) \right]^b$ on $L^2(\mathbb{R}, dx)$ is intrinsically ultracontractive if $b > 2$ and intrinsically supercontractive but not ultracontractive if $0 \le b < 2$.*

Remarks. 1. The case $b = 2$ is an example where $g(\delta) \le C_1 \exp(C_2 \delta^{-1})$ and one does not have ultracontractivity.

2. The same methods we use here show if $a = 2$, there is an estimate of the form (6.6.216) only for $\delta \ge \delta_0 > 0$ for some $\delta_0 > 0$ (for $\log \varphi_0 \sim -\frac{1}{4}x^2$) and if $a < 2$ there is none, since $x^{1 + a/2}$ cannot be bounded by x^a.

3. There are results available for nonspherically symmetric potentials, see the Notes.

4. There are results showing that not only is e^{-tA} ultracontractive in case (a), but e^{-tA^α} is if $\alpha > a^{-1} + \frac{1}{2}$; see Problem 12.

Proof. For both parts, we use the asymptotics for decaying solutions contained in Theorem 15.5.15 of Part 2B (and Example 15.5.18 of that part). We'll restrict to $\nu = 1$; for $\nu \ge 2$, one separates variables and returns to the $\nu = 1$ case with an unimportant $C|x|^{-2}$ term. The result we need is that

$$\varphi_0(x) \, b(x)^{-1} \to C \qquad\qquad (6.6.226)$$

a nonzero constant where

$$b(x) = |V(x)|^{-\frac{1}{4}} \exp\left(- \int_{x_0}^x (V(y) - E)^{\frac{1}{2}} \, dy\right) \qquad (6.6.227)$$

(a) By the above (in the sense of the ratio going to one)

$$-\log \varphi_0(x) \sim Dx^{1 + a/2} \qquad\qquad (6.6.228)$$

It follows that

$$-\log \varphi_0 \le \delta |x|^a + g(\delta)_2 \qquad\qquad (6.6.229)$$

$$g(\delta) = E\delta^{-\beta} \quad \beta = (a + 2)/(a - 2) \qquad (6.6.230)$$

This implies intrinsic ulracontractivity by Corollary 6.6.26. Indeed (Problem 12), one gets that if $\alpha > a^{-1} + \frac{1}{2}$, H^α is intrinsically ultracontractive.

(b) By (6.6.226)/(6.6.227), we have

$$-\log \varphi_0 \sim Dx^2 (\log x)^{b/2} \qquad (6.6.231)$$

By an easy calculation, the maximum of $-x^2(\log x)^{b/2} + \delta x^2 (\log x)^b$ occurs at $\log x \sim \delta^{-2/b}$ and that maximum is $O(\exp(D\delta^{-2/b}))$, i.e., (6.6.216) holds with

$$g(\delta) = \exp(E\delta^{-2/b}) \qquad (6.6.232)$$

By Corollary 6.6.26 this proves intrinsic supercontractivity for all $b > 0$ and intrinsic ultracontractivity if $b > 2$.

Let φ_1 be the eigenvector associated to the second eigenvalue. Then, by (6.6.226)/(6.6.227) and $(V(x) - E_1)^{\frac{1}{2}} - (V(x) - E_0)^{\frac{1}{2}} \sim \frac{1}{2}(E_0 - E_1)V(x)^{-\frac{1}{2}}$

$$\varphi_1/\varphi_0 \sim C \exp\left(\tfrac{1}{2}(E_1 - E_0) \int_{x_0}^x V(y)^{-1/2} dy\right) \qquad (6.6.233)$$

is unbounded as $x \to \infty$ if (and only if) $b \le 2$. It follows that H is not intrinsically ultracontractive if $b \le 2$. $\qquad \square$

This completes the discussion of when semigroups are hypercontractive. We end with two applications of hypercontractivity—boundedness and essential self-adjointness of certain perturbations of hypercontractive semigroups.

Lemma 6.6.28 (Young's Inequality). *For $y \ge 0$ and $x \in \mathbb{R}$*

$$xy \le e^x + y \log y - y \qquad (6.6.234)$$

Remark. This and $\frac{x^p}{p} + \frac{y^q}{q} = 1$ are two cases of the general Young's inequality $xy \le f(x) + f^*(y)$ discussed in the Notes to Section 5.3 of Part 1.

Proof. Fix $y > 0$ ($y = 0$ with "$0\log 0$" $= \lim[y \log y] = 0$ is trivial). Then

$$g(x) = e^x - xy \qquad (6.6.235)$$

goes to ∞ as $x \to \pm\infty$, so it has a finite minimum $g'(x) = e^x - y$. This occurs exactly at $x_0 = \log y$ where $g(x_0) = y - y \log y$. Thus $g(x) \ge g(x_0)$ which is (6.6.234). $\qquad \square$

Theorem 6.6.29. *Let e^{-tA} be a hypercontractive semigroup on $L^2(M, d\mu)$ with $\mu(M) = 1$. Let V be a measurable function on M with*

$$\int e^{-\alpha V(m)} \, d\mu(m) < \infty, \quad \text{all } \alpha \ge 0; \ V \in L^1 \qquad (6.6.236)$$

Then, for each $\lambda > 0$, $Q(A) \cap Q(V)$ is dense and

$$\inf_{f \in Q(A) \cap Q(V); \|f\|_2 = 1} \langle f, (A + \lambda V)f \rangle > -\infty \qquad (6.6.237)$$

Proof. Since $V \in L^1$, $L^\infty \subset Q(V)$. Since e^{-tV} is L^p-contractive $e^{-A}L^\infty \subset L^\infty \cap Q(A)$ and is dense in L^1. Hypercontractivity implies a logarithmic Sobolev inequality

$$\int |f(m)|^2 \log|f(m)|^2 \, d\mu(m) \leq C\langle f, Af\rangle + d\langle f, f\rangle + \|f\|^2 \log\|f\|^2 \quad (6.6.238)$$

Taking $x = -\alpha V(m)$, $y = |f(m)|^2$ in (6.6.234) and integrating $d\mu(m)$ yields (for $f \in Q(V)$)

$$-\alpha \int V(m)|f(m)|^2 \, d\mu(m) \leq \int |f(m)|^2 \log|f(m)|^2 \, d\mu(m) - \|f\|_2^2$$

$$+ \int e^{-\alpha V(m)} \, d\mu(m) \quad (6.6.239)$$

$$\leq C\langle f, Af\rangle + (d-1)\|f\|^2 + \|f\|^2 \log\|f\|^2$$

$$+ \int e^{-\alpha V(m)} d\mu(m)$$

Taking $\alpha = C\lambda$, we obtain (6.6.237).   □

Lemma 6.6.30. *Let $f_n \to f$ in L^p with $\sup_n \|f_n\|_q < \infty$ for some q in $[1, \infty]$. Then, $f \in L^q$.*

Proof. By passing to a subsequence, we can suppose that $f_n(w) \to f(w)$ for a.e. w. If $q = \infty$, we see that $\|f\|_\infty \leq \sup_n \|f_n\|_\infty$. Otherwise, by Fatou's lemma

$$\int |f(w)|^q \, d\mu(w) \leq \liminf_n \|f_n\|_q^q < \infty$$

□

Theorem 6.6.31. *Let e^{-tA} be a hypercontractive semigroup. Let V be a multiplication operator so that (6.6.236) holds. If $V \geq 0$ and $V \in L^2$ or if $V \in L^r$ for some $r > 2$, then $D(A) \cap D(V)$ is dense and $A + V$ is essentially self-adjoint on this set.*

Remarks. 1. The proof below that $\sup_n \|e^{-t(A+V_n)}\|_{2,2} < \infty$ for each t provides another proof of Theorem 6.6.29.

2. This proof makes heavy use of the theory of unbounded self-adjoint operators developed in Chapter 7 of Part 4.

Proof. As in the proof of the previous theorem, $e^{-tA}L^\infty \subset D(A) \cap D(V)$ is dense. Define V_n by

$$V_n(w) = \begin{cases} 0 & \text{if } |V_n(w)| > n \\ V(w) & \text{if } |V_n(w)| \leq n \end{cases} \quad (6.6.240)$$

Since A is self-adjoint on $D(A)$ and V_n is bounded, so is

$$A_n = A + V_n \quad (6.6.241)$$

We'll complete the proof by showing

(a) If $V \geq 0$, e^{-sA_n} is a contraction from L^p to L^p for all p. If V is general, $\|e^{-sA_\infty}\|_{p,p}$ is bounded uniformly in n for each $p \in (1, \infty)$ and $s \geq 0$.

(b) Using (a), we'll prove s-lim e^{-sA_n} exists and defines a self-adjoint semigroup e^{-sA_∞} bounded on each L^p, $1 < p < \infty$ and on L^∞ if $V \geq 0$.

(c) We'll prove that $e^{-A_\infty}[L^\infty]$ is a core for A_∞ and lies in $D(A) \cap D(V)$ and $A_\infty = A + V$ on this set.

(d) This will imply our result by the theory of self-adjoint extensions.

We turn first to step (a). (6.6.70) implies for $s > 0$ and $p \in (1, \infty)$, there is $\alpha > 0$ and $C < \infty$ so that for all large n

$$\|e^{-sA/n}\|_{p,p_n(\alpha)} \leq C^{\frac{1}{n}}, \quad p_n(\alpha) \equiv \left(p^{-1} - \frac{\alpha}{n}\right)^{-1} \tag{6.6.242}$$

((6.6.70) has $e^{\alpha t}$ but since $q < q' \Rightarrow \|\cdot\|_{p,q} \leq \|\cdot\|_{p,q'}$ we can obtain (6.6.242)).

By Hölder's inequality, for all m

$$\|e^{-sV_m/n}\|_{p_n(\alpha),p} = \left[\int e^{-(sV_m/n)\frac{n}{\alpha}} \, d\mu\right]^{\frac{\alpha}{n}}$$

$$= \|e^{-sV_m/\alpha}\|_1^{\frac{\alpha}{n}} \tag{6.6.243}$$

It follows that uniformly in large n and m that $\|e^{-sV_m/n} e^{-sA/n}\|_{p,p}^n$ is bounded. By the Trotter product formula (Theorem 7.6.13 of Part 4) for L^2-convergence and the lemma (for L^p-bounds), we conclude for all $s > 0$, $p \in (1, \infty)$, $\sup_m \|e^{-sA_m}\|_{p,p} < \infty$.

If $V \geq 0$, $\|e^{-sA/n} e^{-sV_m/n}\|_{p,p} \leq 1$ so we get that $\|e^{-sA_m}\|_{p,p} \leq 1$ for all $p \in [1, \infty]$. This proves (a).

For (b), we begin with *Duhamel's formula*. If C and D are two self-adjoint operators, each bounded from below then

$$e^{-tC} - e^{-tD} = t \int_0^1 e^{-(1-\alpha)tC}(D - C) e^{-\alpha tD} \tag{6.6.244}$$

in the sense that for $\varphi \in D(D), \psi \in D(C)$, we have

$$\langle \psi, (e^{-tC} - e^{-tD})\varphi \rangle = t \int_0^1 \left[\langle e^{-(1-\alpha)+C}\psi, e^{-\alpha tD} D\varphi \rangle \right.$$
$$\left. - \langle e^{-(1-\alpha)tC} C\psi, e^{-\alpha tD} \varphi \rangle \right] d\alpha \tag{6.6.245}$$

This is easily proven by noting that $\frac{d}{d\alpha}\langle \psi, e^{(1-\alpha)tC} e^{-\alpha tD}\varphi \rangle$ is the integral in (6.6.245) and integrating. From this one easily gets if $\varphi \in L^r$ where

$\frac{1}{r} + \frac{1}{p} = \frac{1}{2}$ (with $V \in L^p$)

$$\|(e^{-tA_n} - e^{-tA_m})\varphi\|_2 \leq t\|V_n - V_m\|_p \sup_{0 \leq s \leq t} \|e^{-sA_m}\|_{r,r} \|\varphi\|_{r,r} \qquad (6.6.246)$$

This implies for such φ's that $e^{-tA}\varphi \to e^{-tA_\infty}\varphi$ for some A_∞ and then for all $\varphi \in L^2$. By the lemma, one gets e^{-tA} maps $L^q \to L^q$ for $1 < q < \infty$. If $V \geq 0$, we can take $r = \infty$, $p = 2$. This finishes step (b).

We turn to step (c). By the spectral theorem, if r is as above, $\mathcal{D} = e^{-A_\infty}[L^r]$ is a core for A_∞, i.e., if $B = A_\infty \upharpoonright \mathcal{D}$, then

$$\overline{B} = A_\infty \qquad (6.6.247)$$

Let $A + V$ be the operator sum with domain $D(A) \cap D(V)$. Since e^{-A_∞} maps L^r to $L^r \subset D(V)$, $\mathcal{D} \subset D(V)$. Since for $\varphi \in L^r$, $(A_n - V_n)e^{-A_n}\varphi \to (A_\infty - V)e^{-A_\infty}\varphi$ (since L^2-convergence and $L^{r'}$-boundedness, for $r < r' < \infty$ implies L^r-convergence and analyticity implies convergence of derivative). Thus, $\mathcal{D} \subset D(A)$ and $A_\infty - V = A$ on $\mathcal{D}$, i.e.,

$$(A + V) \upharpoonright \mathcal{D} = A_\infty \qquad (6.6.248)$$

This proves step (c).

For step (d), (6.6.248) and hermiticity of A and V implies

$$B \subset A + V \subset (A + V)^* \subset \overline{B} = A_\infty$$

We conclude that $A + V = A_\infty \upharpoonright D(A) \cap D(V)$ and since $\mathcal{D}$ is a core, so is $D(A) \cap D(V)$. $\qquad \square$

Notes and Historical Remarks. There is a huge literature on the topics of this section, so much so that there are more than 100,000 hits on Google for searches of "hypercontractive" or "ultracontractive" or "logarithmic Sobolev." The subject has applications to quantum field theory, nonrelativistic quantum mechanics, statistical mechanics, probability theory, computer science, combinatorics, analysis on manifolds, and even to fundamental topology.

Three relevant books are Ané et al. [12], Royer [**703**], and Davies [**196**]. Gross [**338, 340, 198**] has several reviews on the subject. Stroock [**791**] and Guionnet–Zegarlinski [**345**] have reviews focusing on applications to probability while Garban–Steif [**303**] discuss applications to computer science. Garling [**304**] has a chapter on aspects of the subject.

The history of the ideas of this section is complicated with a primary thread that brought them into the mainstream with some earlier, somewhat obscure, precursors—the result is that some results have many names in the literature.

The primary thread concerns some of the earliest papers in the field known as constructive quantum field theory, the attempt to define a mathematically precise theory of interacting quantum fields. The simplest model is $(\varphi^4)_2$, a φ^4 interaction in one space (and one time—hence the subscript 2 for two space-time dimensions). Eventually fields were constructed in a Euclidean formalism, but these earliest results concerned Hamiltonians with spatial cutoff interactions.

In zero space dimensions, the analog is $-\frac{d^2}{dx^2} + x^2 + x^4$ where the interaction x^4 is bounded from below. But for $(\varphi^4)_2$, the intuitively positive φ^4 is infinite in the sense that even with a spatial cutoff $\int_{-\ell}^{\ell} \varphi^4(x)\,dx$ is not in L^2.

One addresses this with the simplest of all renormalizations—$\varphi^4(x)$ is replaced by a so-called Wick-ordered φ^4 denoted $:\varphi^4:$. $V = \int_{-\ell}^{\ell} : \varphi^4(x) : dx$ is now in all L^p, $p < \infty$ (For much more on what these terms mean, see the books of Simon [740] or Glimm–Jaffe [322].). However, it is not bounded from below, so it is an important question whether $H_0 + V$ is, where H_0 is the free quantum field Hamiltonian, an infinite-dimensional analog of $-\frac{d^2}{dx^2} + x^2$.

In Nelson [602], it was proven that $-\frac{d^2}{dx^2} + x^2$ generates an intrinsic L^p-contractive semigroup which is bounded from L^2 to L^4 (i.e., hypercontractive in modern parlance) and this was used to prove boundedness of a suitably cutoff perturbed Hamiltonian via a variant of Theorem 6.6.29 (but proven as we do in the proof of Theorem 6.6.31). Nelson was only able to handle infinite dimensions by restricting to periodic boundary conditions which allow a special trick. In [319], Glimm proved an important result (Problem 13): if e^{-tA} is a Markov semigroup, $\|e^{-TA}\|_{2,4} < \infty$ for some T and A has a mass gap, i.e., $A \upharpoonright \{\mathbb{1}\}^{\perp} \geq m > 0$, then for some T_1, $\|e^{-T_1 A}\|_{2,4} \leq 1$. This let him prove hypercontractivity in general infinite-dimensional situations (see below).

In a follow-up to Nelson's paper, Federbush [262] provided an alternate way of going from hypercontractivity to semi-boundedness—he used the proof we give here for Theorem 6.6.29. In so doing, he proved directly an L^2- to L^p-bound which he differentiated at $t = 0$, and so gave the first L^2-log Sobolev inequality, at least in the context of these developments in constructive quantum field theory. Ironically, he never explicitly wrote down the derivative $\frac{d}{dt}\|e^{-tH} f\|_{p_t}^{p_t}$ so his log-Sobolev inequality is implicit rather than explicit.

After proving semi-boundedness of $H_0 + V$ in this cutoff $(\varphi^4)_2$ model, it is natural to ask about essential self-adjointness. This was first established by Glimm–Jaffe [320] using techniques that not only relied on hypercontractivity to get the semi-boundedness but instead needed additional estimates.

That L^p- to L^q-semigroup bounds sufficed for this result was found by Segal [723, 724, 725].

This theory was codified and extended by Simon–Hoegh-Krohn [758], who, in particular, invented the term "hypercontractive." I remember Ed Nelson complaining to me that since one only supposed boundedness of the L^2- to L^4-map, not contractivity, it should be called "hyperbounded" not "hypercontractive." I explained that the latter term had a ring to it and indeed it stuck and was used even by Nelson.

The next step in the thread is the proof, using probabilistic methods, by Nelson [603] of the optimal hypercontractive estimates for the Ornstein–Uhlenbeck semigroup, i.e., Theorem 6.6.24. Shortly afterwards (this paper and the Simon–Hoegh-Krohn paper had long delays from preprint to publication due to slow refereeing), Gross [335] produced his seminal paper where the term "log Sobolev" first appeared. More importantly, this paper noted that the connection between log Sobolev and hypercontractive inequalities went in both directions, the important fact that for Dirichlet forms, L^2-log Sobolev inequalities implied L^p-inequalities, the Bonami–Gross inequalities (by a proof that appears in Problem 14) and the proof of Nelson's best inequality via the central limit theorem (as we present it).

One reaction to Gross' paper (or rather a talk Gross gave in Princeton before his paper appeared) is that Beckner [50] proved an analog of the Bonami–Gross inequality for $i\alpha B$ instead of αB and via a central limit argument used it to get optimal constants in the Hausdorff–Young inequality. While this important result used an analog of Gross' result, it was motivated by Gross and didn't state any form of what we call the Bonami–Gross inequality. Nevertheless, when Kahn–Kalai–Linial [439] used the Bonami–Gross inequality in a seminal work in computer science, they attributed it to Beckner! As a result, what we call the Bonami–Gross inequality is called *Beckner's inequality* or the *Bonami–Beckner* inequality in the computer science literature.

We pause in our discussion of the quantum theory development of this subject to point out some precursor occurrences. Bonami [93], in a discussion of Fourier analysis on $\mathbb{Z}_2$, proved the Bonami–Gross inequality a few years before Gross. Stam [773] and Blachman [80] have versions of the optimal L^2-log Sobolev inequality for Gauss measure many years before Gross, but they only are mentioned in passing. Carlen [140] has further discussion of their work and its connection to hypercontractivity.

Alternate proofs of the optimal log-Sobolev inequalities for Gauss measures can be found in [702, 2, 846, 244, 219]. The Bonami–Segal lemma is named after Bonami [93] and Segal [725].

Returning to the study of hypercontractivity in the quantum theory community, several authors used Gross' idea to study hypercontractivity for intrinsic anharmonic oscillators. In particular, Rosen [**700**] realized that for potentials growing faster than x^2, one had L^2- to L^4-contractivity for all $t > 0$, a property he named supercontractivity. Rosen's lemma is taken from that paper. Carmona [**144**] proved for x^α, $\alpha < 2$ potentials, one didn't have hypercontractivity (see also the paper of Davies–Simon [**199**] which, in particular, has the proof that intrinsic hypercontractivity implies (6.6.85).)

Davies–Simon [**199**] realized that one sometimes has L^1- to L^∞-contractivity for all t, a property, following a suggestion of Derek Robinson, they called ultracontractivity. Theorems 6.6.5 and 6.6.7 are from their paper. They also discuss variants of Theorem 6.6.27 that don't require spherical symmetry or even exact asymptotic power behavior at ∞. (a) $\Leftrightarrow$ (b) when $\nu > 2$ in Theorem 6.6.2 is due to Varopoulos [**829**] and (a) $\Leftrightarrow$ (c) to Fabes–Stroock [**257**]. All those authors were motivated in part by connections between inequalities and semigroup bounds in the concrete situation of elliptic operators with rough coefficients in deep papers of Nash [**596**] and Moser [**589**].

Davies–Simon [**199**] realized that ultracontractivity was important because it implied pointwise bounds on heat kernels. Davies [**193, 194, 195**] realized how to apply this to $\varphi H \varphi^{-1}$ for a suitable function φ to get very detailed information on the small $x - y$ and t behavior of the heat kernel associated to general second-order elliptic operators on manifolds. This approach is expounded in detail in Davies' book [**196**].

That any sub-Markovian semigroup obeys a sub-Dirichlet inequality (Theorem 6.6.13) is a result of Stroock [**790**] and Varopoulos [**829**]. Bakry–Émery [**30**] has techniques to replace the sub-Dirichlet inequality by studying another operator rather than the semigroup generator.

Because of their uses and intrinsic interest as optimal inequalities, there is considerable literature on extending the Bonami–Gross and optimal Gaussian L^2-log Sobolev inequalities. Diaconis and Saloff-Coste [**220**] study the two-point inequality when the measure is shifted from equal to unequal weights. Gross [**337**] has a paper on Lie groups. Gross [**336**] began the study of log Sobolev inequalities for fermions using noncommutative integration with optimal constants found by Carlen–Lieb [**142**]; see also Olkiewicz–Zegarlinski [**617**]. Biane [**70**] discussed what can be considered the mixed Boson-Fermi case. Janson [**423**] and Kemp [**461**] discuss extensions involving the Gaussian semigroup on holomorphic functions; see also Carlen [**141**], Gross [**339**], Janson [**424**] and Zhou [**873**].

As one of the few inequalities that holds in infinite dimensions, hypercontractivity has important applications in probability theory. It is useful in the theory of large deviations (see Deuschel–Stroock [**218**]). It is also useful in statistical mechanics; see Guionnet–Zegarlinski [**345**]: if one can prove at some temperature that there is a Gibbs state which obeys a log Sobolev inequality, then this is the unique Gibbs state and the Glauber dynamics converges to it exponentially fast.

Kahn–Kalai–Lindal [**439**] discuss applications to classical computer science and Ben-Aroya et al. [**51**] and Montanaro [**581**] to quantum computer science. Otto–Villani [**625**] discuss connections between log Sobolev inequalities and optimal transport. For later related work, see Figalli et al. [**274**] and Indrei–Marcon [**415**].

Log Sobolev inequalities got extra attention when Perelman [**636**] used it as an important step in his proof of the Poincaré conjecture; see also Morgan–Tian [**586**] and Zhang [**872**].

The natural setting for Nelson's best Gaussian hypercontractive estimates in infinite dimensions is as second quantized operators. Let $\mathcal{H}$ be a Hilbert space. $\mathcal{F}_s(\mathcal{H})$, the *boson Fock space* over $\mathcal{H}$, is $\overset{\infty}{\underset{n=0}{\oplus}} \mathcal{S}^n(\mathcal{H})$, the symmetric algebra over $\mathcal{H}$. Here $\mathcal{S}^n(\mathcal{H})$ is the symmetric tensor product discussed in Section 3.8 of Part 1. If $B \in \mathcal{L}(\mathcal{H})$, the bounded operators on $\mathcal{H}$, one defines $\Gamma(B)\colon \mathcal{F}_s(\mathcal{H}) \to \mathcal{F}_s(\mathcal{H})$ by $\Gamma(B) \restriction \mathcal{S}^n(\mathcal{H}) = B \otimes \cdots \otimes B$ (n times). If A is a positive self-adjoint operator on $\mathcal{H}$, $\Gamma(e^{-tA})$ is a one-parameter self-adjoint group on $\mathcal{F}_s(\mathcal{H})$ and its generator is denoted by $d\Gamma(A)$, the second quantization of A, i.e.,

$$\Gamma(e^{-tA}) = e^{-td\Gamma(A)} \tag{6.6.249}$$

There is a natural way to put an infinite-dimensional Gaussian measure on $\mathbb{R}^\infty$. Then, if $A \geq \alpha\mathbb{1}$, $\Gamma(e^{-tA})$ is a contraction of L^p to L^q if and only if $e^{-\alpha} \leq \sqrt{\frac{p-1}{q-1}}$. This way of describing Nelson's result and this formalism is discussed further in Glimm–Jaffe [**322**], Reed-Simon [**673**], and Simon [**740**].

Extension of the Perron–Frobenius theorem to infinite dimensions goes back at least to Jentzsch [**427**] in 1912. The idea of using it to prove uniqueness of the lowest eigenvalue in quantum mechanics is due to Glimm–Jaffe [**321**] who use it to study cutoff $(\varphi^4)_2$ Hamiltonians. For extension to nonrelativistic quantum mechanics and to Fermi field theories, see, for example, Faris [**258**], Faris–Simon [**259**], and Gross [**334**].

Problems

1. (a) If $0 < \gamma < 1$, prove that $A = (-\Delta)^\gamma$ generates a Markov semigroup (*Hint*: Prove first that

$$e^{-w^\gamma} = \int g_\gamma(y)e^{-yw}\,dy \tag{6.6.250}$$

where $g_\gamma > 0$ and use that to prove $e^{-t(-\Delta)^\gamma}$ has a positive integral kernel.)

(b) Prove a Sobolev estimate for $(-\Delta)^\gamma$ and so an estimate of the form (6.6.27) for ν nonintegral.

2. Let $A = -\Delta + 1$ on $L^2(\mathbb{R}^\nu)$. Prove that for any $\mu \geq \nu$, $\|e^{-tA}\|_{1,2} \leq C t^{-\mu/4}$ and use this and Theorem 6.6.2 to prove inhomogeneous Sobolev estimates.

3. Using Theorem 7.6.2 of Part 4, prove that $C_0^\infty(\mathbb{R}^\nu)$ is a form core for the operator A of Theorem 6.6.8 and use that to verify that $Q(A) = \{f \in L^2(\mathbb{R}^\nu, d\mu) \mid \nabla f \in L^2(\mathbb{R}^\nu, d\mu)\}$ with (6.6.109) for all $f \in Q(A)$

4. Using Theorem 7.6.1 of Part 4, prove that C_0^∞ is an operator core of the operator A of Theorem 6.6.8 if φ_0 is C^1 and prove that $D(A)$ is as given in the remark following Theorem 6.6.9.

5. Deduce (6.6.125) from (6.6.124).

6. (a) Let $S = \{S_1, \ldots, S_\ell\}$ be a partition of a probability measure space (M, Σ, μ), i.e., the $S_j \in \Sigma$ are disjoint with union S. Let P_S be the L^2-projection onto the ℓ-dimensional space of functions constant on each S_j. Prove that P_S is a contraction on each L^p.

(b) If C is a bounded map of $L^p(M, d\mu)$ to $L^p(N, d\nu)$ and Q_T are the projections for $L^2(N, d\nu)$, prove that C is a strong limit of suitable $Q_T C P_S$.

(c) Prove that $Q_T C P_S$ is given by an integral kernel and conclude that every C is a limit of operators with integral kernel and not bigger norm.

7. Let e^{-sA} be a Markov semigroup. Prove that $\bigcup_{s>0} e^{-sA}[L^\infty]$ is an L^p-operator core for the L^p-semigroup generator when $1 < p < \infty$.

Remark. This requires the reader to know about semigroup generators of contraction semigroups on Banach spaces. It will help to use the Stein interpolation method to show that if $\varphi \in L^p$, $e^{-sA}\varphi$ is analytic in s in a neighborhood of $(0, \infty)$.

8. Provide the details of the calculation in Example 6.6.18.

9. Let $C : L^p(M, d\mu) \to L^1(M, d\mu) + L^\infty(M, d\mu)$ and $D : L^p(N, d\nu) \to L^1(N, d\nu) + L^\infty(N, d\nu)$. If C is not bounded from L^p to L^q, prove that $C \otimes D$ does not have a bounded extension of L^p to L^q. (*Hint:* $\|(C \otimes D)(f \otimes g)\|_q = \|Cf\|_q \|Dg\|_q$.)

10. Prove that (6.6.205) for $f \geq 0$ in $C_0^\infty(\mathbb{R}^\nu)$ implies it for all $f \in Q(A)$.

11. (a) Prove that for the Gaussian measure on $\mathbb{R}$ with $E(x) = 0$, $E(x^2) = 1$, one has $\|e^{\alpha x}\|_p = e^{\frac{1}{2}\alpha^2 p}$.

(b) Prove (6.6.212).

12. (a) In the representation (6.6.250), prove that for small y, $g_\gamma(y)$ is $O\big(\exp(-cy^{-\beta})\big)$, $\beta = \gamma/1-\gamma$.

(b) Use this to prove that for $\alpha > a^{-1}+1/2$, A^α is ultracontractive where A is the intrinsic generator for $-\Delta + |x|^a$.

13. Let e^{-tA} be a Markov semigroup on $L^2(M, d\mu)$ with $\mu(M) = 1$. Suppose for some T_0, $\|e^{-T_0 A}\|_{2,4} < \infty$. Suppose also that $A \upharpoonright \{\mathbb{1}\}^\perp \geq m > 0$. Let P be the projection of L^2 to multiples of $\mathbb{1}$.

(a) Suppose that $Pf = \mathbb{1}$ and write $f = \mathbb{1} + f_1$ with $f_1 = (1-P)f$. Prove that for some T_1 large, and all $t \geq T_1$, independently of f,

$$\|e^{-tA}f_1\|_4^4 \leq \frac{1}{4}\|f_1\|_2^4 \tag{6.6.251}$$

(b) Let $c = \langle \mathbb{1}, (e^{-tA}f_1)^2 \rangle = \|e^{-tA}f_1\|_2^2$. Prove that

$$\|e^{-tA}f\|_4^4 \leq (1+c)^2 + 8c + 2\|e^{-tA}f_1\|_4^4 \tag{6.6.252}$$

(c) Prove that for some T_2 large, that we have for $t \geq T_2$ and all f that

$$\|e^{-tA}f_1\|_2^4 \leq \frac{1}{2}\|f_1\|_2^4, \quad c = \|e^{-tA}f_1\|_2^2 \leq \frac{1}{5}\|f_1\|_2^2 \tag{6.6.253}$$

(d) Prove that $\|f\|_2^4 = 1 + 2\|f_1\|_2^2 + \|f_1\|_2^4$.

(e) Prove that for $t \geq \max(T_1, T_2)$, we have

$$\|e^{-tA}f\|_4 \leq \|f\|_2 \tag{6.6.254}$$

and that knowing this for f with $Pf = \mathbb{1}$ implies it for all f. Thus, hypercontractivity plus a mass gap implies a contraction from L^2 to L^4 for large t.

Remark. This argument is due to Glimm [**319**].

14. This will provide an alternate proof of the L^2-log Sobolev inequality for the Bonami–Gross semigroup.

(a) Let $f_s(x) = 1 + sx$ on $\{-1, 1\}$ for $0 \leq s \leq 1$. Prove that the inequality we want is equivalent to

$$h(s) \equiv \tfrac{1}{2}\Big[(1+s)^2 \log(1+s) + (1-s)^2 \log(1-s)\Big] - \tfrac{1}{2}(1+s^2)\log(1+s^2) \leq s^2 \tag{6.6.255}$$

(b) Prove that $h(0) = h'(0) = 0$ and that $h''(s) \leq 2$ by direct computation and conclude that (6.6.255) holds.

Remark. This is Gross' original proof [**335**].

15. Suppose A obeys an L^2-log Sobolev estimate with local norm 0, i.e.,

$$\int |f|^2 \log|f| \, d\mu \leq \Gamma\langle f, Af \rangle + \|f\|^2 + \log\|f\|^2 \qquad (6.6.256)$$

(a) Prove that $A \geq 0$.

(b) Prove that $A\mathbb{1} = 0$.

(c) If $\langle g, \mathbb{1} \rangle = 0, g \in L^\infty$ and real and $f_\varepsilon = 1 + \varepsilon\, g$, prove LHS of (6.6.256) $= 2\varepsilon \|g\|^2 + O(\varepsilon^2)$ while RHS $= \varepsilon\Gamma \langle g, Ag \rangle + O(\varepsilon^2)$. Conclude that A has a mass gap, i.e., for $m > 0$

$$A \restriction \{\mathbb{1}\}^\perp \geq m \qquad (6.6.257)$$

Remark. For $0 < \alpha < 2$, the intrinsic semigroup associated to $\frac{d^2}{dx^2} + |x|^\alpha$ has a mass gap but does not obey (6.6.256) so the converse of this result is false.

6.7. Lieb–Thirring and Cwikel–Lieb–Rosenblum Inequalities

This section is intended to be read after Section 7.9 of Part 4 and will heavily use results from that section. If B is a self-adjoint operator with $\sigma_{\mathrm{ess}}(B) \subset [0, \infty)$ and B bounded from below, we use $\{E_j(B)\}_{j=1}^\infty$ as the negative eigenvalues counting multiplicity with $E_1 \leq E_2 \leq E_3 \leq \ldots \leq 0$. We let $N(B) = \dim(P_{(-\infty, 0)}(B))$, the number of negative eigenvalues. Thus, if $N(B) < \infty$, $E_j(B) = 0$ if $j > N(B)$ and if $j \leq N(B)$, E_j is the jth eigenvalue from the bottom. We define the *eigenvalue moment*,

$$S_p(B) = \sum_{j=1}^\infty |E_j(B)|^p, \quad p > 0, \ S_0(B) \equiv N(B) \qquad (6.7.1)$$

Let $p \geq 0$. We say that we have a (p, ν)-*Lieb–Thirring inequality* if for all $V \in L^{p+\nu/2}(\mathbb{R}^\nu)$, we have

$$S_p(-\Delta - V) \leq c_{\nu, p} \int |V(x)|^{p+\nu/2} \, d^\nu x \qquad (6.7.2)$$

for a universal constant $c_{\nu, p}$.

Lieb–Thirring inequalities hold for Schrödinger operators if:

$$p \geq 0, \quad \nu \geq 3 \qquad (6.7.3)$$

$$p > 0, \quad \nu = 2 \qquad (6.7.4)$$

$$p \geq \tfrac{1}{2}, \quad \nu = 1 \qquad (6.7.5)$$

The cases $p = 0$ if $\nu \geq 3$, $p = \frac{1}{2}$ if $\nu = 1$ are called *critical Lieb–Thirring inequalities*; $p = 0$ for $\nu \geq 3$ are also called *Cwikel–Lieb–Rosenblum* (CLR) *inequalities*. They are more subtle—indeed the problems of Section 7.9 of Part 4 have a proof of the noncritical results. It is known (see Problem 5 of Section 7.9 of Part 4) that Lieb–Thirring inequalities fail if $p = 0$ when $\nu = 2$ or $p < \frac{1}{2}$ when $\nu = 1$.

Our goal in this section is to prove that abstract Sobolev inequalities for $\nu \geq 3$ are equivalent to CLR inequalities not only for $-\Delta$ but for general Markov semigroups (thereby adding a fourth equivalence to Theorem 6.6.2 when $\nu > 2$ and e^{-tA} is a Markov semigroup). We'll also prove noncritical Lieb–Thirring inequalities from Gagliardo–Nirenberg inequalities for $1 \leq \nu \leq 2$. Finally, for the special case of $-\frac{d^2}{dx^2}$, we'll prove the $\nu = 1$ critical Lieb–Thirring inequality (i.e., $p = \frac{1}{2}$).

It will be useful to know that an abstract LT inequality for some p_0 implies it for all $p > p_0$; indeed, we don't even need the full LT inequality at p_0 but only a weak one. Clearly, if $N_\alpha(B), \alpha > 0$ is the number of eigenvalues of B in $(-\infty, -\alpha)$, then

$$N_\alpha(B)\,\alpha^p \leq S_p(B) \tag{6.7.6}$$

and, by an integration by parts in a Stieltjes integral, using the function $N_\alpha(B)$ to define a measure with jumps at the negatives of negative eigenvalues

$$S_p(B) = -\int_0^\infty \alpha^p\,dN_\alpha(B) \tag{6.7.7}$$

$$= p\int_0^\infty \alpha^{p-1}\,N_\alpha(B)\,d\alpha \tag{6.7.8}$$

We say $A - V$ obeys a *weak LT inequality* for (p, ν) when $A \geq 0$ is a positive self-adjoint operator on $L^2(M, d\mu)$ and $V \in L^p(M, d\mu)$ if for $V \geq 0$, $V \in L^p \cap L^\infty$

$$N_\alpha(A - V) \leq C\,\alpha^{-p}\int |V(x)|^{p+\nu/2}\,d\mu(x) \tag{6.7.9}$$

By (6.7.6), this is implied by a bound like (6.7.2) but doesn't imply it since the integral (6.7.8) is logarithmically divergent if one only has an α^{-p} bound on $N_\alpha(B)$.

Proposition 6.7.1. *Let $A \geq 0$ be a self-adjoint operator on $L^2(M, d\mu)$. Then*

(a) A CLR bound of the form $(V \geq 0, V \in L^\infty \cap L^p)$

$$N(A - V) \leq c_0 \int |V(x)|^{\nu/2}\,d\mu(x) \tag{6.7.10}$$

implies an LT bound of the form

$$S_p(A - V) \leq c_p \int |V(x)|^{p+\nu/2}\,d\mu(x) \tag{6.7.11}$$

for all $p > 0$.

(b) *A weak LT bound of the form (6.7.9) for some p_0 implies an LT bound of the form (6.7.11) for all $p > p_0$.*

Proof. (a) can be viewed as (b) for $p_0 = 0$, so we need only prove (b). Without loss, we can also suppose $V \geq 0$ since $A - V_+ \leq A - V$ while $\int |V_+(x)|^q \, d\mu(x) \leq \int |V(x)|^p \, d\mu(x)$.

We begin by noting that for any $\alpha > 0$

$$(V - \tfrac{\alpha}{2})_+ - \tfrac{\alpha}{2} \geq V - \alpha \qquad (6.7.12)$$

since we have equality if $V(x) \geq \tfrac{\alpha}{2}$ and inequality if $\tfrac{\alpha}{2} > V(x)$. Noting that $N_\beta(B) = N_0(B + \beta)$, we have

$$
\begin{aligned}
N_\alpha(A - V) &= N_0(A - (V - \alpha)) \\
&\leq N_0(A - (V - \tfrac{\alpha}{2})_+ + \tfrac{\alpha}{2}) \\
&= N_{\frac{\alpha}{2}}(A - (V - \tfrac{\alpha}{2})_+) \qquad (6.7.13)
\end{aligned}
$$

where we used the variational principle and (6.7.12) which says that

$$A - (V - \alpha) \geq A - (V - \tfrac{\alpha}{2})_+ + \tfrac{\alpha}{2} \qquad (6.7.14)$$

Now, suppose we have (6.7.9) for p_0 and let $p > p_0$. Then, by (6.7.8):

$$S_p(A - V) = p \int_0^\infty \alpha^{p-1} N_\alpha(A - V) \, d\alpha \qquad (6.7.15)$$

$$\leq p \int_0^\infty \alpha^{p-1} N_{\frac{\alpha}{2}}(A - (V - \tfrac{\alpha}{2})_+) \, d\alpha \qquad (6.7.16)$$

$$\leq Cp \int_0^\infty \alpha^{p-p_0-1} \Big[\int (V - \tfrac{\alpha}{2})_+^{p_0 + \frac{\nu}{2}} \, d\mu(x) \Big] \qquad (6.7.17)$$

$$= Cp \int d\mu(x) \Big[\int_0^{2V(x)} \alpha^{p-p_0-1} (V(x) - \tfrac{\alpha}{2})^{p_0 + \frac{\nu}{2}} \, d\alpha \Big] \qquad (6.7.18)$$

$$= 2^{p-p_0} Cp \int \Big[\int_0^1 \beta^{p-p_0-1}(1-\beta)^{p_0 + \frac{\nu}{2}} \, d\beta \Big] V(x)^{p_0 + \frac{\nu}{2}} \, d\mu(x) \qquad (6.7.19)$$

Since $\int_0^1 \beta^{p-p_0-1}(1-\beta)^{p_0 + \frac{\nu}{2}} \, d\beta < \infty$, we have (6.7.11). In the above, (6.7.16) comes from (6.7.13), (6.7.17) from (6.7.9), (6.7.18) from noting for any fixed x, only α's with $0 < \alpha < 2V(x)$ contribute and (6.7.19) from the change of variables $\alpha = 2V(x)\beta$. $\qquad \square$

The two main theorems we'll prove are the following:

Theorem 6.7.2. *Let A generate a sub-Markovian semigroup on $L^2(M, d\mu)$. Let $\nu > 2$. Then the following are equivalent:*

(1) *The three equivalent conditions of Theorem 6.6.2.*

(2) *There is a constant C so that for all $V \geq 0$ in $L^\infty \cap L^{\nu/2}$, we have a CLR inequality*

$$N(A - V) \leq C \int V(x)^{\nu/2} \, d\mu(x) \tag{6.7.20}$$

Theorem 6.7.3. *Let A generate a sub-Markovian semigroup on $L^2(M, d\mu)$. Let $\nu > 0$ and $p_0 > 0$ be such that $p_0 + \nu/2 > 1$. Then the following are equivalent:*

(1) *Let*

$$\theta = \nu/(\nu + 2p_0), \quad q = (\nu + 2p_0)/(\tfrac{\nu}{2} + p_0 - 1) \tag{6.7.21}$$

Then $f \in L^2(M, d\mu) \cap Q(A) \Rightarrow f \in L^q(M, d\mu)$ and for some constant C_1, the Gagliardo–Nirenberg type inequality

$$\|f\|_q^2 \leq C_1 \|f\|_2^{2(1-\theta)} q_A(f)^\theta \tag{6.7.22}$$

holds.

(2) *There is a constant $C > 0$ so that for all $V \geq 0$ in $L^\infty \cap L^{\nu/2}$, and for all $\alpha > 0$, we have a weak LT inequality*

$$N_\alpha(A - V) \leq C \, \alpha^{-p_0} \int V(x)^{p_0 + \nu/2} \, d\mu(x) \tag{6.7.23}$$

Remark. For $A = -\Delta$ on $L^2(\mathbb{R}^\nu)$, (6.3.30) (the original Gagliardo–Nirenberg inequality), say $\|f\|_q \leq C_2 \|\nabla f\|_2^\theta \|f\|_2^{(1-\theta)}$ if $q^{-1} = \theta(\tfrac{1}{2} - \tfrac{1}{\nu}) + \tfrac{1}{2}(1 - \theta)$ which is equivalent to the formula for q in (a) if $\theta = \nu/(\nu + 2p_0)$. Theorem 6.3.4 as stated requires $\nu > 2$ but, as we'll see (Example 6.7.6), (6.7.22) holds when $\nu = 1, 2$ so long as $p_0 + \tfrac{\nu}{2} > 1$.

Proof that (2) $\Rightarrow$ (1) in Theorem 6.7.2. We are supposing (6.7.20). Given $\varphi \in L^1 \cap L^\infty$, let

$$V(x) = d \, |\varphi(x)|^{4/(\nu-2)} \quad q \equiv \frac{2\nu}{\nu - 2} \tag{6.7.24}$$

where d is chosen so that

$$C \, d^{\nu/2} \|\varphi\|_q^q = 1 \Leftrightarrow d = C^{-2/\nu} \|\varphi\|_q^{-4/\nu-2} \tag{6.7.25}$$

By this choice, $C \int V(x)^{\nu/2} \, dx = 1$, so for any $\gamma \in (0,1)$, $N(A - \gamma V) \leq \gamma^{\nu/2} < 1 \Rightarrow A - \gamma V \geq 0 \Rightarrow A - V \geq 0$. In particular,

$$\int V(x) \, |\varphi(x)|^2 \, d\mu(x) \leq q_A(\varphi) \tag{6.7.26}$$

Since $2 + \tfrac{4}{\nu-2} = q$, this says

$$C^{-2/\nu} \|\varphi\|_q^q \|\varphi\|_q^{-4/\nu-2} \leq q_A(\varphi) \tag{6.7.27}$$

But $q - \frac{4}{\nu-2} = \frac{2\nu-4}{\nu-2} = 2$. Thus, (6.7.27) says

$$\|\varphi\|_q^2 \le C^{\nu/2} \, q_A(\varphi)$$

which is (6.6.28). $\qquad\qquad\qquad\qquad\qquad\qquad\qquad\qquad\qquad\qquad\square$

We now turn to the other direction for Theorem 6.7.2. By a simple limiting argument, we can suppose that

$$V \in L^1 \cap L^\infty, \quad V(x) > 0 \text{ for a.e. } x \qquad\qquad (6.7.28)$$

Moreover, we can restrict to the case $A \ge \varepsilon \mathbb{1}$ for some $\varepsilon > 0$ so long as we obtain a constant C in (6.7.17) that only depends on ν and the constant in the Sobolev estimate (6.6.28). For if $A \ge 0$ is a sub-Markovian semigroup and (6.6.28) holds with a constant C, then it holds for $A + \varepsilon \mathbb{1}$ with the same constant and this is also sub-Markovian. And (6.7.20) for $N(A + \varepsilon - V)$ for all ε implies it for $\varepsilon = 0$. So we make those two assumptions.

The hypothesis, $A \ge \varepsilon \mathbb{1}$ implies that $q_A(\varphi)^{1/2}$ defines a norm on $\mathcal{H}_{+1}(A)$ equivalent to $\|\cdot\|_{+1}$. We want to note two relations between $q = \frac{2\nu}{\nu-2}$ and $p = \frac{\nu}{2}$:

$$\frac{1}{p} + \frac{2}{q} = 1 \Rightarrow \|V\,\varphi^2\|_1 \le \|V\|_p \, \|\varphi\|_q^2 \qquad\qquad (6.7.29)$$

$$\frac{1}{q} + \frac{1}{2p} + \frac{1}{2} = 1 \Rightarrow \|fgh\|_1 \le \|f\|_q \, \|g\|_{2p} \, \|h\|_2 \qquad\qquad (6.7.30)$$

The key will be to essentially look at $Y = V^{-1/2} A V^{-1/2}$ which is the inverse of the Birman–Schwinger operator. We do this by considering $\mathcal{H}' = L^2(M, V \, d\mu)$. We note that by (6.7.29) and (6.6.28)

$$\|\varphi\|_{\mathcal{H}'}^2 \le C \, \|V\|_p \, q_A(\varphi) \qquad\qquad (6.7.31)$$

which means any $\varphi \in Q(A)$ first lies in $\mathcal{H}'$ since

$$\|\varphi\|_{\mathcal{H}'} \le \|V\|_\infty^{1/2} \, \|\varphi\|_2 \qquad\qquad (6.7.32)$$

and secondly $\varphi \mapsto q_A(\varphi)$ as a map from $Q(A) \subset \mathcal{H}'$ defines a closed form on $\mathcal{H}'$ (since φ_n Cauchy in $[q_A(\varphi)^2 + \|\varphi\|_{\mathcal{H}'}^2] \equiv \|\varphi\|_Y^2$ implies φ_n Cauchy in $\mathcal{H}_{+1}$-norm implies φ_n convergent to $\varphi \in \mathcal{H}_{+1}$ and then φ_n convergent in $\|\cdot\|_Y$ because of (6.7.31)).

Thus, there is a positive self-adjoint Y on $\mathcal{H}'$ so $Q(Y) = \mathcal{H}_{+1} \subset \mathcal{H}'$ and $q_Y(\varphi) = q_A(\varphi)$. Moreover, since the ordering of functions in $L^2(M, V \, d\mu)$ and $L^2(M, d\mu)$ is the same, the two Beurling–Deny criteria for A imply them for Y, i.e., e^{-tY} is a $L^p(M, V \, d\mu)$-contractive semigroup.

Moreover,

$$\|\varphi\|_{L^q(M, V \, d\mu)}^2 \le \|V\|_\infty^{2/q} \, \|\varphi\|_{L^q(M, d\mu)}$$
$$\le C \, q_A(\varphi) = C \, q_Y(\varphi) \qquad\qquad (6.7.33)$$

obeys a Sobolev inequality, so by Theorem 6.6.2, e^{-tY} is ultracontractive with

$$\|e^{-tY}\,\varphi\|_{L^\infty(M,V\,d\mu)} \leq C_1\, t^{-\nu/2}\,\|\varphi\|_{L^1(M,V\,d\mu)} \tag{6.7.34}$$

It follows (from the Dunford–Pettis theorem; see Theorem 4.9.7 of Part 1) that e^{-tY} has a bounded integral kernel and, since $V\,d\mu$ is a finite measure (recall $V \in L^1$), it is Hilbert-Schmidt and so $e^{-tY} = (e^{-tY/2})^2$ is trace class.

Because of (6.7.31), $Y \geq (C\,\|V\|_p)^{-1/2}\,\mathbb{1}$, so Y has discrete spectrum and $Y^{-\beta}$ is bounded for any $\beta > 0$. Thus, $Y^{-\beta}\,e^{-tY}$ is also trace class and by

$$Y^{-\beta}\,e^{-tY} = \Gamma(\beta)^{-1}\int_0^\infty s^{\beta-1}\,e^{-(s+t)Y}\,ds \tag{6.7.35}$$

$Y^{-\beta}\,e^{-tY}$ also has a positive integral kernel $H_\beta(x,y;t)$. Here is the key link to $N(A-V)$.

Proposition 6.7.4. *Let $U : \mathcal{H}' \to \mathcal{H}$ by $U\varphi = V^{1/2}\varphi$. Then $U Y U^{-1}$ is the unbounded inverse of the Birman–Schwinger operator $V^{1/2}A^{-1}V^{1/2}$. In particular,*

$$N(A-V) = \dim P_{(0,1]}(Y) \tag{6.7.36}$$

$$\leq 2e^2\,\mathrm{Tr}((2Y)^{-1}e^{-2Y}) \tag{6.7.37}$$

Remark. Since $A \geq \varepsilon\mathbb{1}$ and $V \in L^\infty$, $K \equiv V^{1/2}A^{-1}V^{1/2}$ is a product of bounded operators. Since $V > 0$ a.e., it is easy to see that $\ker(K) = \{0\}$, so by the spectral theorem, K has an unbounded self-adjoint inverse.

Proof. $U^{-1}\psi \in Q(Y) \Leftrightarrow V^{-1/2}\psi \in Q(A)$ and

$$q_Y(U^{-1}\psi) = q_A(V^{-1/2}\psi) \tag{6.7.38}$$

i.e., $U Y U^{-1} = V^{-1/2}\,A\,V^{-1/2}$ suitably interpreted. It is not hard to confirm that this is the inverse of K.

Since $A \geq \varepsilon\mathbb{1}$, $N(A-V) = N_\varepsilon((A - \varepsilon\mathbb{1}) - V)$, so the strictly negative Birman–Schwinger principle (Theorem 7.9.4 of Part 4) implies the equality (6.7.36). To get (6.7.37), note that if $0 < \mu_1 \leq \mu_2 \leq \ldots$ are the eigenvalues of Y

$$\mathrm{Tr}((2Y)^{-1}e^{-2Y}) = \sum_j (2\mu_j)^{-1}\,e^{-2\mu_j}$$

$$\geq \sum_{j\,|\,\mu_j \leq 1} (2\mu_j)^{-1}\,e^{-2\mu_j}$$

$$\geq \frac{1}{2}e^{-2}\,\sharp(j \mid \mu_j \leq 1)$$

proving (6.7.37). $\square$

Proposition 6.7.5.

$$\mathrm{Tr}((2Y)^{-1}e^{-2Y}) \le C^{\nu}(\tfrac{\nu}{2}-1)^{\frac{\nu}{2}-1}2^{-\nu/2}\int V(x)^{\nu/2}\,d\mu(x) \qquad (6.7.39)$$

where C is the constant in (6.6.28)

Proof. Define

$$h_{\beta}(s) = \mathrm{Tr}(Y^{-\beta}e^{-2sY}) \qquad (6.7.40)$$

Using $Y^{-1}e^{-2sY} = (e^{-sY})(Y^{-1}e^{-sY})$, we see that

$$h_1(s) = \int_X V(x)\,d\mu(x)\Big(\int V(y)d\mu(y)\,H_1(y,x;s)H_0(x,y;s)\Big)$$

$$\le \int_X V(x)d\mu(x)\Big\{\|H_1\|_q\big[\int H_0(x,y;s)^2 V(y)d\mu(y)\big]^{1/q}$$

$$\big[\int H_0(x,y;s)V(y)^{\frac{\nu+2}{4}}\,d\mu(y)\big]^{\frac{\nu}{2}}\Big\} \qquad (6.7.41)$$

where we used (6.7.29) (in the form $\|fgh\|_1 \le \|f\|_q\,\|g\|_q\,\|h\|_p$) with $f = H_1$, $g = V^{\frac{1}{q}}H_0^{\frac{2}{q}}$, $h = V^{1-\frac{1}{q}}H_0^{\frac{1}{p}}$ and the calculation $p(1-\frac{1}{q}) = \frac{\nu}{2}\big[1 - \frac{(\nu-2)}{2\nu}\big] = \frac{\nu+2}{4}$.

Now we use Hölder in the form (6.7.30) with $h = \|H_1\|_q$, f the second factor and g the last factor inside the integral in (6.7.41). We obtain

$$h_1(s) \le A^{\frac{1}{2}}B^{1/\nu}D^{\frac{1}{q}} \qquad (6.7.42)$$

$$A = \int V(x)\Big(\int d\mu(y)\,H_1(x,y;s)^q\Big)^{\frac{2}{q}}\,d\mu(x) \qquad (6.7.43)$$

$$B = \int_x V(x)\Big(\int_x H_0(x,y;s)\,V(y)^{\frac{\nu+2}{4}}\,d\mu(y)\Big)^2\,d\mu(x) \qquad (6.7.44)$$

$$D = \int_x V(x)\Big(\int_x H_0(x,y;s)^2\,V(y)\,d\mu(y)\Big) \qquad (6.7.45)$$

By a Sobolev estimate

$$A = \int V(x)\|H_1(x,\cdot;s)\|_q^2\,d\mu(x)$$

$$\le C\int V(x)\langle H_1(x,\cdot;s), Y H_1(x,\cdot;s)\rangle\,d\mu(x)$$

$$= C\,\mathrm{Tr}(Y(Y^{-1}e^{-sY})(Y^{-1}e^{-sY}))$$

$$= C\,h_1(s) \qquad (6.7.46)$$

Taking into account that

$$\|e^{-sY}f\|_{L^2(N,Vd\mu)}^2 = \int V(x)\Big(\int H_0(x,y;s)\,V(y)f(y)d\mu(y)\Big)^2\,d\mu(x)$$

and writing $V(y)^{(\nu+2)/4} = V(y)V(y)^{(\nu-2)/4}$, we see that

$$B = \|e^{-sY}\, V^{(\nu-2)/4}\|^2_{L^2(M,Vd\mu)}$$

$$\leq \|V^{(\nu-2)/4}\|^2_{L^2(M,Vd\mu)} = \int V^{(\frac{\nu}{2}-1)}(y)\, V(y)\, d\mu(y) \tag{6.7.47}$$

$$= \int V^{\nu/2}(y)\, d\mu(y) \tag{6.7.48}$$

where, in (6.7.47), we used that e^{-sY} is a contraction on $L^2(M,Vd\mu)$.

Finally, we recognize that

$$D = h_0(s) = -\tfrac{1}{2}h_1'(s) \tag{6.7.49}$$

Thus, (6.7.42) says

$$h_1(s) \leq C^{\frac{1}{2}}\, h_1(s)^{\frac{1}{2}}\, \|V\|^{\frac{1}{2}}_{\nu/2}\big(-\tfrac{1}{2}h_1'(s)\big)^{\frac{1}{q}} \tag{6.7.50}$$

or

$$h_1^{q/2} \leq \tfrac{1}{2}\, C^{q/2}\, \|V\|^{q/2}_{\nu/2}\, (-h_1') \tag{6.7.51}$$

Noting that $\frac{q}{2} - 1 = \frac{2}{\nu-2}$, this says

$$1 \leq \big(\frac{\nu - 2}{4}\big)\, C^{q/2}\, \|V\|^{q/2}_{\nu/2}\, (h_1^{1-q/2})' \tag{6.7.52}$$

Since $h_1(s) \to \infty$ a $s \downarrow 0$, and $1 - \frac{q}{2} < 0$, $h_1^{1-q/2} \to 0$ as $s \downarrow 0$, so this implies

$$h_1(1) \leq \big(\tfrac{\nu}{2} - 1\big)^{\frac{\nu}{2}-1}\, 2^{-\frac{\nu}{2}+1}\, C^{\nu/2}\, \|V\|^{\nu/2}_{\nu/2} \tag{6.7.53}$$

using $[\frac{q}{2}/(\frac{q}{2} - 1)] = \nu/2$. This is (6.7.39). $\square$

Proof that (1) $\Rightarrow$ (2) in Theorem 6.7.2. Suppose $A \geq \varepsilon\mathbb{1}$, $V \in L^1 \cap L^\infty$; $V > 0$. By Proposition 6.7.5, we have (6.7.39) and then a CLR bound by Proposition 6.7.4.

By a density argument, we can go from the special V's to any $V \in L^{\nu/2}$ and then let $\varepsilon \downarrow 0$ as explained before. $\square$

Proof of Theorem 6.7.3. We'll reduce this to Theorem 6.7.2! For each $\alpha > 0$, define

$$A_\alpha = \alpha^{-1+\theta}\, (A + \alpha) \tag{6.7.54}$$

where θ is given by (6.7.21). Then, the semigroup for A_α is just $e^{-\alpha^\theta t}\exp(-t\alpha^{-1+\theta}A)$ is clearly a contraction on all L^p's and positivity preserving—thus also sub-Markovian.

Moreover,

$$A_\alpha - \alpha^{-1+\theta}V = \alpha^{-1+\theta}(A - V + \alpha) \tag{6.7.55}$$

so

$$N(A_\alpha - \alpha^{-1+\theta}V) = N(A - V + \alpha) = N_\alpha(A - V) \tag{6.7.56}$$

This implies that (6.7.23) for a given α is equivalent to an ordinary CLR bounded for A_α with

$$\tilde{\nu} = \nu + 2p_0 \tag{6.7.57}$$

Thus,

$$(6.7.23) \text{ for all } \alpha > 0 \Leftrightarrow \text{CLR for } A_\alpha \text{ for all } \alpha > 0 \tag{6.7.58}$$

On the other hand by elementary calculus, (the minimum occurs at $\alpha = (1-\theta)a/(\theta b)$, for all $\theta \in (0,1)$, $a, b > 0$)

$$\min_{\alpha>0} \left(a\alpha^{-1+\theta} + b\alpha^\theta\right) = \theta^{-\theta}(1-\theta)^{-(1-\theta)}a^\theta b^{1-\theta} \tag{6.7.59}$$

Thus, with q given by (6.7.21)

$$(6.7.22) \Leftrightarrow \forall\alpha \, \|f\|_q^2 \leq C\theta^\theta(1-\theta)^{(1-\theta)}\langle f, A_\alpha f\rangle \tag{6.7.60}$$

Notice that

$$\tilde{q} = \frac{2\tilde{\nu}}{\tilde{\nu} - 2} = q \tag{6.7.61}$$

Thus, Theorem 6.7.2 says that

$$\text{RHS of (6.7.60)} \Leftrightarrow \text{CLR for } A_\alpha \text{ for all } \alpha > 0 \tag{6.7.62}$$

(6.7.58), (6.7.60) and (6.7.62) imply the theorem. $\square$

Example 6.7.6. Let $A = -\Delta$ on $L^2(\mathbb{R}^{\nu_0})$ For $\nu_0 \geq 3$, A obeys a homogeneous Sobolev inequality with $\nu = \nu_0$ and $q = 2\nu_0/(\nu_0 - 2)$. Thus, Theorem 6.7.2 implies the CLR bounds for those ν_0.

For $\nu_0 = 1, 2$, the Fourier transform $J_1(x)$ of $(1+k^2)^{-1/2}$ lies in all L^p, $p < \infty$ (even $p = \infty$ for $\nu_0 = 1$). Thus, by Young's inequality, we get inhomogeneous Sobolev bounds for $2 < q < \infty$, $\nu_0 = 1, 2$

$$\|f\|_q^2 \leq C_{q,\nu_0}(\|f\|_2^2 + \|\nabla f\|_2^2) \tag{6.7.63}$$

By scaling and (6.7.63), we get (6.7.22) for q, θ given by (6.7.21) for any $p_0 > 0$. The condition $p_0 + \frac{\nu_0}{2} > 1$ is true for any $p_0 > 0$ if $\nu_0 = 2$ but requires $p_0 > \frac{1}{2}$ if $\nu_0 = 1$. Applying Theorems 6.7.1 and 6.7.3, we get (b) of the theorem below. $\square$

Theorem 6.7.7. Let $A = -\Delta$ on $L^2(\mathbb{R}^{\nu_0})$. Then

(a) For $\nu_0 \geq 3$, we have CLR bounds and LT bounds for all p.
(b) For $\nu_0 = 1, 2$, we have LT bounds for $p_0 > 0$ if $\nu_0 = 2$ and for $p_0 > \frac{1}{2}$ if $\nu_0 = 1$.

That only leaves the one-dimensional critical LT bound, $p_0 = \frac{1}{2}, \nu_0 = 1$ which we'll discuss at the end of the section.

Example 6.7.8 (Fractional Laplacian). On $\mathbb{R}^{\nu_0}$, let $A = (-\Delta)^\gamma$, $0 < \gamma < 1$. Then e^{-tA} is a Markov semigroup (see Problem 1 of Section 6.6). By Theorem 6.3.3, we have a homogeneous Sobolev estimate

$$\|f\|_{q_0}^2 \leq C \langle f, (-\Delta)^\gamma f \rangle \tag{6.7.64}$$

so long as $\frac{\nu_0}{2} > \gamma$ where q_0 is given by

$$q_0 = \frac{2\nu_0}{\nu_0 - 2\gamma} = \frac{2\nu}{\nu - 2} \quad \text{if } \nu = \nu_0/\gamma \tag{6.7.65}$$

Thus, we obtain CLR bounds ($0 < \gamma < 1$)

$$\frac{\nu_0}{2} > \gamma \Rightarrow N((-\Delta)^\gamma - V) \leq c_{\gamma,\nu_0} \int |V(x)|^{\nu_0/2\gamma} d^{\nu_0} x \tag{6.7.66}$$

We leave the LT bounds for $\nu_0 = 1$, $\gamma > \frac{1}{2}$ to Problem 1. $\square$

Example 6.7.9 (Periodic Schrödinger Operators.). Let $A = -\Delta + V_0(x)$ on $L^2(\mathbb{R}^\nu)$ where $V_0(x + e_j) = V_0(x)$, $j = 1, \ldots, \nu_j$, $x \in \mathbb{R}^\nu$ with $\{e_j\}_{j=1}^\nu$ a basis for $\mathbb{R}^\nu$ and $\int_{|x| \leq R} |V_0(x)|^{\nu/2} d^\nu x < \infty$ for all $R < \infty$. It is then known (see the Notes) that $\lambda_- = \inf \sigma(-\Delta + V_0)$ is finite and there exists a real-valued function, φ_0, so that

$$(-\Delta + V)\varphi_0 = \lambda_-\varphi_0, \quad 0 < \inf_x \varphi_0(x_0) < \sup_x \varphi_0(x_0) < 0 \tag{6.7.67}$$

As in Theorem 6.6.8 (extended since φ_0 is not in L^2), $A - \lambda_0$ is unitarily equivalent to an intrinsic semigroup generator A_0 on $L^2(\mathbb{R}^\nu, \varphi_0^2 d^\nu x)$ with

$$\langle f, A_0 f \rangle = \int (\nabla f)^2 \varphi_0^2 d^\nu x \tag{6.7.68}$$

By (6.7.67) and (6.7.68), there are C_1 and C_2 so that

$$C_1 \int (\nabla f)^2 d^\nu x \leq \langle f, A_0 f \rangle \leq C_2 \int (\nabla f)^2 d^\nu x$$

$$C_1 \int |f(x)|^q d^\nu x \leq \int |f(x)|^q \varphi_0^2 d^\nu x \leq C_2 \int |f(x)|^q d^\nu x \tag{6.7.69}$$

Therefore the ordinary Sobolev inequality on $\mathbb{R}^\nu$ for $\nu \geq 3$ and $q = 2\nu/(\nu-2)$ implies one for $L^2(\mathbb{R}^\nu, \varphi_0^2 d^\nu x)$. We conclude that

$$N_{\lambda_-}(-\Delta + V_0 + W) \leq C \int |W(x)|^{\nu/2} d^\nu x \tag{6.7.70}$$

$\square$

Finally, we turn to the proof of the $\nu = 1$, critical LT inequality when $A = -d^2/dx^2$. We need some preliminaries. We define

$$T_n(B) = \sum_{j=1}^n \lambda_j(B) \tag{6.7.71}$$

If $B \geq 0$ is compact, $\mu_1(B) \geq \mu_2(B) \geq \ldots \geq 0$ are its positive eigenvalues (counting multiplicity) with $\mu_j(B) = 0$ if $\dim P_{(0,\infty)}(B) < j$.

Proposition 6.7.10. *We have for any $B \geq 0$ compact that*

$$T_n(B) = \sup\{\mathrm{Tr}(BP) \mid P^* = P, P^2 = P, \mathrm{Tr}(P) = n\} \tag{6.7.72}$$

In particular, $B \mapsto T_n(B)$ is a convex function B.

Proof. By the min-max principle, $\lambda_j(PBP) \leq \lambda_j(B)$ for all j. Since PBP has rank at most n, $\mathrm{Tr}(PBP) = \sum_{j=1}^{n} \lambda_j(PBP) \leq T_n(B)$. If P is the projection on the eigenvectors with eigenvalues $\{\lambda_j(B)\}_{j=1}^{n}$ (if λ_j is degenerate, we pick only enough vectors from the λ_n eigenvalues to get $\dim(\mathrm{Ran}(P)) = n$), then $\mathrm{Tr}(PBP) = T_n(B)$. Thus, we have proven (6.7.72).

Since $T_n(B)$ is a sup of functions linear in B, it is convex by general principles. $\qquad\square$

Proposition 6.7.11. *Let $\{b_j\}_{j=1}^{n}$ be a set of n nonnegative numbers. For $\mu \in [0, 1)$, let*

$$(L_\mu)_{ij} = b_i^{1/2} \mu^{|i-j|} b_j^{1/2} \tag{6.7.73}$$

be the matrix element of an $n \times n$ matrix. Then,

$$0 < \mu < \eta \leq 1 \Rightarrow T_\ell(L_\mu) \leq T_\ell(L_\eta), \quad \text{all } \ell \tag{6.7.74}$$

Proof. For $\mu_1, \ldots, \mu_{n-1} \in [-1, 1]$, define the $n \times n$ matrix

$$(L_{\mu_1,\ldots\mu_{n-1}})_{ij} = \begin{cases} b_i^{1/2} \pi_{\ell=i}^{j-1} \mu_\ell b_j^{1/2} & \text{if } i \leq j \\ b_j^{1/2} \pi_{\ell=j}^{i-1} \mu_\ell b_i^{1/2} & \text{if } i \geq j \end{cases} \tag{6.7.75}$$

Then $L \geq 0$ (Problem 2). Moreover, if $\mu_1, \ldots \mu_{m-1}, \mu_{m+1}, \ldots, \mu_{n-1}$ are fixed, $L_{(\mu_j)}$ is clearly affine in μ_m. Moreover, if

$$(U\varphi)_j = \begin{cases} \varphi_j, & j = 1, \ldots m-1 \\ -\varphi_j, & j = m, \ldots, n \end{cases} \tag{6.7.76}$$

then $UL_{\mu_1,\ldots\mu_m,\ldots,\mu_n}U^{-1} = L_{\mu_1,\ldots-\mu_m,\ldots,\mu_n}$ only flips the sign of μ_m.

Since T_ℓ is unitary invariant, $T_\ell(\mu_1, \ldots, \mu_n)$ is even in μ_m for the other μ's fixed and convex. Such a function is monotone decreasing in μ_m for $\mu_m \in [0, \infty)$. Then

$$\begin{aligned} T_\ell(L_{\mu,\mu,\ldots\mu}) &\leq T_\ell(L_{\eta,\mu,\mu,\ldots\mu}) \\ &\leq T_\ell(L_{\eta,\eta,\mu,\ldots\mu}) \leq T_\ell(L_{\eta,\ldots,\eta}) \end{aligned} \tag{6.7.77}$$

$\qquad\square$

Theorem 6.7.12 ($\nu = 1$, Critical LT Inequality). *If* $V \in L^1(\mathbb{R})$, *then*

$$S_{1/2}\left(-\frac{d^2}{dx^2} + V\right) \le \frac{1}{2} \int |V(x)| \, dx \qquad (6.7.78)$$

Remark. We saw in Theorem 7.9.18 of Part 4 that if $V \le 0$ then $-\frac{d^2}{dx^2} + \lambda V$ has an eigenvalue of size $\left(\frac{\lambda}{2} \int V(x) \, dx\right)^2 + o(\lambda^2)$ so $S_{1/2}(-\frac{d^2}{dx^2} + \lambda V) = \frac{\lambda}{2} \int V(x) dx + o(\lambda)$, i.e., the constant $\frac{1}{2}$ in (6.7.78) is optimal. As an alternative, we note $-\frac{d^2}{dx^2} - \lambda \delta$ makes sense (see Example 7.5.8 of Part 4) and has an eigenvalue at $-\frac{\lambda^2}{4}$ so V's in L^1 near $\delta(x)$ will come close to saturating (6.7.16).

Proof. As usual we can suppose $V \le 0$, continuous and with compact support. The Birman–Schwinger kernel at energy $e = -k^2$ has kernel

$$K_k(x, y) = \frac{V(x)^{1/2} \, e^{-k|x-y|} \, V(y)^{1/2}}{2k} \qquad (6.7.79)$$

We let $L_k = 2k K_k$. If V is supported in $[-n, n]$ with $n \in \mathbb{Z}$, we can let j run from $-nR$ to nR and let $x_j = j/R$ and define the $(2nR + 1) \times (2nR + 1)$ matrix

$$\left(L_k^{(R)}\right)_{ij} = V(x_i)^{1/2} \, e^{-k|x_j - x_i|} V(x_j)^{1/2} \qquad (6.7.80)$$

Then, a simple argument (Problem 3) proves that

$$T_m(L_k^{(R)}) \to T_m(L_k) \qquad (6.7.81)$$

so the last proposition implies that $T_m(L_k)$ is monotone increasing as k decreases.

Now suppose the eigenvalues of $-\frac{d^2}{dx^2} + V$ occur at $-k_1^2 < -k_2^2 < \dots$. That means K_{k_j} has 1 as its jth eigenvalue from the top, so

$$2k_j = T_j(L_{k_j}) - T_{j-1}(L_{k_j}) \qquad (6.7.82)$$

It follows by monotonicity that

$$2k_1 = T_1(L_{k_1}) \le T_1(L_{k_2})$$
$$2k_1 + 2k_2 = T_1(L_{k_1}) + [T_2(L_{k_2}) - T_1(L_{k_2})] \le T_2(L_{k_2}) \qquad (6.7.83)$$

By induction

$$2k_1 + \dots + 2k_\ell \le T_{\ell-1}(L_{k_{\ell-1}}) + 2k_\ell$$
$$\le T_{\ell-1}(L_{k_\ell}) + 2k_\ell = T_\ell(L_{k_\ell}) \qquad (6.7.84)$$

Thus for any M

$$\sum_{j=1}^{M} 2k_j \le T_M(L_{k_M}) \le T_M(L_0) = \int V(x) dx \qquad (6.7.85)$$

since $L_0(x, y) = V(x)^{1/2} V(y)^{1/2}$ is a rank-one operator with the single nonzero eigenvalue $\mathrm{Tr}(L_0) = \int V(x)\, dx$. Taking $M \to \infty$ yields (6.7.78). $\qquad \square$

Notes and Historical Remarks. The first LT inequality was for $\nu = 3$, $p = 1$ which Lieb–Thirring [**534**] needed in their proof of stability of matter. As they proved and we showed in Section 7.9 of Part 4, this inequality is a dual to a Sobolev inequality for multi-particle fermion wave functions. Lieb and Thirring [**535**] shortly thereafter extended the results to all the noncritical cases.

CLR inequalities are named after their independent discovery (and rather different proofs) of Cwikel [**179**], Lieb [**530, 531**], and Rozenbljum [**704**]. Cwickel and Lieb were motivated by the fact that these inequalities were conjectured by Simon [**741**] who noted they would allow the extension of classical limit results from nice V's of compact suppport. Alternate proofs are due to Conlon [**165**], Fefferman [**268**], and Li–Yau [**528**].

Both LT and CLR inequalities are classical phase space quantities and are intimately related to Weyl asymptotics which we discuss in Section 7.5 of Part 4. Indeed, one can use these inequalities to extend the V's for which one has such asymptotics as discussed there.

The approach of Li–Yau was further developed by Blanchard–Stubbe–Rezende [**81**] and Levin–Solomyak [**522**] who, in particular, realized that this proof would work for general sub-Markovian semigroups. This was extended by Frank–Lieb–Seiringer [**284**] who, in particular, realized the Lieb–Thirring version that appears as our Theorem 6.7.3. Our proofs of both Theorems 6.7.2 and 6.7.3 follow this paper.

The idea in the proof of Proposition 6.7.1 goes back to the original papers of Lieb–Thirring and to Aizenman–Lieb [**9**]; see Laptev–Weidl [**510**] for another way that different LT inequalities are related.

One interesting extension of LT inequalities involves also the Hardy inequality $-\Delta \geq \frac{(\nu-2)^2}{4} r^{-2}$ and proves LT inequalities for $-\Delta - \frac{(\nu-2)^2}{4} r^{-2} - V$ when $\nu \geq 3$. This is much like the $-\Delta$ when $\nu = 2$; there is not a CLR inequality but there are LT inequalities for all $p > 0$. These *Hardy–Lieb–Thirring inequalities* go back to Ekholm–Frank [**238**]; see Frank–Lieb–Seiringer [**283**] and Frank [**279**] for further developments.

LT and CLR inequalities for magnetic fields are discussed, for example, in Frank [**279**] and Frank–Lieb–Seirringer [**284**]. The key is diamagnetic inequalities (a subject discussed in Section 7.6 of Part 4, e.g., Theorem 7.6.11).

The critical $\nu = 1$ LT inequality (for $p = \frac{1}{2}$) was first proven by Weidl [**844**]. The approach we follow that yields the optimal constant is due to Hundertmark–Lieb–Thomas [**408**]. For a partially alternative proof of the optimal constant, see Hundertmark–Laptev–Weidl [**407**]. The result was extended to Jacobi matrices by Hundertmark–Simon [**406**] and our proof of Proposition 6.7.11 borrows one of their ideas.

A different approach to Example 6.7.9 is due to Frank–Simon–Weidl [**286**] which also gives critical $p = \frac{1}{2}$ inequalities when $\nu = 1$. In particular, they discuss the bound in (6.7.67).

There is a large literature on LT bounds in gaps; see Frank–Simon [**285**] and references therein.

Frank [**280**] based in part on ideas of Rumin [**708, 709**] has a proof of CLR bounds that does not require the semigroup e^{-tA} be positivity preserving but only a bound on the integral kernel of e^{-tA} of the form $e^{-tA}(x,x) \leq Ct^{-\nu/2}$.

There is a comprehensive book on LT and CLR equations by Frank, Laptev, and Weidl [**281**].

Problems

1. Prove Lieb–Thirring bounds for $\left(-\frac{d^2}{dx^2}\right)^{\gamma} + V$ on $L^2(\mathbb{R})$ for $\frac{1}{2} < \gamma < 1$, $p + \frac{\gamma}{2} > 1$.

2. (a) Let $\mu \in [-1,1]$ and $k \in \{1, 2, \ldots, n-1\}$ and $\tilde{L}_\mu$ be the $n \times n$ matrix

$$\left(\tilde{L}_\mu\right)_{ij} = \begin{cases} \mu, & i \in \{1,\ldots,k\}, j \notin \{1,\ldots,k\} \text{ or vice versa} \\ 1, & i,j \in \{1,\ldots,k\} \text{ or } i,j \in \{k+1,\ldots,n\} \end{cases}$$

Prove that (for $\mu = \pm|\mu|$)

$$\sum a_i a_j (\tilde{L}_\mu)_{ij} = |\mu|(\sum_{\ell=1}^{k} a_\ell \pm \sum_{\ell=k+1}^{n} a_l)^2 + (1-|\mu|)\left[(\sum_{\ell=1}^{k} a_\ell)^2 + (\sum_{\ell=k+1}^{n} a_\ell)^2\right]$$

and conclude that $\tilde{L}_\mu \geq 0$.

(b) If $A \geq 0$, $B \geq 0$ for $n \times n$ matrices and $C_{ij} = A_{ij}B_{ij}$ (Schur product), prove that $C \geq 0$. (*Hint:* Reduce to the case A, B each of rank 1.)

(c) Let $\tilde{L}_{\mu_1,\ldots,\mu_{n-1}}$ be the matrix of (6.7.75) with $b_j \equiv 1$, prove that $\tilde{L} \geq 0$.

(d) If B is the diagonal matrix $b_j \delta_{ij}$, prove that with L given by (6.7.75),

$$L = B^{\frac{1}{2}} \tilde{L} B^{\frac{1}{2}}$$

and conclude that $L \geq 0$.

3. Prove (6.7.81). (*Hint:* Consider all operators as acting on $L^2([-R + 1, -R + 1])$ by letting $L^{(R)}$ act on the space of functions constant on $[x_j - \frac{1}{2R}, x_j + \frac{1}{2R}]$.)

6.8. Restriction to Submanifolds

In this section, we'll ask about when $\widehat{f}$ has a restriction to a bounded hypersurface, $S \subset \mathbb{R}^\nu$. While we'll focus on the case $\dim(S) = \nu - 1$, similar ideas work in other codimensions (See Problems 1, 2). This is connected to Sobolev embedding theorems: saying $f \in H^{s,2}$ is continuous if $s > \frac{\nu}{2}$ is a statement about restrictions to submanifolds of codimension ν, i.e., points. One interpretation of Theorem 6.8.2 below is that $f \in H^{s,2}$ with $s > \frac{1}{2}$ can be restricted to codimension one smooth compact submanifolds. The first major theme of this section is that if $(1 + |x|^2)^{\frac{s}{2}} f \in L^2(\mathbb{R}^\nu, d^\nu x)$ and $s > \frac{1}{2}$, then $\widehat{f} \restriction M$ lies in L^2. This is essentially the best one can do if M has flat pieces. But if M is curved, we'll prove more: we'll discuss the Tomas–Stein theorem that if $f \in L^p$, $1 \le p \le 2(\nu + 1)(\nu + 3)^{-1}$, then $\widehat{f}$ has an L^2-restriction to the unit sphere.

Later in this section and in the Notes, we'll discuss applications of which there are several. Except for the warmup case of a hyperplane, our M's will be compact although there are examples of interest in applications that are not compact, such as the light cone, $\{(\vec{x}, t) \in \mathbb{R}^4 \mid |\vec{x}| = t; t > 0\}$ and, as we'll see, parabaloids.

We start with $M = \mathbb{R}^{\nu-1} \subset \mathbb{R}^\nu$, say the coordinate plane $k_\nu = c$. We are heading towards

Theorem 6.8.1. *For any $s > \frac{1}{2}$, there is C_s so that for all c and all $f \in \mathcal{S}(\mathbb{R}^\nu)$,*

$$\int_{k_\nu = c} |\widehat{f}(k)|^2 \, d^{\nu-1}k \le C_s^2 \int (1 + |x|^2)^s |f(x)|^2 \, d^\nu x \qquad (6.8.1)$$

Remark. Once one has this, it follows (by uniform limits of continuous functions) that if the integral on the right is finite, then $k_\nu \to \widehat{f}(\cdot, k_\nu)$ defines a continuous map of $\mathbb{R}$ to $L^2(\mathbb{R}^{\nu-1}, d^{\nu-1}k)$. In fact (Problem 3), it is Hölder continuous of any order $\alpha < \frac{1}{2} - s$. This theorem can be restated as a result about $H^{s,2}$ (see Corollary 6.8.3 below).

Proof. We will use duality. Let $d\sigma = d^{\nu-1}k$ on the hyperplane $M = \mathbb{R}^{\nu-1}$ with $k = (k_1, \ldots, k_{\nu-1}, c)$. Let $\widetilde{gd\sigma}$ be the function on $\mathbb{R}^\nu$ given by

$$\widetilde{gd\sigma}(x) = (2\pi)^{-\nu/2} \int_{\mathbb{R}^{\nu-1}, k_\nu = c} e^{-ik \cdot x} g(k) \, d\sigma(k_1, \ldots, k_{\nu-1}) \qquad (6.8.2)$$

We'll prove that

$$\int (1 + x^2)^{-s} |\widetilde{gd\sigma}(x)|^2 \, d^\nu x \le C_s^2 \int_{\mathbb{R}^{\nu-1}} |g(k)|^2 d\sigma \qquad (6.8.3)$$

Since

$$\langle \hat{f} \restriction M, g \rangle_{L^2(\mathbb{R}^{\nu-1}, d\sigma)} = \langle \hat{f}, g d\sigma \rangle_{\mathcal{S}, \mathcal{S}' \text{pairing}}$$

$$= \langle f, \widetilde{g d\sigma} \rangle_{\mathcal{S}, \mathcal{S}' \text{pairing}} \qquad (6.8.4)$$

(6.8.3) implies (6.8.1).

By the Plancherel theorem:

$$\int_{x_\nu = \alpha} |\widetilde{g \, d\sigma}(x_1, \ldots, x_\nu)|^2 \, d^{\nu-1}x = \int |e^{-ic\alpha} g(k_1, \ldots, k_{\nu-1})|^2 \, d^{\nu-1}k$$

$$= \int_{\mathbb{R}^{\nu-1}} |g(k_1, \ldots, k_{\nu-1})|^2 \, d^{\nu-1}k \qquad (6.8.5)$$

We see that

$$\text{LHS of } (6.8.3) \le \int (1 + |x_\nu|^2)^{-s} |\widetilde{g d\sigma}(x)|^2 \, d^\nu x$$

$$= \left[\int_{-\infty}^{\infty} (1 + |y|^2)^{-s} dy \right] \int |g(k_1, \ldots, k_{\nu-1})|^2 d^{\nu-1}k \qquad (6.8.6)$$

which is (6.8.3) with $C_s = \left[\int_{-\infty}^{\infty} (1 + y^2)^{-s} dy \right]^{\frac{1}{2}}$. $\square$

We now want to consider regular hypersurfaces as defined in Section 1.4. $F : \mathbb{R}^\nu \to \mathbb{R}$ is a C^k-function and

$$S = \{x \mid F(x) = 0\} \qquad (6.8.7)$$

where $\vec{\nabla} F(x) \ne 0$ for all of S. If $x^{(0)} \in S$, if $\frac{\partial F}{\partial x_\nu}(x^{(0)}) \ne 0$, the implicit function theorem says there is an open neighborhood N of $x^{(0)}$ and $h : \{y \in \mathbb{R}^{\nu-1} \mid |y| < \delta\} \to \mathbb{R}$, a C^{k-1}-function so that

$$N \cap S = \{(x_1^{(0)} + y_1, \ldots x_{\nu-1}^{(0)} + y_{\nu-1}), h(y_1, \ldots, y_{\nu-1}) \mid |y| < \delta\}$$

We can use $y_1, \ldots, y_{\nu-1}$ as local coordinates for S near $x^{(0)}$. Repeating this (using x_j with $\frac{\partial F}{\partial x_j} \ne 0$ at points with $\frac{\partial F}{\partial x_\nu} = 0$), we see that S is a C^{k-1}-manifold.

Since $\nabla_y F(x^{(0)} + y, h(y)) = 0$, we see that on N

$$(\nabla_x F)(x_1, \ldots, x_\nu) = (-\nabla_y h, 1) \frac{\partial F}{\partial x_\nu} \qquad (6.8.8)$$

so the unit normal to S at x_0 is given by

$$\vec{n} = (-\nabla_y h, 1) / (1 + |\nabla_y h|^2)^{\frac{1}{2}} \qquad (6.8.9)$$

The surface area, $d\sigma$, induced by the Euclidean metric has $d\sigma \wedge n = dx_1 \wedge \ldots \wedge dx_n$, so by (6.8.9), we have

$$d\sigma(y) = (1 + |\nabla_y h|^2) \, d^{\nu-1}y \qquad (6.8.10)$$

Theorem 6.8.2. *Let S_0 be a C^k-hypersurface, $k \geq 2$, and $S = U \cap S_0$ where U is a bounded open subset of $\mathbb{R}^\nu$. For any $f \in L^2(S, d\sigma)$, $f d\sigma$ is a tempered distribution. For any $s > \frac{1}{2}$, there is a C_s so that*

$$\int (1 + |x|^2)^{-s} |(f d\sigma)^\vee(x)|^2 d^\nu x \leq C_s^2 \|f\|_{L^2(S, d\sigma)}^2 \tag{6.8.11}$$

Moreover, for any $f \in \mathcal{S}(\mathbb{R}^\nu)$, we have that

$$\int |\hat{f}(k)|^2 \, d\sigma(k) \leq C_s^2 \int (1 + |x|^2)^s |f(x)|^2 \, d^\nu x \tag{6.8.12}$$

Remarks. 1. If S_0 is compact, we can pick $S = S_0$ which is the most common example.

2. If we let $S_t = \{x \mid F(x) = t\}$ and $\tilde{S}_t = S_t \cap U$ and one picks local coordinates $(y_1, \ldots, y_{\nu-1}, t)$, one can show (Problem 4) that $f \restriction \tilde{S}_t$ is Hölder continuous in $L^2(d\sigma)$ of any order $\alpha < s - \frac{1}{2}$.

Proof. It is easy to see that $f d\sigma$ is a tempered distribution since $\int |g(k) f(k)| \, d\sigma(k) \leq C \int |g(k)|^2 d\sigma \leq C \|g\|_\infty^2 \int_S d\sigma$.

Since $\overline{S}$ is compact, $\min_{x \in S} |\nabla F| = A > 0$, so at any $x \in S$, there is a j with $|\frac{\partial F}{\partial x_j}| \geq A \nu^{-\frac{1}{2}}$.

By writing $d\sigma = \sum_{j=1}^N \chi_j \, d\sigma$ where χ_j is the characteristic function of a subset of the points where $|\frac{\partial F}{\partial x_j}| \geq A \nu^{-\frac{1}{2}}$, we can suppose (by also renumbering coordinates) that $|\frac{\partial F}{\partial x_\nu}| \geq A \nu^{-\frac{1}{2}}$ on S and $y_1, \ldots, y_{\nu-1}$ are global coordinates on S. Then,

$$(f d\sigma)^\vee(x) = (2\pi)^{-\nu/2}$$
$$\int e^{i(x_1 k_1 + \ldots + x_{\nu-1} k_{\nu-1})} \, e^{i x_\nu \cdot h(k_1, \ldots, k_{\nu-1})} f(k_1, \ldots, k_{\nu-1}) J(k_1, \ldots, k_{\nu-1}) d^{\nu-1} k \tag{6.8.13}$$

where $J = (1 + |\nabla h|^2)^{\frac{1}{2}}$. Thus, by the Plancherel theorem, for each x_ν,

$$\int |((f d\sigma)^\vee(x_1, \ldots, x_\nu)|^2 \, d^{\nu-1} x \leq \int |f(k_1, \ldots, k_{\nu-1}) J(k_1, \ldots, k_{\nu-1})|^2 d^{\nu-1} k$$
$$\leq \|J\|_\infty \|f\|_{L^2(S, d\sigma)}^2 \tag{6.8.14}$$

Since $(1 + |x|^2)^{-s} \leq (1 + |x_\nu|^2)^{-s}$ and $\int (1 + |x_\nu|^2)^{-s} dx_\nu < \infty$, we have (6.8.11) (originally for each χ_j piece but then by summing up the pieces for all of S).

As in the flat case, (6.8.11) implies (6.8.12) by duality. $\qquad \square$

Corollary 6.8.3. *Under the hypotheses of Theorem 6.8.2, the restriction map, $g \mapsto g \restriction S$, defined on $\mathcal{S}(\mathbb{R}^\nu)$ extends to a bounded map of $H^{s,2}$ to $L^2(S, d\sigma)$ for any $s > \frac{1}{2}$.*

If $f \in L^2(\mathbb{R}^\nu, (1 + |x|^2)^{-s} d^\nu x)$, then $f \in L^p(\mathbb{R}^\nu, d^\nu x)$ so long as $2 \geq p > 2\nu(\nu + 2s)^{-1}$ so one might hope the restriction of the Fourier transform to S lies in L^2 (or more generally some L^q) for f's in some L^p. By duality, this is asking if $(f d\sigma)^\vee$ lies in some $L^{p'}$ for p' dual to p.

For general S, this cannot be (for $p > 1$, i.e., $p' < \infty$) because if S is a bounded open subset of $\{k \mid k_\nu = 0\}$, $(f d\sigma)^\vee$ is x_ν-independent and so in no L^p, $p < \infty$, even if $f \in C_0^\infty$. In Example 15.3.12 of Part 2B, we saw that for the sphere, $|\breve{\sigma}(x)| \leq C |x|^{-(\nu-1)/2}$ and more generally (see Theorem 15.3.13 of Part 2B) if S is everywhere curved, such a bound holds for $(g d\sigma)^\vee$ if $g \in C_0^\infty(\mathbb{R}^\nu)$. We are heading towards L^p to L^2-results for $f \mapsto \hat{f} \upharpoonright S$ precisely in this curved case.

We call a C^∞-surface everywhere curved if $\det\left(\frac{\partial^2 h}{\partial x_i \partial x_j}\right) > 0$ at each point of S (after some coordinate with $\frac{\partial F}{\partial x_\ell} \neq 0$ is singled out). Theorem 15.3.13 of Part 2B says that if $d\tau = g d\sigma$ where $d\sigma$ is the surface area of an everywhere curved surface in $\mathbb{R}^\nu$ and $g \in C_0^\infty(\mathbb{R}^\nu)$ is positive, then

$$|\hat{\tau}(k)| \leq C \left(1 + |k|\right)^{-(\nu-1)/2} \tag{6.8.15}$$

This came from a use of the stationary phase method. The curvature requirement implied at the point of stationary phase, that the integral was essentially a Gaussian, so we get $|k|^{-\frac{1}{2}}$ from each of the $\nu - 1$ hypersurface coordinates.

The other fact that we'll need is that for some $C > 0$ and all $x_0 \in \mathbb{R}^\nu$ and $R > 0$, we have that since the surface is $\nu - 1$ dimensional,

$$\tau(\{x \mid |x - x_0| \leq R\}) \leq C R^{\nu-1} \tag{6.8.16}$$

We are heading towards a proof of the following:

Theorem 6.8.4 (Tomas–Stein Theorem). *Let τ be a finite measure on $\mathbb{R}^\nu$ of compact support obeying* (6.8.15) *and* (6.8.16). *Let $1 \leq p \leq \frac{2\nu+2}{\nu+3}$. Then there exists C so that for all $f \in \mathcal{S}(\mathbb{R}^\nu)$,*

$$\|\hat{f}\|_{L^2(d\tau)} \leq C \|f\|_{L^p(\mathbb{R}^\nu)} \tag{6.8.17}$$

Remarks. 1. The $L^2(d\tau)$-norm involves $f \upharpoonright \text{supp}(d\tau)$ and this support has measure zero.

2. Once one has this for $f \in \mathcal{S}(\mathbb{R}^\nu)$, we can extend to $f \in L^p(\mathbb{R}^\nu)$.

3. In Example 6.8.6 below, we'll show that for the case $d\tau = $ area measure on $S^{\nu-1}$, the estimate (6.8.17) fails for $p > (2\nu + 2)/(\nu + 3)$, i.e., this is optimal (and the argument works for any surface with positive curvature).

We'll prove this for $p < \frac{2\nu+2}{\nu+3}$ and leave the discussion of the borderline case to the references in the Notes.

We'll begin with:

Lemma 6.8.5. *Let $p < 2$ and $p' = p/(p-1)$ the dual index. Let τ be a finite measure on $\mathbb{R}^\nu$. Then the following are equivalent:*

(1) $\|(f\tau)^\vee\|_{L^{p'}(\mathbb{R}^\nu)} \leq C \|f\|_{L^2(\tau)}$ *for all* $f \in L^2(\tau)$. $\qquad$ (6.8.18)

(2) $\|\widehat{g}\|_{L^2(\tau)} \leq C \|g\|_{L^p(\mathbb{R}^\nu)}$ *for all* $g \in \mathcal{S}(\mathbb{R}^\nu)$. $\qquad$ (6.8.19)

(3) $\|\check{\tau} * h\|_{L^{p'}(\mathbb{R}^\nu)} \leq (2\pi)^{\nu/2} C^2 \|h\|_{L^p(\mathbb{R}^\nu)}$ *for all* $h \in \mathcal{S}(\mathbb{R}^\nu)$. $\qquad$ (6.8.20)

Remarks. 1. The equivalence of (1) and (2) holds if L^2 in (6.8.18) is replaced by $L^{q'}$ and L^2 in (6.8.19) by L^q. But the equivalence to (3) requires $q = 2$, i.e., the reliance of our proof on (6.8.20) restricts us to estimates of $\widehat{f}$ in a $L^2(d\tau)$-norm.

2. (1) $\Leftrightarrow$ (2) is essentially $\|T^*\| = \|T\|$ while (2) $\Leftrightarrow$ (3) is $\|T^*T\| = \|T\|^2$.

3. $1 \leq p \leq \dfrac{2\nu+2}{\nu+3} \Leftrightarrow \dfrac{2\nu+2}{\nu-1} \leq p' \leq \infty$. $\qquad$ (6.8.21)

Proof. The equivalence of (1) and (2) is duality as we've seen. To see the equivalence of (2) and (3), we note that for $h, g \in \mathcal{S}(\mathbb{R}^\nu)$ (with $f^\flat(x) = f(-x)$)

$$\langle \widehat{h}, \widehat{g} \rangle_{L^2(\tau)} = (\widehat{g}d\tau)(\overline{\widehat{h}}) \qquad (6.8.22)$$

$$= (\widehat{g}\tau)^\vee(\overline{h^\flat}) \qquad (6.8.23)$$

$$= (2\pi)^{-\nu/2}(g * \check{\tau})(\overline{h^\flat}) \qquad (6.8.24)$$

where (6.8.23) uses the definition of $\vee$ on distributions and $(\overline{\widehat{h}})^\vee = \overline{h^\flat}$ and (6.8.24) uses $(fg)^\vee = (2\pi)^{-\nu/2}\check{f} * \check{g}$.

If (6.8.20) holds, take $h = g$ in (6.8.24) to see

$$\|\widehat{g}\|^2_{L^2(\tau)} \leq (2\pi)^{-\nu/2} \|g * \check{\tau}\|_{p'} \|g\|_p$$
$$\leq C^2 \|g\|^2_p$$

proving (6.8.19).

Conversely, if (6.8.19) holds, for any $g \in L^p$,

$$\|g * \check{\tau}\|_{p'} = \sup \left\{ |g * \check{\tau}(\overline{h^\flat})| \mid \|h\|_p = 1 \right\}$$

$$= (2\pi)^{\nu/2} \sup_{\|\tau\|_h = 1} \langle \widehat{h}, \widehat{g} \rangle_{L^2(\tau)}$$

$$\leq (2\pi)^{\nu/2} C^2 \sup_{\|h\|_p=1} \left(\|h\|_p \|g\|_p \right)$$

$$= (2\pi)^{\nu/2} C^2 \|g\|_p$$

proving (6.8.20). $\qquad \square$

(6.8.15) and (6.8.20) suggest the HLS inequalities might be useful—they give partial but not full results. (6.8.20) is equivalent to

$$\int |g(x)| \, |\check{\tau}(x - y) \, h(y)| \, d^\nu x \, d^\nu y \le (2\pi)^{\nu/2} \, C^2 \, \|g\|_p \, \|h\|_p \qquad (6.8.25)$$

(6.8.15) says that $\hat{\tau} \in L^\infty$. Thus, since $L^1 * L^\infty \subset L^\infty$, (6.8.25) holds for $p = 1$. The HLS inequalities and $|\hat{\tau}(k)| \le C \, |k|^{-(\nu-1)/2}$ says (6.8.25) holds if

$$\frac{2}{p} + \frac{(\nu - 1)}{2\nu} = 2 \Rightarrow p = \frac{4\nu}{(3\nu + 1)} \qquad (6.8.26)$$

We thus see (using interpolation) that HLS (and Young) imply Tomas–Stein for

$$1 \le p \le \frac{4\nu}{(3\nu + 1)} \qquad (6.8.27)$$

To get the full blown result, we have to use the fact that $\check{\tau}$ oscillates as well as decays. Fortunately, the estimate (6.8.16) will capture what we need from the oscillations. Remarkably, the full estimate will come from interpolation between $L^1 \to L^\infty$ and $L^2 \to L^2$. Since $\check{\tau}*$ is unbounded from $L^2 \to L^2$ ($\check{g}*$ is bounded from L^2 to L^2 if and only if multiplication by g is bounded from L^2 to L^2 if and only if g is bounded) the idea of this interpolation might seem far fetched.

We'll decompose $\check{\tau} = K_0 + \sum_{n=1}^\infty K_n$ and estimate K_n* as a map of L^1 to L^∞ and L^2 to L^2 as

$$\|K_n * f\|_\infty \le C \, 2^{-\alpha n} \, \|f\|_1, \quad \|K_n * f\|_2 \le C \, 2^{\beta n} \, \|f\|_2 \qquad (6.8.28)$$

with $\alpha, \beta > 0$. The sum of the L^2-estimates diverge but interpolation will yield convergent estimates for a range of p's that yields all but the edge of the Tomas–Stein theorem.

The decomposition will be a Littlewood–Paley $2^{n-1} \le |x| \le 2^{n+1}$ type that has been so successful in Sections 6.4 and 6.5. We construct φ_n, as in the proof of Theorem 6.4.9, that obeys:

(i) $\displaystyle\sum_{n=-\infty}^\infty \varphi_n(x) = 1, \quad x \in \mathbb{R}^\nu \setminus \{0\}.$ $\qquad (6.8.29)$

(ii) $\varphi_n(x) = \varphi_0(2^{-n} x).$ $\qquad (6.8.30)$

(iii) $0 \le \varphi_n \le 1, \quad \varphi_n \in C_0^\infty(\mathbb{R}^\nu), \quad \operatorname{supp} \varphi_n \subset \{x \mid 2^{n-1} < |x| < 2^{n+1}\}.$
$\qquad (6.8.31)$

We define

$$K_n(x) = \varphi_n(x)\check{\tau}(x), \, n = 1, 2, \ldots, \quad K_0(x) = \left[1 - \sum_{n=1}^\infty K_n(x)\right]\check{\tau}(x) \quad (6.8.32)$$

Proof of Theorem 6.8.4 (if $p < \frac{2\nu+2}{\nu+3}$). We will prove for $n \geq 1$ that

$$\|K_n * f\|_\infty \leq C \, 2^{-(\nu-1)n/2}\|f\|_1 \tag{6.8.33}$$

$$\|K_n * f\|_2 \leq C \, 2^n \, \|f\|_2 \tag{6.8.34}$$

$K_0 \in C_0^\infty(\mathbb{R}^\nu)$, so L^p to L^q ($p \leq q$) for K_0* are trivial. By the Riesz–Thorin theorem for $1 < p < 2$ and $\frac{\theta}{2} = \frac{1}{p'} = 1 - \frac{1}{p}$, we have that

$$\|K_n * f\|_{p'} \leq C \, 2^{n\alpha(\theta)}\|f\|_p \tag{6.8.35}$$

with $\alpha(\theta) = \theta - (1-\theta)(\frac{\nu-1}{2})$. This implies that

$$\alpha(\theta) < 0 \Leftrightarrow \theta < \frac{\nu-1}{\nu+1} \Leftrightarrow \frac{1}{p} > 1 - \frac{1}{2}\left[\frac{\nu-1}{\nu+1}\right] = \frac{\nu+3}{2\nu+2} \tag{6.8.36}$$

Since $\sum_{n=1}^\infty 2^{n\alpha(\theta)} < \infty \Leftrightarrow \alpha(\theta) < 0$, we have the claimed result.

We'll obtain these bounds by using

$$\|K * f\|_\infty \leq \|K\|_\infty \|f\|_1, \quad \|K * f\|_2 \leq (2\pi)^{\nu/2} \|\widehat{K}\|_\infty \|f\|_2 \tag{6.8.37}$$

The first is an elementary (indeed trivial) case of Young's inequality and the second follows from the Plancherel theorem and $\widehat{K * f} = (2\pi)^{\nu/2}\widehat{K}\widehat{f}$. In fact, one essentially has equality in these estimates—not for individual estimates but for operator norms (Problems 5, 6)

(6.8.33) is an immediate consequence of (6.8.37) and (6.8.15) which implies

$$\sup_x |K_n(x)| \leq \sup_{|x| \geq 2^{n-1}} |\check{\tau}(x)| \leq C \, 2^{(\nu-1)/2}\big[2^{-(\nu-1)n/2}\big] \tag{6.8.38}$$

The estimate on $\|\widehat{K}_n\|_\infty$ is only a little more complicated. By $\varphi_n(x) = \varphi(2^{-n}x_0)$, we have

$$\widehat{K}_n(k) = (2\pi)^{-\nu/2} \int \widehat{\varphi}_n(k - \ell)d\tau(\ell) \tag{6.8.39}$$

$$= (2\pi)^{-\nu/2} \, 2^{n\nu} \int \widehat{\varphi}_0\big(2^n(k-\ell)\big) \, d\tau(\ell) \tag{6.8.40}$$

so, since $\widehat{\varphi} \in \mathcal{S}(\mathbb{R}^\nu)$ for a constant C,

$$|\widehat{K}_n(k)| \leq C \, 2^{n\nu} \int \frac{1}{\big(1 + 2^n|k-\ell|\big)^{\nu+1}} \, d\tau(\ell) \tag{6.8.41}$$

The intuition is clear; by scaling $2^{n\nu}/(1 + 2^n|k|)^\nu$ has an n-independent L^1-norm, so the integral on the right of (6.8.41) as a function of k has an L^1-norm independent of k. Because of the rapid falloff of $(1 + |x|)^{-(\nu+1)}$, this L^1-function is concentrated in an annulus of thickness 2^{-n}, so the L^∞-norm has to be of order 2^n. Here is a formal proof (we use "C" for an n-independent constant that can change from equation to equation):

$$|\widehat{K_n}(k)| \le C\, 2^{n\nu} \sum_{j=-\infty}^{\infty} \tau(\{\ell \mid 2^{j-1} < |k-\ell| \le 2^j\})(1+2^{n+j-1})^{-(\nu+1)}$$
$$(6.8.42)$$

$$\le C\, 2^{n\nu} \sum_{j=-\infty}^{\infty} \tau(\{\ell \mid |k-\ell| \le 2^j\}) \min(1, 2^{-n-j})^{\nu+1} \qquad (6.8.43)$$

$$\le C\, 2^{n\nu}\left[\sum_{j=-n}^{\infty} 2^{-(\nu-1)j}\, 2^{-(n+j)(\nu+1)} + \sum_{j=-\infty}^{-n-1} 2^{(\nu-1)j}\right] \qquad (6.8.44)$$

$$\le C\, 2^{n\nu}\left[2^{-(\nu-1)n} + 2^{-(\nu-1)n}\right] = C\, 2^n \qquad (6.8.45)$$

which, by (6.8.37) implies (6.8.34)

In the above, (6.8.42) follows by breaking the integral in (6.8.39) into contributions of annulli about k, (6.8.43) bounds the annular volume by that of a sphere and uses $(1+2^a)^{-1} \le \min(1, 2^{-a})$ (together with absorbing $2^{\nu+1}$ into C). (6.8.44) follows from (6.8.16). We get the inequality in (6.8.45) by noting that each sum is a geometric series with n-independent ratios (2^{-2} and $2^{-(\nu-1)}$, respectively) so bounded by the multiples of the leading term which are $2^{-n(\nu-1)}$ and $2^{-(n+1)(\nu-1)}$. $\square$

Example 6.8.6 (Knapp's Counterexample). We will show that (6.8.19) fails if $p > \frac{2\nu+2}{\nu+3}$. Pick $\varphi \in \mathcal{S}(\mathbb{R}^\nu)$ with $\widehat{\varphi} \ge 0$ and $\widehat{\varphi}(k) \ge 1$ for $|k| \le \sqrt{2}$. Define

$$g_N(x_1,\ldots,x_{\nu-1},x_\nu) \equiv e^{ix_\nu N^2} \varphi\left(\frac{x_1}{N}, \ldots, \frac{x_{\nu-1}}{N}, \frac{x_\nu}{N^2}\right) \qquad (6.8.46)$$

Clearly, by scaling

$$\|g_N\|_p = \|\varphi\|_p\, N^{(\nu+1)/p} \qquad (6.8.47)$$

and

$$\widehat{g}_N(k_1,\ldots,k_\nu) = N^{\nu+1}\, \widehat{\varphi}(Nk_1,\ldots,Nk_{\nu-1}, N^2(k_\nu-1)) \qquad (6.8.48)$$

φ was precisely chosen so that $\widehat{g}_N$ is concentrated near the point $e = (0,0,\ldots,0,1)$.

Let C_N be the spherical cap near e:

$$C_N = \{k \in \mathbb{R}^\nu \mid k_1^2 + \ldots + k_{\nu-1}^2 \le N^{-2}; k_\nu > 0; |k| = 1\} \qquad (6.8.49)$$

If $k \in C_N$, then $1 \geq k_\nu \geq (1 - N^{-2})^{\frac{1}{2}} \geq 1 - N^{-2}$, i.e., $|k_\nu - 1| \leq N^{-2}$. Thus,

$$k \in C_N \Rightarrow |(Nk_1, \ldots, Nk_{\nu-1}, N^2(k_\nu - 1))| \leq \left[\frac{N^2}{N^2} + \frac{N^4}{N^4}\right]^{\frac{1}{2}} \leq \sqrt{2} \quad (6.8.50)$$

Thus,

$$k \in C_N \Rightarrow |\widehat{g}_N(k)| \geq N^{\nu+1} \quad (6.8.51)$$

so, with $d\sigma$ the Euclidean measure on the unit sphere,

$$\|\widehat{g}_N\|_{L^q(d\sigma)} \geq C^{\frac{1}{q}} N^{\nu+1} N^{-(\nu-1)/q} \quad (6.8.52)$$

since $\sigma(C_N) \sim C N^{(\nu-1)}$ for a suitable constant C and $N \to \infty$.

In order for (6.8.19) to hold, we must have as $N \to \infty$ that $N^{(\nu+1)/p} \geq C N^{(\nu+1)} N^{-(\nu-1)/q}$. i.e.,

$$\frac{1}{p} \geq 1 - \left(\frac{\nu-1}{\nu+1}\right)\frac{1}{q} \quad (6.8.53)$$

If this fails, then (6.8.19) fails for N large.

If $q = 2$, (6.8.53) is equivalent to $p \leq \frac{2\nu+2}{\nu+3}$. $\qquad \square$

The Tomas–Stein theorem has numerous applications. We illustrate by using it to prove an estimate on free Schrödinger equation dynamics known as Strichartz estimates. In the Notes, we'll discuss applications to scattering and spectral theory, to Hausdorff dimension theory, and to the theory of Bochner–Riesz means.

In Section 6.9 of Part 1, we discussed the free Schrödinger equation

$$i\frac{\partial u}{\partial t} = -\Delta u, \quad u(x, t = 0) = f(x) \quad (6.8.54)$$

$u(x, t)$ a function in $\mathbb{R}^{\nu+1}$. We proved that this was solvable if $f \in \mathcal{S}(\mathbb{R}^\nu)$ and the solution obeys

$$\int |u(x, t)|^2 \, d^\nu x = \int |f(x)|^2 \, d^\nu x \quad (6.8.55)$$

(for good physical reason in the theory of nonrelativistic quantum mechanics).

This allows one to extend $f \mapsto (u(\,\cdot\,, t))$ to a map of L^2 to L^2; indeed the natural context, as explained in Section 7.3 of Part 4, is to use the theory of Hilbert space unitary groups and write $u(\,\cdot\,, t) = (e^{it\Delta} f)(\,\cdot\,)$. We also showed that in a weak sense, as $t \to \pm\infty$,

$$u(x, t) \sim (2t)^{-\nu/2} e^{ix^2/4t} \widehat{f}\left(\frac{x}{2t}\right) \quad (6.8.56)$$

The decaying power of t is just such as to preserve the L^2-norm which does not decay in t. But if $\widehat{f} \in L^p$ and $p > 2$, the p-norm in the right side of

(6.8.56) goes to zero. And if $p > 2 + \frac{2}{\nu}$, $\int |\text{RHS of } (6.8.56)|^p \, d^\nu x \, dt < \infty$. This suggests that maybe we can bound an L^p-norm in space $\times$ time by an initial L^2-norm. As our proof will show, scaling determines the only possible p.

Theorem 6.8.7 (Strichartz Estimate). *There is a constant C only depending on ν so that for $f \in \mathcal{S}(\mathbb{R}^\nu)$, the solution u of (6.8.54) obeys*

$$\|u\|_{L^q(\mathbb{R}^{\nu+1})} \leq C \, \|f\|_{L^2(\mathbb{R}^\nu)} \tag{6.8.57}$$

where

$$q = 2 + 4\nu^{-1} \tag{6.8.58}$$

Remarks. 1. This implies that for any $f \in L^2$, $u = e^{it\Delta} f \in L^q$ for a.e. t and (6.8.55) holds for u.

2. The proof will rely on the borderline Tomas–Stein result we haven't proven!

Proof. The intuition is that the $(\nu + 1)$-dimensional Fourier transform of the Schrödinger equation is $\left[k_{\nu+1} + \sum_{j=1}^{\nu} k_j^2 \right] \widehat{u}(k) = 0$ which suggests that u is the Fourier transform of a function on the paraboloid $k_{\nu+1} + \sum_{j=1}^{\nu} k_j^2 = 0$ so we can hope to write it $(\widehat{f}\mu)^\vee$ and use (6.8.18). The paraboloid is not compact, so we'll need to also restrict to $|\sum_{j=1}^{\nu} k_j^2| \leq 1$ and then get general f by scaling. The q will be the exact value consistent with scaling. Here are the details.

u can be expressed by Fourier and inverse Fourier transforms (see (6.9.49) in Part 1):

$$u(x, t) = (2\pi)^{-\nu/2} \int \widehat{f}(k) e^{i(k \cdot x - |k|^2 t)} \, d^\nu k \tag{6.8.59}$$

Consider $f \in \mathcal{S}(\mathbb{R}^\nu)$ with supp $\widehat{f} \subset \{k \mid |k| \leq 1\}$. We'll use κ for an $\mathbb{R}^{\nu+1}$-variable with $|\kappa|^{(\nu)} \equiv \left(\sum_{j=1}^{\nu} \kappa_j^2 \right)^{\frac{1}{2}}$. Let $\varphi \in C_0^\infty(\mathbb{R}^{\nu+1})$ have $\varphi \geq 0$ and $\varphi(\kappa) \equiv 1$ if $|\kappa|^{(\nu)} \leq 1$, $|\kappa_{\nu+1}| \leq 1$. Define the measure μ on $\mathbb{R}^{\nu+1}$ by

$$\int g(\kappa) d\mu(\kappa) = \int \varphi(\kappa) g(\kappa_1, \ldots, \kappa_\nu, -(|\kappa|^{(\nu)})^2) \, d^{\nu-1}\kappa \tag{6.8.60}$$

so $d\mu = \widetilde{\varphi} \, d\sigma$ with σ the natural measure on the paraboloid $\kappa_{\nu+1} = -[|\kappa|^{(\nu)}]^2$ and $\varphi \in C_0^\infty$. Notice that $\kappa = (\kappa_1, \ldots, \kappa_\nu)$ are global coordinates for the paraboloid and (6.8.59) says that

$$u = (2\pi)^{\frac{1}{2}} \left(\widehat{f}\varphi \, d\mu \right)^\vee \tag{6.8.61}$$

where $\widehat{}$ in $\widehat{f}$ is a ν-dimensional Fourier transform and $\vee$ is a $(\nu + 1)$-dimensional inverse Fourier transform (hence the extra $(2\pi)^{\frac{1}{2}}$).

Since we are in $\nu + 1$ dimensions and the critical p' in that dimension by (6.8.21) is $\frac{2(\nu+1)+2}{(\nu-1)+1} = 2 + \frac{4}{\nu}$, we have

$$\text{supp } \widehat{f} \subset \{k \mid |k| \leq 1\}, \quad q \geq 2 + \frac{4}{\nu} \Rightarrow \|u\|_{L^q(\mathbb{R}^{\nu+1})} \leq C_q \|f\|_{L^2(\mathbb{R}^\nu)} \quad (6.8.62)$$

where we used

$$\int |\widehat{f}(\kappa_1, \ldots, \kappa_\nu)|^2 \, (\varphi d\mu)(\kappa_1, \ldots, \kappa_{\nu+1}) = \int |\widehat{f}(k)|^2 \, d^\nu k \quad (6.8.63)$$

since $\widehat{f}(\kappa_1, \ldots, \kappa_\nu)$ is $\kappa_{\nu+1}$ independent.

For any f with supp $\widehat{f} \subset \{k \mid |k| \leq R\}$, let

$$f_R(x) = f\left(\frac{x}{R}\right) \quad (6.8.64)$$

so $\widehat{f_R}(k) = R^\nu \widehat{f}(Rk)$ is supported in $\{k \mid |k| \leq 1\}$. Moreover, by scaling x and k, we see that u_R the solution with initial condition f_R has

$$u_R(x, t) = u\left(\frac{x}{R}, \frac{t}{R^2}\right) \quad (6.8.65)$$

It follows that

$$\|u_R\|_{L^q(\mathbb{R}^{\nu+1})} = R^{(\nu+2)/q} \|u\|_{L^q(\mathbb{R}^{\nu+1})}, \quad \|f_R\|_{L^2(\mathbb{R}^\nu)} = R^{\nu/2} \|f\|_{L^2(\mathbb{R}^\nu)} \quad (6.8.66)$$

Thus, since f_R, u_R obey (6.8.62), we have

$$\|u\|_{L^q} = R^{-(\nu+2)/q} \|u_R\|_q \leq C \, R^{-(\nu+2)/q} \|f_R\|_2$$
$$= C \, R^{-(\nu+2)/q} R^{\nu/2} \|f\|_2 \quad (6.8.67)$$

If $-\frac{\nu+2}{q} + \frac{\nu}{2} = 0$, i.e., $q = 2 + \frac{4}{\nu}$, we see that (u, f) also obey (6.8.62). Since the set of $f \in \mathcal{S}$ with $\widehat{f}$ supported in some ball is dense in $\mathcal{S}$, we have the claimed estimate. $\qquad \square$

Remarks. 1. It is no coincidence that (6.8.65) looks like the Knapp scaling!

2. The scaling argument shows if (6.8.57) holds for some q and all $f \in \mathcal{S}$, then q is given by (6.8.56).

Notes and Historical Remarks. Sobolev spaces' restriction to submanifolds is a natural extension of Sobolev embedding theorems. Ideas close to Theorem 6.8.2 first appeared in work of Gagliardo [**300**] and Aronszajn–Smith [**16**].

As Stein [**779**] remarks, given that Riemann (in his famous work on trigonometric series that defined the Riemann integral and had his version of the Riemann–Lebesgue lemma) already used stationary phase ideas to study asymptotics of some Fourier integrals, it is surprising that the realization that curved surfaces had L^p-restriction properties only dates to the later 1960's when Elias Stein (1931–) began to realize that the decay of $\hat{\sigma}$ for the measure on the sphere should have consequences in Fourier analysis. He suggested the study of this as a thesis problem to his student Charles Fefferman (1949–) whose results appeared in a paper [**263**]. Some of these results have come to be attributed to Fefferman–Stein including the restriction result (6.8.27) that comes from the HLS inequalities. This paper also has a result interpolating a Littlewood–Paley decomposition between convergent estimates and divergent estimates but by not taking a smooth φ (but rather sharp cutoffs), it doesn't obtain what Tomas later did.

The Tomas–Stein theorem in the nonborderline case, i.e., $1 < p < \frac{2\nu+2}{\nu+3}$ appeared in Tomas [**807**]. This remarkable paper, based on Tomas' Ph.D. thesis, is only two pages, half of it with historical remarks: the mathematical part states the theorem, gives the decomposition and states without details the estimates (6.8.33)/(6.8.34). The paper also states that Stein has informed him that he has been able to prove the $p = \frac{2\nu+2}{\nu+3}$ borderline case using the explicit Bessel function in $\hat{\sigma}$ (see (15.3.46) of Part 2B), making the index of the Bessel function a complex variable and using a Stein interpolation-type argument. Stein never wrote a paper on his proof although ten years later he did sketch the result in published Lecture Notes [**779**]. In the meantime, other sketches appeared in Tomas [**808**] for the case of a sphere and Strichartz [**788**] for general curved surfaces. For textbook presentations of this subtle result see Grafakos [**329**], Muscalu–Schlag [**592**], Stein [**780**], Stein-Shakarchi [**781**], or Tao's online lectures notes. [**796**].

The elegant and simple scaling idea, Example 6.8.6, which works for any curved surface is due to Knapp who never published it—it appears in print in Strichartz' paper [**788**] attributed to Knapp and used with permission. The idea has since been reused by many other in related contexts. This work was centered at Cornell where Knapp and Strichartz were professors and Tomas, a student of Knapp. But there are also close connections to Princeton, where Stein and Fefferman were. Strichartz was a student of Stein and Knapp of Bochner who was also at Princeton.

It is a theorem of Iosevich–Lu [**418**] that if $d\sigma$ is the surface measure of a smooth hypersurface and the endpoint Tomas–Stein bound holds, then σ has a nonvanishing Gaussian curvature. There is some literature on situations where the Gaussian curvature is not strictly positive, for example if $\frac{\partial^2 h}{\partial x_i \partial x_j}$ might not be invertible at every point but has at least k nonzero eigenvalues. See, for instance, Greenleaf [**332**], Bourgain [**104**], or Lee–Vargas [**517**].

Space–time estimates for linear PDE's have been a useful tool in the study of nonlinear perturbations of those equations. This idea goes back at least to Morawetz–Strauss [**584**]. Their norm was not a simple L^p-norm but mixed L^∞ in time or space and L^p in space involving derivatives also. It didn't directly imply decay but could be used in proving decay. Their paper showed the usefulness in controlling such space–time norms.

It was Segal [**726**] who had the first L^p space–time bounds—for $\left(\frac{\partial}{\partial t^2} - \Delta\right)u = 0$ in $\mathbb{R}^{3+1}$—and he obtained it via a version of restriction theorems. It was Strichartz [**788**] who found general ν bounds in a simple and elegant paper that resulted in his name being associated to the estimates. His result and proof were very close to the one we use in Theorem 6.8.7. For applications of these types of bounds to nonlinear PDEs (for which there are dozens of papers), see Ginibre–Velo [**314**], Keel–Tao [**458**], or Bourgain [**106**].

Another application of restriction theorems is to spectral and scattering theory for $H = -\Delta + V$ for V's decaying sufficiently rapidly as $|x| \to \infty$. One wants to control $\langle f, (-\Delta + V - z)^{-1} f \rangle$ as $\operatorname{Im} z \downarrow 0$ for $\operatorname{Re} z > 0$. When $V = 0$, if $z = \kappa^2 + i\varepsilon$ as $\varepsilon \downarrow 0$, we have that

$$\lim_{\varepsilon \downarrow 0} \langle f, (-\Delta - z)^{-1} f \rangle = \lim_{\varepsilon \downarrow 0} \int \frac{|\widehat{f}(k)|^2 d^\nu k}{k^2 - \kappa^2 - i\varepsilon} \qquad (6.8.68)$$

whose imaginary part is formally $\pi \int |\widehat{f}(k)|^2 \delta(k^2 - \kappa) \, d^\nu k$ so one sees the relevance of restriction to the sphere. Agmon [**4, 5**] had the idea of using Theorem 6.8.2 and looking at $(-\Delta - z)^{-1}$ as a map of $L_s^2 \to L_{-s}^2$ for $s > \frac{1}{2}$. For such operators, he showed that $(-\Delta - z)^{-1}$ has continuous extension above (and below) the cut on $[0, \infty)$ with boundary values that are Hölder continuous. If now $|V(x)| \leq C \left(1 + |x|\right)^{-1-\varepsilon}$ for some $\varepsilon > 0$, then $(-\Delta - z)^{-1}V$ is a compact operator from $L_{-s}^2 \to L_{-s}^2$ for $s = \frac{1}{2}(1 + \frac{\varepsilon}{2})$ and one can use the Fredholm theory of inversion of $1 - C$ with C compact to prove (with some extra arguments relying on restriction theorems) that $1 - (-\Delta - z)^{-1}V$ is invertible in a one-sided neighborhood of $(0, \infty)$ including up to the edges. Using

$$(-\Delta + V - z)^{-1} = \left(1 - (-\Delta - z)^{-1}V\right)^{-1} (-\Delta - z)^{-1}$$

he obtained enough information on $-\Delta + V$ for a complete analysis including pure a.c. spectrum on $(0, \infty)$. It is known that there are examples for any $\varepsilon > 0$, with $|V_\varepsilon(x)| \leq (1 + x)^{-1+\varepsilon}$ for which $-\Delta + V_\epsilon$ has some non-a.c. spectrum so this result is optimal.

Agmon–Hörmander [**6**] extended this analysis replacing $-\Delta$ by a larger class of constant coefficient PDE's. For monograph presentations of Agmon's original work, see Hörmander [**397**], Volume 2 or Reed–Simon [**674**]. Schlag and collaborators [**326, 416**] found some $V \in L^q$-results using the Tomas–Stein theorem instead of L_s^2.

The restriction method has been extended to other settings. Mockenhaupt [**578**] proves restriction results to certain Cantor sets in $\mathbb{R}$ and Bourgain [**105**] and Green [**330**] to certain discrete sets.

Recall that the Bochner–Riesz means (discussed briefly in Section 6.4) are defined by

$$(S_R^\delta f)(x) = \left[(1 - \frac{|k|^2}{R^2})^\delta \chi_{|k| \le R}\, \widehat{f}\,\right]^{\vee}(x) \qquad (6.8.69)$$

$\delta = 0$ is essentially a "partial sum" of the Fourier expansion and $\delta = 1$ a Cesàro average. There are close connections between these means and the restriction map explored initially by Fefferman in three papers [263, 264, 267].

The *Bochner–Riesz conjecture* says that for $\delta > 0$ and $1 \le p \le \infty$, S_1^δ is bounded on L^p (equivalently $S_R^\delta f$ converges to f as $R \to \infty$) if and only if

$$\left|\frac{1}{p} - \frac{1}{2}\right| < \frac{2\delta + 1}{2\nu} \qquad (6.8.70)$$

Herz [382] proved (6.8.70) was necessary for the boundedness. Fefferman [263, 267] proved that if $f \mapsto \widehat{f} \restriction S^{\nu-1}$ is bounded from L^{p_0} to L^2, then the Bochner–Riesz conjecture holds for all p in $[p_0', p]$ that also obey (6.8.70). This and the Tomas–Stein theorem imply that if $\delta > \frac{\nu-1}{2(\nu+1)}$, we have the Bochner–Riesz conjecture in the optimal range.

For a monograph on this theme, see Davis–Chang [203] and for some of the many papers on the subject, see [138, 154, 580, 797].

The *restriction conjecture* specifies the exact set of (p, q) for which one has a bound $\|\widehat{f} \restriction S^{\nu-1}\|_{L^q(S^{\nu-1})} \le C \|f\|_{L^p(\mathbb{R}^\nu)}$. The conjecture is that this holds if and only if

$$p' \ge \frac{\nu + 1}{\nu - 1}\, q, \quad p < \frac{2\nu}{\nu + 1} \qquad (6.8.71)$$

For $q = 2$, the first condition is stronger so this conjecture is exactly the Tomas–Stein theorem. It is known that for all p, q, (6.8.71) are necessary conditions. For $\nu = 2$, it is a result of Zygmund [879] that this conjecture holds but for $\nu > 2$, there are only partial results (see, for example, the Historical Notes to Chapter 10 of [329] for a summary of results up to 2009).

The difficulty of the Bochner–Riesz and restriction conjectures is seen by their connection to the Kakeya conjecture. In 1917, the Japanese mathematician, Kakeya [441] asked the question of what was the set of minimal area in the plane through which we could rotate a unit length needle by $360°$. In 1919, Besicovitch [62], then displaced by the Russian civil war (and later a professor at Cambridge—he was the successor in the chair that Littelwood held) and unaware of Kakeya's question constructed sets, E, in $\mathbb{R}^2$ of measure zero so that for every $w \in S^1$, there is x_0 in E with $x_0 + tw \in E$, $0 \le t \le 1$. In 1927 [63] he showed that his construction could also prove that there were sets of arbitrarily small Lebesgue measure which have Kakeya's property. Below, we will follow current standard parlance and use "Kakeya set" and "Kakeya conjecture" even though it would be more appropriate to use Besicovitch's name (as has occasionally been done, e.g., Bourgain [104]).

Fefferman's proof [264] that the operator S_R of (5.8.34) is unbounded used Kakeya sets but except for a few papers in the 1970's (e.g., Cordoba [170]), the subject was dormant until Bourgain's 1992 paper [104] that showed information on the dimension of Kakeya sets implied results on the Bochner–Riesz and restriction problems. One defines a Kakeya maximal function as a sup of averages over tubes of cross-section $\delta^{\nu-1}$. Bourgain conjectured a bound on this maximal function by δ^ε for all $\varepsilon > 0$ and showed it would follow from the restriction conjecture.

Since then, these ideas have been a dominant theme in harmonic analysis. Related is the Kakeya conjecture that any Kakeya set in $\mathbb{R}^\nu$ must have Hausdorff dimension ν. The best lower bounds on this dimension are due to Wolff [861] and Katz–Tao [452].

Major subthemes can be found in ideas from combinatorics, especially geometric combinatorics (where Wolff [862, 863] was a pioneer in making the link to harmonic analysis) and arithmetic combinatorics (where Bourgain, Katz and Tao [451, 107, 108] were pioneers in the link). The synthesis has important spinoff in number theory. It is known that Bochner–Riesz conjecture $\Rightarrow$ restriction conjecture $\Rightarrow$ Kakeya maximal function conjecture$\Rightarrow$ Kakeya dimension conjecture and there are some implications in the opposite directions.

In [862], Wolff found an analogy of the Kakeya conjecture for finite fields proven by Dvir [236]. For a review of these connections between Kakeya problems and the restriction problem, see Łaba [500].

Problems

1. Extend (6.8.1) to $k_{\nu-\ell+1} = c_1, \ldots,\ k_\nu = c_\ell$ (i.e., codimension ℓ hyperplane) and $s > \ell/2$.

2. Extend Theorem 6.8.4 to τ's living on smooth compact surfaces of codimension ℓ with positive Gaussian curvature. $1 \le p < \frac{2\nu+2}{2\nu+3}$ has to be replaced by $1 \le p < \frac{2(\nu-\ell)+4}{(\nu-\ell)+4}$.

3. In the context of Theorem 6.8.1, prove that as an $L^2(\mathbb{R}^{\nu-1}, d^{\nu-1}x)$-valued function when the integral on the right of (6.8.1) is finite, one has $c \mapsto \widehat{f}(k_1, \ldots, k_{\nu-1}, c)$ is Hölder continuous of any order $\alpha < s - \frac{1}{2}$.

4. Extend the results of Problem 3 to embedded hypersurfaces.

5. Let $K \in L^\infty(\mathbb{R}^\nu)$. Prove that the norm of $f \mapsto K * f$ as a map of L^1 to L^∞ is exactly $\|K\|_\infty$. (*Hint*: If $|\{x \mid \text{Re } f(x) > \alpha\}| \ne 0$ prove that with f the characteristic function of that set, $\|K * f\|_\infty \ge \alpha \|f\|_1$.)

6. (a) Let $Q \in L^\infty(\mathbb{R}^\nu)$. Prove that the norm of $f \mapsto Qf$ as a map of $L^p(\mathbb{R}^\nu)$ to $L^p(\mathbb{R}^\nu))$ is $\|Q\|_\infty$.

(b) If $\widehat{K} \in L^\infty(\mathbb{R}^\nu)$, prove that the norm of $f \mapsto K * f$ as a map of L^2 to L^2 is $\|\widehat{K}\|_\infty$.

6.9. Tauberian Theorems

Our main goal in this section is to prove

Theorem 6.9.1. *Let μ be a (positive) measure on $\mathbb{R}$ and $\alpha \in (0, 2)$. Suppose that*

$$\int \frac{d\mu(x)}{x^2 + 1} < \infty \qquad (6.9.1)$$

Then

$$\lim_{R \to \infty} \frac{\mu([-R, R])}{R^\alpha} = a \qquad (6.9.2)$$

if and only if

$$\lim_{y \to \infty} y^{2-\alpha} \int \frac{d\mu(x)}{x^2 + y^2} = \frac{a(\alpha\pi/2)}{\sin(\pi\alpha/2)} \qquad (6.9.3)$$

As we'll see, $(6.9.2) \Rightarrow (6.9.3)$ (called an Abelian theorem) is easy (and doesn't require μ to be positive) and $(6.9.3) \Rightarrow (6.9.2)$ (called a Tauberian theorem) is more subtle. It allows one to go from information on $\operatorname{Im} \int \frac{d\mu(x)}{x-z}$ as $\operatorname{Im} z \to \infty$ to information on $\mathbb{R}$.

Abelian/Tauberian pairs are also discussed in Part 2B (see Sections 13.0 and 13.5 and their Notes) and Part 4 (see Chapter 6, especially Section 6.11 and its Notes). Problem 4–6 of Section 6.11 of Part 4 has considerable overlap with this section but we feel that a discussion of this theme is a part of understanding of inequalities.

Here is the Abelian direction:

Theorem 6.9.2. *Let f be a C^1 even function on $\mathbb{R}$ with $f(x) \to 0$ as $|x| \to \infty$ and f monotone decreasing on $[0, \infty)$. Let $\alpha > 0$ with*

$$\int_0^\infty x^\alpha |f'(x)| \, dx < \infty \qquad (6.9.4)$$

Let μ be a signed Borel measure with

$$\lim_{R \to \infty} R^{-\alpha} \mu([-R, R]) = a \qquad (6.9.5)$$

and

$$\sup_{R \geq 0} R^{-\alpha} |\mu([-R, R])| < \infty \qquad (6.9.6)$$

Then,

$$\lim_{y \to \infty} y^{-\alpha} \int f\left(\frac{x}{y}\right) d\mu(x) = a \int_0^\infty \beta^\alpha |f'(\beta)| \, d\beta \qquad (6.9.7)$$

Remarks. 1. One can drop (6.9.6) by the following argument. If $\mu = \mu_+ + \mu_-$ where μ_- is supported on $[-1, 1]$ and μ_+ on $\mathbb{R} \setminus [-1, 1]$, we have (6.9.6) for μ_+ since (6.9.5) holds and $\mu_+([-R, R]) = 0$ for $R < 1$. And since $\alpha > 0$, the contribution of μ_- to (6.9.5) or (6.9.7) is negligible. We include (6.9.6) only for simplicity of exposition.

2. If $f(x) = (1 + x^2)^{-1}$, then $-f'(x) = (2x)(1 + x^2)^{-2}$ and (6.9.4) holds for $0 < \alpha < 2$. By (5.7.30) of Part 2A,

$$\int_0^\infty \beta^\alpha (-f'(\beta)) \, d\beta = \alpha \int_0^\infty \beta^{\alpha-1} f(\beta) \, d\beta$$

$$= \frac{\alpha}{2} \int_0^\infty \frac{u^{\frac{1}{2}\alpha - 1}}{1 + u} \, du = \frac{(\pi\alpha)/2}{\sin(\pi\alpha/2)} \tag{6.9.8}$$

showing that (6.9.7) for this f is (6.9.3) and so this theorem includes (6.9.2) $\Rightarrow$ (6.9.3).

Proof. Define

$$F(y) = \int f\left(\frac{x}{y}\right) d\mu(x), \quad X(y) = \mu([-y, y]) \tag{6.9.9}$$

The wedding-cake representation (2.2.17) in this case says that

$$f(x) = \int_0^\infty g(\beta)\chi_{[-\beta,\beta]}(x) \, d\beta \tag{6.9.10}$$

where $g(\beta) = -f'(\beta)$, so then (6.9.10) says that

$$f(x) = \int_{|x|}^\infty (-f'(\beta)) \, d\beta \tag{6.9.11}$$

which is obvious.

(6.9.10) implies

$$y^{-\alpha} F(y) = y^{-\alpha} \int_0^\infty [-f'(\beta)] \, X(\beta y) \, d\beta$$

$$= \int_0^\infty [-f'(\beta)] \, \beta^\alpha \left[(\beta y)^{-1} X(\beta y)\right] d\beta \tag{6.9.12}$$

By the hypotheses and the dominated convergence theorem, this converges to the right side of (6.9.7). $\qquad\square$

Example 6.9.3. To get a signed measure where we can do explicit calculations, we take one for which (6.9.1) only involves conditionally convergent integrals. We take

$$d\mu(x) = x \sin(x) \, dx \tag{6.9.13}$$

By taking $\frac{d}{db}$ derivatives in (5.7.11)

$$\int_{-\infty}^\infty \frac{d\mu(x)}{x^2 + y^2} = \pi e^{-y} \tag{6.9.14}$$

where again this is integrated as $\lim_{R\to\infty} \int_{-R}^{R} \dots$. Thus, (6.9.3) holds for $a = 0$, any $\alpha \in (0,2)$. By

$$\sup_{|R|\leq M} |\mu([-R,R])| = O(M) \tag{6.9.15}$$

so for $\alpha \in (0,1)$, $\mu([-R,R])$ $R^{-\alpha}$ is unbounded and not 0 as it would be if (6.9.2) holds. This shows that without limiting sign cancellations (i.e., μ positive), one can't hope for the Tauberian part of Theorem 6.9.1. $\qquad\square$

Proof of (6.9.3) $\Rightarrow$ (6.9.2) **in Theorem 6.9.1.** Define for $y \geq 1$, μ_y by

$$\int f(x) d\mu_y(x) = y^{-\alpha} \int f\left(\frac{x}{y}\right) d\mu(x) \tag{6.9.16}$$

Let γ be the measure

$$d\gamma(x) = x^{\alpha-1} dx \tag{6.9.17}$$

which has the property that $\gamma_y = \gamma$; thus $c\gamma$ is a natural possible limit point for μ_y. Indeed, we'll first show that if (6.9.3) holds, then weakly (on even functions) as finite measures on $\mathbb{R} \cup \{\infty\}$,

$$(1+x^2)^{-1} d\mu_y(x) \to \left(\frac{a\alpha}{2}\right)(1+x^2)^{-1}d\gamma \tag{6.9.18}$$

i.e., for g even and bounded continuous on $\mathbb{R} \cup \{\infty\}$

$$\int g(x)(1+x^2)^{-1} d\mu_y(x) \to \frac{a\alpha}{2} \int g(x)(1+x^2)^{-1}d\gamma \tag{6.9.19}$$

(6.9.3) implies that (6.9.19) holds for $g = 1$. Since $d\mu_{yw} = (d\mu_y)_w$, we see that by scaling, (6.9.19) holds for

$$g(x) = \frac{1+x^2}{w^2+x^2} \tag{6.9.20}$$

Let

$$F_y(u) = \int \frac{d\mu_y(x)}{u+x^2}, \quad F_\infty(u) = \frac{a\alpha}{2} \int \frac{d\gamma(x)}{u+x^2} \tag{6.9.21}$$

By a standard Morera theorem argument (see Theorem 3.1.6 of Part 2A), F_y and F_∞ are analytic in $\{u \mid \operatorname{Re} u > 0\}$ and uniformly bounded in $\{u \mid \operatorname{Re} u > \varepsilon\}$ (since $\sup_{\operatorname{Re} u > 0, x \in \mathbb{R}} \left|\frac{1+x^2}{u+x^2}\right| < \infty$). $F_y(u) \to F_\infty(u)$ for $u \in (0,\infty)$ by (6.9.20) above, so, by Vitali's theorem (see Theorem 6.2.8 of Part 2A), they converge uniformly for $u \in \mathbb{H}_+$ and thus $F_y^{(k)}(1) \to F_\infty^{(k)}(1)$, i.e., for each $k \geq 1$, (6.9.19) holds for $g(x) = (1+x^2)^{-k}$. The polynomials in $(1+x^2)^{-1}$ are dense in the even functions in $C(\mathbb{R} \cup \{\infty\})$, so (6.9.19) holds for all such g's. Since γ has no pure points, by a general argument (see Theorem 4.5.7 of Part 1),

$$\mu_y([-1,1]) \to \left(\frac{a\alpha}{2}\right)\gamma([-1,1]) \tag{6.9.22}$$

which is (6.9.2). $\qquad\square$

The same argument works for other basic functions in addition to $(1 + x^2)^{-1}$ on $\mathbb{R}$. For example (Problem 1):

Theorem 6.9.4. *Let μ be a (positive) Borel measure on $[0, \infty)$ so that $\int e^{-tx} d\mu(x) < \infty$ for all $t > 0$. Then for $\gamma \geq 0$, $D \geq 0$, we have*

$$\lim_{t \downarrow 0} t^\gamma \int e^{-tx} d\mu(x) = D \qquad (6.9.23)$$

if and only if

$$\lim_{a \to \infty} a^{-\gamma} \mu([0, a)) = D \, \Gamma(\gamma \pm 1)^{-1} \qquad (6.9.24)$$

Notes and Historical Remarks. There are extensive notes on Tauberian theorems in the Notes to Section 6.11 of Part 4. The underlying method of this section goes back to Karamata [**445**] who, in particular, had the idea of using polynomial approximation. The exact approach we use here I learned from Aizenman and reported on in [**747**].

Problems

1. Prove Theorem 6.9.4.

Bibliography

[1] H. Abels, *Pseudodifferential and Singular Integral Operators. An Introduction with Applications*, De Gruyter Graduate Lectures, De Gruyter, Berlin, 2012. (Cited on 367, 603.)

[2] R. A. Adams and F. H. Clarke, *Gross's logarithmic Sobolev inequality: a simple proof*, Amer. J. Math. **101** (1979), 1265–1269. (Cited on 652.)

[3] R. A. Adams and J. J. F Fournier, *Sobolev Spaces*, second edition, Pure and Applied Mathematics, Elsevier/Academic Press, Amsterdam, 2003. (Cited on 583.)

[4] S. Agmon, *Spectral properties of Schrödinger operators*, In Actes du Congrès International des Mathématiciens (Nice, 1970), Tome 2, pp. 679–683. Gauthier-Villars, Paris, 1971. (Cited on 683.)

[5] S. Agmon, *Spectral properties of Schrödinger operators and scattering theory*, Lezioni Fermiane, Classe di Scienze, Accademia Nazionale dei Lincei, Scuola Normale Superiore di Pisa, Pisa, 1975. (Cited on 683.)

[6] S. Agmon and L. Hörmander, *Asymptotic properties of solutions of differential equations with simple characteristics*, J. Anal. Math. 30 (1976), 1–38. (Cited on 683.)

[7] L. V. Ahlfors, *Development of the theory of conformal mapping and Riemann surfaces through a century*, In Contributions to the Theory of Riemann Surfaces, pp. 3–13, Annals of Mathematics Studies, Princeton University Press, Princeton, NJ, 1953. (Cited on 298.)

[8] H. Aikawa and M. Essén, *Potential Theory–Selected Topics*, Lecture Notes in Mathematics, Springer-Verlag, Berlin, 1996. (Cited on 177.)

[9] M. Aizenman and E. H. Lieb, *On semiclassical bounds for eigenvalues of Schrödinger operators*, Phys. Lett. A **66** (1978), 427–429. (Cited on 669.)

[10] M. Aizenman and S. Warzel, *Random Operators: Disorder Effects on Quantum Spectra and Dynamics*, American Mathematical Society, Providence, RI, to appear. (Cited on 513.)

[11] W. O. Amrein and A. M. Berthier, *On support properties of L_p-functions and their Fourier transforms*, J. Funct. Anal. **24** (1977), 258–267. (Cited on 337.)

[12] C. Ané et al., *Sur les inégalités de Sobolev logarithmiques*, Panoramas et Synthèses. Société Mathématique de France, Paris, 2000. (Cited on 650.)

[13] D. Armitage and S. J. Gardiner, *Classical Potential Theory*, Springer Monographs in Mathematics, Springer-Verlag London, London, 2001. (Cited on 177.)

[14] V. I. Arnol'd, *Geometrical Methods in the Theory of Ordinary Differential Equations*, 2nd edition, Grundlehren der Mathematischen Wissenschaften, Springer-Verlag, New York, 1988. (Cited on 99.)

[15] V. I. Arnol'd and A. Avez, *Ergodic Problems of Classical Mechanics*, W. A. Benjamin, New York-Amsterdam 1968. (Cited on 79, 97.)

[16] N. Aronszajn and K. T. Smith, *Theory of Bessel Potentials. I*, Ann. Inst. Fourier Grenoble **11** (1961), 385–475. (Cited on 276, 681.)

[17] E. Artin, *Ein mechanisches System mit quasiergodischen Bahnen*, Hamb. Math. Abh. **3** (1924), 170–177. (Cited on 125.)

[18] E. W. Aslaksen and J. R. Klauder, *Continuous representation theory using the affine group*, J. Math. Phys. **10** (1969), 2267–2275. (Cited on 387.)

[19] T. Aubin, *Problèmes isopérimétriques et espaces de Sobolev*, J. Diff. Geom. **11** (1976), 573–598. (Cited on 582.)

[20] S. Aubry, *Metal insulator transition in one-dimensional deformable lattices*, in Bifurcation Phenomena in Mathematical Physics and Related Topics (C. Bardos and D. Bessis, eds.), pp. 163–184, NATO Advanced Study Institute Series, Ser. C: Mathematical and Physical Sciences, D. Reidel Publishing, Dordrecht–Boston, 1980. (Cited on 296.)

[21] A. Avez, *Ergodic Theory of Dynamical Systems, Vol. 1*, Mimeographed lecture notes, University of Minnesota Institute of Technology, Minneapolis, 1966. (Cited on 99.)

[22] A. Avila and J. Bochi, *On the subadditive ergodic theorem*, preprint. Available online at http://www.mat.uc.cl/~jairo.bochi/docs/kingbirk.pdf. (Cited on 145.)

[23] A. Avila, Y. Last, and B. Simon, *Bulk universality and clock spacing of zeros for ergodic Jacobi matrices with a.c. spectrum*, Anal. PDE **3** (2010), 81–108. (Cited on 292.)

[24] J. Avron and B. Simon, *Almost periodic Schrödinger operators, II. The integrated density of states*, Duke Math. J. **50** (1983), 369–391. (Cited on 291, 296.)

[25] S. Axler, P. Bourdon, and W. Ramey, *Harmonic Function Theory*, 2nd edition, Graduate Texts in Mathematics, Springer-Verlag, New York, 2001. (Cited on 177.)

[26] K. I. Babenko, *A problem of Gauss*, Dokl. Akad. Nauk. SSSR **238** (1978), 1021–1024. (Cited on 125.)

[27] H. Bacry, A. Grossmann, and J. Zak, *Proof of the completeness of lattice states in the kq-representation*, Phys. Rev. B. **12**, (1975), 1118–1120. (Cited on 401, 402.)

[28] L. Báez-Duarte, *Another look at the martingale theorem*, J. Math. Anal. Appl. **23** (1968), 551–557. (Cited on 161.)

[29] L. Baggett, *Processing a radar signal and representations of the discrete Heisenberg group*, Colloq. Math. **60/61** (1990), 195–203. (Cited on 403.)

[30] D. Bakry and M. Émery, *Diffusions Hypercontractive*, In Séminaire de Probabilité, XIX, pp. 179–206, Lecture Notes in Math., Springer, Berlin, 1985. (Cited on 653.)

[31] R. Balian, *Un principe d'incertitude fort en théorie du signal ou en mécanique quantique*, C. R. Acad. Sci. Paris, **292** (1981), 1357—1362. (Cited on 402.)

[32] S. Banach, *Sur le théorème de M. Vitali*, Fund. Math. **5** (1924), 130–136. (Cited on 49.)

[33] S. Banach, *Sur une classe de fonctions d'ensemble*, Fund. Math. **6** (1924), 170–188. (Cited on 49.)

[34] S. Banach, *Sur la convergence presque partout de fonctionnelles linéaires*, Bull. Soc. Math. France (2) **50** (1926), 27–32, 36–43. (Cited on 24, 25, 46.)

[35] R. Bañuelos and B. Davis *Donald Burkholder's work in martingales and analysis*, In [**202**]. (Cited on 162.)

[36] V. Bargmann, *Irreducible unitary representations of the Lorentz group*, Ann. Math. (2) **48** (1947), 568–640. (Cited on 386.)

[37] V. Bargmann, *On a Hilbert space of analytic functions and associated integral transform I*, Comm. Pure Appl. Math. **14** (1961), 187–214. (Cited on 385.)

[38] V. Bargmann, P. Butera, L. Girardello, and J. R. Klauder, *On the completeness of the coherent states*, Rep. Math. Phys. **2** (1971), 221–228. (Cited on 401.)

[39] N. K. Bari, *Sur les bases dans l'espace de Hilbert*, Dokl. Akad. Nauk. SSSR, **54** (1946), 379–382. (Cited on 401.)

[40] N. K. Bari, *Biorthogonal systems and bases in Hilbert space*, In Mathematics. Vol. IV, Uch. Zap. Mosk. Gos. Univ., pp. 69–107, Moscow Univ. Press, Moscow, 1951. (Cited on 401, 406.)

[41] A. O. Barut and L. Girardello, *New coherent states associated with non compact groups*, Comm. Math. Phys. **21** (1971), 41. (Cited on 386.)

[42] G. Battle, *Heisenberg proof of the Balian–Low theorem*, Lett. Math. Phys. **15** (1988), 175–177. (Cited on 402, 405.)

[43] G. Battle, *Phase space localization theorem for ondelettes*, J. Math. Phys. **30** (1989), 2195–2196. (Cited on 434.)

[44] G. Battle, *Wavelets and Renormalization*, Series in Approximations and Decompositions, World Scientific Publishing Co., Inc., River Edge, NJ, 1999. (Cited on 433.)

[45] H. Bauer, *Axiomatische Behandlung des Dirichletschen Problems für elliptische und parabolische Differentialgleichungen*, Math. Ann. **146** (1962), 1–59. (Cited on 177.)

[46] H. Bauer, *Weiterführung einer axiomatischen Potentialtheorie ohne Kern (Existenz von Potentialen)*, Z. Wahrscheinlichkeit **1** (1962/1963), 197–229. (Cited on 177.)

[47] H. Bauer, *Harmonische Räume und ihre Potentialtheorie*, Lecture Notes in Math., Springer-Verlag, Berlin-New York, 1966. (Cited on 177, 276.)

[48] R. Beals, *A general calculus of pseudodifferential operators*, Duke Math. J. **42** (1975), 1–42. (Cited on 614.)

[49] A. F. Beardon, *The Geometry of Discrete Groups*, corrected reprint of the 1983 original, Graduate Texts in Mathematics, Springer-Verlag, New York, 1995. (Cited on 127.)

[50] W. Beckner, *Inequalities in Fourier analysis*, Ann. Math. (2) **102** (1975), 159–182. (Cited on 335, 652.)

[51] A. Ben-Aroya, O. Regev and R. de Wolf, *A hypercontractive inequality for matrix-valued functions with applications to quantum computing*, In Proc. 49th FOCS, pp. 477–486, IEEE Computer Society Press, Los Alamitos, CA, 2008. (Cited on 654.)

[52] J. J. Benedetto, *Uncertainty principle inequalities and spectrum estimation*, in Recent Advances in Fourier Analysis and Its Applications (J. S. Bymes and J. L. Byrnes, eds.), pp. 143–182, NATO ASI Series, Kluwer Acad. Publ., Dordrecht, 1990. (Cited on 333.)

[53] M. Benedicks, *On Fourier transforms of functions supported on sets of finite Lebesgue measure*, J. Math. Anal. Appl. **106** (1985), 180–183. (Cited on 337.)

[54] F. Benford, *The law of anomalous numbers*, Proc. Amer. Philos. Soc. **78** (1938), 551–572. (Cited on 99.)

[55] C. Bennett, R. A. DeVore, and R. Sharpley, *Weak-L^1 and BMO*, Ann. Math. **113** (1981), 601–611. (Cited on 534.)

[56] C. Bennett and R. Sharpley, *Interpolation of Operators*, Pure and Applied Mathematics, Academic Press, Inc., Boston, MA, 1988. (Cited on 556, 583.)

[57] Ju. M. Berezanskiĭ, *Expansions in Eigenfunctions of Selfadjoint Operators* (Translated from the Russian by R. Bolstein, J. M. Danskin, J. Rovnyak and L. Shulman), Translations of Mathematical Monographs, American Mathematical Society, Providence, RI. 1968; Russian Original 1965. (Cited on 292.)

[58] F. A. Berezin, *Covariant and contravariant symbols of operators*, (Russian), Izv. Akad. Nauk SSSR Ser. Mat. **36** (1972), 1134–1167; English translation, Math. USSR-Izv. **6** (1973), 1117–1151. (Cited on 386.)

[59] F. A. Berezin, *The Method of Second Quantization*, Academic Press, New York, 1966; translated from Russian. Revised (augmented) second edition, Kluwer, Boston, 1989. (Cited on 402.)

[60] J. Bergh and J. Löfström, *Interpolation Spaces. An Introduction*, Grundlehren der Mathematischen Wissenschaften, Springer-Verlag, Berlin–New York, 1976. (Cited on 556, 583.)

[61] S. Bernstein, *Sur l'ordre de la meilleure approximation des fonctions continues par des polynômes de degré donné*, Mem. Cl. Sci. Acad. Roy. Belg. **4** (1912), 1–103. (Cited on 291.)

[62] A. S. Besicovitch, *Sur deux questions d'integrabilite des fonctions*, J. Soc. Phys. Math. **2** (1919), 105–123. (Cited on 684.)

[63] A. S. Besicovitch, *On Kakeya's problem and a similar one*, Math. Z. **27** (1927), 312–320. (Cited on 684.)

[64] A. S. Besicovitch, *A general form of the covering principle and relative differentiation of additive functions (I), (II)*, Proc. Cambridge Philos. Soc. **41** (1945), 103–110; **42** (1946), 1–10. (Cited on 50.)

[65] O. V. Besov, *On a certain family of functional spaces. Embedding and extension theorems*, Dokl. Akad. Nauk SSSR **126** (1959), 1163–1165. (Cited on 583.)

[66] A. Beurling *On two problems concerning linear transformations in Hilbert space*, Acta Math. **81** (1948), 239–255. (Cited on 470, 517.)

[67] A. Beurling and J. Deny, *Espaces de Dirichlet. I. Le cas élémentaire*, Acta Math. **99** (1958), 203–224. (Cited on 177, 276.)

[68] A. Beurling and J. Deny, *Dirichlet spaces*, Proc. Nat. Acad. Sci. USA, **45** (1959), 208–215. (Cited on 276.)

[69] I. Bialynicki-Birula and J. Mycielski, *Uncertainty relations for information entropy in wave mechanics*, Comm. Math. Phys. **44** (1975), 129–132. (Cited on 335.)

[70] P. Biane, *Free hypercontractivity*, Comm. Math. Phys. **184** (1997), 457–474. (Cited on 653.)

[71] L. Bienvenu, G. Shafer, and A. Shen, *On the history of martingales in the study of randomness*, J. Electron. Hist. Probab. Stat. **5** (2009), 40 pp. (Cited on 160.)

[72] P. Billingsley, *Ergodic Theory and Information*, reprint of the 1965 original, Robert E. Krieger Publishing, Huntington, NY, 1978. (Cited on 79, 125.)

[73] G. D. Birkhoff, *A theorem on theories of orthogonal functions with an application to Sturm–Liouville theories*, Proc. Nat. Acad. Sci. USA., **3** (1917), 656–659. (Cited on 406.)

[74] G. D. Birkhoff, *Surface transformations and their dynamical applications*, Acta Math. **43** (1920), 1–119. (Cited on 80.)

[75] G. D. Birkhoff, *Proof of the ergodic theorem*, Proc. Nat. Acad. Sci. USA **17** (1931), 656–660. (Cited on 79, 81.)

[76] G. D. Birkhoff and B. O. Koopman, *Recent contributions to the ergodic theory*, Proc. Nat. Acad. Sci. USA **18** (1932), 279–282. (Cited on 79, 81.)

[77] G. D. Birkhoff and P. A. Smith, *Structure analysis of surface transformations*, J. Math. Pures Appl. **7** (1928), 345–380. (Cited on 80.)

[78] G. Birkhoff and G-C. Rota, *On the completeness of Sturm–Liouville expansions*, Amer. Math. Monthly **67** (1960), 835Ü-841. (Cited on 406.)

[79] E. Bishop, *Foundations of Constructive Analysis*, McGraw–Hill, New York-Toronto-London, 1967. (Cited on 84.)

[80] N. M. Blachman, *The convolution inequality for entropy powers*, IEEE Trans. Inform. Theory **2** (1965), 267–271. (Cited on 652.)

[81] Ph. Blanchard, J. Stubbe, and J. Rezende, *New estimates on the number of bound states of Schrödinger operators*, Lett. Math. Phys. **14** (1987), 215–225. (Cited on 669.)

[82] C. Blatter, *Wavelets, A Primer*, A K Peters, Ltd., Natick, MA, 1998. (Cited on 433.)

[83] J. Bliedtner and W. Hansen, *Potential Theory. An Analytic and Probabilistic Approach to Balayage*, Universitext, Springer-Verlag, Berlin, 1986. (Cited on 177.)

[84] F. Bloch, *Nuclear Induction*, Phys. Rev. **70** (1946), 460–474. (Cited on 386, 387.)

[85] R. M. Blumenthal and R. K. Getoor, *Markov Processes and Potential Theory*, Pure and Applied Mathematics, Academic Press, New York–London, 1968. (Cited on 177.)

[86] M. Bôcher, *Singular points of functions which satisfy partial differential equations of the elliptic type*, Bull. Amer. Math. Soc. (2) **9**, (1903), 455–465. (Cited on 197.)

[87] S. Bochner, *Summation of multiple Fourier series by spherical means*, Trans. Amer. Math. Soc. **40** (1936), 175–207. (Cited on 603.)

[88] A. Boggess and F. Narcowich, *A First Course in Wavelets with Fourier Analysis*, Second edition, John Wiley & Sons, Inc., Hoboken, NJ, 2009. (Cited on 433.)

[89] P. Bohl, *Über ein in der Theorie der säkularen Störungen vorkommendes Problem*, J. Reine Angew. Math. **135** (1909), 189–283. (Cited on 98.)

[90] H. A. Bohr, *Collected Mathematical Works*, E. Følner and B. Jessen, eds., Dansk Matematisk Forening, Copenhagen, 1952. (Cited on 19.)

[91] L. Boltzmann, *Über das Warme Gleichgewicht zwischen mehratomigen Gasmolektilen*, In Wissenschaftliche Abhandlung, Volume 1, pp. 237–258, Cambridge, Cambridge University Press, 2012; original publication, Wien. Ber. **63** (1871), 397–418. (Cited on 79.)

[92] L. Boltzmann, *Über die Eigenschaften monozyklischer und damit verwandter Systeme*, In Wissenschaftliche Abhandlung, Volume 3, pp. 122-152, Cambridge, Cambridge University Press, 2012; original publication, J. Reine Angew. Math. (Crelle's J.) **98**, (1884/1885), 68–94. (Cited on 79, 80.)

[93] A. Bonami, *Étude des coefficients de Fourier des fonctions de $L^p(G)$*, Ann. Inst. Fourier Grenoble **20** (1970), 335–402. (Cited on 652.)

[94] A. Bonami and B. Demange, *A survey on uncertainty principles related to quadratic forms*, Collect. Math. **57** (2006), 1–36. (Cited on 337.)

[95] G. Boole, *On the comparison of transcendents, with certain applications to the theory of definite integrals*, Philos. Trans. Roy. Soc. London **147** (1857), 745–803. (Cited on 513.)

[96] M. Boon and J. Zak, *Amplitudes on von Neumann lattices*, J. Math. Phys. **22** (1981), 1090–1099. (Cited on 402.)

[97] M. Boon, J. Zak, and I. J. Zucker, *Rational von Neumann lattices*, J. Math. Phys. **24** (1983), 316–323. (Cited on 402.)

[98] E. Borel, *Les probabilités dénombrables et leurs applications arithmétiques*, Rend. Circ. Mat. Palermo **27** (1909), 247–271. (Cited on 97.)

[99] E. Borel, *Sur les chiffres décimaux de $\sqrt{2}$ et divers problèmes de probabilités en chaîne*, C. R. Acad. Sci. Paris **230** (1950), 591–593. (Cited on 97.)

[100] W. Bosma, H. Jager, and F. Wiedijk, *Some metrical observations on the approximation by continued fractions*, Indag. Math. Nederl. Akad. Wetensch. **45** (1983), 281–299. (Cited on 125.)

[101] G. Bouligand, *Sur le probleme de Dirichlet*, Ann. Soc. Polonaise Math. **4** (1926), 59–112. (Cited on 231.)

[102] J. Bourgain, *Estimations de certaines fonctions maximales*, C. R. Acad. Sci. Paris **301** (1985), 499–502. (Cited on 49.)

[103] J. Bourgain, *An approach to pointwise ergodic theorems*, In Geometric Aspects of Functional Analysis (1986/87), pp. 204–223, Lecture Notes in Math., Springer, Berlin, 1988. (Cited on 84, 85.)

[104] J. Bourgain, *Besicovitch type maximal operators and applications to Fourier analysis*, Geom. Funct. Anal. **1** (1991), 147–187. (Cited on 682, 685.)

[105] J. Bourgain, *Fourier transform restriction phenomena for certain lattice subsets and applications to nonlinear evolution equations. I. Schrödinger equations*, Geom. Funct. Anal. **3** (1993), 107–156. (Cited on 683.)

[106] J. Bourgain, *Global Solutions of Nonlinear Schrödinger Equations*, American Mathematical Society Colloquium Publications, American Mathematical Society, Providence, RI, 1999. (Cited on 683.)

[107] J. Bourgain, *On the Erdős–Volkmann and Katz–Tao ring conjectures*, Geom. Funct. Anal. **13** (2003), 334–365. (Cited on 685.)

[108] J. Bourgain, N. Katz, and T. Tao, *A sum-product estimate in finite fields, and applications*, Geom. Funct. Anal. **14** (2004), 27–57. (Cited on 685.)

[109] R. Bowen and C. Series, *Markov maps associated with Fuchsian groups*, Publ. Math. Inst. Hautes Etudes Sci. **50** (1979), 153–170. (Cited on 126.)

[110] H. J. Brascamp, E. H. Lieb, and J. M. Luttinger, *A general rearrangement inequality for multiple integrals*, J. Funct. Anal. **17** (1974), 227–237. (Cited on 563.)

[111] O. Bratteli and P. Jorgensen, *Wavelets Through a Looking Glass, The World of the Spectrum*, Applied and Numerical Harmonic Analysis, Birkhäuser Boston, Inc., Boston, MA, 2002. (Cited on 433.)

[112] M. Brelot, *Sur le potentiel et les suites de fonctions sous-harmoniques*, C. R. Acad. Sci. Paris **207** (1938), 836–838. (Cited on 274.)

[113] M. Brelot, *Familles de Perron et problème de Dirichlet*, Acta Sci. Math. Szeged **9** (1938–40), 133–153. (Cited on 231.)

[114] M. Brelot, *Points irréguliers et transformations continues en théorie du potentiel*, J. Math. Pures Appl. **19** (1940), 319–337. (Cited on 276.)

[115] M. Brelot, *Sur la théorie autonome des fonctions sous-harmoniques*, Bull. Sci. Math. **65** (1941), 72–98. (Cited on 274.)

[116] M. Brelot, *Sur les ensembles effilés*, Bull. Sci. Math. **68** (1944), 12–36. (Cited on 276.)

[117] M. Brelot, *Lectures on Potential Theory*, Lectures on Mathematics, Tata Institute of Fundamental Research, Bombay, 1960. (Cited on 177, 276.)

[118] M. Brelot, *On Topologies and Boundaries in Potential Theory*, Lecture Notes in Math., Springer-Verlag, Berlin–New York, 1971. (Cited on 177.)

[119] M. Brelot, *Les étapes et les aspects multiples de la théorie du potentiel*, Enseign. Math. **18** (1972), 1–36. (Cited on 276.)

[120] H. Brezis and M. Marcus, *Hardy's inequalities revisited*,Ann. Scuola Norm. Sup. Pisa **25** (1997), 217–237. (Cited on 336.)

[121] J. R. Brown, *Ergodic Theory and Topological Dynamics*, Pure and Applied Mathematics, Academic Press, New York–London, 1976. (Cited on 65.)

[122] D. L. Burkholder, *Maximal inequalities as necessary conditions for almost everywhere convergence*, Z. Wahrscheinlichkeit **3** (1964), 75–88. (Cited on 25.)

[123] D. L. Burkholder, *Explorations in martingale theory and its applications*, In École d'Été de Probabilités de Saint-Flour XIXŬ1989, pp. 1–66, Lecture Notes in Math., Springer, Berlin, 1991. (Cited on 162.)

[124] D. L. Burkholder, R. F. Gundy, and M. L. Silverstein, *A maximal function characterization of the class H^p*, Trans. Amer. Math. Soc. **157** (1971), 137–153. (Cited on 162.)

[125] H. Busemann and W. Feller, *Zur Differentiation der Lebesgueschen Integrale*, Fund. Math. **22** (1934), 226–256. (Cited on 48.)

[126] A.-P. Calderón, *A general ergodic theorem*, Ann. Math. (2) **58** (1953), 182–191. (Cited on 83.)

[127] A.-P. Calderón, *Lebesque spaces of differentiable functions and distributions*, In Partial Differential Equations, C.B. Morrey (ed.), pp. 33–49, American Mathematical Society, Providence, RI., 1961. (Cited on 276.)

[128] A.-P. Calderón, *Intermediate spaces and interpolation*, Studia Math. **1** (1963), 31–34. (Cited on 36.)

[129] A.-P. Calderón, *Intermediate spaces and interpolation, the complex method*, Studia Math. **24** (1964), 113–190. (Cited on 387.)

[130] A.-P. Calderón, *Spaces between L^1 and L^∞ and the theorem of Marcinkiewicz*, Studia Math. **26** (1966), 273–299. (Cited on 36.)

[131] A.-P. Calderón, *Ergodic theory and translation-invariant operators*, Proc. Nat. Acad. Sci. USA **59** (1968), 349–353. (Cited on 83, 542.)

[132] A.-P. Calderón and R. Vaillancourt, *On the boundedness of pseudo-differential operators*, J. Math. Soc. Japan **23** (1971), 374–378. (Cited on 614.)

[133] A.-P. Calderón and R. Vaillancourt, *A class of bounded pseudo-differential operators*, Proc. Nat. Acad. Sci. USA **69** (1972), 1185–1187. (Cited on 614.)

[134] P. Calderón and A. Zygmund, *On the existence of certain singular integrals*, Acta Math. **88** (1952) 85–139. (Cited on 601.)

[135] J. W. Calkin,*Functions of several variables and absolute continuity, I*, Duke Math. J. **6** (1940), 170–186. (Cited on 581.)

[136] J. J. Callahan, *Advanced Calculus: A Geometric View*, Undergraduate Texts in Mathematics, Springer, New York, 2010. (Cited on 17.)

[137] E. J. Candès, J. Romberg, and T. Tao, *Robust uncertainty principles: exact signal reconstruction from highly incomplete frequency information*, IEEE Trans. Inform. Theory **52** (2006), 489–509. (Cited on 339.)

[138] A. Carbery, *The boundedness of the maximal Bochner–Riesz operator on $L^4(\mathbb{R}^2)$*, Duke Math. J. **50** (1983), 409–416. (Cited on 684.)

[139] A. L. Carey, *Square-integrable representations of non-unimodular groups*, Bull. Aust. Math. Soc. **15** (1976), 1–12. (Cited on 387.)

[140] E. A. Carlen, *Superadditivity of Fisher's information and logarithmic Sobolev inequalities*, J. Funct. Anal. **101** (1991), 194–211. (Cited on 652.)

[141] E. A. Carlen, *Some integral identities and inequalities for entire functions and their application to the coherent state transform*, J. Funct. Anal. **97** (1991), 231–249. (Cited on 653.)

[142] E. A. Carlen and E. Lieb, *Optimal hypercontractivity for Fermi fields and related noncommutative integration inequalities*, Comm. Math. Phys. **155** (1993), 27–46. (Cited on 653.)

[143] L. Carleson, *On convergence and growth of partial sums of Fourier series*, Acta Math. **116** (1966), 135–157. (Cited on 172.)

[144] R. Carmona, *Regularity properties of Schrödinger and Dirichlet semigroups*, J. Funct. Anal. **54** (1979), 259–296. (Cited on 653.)

[145] R. Carmona and J. Lacroix, *Spectral Theory of Random Schrödinger Operators*, Probability and Its Applications, Birkhäuser, Boston, 1990. (Cited on 294.)

[146] H. Cartan, *Théorie du potentiel newtonien: énergie, capacité, suites de potentiels*, Bull. Soc. Math. France **73** (1945), 74–106. (Cited on 274.)

[147] H. Cartan and J. Deny, *Le principe du maximum en théorie du potentiel et la notion de fonction surharmonique*, Acta Sci. Math. Szeged **12** (1950), 81–100. (Cited on 276.)

[148] R. V. Chacon and D. S. Ornstein, *A general ergodic theorem*, Illinois J. Math. **4** (1960), 153–160. (Cited on 86.)

[149] D. G. Champernowne, *The construction of decimals normal in the scale of ten*, J. London Math. Soc. **8** (1933), 254–260. (Cited on 97.)

[150] J.-Y. Chemin and C.-J. Xu, *Inclusions de Sobolev en calcul de Weyl–Hörmander et champs de vecteurs sous-elliptiques*, Ann. Sci. Ecole Norm. Sup. (4) **30** (1997), 719–751. (Cited on 585.)

[151] Y. Cho, E. Koh, and S. Lee, *A maximal inequality for filtration on some function spaces*, Osaka J. Math. **41** (2004), 267–276. (Cited on 172.)

[152] G. Choquet, *Sur les fondements de la théorie fine du potentiel*, C. R. Acad. Sci. Paris **244** (1957), 1606–1609. (Cited on 274.)

[153] V. Chousionis, and X. Tolsa, *The T1 Theorem*, Notes of a short course, Universitat Autònoma de Barcelona, 2012; available at `http://mat.uab.es/~xtolsa/t1.pdf`. (Cited on 603.)

[154] M. Christ, *Weak type endpoint bounds for Bochner–Riesz multipliers*, Rev. Mat. Iberoam. **3** (1987), 25–31. (Cited on 684.)

[155] M. Christ, *Lectures on Singular Integral Operators*, CBMS Regional Conference Series in Mathematics, American Mathematical Society, Providence, RI, 1991. (Cited on 603.)

[156] M. Christ and A. Kiselev, *Absolutely continuous spectrum for one-dimensional Schrödinger operators with slowly decaying potentials: Some optimal results*, J. Amer. Math. Soc. **11** (1998), 771–797. (Cited on 172.)

[157] M. Christ and A. Kiselev, *Maximal functions associated to filtrations*, J. Funct. Anal. **179** (2001), 409–425. (Cited on 172.)

[158] O. Christensen, *Frames and Bases, An Introductory Course*, Applied and Numerical Harmonic Analysis, Birkhäuser Boston, Inc., Boston, MA, 2008. (Cited on 401.)

[159] E. B. Christoffel, *Über die Gaussische Quadratur und eine Verallgemeinerung derselben*, J. Reine Angew. Math. **55** (1858), 61–82. (Cited on 291.)

[160] K. L. Chung, *A Course in Probability Theory*, Academic Press, New York, 1974. (Cited on 155.)

[161] J. A. Cima, A. L. Matheson, and W. T. Ross, *The Cauchy Transform*, Mathematical Surveys and Monographs, American Mathematical Society, Providence, RI, 2006. (Cited on 489.)

[162] R. R. Coifman, *A real variable characterization of H^p*, Studia Math. **51** (1974), 269–274. (Cited on 534.)

[163] R. R. Coifman and Y. Meyer, *Au delà des opérateurs pseudo-différentiels*, Astérisque, Société Mathématique de France, Paris, 1978 (Cited on 614.)

[164] R. R. Coifman and G. Weiss, *Extensions of Hardy-spaces and their use in analysis*, Bull. Amer. Math. Soc. **83** (1977) 569–645. (Cited on 535.)

[165] J. G. Conlon, *A new proof of the Cwikel-Lieb-Rosenbljum bound*. Rocky Mountain J. Math. **15** (1985), 117–122. (Cited on 669.)

[166] C. Constantinescu and A. Cornea, *Potential Theory on Harmonic Spaces*, Die Grundlehren der mathematischen Wissenschaften, Springer-Verlag, New York–Heidelberg, 1972. (Cited on 177.)

[167] A. H. Copeland and P. Erdős, *Note on normal numbers*, Bull. Amer. Math. Soc. **52** (1946), 857–860. (Cited on 97.)

[168] H. O. Cordes, *On compactness of commutators of multiplications and convolutions, and boundedness of pseudodifferential operators*, J. Funct. Anal. **18** (1975), 115–131. (Cited on 614.)

[169] H. O. Cordes, *Elliptic Pseudodifferential Operators—An Abstract Theory*, Lecture Notes in Math., Springer, Berlin, 1979. (Cited on 367.)

[170] A. Cordoba, *The Kakeya maximal function and the spherical summation multipliers*, Amer. J. Math. **99** (1977), 1–22. (Cited on 685.)

[171] A. Cordoba and R. Fefferman, *A geometric proof of the strong maximal theorem*, Ann. Math. (2) **102** (1975), 95–100. (Cited on 48.)

[172] M. Cotlar, *A combinatorial inequality and its applications to L^2-spaces*, Rev. Mat. Cuyana **1** (1955), 41–55. (Cited on 613.)

[173] M. Cotlar, *A unified theory of Hilbert transforms and ergodic theorems*, Rev. Mat. Cuyana **1** (1956), 105–167. (Cited on 83, 542.)

[174] R. Courant and F. John, *Introduction to Calculus and Analysis. Vol. II*, with the assistance of Albert A. Blank and Alan Solomon, Reprint of the 1974 edition, Springer-Verlag, New York, 1989. (Cited on 17.)

[175] W. Craig, *The trace formula for Schröodinger operators on the line*, Comm. Math. Phys. **126** (1989), 379–407. (Cited on 297.)

[176] W. Craig and B. Simon, *Subharmonicity of the Lyapunov index*, Duke Math. J. **50** (1983), 551–560. (Cited on 291, 295, 297.)

[177] P. Crépel, *Quelques matériaux pour l'histoire de la théorie des martingales (1920–1940)*, Publication des séminaires de mathématiques, Université de Rennes, 1984. (Cited on 160.)

[178] H. T. Croft, *Some problems*, Eureka **31** (1968), 18–19. (Cited on 49, 50.)

[179] M. Cwikel, *Weak type estimates for singular values and the number of bound states of Schrödinger operators*, Ann. Math. (2) **106** (1977), 93–100. (Cited on 669.)

[180] M. Cwikel, Y. Sagher, and P. Shvartsman, *A new look at the John–Nirenberg and John–Strömberg theorems for BMO*, J. Funct. Anal. **263** (2012), 129–166. (Cited on 534.)

[181] H. L. Cycon, R. G. Froese, W. Kirsch, and B. Simon, *Schrödinger Operators With Application to Quantum Mechanics and Global Geometry*, Texts and Monographs in Physics, Springer, Berlin, 1987. (Cited on 294.)

[182] B. E. J. Dahlberg, *Estimates of harmonic measure*, Arch. Ration. Mech. Anal. **65** (1977), 275–288. (Cited on 274.)

[183] K. Dajani and C. Kraaikamp, *Ergodic Theory of Numbers*, Carus Mathematical Monographs, Mathematical Association of America, Washington, DC, 2002. (Cited on 123.)

[184] D. Damanik, R. Killip, and B. Simon, *Perturbations of orthogonal polynomials with periodic recursion coefficients*, Ann. Math. **171** (2010), 1931–2010. (Cited on 293.)

[185] S. B. Damelin and W. Willard, Jr., *The Mathematics of Signal Processing*, Cambridge Texts in Applied Mathematics, Cambridge University Press, Cambridge, 2012. (Cited on 401.)

[186] G. Darboux, *Mémoire sur l'approximation des fonctions de très-grands nombres, et sur une classe étendue de développements en série*, Liouville J. (3) **4** (1878), 5–56; 377–416. (Cited on 291.)

[187] I. Daubechies, *Orthonormal bases of compactly supported wavelets*, Comm. Pure Appl. Math. **41** (1988), 909–996. (Cited on 434.)

[188] I. Daubechies, *The wavelet transform, time-frequency localization and signal analysis*, IEEE Trans. Inform. Theory **36** (1990), 961–1005. (Cited on 403.)

[189] I. Daubechies, *Ten Lectures on Wavelets*, CBMS-NSF Regional Conference Series in Applied Mathematics, Society for Industrial and Applied Mathematics (SIAM), Philadelphia, PA, 1992. (Cited on 433.)

[190] I. Daubechies and A. Grossmann, *Frames in the Bargmann space of entire functions*, Comm. Pure Appl. Math. **41** (1988), 151–164. (Cited on 401.)

[191] I. Daubechies and A. J. E. M Janssen, *Two theorems on lattice expansions*, IEEE Trans. Inform. Theory **39** (1993), 3–6. (Cited on 403.)

[192] G. David and J-L. Journé, *A boundedness criterion for generalized Calderón-Zygmund operators*, Ann. Math. (2) **120** (1984), 371–397. (Cited on 602.)

[193] E. B. Davies, *Explicit constants for Gaussian upper bounds on heat kernels*, Amer. J. Math. **109** (1987), 319–334. (Cited on 653.)

[194] E. B. Davies, *Heat kernel bounds for second order elliptic operators on Riemannian manifolds*, Amer. J. Math. **109** (1987), 545–570. (Cited on 653.)

[195] E. B. Davies, *Gaussian upper bounds for heat kernels of some second order operators on Riemannian manifolds*, J. Funct. Anal. **80** (1988), 16–32. (Cited on 653.)

[196] E. B. Davies, *Heat Kernels and Spectral Theory*, Cambridge University Press, Cambridge, UK, 1989. (Cited on 622, 650, 653.)

[197] E. B. Davies, *A review of Hardy inequalities*, In The Maz'ya Anniversary Collection, Vol. 2, pp. 55–67, Operator Theory Advances and Applications, Birkhauser, Basel, 1999. (Cited on 336.)

[198] E. B. Davies, L. Gross, and B. Simon, *Hypercontractivity: a bibliographic review*, In Ideas and Methods in Quantum and Statistical Physics, S. Albeverio, H. Holden, J. Fenstad, and T. Lindstrom (eds.), Volume 2, pp. 370–389, Cambridge University Press, Cambridge, 1992. (Cited on 650.)

[199] E. B. Davies and B. Simon, *Ultracontractivity and the heat kernel for Schrödinger operators and Dirichlet Laplacians*, J. Funct. Anal. **59** (1984), 335–395. (Cited on 653.)

[200] B. Davis, *On the distributions of conjugate functions of nonnegative measures*, Duke Math. J. **40** (1973), 695–700. (Cited on 514.)

[201] B. Davis, *On the weak type* $(1, 1)$ *inequality for conjugate functions*, Proc. Amer. Math. Soc. **44** (1974), 307–311. (Cited on 514.)

[202] B. Davis and R. Song (eds.), *Selected Works of Donald L. Burkholder*, Selected Works in Probability and Statistics, Springer, New York, 2011. (Cited on 162, 693.)

[203] K. M. Davis and Y-C. Chang, *Lectures on Bochner–Riesz means*, London Mathematical Society Lecture Note Series, Cambridge University Press, Cambridge, 1987. (Cited on 684.)

[204] N. G. de Bruijn and K. A. Post, *A remark on uniformly distributed sequences and Riemann integrability*, Indag. Math. Nederl. Akad. Wetensch. **30** (1968), 149–150. (Cited on 99.)

[205] M. de Guzmán, *Real Variable Methods in Fourier Analysis*, North-Holland Mathematics Studies, Notas de Matemática, North-Holland Publishing, Amsterdam–New York, 1981. (Cited on 25.)

[206] C. de la Vallée Poussin, *Sur l'intégrale de Lebesgue*, Trans. Amer. Math. Soc. **16** (1915), 435–501. (Cited on 64.)

[207] K. de Leeuw, and W. Rudin, *Extreme points and extremum problems in* H_1, Pacific J. Math. **8** (1958) 467–485. (Cited on 472.)

[208] C. Dellacherie and P.-A. Meyer, *Probabilities and Potential. B. Theory of Martingales*, North–Holland Mathematics Studies, North–Holland, Amsterdam, 1982. (Cited on 161.)

[209] M. Del Pino and J. Dolbeault, *Best constants for Gagliardo–Nirenberg inequalities and applications to nonlinear diffusions*, J. Math. Pures Appl. **81** (2002), 847–875. (Cited on 582.)

[210] R. del Rio, S. Jitomirskaya, Y. Last and B. Simon, *Operators with singular continuous spectrum, IV. Hausdorff dimensions, rank one perturbations, and localization*, J. Anal. Math. **69** (1996), 153–200. (Cited on 514.)

[211] F. Demengel and G. Demengel, *Functional Spaces for the Theory of Elliptic Partial Differential Equations*, Translated from the 2007 French original by Reinie Erné, Universitext. Springer, London; EDP Sciences, Les Ulis, 2012. (Cited on 583.)

[212] S. A. Denisov, *On Rakhmanov's theorem for Jacobi matrices*, Proc. Amer. Math. Soc. **132** (2004), 847–852. (Cited on 293.)

[213] A. Denjoy, *Sur les courbes définies par les équations différentielles à la surface du tore*, J. Math. Pures Appl. **11** (1932), 333–375 (Cited on 99.)

[214] J. Denker, *Coherent States*, available online at http://www.av8n.com/physics/coherent-states.htm, 2012. (Cited on 373.)

[215] J. Deny, *Sur les infinis d'un potentiel*, C. R. Acad. Sci. Paris **224** (1947), 524–525. (Cited on 274.)

[216] J. Deny, *Familles fondamentales. Noyaux associés*, Ann. Inst. Fourier Grenoble **3** (1951), 73–101. (Cited on 276.)

[217] Y. Derriennic, *Sur le théorème ergodique sous-additif*, C. R. Acad. Sci. Paris **281** (1975), 985–988. (Cited on 145.)

[218] J-D. Deuschel and D. W. Stroock, *Large Deviations*, Pure and Applied Mathematics, Academic Press, Boston, MA, 1989. (Cited on 654.)

[219] J-D. Deuschel and D. W. Stroock, *Hypercontractivity and spectral gap of symmetric diffusions with applications to the stochastic Ising models*, J. Funct. Anal. **92** (1990), 30–48. (Cited on 652.)

[220] P. Diaconis and L. Saloff-Coste, *Logarithmic Sobolev inequalities for finite Markov chains*, Ann. Appl. Probab. **6** (1996), 695–750. (Cited on 653.)

[221] E. DiBenedetto, *Real Analysis*, Birkhäuser Advanced Texts: Basel Textbooks, Birkhäuser, Boston, MA, 2002. (Cited on 50.)

[222] W. Doeblin, *Remarques sur la théorie métrique des fractions continues*, Compositio Math. **7** (1940), 353–371. (Cited on 124.)

[223] D. Donoho, P. Stark, *A note on rearrangements and spectral concentration*, IEEE Trans. Inform. Theory **39** (1993), 257–260. (Cited on 339.)

[224] J. L. Doob, *Regularity properties of certain families of chance variables*, Trans. Amer. Math. Soc. **47** (1940), 455–486. (Cited on 84, 160, 165.)

[225] J. L. Doob, *Stochastic Processes*, reprint of the 1953 original, Wiley Classics Library, John Wiley & Sons, New York, 1990. (Cited on 160, 161.)

[226] J. L. Doob, *Semimartingales and subharmonic functions*, Trans. Amer. Math. Soc. **77** (1954), 86–121. (Cited on 276.)

[227] J. L. Doob, *Probability methods applied to the first boundary value problem*, In Proc. 3rd Berkeley Symposium on Mathematical Statistics and Probability, 1954-1955, Vol. II, pp. 49-80, University of California Press, Berkeley and Los Angeles, 1956. (Cited on 177.)

[228] J. L. Doob, *Classical Potential Theory and Its Probabilistic Counterpart*, Grundlehren der Mathematischen Wissenschaften, Springer-Verlag, New York, 1984. (Cited on 177.)

[229] R. J. Duffin and A. C. Schaeffer, *A class of nonharmonic Fourier series*, Trans. Amer. Math. Soc. **72** (1952), 341–366. (Cited on 401, 403.)

[230] M. Duflo and C. Calvin, *On the regular representation of a nonunimodular locally compact group*, J. Funct. Anal. **21** (1976), 209–243. (Cited on 387.)

[231] J. J. Duistermaat, *Fourier Integral Operators*. Progress in Mathematics, Birkhäuser Boston, Inc., Boston, MA, 1996. (Cited on 367.)

[232] J. J. Duistermaat and L. Hörmander, *Fourier integral operators. II*, Acta Math. **128** (1972), 183–269. (Cited on 350, 367.)

[233] N. Dunford and J. T. Schwartz, *Convergence almost everywhere of operator averages*, J. Ration. Mech. Anal. **5** (1956), 129–178. (Cited on 86.)

[234] P. L. Duren, *Theory of H^p Spaces*, Pure and Applied Mathematics, Academic Press, New York–London, 1970. (Cited on 439, 464, 513.)

[235] R. Durrett, *Probability: Theory and Examples*, 4th edition, Cambridge Series in Statistical and Probabilistic Mathematics, Cambridge University Press, Cambridge, 2010. (Cited on 162, 163.)

[236] Z. Dvir, *On the size of Kakeya sets in finite fields*, J. Amer. Math. Soc. **22** (2009), 1093–1097. (Cited on 685.)

[237] H. Dym and H. P. McKean, *Fourier Series and Integrals*, Probability and Mathematical Statistics, Academic Press, New York–London, 1972. (Cited on 337.)

[238] T. Ekholm and R. L. Frank, *On Lieb–Thirring inequalities for Schrödinger operators with virtual level*, Comm. Math. Phys. **264** (2006), 725–740. (Cited on 669.)

[239] Y. V. Egorov, *Canonical transformations and pseudodifferential operators* (Russian), Trudy. Moskov. Mat. Obsc. **24** (1971), 3–28. (Cited on 368.)

[240] Y. V. Egorov, and B.-W. Schulze, *Pseudo-Differential Operators, Singularities, Applications*, Operator Theory: Advances and Applications, 93. Birkhäuser Verlag, Basel, 1997. (Cited on 367.)

[241] P. Ehrenfest and T. Ehrenfest *Begriffliche Grundlagen der statistischen Auffassung in der Mechanik*, B. G. Teubner, Leipzig, 1911. (Cited on 80.)

[242] M. Einsiedler and T. Ward, *Ergodic Theory With a View Towards Number Theory*, Graduate Texts in Mathematics, Springer-Verlag, London, 2011. (Cited on 79, 123, 126.)

[243] R. W. Emerson, *Self-Reliance*, In Essays: First Series, Houghton, Mifflin and Company, Boston, MA, 1883; available online at `http://en.wikisource.org/wiki/Essays:_First_Series/Self-Reliance` (self-reliance only) and `https://archive.org/details/essaysfirstseconx00emer` (1883 collection of First and Second Series). (Cited on 1.)

[244] M. Émery and J. E. Yukich, *A simple proof of the logarithmic Sobolev inequality on the circle*, In Séminaire de Probabilités, XXI, pp. 173–175, Lecture Notes in Math., Springer, Berlin, 1987. (Cited on 652.)

[245] P. Erdős and P. Turán, *On interpolation, III. Interpolatory theory of polynomials*, Ann. Math. (2), **41** (1940), 510–553. (Cited on 291, 292.)

[246] A. E. Eremenko, *A Picard type theorem for holomorphic curves*, Periodica Math. Hungarica **38** (1999), 39–42. (Cited on 218.)

[247] A. Eremenko and J. L. Lewis, *Uniform limits of certain A-harmonic functions with applications to quasiregular mappings*, Ann. Acad. Sci. Fenn. A **16** (1991), 361–375. (Cited on 218.)

[248] A. E. Eremenko and M. L. Sodin, *On meromorphic functions of finite order with maximal deficiency sum*, Teor. Funktsiǐ Funktsional. Anal. i Prilozhen. **55** (1991), 84–95; translation in J. Soviet Math. **59** (1992), 643–651. (Cited on 218.)

[249] A. E. Eremenko and M. L. Sodin, *On the distribution of values of meromorphic functions of finite order*, Dokl. Akad. Nauk USSR **316** (1991) 538–541; translation in Soviet Math. Dokl. **43** (1991), 128–131. (Cited on 218.)

[250] A. Eremenko and M. Sodin, *Value distribution of meromorphic functions and meromorphic curves from the point of view of potential theory*, Algebra i Analiz **3** (1991), 131–164; translation in St. Petersburg Math. J. **3** (1991), 109–136. (Cited on 218.)

[251] G. I. Eskin, *The Cauchy problem for hyperbolic convolution equations*, (Russian) Mat. Sb. (N.S.) **74** (1967), 262–297. (Cited on 368.)

[252] G. I. Eskin, *Lectures on Linear Partial Differential Equations*, Graduate Studies in Mathematics, American Mathematical Society Providence, RI, 2011. (Cited on 367.)

[253] M. Essén, *A superharmonic proof of the M. Riesz conjugate function theorem*, Ark. Mat. **22** (1984) 241–249. (Cited on 488.)

[254] G. C. Evans, *On potentials of positive mass. I*, Trans. Amer. Math. Soc. **37** (1935), 226–253. (Cited on 274.)

[255] G. C. Evans, *Application of Poincaré's sweeping-out process*, Proc. Nat. Acad. Sci. USA, **19** (1933), 457–461. (Cited on 274.)

[256] G. Faber, *Über Tschebyscheffsche Polynome*, J. Reine Angew. Math. **150** (1919), 79–106. (Cited on 291.)

[257] E. B. Fabes and D. W. Stroock, *A new proof of Moser's parabolic Harnack inequality via the old ideas of Nash*, Arch. Ration. Mech. Anal. **22** (1986), 327–338. (Cited on 653.)

[258] W. G. Faris, *Invariant cones and uniqueness of the ground state for fermion systems*, J. Math. Phys. **13** (1972), 1285–1290. (Cited on 654.)

[259] W. G. Faris and B. Simon, *Degenerate and non-degenerate ground states for Schrödinger operators*, Duke Math. J. **42** (1975), 559–567. (Cited on 654.)

[260] H. M. Farkas and I. Kra, *Riemann Surfaces*, Graduate Texts in Mathematics, Springer, New York–Berlin, 1980. (Cited on 316.)

[261] P. Fatou, *Séries trigonométriques et séries de Taylor*, Acta Math. **30** (1906), 335–340. (Cited on 59, 457.)

[262] P. Federbush, *A partially alternate derivation of a result of Nelson*, J. Math. Phys. **10** (1969), 50–52. (Cited on 651.)

[263] C. Fefferman, *Inequalities for strongly singular convolution operators*, Acta Math. **124** (1970), 9–36. (Cited on 498, 682, 684.)

[264] C. Fefferman, *The multiplier problem for the ball*, Ann. Math. (2), **94** (1971), 330–336. (Cited on 498, 603, 684, 685.)

[265] C. Fefferman, *Characterizations of bounded mean oscillation*, Bull. Amer. Math. Soc. **77** (1971), 587–588. (Cited on 534.)

[266] C. Fefferman, *Pointwise convergence of Fourier series*, Ann. Math. (2) **98** (1973), 551–571. (Cited on 172.)

[267] C. Fefferman, *A note on spherical summation multipliers*, Israel J. Math. **15** (1973), 44–52. (Cited on 684.)

[268] C. Fefferman, *The uncertainty principle*, Bull. Amer. Math. Soc. **9** (1983), 129–206. (Cited on 336, 669.)

[269] C. Fefferman and D. H. Phong, *The uncertainty principle and sharp Gårding inequalities*, Comm. Pure Appl. Math. **34** (1981) 285–331. (Cited on 336.)

[270] C. Fefferman and E. M. Stein, *Some maximal inequalities*, Amer. J. Math. **93** (1971), 107–115. (Cited on 48.)

[271] C. Fefferman and E. M. Stein, H^p *spaces of several variables*, Acta Math. **129** (1972), 137–193. (Cited on 514, 534.)

[272] H. G. Feichtinger and T. Strohmer, eds., *Gabor Analysis and Algorithms, Theory and Applications*, Applied and Numerical Harmonic Analysis, Birkhäuser Boston, Inc., Boston, MA, 1998. (Cited on 390.)

[273] L. Fejér, *Sur les fonctions bornées et intègrables*, C. R. Acad. Sci. Paris **131** (1900), 984–987. (Cited on 434.)

[274] A. Figalli, F. Maggi, and A. Pratelli, *Sharp stability theorems for the anisotropic Sobolev and log-Sobolev inequalities on functions of bounded variation*, Adv. in Math. **242** (2013), 80–101. (Cited on 654.)

[275] E. Findley, *Universality for locally Szegő measures*, J. Approx. Theory **155** (2008), 136–154. (Cited on 292.)

[276] P. Flandrin, *Time-Frequency/Time-Scale Analysis*, with a preface by Y. Meyer, translated from the French by J. Stöckler, Wavelet Analysis and its Applications, Academic Press, Inc., San Diego, CA, 1999. (Cited on 433.)

[277] G. B. Folland and A. Sitaram, *The uncertainty principle: a mathematical survey*, J. Fourier Anal. Appl. **3** (1997) 207–238. (Cited on 333, 338, 342.)

[278] L. R. Ford, *Automorphic Functions*, 2nd ed., Chelsea, New York, 1951. (Cited on 127.)

[279] R. L. Frank, *A simple proof of Hardy–Lieb–Thirring inequalities*, Comm. Math. Phys. **290** (2009), 789–800. (Cited on 669.)

[280] R. L. Frank, *Cwikel's theorem and the CLR inequality*, J. Spectr. Theory **4**, (2014), 1–21. (Cited on 670.)

[281] R. L. Frank, A. Laptev, and T. Weidl, *Lieb–Thirring Inequalities*, book in preparation. (Cited on 670.)

[282] R. L. Frank and E. H. Lieb, *A new, rearrangement-free proof of the sharp Hardy–Littlewood–Sobolev inequality*, In Spectral Theory, Function Spaces and Inequalities, pp. 55–67, Operator Theory Advances and Applications, Birkhäuser/Springer Basel AG, Basel, 2012. (Cited on 564.)

[283] R. L. Frank, E. Lieb, and R. Seiringer, *Hardy–Lieb–Thirring inequalities for fractional Schrödinger operators*, J. Amer. Math. Soc. **21** (2008), 925–950. (Cited on 669.)

[284] R. L. Frank, E. Lieb, and R. Seiringer, *Equivalence of Sobolev inequalities and Lieb–Thirring inequalities*, In XVIth International Congress on Mathematical Physics, pp. 523–535, World Scientific, Hackensack, NJ, 2010. (Cited on 669.)

[285] R. L. Frank and B. Simon, *Critical Lieb–Thirring bounds in gaps and the generalized Nevai conjecture for finite gap Jacobi matrices*, Duke Math. J. **157** (2011), 461–493. (Cited on 670.)

[286] R. L. Frank, B. Simon, and T. Weidl, *Eigenvalue bounds for perturbations of Schrödinger operators and Jacobi matrices with regular ground states*, Comm. Math. Phys. **282** (2008), 199–208. (Cited on 670.)

[287] M. Fréchet, *Sur divers modes de convergence*, Enseign. Math. **22** (1922), 63. (Cited on 40.)

[288] G. Freud, *Orthogonal Polynomials*, Pergamon Press, Oxford, UK, 1971. (Cited on 292.)

[289] K. O. Friedrichs, *Eine invariante Formulierung des Newtonschen Gravititationsgesetzes und des Grenzüberganges vom Einsteinschen zum Newtonschen Gesetz*, Math. Ann. **98** (1927), 566–575. (Cited on 581.)

[290] K. O. Friedrichs and P. D. Lax, *Boundary value problems for first order operators*, Comm. Pure Appl. Math. **18** (1965), 355–388. (Cited on 367.)

[291] B. Fristedt and L. Gray, *A Modern Approach to Probability Theory*, Probability and its Applications, Birkhäuser, Boston, 1997. (Cited on 162.)

[292] O. Frostman, *Potentiel d'équilibre et capacité des ensembles avec quelques applications à la théorie des fonctions*, Thesis, Lunds Univ. Mat. Sem., **3** (1935), 1-118. (Cited on 274, 276.)

[293] M. Fukushima, *Dirichlet Forms and Markov Processes*, North-Holland Mathematical Library, North-Holland Publishing Co., Amsterdam–New York; Kodansha, Ltd., Tokyo, 1980. (Cited on 177, 276.)

[294] Z. Füredi and P. A. Loeb, *On the best constant for the Besicovitch covering theorem*, Proc. Amer. Math. Soc. **121** (1994), 1063–1073. (Cited on 50.)

[295] H. Furstenberg, *Strict ergodicity and transformation of the torus*, Amer. J. Math. **83** (1961), 573–601. (Cited on 84.)

[296] H. Furstenberg, *Noncommuting random products*, Trans. Amer. Math. Soc. **108** (1963), 377–428. (Cited on 146.)

[297] H. Furstenberg, *Recurrence in Ergodic Theory and Combinatorial Number Theory*, M. B. Porter Lectures, Princeton University Press, Princeton, NJ, 1981. (Cited on 123.)

[298] H. Furstenberg and H. Kesten, *Products of random matrices*, Ann. Math. Stat. **31** (1960), 457—469. (Cited on 145.)

[299] D. Gabor, *Theory of communication*, J. Inst. Elec. Engr. **93** (1946), 429–457. (Cited on 334, 386, 401.)

[300] E. Gagliardo, *Caratterizzazioni delle tracce sulla frontiera relative ad alcune classi di funzioni in n variabili*, Rend. Sem. Mat. Univ. Padova **27** (1957), 284–305. (Cited on 681.)

[301] E. Gagliardo, *Proprietà di alcune classi di funzioni in più variabili*, Ric. Mat. **7** (1958), 102–137. (Cited on 582.)

[302] T. W. Gamelin, *Complex Analysis*, Undergraduate Texts in Mathematics, Springer-Verlag, New York, 2001. (Cited on 316.)

[303] C. Garban and J. Steif, *Noise sensitivity and percolation*, In Probability and Statistical Physics in Two and More Dimensions, pp. 49–154, Clay Math. Proc., American Mathematical Society, Providence, RI, 2012. (Cited on 650.)

[304] D. J. H. Garling, *Inequalities: A Journey into Linear Analysis*, Cambridge University Press, Cambridge, 2007. (Cited on 161, 166, 547, 601, 650.)

[305] J. B. Garnett, *Bounded Analytic Functions*, revised first edition, Graduate Texts in Mathematics, Springer, New York, 2007. (Cited on 439, 534.)

[306] J. B. Garnett and D. E. Marshall, *Harmonic Measure*, New Mathematical Monographs, Cambridge University Press, Cambridge, 2005. (Cited on 274.)

[307] A. M. Garsia, *A simple proof of E. Hopf's maximal ergodic theorem*, J. Math. Mech. **14** (1965), 381–382. (Cited on 83, 91.)

[308] A. M. Garsia, *Topics in Almost Everywhere Convergence*, Lectures in Advanced Mathematics, Markham Publishing, Chicago, IL, 1970. (Cited on 49, 50, 86, 161.)

[309] C. F. Gauss, *Allgemeine Lehrsätze in Beziehung auf die im verkehrten Verhältnisse des Quadrats der Entfernung wirkenden Anziehungs- und Abstossungskräfte*, Resultate aus den Beobachtungen des magnetischen Vereins, 1839. (Cited on 197.)

[310] I. M. Gel'fand, *Expansion in characteristic functions of an equation with periodic coefficients*, (Russian) Dokl. Akad. Nauk. SSSR (N.S.) **73**, (1950), 1117–1120. (Cited on 402.)

[311] R. Gençay, F. Selçuk, and B. Whitcher, *An Introduction to Wavelets and Other Filtering Methods in Finance and Economics*, Academic Press, Inc., San Diego, CA, 2002. (Cited on 433.)

[312] D. Gilbarg and N. S. Trudinger, *Elliptic Partial Differential Equations of Second Order*, reprint of the 1998 edition, Classics in Mathematics, Springer-Verlag, Berlin, 2001. (Cited on 177, 276.)

[313] R. Gilmore, *Geometry of symmetrized states*, Ann. Phys. **74** (1972), 391–463. (Cited on 386.)

[314] J. Ginibre and G. Velo, *On the global Cauchy problem for some nonlinear Schrödinger equations*, Ann. Inst. H. Poincaré Anal. Non Linéaire **1** (1984), 309–323. (Cited on 683.)

[315] E. Glasner and B. Weiss, *On the interplay between measurable and topological dynamics*, In Handbook of Dynamical Systems, pp. 597-648, Elsevier, Amsterdam, 2006. (Cited on 99.)

[316] R. J. Glauber, *Photon correlations*, Phys. Rev. Lett. **10** (1963), 84–86. (Cited on 385.)

[317] R. J. Glauber, *The quantum theory of optical coherence*, Phys. Rev. (2) **130** (1963), 2529–2539. (Cited on 385.)

[318] R. J. Glauber, *Coherent and incoherent states of radiation field*, Phys. Rev. **131** (1963), 2766–2788. (Cited on 386.)

[319] J. Glimm, *Boson fields with nonlinear self-interaction in two dimensions*, Comm. Math. Phys. **8** (1968), 12–25. (Cited on 651, 656.)

[320] J. Glimm and A. Jaffe, *A $\lambda\varphi^4$ quantum field without cutoffs, I*, Phys. Rev. (2) **176** (1968), 1945–1951. (Cited on 651.)

[321] J. Glimm and A. Jaffe, *The $\lambda\varphi^4$ quantum field theory without cutoffs, II. The field operators and the approximate vacuum*, Ann. Math. (2) **91** (1970), 362–401. (Cited on 654.)

[322] J. Glimm and A. Jaffe, *Quantum Physics: A Functional Integral Point of View*, Springer-Verlag, New York–Berlin, 1981. (Cited on 651, 654.)

[323] R. Godement, *Sur les relations d'orthogonalité de V. Bargmann. II. Démonstration générale*, C. R. Acad. Sci. Paris **225** (1947), 657–659. (Cited on 386.)

[324] I. Gohberg and N. Krupnik, *One-Dimensional Linear Singular Integral Equations. I, Introduction*, translated from the 1979 German translation by Bernd Luderer and Steffen Roch and revised by the authors, Operator Theory: Advances and Applications, Birkhäuser Verlag, Basel, 1992. (Cited on 603.)

[325] I. Gohberg and N. Krupnik, *One-Dimensional Linear Singular Integral Equations. II, General theory and applications*, translated from the 1979 German translation by S. Roch and revised by the authors, Operator Theory: Advances and Applications, Birkhäuser Verlag, Basel, 1992. (Cited on 603.)

[326] M. Goldberg and W. Schlag, *A limiting absorption principle for the three-dimensional Schrödinger equation with L^p potentials*, Int. Math. Res. Notices **75** (2004), 4049–4071. (Cited on 683.)

[327] A. Gordon, *The point spectrum of the one-dimensional Schrödinger operator*, Usp. Math. Nauk. **31** (1976), 257–258. (Cited on 296.)

[328] L. Grafakos, *Classical Fourier Analysis*, 2nd edition, Graduate Texts in Mathematics, Springer, New York, 2008. (Cited on 603.)

[329] L. Grafakos, *Modern Fourier Analysis*, 2nd edition, Graduate Texts in Mathematics, Springer, New York, 2009. (Cited on 534, 535, 603, 682, 684.)

[330] B. Green, *Roth's theorem in the primes*, Ann. Math. (2) **161** (2005), 1609–1636. (Cited on 683.)

[331] G. Green, *An Essay on the Application of Mathematical Analysis to the Theories of Electricity and Magnetism*, Nottingham, 1828; available online at `http://arxiv.org/abs/0807.0088`. (Cited on 197.)

[332] A. Greenleaf, *Principal curvature and harmonic analysis*, Indiana Univ. Math. J. **30** (1981), 519–537. (Cited on 682.)

[333] K. Gröchenig, *Aspects of Gabor analysis on locally compact abelian groups*, In Gabor Analysis and Algorithms, pp. 211–231, Appl. Numer. Harmon. Anal., Birkhäuser Boston, Inc., Boston, MA, 1998. (Cited on 403.)

[334] L. Gross, *Existence and uniqueness of physical ground states*, J. Funct. Anal. **10** (1972), 52–109. (Cited on 654.)

[335] L. Gross, *Logarithmic Sobolev inequalities*, Amer. J. Math. **897** (1975), 1061–1083. (Cited on 652, 656.)

[336] L. Gross, *Hypercontractivity and logarithmic Sobolev inequalities for the Clifford Dirichlet form*, Duke Math. J. **42** (1975), no. 3, 383–396. (Cited on 653.)

[337] L. Gross, *Logarithmic Sobolev inequalities on Lie groups*, Illinois J. Math. **36** (1992), 447–490. (Cited on 653.)

[338] L. Gross, *Logarithmic Sobolev inequalities and contractivity properties of semigroups*, In *Dirichelt Forms*, pp. 54–88, Lecture Notes in Math., Springer, Berlin, 1993. (Cited on 650.)

[339] L. Gross, *Hypercontractivity over complex manifolds*, Acta Math. **182** (1999), 159–206. (Cited on 653.)

[340] L. Gross, *Hypercontractivity, logarithmic Sobolev inequalities, and applications: a survey of surveys*, In Diffusion, Quantum theory, and Radically Elementary Mathematics, edited by W. G. Faris, pp. 45–73, Mathematical Notes, Princeton University Press, Princeton, NJ, 2006. (Cited on 650.)

[341] A. Grossmann, G. Loupias, and E. M. Stein, *An algebra of pseudodifferential operators and quantum mechanics in phase space*, Ann. Inst. Fourier Grenoble **18** (1969), 343–368. (Cited on 368.)

[342] A. Grossmann and J. Morlet, *Decomposition of Hardy functions into square integrable wavelets of constant shape*, SIAM J. Math. Anal. **15** (1984), 723–736. (Cited on 387.)

[343] A. Grossmann, J. Morlet, and T. Paul, *Transforms associated to square integrable group representations. I. General results*, J. Math. Phys. **26** (1985), 2473–2479. (Cited on 386.)

[344] J. Guckenheimer and P. Holmes, *Nonlinear Oscillations, Dynamical Systems, and Bifurcations of Vector Fields*, Applied Mathematical Sciences, Springer-Verlag, New York, 1983. (Cited on 99.)

[345] A. Guionnet, and B. Zegarlinski, *Lectures on logarithmic Sobolev inequalities*, In Séminaire de Probabilités, XXXVI, pp. 1–134, Lecture Notes in Math., Springer, Berlin, 2003. (Cited on 622, 650, 654.)

[346] A. Haar, *Zur Theorie der orthogonalen Funktionensysteme*, Mathematische Annalen **69** (1910), 331–371. (English translation from [**374**] available at https://www.uni-hohenheim.de/~gzim/Publications/haar.pdf). (Cited on 434.)

[347] J. Hadamard, *Mémoire sur le problème d'analyse relatif à l'équilibre des plaques élastiques encastrées*, In Œuvres de Jacques Hadamard. Tome II, pp. 515–631, Éditions du Centre National de la Recherche Scientifique, Paris, 1968. Original publication Mém. Sav. étrang. (2) **33**, 1908. (Cited on 582.)

[348] P. R. Halmos, *Lectures on Ergodic Theory*, Chelsea Publishing, New York, 1960. (Cited on 79.)

[349] P. R. Halmos, *Shifts on Hilbert spaces*, J. Reine Angew. Math. **208** (1961) 102–112. (Cited on 517.)

[350] J. M. Hammersley and D. J. A. Welsh, *First-passage percolation, subadditive processes, stochastic networks and generalized renewal theory*, In Proc. Internat. Res. Semin., Berkeley, CA, pp. 61–110, Springer-Verlag, New York, 1965. (Cited on 145.)

[351] Q. Han and F. Lin, *Elliptic Partial Differential Equations*, Courant Lecture Notes in Mathematics, Courant Institute of Mathematical Sciences, New York; American Mathematical Society, Providence, RI, 1997. (Cited on 177.)

[352] G. H. Hardy, *The mean value of the modulus of an analytic function*, Proc. London Math. Soc. (2) **14** (1915), 269–277. (Cited on 213, 444.)

[353] G. H. Hardy, *Notes on some points in the integral calculus, XLI. On the convergence of certain integrals and series*, Messenger of Math. **45** (1915), 163–166. (Cited on 557.)

[354] G. H. Hardy, *Note on a Theorem of Hilbert*, Math. Z. **6** (1920), 314–317. (Cited on 46, 335, 557.)

[355] G. H. Hardy, *Notes on some points in the integral calculus, LX. An inequality between integrals*, Messenger of Math. **54** (1925), 150–156. (Cited on 557, 559.)

[356] G. H. Hardy, *A theorem concerning Fourier transforms*, J. London Math. Soc. **8** (1933), 227– 231. (Cited on 337.)

[357] G. H. Hardy and J. E. Littlewood, *Some problems of diophantine approximation: the series $\Sigma e(\lambda_n)$ and the distribution of the points $(\lambda_n \alpha)$*, Proc. Nat. Acad. Sci. USA **3** (1916), 84–88. (Cited on 98.)

[358] G. H. Hardy and J. E. Littlewood, *Some new properties of Fourier constants*, Math. Ann. **97** (1926), 159–209. (Cited on 458.)

[359] G. H. Hardy and J. E. Littlewood, *Some properties of fractional integrals (1)*, Math. Z. **27** (1928), 565–606. (Cited on 562.)

[360] G. H. Hardy and J. E. Littlewood, *A maximal theorem with function-theoretic applications*, Acta Math. **54** (1930), 81–116. (Cited on 36, 46, 52.)

[361] G. H. Hardy and J. E. Littlewood, *On certain inequalities connected with the calculus of variations*, J. London Math. Soc. **5** (1930), 34–39. (Cited on 562.)

[362] G. H. Hardy and J. E. Littlewood, *Some properties of conjugate funtions*, J. Reine Angew. Math. **167** (1932), 405–423. (Cited on 464.)

[363] G. H. Hardy, J. E. Littlewood, and G. Pólya, *Inequalities*, Cambridge University Press, Cambridge, 1934. (Cited on 36, 488, 557, 564.)

[364] A. Harnack,*Die Grundlagen der Theorie des logarithmischen Potentiales und der eindeutigen Potentialfunktion in der Ebene*, Teubner, Leipzig, 1887. (Cited on 198.)

[365] D. Haroske and H. Triebel, *Distributions, Sobolev Spaces, Elliptic Equations*, EMS Textbooks in Mathematics, European Matthematics Society, Zürich, Switzerland, 2007. (Cited on 583.)

[366] P. Hartman, *On the ergodic theorems*, Amer. J. Math. **69** (1947), 193–199. (Cited on 83.)

[367] P. Hartman, *On completely continuous Hankel matrices*, Proc. Amer. Math. Soc. **9** (1958), 862–866. (Cited on 536.)

[368] F. Hartogs, *Zur Theorie der analytischen Funktion mehrerer unabhängiger Veränderlichen, insbesondere über die Darstellung derselben durch Reihen, welche nach Potenzen einer Veränderlichen fortschreiten*, Math. Ann. **62** (1906), 1–88. (Cited on 213.)

[369] V. Havin and B. Joricke, *The Uncertainty Principle in Harmonic Analysis*, Springer, Berlin, 1994. (Cited on 333.)

[370] W. K. Hayman and P. B. Kennedy, *Subharmonic Functions. Vol. I*, London Mathematical Society Monographs, Academic Press, London–New York, 1976. (Cited on 177, 202, 253.)

[371] G. A. Hedlund, *A metrically transitive group defined by the modular groups*, Amer. J. Math. **57** (1935), 668–678. (Cited on 125.)

[372] C. Heil, *History and evolution of the density theorem for Gabor frames*, J. Fourier Anal. Appl. **13** (2007), 113–166. (Cited on 401, 403.)

[373] C. Heil, *A Basis Theory Primer*, Expanded edition, Applied and Numerical Harmonic Analysis, Birkhäuser/Springer, New York, 2011. (Cited on 401.)

[374] C. Heil and D. F. Walnut, eds., *Fundamental Papers in Wavelet Theory*, Princeton University Press, Princeton, NJ, 2006. (Cited on 434, 708.)

[375] W. Heisenberg, *Über den anschaulichen Inhalt der quantentheoretischen Kinematic und Mechanik*, Z. Phys. **43** (1927), 172–198. (Cited on 333.)

[376] L. Helms, *Potential Theory*, Universitext, Springer-Verlag, London, 2009. (Cited on 177.)

[377] D. Hensley, *Continued Fractions*, World Scientific, Hackensack, NJ, 2006. (Cited on 123, 125.)

[378] D. Herbert and R. Jones, *Localized states in disordered systems*, J. Phys. C **4** (1971), 1145–1161. (Cited on 291.)

[379] I. W. Herbst, *Spectral theory of the operator* $(p^2 + m^2)^{1/2} - Ze^2/r$, Comm. Math. Phys. **53** (1977), 285–294. (Cited on 564.)

[380] G. Herglotz, *Über Potenzreihen mit positivem, reellem Teil im Einheitskreis*, Ber. Ver. Ges. Wiss. Leipzig **63** (1911), 501–511. (Cited on 463.)

[381] E. Hernández and G. Weiss, *A First Course on Wavelets*, with a foreword by Y. Meyer, Studies in Advanced Mathematics, CRC Press, Boca Raton, FL, 1996. (Cited on 433.)

[382] C. Herz, *On the mean inversion of Fourier and Hankel transforms*, Proc. Nat. Acad. Sci. USA **40** (1954), 996–999. (Cited on 684.)

[383] D. Hilbert, *Grundzüge einer allgemeinen Theorie der linearen Integralgleichungen, I–VI*, Nachr. Kön. Ges. Wiss. Göttingen (1904), 49–91; (1904), 213–259; (1905), 307–338; (1906), 157–227; (1906), 439–480; (1910), 355–419. (Cited on 487.)

[384] D. Hilbert, *Zur Theorie der konformen Abbildungen*, Nachr. Kön. Ges. Wiss. Göttingen (1909), 314–323. (Cited on 316.)

[385] I. I. Hirschman, *A note on entropy*, Amer. J. Math. **79** (1957) 152–156. (Cited on 335.)

[386] B. Hollenbeck and I. Verbitsky, *Best constants for the Riesz projection*, J. Funct. Anal. **175** (2000) 370–392. (Cited on 489.)

[387] E. Hopf, *On the time average theorem in dynamics*, Proc. Nat. Acad. Sci. USA **18** (1932), 93–100. (Cited on 82.)

[388] E. Hopf, *Statistik der geodätischen Linien in Mannigfaltigkeiten negativer Krümmung*, Ber. Ver. Ges. wiss. Leipzig **91** (1939), 261–304. (Cited on 125.)

[389] E. Hopf, *The general temporally discrete Markoff process*, J. Ration. Mech. Anal. **3** (1954), 13–45. (Cited on 83, 89, 91.)

[390] E. Hopf, *Ergodic theory and the geodesic flow on surfaces of constant negative curvature*, Bull. Amer. Math. Soc. **77** (1971), 863–877. (Cited on 125.)

[391] L. Hörmander, *Estimates for translation invariant operators in L^p spaces*, Acta Math. **104** (1960), 93–140. (Cited on 603.)

[392] L. Hörmander, *Pseudo-differential operators*, Comm. Pure Appl. Math. **18** (1965), 501–517. (Cited on 367.)

[393] L. Hörmander, *Linear Partial Differential Operators*, Die Grundlehren der mathematischen Wissenschaften, Academic Press, New York; Springer-Verlag, Berlin–Göttingen–Heidelberg, 1963. (Cited on 366.)

[394] L. Hörmander, *Pseudo-differential operators and hypoelliptic equations*, In Singular Integrals (Proc. Sympos. Pure Math., Vol. X, Chicago, Ill., 1966), A. Calderón, ed., pp. 138–183, American Mathematical Society, Providence, RI, 1967. (Cited on 367.)

[395] L. Hörmander, *Hypoelliptic second order differential equations*, Acta Math. **119** (1967), 147–171. (Cited on 370.)

[396] L. Hörmander, *Fourier integral operators. I*, Acta Math. **127** (1971), 79–183. (Cited on 350, 367.)

[397] L. Hörmander, *The Analysis of Linear Partial Differential Operators. I. Distribution Theory and Fourier Analysis, II. Differential Operators with Constant Coefficients*, Grundlehren der Mathematischen Wissenschaften, Springer-Verlag, Berlin, 1983; *III. Pseudodifferential Operators, IV. Fourier Integral Operators*, Grundlehren der Mathematischen Wissenschaften, Springer-Verlag, Berlin, 1985. (Cited on 366, 683.)

[398] L. Hörmander, *Lectures on Nonlinear Hyperbolic Differential Equations*, Mathematics & Applications, Springer-Verlag, Berlin, 1997. (Cited on 613.)

[399] L. Hörmander, *Notions of Convexity*, reprint of the 1994 edition, Modern Birkhäuser Classics, Birkhäuser Boston, Inc., Boston, 2007. (Cited on 213.)

[400] J. Horváth, *Encounters with Mischa Cotlar*, Notices Amer. Math. Soc. **56** (2009), 616–620. (Cited on 614.)

[401] R. Howe, *Quantum mechanics and partial differential equations*, J. Funct. Anal. **38** (1980), 18–254. (Cited on 368, 614.)

[402] R. Howe, *On the role of the Heisenberg group in harmonic analysis*, Bull. Amer. Math. Soc. **3** (1980), 821–843. (Cited on 336.)

[403] R. Howe, *The oscillator semigroup*, In The Mathematical Heritage of Hermann Weyl (Durham, NC, 1987), pp. 61–132, Proc. Sympos. Pure Math., Amererican Mathematical Society, Providence, RI, 1988. (Cited on 368, 614.)

[404] S. V. Hruščev and S. A. Vinogradov, *Free interpolation in the space of uniformly convergent Taylor series*, In Complex Analysis and Spectral Theory (Leningrad, 1979/1980), pp. 171–213, Lecture Notes in Math., Springer, Berlin–New York, 1981. (Cited on 514.)

[405] B. B. Hubbard, *The World According to Wavelets, The Story of a Mathematical Technique in the Making*, Second edition. A K Peters, Ltd., Wellesley, MA, 1998. (Cited on 434.)

[406] D. Hundertmark and B. Simon, *Lieb–Thirring inequalities for Jacobi matrices*, J. Approx. Theory **118** (2002), 106–130. (Cited on 670.)

[407] D. Hundertmark, A. Laptev, and T. Weidl, *New bounds on the Lieb–Thirring constants*, Invent. Math. **140** (2000), 693-Ü704. (Cited on 670.)

[408] D. Hundertmark, E. H. Lieb, and L. E. Thomas, *A sharp bound for an eigenvalue moment of the one-dimensional Schrödinger operator*, Adv. Theor. Math. Phys. **2** (1998), 719–731. (Cited on 669.)

[409] G. A. Hunt, *Markov processes and potentials, I–III*, Illinois J. Math. **1** (1957), 44–93; 316–362; **2** (1958), 151–213. (Cited on 177, 276.)

[410] R. A. Hunt, *An extension of the Marcinkiewicz interpolation theorem to Lorentz spaces*, Bull. Amer. Math. Soc. **70** (1964), 803–807. (Cited on 556.)

[411] R. A. Hunt, *On $L(p,q)$ spaces*, Enseign. Math. (2) **12** (1966), 249–276. (Cited on 556.)

[412] R. A. Hunt, *On the convergence of Fourier series*, In Proc. Orthogonal Expansions and their Continuous Analogues, (Edwardsville, Ill., 1967), pp. 235–255, Southern Illinois University Press, Carbondale, IL, 1968. (Cited on 172.)

[413] K. Husimi, *Some Formal Properties of the Density Matrix*, Proc. Phys. Math. Soc. Japan **22** (1940), 264–314. (Cited on 386.)

[414] I. L. Hwang, *The L^2-boundedness of pseudodifferential operators*, Trans. Amer. Math. Soc. **302** (1987), 55–76. (Cited on 614.)

[415] E. Indrei and D. Marcon, *A quantitative log-Sobolev inequality for a two parameter family of functions*, Int. Math. Res. Notices **20** (2014), 5563–5580. (Cited on 654.)

[416] A. Ionescu and W. Schlag, *Agmon–Kato–Kuroda theorems for a large class of perturbations*, Duke Math. J. **131** (2006), 397–440. (Cited on 683.)

[417] A. Ionescu Tulcea and C. Ionescu Tulcea, *Abstract ergodic theorems*, Trans. Amer. Math. Soc. **107** (1963), 107–124. (Cited on 161.)

[418] A. Iosevich and G. Lu, *Sharpness results and Knapp's homogeneity argument*, Canad. Math. Bull. **43** (2000), , 63–68. (Cited on 682.)

[419] M. Iosifescu and C. Kraaikamp, *Metrical Theory of Continued Fractions*, Mathematics and its Applications, Kluwer, Dordrecht, 2002. (Cited on 123.)

[420] R. Isaac, *A proof of the martingale convergence theorem*, Proc. Amer. Math. Soc. **16** (1965), 842–844. (Cited on 161.)

[421] K. Ishii, *Localization of eigenstates and transport phenomena in the one-dimensional disordered system*, Supp. Prog. Theor. Phys. **53** (1973), 77–138. (Cited on 294.)

[422] S. Izu and J. Lakey, *Time-frequency localization and sampling of multiband signals*, Acta Appl. Math. **107** (2009), 399–435. (Cited on 337, 339.)

[423] S. Janson, *On hypercontractivity for multipliers on orthogonal polynomials*, Ark. Mat. **21** (1983), 97–110. (Cited on 653.)

[424] S. Janson, *On complex hypercontractivity*, J. Funct. Anal. **151** (1997), 270–280. (Cited on 653.)

[425] A. J. E. M. Janssen, *Gabor representation of generalized functions*, J. Math. Anal. Appl. **83** (1981), 377–394. (Cited on 402.)

[426] A. J. E. M. Janssen, *Bargmann transform, Zak transform, and coherent states*, J. Math. Phys. **23** (1982), 720–731. (Cited on 402.)

[427] R. Jentzsch, *Über Integralgleichungen mit positiven Kern*, J. Reine Angew. Math. **141**, (1912), 235–244. (Cited on 654.)

[428] M. Jerison, *Martingale formulation of ergodic theorems*, Proc. Amer. Math. Soc. **10** (1959), 531–539. (Cited on 161.)

[429] B. Jessen, J. Marcinkiewicz, and A. Zygmund, *Note on the differentiability of multiple integrals*, Fund. Math. **25** (1935), 217–234. (Cited on 48.)

[430] S. Jitomirskaya, *Metal-insulator transition for the almost Mathieu operator*, Ann. Math. **150** (1999), 1159–1175. (Cited on 296.)

[431] S. Jitomirskaya, *Ergodic Schrödinger operators (on one foot)*, In Spectral Theory and Mathematical Physics: A Festschrift in Honor of Barry Simon's 60th Birthday, (F. Gesztesy et al., eds.), pp. 613–647, Proc. Symp. Pure Math., American Mathematical Society, Providence, RI, 2007. (Cited on 294.)

[432] S. Jitomirskaya and B. Simon, *Operators with singular continuous spectrum: III. Almost periodic Schrödinger operators*, Comm. Math. Phys. **165** (1994), 201–205. (Cited on 296.)

[433] F. John, *Rotation and strain*, Comm. Pure Appl. Math. **14** (1961), 391–413. (Cited on 534.)

[434] F. John and L. Nirenberg, *On functions of bounded mean oscillation*, Comm. Pure Appl. Math. **14** (1961), 415–426. (Cited on 534.)

[435] R. L. Jones, R. Kaufman, J. M. Rosenblatt, and M. Wierdl, *Oscillation in ergodic theory*, Ergodic Theory Dynam. Systems **18** (1998), 889–935. (Cited on 84.)

[436] M. Kac, *On the notion of recurrence in discrete stochastic processes*, Bull. Amer. Math. Soc. **53** (1947), 1002–1010. (Cited on 85.)

[437] M. Ĭ. Kadec, *The exact value of the Paley–Wiener constant*, Dokl. Akad. Nauk. SSSR **155** (1964), 1253–1254. (Cited on 406.)

[438] J-P. Kahane, *Jacques Hadamard*, Math. Intelligencer **13** (1991), 23–29. (Cited on 437.)

[439] J. Kahn, G. Kalai, and N. Linial, *The influence of variables on Boolean functions*, In 29th Annual Symposium on Foundations of Computer Science, pp. 68–80, IEEE Computer Society Press, Los Alamitos, CA, 1988. (Cited on 652, 654.)

[440] G. Kaiser, *A Friendly Guide to Wavelets*, Reprint of the 1994 edition, Modern Birkhäuser Classics. Birkhäuser/Springer, New York, 2011. (Cited on 433.)

[441] S. Kakeya, *Some problems on maximum and minimum regarding ovals*, Tohoku Science Reports **6** (1917), 71–88. (Cited on 684.)

[442] S. Kalikow and R. McCutcheon, *An Outline of Ergodic Theory*, Cambridge Studies in Advanced Mathematics, Cambridge University Press, Cambridge, 2010. (Cited on 79, 97.)

[443] S. Kalikow and B. Weiss, *Fluctuations of ergodic averages* Illinois J. Math. **43** (1999), 480–488. (Cited on 84.)

[444] T. Kamae, *A simple proof of the ergodic theorem using nonstandard analysis*, Israel J. Math. **42** (1982), 284–290. (Cited on 145.)

[445] J. Karamata, *Über die Hardy–Littlewoodschen Umkehrungen des Abelschen Stetigkeitssatzes*, Math. Z. **32** (1930), 319–320. (Cited on 689.)

[446] I. Karatzas and S. Shreve, *Brownian Motion and Stochastic Calculus*, 2nd edition, Graduate Texts in Mathematics, Springer-Verlag, New York, 1991. (Cited on 161.)

[447] T. Kato, *Boundedness of some pseudo-differential operators*, Osaka J. Math. **13** (1976), 1–9. (Cited on 614.)

[448] T. Kato, *Perturbation Theory for Linear Operators*, 2nd edition, Grundlehren der Mathematischen Wissenschaften, Springer-Verlag, Berlin–New York, 1976. (Cited on 337.)

[449] A. Katok and B. Hasselblatt, *Introduction to the Modern Theory of Dynamical Systems*, Encyclopedia of Mathematics and its Applications, Cambridge University Press, Cambridge, 1995. (Cited on 84.)

[450] S. Katok, *Fuchsian Groups*, University of Chicago Press, Chicago, 1992. (Cited on 127.)

[451] N. H. Katz and T. Tao, *Some connections between Falconer's distance set conjecture and sets of Furstenburg type*, New York J. Math. **7** (2001), 149–187. (Cited on 685.)

[452] N. H. Katz and T. Tao, *New bounds for Kakeya problems*, J. Anal. Math. **87** (2002), 231–263. (Cited on 685.)

[453] Y. Katznelson, *An Introduction to Harmonic Analysis*, 3rd edition, Cambridge Mathematical Library, Cambridge University Press, Cambridge, 2004. (Cited on 439.)

[454] Y. Katznelson and B. Weiss, *A simple proof of some ergodic theorems*, Israel J. Math. **42** (1982), 291–296. (Cited on 145.)

[455] B. Kawohl, *Rearrangements and Convexity of Level Sets in PDE*, Lecture Notes in Mathematics, Springer-Verlag, Berlin, 1985. (Cited on 36.)

[456] M. Keane, *A continued fraction titbit*, Fractals **3** (1995), 641–650. (Cited on 123, 124.)

[457] M. Keane and K. Petersen, *Easy and nearly simultaneous proofs of the ergodic theorem and maximal ergodic theorem*, In Dynamics & Stochastics, pp. 248–251, IMS Lecture Notes Monogr. Ser., Institute for Mathematical Statistics, Beachwood, OH, 2006. (Cited on 83, 145.)

[458] M. Keel and T. Tao, *Endpoint Strichartz estimates*, Amer. J. Math. **120** (1998), 955–980. (Cited on 683.)

[459] W. Keller, *Wavelets in Geodesy and Geodynamics*, Walter de Gruyter & Co., Berlin, 2004. (Cited on 433.)

[460] O. D. Kellogg, *Foundations of Potential theory*, reprint of the 1st edition 1929, Die Grundlehren der Mathematischen Wissenschaften, Springer-Verlag, Berlin-New York, 1967. (Cited on 177, 274.)

[461] T. Kemp, *Hypercontractivity in non-commutative holomorphic spaces*, Comm. Math. Phys. **259** (2005), 615–637. (Cited on 653.)

[462] E. H. Kennard, *Zur Quantenmechanik einfacher Bewegungstypen*, Z. Phys. **44** (1927), 326–352. (Cited on 334.)

[463] S. Kesavan, *Symmetrization & Applications*, Series in Analysis, World Scientific Publishing Co., Hackensack, NJ, 2006. (Cited on 36.)

[464] A. Khinchin, *Zu Birkhoffs Lösung des Ergodenproblems*, Math. Ann. **107** (1933), 485–488. (Cited on 83.)

[465] A. Khinchin, *Eine Verschärfung des Poincaréschen "Wiederkehrsatzes"*, Compositio Math. **1** (1934), 177–179. (Cited on 90.)

[466] A. Khinchin, *Fourierkoeffizienten langs einer Bahn im Phasenraum*, Rec. Math., Moscow **41** (1934), 14–15. (Cited on 83.)

[467] A. Khinchin, *Metrische Kettenbruchprobleme*, Compos. Math. **1** (1935), 361–382. (Cited on 124.)

[468] A. Khinchin, *Continued Fractions*, reprint of the 1964 translation, Dover Publications, Mineola, NY, 1997; first Russian edition 1936. (Cited on 123, 124.)

[469] J. L. King, *Three problems in search of a measure*, Amer. Math. Monthly **101** (1994), 609–628. (Cited on 99.)

[470] J. F. C. Kingman, *The ergodic theory of subadditive stochastic processes*, J. R. Stat. Soc. Ser. B. Stat. Methodol. **30** (1968), 499–510. (Cited on 145.)

[471] A. Kiselev, *An interpolation theorem related to the a.e. convergence of integral operators*, Proc. Amer. Math. Soc. **127** (1999), 1781–1788. (Cited on 172.)

[472] J. R. Klauder, *The action option and a Feynman quantization of spinor fields in terms of ordinary c-numbers*, Ann. Phys. **11** (1960), 123–168. (Cited on 385.)

[473] J. R. Klauder, *Continuous-representation theory. I. Postulates of continuous-representation theory*, J. Math. Phys. **4** (1963), 1055–1058. (Cited on 385.)

[474] J. R. Klauder, *Continuous-representation theory. II. Generalized relation between quantum and classical dynamics*, J. Math. Phys. **4** (1963), 1058–1073. (Cited on 385.)

[475] A. W. Knapp and E. M. Stein, *Intertwining operators for semisimple groups*, Ann. Math. (2) **93** (1971), 489–578. (Cited on 613.)

[476] K. Knopp, *Mengentheoretische Behandlung einiger Probleme der diophantischen Approximationen und der transfiniten Wahrscheinlichkeiten*, Math. Ann. **95** (1926), 409–426. (Cited on 125, 131.)

[477] P. Koebe, *Herleitung der partiellen Differentialgleichung der Potentialfunktion aus deren Integraleigenschaft*, Sitzungsber. Berl. Math. Ges. **5** (1906), 39–42. (Cited on 197.)

[478] P. Koebe, *Über die Uniformisierung beliebiger analytischer Kurven, Erste Mitteilung, Zweite Mitteilung*, Nachr. Königl. Ges. Wiss. Göttingen, Math.-phys. Klasse (1907), 191–210; 633–669. (Cited on 316.)

[479] J. J. Kohn and L. Nirenberg, *An algebra of pseudo-differential operators*, Comm. Pure Appl. Math. **18** (1965), 269–305. (Cited on 367.)

[480] A. Kolmogoroff *Sur les fonctions harmoniques conjuguées et les séries de Fourier*, Fund. Math. **7** (1925), 24–29. (*Note*: His name is usually translated Kolmogorov.) (Cited on 35, 463, 488.)

[481] A. Kolmogoroff, *Über die Summen durch den Zufall bestimmter unabhängiger Grössen*, Math. Ann. **99** (1928), 309–319; Math. Ann. **102** (1929), 484–488. (Cited on 162.)

[482] A. Kolmogorov, *Sur la loi forte des grands nombres*, C. R. Acad. Sci. Paris **191** (1930), 910–912. (Cited on 167.)

[483] A. Kolmogorov, *Grundbegriffe der Wahrscheinlichkeitsrechnung*, Springer, Berlin, 1933. (Cited on 79, 167.)

[484] V. I. Kondrachov, *Sur certaines propriétés fonctions dans l'espace L^p*, C. R. (Doklady) Acad. Sci. URSS (N.S.) **48** (1945), 535–538. (Cited on 582.)

[485] B. O. Koopman, *Hamiltonian systems and Hilbert space*, Proc. Nat. Acad. Sci. USA **17** (1931), 315–318. (Cited on 80.)

[486] P. Koosis, *Introduction to H^p Spaces*, 2nd edition, Cambridge Tracts in Mathematics, Cambridge University Press, Cambridge, 1998. (Cited on 439, 513, 534.)

[487] S. Kotani, *Ljapunov indices determine absolutely continuous spectra of stationary random one-dimensional Schrödinger operators*, In Stochastic Analysis (Katata/Kyoto, 1982), pp. 225-247, North-Holland Math. Library, North-Holland, Amsterdam, 1984. (Cited on 296.)

[488] S. G. Krantz, *Mathematical anecdotes*, Math. Intelligencer **12** (1990), 32–38. (Cited on 47.)

[489] M. Krasnosel'skiĭ and Ya. Rutickiĭ, *Convex Functions and Orlicz Spaces*, P. Noordhoff, Groningen, 1961. (Cited on 36.)

[490] S. G. Krein and E. M. Semenov, *A scale of spaces*, (Russian) Dokl. Akad. Nauk. SSSR **138** (1961), 763–766. (Cited on 556.)

[491] U. Krengel, *Ergodic Theorems*, de Gruyter Studies in Mathematics, Walter de Gruyter, Berlin, 1985. (Cited on 79, 145.)

[492] L. Kronecker, *Näherungsweise ganzzahlige Auflösung linearer Gleichungen*, Berl. Ber. (1884), 1179–1193. (Cited on 98.)

[493] A. Kufner, L. Maligranda, and L.-E. Persson, *The prehistory of the Hardy inequality*, Amer. Math. Monthly **113** (2006), 715–732. (Cited on 557.)

[494] A. Kufner, L. Maligranda, and L.-E. Persson, *The Hardy Inequality. About Its History and Some Related Results*, Vydavatelský Servis, Plzen, 2007. (Cited on 336.)

[495] A. Kufner and L.-E. Persson, *Weighted Inequalities of Hardy Type*, World Scientific Publishing Co., Inc., River Edge, NJ, 2003. (Cited on 557.)

[496] L. Kuipers and H. Niederreiter, *Uniform Distribution of Sequences*, Pure and Applied Mathematics, Wiley-Interscience, New York–London–Sydney, 1974. (Cited on 123.)

[497] H. Kumano-go, *Pseudodifferential Operators*, Translated from the Japanese by the author, Rémi Vaillancourt, and Michihiro Nagase, MIT Press, Cambridge, MA–London, 1981. (Cited on 367.)

[498] K. Kuttler, *Modern Analysis*, CRC Press, Boca Raton, FL, 1998. (Cited on 50.)

[499] R. Kuzmin, *Sur un problème de Gauss*, C. R. (Doklady) Acad. Sci. URSS (N.S.) **1928** (1928), 375–380. (Cited on 124.)

[500] I. Łaba, *From harmonic analysis to arithmetic combinatorics*, Bull. Amer. Math. Soc. (N.S.) **45** (2008), 77–115. (Cited on 685.)

[501] M. Lacey and C. Thiele, *A proof of boundedness of the Carleson operator*, Math. Res. Lett. **7** (2000), 361–370. (Cited on 172.)

[502] E. Landau, *A note on a theorem concerning series of positive terms: Extract from a letter of Prof. E. Landau to Prof. I. Schur*, J. London Math. Soc. **1** (1926), 38–39; based on letters from Landau to Hardy dated 6/21/1921 and to Schur dated 6/22/1921. (Cited on 557.)

[503] H. J. Landau, *On the density of phase-space expansions*, IEEE Trans. Inf. Theory **39** (1993) 1152–1156. (Cited on 338.)

[504] H. J. Landau and H. O. Pollak, *Prolate spheroidal wave functions, Fourier analysis and uncertainty II*, Bell Syst. Tech. J. **40** (1961) 65–84. (Cited on 337, 338.)

[505] H. J. Landau and H. O. Pollak, *Prolate spheroidal wave functions, Fourier analysis and uncertainty III: the dimension of the space of essentially time- and band-limited signals*, Bell Syst. Tech. J. **41** (1962) 1295–1336. (Cited on 338.)

[506] H. J. Landau and H. Widom, *Eigenvalue distribution of time and frequency limiting*, J. Math. Anal. Appl. **77** (1980) 469–481. (Cited on 338.)

[507] N. S. Landkof, *Foundations of Modern Potential Theory*, Die Grundlehren der mathematischen Wissenschaften, Springer-Verlag, New York–Heidelberg, 1972. (Cited on 177, 276.)

[508] S. Laplace, *Théorie des attractions des sphéroïdes et de la figure des planètes*, Mem. Acad. Sci. Paris **10** (1785) 113–196. (Cited on 249.)

[509] A. Laptev and M. Solomyak, *On the negative spectrum of the two-dimensional Schrödinger operator with radial potential.* Comm. Math. Phys. **314:1** (2012), 229–241. (Cited on 340.)

[510] A. Laptev and T. Weidl, *Sharp Lieb–Thirring inequalities in high dimensions*, Acta Math. **184** (2000), 87–111. (Cited on 669.)

[511] Y. Last and B. Simon, *Eigenfunctions, transfer matrices, and absolutely continuous spectrum of one-dimensional Schrödinger operators*, Invent. Math. **135** (1999), 329–367. (Cited on 292.)

[512] Y. Last and B. Simon, *Fine structure of the zeros of orthogonal polynomials, IV. A priori bounds and clock behavior*, Comm. Pure Appl. Math. **61** (2008), 486–538. (Cited on 292.)

[513] H. Lebesgue, *Leçons sur l'Intégration et la Recherche des Fonctions Primitives*, Gauthier–Villars, Paris, 1904; available online at http://www.archive.org/details/LeconsSurLintegration. (Cited on 59.)

[514] H. Lebesgue, *Sur l'intégration des fonctions discontinues*, Ann. Sci. Ecole Norm. Sup. (3) **27** (1910), 361–450. (Cited on 59.)

[515] H. Lebesgue, *Sur des cas d'impossibilité du problème de Dirichlet ordinaire*, C.R. Séances Soc. Math. France **41** (1913), 17–17. (Cited on 207, 231.)

[516] H. Lebesgue, *Conditions de régularité, conditions d'irrégularité, conditions d'impossibilité dans le problème de Dirichlet*, C. R. Acad. Sci. Paris **178** (1924), 349–354. (Cited on 231.)

[517] S. Lee and A. Vargas, *Restriction estimates for some surfaces with vanishing curvatures*, J. Funct. Anal. **258** (2010), 2884–2909. (Cited on 682.)

[518] A. M. Legendre, *Recherches sur l'attraction des sphéroïdes homogènes*, Mem. Acad. Sci. Paris **10** (1785), 411–435 [Note: Legendre submitted his findings to the Academy in 1782, but they were published in 1785.]; available on-line (in French) at `http://edocs.ub.uni-frankfurt.de/volltexte/2007/3757/pdf/A009566090.pdf`. (Cited on 249.)

[519] P. G. Lemarié, *Ondelettes à localisation exponentielle*, J. Math. Pures Appl. (9) **67** (1988), 227–236. (Cited on 434.)

[520] A. Lenard, *The numerical range of a pair of projections*, J. Funct. Anal. **10** (1972), 410–423. (Cited on 344.)

[521] G. Leoni, *A First Course in Sobolev Spaces*, Graduate Studies in Mathematics, American Mathematical Society, Providence, RI, 2009. (Cited on 583.)

[522] D. Levin, and M. Solomyak, *The Rozenblum–Lieb–Cwikel inequality for Markov generators*, J. Anal. Math. **71** (1997), 173–193. (Cited on 669.)

[523] E. Levin and D. Lubinsky, *Applications of universality limits to zeros and reproducing kernels of orthogonal polynomials*, J. Approx. Theory **150** (2008), 69–95. (Cited on 292.)

[524] P. Lévy, *Sur la loi de probabilité dont dépendent les quotients complets at incomplets d'une fraction continue*, Bull. Soc. Math. France **55** (1929), 867–870. (Cited on 124.)

[525] P. Lévy, *Sur le développement en fraction continue d'un nombre choisi au hasard*, Compositio Math. **3** (1936), 286–303. (Cited on 124.)

[526] P. Lévy, *Theorie de l'addition des variables aleatoires*, Gauthier–Villars, Paris, 1937. (Cited on 162.)

[527] J. L. Lewis, *Picard's theorem and Rickman's theorem by way of Harnack's inequality*, Proc. Amer. Math. Soc. **122** (1994), 199–206. (Cited on 218.)

[528] P. Li and S. T. Yau, *On the Schrödinger equation and the eigenvalue problem*, Comm. Math. Phys. **88** (1983),309–318. (Cited on 669.)

[529] E. H. Lieb, *The classical limit of quantum spin systems*, Comm. Math. Phys. **31** (1973), 327–340. (Cited on 386.)

[530] E. H. Lieb, *Bounds on the eigenvalues of the Laplace and Schrödinger operators*, Bull. Amer. Math. Soc. **82** (1976), 751–753. (Cited on 669.)

[531] E. H. Lieb, *The number of bound states of one-body Schrödinger operators and the Weyl problem. Geometry of the Laplace operator*, In Proc. Sympos. Pure Math., XXXVI, pp. 241–252, American Mathematicl Society, Providence, RI, 1980. (Cited on 669.)

[532] E. H. Lieb, *Sharp constants in the Hardy–Littlewood–Sobolev and related inequalities*, Ann. Math. (2) **118** (1983), 349–374. (Cited on 564.)

[533] E. H. Lieb and M. Loss, *Analysis*, Second edition, Graduate Studies in Mathematics, American Mathematical Society, Providence, RI, 2001. (Cited on 36, 275, 564.)

[534] E. H. Lieb and W. Thirring, *Bound for the kinetic energy of fermions which proves the stability of matter*, Phys. Rev. Lett. **35** (1975), 687–689; Errata **35** (1975), 1116. (Cited on 669.)

[535] E. H. Lieb and W. Thirring, *Inequalities for the moments of the eigenvalues of the Schrödinger Hamiltonian and their relation to Sobolev inequalities*, In Studies in Mathematical Physics: Essays in Honor of Valentine Bargmann (E. Lieb, B. Simon, and A. Wightman, eds.), pp. 269–303, Princeton University Press, Princeton, NJ, 1976. (Cited on 669.)

[536] T. M. Liggett, *An improved subadditive ergodic theorem*, Ann. Probab. **13** (1985), 1279–1285. (Cited on 145.)

[537] T. M. Liggett, *Continuous Time Markov Processes. An Introduction*, Graduate Studies in Mathematics, American Mathematical Society, Providence, RI, 2010. (Cited on 161.)

[538] D. Lindley, *Degrees Kelvin: A Tale of Genius, Invention, and Tragedy*, Joseph Henry Press, Washington, D.C., 2004. (Cited on 250.)

[539] J.-L. Lions and J. Peetre, *Sur une classe d'espaces d'interpolation*, Publ. Math. Inst. Hautes Etudes Sci. **19** (1964), 5–68. (Cited on 556.)

[540] J. E. Littlewood, *On inequalities in the theory of functions*, Proc. London Math. Soc. (2) **23** (1925), 481–519. (Cited on 464.)

[541] J. E. Littlewood, Mathematical notes (6): On the definition of a subharmonic function, J. London Math. Soc. **s1-2** (1927), 189–192. (Cited on 213.)

[542] J. E. Littlewood and R. E. A. C. Paley, *Theorems on Fourier Series and Power Series, I, II, III*, J. London Math. Soc. **6**, (1931), 230–233; Proc. London Math. Soc. **42**, (1937), 52–89; Proc. London Math. Soc. **43**, (1938), 105–126. (Cited on 603.)

[543] P. I. Lizorkin, *Properties of functions in the spaces* $\Delta_{p,\theta}^r$ (Russian), Studies in the Theory of Differentiable Functions of Several Variables and Its Applications, V., Trudy Mat. Inst. Steklov **131** (1974), 158–181. (Cited on 583.)

[544] L. H. Loomis, *A note on the Hilbert transform*, Bull. Amer. Math. Soc. **52** (1946), 1082–1086. (Cited on 513.)

[545] F. López Safont, *Introduction to Lorentz Spaces*, unpublished, available online at `http://hdl.handle.net/2445/32389`, 2012. (Cited on 556.)

[546] G. G. Lorentz, *Some new functional spaces*, Ann. Math. (2) **51** (1950), 37–55. (Cited on 36, 37, 556.)

[547] G. G. Lorentz, *On the theory of spaces*, Pacific J. Math. **1** (1951), 411–429. (Cited on 36, 37.)

[548] F. E. Low, *Complete sets of wave packets*, In A Passion for Physics—Essays in Honor of Geoffrey Chew, C. DeTar et al. (eds.), pp. 17–22, World Scientific, Singapore, 1985. (Cited on 402.)

[549] D. S. Lubinsky, *Universality limits in the bulk for arbitrary measures on compact sets*, J. Anal. Math. **106** (2008), 373–394. (Cited on 292.)

[550] D. S. Lubinsky, *A new approach to universality limits involving orthogonal polynomials*, Ann. Math. **170** (2009), 915–939. (Cited on 292.)

[551] Y. I. Lyubarskii, *Frames in the Bargmann space of entire functions. Entire and subharmonic functions*, In Adv. Soviet Math., pp. 167–180, American Mathematical Society, Providence, RI, 1992. (Cited on 401.)

[552] T. M. MacRobert, *Spherical Harmonics. An Elementary Treatise on Harmonic Functions with Applications*, 2nd edition, Methuen & Co., London, 1947. (Cited on 177.)

[553] N. G. Makarov, *On the distortion of boundary sets under conformal maps*, Proc. London Math. Soc. (3) **52** (1985), 369–384. (Cited on 274.)

[554] N. G. Makarov, *Fine structure of harmonic measure* (Russian), Algebra i Analiz **10** (1998), 1–62; translation in St. Petersburg Math. J. **10** (1999), 217–268. (Cited on 274.)

[555] S. Mallat, *Multiresolution approximations and wavelet orthonormal bases of $L^2(\mathbb{R})$*, Trans. Amer. Math. Soc. **315** (1989), 69–87. (Cited on 434.)

[556] S. Mallat, *A Wavelet Tour of Signal Processing*, Academic Press, Inc., San Diego, CA, 1998. (Cited on 433.)

[557] R. Mansuy, *The origins of the word "martingale"*, J. Electron. Hist. Probab. Stat. **5** (2009), 1–10. (Cited on 160.)

[558] J. Marcinkiewicz, *Sur les multiplicateurs des séries de Fourier*, Studia Math. **8** (1939), 78–91. (Cited on 603.)

[559] J. Marcinkiewicz, *Sur l'interpolation d'operateurs*, C. R. Acad. Sci. Paris **208** (1939), 1272–1273. (Cited on 556.)

[560] R. S. Martin, *Minimal positive harmonic functions*, Trans. Amer. Math. Soc. **49** (1941), 137–172. (Cited on 276.)

[561] V. P. Maslov, *Perturbation Theory and Asymptotic Methods*, M. G. U., Moscow, 1965. (Russian) (Cited on 368.)

[562] V. Maz'ya, *Sobolev Spaces with Applications to Elliptic Partial Differential Equations*, Second, revised and augmented edition, Grundlehren der Mathematischen Wissenschaften, Springer, Heidelberg, 2011. (Cited on 583.)

[563] A. D. Melas, *The best constant for the centered Hardy–Littlewood maximal inequality*, Ann. Math. (2) **157** (2003), 647–688. (Cited on 49.)

[564] D. Menshov, *Sur les series de fonctions orthogonales*, Fund. Math. **10** (1927), 375–420. (Cited on 172.)

[565] P.-A. Meyer, *A decomposition theorem for supermartingales*, Illinois J. Math. **6** (1962), 193–205. (Cited on 161.)

[566] P.-A. Meyer, *Brelot's axiomatic theory of the Dirichlet problem and Hunt's theory*, Ann. Inst. Fourier Grenoble **13** (1963), 357–372. (Cited on 177.)

[567] P.-A. Meyer, *Probability and Potentials*, Blaisdell Publishing, Waltham, MA, 1966. (Cited on 161, 177, 276.)

[568] Y. Meyer, *Principe d'incertitude, bases hilbertiennes et algèbres d'opérateurs*, In Séminaire Bourbaki, Vol. 1985/86, Astérisque **145–146** (1987), 209–223. (Cited on 434.)

[569] Y. Meyer, *Ondelettes et opérateurs, I.* (French) Actualités Mathématiques, Hermann, Paris, 1990. (Cited on 433.)

[570] Y. Meyer, *Ondelettes et opérateurs, II.* (French) Actualités Mathématiques, Hermann, Paris, 1990. (Cited on 433.)

[571] Y. Meyer, *Wavelets and Operators*, translated from the 1990 French original by D. H. Salinger, Cambridge Studies in Advanced Mathematics, Cambridge University Press, Cambridge, 1992. (Cited on 433.)

[572] Y. Meyer, *Wavelets, Vibrations and Scalings*, with a preface in French by the author, CRM Monograph Series, American Mathematical Society, Providence, RI, 1998. (Cited on 433.)

[573] Y. Meyer and R. R. Coifman, *Ondelettes et opérateurs, III, Opérateurs multilinéaires*, Actualités Mathématiques, Hermann, Paris, 1991. (Cited on 433.)

[574] Y. Meyer and R. R. Coifman, *Wavelets, Calderón–Zygmund and Multilinear Operators*, translated from the 1990 and 1991 French originals by D. H. Salinger, Cambridge Studies in Advanced Mathematics, Cambridge University Press, Cambridge, 1997. (Cited on 433.)

[575] S. G. Michlin, *On the multipliers of Fourier integrals* (Russian), Dokl. Akad. Nauk. SSSR (N.S.) **109** (1956), 701–703. (Cited on 603.)

[576] S. G. Michlin and S. Prössdorf, *Singuläre Integraloperatoren*, Mathematische Lehrbücher und Monographien, II. Abteilung: Mathematische Monographien, Akademie-Verlag, Berlin, 1980. (Cited on 603.)

[577] Y. Mizuta, *Potential Theory in Euclidean Spaces*, Mathematical Sciences and Applications, Gakkōtosho, Tokyo, 1996. (Cited on 177.)

[578] G. Mockenhaupt, *Salem sets and restriction properties of Fourier transforms*, Geom. Funct. Anal. **10** (2000), 1579–1587. (Cited on 683.)

[579] G. Mockenhaupt, A. Seeger, and C. D. Sogge, *Wave front sets, local smoothing and Bourgain's circular maximal theorem*, Ann. Math. (2) **136** (1992), 207–218. (Cited on 49.)

[580] G. Mockenhaupt, A. Seeger, and C. D. Sogge, *Local smoothing of Fourier integral operators and Carleson-Sjölin estimates*, J. Amer. Math. Soc. **6** (1993), 65–130. (Cited on 684.)

[581] A. Montanaro, *Some applications of hypercontractive inequalities in quantum information theory*, J. Math. Phys. **53** (2012),122–206. (Cited on 654.)

[582] H. L. Montgomery, *Ten Lectures on the Interface Between Analytic Number Theory and Harmonic Analysis*, CBMS Regional Conference Series in Mathematics, American Mathematical Society, Providence, RI, 1994. (Cited on 123, 129.)

[583] S. Montiel, and A. Ros, *Curves and Surfaces*, Second edition, translated from the 1998 Spanish original by Montiel and edited by Donald Babbitt, Graduate Studies in Mathematics, American Mathematical Society, Providence, RI; Real Sociedad Matemática Española, Madrid, 2009. (Cited on 17.)

[584] C. S. Morawetz and W. A. Strauss, *Decay and scattering of solutions of a nonlinear relativistic wave equation*, Comm. Pure Appl. Math. **25** (1972), 1–31. (Cited on 682.)

[585] G .W. Morgan, *A note on Fourier transforms*, J. London Math. Soc. **9** (1934) 187–192. (Cited on 337.)

[586] J. Morgan and G. Tian, *Ricci Flow and the Poincaré Conjecture*, Clay Mathematics Monographs, American Mathematical Society, Providence, RI; Clay Mathematics Institute, Cambridge, MA, 2007. (Cited on 654.)

[587] Ch. B. Morrey, Jr., *Functions of several variables and absolute continuity, II*. Duke Math. J. **6** (1940), 187–215. (Cited on 581.)

[588] A. P. Morse, *Perfect blankets*, Trans. Amer. Math. Soc. **61** (1947), 418–442. (Cited on 50.)

[589] J. Moser, *On Harnack's theorem for elliptic differential equations*, Comm. Pure Appl. Math. **14** (1961), 577–591. (Cited on 653.)

[590] J. E. Moyal, *Quantum mechanics as a statistical theory*, Proc. Cambridge Philos. Soc. **45** (1949), 99–124. (Cited on 370, 386.)

[591] R. F. Muirhead, *Some methods applicable to identities and inequalities of symmetric algebraic functions of n letters*, Proc. Edinburgh Math. Soc. **21** (1902), 144–162. (Cited on 36.)

[592] C. Muscalu and W. Schlag, *Classical and Multilinear Harmonic Analysis. Vol. I*, Cambridge Studies in Advanced Mathematics, Cambridge University Press, Cambridge, 2013. (Cited on 682.)

[593] N. I. Muskhelishvili, *Singular Integral Equations*, translated from the original 1946 Russian publication, Springer, New York, 1977. (Cited on 603.)

[594] M. G. Nadkarni, *Basic Ergodic Theory*, 2nd edition, Birkhäuser Advanced Texts, Birkhäuser Verlag, Basel, 1998. (Cited on 79.)

[595] A-H. Najmi, *Wavelets, A Concise Guide*, Johns Hopkins University Press, Baltimore, MD, 2012. (Cited on 433.)

[596] J. Nash, *Continuity of solutions of parabolic and elliptic equations*, Amer. J. Math. **80** (1958), 931–954. (Cited on 582, 653.)

[597] G. P. Nason, *Wavelet Methods in Statistics with R*, U. Springer, New York, 2008. (Cited on 433.)

[598] J. Naumann, *Notes on the prehistory of Sobolev spaces*, unpublished 2002 preprint of Humbolt University available at `edoc.hu-berlin.de/series/mathematik-preprints/2002-2/PDF/2.pdf`; remarkably, while the English language original seems to be unpublished, there is a published Portuguese translation: Bol. Soc. Math. Stud. **63** (2010), 13–55. (Cited on 581.)

[599] F .L. Nazarov, *Local estimates for exponential polynomials and their applications to inequalities of the uncertainty principle type*, Algebra i Analiz **5** (1993) 3–66. (Cited on 337.)

[600] Z. Nehari, *On bounded bilinear forms*, Ann. Math. (2) **65** (1957), 153–162. (Cited on 535.)

[601] E. Nelson, *A proof of Liouville's theorem*, Proc. Amer. Math. Soc. **12**, (1961), 995. (Cited on 197.)

[602] E. Nelson, *A quartic interaction in two dimensions*, In Mathematical Theory of Elementary Particles, edited by R. Goodman and I. Segal, pp. 69–73, M.I.T. Press. Cambridge, MA, 1966. (Cited on 651.)

[603] E. Nelson, *The free Markoff field*, J. Funct. Anal. **12** (1973), 211–227. (Cited on 652.)

[604] I. Netuka and J. Veselý, *Mean value property and harmonic functions*, In Classical and Modern Potential Theory and Applications (Chateau de Bonas, 1993), pp. 359–398, NATO Adv. Sci. Inst. Ser. C Math. Phys. Sci., Kluwer Acad. Publ., Dordrecht, 1994. (Cited on 197.)

[605] C. Neumann, *Untersuchungen über das Logarithmische und Newton'sche potential*, B. G. Teubner, Leipzig, Germany, 1877; available online at `http://gallica.bnf.fr/ark:/12148/bpt6k996161`. (Cited on 275.)

[606] F. Nevanlinna and R. Nevanlinna, *Über die Eigenschaften analytischer Funktionen in der Umgebung einer singulären Stelle oder Linie*, Acta Soc. Fennicae **50**, 46 S., 1922. (Cited on 444, 457.)

[607] R. Nevanlinna, *Über beschränkte analytische Funktionen die in gegebene Punkten vorgeschriebene Werte annehmen*, Ann. Acad. Sci. Fenn. A **13** (1919), 1–71. (Cited on 513.)

[608] R. Nevanlinna, *Das harmonische Mass von Punktmengen und seine Anwendung in der Funktionentheorie*, In C. R.Škand. Mat.-Kongress, Stockholm, pp. 116–133, Håkon Ohlssons Boktryckeri, Lund, 1935. (Cited on 197.)

[609] J. Neveu, *Discrete-Parameter Martingales*, translated from the French by T. P. Speed, Revised edition, North-Holland Mathematical Library, North-Holland Publishing Co., Amsterdam–Oxford; American Elsevier Publishing Co., Inc., New York, 1975. (Cited on 161.)

[610] S. Newcomb, *Note on the frequency of use of the different digits in natural numbers*, Amer. J. Math. **4** (1881), 39–40. (Cited on 100.)

[611] Y. Nievergelt, *Wavelets Made Easy*, 2nd corrected printing of the 1999 original, Modern Birkhäuser Classics, Birkháuser/Springer, New York, 2013. (Cited on 433.)

[612] M. J. Nigrini *I've Got Your Number*, Journal of Accountancy, 1999; available online at `http://www.journalofaccountancy.com/Issues/1999/May/nigrini` (Cited on 100.)

[613] O. Nikodym, *Sur une classe de fonctions considérée dans l'étude du problème de Dirichlet*, Fund. Math. **21** (1933), 129–150. (Cited on 581.)

[614] L. Nirenberg, *On elliptic partial differential equations*, Ann. Scuola Norm. Sup. Pisa (3) **13** (1959), 115–162. (Cited on 582.)

[615] L. Nirenberg, *Pseudo-differential operators*, In Global Analysis, Proc. Sympos. Pure Math., Vol. XVI, S. Chern and S. Smale, eds., pp. 149–167, American Mathematical Society, Providence, R.I, 1970. (Cited on 352.)

[616] J. A. Oguntuase and L.-E. Persson, *Hardy type inequalities via convexity—the journey so far*, Aust. J. Math. Anal. Appl. **7** (2010), 1–19. (Cited on 557.)

[617] R. Olkiewicz and B. Zegarlinski, *Hypercontractivity in noncommutative L^p spaces*, J. Funct. Anal. **161** (1999), 246–285. (Cited on 653.)

[618] R. O'Neil, *Convolution operators and $L(p,q)$ spaces*, Duke Math. J'**30** (1963), 129–142. (Cited on 557.)

[619] B. Opic and A. Kufner, *Hardy-type Inequalities*, Pitman Research Notes in Mathematics Series, Longman Scientific & Technical, Harlow, 1990. (Cited on 336, 557.)

[620] W. Orlicz, *Über eine gewisse Klasse von Räumen vom Typus B*, Bull. Intern. de l'Acad. Pol., série A, Cracow **8/9** (1932), 207–220. (Cited on 36.)

[621] D. S. Ornstein, *Bernoulli shifts with the same entropy are isomorphic*, Adv. in Math. **4** (1970), 337–352. (Cited on 97.)

[622] D. S. Ornstein and B. Weiss, *The Shannon–McMillan–Breiman theorem for a class of amenable groups*, Israel J. Math. **44** (1983), 53–60. (Cited on 83.)

[623] J. Ortega-Cerdà and K. Seip, *Fourier frames*, Ann. Math. (2) **155** (2002), 789–806. (Cited on 406.)

[624] V. I. Oseledec, *A multiplicative ergodic theorem. Lyapunov characteristic numbers for dynamical systems*, Trans. Moscow Math. Soc. **19** (1968), 197–221. (Cited on 145.)

[625] F. Otto and C. Villani, *Generalization of an inequality by Talagrand and links with the logarithmic Sobolev inequality*, J. Funct. Anal. **173** (2000), 361–400. (Cited on 654.)

[626] R. S. Palais, *Seminar on the Atiyah-Singer index theorem*, with contributions by M. F. Atiyah, A. Borel, E. E. Floyd, R. T. Seeley, W. Shih and R. Solovay, Annals of Mathematics Studies, Princeton University Press, Princeton, NJ, 1965. (Cited on 367.)

[627] R. E. A. C. Paley and N. Wiener, *Fourier Transforms in the Complex Domain*, reprint of the 1934 original, American Mathematical Society Colloquium Publications, American Mathematical Society, Providence, RI, 1987. (Cited on 406.)

[628] R. E. A. C. Paley and A. Zygmund, *A note on analytic functions in the unit circle*, Proc. Cambridge Philos. Soc. **28** (1932), 266–272. (Cited on 464.)

[629] W. Parry, *Ergodic properties of affine transformations and flows on nilmanifolds*, Amer. J. Math. **91** (1969), 757–771. (Cited on 123.)

[630] W. Parry, *Topics in Ergodic Theory*, reprint of the 1981 original, Cambridge Tracts in Mathematics, Cambridge University Press, Cambridge, 2004. (Cited on 79.)

[631] L. A. Pastur, *Spectral properties of disordered systems in the one-body approximation*, Comm. Math. Phys. **75** (1980), 179–196. (Cited on 294.)

[632] L. A. Pastur and A. Figotin, *Spectra of Random and Almost-Periodic Operators*, Grundlehren der Mathematischen Wissenschaften, Springer, Berlin, 1992. (Cited on 294.)

[633] J. Peetre, *Nouvelles propriétés d'espaces d'interpolation*, C. R. Acad. Sci. Paris **256** (1963), 1424–1426. (Cited on 556.)

[634] V. V. Peller, *Hankel Operators and Their Applications*, Springer Monographs in Mathematics, Springer-Verlag, New York, 2003. (Cited on 536.)

[635] D. B. Percival and A. T. Walden, *Wavelet Methods for Time Series Analysis*, Cambridge Series in Statistical and Probabilistic Mathematics, Cambridge University Press, Cambridge, 2000. (Cited on 433.)

[636] G. Perelman, *The entropy formula for the Ricci flow and its geometric applications*, arXiv: math/0211159, 2008. (This paper was never published.) (Cited on 654.)

[637] A. M. Perelomov, *Remark on the completeness of the coherent state system*, Teoret. Mat. Fiz. **6** (1971), 213–224; English translation, Theoret. and Math. Phys. **6** (1971), 156–164. (Cited on 401.)

[638] A. M. Perelomov, *Coherent states for arbitrary Lie group*, Comm. Math. Phys. **26** (1972), 222–236. (Cited on 386.)

[639] O. Perron, *Eine neue Behandlung der ersten Randwertaufgabe für $\Delta u = 0$*, Math. Z. **18** (1923), 42–54. (Cited on 212, 231.)

[640] K. Petersen, *Ergodic Theory*, corrected reprint of the 1983 original, Cambridge Studies in Advanced Mathematics, Cambridge University Press, Cambridge, 1989. (Cited on 79.)

[641] W. Philipp, *Some metrical theorems in number theory. II*, Duke Math. J. **37** (1970), 447–458. (Cited on 123.)

[642] J. Phillips, *A note on square-integrable representations*, J. Funct. Anal. **20** (1975), 83–92. (Cited on 387.)

[643] R. Phillips, *Reminiscences about the 1930s*, Math. Intelligencer **16** (1994), 6–8. (Cited on 83.)

[644] S. K. Pichorides, *On the best values of the constants in the theorems of M. Riesz, Zygmund and Kolmogorov*, Studia Math. **44** (1972) 165–179. (Cited on 488.)

[645] É. Picard, *Deux théorèmes élémentaires sur les singularités des fonctions harmoniques*, C. R. Acad. Sci. Paris **176** (1923), 933–935. (Cited on 197.)

[646] G. Pick, *Über beschränkungen analytischer Funktionen, welche durch vorgegebene Funktionswerte bewirkt werden*, Math. Ann. **77** (1916), 7–23. (Cited on 513.)

[647] M. A. Pinsky, *Introduction to Fourier Analysis and Wavelets*, reprint of the 2002 original, Graduate Studies in Mathematics, American Mathematical Society, Providence, RI, 2009. (Cited on 433.)

[648] B. A. Plamenevskii, *Algebras of Pseudodifferential Operators*, translated from the Russian by R. A. M. Hoksbergen, Mathematics and its Applications (Soviet Series), Kluwer Academic Publishers Group, Dordrecht, 1989. (Cited on 367.)

[649] J. Plemelj, *Ein Ergänzungssatz zur Cauchyschen Integraldarstellung analytischer Funktionen, Randwerte betreffend*, Monatsch. für Math. u. Phys. **19** (1908), 205–210. (Cited on 489.)

[650] A. Plessner, *Zur Theorie der konjugierten trigonometrischen Reihen*, Mitt. Math. Sem. Giessen, Giessen Seminar, 1923. (Cited on 463.)

[651] H. Poincaré, *Sur le probleme des trois corps et les equations de la dynamique*, Acta Math. **13** (1890), 1–270. (Cited on 80, 85.)

[652] H. Poincaré, *Sur les equations aux dérivées partielles de la physique mathématique*, Amer. J. Math. **12** (1890), 211–294. (Cited on 231, 275, 581.)

[653] H. Poincaré, *Sur la theorie cinetique des gaz*, Revue generale des Sciences pures et appliquees, **5** (1894), 513–521. (Cited on 80.)

[654] H. Poincaré, *Théorie du potentiel Newtonien*, Leçons professées à la Sorbonne pendant le premier semestre, 1894–1895. (Cited on 231.)

[655] H. Poincaré, *Les méthodes nouvelles de la mécanique céleste. Tome III. Invariant intégraux. Solutions périodiques du deuxième genre. Solutions doublement asymptotiques*, reprint of the 1899 original, Les Grands Classiques Gauthier–Villars, Bibliothèque Scientifique Albert Blanchard, Paris, 1987; *New methods of celestial mechanics. Vol. 3. Integral invariants and asymptotic properties of certain solutions*, revised reprint of the 1967 English translation, History of Modern Physics and Astronomy, American Institute of Physics, New York, 1993; available online at `http://www.archive.org/details/lesmthodesnouv003poin`. (Cited on 80, 85.)

[656] H. Poincaré, *Theorie du potentiel Newtonien*, Georges Carré et C. Naud, Paris, 1899. (Cited on 212, 275.)

[657] H. Poincaré, *Sur l'uniformisation des fonctions analytiques*, Acta Math. **31** (1908), 1–63. (Cited on 316.)

[658] S. D. Poisson, *Sur l'intégrale de l'équation relative aux vibrations des surfaces élastiques et au mouvement des ondes*, Bull. Soc. Philomathique de Paris (1818) 125–128. (Cited on 197.)

[659] S. D. Poisson, *Memoire sur l' integration de quelques equations linéaires aux differences partielles, et particulierement de l'équation generale du mouvement des fluides élastique*, Mémoires Acad. Sci. Inst. France Sci. **3** (1819) 121–176. (Cited on 197.)

[660] M. P. Pollicott, *Lectures on Ergodic Theory and Pesin Theory on Compact Manifolds*, London Mathematical Society Lecture Note Series, Cambridge University Press, Cambridge, 1993. (Cited on 79.)

[661] A. Poltoratski, *On the boundary behavior of pseudocontinuable functions*, St. Petersburg Math. J. **5** (1994), 389–406. (Cited on 64.)

[662] A. Poltoratski, *On the distributions of boundary values of Cauchy integrals*, Proc. Amer. Math. Soc. **124** (1996), 2455–2463. (Cited on 514.)

[663] A. Poltoratski, B. Simon, and M. Zinchenko, *The Hilbert transform of a measure*, J. Anal. Math. **111** (2010), 247–265. (Cited on 514.)

[664] S. C. Port and C. J. Stone, *Brownian Motion and Classical Potential Theory*, Probability and Mathematical Statistics, Academic Press, New York–London, 1978. (Cited on 177.)

[665] I. I. Privalov, *Sur les fonctions conjuguées*, Bull. Soc. Math. France, **44** (1916), 100–103. (Cited on 489.)

[666] M. Quéfflec, *Old and new results on normality*, In Dynamics & Stochastics: Festschrift in Honor of M. S. Keane, pp. 225–236, IMS Lecture Notes Monograph Series, Institute of Mathematical Statistics, Beachwood, OH, 2006. (Cited on 97.)

[667] H. Rademacher, *Einige Sätze über Reihen von allgemeinen Orthogonalfunktionen*, Math. Ann. **87** (1922), 112–138. (Cited on 409.)

[668] M. S. Raghunathan, *A proof of Oseledec's multiplicative ergodic theorem*, Israel J. Math. **32** (1979), 356–362. (Cited on 145.)

[669] E. A. Rakhmanov, *On the asymptotics of the ratio of orthogonal polynomials, I, II*, Mat. Sb. **32** (1977), 199–213; **46** (1983), 105–117. (Cited on 292.)

[670] T. Ransford, *Potential Theory in the Complex Plane*, London Mathematical Society Student Texts, Cambridge University Press, Cambridge, 1995. (Cited on 177, 274.)

[671] M. M. Rao, *Abstract martingales and ergodic theory*, In Multivariate Analysis, III (Proc. Third Internat. Sympos., Wright State Univ., Dayton, Ohio, 1972), P. Krishnaiak, ed., pp. 45–60, Academic Press, New York, 1973. (Cited on 161.)

[672] G. Rauzy, *Propriétés statistiques de suites arithmétiques*, Le Mathématicien, Collection SUP, Presses Universitaires de France, Paris, 1976. (Cited on 123.)

[673] M. Reed and B. Simon, *Methods of Modern Mathematical Physics, II. Fourier Analysis, Self-Adjointness*, Academic Press, New York, 1975. (Cited on 654.)

[674] M. Reed and B. Simon, *Methods of Modern Mathematical Physics, IV: Analysis of Operators*, Academic Press, New York, 1977. (Cited on 683.)

[675] N. Reingold, *Refugee mathematicians in the United States of America, 1933–1941: reception and reaction*, In A Century of Mathematics in America, Part I, pp. 175–200, History of Mathematics, American Mathematical Society, Providence, RI, 1988. (Cited on 83.)

[676] F. Rellich, *Ein Satz über mittlere Konvergenz*, Nachr. Kön. Ges. Wiss. Göttingen, Mathematisch-Physikalische Klasse (1930), 30–35. (Cited on 582.)

[677] C. Remling, *The absolutely continuous spectrum of Jacobi matrices*, Ann. Math. (2) **174** (2011), 125–171. (Cited on 293.)

[678] S. Rickman, *On the number of omitted values of entire quasiregular mappings*, J. Anal. Math. **37** (1980), 100–117. (Cited on 218.)

[679] G. F. B. Riemann, *Grundlagen für eine allgemeine Theorie der Functionen einer veränderlichen complexen Grösse*, inaugural dissertation (1851), Göttingen; Werke, 5–43; available online in German at `http://www.maths.tcd.ie/pub/HistMath/People/Riemann/Grund/`. (Cited on 197.)

[680] F. Riesz, *Sur certains systèmes singuliers d'équations integrales*, Ann. Sci. Ecole Norm. Sup. (3) **28** (1911), 33–62. (Cited on 463.)

[681] F. Riesz, *Über ein Problem des Herrn Carathéodory*, J. Reine Angew. Math. **146** (1916), 83–87. (Cited on 434.)

[682] F. Riesz, *Sur les valeurs moyennes du module des fonctions harmoniques et des fonctions analytiques*, Acta Litt. ac Scient. Univ. Hung. **1**, (1922), 27–32. (Cited on 444.)

[683] F. Riesz, *Über die Randwerte einer analytischen Funktion*, Math. Z. **18** (1923), 87–95. (Cited on 457.)

[684] F. Riesz, *Sur les fonctions subharmoniques et leur rapport à la théorie du potentiel*, Acta Math. **48** (1926), 329–343. (Cited on 212.)

[685] F. Riesz, *Sur la convergence en moyenne*, Acta Sci. Math. Szeged **4** (1928), 58–64. (Cited on 457.)

[686] F. Riesz, *Sur les fonctions subharmoniques et leur rapport à la théorie du potentiel. II*, Acta Math. **54** (1930), 321–360. (Cited on 212, 213.)

[687] F. Riesz, *Sur une inégalité integral*, J. London Math. Soc. **5** (1930), 162–168. (Cited on 562.)

[688] F. Riesz, *Sur un théorème de maximum de MM. Hardy et Littlewood*, J. London Math. Soc. **7** (1932), 10–13. (Cited on 46, 51.)

[689] F. Riesz, *Sur l'existence de la dérivée des fonctions monotones et sur quelques problèmes qui s'y rattachnet*, Acta Sci. Math. Szeged **5** (1932), 208–221. (Cited on 46, 59.)

[690] F. Riesz and M. Riesz, *Über die Randwerte einer analytischen Funktion*, In Quatrième congrès des math. scand. pp. 27–44, Stockholm, 1916. (Cited on 274, 457.)

[691] M. Riesz, *Sur la sommation des séries de Fourier*, Acta Litt. ac Scient. Univ. Hung. **1** (1923), 104–113. (Cited on 603.)

[692] M. Riesz, *Sur les fonctions conjuguées*, Math. Z. **27** (1928), 218–244. (Cited on 488, 492, 497.)

[693] M. Riesz, *Intégrales de Riemann-Liouville et potentiels*, Acata Sci. Math. Szeged, **9** (1938), 1–42. (Cited on 276.)

[694] H. P. Robertson, *The uncertainty principle*, Phys. Rev. **34** (1929) 163–64. (Cited on 334.)

[695] G. Robin, *Sur la distribution de l'électricité à la surface des conducteurs fermés des conducteurs ouverts*, Ann. Sci. Ecole Norm. Sup. (3), (1886), 3–58. (Cited on 274.)

[696] A. M. Rockett and P. Szüsz, *Continued Fractions*, World Scientific, River Edge, NJ, 1992. (Cited on 123.)

[697] C. A. Rogers, *Two integral inequalities*, J. London Math. Soc. **31** (1956), 235–238. (Cited on 564.)

[698] C. A. Rogers, *The number of lattice points in a set*, Proc. London Math. Soc. (3) **6** (1956), 305–320. (Cited on 564.)

[699] C. A. Rogers, *A single integral inequality*, J. London Math. Soc. **32** (1957), 102–108. (Cited on 564.)

[700] J. Rosen, *Sobolev inequalities for weight spaces and supercontractivity*, Trans. Amer. Math. Soc. **222** (1976), 367–376. (Cited on 653.)

[701] G-C. Rota, *Une théorie unifiée des martingales et des moyennes ergodiques*, C. R. Acad. Sci. Paris **252** (1961), 2064–2066. (Cited on 161.)

[702] O. S. Rothaus, *Lower bounds for eigenvalues of regular Sturm-Liouville operators and the logarithmic Sobolev inequality*, Duke Math. J. **45** (1978), 351–362. (Cited on 652.)

[703] G. Royer, *An Initiation to Logarithmic Sobolev Inequalities*, translated from the 1999 French original by Donald Babbitt, SMF/AMS Texts and Monographs, American Mathematical Society, Providence, RI; Société Mathématique de France, Paris, 2007. (Cited on 650.)

[704] G. V. Rozenbljum, *Distribution of the discrete spectrum of singular differential operators*, Dokl. Akad. Nauk. SSSR (N.S.) **202** (1972), 1012–1015. (Cited on 669.)

[705] D. K. Ruch and P. J. Van Fleet, *Wavelet Theory, An Elementary Approach with Applications*, John Wiley & Sons, Inc., Hoboken, NJ, 2009. (Cited on 433.)

[706] W. Rudin, *Real and Complex Analysis*, 3rd edition, McGraw-Hill, New York, 1987. (Cited on 439.)

[707] D. Ruelle, *Ergodic theory of differentiable dynamical systems*, Publ. Math. Inst. Hautes Etudes Sci. **50** (1979), 27–58. (Cited on 145.)

[708] M. Rumin, *Spectral density and Sobolev inequalities for pure and mixed states*, Geom. Funct. Anal. **20** (2010), 817–844. (Cited on 670.)

[709] M. Rumin, *Balanced distribution-energy inequalities and related entropy bounds*, Duke Math. J. **160** (2011), 567–597. (Cited on 670.)

[710] T. Runst and W. Sickel, *Sobolev Spaces of Fractional Order, Nemytskij Operators, and Nonlinear Partial Differential Equations*, de Gruyter, Berlin, 1996. (Cited on 583.)

[711] C. Ryll-Nardzewski, *On the ergodic theorems. II. Ergodic theory of continued fractions*, Studia Math. **12** (1951), 74–79. (Cited on 124.)

[712] C. Sadosky, *Interpolation of Operators and Singular Integrals. An Introduction to Harmonic Analysis*, Monographs and Textbooks in Pure and Applied Mathematics, Marcel Dekker, Inc., New York, 1979. (Cited on 603.)

[713] C. Sadosky, *On the life and work of Mischa Cotlar*, 10th New Mexico Analysis Seminar, 2007; available at `http://www.math.unm.edu/conferences/10thAnalysis/resources/cotlar/sadosky.pdf`. (Cited on 614.)

[714] C. Sadosky, *An Afternoon in Honor to Mischa Cotlar*, in connection with 10th New Mexico Analysis Seminar, 2007, held October 12, 2007; available at `http://www.math.unm.edu/conferences/10thAnalysis/testimonies.html`. (Cited on 614.)

[715] X. Saint-Raymond, *Elementary Introduction to the Theory of Pseudodifferential Operators*, Studies in Advanced Mathematics, CRC Press, Boca Raton, FL, 1991. (Cited on 367.)

[716] S. Saks, *Theory of the Integral*, 2nd revised edition, Dover Publications, New York, 1964; original publication in 1937. (Cited on 64.)

[717] D. Sarason, *Functions of vanishing mean oscillation*, Trans. Amer. Math. Soc. **207** (1975), 391–405. (Cited on 534.)

[718] M. Sato, *Hyperfunctions and partial differential equations*, In Proc. Internat. Conference on Functional Analysis and Related Topics (Tokyo, 1969), pp. 91–94, University of Tokyo Press, Tokyo, 1970. (Cited on 350.)

[719] E. Schrödinger, *Zum Heisenbergschen unschärfeprinzip*, Sitzungsber. Preuss. Akad. Wiss. Physikalisch-mathematische Klasse **14** (1930) 296–303. (Cited on 334.)

[720] I. Schur, *Bemerkungen zur Theorie der beschränkten Bilinearformen mit unendlich vielen Veränderlichen*, J. Reine Angew. Math. **140** (1911) 1–28. (Cited on 488.)

[721] R. T. Seeley, *Refinement of the functional calculus of Calderón and Zygmund*, Indag. Math. **27** (1965), 521–531. (Cited on 367.)

[722] I. E. Segal, *Mathematical characterization of the physical vacuum for a linear Bose–Einstein field. (Foundations of the dynamics of infinite systems. III)*, Illinois J. Math. **6** (1962), 500Ü-523. (Cited on 385.)

[723] I. E. Segal, *Notes towards the construction of nonlinear relativistic quantum fields, III: Proper ties of the C^*-dynamics for a certain class of interactions*, Bull. Amer. Math. Soc. **75** (1969), 1390–1395. (Cited on 652.)

[724] I. E. Segal, *Construction of nonlinear local quantum processes, I*, Ann. Math. **91** (1970), 462–481. (Cited on 652.)

[725] I. E. Segal, *Construction of nonlinear local quantum processes, II*, Invent. Math. **14** (1971), 211–241. (Cited on 652.)

[726] I. E. Segal, *Space-time decay for solutions of wave equations*, Advances in Math. **22** (1976), 305–311. (Cited on 683.)

[727] K. Seip, *Density theorems for sampling and interpolation in the Bargmann–Fock space. I*, J. Reine Angew. Math. **429** (1992), 91–106. (Cited on 401.)

[728] K. Seip, *Interpolation and Sampling in Spaces of Analytic Functions*, University Lecture Series, American Mathematical Society, Providence, RI, 2004. (Cited on 401.)

[729] K. Seip and R. Wallstén, *Density theorems for sampling and interpolation in the Bargmann-Fock space. II*, J. Reine Angew. Math. **429** (1992), 107–113. (Cited on 401.)

[730] C. Series, *On coding geodesics with continued fractions*, In Ergodic Theory (Sem., Les Plans-sur-Bex, 1980), P. de la Harpe, ed., pp. 67–76, Monograph. Enseign. Math. **29**, Univ. Genève, Geneva, 1981. (Cited on 126.)

[731] C. Series, *The infinite word problem and limit sets in Fuchsian groups*, Ergodic Theory Dynam. Systems **1** (1981), 337–360. (Cited on 126.)

[732] C. Series, *Non-Euclidean geometry, continued fractions, and ergodic theory*, Math. Intelligencer **4** (1982), 24–31. (Cited on 126.)

[733] C. Series, *The modular surface and continued fractions*, J. London Math. Soc. (2) **31** (1985), 69–80. (Cited on 126.)

[734] C. Shannon, *A mathematical theory of communication*, Bell Syst. Tech. J. **27** (1948), 379–423, 623–656. (Cited on 334.)

[735] P. B. Shelley, *Queen Mab, A Philosophical Poem*, published privately by the author in 1813, available online at `https://archive.org/details/queenmabwithnote00shel`. (Cited on 319.)

[736] M. A. Shubin, *Pseudodifferential Operators and Spectral Theory*, translated from the Russian by S. I. Andersson, Springer Series in Soviet Mathematics, Springer-Verlag, Berlin, 1987. (Cited on 367, 368, 371.)

[737] W. Sierpinski, *Un théorème sur les nombres irrationnels*, Bull. Acad. Sci. Cracovie **2** (1909), 725–727. (Cited on 98.)

[738] W. Sierpinski, *Démonstration élémentaire d'un théorme de M. Borel sur les nombres absolutment normaux et détermination effective d'un tel nombre*, Bull. Soc. Math. France **45** (1917), 125–144. (Cited on 97.)

[739] C. E. Silva, *Invitation to Ergodic Theory*, Student Mathematical Library, American Mathematical Society, Providence, RI, 2008. (Cited on 79.)

[740] B. Simon, *The $P(\varphi)_2$ Euclidean (Quantum) Field Theory*. Princeton Series in Physics, Princeton University Press, Princeton, NJ, 1974. (Cited on 651, 654.)

[741] B. Simon, *Analysis with weak trace ideals and the number of bound states of Schrödinger operators*, Trans. Amer. Math. Soc. **224** (1976), 367–380. (Cited on 669.)

[742] B. Simon, *The bound state of weakly coupled Schrödinger operators in one and two dimensions*, Ann. Phys. **97** (1976), 279–288. (Cited on 340.)

[743] B. Simon, *The classical limit of quantum partition functions*, Comm. Math. Phys. **71** (1980), 247–276. (Cited on 386.)

[744] B. Simon, *Schrödinger semigroups*, Bull. Amer. Math. Soc. **7** (1982), 447–526. (Cited on 292.)

[745] B. Simon, *Kotani theory for one dimensional stochastic Jacobi matrices*, Comm. Math. Phys. **89** (1983), 227–234. (Cited on 296.)

[746] B. Simon, *Representations of Finite and Compact Groups*, Graduate Studies in Mathematics, American Mathematical Society, Providence, RI, 1996. (Cited on 250, 336, 387.)

[747] B. Simon, *Functional Integration and Quantum Physics*, 2nd edition, AMS Chelsea Publishing, Providence, RI, 2005. (Cited on 689.)

[748] B. Simon, *Orthogonal Polynomials on the Unit Circle, Part 1: Classical Theory*, AMS Colloquium Series, American Mathematical Society, Providence, RI, 2005. (Cited on 291, 472.)

[749] B. Simon, *Orthogonal Polynomials on the Unit Circle, Part 2: Spectral Theory*, AMS Colloquium Series, American Mathematical Society, Providence, RI, 2005. (Cited on 146, 294.)

[750] B. Simon, *Fine structure of the zeros of orthogonal polynomials, I. A tale of two pictures*, Electron. Trans. Numer. Anal. **25** (2006), 328–368. (Cited on 292.)

[751] B. Simon, *Equilibrium measures and capacities in spectral theory*, Inverse Probl. Imaging **1** (2007), 713–772. (Cited on 291, 292, 293, 294.)

[752] B. Simon, *The Christoffel–Darboux kernel*, In Perspectives in PDE, Harmonic Analysis and Applications, Proc. Sympos. Pure Math., D. Mitrea and M. Mitrea, eds., pp. 295–335, American Mathematical Society, Providence, RI, 2008. (Cited on 291.)

[753] B. Simon, *Two extensions of Lubinsky's universality theorem*, J. Anal. Math. **105** (2008), 345–362. (Cited on 292.)

[754] B. Simon, *Weak convergence of CD kernels and applications*, Duke Math. J. **146** (2009), 305–330. (Cited on 291.)

[755] B. Simon, *Regularity and the Cesàro–Nevai class*, J. Approx. Theory **156** (2009), 142–153. (Cited on 293.)

[756] B. Simon, *Convexity: An Analytical Viewpoint*, Cambridge University Press, Cambridge, 2011. (Cited on 36, 563.)

[757] B. Simon, *Szegő's Theorem and Its Descendants: Spectral Theory for L^2 Perturbations of Orthogonal Polynomials*, Princeton University Press, Princeton, NJ, 2011. (Cited on 127, 197, 293, 295.)

[758] B. Simon and R. Hoegh-Krohn, *Hypercontractive semi-groups and two dimensional self-coupled Bose fields*, J. Funct. Anal. **9** (1972), 121–180. (Cited on 652.)

[759] Ya. G. Sinai, *Introduction to Ergodic Theory*, Mathematical Notes, Princeton University Press, Princeton, NJ, 1976. (Cited on 79.)

[760] D. Slepian, *Prolate spheroidal wave functions, Fourier analysis and uncertainty IV: extensions to many dimensions; generalized prolate spheroidal wave functions*, Bell Syst. Tech. J. **43** (1964), 3009–3057. (Cited on 338.)

[761] D. Slepian and H. O. Pollak, *Prolate spheroidal wave functions, Fourier analysis and uncertainty I.*, Bell Syst. Tech. J. **40** (1961), 43–63. (Cited on 338.)

[762] V. I. Smirnov, *Sur les valeurs limites des fonctions, réguliéres à l'intérieur d'un cercle*, J. Soc. Phys.-Math. Leningrad bf 2 (1929), 22–37. (Cited on 470.)

[763] C. Smith and M. N. Wise, *Energy and Empire: A Biographical Study of Lord Kelvin*, Cambridge University Press, Cambridge, UK, 1989. (Cited on 250.)

[764] S. L. Sobolev, *Le problème de Cauchy dans l'espace des fonctionelles*, Dokl. Akad. Nauk. SSSR, **8** (1935), 291–294. (Cited on 582.)

[765] S. L. Sobolev, *On some estimates related to families of functions having derivatives which are square integrable*, Dokl. Akad. Nauk. SSSR, **10** (1936), 267–270. (Cited on 582.)

[766] S. L. Sobolev, *On a theorem in functional analysis* (Russian), Dokl. Akad. Nauk. SSSR **20** (1938), 5–9. (Cited on 582.)

[767] S. L. Sobolev, *On a theorem of functional analysis*, Mat. Sb. (N.S.) **4** (1938), 471–479. Amer. Math. Soc. Transl. Ser. 2, **34** (1963), 39–68. (Cited on 562.)

[768] S. L. Sobolev, *Some Applications of Functional Analysis in Mathematical Physics*, Izdat. Leningrad. Gos. Univ., Leningrad, 1950; translated from the Russian by F. E. Browder, Translations of Mathematical Monographs, American Mathematical Society, Providence, RI, 1963. (Cited on 582.)

[769] C. D. Sogge, *Fourier Integrals in Classical Analysis*, Cambridge University Press, Cambridge, UK, 1993. (Cited on 564.)

[770] S. Spanne, *Sur l'interpolation entre les espaces $L_p^k F$*, Ann. Scuola Norm. Sup. Pisa **3** (1966), 625–648. (Cited on 489.)

[771] F. Spitzer, *Principles of Random Walks*, 2nd edition, Graduate Texts in Mathematics, Springer-Verlag, New York–Heidelberg, 1976. (Cited on 164.)

[772] H. Stahl and V. Totik, *General Orthogonal Polynomials*, In Encyclopedia of Mathematics and its Applications, Cambridge University Press, Cambridge, 1992. (Cited on 291, 293.)

[773] A. J. Stam, *Some inequalities satisfied by the quantities of information of Fisher and Shannon*, Inform. and Control **2** (1959), 101–112. (Cited on 652.)

[774] M. J. Steele, *Kingman's subadditive ergodic theorem*, Ann. Inst. H. Poincaré Probab. Statist. **25** (1989), 93–98. (Cited on 145.)

[775] E. M. Stein, *On limits of seqences of operators*, Ann. Math. (2) **74** (1961), 140–170. (Cited on 25.)

[776] E. M. Stein, *Singular integrals, harmonic functions, and differentiability properties of functions of several variables*, In Singular Integrals (Proc. Sympos. Pure Math., Chicago, Ill., 1966), A. Calderón, ed., pp. 316–335, American Mathematical Society Providence, RI, 1967. (Cited on 489.)

[777] E. M. Stein, *Singular Integrals and Differentiability Properties of Functions*, Princeton Mathematical Series, Princeton University Press, Princeton, NJ, 1970. (Cited on 603.)

[778] E. M. Stein, *Maximal functions. I. Spherical means*, Proc. Nat. Acad. Sci. USA **73** (1976), 2174–2175. (Cited on 49.)

[779] E. M. Stein, *Oscillatory integrals in Fourier analysis*, In Beijing Lectures in Harmonic Analysis (Beijing, 1984), pp. 307–355, Annals of Math. Studies, Princeton University Press, Princeton, NJ, 1986. (Cited on 681, 682.)

[780] E. M. Stein, *Harmonic Analysis: Real-Variable Methods, Orthogonality, and Oscillatory Integrals*, with the assistance of Timothy S. Murphy. Princeton Mathematical Series, Monographs in Harmonic Analysis, III, Princeton University Press, Princeton, NJ, 1993. (Cited on 601, 682.)

[781] E. M. Stein and R. Shakarchi, *Functional Analysis, Introduction to Further Topics in Analysis*, Princeton Lectures in Analysis, Princeton University Press, Princeton, NJ, 2011. (Cited on 487, 682.)

[782] E. M. Stein and J.-O. Strömberg, *Behavior of maximal functions in $\mathbb{R}^n$ for large n*, Ark. Mat. **21** (1983), 259–269. (Cited on 49.)

[783] E. M. Stein and G. Weiss, *Fractional integrals in n-dimensional Euclidean space*, J. Math. Mech. **7** (1958), 503–514. (Cited on 563, 564.)

[784] E. M. Stein and G. Wiess, *An extension of a theorem of Marcinkiewicz and some of its applications*, J. Math. Mech. **8** (1959), 263–284. (Cited on 513.)

[785] E. M. Stein and G. Weiss, *Introduction to Fourier Analysis on Euclidean Spaces*, Princeton Mathematical Series, Princeton University Press, Princeton, NJ, 1971. (Cited on 251.)

[786] P. Stein, *On a theorem of M. Riesz*, J. London Math. Soc. **8** (1933) 242–247. (Cited on 488, 492.)

[787] M. H. Stone, *Linear Transformations in Hilbert Space*, reprint of the 1932 original, American Mathematical Society Colloquium Publications, American Mathematical Society, Providence, RI, 1990. (Cited on 336.)

[788] R. S. Strichartz, *Restrictions of Fourier transforms to quadratic surfaces and decay of solutions of wave equations*, Duke Math. J. **44** (1977), 705–714. (Cited on 682, 683.)

[789] J.-O. Strömberg, *A modified Franklin system and higher-order spline systems on R^n as unconditional bases for Hardy spaces*, In Conference on Harmonic Analysis in Honor of Antoni Zygmund, Vol. II (Chicago, Ill., 1981), W. Beckner et al., eds., pp. 475–494, Wadsworth Math. Ser., Wadsworth, Belmont, CA, 1983. (Cited on 434.)

[790] D. W. Stroock, *An Introduction to the Theory of Large Deviations*, Universitext, Springer-Verlag, New York, 1984. (Cited on 653.)

[791] D. Stroock, *Logarithmic Sobolev inequalities for Gibbs states*, In Dirichlet Forms, pp. 194–228, Lecture Notes in Math., Springer, Berlin, 1993. (Cited on 650.)

[792] E. C. G. Sudarshan, *Equivalence of semiclassical and quantum mechanical descriptions of statistical light beams*, Phys. Rev. Lett. **10** (1963), 277–279. (Cited on 385, 386.)

[793] J. M. Sullivan, *Sphere packings give an explicit bound for the Besicovitch covering theorem*, J. Geom. Anal. **4** (1994), 219–231. (Cited on 50.)

[794] G. Szegő, *Bemerkungen zu einer Arbeit von Herrn M. Fekete: Über die Verteilung der Wurzeln bei gewissen algebraischen Gleichungen mit ganzzahligen Koeffizienten*, Math. Z. **21** (1924), 203–208. (Cited on 291.)

[795] G. Talenti, *Best constant in Sobolev inequality*, Ann. Mat. Pura Appl. (IV) **110** (1976), 353–372. (Cited on 582.)

[796] T. Tao, *Restriction Theorems and Applications*, unpublished Lecture Notes for Math 254B, UCLA, 1999; available at `http://www.math.ucla.edu/~tao/254b.1.99s/`. (Cited on 682.)

[797] T. Tao, *The Bochner–Riesz conjecture implies the restriction conjecture*, Duke Math. J. **96** (1999), 363–375. (Cited on 684.)

[798] T. Tao, *An uncertainty principle for cyclic groups of prime order*, Math. Res. Lett. **12** (2005), 121–127. (Cited on 339.)

[799] T. Tao, *Tricks Wiki article: The tensor power trick*, 2008; available at `https://terrytao.wordpress.com/2008/08/25/`. (Cited on 556.)

[800] T. Tao, *Interpolation of L^p spaces*, Notes for Math 245C, 2009; available at `http://terrytao.wordpress.com/2009/03/30/245c`. (Cited on 557.)

[801] T. Tao, *Stein's spherical maximal theorem*; available at `http://terrytao.wordpress.com/2011/05/21/`. (Cited on 49.)

[802] L. Tartar, *An Introduction to Sobolev Spaces and Interpolation Spaces*, Lecture Notes of the Unione Matematica Italiana, Springer, Berlin; UMI, Bologna, 2007. (Cited on 583.)

[803] M. Taylor, *Pseudodifferential Operators and Nonlinear PDE*, Progress in Mathematics, Birkhäuser Boston, Inc., Boston, MA, 1991. (Cited on 367, 371.)

[804] S. P. Thompson, *The Life of Lord Kelvin*, Chelsea Publishing, New York, 2004; original edition, Macmillan, London, 1910. (Cited on 250.)

[805] W. Thomson and P. G. Tait, *Treatise on Natural Philosophy*, Oxford University Press, Oxford, 1867. (Note: Thomson was later Lord Kelvin.) (Cited on 196, 250.)

[806] D. J. Thouless, *Electrons in disordered systems and the theory of localization*, Phys. Rep. **13** (1974), 93. (Cited on 291.)

[807] P. A. Tomas, *A restriction theorem for the Fourier transform*, Bull. Amer. Math. Soc. **81** (1975), 477–478. (Cited on 682.)

[808] P. A. Tomas, *Restriction theorems for the Fourier transform*, In Harmonic Analysis in Euclidean Spaces (Proc. Sympos. Pure Math., Williams Coll., Williamstown, Mass., 1978), Part 1, G. Weiss and S. Wainger, eds., pp. 111–114, Proc. Sympos. Pure Math., American Mathematical Society, Providence, RI, 1979. (Cited on 682.)

[809] L. Tonelli, *L'estremo assoluto degli integrali doppi*, Ann. Scuola Norm. Sup. Pisa (2) bf 2 (1933), 89–130. (Cited on 581.)

[810] V. Totik, *Universality and fine zero spacing on general sets*, Ark. Mat. **47** (2009), 361–391. (Cited on 292.)

[811] F. Trèves, *Introduction to Pseudodifferential and Fourier integral operators. Vol. 1. Pseudodifferential operators*, The University Series in Mathematics, Plenum Press, New York–London, 1980. (Cited on 367.)

[812] F. Trèves, *Introduction to Pseudodifferential and Fourier Integral Operators. Vol. 2. Fourier integral operators*, The University Series in Mathematics, Plenum Press, New York–London, 1980. (Cited on 367.)

[813] H. Triebel, *Spaces of distributions of Besov type on Euclidean n-space. Duality, interpolation*, Ark. Mat. **11** (1973), 13–64. (Cited on 583.)

[814] H. Triebel, *A remark on embedding theorems for Banach spaces of distributions*, Ark. Mat. **11** (1973), 65–74. (Cited on 583.)

[815] H. Triebel, *Theory of Function Spaces*, Monographs in Mathematics, Birkhäuser Verlag, Basel, 1983. (Cited on 557.)

[816] H. Triebel, *Interpolation Theory, Function Spaces, Differential Operators*, second edition, Johann Ambrosius Barth Verlag, Heidelberg, 1995. (Cited on 583.)

[817] H. Triebel, *The Structure of Functions*, Monographs in Mathematics, Birkhäuser Verlag, Basel, 2001. (Cited on 557.)

[818] M. Tsuji, *Potential Theory in Modern Function Theory*, reprinting of the 1959 original, Chelsea Publishing, New York, 1975. (Cited on 177, 274.)

[819] J. L. Ullman, *On the regular behaviour of orthogonal polynomials*, Proc. London Math. Soc. (3) **24** (1972), 119–148. (Cited on 291.)

[820] J. L. Ullman, *Orthogonal polynomials for general measures, I*, In Rational Approximation and Interpolation (Tampa, FL, 1983), P. R. Gravis-Morris et al., eds., pp. 524–528, Lecture Notes in Math., Springer, Berlin, 1984. (Cited on 291.)

[821] J. L. Ullman, *Orthogonal polynomials for general measures, II*, In Orthogonal Polynomials and Applications (Bar-le-Duc, 1984), C. Brezinski et al., eds., pp. 247–254, Lecture Notes in Math., Springer, Berlin, 1985. (Cited on 291.)

[822] J. L. Ullman and M. F. Wyneken, *Weak limits of zeros of orthogonal polynomials*, Constr. Approx. **2** (1986), 339–347. (Cited on 291.)

[823] J. L. Ullman, M. F. Wyneken, and L. Ziegler, *Norm oscillatory weight measures*, J. Approx. Theory **46** (1986), 204–212. (Cited on 291.)

[824] A. Unterberger and J. Bokobza, *Les opérateurs de Calderón–Zygmund précisés*, C. R. Acad. Sci. Paris **259** (1964), 1612–1614. (Cited on 367.)

[825] A. Unterberger and J. Bokobza, *Les opérateurs de Calderón–Zygmund précisés*, C. R. Acad. Sci. Paris **260** (1965), 34–37. (Cited on 367.)

[826] W. Van Assche, *Invariant zero behaviour for orthogonal polynomials on compact sets of the real line*, Bull. Soc. Math. Belg. Ser. B **38** (1986), 1–13. (Cited on 292.)

[827] J. C. van den Berg, ed., *Wavelets in Physics*, Cambridge University Press, Cambridge, 1999. (Cited on 433.)

[828] J. G. van der Corput, *Diophantische Ungleichungen. I. Zur Gleichverteilung Modulo Eins*, Acta Math. **56** (1931), 373–456. (Cited on 123.)

[829] N. Th. Varopoulos, *Hardy–Littlewood theory for semigroups*, J. Funct. Anal. **63**, (1985), 240–260. (Cited on 653.)

[830] F. Vasilesco, *Sur la continuité du potentiel à travers les masses, et la démonstration d'un lemma de Kellogg*, C. R. Acad. Sci. Paris **200** (1935), 1173–1174. (Cited on 274.)

[831] B. Vidakovic, *Statistical Modeling by Wavelets*, Wiley Series in Probability and Statistics: Applied Probability and Statistics, A Wiley-Interscience Publication, John Wiley & Sons, Inc., New York, 1999. (Cited on 433.)

[832] B. Vidakovic and P. Mueller, *Wavelets for Kids: A Tutorial Introduction*, Duke University Technical Report, 1991; available online at http://gtwavelet.bme.gatech.edu/wp/kidsA.pdf. (Cited on 407, 433.)

[833] J. Ville, *Étude critique de la notion de collectif*, Gauthier–Villars, Paris, 1939. (Cited on 160.)

[834] J. Ville, *Théorie et applications de la notion de signal analytique*, Câbles et Transmission **2** (1948), 61–74. (Cited on 370.)

[835] G. Vitali, *Sui gruppi di punti e sulle funzioni di variabili reali*, Torino Atti **43** (1908), 229–246. (Cited on 49, 59.)

[836] J. von Neumann, *Die eindeutigkeit der Schroderschen Operatoren*, Math. Ann. **104** (1931), 570–578. (Cited on 336.)

[837] J. von Neumann, *Proof of the quasi-ergodic hypothesis*, Proc. Nat. Acad. Sci. USA **18** (1932), 70–82. (Cited on 80.)

[838] J. von Neumann, *Physical applications of the ergodic hypothesis*, Proc. Nat. Acad. Sci. USA **18** (1932), 263–266. (Cited on 82.)

[839] J. von Neumann, *Mathematical Foundations of Quantum Mechanics*, translated by Robert T. Beyer, Princeton University Press, Princeton, 1955. (Translation of the author's Mathematische Grundlagen der Quantenmechanik, Springer, Berlin, 1932). (Cited on 401.)

[840] D. F. Walnut, *An Introduction to Wavelet Analysis*, Applied and Numerical Harmonic Analysis, Birkhäuser Boston, Inc., Boston, MA, 2002. (Cited on 433.)

[841] J. L. Walsh, *Interpolation and Approximation by Rational Functions in the Complex Domain*, fourth edition, American Mathematical Society Colloquium Publications, American Mathematical Society, Providence, RI, 1965; first edition, 1935. (Cited on 291.)

[842] P. Walters, *An Introduction to Ergodic Theory*, Graduate Texts in Mathematics, Springer-Verlag, New York–Berlin, 1982. (Cited on 79.)

[843] C. Watson, *William Thomson, Lord Kelvin (1824–1907)*, In Some Nineteenth Century British Scientists, pp. 96–153, Oxford University Press, Oxford, UK, 1969. (Cited on 250.)

[844] T. Weidl, *On the Lieb–Thirring constants $L_{\gamma,1}$ for $\gamma \geq 1/2$*, Comm. Math. Phys. **178** (1996), 135–146. (Cited on 669.)

[845] A. Weil, *Sur les espaces à structure uniforme et sur la topologie générale*, Act. Sci. Ind. **551**, Paris, 1937. (Cited on 402.)

[846] F. B. Weissler, *Logarithmic Sobolev inequalities and hypercontractive estimates on the circle*, J. Funct. Anal. **37** (1980), 218–234. (Cited on 652.)

[847] J. Wermer, *Potential Theory*, 2nd edition, Lecture Notes in Mathematics, Springer, Berlin, 1981. (Cited on 177.)

[848] H. Weyl, *Singuläre Integralgleichungen mit besonderer Berücksichtigung des Fourierschen Integraltheorems*, PhD thesis, Universität Göttingen, 1908. (Cited on 488.)

[849] H. Weyl, *Die Gibbssche Erscheinung in der Theorie der Kugelfunktionen*, Rend. Circ. Mat. Palermo **29** (1910), 308–323. (Cited on 98.)

[850] H. Weyl, *Über die Gleichverteilung von Zahlen mod. Eins*, Math. Ann. **77** (1916), 313–352. (Cited on 98, 122.)

[851] H. Weyl, *Gruppentheorie und Quantenmechanik*, S. Hirzel, Leipzig, 1928; Revised English edition: *The Theory of Groups and Quantum Mechanics*, Methuen, London, 1931; reprinted by Dover, New York, 1950. (Cited on 334, 336, 368.)

[852] H. Widom, *Polynomials associated with measures in the complex plane*, J. Math. Mech. **16** (1967), 997–1013. (Cited on 292.)

[853] N. Wiener, *Certain notions in potential theory*, J. Math. Phys. **3** (1924), 24–51. (Cited on 231.)

[854] N. Wiener, *The Dirichlet problem*, J. Math. Phys. **3** (1924), 127–146. (Cited on 231.)

[855] N. Wiener, *Note on paper of O. Perron*, J. Math. Phys. **4** (1925), 21–32. (Cited on 231.)

[856] N. Wiener, *The ergodic theorem*, Duke Math. J. **5** (1939), 1–18. (Cited on 49, 83, 84.)

[857] N. Wiener, *I Am a Mathematician*, MIT Press, Cambridge, 1956. (Cited on 334.)

[858] E. P. Wigner, *On the quantum correction for thermodynamic equilibrium*, Phys. Rev. **40** (1932), 749–759. (Cited on 370.)

[859] I. Wik, *On John and Nirenberg's theorem*, Ark. Mat. **28** (1990), 193–200. (Cited on 534.)

[860] E. Wirsing, *On the theorem of Gauss–Kusmin–Lévy and a Frobenius-type theorem for function spaces*, Acta Arith. **24** (1973/74), 507–528. (Cited on 125.)

[861] T. Wolff, *An improved bound for Kakeya type maximal functions*, Rev. Mat. Iberoam. **11** (1995), 651–674. (Cited on 685.)

[862] T. Wolff, *Recent work connected with the Kakeya problem*, In Prospects in Mathematics (Princeton, NJ, 1996), H. Rossi, ed., pp. 129–162, American Mathematical Society, Providence, RI, 1999. (Cited on 685.)

[863] T. Wolff, *Local smoothing type estimates on L^p for large p*, Geom. Funct. Anal. **10** (2000), 1237–1288. (Cited on 685.)

[864] M. F. Wyneken, *On norm and zero asymptotics of orthogonal polynomials for general measures, I*, Rocky Mountain J. Math. **19** (1989), 405–413. (Cited on 293.)

[865] K. Yosida and S. Kakutani, *Birkhoff's ergodic theorem and the maximal ergodic theorem*, Proc. Imp. Acad. Tokyo **15** (1939), 165–168. (Cited on 83.)

[866] R. M. Young, *An Introduction to Nonharmonic Fourier Series*, revised first edition, Academic Press, Inc., San Diego, CA, 2001. (Cited on 401, 406.)

[867] A. C. Zaanen, *Linear analysis. Measure and Integral, Banach and Hilbert Space, Linear Integral Equations*, Interscience Publishers, New York; North-Holland, Amsterdam; P. Noordhoff, Groningen, 1953. (Cited on 36.)

[868] J. Zak, *Finite translations in solid-state physics*, Phys. Rev. Lett. **19** (1967), 1385–1387. (Cited on 402.)

[869] J. Zak, *Dynamics of electrons in solids in external fields*, Phys. Rev. **168** (1968), 686–695. (Cited on 402.)

[870] J. Zak, *The kq-representation in the dynamics of electrons in solids*, Solid State Phys. **27** (1972), 1–62. (Cited on 402.)

[871] S. Zaremba, *Sur le principe de Dirichlet*, Acta Math. **34** (1911), 293–316. (Cited on 231.)

[872] Q. S. Zhang, *Sobolev Inequalities, Heat Kernels under Ricci Flow, and the Poincaré Conjecture*, CRC Press, Boca Raton, FL, 2010. (Cited on 654.)

[873] Z.-F. Zhou, *The contractivity of the free Hamiltonian semigroup in the L^p space of entire functions*, J./ Funct./ Anal./ **96** (1991), 407–425. (Cited on 653.)

[874] J. D. Zund, *George David Birkhoff and John von Neumann: a question of priority and the ergodic theorems, 1931–1932*, Hist. Math. **29** (2002), 138–156. (Cited on 81.)

[875] A. Zygmund, *Sur les fonctions conjuguées*, Fund. Math. **13** (1929), 284–303. (Cited on 36, 463.)

[876] A. Zygmund, *A remark on Fourier transforms*, Proc. Cambridge Philos. Soc. **32** (1936), 321–327. (Cited on 172.)

[877] A. Zygmund, *Smooth functions*, Duke Math. J. **12** (1945), 47–76. (Cited on 489.)

[878] A. Zygmund, *On a theorem of Marcinkiewicz concerning interpolation of operations*, J. Math. Pures Appl. (9) **35** (1956), 223–248. (Cited on 556.)

[879] A. Zygmund, *On Fourier coefficients and transforms of functions of two variables*, Studia Math. **50** (1974), 189–201. (Cited on 684.)

[880] A. Zygmund, *Trigonometric Series. Vol. I, II*, 3rd edition, Cambridge Mathematical Library, Cambridge University Press, Cambridge, 2002. (Cited on 463.)

Symbol Index

Subject Index

Author Index

Index of Capsule Biographies